540
LEW

D1351475

Gorseinon College
Learning Resource Centre
(TreUchaf) Telephone: (01792) 890808
This book is **YOUR RESPONSIBILITY** and is due for return/renewal on or before the last date shown.

CLASS NO. 540 LEW **ACC. NO.** GC000818

9 NOV 2001
27 MAR 2009
10 JAN 2014
13 MAY 2016

RETURN OR RENEW - DON'T PAY FINES

3483
11.95

Advancing Chemistry

Advancing Chemistry

Michael Lewis
Guy Waller

Radley College

GORSEINON COLLEGE
LIBRARY

Oxford University Press

874297

Oxford University Press, Walton Street, Oxford OX2 6DP

Oxford New York Toronto
Delhi Bombay Calcutta Madras Karachi
Petaling Jaya Singapore Hong Kong Tokyo
Nairobi Dar es Salaam Cape Town
Melbourne Auckland

and associated companies in
Beirut Berlin Ibadan Nicosia

Oxford is a trade mark of Oxford University Press

© M. Lewis, G. Waller 1982
First published 1982
Reprinted 1982, 1984, 1986, 1988

All rights reserved. No part of this publication may be reproduced, stored in a retrieval system, or transmitted, in any form, or by any means electronic, mechanical, photocopying, recording, or otherwise, without the prior permission of Oxford University Press

This book is sold subject to the condition that it shall not, by way of trade or otherwise, be lent, re-sold, hired or otherwise, circulated without the publisher's prior consent in any form of binding or cover other than that in which it is published and without a similar condition including this condition being imposed on the subsequent purchaser

British Library Cataloguing in Publication Data

Lewis, Michael
Advancing Chemistry.
1. Chemistry
I. Title II. Waller, Guy
540 QD33

ISBN 0 19 914083 9

Typeset by MHL Typesetting Limited, Coventry.
Printed in Great Britain
at the University Printing House, Oxford
by David Stanford
Printer to the University

Preface

Advancing Chemistry has been written to provide a coherent view of chemistry for A-level candidates and students in colleges whatever the syllabus followed.

Chemistry syllabuses in the last few years have been influenced by three interrelated factors: the encouragement of practical skills and the introduction of an element of discovery into the study of science, the realization of the importance of the sociological and economic implications of advances in the subject, and an appreciation of the student's difficulties in absorbing the many seemingly unrelated facts. In this last respect, the authors believe that the theoretical aspects of chemistry, including a study of reaction mechanisms in some detail, are of great assistance in providing a framework for the subject as a whole. The text of *Advancing Chemistry* has been written on the basis of this premise with the other two factors being considered from the firm foundation provided.

The text covers physical, organic and inorganic chemistry, and is divided into four parts, each of which builds on the student's existing body of knowledge. Part 1 of the book concerns the physical structure of matter. It details the properties of the nucleus, the atom and the electron, and outlines the nature of the various types of bonding. It concludes with an account of the equilibria between the different phases of matter. Part 2 discusses the energy changes and kinetics of chemical reactions before looking at the likely consequences of equilibrium in acid-base reactions, solubility, redox and electrochemistry. Part 3 looks again at the atomic structure of carbon before categorizing and comparing the many facets of organic chemistry in terms of structure, bonding and reaction mechanism. It concludes with a consideration of some of the stereochemical factors affecting reactions. Part 4 examines the bonding and structure of inorganic systems under the headings of their acid-base, redox and complexing properties. It relates the properties of the elements and their compounds to their position in the periodic table and to their atomic structure. The three appendices give details of diagnostic tests for organic and inorganic compounds, of the major reaction routes of organic chemistry, and of a range of laboratory methods for the preparation of inorganic compounds. The book concludes with numerical answers to the questions at the end of each chapter, and a comprehensive index arranged in three parts for ease of use.

The authors hope that the major themes that recur throughout this book will lead to a gradual breakdown of the boundaries between the different branches of the subject and that in consequence each part of the text will reinforce the total knowledge of the student.

The authors would like to express their sincere thanks to John Harrington of Bradford Grammar School for his thorough and exacting advice on all aspects of the text, to MHL Typesetters for their enthusiastic interest and attention to detail, to the staff of Oxford University Press for their patience and encouragement, and to their wives for unflinching support at all times.

Radley 1981

Michael Lewis
Guy Waller

Preface

Contents

Part 1 Physical structure

Part 2 Chemical energy and equilibrium

Part 3 Organic chemistry

Part 1

Physical structure

Chapter 1
Matter

1.1 The three phases

Characteristic properties

Chemistry is the study of matter and its interactions with energy. Anything that has mass and therefore can be weighed is classified as matter. There are three recognizably distinct states of matter: solid, liquid, and gas.

The term *phase* is often used to describe a state of matter. For example, iron at room temperature is in the solid phase, mercury is in the liquid phase, and argon is in the gas phase.

The table below shows the characteristic physical properties of each of the three phases.

solids	*liquids*	*gases*
1. have a fixed shape	1. have no fixed shape and are mobile	1. have no fixed shape or fixed volume
2. have a fixed volume	2. are difficult to compress and so have approximately a fixed volume	2. occupy the volume of their container
3. are usually the most dense phase	3. are almost as dense as their solid phase	3. are far less dense than solids or liquids

It is possible for a system to be in a single state of matter, but for there to be more than one phase present. For example, water and mercury exist in the liquid state of matter. There are however, two liquid phases present because the liquids are immiscible.

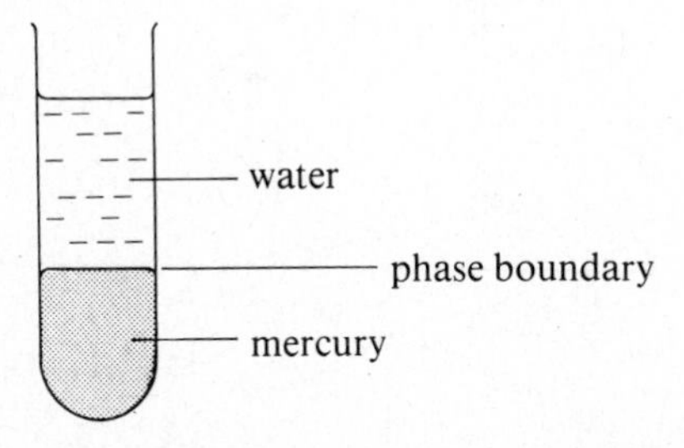

Phase changes

A single pure substance can exist in any one of the three phases. The actual phase in which it is found depends on the external conditions of temperature and pressure.

A change of phase from solid to liquid occurs at a clearly defined temperature for a particular substance. This temperature is called the melting point (m.p.); it varies only a little when the external pressure is changed.

By contrast, the phase change from liquid to gas occurs throughout the whole range of temperature in which the liquid phase exists; it is known as evaporation. The rate of liquid evaporation is increased by

a) increasing the surface area of the liquid,
b) causing convection currents across the surface of the liquid,
c) increasing the temperature,
d) decreasing the pressure.

A liquid reaches its boiling point (b.p.) at the temperature where bubbles of the gas phase are seen to form within the body of the liquid phase. An evaporating liquid gives rise to a vapour pressure due to the gas evaporating from its surface. A boiling liquid exerts a vapour pressure equal to that of the external pressure. The boiling point of a liquid is therefore considerably affected by the external pressure.

The structure of matter

fact	*theory*
Crystals have regular shapes and are hard.	The particles of matter are packed together in an ordered and regular way. Each particle attracts its neighbours strongly. 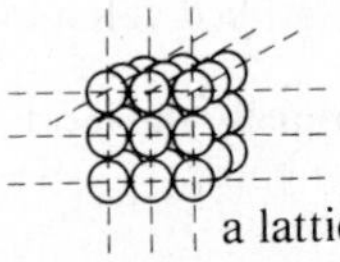 a lattice
A solid melts to give a liquid. A liquid needs to be contained, but occupies almost the same volume as its solid phase	The lattice of particles starts to break up to form small groups of particles that are able to move past one another 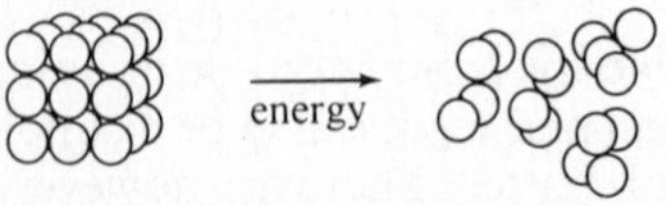
Gases occupy the volume of the container in which they are placed. They can easily be compressed and are much less dense than either the liquid phase or the solid phase.	The particles of matter are far apart and are in constant motion 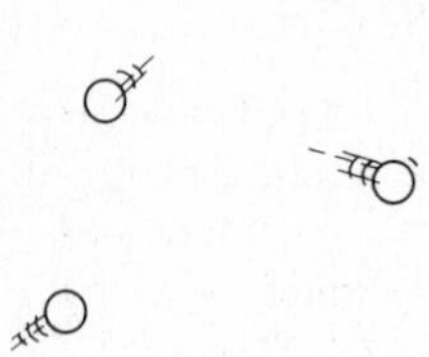

The physical properties of matter and the characteristics of the three phases are all observable facts; they can be measured and validated by repeated experimental observation. However, it is with far less certainty that one can approach questions concerning the actual make up of matter. Nonetheless, all scientists believe that matter is made from infinitely small particles that can neither be seen nor detected in a direct way. The widespread acceptance of this idea contrasts sharply with the acceptance of a particular physical property. The physical property that salt dissolves in water to give a mixture in the liquid phase is an observable *fact*, but the structure of matter is a problem of *theory*.

A theory can continue to find acceptance only so long as it can explain the observed facts. As soon as a set of facts emerges that cannot be explained by the theory, then the theory must either be modified or scrapped.

The 'particle' theory of the structure of matter has been able to account for so much of the experimental evidence that its position is unlikely seriously to be challenged. Some simple examples of its application are shown in the table at the bottom of the previous page.

1.2 Particles of matter

It is possible to think of particles of matter in terms of billiard balls or other similar solid spheres. In order to account for the various properties of matter, it is necessary to assume that the particles have the following characteristics.

1. The particles can possess kinetic energy which may be translational, vibrational or rotational.

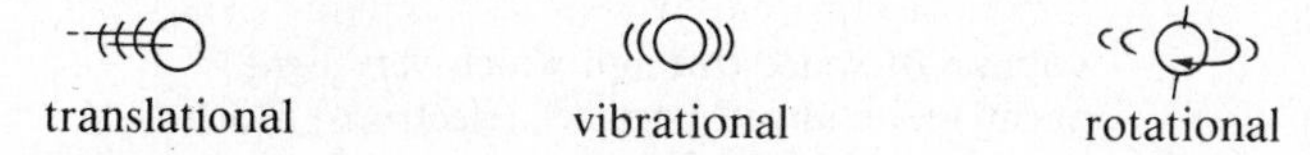

2. The particles can attract one another.
3. The particles can collide and bounce off one another.

However, the model of the billiard balls is found to need significant alteration in the light of the experimental evidence provided by the scattering experiments of Lord Rutherford and the work of Aston on the mass spectrometer. This evidence dates back to the early part of the twentieth century and forms the basis of our understanding of the nature of particles of matter.

Experimental evidence

Rutherford and his co-workers Geiger and Marsden aimed a beam of α-radiation at a gold foil target. Radioactivity had been recently discovered, and the reported properties of α-radiation suggested that the radiation was a stream of positively charged particles whose mass greatly exceeded the mass of an electron. We now know that an α-particle is the nucleus of a helium atom ^4_2He.

They had expected that the α-particles would be scattered as a result of collisions with the gold atoms in the target. From the scattering pattern they had hoped to obtain information about the proportions of the atoms of gold. The results were rather surprising.

1. Nearly all the α-particles passed straight through the target as if it was not there.
2. A few α-particles were scattered at all angles around the target from 0° to 360°.

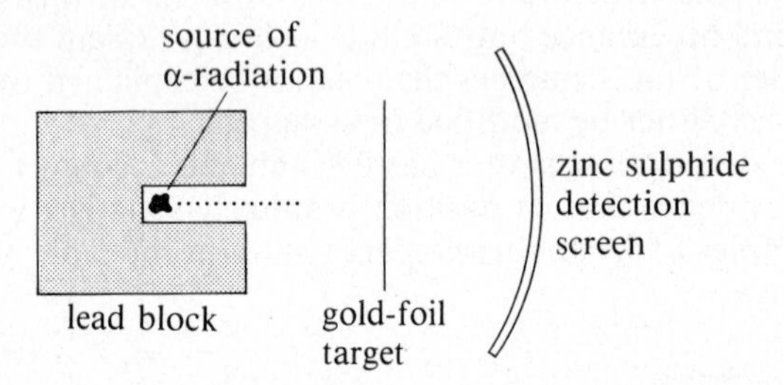

Rutherford came to three important conclusions.

1. Nearly all the volume of an atom is empty space.
2. The mass of an atom is concentrated in a tiny volume, the nucleus, at the centre of the atom.
3. The nucleus of an atom has the same polarity of charge as an α-particle.

The first conclusion explains how most of the α-particles pass through undetected. The last two conclusions explain the severity with which some of the particles are deflected. The model of an atom that is derived from these conclusions is shown below.

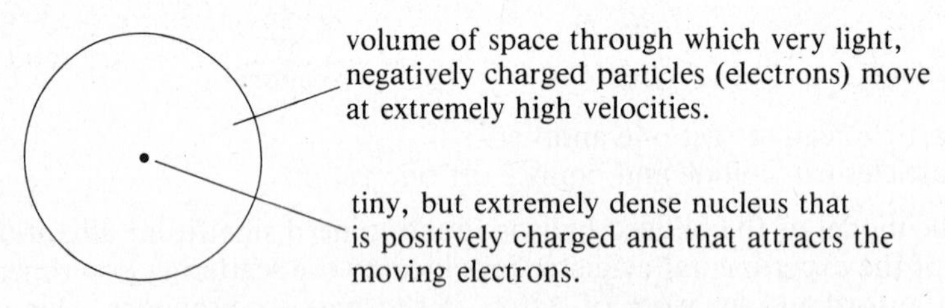

Rutherford's model of the atom is able to account for the difference between atoms of two different elements: the magnitude of the charge on the nucleus determines the identity of the atom. However further modifications were needed.

Aston developed the mass spectrometer and was able to show that a sample of one pure element contained atoms of different masses. It became necessary to suggest that the nucleus of the atom also contained some heavy neutral particles and that the number of these neutral particles was not a constant for any one element. Firm evidence of these particles, called neutrons, was not discovered until 1932 when a beam of neutral radiation was detected from a beryllium target that was bombarded with high-energy α-particles.

$${}^{4}_{2}\mathrm{He} + {}^{9}_{4}\mathrm{Be} \longrightarrow {}^{12}_{6}\mathrm{C} + {}^{1}_{0}\mathrm{n}$$

The modern version of the mass spectrometer is described in the next paragraph. Its earliest forerunner helped to establish the planetary model of the atom.

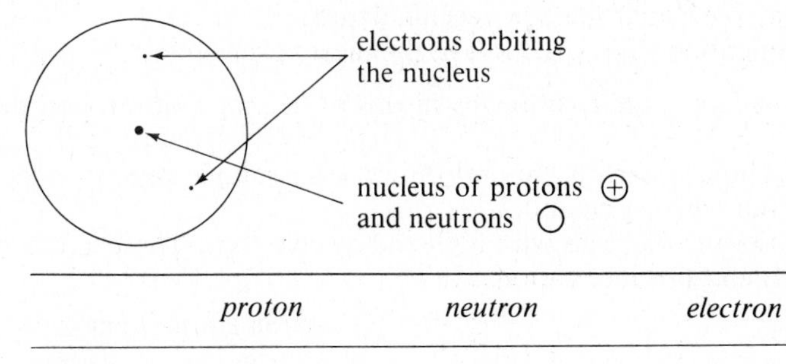

	proton		*neutron*		*electron*
charge	⊕		○		⊖
relative mass	1	:	1	:	$\frac{1}{1860}$

> The atomic number Z of an atom is the number of protons in the nucleus.
>
> The mass number A of an atom is the number of protons and neutrons in the nucleus.
>
> Isotopes of an element have the same atomic number, but different mass numbers, e.g. $^{35}_{17}Cl$ and $^{37}_{17}Cl$ are both isotopes of chlorine.

A pure sample of an element is usually a mixture of isotopes. Although each atom is an atom of the same element, atoms of different mass can be present because the number of neutrons in each nucleus can be different. In this way, the results of Aston's mass spectrometry can be explained.

The mass spectrometer

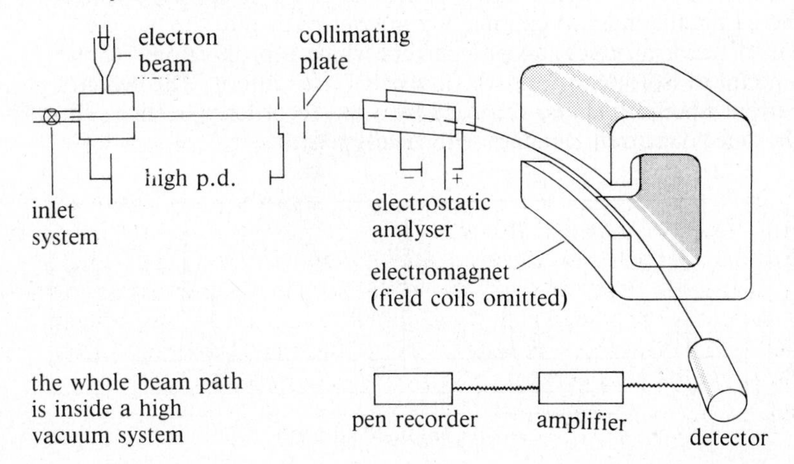

In this instrument a beam of positively charged particles (cations) is deflected in a vacuum by a magnetic field. The degree of deflection is proportional to the mass of the particle. The setting of the field strength is adjusted to allow particles of only one particular mass to reach a detector.

The following functions are carried out in the mass spectrometer.

1. Vaporize the solid or liquid sample by means of heating coils around the inlet system.
2. Evacuate the whole apparatus and introduce some sample vapour at low pressure into the ionizing chamber.
3. Bombard the sample particles with high energy electrons. These knock out other electrons and produce cations.

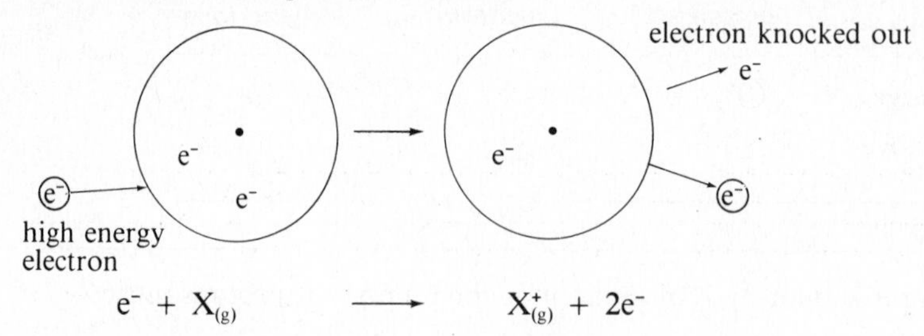

$$e^- + X_{(g)} \longrightarrow X^+_{(g)} + 2e^-$$

and sometimes two electrons are knocked out:

$$e^- + X_{(g)} \longrightarrow X^{2+}_{(g)} + 3e^-$$

4. Accelerate the cations out of the ionizing chamber by applying a high potential difference across the system.
5. Focus and collimate the beam of cations.
6. Pass the beam of cations through two plates at high potential difference. This 'electrostatic analyser' deflects only single-charged cations of a narrow energy range into the magnetic analyser. Other cations are either deflected too much or too little.
7. Adjust the electromagnet to allow cations of one mass only to be deflected into the detector.
8. Slowly and steadily increase the field strength of the magnet by changing the current in the field coils. This has the effect of correspondingly increasing the mass of cations that are deflected into the detector.
9. The detector produces a small current which is proportional to the amount of isotope cations which are deflected into it. The detector current is amplified and recorded by a pen recorder. The chart shows how the detector current changes with field strength.

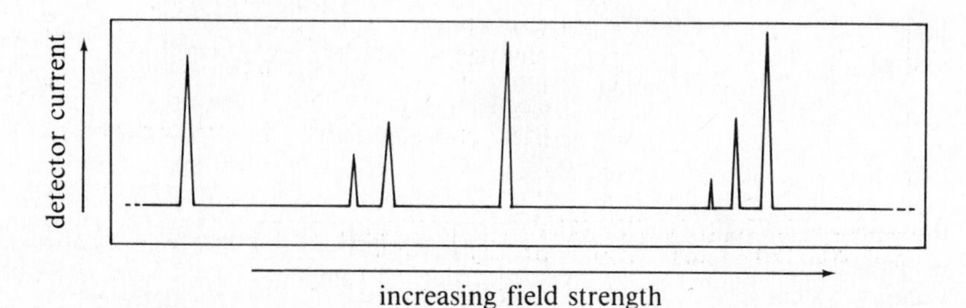

The strength of the magnetic field is directly proportional to the mass of particle being detected. The visual record is calibrated by analysing a substance of known isotopic mass to convert the field strength into mass number A; the visual record is called a mass spectrum. The mass spectrum for neon is shown below. It shows peaks at 20 for ^{20}Ne and at 22 for ^{22}Ne.

atomic mass spectrum for neon

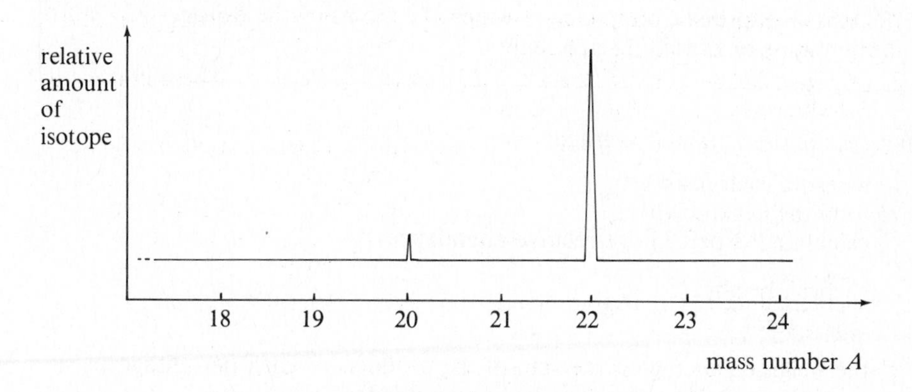

Among the uses of mass spectra are calculation of relative atomic mass, recognition of unknown substances from their characteristic spectra, and very accurate measurement of molecular mass.

1. Calculating relative atomic mass

A simple atomic mass spectrum shows the proportions of all the isotopes present in a sample of an element. The mass spectrum of lead has four main peaks.

atomic mass spectrum for lead

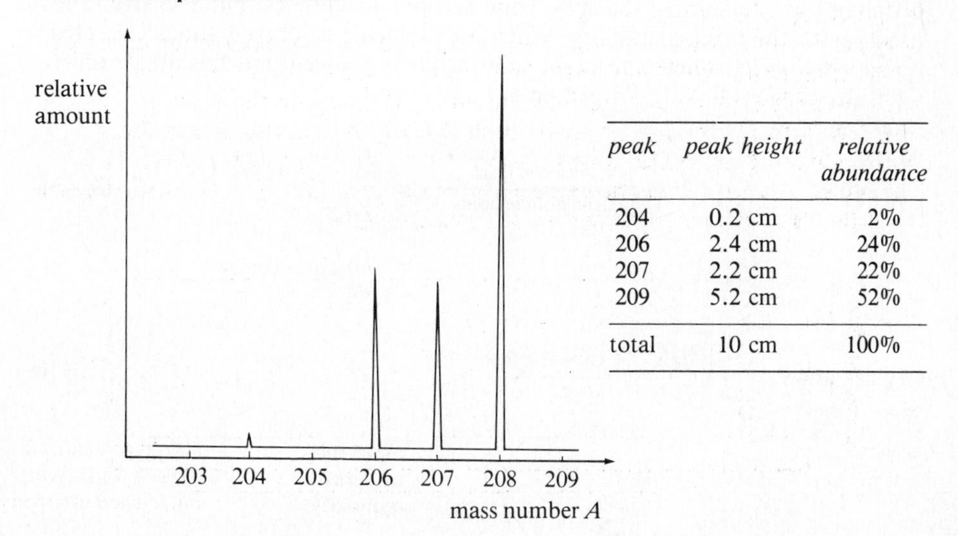

peak	*peak height*	*relative abundance*
204	0.2 cm	2%
206	2.4 cm	24%
207	2.2 cm	22%
209	5.2 cm	52%
total	10 cm	100%

The *relative atomic mass* of an element can be found from the atomic mass spectrum.

> The relative atomic mass (R.A.M.) of an element is the weighted average of the masses of the atoms in a naturally occurring sample of the element.
>
> R.A.M. is expressed on a scale in which the atoms of the isotope ${}^{12}_{6}C$ have a mass of exactly 12 mass units.

To calculate the weighted average:

a) measure each peak height;
b) total the peak heights;
c) calculate the percentage relative abundance:

$$\frac{\text{peak height}}{\text{total}} \times 100;$$

d) the weighted average is the sum of the products of each percentage abundance and its corresponding mass number.

For example for lead:

$$\text{R.A.M.} = \left(\tfrac{2}{100} \times 204\right) + \left(\tfrac{24}{100} \times 206\right) + \left(\tfrac{22}{100} \times 207\right) + \left(\tfrac{52}{100} \times 209\right)$$

$$= 4.08 + 49.44 + 45.54 + 108.16 = 207.22$$

2. Recognition of unknown substances from their mass spectra
Molecular compounds have much more complicated mass spectra. There are different peaks resulting not only from isotopic differences, but also from the break up of the original cations. When the molecule is ionized and accelerated through the spectrometer, it might split into two before it reaches the detector. Consider carbon dioxide for example.

ionization — electron lost from double bond — molecular cation — $A = 44$

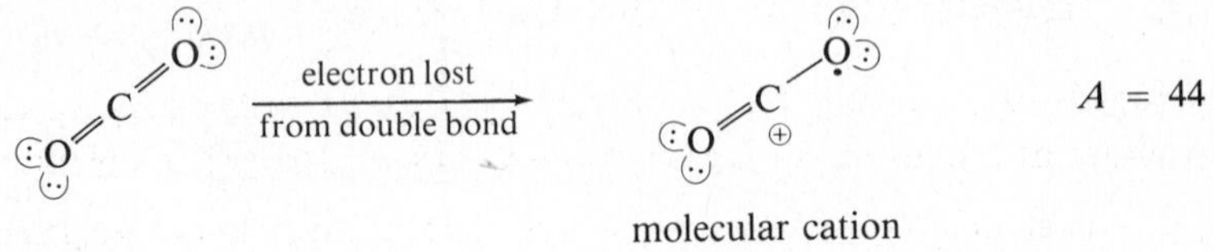

fragmentation — carbon-oxygen bond breaks — molecule of carbon monoxide — an oxygen cation: oxygen with five outer-shell electrons — $A = 16$

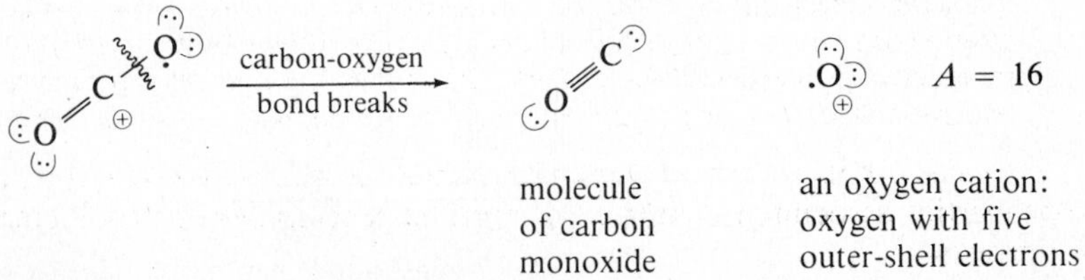

The mass spectrum for carbon dioxide shows a large peak at 16. This strongly suggests that the molecular cation does break up to give oxygen cations. Other peaks are produced from different fragmentation processes.

simplified mass spectrum for carbon dioxide

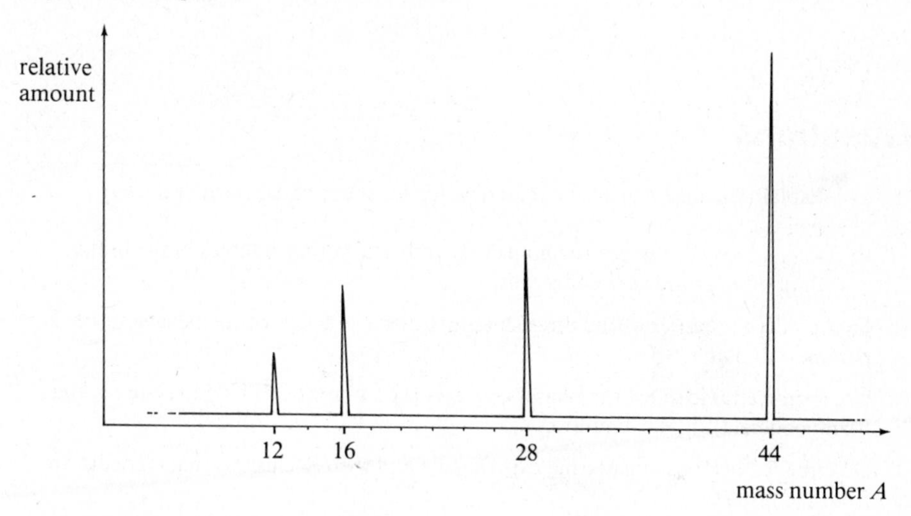

For example,

$$CO_{2(g)}^{+} \longrightarrow CO_{(g)}^{+} + O_{(g)} \qquad A = 28$$

$$CO_{(g)}^{+} \longrightarrow C_{(g)}^{+} + O_{(g)} \qquad A = 12$$

The fragmentation pattern of a particular compound is rather like a finger print. By collecting a large data bank of the mass spectra of known compounds, comparison of an unknown compound's mass spectrum can be made with the data bank. This often establishes its identity without needing any further evidence.

1.3 Electrons and the nucleus

In 1922 Aston was able to summarize the thinking of his time about the structure of the atom in his famous monograph *Isotopes*. He wrote:

> . . . an atom of matter consists of a central massive nucleus carrying a positive charge which is surrounded, at distances relatively great compared with its diameter, by 'planetary' electrons. The central nucleus contains all the positive electricity in the atom, and therefore practically all its mass. The weight of the atom and its radioactive properties are associated with the nucleus; its chemical properties and (emission) spectrum, on the other hand, are properties of its planetary electrons . . .

The clear distinction made between properties associated with the nucleus and properties associated with the electrons has become even clearer since Aston's remarks.

In this book, chapter 2 outlines the chemistry of the nucleus; chapters 3 to 8 are concerned with the 'planetary' electrons. We shall see that the development of electronic theory since 1922 has been enormous. Nearly all chemical properties, structure, and reactivity are now discussed in terms of the behaviour of electrons.

Questions

1. a) Explain the meaning of the following terms: state; phase; evaporation; boiling.
   b) Explain why car tyres are pumped up with air, but air bubbles in the brake fluid of a car are very dangerous.
2. Explain the properties of the three states of matter in terms of the behaviour of the particles of matter.
3. Use a simple table to list the phases or states present in the following systems: fog; foam; concrete; smoke; emulsion.
4. Describe Rutherford's scattering experiments and the conclusions that were drawn from the results.
5. Draw and describe a mass spectrometer, listing its important functions and uses.
6. The mass spectrum for hydrogen chloride is shown below.
   a) Identify the ions producing each peak.
   b) Use this mass spectrum to calculate the relative abundance of the two common isotopes of chlorine.

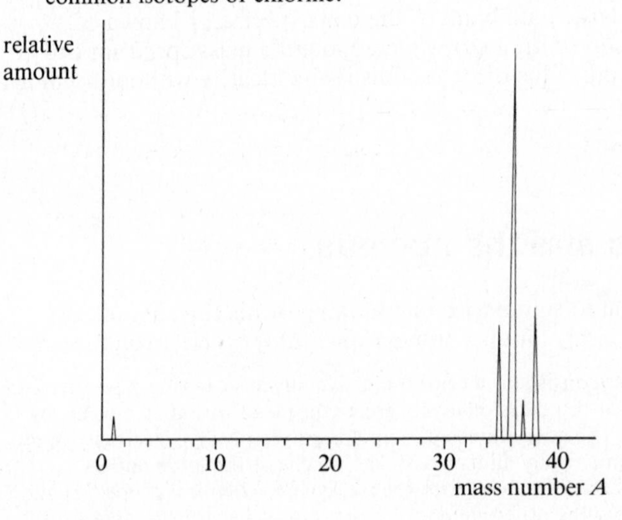

7. The mass spectrum for a compound of hydrogen and oxygen is shown below.
   a) Identify the ions producing each peak.
   b) Identify the compound.

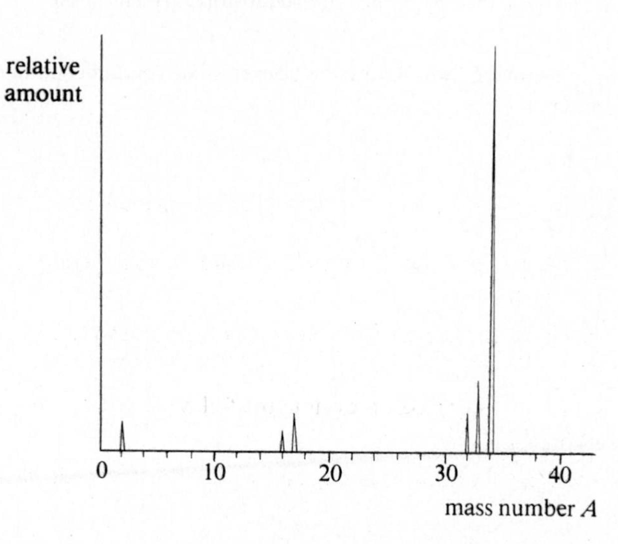

8. a) Calculate the relative atomic mass of boron from the following mass spectrograph data.

	mass	*height of peak*
peak 1	10	10
peak 2	11	40

   b) The mass spectrum data for naturally occurring barium are given in the table below.

	mass	*height of peak*
peak 1	134	1
peak 2	135	2.5
peak 3	136	3.5
peak 4	137	5.5
peak 5	138	33

How many naturally occurring isotopes of barium are there? Calculate the R.A.M. of barium from these data.

9. a) Naturally occurring chlorine consists of 75% of the isotope ^{35}Cl and 25% of the isotope ^{37}Cl. Calculate the R.A.M. of chlorine.
   b) A sample of copper contains the two isotopes ^{63}Cu and ^{65}Cu. If the R.A.M. of copper is 63.5, calculate the percentage of each isotope present in the sample.

10. Explain in simple terms to someone who knows no science what you understand the structure of the atom to be.

Chapter 2

Nuclear chemistry

2.1 The structure of the nucleus

Nuclear binding force

The particles in the nucleus are protons and neutrons; these are known collectively as nucleons. They are packed together in a tiny volume compared with that of the atom. The electrostatic forces of repulsion within the nucleus are extremely high, and yet most nuclei do not split apart. They are held together by an even more powerful attractive force between the nucleons called the *nuclear binding force.* Little is known about the nature of this force, but it is not electrostatic. It appears to depend on the fourth power of the distance between nucleons; compare this with the 'inverse square law' of electrostatic force.

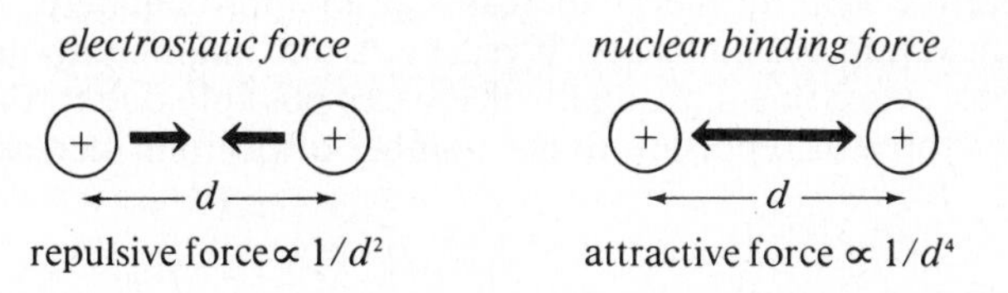

The nuclear binding force is incredibly strong when the distance between nucleons is very small. However the magnitude of the force falls off rapidly as the distance increases.

Recent high-energy particle experiments have been carried out in which protons and neutrons are accelerated towards an atomic nucleus. The results of the experiments indicate that the energy field around the nucleus resembles a hole with a raised rim as shown on the graph below. When a positive particle approaches the nucleus, it experiences stronger and stronger electrostatic forces

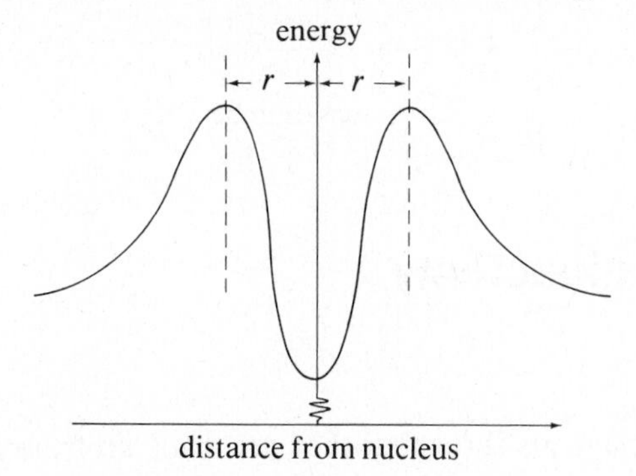

of repulsion. At a distance r from the nucleus, the nuclear binding force overcomes the electrostatic force of repulsion. Once inside distance r, the positive particle 'tumbles' into the energy hole.

The nucleons in the nucleus do not all have the same energy, but the energy required to break a nucleon free from the nucleus is enormous. The nuclear binding energy per mole of nuclei is in the order of 10 to 100 MJ.

Mass deficit

An insight into the possible nature of the nuclear binding force came with the development of a more precise form of mass spectrometer. It was discovered that the measured mass of an atomic nucleus is always slightly less than the sum of the masses of the nucleons making it up. It is as though some of the mass gets lost as the protons and neutrons are bound together to form an atomic nucleus.

This loss in mass, or *mass deficit*, can be accounted for by assuming that it is converted into energy. The amount of energy produced by converting the mass deficits of a mole of nuclei is then equivalent to the nuclear binding energy. The relationship between mass and energy is given by Einstein's famous equation,

$$E = mc^2 \qquad \text{where } c \text{ is the velocity of light.}$$

Calculations show that the binding energy increases at an approximately constant rate as the mass number increases. It reaches a maximum at about $A = 55$ (e.g. iron $^{55}_{26}Fe$). After this it gradually decreases possibly due to the increasing electrostatic force of repulsion as the number of protons increases in the nucleus.

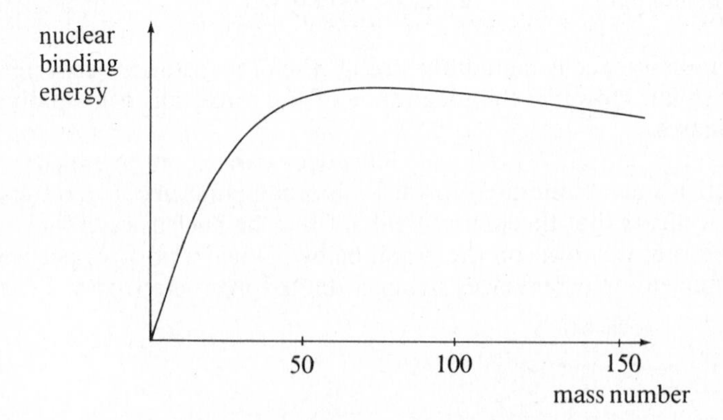

2.2 Isotopes and radioactivity

Isotopic stability

The mass spectra of most pure elements show that they are made up from a mixture of isotopes: atoms of the same atomic number but different mass number. For example, the spectrum of magnesium is shown below.

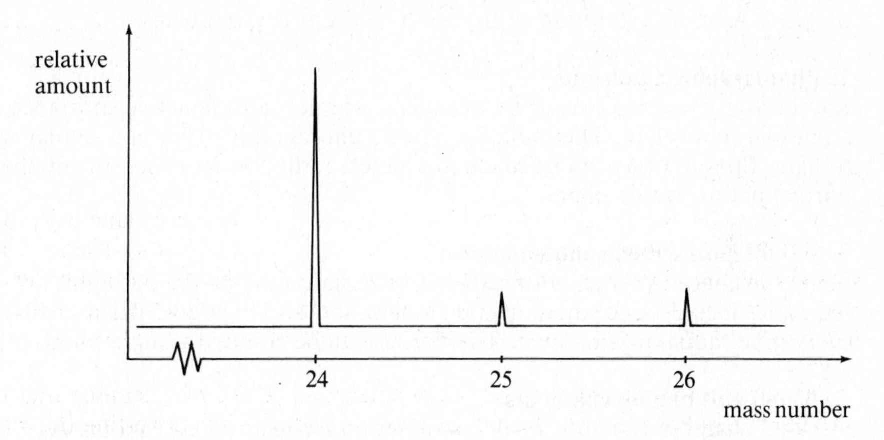

Many isotopes occur naturally, but the number of them varies from one element to another. For example, tin has ten naturally occurring isotopes but iodine has only one, $^{127}_{53}I$. However, 'artificial' isotopes can be made by bombarding others with either neutrons or protons in an atomic reactor; nuclear reactions of this sort are described on page 23. Both natural and artificial isotopes are sometimes found to emit radiation; during the process the isotope itself changes into a different isotope. The term *radioactive decay* is used to describe this effect; isotopes that undergo spontaneous radioactive decay are called *unstable* or *radioactive*.

The stable isotopes of each element that occur naturally are shown on the following graph of atomic number Z against mass number A.

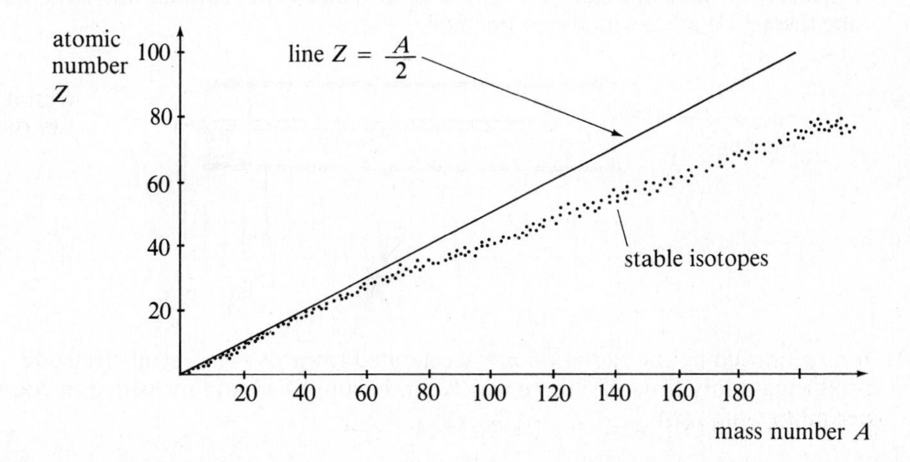

Increasingly more neutrons are needed to produce a stable nucleus as the mass of the nucleus increases; for example, $^{201}_{80}Hg$ has 41 more neutrons than protons.

Detection of radiation

The radiation emitted from a radioactive isotope can be detected in a number of ways. A brief description of the main methods is given below.

1. Photographic emulsions

Radioactivity was discovered by Becquerel who left a radioactive substance on a photographic plate. The radiation affects photographic film in a similar way to light. Special films can be made that detect radiation more sensitively than a normal photographic plate.

2. Scintillation screens and counters

Some substances interact with radioactive radiation by giving out light. Examples include zinc sulphide and caesium iodide which give flashes of light every time radiation hits them. The flashes can be counted using a photo cell.

3. Cloud and bubble chambers

A cloud chamber is a very small container of supersaturated vapour. As radiation passes through the vapour, it produces ions along its path. The ionization increases some of the inter-particle forces so that small droplets of liquid form along the pathway of radiation. A track of tiny droplets shows the presence of radioactive radiation.

A bubble chamber is a small container of superheated liquid. Radiation causes bubbles of vapour to form along its pathway.

4. Ionization chambers and Geiger-Müller tubes

An ionization chamber contains a small volume of gas at low pressure between a pair of electrodes. Radiation entering the chamber causes ionization. These ions then move to the electrodes to constitute a small electric current.

A Geiger-Müller tube is a special application of the idea of an ionization chamber. It is a cylindrical tube with a central electrode. Radiation enters the tube through a mica window at one end.

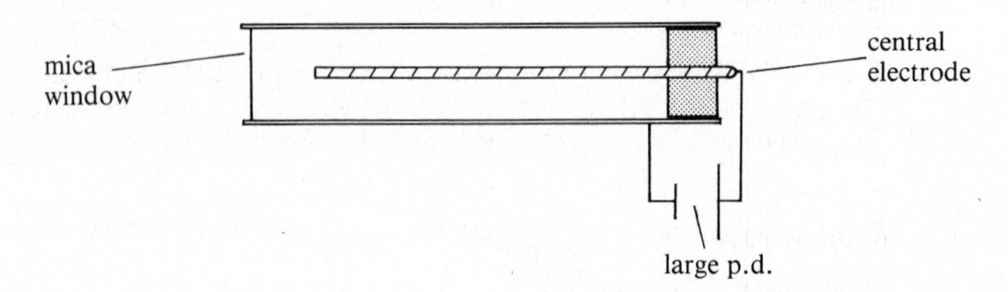

Ions produced by the radiation are accelerated towards the central electrode producing a tiny flow of charge which can be amplified and measured in counts per minute (c.p.m.).

Types of radiation

The stability of a nucleus is related to the number of neutrons and protons present. A nucleus with an even number of protons and neutrons is usually more stable than one with an odd number. There are certain proton numbers that seem to make a nucleus particularly stable; these are 2, 8, 20, 50, and 82. It has been suggested that the proton structure inside a nucleus has some similarity to the electronic shell structure in an atom.

Unstable isotopes usually become more stable by one of the following:

a) reducing their atomic number,
b) reducing their mass number,
c) reducing both.

During any of these three processes, there is an emission of radiation which can be one of several types. The main types are alpha-radiation (α), beta-radiation (β), and gamma-radiation (γ).

α-decay

α-particles are normally emitted by heavy nuclei. The nucleus reduces both its mass and its atomic number by expelling a helium nucleus (two protons and two neutrons).

$$^{238}_{92}U \longrightarrow ^{234}_{90}Th + ^{4}_{2}He$$

α-decay results in

a) reduction in mass number by 4
b) reduction in atomic number by 2.

β-decay

Three forms of beta-decay are known. All of these involve an interaction between neutrons and protons.

1. β^--decay: a neutron becomes a proton and an electron inside the nucleus; the electron is expelled: For example,

$$^{210}_{83}Bi \longrightarrow ^{210}_{84}Po + ^{0}_{-1}e$$

β^--decay results in

a) no change in mass number
b) increase in atomic number by 1

2. β^+-decay: a proton becomes a neutron and a positron inside the nucleus. The positron (a particle with the same properties as an electron except that it is positively charged) is expelled: For example,

$$^{30}_{15}P \longrightarrow ^{30}_{14}Si + ^{0}_{+1}e$$

β^+-decay results in

a) no change in mass number
b) decrease in atomic number by 1

3. K-capture: an inner-shell (K-shell) electron is captured by the nucleus; it converts a proton to a neutron. For example,

$$^{40}_{19}K + ^{0}_{-1}e \longrightarrow ^{40}_{18}Ar$$

K-capture results in

a) no change in mass number
b) decrease in atomic number by 1

Of these three types, β^--decay is the most common.

γ-decay

After emitting α- or β-rays, a nucleus is often left in a high-energy state. It may possess energy due to rotation or vibration of the particles within its structure. The nucleus can reduce its energy by giving out high-energy radiation in the form of γ-rays. These γ-rays are electromagnetic radiation of very short wavelength ($\lambda = 10^{-10}$ to 10^{-13}m) which have similar properties to X-rays. γ-decay results in

a) no change in mass number
b) no change in atomic number

For example,

$${}^{80}_{35}Br^* \longrightarrow {}^{80}_{35}Br + {}^{0}_{0}\gamma$$

where Br* is an 'excited' bromine nucleus.

Properties of α-, β-, and γ-rays

The penetrating power of each type of radiation can be investigated using the apparatus shown below. The effect of a magnetic field can also be studied.

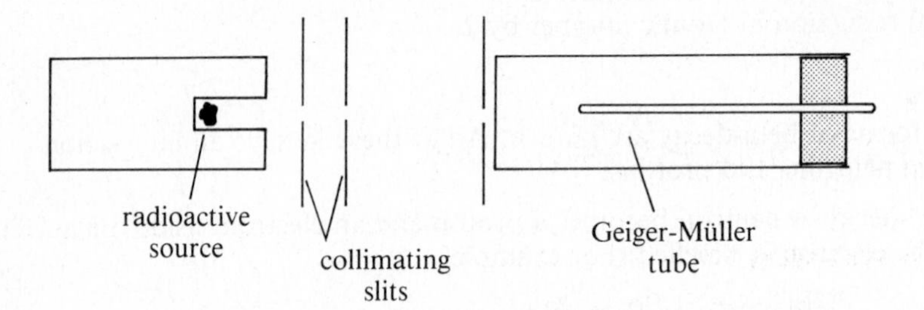

Using α-, β-, and γ-emitting sources separately, the following table of results is produced.

	radiation		
property	*α-rays*	*β-rays*	*γ-rays*
nature of the radiation	helium nuclei ${}^{4}_{2}He$	electrons ${}^{0}_{-1}e$	electromagnetic radiation
distance travelled through air	a few centimetres	a few metres	a long distance
radiation is stopped by	thin foil	a few mm of metal	4 or 5 cm of lead
effect of electric or magnetic field	deflected	strongly deflected in opposite direction to α-rays	undeflected

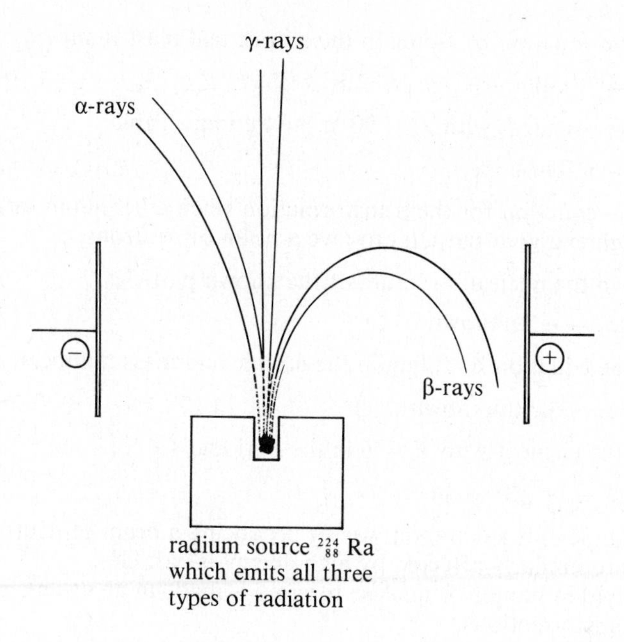

2.3 Nuclear transformations

Nuclear equations

The transformation of a nucleus either by expelling a particle or by taking in a particle can be represented by a nuclear equation. The symbols for the particles are written down with their correct mass numbers and atomic numbers. The equation is balanced when both the total of the mass numbers on each side of the equation are equal and the total of the atomic numbers on each side are equal.

Example a) Write an equation for the α-decay of $^{212}_{83}Bi$

i) Write down the isotope and the α-particle.

$$^{212}_{83}Bi \longrightarrow \text{unknown} + ^{4}_{2}He$$

ii) Balance the equation by filling in the atomic and mass numbers.

$$^{212}_{83}Bi \longrightarrow ^{208}_{81}\text{unknown} + ^{4}_{2}He \qquad \begin{matrix} 212 = 208 + 4 \\ 83 = 81 + 2 \end{matrix}$$

iii) Look up the element with $Z = 81$ in the Periodic Table.

$$^{212}_{83}Bi \longrightarrow ^{208}_{81}Tl + ^{4}_{2}He$$

Example b) Write an equation for the β-decay of $^{227}_{89}Ac$

i) Write down the isotope and the β-particle.

$$^{227}_{89}Ac \longrightarrow \text{unknown} + ^{0}_{-1}e$$

ii) Balance the equation by filling in the atomic and mass numbers.

$$^{227}_{89}\text{Ac} \longrightarrow {}^{227}_{90}\text{unknown} + {}^{0}_{-1}\text{e} \qquad \begin{matrix}227 = 227 + 0\\ 89 = 90 - 1\end{matrix}$$

iii) Look up the element with $Z = 90$ in the Periodic Table.

$$^{227}_{89}\text{Ac} \longrightarrow {}^{227}_{90}\text{Th} + {}^{0}_{-1}\text{e}$$

Example c) Write an equation for the transformation when a beryllium target is bombarded with high-energy α-particles to give a beam of neutrons.

i) Write down the nuclear structure of the known particles.

$$^{9}_{4}\text{Be} + {}^{4}_{2}\text{He} \longrightarrow \text{unknown} + {}^{1}_{0}\text{n}$$

ii) Balance the equation by filling in the atomic and mass numbers.

$$^{9}_{4}\text{Be} + {}^{4}_{2}\text{He} \longrightarrow {}^{12}_{6}\text{unknown} + {}^{1}_{0}\text{n} \qquad \begin{matrix}9 + 4 = 12 + 1\\ 4 + 2 = 6 + 0\end{matrix}$$

iii) Look up the element with $Z = 6$ in the Periodic Table.

$$^{9}_{4}\text{Be} + {}^{4}_{2}\text{He} \longrightarrow {}^{12}_{6}\text{C} + {}^{1}_{0}\text{n}$$

The method in example c) is a common way of producing a beam of neutrons. Other light nuclei are similarly affected by high-energy α-particles.

The plutonium used in bombs or nuclear reactors is made in an atomic pile by the following transformations.

$$^{238}_{92}\text{U} + {}^{1}_{0}\text{n} \longrightarrow \underset{\text{a }\beta\text{-emitter}}{^{239}_{92}\text{U}} \longrightarrow \underset{\text{a }\beta\text{-emitter}}{^{239}_{93}\text{Np}} + {}^{0}_{-1}\text{e}$$

$$^{239}_{93}\text{Np} \longrightarrow {}^{239}_{94}\text{Pu} + {}^{0}_{-1}\text{e}$$

Displacement laws and decay series

The term 'displacement law' is sometimes used to summarize the change that occurs to a nucleus when it undergoes α- or β-decay; α-decay always results in the decrease of the atomic number of a nucleus by 2. This means that the group number of the new nucleus is two less than that of the starting nucleus. A change of minus two in group number is the *displacement law for α-decay*. For example,

a group VII atom becomes a group V atom;
a group I atom becomes a group VII atom.

β-decay always results in the increase of the atomic number of a nucleus by 1. A change of plus one in group number is the *displacement law for β-decay*. For example,

a group VII atom becomes a group VIII atom;
a group VIII atom becomes a group I atom.

Some of the naturally-occurring unstable isotopes decay to give product isotopes which are themselves unstable. Such a product isotope, called a 'daughter' isotope, may itself generate another radioactive isotope which decays to a third isotope. A whole series of products can result from the decay of a single isotope.

A single series of this sort is known as a *decay series*. There are four naturally-occurring decay series. One starts with the uranium isotope $^{235}_{92}\text{U}$ and finishes at a stable lead isotope that has ten fewer protons than its parent isotope.

the uranium decay series

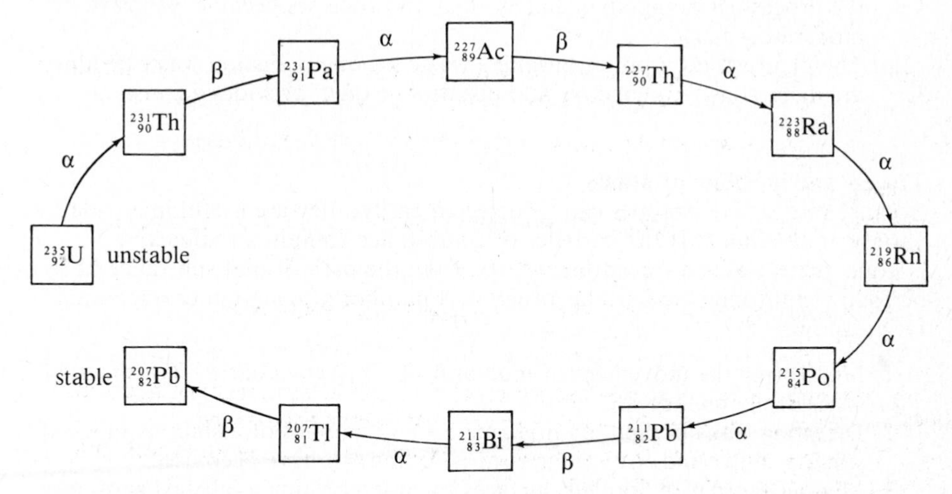

Uses of isotopes

Generating power

With the growing shortage of fossil fuel and the ever-increasing demand for energy, the use of radioactive isotopes for generating power is becoming more widespread. There are two main methods of generating power using isotopes.

1. Fission. The bombardment of certain heavy nuclei by neutrons causes them to break up into two or more fragments. The sum total of the mass of these fragments never quite equals that of the starting nucleus. The loss results from conversion of some of the mass into energy during the fission process. Quite often, further neutrons are produced by the break up; these then trigger the fission of other nuclei producing a chain reaction. For example,

$$^{1}_{0}n + ^{235}_{92}U \longrightarrow ^{100}_{42}Mo + ^{124}_{50}Sn + \text{neutrons} + \text{energy}$$

In a fission reactor, the aim is to provide just enough neutrons to sustain the reaction at a constant and controllable rate. The rate of the reaction can be controlled by inserting *moderators* into the reactor core. A moderator is a substance that slows down the high-energy neutrons in the reactor. The most common moderator is graphite.

2. Fusion. This is the opposite of fission. Two nuclei are fused together to form a single new one. So far, no successful fusion reaction has yet been harnessed to generate power, but the process offers much potential. Efforts have been concentrated around the fusion reaction that takes place in the Sun.

$$^{2}_{1}H + ^{3}_{1}H \longrightarrow ^{4}_{2}He + ^{1}_{0}n + \text{energy}$$

The conditions required to bring about the fusion of a deuterium nucleus $^{2}_{1}H$ and a tritium nucleus $^{3}_{1}H$ include very high temperatures indeed, and a certain critical gas density. The advantages of any future commercial fusion plant over a fission plant are as follows.

a) The plant would be 'fail safe': any failure of the system would result in the process stopping immediately. Fission processes become explosive if moderators fail.
b) The plant would produce harmless waste products. Fission waste products are dangerously radioactive and disposal of them provides a constant problem.

Tracers and labelling of atoms

Because radioactive isotopes can be detected easily, they are useful in a wide variety of situations. If the particles of a substance contain a radioactive isotope, often called a *tracer* or *labelled atom*, the path of that substance can be followed through a particular process. A number of common examples are given below.

1. Monitoring the movement of mud and silt in river estuaries. Tracers used successfully include ${}^{46}_{21}Sc$ and ${}^{51}_{24}Cr$.
2. Detecting leaks in pipes by injecting a small amount of a suitably labelled soluble compound. Compounds of ${}^{24}_{11}Na$ and ${}^{131}_{53}I$ have been used.
3. Measuring wear on moving surfaces by incorporating a labelled substance in the moving parts. For example, piston rings containing ${}^{59}_{26}Fe$ have been used and the activity of the lubricant tested after running the engine for periods of time.
4. Measuring the rate of certain biological processes by including small quantities of labelled compounds as raw materials. The rate of uptake of radioactive material in the organism is then a measure of the rate of the actual process: ${}^{32}_{15}P$ has been used in fertilizers and information about the respiration of plants has been obtained by using ${}^{14}_{6}CO_2$.
5. Determining the actual pathway of a particular reaction by introducing labelled atoms into the particles of one of the reactants. After separating the products, the position of the labelled atoms can then be detected. This often shows exactly how the bonds in the reactant particles have broken, and sometimes in what sequence they have broken. For example, when 'labelled' water reacts with the ester ethyl ethanoate, it is found that the ethanoic acid is labelled and not the ethanol.

$$H{-}{}^{18}OH + CH_3{-}C({=}O){-}O{-}C_2H_5 \longrightarrow H{-}O{-}C_2H_5 + CH_3{-}C({=}O){-}{}^{18}OH$$

In this case, the oxygen-18 isotope is not radioactive, but is detected using a mass spectrometer.

Radioactive tracers have been put to a number of other uses, including the treatment of malignant tumours, sterilizing medical equipment, producing mutations in plant species, measuring the thickness of metal in mills, monitoring the level of liquid in closed containers such as a fire extinguisher, or radio-dating (as explained in the next section).

2.4 Rate of radioactive decay

Decay constant and half-life

In a sample of a radioactive isotope, the probability that any nucleus will undergo transformation within a given time depends only on the nature of that isotope. The more unstable the isotope, the more chance there is that it will decay rapidly. This probability is independent of such external factors as pressure, temperature or chemical composition.

If there are N radioactive atoms present in a sample, then the rate at which radioactive decay takes place is proportional to this number N. The more radioactive atoms there are, the more radiation is detected per unit time. This may be shown mathematically as follows.

$$-\frac{dN}{dt} \propto N \quad \text{or} \quad -\frac{dN}{dt} = \lambda N$$

where $-\frac{dN}{dt}$ means the rate of decrease in the number of radioactive atoms left in the sample, and where λ is a proportionality constant known as the *decay constant.*

Integrating the last equation, we obtain:

$$\log_e \left[\frac{N_0}{N_t}\right] = \lambda t$$

where N_0 = original activity, and N_t = activity after a time lapse t.

The integrated form of the rate equation reveals an important characteristic of decay processes. If the time taken for the activity to drop to half its original level is $t = t_{1/2}$, we can write:

$$t = t_{1/2} \text{ when } N_t = \frac{N_0}{2}$$

and putting these conditions into the integrated rate equation,

$$\log_e \left[\frac{N_0}{\frac{1}{2}N_0}\right] = \lambda t_{1/2}$$

which implies that

$$\log_e 2 = \lambda t_{1/2} \quad \text{or} \quad t_{1/2} = \frac{\log_e 2}{\lambda}$$

where $t_{1/2}$ is known as the *half-life* of the process; it has a constant value.

> The half-life $t_{1/2}$ of an isotope is the time taken for its activity to drop to half its original value
>
> i.e. $t = t_{1/2}$ when $N_t = \frac{N_0}{2}$
>
> $$t_{1/2} = \frac{\log_e 2}{\lambda}; \quad \lambda = \frac{\log_e 2}{t_{1/2}}$$
>
> where λ is the decay constant.

The activity of the sample is measured with a Geiger-Müller tube in counts per minute. The rate of the process decreases with time so that the half-life is always a constant. The graph below illustrates this: the half-lives 90—45 c.p.m., 40—20 c.p.m., and 15—7.5 c.p.m. are shown on the graph to be the same (10 min).

a radioactive decay curve

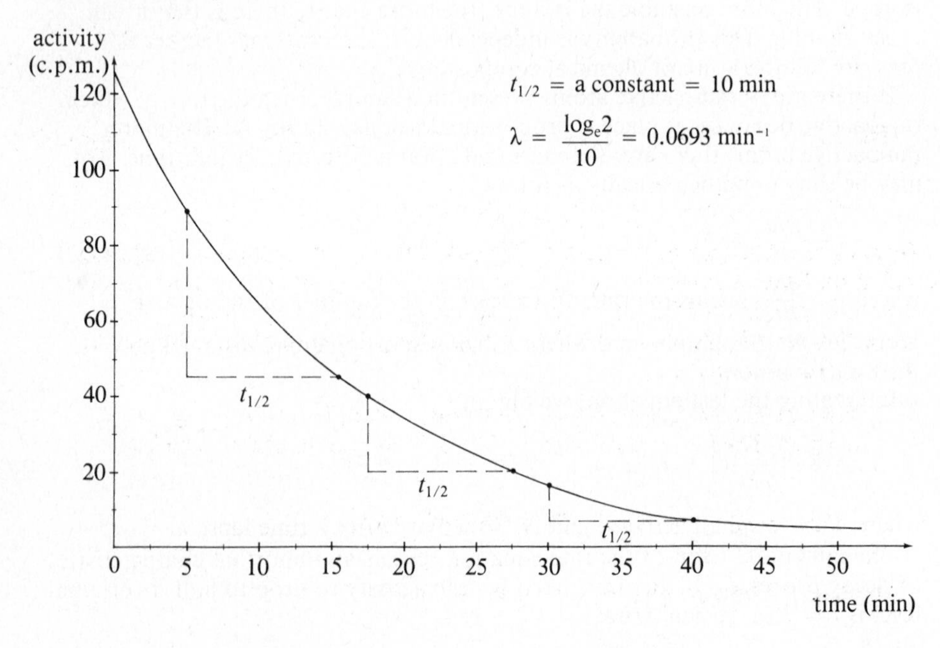

Radioactive decay is one example of a first-order change; these are described in detail in section 15.4.

Sometimes a different shape of decay curve is given by a radioactive sample. This can only happen when one of the decay products is also radioactive. The detection equipment then shows the value of both counts as a single measurement.

Radio-dating

The age of certain substances can be estimated by determining the relative abundance of different isotopes present in the sample. The two most common isotopes that are analysed are

1. ^{40}K in rock samples. The ratio of ^{40}K to ^{39}K slowly changes with time because ^{40}K decays to ^{40}Ar. The value of the ratio $^{40}K : {}^{39}K$ is used to estimate the age of the rock.
2. ^{14}C in organic samples. The level of radioactivity per gram of carbon is measured and compared with that obtained from a sample of live organic matter. The fall-off in activity gives an estimate of the time interval since the object being dated was part of a live organism. It can be used to date samples such as wooden objects or bones.

The second of these methods is called *carbon dating*; it relies on two major assumptions.

a) ^{14}C has been formed and continues to be formed at a constant rate in the upper atmosphere due to cosmic-ray bombardment of nitrogen.

$$\underset{\text{cosmic rays}}{{}^{1}_{0}n} \quad + \quad {}^{14}_{7}N \longrightarrow {}^{14}_{6}C + {}^{1}_{1}H$$

b) The ^{14}C is absorbed into all organic matter in the form of $^{14}CO_2$. The rate of absorption is a constant while the organism is alive, but drops to zero when it dies.

The level of radioactivity in a sample of organic matter is due to the β^--decay of ^{14}C.

$${}^{14}_{6}C \longrightarrow {}^{14}_{7}N + {}^{0}_{-1}e \qquad t_{1/2} = 5568 \text{ years}$$

By analysing the carbon content of the sample and subtracting the background radiation level, a value for the activity per gram of carbon can be obtained. For a live sample this is 15.3 c.p.m. per gram of carbon. For a fragment of the Dead Sea scrolls, the value was 12.0 c.p.m. per gram of carbon. The scrolls were dated as follows.

1. Assume the initial activity of the scrolls was 15.3 c.p.m. per gram of carbon. Then $N_t = 12.0$ and $N_o = 15.3$ where t = age of scrolls.

$$\log_e \left[\frac{N_0}{N_t}\right] = \lambda t \qquad \text{first-order rate equation}$$

2. Find λ from the relationship $\lambda = \dfrac{\log_e 2}{t_{1/2}}$

$$\lambda = \frac{0.693}{5568} \text{ yr}^{-1}$$

3. Put the values in the first-order rate equation.

$$\log_e \left[\frac{N_0}{N_t}\right] = \lambda t = \left(\frac{0.693}{5568}\right) t$$

$$= \log_e \left[\frac{15.3}{12.0}\right] = 0.243$$

$$\text{Therefore } t = \left(\frac{0.243 \times 5568}{0.693}\right) = 1950 \text{ years}$$

The Dead Sea scrolls are approximately 2000 years old.

Questions

1. Explain the meaning of the following terms: radioactive decay; half-life; decay constant; first-order reaction; tracer.

2. Discuss the effect of a 100°C rise in temperature on the rate of:
   a) a typical chemical reaction; b) a nuclear reaction.

3. Compare and contrast the three different types of radioactive decay.

4. Write equations for the following nuclear reactions:
   a) β-decay of $^{231}_{90}Th$
   b) α-decay of $^{235}_{92}U$
   c) α-decay of $^{223}_{88}Ra$
   d) β-decay of $^{228}_{88}Ra$
   e) neutron bombardment of $^{55}_{25}Mn$
   f) neutron bombardment of $^{96}_{42}Mo$
   g) α-bombardment of $^{27}_{13}Al$

5. Explain how the isotope $^{14}_{6}C$ can be used in the dating of old carbon-containing substances.

6. Describe the use of four different radioactive isotopes as tracers.

7. The processes below represent a series of steps in one of the naturally-occurring radioactive decay series.

   $$U \xrightarrow{\alpha} V \xrightarrow{\alpha} W \xrightarrow{\beta} X \xrightarrow{\alpha} Y \xrightarrow{\beta} Z$$

   The letter U stands for the isotope $^{221}_{87}Fr$; the element francium is in group I.
   a) Work out and write down the group numbers of the elements represented by the letters V, W, X, Y, and Z.
   b) Work out and write down the atomic numbers and hence the correct symbols for these elements.
   c) Work out and write down the mass numbers of the isotopes represented by the letters V, W, X, Y, and Z.

8. The disintegration rates at 25°C for a sample of radioactive sodium-24 at various times are given below.

Time in hours	0	2	5	10	20	30
Rate of disintegration (counts s^{-1})	670	610	530	421	267	168

   Using all the data, calculate:
   a) the half-life of sodium-24 b) the value of λ, the decay constant

9. A sample of atmospheric carbon dioxide was monitored for carbon-14 with a Geiger-Müller counter; it gave 128 c.p.m.
   Given that the half-life of carbon-14 is 5568 years, carefully draw a graph showing how the radioactivity will gradually decrease in this sample. Either by using your graph or by calculation, find the age of:
   a) a parchment document if an equal volume of carbon dioxide from the paper produced 100.4 c.p.m.
   b) articles in the prehistoric Lascaux caves in France if an equal volume of carbon dioxide from them produced 18.8 c.p.m.

10. Discuss the advantages and disadvantages of generating power from nuclear reactions.

Chapter 3

The electronic structure of atoms

3.1 Electronic energy levels

The experimental evidence: spectra

Some ten years before Rutherford's work, J.J. Thomson had shown that gaseous matter could be made to interact with electrical energy in a discharge tube containing gas at low pressure.

a discharge tube

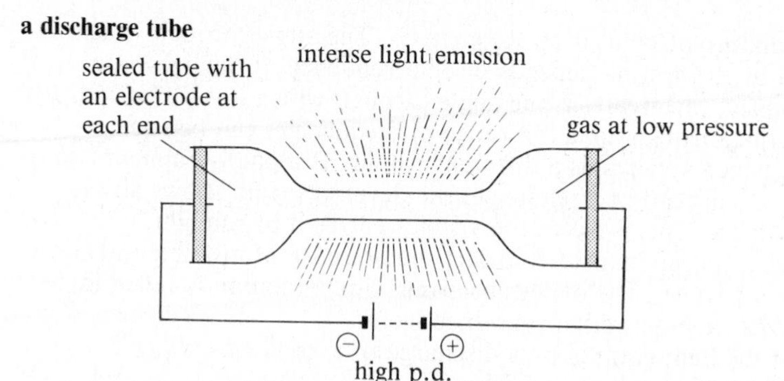

The significance of the results only began to become fully clear after Rutherford's proposal of the planetary model of the atom. The model answered two questions:

a) How can matter interact with electrical energy?
b) How can light be emitted from the electrically excited gas?

1. An orbiting electron has kinetic energy and potential energy.

2. An orbiting electron takes in electrical energy by increasing both its potential energy and its kinetic energy.

3. The resulting atom is said to be 'excited'.

4. The electron returns to a lower energy state. The extra energy is given out as light.

The light emitted from a discharge tube may be analysed by splitting it up into its component frequencies, as for example in the analysis of sunlight ('white' light).

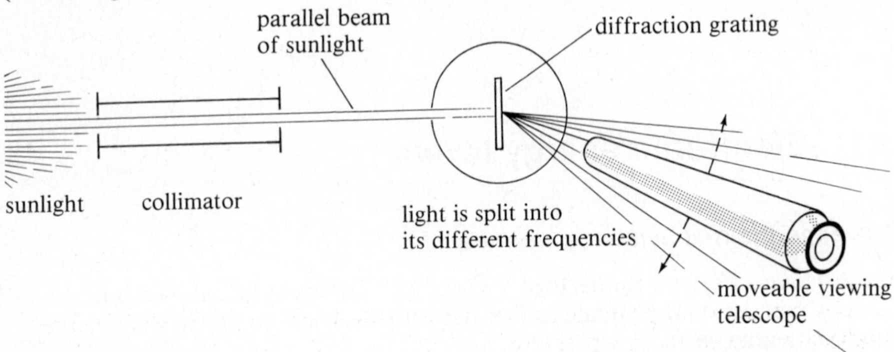

White light is a mixture of light of all frequencies. The angle through which light is diffracted by the grating depends on the frequency of the light: the smaller the frequency, the greater is the angle through which it is diffracted. By moving the telescope through an arc, a range of frequencies can be seen because each frequency appears as a different colour. If a photographic plate is placed round the arc instead of a telescope, the developed film shows all the colours of the rainbow. This is called the visible spectrum of sunlight.

Sensitive instruments can be used to detect the presence of invisible radiation beyond the red or the blue parts of the spectrum. This radiation is called infra-red radiation and ultra-violet radiation respectively.

The analysis of the light emitted by a discharge tube produces rather different results from the analysis of sunlight.

1. The spectrum is *discontinuous*; it appears as a series of coloured lines with darkness between them.
2. The lines on the spectrum *converge*: they gradually get closer and closer until they merge into one another for a short region of continuous light called a *continuum*.
3. The pattern of emissions from a particular element is unique to that element. The more electrons there are in an atom, the more complex is the pattern produced.

For example, the visible part of the spectrum for hydrogen has only three converging lines.

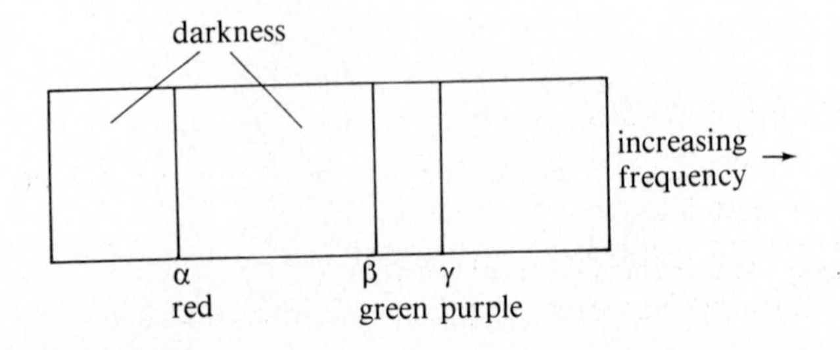

However, there are emissions in both the ultra-violet region and the infra-red region of the hydrogen spectrum. The diagram below shows the series of lines in the far ultra-violet region.

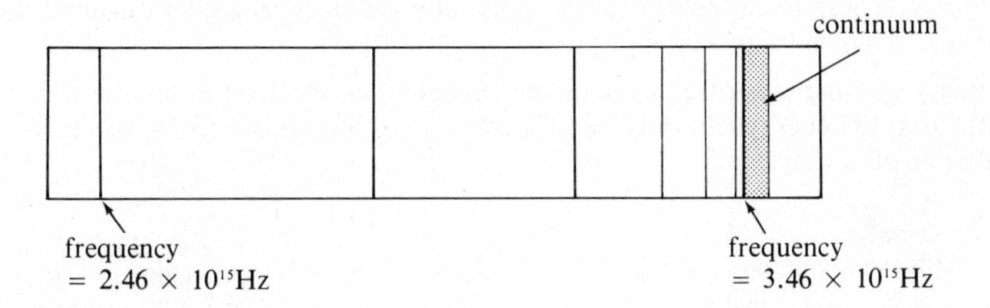

A spectrum that shows all the emissions from a sample of an element is called an *atomic emission spectrum*. Most of the spectrum is likely to be in the invisible infra-red or ultra-violet region.

It is possible to record a spectrum by other means. These usually depend on alternative ways of supplying energy to the sample.

a) The spectrum of a metallic element can be observed by striking an electric arc between two electrodes made of the element.
b) A flame colour is produced by putting a sample of an element or compound into the ionizing part of a Bunsen flame: the light from this can be analysed.
c) An *absorption spectrum* can be recorded by analysing a beam of intense white light that has been passed through a gaseous sample of an element. The emerging light is found to have certain frequencies missing. These frequencies are absorbed by the sample and correspond exactly to the frequencies that would be detected if an emission spectrum was recorded. The absorption spectrum is like a photographic negative of an emission spectrum.

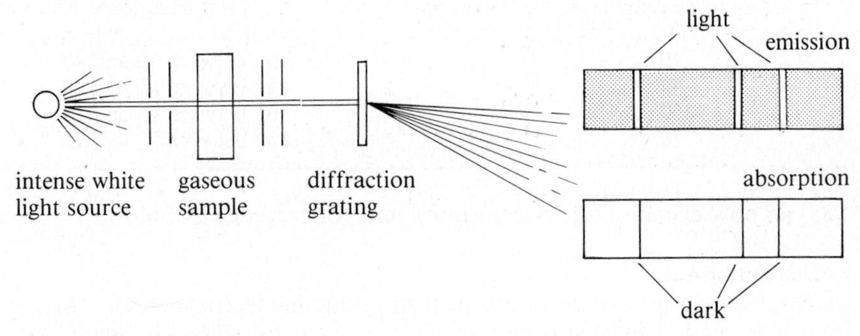

Explaining the evidence: energy levels

Rutherford's model of the atom is able to explain how matter can interact with energy, but it cannot explain:

1. why an emission spectrum is discontinuous,
2. why the lines converge and eventually reach a continuum.

It becomes necessary to make one important change to the model as it stands to account for this new evidence:

> An electron in an atom can only take up certain specific fixed values of energy. These fixed 'energy levels' gradually get closer together as the energy of the electron increases.

The term *quantized* is often used of the energy of an electron in an atom. This means that the energy can only be of a certain fixed *quantity* ; it is easiest to show this on a diagram.

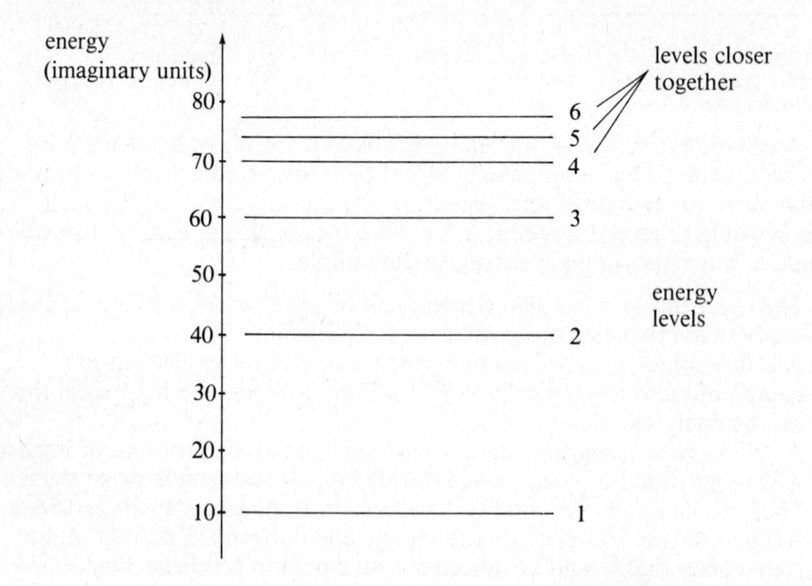

allowed values of energy	*energy level*	*forbidden values of energy*
10	1	any between 10 and 40
40	2	any between 40 and 60
60	3	any between 60 and 70
70	4	any between 70 and 75

We can now explain both discontinuity and convergence as follows.

1. Discontinuity

Light is emitted as a result of an electron taking in electrical energy. In returning to a less excited state, the electron loses the excess energy as emitted light. Assuming fixed energy levels, an electron can only change from one level to another. The values of these energy levels are constant, and so the transition between them must also be of a particular value.

For example, using the simplified energy diagram above, an electron in level 3 can only return to level 2 or level 1. So it can only give out particular amounts of light energy.

3 to 2 ≡ 60 to 40, so 20 units of light energy are given out.
3 to 1 ≡ 60 to 10, so 50 units of light energy are given out.

No other value is possible because the electron can only have either 10, 40, or 60 units of energy, depending on whether it is in level 1, 2, or 3.

A change in electronic energy level is called a *transition*. The frequency of emitted light is related to the difference in the energy levels; this is summarized by the equation

$$\Delta E = h\nu$$

where ΔE is the difference between the energy levels involved in the transition; ν is the frequency of the emitted light; h is Planck's constant.

2. Convergence

Consider the simplified energy diagram again. A series of emissions is produced by electrons returning to level 1 from higher levels.

transition $\Delta E =$	*energy of emission*	*spectrum produced*
$E_6 \rightarrow E_1$	78 to 10 = 68	
$E_5 \rightarrow E_1$	75 to 10 = 65	
$E_4 \rightarrow E_1$	70 to 10 = 60	
$E_3 \rightarrow E_1$	60 to 10 = 50	
$E_2 \rightarrow E_1$	40 to 10 = 30	

lines converging

$\nu = \frac{30}{h}$ $\frac{50}{h}$ $\frac{60}{h}$ $\frac{65}{h}$ $\frac{68}{h}$

increasing frequency ⟶

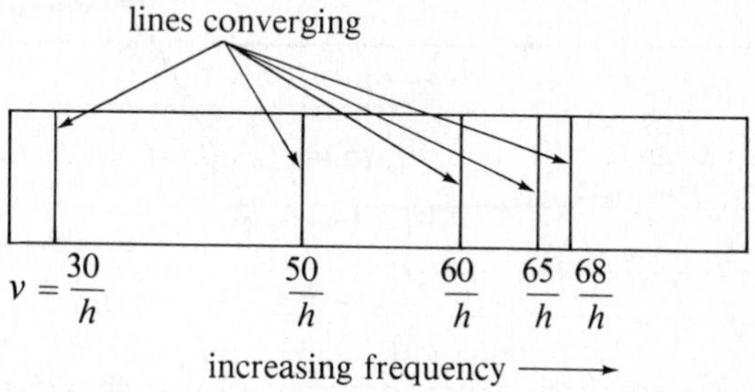

The equation $\Delta E = h\nu$ is used to obtain the emission frequencies.

The converging pattern arises directly out of the assumption that electronic energy levels gradually get closer together as the energy of the electrons increases.

Ionization energy: the continuum

If an electron receives sufficient energy, it can escape altogether from the atom. This results in the production of a free electron and a cation. Once the electron is away from the influence of the nucleus, its energy no longer appears to be quantized. When a free electron of this sort collides with an ion and returns to a particular energy level, the value of the transition is not fixed. It depends on the actual energy of the free electron so that a continuous range of transitions is possible. These appear as the continuum on the spectrum where there are no sharply defined frequencies: each one merges into the next.

The full spectrum for hydrogen contains more than one continuum. There are at least three readily observable continua or series.

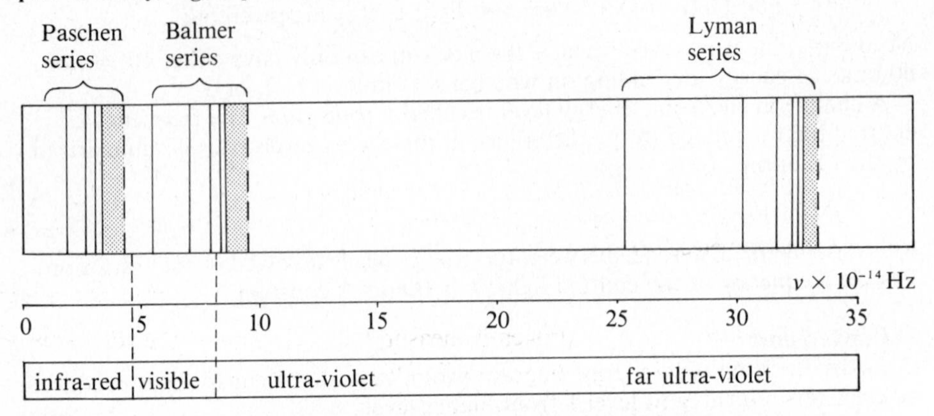

The existence of these series can be explained by considering the possible energy levels to which high energy electrons can return during transitions. If the electrons return to the second level, one of the series is produced; if they return to the first level, a different series results.

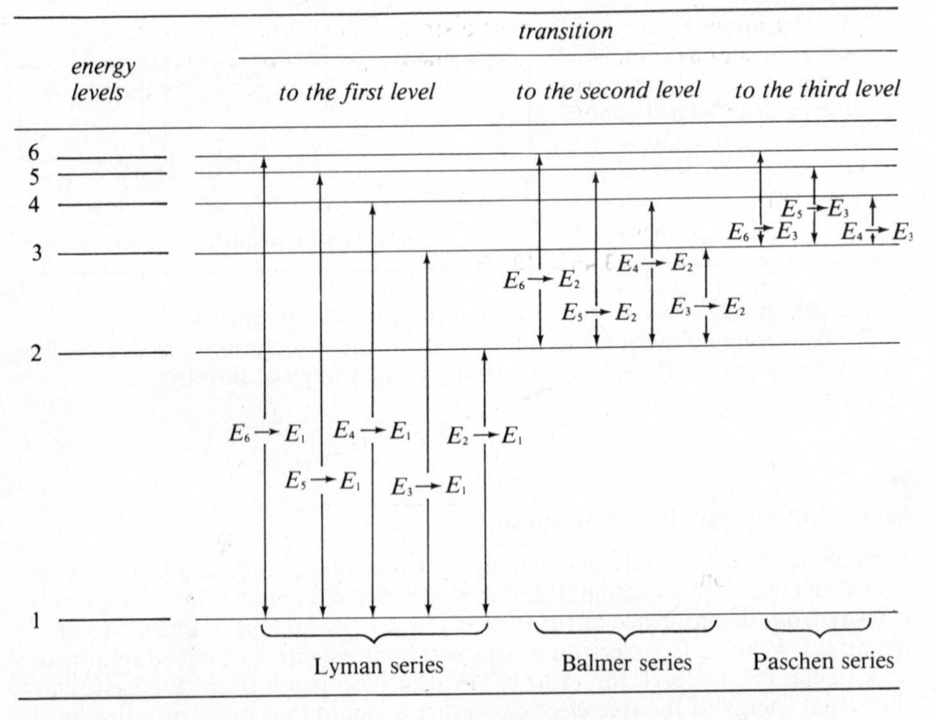

Each series converges towards its own continuum. There is a particular reason for determining the frequency at the start of the Lyman series continuum (transitions to the first level). This emission corresponds to the energy difference between the first energy level of the electron and the energy state where it can break free from the nucleus. It represents the energy required to ionize an atom by electron loss. When this energy is summed over Avogadro's number of atoms, it is called the first ionization energy of the element.

$$X_{(g)} \longrightarrow X^{+}_{(g)} + e^{-}$$

> The first ionization energy (I.E.) of an element is the energy needed to convert a mole of gaseous atoms into a mole of gaseous cations, each atom losing one electron.

The ionization energy of hydrogen is measured by determining the frequency at the start of the Lyman series continuum. This is done as follows.

1. Pick about five successive spectral lines from the Lyman series: measure their frequencies.
2. Let their values be ν_1, ν_2, ν_3, ν_4 and ν_5.
3. Calculate the differences between successive frequencies ($\Delta\nu$), i.e. ($\nu_5 - \nu_4$), ($\nu_4 - \nu_3$), ($\nu_3 - \nu_2$), etc.
4. Plot $\Delta\nu$ against ν.
5. Extrapolate the curve (almost a straight line) back to $\Delta\nu = 0$. This corresponds to the value where one frequency merges into the next: the start of the continuum.
6. Read off the value of ν at $\Delta\nu = 0$.

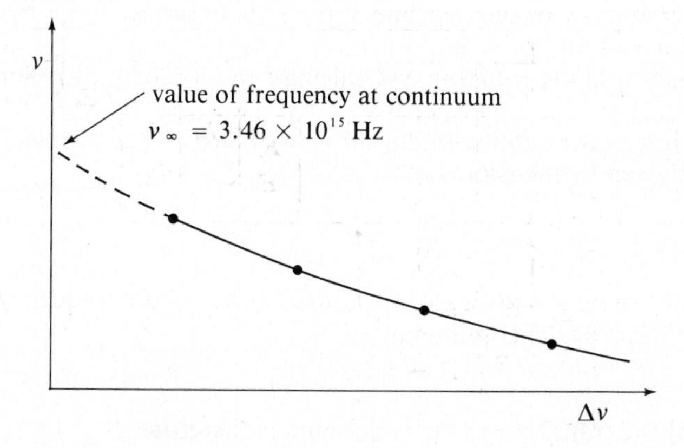

Bohr's equation gives the ionization energy for one atom.

$$\Delta E = h\nu$$

where ΔE = the energy difference between the first level and that required by an electron to escape,

and ν = the frequency at the start of the continuum (3.46×10^{15} Hz).

Multiplying by Avogadro's number gives the ionization energy per mole of atoms.

$$\text{I.E.} = (6.226 \times 10^{-34}) \times (3.46 \times 10^{15}) \times (6.023 \times 10^{23})\ \text{J mol}^{-1}$$
$$= 1300\ \text{kJ mol}^{-1}$$

The Rydberg constant

Sometimes it is not necessary to know as many as four or five spectral frequencies to calculate the ionization energy of an element. It is found that the spacing of the lines in a series obeys a relationship of the form shown below (for hydrogen).

$$\frac{1}{\lambda} = R_H\left(\frac{1}{n^2} - \frac{1}{m^2}\right)$$

where λ = the wavelength of a line in the nth series ($n = 1, 2, 3, \ldots$)
R_H = a constant called the Rydberg constant for hydrogen
m = a whole number greater than n ($m = n + 1, n + 2, n + 3, \ldots$)

For example, for the Lyman series, $n = 1$,

$$\frac{1}{\lambda} = R_H\left(\frac{1}{1^2} - \frac{1}{m^2}\right) \quad \text{where } m = 2, 3, 4, \ldots \infty$$

For the Balmer (visible) series, $n = 2$,

$$\frac{1}{\lambda} = R_H\left(\frac{1}{2^2} - \frac{1}{m^2}\right) \quad \text{where } m = 3, 4, 5, \ldots \infty$$

The Lyman equation can be used to calculate the ionization energy by the following method.

1. Measure the frequency of the first line in the Lyman series.
2. For this line, $n = 1$ and $m = 2$.
3. Put these values into the equation and calculate the Rydberg constant for hydrogen.
4. The spectral line at the continuum has $m = \infty$; therefore its wavelength has the value given by the expression

$$\frac{1}{\lambda} = R_H\left(\frac{1}{1^2} - \frac{1}{\infty^2}\right) \quad \Rightarrow \quad \frac{1}{\lambda} = R_H$$

5. Since $\lambda = c/\nu$, where c = the velocity of light and ν = the frequency of the line at the start of the continuum.

$$\Rightarrow \nu = cR_H$$

6. The energy of the transition at the continuum is therefore

$$\Delta E = h\nu$$
$$= hcR_H$$

7. Since h and c are known constants, and R_H has just been determined, the energy of the transition is found.
8. Multiply by Avogadro's number to give the ionization energy per mole of atoms.

The same method could be based on the second or third line in the series (m is 3 or 4). It is not necessary to use the first line, but it is necessary either to know which line is being measured or to know the values of two consecutive lines in order to be able to obtain a value for R_H.

3.2 Orbitals and shells

Quantum mechanics

At the same time that the experimental research workers were investigating electronic structure, theoretical researchers were trying to construct mathematical equations that would adequately reflect the experimental evidence. Classical mechanics suggested that it should be possible to work out an equation of motion for a moving electron like that which describes the motion of an object moving on the surface of a sphere, shown below.

equation of motion:

$$x^2 + y^2 + z^2 = a^2$$

radius = a

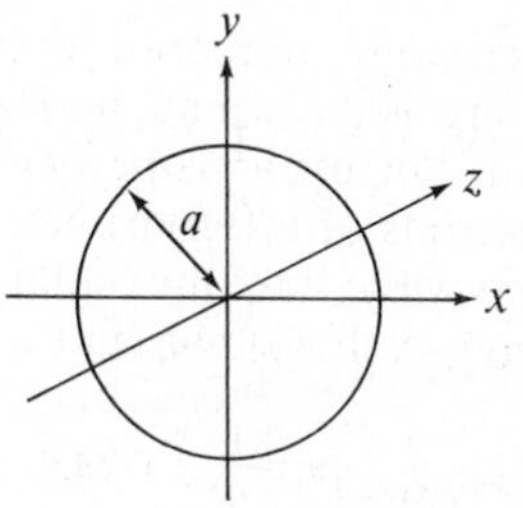

When this approach was applied to the problems of electronic motion, it proved a failure. In the same way that the simple planetary model was found to be inadequate, the mathematics that cope with planetary motion could not explain why only certain energy levels were allowed.

A fresh approach was needed. The most promising solution is as hard to visualize as the concept of fixed energy levels: it assumes that a moving electron has the properties of wave motion. The mathematical techniques that have evolved from this approach have been called 'wave mechanics', or more commonly *quantum mechanics*. They are able to generate equations that predict the quantization of electronic energy. Fortunately it is not necessary for chemists to have to understand the mathematics of quantum mechanics: the solutions provided by the techniques can readily be grasped, and these are the important issues.

The idea of an electron behaving as a moving particle is modified: it should be thought of as a charge cloud of variable density whose limits are described by the wave equation. This equation gives the probability of finding an electron at any point in space around the nucleus. A volume of space in which there is a high probability of finding electron density is called an *orbital*. It is important to be clear that an orbital is a volume of space: it should not be confused with the term 'orbit'.

The solution of the wave equation indicates that each orbital corresponds to a particular energy level. An electronic transition therefore involves an electron moving from an orbital of one energy level to another orbital of a different energy level. A fuller discussion of quantum mechanics is found in *Physical Chemistry* by P.W. Atkins (O.U.P.).

There are four major conclusions that derive directly from the solutions of the wave equation.

1. There exist a number of different types of orbital which differ in shape from one another.

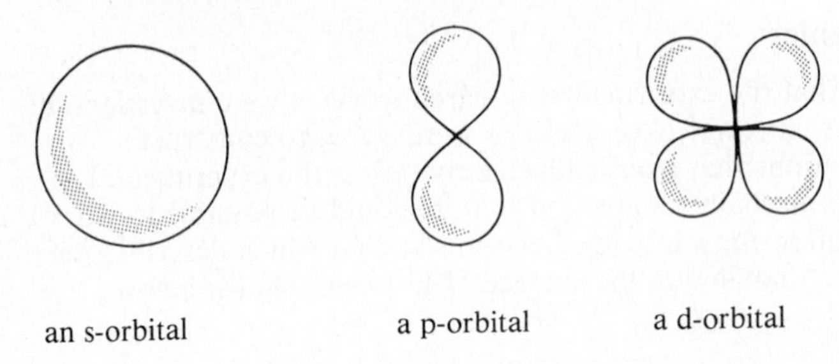

an s-orbital a p-orbital a d-orbital

2. The maximum number of electrons that can be fitted into a single orbital is two: this is known as the *Pauli exclusion principle*.
Electrons possess a property that can be interpreted as *spin*, and so the two electrons of a full orbital make up a 'spin pair'. One spins in one direction while the other spins in the opposite direction. On a diagram, spin is represented by a half arrow.

a full orbital ⇅ a half-full orbital ↿

3. The orbitals occur in a series of clusters called *shells* whose distance from the nucleus gradually increases. The shells are numbered going outwards from the nucleus; they are sometimes represented by letters K, L, M, etc.

nucleus 1st shell K 2nd shell L 3rd shell M

4. The number of orbitals that can be fitted into each shell follows a fixed pattern, as shown in the table below.

shell	*total number of orbitals*	*number of each type of orbital*		
		s	p	d
1	1	1		
2	4	1	3	
3	9	1	3	5

The table shows that:

a) there are n^2 orbitals in the nth shell,
b) the number of each type of orbital present in a shell follows the sequence of odd numbers:

s, p, d, f, . . . 1, 3, 5, 7, . . .

For example, the shapes of the three p-orbitals, p_x, p_y, and p_z, are as shown below.

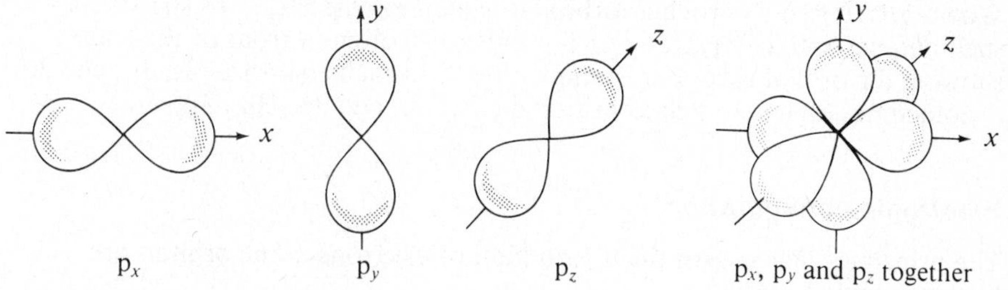

p_x p_y p_z p_x, p_y and p_z together

Shells, orbitals and subshells

An *orbital* is a volume of space in which there is a high probability of finding an electron. An orbital has a fixed energy level.

A *shell* is a group of orbitals whose radial distribution from the nucleus is approximately equal. The energy level of the orbitals in a shell are not necessarily equal.

The three p-orbitals of the second shell (each containing a pair of electrons when full) are all at the same energy level. These p-orbitals therefore represent a separate *subshell* of the second shell. The second shell can contain two subshells altogether, one containing an s-orbital and one containing three p-orbitals: a total of 2^2 (=4) orbitals in the second shell.

The third shell contains three subshells:

a) A subshell containing one s-orbital,
b) A subshell containing three p-orbitals,
c) A subshell containing five d-orbitals;

So the third shell contains three subshells, but 3^2 (=9) orbitals. The orbitals of a subshell have the same energy but different orientations in space as shown below.

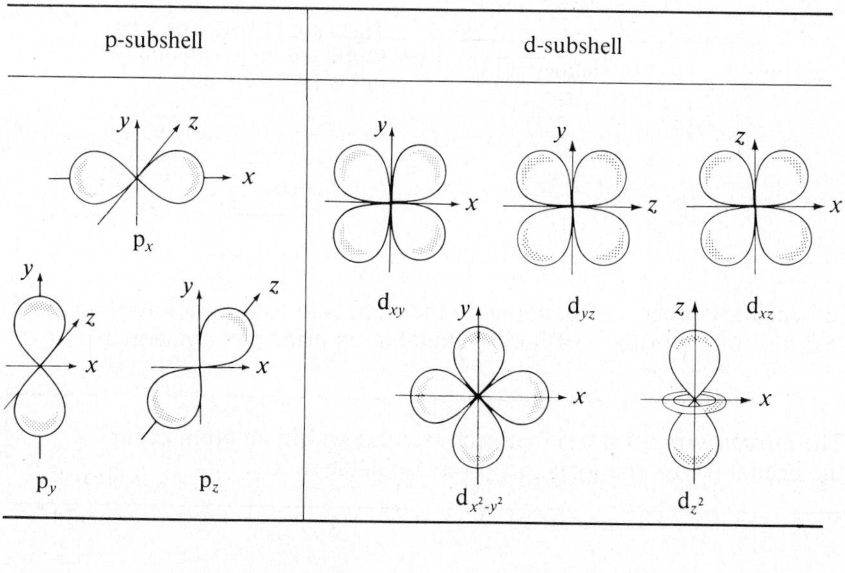

p-subshell	d-subshell
p_x, p_y, p_z	d_{xy}, d_{yz}, d_{xz}, $d_{x^2-y^2}$, d_{z^2}

When referring to a particular orbital it is therefore necessary to specify its shell number and its type. The shell number is written in front of the letter showing the orbital type. For example, the third shell comprises 3s, 3p, and 3d.

Sometimes each sub-shell is written out in full, e.g. $3p_x$, $3p_y$, $3p_z$.

Electronic configuration

The principles that govern the distribution of electrons in the orbitals are largely common sense.

1. Electrons are fitted successively into the orbitals of lowest available energy. An atom whose electrons occupy the orbitals of lowest available energy is said to be in its *ground state*. The order of orbital energies can be remembered from the pattern shown.

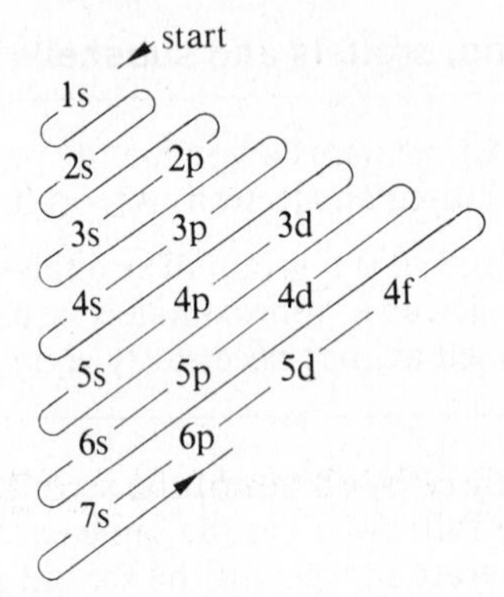

2. *The Hund principle*. When filling a sub-shell, there is less electron repulsion if each of the orbitals is half-filled first before any single one is completely filled.

So, in assigning the electrons of a sodium atom to particular orbitals, start at the 1s-orbital as indicated by the energy diagram and carry on until all the electrons have been assigned.

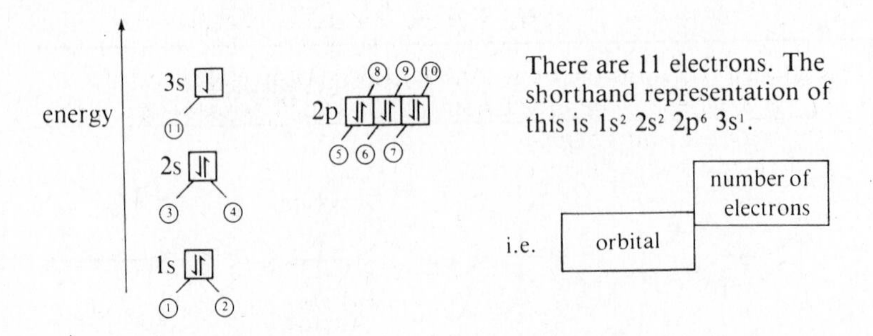

There are 11 electrons. The shorthand representation of this is $1s^2\ 2s^2\ 2p^6\ 3s^1$.

The imaginary process of building up the electronic configuration of an atom is called the aufbau prinzep (meaning 'building-up principle' in German).

The *aufbau prinzep* states that each electron within an atom occupies the orbital whose energy is the lowest available to it.

The principle can be applied to the electrons in any atom: the shorthand way of recording the result is called the *electronic configuration* of the atom. Two more examples are given below.

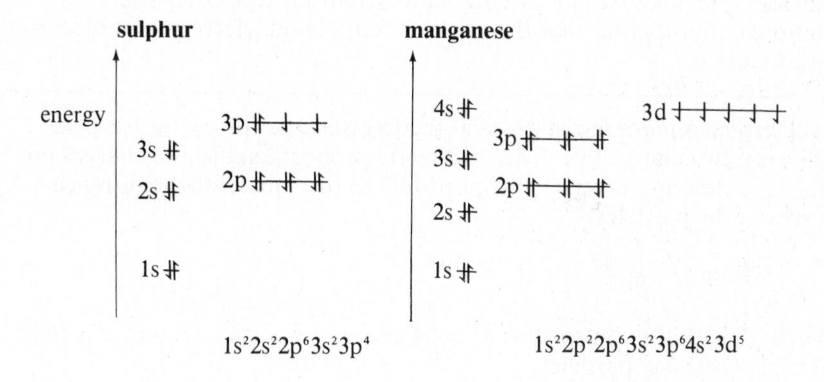

3.3 Ionization energies and electron affinities

Successive ionization energies

Support for the theoretical model of electronic shells is provided by the measured values of successive ionization energies. In the same way that an atom can be ionized by losing an electron, the resulting ion can be further ionized.

$$X_{(g)} \longrightarrow X^+_{(g)} + e^- \quad \Delta H_1 = \text{first ionization energy}$$
$$X^+_{(g)} \longrightarrow X^{2+}_{(g)} + e^- \quad \Delta H_2 = \text{second ionization energy}$$
$$X^{2+}_{(g)} \longrightarrow X^{3+}_{(g)} + e^- \quad \Delta H_3 = \text{third ionization energy}$$

The process could be continued until there are no more electrons left surrounding the nucleus. The graph below shows successive values of the ionization energy for potassium.

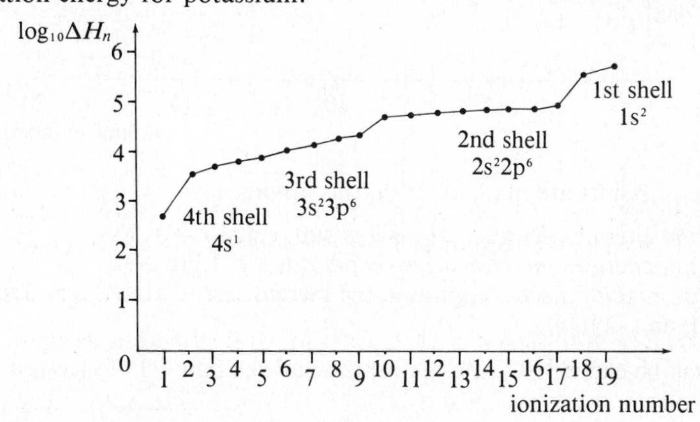

The graph shows clear discontinuities between ionization numbers 1 and 2, 9 and 10, and 17 and 18. This agrees with the theory that the electrons are grouped in shells that have the structure 2, 8, 8, 1. The closer an electron is to the nucleus, the more energy is required to remove it from the atom.

The force of attraction that the nucleus exerts on an electron is subject to the inverse square law.

The *inverse square law* states that the electrostatic force of attraction between two charged particles is directly proportional to the charges on the particles, and inversely proportional to the square of the distance between the particles.

$$F \propto \frac{q_a \times q_b}{d^2}$$

where q_a is charge on particle a; q_b is charge on particle b; and d is the distance between particles.

Variation in first ionization energy

A graph of the first ionization energies for each element against atomic number also provides evidence for the successive build-up of electronic shells.

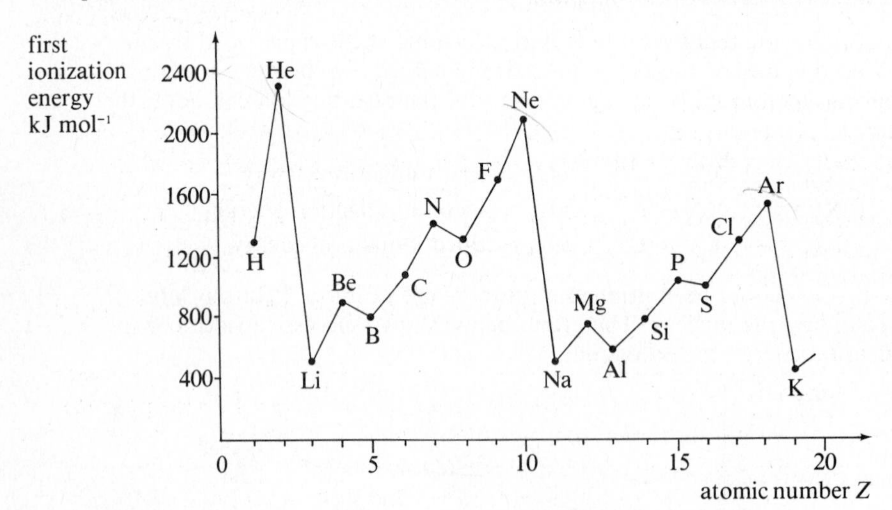

Some important points are made clear on this graph.

1. Ionization energies *decrease down a group*, e.g. Li, Na, K.
2. Ionization energies *increase across a period*, e.g. Li to Ne.
3. There are *regular discontinuities* in the increase trend across a period, e.g. Be to B, and N to O.

Each point can be explained using the orbital and shell model of electronic structure.

1. Decrease down a group

Two atoms in the same group have the same number of outer-shell electrons, but they differ in the number of protons in their nuclei and in the number of shielding or screening inner-shell electrons.

For example, lithium and sodium have one outer-shell electron each. If it is assumed that the inner-shell electrons provide 100% screening, the table below shows the core charge attracting the outer-shell electron in the right-hand column.

atom	*nucleus*	*screening electrons*	*outer-shell electrons*	*core charge attracting the outer-shell electron*
$_3Li$	(+3)	−2	•	+3 −2 = +1
$_{11}Na$	(+11)	−2 −8	•	+11 −8 −2 = +1

The same net charge attracts the outer-shell electrons to the core in both cases but, going down the group, the outer shell is increasingly further from the nucleus. From the inverse square law, this means that the force of attraction between the outer electron and the core becomes steadily less: consequently less energy is needed to remove the electron.

As we shall see later, inner-shell electrons do not completely screen the outer shell from the nucleus, but the effects are not sufficiently marked to upset the trend discussed above.

2. Increase across a period

Atoms of elements in the same period all have their outer-shell electrons in the same shell. They differ in the number of protons in the nucleus and in the number of electrons in the outer shell. Going across a period, the positive charge on the nucleus increases, but the distance of an outer-shell electron from the nucleus remains approximately constant. More energy is therefore needed to remove an electron as the nuclear charge increases.

The table below illustrates this, again assuming 100% screening by the inner-shell electrons.

atom	*nucleus*	*screening electrons*	*outer-shell electrons*	*core charge attracting the outer-shell electrons*
$_3Li$	(3+)	−2	•	+3 −2 = +1
$_4Be$	(4+)	−2	•	+4 −2 = +2

3. The regular discontinuities

The discontinuities can be explained by looking at the orbitals from which electrons are lost. The breaks come at the third and sixth atoms in the periods, i.e. $s^2p^1 \rightarrow s^2$ and $s^2p^4 \rightarrow s^2p^3$ require less energy than expected.

First compare the ionization of the second and third atoms in a period.

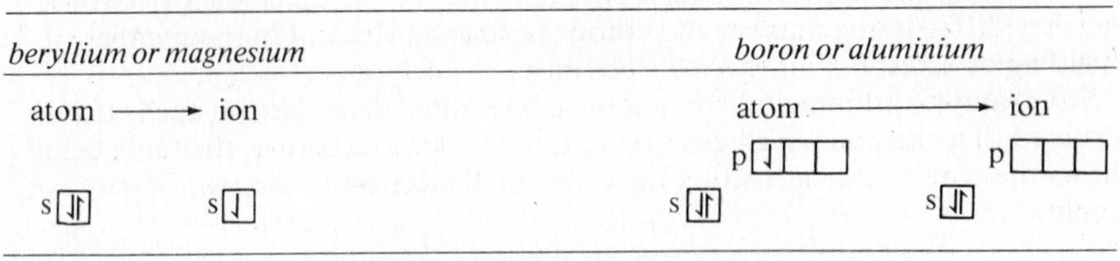

beryllium or magnesium		*boron or aluminium*	
atom →	ion	atom →	ion
		p [↓] [ ] [ ]	p [ ] [ ] [ ]
s [↓↑]	s [↓]	s [↓↑]	s [↓↑]

The electron lost by boron or aluminium comes from a p-orbital. This electron is at a higher energy already than those in the corresponding s-orbital, and therefore less energy is required to remove it.

Similarly compare the ionization of the fifth and sixth atoms in a period.

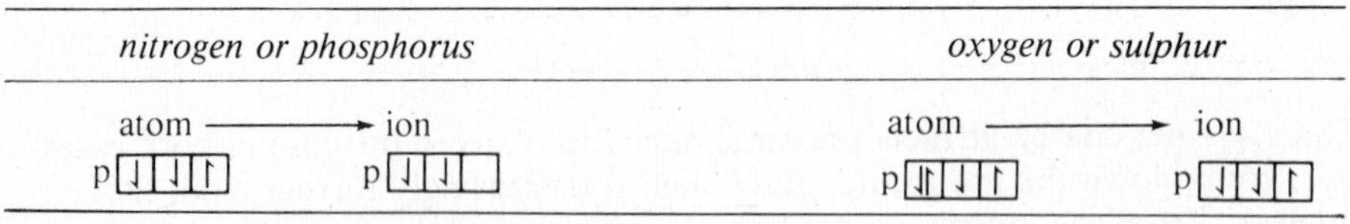

nitrogen or phosphorus		*oxygen or sulphur*	
atom →	ion	atom →	ion
p [↓] [↓] [↑]	p [↓] [↓] [ ]	p [↓↑] [↓] [↑]	p [↓] [↓] [↑]

The electron lost by oxygen or sulphur comes from a paired p-orbital. This electron is at a higher energy than the p-electrons of the nitrogen or phosphorus atom because of the repulsion effects (the Hund principle — page 40). Less energy is required to remove it in spite of the increase in nuclear charge.

Electron affinity and electronegativity

An ionization energy is a measure of the attractive force that a nucleus has for its own outer-shell electrons. There is another similar energy term that is concerned with the force of attraction that a nucleus can exert on other electrons external to the atom. It is called the *electron affinity* (E.A.) of the element; like ionization energy, there are successive electron affinities for any element.

> The *first-electron affinity* of an element is the energy change when one mole of gaseous atoms is converted into one mole of gaseous anions by gaining one electron per atom.
>
> $X_{(g)} + e^- \rightarrow X^-_{(g)}$ $\quad \Delta H$ = first-electron affinity

Most first-electron affinities are exothermic. A neutral atom has some attraction for an electron because of the incomplete screening of the nucleus by its own electrons. Second and subsequent electron affinities are all endothermic because the process involves the bringing together of two negatively charged particles.

A third concept is commonly used to describe the ability of an atom to attract electrons to itself; it is called electronegativity. The electronegativity of an element is not an energy term and it has no units.

> The *electronegativity* of an element is defined as a numerical value for the attracting power that an atom of the element has for bonding and non-bonding electrons in its outer shell.

The most widely used scale is that worked out by Pauling; values for the first twenty elements are shown below.

			$H_{2.1}$				He
$Li_{1.0}$	$Be_{1.5}$	$B_{2.0}$	$C_{2.5}$	$N_{3.0}$	$O_{3.5}$	$F_{4.0}$	Ne
$Na_{0.9}$	$Mg_{1.2}$	$Al_{1.5}$	$Si_{1.8}$	$P_{2.1}$	$S_{2.5}$	$Cl_{3.0}$	Ar
$K_{0.8}$	$Ca_{1.0}$						

The higher the value, the stronger is the attractive force exerted; values range from 0.7 to 4.0.

3.4 Types of atom

Metallic atoms

Metallic atoms are characterized by their low degree of outer-shell electronic control. Metals have low ionization energies and electronegativity values usually less than 2.0. In the Periodic Table three blocks of metals can be identified.

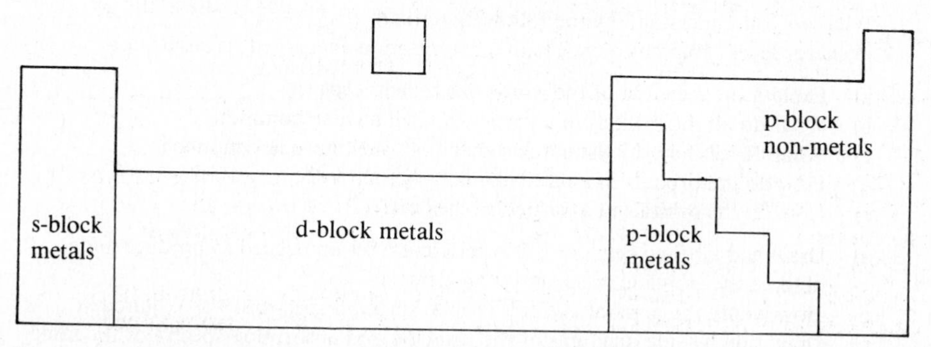

In any particular period, a metallic atom is likely to be larger than a non-metallic atom. The low electronic control leads to the outer shell being spread further from the nucleus.

Non-metallic atoms

Non-metallic atoms are characterized by a high degree of electronic control. They have electronegativity values greater than 2.0 and appear as a block on the right of the Periodic Table. Non-metallic atoms are smaller than metallic atoms of the same period.

	Li	Be	B	C	N	O	F
covalent radius m $\times 10^{-10}$	1.23	1.06	0.88	0.77	0.70	0.66	0.64

approximate relative atomic sizes

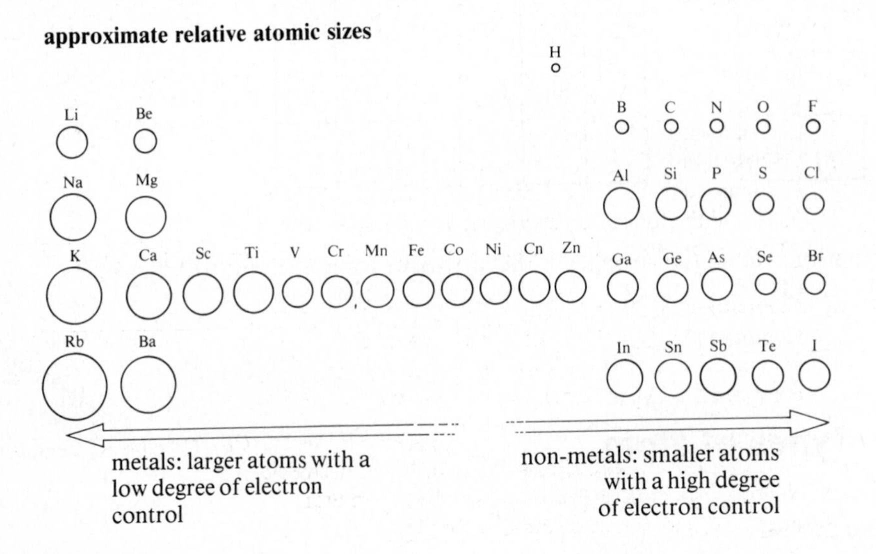

Questions

1. What do you understand by the following terms?
   a) energy level b) convergence limit c) ionization energy d) ground state

2. a) Explain the meaning of the words *orbital* and *shell*.
   b) What do all the orbitals in a particular shell have in common?
   c) What do all the orbitals in a particular sub-shell have in common?
   d) How do the orbitals in a particular sub-shell differ?
   e) How do the orbitals in a particular shell differ?

3. a) Draw and label a sketch of the experimental set up needed to produce and observe the emission spectrum of an element.
   b) How would the apparatus differ if an absorption spectrum was required?
   c) Draw side-by-side diagrams of the emission and absorption spectra of the same element.
   d) What are the two main differences between the spectrum of white light and the emission spectrum of an element?

4. a) Sketch the visible emission spectrum for the element hydrogen, labelling the colour of the lines.
   b) Using this spectrum as your example, explain what conclusions can be drawn about the electronic structure of atoms from the observation of atomic emission spectra.

5. a) Three rules or principles are used in working out the electronic structure of the atom of an element in its ground state. Name each of these principles and explain what they mean.
   b) Write down the electronic configuration of the atoms whose elements have the following atomic numbers:
   i) 1; ii) 7; iii) 21; iv) 23; v) 28; vi) 31; vii) 45; viii) 53;
   c) Name the elements whose atoms have the following electronic structures:
   i) $1s^2\ 2s^2\ 2p^6\ 3s^1$ ii) $1s^2\ 2s^2$ iii) $1s^2\ 2s^2\ 2p^6\ 3s^2\ 3p^6\ 4s^1$ iv) $1s^2\ 2s^2\ 2p^6\ 3s^1$
   v) $1s^2\ 2s^2\ 2p^2$

6. a) Write down the electronic configuration of the atoms of the elements in group V.
   i) Compare their outer shells.
   ii) Compare the number of shells each atom has.
   iii) Compare the size of the atoms going down the group.
   iv) How will the ionization energy change going down the group?
   b) Write down the electronic configuration of the atoms of the elements in the second period Li → Ne.
   i) Compare the number of electrons in the outer shell.
   ii) Compare the number of shells each atom has.
   iii) Compare the size of the atoms going across the period.
   iv) How will the ionization energy change going across the period?

7. The sulphide ion S^{2-} has 18 electrons. So does the chloride ion Cl^-. Particles with the same total number of electrons are said to be isoelectronic.
   a) Write down five ions isoelectronic with the neon atom.
   b) Put these particles in increasing order of size.

8. Carefully sketch a graph of the first ionization energies for the first 20 elements against mass number.
   a) Mark in on the graph different periods.
   b) Explain the changes in ionization energies across the second period.
   c) Explain the changes in ionization energies down a group. Carefully sketch a graph of the 11 successive ionization energies of sodium against ionization number using a log scale for the ionization.
   d) Mark in the different shells of electrons.
   e) Explain what this graph tells about the distribution of electrons around a sodium nucleus.

9. Compare clearly and concisely the characteristics of a typical metallic atom and a typical non-metallic atom.

Chapter 4

The attraction between atoms

4.1 The electrostatic nature of the forces

The interaction of two atoms

In chapters 2 and 3, the forces between the particles present in an isolated atom are discussed. In this chapter, we are concerned with the forces that result from the interaction of one atom with another.

First consider a hypothetical situation in which two isolated atoms are slowly brought together. The positive and negative parts of each atom begin to interact with one another. In the diagram below, only one electron is shown to be present in each charge cloud.

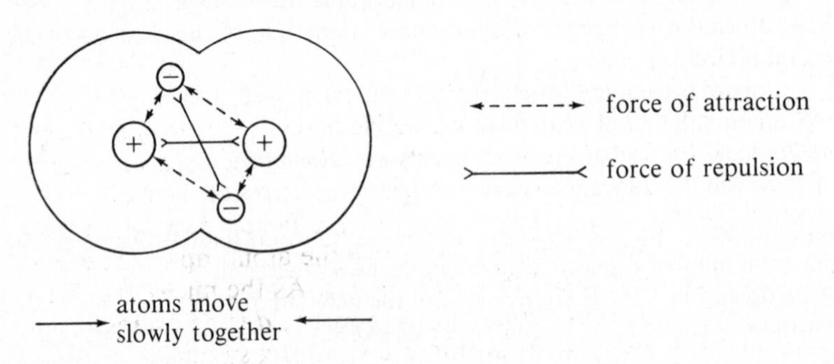

Two possible outcomes can result from the interaction.

a) The forces of repulsion outweigh the forces of attraction.

Here work must be done to move the atoms together. So the potential energy of the two-atom system increases sharply as the two atoms are brought closer together.

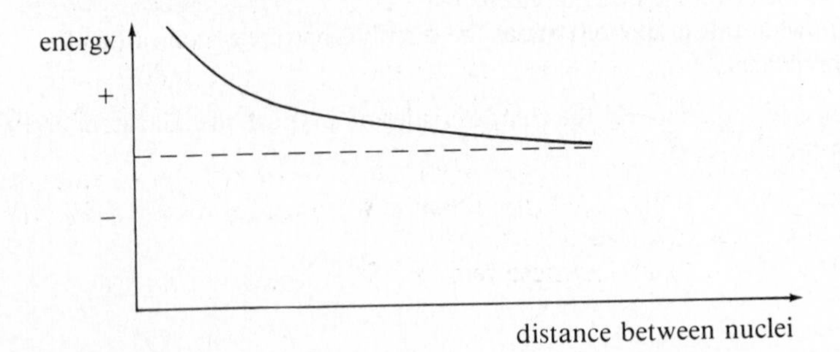

b) The forces of attraction outweigh the forces of repulsion.

As the atoms come closer together, the potential energy of the system must fall to a minimum and then rapidly increase as the distance between the nuclei becomes very small. The drop in potential energy corresponds to the amount of work that would have to be done to break the atoms completely apart.

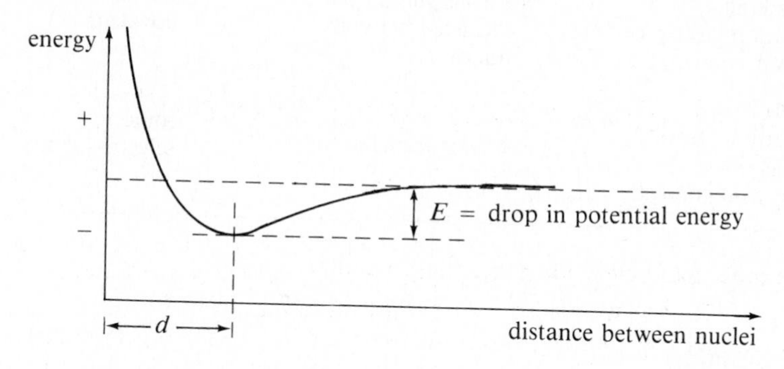

d = the distance between nuclei at the minimum potential energy
E = the drop in potential energy caused by the reaction

The chemical bond and bond energy

A *chemical bond* exists whenever the attractive forces between nuclei and electrons outweigh the repulsive forces between the nuclei themselves and between the electrons themselves.

The potential energy drop *E* is a measure of the *bond energy*; it is the amount of work that must be supplied to force the atoms apart. The internuclear distance *d* is often called the *bond length*. As the nuclei are likely to be in a state of vibration about some mean position, *d* is an average value which depends on the total energy within the vibrating system.

Two different bonding possibilities can be isolated from the general pattern.

1. The two nuclei have approximately equal forces of attraction for the bonding electrons.
2. The attraction of one nucleus for the bonding electrons is much greater than that of the other nucleus.

The first case results in the *sharing* of electrons between the nuclei. If the two atoms are large with a low degree of electron control (i.e. they are metallic atoms), the shared electrons are *delocalized* between many nuclei: this is known as *metallic bonding*. If the atoms are small with a high degree of electron control (i.e. they are non-metallic atoms), the electrons are *localized* between two nuclei: this is called *covalency*.

The second case leads to the transfer of an electron or electrons from one atom to another: this is called *electrovalency*. Two oppositely charged particles are produced and these are called *ions*.

The table overleaf summarizes these results.

types of atoms	*electron situation*	*bonding*
both metallic (large with low degree of electron control)	electrons shared but delocalized	metallic
both non-metallic (small with high degree of electron control)	electrons shared but localized between nuclei	covalent
one metallic (low electron control) one non-metallic (high electron control)	electrons transferred	ionic (electrovalent)

The three diagrams below illustrate these results.

Metallic bonding

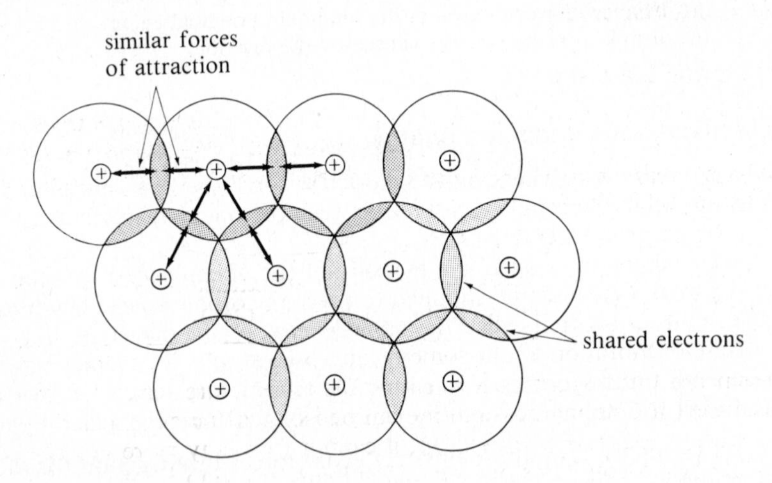

Covalent bonding

A group of atoms that are covalently bonded together is called a *molecule*.

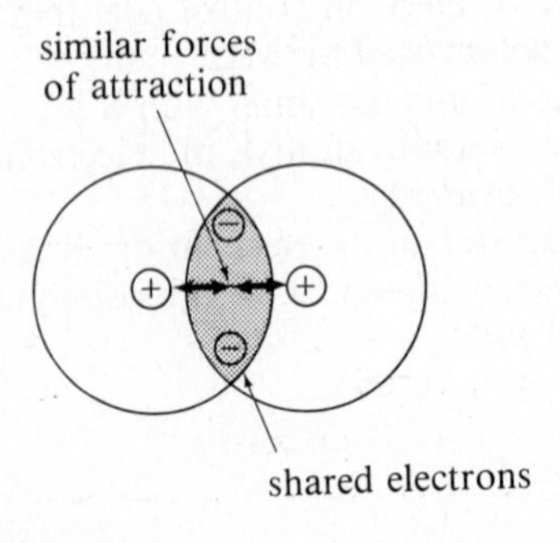

Ionic bonding

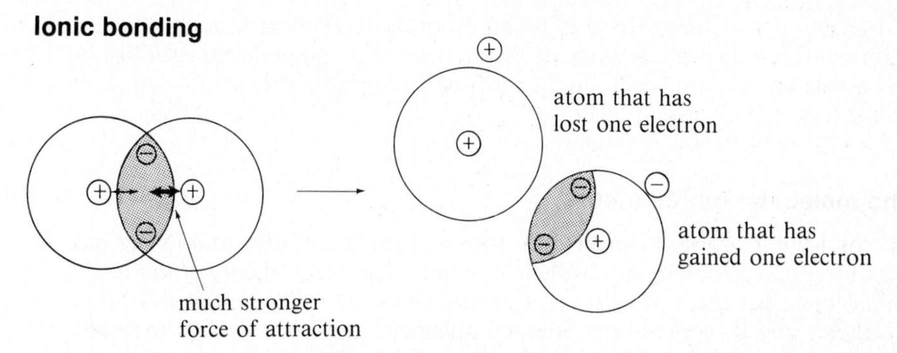

Metallic bonding is described in detail in Chapter 5. The rest of this chapter explores the possibilities of covalency and electrovalency in more detail.

4.2 Covalency

The valence-bond theory

When two atoms interact with each other, it is their outer shells that come into contact. The valence-bond theory (V.B. theory) describes the outer shell of an atom as its 'valence shell' because it is the electrons in this shell that are involved in the formation of bonds.

> A *covalent bond* is the electrostatic force of attraction between two nuclei and a shared pair of electrons between them.

An atom can form as many covalent bonds as there are spaces left in its valence shell for electrons. Once the valence shell of an atom is full or all its valence electrons are used up, it cannot form any more bonds.
The table below compares oxygen and sulphur in group VI.

oxygen	*sulphur*
There are 6 valence-shell electrons in the outer shell whose maximum is 8. There are only two spaces. So oxygen forms two bonds only.	There are 6 valence-shell electrons in the outer shell whose maximum is 18. There are more than 6 spaces. So sulphur can form as many as six bonds.
H_2O structures: :O(·×)H with (·×)H, or H–O–H	O=S(=O)=O (6 bonds), O=S=O (4 bonds), H–S–H (2 bonds)

The number of bonds formed by an atom of an element in a particular compound is called the *valency* of the element. For example, oxygen always has a valency of two, while sulphur exhibits a variable valency of two, four, or six.

The molecular orbital theory

The molecular orbital theory (M.O. theory) is an alternative and rather more exacting way of looking at covalent bonding. The M.O. theory is *not* a replacement for the V.B. theory: it concentrates on a different aspect of covalency and its conclusions often complement those of the V.B. theory. Both theories are used in later parts of this book. The choice of which model to pick is governed by the bonding features that need emphasis.

Consider again the approach of two isolated atoms. As they get closer together, the orbitals of the outer shell of one atom start to overlap with the orbitals of the other atom. An electron in the region of the overlap comes under the influence of both nuclei and hence cannot be said to be in an 'atomic' orbital any longer: instead it is in a 'molecular' orbital. The principles that govern the formation of molecular orbitals are given below; the general principle is called the linear combination of atomic orbitals (l.c.a.o.).

1. The overlap of N atomic orbitals produces N molecular orbitals, e.g. 3 atomic orbitals overlap to give 3 molecular orbitals.
2. The average energy of the M.O.'s produced by l.c.a.o. is the same as the average energy of the overlapping A.O.'s.
3. The M.O.'s are filled with electrons in accordance with the aufbau and Hund principles (just like atomic orbitals).

These three principles are best illustrated by examples.

Example a) two hydrogen atoms, $1s^1$ and $1s^1$.
The two 1s-orbitals overlap and produce two molecular orbitals. Their distributions are shown below.

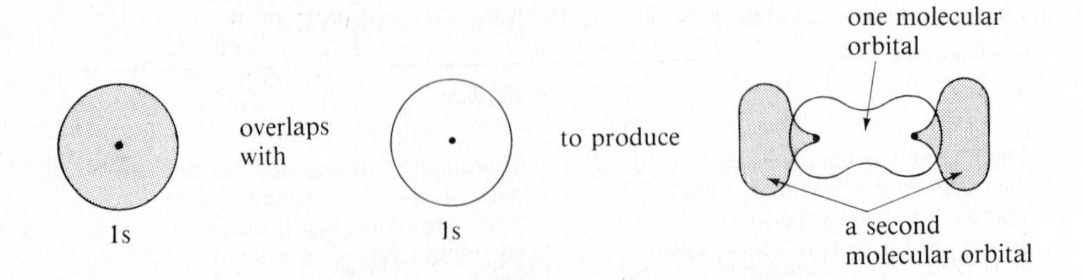

The two molecular orbitals are different both in character and in energy. One is split into two volumes spreading out and away from the nuclei; the other covers mainly the volume of space between the nuclei. Their average energy equals the energy of the 1s-atomic orbitals.

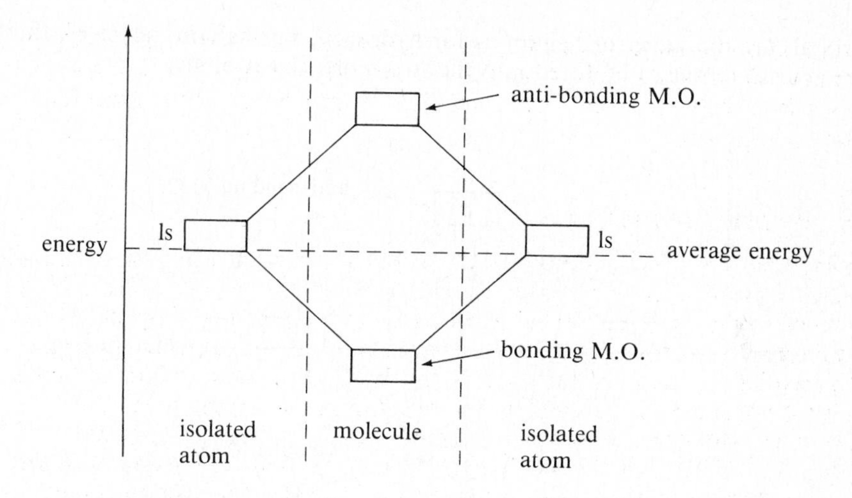

The bonding M.O. is the one that occupies the volume of space between the nuclei. Electron density in this orbital leads to a bonding effect. The split-volume M.O. is anti-bonding because it leaves the two atomic cores facing each other with no shielding electrons between them. This results in repulsion and the tendency for the system to revert to separate atoms.

The M.O.'s are filled according to the aufbau principle. For a hydrogen molecule, there are two electrons which both go in the bonding M.O.

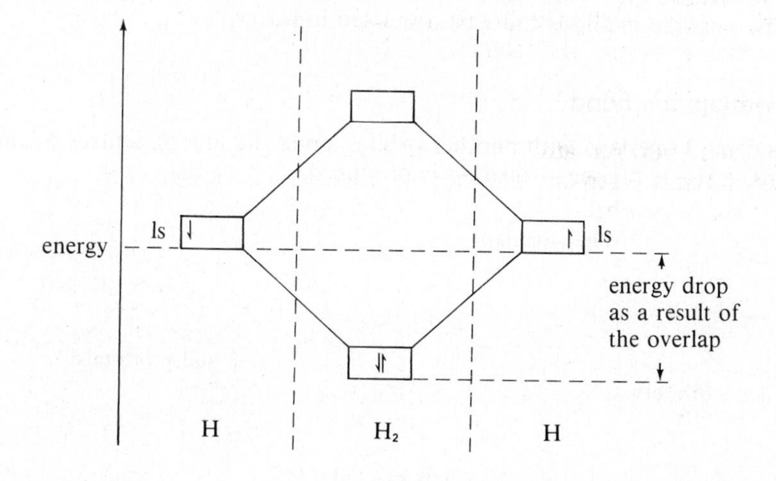

The overlap and subsequent filling of a bonding M.O. leads to a drop in the system's potential energy which is equivalent to that shown in the graph on page 49. The bond energy must be supplied to the electrons in a bonding M.O. to excite them up to the energy level of the separate atomic orbitals.

Example b) two helium atoms, $1s^2$ and $1s^2$.
Exactly the same molecular orbitals are produced because the two atomic

orbitals are the same for helium as for hydrogen. For helium, however, there are four electrons to be fitted into the two molecular orbitals.

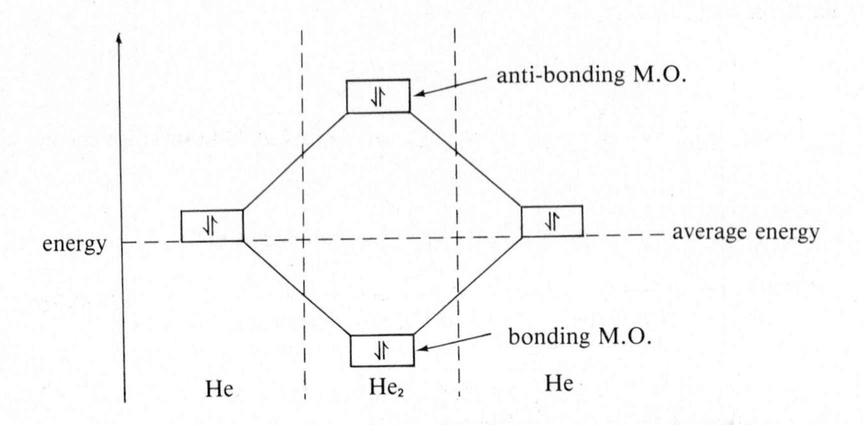

Both the bonding and the anti-bonding orbitals are occupied. The net effect is *non-bonding*: there is no net energy drop as a result of the distribution of electrons in the M.O.'s formed. In other words, He_2 is unlikely to exist whereas H_2 is very likely to be formed.

These two simple examples emphasize the general principles of M.O. theory. Since there is more than one way in which two atomic orbitals can overlap, each of the possible methods must be discussed in turn.

Single overlap: a σ-bond

When an orbital overlaps with another orbital along the line of centres of the two atoms, a single region of overlap is produced.

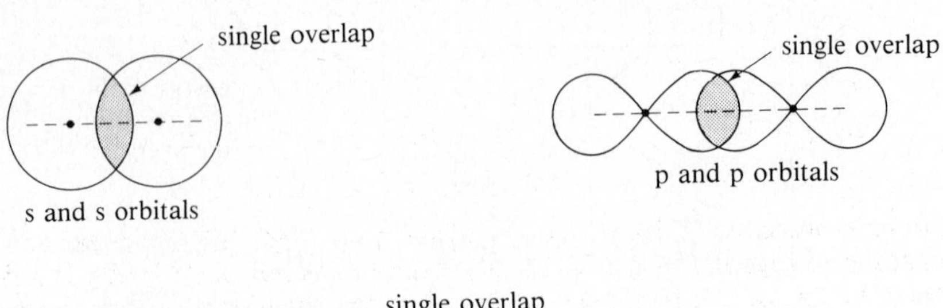

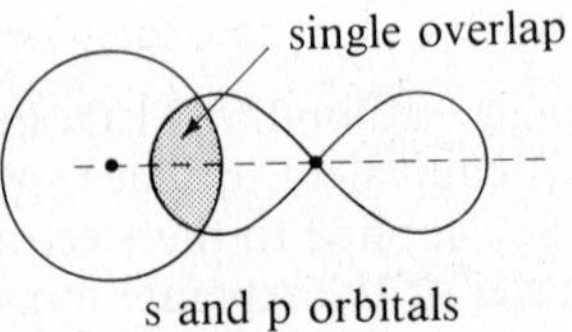

A single overlap of this sort results in the production of a bonding-M.O. which fills the volume between the nuclei. The presence of electrons in this orbital leads to high electron density between the nuclei which is called a *sigma bond* (σ-bond): 'σ' stands for single overlap and a σ-bond is the equivalent of a single bond as understood by V.B. theory. For example, the hydrogen molecule contains a σ-bond.

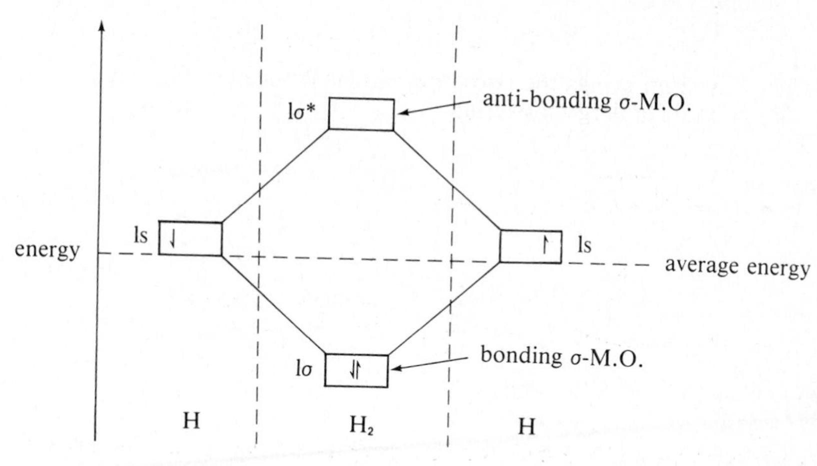

Double overlap: a π-bond

There is another way in which p-orbitals and d-orbitals are able to overlap: these multi-lobed orbitals can overlap sideways on as well as end-to-end. If they overlap sideways on, the overlap is in two separate volumes *not* along the line of centres.

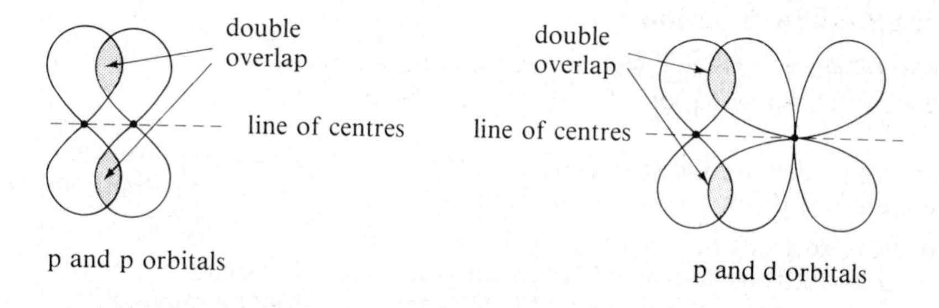

p and p orbitals

p and d orbitals

A double overlap of this sort can only take place when the two nuclei are already held together by a σ-bond. It produces a bonding-M.O. above and below the line of centres of the atoms. The presence of electrons in this orbital leads to high electron density in two separate volumes above and below the σ-bond. This is called a *pi bond* (π-bond).

Example: π-bonding in the nitrogen molecule

1. The electronic configuration of a nitrogen atom is $1s^2\ 2s^2\ 2p^3$. There are three unpaired electrons in the 2p sub-shell of each atom.
2. When the two nitrogen atoms approach, their $2p_x$-orbitals overlap and a σ-bond forms (two electrons in the bonding σ-M.O.).

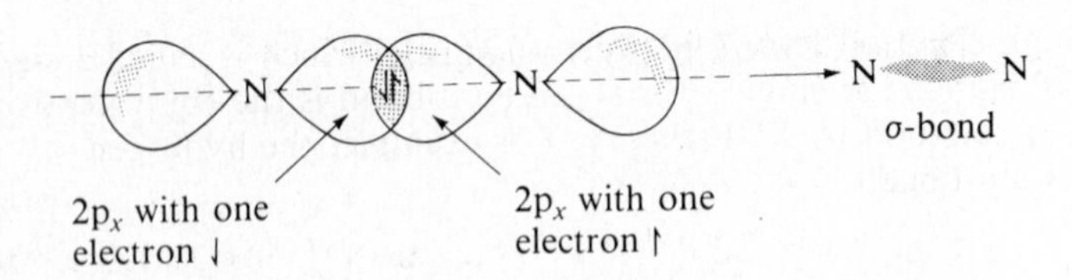

3. The σ-overlap brings the two atoms so close together that their $2p_y$- orbitals undergo π-overlap.

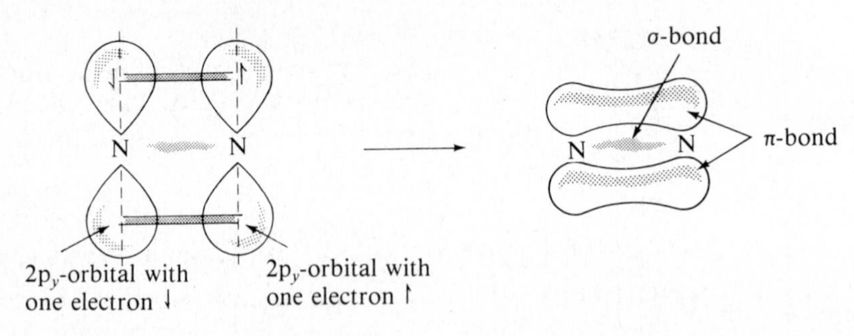

4. The $2p_z$-orbitals also undergo π-overlap to produce a second π-bond.

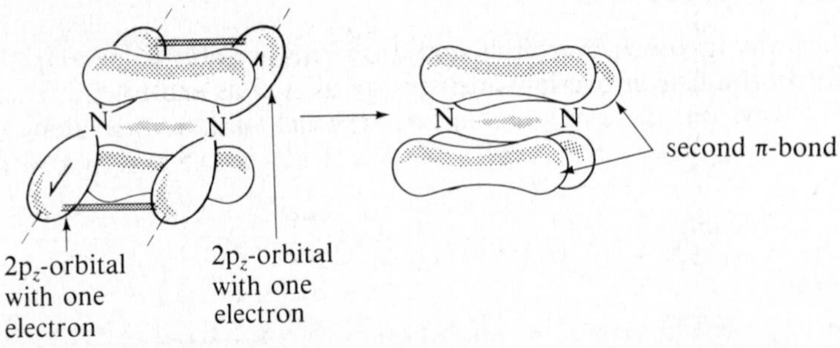

Notice that each π-bond is in two separate volumes surrounding the σ-bond. There are five distinct volumes of electron density in the nitrogen molecule.

a) the σ-electrons between the nuclei: a σ-bond,
b) the π-electrons above and below the σ-electrons: a π-bond,
c) the π-electrons in front of and behind the σ-electrons: another π-bond.

Compare the M.O. model of N_2 with the V.B. picture of a triple-bonded molecule.

:N (:x)(:x)(:x) N ⁑ or (:N≡N:)

The 'triple bond' is unable to give any information about the distribution of the electrons in the molecule.

The energy diagram for the M.O. model of N_2 is shown below.

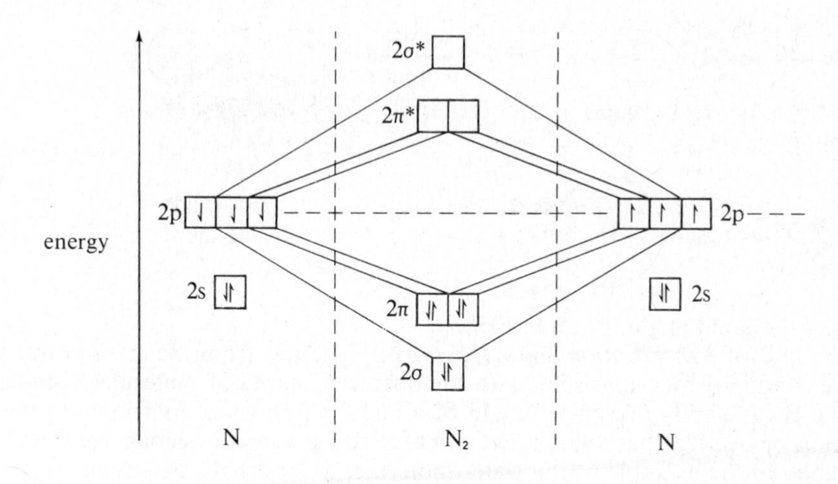

Only the outer shell is put in the diagram. There is one pair of electrons left on each atom that takes no part in the bonding. This is called a lone pair of electrons.

A *lone pair* is an outer-shell pair of electrons under the influence of only one nucleus. In this book a lone pair is shown: (:)

Molecular geometry and hybridization

The ideas of orbital overlap are summarized below.

1. Atomic orbitals overlap to form molecular orbitals.
2. The filling of the molecular orbitals can lead to a drop in the electronic energy-level of the system, i.e. a 'bond' forms.
3. There are three types of electron-pairs:

   a) σ-pairs which are trapped between the nuclei,
   b) π-pairs which are above and below the σ-electrons,
   c) lone pairs which are under the influence of one nucleus.

It should be possible to predict the shapes of molecules from the overlap patterns. As far as the theory has been developed so far, a water molecule would be expected to have a bond angle of 90° as shown by the diagram on page 58.

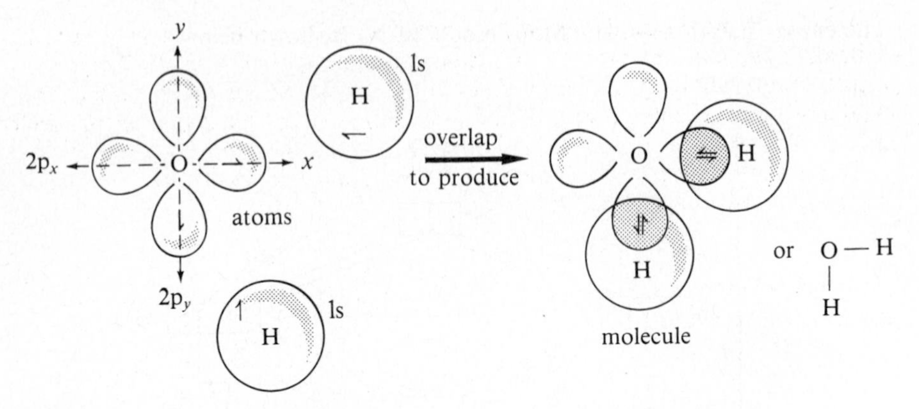

In fact the bond angle is nearer 105°.

The idea of *hybridization* and *hybrid orbitals* comes from the attempt to make overlap theory consistent with the observed shapes of molecules such as water. By assuming that the strength of a bond is directly proportional to the amount of overlap that takes place, a more critical view of overlap becomes possible. In the example of the water molecule, at least half the volume of the 2p-orbitals of the oxygen atom is unavailable as far as bonding is concerned. This suggests that the approach may need modification.

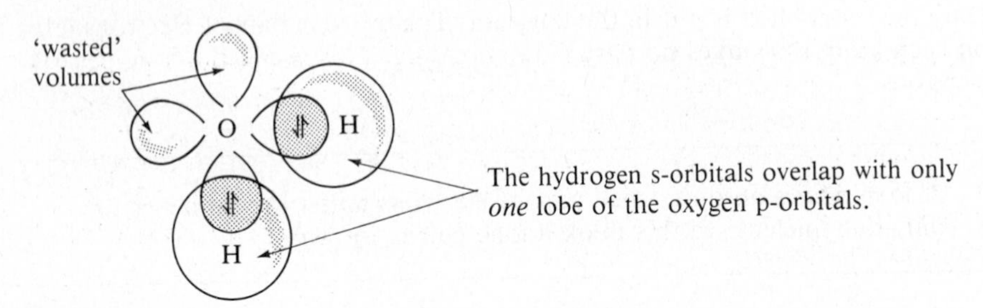

The existence of s-, p- and d-orbitals is derived from the solution of lowest energy to the equation of motion for an electron in an isolated atom. However, it is possible to find a less favourable solution which defines an alternative set of orbitals. These orbitals have higher energy levels than those of the s-, p- and d-orbitals of the corresponding shell and are known as *hybrid* orbitals because their properties are interpreted as a mixture (or hybrid) of s-, p- and d-character.

For example, the shape of an sp-hybrid orbital is:

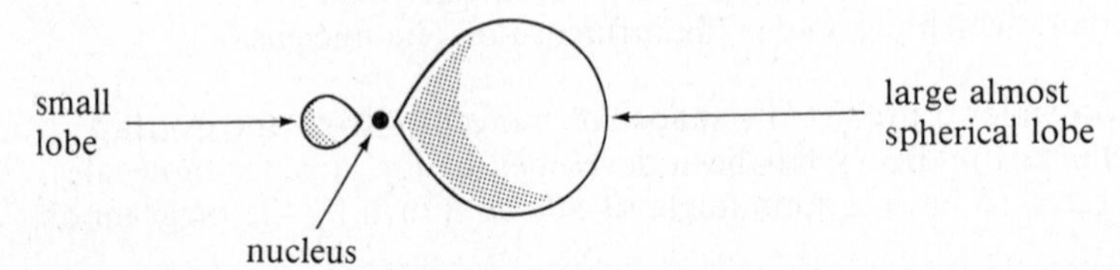

There is less wasted volume when an sp-orbital overlaps with another atomic orbital. Only the smaller lobe is left as wasted volume.

A set of hybrid orbitals becomes occupied in preference to the pure atomic orbitals if the amount of overlap increases appreciably as a result. The increased overlap causes a larger drop in the energy of the system so that the whole process becomes energetically favourable despite the extra energy needed initially for the hybrid orbitals to be filled.

These ideas are easier to understand when applied to the example of the water molecule considered earlier.

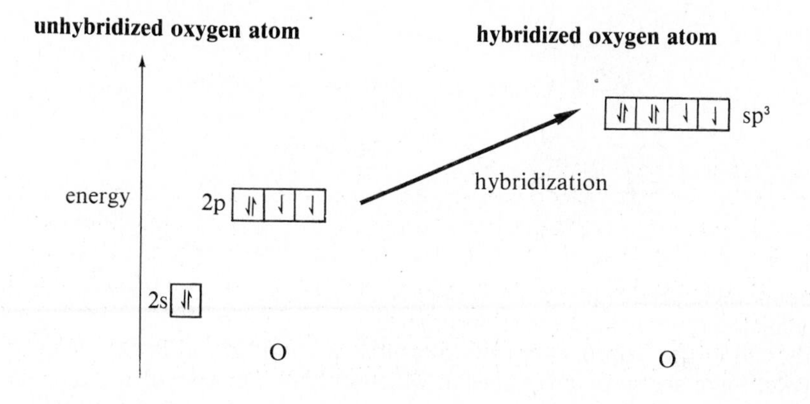

The hybrid orbitals here are called sp^3 because they derive from one s-orbital and three p-orbitals. They are of higher energy than the 2s- and 2p-orbitals and so are not occupied in an isolated oxygen atom. They have better overlapping properties than p-orbitals, however, and therefore it can become favourable for them to be occupied under bonding conditions. The M.O.'s formed from the overlap of the hydrogen 1s-orbital in the two possible cases are compared below.

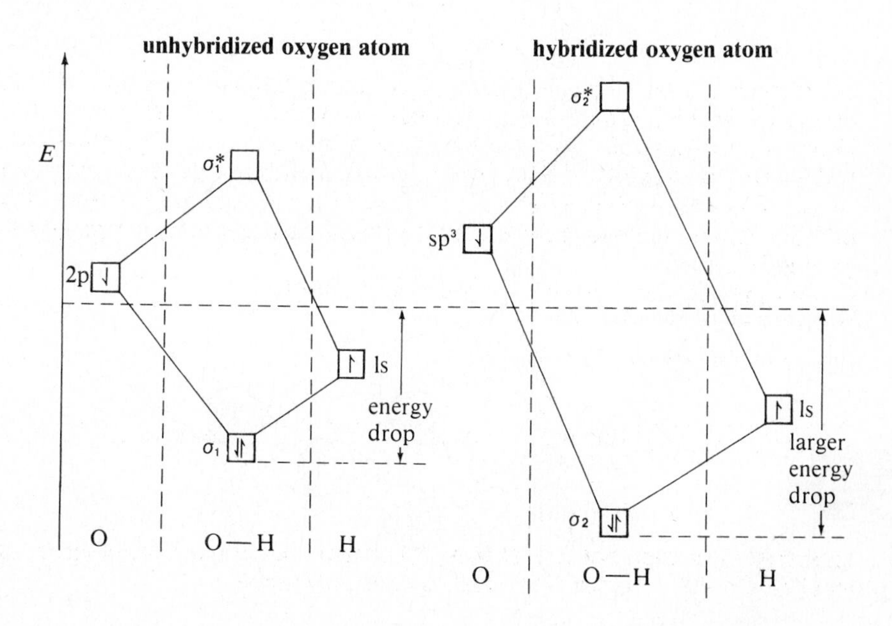

Even though the sp³-orbital is of higher energy than the 2p-orbital, its occupation leads to a lower energy M.O. because σ_2 is of lower energy than σ_1. This is because there is a much greater energy drop due to the more extensive overlapping of the sp³-orbital with the hydrogen 1s-orbital.

There are four equivalent sp³-orbitals in a sub-shell and these are distributed symmetrically around the nucleus. Shown below is the hybridized picture of a water molecule.

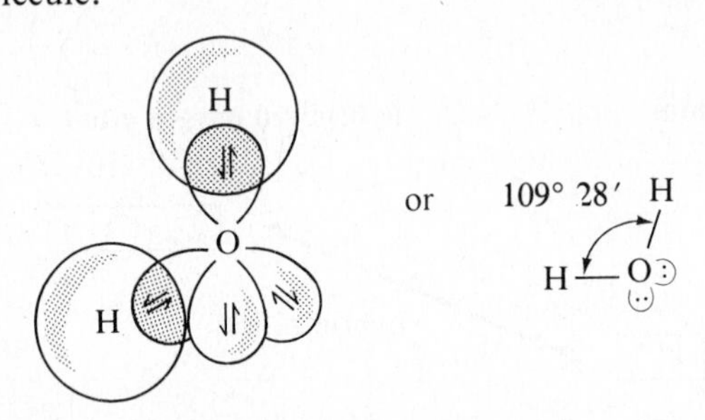

The bond angle predicted by this model is clearly a great improvement on the 90° predicted by the simple overlap model.

Although in this approach all the p-orbitals are hybridized in the water molecule, there are some molecules in which only one or two of the p-orbitals become hybridized. In these molecules, the unhybridized p-orbitals usually undergo π-overlap to produce a π-bond. A simple example is an ethene molecule: $\mathrm{H_2C{=}CH_2}$

looking 'sideways' at the molecule (note $2p_y$-orbitals unhybridized)

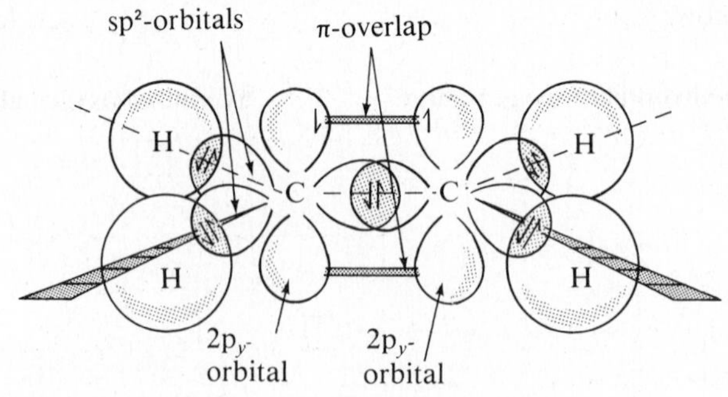

view from above the molecule

the $2p_y$-orbitals are not shown

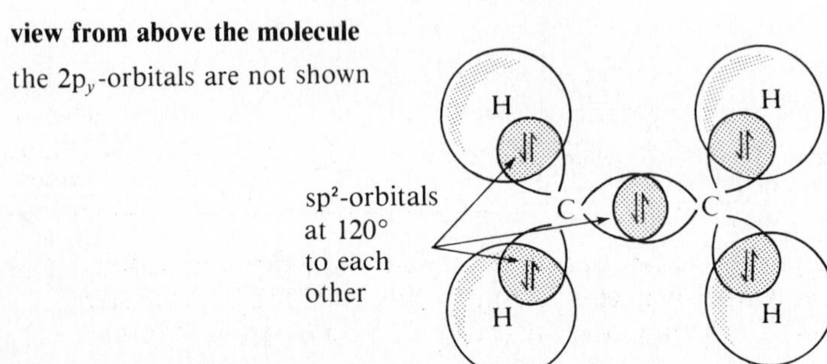

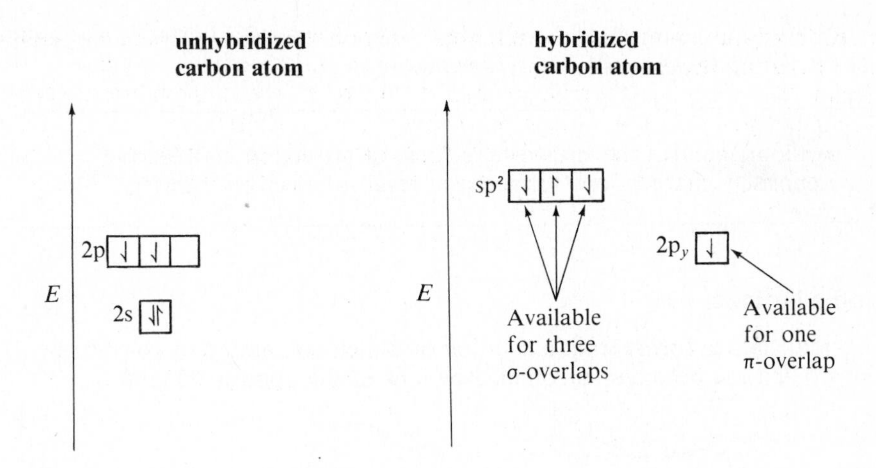

4.3 Electrovalency

Electron transfer and ions

Of the two bonding possibilities outlined on page 49, the second is the easier to visualize. When the attraction of one nucleus for a bonding pair of electrons greatly exceeds the other's, electron transfer occurs and ions form.

Metal atoms have low electronegativities and electron-deficient valence shells. They tend to lose their electrons to non-metal atoms which have higher electronegativities and valence shells that are almost full. The number of electrons lost by a metallic atom or gained by a non-metallic atom is known as its *electrovalency* or *valency* for short.

The ions that form after electron transfer usually have the electronic configuration of noble gas atoms; d-block ions are the most notable exceptions.

metals	Na	$1s^22s^22p^63s^1$	loses 1 electron	$1s^22s^22p^6$	Na^+	0.95	ionic radius $m \times 10^{-10}$
	Mg	$1s^22s^22p^63s^2$	loses 2 electrons	$1s^22s^22p^6$	Mg^{2+}	0.65	
	Al	$1s^22s^22p^63s^23p^1$	loses 3 electrons	$1s^22s^22p^6$	Al^{3+}	0.50	
non-metals	F	$1s^22s^22p^5$	gains 1 electron	$1s^22s^22p^6$	F^-	1.36	ionic radius $m \times 10^{-10}$
	O	$1s^22s^22p^4$	gains 2 electrons	$1s^22s^22p^6$	O^{2-}	1.40	
	N	$1s^22s^22p^3$	gains 3 electrons	$1s^22s^22p^6$	N^{3-}	1.71	

All the ions shown above are *isoelectronic*: they have the same electronic configuration which is that of a neon atom. Notice the trend in ionic sizes, however; it reflects the changing nuclear charge from nitrogen +7 to aluminium +13.

Cations and anions attract each other because they are oppositely charged. The resulting bond is called an *electrovalent* or *ionic* bond.

> An ionic bond is the electrostatic force of attraction between two oppositely charged ions formed as a result of electron transfer.

Ionic lattices

An ionic lattice forms from a solution or a melt containing its constituent ions. Each ion becomes surrounded by ions of the opposite charge:

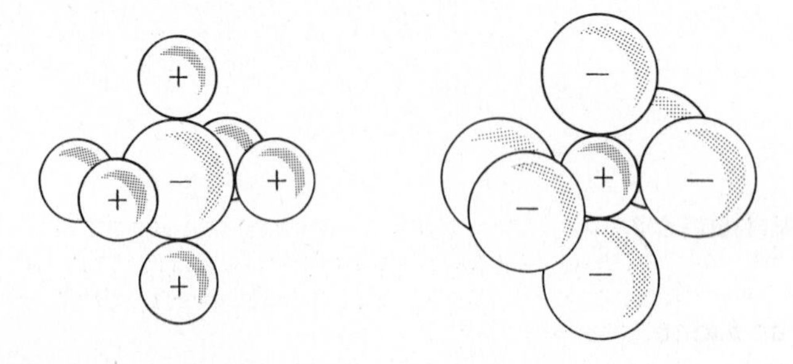

The process leads to the building up of a complete lattice. A *lattice* is an ordered arrangement of particles in three dimensions.

A lattice of cations and anions

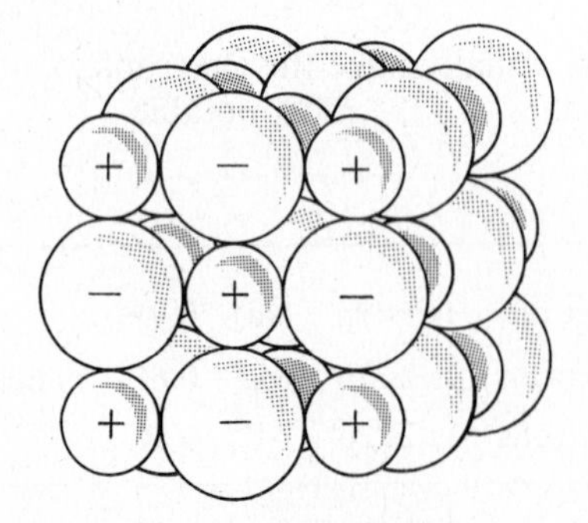

There are some important differences between the make-up of an ionic and a molecular compound:

1. The molecular compound can exist as discrete compound particles (molecules). There are no discrete compound particles in an ionic lattice.
2. The bonds in a molecular compound are specific and directional and are a function of the covalency; but the bonds in an ionic lattice are non-directional and are a function of the size and charge of the ions.

4.4 Factors affecting bond type

The range of bond type

The covalent and ionic models of bonding represent the extremes of the general case introduced on page 49. There is in fact a continuous range of bond types between these extremes.

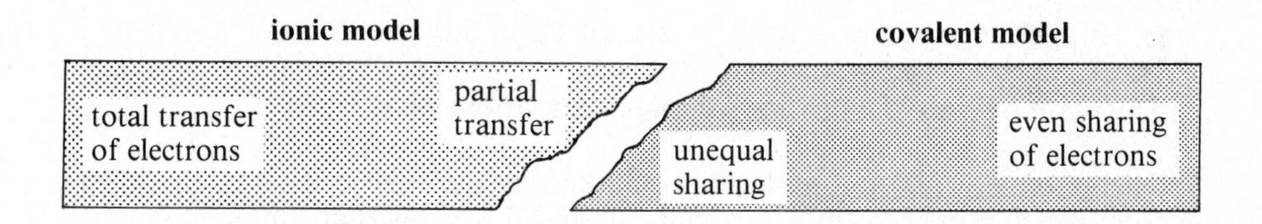

In the majority of cases, atoms neither share their electrons absolutely equally nor totally transfer them from one to another. It is usually easy to decide which of the two extremes is nearer for a particular bond. The V.B. principles of bonding provide the best guide.

Here are two examples in which electron-density maps illustrate the electron distribution in two bonds:

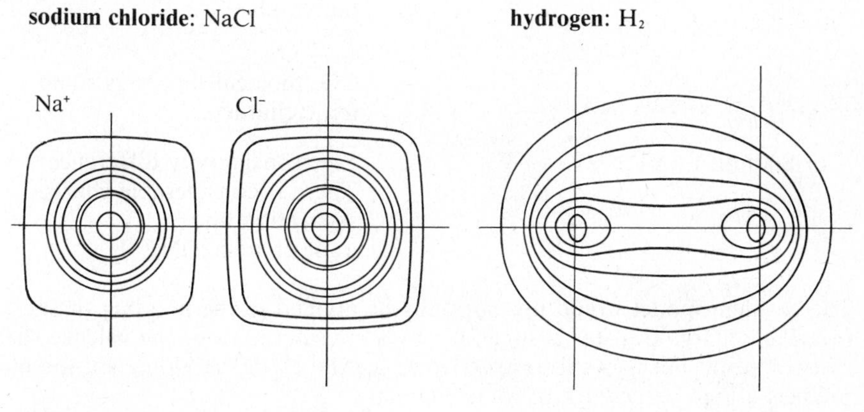

Here the electron density is high around each nucleus but falls away to a very low value between the ions.

Here the electron density is high around the nuclei, but there is a region of fairly high electron density *between* the nuclei.

The covalent model

Electronegativity is the key idea that helps to decide the nature of a bond using the covalent model. Electronegativity is defined on page 45 as a numerical value for the electron-attracting power that a particular atom has for its bonding and lone pairs of electrons. The following steps are made:

1. The bond between two atoms is assumed to be covalent in character.
2. If the electronegativity difference between the atoms is zero, then the electrons are shared absolutely equally. In other words, the bond has 100% covalent character or 0% ionic character.
3. If the electronegativity difference between the atoms is more than zero, then the electrons are *not* shared equally. One atom takes more control of the electrons and a charge separation occurs along the bond. For example

In HCl, electronegativity of H = 2.1, of Cl = 3.0.

Thus, a molecule of HCl has a charge separation: $\overset{\delta+}{H}—\overset{\delta-}{Cl}$

Charge separation on a molecule is marked by the signs $\delta+$ and $\delta-$. δ is a symbol which means 'a little', so that $\delta+$ and $\delta-$ show that the bond possesses some ionic character.

A few examples illustrate these ideas:

a) Hydrogen $\overset{2.1}{H}—\overset{2.1}{H}$

Electronegativity difference = 0
Therefore there is no charge separation
and 0% ionic character.

b) Carbon dioxide $\overset{3.5}{O}=\overset{2.5}{C}=\overset{3.5}{O}$

Electronegativity difference = 1.0.
There is a charge separation along the bonds:
$\overset{\delta+}{C}—\overset{\delta-}{O}$.
CO_2 molecules possess some ionic character.

c) Sodium fluoride $\overset{0.9}{Na}—\overset{4.0}{F}$

Electronegativity difference = 3.1.
There is considerable charge separation and
ionic character $Na^{+}\,^{-}F$

The covalent model would not normally be applied to the bonding of sodium fluoride. Charge transfer is suggested by a consideration of the valence shells of each atom, but it is still instructive to see the effects of doing so: the model predicts a high percentage of ionic character.

The ionic model

In any argument where two extreme positions exist, it is possible to start at either extreme and move towards the other. When using the covalent model, it is the percentage of ionic character in the bond that is being assessed. An alternative approach is to start with an assumption of charge transfer and then assess the degree of covalency between the ions:

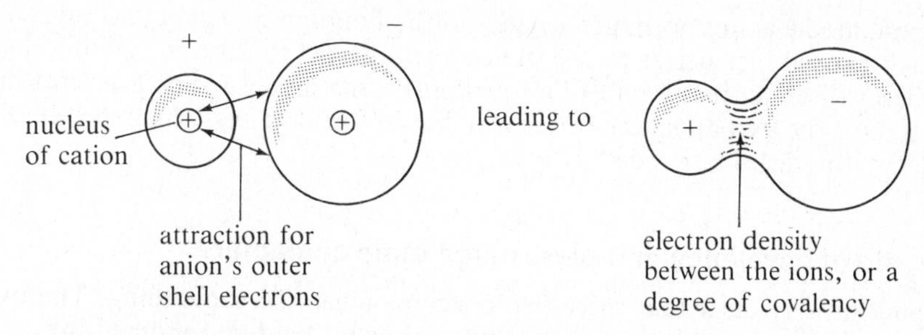

The distortion of an anion caused by the positive charge is called *polarization*. A cation possesses a *polarizing power* and an anion a *degree of polarizability*. The extent to which polarization occurs clearly depends on these two factors. It can be worked out largely by common sense and use of the inverse-square law.

Polarization increases with:

1. a decrease in the distance between the cation's nucleus and the anion's outer shell (stronger attractive force)
2. an increase in the distance between the anion's nucleus and its outer shell (weaker control of electrons).
3. an increase in the charge of either ion.

Fajans summarized these conclusions by stating that all compounds possess a degree of covalency.

> High degrees of covalency exist when:
> the likely *cation* formed from the bonding atoms is *small*,
> the likely *anion* formed from the bonding atoms is *large*,
> the *charges* of the likely ions are *high*.

These ideas are illustrated in the three examples below which should be compared with examples of the use of the covalent model opposite.

a) Sodium fluoride

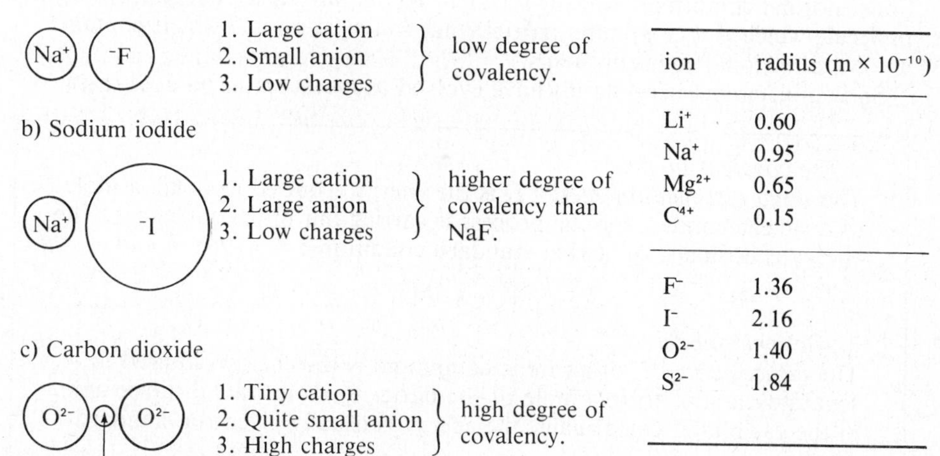

ion	radius ($m \times 10^{-10}$)
Li^+	0.60
Na^+	0.95
Mg^{2+}	0.65
C^{4+}	0.15
F^-	1.36
I^-	2.16
O^{2-}	1.40
S^{2-}	1.84

The ionic model is not normally applied to the bonding of carbon dioxide. But just as it proved instructive to use the covalent model to discuss the bonding of sodium fluoride, it is interesting to use the ionic model for carbon dioxide. Once again, the two approaches coincide because it predicts a high degree of covalency for carbon dioxide.

A degree of covalency or a percentage ionic character?

It should now be clear that these two concepts mean the same thing. The two bonding models offer alternative starting points in the description of the nature of a bond between two atoms.

A helpful analogy is the problem of describing the colour grey. It is either said to be white with black added to it, or black with white added to it. So it is with bonding as well; there are few black or white cases: most are grey and the best approach is to consider them as a modified form of one extreme or the other.

degree of covalency	*percentage ionic character*
Use the *ionic model* 1. Assume complete charge transfer 2. Estimate polarization using Fajans' Rules	Use the *covalent model* 1. Assume electron-sharing 2. Estimate the charge separation using difference in electronegativity

The strength of bonds

The two models have significantly different approaches to the question of bond strength. These arise from the important distinctions made about what constitutes a compound particle. On page 62 it was seen that the ionic model cannot incorporate the idea of a bond between only two atoms. A whole lattice is the smallest particle of an ionic compound that can exist on its own under normal conditions. But the covalent model introduces the concept of a molecule which is a compound particle that can exist on its own. It contains its own directional covalent bonds.

So two different energy terms have evolved as measures of bond strength:

1 The covalent model
The *bond dissociation energy D* is the energy required to break a mole of covalent bonds when the process is carried out in the gas phase. The energy is normally quoted at standard conditions: T = 298 K and P = 1 atm.

2 The ionic model
The *lattice energy U* of an ionic compound is the energy required to separate the ions from a mole of the lattice to an infinite distance apart in the gas phase. Once again, the energy is usually given for standard conditions.

These important terms are often used to provide evidence for the nature of the bonding within a compound. They are discussed in much more detail in chapter 14.

Questions

1. Explain the meaning of the terms:
   a) bond energy b) bond length c) molecular orbital d) hybridization.
2. Distinguish clearly between:
   a) electrovalency and covalency
   b) sigma pair, pi pair and lone pair
   c) polarizing power and degree of polarizability
   d) degree of covalency and percentage ionic character.
3. a) List the electrostatic forces occurring between two atoms next to each other.
   b) Describe three different situations resulting from the effects of these forces.
4. Using first the valence bond theory and then the molecular orbital theory explain why
   a) hydrogen is found as the molecule H_2.
   b) helium is an atomic gas He.
   c) the species HHe^+ can be detected in certain gas discharge tubes.
5. a) Draw two clumps of ions like those on page 62, one representing a sodium ion surrounded by chloride ions, and the other a chloride ion surrounded by sodium ions.
   b) What is the electrovalency of each ion?
   c) How many near neighbours has each ion?
   d) How many ionic bonds does each ion make in the lattice?
6. a) Draw a water molecule in two ways:
      i) to show the overlap of orbitals and how the orbitals are occupied by electrons,
      ii) to show clearly the bonding, the lone pairs and the bond angle.
   b) What is the covalency of each atom in the molecule?
   c) How many near neighbours has each atom got in the molecule?
   d) How many covalent bonds does each atom make in the molecule?
7. Applying the covalent model, discuss the bond type in the following:
   a) Cl_2, b) HF, c) SO_2, d) NaH, e) KBr, f) NaOH, g) MgO.
8. a) The ionic model approach to bonding often involves applying what are known as Fajans' Rules. Explain these clearly.
   b) Apply the ionic model to the bond type in the following substances: i) KF, ii) LiI, iii) CaO, iv) CS_2, v) $AlCl_3$.
9. Distinguish between the terms *bond strength* and *lattice energy* and explain clearly when it is correct and appropriate to use each of them.
10. Discuss the statement 'All bonds are intermediate in type; it is not possible to have entirely ionic or purely covalent bonds.'

Chapter 5

Metallic lattices

5.1 Physical properties

Metals are readily recognizable from their characteristic physical properties. The three most important ones are:

1. their ductility and malleability
2. their high thermal and electrical conductivity
3. their shiny surfaces

Ductility and malleability

Metals can be bent, stretched and deformed into different shapes without breaking. When a metal is being deformed under tension, its behaviour is described as *ductile*. When it is being deformed under compression, its behaviour is described as *malleable*.

In general, the more malleable and ductile a metal is, the less is its mechanical strength. Copper and lead can both be deformed easily but neither will support heavy loads. High-strength metals such as hard steels are not very malleable or ductile.

Metals become increasingly less malleable and ductile the more they are deformed.

Conductivity

Metals conduct both heat and electricity very well compared with non-metallic substances. On average they conduct heat ten to a hundred times better and electricity 10^{10} to 10^{11} times better than non-metallic substances. However, a metal's conductivity decreases with increasing temperature. The graphs below

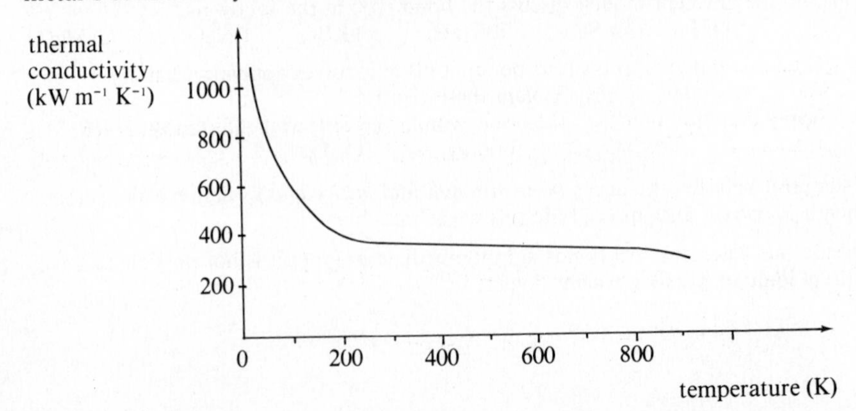

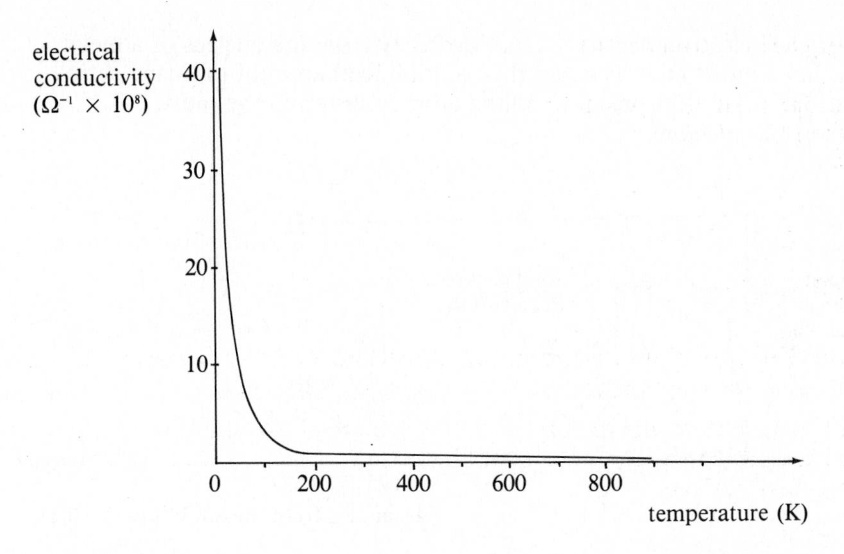

show the thermal and the electrical conductivity of the metal copper at various temperatures.

Surface lustre and colour

Another well-known characteristic of metals is their surface lustre or sheen. The clean surface of nearly all metals is shiny and reflects light giving the metal a silvery colour. Two important exceptions are copper and gold with their distinctive colours.

5.2 Metal structure

A satisfactory theory of metallic bonding must be able to explain the physical properties of the bulk metal. All pure metals are elements and contain only one type of atom in their lattice. Before attempting to describe a metallic lattice, a revision of the nature of a metal atom is necessary.

Metal atoms

A metal atom has an electron-deficient outer shell that is held weakly to the nucleus. The outer shell is shielded from the nucleus by the screening effect of the inner-shell electrons.

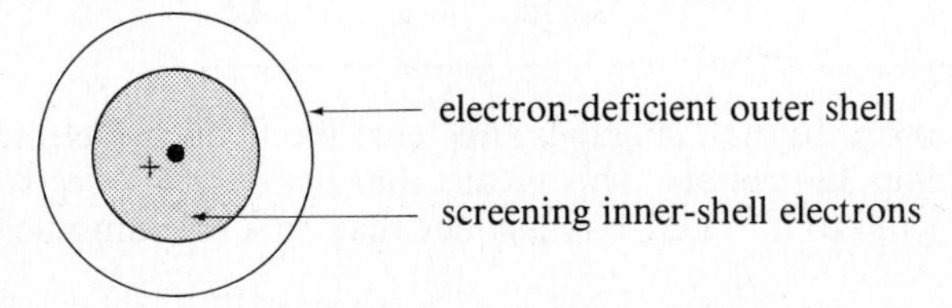

The outer-shell electron density spreads diffusely from the nucleus of a metal atom so that a metal atom is larger than a non-metal atom of the same period. The distribution of electrons in a sodium atom is shown in the radial distribution curve below.

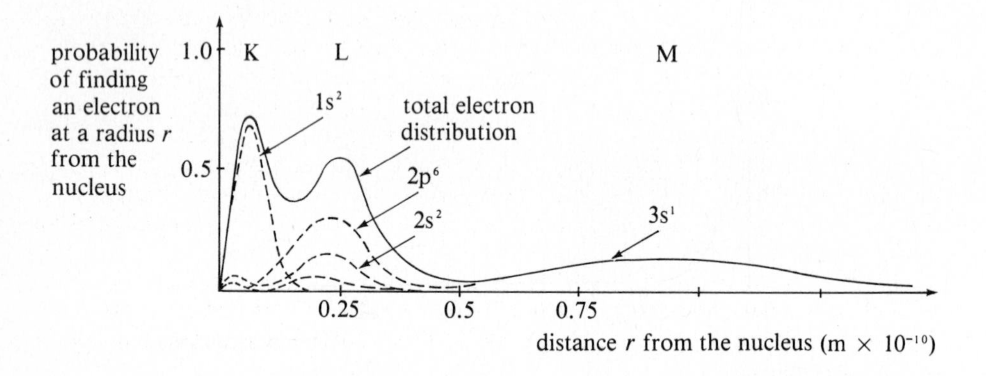

Sodium is an example of an s-block element. There are three blocks of metallic elements in the Periodic Table.

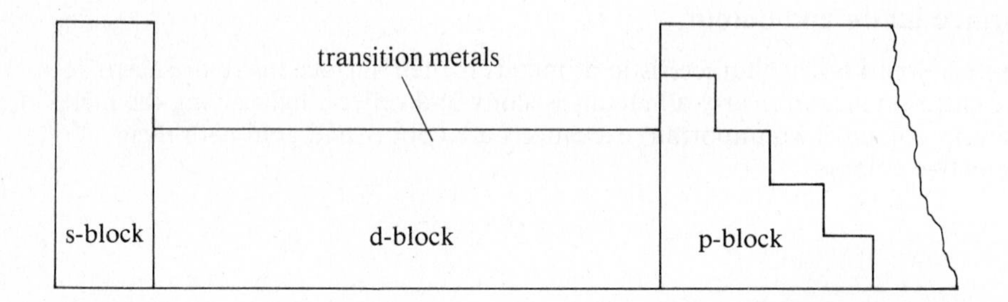

The shielding of the outer shell from the nucleus is at a maximum for the s-block metals. The degree of shielding depends on the sub-shells that the inner electrons occupy: p-electrons shield less well than s-electrons, and d-electrons even less than p-electrons. Compare an atom of calcium with an atom of manganese.

atom	*nuclear charge*	*shielding electrons*	*outer-shell electrons*
Ca	+20	$1s^2 2s^2 2p^6 3s^2 3p^6$	$4s^2$
Mn	+25	$1s^2 2s^2 2p^6 3s^2 3p^6 3d^5$	$4s^2$

There are five more protons in the manganese nucleus; the five extra electrons are in the poorly shielding 3d-orbitals. This means that the outer 4s-electrons are more strongly attracted to a manganese nucleus than to a calcium nucleus.

Generally speaking, s-block atoms are therefore:

1. less electronegative
2. larger

than the transition metal atoms of the same period.

The p-block metals owe their metallic character entirely to the poor shielding effects of the inner-shell electrons. The valence shells of these atoms are not at first sight electron-deficient. For example, lead's outer-shell configuration is $6s^2 6p^2$ which might result in it being non-metallic like carbon, $2s^2 2p^2$. However, the 6s-electrons experience such poor shielding that they almost become a part of the inner shells. This leaves the 6p-electrons to all intents and purposes as the only 'outer-shell' electrons in a lead atom. These p-electrons are themselves shielded by the withdrawn 6s-electrons so that they are less tightly bound to the nucleus and can give the atom metallic character.

The withdrawal of the outer-shell s-electrons into the atomic core of the p-block atoms is called the *inert pair effect*. The full scope of this effect is discussed in Chapter 43. It is clearly shown in the radial distribution curve for lead shown below.

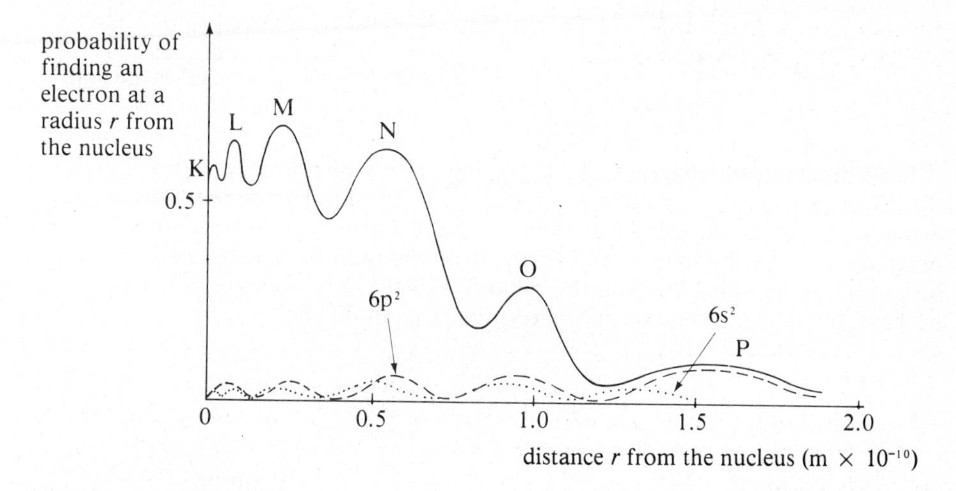

Overlap of orbitals

When metal atoms group together to form a lattice, multiple overlap of their diffuse outer-shell orbitals takes place. This leads to the formation of *bands* of molecular orbitals whose energy ranges are almost continuous. The easiest way of visualizing the formation of an energy bond is first to imagine overlap in only one direction. For example, when two s-orbitals overlap, they produce two molecular orbitals (e.g. see page 52).

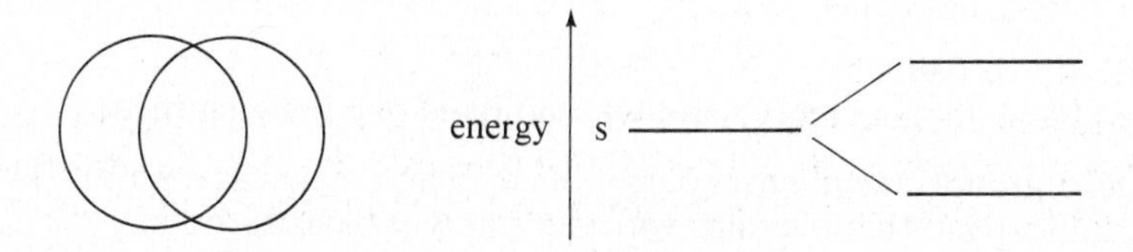

If a *third* atom's s-orbital now overlaps with the first two, *three* molecular orbitals are produced:

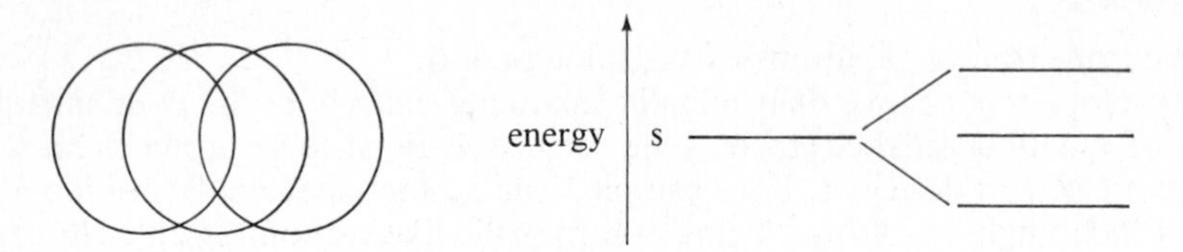

If a *fourth* atom's s-orbital also overlaps with the group, *four* molecular orbitals result; this process can be extended until such a large number of overlaps result that the molecular orbitals produced have energies that form a continuous band.

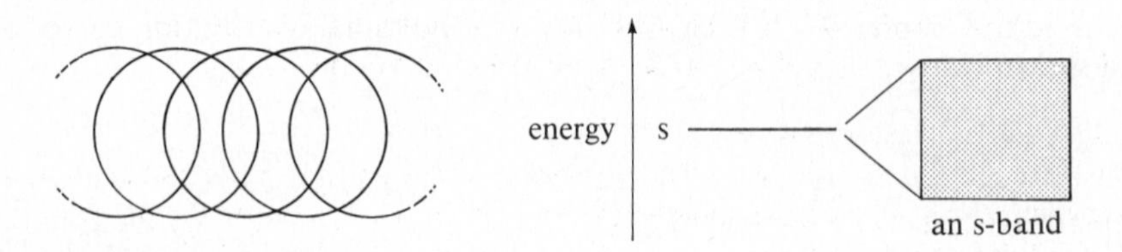

For simplicity, the diagrams show overlap in only one direction. In a lattice the effect extends throughout the three dimensions and happens with other orbitals as well. In the same way that an s-band forms from the multiple overlap of s-orbitals, so a p-band forms from the multiple overlap of p-orbitals. In an s-block metal, the p-band is likely to be completely empty because an s-block atom has no p-electrons in its outer shell.

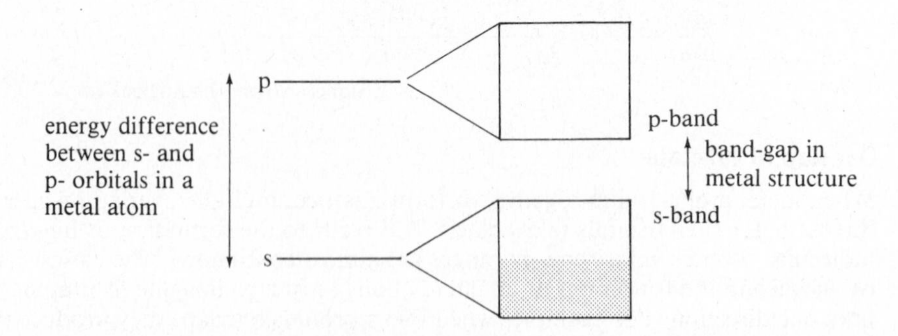

A typical structure has either

1. a partly-filled band, or
2. a filled band and an empty band with no band gap between them.

For example, a group I metal has a partly filled s-band, whereas a group II metal has a full s-band that overlaps with the empty p-band above it.

group I metal	*group II metal*
p-band s-band	p-band s-band

Metal lattices

The band model of the electronic structure of a metal lattice is able to account for the bulk properties of the metal, namely mechanical properties, conductivity and surface appearance. We shall consider each in turn.

1. Mechanical properties

A band structure arises from the multiple overlap of the atomic orbitals of a very large number of atoms. The outer electrons from each atom therefore come under the influence of a very large number of other atoms and so are free to move from one atom to another through the lattice. The term *'delocalized'* is often used to describe their condition. A metal lattice can therefore be viewed as an array of cations held together by a cloud of delocalized outer-shell electrons.

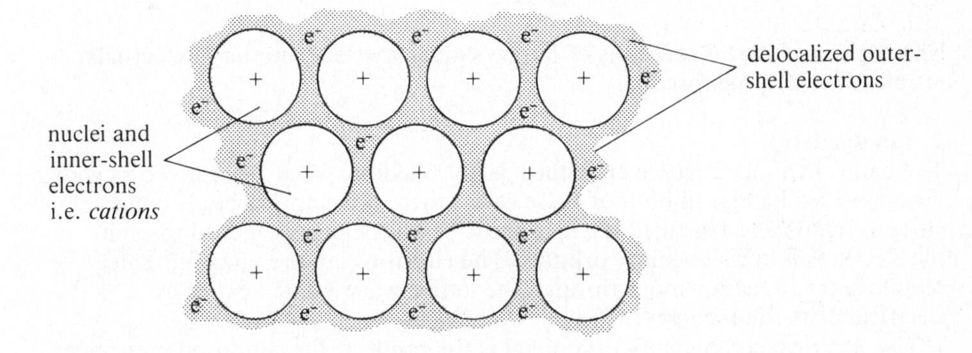

A *metallic bond* is the electrostatic force of attraction that two neighbouring cations have for the delocalized electrons between them.

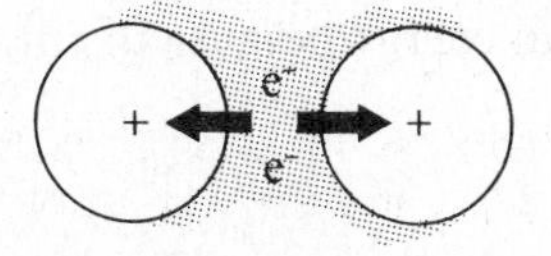

When stress is applied to the metal lattice, adjacent layers of cations can slide over one another.

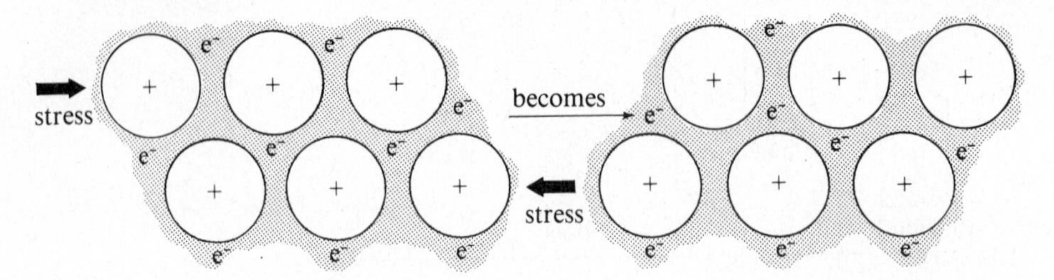

On a large scale, an enormous number of atomic slips like this account for the ductility and malleability of the bulk metal.

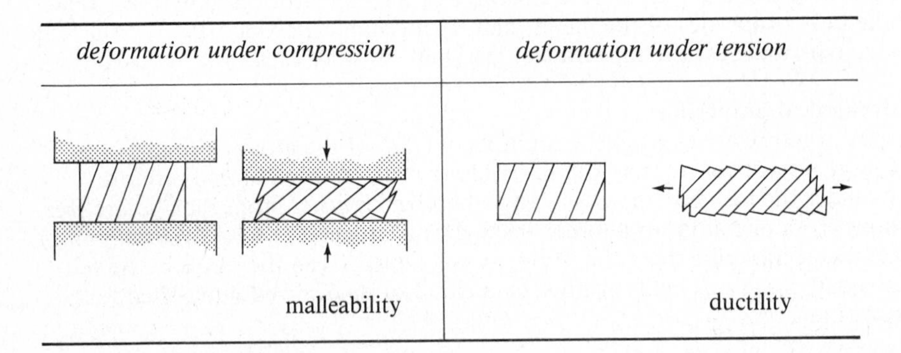

The steps produced by sections of lattice sliding past one another are actually visible through a microscope.

2. Conductivity
The bands in a metal lattice are either partly filled or, when filled, overlap with other unfilled bands. In both of these cases, there are empty orbitals lying very close to the uppermost filled level. It needs very little energy indeed to excite the electrons into these empty orbitals. The electrons are therefore extremely mobile and can readily move through the lattice when excited either by electrical or thermal energy.

The electrical conductivity of a metal is the result of the concerted movement of delocalized electrons through the lattice. The thermal conductivity of a metal is the result of the exchange of kinetic energy between the delocalized electrons in the lattice. In both cases, the upper energy levels of the unfilled band are taken up by the excited electrons. On page 68 the sharp drop in conductivity as temperature increases is described. This can be readily explained. At higher temperatures, the electrons are likely to occupy the higher energy levels in the band. This decreases their ability so easily to take in energy because the higher energy levels are now already populated.

3. Surface appearance
As light falls onto a metal surface, it excites the electrons into unfilled levels of

the band. Because a whole range of electron transitions is possible, a whole range of frequencies is absorbed.

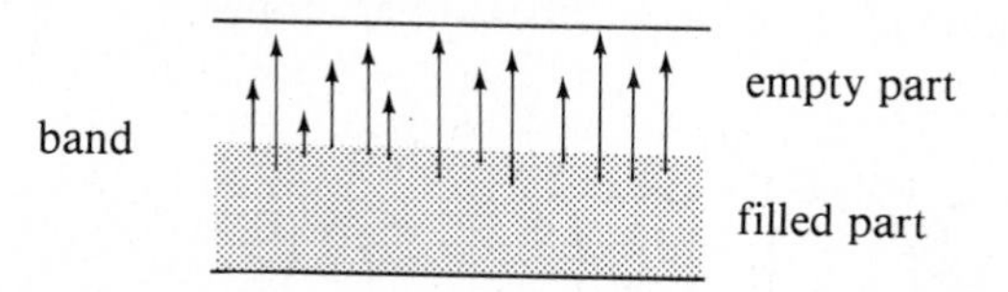

The electrons return to lower energy states by giving back the absorbed energy as light again. In this way, all the light is reflected and this gives the metal its silvery mirror-like appearance.

Most freshly polished metals are silvery in appearance, but copper and gold are coloured. Copper, silver and gold all have the electronic configuration $d^{10}s^1$. There is a small energy gap between the filled d-band and the level of the empty part of the s-band.

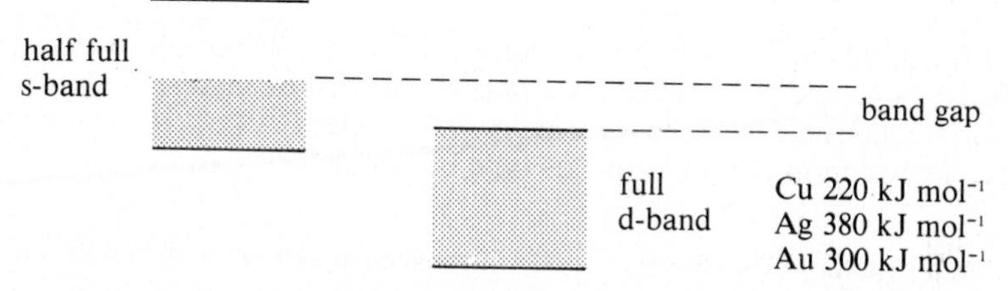

To be excited into the s-band, d-electrons must absorb energy greater than the size of the band gap. For copper, this energy corresponds to the frequencies of green and blue light, leaving only red unabsorbed. For gold with a larger gap, blue and violet light are absorbed, leaving yellow light. The band gap of silver is so large that the visible spectrum is unchanged.

5.3 Lattice types

Ideal lattices

The type of lattice arrangement adopted by a metal depends on a number of factors:

a) the size of the atoms
b) the number of delocalized electrons per atom
c) the degree of shielding of the inner shells.

A particular metal's lattice is the arrangement of lowest possible potential energy for the atoms concerned. There are three major types of lattice structure. Two of these involve the *close-packing* of the atoms; the third involves *cubic packing* of the atoms:

1) close-packing: hexagonal close-packing (h.c.p.)
cubic close-packing (c.c.p.)
2) cubic packing: body-centred cubic packing (b.c.c.)

1. Close-packing
Close-packed atoms are grouped together as tightly as possible. Each atom in a close-packed layer is surrounded by a hexagon of six other atoms.

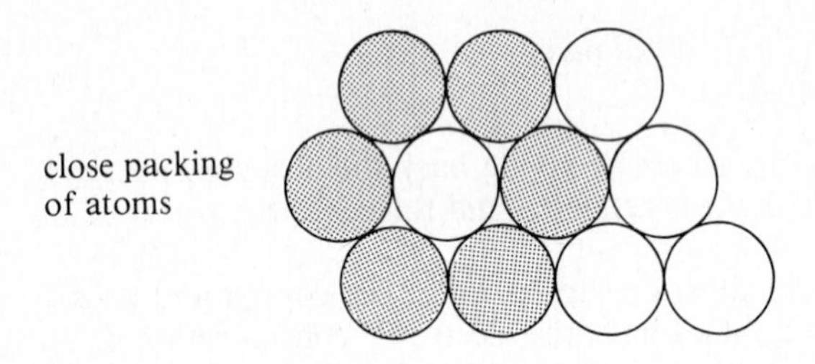

Packing the atoms in this way gives a lattice in which 74% of all the available space is used up. In a **hexagonal close-packed** lattice, the layers of atoms are stacked in an alternating sequence usually called ABABA packing. Magnesium and zinc are common metals with this type of lattice.

layers viewed from above

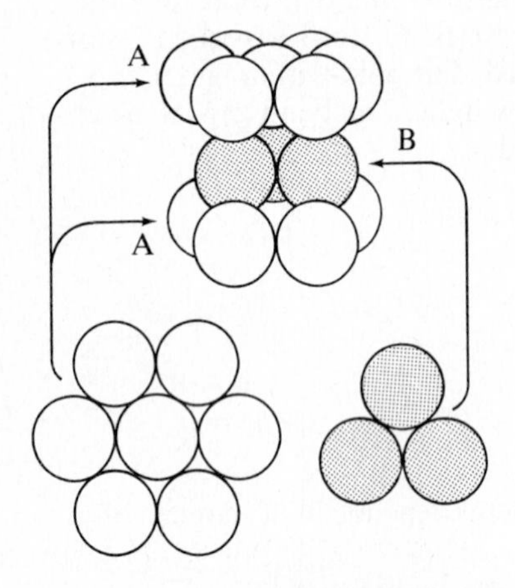

arrangement viewed as if spaced out

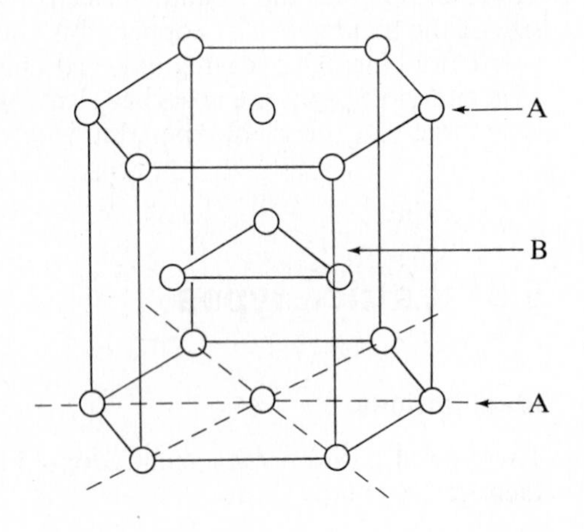

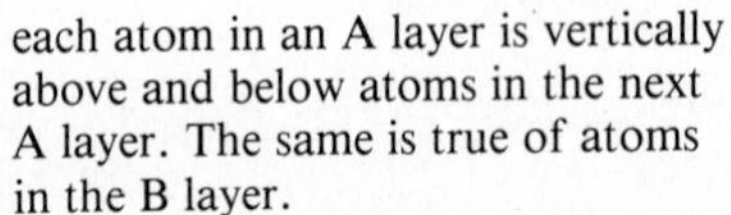
each atom in an A layer is vertically above and below atoms in the next A layer. The same is true of atoms in the B layer.

Another packing sequence is possible, however. Instead of repeating an A layer on top of an AB structure, the next layer may be sited so that its atoms are not vertically above those of the bottom A layer. The arrangement becomes ABCABC packing, known as **cubic close-packing.**

arrangement viewed as if spread out.

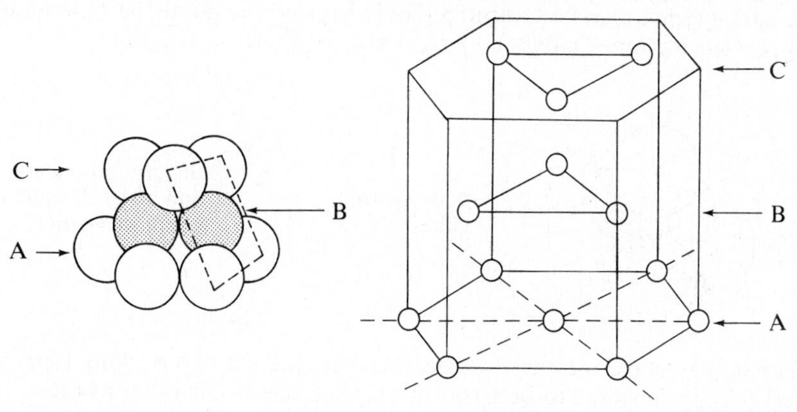

Each atom in the B layer is vertically over the centre of a triangle formed by three atoms in the A layer. The atoms in the C layer are vertically over the centres of a different set of triangles.

Cubic close-packing is often called face-centred cubic packing because, if viewed from a particular angle, the atoms can be considered as being at the corners of a cube with another atom in the centre of each face of the cube. One such face is indicated above on the left, and the close-packed layers in a face-centred cube are indicated in the diagram below.

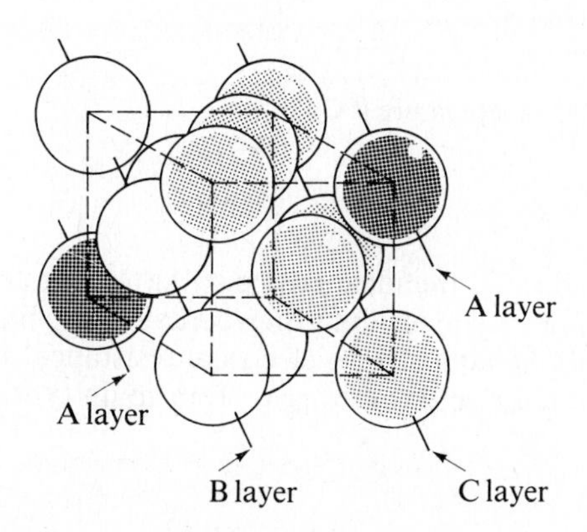

Face-centred cubic metals include calcium, nickel, copper, silver and gold.

2. Cubic packing

Atoms that are cubic-packed are less efficiently arranged than those that are close-packed:

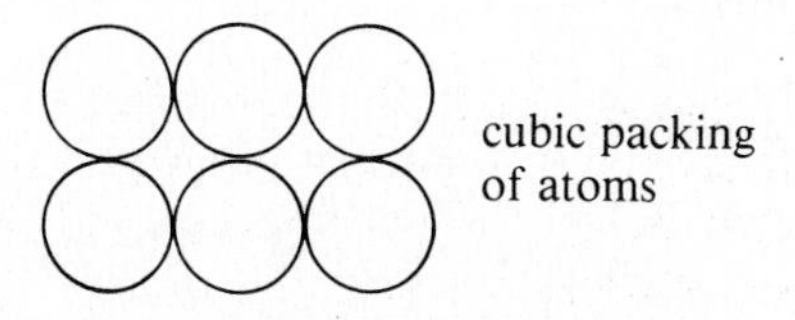

Only 68% of the available space is taken up by this form of packing. It leads to a lattice type called body-centred cubic because the atoms occupy the corners and centre of a lattice cube.

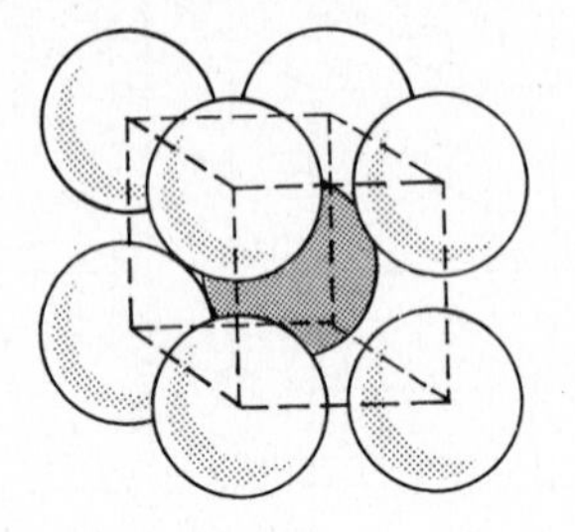

body-centred cube

the group I metals and many transition metals adopt this lattice.

For some metals, the energy difference between the two packing types is small enough for there to be a transition from one to the other as the temperature changes. A well-known example is iron:

the allotropy of iron

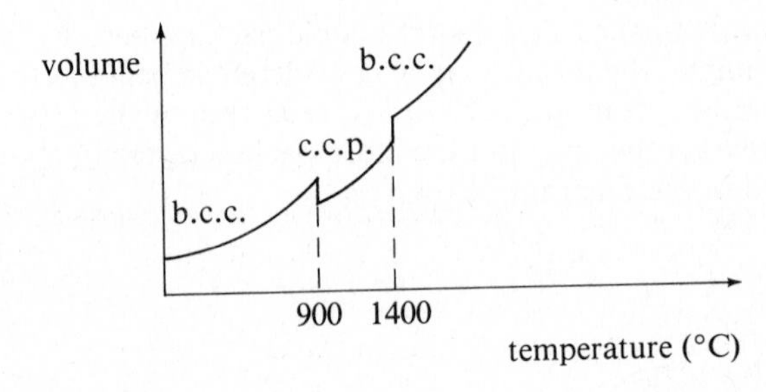

Real lattices

Calculations suggest that if metals adopted the ideal lattice structures perfectly, they would be at least a thousand times stronger than they actually are. Band theory also suggests that there should be almost zero electrical resistance. Real lattices in fact have imperfections in them and these are of four main types:

1. vacancies
2. dislocations
3. grain boundaries
4. impurities.

1. Vacancies

As a lattice forms, sometimes a gap is left. The surrounding atoms are bonded so tightly that they do not fall into the hole. The number of vacancies in the lattice increases with temperature until near the melting point there is one vacancy for approximately every 10^4 atoms.

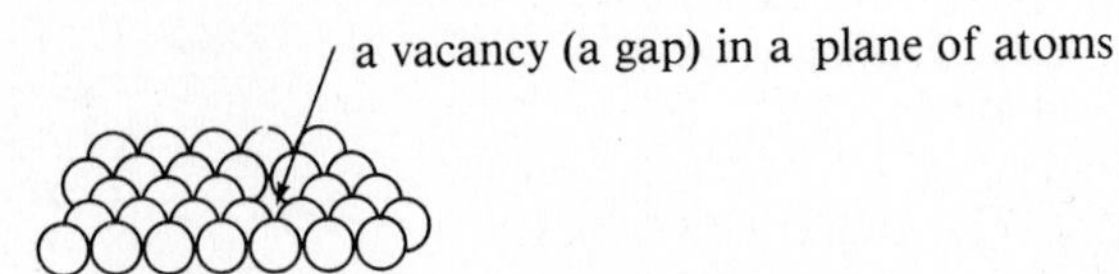

2. Dislocations
A dislocation is the name given to a line of atoms that suddenly stops in the middle of the lattice. At this point there is a region of disorder very like that caused by a vacancy. The major difference is that a vacancy is a single-point defect whereas a dislocation runs through the lattice (at right angles to the plane drawn here) for thousands of atoms distance.

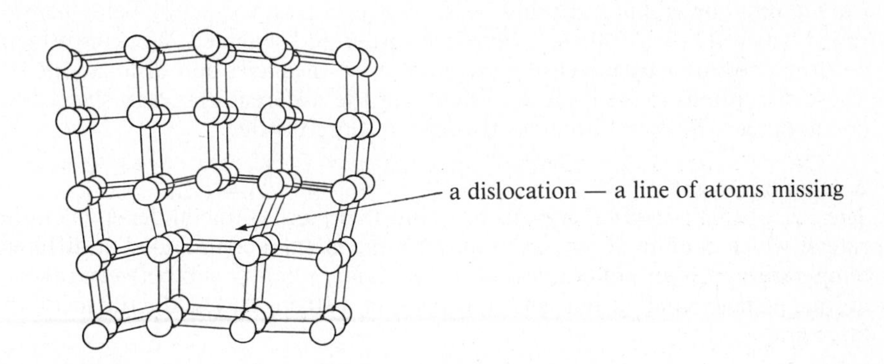

3. Grain boundaries
When a molten metal solidifies, crystals start growing at many points in the melt. When two growing crystals meet, their two lattices normally do not have the same orientation and so a grain boundary results.

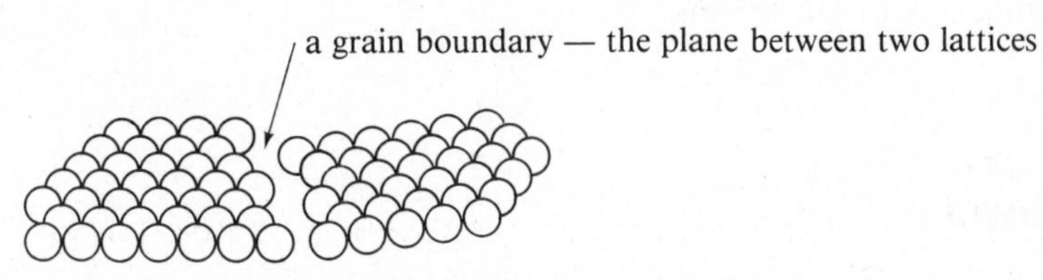

The three disruptions so far discussed can be summarized as follows:

1. vacancies — point disruptions
2. dislocations — linear disruptions
3. grain boundaries — planes of disruption.

4. Impurities
Adding an impurity to a metal introduces a different source of atoms to the lattice. These impurity atoms disrupt the lattice because they have a different size and electron availability. Sometimes impurities are deliberately added to alter the properties of a particular metal lattice. For example, carbon is added to iron to make steel.

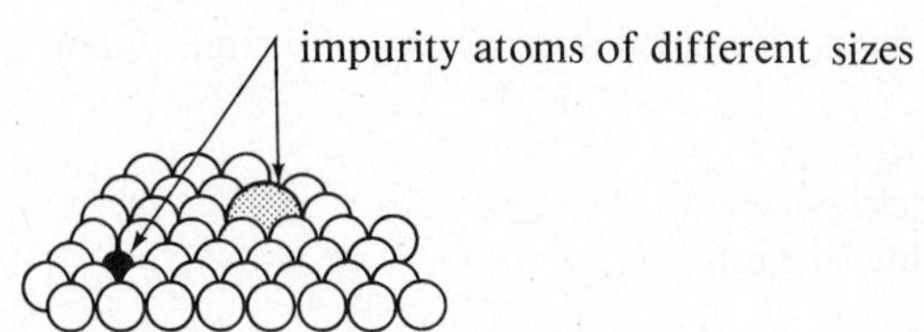

5.4 Metallurgy

The study of the properties of metals is called metallurgy. It is particularly concerned with the use to which metals can be put and therefore with the ways in which metal properties can be adapted. There are three main methods of altering the characteristics of a metal:

1. Work hardening
Pure aluminium is quite soft and bends easily. If it is worked by being bent or rolled between heavy rollers, it becomes harder and stronger. Working a metal deforms the lattice by increasing the number of vacancies and dislocations in it. These disruptions make it more difficult for the different planes to slide over one another. The metal becomes harder but more brittle.

2. Heat treatment
The size of the individual crystals or grains in a piece of metal depends on the rate at which cooling occurs and how long the metal is maintained at different temperatures. Grain boundaries, like dislocations, inhibit slip between the atomic planes. Small grains lead in general to a stronger, less ductile metal than large grains.

3. Alloying
The addition of another component to a metal can radically alter its properties. Tin can be made to melt at a much lower temperature by adding lead (solder); copper can be made much harder by adding zinc (brass); iron can be converted into a wide range of steels by the addition of varying amounts of carbon, silicon, manganese or chromium.

Questions

1. With the aid of diagrams describe:
   a) a typical metal atom
   b) a typical metallic lattice
   c) the formation of a metallic bond.
2. a) List the properties of a typical metal.
   b) Account for these properties in terms of the structure of a metal.
3. What do you understand by the following terms?
   a) Delocalization of electrons
   b) The inert pair effect
   c) The formation of bands
   d) The shielding effect of electrons from different sub-shells.
4. Using the diagrams, describe the arrangement and packing of metallic atoms in the following lattice types:
   a) hexagonal close-packing (h.c.p.)
   b) face-centred cubic/cubic close-packing (f.c.c. or c.c.p.)
   c) body-centred cubic-packing (b.c.c.)

5. a) Explain why most untarnished metals are silvery in appearance while copper and gold are coloured.
   b) Account for the colour of brass, an alloy of copper and zinc.
6. In what ways do real lattices differ from the perfect or ideal structures described in question 4?
7. How can the properties of a metal be modified to adapt it for a new use? Choose three metals widely used in everyday life, and explain why each is chosen for that use.

Chapter 6

Ionic lattices

6.1 Physical properties

The range of bond types that can exist between two atoms was described on page 61 in Chapter 4. Ionic bonding is at one extreme of this range; covalent bonding is at the other. A high percentage of ionic character results when:

1. there is a large difference in electronegativity between the elements making up the compound
2. the likely cation produced is large and of low charge
3. the likely anion produced is small and of low charge.

These conditions are best met by a compound made from elements at the bottom of group I and at the top of group VII. In the discussion that follows, the alkali metal halides in general should be viewed as typical ionic compounds (although lithium iodide possesses a high degree of covalency).

Mechanical strength

Ionic compounds are solids at room temperature. They are harder than most organic solids but tend to be less hard than many metals. Unlike metals they are not ductile or malleable: when they are deformed, they suddenly snap or break. They are *brittle*. Most typical ionic substances have high melting and boiling points which indicate the strength of the ionic bonding.

compounds	*m.p.*(°C)	*b.p.* (°C)
sodium chloride	801	1465
potassium chloride	770	1407
magnesium chloride	712	1418
magnesium oxide	2640	3043

Conductivity

Ionic solids are insulators but when molten or dissolved in water, they become conductors. The passage of current causes decomposition. In the molten phase, the pure elements are produced. In aqueous solution, more often than not, it is the water that is decomposed to hydrogen and oxygen. The decomposition is called electrolysis and the conducting liquid is an electrolyte. Both are described in detail in Chapter 20.

An example of the passage of electricity through a molten electrolyte is shown below:

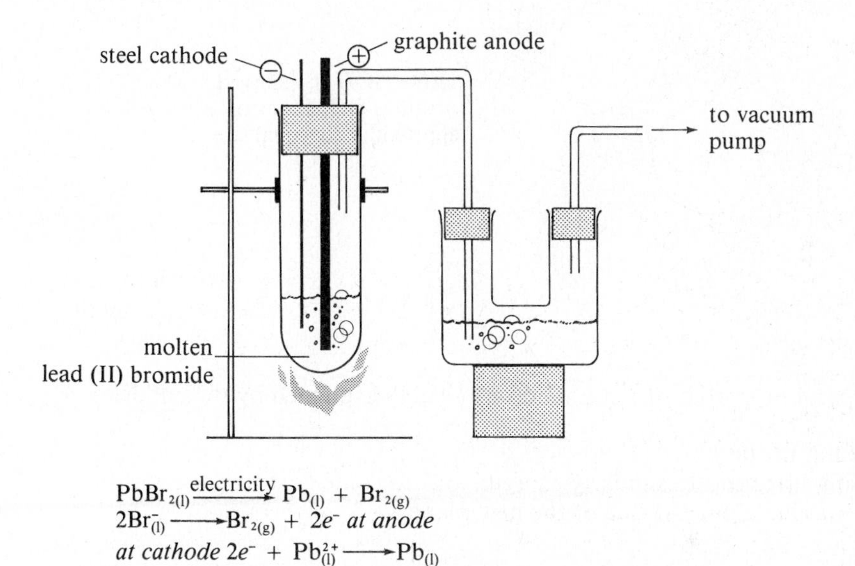

$PbBr_{2(l)} \xrightarrow{\text{electricity}} Pb_{(l)} + Br_{2(g)}$
$2Br^-_{(l)} \longrightarrow Br_{2(g)} + 2e^-$ *at anode*
at cathode $2e^- + Pb^{2+}_{(l)} \longrightarrow Pb_{(l)}$

A globule of lead is produced at the bottom of the tube and the water in the wash bottle is turned brown by the bromine dissolving in it.

Solubility

Ionic solids show a wide range of solubility in water. The majority of them dissolve appreciably, however, and they are all much more soluble in water than in any organic solvents. Only a few classes of molecular solids dissolve in water, so it is likely that a solid dissolving in water is ionic.

6.2 Ionic structure

Lattice of ions

In a metal lattice, all the ions are of the same size and are held together by the delocalized electrons. In an ionic lattice, cations are smaller than anions and there are sometimes two cations for every anion or *vice versa*, e.g.

Na_2O	$2Na^+$	O^{2-}
$MgCl_2$	Mg^{2+}	$2Cl^-$

The lattice that forms must take these factors into account. It is held together by electrostatic attraction between the ions.

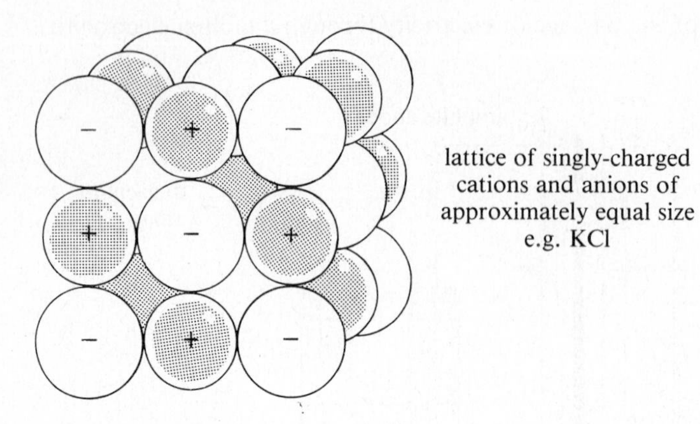

The physical properties of ionic solids are readily explained by this model:

1. Hard but brittle
A shearing force causes similarly charged ions to come next to one another. Fracture occurs along the line of the new plane.

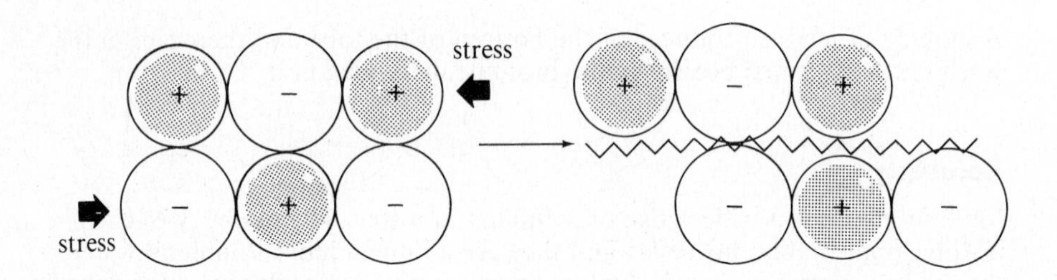

2. Conductivity
In the molten phase, the ions are free from the lattice. They can migrate to the electrodes where:

i) anions give up electrons at the anode
ii) the electrons flow through the circuit: conduction
iii) cations receive the electrons at the cathode.

Chemical change takes place because the ions are converted to atoms or molecules.

3. Solubility in water
Water molecules are polar as a result of charge separation along the oxygen–hydrogen bonds. The poles of the water molecules interact with the ions of the lattice so that each ion becomes surrounded by water molecules. This weakens the forces of attraction between the ions, and the lattice breaks up.

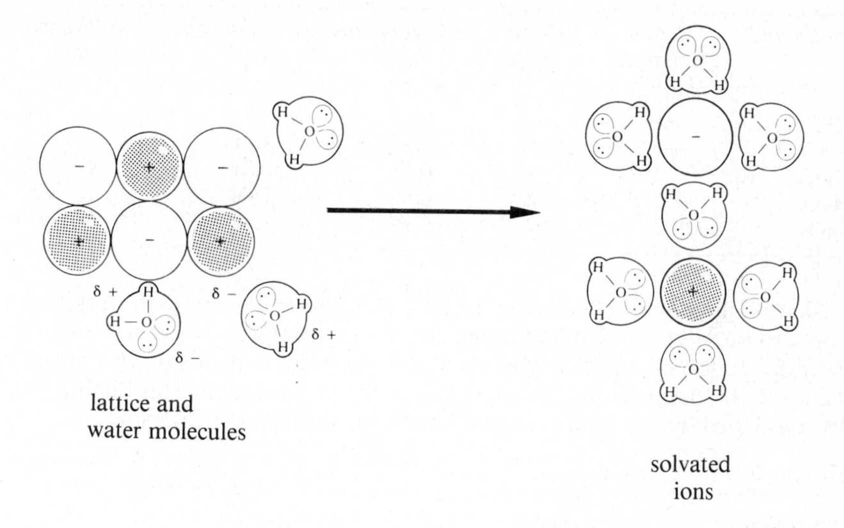

e.g. $NaCl_{(s)} \xrightarrow{\text{water}} Na^+_{(aq)} + Cl^-_{(aq)}$

Lattice energies

The lattice energy of a particular ionic compound is the energy required to separate a mole of the compound into its constituent ions in the gas phase so that the ions are an infinite distance apart. The lattice energy is normally quoted at standard conditions of temperature and pressure, 298 K and 1 atmosphere, e.g.

$$NaCl_{(s)} \rightarrow Na^+_{(g)} + Cl^+_{(g)}: \quad \Delta H^{\ominus} = 771 \text{ kJ mol}^{-1}$$

Experimental values are obtained by means of a suitable Born – Haber cycle (this technique is outlined in Chapter 14). Theoretical values can be calculated from an assumption that the ions are hard spheres of particular size and charge. The theoretical values very often do not quite agree with the experimental ones: this generally indicates that the compound has a marked degree of covalency so that the experimental value is almost a bond dissociation energy rather than a lattice energy.

The inverse square law states that the force of attraction F between two charged particles depends inversely on the square of the distance d between them and directly on the value q^+ and q^- of the two charges

$$F \propto \frac{q^+ \cdot q^-}{d^2}$$

Therefore, a lattice energy increases as the ionic radii decrease, or as the ionic charges increase. This is shown by the figures in the following table.

compound	*cation charge*	*anion charge*	*cation size* m $^{3}10^{-10}$	*anion size* m$^{3}10^{-10}$	*lattice energy* kJ mol^{-1}
LiCl	+1	−1	0.60	1.81	846
NaCl	+1	−1	0.95	1.81	771
$MgCl_2$	+2	−1	0.65	1.81	2493
MgO	+2	−2	0.65	1.40	3889
CaO	+2	−2	0.99	1.40	3513

Melting points are often taken to be a measure of the strength of a lattice. This can be rather misleading because a lattice energy is concerned with the *total* disruption of a lattice whereas a melting point depends on the nature of the particles that exist in the liquid phase of the substance. This distinction is demonstrated by the figures shown below for three metal chlorides.

compound	*lattice energy* kJ mol^{-1}	*melting point* °C	
NaCl	771	801	
$MgCl_2$	2493	712	
$AlCl_3$	5103	183	(sublimes)

The lattice energies follow the expected trend suggested by the inverse square law but the melting points are in exactly the reverse order. An explanation for these figures can be found by considering the nature of the bonding in the three compounds. According to Fajans' rules (see page 65 again) the degree of covalency in an ionic compound is highest when the cation is small, the anion is large, or the ions have a high charge, as illustrated in the table below.

lattice	*cation size* m × 10^{-10}	*cation charge*	*degree of covalency*
NaCl	0.95	+1	
$MgCl_2$	0.65	+2	
$AlCl_3$	0.50	+3	

When sodium chloride melts, the particles in the liquid phase are likely to be discrete ions because the degree of covalency is low. When aluminium chloride sublimes, however, Al_2Cl_6 molecules are produced in the gas phase ($2AlCl_{3(s)} \rightleftharpoons Al_2Cl_{6(g)}$). The aluminium ions polarize the chloride ions so extensively that the lattice can almost be considered to be a molecular lattice of Al_2Cl_6 molecules. The force of attraction between neutral molecules is much weaker than the electrostatic force of attraction between ions.

'molecules' of Al_2Cl_6

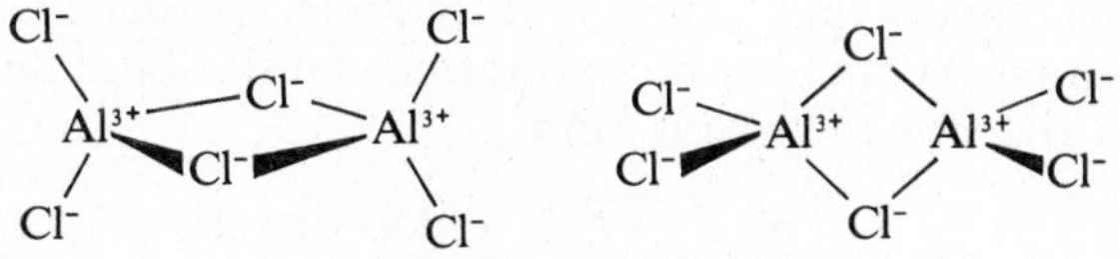

Magnesium chloride represents a halfway stage between the two cases. Liquid magnesium chloride contains an appreciable concentration of $MgCl_2$ molecules.

The lattice energy of a compound is a measure of the energy required to separate it into its individual component particles. The melting point of a compound is a measure of the kinetic energy needed to separate clumps of particles from each other. Within the clumps, the bonds are still intact and have not yet been broken. So melting involves only a partial break up of the lattice.

6.3 Lattice types

Co-ordination number

In an ionic lattice, each ion is surrounded closely by a number of ions of the opposite charge. The number of near-neighbour ions in the arrangement is called the *co-ordination number* of the lattice. The co-ordination number depends on the relative sizes of the ions in the lattice. A large ion can fit more ions around it than a small ion.

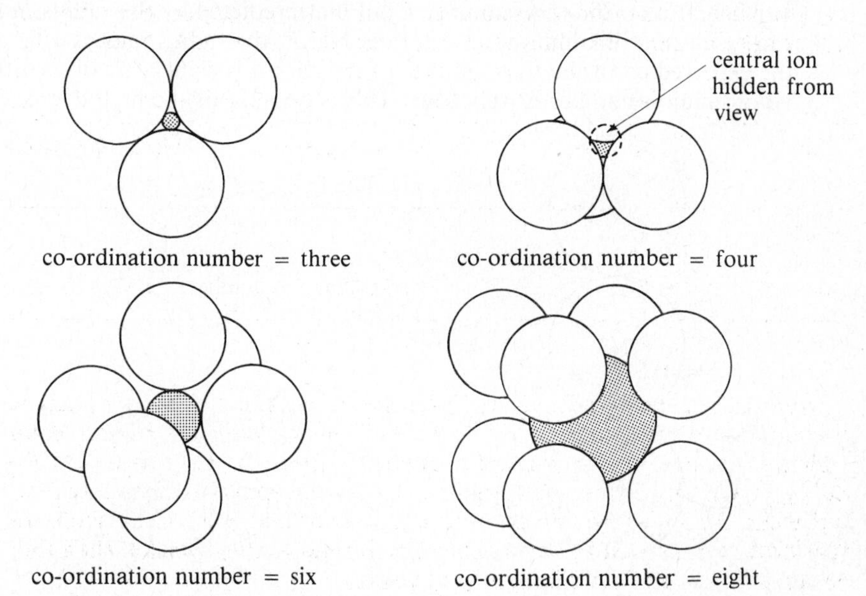

co-ordination number = three

co-ordination number = four

co-ordination number = six

co-ordination number = eight

Radius ratio

Cations are almost always smaller than anions and so the factor that determines lattice structure is the number of anions that can fit round the cation. A useful guide is given by the *radius ratio*:

$$\text{Radius ratio} = \frac{r_c}{r_a} \qquad \begin{aligned} r_c &= \text{radius of cation} \\ r_a &= \text{radius of anion} \end{aligned}$$

The value of the radius ratio determines what the co-ordination number of the cation is likely to be.

radius ratio	*co-ordination number*
Less than 0.23	3
0.23 to 0.42	4
0.42 to 0.73	6
More than 0.73	8

In a real example of a lattice structure, two other factors apart from the radius ratio affect the arrangement of the ions:

1. The *stoichiometry of the compound*, e.g. there are twice as many anions as cations in $MgCl_2$ and so there are holes in the lattice structure to preserve this number ratio.
2. The *degree of covalency of the compound*. Covalent bonds are directional. So in cases where there is a high degree of covalent character, a different co-ordination number from that predicted by the radius ratio can result, e.g. in ammonium fluoride, NH_4F, the radius ratio is 1.01 and the expected co-ordination number of the cation is eight. The observed co-ordination number is only four. This is because of the high degree of

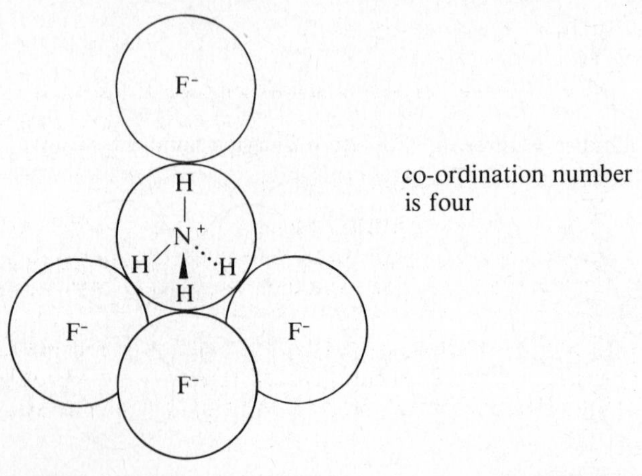

co-ordination number is four

covalency between the hydrogen atoms of the ammonium ion and the surrounding fluoride ions.

Some examples of lattices that agree with the predictions of the radius ratio are shown below.

1. Beryllium chloride
Although the co-ordination number of beryllium in a lattice of beryllium chloride is four, in the vapour phase a high proportion of the chloride contains three co-ordinated beryllium.

radius ratio = 0.17 cation co-ordination number = 3

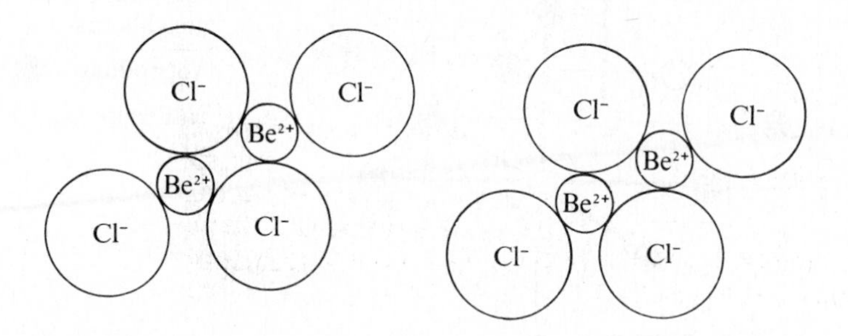

'molecules' of Be_2Cl_4 in the vapour of beryllium chloride; each beryllium ion fits between three chloride ions

2. Zinc sulphide

radius ratio = 0.40 cation co-ordination number = 4
anion co-ordination number = 4
described as 4:4 co-ordinated

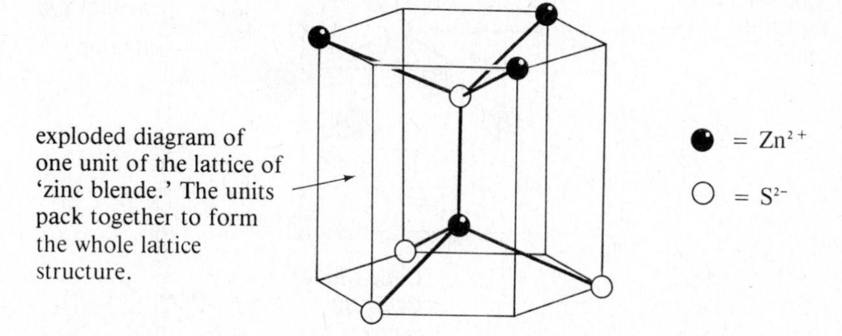

exploded diagram of one unit of the lattice of 'zinc blende.' The units pack together to form the whole lattice structure.

3. Sodium chloride

radius ratio = 0.52 cation co-ordination number = 6
anion co-ordination number = 6
described as 6:6 co-ordinated

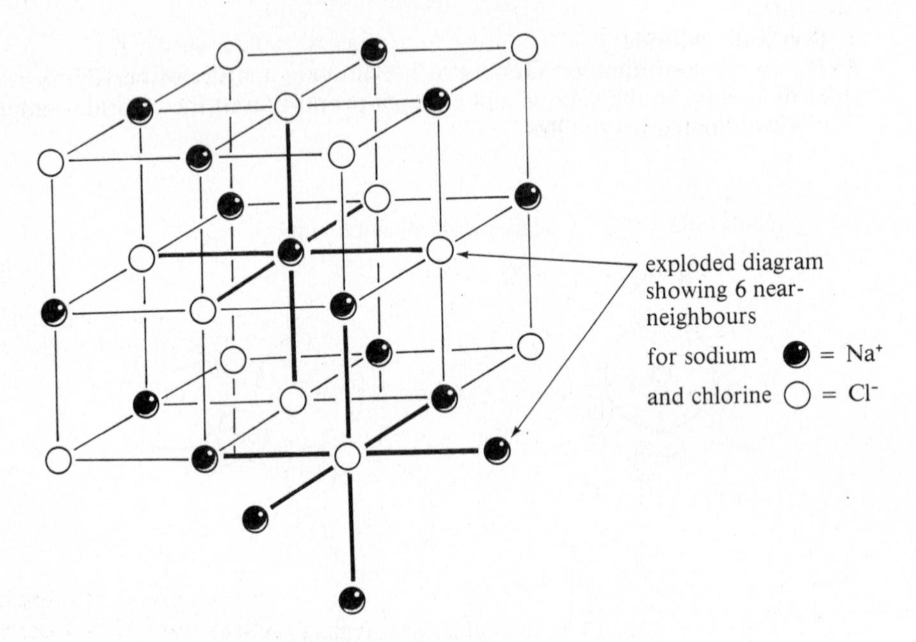

4. Caesium chloride

radius ratio = 0.93 cation co-ordination number = 8
anion co-ordination number = 8
described as 8:8 co-ordinated

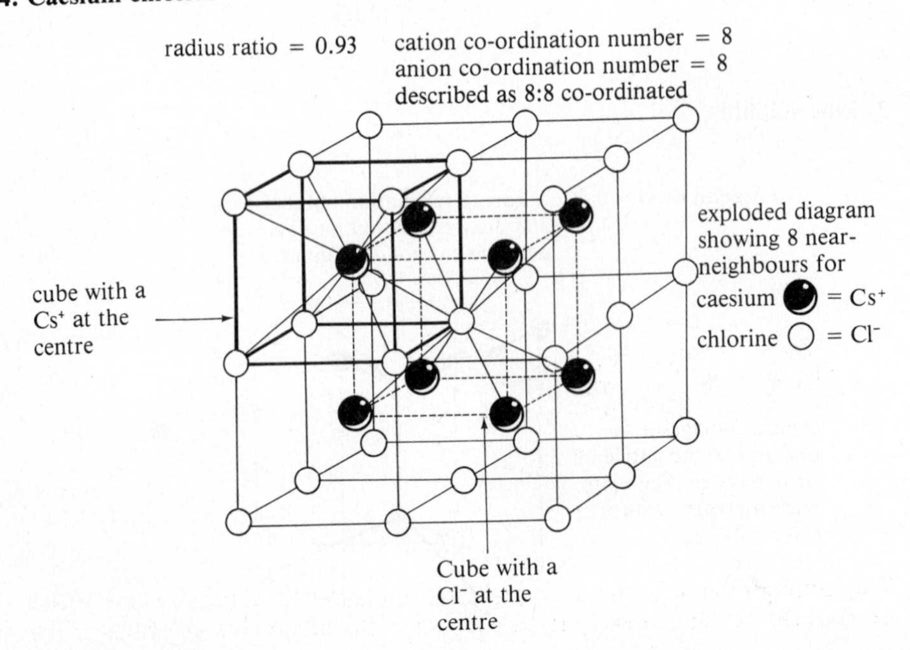

Layer lattices

A layer lattice is one in which there are layers of anions packed on top of each other. This comes about only in lattices where there is a stoichiometric ratio of cations:anions greater than one and where there is a high degree of covalent character.

A good example of a lattice like this is aluminium chloride. A sideways view of a small portion of the lattice is shown below; notice the layers of chloride ions with small aluminium ions sandwiched between alternate layers. The lattice contains double or 'sandwich' layers.

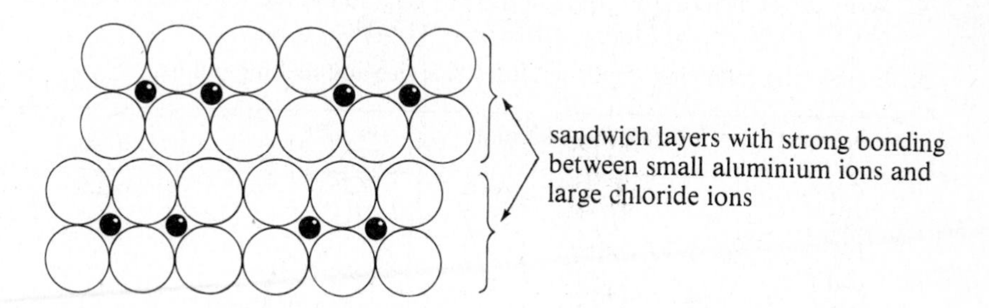

There are only weak forces between the sandwich layers because of the absence of aluminium ions between them. The Al_2Cl_6 molecules mentioned on page 87 can clearly be seen in the above diagram. A sandwich layer is shown below 'opened up'.

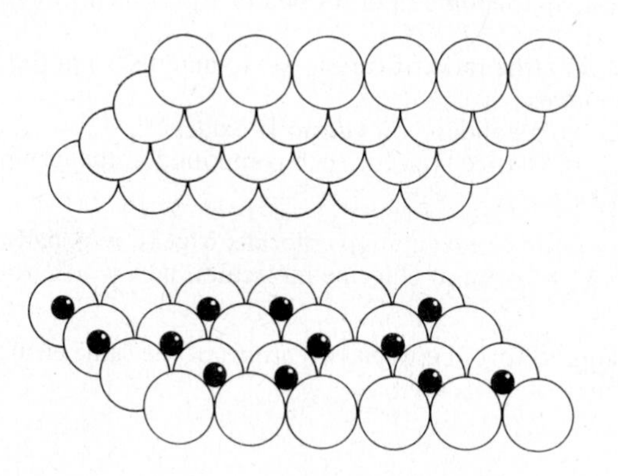

The cadmium iodide lattice is very similar but with continuous rows of metal ions instead of the gap every third position in the aluminium chloride lattice.

Questions

1. a) Describe the properties of a typical ionic substance.
   b) Account for these properties in terms of the structure of the compound.
2. Ammonium chloride dissolves in liquid ammonia yielding ammonium ions and chloride ions. Draw diagrams to show the orientation of the ammonia molecules around each of these ions.
3. Draw clear diagrams of the sodium chloride lattice and the caesium chloride lattice and use them to explain the fact that the sodium chloride lattice is described as 6:6 co-ordinated while the caesium chloride lattice is described as 8:8 co-ordinated.
4. Write down the electron configuration of each ion in the following substances:
   a) KF b) Na_2O c) $MgBr_2$ d) NH_4Cl e) LiH.
5. Explain the following trends in lattice energies and melting points.

substance	*lattice energy* kJ mol^{-1}	*m.p.*°C.
CsI	585	626
CsF	716	715
LiF	1022	870
CaF_2	2602	1392
$AlCl_3$	5103	183 (sublimes)

6. Account for the following data comparing sodium with aluminium chloride.

property	*solubility in water*	*solubility in benzene*	*conductivity when molten*	*conductivity in aqueous sol.*
sodium chloride	high	low	high	high
aluminium chloride	high	higher	lower	high

7. Using aluminium chloride as your example, draw and describe a layer lattice.
8. a) Using your data book, look up the ionic radii formed by the elements in groups I, II, VI and VII.
   b) Now calculate the radius ratio (the ratio of cation size to anion size) in the following groups of compounds.
   i) Group I halides ii) Group II halides iii) Group II oxides
   c) Use these results to predict the lattice type for each compound in the groups above.
9. If a student argued that the structure of magnesium chloride, $MgCl_2$, was made up of magnesium atoms packed into a lattice of chlorine molecules, how would you convince him that he was wrong?
10. Is it impossible for an ionic bond to form between two atoms of the same element?

Chapter 7

Molecular lattices

7.1 Physical properties

The range of bond types that can exist between two atoms is described on page 63 in Chapter 4. Covalent bonding is at one extreme of this range; ionic bonding is at the other. A high degree of covalency results when:

1. there is a small difference in electronegativity between the elements making up the compound
2. the possible cation produced is small and highly charged
3. the possible anion produced is large and highly charged.

These conditions are best met by a compound made from two non-metallic elements. Although some metal/non-metal compounds have appreciable covalent character (e.g. $AlCl_3$), they are not typical.

Covalent bonding between atoms leads to the formation of either simple molecules or macromolecules:

simple molecules H—H CCl_4

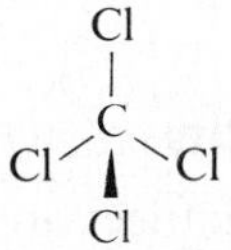

macromolecules 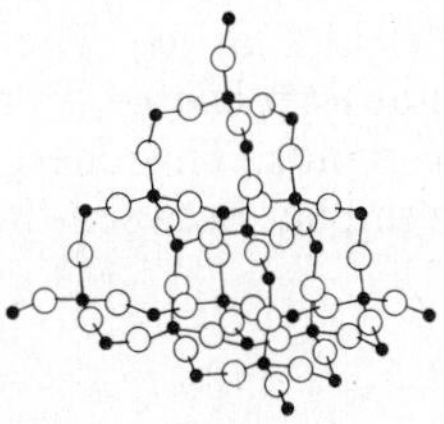 silica 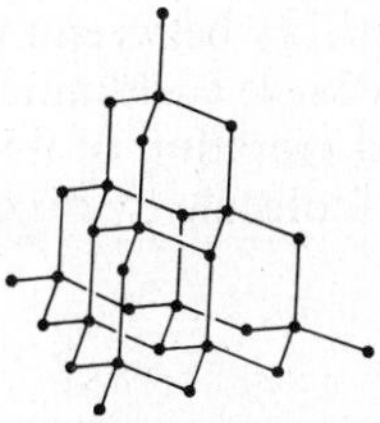 diamond

In this chapter, simple molecules only are discussed. Chapter 8 describes the properties of macromolecular lattices.

Mechanical strength

At room temperature, many molecular substances exist in the liquid or gas phase: a molecular lattice is far more easily disrupted than a metallic or ionic lattice. Compare the fixed points of some typical ionic, metallic and molecular substances.

type	*substance*	*m.p.* (°C)	*b.p.* (°C)
molecular	oxygen: O_2	−218	−183
	bromine: Br_2	−7	58.8
	ethanol: C_2H_5OH	−117	78.5
	sulphur: S_8	119	445
metallic	magnesium: Mg	650	1110
	iron: Fe	1535	3000
ionic	sodium chloride :NaCl	801	1465
	magnesium oxide :MgO	2640	3043

When the temperature is low enough for a molecular lattice to form, the solid produced is both weak and brittle. There are a few exceptions to this generalization, but it is true for the vast majority of simple molecular compounds. The force of attraction between neutral molecules is considerably weaker than that between either:

1. cations and delocalized electrons (metals),
2. cations and anions (ionic compounds).

The nature of inter-molecular forces is discussed in section 7.3.

Conductivity

Molecular lattices are both thermal and electrical insulators. There are no charged particles free to move through the lattice and so electrical conductivity is extremely small. Thermal conductivity depends on the rate of transfer of kinetic energy between the particles in a lattice. The rate is slow in a molecular lattice because each molecule can absorb energy separately in the form of increased vibration of its covalent bonds. This energy is only transferred to another molecule by means of an impact between the two molecules in the lattice.

Solubility

Many molecular solids that are insoluble in water dissolve well in the liquid phase of other molecular substances. These non-aqueous solvents are usually organic in nature: in other words they are molecular compounds of carbon. Some common examples of non-aqueous solvents are given in the table at the top of page 95.

As a general rule, molecular substances dissolve better in non-aqueous solvents than in water. Water is a strongly 'polar' solvent and the neutral molecules of many molecular lattices are unable to disrupt the forces of attraction between the oppositely charged dipoles of the water molecules. This prevents mixing of the two types of molecule.

solvent	*common name*	*formula*	*liquid temperature range* (°C)
ethanol	alcohol	C_2H_5OH	−117.0 ⟶ 78.2
ethoxy-ethane	ether	$C_2H_5.O.C_2H_5$	−116.0 ⟶ 34.5
propanone	acetone	$CH_3.CO.CH_3$	−95.4 ⟶ 56.2
benzene	benzene	C_6H_6	+5.5 ⟶ 80.1
methyl benzene	toluene	$C_6H_5CH_3$	−95.0 ⟶ 111.0

7.2 Molecular structure

Pairs of electrons

There are three types of pairs of electrons found in the structure of molecules. These are described in section 4.2. Their main features are summarized below.

1. A σ-pair

A pair of electrons in an orbital along the line of centres of two bonding atoms is called a σ-pair. A σ-pair is the same as a 'single' covalent bond.

σ-pairs

2. A π-pair

A pair of electrons in an orbital above and below the line of centres of two bonding atoms is called a π-pair. π-bonding only happens between two atoms that are already bonded together by a σ-pair. The combination of a σ-pair and a π-pair is the same as a 'double' covalent bond.

π-pair

σ-pairs

or

'double' bond
(1σ-pair and 1 π-pair)

3. A lone pair

An outer-shell pair of electrons under the influence of only one atom's nucleus is called a lone pair. The number of lone pairs of a bonded atom varies with

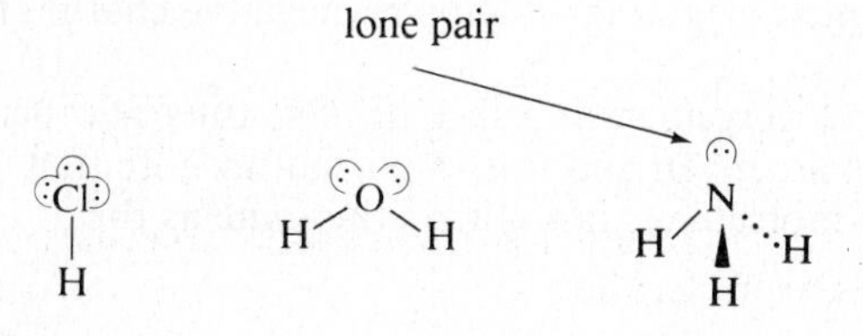

the group number of the atom. Halogen atoms usually have three, oxygen atoms two and nitrogen atoms one.

Molecular ions

An atom becomes an ion by transferring electrons to or from its valence shell. A molecule can undergo a similar process, and the particle formed is then a *molecular ion*. A molecular cation is most commonly produced as a result of proton transfer rather than electron transfer. Proton transfer is the characteristic of acidic particles and is described in detail in section 17.1. Some common molecular ions are listed below.

molecular anions	*molecular cations*
hydroxide: OH^-	hydroxonium: H_3O^+
carbonate: CO_3^{2-}	ammonium: NH_4^+
nitrate (V): NO_3^-	
sulphate (VI): SO_4^{2-}	

Molecular shape

The shape of a molecule can be predicted by two different methods:

1. by considering the distribution of the atomic orbitals of the central atom
2. by considering the effects of repulsion between the pairs of electrons in the valence shell of the central atom.

The first of these two methods is the more precise but is also the more complex. It is concerned with working out the state of hybridization of the central atom. This then determines the arrangement of the likely bonds formed by overlap of the hybrid orbitals with the atomic orbitals of the neighbouring atoms. A discussion of the principles of hybridization and orbital overlap can be found on page 57. In this section, only the second of the two methods will be described.

Electron-pair repulsion

The arrangement of electron pairs (either bonding or non-bonding) around an atom depends on the number of pairs.

Each electron pair occupies a localized molecular orbital and these orbitals are arranged so that they are a maximum distance apart. This is a direct result of the Pauli exclusion principle. Each orbital excludes any others from the volume of space that it occupies and so it appears as though the orbitals are repelling each other. As an occupied orbital is a region of negative charge, this is not unexpected behaviour.

The following table shows the arrangement in space of increasing numbers of electron pairs. Some examples are given and it is important to note that the description of the shape of a molecule is not always the same as the orbital arrangement because, while the total number of electron pairs controls the arrangement, shape refers only to the placing of atomic centres. For the sake of completion, the hybridization of the orbitals is given also.

no of charge clouds	*hybridization*	*arrangement of charge clouds*	*shape of the molecule*	*example*
2	sp	linear 	linear 	BeH_2
3	sp^2	planar trigonal 	planar trigonal 	BF_3
			bent 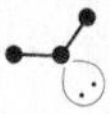	$SnCl_{2(g)}$
4	sp^3	tetrahedral 	tetrahedral 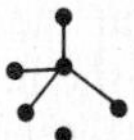	CH_4
			trigonal pyramidal 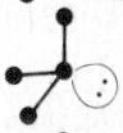	NH_3
			bent 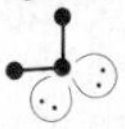	H_2O
5	sp^3d	trigonal bipyramidal 	trigonal bipyramidal 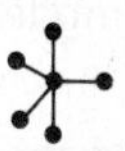	PCl_5
			distorted tetrahedral 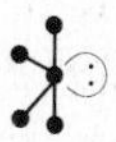	SF_4
			T-shaped 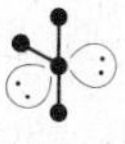	ClF_3
			linear 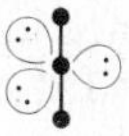	I_3^-
6	sp^3d^2	octahedral 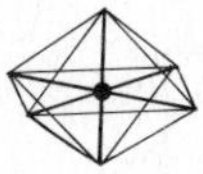	octahedral 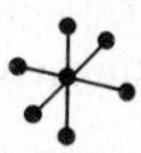	SF_6
			square pyramidal 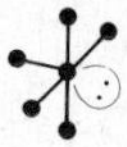	IF_5
			square planar 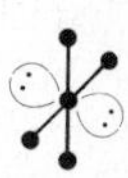	ICl_4

Bond angles

Two particles may have the same distribution of charge clouds but the angles separating their bonding pairs are often not the same. For example, a water molecule and an ammonia molecule both have four charge clouds around the central atom. However, the angle between the bonding hydrogen atoms is not the same in the two molecules.

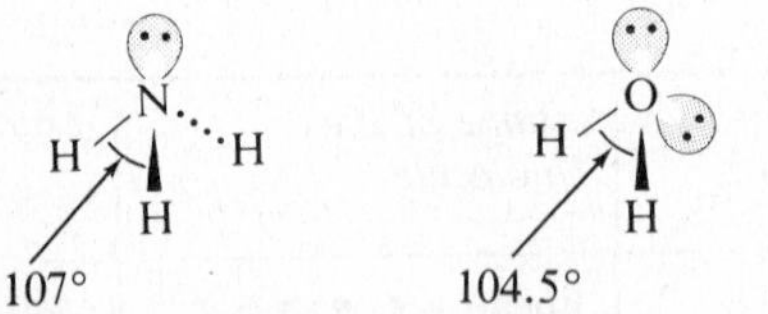

NH_3: 107° H_2O: 104.5°

Three refinements of the simple electron-pair repulsion theory are commonly included to account for variations of this sort:

1. Lone pairs repel more than bonding pairs. A lone pair takes up more space at the surface of an atom than does a bonding pair because it is under the influence of only one nucleus. This idea can be used to explain the difference in the bond angles above: an oxygen atom has two lone pairs whereas a nitrogen atom has only one. It can also be used to explain the observation that lone pairs always occupy the equatorial positions in a trigonal bipyramidal arrangement.
   There is more space available in the equatorial positions to accommodate the bulkier lone pairs.

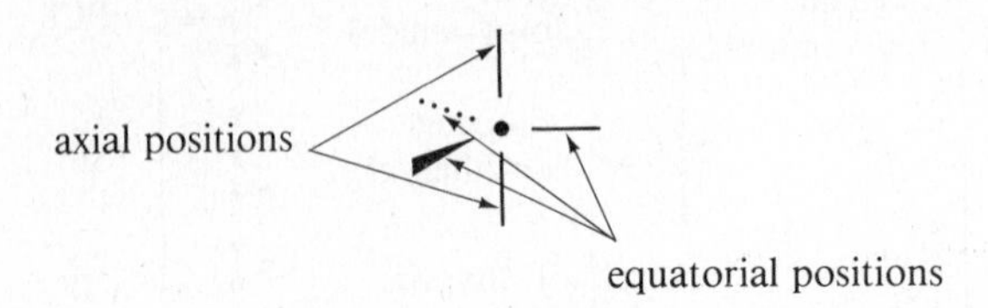

2. The space taken up by a bonding pair depends on the electronegativity of the bonded atoms. A bonding pair is distorted into a thinner volume the more electronegative the bonding atom. Compare the bond angles in the molecules of the hydride and fluoride of nitrogen; fluorine is more electronegative than hydrogen.

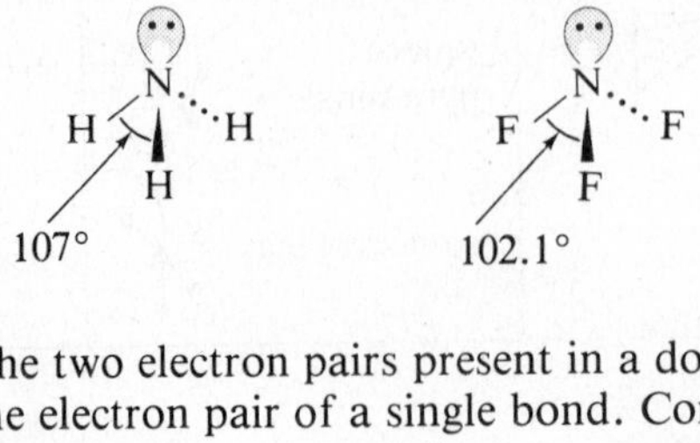

NH_3: 107° NF_3: 102.1°

3. The two electron pairs present in a double bond take up more space than the electron pair of a single bond. Compare the bond angles below:

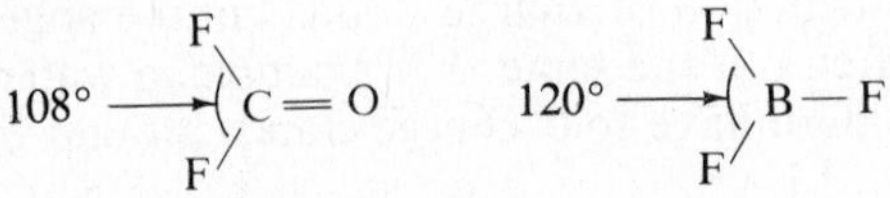

$F_2C{=}O$: 108° BF_3: 120°

Predicting shapes
Sidgwick and Powell originally put forward the theory for predicting a particle's shape by electron-pair repulsion. It was later refined by Nyholm and Gillespie and is often summarized by the following rules, which are named after one (or both) of these pairs of workers.

1. Electron pairs maximize their distance apart.
2. The space taken up by an electron pair depends on the electronegativity of the atom concerned.
3. The space taken up by electron pairs is in the following order:
   lone pair > double-bonding pair > single-bonding pair.

For a particle that has no double bonds (which usually means one with no oxygen atoms), e.g. SF_6, its shape can be predicted by the method set out below:

	method	
i)	count the total number of outer-shell electrons in the central atom	S has 6
ii)	add one if the particle is negatively charged or subtract one if it is positively charged (i.e. a molecular anion or cation)	SF_6 is neutral: 0
iii)	add one for each bonding atom	$6 \times F = 6$
iv)	divide the total by two, which gives the number of *pairs* of electrons	$\frac{6+6}{2} = \frac{12}{2} = 6$ pairs
v)	assume these pairs are placed symmetrically around the central atom	(S with six bonds)
vi)	count the number of bonded atoms: the extra pairs left over are lone pairs	none left over; SF_6 (S bonded to six F) octahedral

Here are four other examples: a) BF_3, b) PCl_4^+, c) BrF_3 d) I_3^-.

a) BF_3 i) 3
ii) 0
iii) 3
iv) $6/2 = 3$
v) (B with three bonds)
vi) Three bonds, therefore no lone pairs ⇒ BF_3 (B bonded to three F) trigonal planar

b) PCl_4^+ i) 5

ii) −1: there is a positive charge (it is a molecular cation).

iii) 4

iv) 8/2 = 4

v)

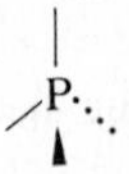

vi) Four bonds, therefore no lone pairs ⇒ 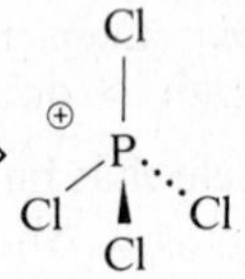 tetrahedral

c) BrF_3 i) 7

ii) 0

iii) 3

iv) 10/2 = 5

v) 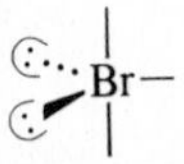 trigonal bipyramidal distribution of electrons

vi) Three bonds, therefore two lone pairs that go in equatorial sites ⇒ 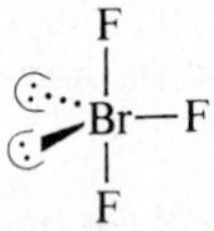 T-shaped

d) I_3^- i) 7

ii) 1 : there is a negative charge (it is a molecular anion).

iii) 2 : one central iodine atom bonded to two others.

iv) 10/2 = 5

v)

vi) Two bonds, therefore three lone pairs that go in equatorial sites ⇒ 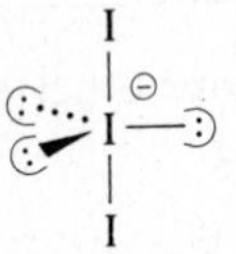 linear

For particles that *have* double bonds, the rules are applied a little differently. It is usually fair to assume that this concerns only those particles containing oxygen atoms. In oxy-molecules and oxy-anions, there are three types of bonded oxygen atom likely.

1. A σ- and π-bonded oxygen atom $=\ddot{O}$ e.g. $O=C=O$

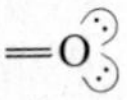

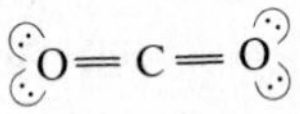

The experimental evidence suggests that all of the bonds have the same amount of double-bond character. **A, B** and **C** must therefore be seen as extreme forms of the actual structure, and are known as *canonical forms*. They can be compared with the way that the colour grey is often described in terms of the two extremes, black and white. Arguments of this sort are known as *resonance*. Resonance identifies all the extreme positions that are possible in a particular issue so that the truth appears to owe a little to each extreme. As applied to the theory of chemical structure, the bonding in a particle is represented by a number of canonical forms. Each canonical form is a different valence-bond structural drawing of the same particle, like **A, B** and **C** above. The actual structure of the particle is called the *resonance hybrid* of the canonical forms. The idea of resonance can be applied to *any* particle whose structure cannot be represented accurately by a single valence-bond diagram.

For example, resonance helps to describe the bonding in:

1. organic molecules, such as the benzene molecule, see page 218.
2. oxy-anions, such as the sulphate (VI) and nitrate (V) ions.

Benzene molecule

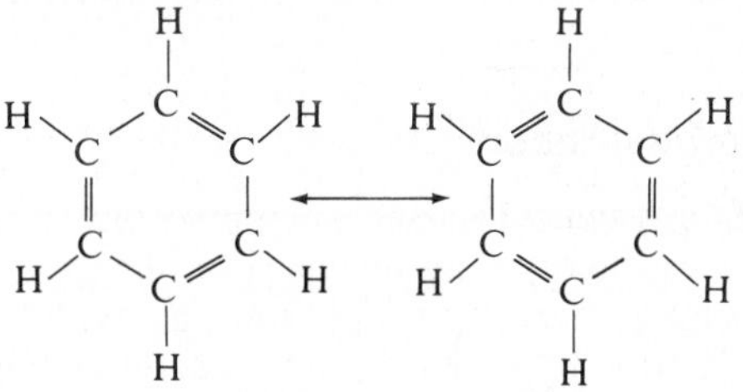

Sulphate (VI) ion, SO_4^{2-}

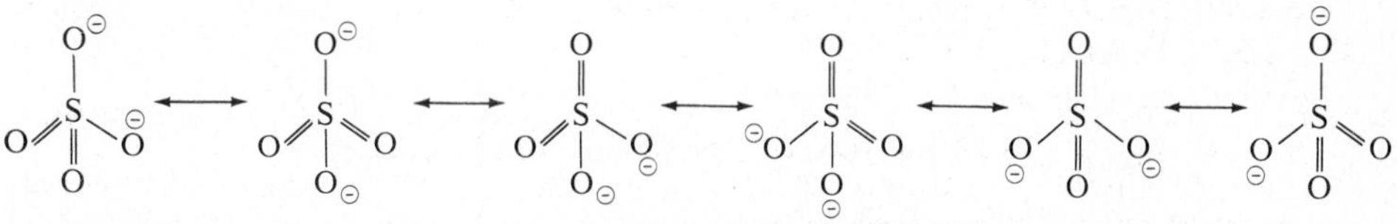

Six different canonical forms of the sulphate (VI) ion can be drawn. When comparing the stability of two similar structures, the number of canonical forms possible for each structure is a useful guide. In general the particle with the greater number of canonical forms is the more stable. For example, the relative stability of the sulphate (VI) ion and of the thiosulphate (VI) ion can be explained in part by the fact that only three significant canonical forms can be drawn for $S_2O_3^{2-}$:

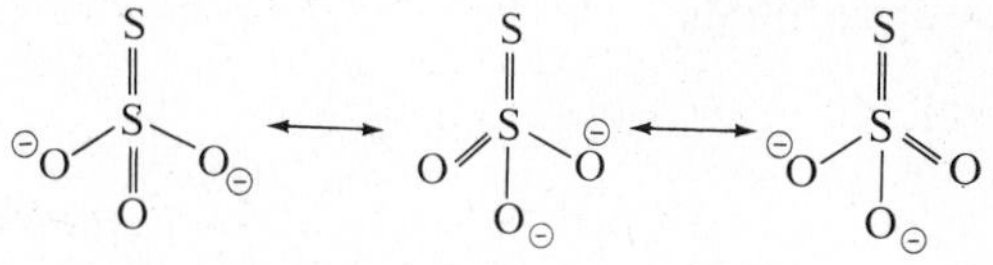

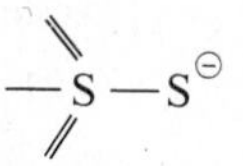

The canonical forms containing $-S-S^{\ominus}$ are discounted because the negative charge is unlikely to be localized on the sulphur atom. The sulphur atom competes for the negative charge with three oxygen atoms which are all more electronegative than it is.

Nitrate (V) ion, NO_3^-

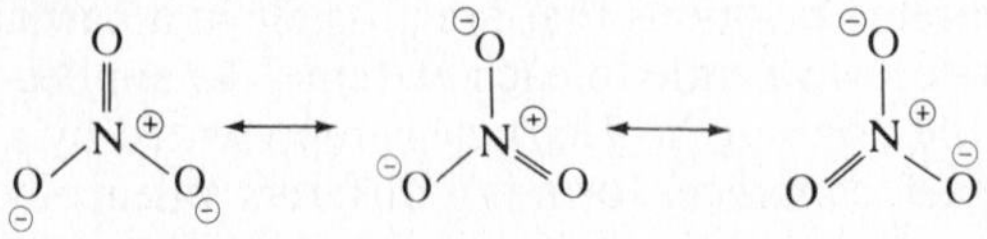

A nitrogen atom that has formed four covalent bonds is always positively charged, for example

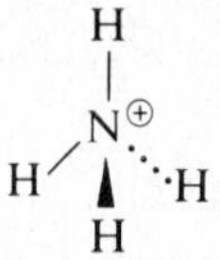

The lone pair has been used to form an extra bond: a 'dative bond'. The nitrogen atom controls only its half share (on average) of this new bond, and so is shown positive because it has effectively lost control of one electron as a result.

Resonance and the molecular orbital theory

There are two important theories of covalency; resonance applies only to the valence bond theory. Molecular orbital theory does not need to rely on the ideas of resonance because a structure is shown to result from the overlap of the atomic orbitals of the separate atoms. In the case of the carbonate ion described on page 102, there is π-overlap between all the $2p_y$ orbitals of the atoms.

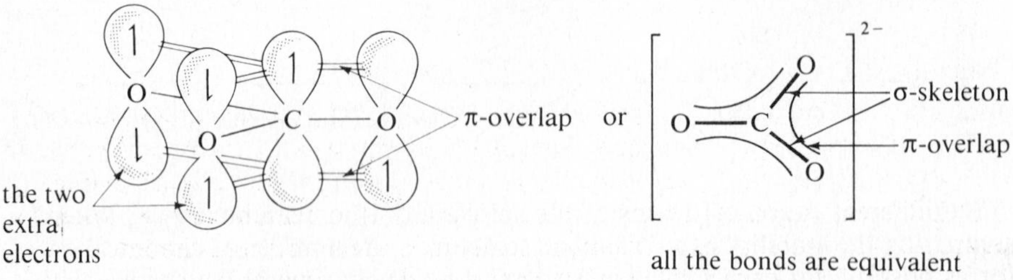

π-overlap or
σ-skeleton
π-overlap
the two extra electrons
all the bonds are equivalent

The same pattern repeats itself in the nitrate (V) ion. The structures are identical except that a nitrogen nucleus has one more proton than a carbon nucleus. Both structures have a total of six electrons in the π-molecular orbitals produced by the overlap of four $2p_y$ orbitals

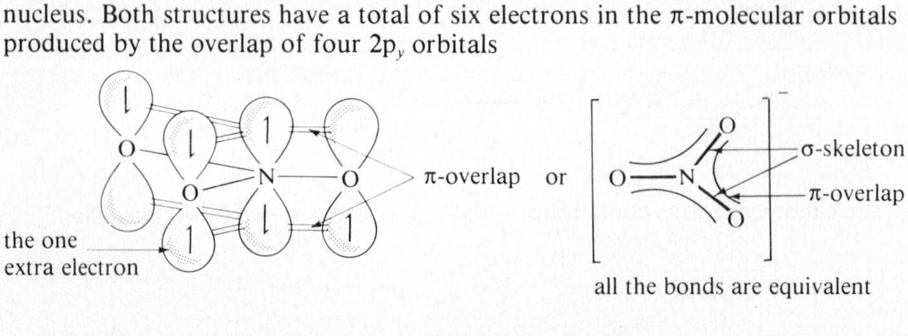

π-overlap or
σ-skeleton
π-overlap
the one extra electron
all the bonds are equivalent

The orbital drawing for the sulphate (VI) ion is more complicated. Its structure is usually shown like this:

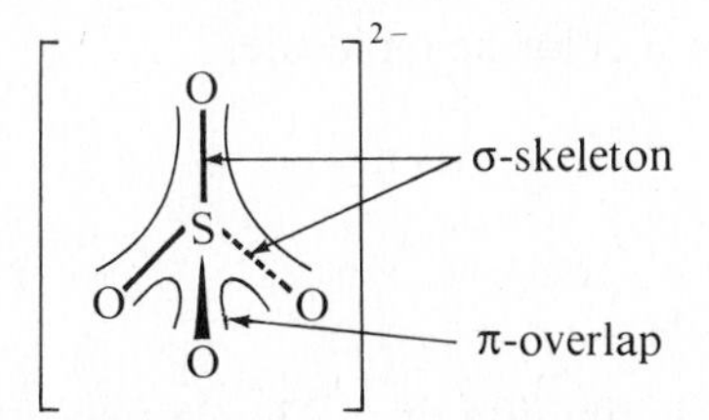

Again all the bonds are equivalent.

In a benzene molecule, the $2p_y$ orbitals of all six carbon atoms overlap to produce π-electrons above and below the σ-skeleton:

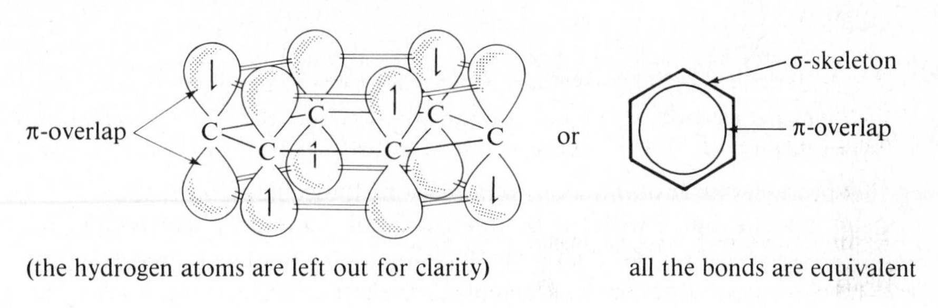

(the hydrogen atoms are left out for clarity) all the bonds are equivalent

7.3 Inter-molecular forces

Van der Waals' forces

Molecular elements and even the atomic noble gases can be liquefied and then solidified by cooling. This implies that there are forces of attraction even between particles that possess no charge separation. For example, an iodine molecule has no charge separation: both iodine atoms attract the bonding pair equally. Yet iodine is a solid at room temperature, which suggests that there are quite strong forces of attraction between the neutral molecules.

A clue to the nature of these forces is provided by the figures in the table below.

molecule	*number of electrons in the molecule*	*b.p. of the liquid system* (°C)
H_2	2	−253
N_2	14	−196
O_2	16	−183
Cl_2	34	− 35
I_2	126	+185

The number of electrons in each molecule appears to affect the force of attraction of one molecule for the next. The stronger the attractive force, the higher is the liquid's boiling point.

Consider a picture of the charge cloud of a chlorine molecule.

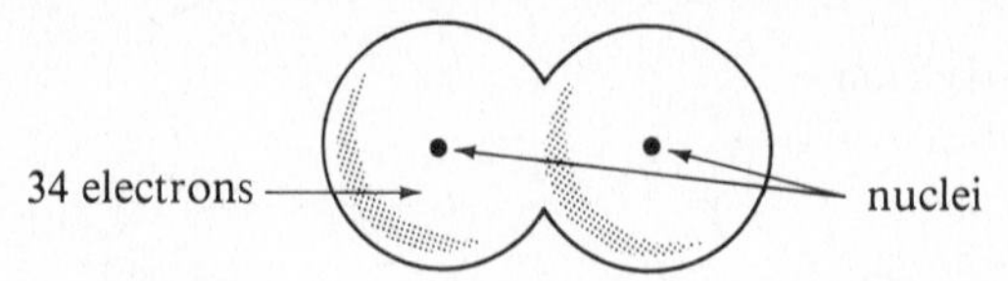

It is quite possible at any moment for the distribution of the 34 electrons to be uneven through the molecule. By chance there may be momentarily a greater electron density at one end of the molecule than the other:

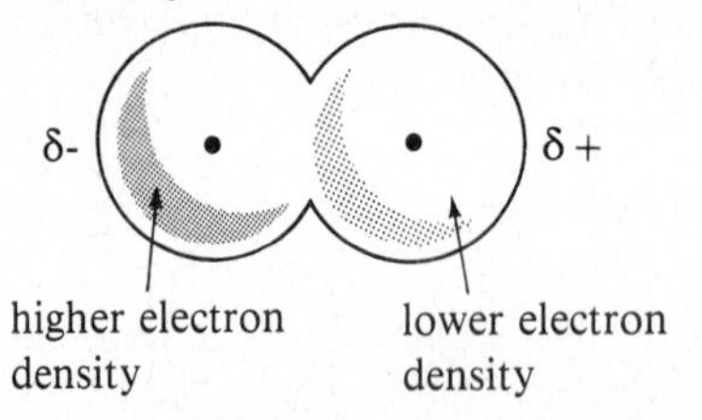

This produces an *instantaneous dipole* in the molecule. Contrast an instantaneous dipole with the permanent dipole caused by a difference in electronegativity e.g. $\overset{\delta+}{H} - \overset{\delta-}{Cl}$. The instantaneous dipole has the effect of inducing a second dipole on a neighbouring molecule:

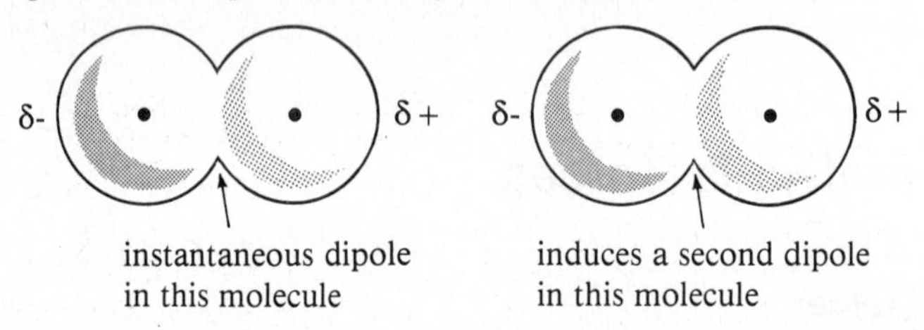

> The weak force of attraction between the neighbouring oppositely charged ends of two instantaneous dipoles is called a *van der Waals' force.*

The greater the number of electrons in a particle, the more likely the electrons are to be momentarily localized at one part of the particle in preference to another. The strength of the van der Waals' force therefore increases with electron number.

Dipole-dipole forces

When a molecule contains atoms of different elements, the bonding electrons between the atoms will not be evenly shared. The atom of the element with the higher electronegativity attracts the bonding electrons more strongly. For example, in hydrogen chloride:

	H $\overset{x}{o}$ Cl	➡	H $\overset{x}{o}$Cl
electronegativity	2.1 3.0		

The hydrogen atom becomes *electron deficient* and the chlorine atom becomes *electron excessive*. This results in *charge separation* along the bond shown like this:

$$\overset{\delta+}{\mathrm{H}}—\overset{\delta-}{\mathrm{Cl}}$$

A *permanent dipole* exists in the molecule.

The permanent dipoles produced as a result of the charge separation lead to two types of molecule:

1. A molecule with a dipole moment. Here the charge separation is not symmetric through the molecule and there are recognizable positive and negative ends to the molecule, e.g.

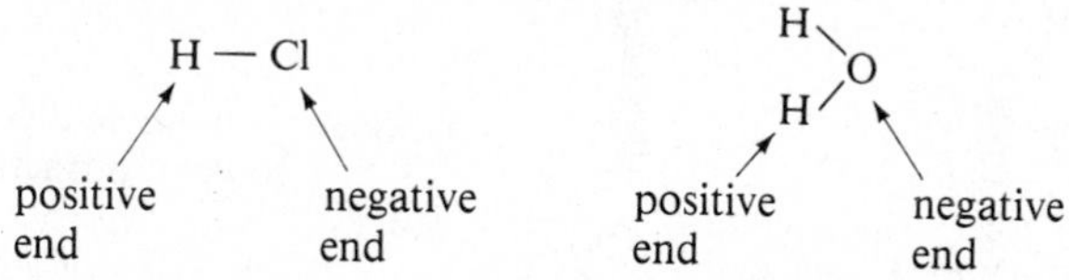

2. A molecule with more than one dipole but *no* overall dipole moment. Here the charge separation is symmetric throughout the molecule, so that the effects of the separate dipoles work against one another, e.g.

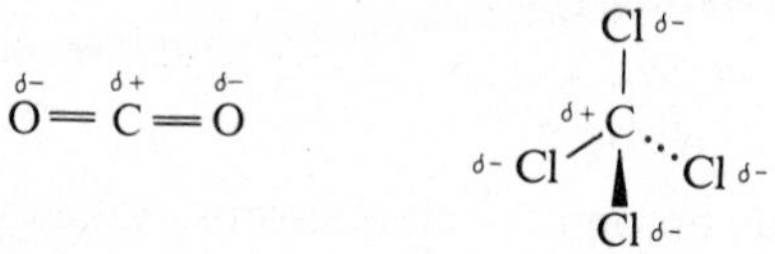

no negative and positive ends

The centres of the molecules are electron deficient and their outside parts are electron excessive, but there are no positive and negative ends. Molecules of this sort do not line up in any particular orientation when an electric field is applied to the system. A dipole moment exists when a charge of $+q$ is separated from a charge of $-q$ by a distance d across the whole length of the molecule. The size of the dipole moment $= q.d$ and is given in units of Debyes (D): 1D $= 3.338 \times 10^{-30}$ coulomb metres.

molecule	⊕ *dipoles* ⊖	*dipole moment* (D)
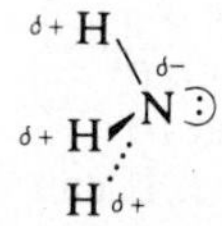	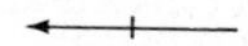	1.48
$\overset{\delta+}{\mathrm{H}}—\overset{\delta-}{\mathrm{Cl}}$	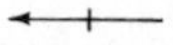	1.05
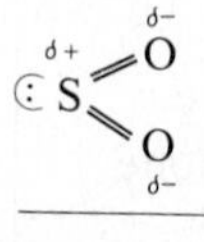		1.63
$\overset{\delta-}{\mathrm{O}}=\overset{\delta+}{\mathrm{C}}=\overset{\delta-}{\mathrm{O}}$	 cancelling	0.00
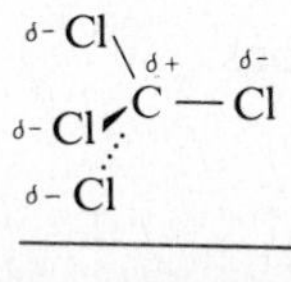	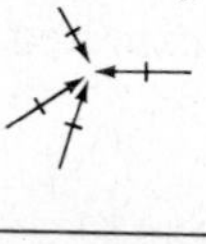	0.00

Two neighbouring molecules attract each other as a result of the force of attraction between their dipoles, e.g. by simple van der Waals' considerations carbon dioxide might be expected to boil at a temperature of about −120°C. The molecule has 22 electrons and this number lies between that of O_2 (16) and Cl_2 (34) whose b.p's are −183°C and −35°C respectively. In fact, carbon dioxide changes phase at a temperature some 40°C higher than that predicted by electron number. Its lattice is held together by more than just simple van der Waals' forces: there are attractive forces between the opposing dipoles.

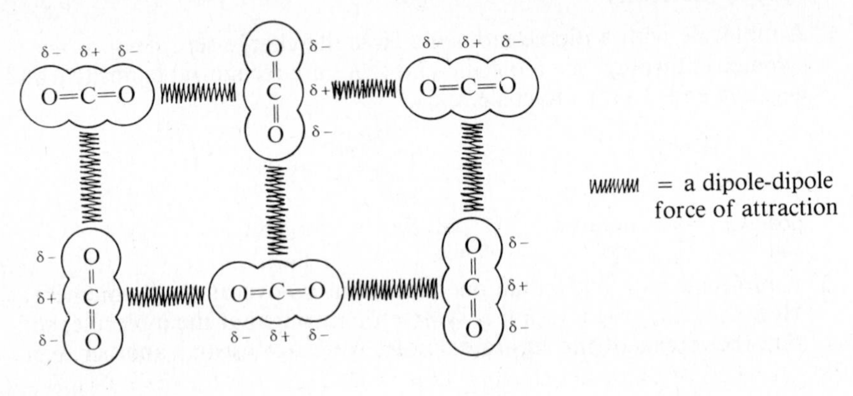

Hydrogen bonding

A hydrogen atom is unique because it has no inner-shell electrons. When a hydrogen atom bonds to a more electronegative atom, the resulting molecule has in effect a covalently bonded but poorly shielded proton in its structure, e.g. a water molecule

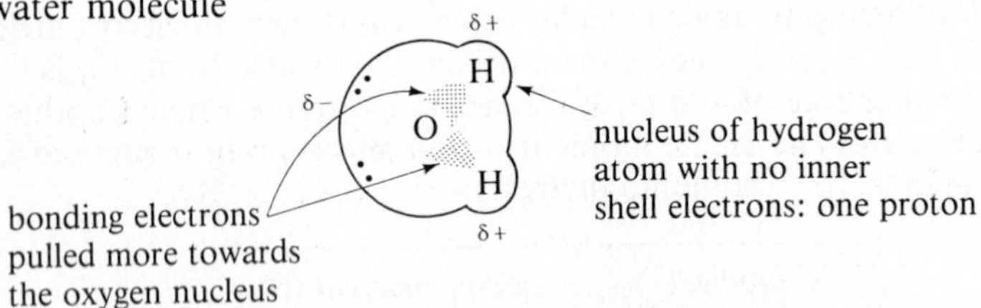

The poorly shielded proton in one molecule is attracted to a lone pair of electrons in a neighbouring molecule producing a particularly strong type of dipole-dipole force. This is at its most marked when hydrogen is bonded to nitrogen, oxygen or fluorine and the force is known as a *hydrogen bond*. It can be formally defined as follows.

> A hydrogen bond is the electrostatic force of attraction between the poorly shielded proton of a hydrogen atom bonded to a small highly electronegative atom such as nitrogen, oxygen or fluorine, and the lone pair of a neighbouring molecule.

Hydrogen bonding leads to some otherwise unexpected observations.

Consider the boiling points of the hydrides shown below:

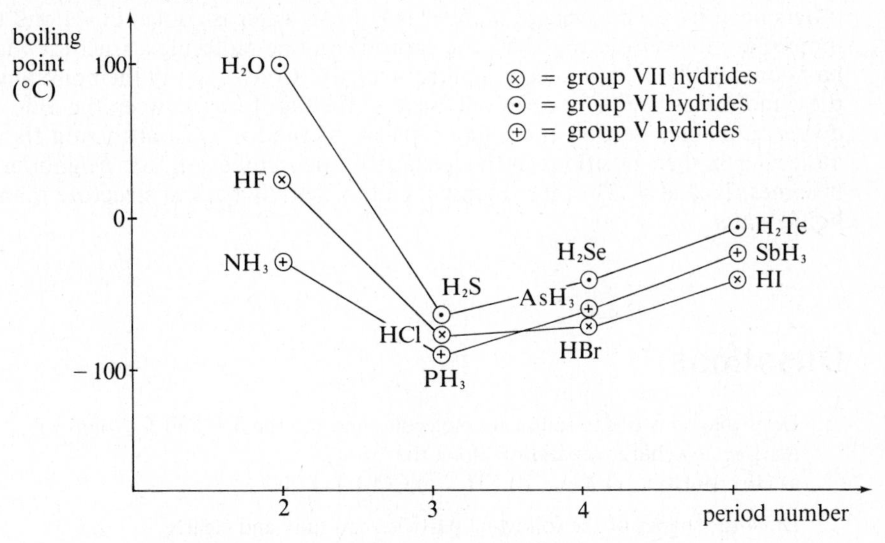

The hydrides in period 2 all have much higher boiling points than are expected from the trend of decreasing boiling point with number of electrons per molecule. Hydrogen bonding between these molecules is at a maximum because the atoms in period 2 are:

1. The most electronegative, and therefore cause the least shielding of the protons.
2. The smallest, and therefore the lone pairs are most strongly attracted to neighbouring unshielded protons.

Ice

The structure of ice is determined by the strong hydrogen bonds formed between neighbouring water molecules. It is interesting to note that solid water is less dense than liquid water. This can be explained as follows:

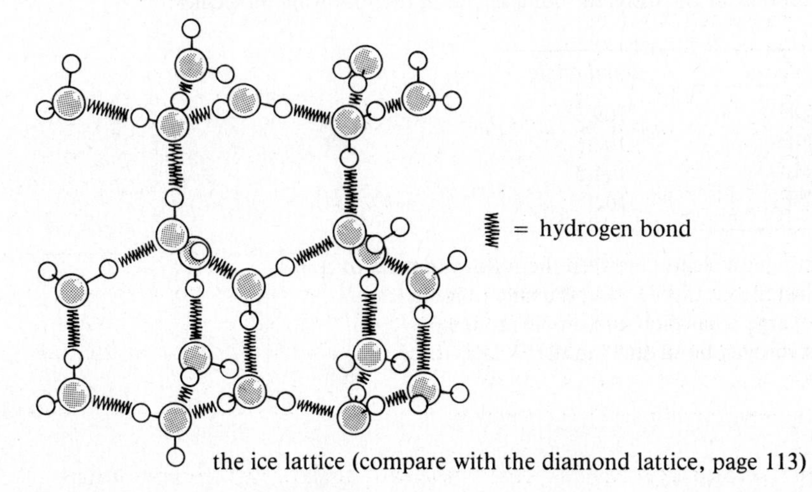

the ice lattice (compare with the diamond lattice, page 113)

In the liquid phase of water, some molecules have four near-neighbours while others have five. The average number is 4.4. As water is cooled down and the molecules move closer together, the protons on one molecule attract the lone pairs on neighbouring molecules more strongly. Consequently the bond angle in these molecules begins to open out because the repulsion between the lone pairs decreases as they become more like σ-pairs. As the forces of attraction fix the molecules in their positions in the ice lattice, the number of near neighbours becomes fixed at 4. Thus ice represents a less densely packed structure than liquid water.

Questions

1. Draw diagrams of the following molecules and use the δ + and δ – signs to mark in any charge separation along the bonds.
   a) HF b) H_2O c) NH_3 d) CH_4 e) CO_2 f) $CHCl_3$
2. Draw the shapes of the following particles carefully and clearly.
   a) SO_2 b) CS_2 c) O_3 d) BCl_3 e) CCl_4 f) SF_4 g) PCl_5 h) PCl_4^+
   i) PCl_6^- j) IF_7
3. By carefully drawing diagrams showing their shapes, compare the dipole moments of the following pairs of particles:
   a) CCl_4 and SCl_4 b) BF_3 and NF_3 c) CO_2 and SO_2 d) PF_5 and IF_5
   e) ICl_4^+ and ICl_4^-
4. Using the sulphate ion as your example, explain clearly the meaning of the following terms:
   a) resonance b) canonical form c) resonance hybrid.
5. Draw diagrams which show the differences in shape and in charge distribution for the following ions:
   a) CO_3^{2-} b) NO_3^- c) SO_3^{2-} d) ClO_3^-
6. Account for the different bond angles in the following molecules:

molecule	*bond angle*
CH_4	109.5°
NH_3	107°
H_2O	104.5°
NF_3	102°

7. Distinguish clearly between the following pairs of terms:
   a) instaneous dipole and permanent dipole
   b) charge separation and dipole moment
   c) hydrogen bond and van der Waals' force.

8. Account for the difference in boiling point of the following substances.

substance	*boiling point* °C
Ne	−246
O_2	−183
I_2	183
C_2H_6	−88
C_2H_5OH	78.5
CH_3CO_2H	117.5

9. a) Phosphorus (V) chloride decomposes on heating.
   $PCl_5 \rightarrow PCl_3 + Cl_2$.
   Draw the structure of PCl_5 and then explain whether it is from the axial or equatorial positions that the chlorines break away.
   b) Experimental evidence indicates that solid phosphorus (V) chloride consists of ions: the cation is tetrahedral in shape while the anion is octahedral. However PCl_5 is a very poor conductor in both solid and liquid states. Account for these facts.

10. Using appropriate examples, outline the factors that influence the shape of a covalent molecule.

Chapter 8

Macromolecular lattices

8.1 Physical properties

Macromolecular substances have quite distinct properties. These contrast sharply with those of molecular substances despite the fact that both types have covalently bonded particles. These characteristic properties are:

1. Very high fixed points
Macromolecular substances have some of the highest fixed points of all, as the values below show.

substance	*m.p.* (°C)	*b.p.* (°C)
boron	2030	
carbon: graphite	3730	4830
silicon dioxide	1700	

2. Hardness and strength
Macromolecular substances include the hardest of all known materials. Both diamond and boron nitride, the two hardest solids, are macromolecular in structure. However, they also include flaky materials like mica, talc and slate, and fibrous substances such as asbestos, nylon and silk. These differences in physical properties directly reflect the structure of the macromolecules as the later sections of the chapter explain.

3. Insolubility
Macromolecular substances do not dissolve in polar or non-polar solvents.

8.2 Macromolecular structure

The hardness, insolubility and high fixed points of macromolecular substances result from the extended covalent bonding present in the lattice. A molecular substance contains discrete and separate covalently bonded particles in a lattice held together by weak dipole forces; a macromolecular substance contains atoms that are covalently bonded throughout the lattice. The bonding can extend in all three dimensions (as in the case of diamond), or only in two dimensions (as in the case of graphite).

In any process that involves the disruption of the lattice, huge amounts of energy are required to overcome the covalent bonds. A perfect diamond is in fact one single giant molecule.

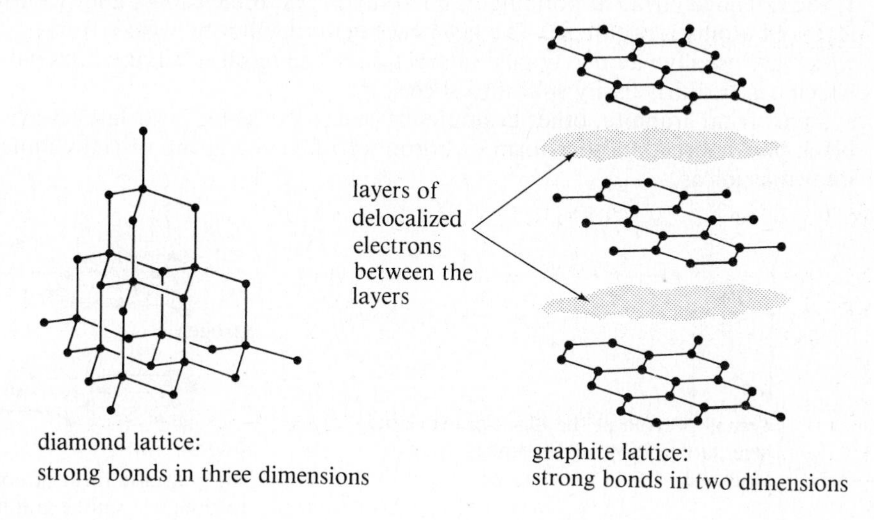

diamond lattice:
strong bonds in three dimensions

graphite lattice:
strong bonds in two dimensions

8.3 Lattice types

Extended covalent bonding in three dimensions

If the network of covalent bonding extends in all three dimensions, the macromolecular substance is an extremely hard solid with a very high melting point. Apart from diamond, two other examples of this type of lattice can be seen in silicon dioxide $(SiO_2)_n$ or quartz, and in boron nitride $(BN)_n$. Boron nitride is isoelectronic with carbon because each boron atom has three electrons while each nitrogen atom has five.

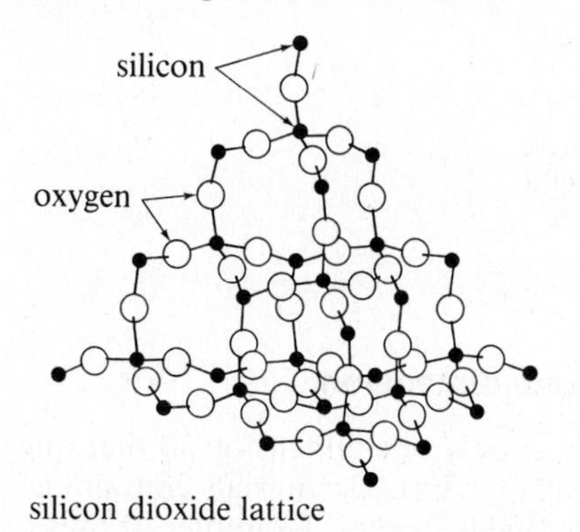

silicon dioxide lattice

there are several forms of quartz which differ mainly in the size of the Si-O-Si bond angle.

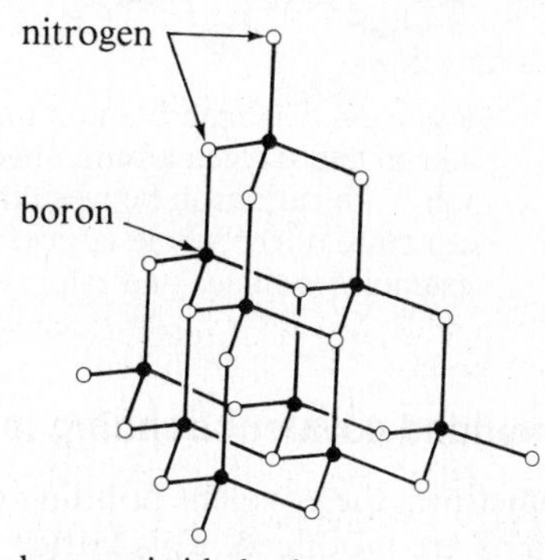

boron nitride lattice

this is the diamond form of the compound

Extended covalent bonding in two dimensions

If the extended covalent bonding occurs only in two dimensions, a lattice with layers of atoms is produced. The layers are held together by weaker forces which are usually van der Waals' interactions. The result is a flaky material which can be fairly easily split into sheets.

Apart from graphite, other examples of macromolecular layer lattices are black phosphorus, another form of boron nitride and a group of flaky minerals known as micas.

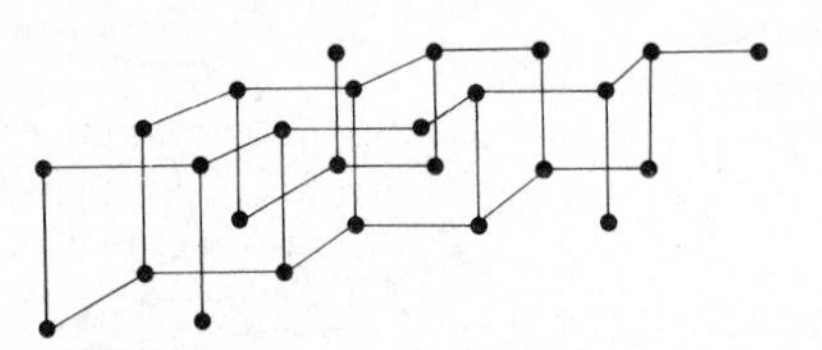

a small portion of the black phosphorus layer lattice showing the linked zig-zag structure.

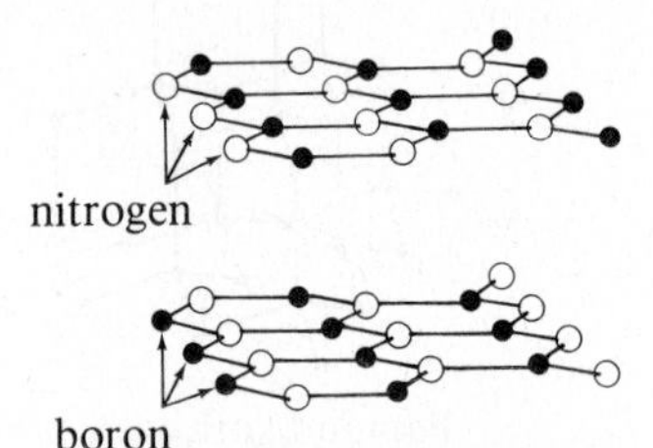

the graphitic form of boron nitride; it is unlike graphite in that it is an insulator.

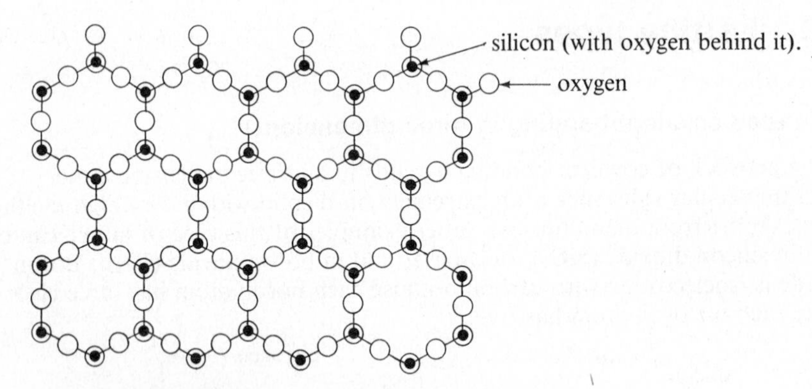

a portion of a macro-anion made up of silicon and oxygen atoms; sheets of these ions with cations in between make up the structure of the whole lattice; common examples are mica and talc.

Extended covalent bonding in only one dimension

Sometimes the covalent bonding extends in only one dimension so that huge chains are produced. This lattice type leads to fibrous materials containing giant molecules held together by van der Waals' forces. Examples of these include: silicate asbestos, organic polymers such as proteins and carbohydrates, and synthetic polymers such as polythene.

Parts of silicate chains: two forms of asbestos

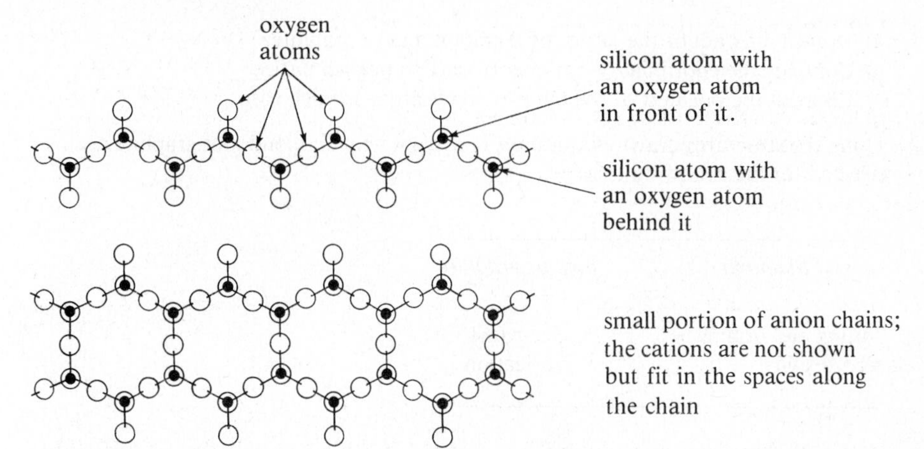

Part of a protein chain

R represents various short carbon side chains

Part of a starch molecule: a carbohydrate

Part of a hydrocarbon chain: polythene

Questions

1. a) Draw a diagram of the lattice of diamond and of graphite.
   b) Compare the coordination number of carbon in each lattice.
   c) Contrast the position of the outer-shell electrons in each case.

2. Using the structures drawn in question 1, explain why diamond and graphite are selected for the following uses:

uses of diamond	*uses of graphite*
cutting and drilling	electrodes
record styli	lubrication

3. Carbon dioxide and silicon dioxide are both covalently bonded substances but with very different properties.
   a) Look up and list their fixed points.
   b) Compare their solubilities in water.
   c) Draw small portions of their solid lattices.
   d) Use these diagrams to account for the difference in the properties of the two oxides.

4. To what extent is it possible to relate the physical properties of a macromolecular substance to its lattice structure?

5. Is it possible to produce a list of typical macromolecular properties? If so list them and account for them.

Chapter 9
Phase changes

9.1 Phase equilibria

States and phases

Matter can exist in one of three states: solid, liquid, or gas. But within each state of matter there can be regions containing different substances. For example, concrete contains pebbles and grains of sand in a matrix of cement. Similarly, a mixture of oil and water contains two distinct regions which are known as phases.

> A *phase* is any homogeneous part of a system which remains physically distinct from the rest of the system.

In each of the two examples above, the phases are in the same state. The pebbles are separated from the sand and the cement by a *phase boundary*. However much the oil and water are mixed, there remains a distinct boundary or interface between the two liquids; the system consists of two phases in one state of matter.

Although in these examples the different phases are in the same state, this need not always be so. A half-full bottle of water contains a liquid phase and a vapour phase; these are separated from each other by the surface of the liquid which is the phase boundary.

Phase diagrams

Pure substances can change phase if the conditions of temperature or pressure change. A representation of the conditions under which each phase exists is called a *phase diagram*. The phase diagram for carbon dioxide shown below illustrates the typical characteristics.

the phase diagram for carbon dioxide

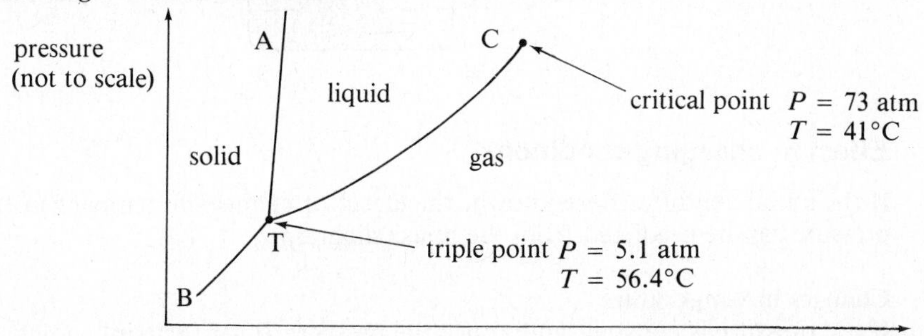

The diagram is divided into three areas by curves which form a Y-shape that slopes to the right. Lines BT and TC are two separate curves that meet at T. The three areas are labelled solid, liquid, and gas. Variation of temperature or pressure within any one of these areas produces no change of phase.

The curve between each area shows the conditions under which the two phases are in equilibrium. So a point on AT shows both the temperature and pressure under which a mixture of solid and liquid remains indefinitely in equilibrium. This is because under these conditions, solid melts to liquid at exactly the same rate as liquid freezes to solid. Because these two opposing processes occur at exactly the same rate, they cancel out the effect of each other. For every gram of solid that melts, a gram of liquid freezes. This is an example of dynamic equilibrium; it is shown using an equilibrium sign:

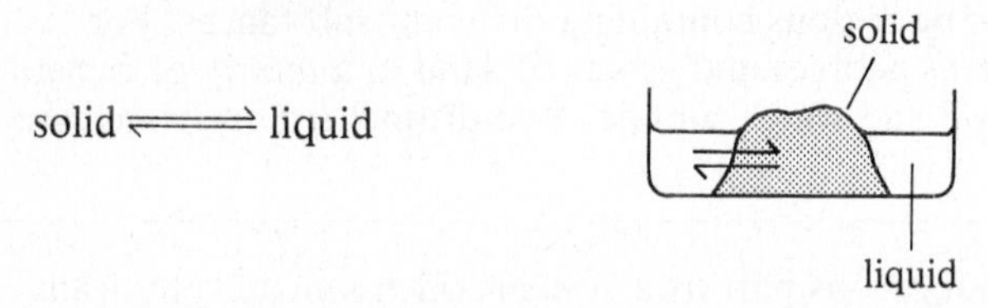

Curve TC represents the range of conditions under which liquid and gas phases are in equilibrium in an isolated system.

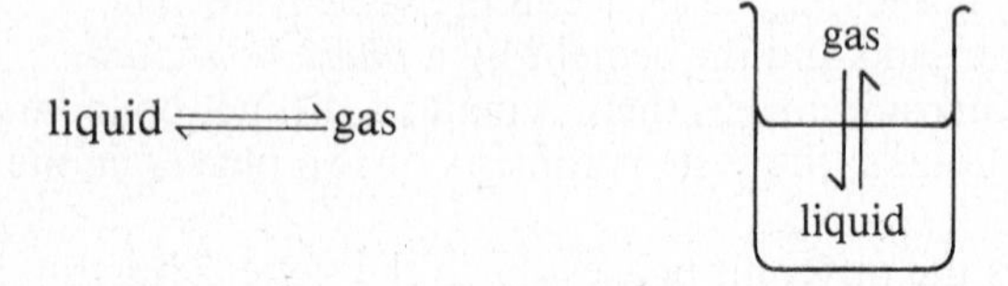

Curve BT represents the range of conditions under which solid and gas phases are in equilibrium in an isolated system.

The point T is called the triple point. This defines the unique conditions under which all three phases are in equilibrium with each other in an isolated system.

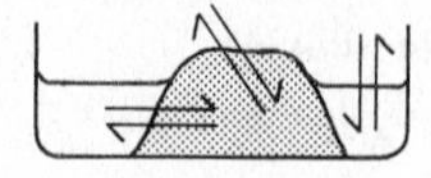

Effect of changing conditions

If the initial conditions are known, the effect of changes in temperature and in pressure can be predicted from the phase diagram.

Changes in temperature

If the pressure is constant and above the pressure P_T of the triple point, an increase in temperature causes first melting at a temperature T_m and then boiling at a temperature T_b. The temperature at which each phase change occurs is found from the point where the line of constant pressure P_1 crosses the phase line.

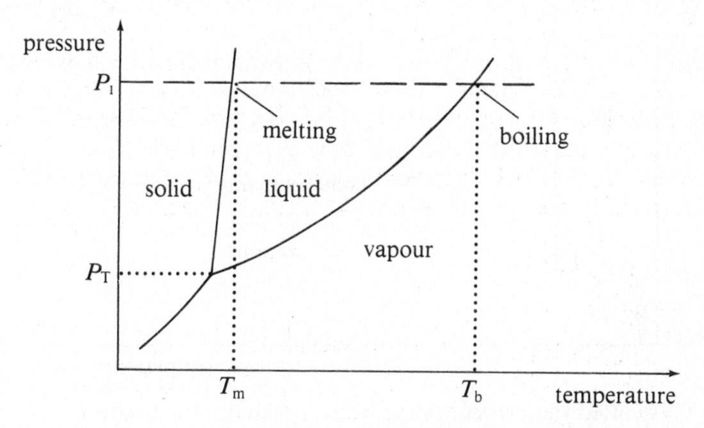

A decrease in temperature under these conditions will produce first condensation and then freezing. An increase in temperature at a higher constant pressure results in melting and boiling at higher temperatures. These temperatures at which the phase changes occur are shown by the points at which the new constant pressure line cuts the phase line.

If the pressure is constant and below the pressure of the triple point, an increase in temperature causes a phase change from solid direct to vapour; this type of phase change is called *sublimation*. For carbon dioxide P_T is 5.1 atmospheres. So at atmospheric pressure carbon dioxide ('dry ice') sublimes at −78°C because the vapour pressure of carbon dioxide is well below 5.1 atmospheres.

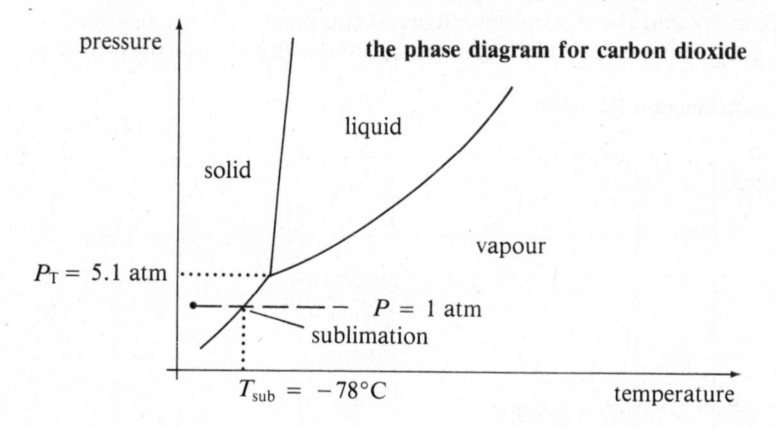

Changes in pressure

The general shape of the phase diagram shows that an increase in pressure at constant temperature usually produces only a change of phase from gas to liquid. However, if the temperature is just above that of the triple point, the substance may go through all three phases.

An increase in pressure at constant temperature on the liquid phase produces solidification; increase in pressure on the gas phase produces condensation. A decrease in pressure produces melting and boiling respectively.

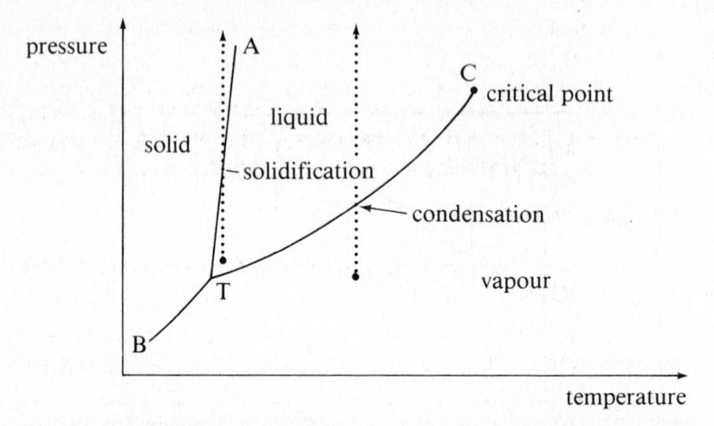

The point C is called the *critical point*; this represents the highest temperature at which the gas can be compressed to a liquid by change of pressure alone. Above the critical temperature the phase is called a gas; its behaviour approximates to the ideal behaviour of gases described in the next chapter. Below the critical temperature the phase should properly be called a vapour.

The phase diagram for water

The hydrogen bonding of water results in a number of abnormal properties. One of these is the diamond-like structure of ice (see page 109) in which each water molecule is bonded to four neighbouring molecules. As the temperature rises, the ice melts and the disruptive effects of the kinetic energy begin to overcome the ordering effect of the hydrogen bonds. But the strength of these

the phase diagram for water

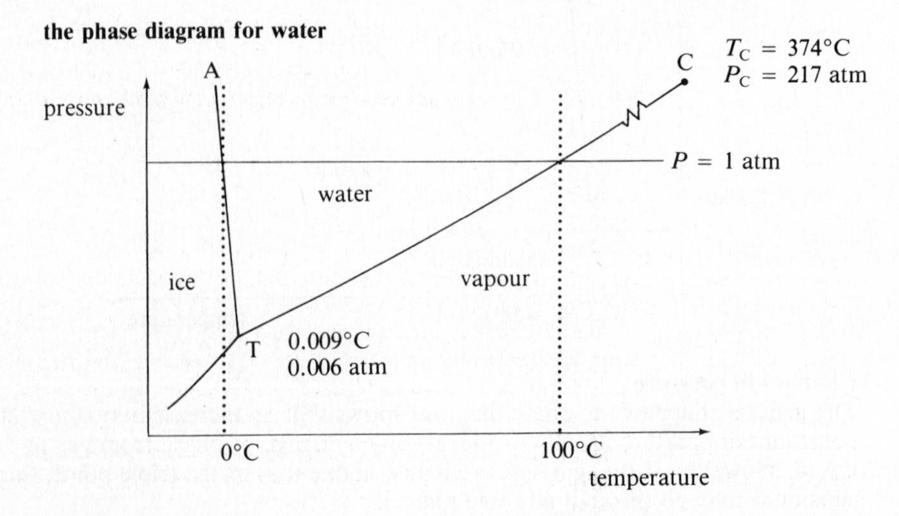

bonds causes the number of neighbouring molecules to increase to an average of 4.4 in the liquid, so that cold water is denser than ice and the ice floats.

This also leads to an important difference in the phase diagram for water. Instead of sloping to the right, the line TA between solid and liquid slopes slightly to the left. Compare with the phase diagram for carbon dioxide.

Unlike carbon dioxide and nearly all other solids, ice can be caused to melt by an increase in pressure at constant temperature. The line of increasing pressure crosses the phase line TA from solid to liquid.

A second major difference between the phase diagram for water and that for carbon dioxide is that the pressure at the triple point is well below 1 atm. Ice does not sublime under most atmosphere conditions.

Rapid phase changes and metastability

A phase diagram shows the conditions of temperature and pressure under which different phases exist. If no phase line is crossed, the particular phase defined by the diagram exists indefinitely without changing.

However, sometimes the conditions are changed so rapidly that a phase line is crossed without any visible change occurring. This happens because the rearrangement of particles during a phase change takes a finite time.

When a liquid is rapidly heated, its temperature may rise several degrees above its boiling point before it begins to boil; it is then said to be *superheated*. Once it does boil, the temperature falls back to the boiling point as the latent heat is absorbed by the system. Similarly, a liquid can be cooled several degrees below its melting point before it begins to freeze: the liquid is *supercooled*.

A phase existing under conditions in which it would normally have changed into another phase if it had had time to do so is said to be *metastable*. The change may eventually occur as predicted if given enough time. Pecked lines representing the behaviour of metastable phases are sometimes shown on a phase diagram.

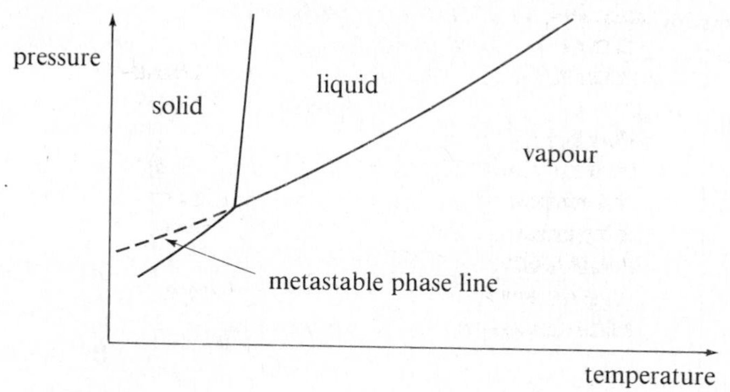

9.2 Allotropy

A number of pure substances are made up of particles that can adopt more than one lattice arrangement. Such substances can exist in more than one phase (or form) in the same state. An element that can exist in this way is said to display *allotropy*; for example, solid carbon has the allotropes diamond and graphite. Compounds also exist in different forms — the phenomenon is then called *polymorphism*. For example, calcium carbonate is found naturally in several forms or polymorphs.

Allotropic phase diagrams

The phase diagram for an element that has two allotropes consists of two of the basic Y-shaped curves.

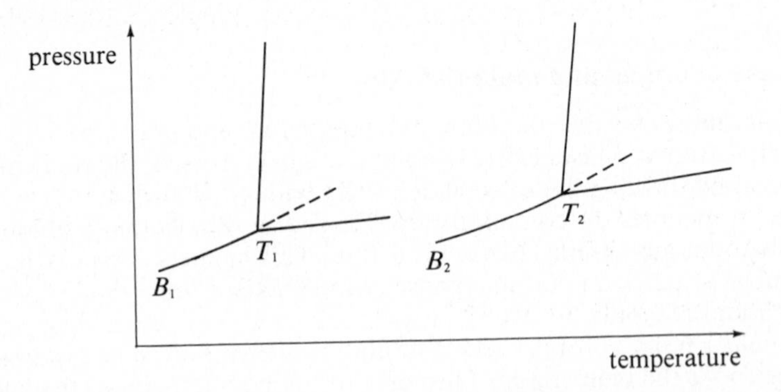

These curves can meet in one of two ways: either B_1T_1 has the steeper gradient or B_2T_2 has the steeper gradient. In the first case, either form of the element is stable under the right conditions and the element is said to display *enantiotropy*. In the second case, one form of the element is stable under all conditions while the other is metastable, a phenomenon known as *monotropy*.

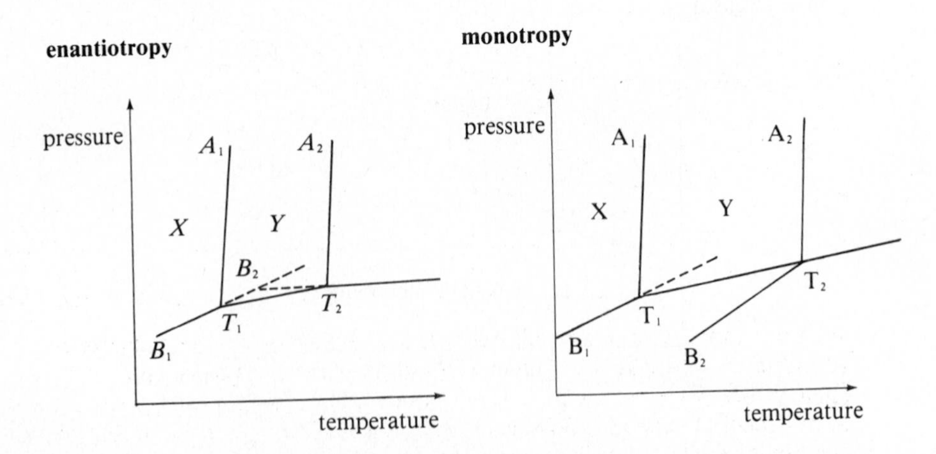

In the left-hand diagram, solid X is stable to the left of A_1T_1 but solid Y is stable between A_1T_1 and A_2T_2; the change is reversible but usually slow. In the right-hand diagram, solid X is metastable.

Enantiotropy

> *Enantiotropy* is defined as the existence of an element in more than one crystalline form, each of which is stable over a particular range of conditions. The different forms are called *enantiotropes*.

Tin

At atmospheric pressure the transition temperature for tin is 13.2°C. Above this temperature white metallic tin is the stable form. Below this temperature the stable form is known as grey tin: the atoms are arranged in a diamond type of lattice, but the change is slow.

$$\text{grey tin} \xrightleftharpoons{13.2°C} \text{white tin}$$

Sulphur

The phase diagram for sulphur shows the conditions under which the two crystalline forms of sulphur are stable. Notice that every phase line in the diagram slopes to the right.

enantiotropic phase diagram for sulphur

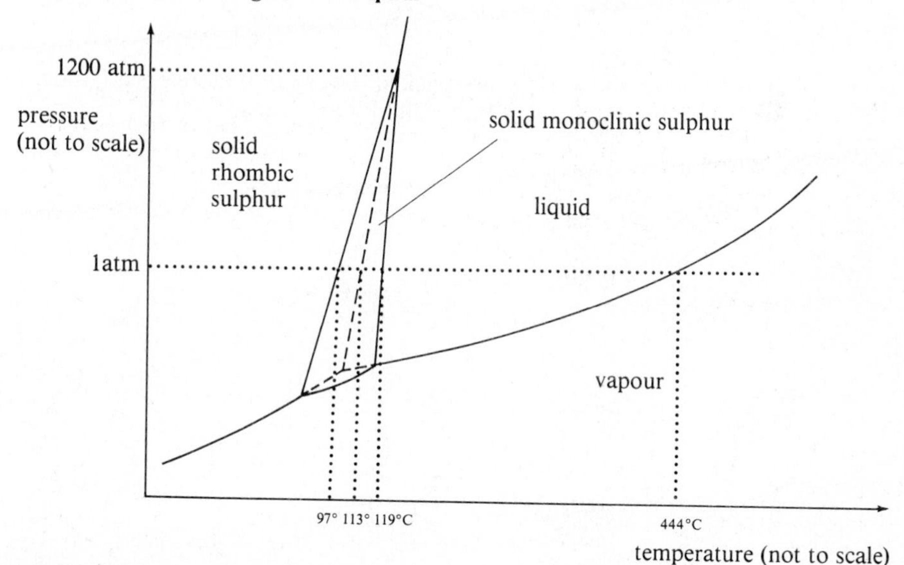

At 1 atmosphere pressure sulphur exists in the rhombic crystalline form (α-sulphur) below 97°C. Above 97°C sulphur exists as monoclinic sulphur (β-sulphur). When rhombic sulphur is heated at 1 atmosphere pressure, it should change into monoclinic sulphur at 97°C, but the rate of change from one phase to the other is slow. So the rhombic form becomes metastable and melts at 113°C. This is shown by the line at 1 atmosphere pressure crossing the broken metastable phase line at 113°C.

Monoclinic sulphur is the stable form above 97°C. It has a more open structure and does not melt until 119°C because the more open lattice can absorb more kinetic energy before being disrupted enough to melt.

When liquid sulphur is cooled, the monoclinic form of the solid is normally produced because this is the first solid phase to be encountered. However, this will gradually change to the rhombic form unless it is kept above 97°C.

Rhombic sulphur exists as S_8 rings arranged in a densely packed lattice.

rhombic sulphur

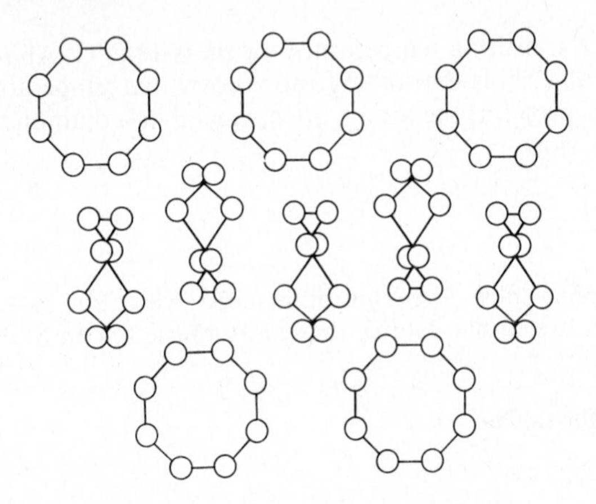

In monoclinic sulphur, the S_8 rings are packed loosely in a more open lattice. Each S_8 ring has enough space to vibrate slightly as shown.

monoclinic sulphur

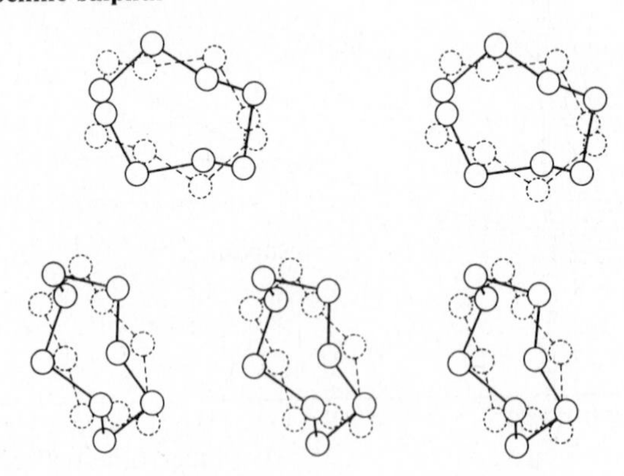

Iron

The metal exists in at least three different enantiotropic forms. The structure of the crystals is normally modified by the presence of carbon atoms arranged in the interstices of the lattice. Pure iron exists as α-iron up to 766°C; it is soft and magnetic but becomes non-magnetic above this temperature. From 900°C to 1400°C the stable form is γ-iron which is miscible with many alloying elements to form hard isomorphous solid solutions. Above 1400°C δ-iron is the stable form; it resembles the non-magnetic enantiotrope in structure and properties.

Monotropy

> *Monotropy* is the existence of an element in more than one crystalline form, one of which is always more stable under all conditions while the other is always metastable. The different crystalline forms are called *monotropes.*

Carbon

Diamond is unstable with respect to graphite, but the conversion of diamond to graphite involves breaking some of the carbon-carbon bonds in the diamond lattice. As this hardly ever happens under normal conditions, diamonds can almost be said to last for ever.

Phosphorus

The phase diagram shows that the element exists as two monotropes. Red phosphorus is the crystalline form that is stable. If it is heated at 1 atmosphere pressure, the phosphorus sublimes at 417°C where the phase line BT is crossed; a stable liquid phase can only be produced at pressures above 43.1 atm which is the pressure at the triple point T.

monotropic phase diagram for phosphorus

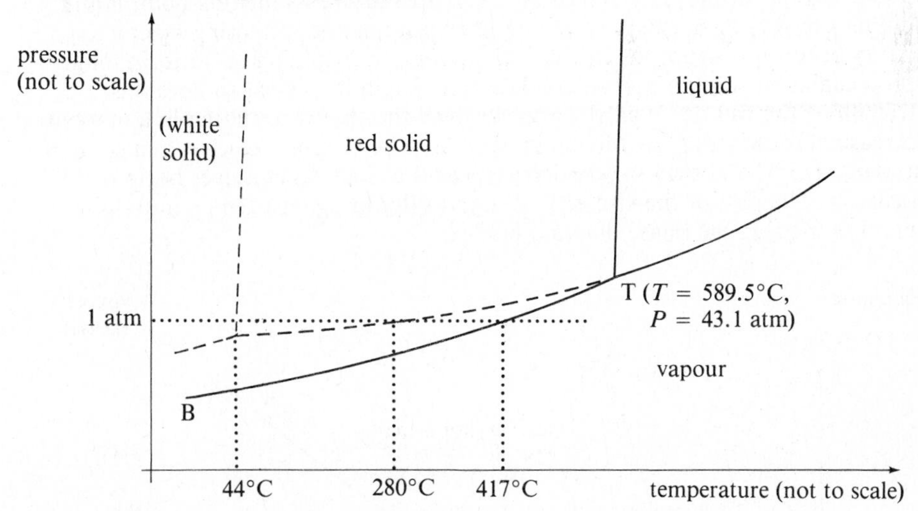

White phosphorus is the metastable monotrope of the element. It melts at 44°C and boils at 280°C where the two metastable phase lines are cut by the line at 1 atmosphere pressure.

White phosphorus (sometimes called yellow phosphorus) consists of tetrahedral P_4 units which are all joined by bonds at an angle of 60°. The P_4 molecules are held together in a lattice by weak van der Waals' forces. Melting only involves overcoming these weak forces.

Red phosphorus consists of long chains of phosphorus atoms held together by covalent bonds, some of which are at angles greater than 60°. When red phosphorus sublimes, these covalent bonds have to be broken.

P_4 molecule of white phosphorus

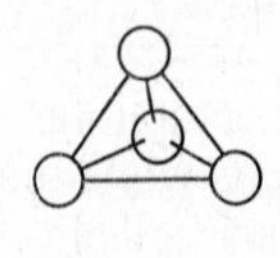

chain of atoms in red phosphorus

Dynamic allotropy

Covalent bonds may be broken if there is sufficient kinetic energy in the system. A broken bond leaves unpaired electrons which may result in the formation of new bonds and new molecules.

> Dynamic allotropy is the existence in the same phase of two molecular forms of the element which are always in equilibrium with each other.

Like all equilibrium systems, the relative concentrations of each dynamic allotrope depend on the temperature of the system.

Sulphur

Liquid sulphur displays dynamic allotropy. Just above its melting point liquid sulphur consists of S_8 rings. At higher temperatures the increasing kinetic energy begins to disrupt the S_8 rings; these then start joining up to form long spiral chains of sulphur atoms. As the chain length increases, so does the strength of the van der Waals' forces between the chains, resulting in a marked increase in the viscosity of the liquid sulphur. The viscosity reaches a maximum at about 175°C and then slowly decreases as the chains themselves begin to break up. The plot of the viscosity of liquid sulphur against temperature is superimposed on the phase diagram below.

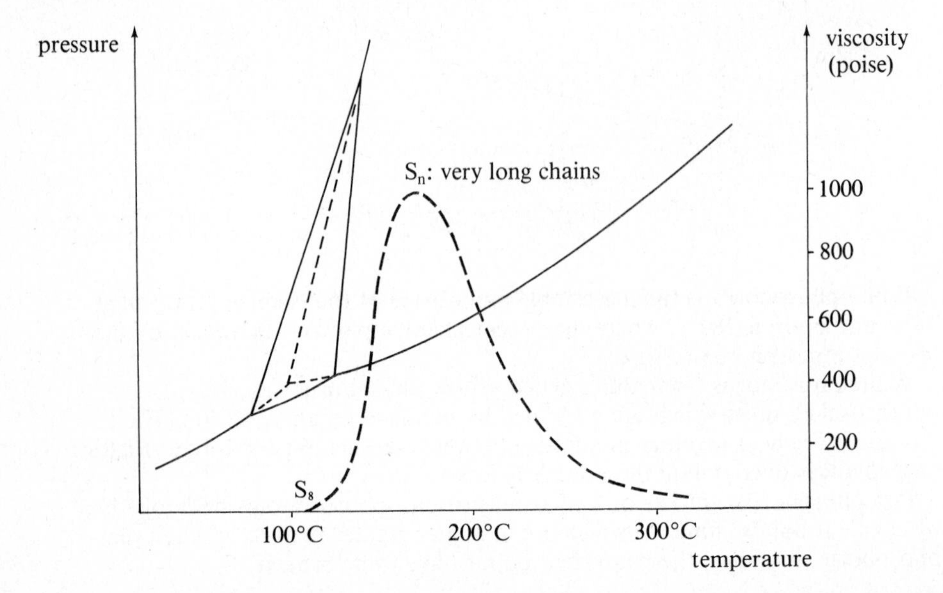

The colour of the liquid sulphur also becomes darker as the liquid becomes more viscous. It starts a golden yellow and becomes a darker red until it is almost black. This is caused by the unpaired electrons at the ends of the chains.

If the thick viscous liquid sulphur at about 175°C is suddenly cooled by pouring it into cold water, a metastable form of liquid sulphur called *plastic sulphur* is formed. This rubbery solid changes back to rhombic sulphur on standing, particularly if it is worked between the hands.

Questions

1. What do you understand by the following terms?
   a) Phase b) Phase diagram c) Triple point d) Sublimation e) Metastable state.

2. Explain the following: when pure sulphur is heated very slowly, it melts at 119°C; but if it is rapidly heated, it melts at 112°C. However liquid sulphur always freezes at 119°C at atmospheric pressure.

3. a) Draw the molecules making up the two forms of phosphorus. Use a data book to mark in the bond angles. Which allotrope is more stable? Which has the larger bond angles?
   b) Draw the lattices of the two forms of carbon. Mark in the bond angles. Which allotrope is more stable? Which has the larger bond angle?
   c) Comment on these results.

4. Use examples to explain the difference between:
   a) Monotropy b) Enantiotropy c) Dynamic allotropy.

5. Explain why increasing the pressure on liquid carbon dioxide makes it solidify, while increasing the pressure on ice makes it melt.

6. Look back at the chapter on metallic lattices and find the reference to the allotropy of iron on page 78. Explain which form of allotropy it is.

7. Explain the following:
   a) Liquid sulphur thickens and darkens on heating.
   b) Ice is less dense than water.
   c) Diamond is less stable than graphite but does not appear to change into it.

8. Describe and explain the effect of heat on sulphur.

9. Explain why, when liquid phosphorus freezes, white phosphorus always forms irrespective of whether it was red or white phosphorus that was originally melted.

10. Explain why some elements exhibit allotropy, but others do not.

Chapter 10
The gas phase

10.1 Properties of gases

Unlike solids and liquids, the volume occupied by a given mass of gas depends on the conditions applied. Changes in either pressure or temperature will affect the volume of a gas. For this reason the conditions under which the volume is measured should always be quoted. *Standard conditions of temperature and pressure* (s.t.p.) are by convention 273 K (0°C) and 1 atmosphere.

The relationships between the volume of a given mass of gas and the prevailing conditions are known as the *gas laws.*

Boyle's law

The relationship between pressure and volume is known as Boyle's law.

Boyle's law states that, at constant temperature, the volume of a given mass of gas is inversely proportional to the applied pressure.

$V \propto 1/P$ which becomes $PV = \text{constant}$

or $P_1V_1 = P_2V_2$

Boyle's law can be represented graphically in several ways.

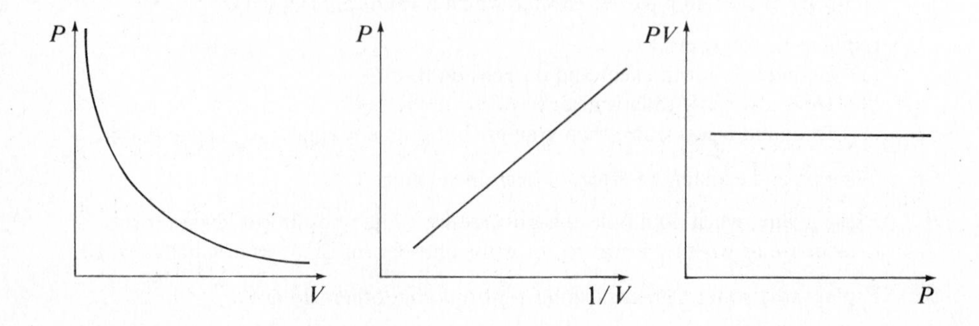

Charles' law

The relationship between temperature and volume is known as Charles' law.

Charles' law states that, at constant pressure, the volume of a given mass of gas is proportional to its temperature expressed in Kelvin.

$V \propto T$ which becomes $\frac{V}{T}$ = constant

or $\frac{V_1}{T_1} = \frac{V_2}{T_2}$

Graphically Charles' law looks like this on the Celsius scale.

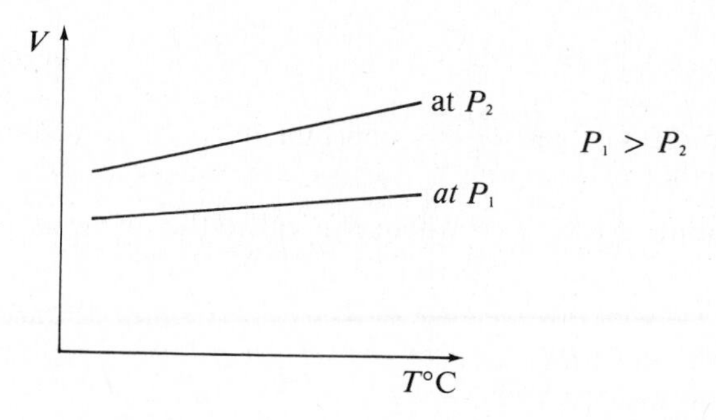

When this graph is plotted in Kelvin, we see that the lines, when extrapolated, go through *absolute zero*.

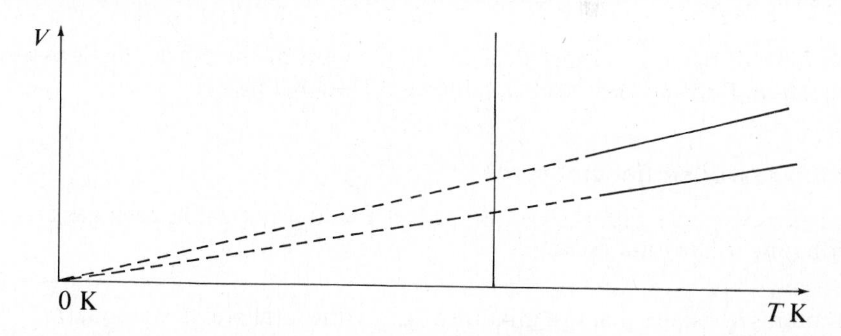

The equation of state or the ideal gas equation

Using Boyle's law and Charles' law, it is possible to derive an equation relating all three functions of volume, temperature and pressure.

Take a gas of volume V_1 under conditions P_1 and T_1. How will these values be related to the new volume V_2 under conditions P_2 and T_2?

First apply a Boyle's law change, keeping the temperature constant, i.e. initially P_1, V_1, T_1 and finally P_2, V' and T_1. Then according to Boyle's law

$$P_1 V_1 = P_2 V'$$

Now apply a Charles' law change, keeping the pressure constant, i.e. initially P_2, V', T_1 and finally P_2, V_2, T_2. Then according to Charles' law

$$\frac{V'}{T_1} = \frac{V_2}{T_2}$$

Rearrange this equation:

$$V' = \frac{V_2 T_1}{T_2}$$

and substitute the value for V' in the Boyle's law equation:

$$P_1 V_1 = \frac{P_2 V_2 T_1}{T_2}$$

Dividing both sides by T_1 gives

$$\frac{P_1 V_1}{T_1} = \frac{P_2 V_2}{T_2}$$

which means that for a given mass of gas $\frac{PV}{T}$ = constant.

For 1 mole of a gas this becomes $PV/T = R$ where R is called the *universal gas constant*.

For n moles of gas, $PV = nRT$
This is known as the *ideal gas equation*.
In the form $\frac{P_1 V_1}{T_1} = \frac{P_2 V_2}{T_2}$ it is known as the *equation of state*.

The value of the gas constant depends on the units of the other functions in the equation. In SI units R has the value of 8.314 J K^{-1} mol^{-1}.

Dalton's law of partial pressures

All gases mix readily with each other, and in a mixture of gases, each gas is contributing to the total pressure.

Dalton's law states that in a mixture of gases, the total pressure is equal to the sum of the partial pressures of all the gases making up the mixture.

$$P_{total} = p_A + p_B + p_C + \ldots$$

The *partial pressure* is defined as the pressure that a gas in a mixture would exert if it alone occupied the container. It can be shown that:

the partial presure p_A of gas A = $\frac{\text{no of moles of gas A}}{\text{total no of moles of gas}} \times P_{total}$

So in a sample of air at 1 atmosphere; assuming 80% N_2, 20% O_2,

partial pressure P_{N2} of N_2 = $\frac{80}{100} \times 1 = 0.8$ atm

partial pressure P_{O2} of O_2 = $\frac{20}{100} \times 1 = 0.2$ atm

The *molar concentration* of a gas at any temperature can be calculated from its partial pressure by using the ideal gas equation:

$$\frac{n}{V} = \frac{P}{RT}$$

Example Find the molar concentration of oxygen in the air (20% O_2) at 298 K and 1 atm (1.01×10^5 Nm^{-2}).

$$\text{molar concentration} = \frac{n}{V} = \frac{P}{RT}$$

$$= \frac{0.2 \times 1.01 \times 10^5}{8.314 \times 298} = 8.15 \text{ mol m}^{-3}$$

Graham's law of diffusion

The particles in a gas are so small that they can escape through tiny holes and through porous substances. The passage of a gas through a small hole is called *effusion* and its passage through a porous substance is called *diffusion*.

A series of simple experiments using gas syringes in the way shown below reveals a relationship between the rate of escape of the gas and its relative molar mass.

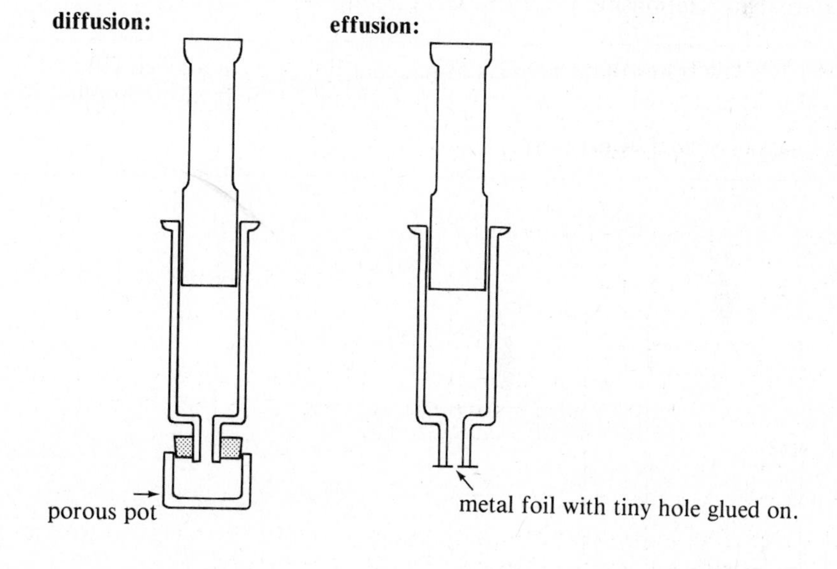

This is known as *Graham's law of diffusion*.

> Graham's law states that the rate of diffusion of a gas at a particular temperature and pressure is inversely proportional to the square root of its density
>
> $$\text{rate of diffusion} \propto \frac{1}{\sqrt{\text{density}}}$$

Since the density of a gas is directly proportional to its relative molecular mass, Graham's law can be written in the form:

$$\text{rate of diffusion of gas} \propto \frac{1}{\sqrt{\text{R.M.M. of gas}}}$$

for hydrogen R.M.M. = 2.
oxygen R.M.M. = 32.

$$\frac{\text{rate of diffusion of hydrogen}}{\text{rate of diffusion of oxygen}} = \sqrt{\frac{\text{R.M.M. of oxygen}}{\text{R.M.M. of hydrogen}}}$$

$$= \sqrt{\frac{32}{2}} = \frac{4}{1}$$

So hydrogen will escape four times more rapidly than oxygen.

Henry's law

A gas above a liquid dissolves in the liquid to some extent. For gases that do not react with the liquid they are in contact with, it is found that there is a relationship between the pressure of the gas and the amount of gas that dissolves. This relationship is called *Henry's law*.

> Henry's law states that the mass of gas that dissolves in a given volume of liquid at a fixed temperature is proportional to the pressure applied.
>
> mass dissolved $\propto$ pressure.

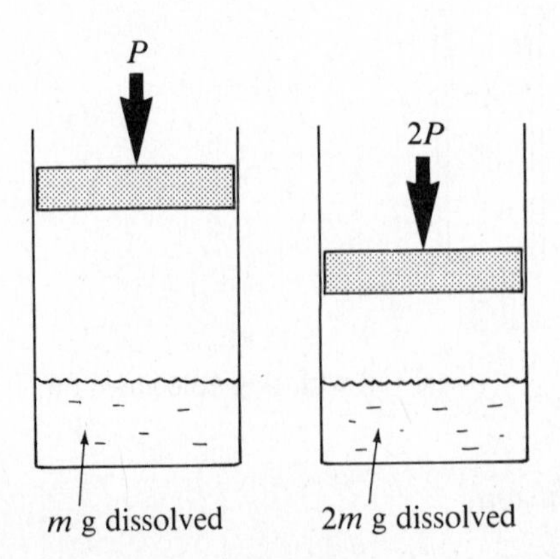

The greater the applied pressure, the more gas will dissolve. A well known example of this is the dissolving of nitrogen in a diver's blood under the sea. If the diver surfaces too quickly, the sudden drop in pressure allows the nitrogen to come out of solution as tiny bubbles. This is known to divers as 'the bends'.

10.2 Ideal gases

The kinetic theory of gases

The particles in a gas are very far apart (between 1000 and 10 000 times their own diameters) and are moving at about the speed of a bullet. So it is possible to produce a theoretical model of a gas using fairly simple mathematics. This mathematical treatment of gas particles is known as the *kinetic theory of gases* and the hypothetical system it describes is called an *ideal gas*.

The kinetic theory makes certain assumptions or postulates, as they are sometimes called, about the behaviour of gas particles. These assumptions are:

1. A gas is made up of small particles which are a very long way apart compared to their size. So the total volume of the particles is very small compared to the volume of the gas as a whole.
2. The particles do not attract each other at all.
3. The gas particles are in continuous random motion. This means that there will be a range of particle speeds. Even if they all started with the same speed, the results of collisions between particles will slow some particles down while accelerating others. The distribution of particle speeds can be calculated purely from the laws of probability. This was first done by Maxwell and Boltzmann and it is often called a Maxwell-Boltzmann distribution. The distribution at two different temperatures is shown below for oxygen molecules.

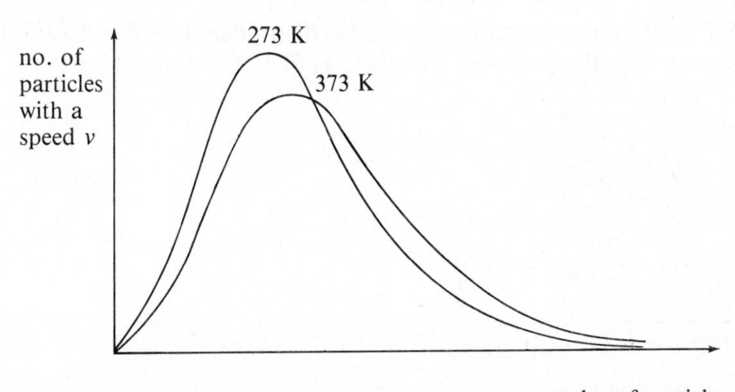

notice that the increase in temperature has:
a) increased the average speed
b) increased the range of speeds in the system

4. The average kinetic energy of the particles is proportional to the temperature of the system. The particles have energy because they are moving, and this kinetic energy is a function of their mass and speed.

$$\text{kinetic energy} = \tfrac{1}{2}mv^2$$

Because there is a range of particle speeds, there will be a range of particle energies. As energy is proportional to the square of the particle speed, a plot of speed squared gives a distribution of the energies in a gas system.

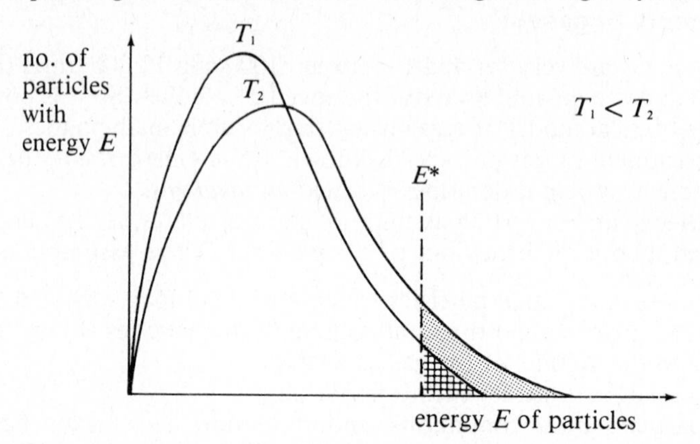

This Maxwell-Boltzmann distribution of energies can be expressed in equation form:

$$N = N_t e^{-E^*/kT}$$

where N is the number of particles with an energy greater than E^* (represented by the shaded portion under the graph to the right of E^*), N_t is the total number of particles, k is the Boltzmann constant and T is the kelvin temperature.

Starting with these assumptions, it is possible to derive a relationship for the pressure exerted by the particles. Consider a cube of gas of side = l.

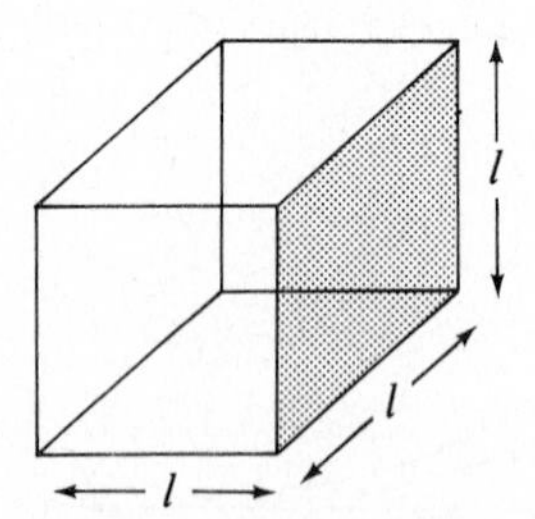

The pressure on the faces of the cube is due to the particles hitting the faces and bouncing off.

Take a single particle moving with velocity c. This velocity can be resolved into three components x, y and z parallel to the faces of the cube.

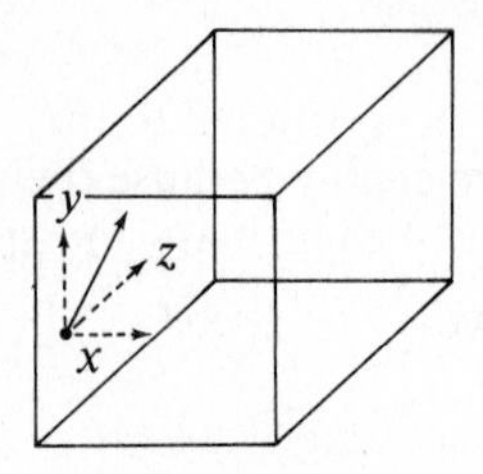

velocity in x-direction = x
velocity in y-direction = y
velocity in z-direction = z
and $c^2 = x^2 + y^2 + z^2$

The particle will hit a wall of the cube every time it travels a distance l parallel to the sides of the cube. So in the x-direction the number of collisions will be x/l per second.

Now if the particle has mass m, then its momentum in the x-direction is mx. If the collisions with the wall are completely elastic, the momentum after the collision will be $-mx$. So for each collision the change of momentum is $2mx$. For all the collisions for the one particle, the rate of change in momentum in the x-direction is:

$$2mx \times \frac{x}{l} = \frac{2mx^2}{l}$$

This is also true in the other two directions, so the total rate of change in momentum for the particle is:

$$\frac{2m}{l}(x^2 + y^2 + z^2) = \frac{2mc^2}{l}$$

Now if the total number of particles in the system is N and the average of their squared velocities is $\overline{c^2}$, the total rate of change of momentum at the walls of the cube is

$$\frac{2Nm\overline{c^2}}{l}$$

According to Newton the rate of change of momentum is equal to the *force* on the walls of the cube. But force/area = pressure and the total area of the six walls of the cube is $6(l^2)$. So the pressure P exerted by the gas is given by:

$$P = \frac{\text{force}}{\text{area}} = \frac{2Nm\overline{c^2}}{l \times 6l^2} = \frac{1}{3}\frac{Nm\overline{c^2}}{l^3}$$

but l^3 is the volume V of the cube, so

$$P = \frac{1}{3}\frac{Nm\overline{c^2}}{V} \text{ which becomes } PV = \tfrac{1}{3}Nm\overline{c^2}$$

$$\text{Since } \frac{Nm}{V} = \frac{\text{total mass}}{\text{volume}} = \text{density of the gas } \rho$$

we can write $P = \frac{1}{3}\rho\overline{c^2}$

The equation $PV = \frac{1}{3}Nm\overline{c^2}$ is at the heart of the kinetic theory. The experimentally derived gas laws can be predicted from this theoretical equation.

Kinetic theory and the gas laws

Charles' law

Charles' law refers to a given mass of gas at constant pressure. So in the equation $PV = \frac{1}{3}Nm\overline{c^2}$, P, m and N are all constants. This leaves

$$V \propto \overline{c^2}.$$

As temperature is a measure of the kinetic energy of the gas $T \propto \frac{1}{2}m\overline{c^2}$ and m is constant, then

$$T \propto \overline{c^2}.$$

So it follows that $V \propto T$ which is Charles' law.

Boyle's law

Boyle's law refers to a given mass of gas at constant temperature. So in the equation $PV = \frac{1}{3} mN\overline{c^2}$, if the mass is fixed then mN is constant, and if the temperature is constant than the value of $\overline{c^2}$ is constant so the equation becomes PV = constant which is Boyle's law.

Graham's law

Two different gases at the same temperature have the same kinetic energy, e.g. for oxygen and hydrogen

$$\frac{1}{2} m_O\overline{c_O^2} = \frac{1}{2} m_H\overline{c_H^2}$$

rearranging this equation it becomes

$$\frac{m_H}{m_O} = \frac{\overline{c^2}_O}{\overline{c_H^2}}$$

$$\text{or } \frac{c'_O}{c'_H} = \sqrt{\frac{m_H}{m_O}}$$

where c' is the root mean square velocity of the molecules. $c' = \sqrt{\overline{c^2}}$

This is a statement of Graham's law because the root mean square velocity of the particles is proportional to the rate at which they will escape.

Avogadro's hypothesis

If equal volumes of two different gases under the same conditions are taken, then for each gas

$$PV = \frac{1}{3} Nm\overline{c^2}.$$

They are under the same conditions so PV is the same for both and so is $m\overline{c^2}$ which is proportional to the temperature.

It follows therefore that $N_1 = N_2$, i.e. there are the same number of particles of each gas. This is in agreement with *Avogadro's hypothesis*.

> Avogadro's hypothesis states that, under identical conditions, equal volumes of all gases contain equal numbers of molecules.

10.3 Real gases

Deviations from ideal behaviour

More careful study of real gases shows that in fact they do not obey Boyle's law exactly and that the kinetic theory model of an ideal gas is an over-simplification. Not only do real gases deviate from the predictions made for an ideal gas, but the deviations depend on the conditions.

Here for example are the plots for a real gas: nitrogen.

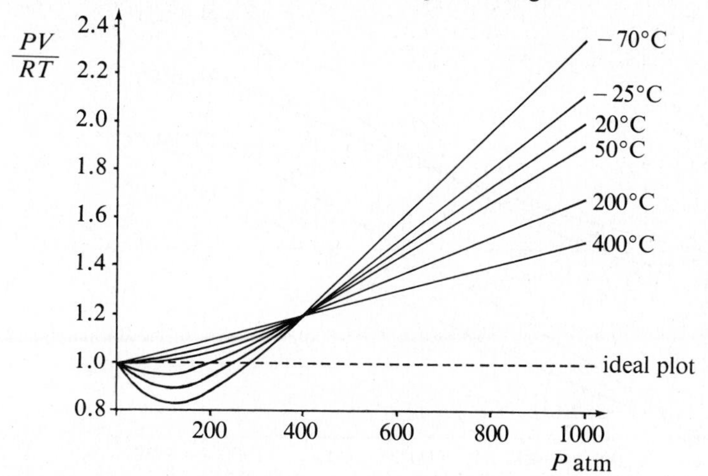

These curves show that a real gas deviates more from ideal behaviour at:

1. high pressures
2. low temperatures

The low temperature deviations were investigated further by Andrews in some famous experiments on carbon dioxide. His results were plotted in the form of P against V curves. Compare these with the ideal case on p. 128.

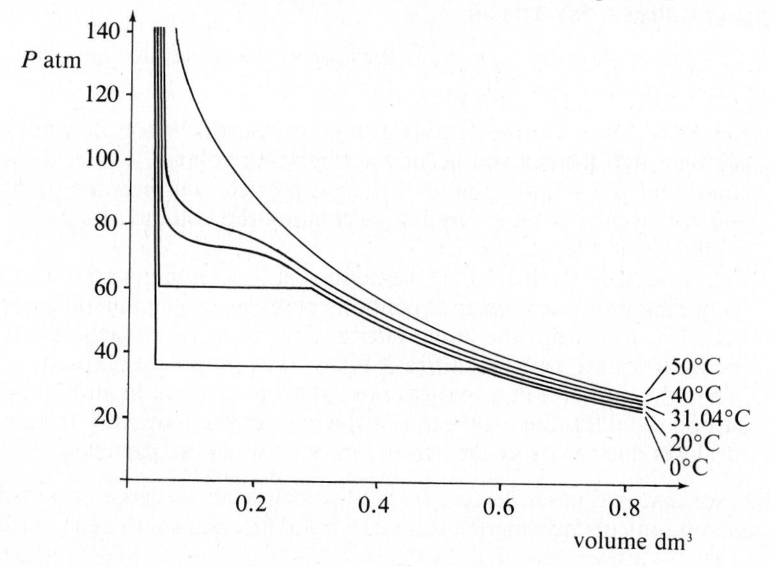

Only the highest temperature plot approaches ideal behaviour; all the others deviate considerably.

The horizontal portions of the two lower curves indicate that the pressure increase has made the gas into a liquid. The highest temperature at which this can occur is called the *critical temperature.*

Just as the deviations in the behaviour of a real gas vary from the ideal model with changing conditions, so they vary from one gas to another as the curves below indicate.

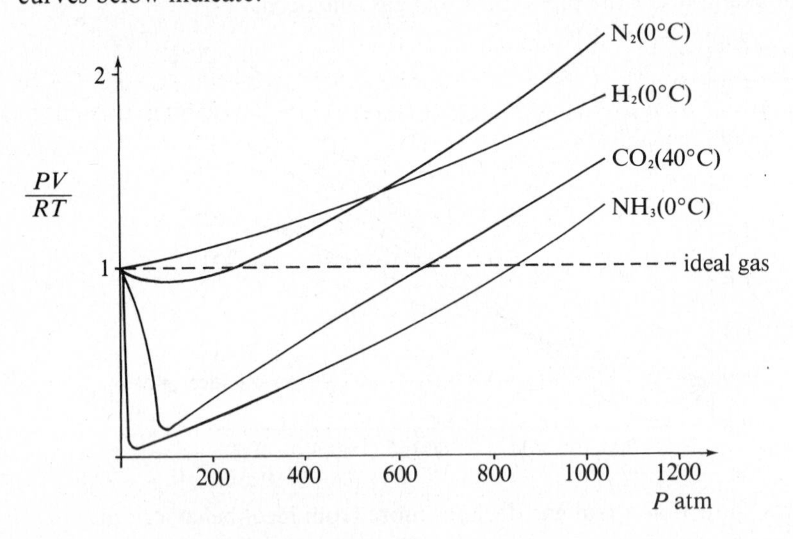

These curves clearly show that some gases are much less ideal in their behaviour than others. Carbon dioxide and ammonia both deviate much more than hydrogen and nitrogen.

Interpreting these deviations

The sets of curves showing how the behaviour of real gases deviate from the ideal model emphasize three points.

1. Real gases deviate considerably at high pressures. When the pressures are very high the gas will occupy a very small volume. Under these conditions the actual volume of the gas particles has become significant and can no longer be ignored in calculating the volume of space available for a single particle.
2. Real gases deviate from ideal behaviour at low temperatures. Just above its boiling point, a substance contains particles whose kinetic energy is sufficient to disrupt the liquid lattice. The forces of attraction between the particles are still a significant factor that cannot be ignored.
3. When the particles in a real gas are polar (as they are in ammonia or carbon dioxide), the properties of the gas deviate markedly from of an ideal gas due to strong attractive forces between the particles.

We thus see that real gases deviate from ideal behaviour because two of the basic assumptions of the kinetic theory are not entirely true: the gas particles do *not* have negligible volume, and there *are* attractive forces between the particles.

Van der Waals' equation

There have been many attempts to modify the ideal gas equation

$$PV = RT$$

so that it accounts more accurately for the behaviour of real gases. The one below was produced by van der Waals.

If the behaviour of real gases reflects the fact that two of the Kinetic Theory assumptions are wrong, then isotherms might be the result of two conflicting factors. This is illustrated below.

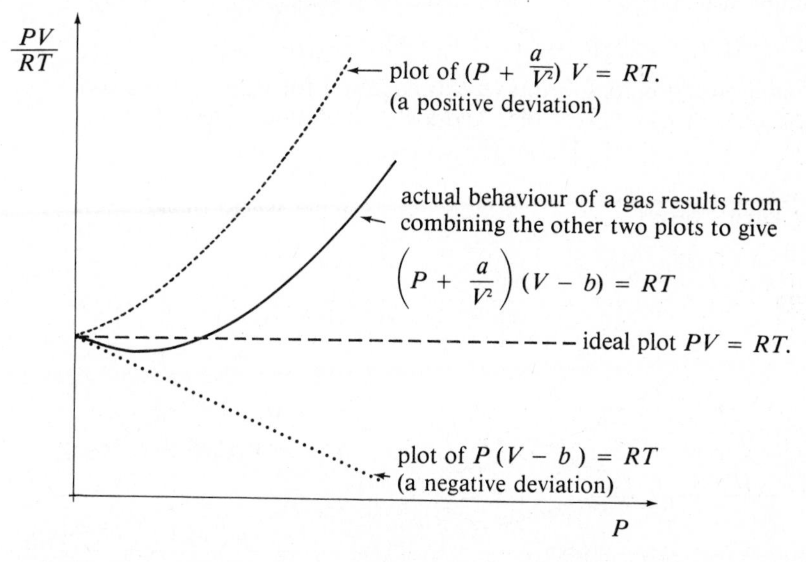

The negative deviation is a result of the gas particles actually having a definite volume. Therefore the space available for any individual particle is not V but $V - b$, where b is a constant for that gas. This would produce a linear equation of the form $P(V - b) = RT$.

The positive deviation is a result of attractive forces between the particles. A particle arriving at the wall of the container is going more slowly in a real gas than in an ideal gas because the other particles are pulling it back. A factor must be added to the pressure to allow for the result of this force. The drop in pressure depends on

1. the number of particles at the surface
2. the number of particles exerting a pull from within.

At constant temperature for a fixed amount of gas, both these factors are inversely proportional to the volume of the container. Hence a factor a/V^2 must be added, producing an equation of the form

$$\left(P + \frac{a}{V^2}\right)V = RT$$

When these two equations are combined, the following equation is produced

$$\left(P + \frac{a}{V^2}\right)(V - b) = RT$$

It is known as the *van der Waals'* equation and it agrees fairly well with the behaviour of real gases.

The two constants a and b vary from gas to gas and with temperature; their values are given in some data books.

Van der Waals' equation and the critical point

If the van der Waals' equation is rearranged it becomes:

$$PV^3 - (RT + Pb)V^2 + aV - ab = 0.$$

This is a cubic equation in V. Plotting this equation for values of P and V produces a series of curves very like Andrew's experimental curves.

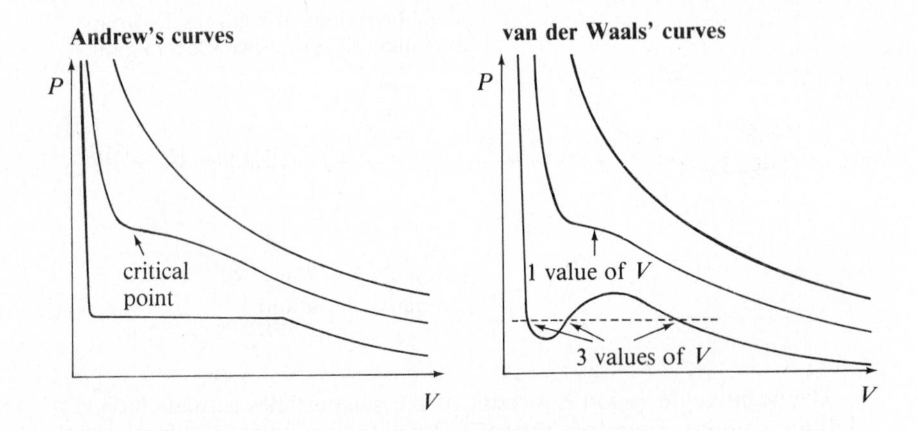

The main difference is that the straight-line portion of the lower Andrew's curves representing the change of state from gas to liquid is replaced by a portion with a double curve in it. This gives three values of V for one value of P. As the temperature rises, the three values of V get closer and closer together. When the three values of V have become a single point, this defines the critical point.

Solving the van der Waals' equation for these conditions gives relationships between the van der Waals' constants a and b and the critical conditions.

the critical pressure $P_c = a/27b^2$

the critical temperature $T_c = 8a/27Rb$.

the critical volume for one mole of gas $V_c = 3b$.

10.4 Analysing gases

Finding the R.M.M. of a gas

The relative molar mass of a gas or of a volatile liquid can now be very accurately measured using a mass spectrometer. However it can also be found fairly simply in the school laboratory. The method rather elegantly relies on Avogadro's hypothesis and is described below.

A pair of gas syringes are sealed with rubber caps and placed side by side in an oven and heated. A small sample of a known volatile liquid is drawn into a hypodermic and weighed. The sample is then injected through the rubber cap into one of the gas syringes. A sample of the liquid of unknown R.M.M. is drawn into a second hypodermic and also weighed. This sample is then injected into the second gas syringe. The volume of both gas syringes is now measured. Specimen results and the calculation of the R.M.M. are shown below.

Known liquid: Propanone CH_3COCH_3.
R.M.M. = 58
Mass of propanone injected = 0.155g
Volume of propanone gas syringe = 60 cm^3

Unknown liquid X.
R.M.M. = ?
Mass of X injected = 0.230 g
Volume of X gas syringe = 43 cm^3

Calculation: amount of propanone $= \frac{0.155}{58} = 2.67 \times 10^{-3}$ mol.

This occupies 60 cm^3 in the hot gas syringe. Because the two gas syringes are side by side under the same conditions, Avogadro's hypothesis can be applied.

$$60 \text{ cm}^3 \text{ is the volume of } 2.67 \times 10^{-3} \text{ mol}$$

$$\therefore 43 \text{ cm}^3 \text{ is the volume of } 2.67 \times 10^{-3} \times \frac{43}{60} \text{ mol.}$$

$$= 1.915 \times 10^{-3} \text{ mol}$$

So the second gas syringe contained 1.915×10^{-3} mol of X. This weighs 0.230 g.

$$1.915 \times 10^{-3} \text{ mol weighs } 0.230 \text{ g}$$

$$\text{then 1 mol weighs } \frac{0.230}{1.915 \times 10^{-3}} = 120.1 \text{ g}$$

So the R.M.M. of X is found to be 120.

Gas chromatography

Gas chromatography is a relatively modern technique (the first reported account was in 1941) which has rapidly grown in importance. There are now very many applications for gas chromatography in fields ranging from medicine and pure research to forensic and industrial sample monitoring.

Gas chromatography involves injecting a small sample (as little as 2—3 microlitres) into a stream of gas which flows through a tube containing some powdered material. As the sample is carried through the tube it is separated into its various components which are detected at the far end as they emerge one at a time.

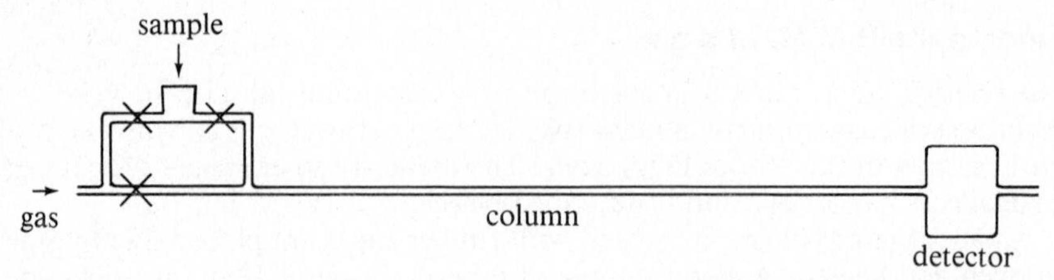

The sample particles are carried along in the stream of the carrier gas until they meet the solid packed in the column. On contact with the solid surface they become more or less strongly adsorbed onto the surface. As more gas comes through, the sample particles are desorbed from the surface and carried a little further into the column before being re-adsorbed onto a new piece of solid. By this process of adsorption and desorption, the sample is gradually carried through the column. Each component in the sample will be adsorbed and desorbed at a different rate, and so the components gradually become separated in the column.

The carrier gas is usually chosen for its inertness; some that are frequently used are He, Ar, H_2, and N_2.

The powdered material in the column is one that adsorbs substances onto its surface quite strongly. Charcoal, aluminium oxide and silicon dioxide are all used commonly, and they are often said to be *activated* before use. This consists of heating them to a temperature at which some disruption in their lattice structure occurs, creating what are known as active sites on the surface. They are then suddenly cooled leaving the surface covered with these sites onto which other particles will bond.

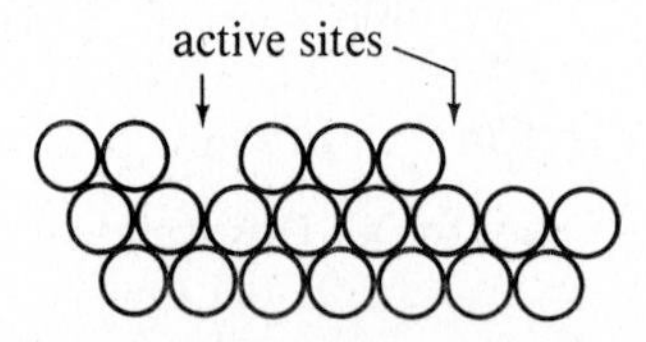

The tube (made of metal, glass or silica) is usually about 0.5 cm in diameter and between 3 and 30 m long. To keep the apparatus compact, the tubes are usually looped or coiled.

There are various detection methods. Some gas columns are connected direct to a spectrometer. Two other forms of detection are commonly used.

1. Flame ionizer.
   Hydrogen carrier gas is burnt between a pair of electrodes. The conductivity of the flame depends on the composition of the gas and changes as the carrier gas brings out one of the components.

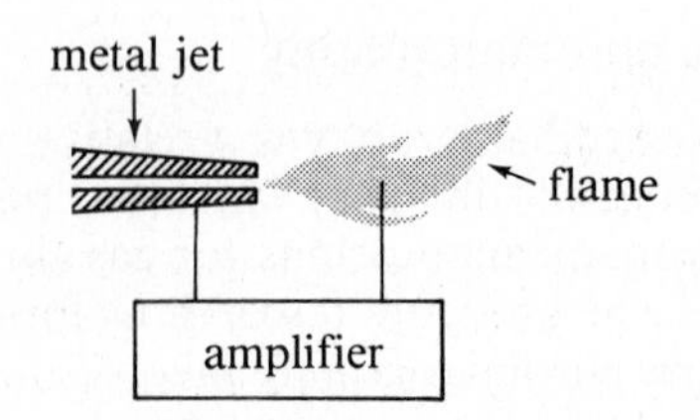

2. Thermal conductivity.
The carrier gas flows over either a heated wire or a thermistor cooling it. The rate of cooling varies with the composition of the gas.

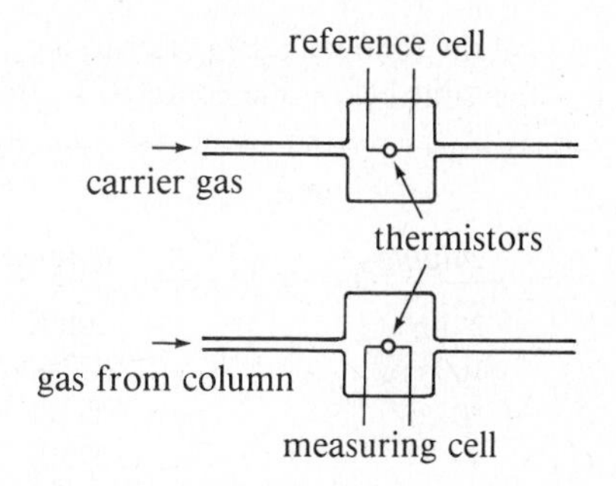

Gas chromatography has several important advantages over other forms of chromatography. In particular:

i) it can deal with really tiny samples,
ii) it is able to separate even the most similar mixtures well. This is because a gas column can be much longer than a liquid one due to the much lower viscosity of gases.

The behaviour of the gases within a column cannot be predicted accurately because of the change in pressure along the tube. The apparatus always has to be calibrated with samples of known composition.

Questions

1. Explain the meaning of:
a) s.t.p. b) an ideal gas c) partial pressure d) diffusion e) Avogadro's hypothesis.

2. a) What are the assumptions of the Kinetic Theory?
b) Explain why some of these assumptions are not correct for real gases.
c) Describe the conditions under which a real gas most closely approaches ideal behaviour and explain why these conditions favour ideal behaviour.

3. All the following data refer to a given mass of gas at constant temperature. Draw out the table below and complete it, filling in all the gaps.

initial		*final*	
pressure	*volume*	*pressure*	*volume*
1.6 atm	250 cm^3	1.0 atm	
0.01 atm	2 dm^3		100 cm^3
740 torr	50 cm^3	770 torr	
140 kN m^{-2}	7.5 l	101.3 kN m^{-2}	
750 mm of Hg	150 cm^3	760 mm of Hg	

4. All the following data refer to a given mass of gas at constant pressure. Draw out the table below and complete it, filling in all the gaps.

initial		*final*	
volume	*temperature*	*volume*	*temperature*
2 dm^3	200 K		273 K
100 cm^3	273 K	147 cm^3	
10 cm^3	25°C	10.8 cm^3	
4 l	273 K		100°C
1.5 dm^3	300°C		500 K

5. All the following data refer to a given mass of gas. Draw out the table below and complete it filling in all the gaps.

initial			*final*		
pressure	*volume*	*temperature*	*pressure*	*volume*	*temperature*
1 atm	25 cm^3	200 K		22.8 cm^3	273 K
750 mm Hg	2 dm^3	23°C	760 mm Hg		100°C
100 kN m^{-2}	400 cm^3	500°C		138 cm^3	100°C
700 torr	3 l	300 K	760 torr	2.51 l	
0.5 atm	22.4 dm^3	273 K	1.1 atm		25°C
96 kN m^{-2}	150 cm^3	0°C	101 kN m^{-2}		20°C
101325 N m^{-2}	1.01 dm^3	100°C	96 kN m^{-2}	0.923 dm^3	
100 mm Hg	1.3 l	288 K	760 mm Hg	0.416 l	

6. A gas syringe contained 0.275g of ethanoic acid (CH_3CO_2H). At 127°C this sample occupied 75 cm^3. Another syringe contained 0.0995g of ethanoic acid. This sample occupied the same volume as the other one at 277°C. Both were at 1 atm.

   Calculate the R.M.M. of ethanoic acid at each temperature. Comment on the results.

7. Two glass bulbs are connected by a thin capillary tube with a tap in it.

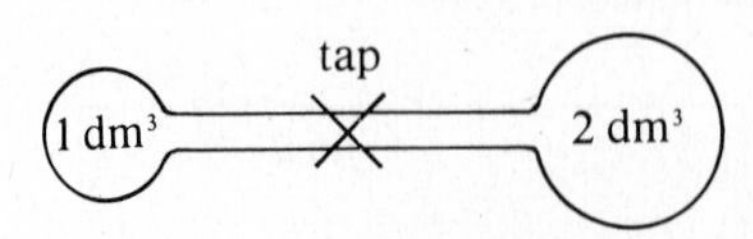

   One bulb of volume 1 dm^3 contains gas at 10 atm pressure. The other of volume 2 dm^3 contains gas at 8 atm pressure.

   What will be the pressure in the whole system if the tap is opened?

8. Two pieces of cotton wool, one dipped into a concentrated solution of ammonia, the other into a concentrated solution of hydrogen chloride, are simultaneously held at opposite ends of a glass tube 150 cm long.

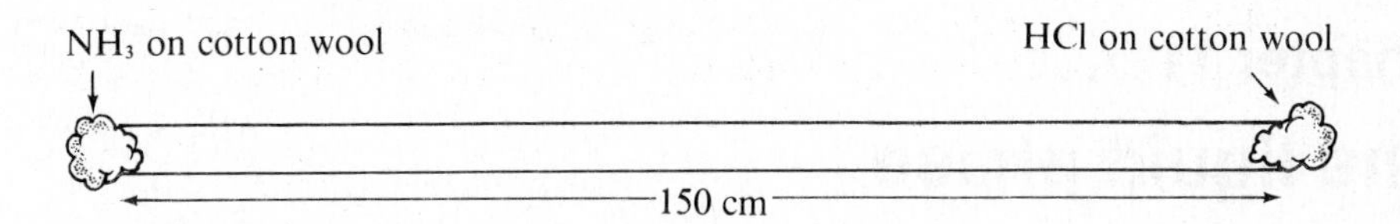

Molecules of the two gases leave the cotton wool and diffuse down the tube until they meet. White rings of solid ammonium chloride form on the tube at the point where they do meet.

a) Which law is concerned with this sort of process?
b) Which end will the rings be nearer?
c) Calculate the distance of the rings from the ammonia end.
d) If hydrogen chloride was replaced by hydrogen bromide, how far would the rings be from the ammonia end?

9. If real gas behaviour deviates so much from the ideal model of the Kinetic Theory, what is the point in learning about ideal gases? Can it serve any useful purpose?

Chapter 11
The liquid phase

11.1 Liquid properties

The physical properties of a typical liquid are listed and discussed below.

Density similar to solids

Heating a solid causes it to expand steadily. When it melts, this expansion continues without any sudden discontinuity. This is in marked contrast to the change in volume of about one thousand times that occurs when a liquid boils. If solids and liquids have roughly similar densities, it implies that the packing of particles in the two states is about the same in distance, if not in order.

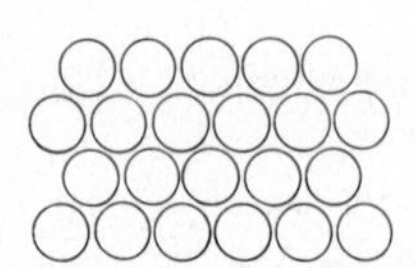

solid lattice: particles close together in a lattice which is ordered over long distances

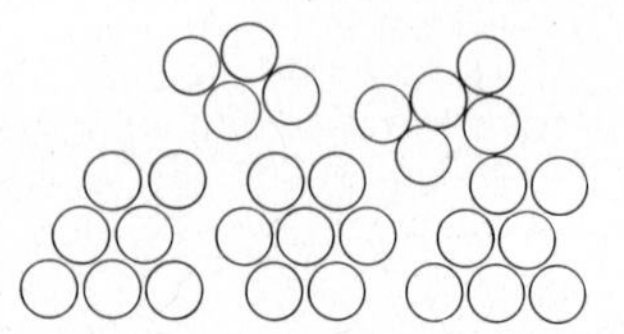

liquid lattice: particles close together in clumps which are fairly ordered but there is no long range order

Because the particles are fairly tightly packed together in a liquid, liquids are almost incompressible. This is obviously very different from gases.

No definite shape

Although liquids are like solids in having a definite volume almost independent of pressure, liquids have no definite shape. Like gases, they take on the shape of the container and can flow if a force is applied. Liquids have no definite shape because in the liquid state there is enough kinetic energy in the system for some of the attractive forces to be overcome. As a result, whole clumps of particles can slide past each other easily. That there are still significant attractive forces between the particles is shown by the way liquids form drops (or large globules in weightless conditions).

Liquids evaporate and boil

All liquids evaporate when left open to the atmosphere. The process of evaporation results in a cooling of the liquid, and is accelerated by a large surface area, a high temperature and a low external pressure.

The rate of evaporation increases with temperature until the surface of the liquid becomes agitated and bubbles are seen forming in the body of the liquid. The liquid is then said to boil. Once the boiling point is reached, the temperature remains constant (no matter how much energy is supplied) until all the liquid has changed into the vapour phase.

The Kinetic Theory can also be applied to the liquid state to account for these observations.

Like the particles in a gas, those in a liquid are all thought to be moving randomly. However in a liquid each particle is in a cage consisting of the surrounding particles, and so both the actual particle speed and the range of speeds are much smaller.

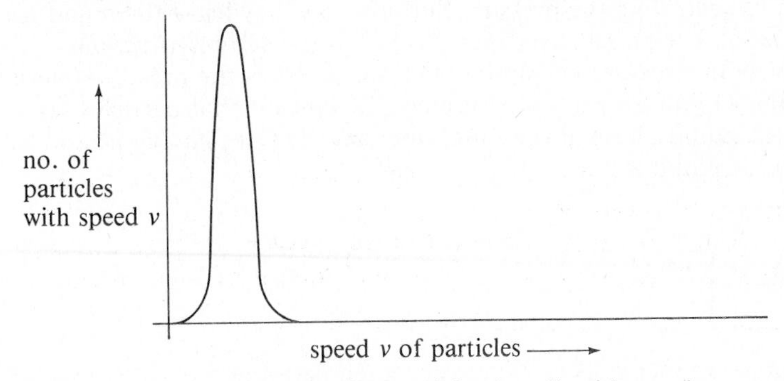

Unlike the particles in a gas, the particles in a liquid are close together and so there are definite attractive forces between them.

In spite of these attractive forces, some particles will be able to escape from the liquid. The particles that do escape are those which, as the result of a series of collisions, reach the speed needed to overcome the attractive forces pulling them back.

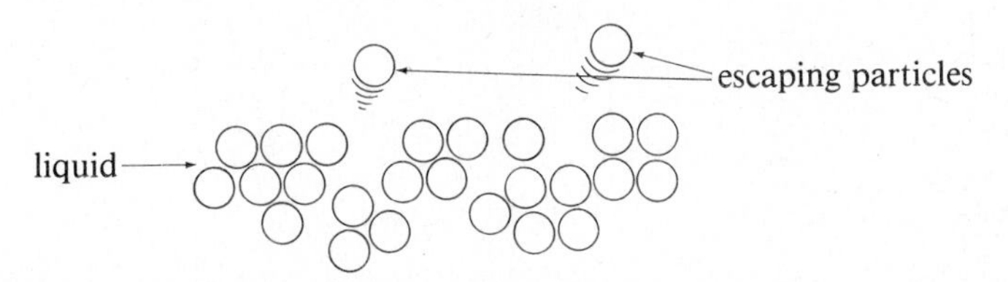

The minimum speed needed to escape from the liquid is named the *escape speed*; it is marked on the graphs below of particle speed against number of particles for various temperatures.

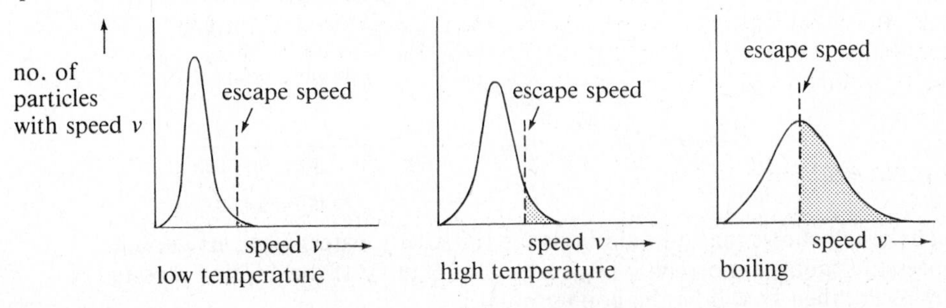

The graphs show that the escape speed does not change with temperature, but the proportion of particles with at least the escape speed does change, as shown by the shaded area under the curves. At low temperatures very few particles have the escape speed and evaporation is slow. At high temperatures a noticeable fraction of the particles has at least the escape speed and evaporation is rapid. The liquid boils when the average particle speed (shown by the peak of the curve) exceeds the escape speed.

The escape speed is unique and characteristic to each liquid. It depends on the nature of the particles making up the liquid and the attractive forces between them.

A measure of the escape speed of the particles at a particular temperature can be found by recording the pressure they exert as they leave the liquid for the vapour phase. The pressure measured is called the *saturated vapour pressure* (or vapour pressure for short), and is defined as the pressure over the liquid when the vapour above it is saturated. This means that particles are returning to the liquid phase at the same rate that they are leaving it, and the two phases are in equilibrium.

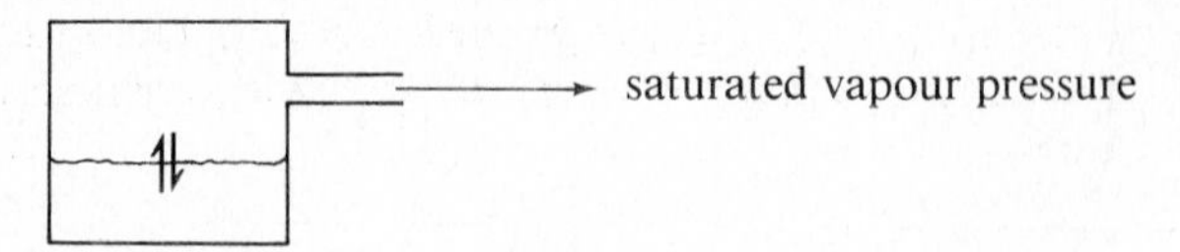

The vapour pressure increases with increasing temperature.

The change in vapour pressure with temperature for water is shown on this graph.

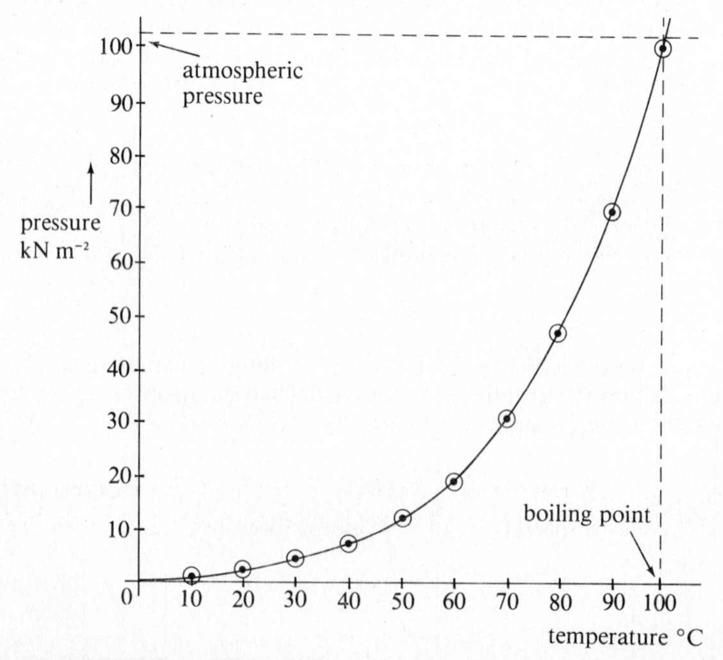

The graph shows that a liquid boils (at 100°C for water) when its vapour pressure equals atmospheric pressure (101 kN m^{-2}). If the applied pressure is reduced, then so will be the boiling point.

The ability to dissolve solutes

Gases
These dissolve in liquids, some reacting such as ammonia or hydrogen chloride with water, others obeying Henry's law. The solubility of gases decreases with increasing temperature.

Liquids
Many liquids dissolve completely in each other in all proportions and are said to be *miscible*. Alcohol and water are an example. Others are almost completely insoluble in each other and are said to be *immiscible*. They form two separate layers when mixed such as oil and water. Between these two extremes are some *partially miscible* mixtures such as phenol and water or nicotine and water. These mixtures either form a single, homogeneous layer or they separate into two layers (or phases) depending on the conditions.

Solids
These may or may not dissolve depending on the nature of the forces acting between solute and solvent particles. This is discussed in detail in the next section. In general the solubility of solids increases with increasing temperature.

Solvent properties

Liquids can be divided into two broad classes according to their solvent properties. The two classes of solvent are called *polar* and *non-polar* solvents. Like any attempt at classification the division is not clear cut but has some overlap.

polar solvents			**non-polar solvents**	
H_2O	CH_3CH_2OH	CH_3COCH_3	$CHCl_3$	CCl_4
	CH_3COOH		CH_3CCl_3	C_6H_6
NH_3			$CH_3CH_2OCH_2CH_3$	C_6H_{14}

As the name implies, polar solvents are those whose particles have a dipole moment as the result of charge separation in an asymmetric bond structure, e.g.

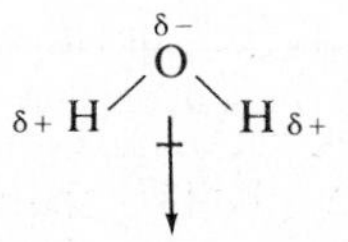

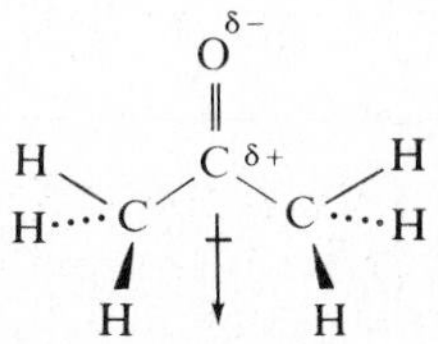

The presence or absence of dipole moments affects the way in which the solvent particles interact with each other and with the solute particles.

When considering the dissolving of any substance, it is possible to divide up the process into two steps:

i) the disruption of solute and solvent lattices
ii) the interaction between solute and solvent particles.

Put another way, existing bonds in the solute and solvent have to be broken and then new ones form in the solution. For example:

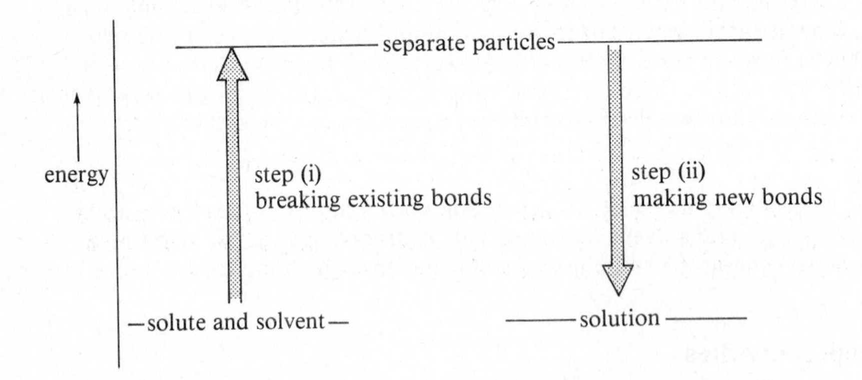

This is dealt with at greater length in the chapter on energy, but at this stage it is only necessary to realize that it is the relative sizes of the energy changes involved in these two theoretical steps which control solubility. For example;

salt dissolves in water but not in benzene.
wax does not dissolve in water but does dissolve in benzene.

These facts are explained like this:

sodium chloride in water	*sodium chloride in benzene*
Step i) consists of disruption of the strong bonds in the ionic salt lattice and some of the hydrogen bonds in water.	Step i) consists of disruption of the strong bonds in the ionic salt lattice and the weak van der Waals' forces between benzene molecules.
Step ii) involves making strong bonds between the sodium and chloride ions and the dipoles of water molecules as the ions are dissolved.	Step ii) would only result in very weak forces of attraction between non-polar benzene molecules and the ions.
These two steps have similar energies: the process is favourable and salt dissolves	The energy changes in step ii) are nothing like those in step i). The process does not occur — salt is insoluble.

wax in water	*wax in benzene*
Step i) consists of disruption of the weak van der Waals' forces in the molecular wax lattice, and the hydrogen bonds in water.	Step i) consists of disruption of the weak forces in the wax lattice and even weaker van der Waals' forces in benzene.
Step ii) would only result in very weak forces of attraction between water and wax molecules.	Step ii) results in the formation of more very weak forces of attraction between benzene and wax molecules.
Although the energy changes in step i) are small, those in step ii) are even smaller. The process does not occur — wax is insoluble.	Both these steps involve similarly small energy changes — wax dissolves.

It is possible as a result of many observations to make a general statement which can be used as a rough guide when considering the solubility of solids.

Solids with a high degree of ionic character are likely to be more soluble in polar than in non-polar solvents, and solids that are largely covalent are likely to be more soluble in non-polar than in polar solvents.

Colloids

By definition a solution is homogeneous. It is a one-phase system in which the solute particles are intimately mixed with the solvent particles. The solute particle size in a solution ranges from that of individual ions (1×10^{-12}m), to that of the largest molecules such as proteins (5×10^{-9} m).

A suspension is defined as a heterogeneous, two-phase system. It consists of a suspended phase of visible particles spread through a solvent medium. The smallest suspended particles which can be seen are about 2×10^{-7}m in diameter.

Between these two limits are mixtures called *colloids*.

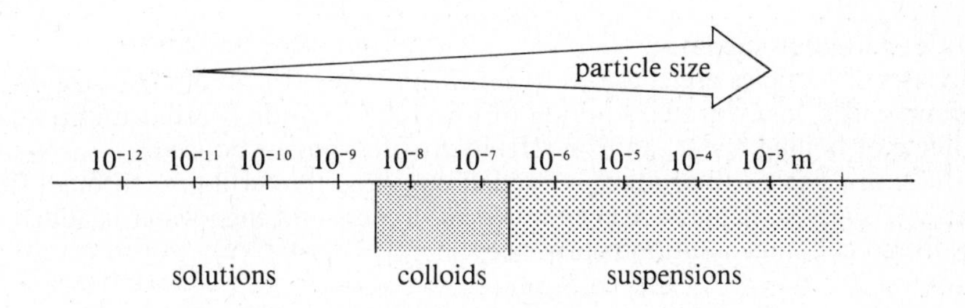

So colloids contain particles larger than molecules but not large enough to be seen optically. These particles called the *disperse phase* are spread through a substance called the *disperse medium*. If the colloidal mixture is liquid, the whole system is called a *sol*. If it is semi-rigid, it is called a *gel*.

Many colloids are found in everyday life. Examples are: jellies and mousses, foams, smokes, aerosol sprays. Colloids can be recognized by their optical properties. Although the particles are too small to be seen individually, they will scatter light. When a beam of light is focused into a colloid, the converging beam shows up as a cone of light known as *Tyndall's cone.*

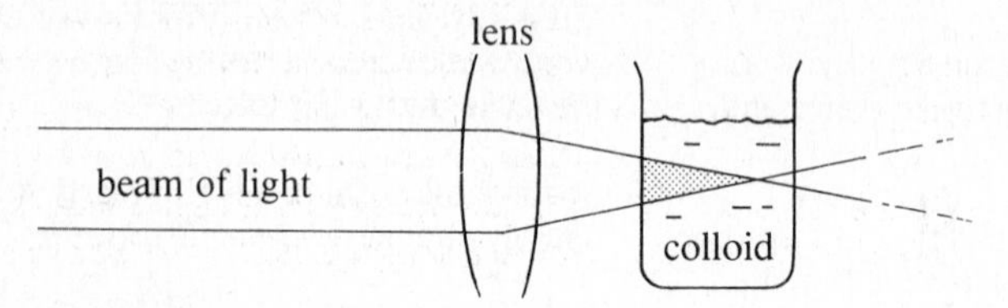

A similar effect is produced by a stream of sunlight in a dusty room, or a projector beam in a smoky cinema.

Colloids are usually divided into two classes. One type is colloidal because the size of the particles of its disperse phase naturally fall into the colloidal size range. These colloids are called *lyophilic colloids* and are made by adding a suitable solvent to a substance whose molecules are very large, e.g.

add water to gelatin or starch
add benzene to rubber.

These colloids are stable and are not precipitated if an electrolyte is added to them. If they are dried by heat, they can usually be reconstituted by adding the solvent again.

The other type of colloid is called a *lyophobic colloid.* It is prepared in the disperse medium by some method that happens to produce particles of colloidal size. Lyophobic colloids can be thought of as artifical colloids and there are two general ways of making them:

i) Dispersion methods
These involve breaking up large portions of solid into particles of colloidal size. For example, if an arc is sparked between two silver electrodes under water, colloidal-sized particles of silver are formed, so producing a silver sol.

ii) Condensation methods
Here small particles are clumped together until they reach colloidal size. For example, if a concentrated solution of iron (III) chloride is dripped into a large volume of boiling water, an iron (III) hydroxide sol is produced.

Both these sols will produce a Tyndall cone and they will pass through filter paper. However, if they are evaporated to dryness and then water is added again, no colloidal solution is produced.

11.2 Ideal solutions with non-volatile solutes

Ideal solutions

When a solute is dissolved in a liquid, new particles are introduced among the original particles of the liquid. In the Kinetic Theory model of a solution, the addition of a solute changes the population of any particular type of particle, but *not* the forces on it. So in a solution of this sort, the forces acting on a particle are the same whatever type of particle there is around it. A solution of this sort is known as an *ideal solution*.

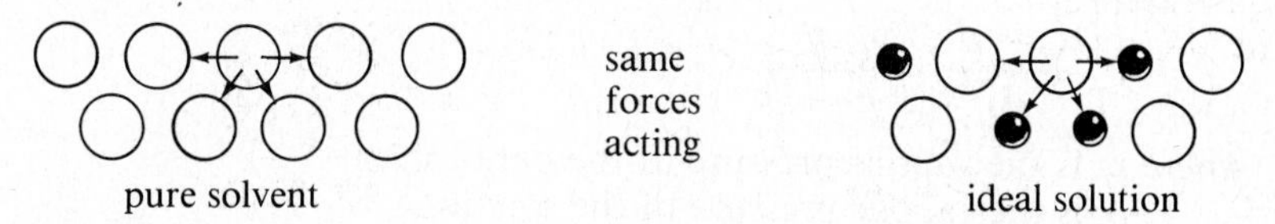

In an ideal gas there are no forces between the particles; in an ideal solution there are constant forces between the particles whatever their identity.

If this ideal situation in a solution is true, then the escaping tendency of any particle will depend only on the statistical chance of the particle occupying a position on the liquid's surface when it has the escape speed. This in turn depends on the concentration of the solution.

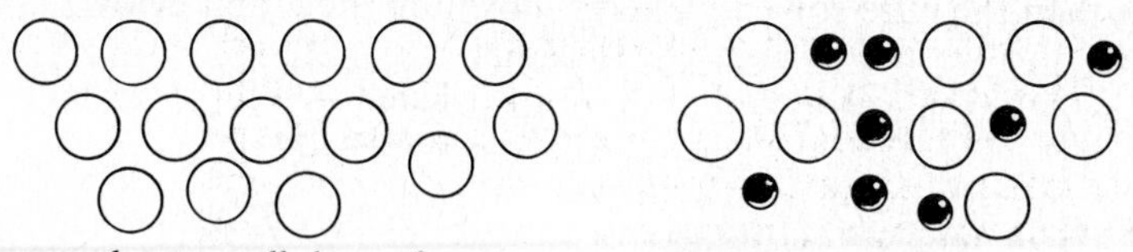

pure solvent — all the surface positions occupied by solvent particles.
50% by mole fraction solvent — only half the surface positions occupied by solvent particles

Non-volatile solutes

If the solute particles show no escaping tendency, the solute is described as being *non-volatile*. Sugar or salt added to water are examples of non-volatile solutes. In these solutions, the solute particles contribute no vapour pressure of their own, but they produce a reduction in the solvent vapour pressure because they are occupying surface positions.

In an ideal solution, therefore, the vapour pressure of the solvent at a given temperature is proportional to the mole fraction of solvent in the solution.

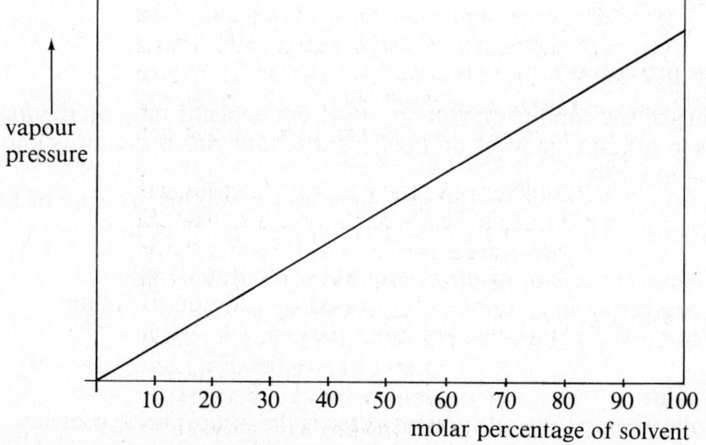

This relationship between the lowering of the vapour pressure of a solvent and its mole fraction or mole percentage is known as *Raoult's law*.

Raoult's law states that, at constant temperature, the relative lowering in vapour pressure of a solvent is equal to the mole fraction of solute in the solution.

$$\frac{p_o - p}{p_o} = \frac{n}{n + n_s}$$

where p_o is the vapour pressure of the pure solvent
p is the vapour pressure of the solution
n is the number of moles of solute
n_s is the number of moles of solvent

All this refers to the solution at one particular temperature. The vapour pressure of the solvent at other temperatures will be correspondingly lowered. A solution that contains 10% by moles of solute always has a vapour pressure 10% lower than the pure solvent. This is shown on the graph below.

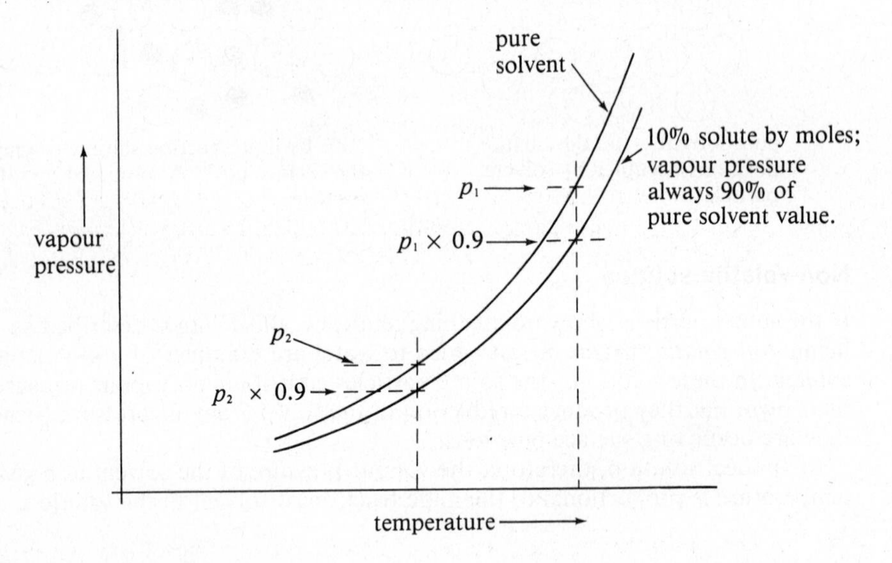

Colligative properties

The lowering of the vapour pressure of a solvent depends only on the *number* of particles in solution and not on their identity. For this reason it is known as a *colligative property*.

There are three other properties all related to the vapour pressure of the solution. Therefore they too are colligative properties. They are:

1. the elevation in boiling point caused by addition of solute
2. the depression in freezing point caused by addition of solute
3. the osmotic pressure of a solution.

1. Boiling point

A liquid boils when its vapour pressure equals the atmospheric pressure. When the vapour pressure of a liquid changes due to addition of solute, its boiling point also changes.

A magnified portion of the vapour pressure curves shows this.

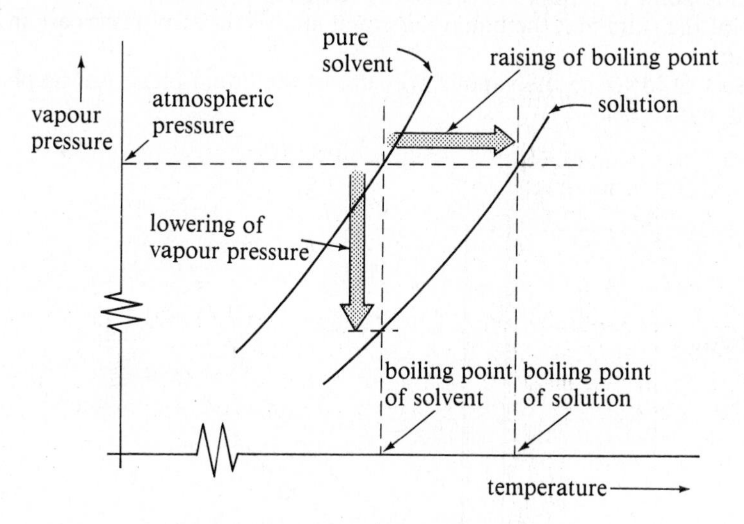

The following points arise from these graphs:

i) Adding a solute *raises* the boiling point of the liquid.
ii) In agreement with Raoult's law, the two vapour pressure curves are nearly parallel. This means that a particular lowering of the vapour pressure produces a related raising of the boiling point and an analogue to the Raoult's law equation can be written:

$$\frac{T - T_0}{T_0} \propto \frac{n}{n + n_s}$$

where T_0 is the boiling point of the pure solvent
T is the boiling point of the solution
n is the number of moles of solute
n_s is the number of moles of solvent

The change in the boiling point of a solvent for addition of any solute is unique to that solvent. Tables have been produced showing how the boiling point of 1000 g of solvent changes when one mole of solute is added. These figures are called *ebullioscopic constants* or *boiling-point constants*.

The constants for some common solvents are listed below.

solvent	*boiling-point constant* (°C mol^{-1} 1000 g^{-1})
water	0.52
ethanol	1.15
propanone	1.72
benzene	2.57

These figures mean that if 1 mole of *any* solute is added to 1000 g of water, its boiling point is raised by 0.52°C and it boils at 100.52°C. If 1 mole of a solute is added to 1000 g of propanone, its boiling point is raised by 1.72°C.

2. Freezing point
The freezing point of a liquid is defined as the temperature at which the vapour pressures of the solid and the liquid are equal and so the two phases are in equilibrium.

The effect of lowering the vapour pressure of the liquid phase on the phase diagram is shown below.

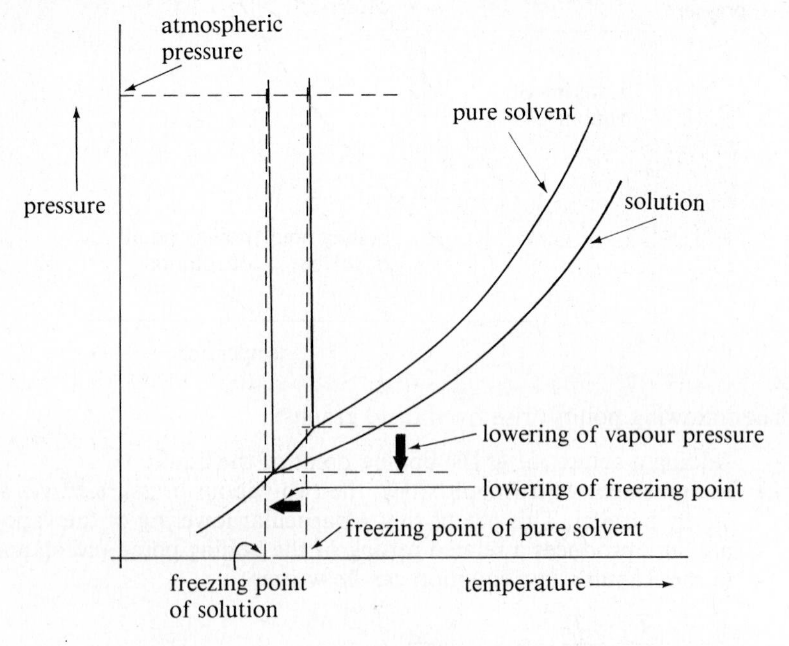

From the phase diagram it can be seen that:

i) Adding a solute *lowers* the freezing point of the liquid.
ii) Like the boiling point, a change in the vapour pressure of the liquid produces a related lowering of the freezing point. Another Raoult's law analogue can be written:

$$\frac{T_0' - T'}{T_0'} \propto \frac{n}{n + n_s}$$

where T_0' is the freezing point of the pure solvent
T' is the freezing point of the solution
n is the number of moles of solute
n_s is the number of moles of solvent

Like boiling-point figures, tables have been produced of *cryoscopic* or *freezing-point* constants. Some typical ones are given below.

solvent	*freezing-point constant* (°C mol^{-1} 1000 g^{-1})
water	1.85
ethanoic acid	3.90
benzene	5.12

These values mean that if 1 mole of a solute is added to 1000 g of water, its freezing point is lowered by 1.85° and it freezes at −1.85°C.

3. Osmotic pressure

When two solutions of different concentrations are separated by a certain type of membrane, it is found that *solvent* flows through the membrane from the dilute solution to the more concentrated solution until the two concentrations are equal. This process is known as *osmosis*.

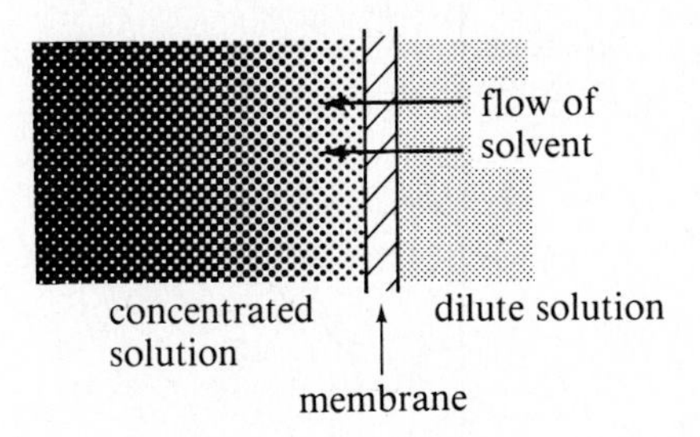

A membrane which allows the passage of solvent only through it is called a *semi-permeable membrane*.

The flow of solvent across the membrane is related to the vapour pressures of the two solutions it separates because it is a direct consequence of the number of collisions by solvent particles on each side of the membrane.

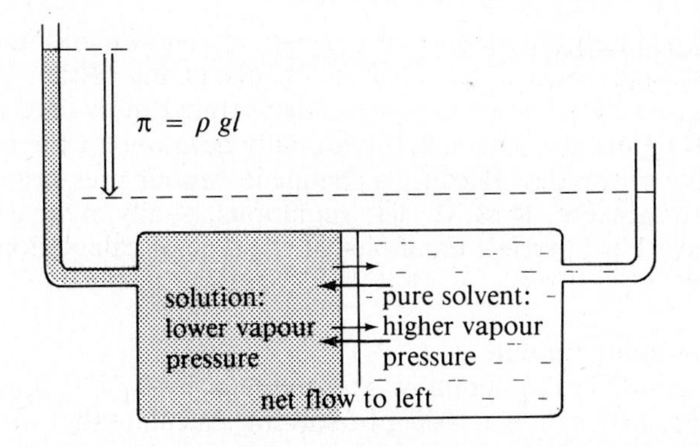

This net flow of solvent in one direction creates a pressure difference across the membrane. This pressure is called the *osmotic pressure*, symbol π, and can be measured by applying an opposing pressure just large enough to stop the flow of solvent.

> Osmotic pressure is the pressure that must be applied to a solution, when separated from a more dilute solution by a semi-permeable membrane, in order to prevent the inflow of solvent.

Osmotic pressure π increases with the kelvin temperature T of the system and with the concentration difference c across the membrane.

$\pi \propto T$ and $\pi \propto c \Rightarrow \pi \propto \frac{1}{V}$ where V is the volume of the solution.

Therefore $\pi \propto \frac{T}{V}$ or $\pi V \propto T$

If n moles of solute are dissolved in the volume V

$\pi V = nRT$ where R is the gas constant

This is analogous to the ideal gas equation.

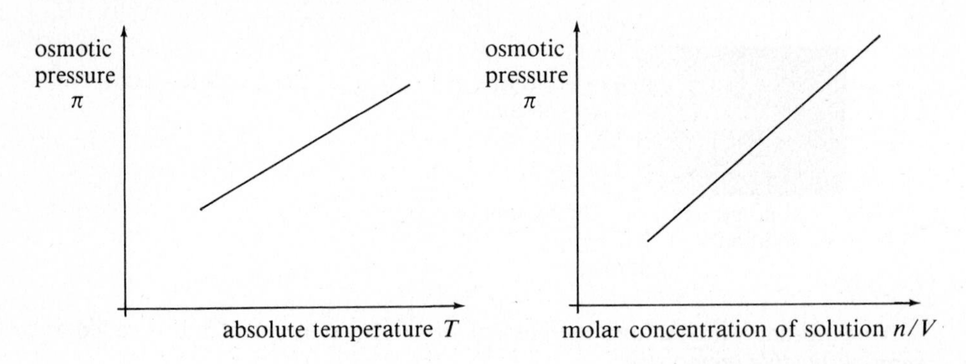

Finding the R.M.M. of solutes using Raoult's law

Colligative properties depend on the *number* of particles in solution and not on their identity. For this reason a measurement of the change in any of the colligative properties caused by a known mass of added solute can be used to calculate the R.M.M. of the solute. This has traditionally been one of the main applications of colligative properties. Because a change in vapour pressure is in practice more difficult to measure, R.M.M. determinations usually involve one of the other three colligative properties. Examples of the type of calculation involved are given below.

R.M.M. from changes in boiling point

4.5 g of a substance raised the boiling point of 56 g of water by 0.71°C. Given the boiling-point constant is 0.52°C per 1000 g of solvent, calculate the R.M.M.of the substance.

4.5 g in 56 g of water changed the b.p. by 0.71°C.

thus 4.5 g in 1000 g of water would change the b.p. by

$$\left[\frac{0.71 \times 56}{1000}\right] = 0.04°\text{C}$$

So to change the b.p. of 1000 g of water by 0.52°C, the boiling-point constant, requires

$$\left[\frac{4.5 \times 0.52}{0.04}\right] = 58.5\ \text{g}$$

This must be 1 mole, so the R.M.M. of the substance is 58.5.

R.M.M. from changes in freezing point

When 1.2 g of a substance is dissolved in 25 g of benzene, the solution freezes at 2.73°C. Calculate the R.M.M. of the substance given the freezing point of pure benzene is 5.4°C and its freezing-point constant is 5.12°C per 1000 g.

Change in freezing point = 5.4 − 2.73 = 2.67°C.
So 1.2 g in 25g of benzene changes the f.p. by 2.67°C
thus 1.2 g in 1000 g of benzene would change the f.p. by

$$\left[\frac{2.67 \times 25}{1000}\right] = 0.067°\text{C}$$

So to change the freezing point of benzene by 5.12°C, the freezing-point constant, requires

$$\left[\frac{1.2 \times 5.12}{0.067}\right] = 91.7\text{g}$$

This must be 1 mole, so the R.M.M. = 91.7.

R.M.M. from osmotic pressure measurements

0.25 g of starch in 120 cm³ of ethanol exerts an osmotic pressure of 8.4×10^{-4} atmospheres at 30°C. Find the apparent R.M.M. of starch.

The osmotic pressure equation $\pi V = nRT$ means that 1 mole of solute dissolved in 22.4 dm³ of solution will produce an osmotic pressure of 1 atmosphere at 273 K. The problem can be solved by working to this statement.

$$\pi \text{ at } 273\ K = \left[\frac{8.4 \times 10^{-4} \times 273}{303}\right] = 7.57 \times 10^{-4} \text{ atm}$$

0.25 g in 120 cm³ produces an osmotic pressure of 7.57×10^{-4} atm thus 0.25 g in 22 400 cm³ would give a pressure of

$$7.57 \times 10^{-4} \times \left[\frac{120}{22\,400}\right] = 4.05 \times 10^{-6} \text{ atm}$$

To produce an osmotic pressure of 1 atmosphere would require

$$\left[\frac{0.25 \times 1}{4.05 \times 10^{-6}}\right] = 61\,700 \text{ g of solute in } 22\,400 \text{ cm}^3 \text{ of solution}$$

This must be 1 mole, so the R.M.M. of the substance is 61 700.

Anomalous values for the R.M.M.

Unexpected or anomalous results are sometimes produced in these calculations of the relative molar mass. These arise because the solute either splits up into more particles than expected, or else it associates and so produces fewer particles. For example, ionic solids dissociate into their component ions in water:

$$Na_2SO_{4(s)} + H_2O_{(l)} \longrightarrow 2Na^+_{(aq)} + SO^{2-}_{4(aq)}$$

1 mole of sodium sulphate produces 3 moles of particles and so all the colligative property changes are affected three times as much as expected.

Conversely, carboxylic acids link up as the result of hydrogen bonding and form dimers in organic solvents:

$$2CH_3COOH_{(l)} \xrightarrow{\text{benzene}} CH_3C\begin{matrix} \diagup\!\!\!\!= O \cdots H - O \diagdown \\ \diagdown O - H \cdots O =\!\!\!\!\diagup \end{matrix}CCH_3$$

So 1 mole of ethanoic acid only produces half a mole of particles.

These anomalous values emphasize again the nature of colligative properties. They are related only to the *number* of particles and not to their nature.

Practical methods of measuring changes in colligative properties

The practical difficulties in finding a relative molar mass by measuring a change in a colligative property are considerable. Raoult's law is only strictly obeyed in fairly dilute solutions and consequently the changes in vapour pressure, freezing point, or boiling point to be measured are tiny.

Freezing point and boiling point changes are the two colligative properties that are most easily determined, and the methods used illustrate how the difficulties are more or less overcome. The problems are:

i) Measuring accurately a very small temperature change. This is solved by using a special thermometer; it need only have a very small range because the changes in fixed points are small for dilute solutions, but it must be very sensitive. This is achieved by having a very large bulb of mercury and a very fine bore. The most common type is called a Beckmann thermometer.

ii) Avoiding superheating or supercooling. The changes in fixed points are only fractions of a degree, so it is essential to avoid an error of several degrees produced as the result of rapid heating or cooling.

Two typical methods are described below.

Landsberger's boiling point apparatus

This avoids superheating by boiling the liquid with its own vapour

The apparatus is set up as shown at the top of p. 161.Pure solvent is put into the inner tube. Vapour from the boiling solvent in the flask on the left is passed into the solvent in the inner tube. The latent heat given out as the vapour condenses is used to heat the solvent to its boiling point, but it cannot heat it any higher because this would involve more energy than is available from the latent heat. So superheating is avoided. The highest temperature reached is recorded as the boiling point of the pure solvent. A weighed pellet of solute is now dropped into the inner tube and the resulting solution is boiled by passing more vapour through it. A reading of the boiling point is recorded and immediately the supply of vapour is stopped. The volume of the solution can now be measured using the graduations on the inner tube.

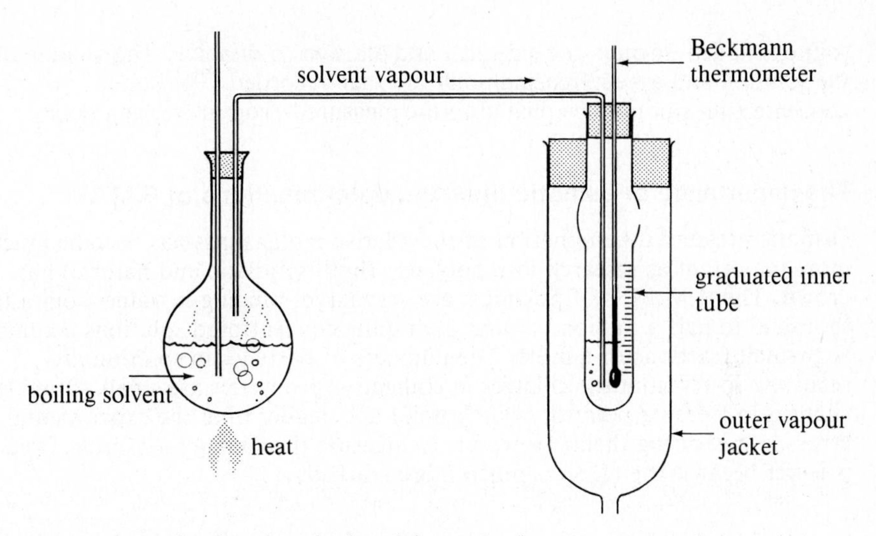

The concentration of solute is calculated from the known mass and the measured volume. It can now be related to the increase in boiling point.

Beckmann's freezing-point apparatus
In this apparatus supercooling is deliberately induced.

If a solvent or solution is below its freezing point, freezing can be initiated by scratching the side of the container with a glass rod. Once freezing starts, the latent heat given out raises the temperature of the mixture until the freezing point is reached.

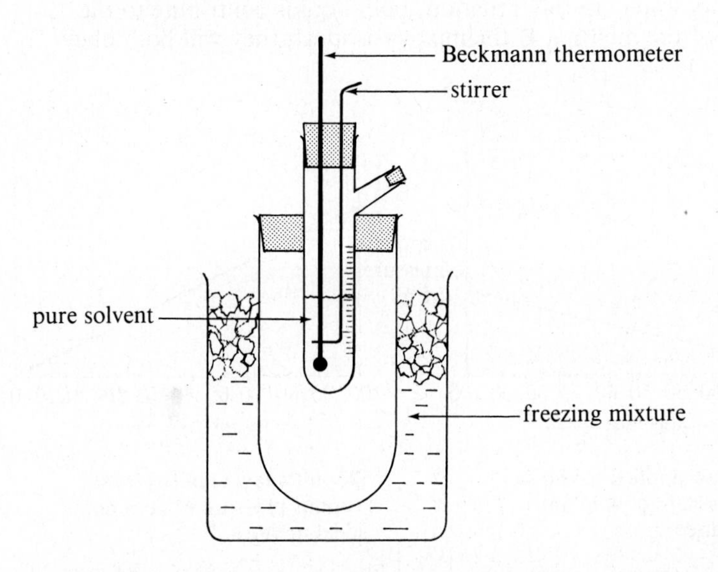

A weighed sample of pure solvent is put into the inner tube and cooled below its freezing point. Freezing is initiated by agitating the stirrer and the highest temperature reached is recorded. This is the freezing point. A weighed pellet of solute is added through the side arm and allowed to dissolve. The volume of the solution and a new freezing point are then recorded. The solute concentration can now be related to the measured drop in freezing point.

The importance of osmotic pressure determinations of R.M.M.

Osmotic pressure determination of the relative molar mass has become much more important as research into polymers (both synthetic and natural) has grown. The R.M.M.'s of polymers are very large, ranging in value from a few thousand to half a million, so that even quite concentrated solutions contain very small fractions of a mole. The numbers of particles in solution are relatively so few that the changes in colligative properties are small. Often the changes in freezing point or boiling point are smaller than the experimental errors in measuring them. However, to measure the change in osmotic pressure is easier because the effect is much more marked.

11.3 Ideal solutions with volatile solutes

So far all the solutions discussed have contained involatile solutes. These solutes have such a low vapour pressure that it can be ignored, and the only vapour pressure present in the system is that of the solvent.

It is perfectly possible to have a solution of one liquid in another, for example alcohol in water. In this situation, both liquids contribute to the vapour pressure of the mixture. If the mixture is ideal, they will both obey Raoult's law.

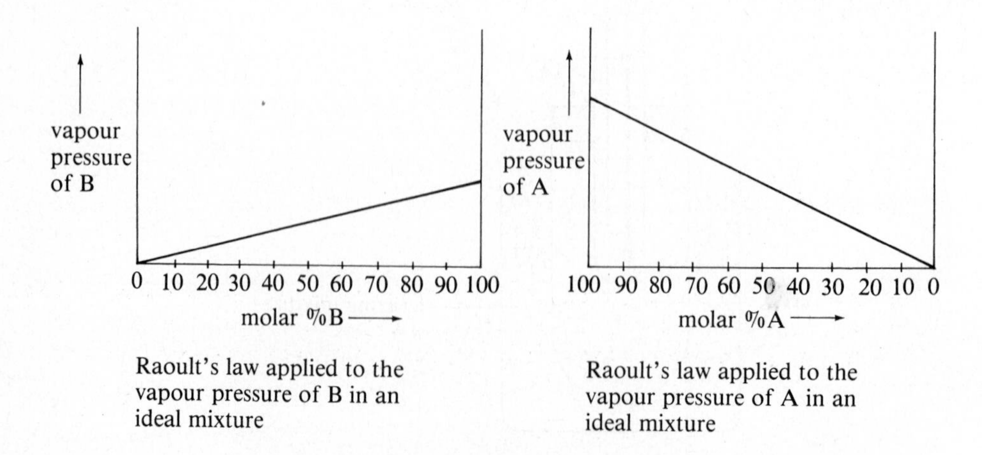

Raoult's law applied to the vapour pressure of B in an ideal mixture

Raoult's law applied to the vapour pressure of A in an ideal mixture

These two graphs can be combined into a single diagram, and if the two vapours behave like ideal *gases*, Dalton's law of partial pressures can be applied to find the total vapour pressure of the mixture.

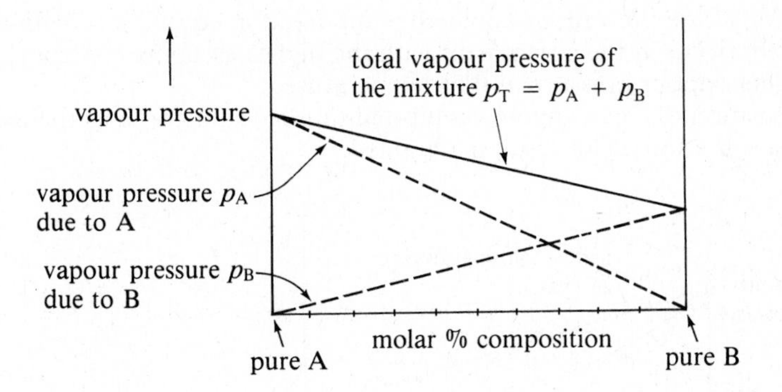

Boiling a nearly ideal mixture

Most mixtures contain one component of higher vapour pressure and one of lower vapour pressure. Vapour pressure is related to the escaping tendency of the two types of particles.

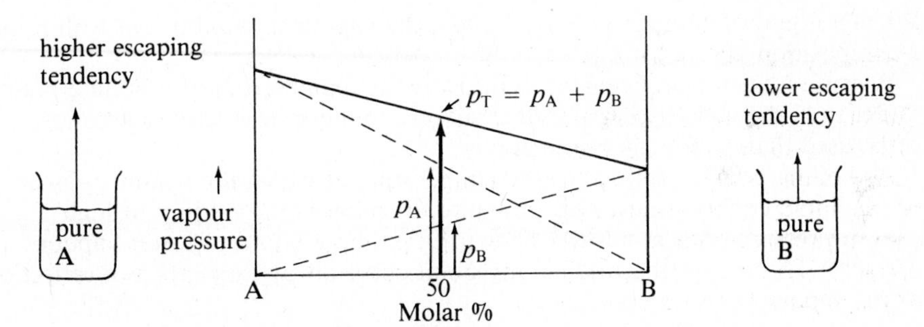

What happens if a 50/50 mixture of A and B is boiled?

p_A p_B

50% A
50% B

heat

The diagram indicates that the vapour above the 50/50 liquid mixture is richer in A because its particles have a greater escaping tendency.

In fact the composition of the vapour can be calculated as follows.

$$\text{mole \% A} = \frac{p_A}{p_A + p_B} \approx 70\%$$

$$\text{mole \% B} = \frac{p_B}{p_A + p_B} \approx 30\%$$

This argument can be applied to the boiling of a whole range of mixtures of

A and B. In every case the vapour above the liquid and in equilibrium with it will be relatively richer in the component with the higher escaping tendency, i.e. the one of higher vapour pressure at that temperature.

If the composition of these vapours is plotted on the same graph as the total vapour pressure, the following result is obtained.

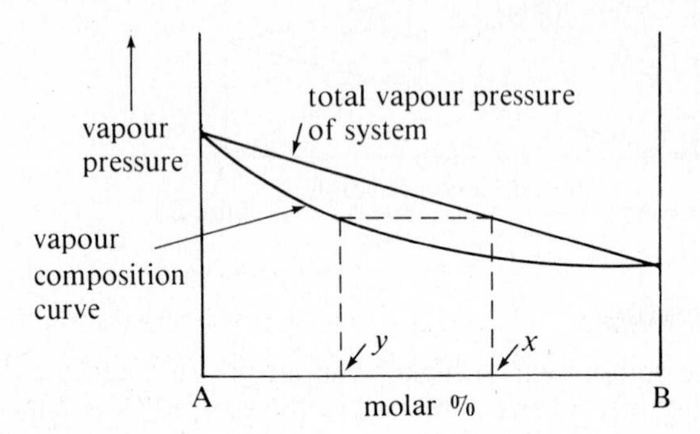

When a liquid of composition x is boiled, the vapour in equilibrium with it has a composition y.

Because it is more convenient to measure the temperature of a boiling point rather than the vapour pressure of a mixture, boiling-point curves are more often used than vapour-pressure curves.

As boiling point is defined as the temperature at which the vapour pressure of the liquid equals atmospheric pressure, it follows that a liquid of high vapour pressure boils at a lower temperature than a liquid of lower vapour pressure. This means that the boiling-point curve always appears as a reflection of the vapour pressure curve.

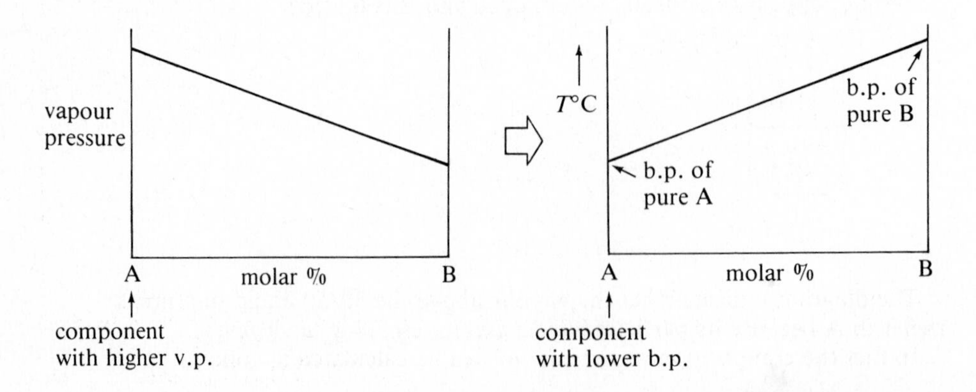

The vapour-composition curve can also be plotted on a boiling-point curve. It shows that the composition of the vapour above a boiling liquid is always relatively richer in the component that boils most readily, i.e. the one of lower boiling point. For example, by boiling a liquid of composition x a vapour of composition y is driven off, as represented by the graph on page 165.

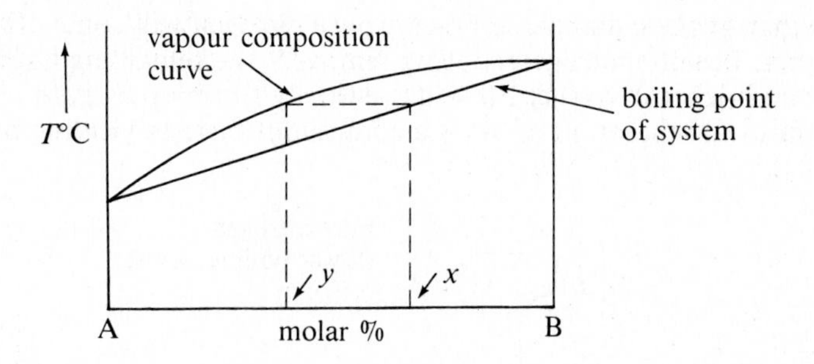

Fractional distillation

The vapour from a boiling liquid mixture usually has a different composition from that of the liquid. If this vapour is removed and condensed, another liquid is produced. This second liquid can itself be boiled, yielding a new vapour and so on. This is the basis of separating liquid mixtures by *fractional distillation*. The process is represented on the experimental curves for the nearly ideal oxygen/nitrogen mixture as shown below.

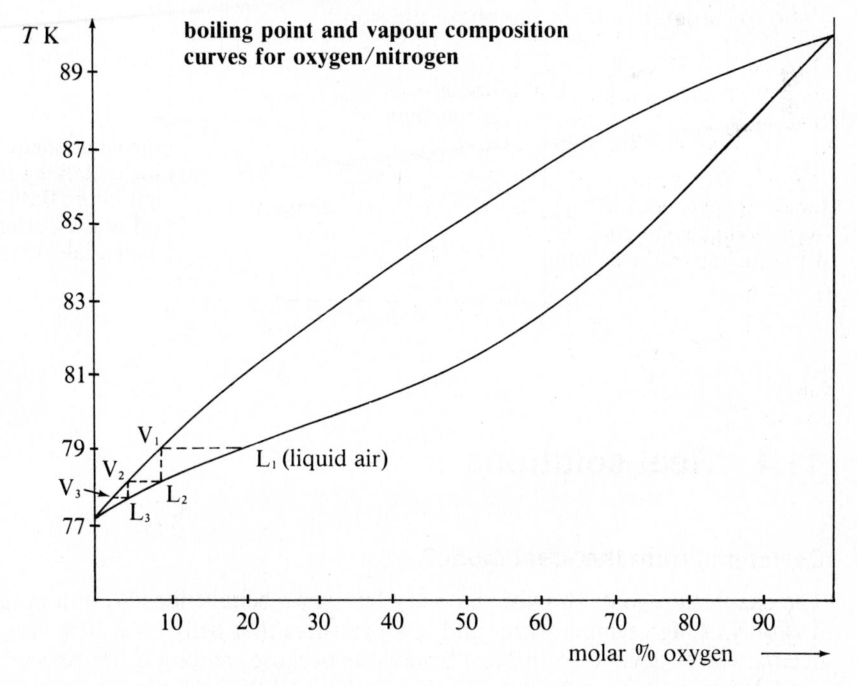

Boiling liquid air L_1 in a fractionating column produces vapour V_1 richer in nitrogen, the lower boiling point component. This vapour is condensed in the column and starts to run back down the column as L_2. But the remaining liquid, now richer in oxygen and therefore with a higher boiling point, produces a second vapour which is hotter. This causes liquid L_2 to boil producing vapour V_2 and the cycle repeats itself again.

The result is that, if the column is efficient, pure nitrogen will come off the top of the column. Because nitrogen is being removed, the remaining liquid becomes richer and richer in oxygen. It will eventually finish up as pure oxygen. The industrial separation of air is a continuous process yielding both pure gases.

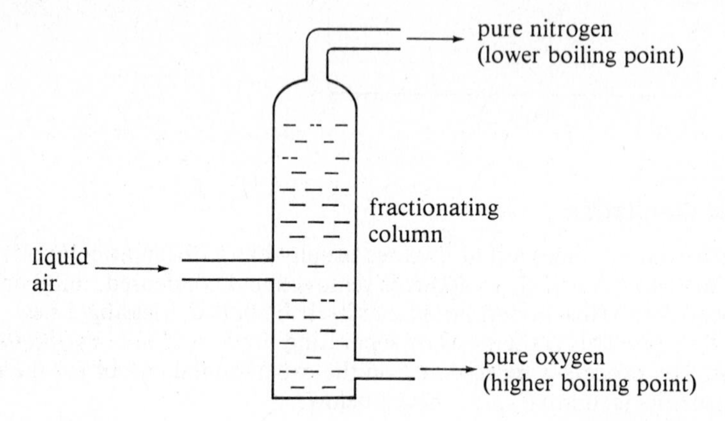

So to summarize from the phase diagram:

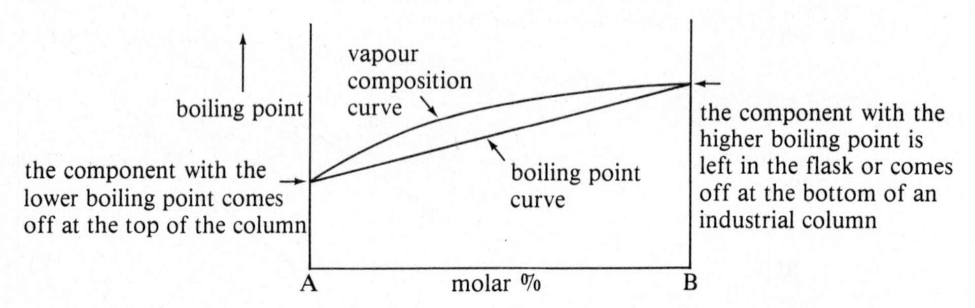

11.4 Real solutions

Deviations from the ideal model

The chapter on gases showed that very few *gases* behave ideally, and even then it is only at high temperatures and low pressures that truly ideal behaviour occurs. Gases deviate from ideal behaviour because, among other reasons, the particles do attract each other. In the ideal model they do not.

Similarly, most *solutions* deviate from the ideal behaviour described by Raoult's law. They deviate not because there are forces between the particles (there always are in the liquid state), but because these forces are not constant. Solvent particles behave differently when surrounded by solute particles from when surrounded by more solvent particles.

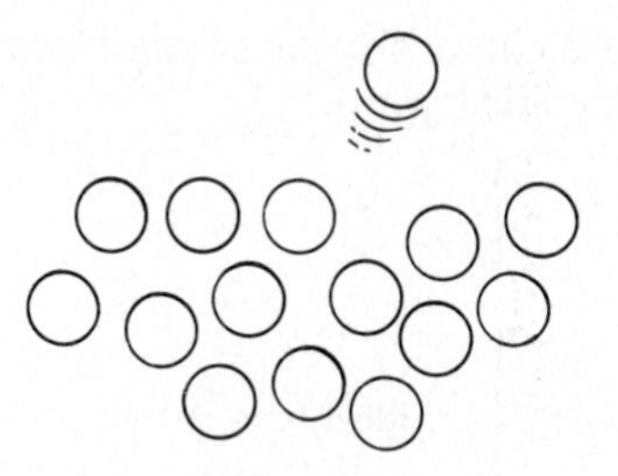

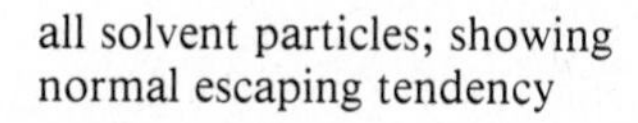

all solvent particles; showing normal escaping tendency

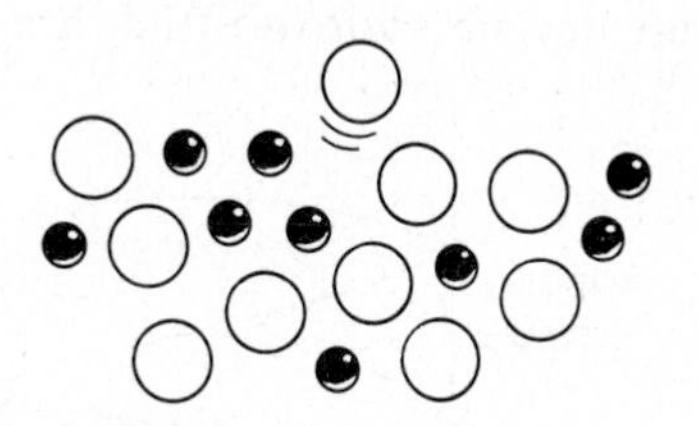

solvent and solute particles; the escaping tendency of the solvent particles is altered.

The escaping tendency of the particles must be changed if the strength of the forces between solute and solvent particles F_{AB}, is different from the strength of the forces between solvent particles F_{AA}, or between solute particles F_{BB}. Depending on the relative strengths of these forces, a real solution can deviate from ideal behaviour in different ways. Real solutions may be divided into three classes.

1. Nearly ideal mixtures

These always consist of two components whose particles are very similar. Because the particles are so similar the forces between them change very little if the particles are mixed.

Nearly ideal solutions are produced from mixtures of oxygen/nitrogen as in liquid air (see page 165), alkanes as in crude oil, and benzene/toluene.

The experimental vapour-pressure curves for the benzene/toluene mixture are ideal within experimental error.

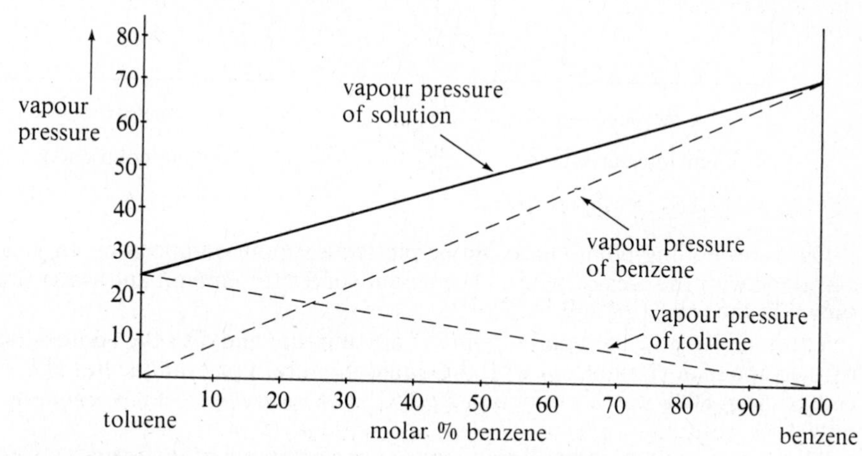

2. Mixtures which deviate positively from Raoult's law

Some mixtures are found to exert a greater vapour pressure than the ideal model. As the vapour pressure is larger than the predicted value they are said to deviate *positively* from Raoult's law. Examples include mixtures of ethanol/water, propanone/water, ethyl ethanoate/ethanol, and ethanol/carbon disulphide. The experimental vapour-pressure curves for the ethanol/water mixture deviate *positively* (they are above the ideal lines).

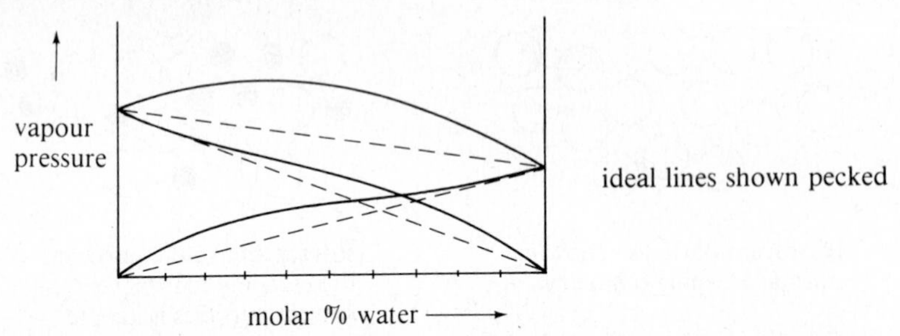

This positive deviation means that the vapour pressure of water and of ethanol is higher in the mixture than would be predicted by Raoult's law. If the vapour pressure is higher, it must mean that the particles have a greater escaping tendency in the mixture than expected. This implies that the force of attraction between a water and an ethanol molecule F_{AB} is less than the force of attraction between two water molecules F_{AA} or between two ethanol molecules F_{BB}. So if $F_{AB} < F_{AA}$ or F_{BB}, a positive deviation results.

Remember from ideal mixtures that the boiling-point curve looks like a reflection of the vapour-pressure curve. If the vapour-pressure curve from a positively deviating mixture has a hump in it, its boiling-point curve has a dip in it.

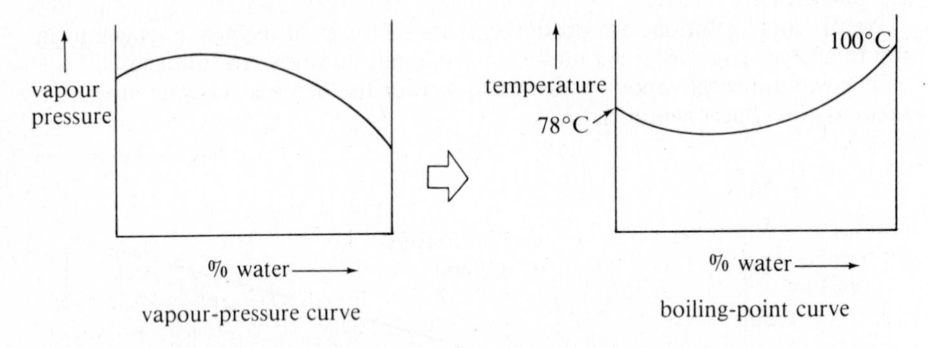

Onto this boiling-point curve can be plotted a vapour composition curve as was done with the ideal mixture. The actual curves for ethanol and water are shown at the top of p. 169.

These curves give the boiling point of any mixture and also the composition of the vapour in equilibrium with the liquid mixture. For example liquid L of composition 80% water will boil at 82.3°C. The vapour above it V contains only 41% water.

The curves show that any liquid mixture has a vapour of different composition in equilibrium above it. There is only one exception and this is defined by the point where the two curves touch. A mixture of this particular composition is called an *azeotrope*.

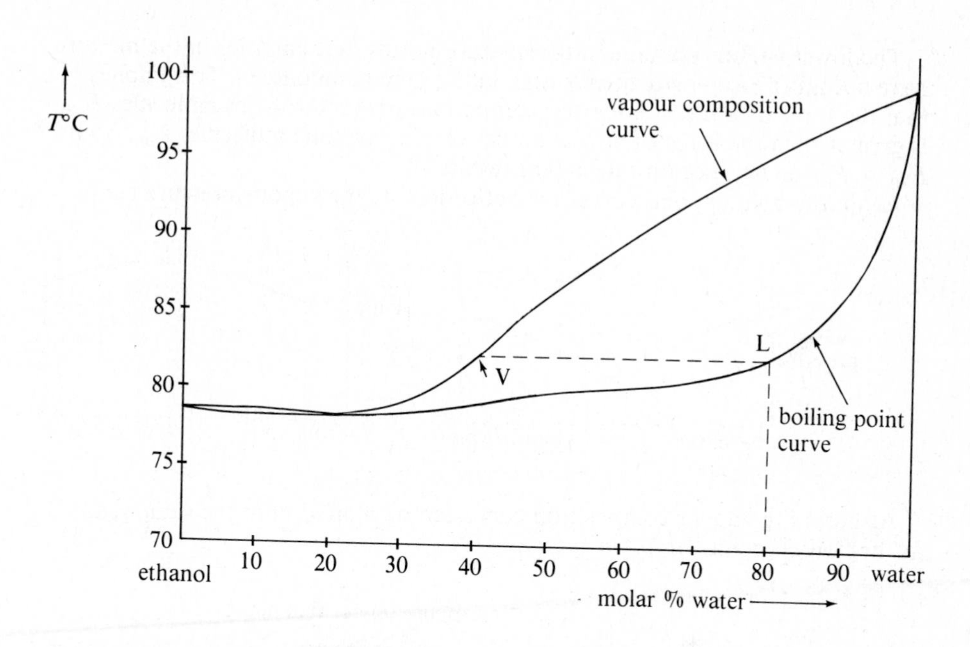

An *azeotrope* is a liquid mixture which distils unchanged at a definite temperature.

The composition of the vapour above the liquid is the same as the composition of the liquid itself. Because an azeotrope distils at a constant temperature with no change in composition, it is sometimes referred to as a *constant boiling mixture*.

3. Mixtures which deviate negatively from Raoult's law

Another set of mixtures is found to exert vapour pressures less than those predicted by Raoult's law, and so they are said to deviate *negatively*. Examples include mixtures of hydrogen halides/water and propanone/trichloromethane (chloroform) whose vapour pressure curves are given below.

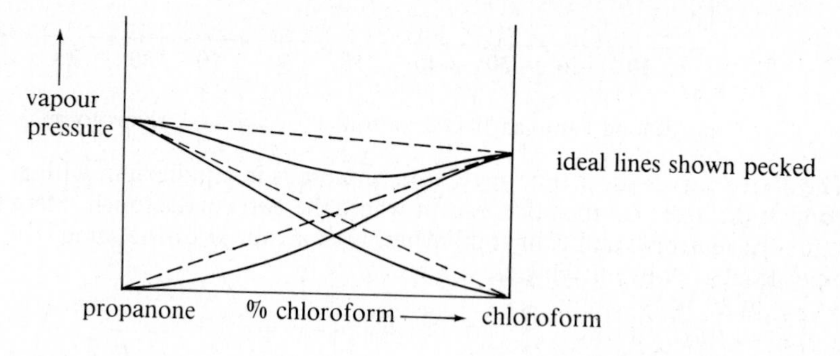

The lower vapour pressure in the mixture means that particles in the mixture have a smaller escaping tendency than in the pure components. This implies that the force of attraction between propanone and chloroform molecules F_{AB} is greater than the force of attraction between propanone molecules F_{BB}. So if $F_{AB} > F_{AA}$ or F_{BB}, negative deviation results.

Again the boiling point curve is a 'reflection' of the vapour pressure curve.

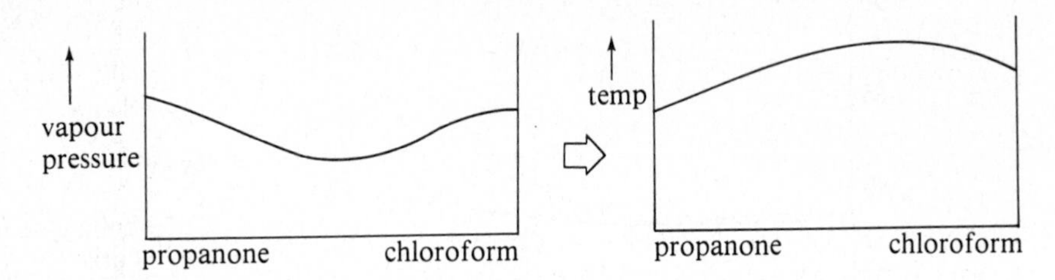

As before, a vapour composition curve can be plotted onto the same graph as the boiling-point curve.

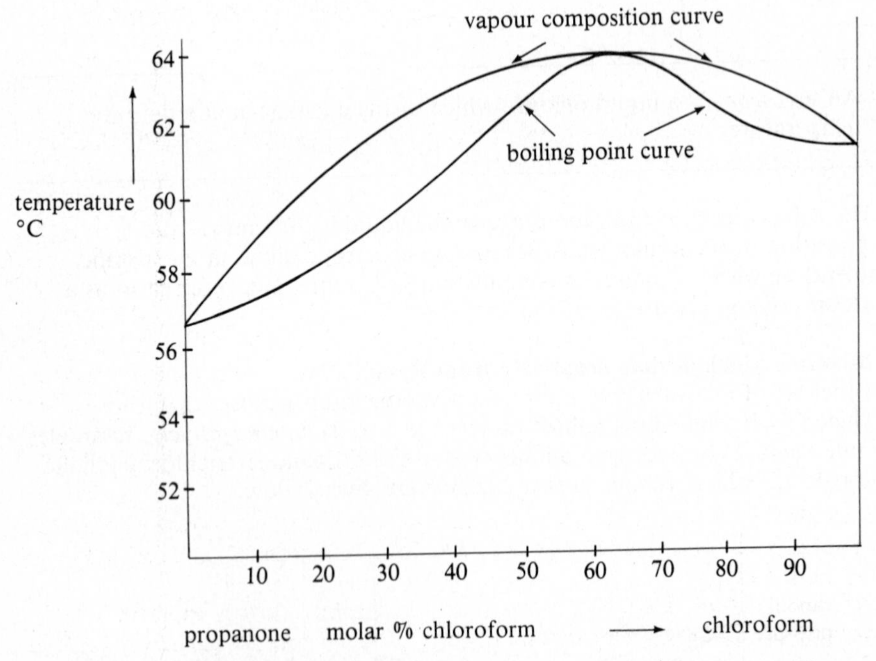

The above curves show that any liquid mixture is in equilibrium with a vapour of different composition except where the two curves touch. Here the mixture is azeotropic and is in equilibrium with a vapour of the same composition. It boils unchanged.

Distilling real solutions

The summary on page 166 describes the distillation of an ideal mixture. The component with the lower boiling point is produced at the top of a distillation column, while the higher boiling point component is collected at the bottom.

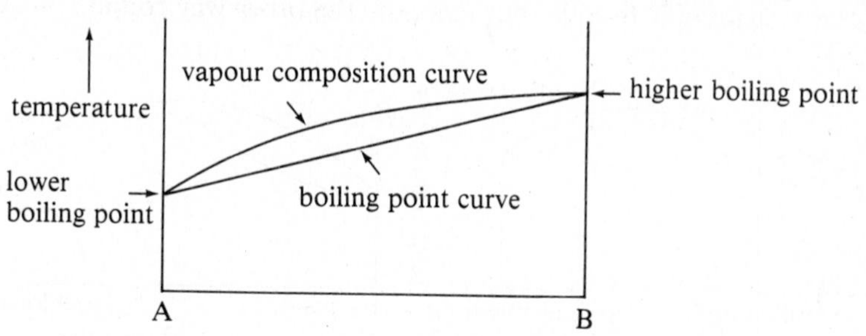

A is produced at the top of the column; B is collected at the bottom.

This is a useful guide in considering the distillation of real solutions which fall into three classes.

1. Solutions whose boiling points have neither a maximum nor a minimum
The deviation in these solutions is sufficiently small for them to be treated as ideal. Fractional distillation produces complete separation of the components as in an ideal mixture.

2. Solutions with a boiling-point minimum
Mixtures of this sort, such as alcohol and water, can be treated as two nearly ideal systems side by side:

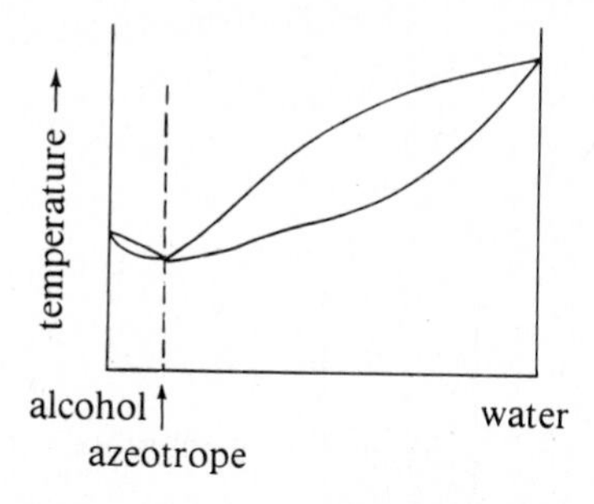

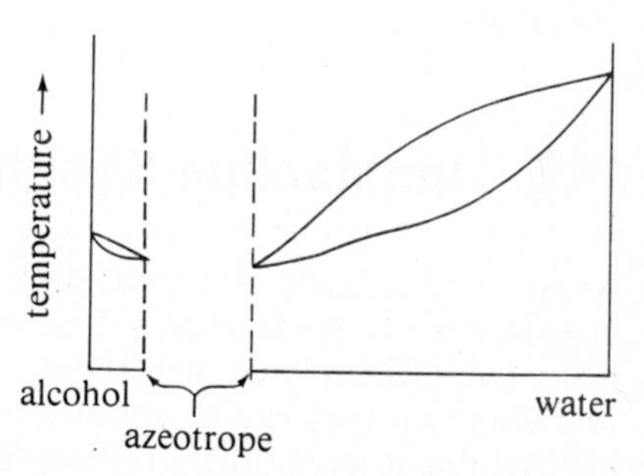

Consider the left-hand diagram as being made up of the two parts of the right-hand diagram pushed together. If this is done, predicting the result of distillation is easy.

Distillation of any alcohol/water mixture to the left of the azeotropic composition will yield the azeotrope at the top of the column because it has the lower boiling point, and pure alcohol will collect in the flask at the bottom.

Distillation of any mixture to the right of the azeotropic composition will again yield the azeotrope at the top of the column, but this time water is left in the flask when distillation is complete.

The second situation is the usual one used to enrich fermented alcohol/water mixtures to make 'spirits'. This diagram shows that no matter how many times the mixture is distilled, pure alcohol will never be produced. To obtain pure alcohol, the remaining water in the azeotrope can be removed by reacting it with calcium metal.

3. Solutions with a boiling-point maximum
Mixtures with a maximum in the boiling-point curve can also be treated as two nearly ideal systems side by side, but this time the other way round.

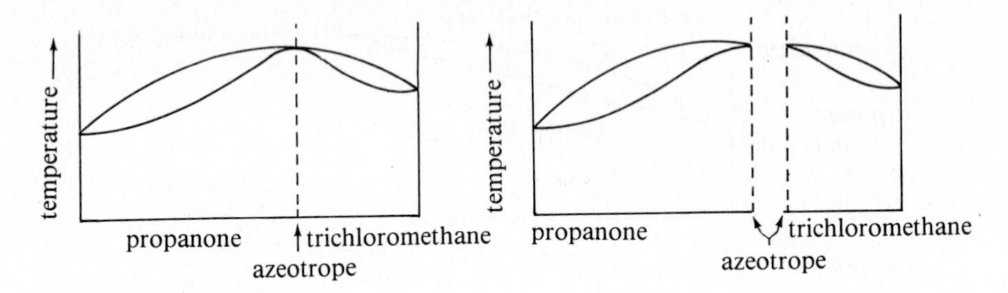

Again consider the left-hand diagram as the two parts of the right-hand diagram pushed together.

Distillation of any mixture of propanone/trichloromethane to the left of the azeotropic composition will yield pure propanone at the top of the column because it has the lower boiling point. The azeotropic mixture will this time collect at the *bottom*.

Distillation of any mixture richer in trichloromethane than the azeotropic composition (and so to the right of the azeotrope) will yield pure trichloromethane at the top of the column and again leave the azeotrope at the bottom.

11.5 Immiscible liquids

Although practically all liquids dissolve in each other to some extent, there are many pairs which have such a low solubility that they can be considered to be insoluble in each other. As described earlier, these *immiscible* liquids form two separate layers when added together.

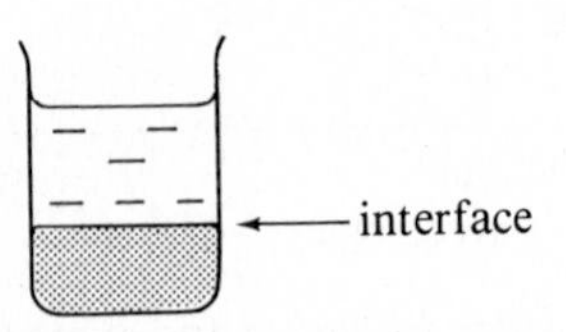

So two immiscible liquids form a two-phase system in the same state. The phase boundary or *interface* is the mutual surface between the two liquids. Examples include mixtures of water/tetrachloromethane, water/ethoxyethane, water/phenylamine, and water/benzene.

Boiling mixtures of immiscible liquids

If an immiscible mixture of two liquids is heated, it boils when the total pressure of the system reaches atmospheric pressure. Boiling is recognized by the formation of bubbles in the liquid: consider the formation of a bubble at the interface of the two liquids as shown on page 173.

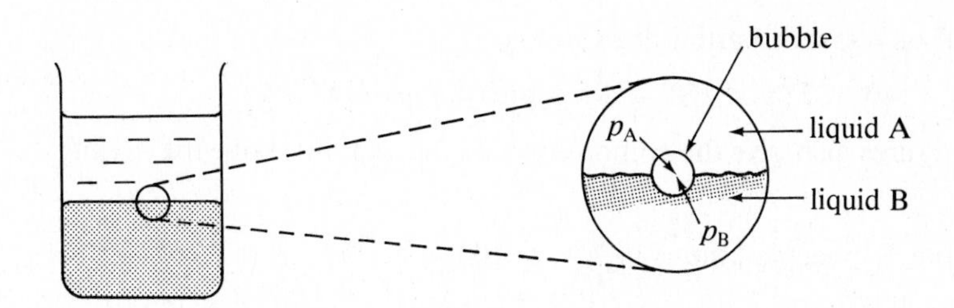

The liquid boils when the pressure p_T in the bubble, made up of the vapour pressure p_A of liquid A and the vapour pressure p_B of liquid B equals atmospheric pressure. Boiling occurs when $p_T = p_A + p_B = p_{atm.}$

The turbulence caused by heating the liquid means that in reality the two liquids are not present in two separate layers, as shown above, but are being vigorously mixed to produce a large area of interface on which bubbles can form. This means that the mixture boils at a lower temperature than either of the component liquids. The boiling point can be predicted by plotting the vapour pressures of the liquids. Here are the curves for phenylamine and water.

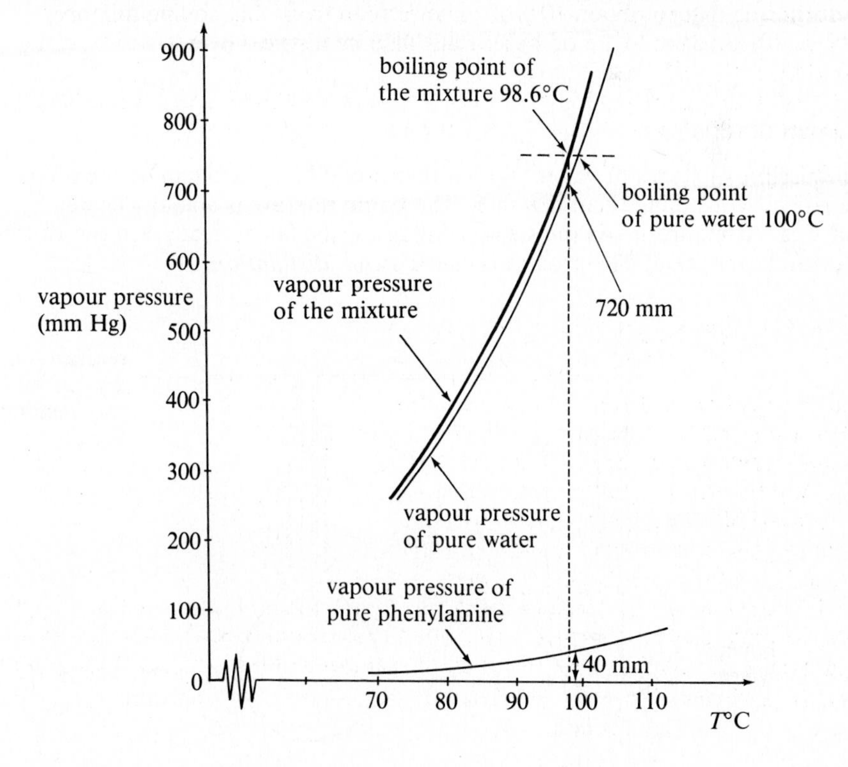

The curves show that the mixture boils at 98.6°C when the sum of the two partial pressures equals atmospheric pressure, i.e.

Vapour pressure of phenylamine = 40 mm
Vapour pressure of water = 720 mm

Using Dalton's law of partial pressures,

$$p_T = 40 + 720 = 760 = \text{atmospheric pressure.}$$

These figures also give the composition of the vapour above the boiling mixture.

$$\text{mole \% phenylamine} = \frac{40}{760} \times 100 = 5.3\%$$

$$\text{mole \% water} = \frac{720}{760} \times 100 = 94.7\%$$

So in every 1 mole of particles above the liquid, 0.053 mole are phenylamine and 0.947 mole are water.

As the R.M.M.'s are phenylamine = 93, water = 18, the proportions by mass are:

$$\text{phenylamine } 0.053 \times 93 = 4.93\text{g}$$
$$\text{water } 0.947 \times 18 = 17.0\text{g}$$

This means that to obtain 10 g of phenylamine from this boiling mixture, $(17 \times 10)/4.93 = 34.5$ g of water must also be distilled over.

Steam distillation

Immiscible mixtures of water and another liquid like phenylamine are often purified by distilling them together. The liquid mixture is actually heated by passing steam into it. As the steam condenses, the latent heat given out causes the mixture to boil. The process is called *steam distillation.*

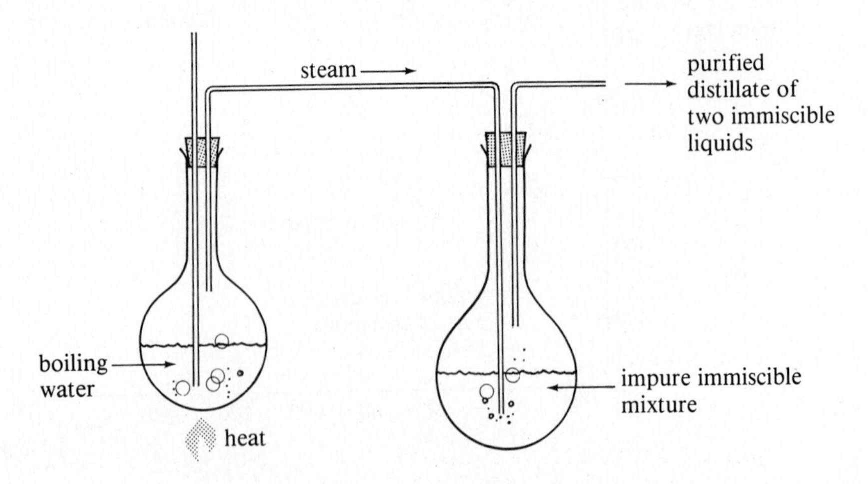

Steam distillation can therefore be used to purify water-immiscible substances like phenylamine by distilling them at much lower temperatures than their own boiling points (phenylamine b.p. = 184.4°C). This avoids the risk of decomposing the substance.

The distribution law

If a solute is added to a mixture of two immiscible liquids, it will dissolve to some extent in both of them. Once in solution, the solute particles at the interface can move from one liquid to another. The rate at which solute particles move across the interface depends on the concentration of particles in solution: the more concentrated the solution, the faster the rate.

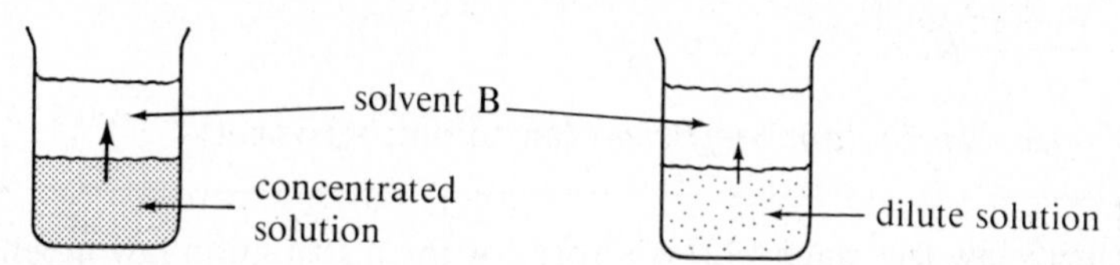

Once solute particles are in both solvents, they will be passing across the interface in opposite directions. When the rate of flow of solute particles in both directions is equal, an equilibrium situation is set up. This will occur when the ratio of the concentration of the solute in the two solvents reaches a particular value.

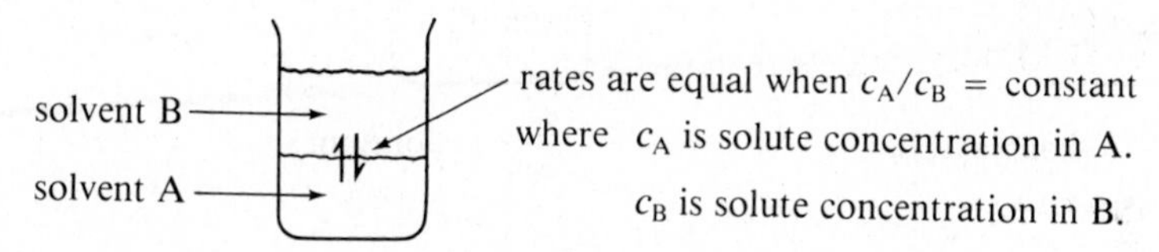

This situation can be reached under a variety of conditions as the following diagrams show.

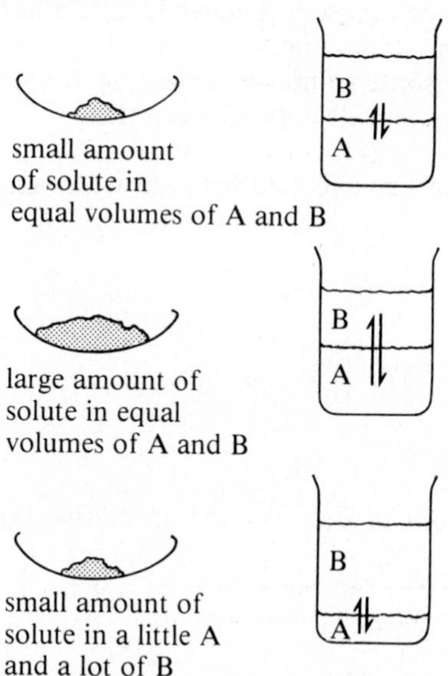

Experiment shows that, irrespective of the amount added or the relative volumes of the two solvents, the solute distributes itself between the two solvents so that the ratio of solute concentrations c_A/c_B is constant.

The distribution law.

A solute added to two immiscible liquids distributes itself between the two liquids in a constant ratio of concentrations irrespective of the amount of solute added, provided the temperature is unchanged.

$$\frac{c_A}{c_B} = K$$

where K is the *distribution coefficient* (partition coefficient).

In fact Henry's law (see page 132) is a form of the distribution law in which a gas distributes itself between a liquid and a gas.

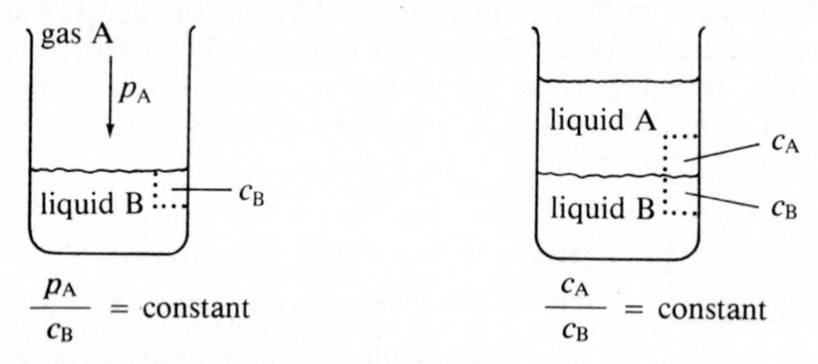

Exceptions to the distribution law

The distribution law only strictly applies for fairly dilute solutions. Once concentrations begin to rise, the two solutions cease to behave ideally and the distribution of solute deviates from the expected value.

However, even at dilute concentrations, some solutes associate or dissociate in one of the solvents so that there are apparent deviations from the distribution law.

For example, consider the distribution of benzoic acid between water and benzene.

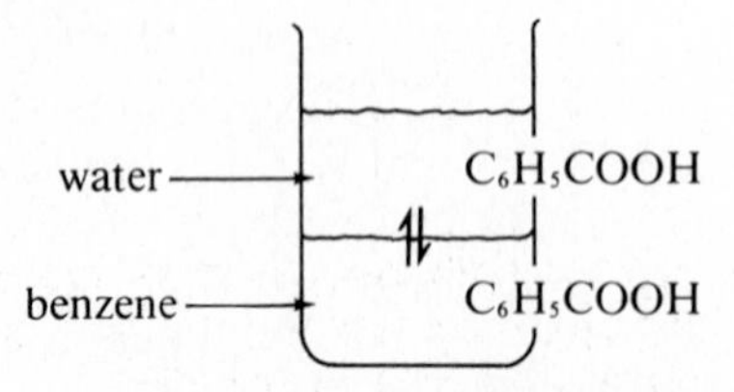

Here are the figures for the concentrations of benzoic acid in the two layers and their ratios.

benzoic acid concentration in mol dm^{-3}		*ratio of concentrations*
in water c_1	*in benzene c_2*	c_1/c_2
0.06	0.483	0.124
0.12	1.92	0.063
0.14	2.63	0.053
0.20	5.29	0.038

These experimental figures show that the ratio of the concentrations is not constant, and so it appears that the distribution law is not obeyed.

This problem arises because other equilibria are occurring which have not been considered.

$$\text{in water} \quad C_6H_5COOH + H_2O \rightleftharpoons C_6H_5COO^-_{(aq)} + H_3O^+_{(aq)}$$

interface ⇅

$$\text{in benzene} \quad C_6H_5COOH \rightleftharpoons (C_6H_5COOH)_2$$

The equilibrium in water can be ignored because benzoic acid is a weak acid ($K_a = 6.3 \times 10^{-5}$) and only a fraction of one percent has reacted with water. However the equilibrium in benzene cannot be ignored. Nearly all the benzoic acid molecules in benzene have paired up as the result of hydrogen bonding producing dimers.

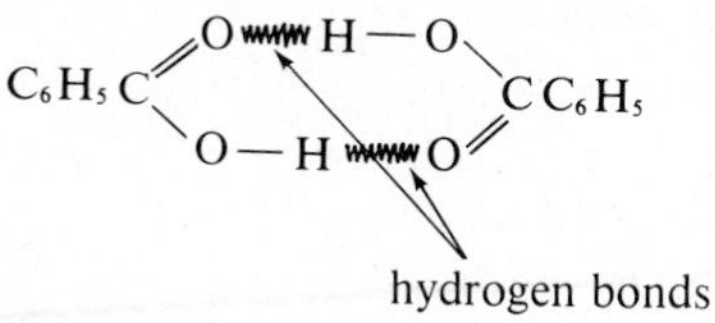

C₆H₅C(=O···H—O)(—O—H···O=)CC₆H₅

hydrogen bonds

Another similar case occurs with the distribution of iodine between water and an organic solvent.

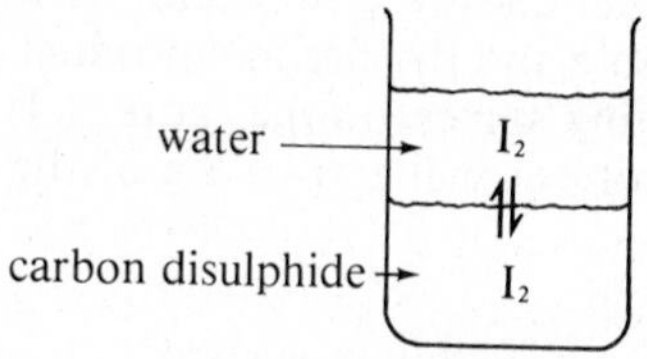

water → I_2

⇅

carbon disulphide → I_2

$$\frac{I_{2(aq)}}{I_{2(CS_2)}} = K = 1.6 \times 10^{-3} \text{ at } 20°C.$$

This system obeys the distribution law until iodide ions are added to the aqueous layer. This introduces a new equilibrium.

$$\text{in water} \quad I_{2(aq)} + I^-_{(aq)} \rightleftharpoons I^-_{3(aq)}$$

interface ⇅

$$\text{in carbon disulphide} \quad I_{2(CS_2)}$$

The new equilibrium in the aqueous layer alters the total amount of iodine in one form or another in the aqueous layer, although the concentration of iodine *molecules* in the water layer remains in agreement with the distribution law.

These two examples help to emphasize a very important point. The distribution law only applies to the same molecular species in the two solvents. It does not refer to different types of particle in each solvent.

Applications of the distribution law

Solvent extraction

At the end of an organic synthesis when the product is in a large volume of aqueous solution, it is often impractical to remove the water by evaporation or boiling. The organic product may be too volatile or unstable to heat.

In these cases one common technique is to find a second immiscible solvent which will easily dissolve the organic substance and which itself is very volatile.

By shaking the two solvents together, the organic product will distribute itself between the two layers according to the distribution law. The second solvent is then run off using a tap funnel and allowed to evaporate (or distilled off) leaving the organic product. In spite of the fire hazard, ether is frequently used as the second, extracting solvent, and the process is called *ether extraction*.

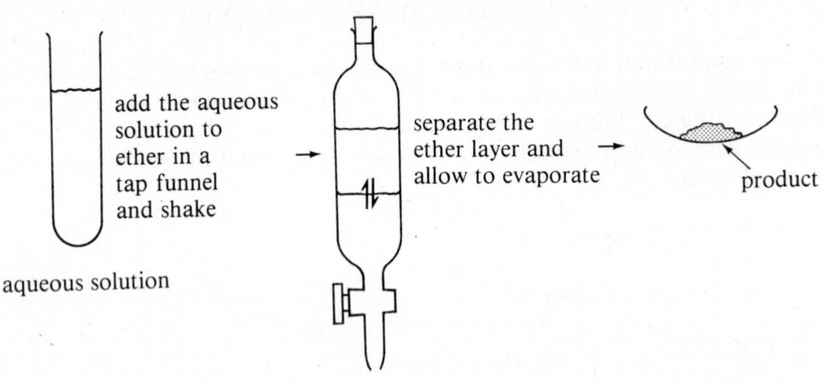

Simple calculation shows that it is better to add the extracting solvent in several small portions rather than in one large lot.

For example, imagine 12 g of organic product dissolved in 50 cm³ of water. There is 50 cm³ of extracting solvent available and the organic product will distribute itself between water and the extracting solvent in the ratio of 1:2.

A single extraction with all the extracting solvent will extract 8 g of the organic product.

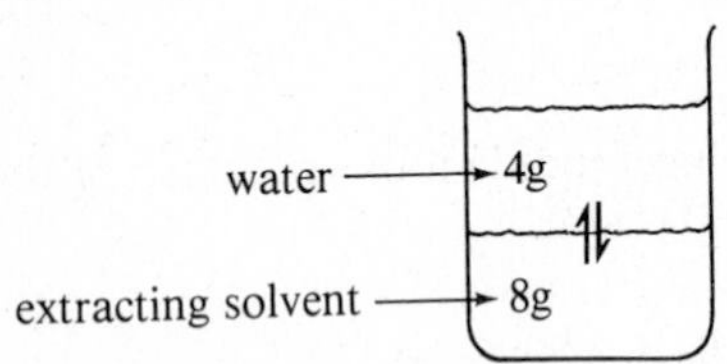

Two successive extractions with 25 cm³ portions of the extracting solvent will extract 6 + 3 = 9 g of the organic product.

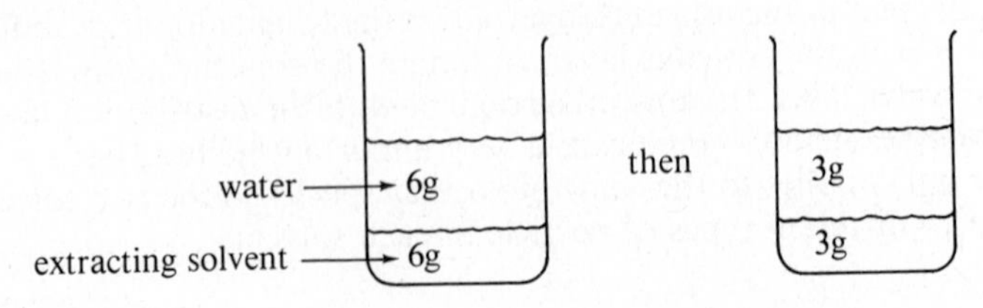

Liquid-phase chromatography

If two successive extractions above are better than one, obviously even more extractions using tiny amounts of the extracting solvent would be even better.

An almost infinite number of extracting steps can be achieved using liquid-phase chromatography. This is analogous to the jump from several successive distillations to a fractional distillation column.

In liquid-phase chromatography, in which there are many solutes, the substances to be separated are dissolved in a minimum volume of some suitable solvent. This is then deposited either as a spot or a drop onto a sheet of absorbent paper or a thin layer of absorbent solid on a plate. Larger volumes are poured into the top of a glass tube or column filled with some absorbent solid. Typical solids used for this are aluminium oxide or charcoal. A second solvent is then allowed to flow across the deposited solution.

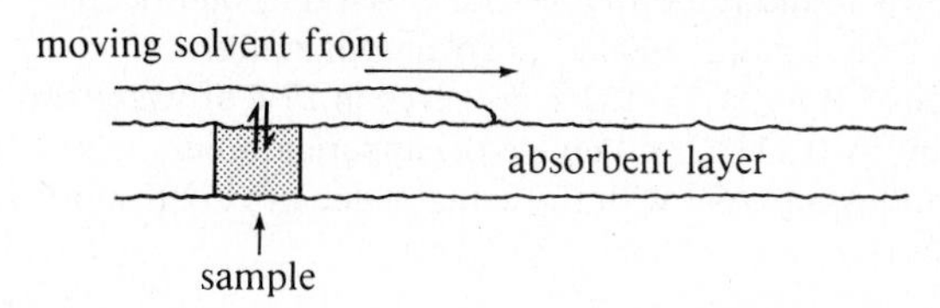

As the solvent flows across the sample, each solute will distribute itself between the initial and second or *eluting* solvent in a characteristic way depending on its distribution coefficient in these two solvents.

Sometimes the same solvent which originally dissolved the sample is used to elute it. In this case the partition is between the solvent and the residual water trapped in the absorbent layer.

Once in the moving solvent, the solutes get carried along and deposited back on the absorbent material a little further along. Here a second partition occurs, and the cycle repeats itself again.

While this form of separation, known as *liquid-phase chromatography*, is largely a partition phenomenon, it also involves surface adsorption onto the stationary solid material.

Paper chromatography and *thin-layer* chromatography (known as T.L.C.) are used for analysis of small samples. *Column* chromatography can be used for small bulk separations as well as for analysis.

Questions

1. Explain the meaning of:
   a) liquid b) ideal solution c) evaporation d) boiling e) azeotrope

2. Explain: a) the meaning of Raoult's law, b) what a colligative property is, c) how relative molar masses of soluble substances can be found.

3. Calculations based on changes in the freezing point.
   a) 0.2 g of a substance lowers the freezing point of 70 g of its aqueous solution by 0.0673°C. Calculate the R.M.M. of the substance given the cryoscopic constant for water is 1.86 K mol^{-1} 1000 g^{-1}.
   b) The freezing point of ethanoic acid was lowered from 16.63°C to 15.18°C when 1.1 g of an organic substance was added to 50 g of the acid. Calculate the R.M.M. of the organic substance given that the freezing-point constant for ethanoic acid is 3.90 K mol^{-1} 1000 g^{-1}.

c) 0.15 g of glucose, $C_6H_{12}O_6$ was dissolved in ethanoic acid. The freezing point was lowered by 5°C. Calculate the mass of ethanoic acid used as solvent given that the freezing-point constant for ethanoic acid is 3.9 K mol^{-1} 1000 g^{-1}.

4. Calculations based on changes in the boiling point.
   a) 30 g of a substance are dissolved in 250 g of ethanol. The boiling point of the solution was 79.67°C. Calculate the R.M.M. of the substance given the boiling point of ethanol is 78.52°C and the boiling-point constant is 1.15 K mol^{-1} 1000 g^{-1}.
   b) A substance of R.M.M. = 58.5 is selected; 2.5 g of it are put in 45 g of water and the boiling point is found to be 100.998°C. Given the ebulioscopic constant of water is 0.52 K mol^{-1} 1000 g^{-1}, explain this result.
   c) 0.76 g of a substance of R.M.M. = 132 is dissolved in 27 g of water and raises its boiling point by 0.333°C. Calculate the apparent boiling-point constant for water and compare it with the actual value of 0.52 K mol^{-1} 1000 g^{-1}. Explain this anomaly.

5. Calculations based on osmotic pressure changes.
   a) 0.4 g of an organic compound in 100 cm^3 of benzene exerts an osmotic pressure of 0.0068 atm at 20°C. Find the R.M.M. of the organic compound.
   b) 0.3 g of insulin in 150 cm^3 of ethoxyethane exerts an osmotic pressure of 6.21 mm of mercury at 25°C. Find the R.M.M. of insulin.
   c) 12 g of cellulose in 200 cm^3 of an organic solvent exerts an osmotic pressure of 0.005 95 atm at 10°C. Find the R.M.M. of cellulose.

6. Explain and describe what happens when a mixture, which deviates negatively from Raoult's law and forms an azeotrope, is fractionally distilled.

7. Given a 50% mixture of aqueous alcohol, describe how you would obtain a sample of pure alcohol.

8. The partition coefficient for iodine between benzene and water is 400,

$$\frac{I_{2(C_6H_6)}}{I_{2(aq)}} = 400.$$

When iodine is distributed between benzene and a 0.125 mol dm^{-3} aqueous solution of KI, the concentration of iodine in benzene is found to be 0.2 mol dm^{-3}. The total iodine concentration in the aqueous layer (not counting iodide ions from the KI) is found to be 0.0465 mol dm^{-3}.
   a) Calculate the concentration of $I^-_{3(aq)}$ in the aqueous layer.
   b) Calculate the value of the equilibrium constant for the reaction

$$I_{2(aq)} + I^-_{(aq)} \rightleftharpoons I^-_{3(aq)}$$

9. Explain the principles behind: a) Solvent extraction b) Liquid-phase chromatography.

10. Discuss the concept of an ideal solution and explain why real solutions deviate from the ideal model.

Chapter 12
The solid phase

12.1 Solid properties

The wide range of solid properties

Unlike gases or liquids, solids have a definite shape and can be subjected to systematic deformation. They therefore have a larger range of properties than the other two states.

Apart from unambiguous properties such as shape, melting point or density, many other terms are loosely used in describing solids: e.g. hardness, strength, malleability and toughness.

It is often difficult to give a concise definition of a solid property.

Many solids retain their shapes but some (like glasses or waxes) gradually deform under gravity. Most solids have a sharp melting point, yet the glasses soften over a range of temperature. In spite of these problems, it is useful to attempt to define certain properties which help to describe and compare solids. These are:

Hardness

This is measured on a comparative scale and there are various methods for expressing the results. They all consist of marking the solid with a tool of known shape and hardness, and measuring the size of indentation. One scale commonly used to express hardness is the Moh scale which compares the hardness of materials with that of diamond.

Strength

This property is usually measured by applying a load to a specially shaped specimen. Different results are produced depending on whether the load is applied in tension or compression. The results can be expressed in terms of the percentage elongation, the reduction in cross-sectional area or as a ratio of the two. Another simple way of expressing tensile strength is in terms of the length of material that will support itself when hanging freely.

Brittleness and ductility or malleability

A material is said to be brittle if under load it deforms very little before breaking. Ductility (or malleability) refers to materials that deform progressively and to a considerable extent under load without fracture.

However even when the properties have been clearly defined and ways of measuring them devised, the problems are not over. The properties are not always constant, but can change depending on the history of the material. The properties of pure metals like aluminium can be changed by stretching or rolling the metal, and the rate of cooling of the liquid when it is cast can radically change the properties of many materials.

In fact the technological stage has been reached where materials can be manufactured to meet almost any requirement. Many of these requirements are conflicting, such as ease of machining of the metal with high resistance to deformation in the finished object.

Relating the bulk properties to the structure and bonding of the particles

All the bulk properties of a solid material are a consequence of the structure and bonding in the solid. Sometimes it is easy to explain the properties in terms of the structure and bonding. On other occasions this rationalization is more difficult. In these cases it is usually an incomplete description of the structure of the solid which causes the problem.

For example, a sharp melting point is interpreted as implying a regular lattice of particles, all bonded uniformly to their neighbours. If the melting point is high (as in magnesium oxide or diamond), it reflects strong bonding between the particles forming the lattice. A low melting point (as in carbon dioxide), suggests weak bonding between the lattice units. A softening range (as in a glass), implies that the lattice is not regular and uniform and so there are bonds of varying lengths and strengths which do not all break at the same temperature.

Uniform properties in all directions (or *isotropy*) implies an extended lattice in three dimensions. Different properties in different directions (or *anisotropy*), implies an extended lattice in only one or two dimensions with bonding of a different type and strength in the others. Examples of anisotropic materials are the flakey substances like graphite and mica having two-dimensional lattices, fibrous substances such as asbestos and fibrous proteins having one-dimensional lattices, varying electrical conductivity in different directions (e.g. graphite and semiconductors), and different refractive indices in different directions (e.g. calcium carbonate as Iceland Spar used in Nicol prisms).

So a clear understanding of the structure of a solid in terms of particles allows the physical properties to be explained and the behaviour of the solid to be predicted under specified conditions.

12.2 Pure solids

Lattice determination in pure solids

The main experimental technique which has led to the present knowledge of lattice structures is *X-ray spectrometry*. The wavelengths of X-ray beams are short enough to interact with the particles and planes of particles in a lattice. By studying the scattering pattern, quite a lot of information can be gained about the structure of the solid.

For instance, rows or planes of atoms can act as obstacles to the X-ray waves which will be disturbed and bent around them. This phenomenon is known as *diffraction* and produces a variety of patterns.

Each gap between a pair of atoms acts as a separate source of waves. The waves spreading out from each gap interfere with each other, sometimes

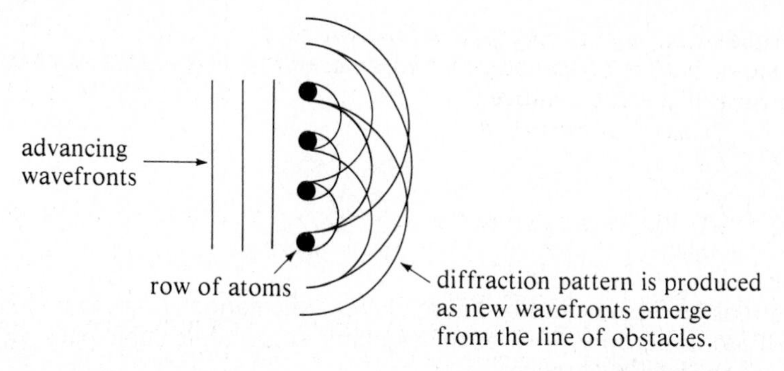

constructively producing bigger waves, and sometimes destructively cancelling each other out. This interference results in the impression that the direction of travel of the waves is being bent into a new path.

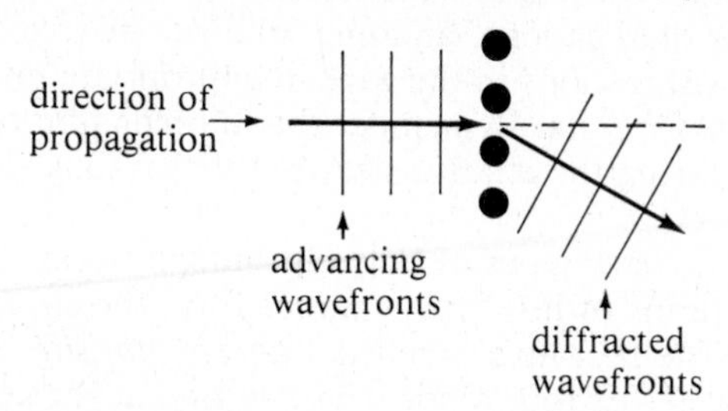

Bragg interpreted the diffraction in terms of the reflection of X-rays from successive planes of particles. For a given wavelength λ, he derived an equation relating the angle θ between the reflected ray and the plane of particles to the distance d between two successive planes.

$$n\lambda = 2d \sin \theta \quad \text{where } n = 1, 2, 3, \ldots \quad \text{(The Bragg equation)}$$

This equation is used to calculate lattice distances between atoms and planes of atoms.

To be able to infer the whole structure from a few experimental results assumes that the whole lattice is uniform. Even in pure solids the actual structure nearly always deviates from the ideal lattice. The lattices of almost all pure solids contain imperfections in the form of either extra particles causing *dislocations*, or gaps in the lattice called *vacancies*. These are discussed on page 79.

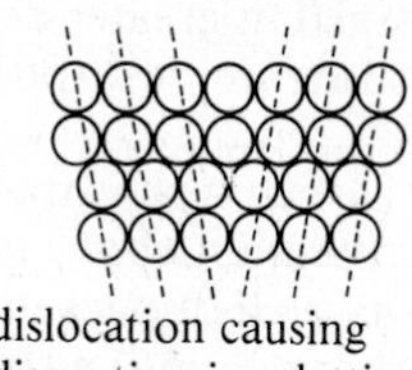
dislocation causing
disruption in a lattice

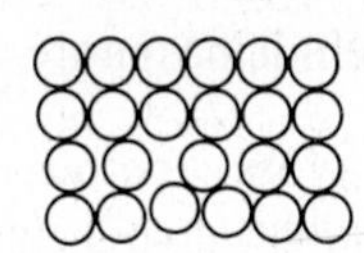
vacancy disturbing
the pattern in a lattice

Lattice types

Although almost an infinite number of lattice types is theoretically possible, in practice only a relatively small number is found commonly. Here is a brief summary of the more important ones.

Metallic solids
Metallic atoms pack together either in *close-packed* lattices or in a more open arrangement called a *cubic* lattice.

If the atoms are assumed to be hard spheres, calculation shows that close packing of atoms occupies 74% of the available space while cubic packing only uses 68% of the available space.

Ionic solids
These are more complicated because there are two different ions to consider. Because anions are usually bigger than cations, an ionic solid can be treated as a close-packed lattice of anions with cations in the gaps in appropriate numbers to satisfy the stoichiometry of the formula. Remember that a prediction of the lattice type can be made by comparing the relative sizes of the two ions — the *radius ratio*, see page 88.

In a close-packed lattice there are two types of hole big enough to take other particles. These are classified in terms of how many anions form the cluster which has the hole at its centre. The first hole is called a *tetrahedral site* because it is at the centre of a cluster of four anions. The second at the centre of a cluster of six anions is called an *octahedral site*.

Lattices can be classified in terms of which and what proportion of these sites are occupied by cations. Each lattice is named after a particular substance displaying that type of lattice.

In the table on p. 185 notice that the anion lattices are divided into two. *Cubic close-packed*, c.c.p., lattices contain a vertical sequence of layers that go ABCABC, etc. *Hexagonal close-packed*, h.c.p., lattices contain a vertical sequence of layers that go ABABAB, etc. This is explained in greater detail on page 76. Notice that zinc sulphide has more than one structure: it displays polymorphism.

In addition to these lattice types, another important lattice type is the caesium chloride lattice. This does not fit into the classification above because caesium ions are so large they do not fit into any gaps in a close-packed array of anions. In caesium chloride, the chloride ions are separated into a simple cubic lattice.

Even more complicated structures are found in *spinels*. These have lattices whose layers alternate between one type and another. On one layer tetrahedral sites may be occupied, while on the next octahedral sites are the ones occupied. Examples are the oxides of the formula M_3O_4.

Lattice type	*Anion lattice*	*Cation sites occupied*
NaCl	c.c.p.	all the octahedral sites
$CdCl_2$	c.c.p.	$\frac{1}{2}$ the octahedral sites
$CrCl_3$	c.c.p.	$\frac{1}{3}$ the octahedral sites
CaF_2	c.c.p.	all the tetrahedral sites
ZnS	c.c.p.	$\frac{1}{2}$ the tetrahedral sites
NiAs	h.c.p.	all the octahedral sites
CdI_2	h.c.p.	$\frac{1}{2}$ the octahedral sites
BiI_3	h.c.p.	$\frac{1}{3}$ the octahedral sites
ZnS	h.c.p.	$\frac{1}{2}$ the tetrahedral sites

12.3 Solid solutions

Alloys and mixtures of molecular solids

In metallic lattices and in molecular lattices the bonding is sufficiently non-specific for large amounts of impurities to be absorbed in the lattice. Obviously the impurities must be added in the liquid state, and most molten metals dissolve each other in all concentrations. Molecular solids such as waxes also dissolve in large quantities of other organic solids. However when the liquid mixture begins to cool and a low-potential energy lattice begins to form, not all the impurity particles will be accepted in the lattice. Several solid phases may form. These can be diagnosed by examining cooling curves for the mixture as it freezes. This process is called *thermal analysis.*

Cooling curve for a pure substance

The typical cooling curve for a pure substance has three portions. Each signifies a different process.

The top section of the graph below represents the liquid cooling. The kinetic energy of the particles decreases, quickly at first, then more slowly (according to Newton's law of cooling). As temperature is a measure of the average kinetic energy of the particles, this fall is reflecting the decrease in kinetic energy.

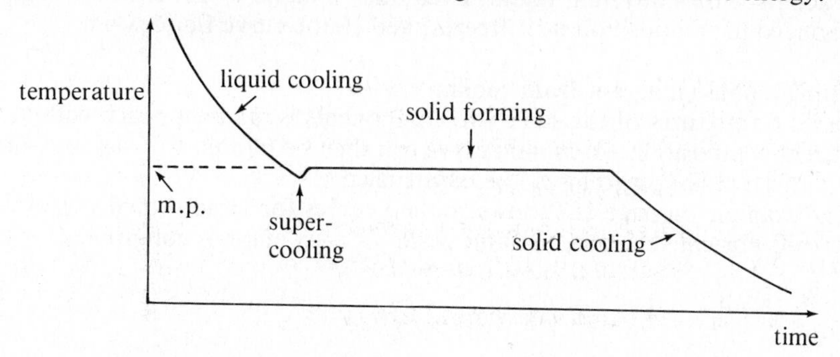

At the melting point, the curve changes gradient sharply because the solid is starting to form. As the particles move together to form a solid lattice, it is their *potential* energy that drops. This does not affect their kinetic energy and so the temperature remains constant.

When all the liquid has frozen, the third portion of the graph begins. This represents solid cooling. The kinetic energy (mostly due to vibration) of the particles in the solid lattice decreases. Again this change in kinetic energy is reflected in a change in temperature.

Cooling curve for a mixture

A cooling curve for a typical mixture differs from the curve above as the following diagram shows.

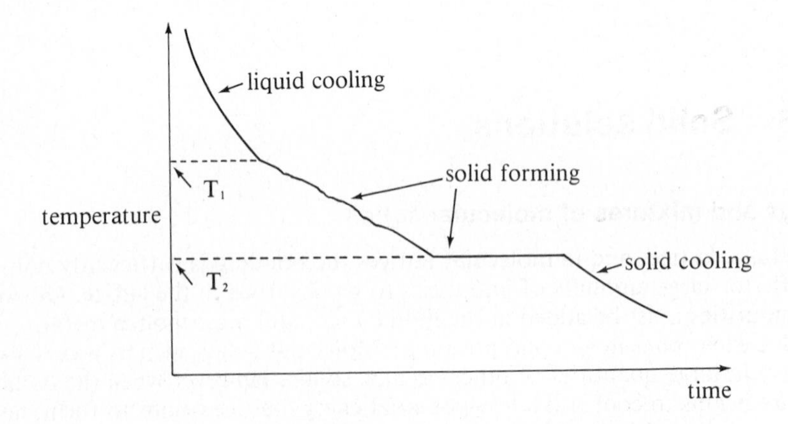

The first section of the curve is similar to that for a pure substance.

However, although there is a change in gradient as the first solid is seen to form at T_1, the temperature continues to fall in a rather irregular way until T_2. The fall in temperature ceases here until all the solid has formed. The temperature then continues to fall in the same way as for a pure substance.

The central portion of this curve is interpreted as representing the formation of a solid, leaving the remaining liquid of a different composition. As the solid forms, the composition of the liquid slowly changes so the gradient of the cooling curve also slowly changes. Each portion represents the cooling characteristic of a different liquid. Eventually a liquid is formed which remains unchanged in composition as it freezes, and so the curve flattens out.

Plotting a phase diagram from cooling curves

A series of mixtures of the same two components is made up, each with a different composition. A cooling curve can then be plotted for each mixture. From this a phase diagram can be constructed.

The example on page 187 shows cooling curves for various mixtures of tin and lead. For convenience, the time scale for each curve is compressed.

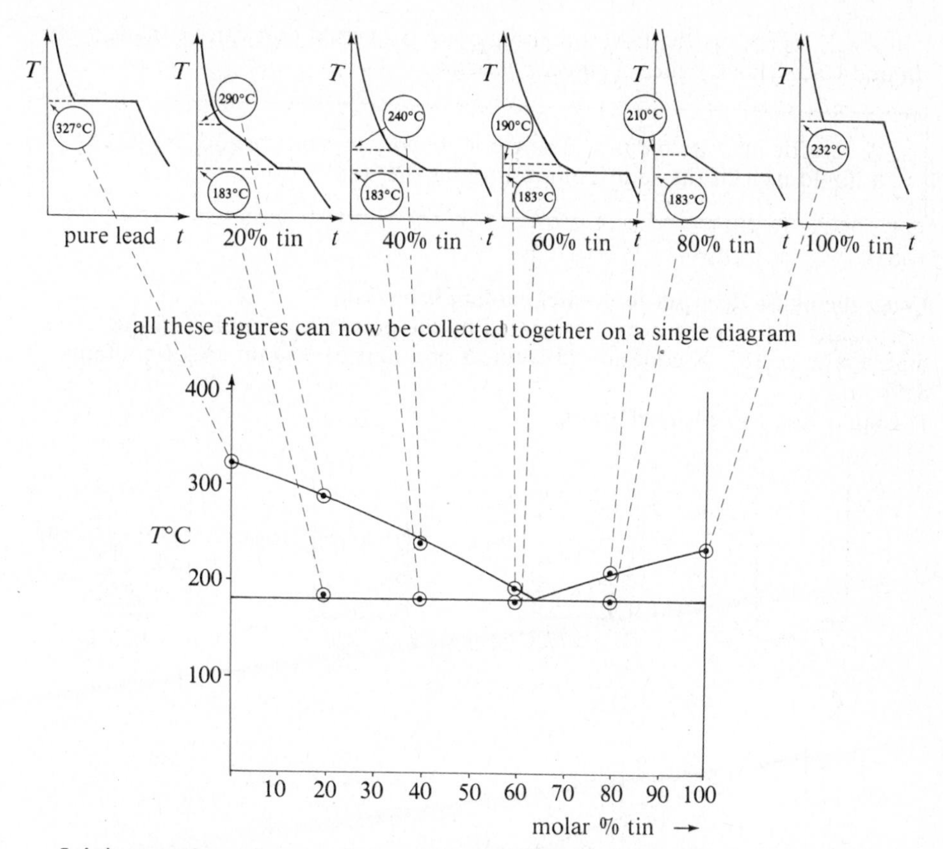

Joining up the points produces the phase diagram above. Like all phase diagrams, the areas on this one represent the conditions under which different phases exist and so they can be labelled as shown below.

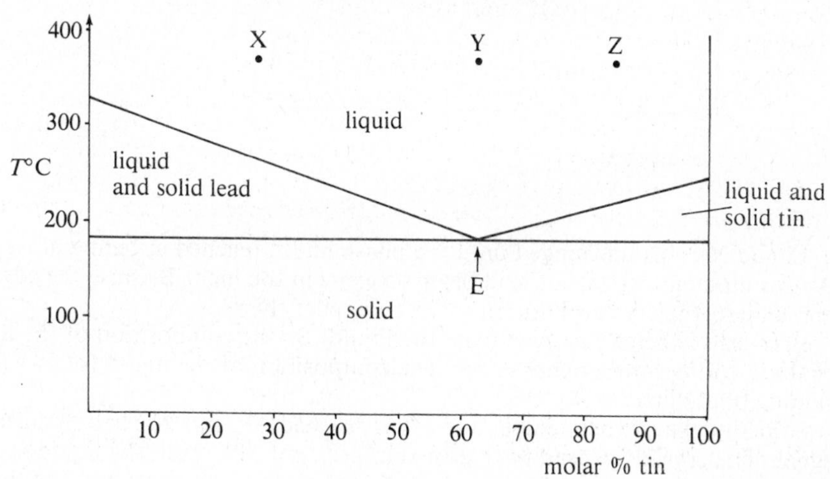

Point E represents the lowest melting point of a solid that can be made from tin and lead. This is called a *eutectic mixture.*

> A eutectic mixture melts at a single, defined temperature and produces a liquid melt of the same composition as the solid.

Using the phase diagram to predict cooling behaviour

Consider the three systems designated by the points X, Y and Z on the diagram on p. 187. X contains 25% tin, Y contains 61.9% tin and Z contains 85% tin.

1. Liquid X is cooled from 350°C

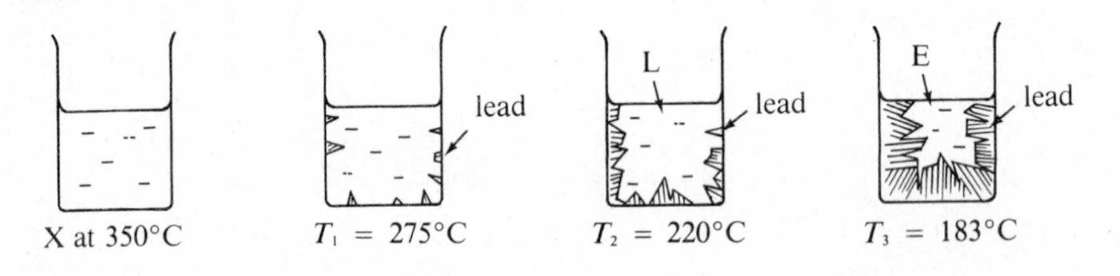

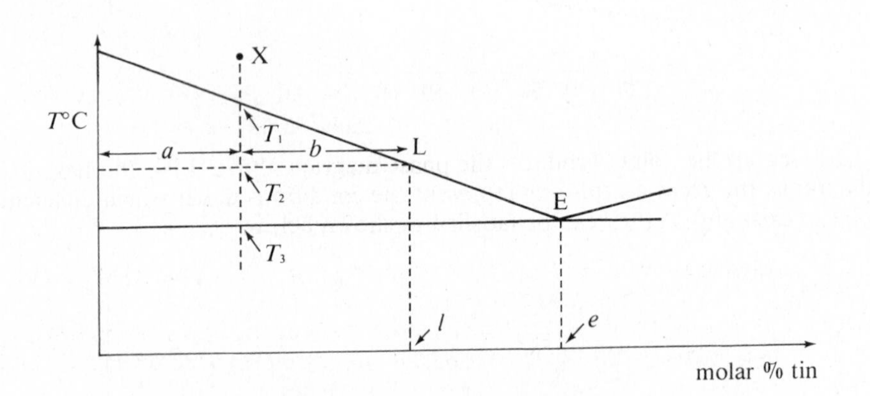

Liquid X cools unchanged until the phase line is reached at temperature T_1. At this moment crystals of lead begin to grow in the melt. Because the edges are coolest, the crystals naturally start to appear there.

Pure lead is being removed from the liquid. So the composition of the liquid changes and becomes richer in tin. The composition of the liquid follows the sloping phase line.

Sometime later when temperature T_2 is reached, the mixture has changed again. The crystals of lead have grown larger, and as a result the liquid is even richer in tin. The composition of the liquid is given by the intercept l of the point L on the composition axis. The relative amounts of lead and of liquid L are given by the lengths of the two portions of the line through point T_2. This line is called a *tie line* and joins the two phases in equilibrium with each other at that temperature.

$$\frac{\text{amount of lead}}{\text{amount of liquid L}} = \frac{b}{a}$$

At temperature T_3, the lead crystals have grown even larger and the liquid composition is so rich in tin that it has reached the eutectic point E and has composition *e*. Now all the remaining liquid freezes out unchanged. It forms a solid mixture consisting of tin and lead in eutectic proportions.

2. Liquid Y is cooled from 350°C.
Here the sequence of events is much simpler. Liquid Y cools unchanged until temperature T_3 is reached. At T_3 solid begins to form consisting of crystals of the *same* composition. The entire liquid freezes at T_3 as though it were a pure substance.

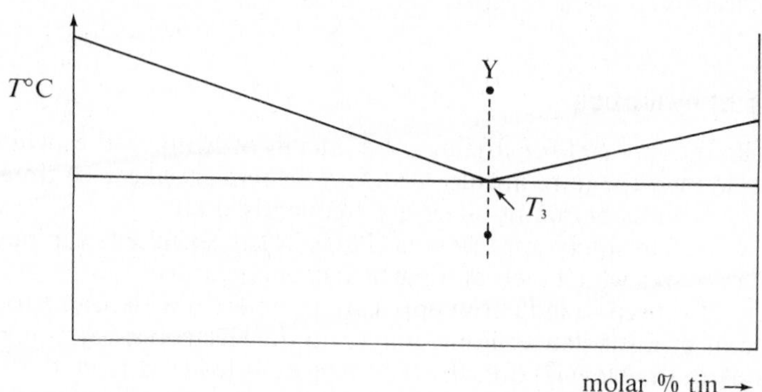

3. Liquid Z is cooled from 350°C.
The composition of Z is to the right of the eutectic point and so when T_1 is reached, crystals of tin begin to grow in the melt.
Because tin is leaving the liquid and crystallizing out, the liquid becomes richer in lead. The liquid composition follows the sloping phase line downwards in the diagram below.

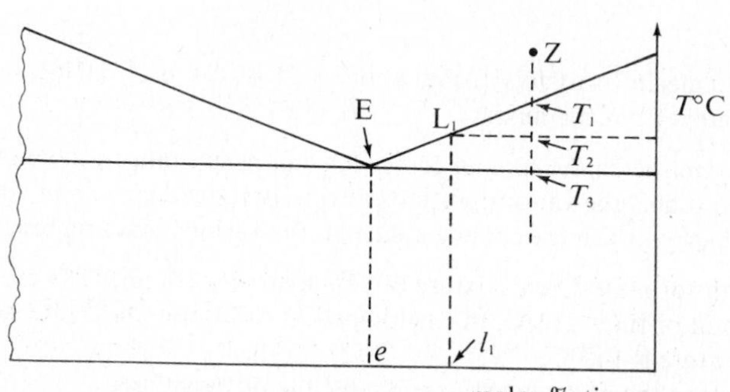

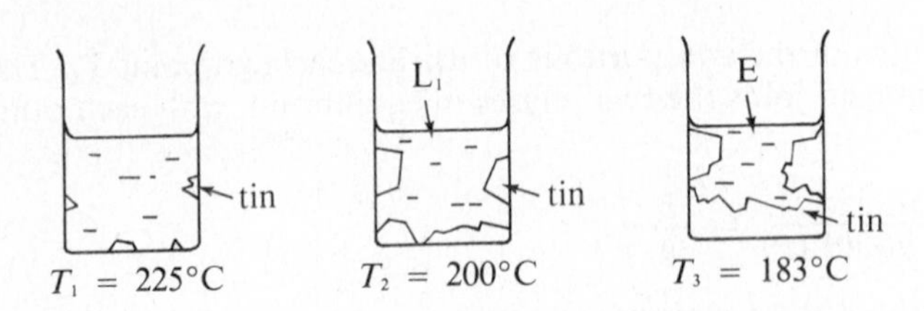

At T_2 the system consists of crystals of tin in a liquid L_1 of composition l_1. At temperature T_3 the liquid reaches eutectic proportions of composition *e*. This remaining liquid freezes out unchanged.

A mixture of eutectic composition is unique in that it freezes out at a single temperature without changing its composition. It can be compared to an azeotrope which boils at a single temperature without changing its composition. Under a microscope the regions of eutectic solid are usually clearly visible because they have a speckled appearance. The crystals of lead and tin have a uniform appearance.

Applications of eutectics

Eutectics are used as *solders* for joining other metals at temperatures which are not so high as to melt the pure metals. Ordinary solders are alloys of tin and lead, but stronger ones containing silver are frequently used.

Eutectics are used in safety applications. Plugs in fire sprinklers can be made from eutectic mixtures which melt at a particular temperature.

Eutectics are also used in industrial applications where it is desirable to lower the melting point of a substance. For example, in the electrolysis of molten sodium chloride in the Downs' cell, the temperature is lowered from the melting point of salt (801°C) to about 570°C by the addition of about 60% calcium chloride. This is important to prevent the sodium (b.p. 892°C) vaporizing off. Similarly in the extraction of aluminium from bauxite, cryolite is added to bring the operating temperature down from 2045°C to a more manageable 970°C.

Questions

1. Describe what is meant by a) anisotropy, b) dislocation, c) close packing, d) tetrahedral site, e) eutectic mixture.

2. On the same graph sketch the cooling curves for a pure substance and a non-eutectic mixture. Label the various portions of the curves and explain in your own words what is happening in the cooling system at the various times represented.

3. Tin and bismuth form a eutectic mixture of composition 42% tin, 58% bismuth. The melting point of tin is 232°C, the melting point of bismuth is 271°C, and the eutectic temperature is 140°C.
   a) Carefully draw the phase diagram and label the different areas.

b) Describe the sequence of changes when a liquid of composition 20% tin, 80% bismuth is cooled.

c) Compare the cooling of the system above with that of a mixture containing 80% tin and 20% bismuth.

4. Explain how the facts needed to construct a eutectic phase diagram are determined, and how they are used to produce the diagram.

5. An X-ray beam of wavelength 1×10^{-10}m is diffracted through an angle of 13° by an array of atoms. Calculate the distance apart of the atoms.

6. An X-ray beam of wavelength 1×10^{-10}m is diffracted by a metal lattice. The atoms in the lattice are 7.1×10^{-10}m apart. Calculate the angle of diffraction.

7. 1 mole of copper weighs 63.5 g. The density of copper is 8.96 g cm^{-3}.
   a) Calculate the volume of 1 mole of copper.
   b) Calculate the length of the side of a cube of 1 mole copper.
   c) Given that 1 mole contains 6×10^{23} atoms, calculate the distance apart of copper atoms.
   d) If an X-ray beam of wavelength 2.28×10^{-10}m is diffracted by the copper, what is the angle of diffraction?

8. Imagine a close-packed lattice made up of hard spheres 1×10^{-10}m in radius.
   a) Calculate the size of the largest sphere that could fit into a tetrahedral site in this array.
   b) Calculate the size of the largest sphere that could fit into an octahedral site in this array.

9. Explain why some solids are hard, while others are flakey like slate or mica, and yet others are fibrous like asbestos.

10. To what extent is it possible to explain the physical properties of solids in terms of their structure and bonding? In your answer use as examples diamond, quartz, copper, sodium, graphite, polythene, glass, salt, and wax.

Part 2

Chemical energy and equilibrium

Chapter 13
Chemical reactions

13.1 Recognizing chemical change

Physical chemistry is the study of both the physical and chemical properties of matter. In Chapters 1 to 12, bonding, structure, and the phases are discussed. These can all be thought of as physical properties because they concern only the composition and arrangement of the particles making up the system. The identity of the particles in the system does not change. However, there are many processes in which this is not so: the particles may break up into smaller particles or join up to form larger ones. Processes of this sort are chemical changes, and the tendency of a system to undergo a particular chemical change can be thought of as a chemical property of the system. In Chapters 13 to 20, the characteristics of chemical change are described.

At an elementary level, we can identify three factors that are common to all chemical changes:

1. A new substance is produced.

   reactants → products

2. There is usually an energy exchange between the system and its surroundings.

   system → surroundings (*exothermic reaction*)   system ← surroundings (*endothermic reaction*)

3. The masses of reactants and products in a particular reaction are in a fixed ratio to one another.

All three factors result from the changes that happen to the particles in the system:

1. A new substance is produced because there are new particles forming in the reaction.
2. Bonds are both broken and made in the process and the energy involved in breaking the bonds is not usually the same as the energy required to make the new bonds.
3. We have seen in earlier chapters how the particles making up pure elements and compounds combine and bond in specific ways. This means that a fixed number of reactant particles is rearranged to produce a fixed number of product particles. When this process is scaled up many times, it is reflected in an identifiable mass relationship between reactants and products. The scale factor most commonly used is the *Avogadro Number* L. This is the scale factor needed to convert atomic mass units to grams and is defined as follows.

The Avogadro Number L is the number of carbon atoms in exactly 12 g of the isotope carbon-12. This number is 6.02×10^{23}.
The amount of any substance containing Avogadro's Number of particles of that substance is called a mole.

The number ratio of reactant particles to product particles in a chemical change is called the *stoichiometry* of the reaction.

Chemical changes are commonly represented by the use of equations. All these factors can be illustrated by the example of methane burning in oxygen.

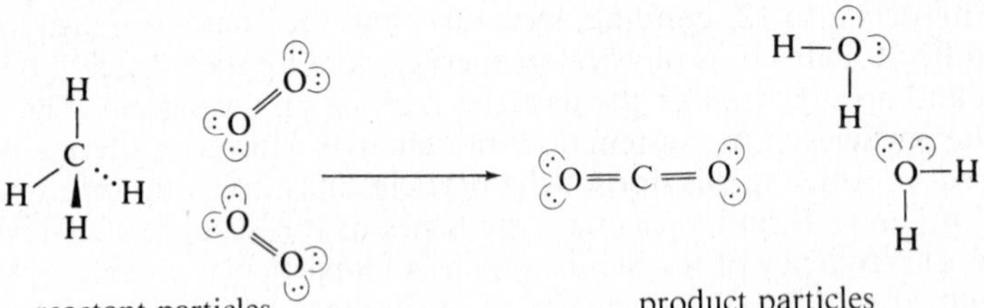

reactant particles

product particles

Equation:

$$CH_{4(g)} + 2O_{2(g)} \longrightarrow CO_{2(g)} + 2H_2O_{(g)}$$

1. *New substances*: carbon dioxide and water.
2. *Energy exchange*: from the system to the surroundings (exothermic). Energy is supplied to break a bond and therefore energy is released when a bond is made. Less energy is needed to break the bonds in methane and oxygen than is given out in forming the new bonds in carbon dioxide and in water.
3. *Fixed ratio*: one methane molecule and two oxygen molecules always produce one carbon dioxide molecule and two water molecules. Measurement of the mass of reactants and of products confirms (on a scale many millions of times greater) that this is the number of molecules involved.

$$\underset{16}{CH_4} : \underset{64}{2O_2} : \underset{44}{CO_2} : \underset{36}{2H_2O} = 4 : 16 : 11 : 9$$

13.2 Chemical equations

A chemical equation is used in two different ways. As has just been described, it can show the reactant and product particles in their reacting number ratio.

Sometimes, however, it is necessary to concentrate on the whole reaction system rather than on the individual reacting particles. By scaling up the particle equation Avogadro's Number of times, a system equation showing molar ratios is produced. For example,

$$SO_2 + \frac{1}{2}O_2 \longrightarrow SO_3$$

does not mean that one molecule of sulphur (IV) oxide combines with half a molecule of oxygen. It implies that 64 grams of sulphur (IV) oxide (*L* molecules) react with 16 grams of oxygen (*L*/2 molecules).

The difference between a particle equation and a system equation is illustrated below by the two examples of a precipitation reaction and an acid neutralization.

1. **Precipitation**

The system equation for the precipitation of silver chloride from a sodium chloride solution is

$$AgNO_3 + NaCl \longrightarrow AgCl \downarrow + NaNO_3$$

However, the particle equation shows only the particles that are changing during the reaction. In this case, they are chloride ions and silver ions.

$$Ag^+_{(aq)} + Cl^-_{(aq)} \longrightarrow AgCl_{(s)}$$

The other particles in the system take no part in the reaction and so are not shown in the equation. Particles of this sort are known as spectators. In the above reaction, $Na^+_{(aq)}$ and $NO^-_{3(aq)}$ are *spectator ions*.

2. **Neutralization**

The system equation for the neutralization of hydrochloric acid by sodium hydroxide is

$$NaOH + HCl \longrightarrow NaCl + H_2O$$

However, the actual particles changing in the reacting system are hydroxonium ions from the acid solution and hydroxide ions from the alkaline solution.

$$HCl_{(g)} + H_2O_{(l)} \longrightarrow H_3O^+_{(aq)} + Cl^-_{(aq)} \qquad \textit{acid}$$

$$NaOH_{(s)} \longrightarrow Na^+_{(aq)} + OH^-_{(aq)} \qquad \textit{alkali}$$

$$H_3O^+_{(aq)} + OH^-_{(aq)} \longrightarrow 2H_2O_{(l)} \qquad \textit{neutralization}$$

The sodium ions and chloride ions are spectators.

Particle equations are most useful when it is necessary to emphasize structural changes that happen as a result of a chemical change. System equations are used more as a description of a chemical reaction or to calculate specific amounts of either a reactant or a product. It is common practice to include 'state symbols' after the formulae of the particles in a particle equation.

$_{(s)}$ for solid
$_{(l)}$ for liquid
$_{(g)}$ for gas
$_{(aq)}$ for an aqueous solution

These symbols describe the immediate environment of the reacting particle. For example, $Na^{+}_{(aq)}$ means a sodium ion surrounded by water molecules:

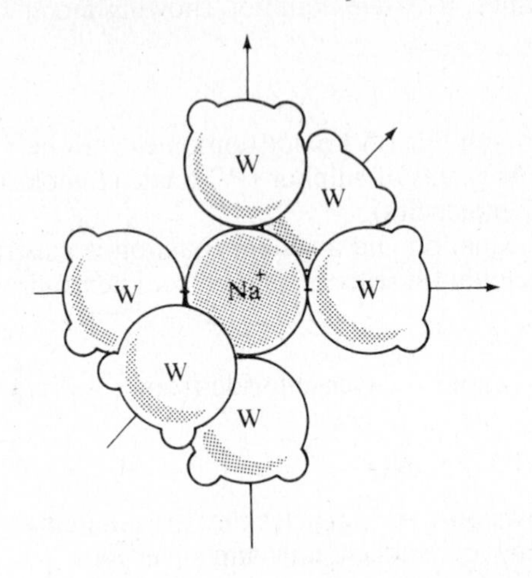

13.3 Using equations in calculations

Mole ratios shown by equations are used in a number of different types of calculations.

1. The masses of reactants and products
The mole ratio given by the equation is converted to a mass ratio. Thus for elements and covalent compounds the relationship is:

R.A.M. in g ⟷ 1 mole of substance
R.M.M. in g ⟷ 1 mole of substance.

For ionic compounds which contain no molecules, the term relative formula mass is used thus:

R.F.M. in g ⟷ 1 mole of substance

Example Calculate the mass of magnesium oxide produced by burning 15 g of magnesium.

equation: $2Mg + O_2 \longrightarrow 2MgO$

molar ratio: 1 mole of Mg produces 1 mole of MgO

mass ratio: 24 g of Mg produces 40 g of MgO

scaling: 15 g of Mg produces $\left(40 \times \frac{15}{24}\right)$g of MgO

Therefore mass produced = 25 g

2. **The concentration of solutions** (using titration)
A solution of one reactant is run from a burette into a fixed volume of a solution of a second reactant. Some method of deciding that reaction is complete, the end-point of the titration, is found and the reaction is carried out. Knowing the two reacting volumes and the concentration of one of them, the concentration of the other can be calculated from the molar ratio shown by the equation.

Example (a) Calculate the concentration of a solution of sulphuric acid given that 20 cm³ of a 0.1M solution of sodium hydroxide was needed to neutralize 8 cm³ of acid.

Note: 1M means 1 mole dissolved in 1 dm³, i.e. 1 mol dm^{-3}; so 0.1M means 0.1 mol dm^{-3}.

equation: $2NaOH + H_2SO_4 \longrightarrow Na_2SO_4 + 2H_2O$

molar ratio: 2 moles of alkali neutralize 1 mole of acid

working: In 20 cm³ of NaOH, there are $\left(\frac{20}{1000} \times 0.1\right)$ moles

thus in 8 cm³ of H_2SO_4, there must be $\left(\frac{1}{2} \times \frac{20}{1000} \times 0.1\right)$ moles

$\Rightarrow$ in 1000 cm³ of H_2SO_4, there are $\left(\frac{1000}{8} \times \frac{1}{2} \times \frac{20}{1000} \times 0.1\right)$ moles

Therefore concentration of solution = 0.125 mol dm^{-3}

Example (b) Calculate the concentration of a solution of iron (II) sulphate (VI) given that 20 cm³ of a 0.05M solution of potassium manganate (VII) were required to oxidize 40 cm³ of the solution.

equation: $10FeSO_4 + 2\,KMnO_4 + 8H_2SO_4 \longrightarrow$
$5Fe_2(SO_4)_3 + K_2SO_4 + 2MnSO_4 + 8H_2O$

molar ratio: 5 moles of $FeSO_4$ react with 1 mole of $KMnO_4$

working: In 20 cm³ of manganate, there are $\left(\frac{20}{1000} \times 0.05\right)$ moles

thus in 40 cm³ of iron (II), there must be $\left(5 \times \frac{20}{1000} \times 0.05\right)$ moles

$\Rightarrow$ in 1000 cm³, there are $\left(\frac{1000}{40} \times 5 \times \frac{20}{1000} \times 0.05\right)$ moles

Therefore concentration of solution = 0.125 mol dm^{-3}

3. **The volume of gases involved in a reaction**
Avogadro's hypothesis states that, under similar conditions, equal volumes of gases contain equal numbers of molecules. Thus reacting volumes can be converted to moles or vice versa. Also experiment shows that, to a first approximation, the molar volume of all gases is the same.

> Under standard conditions of temperature and pressure, 1 mole of any gas occupies 22.4 dm^3.

Example (a) Calculate the volume of carbon dioxide produced by decomposing 10 g of calcium carbonate.

equation: $CaCO_3 \longrightarrow CaO + CO_2$

molar ratio: 1 mole of $CaCO_3$ produces 1 mole of CO_2

mass-volume ratio: 100 g of $CaCO_3$ produces 22.4 dm^3 of CO_2 at s.t.p.

scaling: 10 g of $CaCO_3$ produces $\left(22.4 \times \frac{10}{100}\right)$ dm^3

Therefore volume produced = 2.24 dm^3 at s.t.p.

Example (b) Calculate the volume of oxygen that is needed to oxidize 10 dm^3 of ammonia to nitrogen (II) oxide.

equation: $4NH_{3(g)} + 5O_{2(g)} \longrightarrow 4NO_{(g)} + 6H_2O_{(g)}$

molar ratio: 4 moles of ammonia react with 5 moles of oxygen

volume ratio: 4 volumes of ammonia react with 5 volumes of oxygen: this follows directly from Avogadro's hypothesis.

scaling: 10 dm^3 of ammonia react with $\left(\frac{5}{4} \times 10\right)$ dm^3 of oxygen

Therefore volume needed = 12.5 dm^3

4. The charge flowing in an electrolytic cell

Equations involving electron transfer lead to calculations of charge.

> Avogadro's Number of electrons carry a charge of 96 500 coulombs or 1 Faraday.

Example Calculate the mass of aluminimum produced in a Hall's cell that passes a current of 1000 A for one hour. (R.A.M. for Al = 27)

Note: the charge flowing is given by $Q = it$ where i = current in A and t = time in s.

equation: $Al^{3+} + 3e \longrightarrow Al$ at cathode

molar ratio: 3 moles of electrons produce 1 mole of aluminium

charge-mass ratio: $(3 \times 96\,500)$ coulombs produce 27 g of aluminium

scaling: (1000×3600) coulombs produce

$$\left(27 \times \frac{1000 \times 3600}{3 \times 96\,500}\right) \text{ g of Al}$$

Therefore mass produced = 335.8 g

Questions

1. Calculate the number of moles in:
   a) 36 g of water
   b) 4 g of carbon dioxide
   c) 51 g of ammonia
   d) 32 g of sulphur S_8
   e) 2.1 g of HNO_3
   f) 3 g of hydrogen molecules
   g) 100 g of sodium chloride
   h) 5 dm^3 of air at s.t.p.
   i) 250 cm^3 of hydrogen at s.t.p.
   j) 145 cm^3 of oxygen at 25°C and 730 mm.

2. Calculate the mass in gram of:
   a) 0.5 mol of carbon dioxide
   b) 2.0 mol of sulphur dioxide
   c) 0.1 mol of hydrogen chloride
   d) 0.1 mol of sulphur S_8
   e) 0.25 mol of sulphuric acid
   f) 0.5 mol of $Na_2CO_3 . 10H_2O$
   g) 0.01 mol of PCl_5

3. What mass of each of the following would have to be dissolved in water to make 1 dm^3 of a solution whose concentration was 0.5 mol dm^{-3}?
   a) H_2SO_4
   b) HCl
   c) NaOH
   d) NaCl
   e) Na_2CO_3
   f) $AgNO_3$
   g) NH_4Cl
   h) $(NH_4)_2 SO_4$
   i) $Al_2 (SO_4)_3$

4. What mass of each of the following is needed to make up 250 cm^3 of a solution whose concentration was 0.1 mol dm^{-3}?
   a) $AgNO_3$
   b) HCl
   c) H_2SO_4
   d) NaOH
   e) $KMnO_4$
   f) $Na_2S_2O_3 . 5H_2O$

5. How many moles of each solute are contained in the following solutions?
   a) 500 cm^3 of a 0.1 mol dm^{-3} NaCl solution
   b) 20 cm^3 of a 2.0 mol dm^{-3} HCl solution
   c) 250 cm^3 of a 0.01 mol dm^{-3} NaOH solution

6. How many grams of each solute are contained in the following solutions?
   a) 500 cm^3 of a 1.5 mol dm^{-3} NaOH solution
   b) 100 cm^3 of a 0.1 mol dm^{-3} K_2CO_3 solution
   c) 200 cm^3 of a 0.15 mol dm^{-3} $KMnO_4$ solution

7. a) What is the maximum mass of carbon that can be burnt in 16.0 g of oxygen to give carbon dioxide?
   b) Aluminium reacts with oxygen to form alumina, Al_2O_3. What mass of oxygen is needed to react with 2.7 g of aluminium?

8. a) What volume of oxygen at s.t.p. is needed to burn 6.4 g of sulphur?
   b) When calcium carbonate is heated it decomposes to give calcium oxide and carbon dioxide. What mass of calcium oxide and what volume of carbon dioxide at s.t.p. are produced when 10 g of calcium carbonate are heated?

9. Magnesium is a reactive metal which dissolves in acids. What is the maximum mass of magnesium metal which will dissolve in:
   a) 1 dm^3 of a 0.01 mol dm^{-3} H_2SO_4 solution
   b) 100 cm^3 of a 0.1 mol dm^{-3} H_2SO_4 solution

c) 53 cm^3 of a 2.0 mol dm^{-3} H_2SO_4 solution
d) 10 cm^3 of a 2.0 mol dm^{-3} HCl solution

10. The reaction between a solution of copper (II) chloride and a solution of sodium carbonate can be written as:

$$CuCl_2 + Na_2CO_3 \longrightarrow CuCO_3\downarrow \quad + 2NaCl$$

or $$Cu^{2+}_{(aq)} + CO^{2-}_{3(aq)} \longrightarrow CuCO_{3(s)}$$

If the concentration of the sodium carbonate solution is 2 mol dm^{-3}, calculate:
a) the number of moles of sodium carbonate in 250 cm^3 of solution
b) the mass of this sodium carbonate
c) the volume of a 0.5 mol dm^{-3} solution of copper (II) chloride needed to react completely with the sodium carbonate
d) the mass of copper carbonate precipitated
e) the total volume of liquid involved
f) the concentration in mol dm^{-3} of the resulting sodium chloride solution.

Chapter 14

Energy changes in chemical reactions

14.1 Measuring and recording energy changes

Energy, heat and temperature

A firework sparkler contains iron filings in an oxidizing mixture. When it is lit, the sparkler throws out glowing specks of iron at a temperature of about 700°C. However, there is less chance of being burnt by these sparks than there is of being scalded by a drop of boiling water at 100°C. This comparison emphasizes the difference between the related ideas of energy, heat and temperature: both systems, the iron sparks and the water drop, contain energy which is being transferred in the form of heat because their temperatures are higher than that of their surroundings.

The *energy* of a system is the sum of all the individual kinetic and potential energies of the particles present in the system.

Heat is the term used to describe the process of energy transfer to or from the system.

Temperature is a measure of the average kinetic energy (K.E.) of the particles in the system; a temperature difference then gives an indication of the direction of heat flow.

a 'hot' system

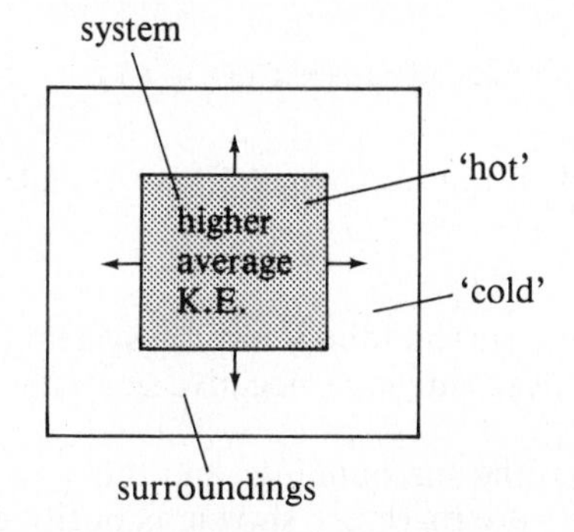

the 'system' particles slow down as the 'surroundings' particles speed up

a 'cold' system

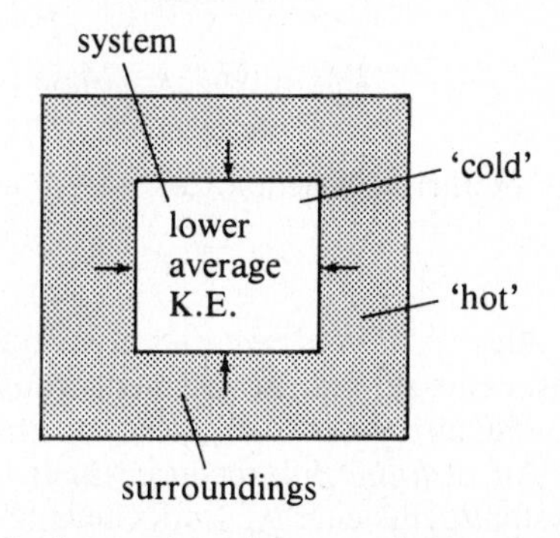

the 'system' particles speed up as the 'surroundings' particles slow down

An iron spark is less likely to cause burning because its mass is far less than that of a water drop. Even though the average energy of the particles is much higher, there are fewer particles present to transfer energy. So these two factors:

a) the number of particles in the system, and
b) the average energy of the particles in the system

when scaled up are equivalent to

a) the mass of the system, and
b) the temperature of the system.

The heat transferred by a hot body can then be written in equation form:

$$q \propto m\Delta T$$

where q is the heat transferred to or from a system of mass m and at a temperature ΔT different from the surroundings.

For a particular substance, a proportionality constant can be included that is called the specific heat capacity of the substance. This is the energy required to raise one kilogram of the substance through one kelvin.

$$q = ms\Delta T$$

where s is the specific heat capacity.

Endothermic and exothermic reactions

One of the three characteristics of a chemical change is that an energy exchange takes place between the reacting system and its surroundings. The exchange can either be to or from the system.

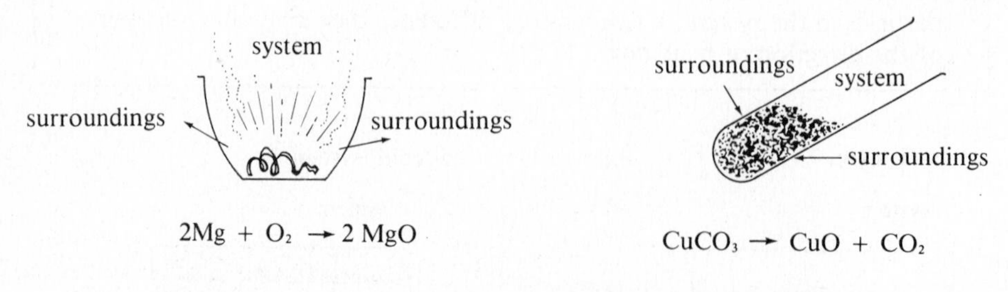

An *exothermic reaction* gives out energy to the surroundings and the system loses energy. Exothermic heat transfers are therefore shown as negative quantities to indicate the loss of energy to the surroundings.

An *endothermic reaction* takes in energy from the surroundings and the system gains energy. Endothermic heat transfers are therefore shown as positive quantities.

Both types of heat transfer are commonly represented on an energy diagram.

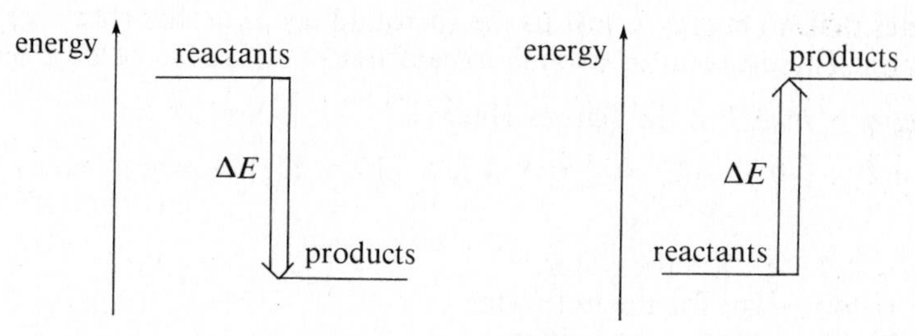

exothermic reaction: energy is given out by the system

$\Delta E = -q$

endothermic reaction: energy is taken in by the system

$\Delta E = +q$

Simple calorimetry

The energy change brought about by a chemical reaction is measured using a *calorimeter*. The temperature difference is recorded and this is converted to an energy change by applying the equation

$$q = ms\Delta T$$

The simplest calorimeter is a polystyrene beaker with a thermometer as a stirrer. It can be used to measure the energy change of a reaction that takes place in solution; for example, the exothermic reaction between an acid and an alkali. The following procedure is carried out.

1. Find the *heat capacity* of the calorimeter and the stirrer by mixing equal volumes of hot and cold water in the calorimeter system.

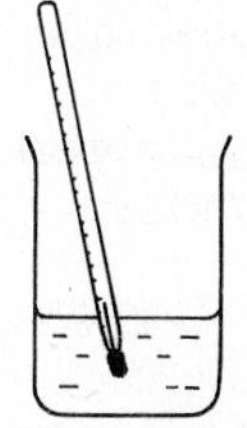

1. add 20 g of cold water at 15°C to calorimeter

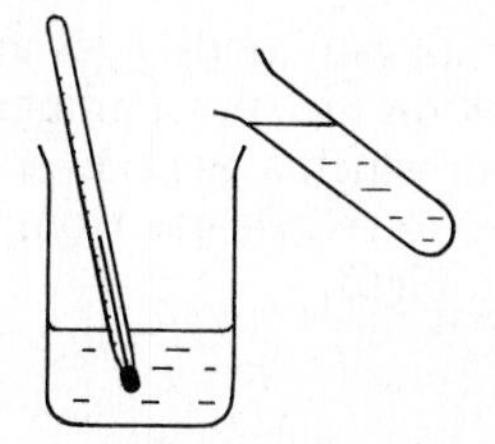

2. add 20 g of hot water at 40°C to the system

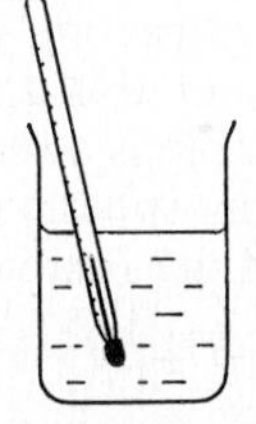

3. stir and record the final temperature, e.g. 26°C

> The first law of thermodynamics states that energy can neither be created nor destroyed but is simply transferred from one form or system to another.

Applying this law, we can write:

energy lost by hot water = energy gained by cold water + energy gained by calorimeter system

This assumes that no energy is lost to the surroundings, i.e. that there are no cooling losses. The problem of cooling in calorimeters is discussed on page 211.

Applying $q = ms\Delta T$ to the figures given:

$\Rightarrow$ $[20 \times 4.2 \times (40 - 26)] = [20 \times 4.2 \times (26 - 15)]$ + energy gained by calorimeter system

because $s = 4.2$ J g^{-1} K^{-1} for water,
$\Delta T = (40 - 26)$ for the hot water
and $\Delta T = (26 - 15)$ for the cold water.

$\Rightarrow$ energy gained by calorimeter system for a temperature rise of 11°C $= (1176 - 924)$ J $= 252$ J

The heat capacity of the calorimeter system is the energy needed to raise the temperature of the thermometer and the calorimeter by 1°C.

$$\Rightarrow \text{heat capacity} = \frac{252}{11} = 22.9 \text{ J °C}^{-1}$$

2. Carry out the reaction between the alkali and the acid in the calorimeter and stir with the thermometer. Record the temperature rise.

If 20 cm³ of alkali and 20 cm³ of acid at 15°C give a salt solution at a temperature of 17°C, the heat released by the reaction is found from $q = ms\Delta T$ as follows (assume $s = 4.2$ J g^{-1} K^{-1} and density of solution is 1 g cm^{-3}).

energy given out by the reaction = heat gained by the 40 cm³ of solution + heat gained by the calorimeter system

$= (40 \times 1 \times 4.2 \times 2) + (22.9 \times 2)$
$= 381.8$ J

Sometimes a reacting system not only undergoes an energy exchange with its surroundings but also does work on the surroundings as well. The most common example is a reaction in which a gas is evolved. The system must work against the atmospheric pressure to release gas from the reaction. A further application of the first law is required.

$$\Delta E = q + w$$

where ΔE is the total energy change, q is the heat exchange and w is the work done by the system.

When a gas is evolved at a constant pressure, the work done is given by the equation

$$w = P(V_1 - V_2)$$

where P is the pressure, V_1 is the initial volume of the system and V_2 its final volume. This means that the equation $\Delta E = q + w$ becomes

$$\Delta E = q - P\Delta V$$

where $\Delta V = (V_2 - V_1)$ or the volume of gas evolved by the reaction.

Example 0.56 g of iron dissolve in 30 cm³ of acid in a calorimeter of heat capacity 22.9 J K^{-1} to give a temperature rise of 5.4 K. Given that a mole of gas under the reaction conditions occupies 24 dm³, find the total energy exchange between system and surroundings. Assume atmospheric pressure is 10^5N m^{-2}

equation: $Fe_{(s)} + 2H_3O^+_{(aq)} \rightarrow Fe^{2+}_{(aq)} + H_{2(g)} + 2H_2O_{(l)}$
molar ratio: 1 mole of iron produces 1 mole of hydrogen
mass-volume ratio: 0.56 g of iron produces 0.24 dm³ of gas under the reaction conditions = 2.4×10^{-4} m³

$$\begin{aligned} \text{Thus } \Delta V &= 2.4 \times 10^{-4} \text{ m}^3 \\ \text{and } P &= 10^5 \text{ N m}^{-2} \\ \Rightarrow w &= P\Delta V = (10^5 \times 2.4 \times 10^{-4}) \text{ N m} \\ &= 24 \text{ J because 1 J} = 1 \text{ N m} \end{aligned}$$

The heat energy q is given *out* to both solution and calorimeter and is therefore negative.

$$\begin{aligned} \Rightarrow q &= ms\Delta T \\ &= -[(30.56 \times 1 \times 4.2 \times 5.4) + (22.9 \times 5.4)] = -816 \text{ J} \\ \text{Now } \Delta E &= q - P\Delta V \\ \Delta E &= -816 - 24 = -840 \text{ J} \end{aligned}$$

14.2 Enthalpy and standard enthalpy changes

The total heat content of a system is called its *enthalpy H*. It can be regarded as the sum of the internal and external energies of the system: $H = U + PV$

1. *The internal energy, U,* is the energy possessed by the system as a result of the motion and spatial arrangement of all the particles making up its atoms and molecules.
2. *The external energy, PV,* is the product of the pressure and volume of the system; it is a measure of the energy the system possesses as a result of the space it occupies.

The enthalpy of a system can be represented in the form of a bar chart. The series of diagrams below shows a system whose total heat content is decreasing

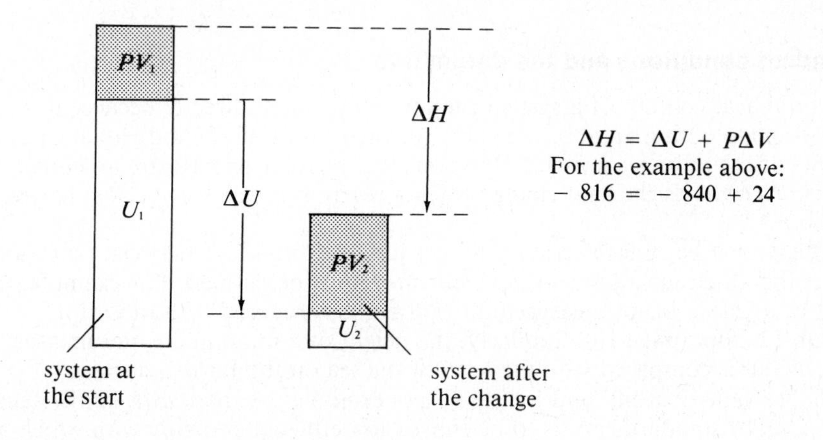

as energy is given out to the surroundings. The system also does work on the surroundings because its external energy increases during the change.

From the bar charts, it can be seen that ΔH is a negative quantity and that ΔU is a larger negative quantity. The smaller decrease in the total heat content occurs because the external energy of the system increases as a result of the change. These relationships are illustrated by the example given on page 207: iron dissolving in acid.

1. The total heat content H decreases as heat energy is given out to the surroundings: $\Delta H = -816$ J.
2. Work is done on the surroundings because the external energy of the system increases: $P\Delta V = 24$ J where $\Delta V = (V_2 - V_1) = 0.24$ dm^3.
3. The internal energy of the system decreases by a larger value than that of the total heat content which takes into account the small increase in the external energy: $\Delta U = -840$ J.

By applying this example more generally, the following important definitions are derived.

> The enthalpy change of a reaction ΔH is the heat change when the reaction is carried out at constant pressure.

i.e. $\Delta H = q = \Delta U + P\Delta V$ at constant pressure.

If no external work is done by the system because there is no volume change, i.e. $\Delta V = 0$ and $P\Delta V = 0$, then $\Delta H = q = \Delta U$, or

> The internal energy change of a reaction ΔU is the heat change when the reaction is carried out at constant volume.

The two definitions are represented by the two equations,

$$\Delta H = q_P \quad \text{and} \quad \Delta U = q_V$$

Standard conditions and the datum line

The total heat content of a system cannot be measured directly because it is impossible to obtain an absolute value for the sum of all the individual energy factors of the particles present. However, it *is* possible to measure an *enthalpy change* ΔH: this is the heat change when a reaction is carried out at constant pressure.

Whenever it becomes necessary to measure the difference between two values rather than their absolute values, a *standard* must be defined. For example, the mass of an atom is not measured directly but is compared with that of a carbon-12 atom (page 10). Similarly, the height of a mountain is not measured directly but is compared with the level of the sea on the Earth's surface.

The 'carbon-12 atom' and 'sea level' are examples of *standards*. In the same way, a set of standards needs to be chosen to define an enthalpy with which all other enthalpies can be compared. These are shown in the table below.

standards chosen	*reasons*
	The enthalpy of a system depends on
1. One mole of ...	1. The number of particles present
2. ... an element in its natural state ...	2. the nature of the particles present
3. ... at 298 K and ...	3. the kinetic energy of the particles
4. ... at 1 atmosphere pressure	4. the space occupied by the particles

The symbol $H^\ominus$ is used to mean *standard enthalpy* and it is the enthalpy of a system compared with the enthalpy of one mole of an element in its natural state at 298 K and one atmosphere pressure. Logically, it therefore follows that for any element $H^\ominus = 0$. When its enthalpy is compared with the set of standards, the difference is zero. This is shown on an *enthalpy diagram* by the *datum line*. The datum line is the line $H^\ominus = 0$ and is the standard enthalpy of one mole of any element under the standard conditions.

For example, solid carbon and gaseous oxygen are on the datum line of the enthalpy diagram shown below.

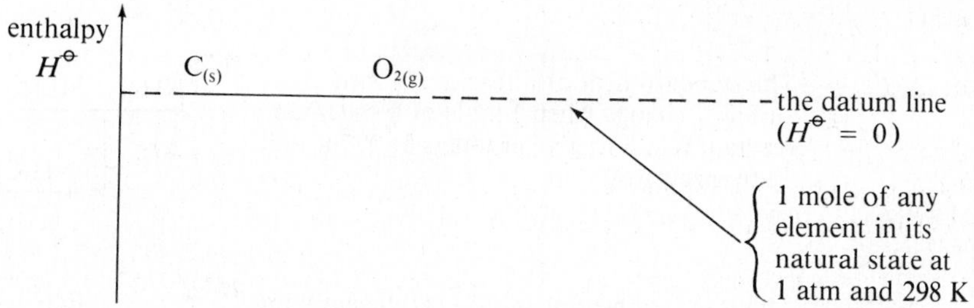

If a system undergoes a change, the enthalpy change can be shown on the enthalpy diagram. For example, the reaction of carbon and oxygen to give carbon dioxide is exothermic.

$$C_{(s)} + O_{2(g)} \longrightarrow CO_{2(g)}$$

The heat content of the system falls as energy is given out to the surroundings. So, on the enthalpy diagram:

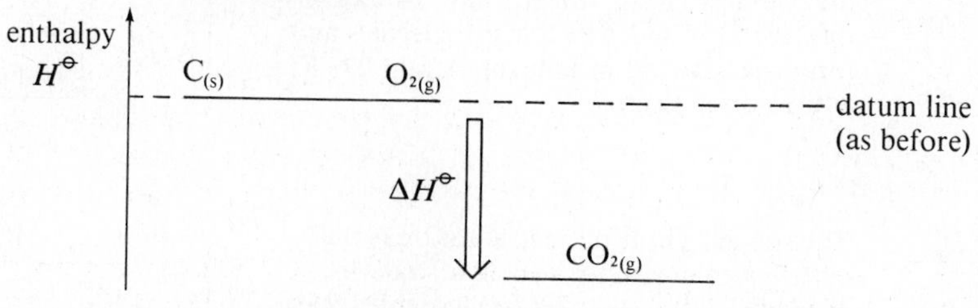

The *standard enthalpy change* $\Delta H^\ominus$, when one mole of carbon dioxide is produced from its elements, is −283 kJ. This can be written after the equation for the chemical change:

$$C_{(s)} + O_{2(g)} \longrightarrow CO_{2(g)} \qquad \Delta H^\ominus = -283 \text{ kJ mol}^{-1}$$

When a mole of compound is being formed from its elements, the standard enthalpy change is known as a *standard heat of formation* $\Delta H_f^\ominus$.

There are a number of different standard heats (or standard enthalpy changes) that are in common use. A list of some of them is given below.

symbol	*definition*	*example*
$\Delta H_f^\ominus$	The standard heat of formation is the enthalpy change when 1 mole of a compound is formed from its elements in their natural states at 298 K and 1 atm pressure.	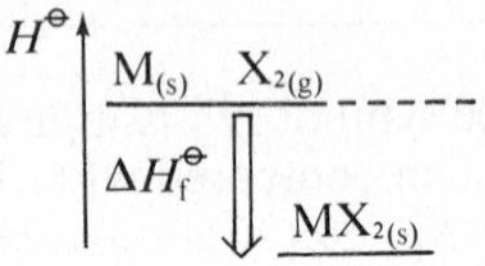 $H^\ominus$; $M_{(s)}$ $X_{2(g)}$; $\Delta H_f^\ominus$; $MX_{2(s)}$
$\Delta H_c^\ominus$	The standard heat of combustion is the enthalpy change when 1 mole of a compound is completely burnt in oxygen at 298 K and 1 atm pressure.	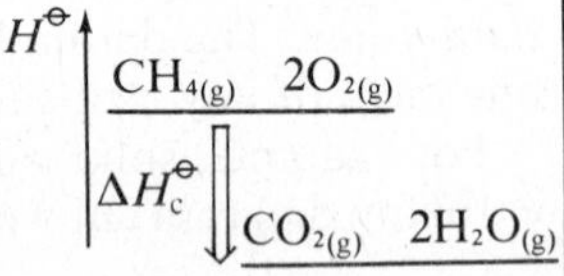 $H^\ominus$; $CH_{4(g)}$ $2O_{2(g)}$; $\Delta H_c^\ominus$; $CO_{2(g)}$ $2H_2O_{(g)}$
$\Delta H_r^\ominus$	The standard heat of a reaction is the enthalpy change when 1 mole of a specified reactant reacts to give products at 298 K and 1 atm pressure.	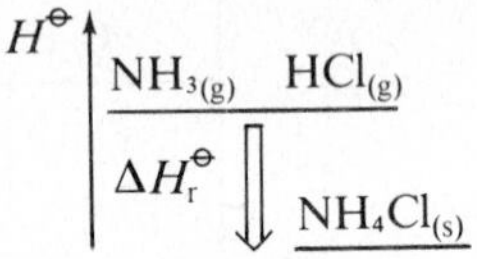 $H^\ominus$; $NH_{3(g)}$ $HCl_{(g)}$; $\Delta H_r^\ominus$; $NH_4Cl_{(s)}$
$\Delta H_{sol}^\ominus$	The standard heat of solution is the enthalpy change when 1 mole of substance is dissolved by solvent such that further dilution produces no measurable heat change at 298 K and 1 atm pressure.	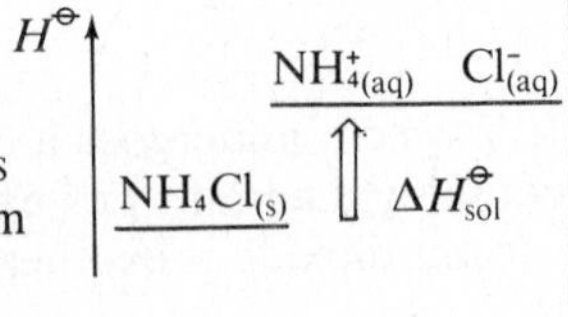 $H^\ominus$; $NH_{4(aq)}^+$ $Cl_{(aq)}^-$; $NH_4Cl_{(s)}$; $\Delta H_{sol}^\ominus$
$\Delta H_{hyd}^\ominus$	The standard heat of hydration of any ion is the enthalpy change when 1 mole of gaseous ions are surrounded by water molecules and form a solution at infinite dilution at 298 K and 1 atm pressure.	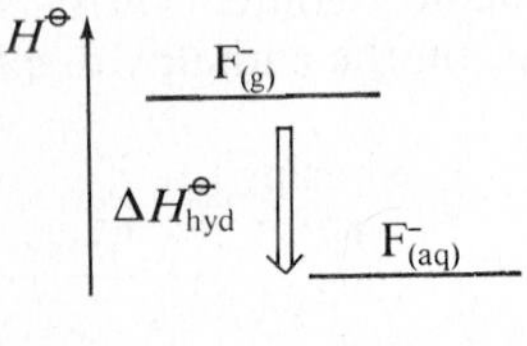 $H^\ominus$; $F_{(g)}^-$; $\Delta H_{hyd}^\ominus$; $F_{(aq)}^-$
$\Delta H_n^\ominus$	The standard heat of neutralization is the enthalpy change when 1 mole of water is produced as a result of the reaction between an acid and an alkali at 298 K and 1 atm pressure.	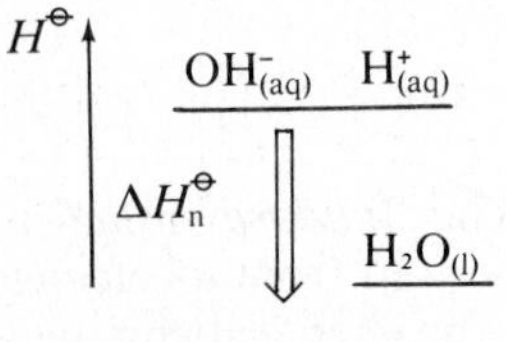 $H^\ominus$; $OH_{(aq)}^-$ $H_{(aq)}^+$; $\Delta H_n^\ominus$; $H_2O_{(l)}$

Determining standard enthalpy changes in the laboratory

For reactions that happen in solution, a simple calorimeter is used. This piece of apparatus is described on page 205. For example, the standard heat of reaction for the displacement of copper from copper sulphate solution by zinc is determined as follows:

$$Zn_{(s)} + Cu^{2+}_{(aq)} \longrightarrow Zn^{2+}_{(aq)} + Cu_{(s)}$$

1. Determine the calorimeter constant as outlined earlier.
2. Add a measured volume of copper sulphate solution to the calorimeter. The number of moles of copper ions can be calculated from the concentration of the solution:

$$20\ \text{cm}^3 \text{ of } 0.1\ \text{M solution} \equiv \left(\frac{20}{1000} \times 0.1\right) = 2 \times 10^{-3} \text{ moles}$$

3. Add powdered zinc stirring with the thermometer until there is a measured excess present. Record the time taken during the addition process and the temperature of the system at regular intervals.
4. Plot a temperature-time curve and extrapolate it to reaction time = zero. The temperature change at this point is the maximum that would have been observed had there been no cooling during the addition process.

corrections made for cooling

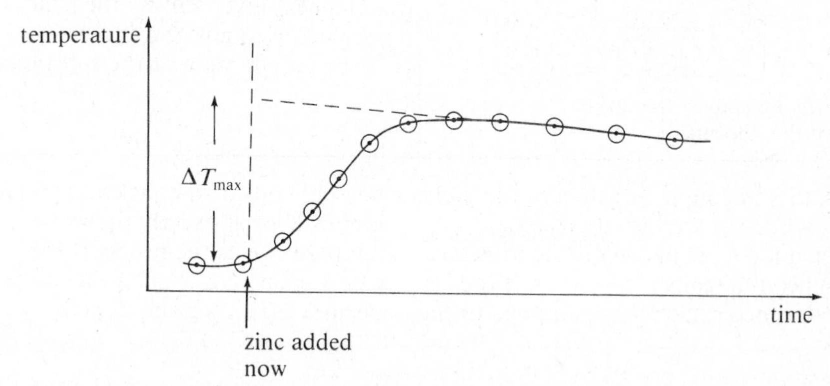

5. Use the relationship $q = (ms\Delta T_{max}) + (\text{calorimeter constant} \times \Delta T_{max})$ to calculate the heat evolved from 2×10^{-3} moles of copper ions.
6. Scale this heat change up to the value obtained for a mole of copper ions and this gives $\Delta H_r^\ominus$.

It is much more difficult to determine the heat of a reaction that does not take place in solution. A special type of device, called a bomb calorimeter, is used to measure heats of combustion. It is described overleaf.

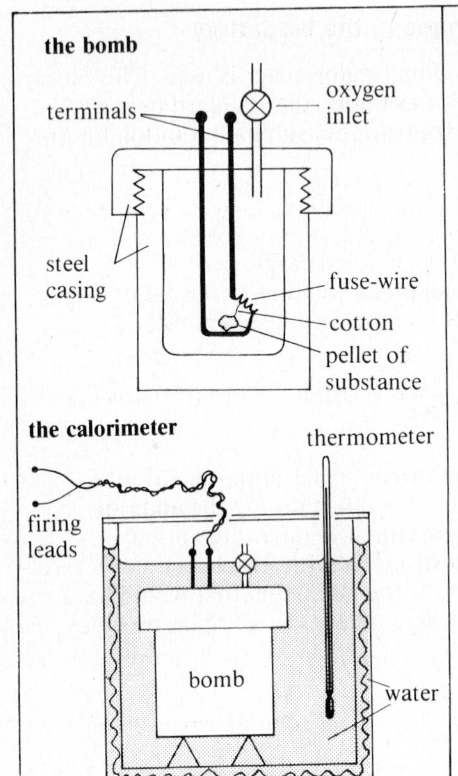

1. A small pellet of known mass (or volume if liquid) is put into the reaction cup of the bomb.
2. A piece of cotton is used to connect it to the fuse wire that shorts the two terminals.
3. About 25 atmospheres of pure oxygen is introduced.
4. The bomb is placed in the calorimeter and the fuse wire is blown by a pulse of high current.
5. The known mass of substance is burnt in the calorimeter.
6. Corrections are made for the energy given out by the burning cotton and fuse wire.
7. The apparatus is calibrated by burning a known mass of a substance whose heat of combustion is also known. The common standard is benzoic acid C_6H_5COOH. $\Delta H_c^\ominus = -3227$ kJ mol^{-1}.
8. The heat given out by the pellet is scaled up to give the amount evolved per mole of the substance.

A thermocouple attached to the jacket is wired via a relay circuit to a thermocouple inside the calorimeter. When the temperature rises in the calorimeter, current is supplied to the heating coils of the jacket. The jacket is thus kept at exactly the *same* temperature as the inside of the calorimeter. No cooling therefore occurs: ΔT_{max} is given directly.

Here are some typical results obtained for glucose $C_6H_{12}O_6$ (R.M.M. = 180).

$$0.90 \text{ g glucose} \Rightarrow \Delta T_{max} = 0.36°C$$
$$\text{while } 1.22 \text{ g benzoic acid} \Rightarrow \Delta T_{max} = 0.82°C$$

For benzoic acid C_6H_5COOH (R.M.M. 122), $\Delta H_c^\ominus = 3227$ kJ mol^{-1} (given)

So 122 g benzoic acid evolve 3227 kJ
$\Rightarrow$ 1.22 g benzoic acid evolve 32.27 kJ
$\therefore$ 0.82°C is the temperature rise for 32.27 kJ evolved
$\therefore$ 0.36°C is the temperature rise for $\left(32.27 \times \dfrac{0.36}{0.82}\right)$ kJ

This energy is evolved by 0.90 g glucose = 0.90/180 mole.

$$\therefore \quad \Delta H_c^{\ominus} = -\left[32.27 \times \frac{0.36}{0.82} \times \frac{180}{0.90}\right] \text{kJ mol}^{-1}$$

$$= -2830 \text{ kJ mol}^{-1} \qquad \text{(accepted value} = -2816 \text{ kJ mol}^{-1}\text{).}$$

Even when the problems of calorimetry are overcome, there are many enthalpies that still cannot be measured directly, because these changes cannot be brought about in the laboratory. For example, it is not possible to form methane under controlled conditions from carbon and hydrogen, so it is not possible to measure $\Delta H_f^{\ominus}$ for methane directly. This is true for many other substances. Enthalpy changes of this sort must be worked out by applying the first law of thermodynamics (page 205) to quantities that *can* be directly measured.

14.3 Applications of the first law

Hess's law

> The total heat change in a chemical reaction is determined only by the initial and final states of the system and is independent of the pathway followed (provided it is only heat energy that is exchanged).

Hess's law is a direct consequence of the first law. Consider a process in which a system undergoes a change from state A to state B and 100 kJ are evolved. If it is possible to find an alternative route from A to B in which only 80 kJ are evolved, then a cycle can be set up as shown below.

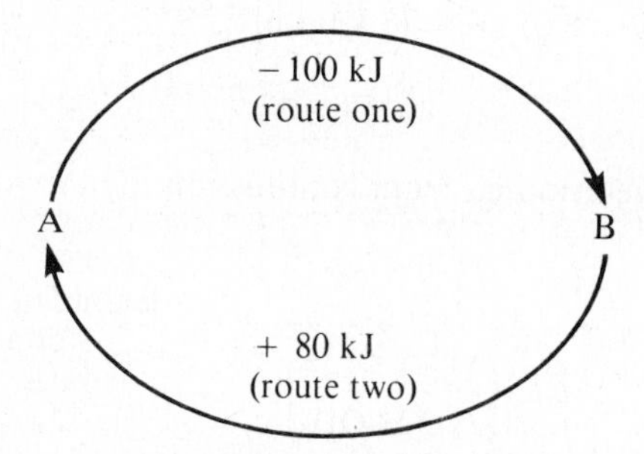

For each complete cycle, 20 kJ are created. This contravenes the first law and it is therefore necessary to assume that both routes have the same enthalpy change.

Hess's law can be applied to the problem of determining a heat of formation such as that of methane (mentioned above).

$$C_{(s)} + 2H_{2(g)} \longrightarrow CH_{4(g)}$$

This reaction cannot be brought about in the laboratory. Its enthalpy change can be calculated from combustion data because the products of burning the reactants (carbon and hydrogen) are the same as the products of burning methane. A bomb calorimeter is used to measure $\Delta H_c^\ominus$ for each substance and all the data are collected on a single enthalpy diagram.

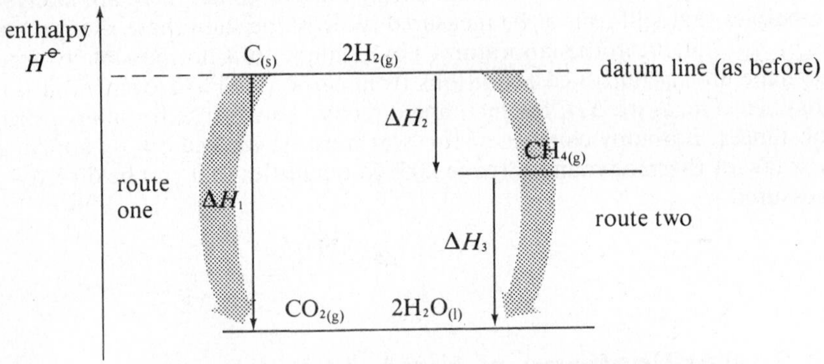

The diagram shows that there are, in theory, two routes possible for the conversion of carbon and hydrogen to their respective oxides.

1. The direct route, whose enthalpy change ΔH_1 is given by the sum of the heat of combustion of carbon and twice the heat of combustion of hydrogen.
   $$\Delta H_1 = (\Delta H_c^\ominus \text{ of } C_{(s)}) + 2\,(\Delta H_c^\ominus \text{ of } H_{2(g)})$$
   $$= -966 \text{ kJ}$$

2. A route via an intermediate stage: methane. The enthalpy change of this route is $(\Delta H_2 + \Delta H_3)$ and it can be seen that $\Delta H_2 = \Delta H_f^\ominus$ of $CH_{4(g)}$ and $\Delta H_3 = \Delta H_c^\ominus$ of $CH_{4(g)}$. Since $\Delta H_c^\ominus$ for $CH_{4(g)}$ is known from calorimetry to be $-$ 890 kJ mol^{-1}, $\Delta H_f^\ominus$ is found by subtraction because, by Hess's law, $\Delta H_1 = \Delta H_2 + \Delta H_3$.

   i.e. $\Delta H_f^\ominus$ for $CH_{4(g)}$ is $(-966 + 890) = -76$ kJ mol^{-1}.

Applications of Hess's law

The heat of formation of ethanol is determined from combustion data by the same procedure.

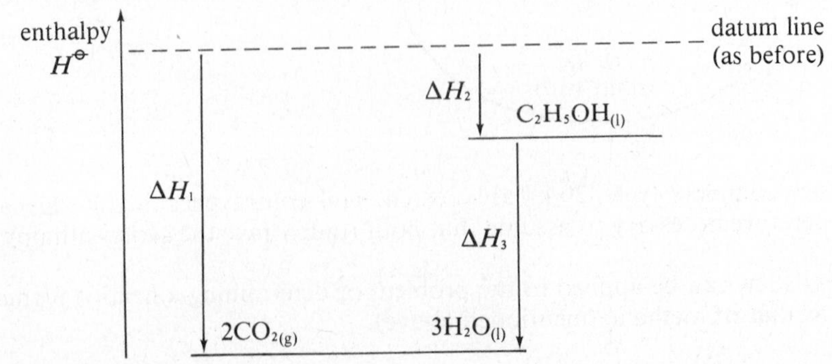

$\Delta H_1 = 2(\Delta H_c^{\ominus}\text{of } C_{(s)}) + 3(\Delta H_c^{\ominus}\text{of } H_{2(g)}) = -1649 \text{ kJ}$

$\Delta H_2 = \Delta H_f^{\ominus}\text{of } C_2H_5OH_{(l)} = (\Delta H_1 - \Delta H_3)$ from the diagram

but $\Delta H_3 = \Delta H_c^{\ominus}\text{of } C_2H_5OH_{(l)} = -1371 \text{ kJ}$

Thus $\Delta H_f^{\ominus}$of ethanol $= -278 \text{ kJ mol}^{-1}$

The heat of any reaction can be determined if the heats of formation of the reactants and products are known. For example, the reaction between ammonia and hydrogen chloride is exothermic.

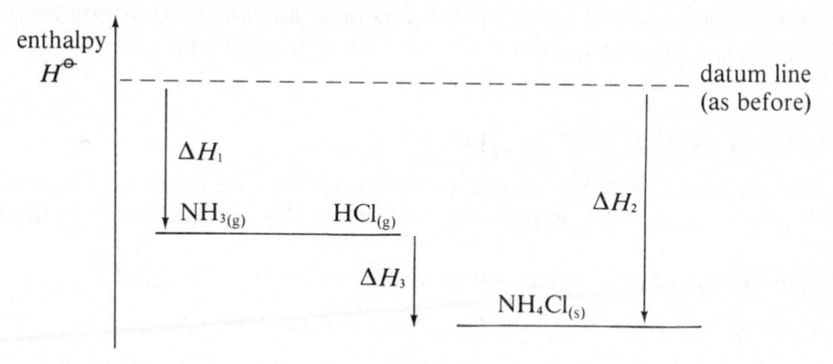

$\Delta H_1 = (\Delta H_f^{\ominus}\text{of } NH_{3(g)}) + (\Delta H_f^{\ominus}\text{of } HCl_{(g)}) = -138 \text{ kJ}$

$\Delta H_2 = \Delta H_f^{\ominus}\text{of } NH_4Cl_{(s)} = -315 \text{ kJ}$

ΔH_3 = the required heat of reaction, $\Delta H_r^{\ominus} = (\Delta H_2 - \Delta H_1)$ from the diagram

Thus $\Delta H_r^{\ominus} = -177 \text{ kJ mol}^{-1}$

The heat of reaction per mole of reactant is sometimes not the same as the overall heat of reaction as expressed by the reaction equation. For example, in the reaction of hydrazine with hydrogen peroxide, the enthalpy change per mole of hydrazine is twice that per mole of peroxide.

$$2H_2O_{2(l)} + N_2H_{4(l)} \longrightarrow 4H_2O_{(g)} + N_{2(g)}$$

The overall enthalpy change is calculated first and then converted to the value for the specified reactant. For example, the heat of reaction per mole of peroxide is found as follows.

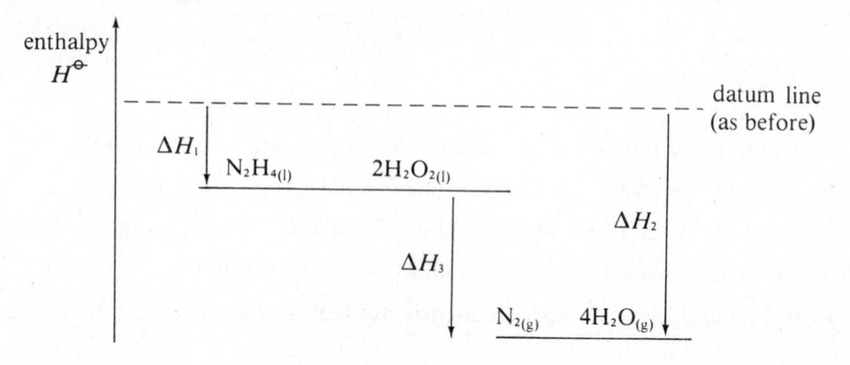

$$\Delta H_1 = (\Delta H_f^\ominus \text{of } N_2H_{4(l)}) + 2(\Delta H_f^\ominus \text{of } H_2O_{2(l)})$$
$$= +50 + 2(-188) = -326 \text{ kJ}$$
$$\Delta H_2 = 4(\Delta H_f^\ominus \text{of } H_2O_{(g)}) = -986 \text{ kJ} \quad (\Delta H_f^\ominus \text{of } N_{2(g)} = 0)$$
ΔH_3 = the heat of reaction for two moles of peroxide
$= (\Delta H_2 - \Delta H_1)$ from the diagram $= -660$ kJ

Thus $\Delta H_r^\ominus = -330 \text{ kJ mol}^{-1}$ of hydrogen peroxide

In the data given above, it can be seen that the standard enthalpy of formation of hydrazine is a positive quantity. Hydrazine is an *endothermic compound* because it has a greater standard enthalpy than its elements.

Born-Haber cycles

Any simple process can be considered to take place as a result of a series of single theoretical steps. For example, salt dissolving in water involves the two steps:

1. Disruption of the lattice into separate ions in the gas phase:

$$NaCl_{(s)} \longrightarrow Na^+_{(g)} + Cl^-_{(g)}$$

2. Solvation of the individual ions by water molecules (hydration)

$$Na^+_{(g)} \xrightarrow{aq} Na^+_{(aq)}$$
$$Cl^-_{(g)} \xrightarrow{aq} Cl^-_{(aq)}$$

Overall:

$$NaCl_{(s)} \xrightarrow{aq} Na^+_{(aq)} + Cl^-_{(aq)}$$

When the energetics of these theoretical steps are shown on an energy diagram, the result is called a *Born-Haber cycle*.

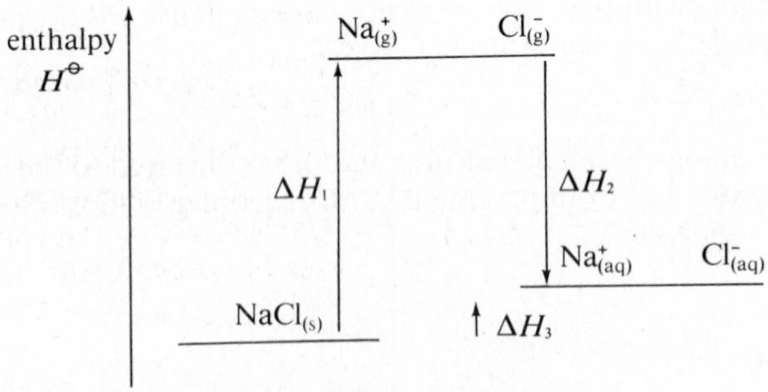

enthalpy $H^\ominus$

$Na^+_{(g)}$ $Cl^-_{(g)}$

ΔH_1 ΔH_2

$Na^+_{(aq)}$ $Cl^-_{(aq)}$

$NaCl_{(s)}$ $\uparrow \Delta H_3$

ΔH_1 = the lattice energy of sodium chloride $= 771 \text{ kJ mol}^{-1}$

$\Delta H_2 = (\Delta H^\ominus_{hyd} \text{ of } Na^+_{(g)}) + (\Delta H^\ominus_{hyd} \text{ of } Cl^-_{(g)}) = -770 \text{ kJ mol}^{-1}$

$\Delta H_3 = \Delta H^\ominus_{sol}$ of sodium chloride $= (\Delta H_1 + \Delta H_2)$ from the diagram

Thus heat of solution of sodium chloride $= +1 \text{ kJ mol}^{-1}$.

Born-Haber cycles are most frequently applied to the formation of compounds.

Example (a) When applied to an *ionic compound* (e.g. salt), the theoretical steps are:

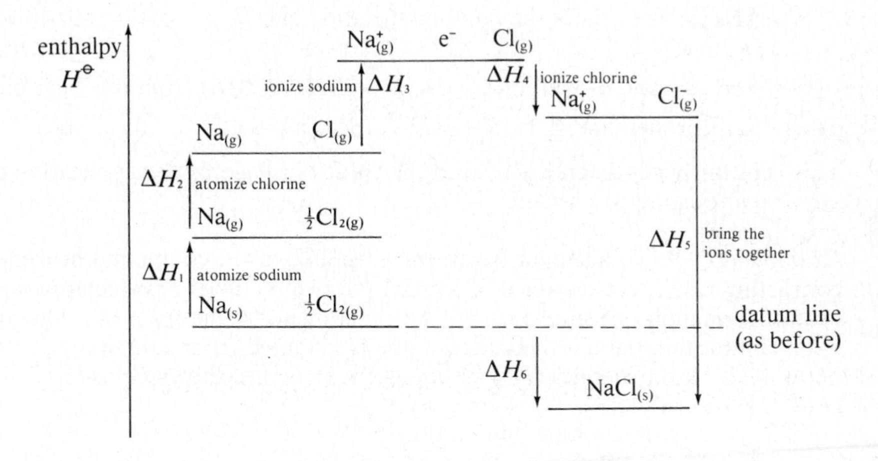

ΔH_1 = $\Delta H^{\ominus}_{at}$ of sodium = 109 kJ

ΔH_2 = half the bond dissociation energy of Cl – Cl = 121 kJ

ΔH_3 = first ionization energy of sodium = 494 kJ

ΔH_4 = first electron affinity of chlorine = – 364 kJ

ΔH_5 = the lattice energy of sodium chloride = – 771 kJ (given out)

ΔH_6 = $\Delta H^{\ominus}_f$ of sodium chloride = (ΔH_1 + ΔH_2 + ΔH_3 + ΔH_4 + ΔH_5) from the diagram.

From the theoretical predictions of the Born—Haber cycle, the heat of formation of salt is given by

$$\Delta H_6 = (109 + 121 + 494 - 364 - 771) = -411 \text{ kJ mol}^{-1}$$

This is in good agreement with the experimentally determined value and so it can be concluded that the ionic model is a sensible bonding model for sodium chloride.

Example (b) When a Born-Haber cycle is applied to the formation of a *covalent compound* (e.g. methane), the theoretical steps are:

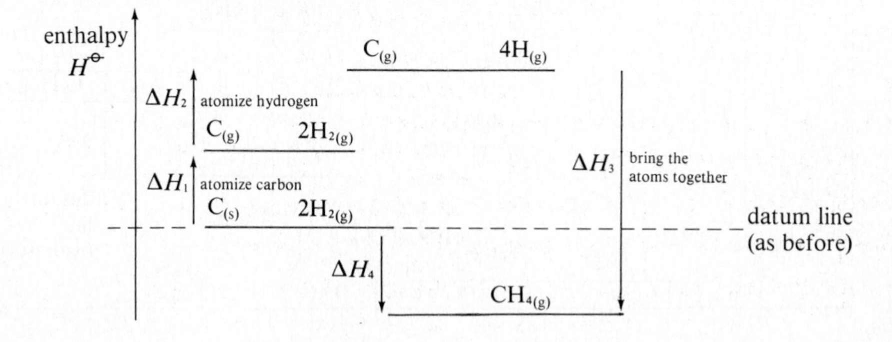

$\Delta H_1 = \Delta H^{\ominus}_{at}$ of carbon $= 715$ kJ

$\Delta H_2 =$ two times the bond dissociation energy H — H $= 872$ kJ

$\Delta H_3 =$ four times the bond dissociation energy C — H $= -1648$ kJ (given out)

$\Delta H_4 = \Delta H^{\ominus}_f$ of methane $= (\Delta H_1 + \Delta H_2 + \Delta H_3)$ from the diagram

Thus $\Delta H^{\ominus}_f$ for methane $= (715 + 872 - 1648)$ kJ mol^{-1} $= -61$ kJ mol^{-1}

This is in fairly good agreement with the value obtained by the application of Hess' law to combustion data.

Example (c) When a similar treatment is applied to the compound benzene, conflicting results are obtained. Compare the two methods of calculating the heat of formation of benzene below. An experimental value for $\Delta H^{\ominus}_f$ is obtained from combustion data; a theoretical value is obtained from data derived from bond dissociation energies ($D^{\ominus}$) assuming the structure shown below

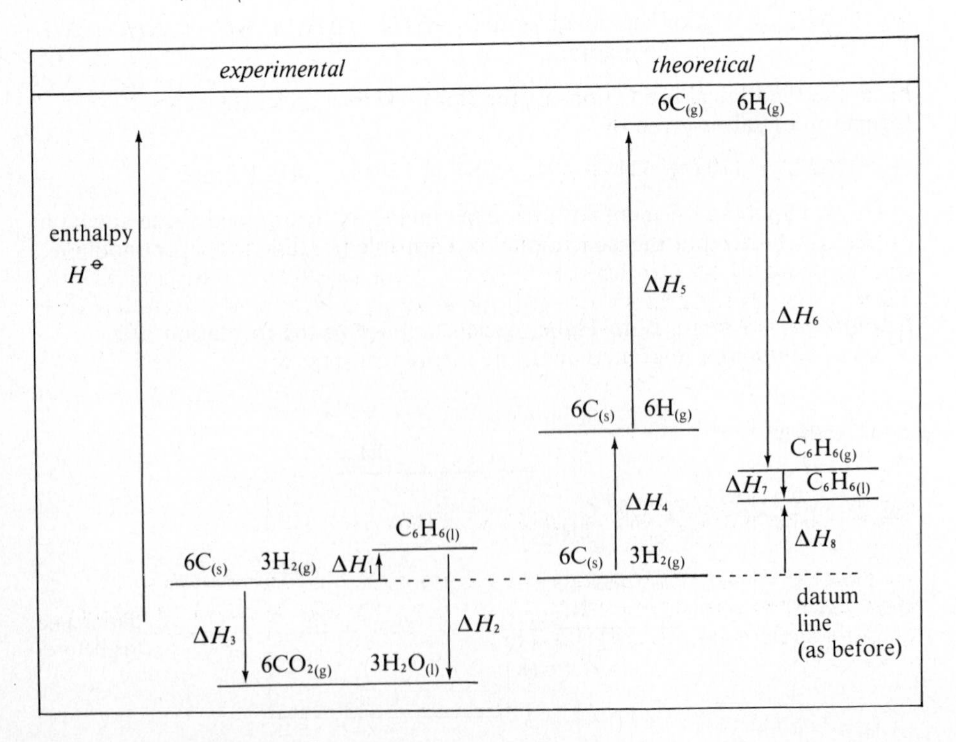

experimental	*theoretical*
$\Delta H_1 = \Delta H_f^\ominus$ of benzene $= (\Delta H_3 - \Delta H_2)$ (from the diagram) $\Delta H_2 = \Delta H_c^\ominus$ of benzene $= -3273$ kJ $\Delta H_3 = 6(\Delta H_c^\ominus$ of carbon) $+ 3\ (\Delta H_c^\ominus$ of hydrogen) $= -3224$ kJ $\therefore \Delta H_1 = (-3224 + 3273)$ kJ $\Rightarrow \Delta H_f^\ominus$ of benzene $= +49$ kJ mol^{-1}	$\Delta H_4 = 3(D^\ominus$ of H—H) $= 1308$ kJ $\Delta H_5 = 6\Delta H_{at}^\ominus$ of carbon $= 4290$ kJ $\Delta H_6 = 6(D^\ominus$ of C—H) $+ 3(D^\ominus$ of C=C) $+ 3\ (D^\ominus$ of C—C) $= -5352$ kJ (given out) $\Delta H_7 =$ latent heat of vaporization $= -34$ kJ $\Delta H_8 = \Delta H_f^\ominus$ of benzene $= (\Delta H_4^\ominus + \Delta H_5^\ominus$ $+ \Delta H_6^\ominus + \Delta H_7^\ominus)$ (from the diagram) $\Delta H_f^\ominus$ of benzene $= +212$ kJ mol^{-1}

The value of $\Delta H_f^\ominus$ determined by experiment is 163 kJ mol^{-1} lower than the value predicted by the bonding model. In other words, the true structure of benzene is *thermodynamically more stable* by 163 kJ mol^{-1} than the structure of the bonding model adopted. This is clear evidence that the structure shown below is an inadequate representation of a benzene molecule.

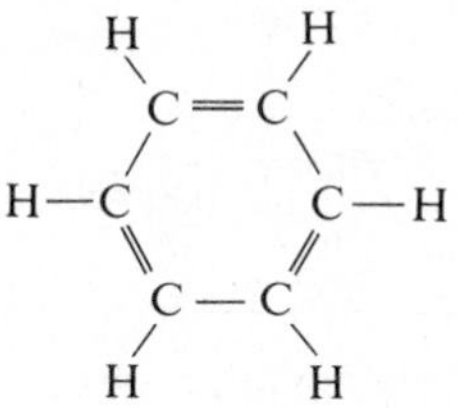

```
      H       H
       \     /
        C═══C
       /     \
  H—C         C—H
      \\     //
        C—C
       /     \
      H       H
```

There are a number of compounds for which conflicting results of this sort can be obtained. The extra bond strength not accounted for by the simple valence bond interpretation is called *resonance energy* or *delocalization energy*. It is discussed in more detail on page 488.

14.4 Spontaneous processes

On page 204, the idea of an endothermic process is first introduced. For example, when sodium nitrate (V) dissolves in water, the system spontaneously takes in energy from the surroundings.

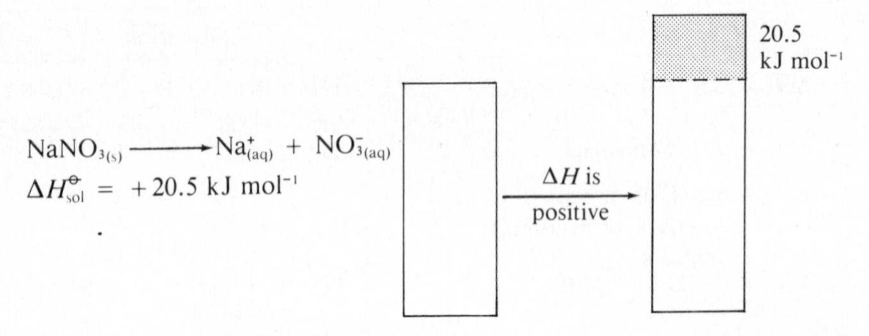

$$NaNO_{3(s)} \longrightarrow Na^+_{(aq)} + NO^-_{3(aq)}$$
$$\Delta H^{\ominus}_{sol} = +20.5 \text{ kJ mol}^{-1}$$

At first sight, this process appears rather strange: it is not a common experience that the energy of a system should suddenly increase (unless some external agent does work on the system). The last section of this chapter seeks to explain the factors that determine the spontaneity of any change. To that end, the spontaneous endothermic change described above can be contrasted with a spontaneous process that is familiar from the physical world: the movement of a ball on the slope of a hill.

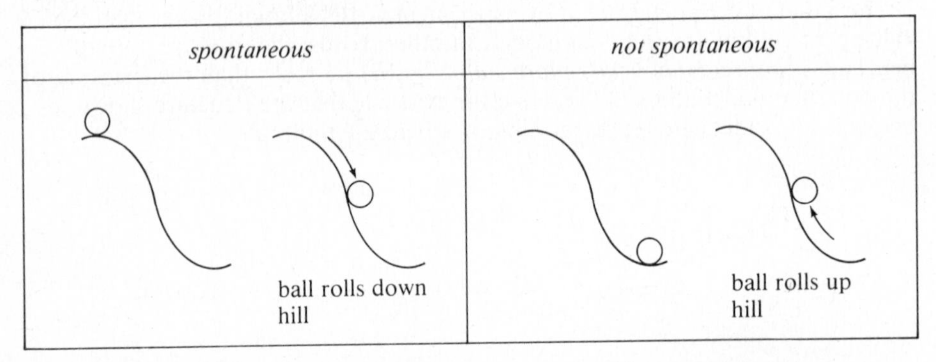

The direction of this spontaneous change is explained by examining the potential energy of the system. The energy of most systems tends to a minimum and the direction of any likely change can be predicted on this basis. But the prediction does not seem to work for the endothermic reaction: spontaneously it increases in enthalpy. Clearly, there is another factor involved in the controlling of spontaneous processes.

A clue to this other factor can be found by considering the behaviour of the particles in a piece of metal that is hot at one end and cold at the other. Common experience is that the temperature quickly becomes equal throughout the system:

spontaneous		*not spontaneous*	
hot: fast-moving particles; cold: slow-moving particles	warm: all particles with approx. the same random energy	warm: all particles with approx. the same random energy	hot: fast-moving particles; cold: slow-moving particles

The opposite process does not happen: an iron bar does not suddenly become very hot at one end and very cold at the other. The random distribution of the particles is unlikely to become ordered so that the fast-moving particles move to one end of the system and the slow-moving ones move to the other. Systems tend towards a state of randomness or lack of order rather than the other way round.

The examples of the ball on the hill and the hot iron bar identify two different tendencies about spontaneous changes:

Systems tend to a minimum in potential energy
Systems tend to maximum disorder

The two tendencies are sometimes in opposition to each other and their effects can be related by the following assumptions,

1. A system possesses a certain amount of energy which is directly associated with the degree of disorder of the particles within the system.
2. This energy (part of the total heat content H) cannot be converted to external energy to do work on the surroundings.

On page 207 the heat content of a system is described as the sum of the internal and external energies. But there is now an alternative way of dividing up the heat content. It can be represented as the sum of the energy that *can* be converted into external work (*free energy*) and the energy unavailable for conversion.

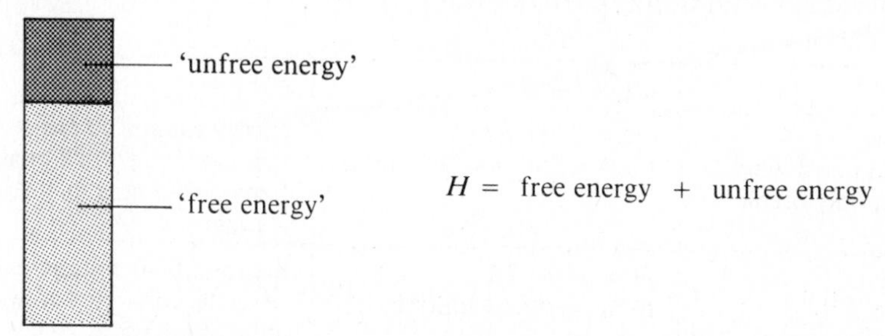

The degree of disorder of a system is temperature dependent. The higher the temperature, the more disorder there is likely to be. The energy associated with this disorder must therefore be given per degree; it is called the *entropy* of the system and is defined by the expression:

$$dS = \frac{dq}{T} \text{ (for reversible conditions)}$$

where dS = the small increase in entropy when a small quantity of heat, dq, is absorbed by the system at a temperature T.

The 'unfree energy' associated with a system's heat content can then be written as

'unfree energy' $= TS$

where S = entropy and T = absolute temperature.

The 'free energy' of the system is also called the Gibbs Function and is given the symbol, G:

$$G = H - TS$$

All the factors that have been discussed in connection with the heat content of a system are summarized in the two bar-charts below:

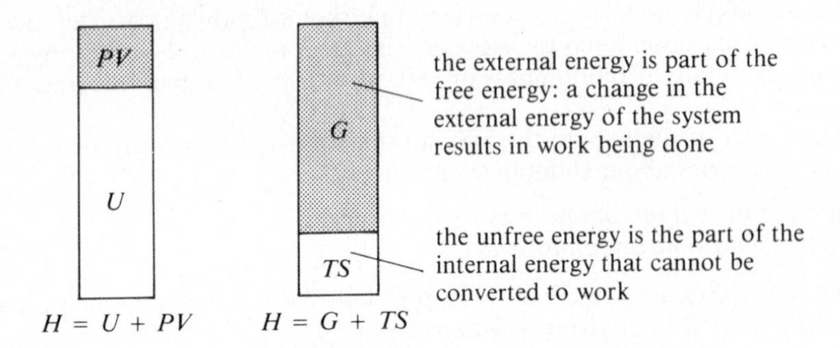
PV
U
G
TS
the external energy is part of the free energy: a change in the external energy of the system results in work being done
the unfree energy is the part of the internal energy that cannot be converted to work
H = U + PV
H = G + TS

These two different ways of dividing up the heat content of a system are really disguised applications of the first two laws of thermodynamics. The first law has already been given; the second is often given in the following form:

> Heat cannot be completely converted into work without some part of the system undergoing change.

Both laws have been put more frivolously:

law	*application*	*reason*
First law: You can never win: you can only break even	$H = U + PV$ total = internal + external	energy cannot be created, so the maximum work possible from a system is its heat content
Second law: You can only break even at a temperature of absolute zero	$H = G + TS$ total = free + unfree	the total heat content can never be converted to work unless $H = G$: i.e. when $TS = 0$.

(There is a third law of thermodynamics which completes a gloomy trio . . . you can never reach absolute zero!)

The problem of the spontaneous endothermic change of page 220 can now be solved using the relationship $H = G + TS$. Again, bar-charts are helpful to make the difference clear (see page 223).

Although ΔH is positive (an increase in total heat content) ΔG is negative. The free energy of the system has decreased. The reason for this should be clear: there is a large increase in unfree energy because entropy has increased, $S_2 > S_1$. When the actual process is examined, it can be seen that the products are considerably more disordered than the reactants, and that therefore it should be expected that $S_2 > S_1$. The reactants are water and a highly structured crystal lattice of ions. The products are water and a random distribution of hydrated ions in solution.

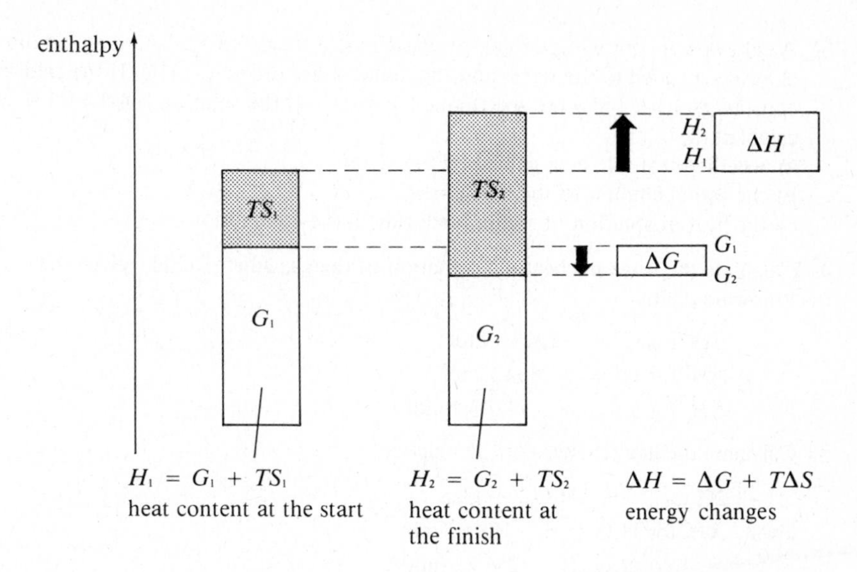

The critical factor in deciding the direction of a change is therefore ΔG. When the free energy change is negative, then the change is spontaneously likely to occur. When the free energy change is positive, it cannot occur spontaneously.

ΔG is negative for all spontaneous changes

Questions

1. 50 cm^3 of hot water at 55°C are added to 50 cm^3 of cold water at 12°C in a calorimeter. The highest temperature reached by the mixture is 31°C. Calculate:
   a) the drop in temperature of the hot water
   b) the heat loss for the hot water
   c) the rise in temperature for the cold water
   d) the heat gained by the cold water
   e) the heat taken in by the calorimeter
   f) the temperature change for the calorimeter
   g) the calorimeter constant.

2. Represent the following reactions on energy diagrams and clearly label them endothermic or exothermic.
   a) $C_{(s)} + O_{2(g)} \longrightarrow CO_{2(g)}$ $\Delta H = -394$ kJ.
   b) $H_2O_{(l)} \longrightarrow H_{2(g)} + \frac{1}{2}O_{2(g)}$ $\Delta H = +286$ kJ.
   c) $N_{2(g)} + 2O_{2(g)} \longrightarrow N_2O_{4(g)}$ $\Delta H = +9.7$ kJ.
   d) $C_{(diamond)} \longrightarrow C_{(graphite)}$ $\Delta H = -1.9$ kJ.

3. A calorimeter containing 25 cm^3 of water is at a steady 18°C. 5.85 g of sodium chloride is added to the water and the temperature drops to 15°C. If the calorimeter constant is 3 J K^{-1} and the specific heat capacity of the solution is 4.2 J g^{-1} K^{-1}; calculate
   a) whether heat is lost or gained by the system
   b) the actual amount of the heat change
   c) the heat of solution of sodium chloride in these conditions.

4. Calculate the standard heat of formation of the gas ethene, C_2H_4, given the following data:

$$\Delta H_c^{\ominus} \text{ for C} = -394 \text{ kJ mol}^{-1}$$
$$\Delta H_c^{\ominus} \text{ for } H_2 = -286 \text{ kJ mol}^{-1}$$
$$\Delta H_c^{\ominus} \text{ for } C_2H_4 = -1428 \text{ kJ mol}^{-1}.$$

5. Calculate the heat change for the reaction

$$2H_2O_{2(l)} \rightarrow 2H_2O_{(l)} + O_{2(g)}$$

given $\Delta H_f^{\ominus}$ for $H_2O_{2(l)} = -188$ kJ mol^{-1}
$H_2O_{(l)} = -286$ kJ mol^{-1}.

6. Calculate the heat change for the reaction

$$2Al_{(s)} + Cr_2O_{3(s)} \rightarrow Al_2O_{3(s)} + 2Cr_{(s)}$$

given $\Delta H_f^{\ominus}$ for $Al_2O_3 = -1669$ kJ mol^{-1}
$\Delta H_f^{\ominus}$ for $Cr_2O_3 = -1128$ kJ mol^{-1}.

7. Calculate the heat change for the reaction

$$Br_{2(g)} + H_{2(g)} \rightarrow 2HBr_{(g)}$$

given the following bond enthalpies:

H – H $\Delta H = +436$ kJ mol^{-1}
Br – Br $\Delta H = +193$ kJ mol^{-1}
H – Br $\Delta H = +366$ kJ mol^{-1}

8. Calculate the heat change for the reaction

$$CH_3OH_{(g)} + HCl_{(g)} \rightarrow CH_3Cl_{(g)} + H_2O_{(g)}$$

given the following bond enthalpies:

C – H $\Delta H = +412$ kJ mol^{-1}
C – O $\Delta H = +360$ kJ mol^{-1}
O – H $\Delta H = +463$ kJ mol^{-1}
H – Cl $\Delta H = +431$ kJ mol^{-1}
C – Cl $\Delta H = +338$ kJ mol^{-1}

9. Using a data book, construct a Born-Haber cycle and hence calculate the lattice energy for a) potassium chloride, b) silver chloride.

   Theoretical values for the lattice energies of these two solids calculated using the electrostatic model are: KCl −692 kJ mol^{-1}; AgCl −768 kJ mol^{-1}. Compare your figures with these and explain any apparent discrepancies.

10. Discuss why certain chemical reactions occur spontaneously while others do not. Use as examples in your answer i) the rusting of iron, ii) the rotting of wood, iii) photosynthesis.

Chapter 15
The rates of chemical reactions

15.1 Measuring rates

The study of the rate of chemical change is known as *kinetics*. It provides an alternative description of a reaction's profile to that given by the *thermodynamics* of the reaction (Chapter 14).

The word 'rate' can be applied to any process in which there is a measurable change taking place during an interval of time. For example, the rate of a man's travel is expressed as the change in his position (measured by the distance moved) in a unit of time. The rate of a chemical change is interpreted for the reaction A → B in one of two ways:

1. the decrease in reactant concentration [A] per unit time, or
2. the increase in product concentration [B] per unit time

(where the square brackets [ ] mean 'concentration in mol dm^{-3} of').

> The rate of a reaction is a measure of the amount of reactants being converted to products in a unit of time.

The symbol d/dt is a useful means of expressing rate: dx/dt ('dx by dt') means the change in x with respect to time t. So, for a chemical change, rate is given by either

1. $\frac{-d[A]}{dt}$ (the negative sign indicates [A] is decreasing), or

2. $\frac{d[B]}{dt}$

When measuring a rate, it is usual to measure the concentration of a reactant or a product at a number of time intervals. This can be done in a variety of ways. Three common methods are described below: measurement of the volume of a gaseous product, of the optical properties, or by titration.

The volume of a gaseous product

The rate of the reaction between calcium carbonate and acid is followed by recording the volume of carbon dioxide at a number of reaction times.

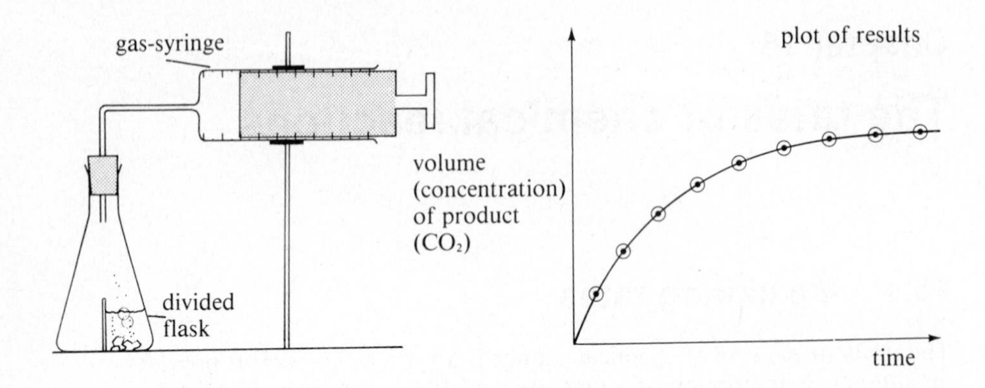

$$CaCO_{3(s)} + 2H_3O^+_{(aq)} \longrightarrow Ca^{2+}_{(aq)} + 3H_2O_{(l)} + CO_{2(g)}$$

A measure of the rate at any particular reaction time is given by the slope of the concentration—time curve.

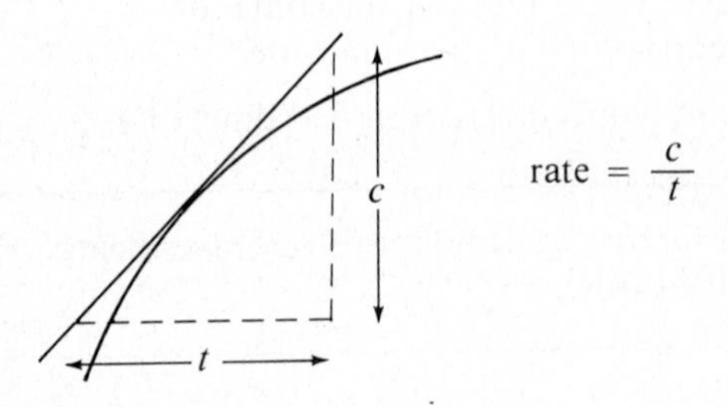

$$\text{rate} = \frac{c}{t}$$

The slope expresses the amount of product being formed in a particular interval of time.

The optical properties of a reactant or a product

Some reactions in solution bring about a change in an optical property of the system. For example,

a) the product may be a different colour from the reactant;
b) the product may be a precipitate, blocking the passage of light through the system;
c) either a reactant or a product may be 'optically active' and so the plane of polarized light shone through the system might be slowly rotated; see page 549.

The reactant concentration at different reaction times is measured by using an instrument that can detect the change in colour, intensity or polarization of light passing through the system (e.g. colorimeters, polarimeters). A plot of reactant concentration against time can be made from the results obtained. The plot has the reverse shape from a curve of product concentration against time because the reactants are being used up.

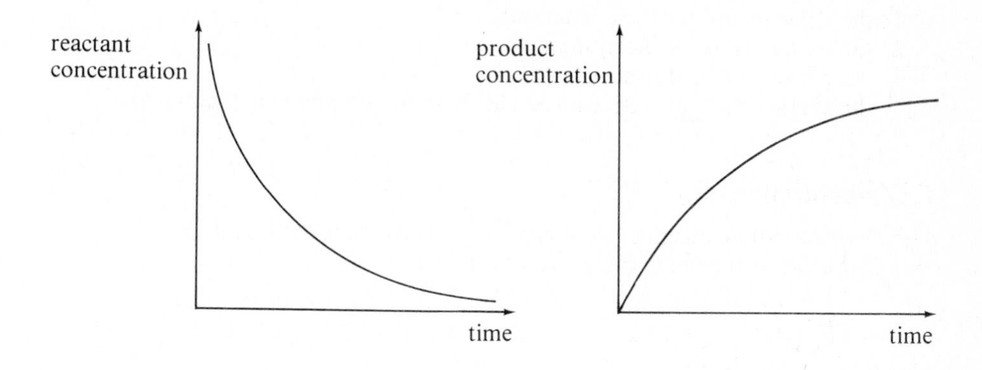

Titrimetric methods

For many reactions in solution, the methods discussed so far cannot be applied. Instead, a small sample must be taken out of the reaction mixture at various times and analysed separately for a particular reactant or product. The major problem with this method is that the reaction continues in the sample volume after it has been removed from the mixture. To overcome this, the sample volume is either

a) added to a large volume of cold solvent to slow the reaction down, or
b) added to a known excess of 'quenching reagent'.

A 'quenching reagent' is a substance that reacts immediately with one of the reactants. By analysing the amount of quenching reagent left in the sample volume after addition, the original amount of reactant can be found. A typical case is described below:

$$\underset{\text{methyl ethanoate}}{CH_3COOCH_3} + \underset{\text{sodium hydroxide}}{NaOH} \longrightarrow \underset{\text{sodium ethanoate}}{CH_3COONa} + \underset{\text{methanol}}{CH_3OH}$$

1. Mix a known amount of the two reactants and start the clock.
2. At a number of different reaction times, use a pipette to remove a small sample from the reaction mixture.
3. Add each sample to a known excess of dilute acid. The acid reacts immediately with the alkali reactant present in the sample.
4. Titrate the quenched sample against standard alkali and find out the amount of acid in excess in each sample.
5. By subtraction, this gives the amount of acid needed to react with the alkali present and therefore the concentration of alkali reactant at each sample time.

15.2 Factors affecting the rate of a reaction

Using any of the methods described in the last section, it is possible to show experimentally that the rate of a reaction depends on:

1. the concentration of the reactants
2. the temperature of the system
3. the presence of catalysts
4. the surface area of reactants or catalysts (heterogeneous reactions).

1. Concentration

The characteristic concentration-time curves show that the larger the concentration of reactant, the greater is the rate:

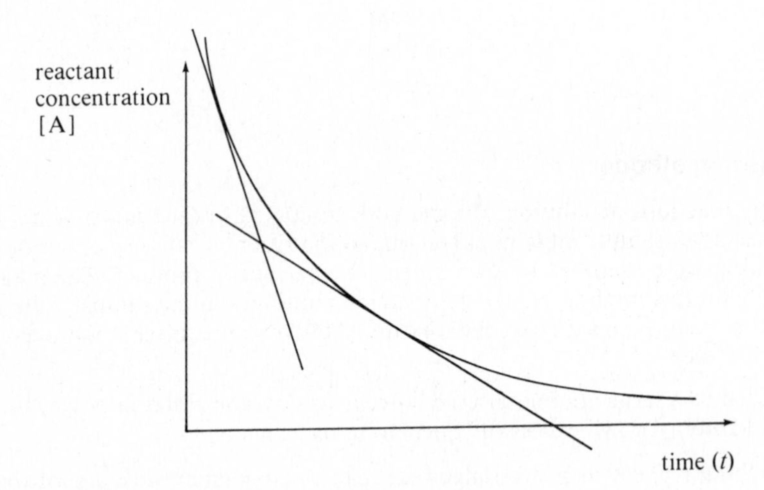

the slope gets flatter as the reaction time increases:

$$\frac{-d[A]}{dt} \text{ is largest at time } t = 0$$

and

$$\frac{-d[A]}{dt} \rightarrow 0 \text{ as } t \rightarrow \infty$$

The relationship between the rate of reaction and a reactant concentration is studied more precisely by the following method:

1. Set up a number of different reaction mixtures so that each mixture has a different known starting concentration of one of the reactants.
2. Keep all the other reactant concentrations constant and keep the temperature constant for each mixture.

3. By some suitable means, determine the initial rate of the reaction in each case.
4. Compare the initial rates of reaction with the initial reactant concentrations in each case.

The rate of the reaction between thiosulphate and acid can be studied by this method. One of the products is sulphur, which is slowly precipitated as a colloidal suspension. The initial rate of a particular reaction mixture is measured by the length of time it takes for sufficient sulphur to be produced so that a mark can no longer be seen through the reaction beaker.

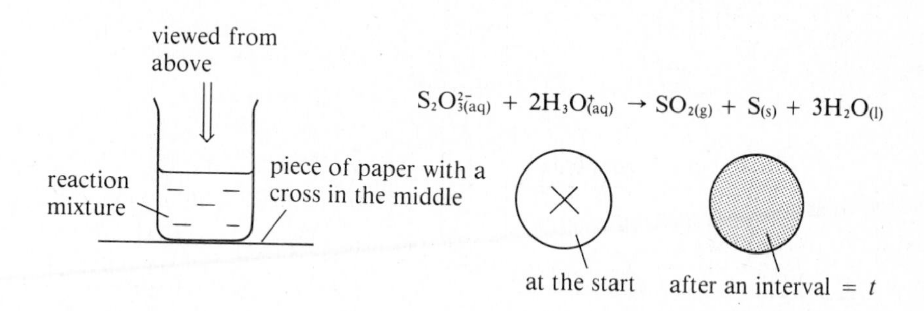

The larger the value of t, the slower is the rate of the reaction.

Clock reactions

Clock reactions are further examples of the application of this method. A typical example is the reaction of iodate, iodide and acid in the presence of a small, measured amount of thiosulphate and starch. The equation for the reaction is

$$IO^-_{3(aq)} + 5I^-_{(aq)} + 6H_3O^+_{(aq)} \longrightarrow 3I_{2(aq)} + 9H_2O_{(l)}$$

1. A number of different reaction mixtures are set up in which the concentration of one reactant (e.g. iodide) is varied while the concentration of the others is held constant.
2. The same small amount of thiosulphate and starch solution is added to each mixture. The thiosulphate is a 'quenching' reagent (see page 227): it reacts with the product, iodine, removing this product from the system

$$\underset{\text{reaction product}}{I_{2(aq)}} + \underset{\text{quenching reagent}}{2S_2O^{2-}_{3(aq)}} \longrightarrow 2I^-_{(aq)} + S_4O^{2-}_{6(aq)}$$

3. When a particular concentration of iodine has been produced, the quenching agent is used up. At this moment, the solution suddenly turns blue because iodine is detected by the added starch for the first time in the reaction mixture (starch and iodine give a blue complex).
4. The rate of reaction is inversely proportional to the time taken for the blue colour to appear (the 'clock time').

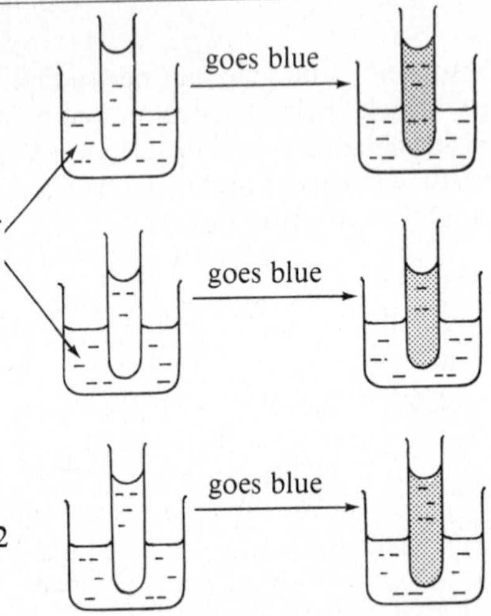

initial concentration (mol dm^{-3}) of potassium iodide	*clock-time t (seconds)*	[I$^-$] (mol dm^{-3})	1/*t* (s^{-1})
0.02 (goes blue)	10	2×10^{-2}	10×10^{-2}
0.01 (goes blue)	20	1×10^{-2}	5×10^{-2}
0.002 (goes blue)	100	0.2×10^{-2}	1×10^{-2}

water bath

The results show that the rate ∝ [I$^-$]. The method can be repeated for each reactant in turn.

Gas-phase reactions

In gas-phase reactions, it is often more convenient to measure the concentration of a reactant or product as a partial pressure (page 130). However, the same general observations are still made: an increase in the partial pressure of a reactant usually leads to an increase in the reaction rate. For many gas-phase reactions, it is possible to follow the rate of reaction by recording the *total* pressure of the system. For example, in the pyrolysis of ethanal.

$$CH_3CHO_{(g)} \longrightarrow CH_{4(g)} + CO_{(g)}$$

one mole of reactant becomes two moles of product. Therefore, at constant volume, since the pressure of a gaseous system is proportional to the number of moles present, the rate of increase of pressure in the above system equals the rate of reaction.

2. The temperature of the system

Raising the temperature of a reaction always leads to an increase in the reaction rate. The increase can by demonstrated experimentally by either of the two reactions discussed on page 229: a sharp decrease in clock-time is observed when the temperature increases.

3. Catalysts

A catalyst is a substance that increases the rate of a chemical reaction but is left chemically unchanged at the end of the reaction.

Catalysts tend to be fairly specific in their action: a substance may catalyse one reaction and not another. Some examples of catalysis are:

a) The increase in rate of decomposition of hydrogen peroxide by adding blood.
b) The increase in rate of oxidation of methanol vapour in the presence of a hot platinum spiral.

In the first example, the catalyst is in the same phase as the reactant: it is a *homogeneous catalyst*. In the second, it is in a different phase: it is a *heterogeneous catalyst*.

Certain substances, while not catalysing the reaction themselves, may increase the effect the catalyst has on the rate of the reaction. These substances are called *promoters*. An example is the addition of metal oxides to the iron used as a catalyst in the Haber process.

4. The surface area of reactants or catalysts (heterogeneous reactions)

The explosion of coal dust is a major hazard of the mining industry. The increased surface area of the solid reactant causes a large increase in the rate of the solid-gas reaction. The effect of increased surface area is easily shown in the laboratory, for example:

a) Compare the rate of evolution of carbon dioxide from lumps of calcium carbonate with that from powdered calcium carbonate.
b) Compare the rate of pyrolysis of ethanal in a silica vessel packed with glass tubing with the rate when the vessel contains only ethanal.

15.3 The collision theory

A successful theory of reaction rates must be able to explain the two major factors that affect rates:

1. The rate of a reaction generally increases with an increase in reactant concentration; but also there are some reactants whose concentrations do not appear to alter the rate of certain reactions.
2. The rate of a reaction increases with an increase in temperature; but the extent of the increase changes from reaction to reaction.

The collision theory is the most widely accepted theory of reaction rate. It developed from the kinetic theory of gases.

Major postulates of the collision theory

The major postulates are:

1. Reactant particles must collide in order to react.
2. Not every collision results in a reaction: there must be a certain minimum energy in the collision to bring about the necessary reorganization of the bonds in the colliding particles.
3. The colliding particles must be correctly orientated with respect to each other in the collision, otherwise the new bonds that are likely to be formed may not have the chance of forming.

Some of these ideas are illustrated below for collisions between molecules of hydrogen and iodine:

$$H_{2(g)} + I_{2(g)} \longrightarrow 2HI_{(g)}$$

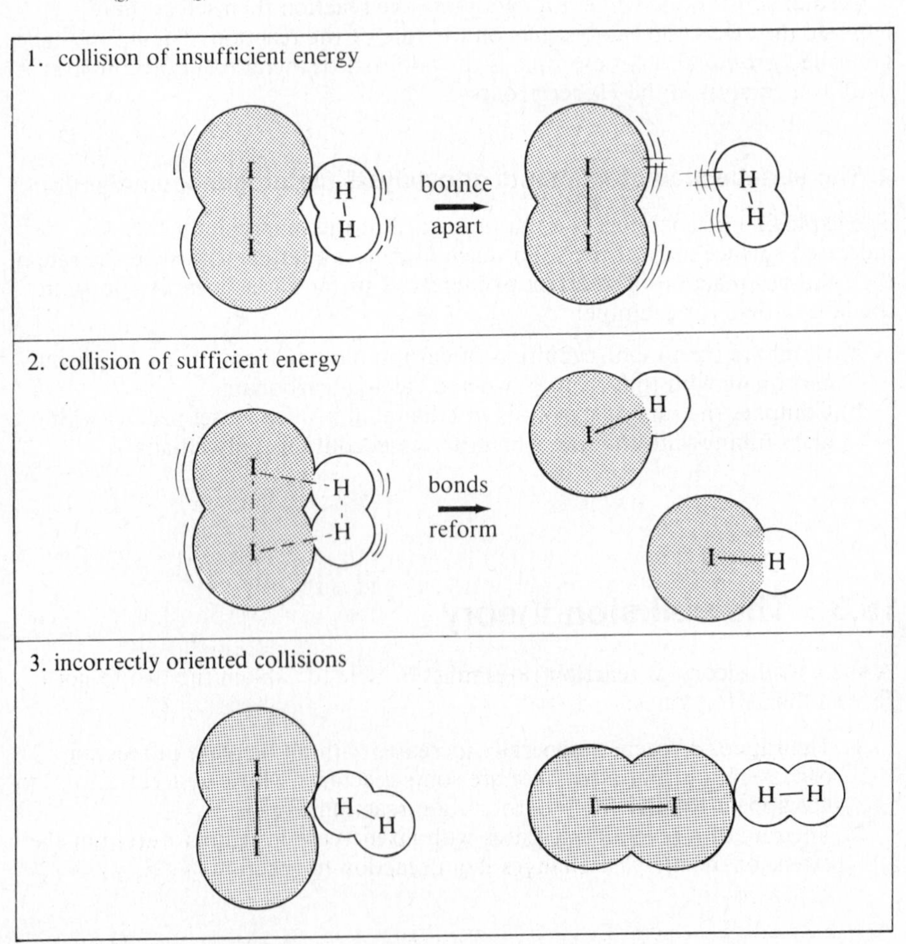

Concentration, temperature and rate

As reactant concentration increases, the frequency of collisions increases and so therefore does the frequency of collisions having sufficient energy to cause reaction.

To explain the temperature dependence of rate, a closer look at the distribution of energy within the system is necessary. Temperature is a measure of the mean kinetic energy of the particles in the system. On page 133, the distribution of molecular speeds in a gas is described. Molecular energies must follow a squared form of this distribution function because K.E. $= \frac{1}{2} mv^2$.

It is therefore likely that the energies of collision between particles will also be governed by this squared function. It is called the *Maxwell-Boltzmann distribution* and is shown below for the collision frequencies (number of collisions per second) at two different temperatures T_1 and T_2. T_1 is greater than T_2 and therefore the mean energy of the collisions is higher.

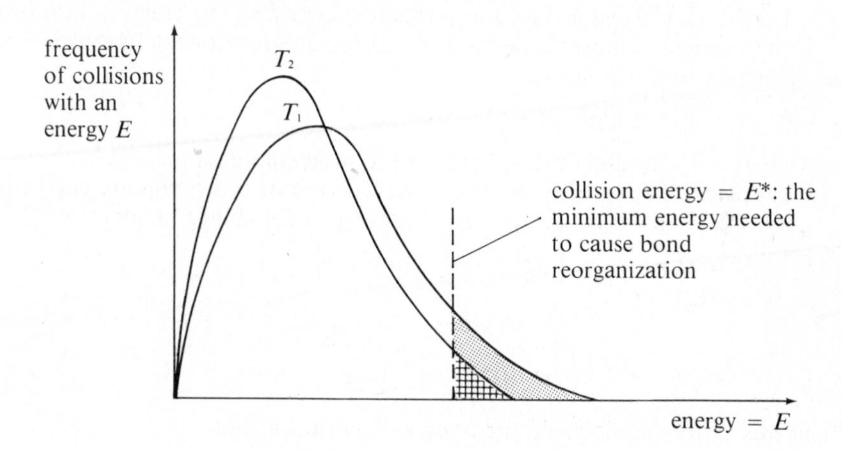

1. The total area enclosed by each curve ≡ the total number of collisions taking place per second.
2. The hatched area represents the fraction of the collisions at the lower temperature that have sufficient energy to lead to reaction.
3. The shaded area represents the fraction of the collisions at the higher temperature that have sufficient energy to lead to reaction.
4. There is a marked increase in area and therefore there is likely to be a marked increase in the rate of the reaction as the temperature increases.

The rate of reaction increases

a) *with reactant concentration*, because the collision frequency increases,
b) *with temperature*, because the fraction of collisions with sufficient energy increases.

The rate constant

A mathematical expression for the rate of a reaction based on the collision theory can be obtained from the assumption below:

$$\text{rate of reaction} = \text{number of effective collisions per second} \quad \text{in a unit volume of the system}$$

If Z = the number of collisions per second between one molecule of reactant A and one molecule of reactant B in a unit volume of the system, N_A = the number of A molecules present in the unit volume and N_B = the number of B molecules present in the unit volume then

$$\text{total collision frequency per unit volume} = ZN_AN_B$$

However only *some* of these collisions are effective:

i) They must possess a certain minimum energy E^*. The fraction that have this energy (or more) is derived from the equation for the Maxwell-Boltzmann distribution:

$$\text{This fraction} = e^{-E^*/kT}$$

where k = the Boltzmann constant, T = absolute temperature.

ii) They must be correctly oriented; the fraction of those that are correctly oriented is given by P, the steric factor (or probability factor) $0 < P < 1$.

$$\text{rate of reaction} = \underset{\text{fraction with correct orientation}}{P} \times \underset{\text{fraction with sufficient energy}}{e^{-E^*/kT}} \times \underset{\text{collision frequency}}{\{Z \times N_AN_B\}}$$

When this expression is given for a *mole* of each reactant

$$N_A \text{ and } N_B \equiv \text{the concentration of the reactants}$$

and $e^{-E^*/kT}$ becomes $e^{-E_A/RT}$, where R = the ideal gas constant and E_A is called the *activation energy* of the reaction, i.e.

$$\text{rate} = PZe^{-E_A/RT} \text{ [concentration factors]}$$

At constant temperature, $PZe^{-E_A/RT}$ is a constant for a particular reaction: it is known as the rate constant for the reaction and is given the symbol k (*not* the same as the Boltzmann constant).

$$\text{rate constant, } k = PZe^{-E_A/RT}$$

This equation was derived experimentally in 1889 by Arrhenius and is named after him. The two factors PZ are given as a single constant A called the pre-exponential factor; all these constants are summarized in the table on the next page.

$k = Ae^{-E_A/RT}$	*the Arrhenius equation*
k = the rate constant	the constant of proportionality that relates the rate of reaction to the concentrations of reactants present at a particular temperature
A = the pre-exponential factor	the constant interpreted by collision theory as the product of the steric factor P and the collision number Z
E_A = the activation energy of the reaction	the minimum collision energy (per mole of collisions) that brings about a reaction between two colliding reactant particles

Activation energy of a reaction

The activation energy of a reaction is often interpreted in a different way by looking at the energy content of the whole system. Consider the hydrogen and iodine reaction again. Three different stages of the reaction can be identified:

I—I H—H	—collision of sufficient energy→	I---H over I---H (dashed bonds between them)	——→	I—H I—H
reactants		transition-state or activated complex		products

The energy of the activated complex is higher than that of the reactants. An activated complex can only be produced as a result of high-energy collisions between the reactant particles. When shown on an energy diagram, an 'energy profile' of the reaction is produced:

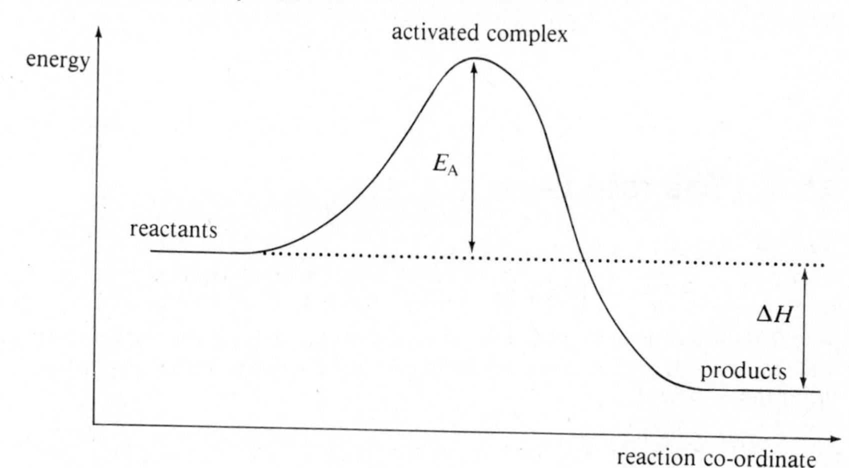

The activation energy represents an energy barrier that the reactants must get over in order to become products.

Catalysts and activation energy
From the Arrhenius equation, $k = Ae^{-E_A/RT}$, it can be seen that a large activation energy leads to a small rate constant and therefore a slow reaction. A catalyst works by lowering the activation energy of a particular reaction. It can do this in one of two main ways:

1. The same activated complex is formed, but the energy required to form it is reduced by the presence of the catalyst, e.g. most heterogeneous catalysts.
2. A different activated complex of lower energy is formed in the presence of the catalyst, e.g. most homogeneous catalysts.

Both these processes are discussed on page 246. Their effect on the reaction profile is the same:

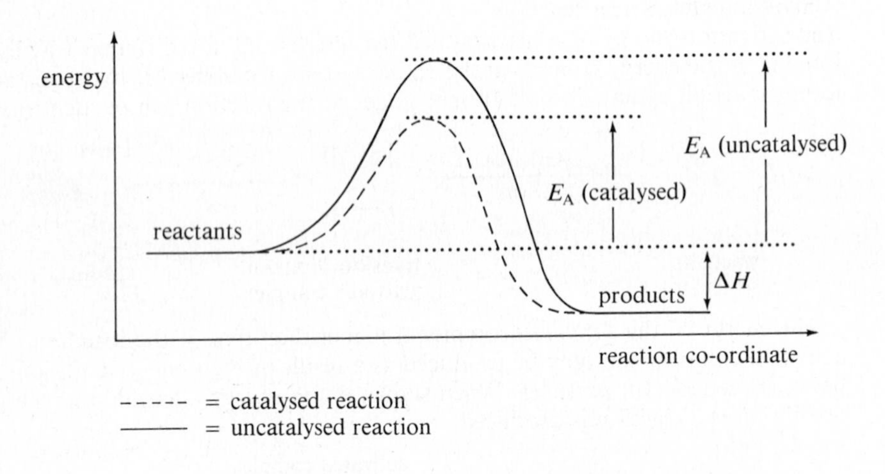

15.4 The rate laws

For the reaction

$$A + B \longrightarrow \text{products}$$

at constant temperature, the rate of reaction depends on the concentrations of the reactants A and B. This dependence can be written in the form of an equation:

$$\text{rate} = k\,[A]^x\,[B]^y$$

An equation of this sort is called the *rate equation* for the reaction and k is the *rate constant* (or the *reaction velocity constant*). In most cases the powers x and y turn out to be 0 or 1. If they are both 1, the equation becomes

$$\text{rate} = k\,[A]\,[B]$$

which means that the rate is directly proportional to the concentration of each reactant.

It is usually not possible to predict the values of these powers. They have to be found by experiment using the methods that are described earlier in the chapter. A special name is given to the sum of these powers: it is called the *order* of the reaction. In the case where x and y both equal 1, the order of the reaction is 2: it is a 'second-order' reaction. Sometimes it is found that the reaction rate remains unaltered even though the concentration of one of the reactants is increased. In a case like this, the power to which the particular reactant concentration is raised in the rate equation is equal to zero, e.g.

$$\text{rate} = k\,[A]^1\,[B]^0$$
$$= k\,[A]^1 \qquad \text{(rate is not altered by changes in [B] : } [B]^0 = 1\text{).}$$

The order of a reaction of this sort = (1 + 0) = 1; it is a 'first-order' reaction.
Some first-order and second-order reactions are given in the table below.

order	*equation*	*rate law*
first	radioactive decay: $^{14}_{6}C \longrightarrow ^{14}_{7}N + ^{0}_{-1}e$	rate = $k\,[^{14}_{6}C]^1$
	$2N_2O_{5(g)} \longrightarrow 4NO_{2(g)} + O_{2(g)}$	rate = $k\,[N_2O_5]^1$
	$CH_3COCH_{3(aq)} + Br_{2(aq)} \longrightarrow CH_3COCH_2Br_{(aq)} + HBr_{(aq)}$	rate = $k\,[CH_3COCH_3]^1\,[Br_2]^0$
second	$H_{2(g)} + I_{2(g)} \longrightarrow 2HI_{(g)}$	rate = $k\,[H_2]^1\,[I_2]^1$
	$3ClO^-_{(aq)} \longrightarrow ClO^-_{3(aq)} + 2Cl^-_{(aq)}$	rate = $k\,[ClO^-]^2$
	$HCOOCH_{3(aq)} + HO^-_{(aq)} \longrightarrow HCOO^-_{(aq)} + CH_3OH_{(aq)}$	rate = $k\,[HCOOCH_3]^1[HO^-]^1$

The order of a reaction is the sum of the powers to which the reactant concentrations are raised in the rate equation.

It is important to remember that the order can be determined only by experiment and is not necessarily related to the stoichiometry of the reaction.

First order reactions

If c is the concentration of the reactants at time = t, the first order reaction rate is given by the expression

$$\frac{-dc}{dt} = kc^1$$

$-dc/dt$ means the decrease in reactant concentration with respect to time (see page 225). This is a differential equation which must be integrated to clear the differentials dc and dt. The integrated form is

$$\log_e\left(\frac{c_0}{c}\right) = kt$$

where c_0 is the reactant concentration when a particular reading is taken (i.e. at time t = 0) and c is the reactant concentration after an interval t has passed.

A reaction whose rate obeys the first-order law gives results that fit these equations.

differential form	*integrated form*
$\frac{-dc}{dt} = kc$	$\log_e\left(\frac{c_0}{c}\right) = kt$
	c_0 = a chosen initial reactant concentration
	c = the reactant concentration after a time t

For example, the graph of reactant concentration against time for a typical first order equation has the shape shown below.

first order plot

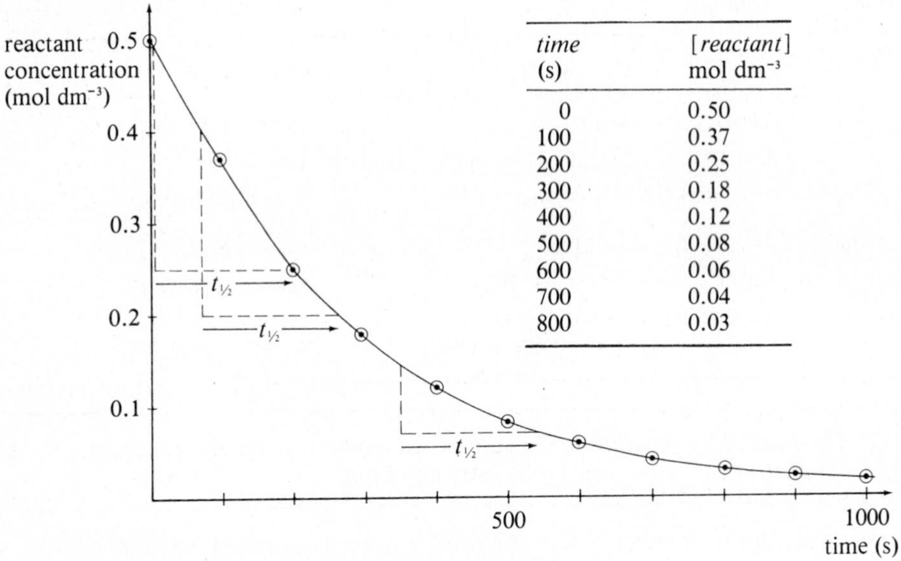

time (s)	[*reactant*] mol dm^{-3}
0	0.50
100	0.37
200	0.25
300	0.18
400	0.12
500	0.08
600	0.06
700	0.04
800	0.03

The curve has a special property: *the 'half-life' is constant.*

A half-life is the time taken for the reactant concentration to fall to half of its initial value. Its symbol is $t_{1/2}$.

The half-life of the reaction illustrated by the above curve is 200 seconds no matter what value of initial reactant concentration is chosen.

This important result can also be derived by substituting the half-life condition into the integrated rate equation. By definition,

$$t = t_{1/2} \text{ when } c = c_0/2$$

and for a first order reaction

$$\log_e\left(\frac{c_0}{c}\right) = kt$$

$$\Rightarrow \log_e\left(\frac{c_0}{c_0/2}\right) = kt_{1/2}$$

$$\Rightarrow \log_e 2 = kt_{1/2}$$

or $$t_{1/2} = \frac{0.693}{k}, \text{ a constant}$$

If the half-life of a first order reaction is known, its rate constant can be calculated from the above relationship. A more rigorous way of showing that a given set of results fits the first order equation is described below.

<table>
<tr><th>procedure</th><th>working</th></tr>
<tr><td>1. tabulate the values of c and t and work out $\log_{10}c$ in each case.
2. plot $\log_{10}c$ against t. If the plot is a straight line, the results fit the first order equation and the gradient is $-k/2.303$.</td><td><table><tr><th>t (s)</th><th>c (mol dm^{-3})</th><th>$\log_{10}c$</th></tr><tr><td>0</td><td>0.50</td><td>−0.30</td></tr><tr><td>100</td><td>0.37</td><td>−0.43</td></tr><tr><td>200</td><td>0.25</td><td>−0.60</td></tr><tr><td>300</td><td>0.18</td><td>−0.74</td></tr><tr><td>400</td><td>0.12</td><td>−0.92</td></tr></table></td></tr>
<tr><td>this follows from the integrated rate equation,
$\log_e\left(\frac{c_0}{c}\right) = kt$
$\log_{10} c = \frac{-k}{2.303} t + \log_{10} c_0$
the above equation is of the form
y = (gradient) x + intercept</td><td>$\log_{10}c$
slope = $\left(\frac{-0.3}{200}\right)$ s^{-1}
−0.5
0.3
200
−1.0
100 200 300 400 500
time (s)</td></tr>
</table>

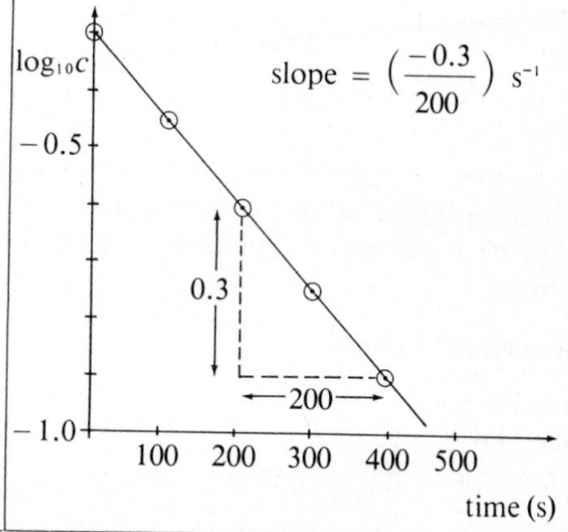

Either the measured constant half-life or the value of the slope of the log plot confirms that the reaction is first order. The rate constant can be worked out in both cases.

from half-life	*from slope*
$kt_{1/2} = \log_e 2$	slope $= -k/2.303 = -0.3/200$
$k = \left(\frac{0.693}{200}\right)$ s^{-1}	$k = \left(\frac{0.3 \times 2.303}{200}\right)$ s^{-1}
$k = 3.47 \times 10^{-3}$ s^{-1}	$k = 3.46 \times 10^{-3}$ s^{-1}

Second order reactions

A second order reaction has a rate equation of the form,

$$\frac{-dc}{dt} = kc^2$$

under the following conditions:

1. The concentrations of the different reactants are equal at the start of the reaction.
2. One mole of one reactant reacts with only one mole of the other reactant (i.e. $A + B \longrightarrow$ products).

This is another differential equation (see page 237) and must be integrated to clear the differentials dc and dt. The integrated form is

$$\frac{1}{c} - \frac{1}{c_0} = kt$$

where c_0 is the reactant concentration when an initial reading is taken, i.e. at time $t = 0$. A reaction whose rate obeys the second-order law gives results that fit these equations.

differential form	*integrated form*
$\frac{-dc}{dt} = kc^2$	$\frac{1}{c} - \frac{1}{c_0} = kt$ c_0 = a chosen initial reactant concentration c = the reactant concentration after a time t

The graph of reactant concentration against time for a second-order reaction appears to have the same shape as the one for a first-order reaction (see page 238).

second-order plot

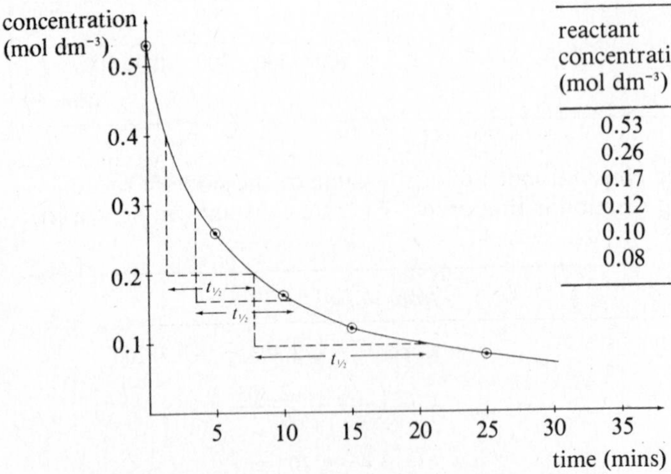

concentration (mol dm^{-3})
0.5
0.4
0.3
0.2
0.1
$t_{1/2}$
$t_{1/2}$
$t_{1/2}$
5 10 15 20 25 30 35
time (mins)

experimental results

reactant concentration (mol dm^{-3})	time (min)
0.53	0
0.26	5
0.17	10
0.12	15
0.10	20
0.08	25

However, a closer inspection shows that $t_{1/2}$ is *not* a constant but that $c_0 t_{1/2}$ is.

The figures in the following table are taken from the graph for various initial concentrations c_0.

c_0 (mol dm^{-3})	$t_{1/2}$ (min)	$c_0\ t_{1/2}$ (mol dm^{-3} min)
0.40	6.0	2.40
0.30	8.0	2.40
0.20	12.0	2.40

By applying the conditions $t = t_{1/2}$ when $c = c_0/2$ (as on page 238), the following equation is obtained for a second-order reaction:

$$kt_{1/2} = \frac{1}{c_0/2} - \frac{1}{c_0}$$

$$\Rightarrow kt_{1/2} = \frac{2}{c_0} - \frac{1}{c_0} = \frac{1}{c_0}$$

$$k = \frac{1}{c_0 t_{1/2}}$$

or $$c_0 t_{1/2} = \frac{1}{k}, \text{ a constant.}$$

Since $c_0 t_{1/2}$ is constant for the above curve, it can be deduced that the reaction is second order. Also, the rate constant can now be calculated:

$$k = \frac{1}{c_0 t_{1/2}} = \frac{1}{2.40} \Rightarrow k = 0.42 \text{ dm}^3 \text{ mol}^{-1} \text{ min}^{-1}$$

It is also possible to apply the same sort of rigorous approach as the one outlined on page 239. Instead of plotting $\log_{10} c$ against t, $1/c$ is plotted against t.

From the slope of the inverse concentration plot, see page 242:

$$k = \left(\frac{5}{12}\right) \text{dm}^3 \text{ mol}^{-1} \text{ min}^{-1}$$

$$= 0.42 \text{ dm}^3 \text{ mol}^{-1} \text{ min}^{-1}$$

The value agrees with the one worked out from the equation $k = 1/c_0 t_{1/2}$.

<table>
<tr><th>procedure</th><th>working</th></tr>
<tr><td>1. tabulate the values of c and t and work out $1/c$ in each case.
2. plot $1/c$ against t. If the plot is a straight line, the results fit the second order equation and the gradient is k.</td><td>

t (min)	c (mol dm^{-3})	$1/c$ (dm^3 mol^{-1})
0	0.53	1.89
5	0.26	3.76
10	0.17	5.89
15	0.12	8.34
20	0.10	10.00

</td></tr>
<tr><td>this follows from the integrated rate equation,
$$\frac{1}{c} - \frac{1}{c_0} = kt$$
$$\frac{1}{c} = kt + \frac{1}{c_0}$$
the above equation is of the form,
y = (gradient) x + intercept</td><td>slope $= \left(\frac{5}{12}\right)$ dm^3 mol^{-1} min^{-1}
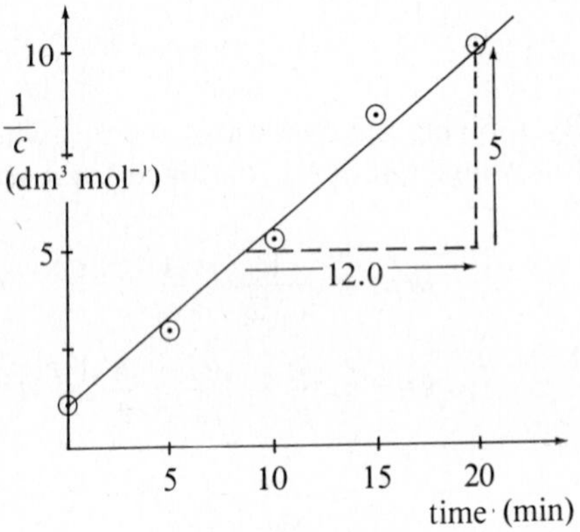
</td></tr>
</table>

Summary of rate equations

There are two principal factors that characterize the rate of a chemical reaction:

i) the rate constant k which is temperature-dependent
ii) the order of the reaction.

When these factors are determined experimentally, the process is known as finding the *kinetics of the reaction*. The procedure for this is summarized below.

1. Measure reactant concentrations c at different reaction times t.
2. Plot reactant concentration against time and calculate a number of half-lives from the curve.
3. If the half-lives are constant, the reaction is first order and can be confirmed as first order by plotting $\log_{10}c$ against t when a straight line of gradient $-k/2.303$ is produced.
4. If the half-lives are not constant, values of $c_0t_{1/2}$ are calculated and these usually turn out to be constant. If they are, the reaction is second-order. A plot of $1/c$ against t can be carried out to check this: a straight line of gradient k is produced.

Activation energy of the reaction
The activation energy of a reaction can be calculated by determining values of the rate constant at different temperatures.

Since $k = Ae^{-E_A/RT}$ (the Arrhenius equation, page 235)

$$\log_e k = \log_e A - \frac{E_A}{RT}$$

thus $$\log_{10} k = \frac{-E_A}{2.303R}\left\{\frac{1}{T}\right\} + \log_{10} A$$

which is of the form,

$$y = (\text{gradient})\,x + \text{intercept}$$

By plotting $\log_{10} k$ against $1/T$ a straight line whose slope is $-E_A/2.303R$ is produced; E_A is calculated from the value of the slope.

15.5 Multi-step reactions

In the reaction

$$A + B \longrightarrow \text{products}$$

simple collision theory suggests that the order of the reaction should be two. The rate is likely to depend on the concentration of A and on the concentration of B, i.e.

$$\text{rate} = k\,[A]\,[B]$$

The term *individual order* is often used to describe the dependence of the rate on a particular reactant concentration. In the above expression, the individual orders are one with respect to A and one with respect to B. The overall order is the sum of the individual orders. For an overall order of two, the reaction is said to be a simple or one-step process and the activated complex is likely to be formed as a result of the collision between an A particle and a B particle.

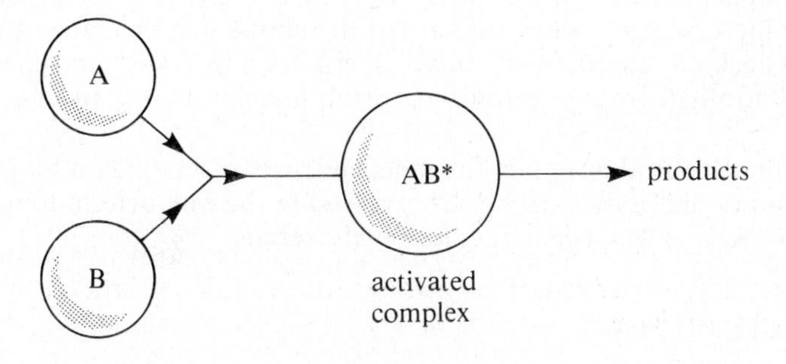

Doubling the concentration of A leads to twice as many collisions between A and B, and therefore twice the rate: rate $\propto [A]^1$. The same is true for B: rate $\propto [B]^1$. Overall, rate $\propto [A]^1 [B]^1$: order = 2.

However, in certain cases, the reaction turns out to have a different order. For example, the iodination of aqueous propanone has an order of one with respect to propanone but is independent of the concentration of iodine: the reaction is zero order with respect to iodine.

$$CH_3\overset{\overset{O}{\|}}{C}CH_3 + I_2 \longrightarrow CH_3\overset{\overset{O}{\|}}{C}CH_2I + HI$$

$$\text{rate} = k\,[CH_3\overset{\overset{O}{\|}}{C}CH_3]^1\,[I_2]^0 \qquad \text{order} = 1$$

When the measured order is not the same as the expected order, the reaction must be taking place in more than one step. The reaction is a complex or multi-step process.

Before the chemical example above is discussed further, consider the rates of some two-step processes that occur in everyday life.

process	*steps*	
shopping at a busy supermarket	1. collecting the items from the shelves	2. checking out at the counter
getting into a football ground	1. passing through the turn-stiles	2. walking to the stands

The time taken to shop at a busy supermarket is often independent of the time taken to collect the necessary items. It may only be necessary to collect one item, but the large queue at the check-out counter makes the whole process take almost as long as the process of doing the whole week's shopping. Similarly, the time taken to get into the stands of a football ground is often independent of the time taken to find a place to watch from, once inside the ground. The long queues to get into the ground control the rate of the whole process. In both these situations it is the slow step which controls the overall rate of the process: this is called the *rate-determining step* and it controls the rate whether it happens at the start of the process or at the end of the process. The overall rate increases only when the rate of the above step increases: for example, more check-out counters in a busy supermarket or fewer people at the turn-stiles of the football ground (a fourth division match rather than first division!).

To return to the chemical example, the independence of the rate on iodine concentration shows that iodine cannot be involved in the rate-determining step. This step must involve propanone, however, because

$$\text{rate} \propto [CH_3\overset{\overset{O}{\|}}{C}CH_3]^1$$

There are two forms of propanone;

'keto' form

'enol' form

and iodine reacts with only one form, the enol form. Under ordinary aqueous conditions, propanone is almost entirely present as its keto-form. The conversion from keto to enol is slow. So the overall reaction happens in the two steps shown below:

1. keto-form $\xrightarrow{\text{slow}}$ enol-form rate $\propto$ [propanone]^1

2. enol-form + iodine $\xrightarrow{\text{fast}}$ products

This process is of the same type as the entry into the football ground: a slow first step that controls the overall rate. In the reaction, *as soon as* some enol-propanone forms, it reacts with iodine to give products. Doubling the concentration of iodine can have little effect on the overall rate because it cannot alter the rate at which the other reactant, the enol-propanone is formed. Doubling the concentration of propanone, however, *does* have an effect: twice the amount of enol-propanone forms and so the reaction goes twice as fast.

The term *molecularity* is also used to classify reactions. In a simple one-step reaction, the molecularity of the reaction is the same as the order of the reaction. In complex processes, however, the two tend to be different. The two terms are illustrated by reference to the reaction above which is *bimolecular* but *first-order* with respect to propanone.

term	*definition*	*example*
order	the order of a reaction is equal to the number of particles taking part in the rate determining step	keto $\longrightarrow$ enol one particle *1st order*
molecularity	the molecularity of a reaction is equal to the total number of particles needed to form the activated complex from which the products are given	*bimolecular* enol + I_2 give the activated complex from which the products form

Homogeneous and heterogeneous catalysis are both processes that are multi-step; they provide further examples to help clarify the terms rate-determining step, individual order, order and molecularity.

Heterogeneous catalysis

The decomposition of ammonia is catalysed by a hot tungsten wire. The process is heterogeneous because the catalyst is in a different phase from the reactant (see page 231).

$$2NH_3 \xrightarrow{W_{(s)}} N_{2(g)} + 3H_{2(g)}$$ (W is the symbol for the element tungsten)

Three likely steps can be put forward for any heterogeneous catalysis.

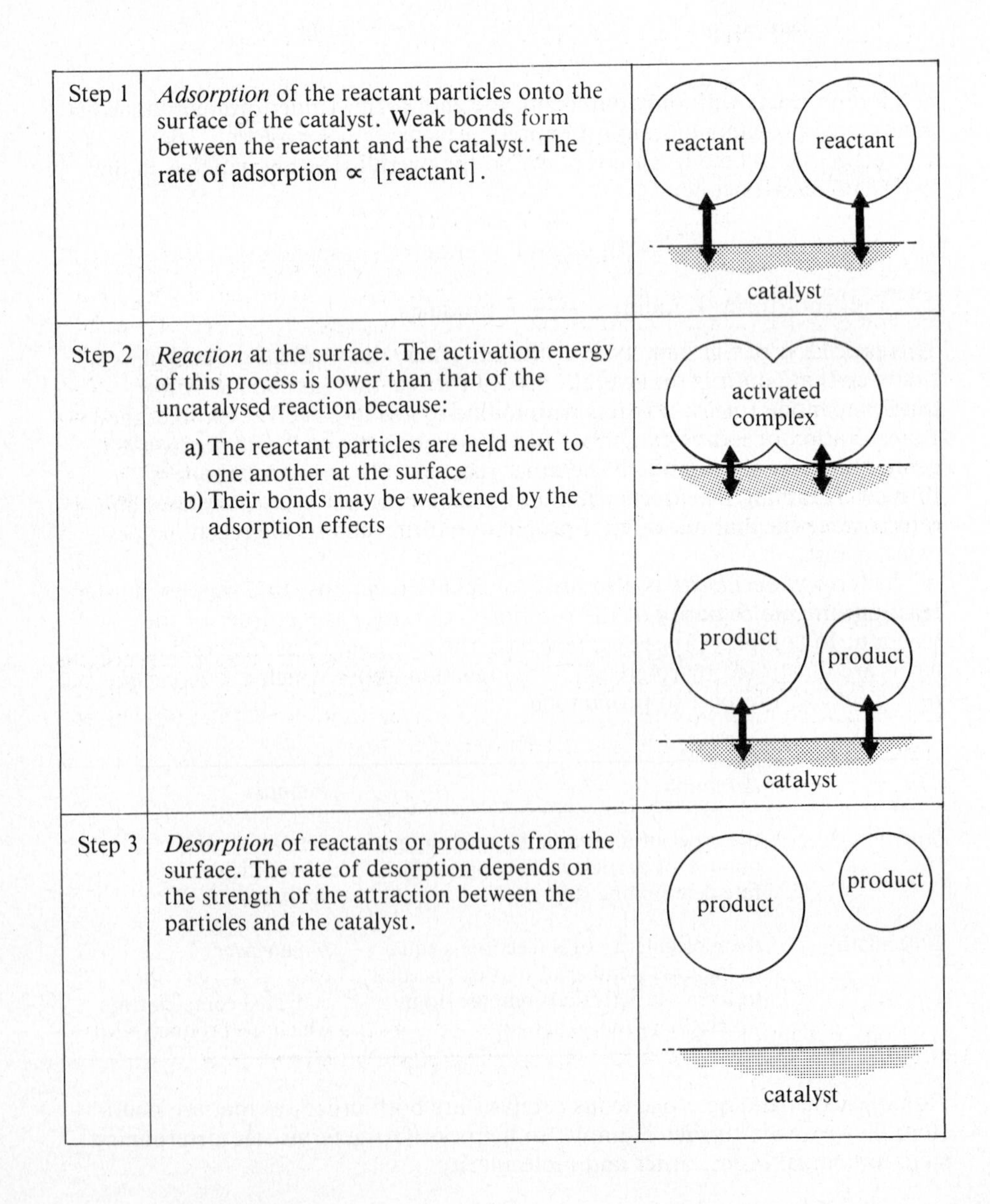

Step 1	*Adsorption* of the reactant particles onto the surface of the catalyst. Weak bonds form between the reactant and the catalyst. The rate of adsorption ∝ [reactant].	
Step 2	*Reaction* at the surface. The activation energy of this process is lower than that of the uncatalysed reaction because: a) The reactant particles are held next to one another at the surface b) Their bonds may be weakened by the adsorption effects	
Step 3	*Desorption* of reactants or products from the surface. The rate of desorption depends on the strength of the attraction between the particles and the catalyst.	

The rate of decomposition is found to be 2×10^{-3} atm s^{-1} at 1030 K when the initial pressure is 0.2 atm of ammonia. The *same* initial rate is found when the initial reactant pressure is raised to 0.3 atm and again to 0.4 atm; the reaction rate is independent of the partial pressure of ammonia, i.e.

$$\text{rate} = k[NH_3]^0 = \text{constant} = 2 \times 10^{-3} \text{ atm s}^{-1}$$

The reaction is *zero* order. The reaction must be a multi-step process in which the rate-determining step does not depend on the concentration of ammonia.

The rates of the three steps are therefore likely to be as follows:

1. *Adsorption* $NH_{3(g)} \xrightarrow[\text{fast}]{W_{(s)}} NH_{3(W)}$ Rate $\propto [NH_3]$
2. *Reaction* $2NH_{3(W)} \xrightarrow[\text{fast}]{} N_{2(W)} + 3H_{2(W)}$ Rate $\propto [NH_3]^2$
3. *Desorption* $N_{2(W)} + 3H_{2(W)} \xrightarrow[\text{slow}]{}$ products + clean surface; $NH_{3(W)} \longrightarrow NH_{3(g)}$ + clean surface Rate is *not* $\propto [NH_3]$

The desorption is rate-determining. The surface is completely covered by either reacting particles or products: as soon as some particles leave the surface, they are immediately replaced by more reactant ammonia molecules. No ammonia can react, however, until a vacancy at the surface appears.

It is interesting to note that the decomposition of phosphine under the same conditions is first order. In this case, the rate of adsorption is rate-determining. There are vacancies at the catalyst surface that are waiting to be filled: their rate of filling is proportional to the partial pressure of phosphine. In both cases, the reaction is *bimolecular* because it takes two molecules to produce the activated complex at the catalyst surface.

Homogeneous catalysis

The iodination of propanone is catalysed by the addition of acid. The process is homogeneous because the catalyst is in the same phase as the reactants. It is found that the reaction is second-order in acidic conditions, but the individual order with respect to iodine is still zero (see page 245).

$$CH_3COCH_3 + H_3O^+ + I_2 \longrightarrow CH_3COCH_2I + H_3O^+ + HI$$

$$\text{rate} = k\,[CH_3COCH_3]^1\,[H_3O^+]^1\,[I_2]^0 \qquad \text{Order} = 2$$

The acid must speed up the rate of the rate-determining step: the conversion of the keto-form to the enol-form. A possible method is shown overleaf.

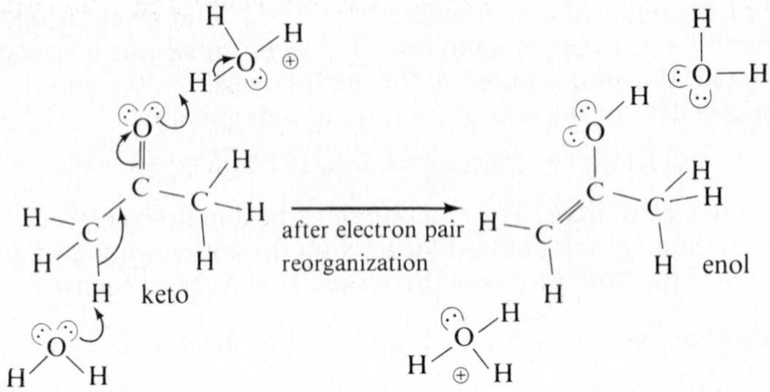

keto

after electron pair reorganization

enol

1. keto + acid $\xrightarrow{\text{slow}}$ enol + acid rate $\propto$ [propanone]^1 [acid]^1
2. enol + iodine $\xrightarrow{\text{fast}}$ products

Homogeneous catalysts sometimes work by providing a different route of lower activation energy from the uncatalysed process. For example, the reaction

$$Fe^{3+}_{(aq)} + V^{3+}_{(aq)} \longrightarrow Fe^{2+}_{(aq)} + V^{4+}_{(aq)}$$

is first-order with respect to both iron (III) ions and vanadium (III) ions. In the presence of copper (II) ions, however, the reaction proceeds much more rapidly and is zero-order with respect to iron (III) ions. An explanation for this is suggested by the following reaction scheme.

Uncatalysed $Fe^{3+}_{(aq)} + V^{3+}_{(aq)} \longrightarrow Fe^{2+}_{(aq)} + V^{4+}_{(aq)}$: rate $\propto [Fe^{3+}]^1 [V^{3+}]^1$

Catalysed Step 1. $Cu^{2+}_{(aq)} + V^{3+}_{(aq)} \xrightarrow{\text{slow}} Cu^{+}_{(aq)} + V^{4+}_{(aq)}$

Step 2. $Cu^{+}_{(aq)} + Fe^{3+}_{(aq)} \xrightarrow{\text{fast}} Cu^{2+}_{(aq)} + Fe^{2+}_{(aq)}$

The combined effect of the two-step process is faster than the uncatalysed process, but the second step is much quicker than the first step. The first step is rate-determining and so the rate of the reaction is governed by the speed of this step, i.e.

$$\text{rate} \propto [V^{3+}]^1 [Cu^{2+}]^1 [Fe^{3+}]^0$$

Chain reactions

In all the multi-step reactions described so far, the rate of the slowest step determines the rate of the overall reaction. The other steps have to wait for the slower step to take place before the overall result is achieved. There is a group of reactions, however, that do not fit this pattern. They are called chain reactions and most of them involve reaction mixtures that contain free atoms or free radicals. The slowest step in the chain is always the production of free radicals but, once one radical is produced, it sets up a whole chain of interactions. A classic example is the reaction of hydrogen and chlorine in the presence of sunlight. The rate of this 'photochemical' process is very fast indeed and is not controlled by the rate of initiation.

Step 1 Initiation
A chlorine molecule is split into two atoms (free radicals) by radiation energy. The process has a high activation energy and is slow

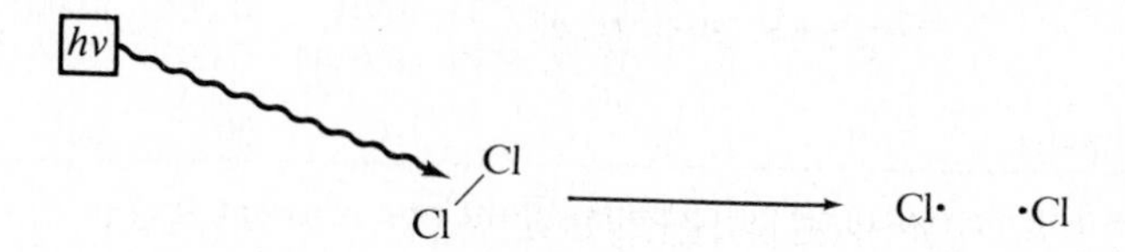

Step 2 Propagation
Radicals collide with molecules and cause new molecules to form. In every case, a new radical is also produced and so a 'chain' of reactions is set up:

Cl· H—H ⟶ Cl—H ·H

then H· Cl—Cl ⟶ H—Cl ·Cl

up to 5 000 000 of these steps happen for every initiation.

this chlorine atom restarts the chain

Step 3. Termination
The chain can only stop when two radicals combine. This happens when two radicals collide and before they break apart again, suffer a further collision with a third particle. In this second collision, some energy is lost from the radicals as kinetic energy.

$$H\cdot \quad \cdot Cl \xrightarrow{+M} H—Cl$$

M could be a chlorine molecule or a particle at the vessel wall.

Questions

1. Explain the meaning of:
   a) rate of reaction, b) order of reaction, c) activation energy d) molecularity of reaction, e) rate-determining step, f) chain reaction.

2. List the factors that affect the rate of a chemical reaction and explain how the collision theory accounts for them.

3. Cyclopropane converts to propene: the kinetics of this reation are followed by measuring the percentage of cyclopropane that remains at various times. From the data below calculate a) the order of reaction and b) the rate constant for the reaction.

% cyclopropane remaining	100	91	79	63	40	25
time in hours	0	2	5	10	20	30

4. Benzoyl chloride reacts with phenylamine in benzene solution. The following data are obtained for reactions at two different temperatures.

[benzoyl chloride] = [phenylamine]	mol dm^{-3}				
at 25°C	0.02	0.0133	0.01	0.0069	0.0051
at 60°C	0.02	0.0074	0.0044	0.0025	0.0017
time in minutes	0	5	10	20	30

From these values calculate a) order of reaction, b) the rate constant at each temperature, c) the activation energy for the reaction.

5. Methoxymethane decomposes at 550°C to give methane, carbon monoxide and hydrogen.

$$CH_3OCH_{3(g)} \longrightarrow CH_{4(g)} + CO_{(g)} + H_{2(g)}$$

This decomposition is carried out in a closed vessel and the total pressure is measured as the reaction proceeds.

total pressure in mm Hg	420	743	954	1054	1198	1258
time in minutes	0	114	219	299	564	∞

From these data, calculate a) order of reaction, b) the rate constant for the reaction.

6. Ethanal decomposes at 800 K according to the reaction

$$CH_3CHO_{(g)} \longrightarrow CH_{4(g)} + CO_{(g)}$$

When this decomposition is carried out in a closed vessel, the total pressure of the system changes as follows:

total pressure in kPa	48.4	58.2	66.2	74.2	80.9	86.2
time in seconds	0	105	242	480	840	1440

Use these data to calculate a) the order of the reaction, b) the rate constant for the reaction.

7. Hydrogen peroxide decomposes to water and oxygen and the following readings were obtained for the concentration of hydrogen peroxide at various times.

[H_2O_2] mol dm^{-3}	0.307	0.247	0.199	0.131	0.082	0.057
time in minutes	0	5	10	20	30	40

Calculate a) the order of reaction, b) the rate constant for the reaction.

8. At 40°C a certain first order reaction was 20% complete in 15 minutes. At 60°C the same reaction was 20% complete in 3 minutes. Use these data to calculate the activation energy for the reaction.

9. Account for the following:
   a) When a mixture of hydrogen and deuterium, D_2, is kept over a nickel catalyst, molecules of HD are detected.
   b) At room temperature hydrogen does not react with chlorine in the dark, but when exposed to sunlight the mixture explodes.
   c) The rate of catalytic decomposition of phosphine, PH_3, on a glass surface is proportional to the pressure of phosphine, but the rate of catalytic decomposition of ammonia on a tungsten surface is independent of the pressure of ammonia.

10. By choosing *three* specific examples, discuss how a study of the kinetics of a reaction can give information about the pathway that the reaction follows.

Chapter 16

Chemical equilibrium

16.1 Systems at equilibrium

When two substances react to produce products, it is possible that the products themselves may then react to give back the reactants again:

$$A + B \longrightarrow C + D$$
$$\text{but also } C + D \longrightarrow A + B$$

Two examples of reactions of this sort are:

Ethanoic acid and ethanol

$$CH_3COOH + C_2H_5OH \longrightarrow CH_3COOC_2H_5 + H_2O$$

the products, ethyl ethanoate and water, themselves react to give ethanoic acid and ethanol:

$$CH_3COOC_2H_5 + H_2O \longrightarrow CH_3COOH + C_2H_5OH$$

Iron (III) chloride and potassium thiocyanate

$$Fe^{3+}_{(aq)} + CNS^-_{(aq)} \longrightarrow \underset{\text{deep red complex}}{[FeCNS]^{2+}_{(aq)}}$$

the product, a deep red-coloured complex, itself decomposes to give iron (III) ions and thiocyanate ions:

$$[FeCNS]^{2+}_{(aq)} \longrightarrow Fe^{3+}_{(aq)} + CNS^-_{(aq)}$$

In both these systems there are two opposing processes taking place. To show this, the sign $\rightleftharpoons$ is used: it means '*is in equilibrium with*' and implies that the rates of the opposing processes are equal, e.g.

$$CH_3COOH + C_2H_5OH \rightleftharpoons CH_3COOC_2H_5 + H_2O$$
$$Fe^{3+}_{(aq)} + CNS^-_{(aq)} \rightleftharpoons [FeCNS]^{2+}_{(aq)}$$

Under these conditions there is no net change in concentration of substances in the system. If 0.1 mol dm^{-3} of reactants turns to products at the same time that another 0.1 mol dm^{-3} of products gives back reactants, then the net change is zero.

Since it is possible to view the substances on both the left and right of the equilibrium equation as reactants, an agreed convention is adopted:

1. The substances on the left of the equation are called reactants
2. The substances on the right of the equation are called products
3. The reaction from left to right is called the forward reaction
4. The reaction from right to left is called the back reaction

$$\underset{\text{reactants}}{A + B} \underset{\text{back}}{\overset{\text{forward}}{\rightleftharpoons}} \underset{\text{products}}{C + D}$$

The dynamic nature of equilibrium

Although there is no change in reactant or product concentration in a system at equilibrium, both forward and back reactions are still proceeding. For this reason, chemical equilibrium is said to be *dynamic*. It is instructive to consider the approach to equilibrium of the iron (III)/thiocyanate reaction.

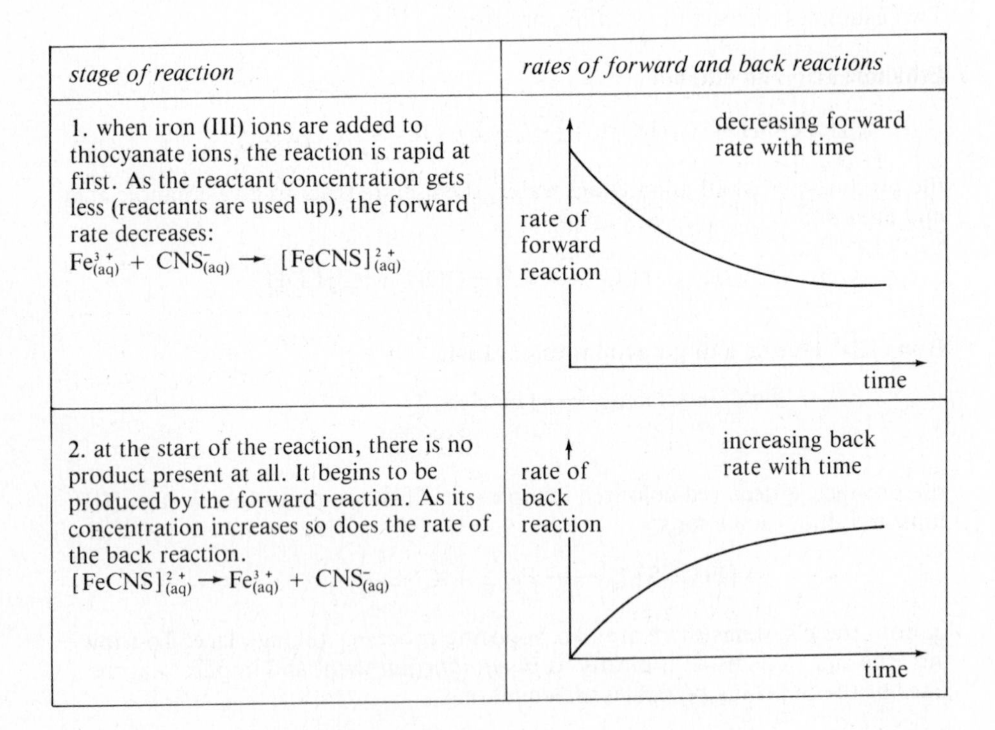

stage of reaction	*rates of forward and back reactions*
1. when iron (III) ions are added to thiocyanate ions, the reaction is rapid at first. As the reactant concentration gets less (reactants are used up), the forward rate decreases: $Fe^{3+}_{(aq)} + CNS^{-}_{(aq)} \rightarrow [FeCNS]^{2+}_{(aq)}$	decreasing forward rate with time; rate of forward reaction; time
2. at the start of the reaction, there is no product present at all. It begins to be produced by the forward reaction. As its concentration increases so does the rate of the back reaction. $[FeCNS]^{2+}_{(aq)} \rightarrow Fe^{3+}_{(aq)} + CNS^{-}_{(aq)}$	increasing back rate with time; rate of back reaction; time

After a particular time interval, the rates of both reactions become equal and the system reaches equilibrium. This can be shown by including both forward and back rates on the same graph:

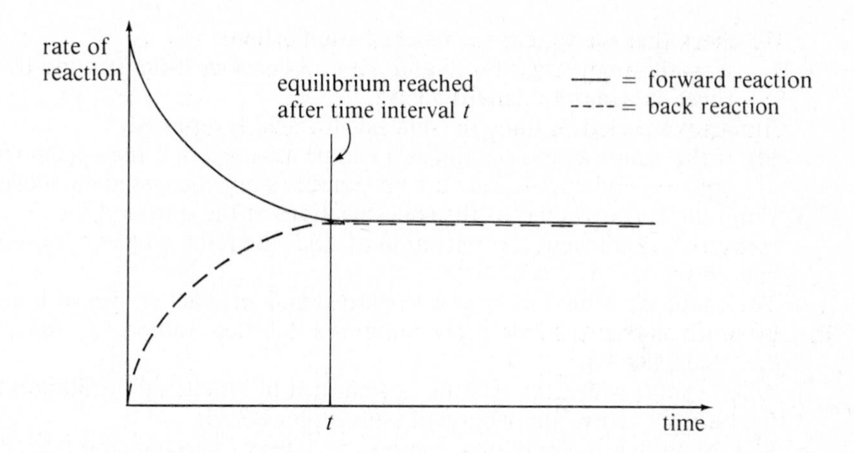

Evidence for the dynamic nature of equilibrium

Since there are no changes in reactant or product concentrations in an equilibrium system, it is fair to question whether anything is happening at all. Perhaps an equilibrium system is simply a very slow reaction.

This theory is readily disproved by increasing the temperature of the system. An increase in temperature causes an increase in reaction rate which it should be possible to detect. Compare the effect of a temperature increase on an equilibrium system with that on a genuine slow reaction.

red-coloured solution at 16°C	*purple-coloured solution at 16°C*
$Fe^{3+}_{(aq)} + CNS^{-}_{(aq)} \rightleftharpoons [FeCNS]^{2+}_{(aq)}$	$2MnO^{-}_{4(aq)} + 5(COOH)_{2(aq)} + 6H^{+}_{(aq)} \rightarrow 2Mn^{2+}_{(aq)} + 10CO_{2(g)} + 8H_2O_{(l)}$
an increase in temperature causes a slight colour change, but the original colour returns when the temperature drops back to its original value.	an increase in temperature causes the ethanedioic acid to reduce the dilute solution of purple-coloured manganate (VII); the colour goes and does not return when the temperature drops back to its original value.

The absence of any marked colour change for the first system can be explained by assuming that there are two opposing processes taking place. The increased temperature leads to an increase in the rate of both and therefore little change is detected.

More direct evidence of the dynamic nature of an equilibrium can be obtained as follows (for the equilibrium between organic acid, alcohol, ester and water):

$$\underset{\text{ethanoic acid}}{CH_3COOH} + \underset{\text{ethanol}}{C_2H_5OH} \rightleftharpoons \underset{\text{ethyl ethanoate}}{CH_3COOC_2H_5} + H_2O$$

1. Measured amounts of acid and alcohol are mixed and left to equilibrate for a long time at constant temperature.

2. To check that the system has reached equilibrium:
   i) a small sample is removed and the acid concentration is found by titration against standard alkali;
   ii) after a period of time, the analysis for acid is repeated;
   iii) if the same value is obtained, it can be assumed that the system is in equilibrium because there are no concentration changes at equilibrium.
3. From the known values of the concentrations at the start and the measured equilibrium concentration of acid, the remaining equilibrium concentrations are calculated.
4. The equilibrium mixture is now discarded and an exact replica of it is set up in which each acid molecule contains a 'labelled' radioactive isotope, e.g. $^{14}CH_3COOH$.
5. After a short while, the mixture is separated by fractional distillation and the radioactivity of the separated ester is monitored.
6. It is found that the ester now contains 'labelled' molecules and this proves that the forward reaction is still proceeding even in an equilibrium mixture.

16.2 The equilibrium law

For a given reaction, there is a fixed relationship between the numerical values of the reactant and product concentrations at equilibrium. For example:

a) In the system, $CH_3COOH + C_2H_5OH \rightleftharpoons CH_3COOC_2H_5 + H_2O$ at equilibrium, it is found that

$$\frac{[CH_3COOC_2H_5]\ [H_2O]}{[CH_3COOH]\ [C_2H_5OH]} = \text{a constant}$$

where [A] is the molar concentration of A at equilibrium.

b) In the system, $H_{2(g)} + I_{2(g)} \rightleftharpoons 2HI_{(g)}$ at equilibrium,

$$\frac{[HI]^2}{[H_2]\ [I_2]} = \text{a constant}$$

c) In the system, $2SO_{2(g)} + O_{2(g)} \rightleftharpoons 2SO_{3(g)}$ at equilibrium,

$$\frac{[SO_3]^2}{[SO_2]^2\ [O_2]} = \text{a constant}$$

These relationships are all examples of the *equilibrium law*.

> For the general reaction
>
> $$aA + bB \rightleftharpoons cC + dD$$
>
> it is found that the expression $\dfrac{[C]^c\ [D]^d}{[A]^a\ [B]^b}$ is a constant.

The constant is called the *equilibrium constant* and is given the symbol K_c when the reactant and product concentrations are given in mol dm^{-3}.

There are some important points to note about the equilibrium law.

The product concentrations are always given on the top line of the expression:

$$\text{reactants} \rightleftharpoons \text{products} \qquad K_c = \frac{[\text{products}]}{[\text{reactants}]}$$

The constant may or may not have units: it depends on the form of the equilibrium expression, e.g.

$$2HI_{(g)} \rightleftharpoons H_{2(g)} + I_{2(g)} \qquad K_c = \frac{[H_2]\,[I_2]}{[HI]^2}$$

$$2SO_{2(g)} + O_{2(g)} \rightleftharpoons 2SO_{3(g)} \qquad K_c = \frac{[SO_3]^2}{[SO_2]^2\,[O_2]}$$

There are no units for the first expression because the concentrations cancel. For the second, the units are $(\text{conc})^{-1} = \text{dm}^3\ \text{mol}^{-1}$.

For gaseous reactions, the concentrations can be given as partial pressures. In this case the equilibrium constant is written as K_p

$$N_{2(g)} + 3H_{2(g)} \rightleftharpoons 2NH_{3(g)} \qquad K_p = \frac{P^2_{NH_3}}{P_{N_2}\,P^3_{H_2}}$$

(partial pressure = mole fraction × total pressure).

Finding the value of an equilibrium constant

The equilibrium between acid, alcohol, ester and water is an equilibrium system that is quite easy to study in the laboratory. To obtain a value of K_c for this system, the following procedure is adopted:

1. Set up a number of reaction flasks with known initial amounts of each component. The reaction is catalysed by acid, and so the same volume of concentrated sulphuric acid is added to each flask. For example, seven flasks might be listed as shown below:

flask	*moles of acid*	*moles of alcohol*	*moles of ester*	*moles of water*	*volume (cm^3) of H_2SO_4*
1	—	—	—	—	1.0
2	1	1	—	—	1.0
3	2	1	—	—	1.0
4	1	1	1	—	1.0
5	—	1	1	1	1.0
6	—	—	1	1	1.0
7	—	—	2	1	1.0

No two flasks have the same initial amounts
Flask 1 is a 'control' flask.

2. Allow each system to reach equilibrium by putting them into a thermostatic bath at 298 K for a week (the reaction is slow). The flasks are stoppered to prevent evaporation losses.
3. When each system has reached equilibrium, the amount of acid present is found by titrating against standard alkali. The control flask shows how much alkali is needed to neutralize the sulphuric acid used as a catalyst; this amount is subtracted from each reading.
4. Using both the known initial amounts and the measured amount of acid at equilibrium, the remaining equilibrium amounts can be calculated from the equation's stoichiometry. For example, for flask 2:

	acid	+ alcohol	⇌ ester	+ water	
initially:	1	1	0	0	moles in V dm^3

By titration, it is found that there are 0.33 moles at equilibrium.

at equilibrium:	0.33	0.33	0.67	0.67	moles in V dm^3

5. Since K_c is given in terms of concentrations, the equilibrium amounts have to be divided by the volume of solution present: e.g. for flask 2,

$$K_c = \frac{[\text{ester}][\text{water}]}{[\text{acid}][\text{alcohol}]} = \frac{(0.67/V)(0.67/V)}{(0.33/V)(0.33/V)}$$

In this case, all the volume factors cancel

$$\Rightarrow K_c = \frac{(0.67)(0.67)}{(0.33)(0.33)} = 4.12$$

K_c is worked out for each set of equilibrium amounts; an average value of K_c is calculated from the results shown below.

at equilibrium

flask	*moles of acid*	*moles of ethanol*	*moles of ester*	*moles of water*	K_c
2	0.33	0.33	0.67	0.67	4.12
3	0.15	0.15	0.85	0.85	4.18
4	0.66	0.66	1.34	1.34	4.12
5	0.15	1.15	0.85	0.85	4.18
6	0.33	0.33	0.67	0.67	4.12
7	0.45	0.45	1.55	0.55	4.20

Heterogeneous equilibria

When the reactants and products of an equilibrium system are not all in the same phase, the equilibrium is *heterogeneous*. A typical case is the decomposition of a metal carbonate in a closed container.

$$CaCO_{3(s)} \rightleftharpoons CaO_{(s)} + CO_{2(g)}$$

An experiment is carried out in which various masses of calcium carbonate are heated to the same temperature in similar containers. The partial pressure of carbon dioxide is found to be the same whatever the mass of the solid.

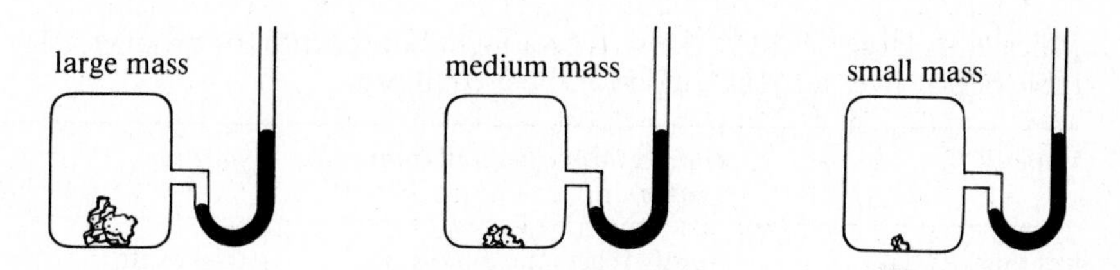

This can be explained by applying the equilibrium law to the heterogeneous equilibrium:

$$K_c = \frac{[CaO_{(s)}]\ [CO_{2(g)}]}{[CaCO_{3(s)}]}$$

The concentration of any solid is the same as its density expressed in mol dm^{-3} (called its mass concentration). Because the density of a solid is constant the above expression reduces to

$$K_c = \text{(a constant)}\ [CO_{2(g)}]$$

By combining the two constants and expressing the concentration of carbon dioxide as a partial pressure, the equation becomes

$$K_P = P_{CO_2}$$

i.e. the partial pressure of carbon dioxide is constant.

In considering any gas-solid equilibria, therefore, the concentration factor of the solid can be taken to be constant providing there is some solid present, e.g.

$$CO_{2(g)} + C_{(s)} \rightleftharpoons 2CO_{(g)}$$

the equilibrium law reduces to $K_P = P^2_{CO}/P_{CO_2}$.

The position of equilibrium

In an equilibrium system, the reactants and products have concentrations whose values remain the same because the rates of forward and back reactions are equal. The values are related by the equilibrium constant K_c. The term *position of equilibrium* is used to indicate whether the system contains a larger proportion of reactants or products:

$$\text{reactants} \rightleftharpoons \text{products}$$

'equilibrium position lies to the *left*' means that the system contains a greater proportion of *reactants*.

'equilibrium position lies to the *right*' means that the system contains a greater proportion of *products*.

The actual size of an equilibrium constant gives a measure of the position of equilibrium. When K_c is very small, there is likely to be a higher proportion of reactants because

$$K_c \text{ varies with } \frac{[\text{product}]}{[\text{reactant}]}$$

When K_c is large, there is likely to be a higher proportion of products. The table below gives a rough guide for most equilibria.

value of K_c	*composition of equilibrium system*	*position*
less than 10^{-2}	mostly reactants; almost no products formed	to the left
between 10^{-2} and 10^{2}	reactants and products in appreciable amounts	central
larger than 10^{2}	mostly products; reaction almost complete	to the right

If the proportion of products in the mixture increases, the position of equilibrium is said to move to the right. Similarly the position of equilibrium is said to move to the left if an increase in the proportion of reactants takes place.

16.3 The effect of changing conditions

There are a number of factors that may affect the position of a particular equilibrium:

1. the concentration of the reactants and of the products
2. the temperature
3. the total pressure
4. the presence of other substances in the system

The effects of each of these factors are discussed in turn during this section. To simplify the discussion, the same equilibrium system is used as an example;

$$N_{2(g)} + 3H_{2(g)} \rightleftharpoons 2NH_{3(g)} \qquad \Delta H_f^\ominus = -46.2 \text{ kJ mol}^{-1}$$

$$\text{At 650 K, } K_c = \frac{[NH_3]^2}{[N_2]\,[H_2]^3} = 9.25 \times 10^{-2} \text{ dm}^6 \text{ mol}^{-2}$$

Reactant and product concentration

If the concentration of a reactant in an equilibrium system is suddenly changed, the rate of the forward reaction alters because rate $\propto$ [reactant]. Similarly, if a product concentration is suddenly changed, the rate of the back reaction alters. In either of these two cases, equilibrium is disrupted because the rates of the two opposing processes are no longer equal. The concentrations of reactants and products change until a new equilibrium position is reached.

Example (a) Add hydrogen at constant volume:

$$N_{2(g)} + 3H_{2(g)} \rightleftharpoons 2NH_{3(g)}$$

i) The forward rate increases with increased $[H_2]$.
ii) Thus the forward rate becomes greater than the back rate.
iii) The concentration of ammonia increases while that of nitrogen decreases until the two rates become equal again.
iv) There is a shift of equilibrium position to the right.

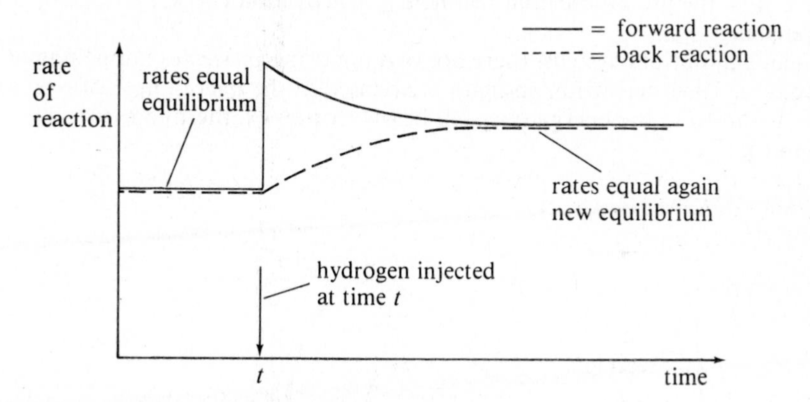

Example (b) Remove ammonia at constant volume:

i) The back rate decreases with decreased $[NH_3]$.
ii) Thus the back rate becomes less than the forward rate.
iii) The concentration of ammonia builds up again at the expense of nitrogen and hydrogen until the two rates are equal once more.
iv) There is a shift of equilibrium position to the right.

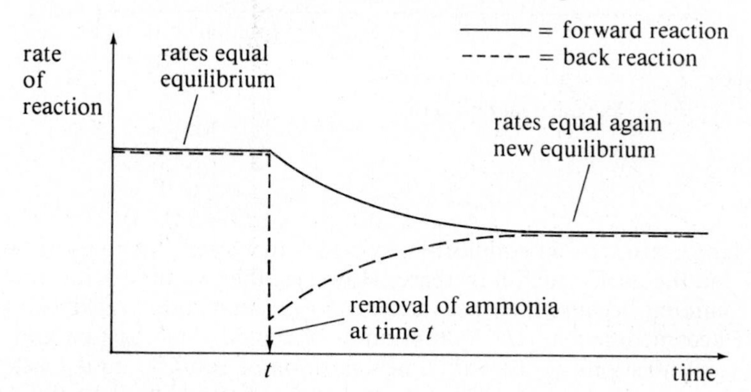

In both the above cases, the value of K_c remains unchanged although the position of equilibrium is changing.

Similarly, the addition of a product or the removal of a reactant from an equilibrium mixture leads to a shift in equilibrium position to the left.

Temperature

An increase in the temperature of any reaction causes an increase in its rate: the rate constant k is temperature dependent as given by the Arrhenius equation (page 234).

$$k = Ae^{-E_A/RT}$$

where A = the pre-exponential factor; E_A = activation energy; T = temperature in kelvins.

In an equilibrium system, there are two opposing processes happening at the same rate. Their activation energies are related to the overall heat of reaction, ΔH, as shown in the energy diagram below (for an exothermic forward reaction):

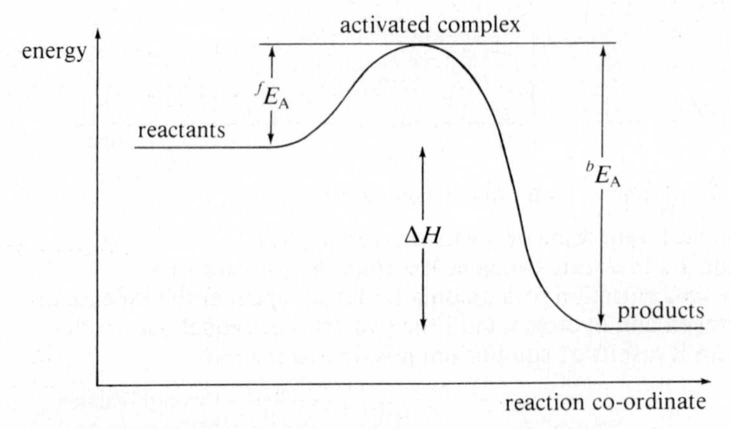

where ^fE_A = forward activation energy
^bE_A = back activation energy

If the temperature of an equilibrium system is increased, the rates of both the forward and the back reaction increase. However, the two do not increase by the same amount because their activation energies are different. The two rates therefore become unequal and equilibrium is disrupted. The reactant and product concentrations adjust until a new position of equilibrium is reached but, in this case, the new equilibrium is no longer governed by the value of K_c at the lower temperature. In other words, *the equilibrium constant changes in value when the temperature changes.*

Example a) An exothermic reaction (${}^fE_A < {}^bE_A$). For the synthesis of ammonia:

1. In an equilibrium system at 500 K, $K_P = 3.55 \times 10^{-2}$ atm^{-2} and the rates of the forward and back reaction are equal.

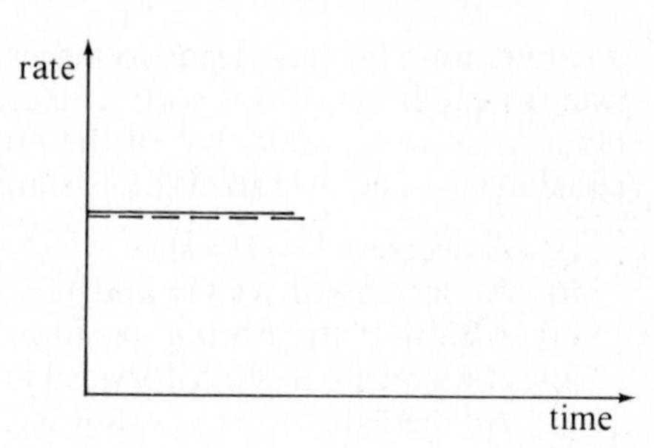

2. When the temperature is increased to 700 K, the rates of *both* reactions are increased. The rate of the back reaction is increased by a larger amount* than the rate of the forward reaction because ${}^bE_A > {}^fE_A$.

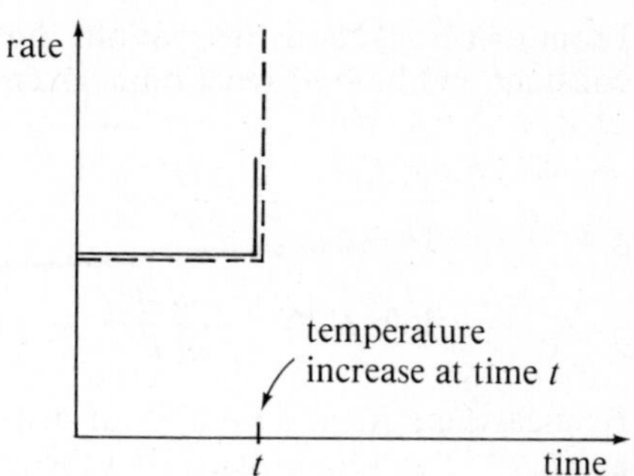

3. The back rate is now greater than the forward rate and so $[NH_3]$ decreases while $[N_2]$ and $[H_2]$ increase until the two rates become equal again. Since

$$K_P = \frac{P^2_{NH_3}}{P_{N_2}\,P^3_{H_2}}$$

its value gets smaller. ($K_P = 7.76 \times 10^{-5}$ atm^{-2} at 700 K).

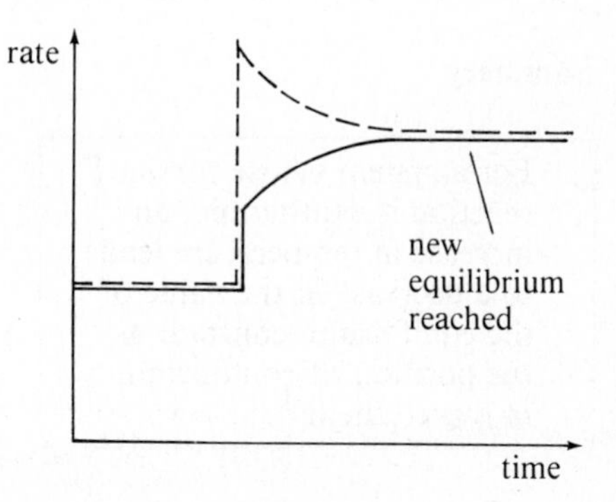

The overall effect of an increase in temperature on the equilibrium system is therefore:

i) A decrease in $[NH_3]$
ii) An increase in $[N_2]$ and $[H_2]$
iii) A shift in equilibrium position to the left
iv) An increase in both forward and back rates
v) A decrease in the value of K_P.

---

*The Arrhenius equation can be used to show that the increase in k caused by an increase in T is largest when E_A is largest.

Example b) An endothermic reaction (${}^fE_A < {}^bE_A$). For the water-gas reaction

$$C_{(s)} + H_2O_{(g)} \rightleftharpoons CO_{(g)} + H_{2(g)} \qquad \Delta H^{\ominus} = +131 \text{ kJ mol}^{-1}$$

a temperature increase leads to a greater increase in forward rate than in back rate for equilibria of this sort. The concentration of the reactants therefore tends to decrease while that of the products increases until the two rates are equal again. The overall effect is summarized below.

i) A decrease in $[H_2O]$
ii) An increase in $[CO]$ and $[H_2]$
iii) A shift in equilibrium position to the right
iv) An increase in both forward and back rates
v) An increase in the value of K_P.

From thermodynamic derivation, the actual relationship between equilibrium constant and heat of reaction is given by the van't Hoff equation:

$$\frac{d(\log_e K_P)}{dT} = \frac{\Delta H^{\ominus}}{RT^2}$$

$$\Rightarrow \quad \log_e K_P = \frac{-\Delta H^{\ominus}}{RT} + \text{a constant (assuming } \Delta H^{\ominus} \neq f(T))$$

By measuring K_P at a number of different temperatures, a plot of $\log_e K_P$ against $1/T$ can be used to find the enthalpy change for the reaction concerned. The plot should give a straight line of slope $-\Delta H^{\ominus}/R$.

Summary

For a system whose forward reaction is exothermic, an increase in temperature leads to a decrease in the value of the equilibrium constant as the position of equilibrium moves to the left.

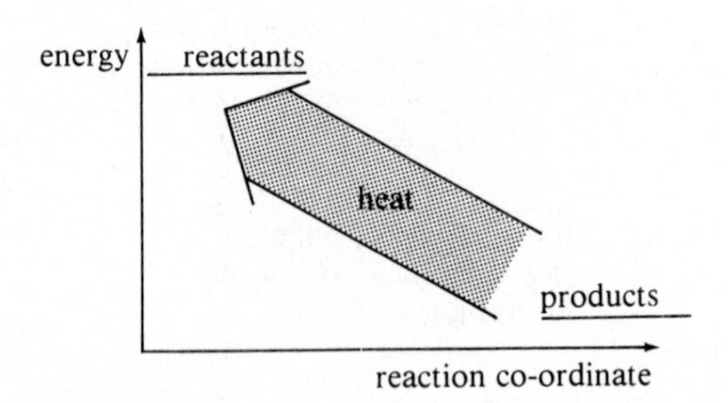

For a system whose forward reaction is endothermic, an increase in temperature leads to an increase in the value of the equilibrium constant as the position of equilibrium moves to the right.

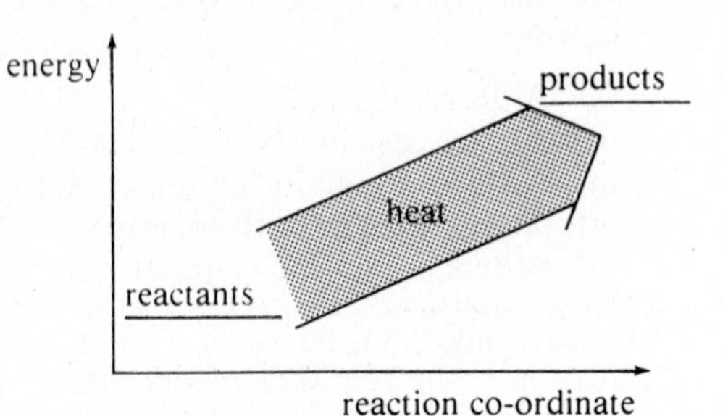

Pressure

Pressure changes can only affect those equilibria that involve gaseous reactants or products. The concentration of a gas is proportional to its partial pressure (see page 230) and partial pressures depend on the total pressure of the system. In the ammonia synthesis,

$$N_{2(g)} + 3H_{2(g)} \rightleftharpoons 2NH_{3(g)}$$

$$P_{NH_3} = x_{NH_3}P_T, \quad P_{N_2} = x_{N_2}P_T \quad \text{and} \quad P_{H_2} = x_{H_2}P_T$$

where P_Z is the partial pressure of Z, P_T = total pressure and x_Z is the mole fraction of Z in the gaseous mixture.

$$\text{Thus } K_P = \frac{(x_{NH_3}\,P_T)^2}{(x_{N_2}\,P_T)(x_{H_2}\,P_T)^3} = \left(\frac{x^2_{NH_3}}{x_{N_2}\,x^3_{H_2}}\right)\frac{1}{P_T^2}$$

Example a) Increase the total pressure. If P_T is increased by a factor of three, in the above expression for K_P, the pressure factor $1/P_T^2$ falls to one-ninth of its initial value. Since the equilibrium constant governs the concentrations of reactants and products in any equilibrium mixture, the mole fractions of reactants and products must alter to cancel out the fall in the pressure factor: K_P is *constant* at constant temperature, i.e.

$$\left(\frac{x^2_{NH_3}}{x_{N_2}x^3_{H_2}}\right)$$

must increase by a factor of nine to cancel out the decrease in $1/P_T^2$. Therefore:

i) $[NH_3]$ increases.
ii) $[N_2]$ and $[H_2]$ decrease.
iii) The position of equilibrium shifts to the right.
iv) The pressure increase favours the side with the least number of moles.

It is worth noticing, therefore, that pressure changes cannot affect the position of an equimolar equilibrium such as the synthesis of hydrogen iodide.

$$H_{2(g)} + I_{2(g)} \rightleftharpoons 2HI_{(g)}$$

$$K_P = \left(\frac{x^2_{HI}}{x_{H_2}\,x_{I_2}}\right)\frac{P_T^2}{P_T^2} \quad \text{i.e. } K_P \neq f(P_T)$$

Example b) The addition of an inert gas. If argon is added at *constant volume* to an equilibrium mixture of nitrogen, hydrogen and ammonia

i) P_T increases, but
ii) each mole fraction decreases by the same factor, and therefore
iii) each partial pressure remains constant: there is no effect on the position of equilibrium.

However, if argon is added at *constant pressure*, there is a dilution effect which alters the forward and back rates in a different way.

i) each mole fraction decreases by the same amount, but
ii) the total pressure remains constant

So in the equilibrium law

$$K_P = \left(\frac{x_{NH_3}^2}{x_{N_2}\, x_{H_2}^3}\right)\frac{1}{P_T^{\,2}}$$

if each mole fraction is decreased by a factor of a half, the expression in brackets increases by a factor of four:

$$\left(\frac{(\frac{1}{2})^2}{(\frac{1}{2})\,(\frac{1}{2})^3}\right) = 4.$$

The composition of the system must now alter so that equilibrium can be established again: to bring this about the expression in brackets decreases by a factor of four.

i) $[NH_3]$ decreases even further.
ii) $[N_2]$ and $[H_2]$ increase (to offset partially the dilution caused by the argon addition).
iii) the equilibrium position shifts to the left.

The overall effect is the opposite of that brought about by increasing the total pressure of the system (page 263).

Catalysts

A catalyst is a substance that increases the rate of a reaction but is itself left chemically unaltered at the end of the process (see pages 246 and 247). When a catalyst is added to an equilibrium system, it speeds up the rate of both the forward and back reactions.

—— = uncatalysed equilibrium
---- = catalysed equilibrium

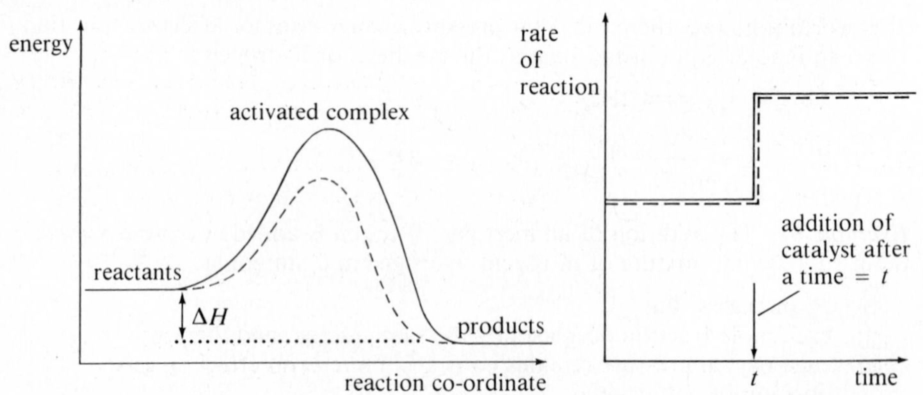

Unlike the effect of temperature, the increase in rate is the same for both forward and back reactions: a catalyst *cannot* alter the value of the equilibrium constant. On page 262, it was seen that the equilibrium constant is related to the overall heat of reaction:

$$\log_e K_P = \frac{-\Delta H^{\ominus}}{RT} + \text{a constant.}$$

Since the heats of both the catalysed and uncatalysed reactions are equal (the catalyst is unchanged at the end and so does not affect the energy state of either reactants or products), K_P is also equal for both processes. The first law of thermodynamics can be used to illustrate that $\Delta H^{\ominus}_{cat} = \Delta H^{\ominus}_{uncat}$.

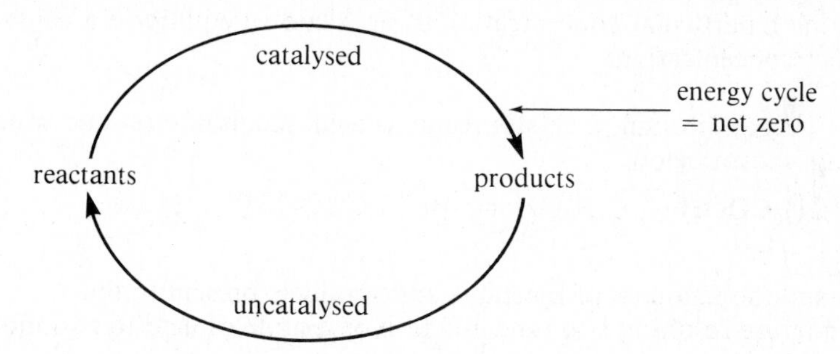

A more exact expression shows that the equilibrium constant is related to the change in free energy rather than enthalpy during a reaction.

$$\log_e K_P = \frac{-\Delta G^{\ominus}}{RT}$$

$\Delta G^{\ominus}$ = the standard free energy change.

Summary: Le Chatelier's principle

The effects of changing the conditions of an equilibrium system were first summarized by Le Chatelier in a principle that bears his name:

> When a change is imposed on an equilibrium system, the system reacts in such a way as to tend to oppose the change.

This principle is *not* an explanation for the processes taking place: it is simply a rule of thumb. Some examples of its application provide a useful summary of the whole of Section 16.3.

changing conditions	*application of Le Chatelier's principle*
increase reactant concentration	equilibrium position shifts to the right (towards products) to tend to oppose the effects of increased reactant concentration
decrease product concentration	equilibrium position shifts to the right (towards products) to tend to oppose the effects of decreased product concentration
increase the temperature (exothermic forward reaction)	equilibrium position shifts to the left as the endothermic back reaction absorbs more heat and tends to oppose the effects of increased temperature
increase the pressure of the equilibrium $N_{2(g)} + 3H_{2(g)} \rightleftharpoons 2NH_{3(g)}$	equilibrium position shifts to the right as the products occupy a smaller volume than the reactants (2 moles c.f. 4 moles) and the smaller volume tends to oppose the effects of increased pressure

16.4 Calculations involving equilibrium constants

There are two principal types:

1. calculating an equilibrium constant from concentration data
2. calculating a particular concentration using a known equilibrium constant and other concentrations.

Example (a) At equilibrium, a vessels contains acid, alcohol, ester and water in the amounts shown below.

$$\underset{1.43}{CH_3COOH} + \underset{0.43}{C_2H_5OH} \rightleftharpoons \underset{1.57}{CH_3COOC_2H_5} + \underset{1.57\ \text{mol}}{H_2O}$$

Calculate K_c and the amounts of reactants and products present in the equilibrium mixture resulting from the addition of a mole of acid to two moles of alcohol.

equation:	CH_3COOH +	C_2H_5OH $\rightleftharpoons$	$CH_3COOC_2H_5$ +	H_2O
equilibrium amounts	1.43	0.43	1.57	1.57 mol in V dm³
equilibrium concentrations	$\frac{1.43}{V}$	$\frac{0.43}{V}$	$\frac{1.57}{V}$	$\frac{1.57}{V}$ mol dm⁻³

$$K_c = \frac{[CH_3COOC_2H_5]\,[H_2O]}{[CH_3COOH]\,[C_2H_5OH]} = \frac{(1.57/V)(1.57/V)}{(1.43/V)(0.43/V)}$$

$$\Rightarrow K_c = 4.00$$

The value of K_c is used to solve the second part. Exactly the same procedure is adopted.

equation:	CH_3COOH +	C_2H_5OH $\rightleftharpoons$	$CH_3COOC_2H_5$ +	H_2O
initial amounts	1	2	0	0 mol in V dm³

Let x moles of acid react; thus x moles of alcohol also react and x moles each of ester and water are formed.

equilibrium amounts	$(1-x)$	$(2-x)$	x	x mol in V dm³
equilibrium concentrations	$\left(\frac{1-x}{V}\right)$	$\left(\frac{2-x}{V}\right)$	$\frac{x}{V}$	$\frac{x}{V}$ mol dm⁻³

$$K_c = \frac{(x/V)(x/V)}{(1-x/V)(2-x/V)} = \frac{x^2}{2-3x+x^2} = 4$$

$$\Rightarrow \quad 3x^2 - 12x + 8 = 0$$

A quadratic equation is solved by using the formula shown below for the general equation of form: $ax^2 + bx + c = 0$

$$x = \frac{-b \pm \sqrt{(b^2 - 4ac)}}{2a} \text{ where } a = 3,\ b = -12 \text{ and } c = 8$$

$$\Rightarrow x = \frac{12 \pm \sqrt{(144 - 96)}}{6} = 2 \pm 1.15$$

$\Rightarrow x = 3.15$ (impossible because there is only one mole of acid at the start)
or $x = 0.85$
The equilibrium amounts are therefore:

$[CH_3COOH] = 0.15$ mol $[CH_3COOC_2H_5] = 0.85$ mol
$[C_2H_5OH] = 1.15$ mol $[H_2O] = 0.85$ mol

Example (b) A 1:3 mixture by volume of nitrogen and hydrogen gas is made up and left to equilibrate at 1.01 MN m^{-2} and 600 K. Calculate K_p given that the equilibrium partial pressure of ammonia is 15%.

equation	$N_{2(g)}$ + $3H_{2(g)}$ $\rightleftharpoons$ $2NH_{3(g)}$			
initial amounts	1 : 3 : 0	by volume		

Avogadro's Law (page 136) states that equal volumes of gases contain equal numbers of molecules at the same temperature and pressure:

gas volumes $\equiv$ molar quantities

Since the reaction uses one volume of nitrogen for every three of hydrogen, the proportion in the equilibrium mixture is still 1:3. If 15% is ammonia (given), then the remaining 85% is split in the ratio 1:3

equilibrium molar ratio:	$N_{2(g)}$ + $3H_{2(g)}$ $\rightleftharpoons$ $2NH_{3(g)}$ 21.25 : 63.75 : 15	molar ratio

From these ratios, the equilibrium partial pressures are calculated.

$P_{N_2} = 0.2125\ P_T$ $P_{H_2} = 0.6375\ P_T$ $P_{NH_3} = 0.15\ P_T$

where P_T = total pressure = 1.01 MN m^{-2} (given)

$$K_p = \frac{P^2_{NH_3}}{P_{N_2}\,P^3_{H_2}} = \frac{(0.15)^2}{(0.2125)(0.6375)^3} \times \left(\frac{1}{1.01}\right)^2$$

$K_p = 0.41$ m^4 $(MN)^{-2}$ [note the units of $(\text{pressure})^{-2}$]

More worked examples involving equilibrium constants and concentrations are given for acid-base equilibria (see page 280).

Questions

1. Describe an experiment that provides evidence for the dynamic nature of equilibrium.

2. Describe the practical determination of the value of the equilibrium constant for an equilibrium system of your choice.

3. Write an expression for K_p for each of the following systems:
   a) $N_2 + O_2 \rightleftharpoons 2\,NO$ b) $Cl_2 + 3F_2 \rightleftharpoons 2\,ClF_3$
   c) $CO + \frac{1}{2}O_2 \rightleftharpoons CO_2$ d) $3O_2 \rightleftharpoons 2O_3$

4. a) For the equilibrium system $A + B \rightleftharpoons C$, the equilibrium concentrations in mol dm^{-3} are: [A] = 0.1; [B] = 0.2; [C] = 0.3. Calculate K_c.
   b) For the equilibrium system $3A + B \rightleftharpoons C + 2D$, the equilibrium concentrations in mol dm^{-3} are [A] = 0.5; [B] = 0.33; [C] = 0.25; [D] = 0.2. Calculate K_c
   c) For the system $A + B \rightleftharpoons C + D$, $K_c = 4$ at a particular temperature. In an equilibrium mixture at the same temperature the known concentrations in mol dm^{-3} are [A] = 5; [B] = 2; [C] = 4. Calculate the equilibrium concentration of D.

5. a) The equilibrium constant K_p for the system $2X \rightleftharpoons 3Y$ is 0.125 atm, and the partial pressure of X $p_X = 8$ atm. Calculate the partial pressure of Y and hence the total pressure of the system.
   b) A 500 cm^3 flask contains 1 mole of X, 1.5 mole of Y and 2.5 mole of Z in equilibrium. Calculate K_c for the equilibrium $X \rightleftharpoons Y + Z$.

6. a) An equilibrium mixture contains 2 mole of bromine, 1.25 mole of hydrogen and 0.5 mole of hydrogen iodide in a 4 dm^3 vessel. Calculate K_c for the reaction
      i) $H_2 + Br_2 \rightleftharpoons 2HBr$
      ii) $\frac{1}{2}H_2 + \frac{1}{2}Br_2 \rightleftharpoons HBr$
      iii) $HBr \rightleftharpoons \frac{1}{2}H_2 + \frac{1}{2}Br_2$.
   b) Some hydrogen bromide is injected into a 2 dm^3 evacuated flask at the same temperature as the system in (a) above. When equilibrium is reached, some of the hydrogen bromide has decomposed yielding 5.0 mole of bromine.
      i) Calculate the equilibrium concentrations of each component in the system.
      ii) Calculate the original mass of hydrogen bromide put into the vessel.

7. When 60 g of ethanoic acid, CH_3COOH, reacts with 46 g of ethanol, CH_3CH_2OH, the equilibrium amount of ethyl ethanoate, $CH_3COOCH_2CH_3$, formed is 58.7 g.
   a) Calculate the value of K_c for the system
      $CH_3COOH + CH_3CH_2OH \rightleftharpoons CH_3COOCH_2CH_3 + H_2O$
   b) What mass of ethyl ethanoate would be formed at the same temperature if 60 g of ethanoic acid are mixed with 23 g of ethanol?

8. For the system:
   $CH_3CH_2COOH + CH_3CH_2OH \rightleftharpoons CH_3CH_2COOCH_2CH_3 + H_2O$
   propanoic acid ethanol ethyl propanoate
   the value of K_c at 50°C is 7.5. Calculate the mass of ethanol that must be mixed with 60 g of propanoic acid to yield 80 g of ethyl propanoate at equilibrium at 50°C.

9. For the system $2SO_2 + O_2 \rightleftharpoons 2SO_3$, $\Delta H = -94.8$ kJ mol^{-1}. Discuss the effect on the system of:
   a) adding more oxygen
   b) increasing the pressure
   c) adding a platinum catalyst
   d) increasing the temperature.
   (For each change you should mention the effect on individual concentrations, equilibrium position, reaction rates and the value of K_c).

10. The Haber process for the manufacture of ammonia involves the equilibrium system $N_2 + 3H_2 \rightleftharpoons 2NH_3$, $\Delta H = -46.2$ kJ mol^{-1}. Predict the reaction conditions that would yield the maximum amount of ammonia.
    Now look up the Haber process and compare your predicted conditions with the actual conditions used in industry. Account for any differences.

Chapter 17

Equilibrium in acid-base reactions

17.1 Acids and proton transfer

Most common acids are compounds of hydrogen and a non-metal, usually with oxygen making up a third element, for example:

compound	*formula*	*name of acid*
hydrogen chloride	HCl	hydrochloric acid
hydrogen nitrate (V)	HNO_3	nitric acid
hydrogen sulphate (VI)	H_2SO_4	sulphuric acid
hydrogen phosphate (V)	H_3PO_4	phosphoric acid

A typical acid has the following properties:

1. In its pure state it is a molecular substance that does not conduct electricity.
2. It dissolves exothermically in water to produce a solution that does conduct electricity.
3. Its aqueous solution
   i) turns litmus and universal indicator red,
   ii) reacts with and dissolves metal oxides that are insoluble in water,
   iii) reacts with most metals liberating hydrogen gas,
   iv) reacts with metal carbonates liberating carbon dioxide gas.

All the acids in the above table show these properties. The fact that their aqueous solutions behave in the same way suggests that their molecules interact with water molecules in the same way. Since the aqueous solutions are all conductors of electricity, the molecular interaction must result in the production of ions.

The theory of *proton transfer* accounts for all these properties; it is often named after Brønsted and Lowry who proposed it at the same time independently of each other. Its principles are outlined below.

1. An acid molecule contains a covalently bonded hydrogen atom that is electron deficient as a result of bonding to a more electronegative non-metal atom, for example:

$\overset{\delta+}{H}—\overset{\delta-}{Cl}$ or $\overset{\delta+}{H}—\overset{\delta-}{O}—S(=O)_2—\overset{\delta-}{O}—\overset{\delta+}{H}$ and in general $\overset{\delta+}{H}—\overset{\delta-}{A}$

2. When an acid molecule collides with a water molecule, the lone pairs of the water molecule are attracted to the electron-deficient hydrogen atom.

$H_2\ddot{O}:$ $\overset{\delta+}{H}—\overset{\delta-}{A}$

3. The attraction between the acid molecule and the water molecule can become so strong that the nucleus of the hydrogen atom (a proton) breaks free from the acid molecule.

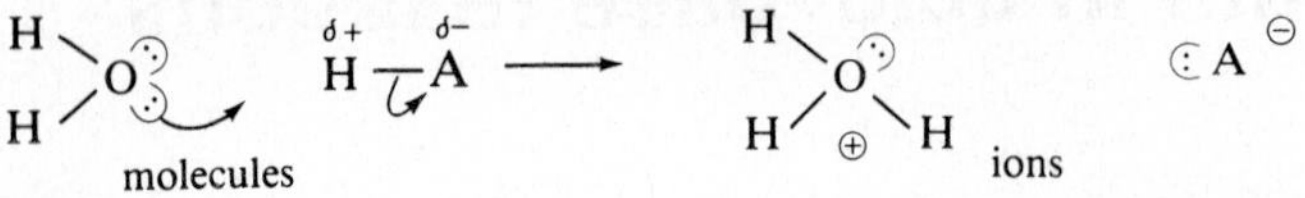

$$H_2O + \overset{\delta+}{H}-\overset{\delta-}{A} \longrightarrow H_3O^{+} + A^{-}$$

molecules → ions

4. When the proton breaks free, ions are produced. No matter what the acid is, the cations are H_3O^+ ions. An H_3O^+ ion is called a *hydroxonium ion* and is a protonated water molecule, for example:

$$HCl_{(g)} + H_2O_{(l)} \longrightarrow H_3O^{+}_{(aq)} + Cl^{-}_{(aq)}$$

$$H_2SO_{4(l)} + 2H_2O_{(l)} \longrightarrow 2H_3O^{+}_{(aq)} + SO^{2-}_{4(aq)}$$

$$H_3PO_{4(l)} + 3H_2O_{(l)} \longrightarrow 3H_3O^{+}_{(aq)} + PO^{3-}_{4\,(aq)}$$

Sulphuric acid is said to be *dibasic* because it can transfer two protons per molecule. Phosphoric acid is *tribasic*.

The protonated water molecules (hydroxonium ions) are responsible for all the acid properties. For example:

Acids react with metal oxides

$$CuO_{(s)} + 2H_3O^{+}_{(aq)} \longrightarrow Cu^{2+}_{(aq)} + 3H_2O_{(l)}$$

All metal oxides contain the oxide ion O^{2-}.

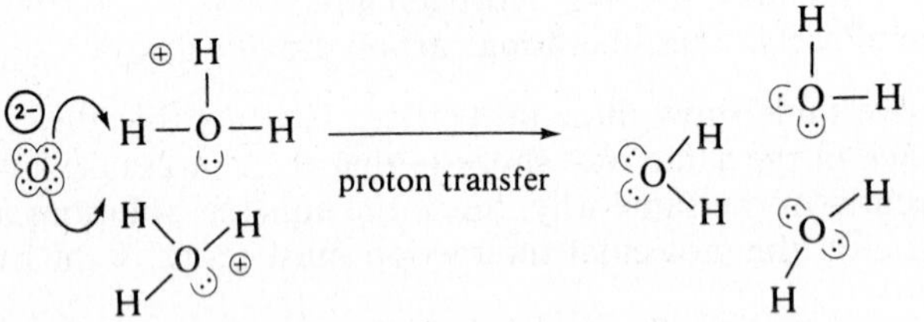

$$O^{2-} + 2H_3O^{+} \xrightarrow{\text{proton transfer}} 3H_2O$$

Acids react with metal carbonates

$$MgCO_{3(s)} + 2H_3O^{+}_{(aq)} \longrightarrow Mg^{2+}_{(aq)} + 3H_2O_{(l)} + CO_{2(g)}$$

All metal carbonates contain the carbonate ion CO_3^{2-}

i) Two protons are transferred

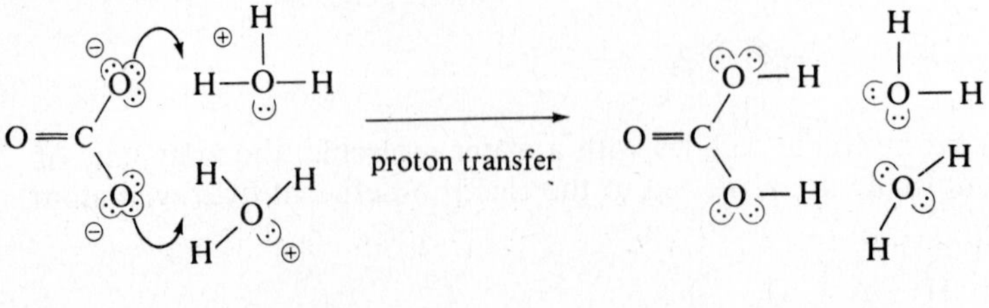

$$CO_3^{2-} + 2H_3O^{+} \xrightarrow{\text{proton transfer}} H_2CO_3 + 2H_2O$$

ii) The molecule H_2CO_3 is unstable and splits into two pieces

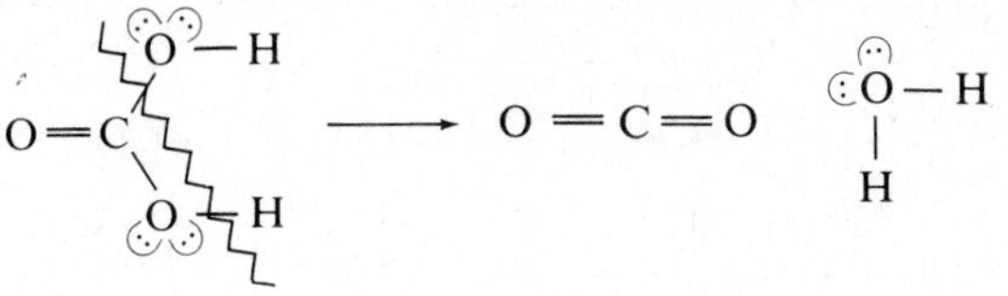

$$O{=}C(OH)_2 \longrightarrow O{=}C{=}O \quad H{-}O{-}H$$

Acids react with the more reactive metals

$$Zn_{(s)} + 2H_3O^+_{(aq)} \longrightarrow Zn^{2+}_{(aq)} + 2H_2O_{(l)} + H_{2(g)}$$

All metals contain delocalized outer-shell electrons in their lattice. These are attracted away from the lattice towards the hydroxonium ions.

$$\text{i.e. } 2e^- + 2H_3O^+_{(aq)} \longrightarrow H_{2(g)} + 2H_2O_{(l)}$$

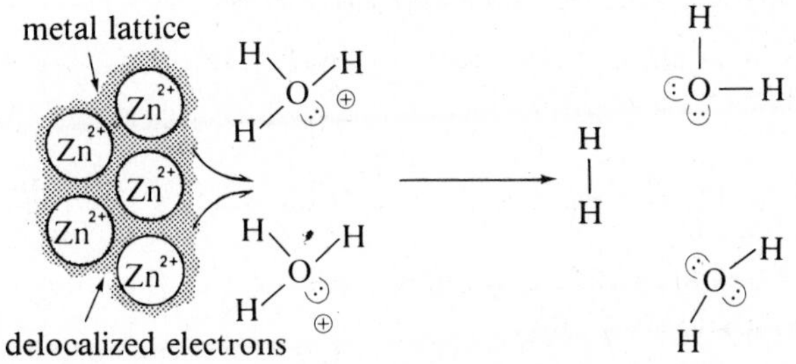

metal lattice

delocalized electrons

So, according to the Brønsted-Lowry theory:

An acid contains particles that act as proton donors

In each case of proton transfer, another particle accepts protons. This other particle is a *basic* particle.

A base contains particles that act as proton acceptors

So, for example, metal oxides and metal carbonates are bases.

Reversibility of proton transfer

When a molecule of HA protonates a water molecule, it acts as an acid, while the water molecule acts as a base.

$$\underset{\text{acid}}{HA_{(aq)}} + \underset{\text{base}}{H_2O_{(l)}} \longrightarrow \underset{\text{lost a proton}}{A^-_{(aq)}} + \underset{\text{gained a proton}}{H_3O^+_{(aq)}}$$

The two particles that are produced as a result of the proton transfer may themselves collide and transfer a proton between them. In this case the hydroxonium ion protonates the anion A^- and so it acts as an acid while the anion acts as a base.

$$A^-_{(aq)} + H_3O^+_{(aq)} \longrightarrow HA_{(aq)} + H_2O_{(l)}$$

base	acid	gained a proton	lost a proton

Both of these two opposing processes are likely to happen at the same time in the mixture and eventually the rates of the two processes will become equal so that equilibrium is reached.

$$HA_{(aq)} + H_2O_{(l)} \rightleftharpoons A^-_{(aq)} + H_3O^+_{(aq)}$$

acid 1	base 2	base 1	acid 2

In this equilibrium system, there are two acids and two bases.

acids	*bases*
HA	A^-
H_3O^+	H_2O

These pairs of acids and bases are called conjugate pairs. Notice that they differ from each other by one proton only.

The conjugate base of the acid HA is A^-.
The conjugate acid of the base H_2O is H_3O^+.

So for example, the conjugate base of HCl is Cl^-; the conjugate base of the acid H_2SO_4 is HSO_4^-. And the conjugate acid of NH_3 is NH_4^+; the conjugate acid of OH^- is H_2O.

When the equilibrium law is applied to the following reaction,

$$HA + H_2O \rightleftharpoons H_3O^+ + A^-$$

the equilibrium constant K_c is given by the expression

$$K_c = \frac{[H_3O^+][A^-]}{[HA][H_2O]}$$

But for dilute solutions, the concentration of water is approximately constant as the table below shows.

solution	*concentration of solute*	*concentration of water*	
	g dm^{-3}	g dm^{-3}	mol dm^{-3}
0.1 M HCl	3.6	997	55.5
1 M HNO_3	63.0	981	54.5
0.5 M H_2SO_4	49.0	984	54.6
1 M NaOH	40.0	1002	55.6
pure water	0	1000	55.5

The value of $[H_2O]$ is always approximately 55 mol dm^{-3}. When this value is put into the equilibrium expression, we get

$$K_c = \frac{[H_3O^+][A^-]}{[HA]} \times \frac{1}{55} \quad \Rightarrow \quad K_c \times 55 = \frac{[H_3O^+][A^-]}{[HA]}$$

A new equilibrium constant K_a is used instead of $K_c \times 55$ and so the equilibrium condition is written as follows.

$$HA_{(aq)} + H_2O_{(l)} \rightleftharpoons H_3O^+_{(aq)} + A^-_{(aq)}; \quad K_a = \frac{[H_3O^+][A^-]}{[HA]}$$

K_a is called the *dissociation constant* for the acid HA.

A similar expression can be derived for the equilibrium condition of a base in water. A base accepts a proton from a water molecule and therefore the equilibrium produced is

$$B_{(aq)} + H_2O_{(l)} \rightleftharpoons HB^+_{(aq)} + OH^-_{(aq)}; \quad K_b = \frac{[HB^+][OH^-]}{[B]}$$

K_b is called the *dissociation constant* for the base B.

In the two equilibria discussed above, water acts first as a base and then as an acid. A compound that is able to act either as an acid or as a base is called an *ampholyte*; its behaviour is described as *amphoteric*.

Water molecules can be both proton donors and proton acceptors depending on the nature of the system. In pure water, there is a tendency for water molecules to undergo proton transfer as shown below.

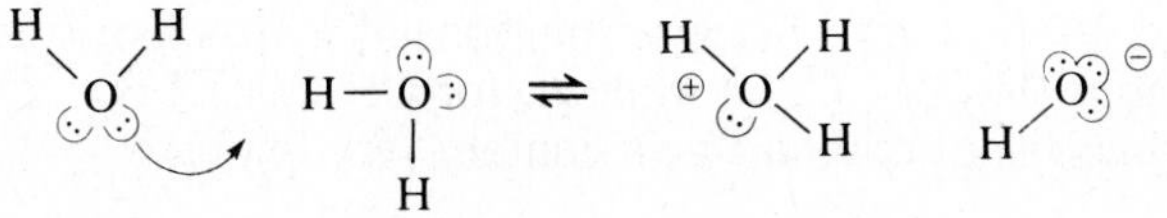

$$H_2\ddot{O}: \;+\; H{-}\ddot{O}{-}H \;\rightleftharpoons\; H_3O^{\oplus} \;+\; H{-}\ddot{O}:^{\ominus}$$

$$\underset{\text{base}}{H_2O_{(l)}} + \underset{\text{acid}}{H_2O_{(l)}} \rightleftharpoons \underset{\text{acid}}{H_3O^+_{(aq)}} + \underset{\text{base}}{OH^-_{(aq)}}; \quad K_c = \frac{[H_3O^+][OH^-]}{[H_2O]^2}$$

But $[H_2O]$ is constant, so a new constant can be written for $K_c\,[H_2O]^2$. It is given the symbol K_w and is the *dissociation constant* for water.

$$K_w = [H_3O^+][OH^-]$$

K_w has the value 1×10^{-14} mol^2 dm^{-6} at 298 K and, like all equilibrium constants, it is constant unless the temperature changes.

There is an interesting relationship between the dissociation constant of an acid and that of its conjugate base.

acid: HA — *conjugate base:* A^-

$$HA_{(aq)} + H_2O_{(l)} \rightleftharpoons A^-_{(aq)} + H_3O^+_{(aq)} \qquad A^-_{(aq)} + H_2O_{(l)} \rightleftharpoons HA_{(aq)} + OH^-_{(aq)}$$

$$K_a = \frac{[H_3O^+][A^-]}{[HA]} \qquad K_b = \frac{[HA][OH^-]}{[A^-]}$$

$$\text{So } K_a \times K_b = \frac{[H_3O^+][A^-]}{[HA]} \times \frac{[HA][OH^-]}{[A^-]}$$

$$= [H_3O^+] \times [OH^-]$$

$$= 10^{-14}\ \text{mol}^2\ \text{dm}^{-6}$$

Therefore $K_aK_b = K_w$ for the dissociation constants of an acid and its conjugate base.

Strength of acids and bases

Just as all equilibrium constants reflect the position of an equilibrium system, so the value of the dissociation constant K_a reflects the extent of protonation in an acidic solution.

The extent of protonation is often referred to in terms of the strength of an acid as shown below.

value of K_a mol dm^{-3}	*position of the equilibrium* $HA_{(aq)} + H_2O_{(l)} \rightleftharpoons H_3O^+_{(aq)} + A^-_{(aq)}$ (molecules ⇌ ions)	*strength of acid*
greater than 10^2	to the right: mainly ions	a strong acid: it shows a strong tendency to protonate water molecules
less than 10^{-2}	to the left: mainly molecules	a weak acid: it shows a weak tendency to protonate water molecules

For example, K_a for HCl is 10^7 mol dm^{-3} because solutions of hydrochloric acid consist almost entirely of ions ($H_3O^+_{(aq)}$, $Cl^-_{(aq)}$) while K_a for CH_3COOH is $10^{-4.8}$ mol dm^{-3} because solutions of ethanoic acid contain very few ions ($H_3O^+_{(aq)}$, $CH_3COO^-_{(aq)}$).

Similarly the dissociation constant K_b for a base indicates the extent of deprotonation in a solution of a base. Strong bases exist mainly as ions, while weak bases consist mainly of molecules.

A strong acid has a weak conjugate base, while a weak acid has a strong conjugate base because the dissociation constants of an acid and its conjugate base are related by the equation

$$K_aK_b = 10^{-14}\ \text{mol}^2\ \text{dm}^{-6}$$

Example a) HCl is a strong acid; its conjugate base is Cl^-

For HCl, $K_a = 10^7$ mol dm^{-3}

$$\text{Thus for } Cl^-,\ K_b = \frac{K_w}{K_a} = \frac{10^{-14}}{10^7} = 10^{-21}\ \text{mol dm}^{-3}$$

With $K_b = 10^{-21}$ mol dm^{-3}, Cl^- is a very weak base indeed.

Example b) H_2O is itself a weak acid; its conjugate base is OH^-

$$\text{For } H_2O\ K_a = \frac{[H_3O^+][OH^-]}{[H_2O]} = 10^{-15.8}\ \text{mol dm}^{-3}$$

$$\text{Thus for } OH^-,\ K_b = \frac{K_w}{K_a} = \frac{10^{-14}}{10^{-15.8}} = 10^{1.8}\ \text{mol dm}^{-3}$$

With $K_b = 10^{1.8}$ mol dm^{-3}, OH^- is a strong base.

The inverse relationship between the strength of any acid and the strength of its conjugate base is shown in the table of dissociation constants below.

	K_a mol dm^{-3}	*acid*	*conjugate base*	K_b mol dm^{-3}	
increasing acid strength	10^{7}	HCl	Cl^-	10^{-21}	increasing base strength
	10^{2}	H_2SO_4	HSO_4^-	10^{-16}	
	$10^{1.8}$	H_3O^+	H_2O	$10^{-15.8}$	
	$10^{-1.8}$	HSO_4^-	SO_4^{2-}	$10^{-12.2}$	
	$10^{-4.8}$	CH_3COOH	CH_3COO^-	$10^{-9.2}$	
	$10^{-6.4}$	H_2CO_3	HCO_3^-	$10^{-7.6}$	
	$10^{-7.0}$	H_2S	HS^-	$10^{-7.0}$	
	$10^{-9.2}$	NH_4^+	NH_3	$10^{-4.8}$	
	$10^{-15.8}$	H_2O	OH^-	$10^{1.8}$	
	10^{-30}	H_2	H^-	10^{16}	

There are two important conclusions to be drawn from this table of acid strengths.

1. *Predicting acid-base reactions.* The positions of the aqueous equilibria can be used as a guide even in non-aqueous conditions. For example if the acid $H_2SO_{4(l)}$ is added to the conjugate base of a weaker acid such as ethanoic acid, the table of strengths predicts that a reaction should occur.

$$H_2SO_{4(l)} + CH_3COO^-Na^+_{(s)} \longrightarrow Na^+HSO^-_{4(s)} + CH_3COOH_{(l)}$$

The opposite reaction is unlikely to happen: ethanoic acid is a weaker acid than sulphuric acid and is unlikely to be able to protonate the very weak base sodium hydrogensulphate.

In aqueous solutions, a similar argument suggests that sulphuric acid is present as $HSO^-_{4(aq)}$ and $H_3O^+_{(aq)}$ ions with a very small concentration of $SO^{2-}_{4(aq)}$ ions; $H_3O^+_{(aq)}$ is a stronger acid than $HSO^-_{4(aq)}$ and the position of equilibrium (below) lies to the left.

$$HSO^-_{4(aq)} + H_2O_{(l)} \rightleftharpoons H_3O^+_{(aq)} + SO^{2-}_{4(aq)}$$

weaker acid weaker base stronger acid stronger base

As a general rule, there will be appreciable reaction between any acid and the conjugate base of another acid having a lower K_a value.

2. *The levelling effect of the solvent.* When the above general rule is applied to the base water, it is found that all acids stronger than H_3O^+ (i.e. those acids with K_a greater than $10^{1.8}$ mol dm^{-3}) react to produce

solutions of $H_3O^+_{(aq)}$ and thus they all appear to be of the same strength. e.g.

$$H_2SO_{4(aq)} + H_2O_{(l)} \rightleftharpoons HSO^-_{4(aq)} + H_3O^+_{(aq)}$$

$$HCl_{(aq)} + H_2O_{(l)} \rightleftharpoons Cl^-_{(aq)} + H_3O^+_{(aq)}$$

The strongest acid present in each equilibrium system is the hydroxonium ion $H_3O^+_{(aq)}$. In other words, the strength of the two acids is *levelled* to that of the protonated form of the solvent water.

A similar effect takes place for bases stronger than OH^- (i.e. those bases with K_b greater than $10^{1.8}$ mol dm^{-3}). The strength of these bases is levelled to the strength of the depronated form of the solvent, i.e. $OH^-_{(aq)}$. Thus for example, the heat of neutralization of any strong acid by a strong base is the same. The reaction taking place is

$$H_3O^+_{(aq)} + OH^-_{(aq)} \longrightarrow 2H_2O_{(l)}$$

The heat of reaction $\Delta H_r^\ominus$ is -57 kJ mol^{-1} regardless of the identity of the acid or the base that was dissolved in the solution.

17.2 Acidity and pH

The acidity of an aqueous solution depends on the presence of hydroxonium ions H_3O^+. The concentration of these ions provides a measure of the relative acidity. For example, in pure water

$$H_2O_{(l)} + H_2O_{(l)} \rightleftharpoons H_3O^+_{(aq)} + OH^-_{(aq)}$$

for every H_3O^+ ion produced, there is one OH^- ion produced, i.e.

$$[H_3O^+] = [OH^-]$$

But we know that $K_w = [H_3O^+]\,[OH^-] = 10^{-14}$ mol^2 dm^{-6} at 298K, thus

$$[H_3O^+]^2 = 10^{-14} \Rightarrow [H_3O^+] = 10^{-7} \text{ mol dm}^{-3}$$

In a solution of hydrogen chloride containing 0.1 mol dm^{-3},

$$HCl_{(g)} + H_2O_{(l)} \rightarrow H_3O^+_{(aq)} + Cl^-_{(aq)}$$

If we assume complete proton transfer by the strong acid HCl:

$$[HCl] = [H_3O^+] \Rightarrow [H_3O^+] = 0.1 = 10^{-1} \text{ mol dm}^{-3}$$

system	$[H_3O^+]$ mol dm^{-3}	*acidity*
pure water	10^{-7}	less acidic
0.1M HCl	10^{-1}	more acidic

A shorthand method of representing the concentration of hydroxonium ions is very often used because its actual value is usually small, e.g. 10^{-7} and 10^{-1} as above. The procedure is as follows.

1. Work out the logarithm to base 10 of the value.
2. Put a minus sign in front of it, i.e. change its sign.

So, for $[H_3O^+] = 10^{-7}$ mol dm^{-3}; $\log_{10}[H_3O^+] = -7; -(-7) = +7$
for $[H_3O^+] = 10^{-1}$ mol dm^{-3}; $\log_{10}[H_3O^+] = -1; -(-1) = +1$

The resulting number is called the pH of the solution.

The pH of pure water at 298 K is therefore 7.
The pH of 0.1M HCl is 1.

$$pH = -\log_{10}[H_3O^+] \quad \text{or} \quad [H_3O^+] = 10^{-pH}$$

The pH of a solution is the negative of the logarithm to base 10 of its hydroxonium ion concentration.

$[H_3O^+]$ mol dm^{-3}	10^1	10^0	10^{-1}	10^{-2} . . .	10^{-7} . . .	10^{-13}	10^{-14}	10^{-15}	10^{-16}
pH	−1	0	1	2 . . .	7 . . .	13	14	15	16

← increasing acidity

← decreasing pH

This means that:

1. Acidic solutions have low pH values (pH < 7 at 298 K).
2. Neutral solutions (at 298 K) have pH $= 7$.
3. Basic solutions have high pH values (pH > 7 at 298 K).

The term pOH is sometimes used; pOH $= -\log_{10}[OH^-]$. It can easily be shown that pH + pOH = 14 at 298 K.

$$K_w = [H_3O^+][OH^-] = 10^{-14} \text{ mol}^2 \text{ dm}^{-6} \text{ at 298 K}$$

$$\log_{10}[H_3O^+][OH^-] = -14$$

$$\log_{10}[H_3O^+] + \log_{10}[OH^-] = -14$$

$$-\log_{10}[H_3O^+] - \log_{10}[OH^-] = 14 \Rightarrow pH + pOH = 14$$

So pOH for pure water = 7 and pOH for 0.1M HCl = 13.

This treatment is also applied to dissociation constants. Thus

$$pK_a = -\log_{10}K_a$$

e.g. pK_a for ethanoic acid is 4.8.

Measuring the pH of a solution

There is a range of naturally occurring dyes whose colours are affected by the concentration of hydroxonium ions present. These substances are called indicators and their colours and colour changes give some guide to the pH of the solution into which they are added. Indicators are discussed in more detail in section 17.5.

In order to obtain an accurate measurement of pH, it is necessary to determine the hydroxonium ion concentration. There are two main methods by which this can be achieved.

The first method employs a *concentration cell* (page 327) in which one half-cell contains a standard hydrogen electrode dipping into molar acid and the other half-cell is a hydrogen electrode dipping into the solution under test. The cell e.m.f. is proportional to the hydroxonium ion concentration of the test solution.

The second method uses a *glass electrode* and high resistance voltmeter: together these make up a 'pH meter'. The major feature of the apparatus is the tip of the electrode which is a special glass membrane whose inside surface is in contact with a standard buffer solution sealed inside the electrode. When the electrode dips into the test solution, the outside surface of the membrane comes into contact with the test solution. This is illustrated in the schematic diagram below.

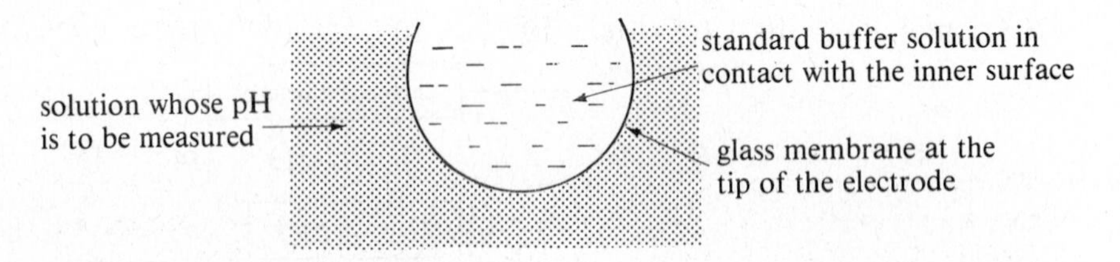

The different concentrations of hydroxonium ions in contact with the two surfaces of the membrane cause a small potential difference to exist between them. The voltmeter records this potential which is proportional to the difference in hydroxonium ion concentration. Since the solution in contact with the inside has a constant $[H_3O^+]$, the potential is proportional to the concentration on the outside.

Calculating the pH of a solution

Usually the calculation of the pH of a solution can be simplified in one of two ways.

1. If the acid or base is strong, we can assume that it is completely ionized.
2. If the acid or base is weak, we can assume that the equilibrium concentration of it is the same as its initial concentration.

For the pH calculations of acids or bases of medium strength, see page 281.

1. Strong acids or bases

Example (a) Calculate the pH of 0.001 mol dm^{-3} HCl

HCl is a strong acid, so assume $[H_3O^+] = [HCl]$

$[H_3O^+] = 10^{-3}$ mol dm$^{-3} \Rightarrow$ pH = 3

Example (b) Calculate the pH of 10 mol dm^{-3} HCl

Again assume $[H_3O^+] = [HCl]$

$[H_3O^+] = 10$ mol dm^{-3} = 10^1 mol dm$^{-3} \Rightarrow$ pH = -1

Example (c) Calculate the pH of 0.5 mol dm^{-3} NaOH
NaOH is a strong base, so assume $[OH^-]$ = [NaOH]
$[OH^-] = 5 \times 10^{-1}$ mol dm^{-3} = $10^{0.3}$ mol dm^{-3}
pOH = 0.3; but pOH + pH = 14 $\Rightarrow$ pH = 13.7

Example (d) Calculate the hydroxonium ion concentration when pH = 5.2
$[H_3O^+] = 10^{-5.2}$ mol dm^{-3} = $10^{0.8} \times 10^{-6}$ = 6.3×10^{-6} mol dm^{-3}

In all these calculations the concentration of hydroxonium ions that results from the dissociation of water has been ignored. It is very small: 10^{-7} mol dm^{-3} at 298 K. However, if the solution under consideration is very dilute, e.g. 10^{-8} mol dm^{-3} HCl, this contribution cannot be ignored. In this case:

$$\text{total } [H_3O^+] = [H_3O^+]_{\text{water}} + [H_3O^+]_{\text{HCl}}$$
$$= 10^{-7} + (0.1 \times 10^{-7})$$
$$= 1.1 \times 10^{-7} \text{ mol dm}^{-3}$$
$$= 10^{0.04} \times 10^{-7} \text{ mol dm}^{-3}$$
$$= 10^{-6.96} \text{ mol dm}^{-3}$$
$$\Rightarrow \text{pH} = 6.96$$

2. Weak acids or bases
These are not fully dissociated in solution. The fraction or percentage of molecules that dissociate into ions is termed the degree of dissociation α.
Assume the volume of solution that contains 1 mole of solute is V dm^3.

	$HA + H_2O$	$\rightleftharpoons$	H_3O^+	$+ A^-$	
initially:	1		0	0	moles in V dm^3
at equilibrium:	$(1 - \alpha)$		α	α	moles in V dm^3

This is because $(\alpha \times 1)$ moles of HA dissociate to produce α moles of H_3O^+ and α moles of A^-: $(1 - \alpha)$ moles of HA are left.

equilibrium concentrations:	$\left(\frac{1-\alpha}{V}\right)$	$\left(\frac{\alpha}{V}\right)$	$\left(\frac{\alpha}{V}\right)$	mol dm^{-3}

These equilibrium concentrations can be put into the equilibrium law:

$$K_a = \frac{[H_3O^+][A^-]}{[HA]} \Rightarrow K_a = \left(\frac{\alpha}{V}\right)\left(\frac{\alpha}{V}\right)\Big/\left(\frac{1-\alpha}{V}\right) = \frac{\alpha^2}{V(1-\alpha)}$$

This expression was first derived by Ostwald for the dissociation of a weak acid in aqueous solution and is known as

$$\text{Ostwald's dilution law: } K_a = \frac{\alpha^2}{V(1-\alpha)}$$

For many weak acids, the proportion of molecules dissociating is very small so $[HA]_{initial} \approx [HA]_{equilibrium} \quad \Rightarrow \quad 1 - \alpha \approx 1$

Ostwald's dilution law then becomes $\boxed{K_a = \dfrac{\alpha^2}{V}}$

Example (a) Find (i) the degree of dissociation, (ii) the pH of a solution of a weak acid ($K_a = 10^{-6}$ mol dm^{-3}) of concentration 0.1 mol dm^{-3}.

As the acid is weak, assume that α moles of ions per mole of acid dissociate.

$$HA \rightleftharpoons H_3O^+ + A^-$$

There is 0.1 mol HA in 1 dm³, thus 1 mol is present in 10 dm³.

	HA	H_3O^+	A^-	
initially:	1	0	0	mol in 10 dm³
at equilibrium:	$1 - \alpha$	α	α	mol in 10 dm³
concentration:	$\dfrac{1-\alpha}{10}$	$\dfrac{\alpha}{10}$	$\dfrac{\alpha}{10}$	mol dm^{-3}

(i) Because the acid is very weak, assume that $1 - \alpha = 1$

$$K_a = \frac{\alpha^2}{V} \Rightarrow \frac{\alpha^2}{10} = 10^{-6}$$
$$\alpha^2 = 10^{-5}$$
$$\alpha = \sqrt{10^{-5}} = 3.16 \times 10^{-3}$$

The degree of dissociation is 3.16×10^{-3} or 0.316%

(ii) $$[H_3O^+] = \frac{\alpha}{10} = \frac{3.16 \times 10^{-3}}{10} = 10^{0.5} \times 10^{-4} = 10^{-3.5}\ \text{mol dm}^{-3}$$

$\Rightarrow$ pH = 3.5

Example (b) Calculate K_a for a weak acid of pH 4.5 in a solution of concentration 0.05 mol dm^{-3}.

$$HA \rightleftharpoons H_3O^+ + A^-$$

There is 0.05 mol HA in 1 dm³, thus 1 mol is present in 20 dm³

	HA	H_3O^+	A^-	
initially:	1	0	0	mol in 20 dm³
at equilibrium:	$1 - \alpha$	α	α	mol in 20 dm³
concentration:	$\dfrac{1-\alpha}{20}$	$\dfrac{\alpha}{20}$	$\dfrac{\alpha}{20}$	mol dm^{-3}

Now $[H_3O^+] = [A^-] = \dfrac{\alpha}{20} = 10^{-4.5}$ mol dm^{-3}

Applying the equilibrium law: $K_a = \dfrac{[H_3O^+][A^-]}{[HA]} = \dfrac{10^{-9}}{(1-\alpha)/20}$

Assume that HA is a weak acid so that $1 - \alpha = 1$

$$\text{Then } K_a = \frac{10^{-9}}{1/20} = 20 \times 10^{-9} = 2.0 \times 10^{-8} \text{ mol dm}^{-3}$$

Example (c) Calculate the pH of a 0.1 mol dm^{-3} solution of ethanoate ions given that $K_a = 1.8 \times 10^{-5}$ mol dm^{-3} for CH_3COOH, the conjugate acid of the ethanoate base.

$$CH_3COO^-_{(aq)} + H_2O \rightleftharpoons CH_3COOH_{(aq)} + OH^-_{(aq)}$$

There is 0.1 mol ethanoate in 1 dm^3, thus 1 mol is present in 10 dm^3,

initially:	1	0	0	mol in 10 dm^3
at equilibrium:	$1 - \alpha$	α	α	mol in 10 dm^3
concentration:	$\frac{1 - \alpha}{10}$	$\frac{\alpha}{10}$	$\frac{\alpha}{10}$	mol dm^{-3}

$$\text{Applying the equilibrium law: } K_b = \frac{[CH_3COOH]\,[OH^-]}{[CH_3COO^-]} = \frac{\alpha^2}{10(1 - \alpha)}$$

If K_a for $CH_3COOH = 1.8 \times 10^{-5}$ mol dm^3

$$\text{then } K_b \text{ for } CH_3COO^- = \frac{K_w}{K_a} = \frac{10^{-14}}{1.8 \times 10^{-5}} = 5.56 \times 10^{-10} \text{ mol dm}^{-3}$$

Because the ethanoate is a weak base, assume that $1 - \alpha = 1$;

$$\text{Thus } 5.56 \times 10^{-10} = \frac{\alpha^2}{10} \Rightarrow \alpha^2 = 5.56 \times 10^{-9}$$

$$\Rightarrow \alpha = 7.46 \times 10^{-5}$$

$$\text{Now } [OH^-] = \frac{\alpha}{10} = 7.46 \times 10^{-6} \text{ mol dm}^{-3} \Rightarrow pOH = 5.12$$

$$\text{but } pH = 14 - pOH \qquad \Rightarrow pH = 8.88$$

Acids of medium strength.
When the strength of the acid is such that $10^2 > K_a > 10^{-4}$ mol dm^{-3}, the assumptions either that the acid is so strong that all the acid dissociates or that the acid is so weak that $(1 - \alpha) = 1$ do not apply.

In such cases a quadratic equation has to be solved to find the degree of dissociation α of the acid from the equilibrium constant K_a and the volume V of a solution of the acid of molar strength.

$$HA \rightleftharpoons H_3O^+ + A^-$$

$$\text{equilibrium concentrations: } \left(\frac{1 - \alpha}{V}\right) \quad \left(\frac{\alpha}{V}\right) \quad \left(\frac{\alpha}{V}\right) \text{ mol dm}^{-3}$$

These equilibrium concentrations are substituted in the equilibrium law

$$K_a = \frac{[H_3O^+]\,[A^-]}{[HA]} = \frac{\alpha^2}{V(1 - \alpha)} \Rightarrow \alpha^2 + VK_a\,\alpha - VK_a = 0$$

A value for α is obtained which is used to find $[H_3O^+] = \alpha/V$ and hence the pH.

17.3 Hydration and hydrolysis

When a solute dissociates into ions in aqueous solution, the ions are surrounded by a sheath of solvent water molecules. Water molecules possess dipoles and the oppositely charged ends are attracted to the ions. For example,

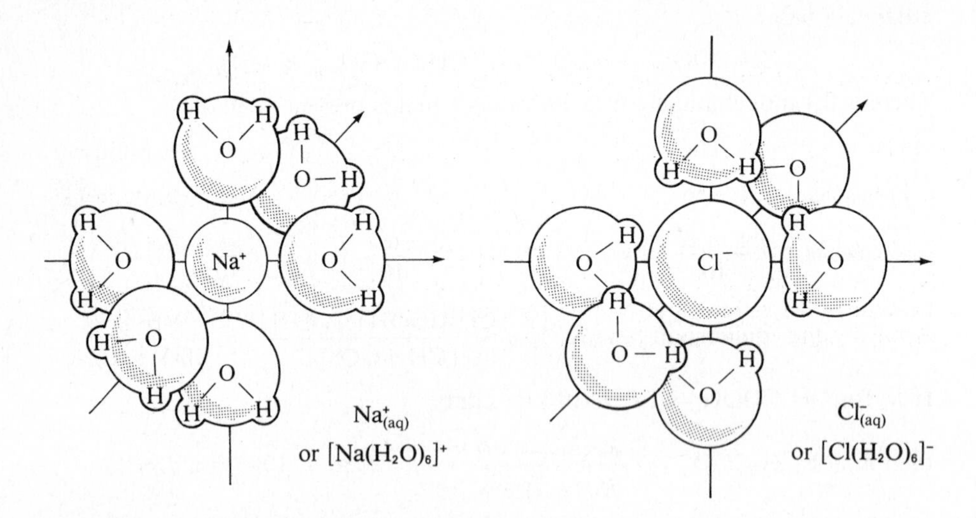

The water molecules are acting as *ligands* in the above two examples: they are coordinated to the central ions by electrostatic forces and there is little charge transferred between the surrounding molecules and each central ion. This process is called *hydration*, or *solvation* for any solvent that may be specified.

Hydration can be recognized by the lack of any reaction (no change in pH) between the solvating water molecules and the ions. In many cases however, the ions do react with the solvent: this process is called *hydrolysis* if the solvent is water, or *solvolysis* for any solvent that may be specified.

Hydrolysis can take place in one of three different ways, all of which produce a change in pH. Examples of each of these ways are given below.

1. When the cation or anion is a proton donor

Example (a) Ammonium chloride in water is acidic because it is hydrated to form ammonium aquo-ions which are then hydrolysed.

$$NH_4Cl_{(s)} \rightarrow NH^+_{4(aq)} + Cl^-_{(aq)} \quad \text{hydration}$$

$$NH^+_{4(aq)} + H_2O_{(l)} \rightleftharpoons NH_{3(aq)} + H_3O^+ \quad \text{hydrolysis}$$

$$K_a = 10^{-9\cdot 2} \text{ mol dm}^{-3}$$

Example (b) Sodium hydrogen sulphate in water is acidic for similar reasons.

$$NaHSO_{4(s)} \rightarrow Na^+_{(aq)} + HSO^-_{4(aq)} \quad \text{hydration}$$

$$HSO^-_{4(aq)} + H_2O_{(l)} \rightleftharpoons SO^{2-}_{4(aq)} + H_3O^+_{(aq)} \quad \text{hydrolysis}$$

$$K_a = 10^{-1\cdot 8} \text{ mol dm}^{-3}$$

2. When the anion is the conjugate base of a weak acid
Example (a) Sodium ethanoate in water is alkaline because it is hydrated to form ethanoate aquo-ions which are then hydrolysed.

$$CH_3COONa_{(s)} \rightarrow CH_3COO^-_{(aq)} + Na^+_{(aq)} \qquad \text{hydration}$$
$$CH_3COO^-_{(aq)} + H_2O_{(l)} \rightleftharpoons CH_3COOH_{(aq)} + OH^-_{(aq)} \qquad \text{hydrolysis}$$
$$K_b = 10^{-9.2} \text{ mol dm}^{-3}$$

Example (b) Potassium fluoride in water is alkaline for similar reasons.

$$KF_{(s)} \rightarrow K^+_{(aq)} + F^-_{(aq)} \qquad \text{hydration}$$
$$F^-_{(aq)} + H_2O_{(l)} \rightleftharpoons HF_{(aq)} + OH^-_{(aq)} \qquad \text{hydrolysis}$$
$$K_b = 10^{-10.75} \text{ mol dm}^{-3}$$

3. When the cation has a high polarizing power
A small or highly charged ion attracts the lone pairs of the coordinating water molecules so strongly that charge transfer can take place by the mechanism shown in the following example of $Al^{3+}_{(aq)}$.

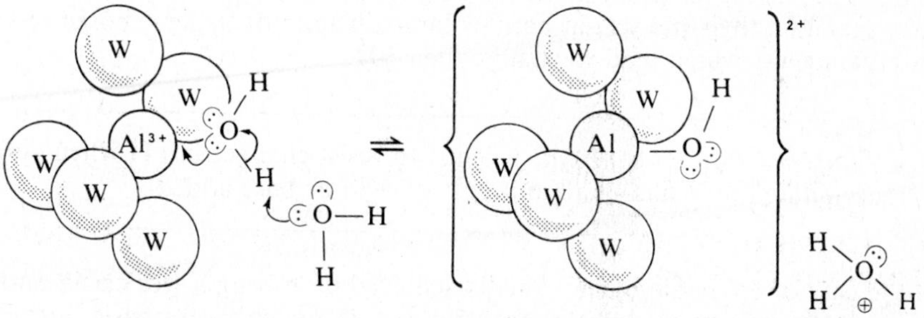

$$[Al(H_2O)_6]^{3+}_{(aq)} + H_2O_{(l)} \rightleftharpoons [Al(H_2O)_5OH]^{2+}_{(aq)} + H_3O^+_{(aq)}$$

The ligand water molecules are extensively polarized by the aluminium ions and therefore their tendency to act as proton donors is greatly increased as a result. It is possible for further proton transfer to occur, e.g.

$$[Al(H_2O)_5OH]^{2+}_{(aq)} + H_2O_{(l)} \rightleftharpoons [Al(H_2O)_4(OH)_2]^+_{(aq)} + H_3O^+_{(aq)}$$

The effect that different cations have on the pH of a solution is shown in the table of acid strengths below.

cation	*size* m 10^{-10}	*positive charge*	*hydrolysis reaction*	K_a mol dm^{-3}	*pH of 0.1M solution*
Na^+	9.5	1	$[Na(H_2O)_6]^+ \rightleftharpoons [Na(H_2O)_3OH] + H_3O^+$	2×10^{-16}	7.0
Li^+	6.0	1	$[Li(H_2O)_6]^+ \rightleftharpoons [Li(H_2O)_5OH] + H_3O^+$	3×10^{-14}	7.0
Mg^{2+}	6.5	2	$[Mg(H_2O)_6]^{2+} \rightleftharpoons [Mg(H_2O)_5OH]^+ + H_3O^+$	4×10^{-12}	6.2
Al^{3+}	5.0	3	$[Al(H_2O)_6]^{3+} \rightleftharpoons [Al(H_2O)_5OH]^{2+} + H_3O^+$	1.6×10^{-6}	3.4
Fe^{3+}	6.4	3	$[Fe(H_2O)_6]^{3+} \rightleftharpoons [Fe(H_2O)_5OH]^{2+} + H_3O^+$	6.3×10^{-3}	1.6

For some compounds both cation and anion hydrolysis takes place in solution; the overall result is neutrality. For example, when ammonium fluoride dissolves in water at 298 K, the pH is almost exactly 7. There is a large number of equilibria that have to be considered in a system of this nature.

$$NH_4F_{(s)} \rightleftharpoons NH^+_{4(aq)} + F^-_{(aq)} \quad \text{hydration}$$

$$\left.\begin{array}{l} NH^+_{4(aq)} + H_2O_{(l)} \rightleftharpoons NH_{3(aq)} + H_3O^+_{(aq)} \\ F^-_{(aq)} + H_2O_{(l)} \rightleftharpoons HF_{(aq)} + OH^-_{(aq)} \end{array}\right\} \text{hydrolysis}$$

$$NH_{3(aq)} + HF_{(aq)} \rightleftharpoons NH^+_{4(aq)} + F^-_{(aq)}$$

$$H_3O^+_{(aq)} + OH^-_{(aq)} \rightleftharpoons H_2O_{(l)} + H_2O_{(l)}$$

17.4 Buffer solutions

In many biological systems the control of pH within a narrow range is vital for the correct balance of the biochemical reactions. For example, healthy human blood has a pH in the range 7.1 – 7.8, while saliva has a pH of 6.8. In the stomach the digestive enzymes require a pH of 1.7 to operate effectively. These solutions and many others like them are buffer solutions: they maintain their pH within narrow limits in spite of being exposed to external agents which tend to change their pH.

> A *buffer solution* is one which tends to resist changes in pH when either it is diluted or it has small amounts of acid or base added.

In the case of an acidic buffer, this is achieved by mixing a weak acid and its conjugate base in appreciable concentrations. In the case of a basic buffer, the components are a weak base and its conjugate acid.

For example, a solution of ethanoic acid and sodium ethanoate constitutes an acidic buffer solution: its pH is below 7. In this solution the equilibrium system is:

$$CH_3COOH_{(aq)} + H_2O_{(l)} \rightleftharpoons CH_3COO^-_{(aq)} + H_3O^+_{(aq)}$$

Ethanoic acid is a weak acid and only dissociates slightly. The degree of dissociation is repressed further by the presence of the ethanoate ions from the added sodium ethanoate. The solution therefore contains a large concentration of undissociated acid.

Sodium ethanoate is a base of medium strength because its conjugate acid, ethanoic acid is weak. Consequently ethanoate ions in the solution tend to accept protons and form ethanoic acid. However the tendency for this to occur is reduced by the presence of the large quantity of undissociated acid.

1. The effect of adding a small amount of acid to the solution
The ethanoate ions react with the hydroxonium ions added and this causes the

position of the buffer equilibrium to shift to the left. The pH change is therefore minimal as long as there is a high enough concentration of ethanoate ions present at the start to react with the hydroxonium ions added.

2. The effect of adding a small amount of base to the solution
The added base reacts with the hydroxonium ions present in the buffer solution. As these are removed, the position of the buffer equilibrium shifts to the right: more undissociated ethanoic acid splits up and goes some way toward replacing the hydroxonium ions. The pH change is therefore again minimal as long as there is a high enough concentration of ethanoic acid molecules present at the start.

For example for any weak acid:

$$HA_{(aq)} + H_2O_{(l)} \rightleftharpoons H^+_{(aq)} + A^-_{(aq)}$$

$$K_a = \frac{[H_3O^+][A^-]}{[HA]}$$

$$\log_{10} K_a = \log_{10} [H_3O^+] + \log_{10} \frac{[A^-]}{[HA]}$$

Since $-\log_{10} K_a = pK_a$ and $-\log_{10} [H_3O^+] = pH$, this expression becomes

$$-pK_a = -\ pH + \log_{10} \frac{[A^-]}{[HA]}$$

$$\text{or } pH = pK_a + \log_{10} \frac{[A^-]}{[HA]}$$

As long as $[A^-] \approx [HA]$, $\log_{10} \frac{[A^-]}{[HA]} \approx \log_{10} 1.0 = 0$. Thus

$$pH \approx pK_a$$

Since K_a is a constant, pH remains approximately constant.

For a basic buffer, a similar treatment can be applied.

$$A^-_{(aq)} + H_2O_{(l)} \rightleftharpoons HA_{(aq)} + OH^-_{(aq)}$$

$$K_b = \frac{[HA][OH^-]}{[A^-]} \Rightarrow pOH = pK_b + \log_{10} \frac{[HA]}{[A^-]}$$

and $pOH \approx pK_b$ when $[A^-] \approx [HA]$.

Notice that the effect of diluting a buffer solution is to change all the concentrations by the same amount. The ratio of $[A^-]$ to $[HA]$ remains the same and so the pH remains constant.

A buffer solution must contain a weak acid (or base) and its conjugate base (or acid) in approximately equal concentration.

Procedure for making up a buffer solution to a given pH

Example a) To make up an acidic buffer of pH = 4:

1. Find from the data book a pK_a value near to the required pH. Here, methanoic acid ($pK_a = 3.75$) would do.
2. Use the acidic buffer expression to calculate the ratio $[A^-] : [HA]$

$$pH = pK_a + \log_{10}\frac{[A^-]}{[HA]}$$

Substituting for pH and pK_a:

$$4 = 3.75 + \log_{10}\frac{[A^-]}{[HA]}$$

$$\Rightarrow \log_{10}\frac{[A^-]}{[HA]} = 0.25$$

$$\Rightarrow \frac{[A^-]}{[HA]} = 1.78$$

Here A^- is $HCOO^-$ from HCOONa, and HA is HCOOH.

3. Make up the buffer solution by taking one mole of methanoic acid and adding 1.78 mole of sodium methanoate for every dm^3 of solution needed.

Example b) To make up a basic buffer of pH = 9.75 (pOH = 4.25)

1. Tables of pK_b show many suitable bases, but let us choose ammonia, NH_3 ($pK_b = 4.75$).
2. Use the basic buffer expression to calculate the ratio $[HA] : [A^-]$.

$$pOH = pK_b + \log_{10}\frac{[HA]}{[A^-]}$$

substituting for pOH and pK_b

$$\Rightarrow 4.25 = 4.75 + \log_{10}\frac{[HA]}{[A^-]}$$

$$\Rightarrow \log_{10}\frac{[HA]}{[A^-]} = -0.5$$

$$\Rightarrow \frac{[HA]}{[A^-]} = 0.316$$

here HA is NH_4^+ and A^- is NH_3.

3. Make up the buffer solution by taking one mole of ammonia and adding 0.316 mole of ammonium chloride for every dm^3 of solution needed.

17.5 Indicators and titrations

An indicator is a substance which changes colour as the pH of the solution in which it is dissolved changes. For example, phenolphthalein is

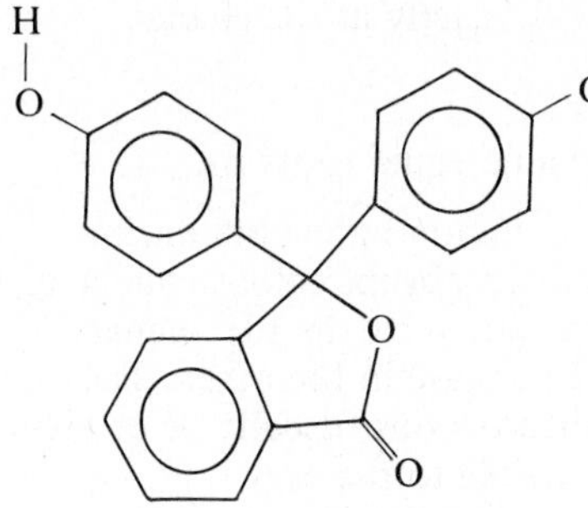

$$H\text{—}O\text{—}(C_{20}H_{13}O_3)$$

> An indicator is a weak acid whose conjugate base has a different colour from that of the acid.

For phenolphthalein:

weak acid (colourless) ⇌ *conjugate base* (pink)

$$H\text{—}O\text{—}(C_{20}H_{13}O_3)_{(aq)} + H_2O_{(l)} \rightleftharpoons H_3O^+_{(aq)} + {}^-O\text{—}(C_{20}H_{13}O_3)_{(aq)}$$

In neutral solution, very little of the weak acid dissociates and so it is colourless. If the above equilibrium is disturbed by the addition of base the deprotonated form of the weak acid is produced and this is coloured pink. The actual relationship between the pH of the solution and the colour of an indicator can be expressed more precisely as follows.

1. If the general formula of an indicator is HIn, where HIn has a colour *A* and its conjugate base In^- has a colour *B*, then

   $$HIn_{(aq)} + H_2O_{(l)} \rightleftharpoons In^-_{(aq)} + H_3O^+_{(aq)}$$

2. The dissociation constant K_{In} for the weakly acidic indicator is

   $$K_{In} = \frac{[In^-][H_3O^+]}{[HIn]} \text{ or } K_{In} = \frac{[In^-]}{[HIn]} \times [H_3O^+]$$

3. The colour of the solution depends on the value of the ratio $[In^-]/[HIn]$. When this is very large, $[In^-] > [HIn]$ and the colour is *B*; when this is very small, $[HIn] > [In^-]$ and the colour is *A*.
4. From the equation, this ratio is related to $[H_3O^+]$:

   $$\frac{[In^-]}{[HIn]} \times [H_3O^+] = \text{a constant}$$

So when the ratio is very large, $[H_3O^+]$ is likely to be small, i.e. at higher pH, the colour is *B*; and when the ratio is very small, $[H_3O^+]$ is likely to be larger, i.e. at lower pH, the colour is *A*. The actual range is determined by the magnitude of the dissociation constant K_{In}.

5. The most interesting situation arises when $[In^-] = [HIn]$. This represents a solution in which the colour is exactly at the change point. Under these conditions:
   $K_{In} = [H_3O^+]$ or $pK_{In} = pH$
   So an indicator changes colour at a pH value equal to its pK_{In}

The list below shows the pH range of some common indicators during which a colour change can be seen. There is not an abrupt colour change at one specific pH value. The shift in equilibrium caused by the pH change produces a gradual decrease in one colour and increase in the other. The concentration of one coloured form of the indicator must usually be at least ten times that of the other for a change to be visible to the eye.

indicator	HIn	In^-	K_{In} mol dm^{-3}	*pH range*
methyl orange	red	yellow	2×10^{-4}	3.1 – 4.4
bromophenol blue	yellow	blue	1×10^{-4}	3.0 – 4.6
methyl red	red	yellow	8×10^{-6}	4.2 – 6.3
bromothymol blue	yellow	blue	1×10^{-7}	6.0 – 7.6
phenol red	yellow	red	1×10^{-8}	6.8 – 8.4
phenolphthalein	colourless	red	5×10^{-10}	8.3 – 10.0

Titrations

A titration is the determination of the volume of a reactant solution that is the exact molar equivalent of a fixed volume of a second reactant solution.

The fixed volume is run into the reaction flask from a pipette; the volume being measured is then added little by little from a burette until the *equivalence point* (or end point) is reached.

At the equivalence point of a titration:

the molar ratio of the two reactants = the molar ratio specified by the reaction equation

The difficulty in a titration is to know at what point exactly the right volume has been added.

The purpose of an acid-base titration is normally to 'standardize' one solution i.e. to calculate its concentration from the results of titrating it against a second solution of known or 'standard' concentration.

For example, the *strong* acid HCl can be standardized against the *strong* base NaOH as follows.

1. Add 20 cm^3 of standard base to a conical flask using a pipette.
2. Dip a pH glass electrode into the flask.
3. Run acid solution into the flask and record the pH after the addition of specific volumes up to about 40 cm^3.
4. Plot the results on a graph.

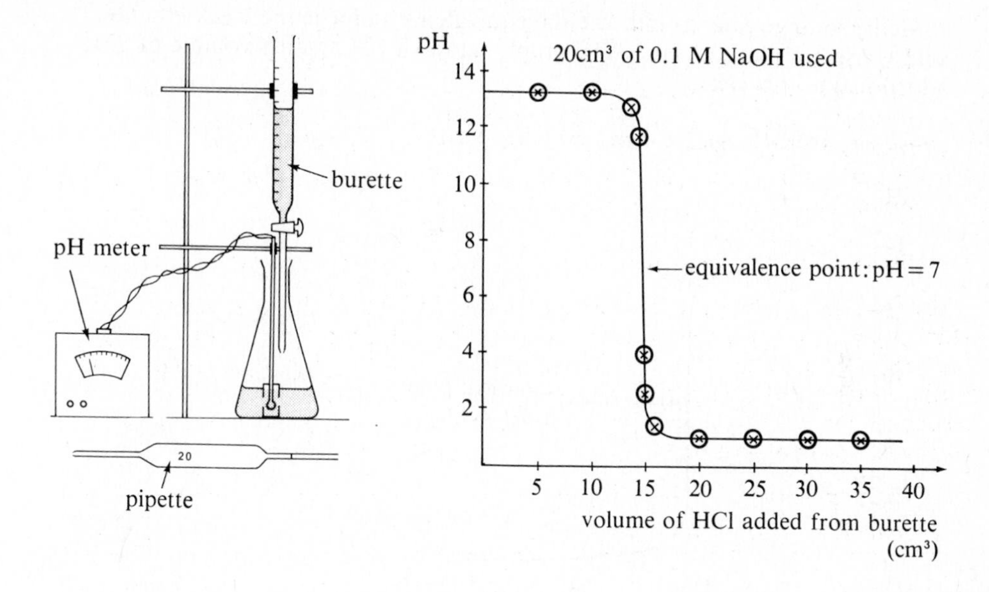

$$NaOH + HCl \rightarrow NaCl + H_2O$$

At the equivalence point, the system contains a solution of sodium chloride and water whose pH is 7. There is a large change in pH (from 11 to 3) for only a very small addition (about 0.2 cm^3) of acid. The mid-point of this pH change is after 15.7 cm^3 of acid have been added, and this is the equivalence point.

The concentration of the acid solution is calculated as follows.

20 cm^3 of 0.1M NaOH $\equiv$ 15.7 cm^3 of HCl ($\equiv$ means is equivalent to)

So at the equivalence point there are ($\frac{20}{1000} \times 0.1$) moles of NaOH and, because 1 mole of HCl reacts with 1 mole of NaOH, there must be the same number of moles of HCl.

Thus 15.7 cm^3 of HCl contains $\frac{20}{1000} \times 0.1$ moles

and concentration of HCl $= \frac{20}{1000} \times 0.1 \times \frac{1000}{15.7} = 0.127 \text{ mol dm}^{-3}$

Use of indicators in titrations

The above titration may be repeated using three drops of methyl orange indicator in the solution instead of a pH meter. At the equivalence point the methyl orange would be yellow because the pH of the solution is greater than 4.4 and it is present as In^-. The colour of methyl orange changes in the pH range 3.1 — 4.4 but, because for this strong acid the pH changes rapidly at the equivalence point, the colour change from yellow to red of this indicator can be used to determine the equivalence point.

Methyl orange *fails* to indicate the equivalence point if the weak ethanoic acid is used in place of HCl. The graph below of pH against volume of acid added makes this clear.

strong base (NaOH) against *weak acid* (CH_3COOH)

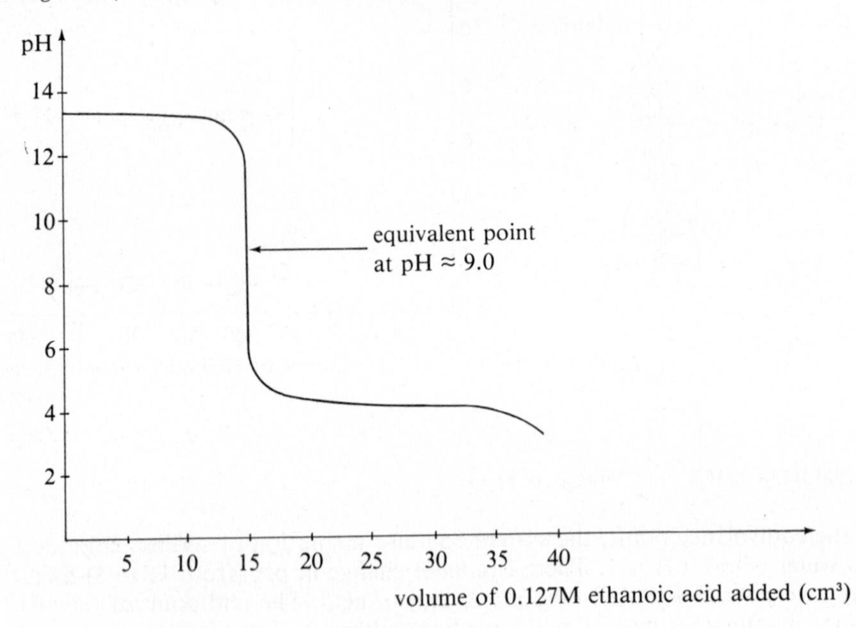

There are three important differences between this graph and the one given by a strong base and a strong acid.

1. The equivalence point is reached when the solution contains sodium ethanoate:

   $$CH_3COOH + NaOH \rightarrow CH_3COONa + H_2O$$

   Ethanoate ions in solution are basic (the conjugate base of a weak acid). So the equivalence point is reached at a pH greater than 7 (in this case pH ≈ 9).

   $$CH_3COO^-_{(aq)} + H_2O_{(l)} \rightleftharpoons CH_3COOH_{(aq)} + OH^-_{(aq)} \qquad pK_b = 9.8$$

2. When a small volume of acid is added close to the equivalence point, the change in pH is not so great.
3. After an excess of acid is added, the final pH is not much below 4.0. This is not surprising because ethanoic acid is a weak acid which is less dissociated.

The change in pH near the equivalence point is from about 11 to about 6. Since methyl orange changes colour in the pH range 3.1—4.4, it cannot register the equivalence point. An indicator whose pK_{In} is about 9 should be chosen instead, e.g. phenolphthalein.

A similar set of conclusions is reached for the titration of a strong acid by a weak base, e.g. hydrochloric acid and ammonia. The equivalence point is in the acid pH range and the initial pH is lower than given by a strong base. For this titration, methyl orange would be suitable, but phenolphthalein would not be suitable.

All four possible titration curves are shown below for comparison. They represent the changes for monobasic acids and bases whose concentrations are equal.

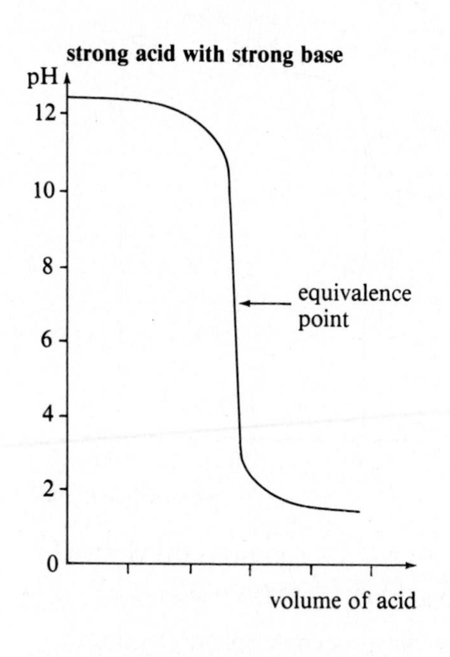

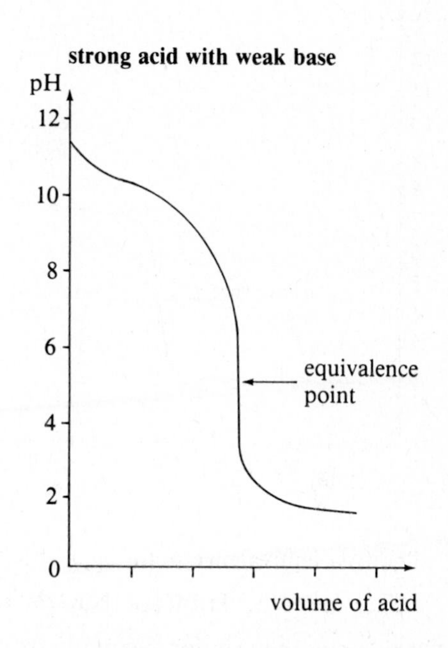

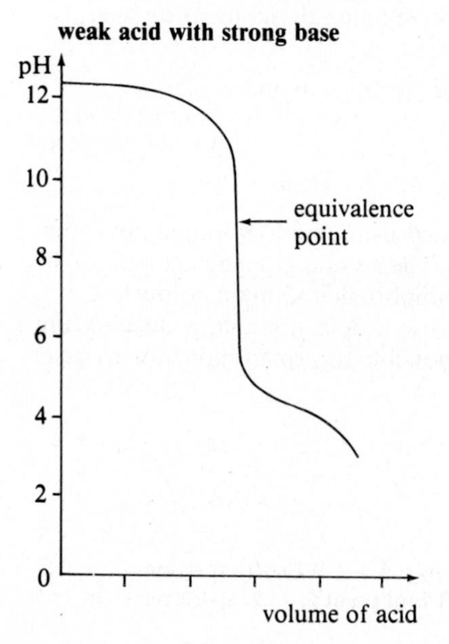

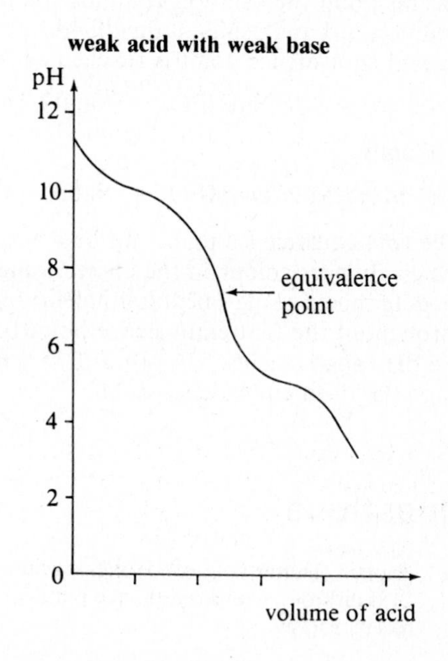

The titration curve for a diabasic acid or a tribasic acid has more than one inflection showing that there is more than one equivalence point. For example, the titration curve for 20 cm^3 of 0.1M H_3PO_4 has the shape shown in the graph below.

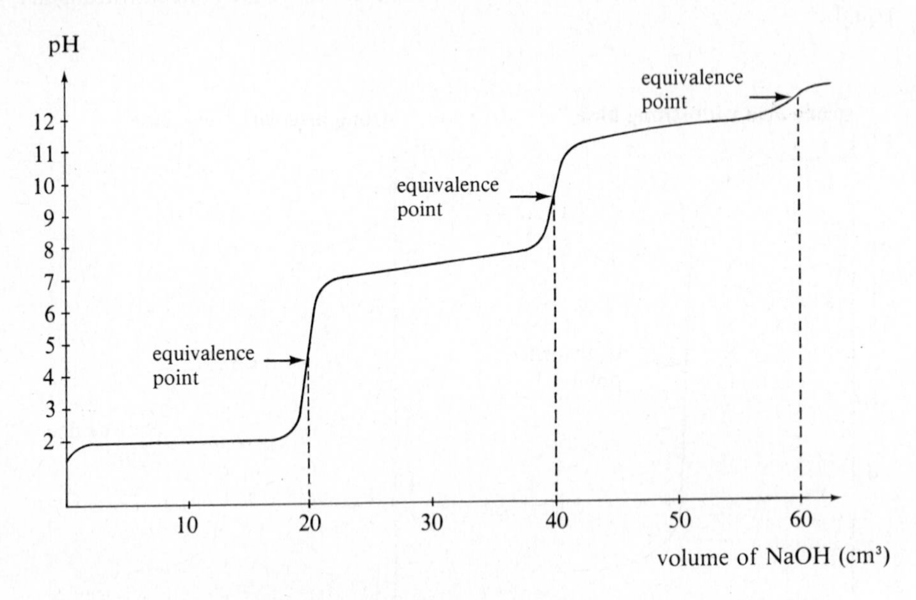

The first equivalence point is at about pH 4.4 and represents the change:

$$H_3PO_4 + NaOH \rightarrow NaH_2PO_4 + H_2O$$

At this point the solution contains sodium dihydrogenphosphate (V). Further addition of base results in the dihydrogenphosphate salt acting as an acid; the second equivalence point is reached as follows:

at pH $\approx$ 9.6: $NaH_2PO_4 + NaOH \rightarrow Na_2HPO_4 + H_2O$

Finally

at pH $\approx$ 12.2: $Na_2HPO_4 + NaOH \rightarrow Na_3PO_4 + H_2O$

The first equivalence point can be determined using methyl orange indicator which changes colour in the correct range. The second equivalence point can be determined using phenolphthalein; phenolphthalein remain colourless throughout the first equivalence point because it does not change to pink until the pH range reaches 8.3—10.0. It is not possible to use an indicator to determine the third equivalence point.

Questions

1. Explain the meaning of: a) Weak acid b) pH c) Buffer solution d) Indicator e) Equivalence point f) End point g) Hydration h) Hydrolysis.

2. a) Outline the Brønsted-Lowry theory of acids and bases.
   b) Complete the following equations and label each species either acid or base.
      i) $H_2O_{(l)} + CO^{2-}_{3(aq)} \rightleftharpoons$
      ii) $H_2O_{(l)} + HCO^{-}_{3(aq)} \rightleftharpoons$
      iii) $HCl_{(g)} + NH_{3(g)} \rightleftharpoons$
      iv) $H_3PO_{4(aq)} + H_2O_{(l)} \rightleftharpoons$
      (v) $H_2PO^{-}_{4(aq)} + H_2O_{(l)} \rightleftharpoons$
   c) List any ampholytes that occur in the equations above.
3. Explain why:
   a) a solution of aluminium chloride in water is acidic
   b) a solution of sodium carbonate in water is alkaline.
4. Carefully sketch the curves showing how the pH of the mixture changes when a 0.1 mol dm^{-3} solution of sodium hydroxide is added to:
   a) 10 cm^3 of 0.1 mol dm^{-3} nitric acid solution
   b) 10 cm^3 of 0.1 mol dm^{-3} propanoic acid solution (pK_a = 4.87)
   c) 10 cm^3 of 0.1 mol dm^{-3} phosphoric (V) acid.
5. Calculate the pH of the following solutions in each case assuming complete ionization.
   a) 1.0 mol dm^{-3} HCl
   b) 0.001 mol dm^{-3} HCl
   c) 10^{-5} mol dm^{-3} HNO_3
   d) 10^{-8} mol dm^{-3} HNO_3
   e) 1.25 mol dm^{-3} HCl
   f) 1.3×10^{-4} mol dm^{-3} HCl
   g) 1.5×10^{-3} mol dm^{-3} H_2SO_4
   h) 10 mol dm^{-3} HCl
   i) 20 mol dm^{-3} HNO_3
   j) 1 mol dm^{-3} NaOH
   k) 10^{-2} mol dm^{-3} NaOH
   l) 1.5×10^{-3} mol dm^{-3} NaOH
   m) 2×10^{-5} mol dm^{-3} $Ca(OH)_2$
   n) 2.3×10^{-6} mol dm^{-3} $Al(OH)_3$
   o) 10 mol dm^{-3} KOH
6. Calculate the hydroxonium ion concentration, $[H_3O^+]$, in mol dm^{-3} of solutions whose pH values are given below. Express your answers in standard form.
   a) 2 b) 6 c) 14 d) 0 e) −1
   f) 0.2 g) 0.629 h) 12.3 i) 5.3 j) 12.68
   k) 9.806 l) 12.89 m) 7.4 n) −0.461 o) −0.111

In questions 7 to 10, K_a is given in mol dm^{-3}.

7. a) Given $K_a = 4.5 \times 10^{-4}$ calculate $[H_3O^+]$ for a 0.02 mol dm^{-3} solution of HNO_2.
   b) Given $K_a = 2.06 \times 10^{-9}$ calculate the pH of a 0.26 mol dm^{-3} solution of HOBr.
   c) Given $K_a = 3.2 \times 10^{-8}$ calculate the pH of a 0.186 mol dm^{-3} solution of HOCl.
8. a) Given $K_a = 1.2 \times 10^{-4}$ calculate the concentration of a solution of HCNO whose pH = 2.65.
   b) A 0.1 mol dm^{-3} solution of an acid has a pH = 5.1. Find K_a for the acid.
   c) A 0.1 mol dm^{-3} solution of an acid has a pH of 2.89. Find K_a for the acid.
9. a) Given that $K_a = 1.8 \times 10^{-5}$, for ethanoic acid, calculate the pH and the degree of dissociation for solutions which were made up to be initially:
      i) 1 mol dm^{-3} ii) 0.1 mol dm^{-3} iii) 0.001 mol dm^{-3}
   b) Comment on the effect dilution has on the degree of dissociation of the acid.
10. a) Calculate the pH of a buffer made by adding 0.225 mol of ethanoic acid to 0.35 mol of sodium ethanoate in 0.6 dm^3 of solution. ($K_a = 1.8 \times 10^{-5}$.)
    b) Calculate the pH of a buffer made by adding 0.15 mol of ammonia to 0.25 mol of ammonium chloride in 0.75 dm^3 of solution. ($K_b = 4 \times 10^{-4}$)
    c) Calculate the mole ratio of ethanoic acid to sodium ethanoate needed to produce a buffer of pH = 4.55.

Chapter 18
Equilibrium and solubility

18.1 Solubility product

When a solid is added to a solvent, it begins to dissolve. This means that particles leave the solid lattice where they are surrounded by other similar particles and move into the liquid phase where they are surrounded by solvent particles. For example, the ionic solid sodium chloride dissociates into ions that become surrounded by water molecules when salt is put into water:

$$NaCl_{(s)} \xrightarrow{(aq)} Na^+_{(aq)} + Cl^-_{(aq)}$$

Initially there are only particles leaving the solid phase. After a short while, however, the concentration of aqueous ions builds up and the reverse process starts to happen. Aqueous ions arrive at the solid lattice surface and become an extension of it:

$$Na^+_{(aq)} + Cl^-_{(aq)} \longrightarrow NaCl_{(s)}$$

When the two opposing rates are equal, the system reaches equilibrium:

$$NaCl_{(s)} \rightleftharpoons Na^+_{(aq)} + Cl^-_{(aq)}$$

A solution which is in contact with solute and which is in equilibrium with it is said to be a *saturated solution*.

The example discussed above concerns an ionic solid dissolving in water. The same principles apply to any solid dissolving in any solvent; for example, iodine dissolves in tetrachloromethane.

$$I_{2(s)} \rightleftharpoons I_{2(CCl_4)}$$

Removal of solvent increases the concentration of the dissolved particles and therefore increases the rate of the back reaction. Equilibrium is disturbed and crystallization takes place until the concentration of the solute has decreased to a level where the rate of dissolving again equals the rate of crystallizing.

Solubility

The mass of any substance that dissolves in a given volume of a solvent is constant for that substance in the particular solvent at a specified temperature. This mass is known as the *solubility* of the substance and is measured in g dm^{-3}, mol dm^{-3} or g per 100 g.

Solids show a wide range of solubilities in water. The factors that affect solubility are discussed on page 150. Here are a few figures that illustrate the range:

type of substance	*substance*	*solubility* (at 293 K) g dm^{-3}
ionic	potassium chloride	4.2×10^2
	sodium chloride	3.6×10^2
	lead (II) chloride	1.1×10^1
	barium sulphate (VI)	2.3×10^{-3}
covalent	glucose	2.0×10^3
	bromine	3.2×10^1

The solubility of a substance can be used to calculate a form of equilibrium constant for the dissolving-crystallizing process taking place in a saturated solution.

For a solute $MX_{(s)}$ that is ionic and soluble in water, we can write:

$$MX_{(s)} \rightleftharpoons M^+_{(aq)} + X^-_{(aq)}$$

$$K_d = \frac{[M^+_{(aq)}]\,[X^-_{(aq)}]}{[MX_{(s)}]}$$

where K_d is a dissociation constant like K_a for an acid (see page 273). Since the concentration $[MX_{(s)}]$ of a solid is a constant, a new equilibrium constant K_{sp} for the solid can be used.

$$K_{sp} = [M^+]\,[X^-]$$

where K_{sp} is called the *solubility product*.

The units of this equilibrium constant depend on the stoichiometry of the solute. This is illustrated by the solubility products shown in the table below.

reaction	K_{sp}	*value at* 298 K	*units*
$AgCl_{(s)} \rightleftharpoons Ag^+_{(aq)} + Cl^-_{(aq)}$	$[Ag^+]\,[Cl^-]$	1.8×10^{-10}	$mol^2\ dm^{-6}$
$CaCO_{3(s)} \rightleftharpoons Ca^{2+}_{(aq)} + CO^{2-}_{3(aq)}$	$[Ca^{2+}]\,[CO_3^{2-}]$	5.1×10^{-9}	$mol^2\ dm^{-6}$
$PbBr_{2(s)} \rightleftharpoons Pb^{2+}_{(aq)} + 2Br^-_{(aq)}$	$[Pb^{2+}]\,[Br^-]^2$	7.9×10^{-5}	$mol^3\ dm^{-9}$
$Fe(OH)_{3(s)} \rightleftharpoons Fe^{3+}_{(aq)} + 3OH^-_{(aq)}$	$[Fe^{3+}]\,[OH^-]^3$	2.0×10^{-39}	$mol^4\ dm^{-12}$

The solubility of a substance can be used to work out the solubility product.

Example (a) The solubility of $BaSO_4$ is 2.33×10^{-3} g dm^{-3} at room temperature (R.M.M. of $BaSO_4$ = 233).

equation	$BaSO_{4(aq)} \rightleftharpoons$	$Ba^{2+}_{(aq)}$	+	$SO^{2-}_{4(aq)}$	
initial amount	$\left(\frac{2.33 \times 10^{-3}}{233}\right)$	0		0	mol in 1 dm^3
equilibrium concentrations:		$\left(\frac{2.33 \times 10^{-3}}{233}\right)$		$\left(\frac{2.33 \times 10^{-3}}{233}\right)$	mol dm^{-3}

$$K_{sp} = [Ba^{2+}]\,[SO_4^{2-}] = (10^{-5})\,(10^{-5})\ \text{mol}^2\ \text{dm}^{-6}$$
$$K_{sp} = 1 \times 10^{-10}\ \text{mol}^2\ \text{dm}^{-6}$$

The reverse procedure is adopted to work out the solubility of a solute from its solubility product.

Example (b) The solubility of iron (III) hydroxide can be calculated from the value of K_{sp} shown in the table on page 295.
Let x moles of iron (III) hydroxide dissolve in 1 dm^3.

equation	$Fe(OH)_{3(s)} \rightleftharpoons$	$Fe^{3+}_{(aq)}$	+	$3OH^-_{(aq)}$	
initial amount	x	0		0	mol in 1 dm^3
equilibrium concentrations:		x		$3x$	mol dm^{-3}

but
$$K_{sp} = [Fe^{3+}]\,[OH^-]^3 = 2 \times 10^{-39}$$
$$= (x)\,(3x)^3$$
$$27x^4 = 2.0 \times 10^{-39}$$
$$x = 9.28 \times 10^{-11}$$
$$\text{solubility} = (9.28 \times 10^{-11} \times 108)\ \text{g dm}^{-3}\ (\text{R.M.M. of } Fe(OH)_3 = 108)$$
$$= 1 \times 10^{-8}\ \text{g dm}^{-3}$$

Example (c) Calculate the solubility of bismuth sulphide given that $K_{sp} = 1 \times 10^{-97}$ mol^5 dm^{-15}.
Let x moles of bismuth sulphide dissolve in 1 dm^3.

equation	$Bi_2S_{3(s)} \rightleftharpoons$	$2Bi^{3+}_{(aq)}$	+	$3S^{2-}_{(aq)}$	
initial amount	x	0		0	mol in 1 dm^3
equilibrium concentrations:		$2x$		$3x$	mol dm^{-3}

but
$$K_{sp} = [Bi^{3+}]^2\,[S^{2-}]^3 = 1 \times 10^{-97}$$
$$= (2x)^2\,(3x)^3$$
$$(4 \times 27)\,x^5 = 1 \times 10^{-97}$$
$$x = 1.5 \times 10^{-20}$$
$$\text{solubility} = (1.5 \times 10^{-20} \times 262)\ \text{g dm}^{-3}\ \ (\text{R.M.M. of } Bi_2S_3 = 262)$$
$$= 4 \times 10^{-18}\ \text{g dm}^{-3}$$

Solubility in the presence of other salts: the common ion effect

The solubility product of a compound is a function of the concentration of the ions making it up. If the compound is to be dissolved in a solution that already contains some of these ions, the solubility of the compound must be affected. Less can dissolve because the resulting total ionic concentrations cannot exceed the solubility product.

Example Given that K_{sp} for iron (II) hydroxide at 298 K is 7.9×10^{-16} mol^3 dm^{-9} and its R.M.M. is 90, calculate its solubility in

i) pure water
ii) 0.1 M iron (II) sulphate (VI) solution
iii) 1 M sodium hydroxide solution

i) $Fe(OH)_{2(s)} \rightleftharpoons Fe^{2+}_{(aq)} + 2OH^-_{(aq)}$ $\qquad K_{sp} = [Fe^{2+}]\,[OH^-]^2$

Let the solubility be x mol dm^{-3} at 298 K, then, at equilibrium,

$$[Fe^{2+}] = x \text{ mol dm}^{-3}$$

$$[OH^-] = 2x \text{ mol dm}^{-3}$$

$$x(2x)^2 = 7.9 \times 10^{-16}$$

$$4x^3 = 7.9 \times 10^{-16} \Rightarrow x = 5.82 \times 10^{-6} \text{ mol dm}^{-3}$$

$$\text{solubility of } Fe(OH)_2 \text{ in water} = 5.24 \times 10^{-4} \text{ g dm}^{-3}$$

ii) In 0.1 M $FeSO_4$, there is already present $[Fe^{2+}] = 0.1$ mol dm^{-3}

Let the new solubility of $Fe(OH)_2$ be y mol dm^{-3}, then, at equilibrium,

$$[Fe^{2+}] = (0.1 + y) \approx 0.1 \text{ mol dm}^{-3} \ (y \text{ is very small})$$

$$[OH^-] = 2y \text{ mol dm}^{-3}$$

$$(0.1)\,(2y^2) = 7.9 \times 10^{-16}$$

$$0.4y^2 = 7.9 \times 10^{-16} \Rightarrow y = 4.44 \times 10^{-8} \text{ mol dm}^{-3}$$

$$\text{solubility of } Fe(OH)_2 \text{ in 0.1 M } FeSO_4 = 4.00 \times 10^{-6} \text{ g dm}^{-3}$$

Iron (II) hydroxide is therefore 0.76% as soluble in the iron (II) salt solution as in pure water.

iii) In 1 M NaOH, there is already present $[OH^-] = 1$ mol dm^{-3}

Let the new solubility of $Fe(OH)_2 = z$ mol dm^{-3}, then, at equilibrium,

$$[Fe^{2+}] = z \text{ mol dm}^{-3}$$

$$[OH^-] = (1 + 2z) \approx 1 \text{ mol dm}^{-3} \ (z \text{ is very small})$$

$$z(1)^2 = 7.9 \times 10^{-16}$$

$$z = 7.9 \times 10^{-16} \text{ mol dm}^{-3}$$

$$\text{solubility of } Fe(OH)_2 \text{ in 1 M NaOH solution} = 7.11 \times 10^{-14} \text{ g dm}^{-3}$$

Iron (II) hydroxide is over ten thousand million times less soluble in the hydroxide solution than in pure water.

> The reduced solubility of a salt in a solution that already contains an ion common to that salt is known as the *common ion effect*.

Using solubility product to predict precipitation

A saturated solution contains dissolved ions in equilibrium with ions present in a solid lattice:

$$MX_{(s)} \rightleftharpoons M^{+}_{(aq)} + X^{-}_{(aq)}$$

It follows therefore that the product of the ionic concentrations $[M^+]$ $[X^-]$ represents the maximum value permitted by the equilibrium condition. It is possible for the ionic product to be *less* than the value of the solubility product. In this case, the rate of the back reaction is slower than that of the forward reaction: more solid must dissolve in order to restore equilibrium. If there is no solid there to dissolve, the system simply remains as a solution containing hydrated ions.

The addition of more hydrated ions by means of a second solution leads to precipitation if the product of the ionic concentrations has a value greater than that of the solubility product.

$$\text{ionic product} > \text{value of the solubility product}$$

If the ionic product is *still* less than the value of K_{sp}, no solid can form because the rate of its dissolving would be greater than the rate of its formation from the solution.

Example 100 cm³ of a solution containing 0.001 mol dm⁻³ of magnesium chloride are poured into 400 cm³ of a solution containing 0.001 mol dm⁻³ of sodium hydroxide. Does precipitation occur at 298 K? (K_{sp} $Mg(OH)_2$ = 1.2×10^{-11} mol³ dm⁻⁹.)

$$Mg(OH)_{2(s)} \rightleftharpoons Mg^{2+}_{(aq)} + 2OH^{-}_{(aq)}$$

In 100 cm³ $Mg^{2+}_{(aq)}$, there is $\left(\frac{100}{1000} \times 0.001\right)\text{mol} = 10^{-4}$ mol

In 400 cm³ $OH^{-}_{(aq)}$, there is $\left(\frac{400}{1000} \times 0.001\right)\text{mol} = 4 \times 10^{-4}$ mol

When these are added together, the total volume = 500 cm³.

$\Rightarrow$ $[Mg^{2+}] = 2 \times 10^{-4}$ mol dm⁻³

$\Rightarrow$ $[OH^-] = 8 \times 10^{-4}$ mol dm⁻³

K_{sp} controls the maximum permitted amounts of the two ions that can exist in solution together:

$$K_{sp} = [Mg^{2+}]\,[OH^-]^2 = 1.2 \times 10^{-11} \text{ mol}^3 \text{ dm}^{-9}$$

In this case, the ionic product,

$$[Mg^{2+}]\,[OH^-]^2 = (2 \times 10^{-4})\,(8 \times 10^{-4})^2 = 1.28 \times 10^{-10} \text{ mol}^3 \text{ dm}^{-9}$$

This exceeds the value of K_{sp} (1.2×10^{-11} mol³ dm⁻⁹) and so precipitation occurs until the ionic product reaches 1.2×10^{-11}.

Variation of solubility with temperature

The value of a solubility product is usually quoted at 298 K. Like other equilibrium constants, it is temperature dependent and usually increases with increasing temperature. This means that a calculation to predict whether or not precipitation occurs is valid only at the temperature for which the solubility product is quoted.

The reason for the general increase with temperature of the solubility of solids in liquids is the growing importance of the entropy term $T\Delta S$ for the process in the equation $\Delta G = \Delta H - T\Delta S$ (see page 223).

18.2 Complex ion formation

> An ion which is surrounded by and bonded to a discrete group of coordinating particles is called a *complex ion*.

A typical example is an ion surrounded by water molecules.

The central ion is sometimes called the *nuclear ion*, while the surrounding particles are called *ligands*. Most ligands have a lone pair of electrons that becomes coordinated to a central cation, and the charge on the resulting complex ion depends on the relative charges of the cation and the ligands. Complex ions may be cationic, neutral or anionic (see page 300 for examples).

When complex ions are formed, the solubility of the ionic compound tends to be higher. For example, copper (II) hydroxide is far more soluble in water containing dissolved ammonia. The ammonia molecules act as ligands to produce a complex that is more soluble than the simple hydrated form.

$[Cu(NH_3)_4]^{2+}$: a deep blue-purple colour in solution.

Another example is the increase in the solubility of iodine in water when potassium iodide is added to the solution. The complex $K^+I_3^-$ forms, and this is more soluble than molecular iodine.

$[I—I—I]^-$: a deep brown colour in solution

(I_3^- can be considered as an I^+ ion complexed by two I^- ligands).

Examples of complex ions

central ion	*ligands*	*formula*	*structure*	*charge*
Al^{3+}	6 H_2O molecules	$[Al(H_2O)_6]^{3+}$	octahedral; W ligands around Al; where (W)— is H_2O:	3+ cationic
Ag^+	2 NH_3 molecules	$[Ag(NH_3)_2]^+$	linear; (A)—Ag—(A); where (A)— is H_3N:	1+ cationic
Pt^{2+}	2 Cl^- ions and 2 NH_3 molecules	$Pt(NH_3)_2Cl_2$	square planar; (A), (Cl)—Pt—(Cl), (A)	neutral
Zn^{2+}	4 OH^- ions	$[Zn(OH)_4]^{2-}$	tetrahedral; (OH), (HO), Zn, (OH), (OH)	2− anionic
Cr^{3+}	6 F^- ions	$[CrF_6]^{3-}$	octahedral; F ligands around Cr	3− anionic

Stability and ligand replacement

When a suspension of cobalt (II) hydroxide in water is shaken with aqueous ammonia, the solubility equilibrium below is disturbed.

$$Co(OH)_{2(s)} \rightleftharpoons Co^{2+}_{(aq)} + 2OH^-_{(aq)} \qquad K_{sp} = [Co^{2+}][OH^-]^2$$

Ammonia molecules collide with the aqueous cobalt (II) ions and a process of *ligand replacement* occurs. This is illustrated in the diagram below where (W) is a water molecule and (A) an ammonia molecule.

First replacement

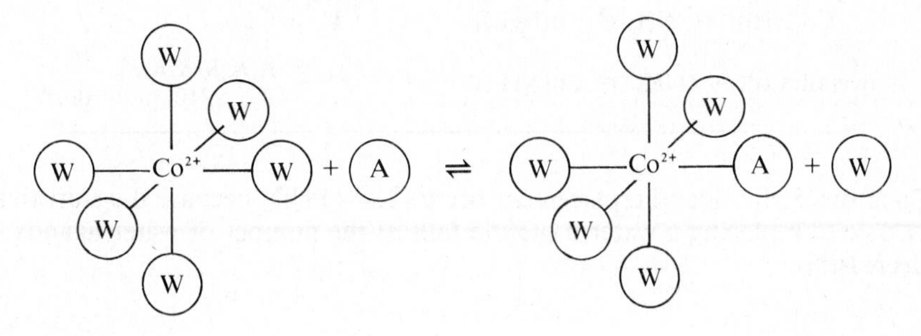

Second replacement

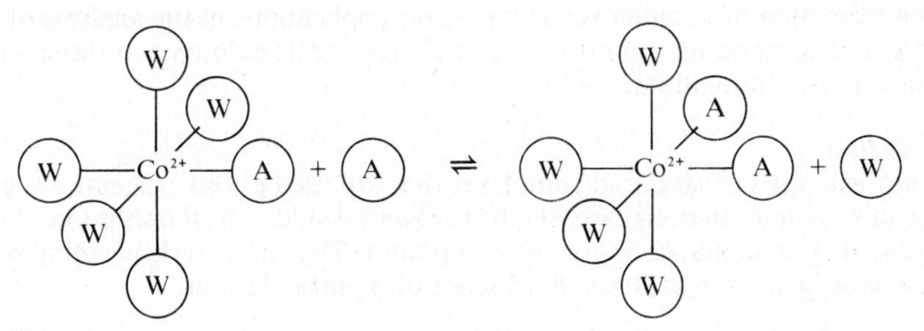

As the aqueous cobalt (II) ions are converted into ammonia-complexed ions, the position of the solubility equilibrium shifts to the right and more solid dissolves. In excess ammonia, $Co(NH_3)_6^{2+}$ is formed.

Ligand replacement is usually explained in terms of the relative stabilities of the complex ions involved. A measure of stability is provided by the dissociation constant of the complex ion in aqueous conditions.

$$Co(NH_3)_6^{2+} \overset{(aq)}{\rightleftharpoons} Co(NH_3)_5^{2+} + NH_3$$

$$K_d = \frac{[Co(NH_3)_5^{2+}][NH_3]}{[Co(NH_3)_6^{2+}]}$$

It is more usual to write the equilibrium the other way round so that a larger value of equilibrium constant indicates a more stable complex ion in solution. When written in this way, the constant is known as a *stability constant* and can take the form of a series of 'stepwise constants' or an overall constant, K_{stab}, for example:

equilibrium	*stepwise constant* (mol^{-1} dm^{3})
$Co^{2+} + NH_3 \rightleftharpoons Co(NH_3)^{2+}$	$K_1 = 100$
$Co(NH_3)^{2+} + NH_3 \rightleftharpoons Co(NH_3)_2^{2+}$	$K_2 = 30$
$Co(NH_3)_2^{2+} + NH_3 \rightleftharpoons Co(NH_3)_3^{2+}$	$K_3 = 10$
$Co(NH_3)_3^{2+} + NH_3 \rightleftharpoons Co(NH_3)_4^{2+}$	$K_4 = 4$
$Co(NH_3)_4^{2+} + NH_3 \rightleftharpoons Co(NH_3)_5^{2+}$	$K_5 = 1$
$Co(NH_3)_5^{2+} + NH_3 \rightleftharpoons Co(NH_3)_6^{2+}$	$K_6 = 0.3$
overall: $Co^{2+} + 6NH_3 \rightleftharpoons Co(NH_3)_6^{2+}$	$K_{stab} = K_1K_2K_3K_4K_5K_6$ $= 3.6 \times 10^4$ mol^{-6} dm^{18}

Each successive ligand replacement occurs less readily because the statistical chance of replacing a water molecule falls as the number of water ligands decreases.

Uses of complex formation in analysis

The formation of complexes has two main applications in the analysis of inorganic compounds: complexes with characteristic colours and those with characteristic solubilities.

1. Colour

Some examples of coloured complexes that are used as test reagents are given in the table below. In each case, the test reagent is added to the solution containing an unknown inorganic compound. The characteristic complex colour indicates the presence or absence of a particular ion.

test reagent	*positive result*	*complex*
aqueous thiocyanate ions: $[N{\equiv}C{-}S]^-$	a blood-red complex indicates $Fe^{3+}_{(aq)}$	$Fe(CNS)^{2+}$
aqueous ammonia NH_3	a deep blue-purple complex indicates $Cu^{2+}_{(aq)}$	$Cu(NH_3)_4^{2+}$
dimethyl glyoxime (butanedione dioxime)	a deep red precipitate indicates $Ni^{2+}_{(aq)}$	$Ni(dmg)_2$

```
HO       OH
  \ ..  .. /
   N     N
   ‖     ‖
   C  —  C
H /        \ /H
 C          C
H/ \       / \H
    H     H
```

The last example brings out further aspects of a ligand's properties. Dimethyl glyoxime is a *bidentate ligand*: each molecule possesses two separate atoms that are co-ordinated to the same nickel atom:

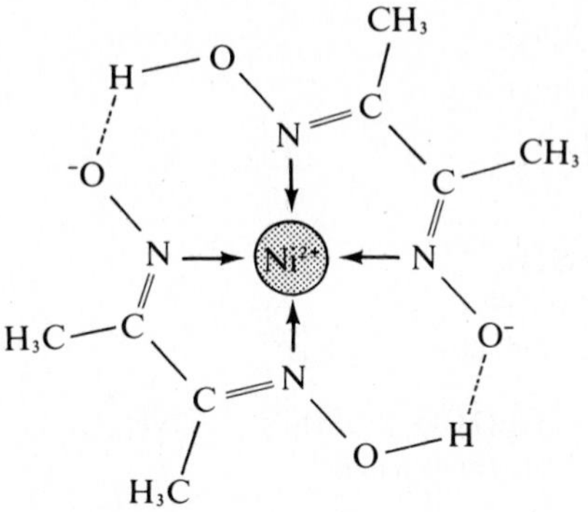

notice that each dmg molecule has lost one proton. The overall charge of Ni^{2+} and $2(dmg)^-$ is therefore neutral.

square planar

Ammonia and thiocyanate ions are *monodentate* ligands. Some other examples of polydentate ligands are:

a) 1,2-diaminoethane ('ethylene diamine') *bidentate*

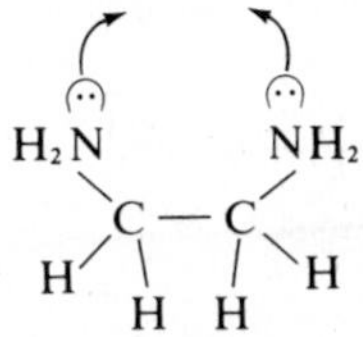

b) edta ('ethylene diamine tetracetic acid') $CH_2.N.(CH_2COO^-)_2$ | $CH_2.N.(CH_2COO^-)_2$ *hexadentate*

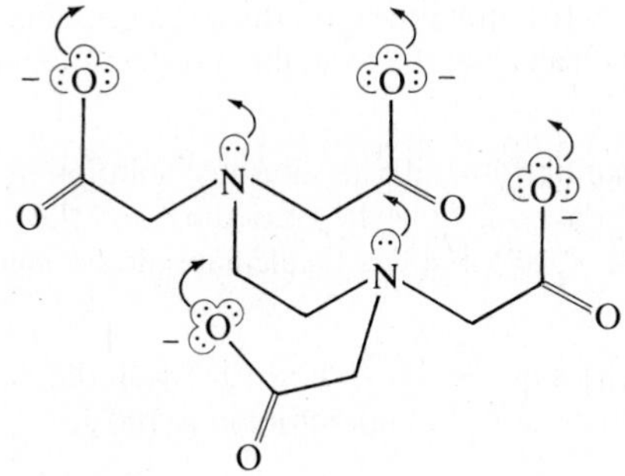

e.g. 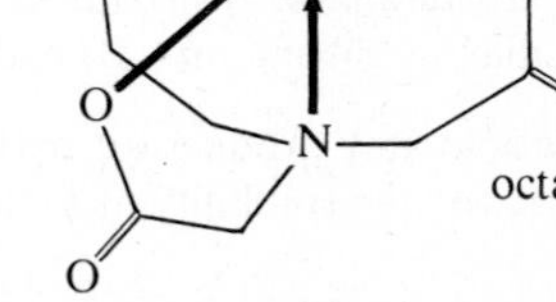 $[Ca(edta)]^{2-}$

octahedral

The term 'chelate compound' is often used to describe a compound containing ions complexed by polydentate ligands. The word chelate comes from the Greek word for crab's claw and provides a strong image of the effect of the polydentate ligand.

2 Solubility

Two examples of soluble complexes that are used to indicate the presence of cations are shown below.

a) Amphoteric hydroxides:

$$Al^{3+}_{(aq)} + 3OH^-_{(aq)} \rightarrow Al(OH)_{3(s)}$$

white precipitate

$\Big\Updownarrow$ excess $OH^-_{(aq)}$

$$Al(OH)_4^-$$

soluble complex

b) Ammonia complexes

$$Cu(H_2O)_6^{2+} + 2NH_{3(aq)} \rightleftharpoons Cu(H_2O)_4(OH)_{2(s)} + 2NH^+_{4(aq)}$$

blue precipitate

$\Big\Updownarrow$ excess $NH_{3(aq)}$

$$[Cu(H_2O)_2(NH_3)_4]^{2+} + 2OH^-_{(aq)}$$

soluble deep blue complex

Questions

1. Write solubility product expressions for each of the following:
   a) $AgBr$ b) $BaSO_4$ c) $PbCl_2$ d) $Al(OH)_3$ e) $Ag_2 CrO_4$ f) $Mg(OH)_2$
2. Solid magnesium carbonate is in equilibrium with its saturated solution in water. If the concentrations of magnesium and carbonate ions are each 3.16×10^{-3} mol dm^{-3}, calculate the solubility product of magnesium carbonate.
3. Solid iron (II) hydroxide is in equilibrium with its saturated solution in water. If the concentration of iron (II) ions is 5.82×10^{-6} mol dm^{-3} and the concentration of hydroxide ions is 1.16×10^{-5} mol dm^{-3}, calculate the solubility product of iron (II) hydroxide.
4. Solid chromium (III) hydroxide is in equilibrium with its saturated solution in water. If the concentration of chromium (III) ions is 2.47×10^{-9} mol dm^{-3} and the concentration of hydroxide ions is 7.40×10^{-9} mol dm^{-3}, calculate the solubility product of chromium (III) hydroxide.
5. The solubility product for barium sulphate at 298 K is 1.3×10^{-10} mol^2 dm^{-6}. Calculate the solubility of barium sulphate at this temperature in g/100 g.
6. Here are some values of solubility products:

compound	K_{sp}	
$AgBr$	5×10^{-13}	$mol^2\ dm^{-6}$
$SrCO_3$	1.1×10^{-10}	$mol^2\ dm^{-6}$
Ag_2CrO_4	1.3×10^{-12}	$mol^3\ dm^{-9}$
$Co(OH)_2$	6.3×10^{-15}	$mol^3\ dm^{-9}$

Put these compounds in order of increasing solubility expressed in mol dm^{-3}.

7. Describe two practical methods of determining solubility products

8. Write stability constant expressions for the following complexes:
   a) $FeCNS^{2+}$ b) $CuCl_4^-$ c) $Cu(NH_3)_4^{2+}$ d) $Fe(CN)_6^{3-}$ e) $Pb(OH)_3^-$ f) $Co(NH_3)_6^{2+}$

9. Silver forms the complex $Ag(NH_3)_2^+$ in aqueous ammonia. In a 1×10^{-3} mol dm^{-3} equilibrium concentration of ammonia, the silver ion concentration is found to be 1×10^{-1} mol dm^{-3} and this is equal to the concentration of the complex ion. Calculate the stability constant for this complex.

10. Discuss the importance of complex ions in:
    a) the analysis of unknown solutions
    b) changing the solubility of compounds

Chapter 19

Equilibrium in redox reactions

In Chapter 17 the ideas of equilibrium in acid-base reactions are discussed. In this chapter, they are extended to oxidation and reduction.

Whereas acid-base reactions are understood in terms of the transfer of protons, oxidation-reduction reactions involve the transfer of electrons. There are some important similarities and differences between these two processes. To help make this clear, the first sections of Chapters 17 and 19 are structured in the same way. Some direct references are also given on occasion.

19.1 Electron transfer

Many metals react with water or dilute solutions of acids. In doing so, the solid metal dissolves and becomes a solution of hydrated metal cations. In order to do this, each metal atom loses electrons:

$$M_{(s)} \longrightarrow M^{n+}_{(aq)} + ne^-$$

The tendency of a metal atom to lose electrons is a measure of its reactivity. Each metal has a different reactivity associated with the tendency of its atoms to lose electrons.

Conversely, many non-metals react with metals or water to form hydrated non-metal anions. In order to do this, each non-metal atom gains electrons, e.g. for the halogens, oxygen or nitrogen

$$X_{2(g)} + (2n)e^- \longrightarrow 2X^{n-}_{(aq)}$$

The process of electron loss is called oxidation. The process of electron gain is called reduction.

Metals tend to be oxidized while non-metals are often reduced.

Reversibility of electron transfer

Although metal atoms tend to lose electrons, there is also some tendency for the metal cations produced to gain electrons. Consider the case of a zinc rod dipping into an aqueous solution of zinc ions: zinc atoms in the metal lattice come into contact with hydrated zinc ions. Two possible electron transfer reactions can take place.

1. A zinc atom may lose two electrons and become a zinc ion. The two electrons are left behind on the electrode making it negatively charged

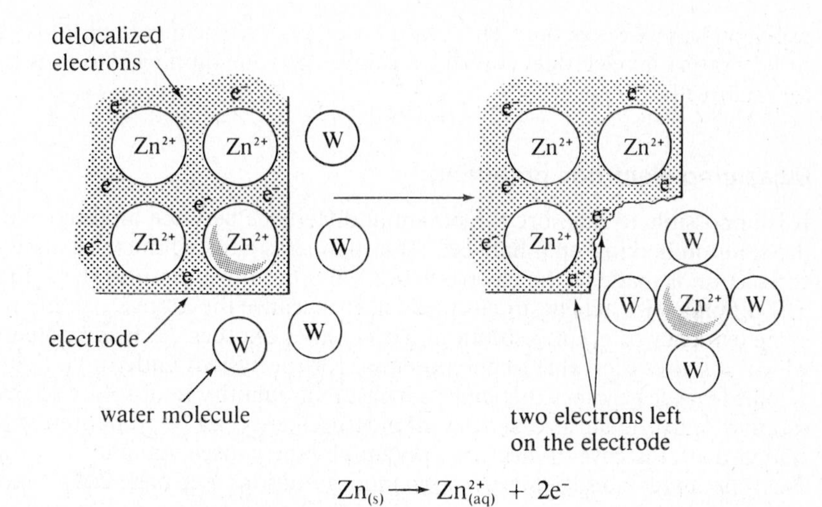

$$Zn_{(s)} \longrightarrow Zn^{2+}_{(aq)} + 2e^-$$

2. A zinc ion may remove two electrons from the lattice of the electrode and become a zinc atom. This leaves the electrode positively charged

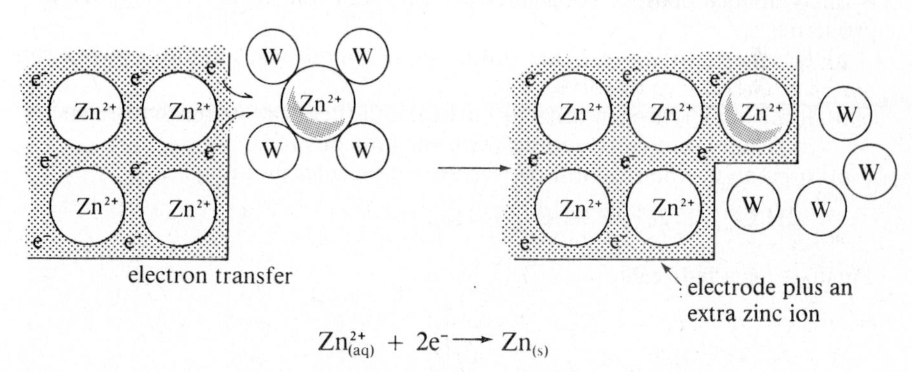

$$Zn^{2+}_{(aq)} + 2e^- \longrightarrow Zn_{(s)}$$

The two tendencies are not equal for any metal, and consequently one of the above processes occurs more readily. The electrode therefore becomes progressively more charged with respect to the solution until the rates of the two opposing processes become equal. Under these conditions, equilibrium exists and the potential difference between the metal electrode and the solution of its ions is called the *electrode potential* of the metal. More generally,

> The electrode potential of an element is the potential difference between an aqueous solution of its ions and the element when it is in equilibrium with that solution.
>
> For a metal, $M^{n+}_{(aq)} + ne^- \rightleftharpoons M_{(s)}$ E = electrode potential

There is a convenient term used to describe the arrangement of an electrode dipping into a solution of ions: *half-cell*. Two different electrodes are likely to be at a different potential and so a flow of electricity would take place if they were connected. Since the term 'cell' is used to describe any apparatus capable of generating an electrical current, a single electrode dipping into ions is known as a half-cell.

Measuring electrode potentials

It is impossible to measure the potential difference between an electrode and the solution making up a half-cell. If a high resistance voltmeter is used, one connection is made to the electrode but the other must dip into the solution. The dipping wire acts as an electrode itself because the atoms forming it have some tendency to go into solution. This creates another electrode potential which obscures the value of the potential for the system under investigation.

Whenever it becomes difficult to measure a quantity in absolute terms, a standard must be defined so that all measurements can be compared with this standard. In the case of electrode potentials, the chosen standard is a hydrogen electrode under standard thermodynamic conditions (see page 209). Since hydrogen is a gas at room temperature, a different design of electrode apparatus is needed. An 'inert electrode' of platinum black is used to bring hydrogen molecules into contact with hydrated hydrogen ions. Platinum black is finely divided platinum deposited on platinum foil. It has three principal properties:

a) It is inert in the sense that, under these conditions, platinum shows almost no tendency to dissolve.
b) It adsorbs hydrogen onto its surface and therefore brings hydrogen molecules into close contact with the aqueous ions.
c) It catalyses the reaction between the molecules and ions.

$$2H_3O^+_{(aq)} + 2e^- \rightleftharpoons 2H_2O_{(l)} + H_{2(g)}$$

The hydrogen half-cell

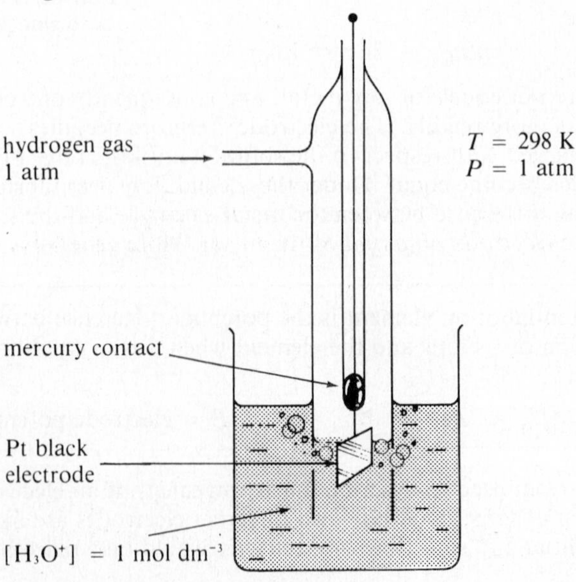

The standard hydrogen half-cell is used to measure the *standard electrode potential* $E^{\ominus}$ of any other half-cell as follows.

1. Set up the two half-cells next to each other and ensure that all the concentrations of the solutions are 1 mol dm^{-3}, that the temperature is 298 K and that the pressure is 1 atm.
2. Connect the two solutions with an electrolyte 'salt bridge'. This provides electrical contact between the two half-cells, but avoids the problem of using a wire dipping into the two solutions: a wire connector creates its own electrode potentials which interfere with the measurement to be made. The most commonly used electrolyte in a salt bridge is saturated potassium chloride solution.
3. Measure the p.d. between the standard hydrogen half-cell and the other half-cell. Either a potentiometer or a high-resistance value voltmeter can be used. For example, for a metal.

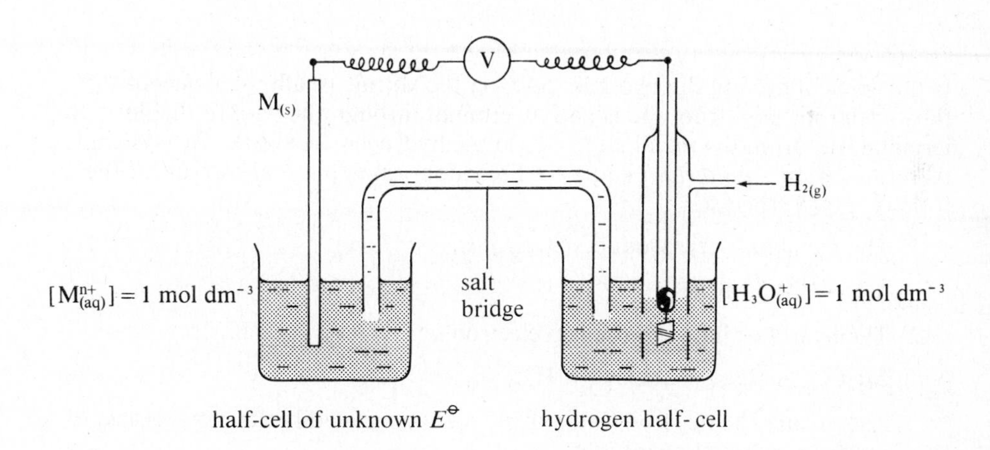

> The standard electrode potential of an element $E^{\ominus}$ is defined as the potential difference between a standard hydrogen half-cell and a half-cell of the element containing solutions of 1 mol dm^{-3} at 298 K and 1 atm.

Standard metal electrodes

When different metal half-cells are set up as shown above, it is usually found that the metal electrode is negatively charged compared with the hydrogen electrode: there is a greater density of electrons on the metal electrode.

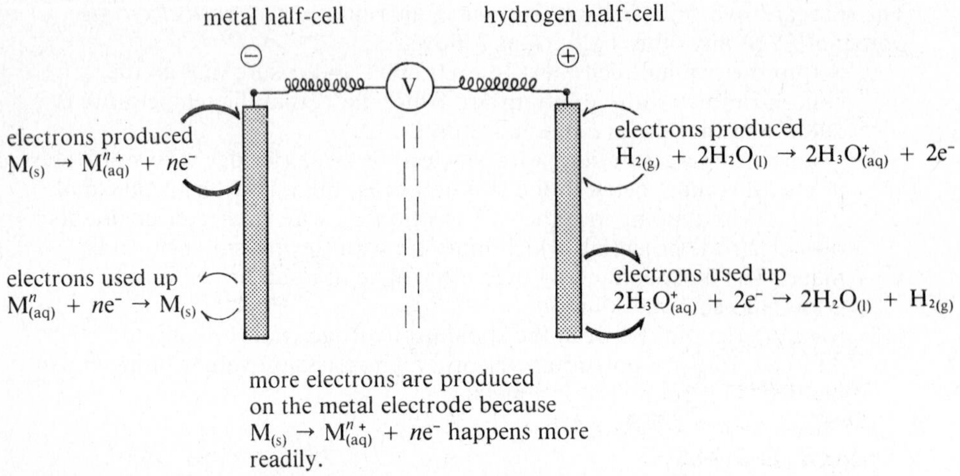

metal half-cell

hydrogen half-cell

⊖ V ⊕

electrons produced
$M_{(s)} \rightarrow M^{n+}_{(aq)} + ne^-$

electrons used up
$M^{n}_{(aq)} + ne^- \rightarrow M_{(s)}$

electrons produced
$H_{2(g)} + 2H_2O_{(l)} \rightarrow 2H_3O^+_{(aq)} + 2e^-$

electrons used up
$2H_3O^+_{(aq)} + 2e^- \rightarrow 2H_2O_{(l)} + H_{2(g)}$

more electrons are produced
on the metal electrode because
$M_{(s)} \rightarrow M^{n+}_{(aq)} + ne^-$ happens more
readily.

If the wire connecting the two half-cells via the voltmeter allows a current to flow, electrons pass from the negative terminal through the wire to the positive terminal, i.e. from the metal electrode to the hydrogen electrode. The system is therefore *not at equilibrium* and a 'cell reaction' takes place as a result of the following two processes.

1. The metal half-cell produces electrons:

   $$M_{(s)} \longrightarrow M^{n+}_{(aq)} + ne^-$$

2. The hydrogen half-cell uses up electrons:

   $$2H_3O^+_{(aq)} + 2e^- \longrightarrow 2H_2O_{(l)} + H_{2(g)}$$

   The greater the negative value of $E^{\ominus}$, the more readily the metal reacts to produce hydrated metal ions.

Some metals have positive $E^{\ominus}$ values. This means that their atoms lose electrons *less* readily than do hydrogen molecules. For example, a copper rod dipping into 1 M copper nitrate solution gives a reading of +0.34 volts. In this cell, electrons flow through the wire from the hydrogen electrode to the copper electrode and the reactions are as follows.

1. The hydrogen half-cell produces electrons:

   $$H_{2(g)} + 2H_2O_{(l)} \longrightarrow 2H_3O^+_{(aq)} + 2e^-$$

2. The copper half-cell uses up electrons:

   $$Cu^{2+}_{(aq)} + 2e^- \longrightarrow Cu_{(s)}$$

A list of the $E^{\ominus}$ values for the metals is called *the electrochemical series*: some of these are shown below.

reaction	$E^{\ominus}$ (V)	*reactivity*
$K^{+}_{(aq)} + e^{-} \rightleftharpoons K_{(s)}$	−2.92	these metals displace hydrogen from dilute acid
$Ca^{2+}_{(aq)} + 2e^{-} \rightleftharpoons Ca_{(s)}$	−2.87	
$Na^{+}_{(aq)} + e^{-} \rightleftharpoons Na_{(s)}$	−2.71	
$Mg^{2+}_{(aq)} + 2e^{-} \rightleftharpoons Mg_{(s)}$	−2.37	
$Al^{3+}_{(aq)} + 3e^{-} \rightleftharpoons Al_{(s)}$	−1.66	
$Zn^{2+}_{(aq)} + 2e^{-} \rightleftharpoons Zn_{(s)}$	−0.76	
$Fe^{2+}_{(aq)} + 2e^{-} \rightleftharpoons Fe_{(s)}$	−0.44	
$Sn^{2+}_{(aq)} + 2e^{-} \rightleftharpoons Sn_{(s)}$	−0.14	
$Pb^{2+}_{(aq)} + 2e^{-} \rightleftharpoons Pb_{(s)}$	−0.13	
$2H_3O^{+}_{(aq)} + 2e^{-} \rightleftharpoons 2H_2O_{(l)} + H_{2(g)}$	0.00	
$Cu^{2+}_{(aq)} + 2e^{-} \rightleftharpoons Cu_{(s)}$	+0.34	these metals do not displace hydrogen
$Ag^{+}_{(aq)} + e^{-} \rightleftharpoons Ag_{(s)}$	+0.80	

Non-metals: the use of inert electrodes

Although non-metal atoms tend to accept electrons and become aqueous anions, there is also a tendency for the anions to lose electrons. The use of an inert electrode dipping into an aqueous solution of anions enables these tendencies to be studied. For example, the standard electrode potential of chlorine is measured by using a cell like the one shown below:

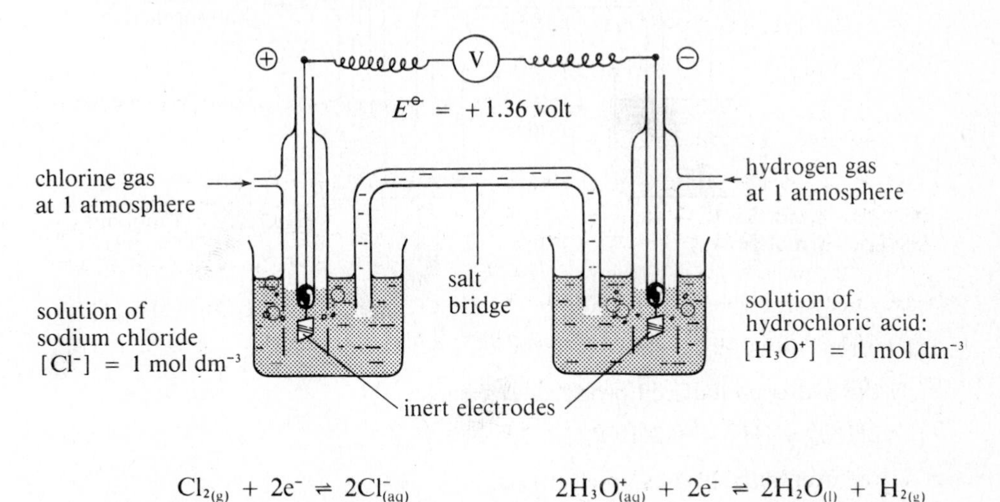

$$Cl_{2(g)} + 2e^{-} \rightleftharpoons 2Cl^{-}_{(aq)} \qquad 2H_3O^{+}_{(aq)} + 2e^{-} \rightleftharpoons 2H_2O_{(l)} + H_{2(g)}$$

It is found that the chlorine electrode is positive compared with the hydrogen electrode. This indicates that there is a higher electron density on the hydrogen electrode or that, if a current flows,

1. the hydrogen half-cell produces electrons:

$$H_{2(g)} + 2H_2O_{(l)} \longrightarrow 2H_3O^+_{(aq)} + 2e^-$$

2. the chlorine half-cell uses up electrons:

$$Cl_{2(g)} + 2e^- \longrightarrow 2Cl^-_{(aq)}$$

Inert electrodes can also be used to investigate the transfer of electrons between particles in solution. For example, iron exists in either of the two oxidation states (II) or (III) in its simple salts. In an aqueous solution containing both $Fe^{2+}_{(aq)}$ and $Fe^{3+}_{(aq)}$ ions, there is a tendency for iron (II) ions to lose electrons and iron (III) ions to gain them:

$$Fe^{2+}_{(aq)} \longrightarrow Fe^{3+}_{(aq)} + e^- \quad \text{and} \quad Fe^{3+}_{(aq)} + e^- \longrightarrow Fe^{2+}_{(aq)}$$

An inert electrode dipping into the solution brings these ions into contact with each other at the electrode surface. The electron density at the electrode surface is affected by the relative tendencies of the two ions to gain or lose electrons. Once again, this can be determined using a hydrogen electrode as a standard reference. Both an iron (II) salt and an iron (III) salt are dissolved to make a solution that is molar with respect to each ion:

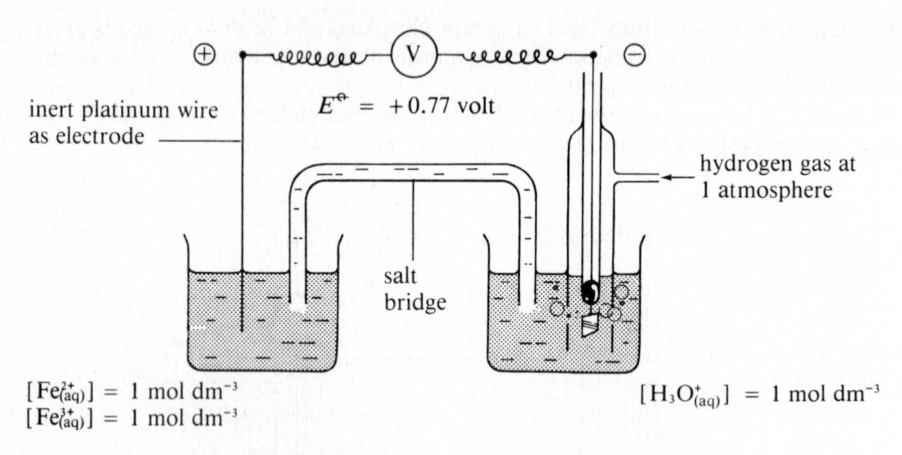

$$Fe^{3+}_{(aq)} + e^- \rightleftharpoons Fe^{2+}_{(aq)} \qquad 2H_3O^+_{(aq)} + 2e^- \rightleftharpoons 2H_2O_{(l)} + H_{2(g)}$$

If a current flows in the above cell,

1. the hydrogen half-cell produces electrons:

$$2H_2O_{(l)} + H_{2(g)} \longrightarrow 2H_3O^+_{(aq)} + 2e^-$$

2. the iron half-cell uses up electrons:

$$Fe^{3+}_{(aq)} + e^- \longrightarrow Fe^{2+}_{(aq)}$$

It is inconvenient to set up a hydrogen half-cell every time a standard electrode potential is to be measured. A reference half-cell, whose standard potential is known can be used instead. The most common one is the *calomel electrode* whose potential relative to the standard hydrogen electrode is +0.33 volts. The electrode is an inert platinum wire dipping into a saturated solution of mercury (I) chloride in contact with liquid mercury.

$$Hg_2Cl_{2(s)} + 2e^- \rightleftharpoons 2Hg_{(l)} + 2Cl^-_{(aq)}$$

$$E^\ominus = +0.33 \text{ volts in } 0.1 \text{ mol dm}^{-3} \text{ KCl}$$

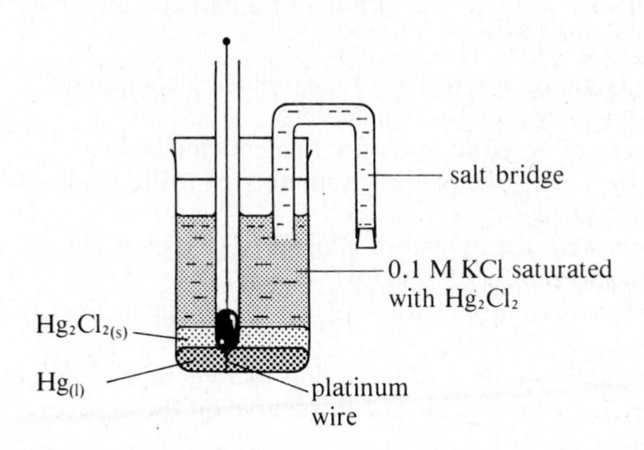

If the calomel electrode is used as a reference half-cell, the reading is 0.33 volts different from that given when a standard hydrogen half-cell is used, e.g. with respect to the calomel electrode.

$$Ag^+_{(aq)} + e^- \rightleftharpoons Ag_{(s)} \qquad E = +0.47 \text{ volt}$$

$$Zn^{2+}_{(aq)} + 2e^- \rightleftharpoons Zn_{(s)} \qquad E = -1.09 \text{ volt}$$

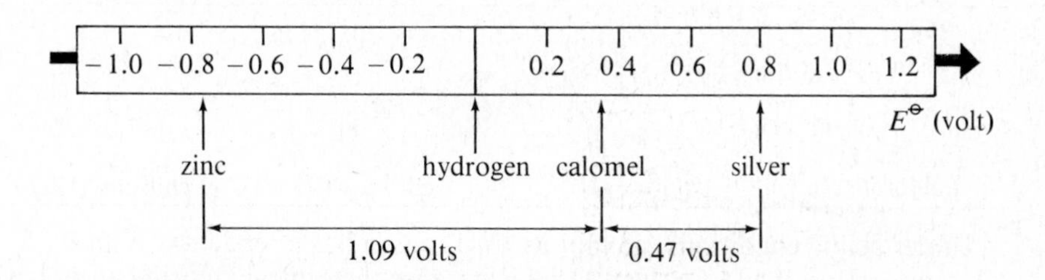

Partial loss and gain of electrons: oxidation number

Electron transfer takes place not only at the surface of an electrode but also as a result of the bonding between atoms. In an ionic compound the transfer of electrons is almost complete whereas in a covalent compound only partial transfer occurs. The electrons of a covalent bond are attracted more strongly by the atom of higher electronegativity (see page 44) and this leads in effect to the partial transfer of an electron from the less electronegative atom to the more electronegative atom.

Each atom present in a molecule or ion can be assigned an *oxidation number* to indicate the electron transfer that has occurred as a result of the bonding.

> The oxidation number of an atom shows the number of electrons over which it has lost or gained control as a result of its bonding. A positive number indicates loss of control; a negative number indicates gain of control.

The oxidation number of each atom in the formula of a particle can be worked out by using the following guide lines.

1. The sum of all the oxidation numbers of the atoms in a compound particle (e.g. CO_2, NaCl, $Mg(NO_3)_2$) equals zero.
2. The oxidation number of an atom in a pure element equals zero.
3. In a compound of hydrogen, the oxidation number of hydrogen is always +1 (except in metal hydrides).
4. In a compound of oxygen, the oxidation number of oxygen is always −2 (except in peroxides and fluorocompounds).
5. The oxidation number of an atomic ion (e.g. Fe^{2+}, Cl^-) equals the value of the charge on the ion.
6. The sum of the oxidation numbers of the atoms making up a molecular ion (e.g. MnO_4^-, NH_4^+) equals the value of the charge on the molecular ion.

Example Chlorine forms the four different oxyanions shown below depending on the number of bonding electrons that each chlorine atom uses. Oxygen is more electronegative than chlorine and therefore takes more than an equal share of the bonding pairs between them.

oxidation number of chlorine atom:

1	3	5	7
chlorate (I)	chlorate (III)	chlorate (V)	chlorate (VII)

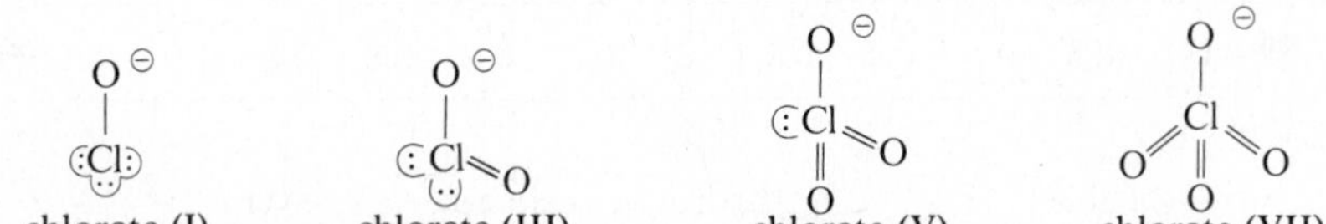

Under acidic conditions, a chlorate (VII) anion has a tendency to *gain* electrons rather than lose them. This is because the central chlorine atom has lost control of seven electrons and has a high electron affinity. The diagram below shows this tendency in action at the surface of an inert electrode dipping into a solution of chlorate (VII) ions.

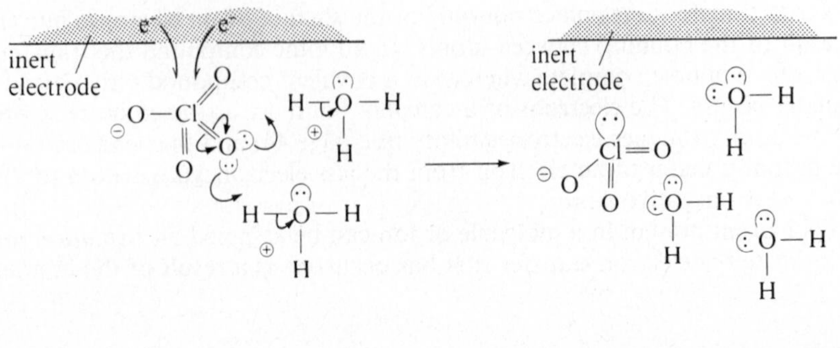

inert electrode

inert electrode

Chlorate (V) ions and water molecules are produced as a result of the electron transfer. The process is reversible, however: a chlorate (V) ion shows some tendency to lose its lone pair of electrons after a collision with a water molecule at the electrode surface. The necessary proton transfer is likely to take place almost instantaneously.

When the inert electrode is at the correct potential with respect to the solution, the two opposing processes take place at the same rate and the system is at equilibrium. The *standard electrode potential* of the system is measured using the same principles discussed earlier.

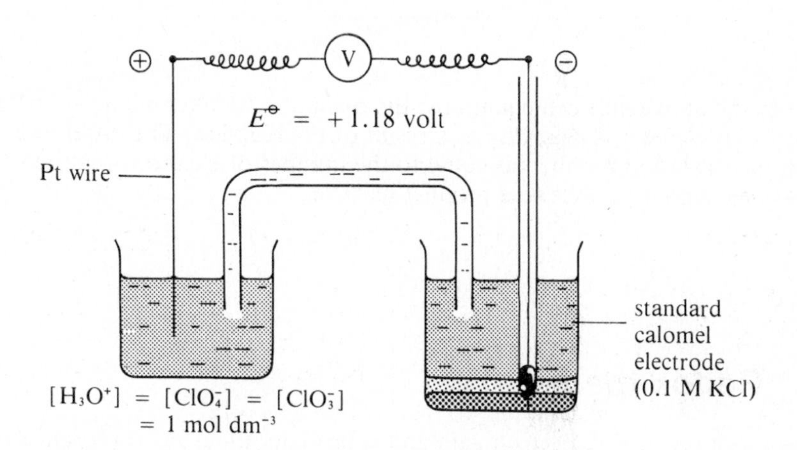

$$\begin{aligned} \text{measured p.d.} &= +0.85 \text{ volt} \\ E^\ominus &= (0.85 + 0.33) \text{ volt} \\ &= +1.18 \text{ volt} \end{aligned}$$

$$ClO^-_{4(aq)} + 2H_3O^+_{(aq)} + 2e^- \rightleftharpoons ClO^-_{3(aq)} + 3H_2O_{(l)}$$

The table below lists some standard electrode potentials that are measured using inert electrodes. It would be good practice to assign oxidation numbers to all the atoms present in the formulas shown in the table. For example, in the first reaction Fe(6) → Fe(3), and in the second O(−1) → O(−2).

reaction (all ions in (aq))	$E^{\ominus}$ (V)
$FeO_4^{2-} + 8H_3O^+ + 3e^- \rightleftharpoons Fe^{3+} + 12H_2O$	+2.20
$H_2O_2 + 2H_3O^+ + 2e^- \rightleftharpoons 4H_2O$	+1.77
$MnO_4^- + 4H_3O^+ + 3e^- \rightleftharpoons MnO_{2(s)} + 6H_2O$	+1.67
$MnO_4^- + 8H_3O^+ + 5e^- \rightleftharpoons Mn^{2+} + 12H_2O$	+1.52
$2BrO_3^- + 12H_3O^+ + 10e^- \rightleftharpoons Br_2 + 18H_2O$	+1.52
$2ClO_3^- + 12H_3O^+ + 10e^- \rightleftharpoons Cl_{2(g)} + 18H_2O$	+1.47
$PbO_{2(s)} + 4H_3O^+ + 2e^- \rightleftharpoons Pb^{2+} + 6H_2O$	+1.47
$Cr_2O_7^{2-} + 14H_3O^+ + 6e^- \rightleftharpoons 2Cr^{3+} + 21H_2O$	+1.33
$MnO_{2(s)} + 4H_3O^+ + 2e^- \rightleftharpoons Mn^{2+} + 6H_2O$	+1.23
$2IO_3^- + 12H_3O^+ + 10e^- \rightleftharpoons I_{2(s)} + 18H_2O$	+1.19
$NO_3^- + 4H_3O^+ + 3e^- \rightleftharpoons NO_{(g)} + 6H_2O$	+0.96
$NO_3^- + 2H_3O^+ + e^- \rightleftharpoons NO_{2(g)} + 3H_2O$	+0.81
$O_{2(g)} + 2H_3O^+ + 2e^- \rightleftharpoons H_2O_2 + 2H_2O$	+0.68
$O_{2(g)} + 2H_2O + 4e^- \rightleftharpoons 4OH^-$	+0.40
$HSO_4^- + 2H_3O^+ + 2e^- \rightleftharpoons HSO_3^- + 3H_2O$	+0.17
$2CO_{2(g)} + 2H_3O^+ + 2e^- \rightleftharpoons H_2C_2O_4$	−0.49

In each of the above half-cell equations, the oxidation number of the atoms of one particular element is changing as a result of the reaction. The difference between the oxidation numbers is equal to the number of electrons required to convert one reactant particle to a product particle.

19.2 Redox theory

Reduction is a process of electron gain and is brought about by a substance whose particles readily lose electrons so that they are donated to other particles in the system. A substance of this sort is called a *reducing agent* or *reductant*. Reactive metals are good examples of reductants because their atoms show a marked tendency to lose electrons and become ions.

$$M_{(s)} \longrightarrow M^{n+}_{(aq)} + ne^-$$

The more readily the particles of a substance lose electrons, the stronger a reductant the substance is.

Conversely, oxidation is a process of electron loss and is brought about by a substance whose particles readily gain electrons so that they are taken from other particles in the system. A substance of this sort is called an *oxidizing agent* or *oxidant*. Reactive non-metals are good examples of oxidants because their atoms show a tendency to accept electrons and become anions:

$$O_{2(g)} + 4e^- \longrightarrow 2O^{2-}$$
$$Cl_{2(g)} + 2e^- \longrightarrow 2Cl^-$$

The reversibility of electron transfer has been outlined in 19.1. A metal atom causes reduction because of its tendency to lose electrons, but then the metal ion formed can cause oxidation because it has some tendency to accept electrons. Similarly a non-metal anion can cause reduction because of its tendency to lose electrons. Every oxidant therefore has its *conjugate reductant* and every reductant its *conjugate oxidant* in the same way that every acid has a conjugate base (see page 272).

$$\underset{\text{reductant}}{M_{(s)}} \underset{\text{gains electrons}}{\overset{\text{loses electrons}}{\rightleftharpoons}} \underset{\text{oxidant}}{M^{n+}_{(aq)}} + ne^-$$

$$\underset{\text{oxidant}}{Cl_2} + 2e^- \underset{\text{loses electrons}}{\overset{\text{gains electrons}}{\rightleftharpoons}} \underset{\text{reductant}}{2Cl^-}$$

By convention a conjugate redox pair is always written with the oxidant on the left and the reductant on the right. A list of these conjugate pairs is therefore the same as the list of standard electrode reactions on pages 311 and 316. For this reason, the electrode potentials are often called *standard redox potentials*. Some further examples not included earlier are given below.

electrode reaction (all ions in (aq)) *oxidant*	*reductant*	*standard redox potential (volt)*
$O_{3(g)} + 2H_3O^+ + 2e^-$	$\rightleftharpoons O_{2(g)} + 3H_2O_{(l)}$	+2.07
$2ClO^- + 4H_3O^+ + 2e^-$	$\rightleftharpoons Cl_{2(g)} + 6H_2O_{(l)}$	+1.64
$2BrO^- + 4H_3O^+ + 2e^-$	$\rightleftharpoons Br_2 + 6H_2O_{(l)}$	+1.59
$2IO^- + 4H_3O^+ + 2e^-$	$\rightleftharpoons I_{2(s)} + 6H_2O_{(l)}$	+1.45
$Cl_{2(g)} + 2e^-$	$\rightleftharpoons 2Cl^-$	+1.36
$Br_2 + 2e^-$	$\rightleftharpoons 2Br^-$	+1.07
$I_{2(s)} + 2e^-$	$\rightleftharpoons 2I^-$	+0.54

Oxidation state

The term 'oxidation state' is close in meaning to that of oxidation number. Whereas an oxidation number refers to the individual atoms of an element and their loss or gain of control of electron density, the oxidation state refers to the element as a whole in the compound. It is a bulk descriptive term and, to distinguish it from oxidation number, the oxidation state of an element in a compound is shown as a roman numeral.

For example, the +VI oxidation state of sulphur includes the following compounds:

sulphur (VI) oxide	SO_3
sulphur (VI) fluoride	SF_6
sodium sulphate (VI)	Na_2SO_4

In each of these compounds, the sulphur atoms are bonded in such a way that they have lost control of six electrons compared with their control in the pure element: the oxidation number of each sulphur atom is +6.

The concept of oxidation state is useful in two main ways:

1. in organizing a study of the inorganic chemistry of an element. It is used extensively in part four of this book.
2. for recognizing a redox reaction from the equation for the reaction. One element goes up in oxidation state while another comes down; in particle terms, this is understood by the transfer of electrons from one particle to another.

Example Consider the four equations shown below.

a) $NH_{3(g)} + HCl_{(g)} \longrightarrow NH_4Cl_{(s)}$

b) $2SO_{2(g)} + O_{2(g)} \longrightarrow 2SO_{3(g)}$

c) $Cr_2O^{2-}_{7(aq)} + 3H_2O_{(l)} \longrightarrow 2CrO^{2-}_{4(aq)} + 2H_3O^+_{(aq)}$

d) $Cr_2O^{2-}_{7(aq)} + 6Fe^{2+}_{(aq)} + 14H_3O^+_{(aq)} \longrightarrow 2Cr^{3+}_{(aq)} + 6Fe^{3+}_{(aq)} + 21H_2O_{(l)}$

Only two are redox changes: b) and d). In reactions a) and c) all the elements keep the same oxidation state throughout, whereas in b) and d) the changes are:

b) S(IV) to S(VI)
while
O(0) to O(−II)

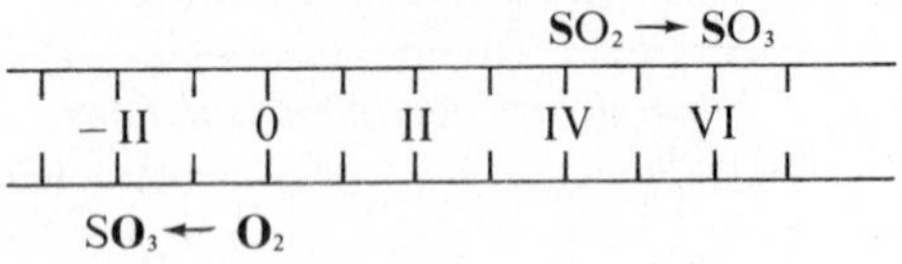

$SO_2 \rightarrow SO_3$

−II 0 II IV VI

$SO_3 \leftarrow O_2$

d) Cr(VI) to Cr(III)
while
Fe(II) to Fe(III)

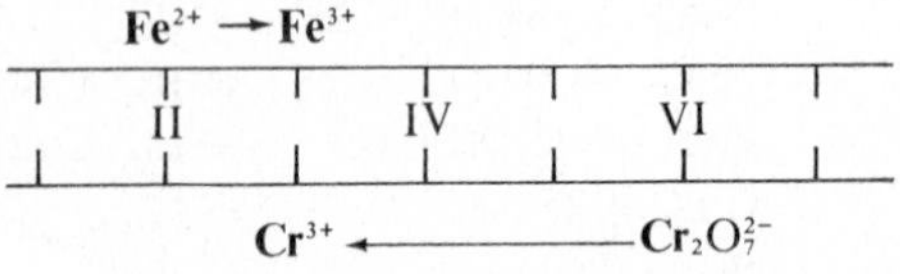

$Fe^{2+} \rightarrow Fe^{3+}$

II IV VI

$Cr^{3+} \longleftarrow Cr_2O_7^{2-}$

Equations a) and c) are, in fact, acid-base reactions. Instead of electron transfer between reactant particles, proton transfer takes place:

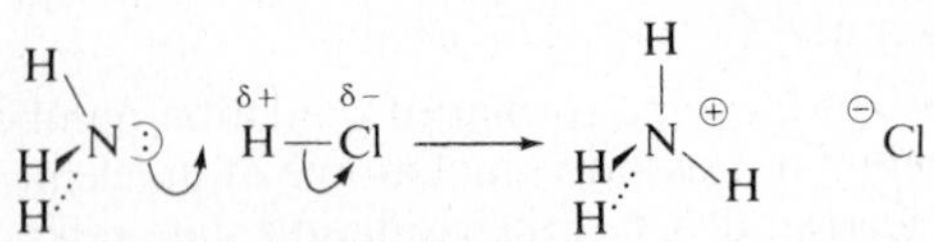

H, H, H–N: + H(δ+)–Cl(δ−) ⟶ H, H, H, H–N⊕ + ⊖Cl

Summary

An oxidant:

1. causes oxidation,
2. contains an element whose oxidation state goes down as a result of the redox reaction,
3. is reduced in the process.

A reductant:

1. causes reduction,
2. contains an element whose oxidation state goes up as a result of the redox process,
3. is oxidized in the process.

Using $E^\ominus$ data to predict redox reactions

The standard redox potential of the electrode reaction between an oxidant and its conjugate reductant is a measure of the oxidizing or reducing strength of the members of the conjugate pair. The greater the positive value of $E^\ominus$, the stronger is the oxidant and the weaker its conjugate reductant (just as a strong acid has a weak conjugate base, see page 274). Conversely, the greater the negative value of $E^\ominus$, the weaker is the oxidant and the stronger is its conjugate reductant.

	redox conjugate pair *oxidant*	⇌	*reductant*		$E^\ominus$ *volt*
	$F_{2(g)}$		$2F^-_{(aq)}$		+2.87
	$H_2O_{2(aq)}$		$2H_2O_{(l)}$		+1.77
increasing	$MnO^-_{4(aq)}$		$Mn^{2+}_{(aq)}$		+1.52
oxidizing	$Cl_{2(g)}$		$2Cl^-_{(aq)}$		+1.36
strength	$Cr_2O^{2-}_{7(aq)}$		$2Cr^{3+}_{(aq)}$		+1.33
	$Fe^{3+}_{(aq)}$		$Fe^{2+}_{(aq)}$		+0.77
	$S_{(s)}$		$H_2S_{(g)}$		+0.14
	$2H_3O^+_{(aq)}$		$H_{2(g)}$		0.00
	$Fe^{2+}_{(aq)}$		$Fe_{(s)}$	increasing	−0.47
	$Zn^{2+}_{(aq)}$		$Zn_{(s)}$	reducing	−0.76
	$Al^{3+}_{(aq)}$		$Al_{(s)}$	strength	−1.66
	$Mg^{2+}_{(aq)}$		$Mg_{(s)}$		−2.37
	$K^+_{(aq)}$		$K_{(s)}$		−2.92

Example 1 Is it likely that potassium manganate (VII) will oxidize sodium chloride solution to chlorine?

A half-cell containing the redox couple MnO_4^-/Mn^{2+} is positive compared with a half-cell containing the couple Cl_2/Cl^-. Given:

MnO_4^-/Mn^{2+} $E^\ominus = +1.52$ volt
Cl_2/Cl^- $E^\ominus = +1.36$ volt

The manganese half-cell is 0.16 volts more positive.

If a cell is set up from these two half-cells (like those described on page 309), electrons flow through the wire towards the manganese half-cell.

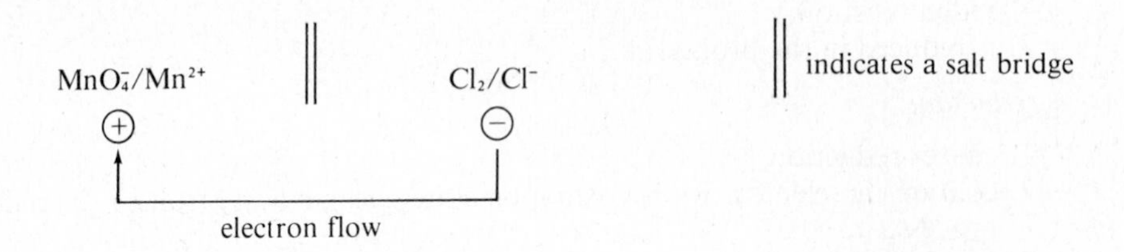

The cell reaction is given by the following two processes.

1. The chlorine half-cell produces electrons:

   $2Cl^-_{(aq)} \rightarrow Cl_{2(g)} + 2e^-$

2. The manganese half-cell uses up electrons:

   $MnO^-_{4(aq)} + 8H_3O^+_{(aq)} + 5e^- \rightarrow Mn^{2+}_{(aq)} + 12H_2O_{(l)}$

In other words, chloride ions and manganate (VII) ions are used up: manganate (VII) is likely to oxidize chloride to chlorine.

Example 2 Is it likely that an acidified iron (II) sulphate solution will reduce a suspension of sulphur to hydrogen sulphide?

Given:

Fe^{3+}/Fe^{2+} $E^\ominus = +0.77$ volt
S/H_2S $E^\ominus = +0.14$ volt

The iron half-cell is more positive (by 0.63 volt) and therefore in a cell set up from these two half-cells, electrons flow through the wire to the iron half-cell. The cell reaction is:

1. The sulphur half-cell producing electrons:

   $H_2S_{(g)} + 2H_2O_{(l)} \rightarrow S_{(s)} + 2H_3O^+_{(aq)} + 2e^-$

2. The iron half-cell using up electrons:

   $Fe^{3+}_{(aq)} + e^- \rightarrow Fe^{2+}_{(aq)}$

The reactants are iron (III) and hydrogen sulphide. In other words, an iron (II) solution is *not* likely to reduce sulphur to hydrogen sulphide.

A useful generalization can be made concerning the list of redox couples shown in decreasing order of $E^\ominus$ values (as on page 319).

> An oxidant is likely to oxidize any reductant whose conjugate oxidant is below the selected oxidant on the list.

Precautions needed in $E^\ominus$ predictions

There are two reasons why a reaction predicted by an inspection of redox couple $E^\ominus$ values might *not* take place.

1. $E^\ominus$ values are all measured under standard conditions: solutions that are molar, pressures of 1 atmosphere and temperature = 298 K. Many redox changes need to be carried out under non-standard conditions and therefore $E^\ominus$ predictions may not be valid.
2. The potential difference between the two half-cells is a measure of the *free energy change* (see page 221) that takes place if the cell reaction occurs. The $E^\ominus$ predictions are therefore based on the relative thermodynamic stabilities of the reactants and products. The thermodynamic stability of a product with respect to its reactants is never a sufficient confirmation that a reaction will happen. For example, water is 287 kJ mol^{-1} more stable than hydrogen and oxygen and yet mixtures of hydrogen and oxygen do not become water — until they are ignited. Ignition supplies the necessary activation energy. At room temperature, the reaction between hydrogen and oxygen is extremely slow.

   The same caution must be applied to $E^\ominus$ predictions. These give information concerning the feasibility of a reaction from an energetics point of view, but they cannot say how fast the reaction is likely to proceed. Under the particular reaction conditions, a *rate factor* might be the dominant one and no apparent reaction take place. A system is often said to be *thermodynamically unstable* but *kinetically stable* when a reaction is predicted from an energetics point of view but is not observed to proceed at a measurable rate.

Writing redox equations

The equation for a redox reaction is worked out from the equations of the two half-cell processes involved: these are known as *half-equations*. A simple example is shown below for the reaction of zinc dust with silver nitrate solution.

1. Work out the two relevant half-equations.

$$Ag^{+}_{(aq)} + e^{-} \rightleftharpoons Ag_{(s)}$$
$$Zn^{2+}_{(aq)} + 2e^{-} \rightleftharpoons Zn_{(s)}$$

2. Balance the half-equations so that the number of electrons produced by the one equals the number used by the other

$$Zn_{(s)} \longrightarrow Zn^{2+}_{(aq)} + 2e^{-} \qquad 2e^{-} + 2Ag^{+}_{(aq)} \longrightarrow 2Ag_{(s)}$$

3. Add the two half-equations together and cancel out any particles that appear on both sides of the equation

$$Zn_{(s)} + 2Ag^{+}_{(aq)} \longrightarrow Zn^{2+}_{(aq)} + 2Ag_{(s)}$$

The simple example given above is easy to work out because the half-equations are so straightforward. This is less true for the reactions that take place between oxysalts in either acidic or alkaline solution. The procedure for working out their half-equations is illustrated below for the couple bromate (V) -bromide.

1. Write down the formula of the oxidant and its conjugate reductant.

$$BrO_3^- \rightleftharpoons Br^-$$

2. Balance the difference in oxidation number by adding the appropriate number of electrons (5 − 6 = −1)

$$\overset{V}{Br}O_3^- + 6e^- \rightleftharpoons \overset{-I}{Br^-}$$

3. Balance the overall charge on each side of the half-equation by adding the appropriate number of H_3O^+ ions (if in acid conditions) or OH^- ions (if in alkaline conditions)

acid $\quad BrO_3^- + 6e^- + 6H_3O^+ \rightleftharpoons Br^-$

alkali $\quad BrO_3^- + 6e^- \rightleftharpoons Br^- + 6OH^-$

4. Balance for hydrogen and oxygen by adding the appropriate number of water molecules

acid $\quad BrO_3^- + 6e^- + 6H_3O^+ \rightleftharpoons Br^- + 9H_2O$

alkali $\quad BrO_3^- + 6e^- + 3H_2O \rightleftharpoons Br^- + 6OH^-$

Two further examples of the whole procedure are given below.

Example 1 Potassium manganate (VII) oxidizes ethanedioic acid to carbon dioxide under acidic conditions at 60°C.

1. Work out the half-equations

(a)	(b)
$\overset{VII}{Mn}O_4^- \rightleftharpoons \overset{II}{Mn^{2+}}$	$2\overset{IV}{C}O_2 \rightleftharpoons (\overset{III}{C}OOH)_2$
$\overset{VII}{Mn}O_4^- + 5e^- \rightleftharpoons \overset{II}{Mn^{2+}}$	$2\overset{IV}{C}O_2 + 2e^- \rightleftharpoons (\overset{III}{C}OOH)_2$
$MnO_4^- + 5e^- + 8H_3O^+ \rightleftharpoons Mn^{2+}$	$2CO_2 + 2e^- + 2H_3O^+ \rightleftharpoons (COOH)_2$
$MnO_4^- + 5e^- + 8H_3O^+ \rightleftharpoons Mn^{2+} + 12H_2O$	$2CO_2 + 2e^- + 2H_3O^+ \rightleftharpoons (COOH)_2 + 2H_2O$

2. Balance the electron transfer

$$2MnO_4^- + 16H_3O^+ + 10e^- \rightarrow 2Mn^{2+} + 24H_2O$$
$$5(COOH)_2 + 10H_2O \rightarrow 10CO_2 + 10H_3O^+ + 10e^-$$

3. Add the half-equations and cancel out the particles appearing on both sides of the equation

$$2MnO_4^- + 5(COOH)_2 + 6H_3O^+ \rightarrow 2Mn^{2+} + 14H_2O + 10CO_2$$

Example 2 Hydrogen peroxide oxidizes chromium (III) hydroxide to a chromate (VI) salt in alkaline conditions

1. Work out the half-equations

(a)

$\overset{-I}{H_2O_2} \rightleftharpoons \overset{-II}{2H_2O}$

$\overset{-I}{H_2O_2} + 2e^- \rightleftharpoons \overset{-II}{2H_2O}$

$H_2O_2 + 2e^- \rightleftharpoons 2H_2O + 2OH^-$

$H_2O_2 + 2e^- \rightleftharpoons 2OH^-$

(b)

$\overset{VI}{CrO_4^{2-}} \rightleftharpoons \overset{III}{Cr(OH)_3}$

$\overset{VI}{CrO_4^{2-}} + 3e^- \rightleftharpoons \overset{III}{Cr(OH)_3}$

$CrO_4^{2-} + 3e^- \rightleftharpoons Cr(OH)_3 + 5OH^-$

$CrO_4^{2-} + 3e^- + 4H_2O \rightleftharpoons Cr(OH)_3 + 5OH^-$

2. Balance the electron transfer

$$3H_2O_2 + 6e^- \longrightarrow 6OH^-$$
$$2Cr(OH)_3 + 10OH^- \longrightarrow 2CrO_4^{2-} + 6e^- + 8H_2O$$

3. Add the half-equations and cancel out the particles appearing on both sides of the equation.

$$3H_2O_2 + 2Cr(OH)_3 + 4OH^- \longrightarrow 2CrO_4^{2-} + 8H_2O$$

Testing for redox reagents in solution

A practical test in the laboratory should have the properties of being quick and easy to carry out and having a rapidly detected result if the test proves positive. A good example is a pH sensitive indicator. An indicator works because it consists of a weak acid whose conjugate base is a different colour (see page 287). In a similar way a redox conjugate pair consisting of differently coloured members can be used to test for oxidants and reductants. These two ideas are contrasted in the table below.

$HIn \rightleftharpoons H_3O^+ + In^-$ (HIn: colour A; In^-: colour B)	oxidant + electrons $\rightleftharpoons$ reductant (oxidant: colour A; reductant: colour B)
addition of acid shifts position of equilibrium to the left: colour A is seen.	addition of an oxidant shifts position of equilibrium to the left: colour A is seen.
addition of base shifts position of equilibrium to the right: colour B is seen	addition of a reductant shifts position of equilibrium to the right: colour B is seen.

These ideas are applied as follows.

Testing for reductants

add	observe	change
acidified manganate (VII)	purple to colourless	$MnO^-_{4(aq)} \rightarrow Mn^{2+}_{(aq)}$
acidified dichromate (VI)	orange to green	$Cr_2O^{2-}_{7(aq)} \rightarrow Cr^{3+}_{(aq)}$

Testing for oxidants

add	observe	change
aqueous iodide ions	pale yellow to brown	$I^-_{(aq)} \rightarrow I_{2(aq)}$
aqueous sulphide ions	colourless to pale yellow precipitate	$S^{2-} \rightarrow S_{(s)}$

19.3 Cells

A cell is made by connecting together two half-cells. The electrical contact is usually brought about by a salt bridge. The characteristic property of any cell is that current flows through a wire that is connected to the electrodes of each half-cell. The potential difference between the two electrodes is called the *electromotive force* (e.m.f.) of the cell when the measurement is made so that no current is being drawn from the cell (on 'open circuit').

The e.m.f. of a cell is given by the difference in the electrode potentials of the two half-cells. Most cells do not operate under standard conditions and the electrode potential of a non-standard half-cell depends on:

1. the redox couple present
2. the concentration of the reactants and products of the electrode process
3. the temperature.

The redox couple present

The standard electrode potentials give *some* indication of the electrode potential expected from a redox couple under non-standard conditions.

For example, the common 'dry cell' that is used as the battery for torches and other everyday appliances has a Zn^{2+}/Zn half-cell. Under the conditions of the cell, the electrode reaction that takes place is:

$$[Zn(NH_3)_4]^{2+} + 2e^- \rightleftharpoons Zn_{(s)} + 4NH_{3(g)}$$

zinc ions complexed
by ammonia molecules

This reaction is different from the one that takes place when a zinc electrode dips into a molar solution of zinc (II) nitrate (V). Under standard conditions, the electrode potentials of the two processes are:

$$[Zn(H_2O)_4]^{2+} + 2e^- \rightleftharpoons Zn_{(s)} + 4H_2O_{(l)} \qquad E^\ominus = -0.76 \text{ volts}$$

$$[Zn(NH_3)_4]^{2+} + 2e^- \rightleftharpoons Zn_{(s)} + 4NH_{3(g)} \qquad E^\ominus = -1.03 \text{ volts}$$

The different ligands around the central ion alter its tendency to gain electrons: when a zinc ion is co-ordinated by the more strongly electron-releasing ligands (ammonia molecules) its tendency to gain electrons is reduced. The back reaction is therefore even more favourable and the $E^\ominus$ is even more negative.

The *effect of ligands* on the ability of an ion to gain or lose electrons is most marked when the ligand is strongly electron-releasing. For example, compare the electrode potentials for iron (III) to iron (II) when the ligands are firstly water molecules and secondly cyanide ions:

$$[Fe(H_2O)_6]^{3+} + e^- \rightleftharpoons [Fe(H_2O)_6]^{2+} \qquad E^\ominus = +0.77 \text{ volts}$$

$$[Fe(CN)_6]^{3-} + e^- \rightleftharpoons [Fe(CN)_6]^{4-} \qquad E^\ominus = +0.36 \text{ volts}$$

In the 'dry cell', the other half-cell is made up from a paste of manganese (IV) oxide with ammonium chloride in contact with an inert carbon rod as electrode. The electrode reaction is not simple:

$$NH_4^+ + MnO_2 + H_2O + e^- \rightleftharpoons Mn(OH)_3 + NH_3$$

If it could be carried out under standard conditions, the expected $E^\ominus$ value of this reaction is approximately +1.0 volts.

The overall e.m.f. of the dry cell can therefore be worked out from the values of the electrode potentials:

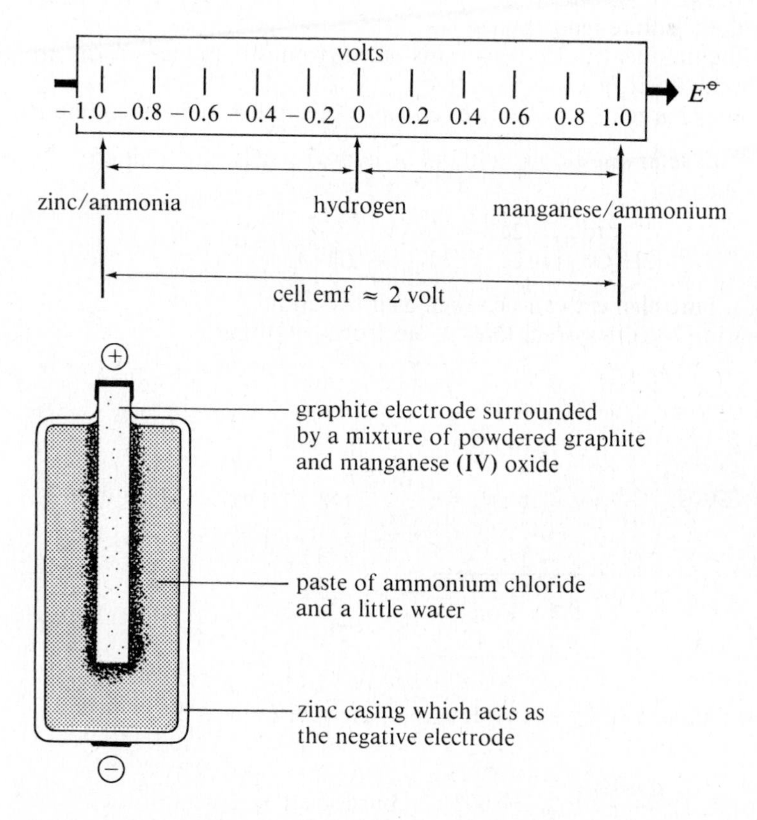

Due to the internal resistance of the cell and the non-standard conditions of the reaction, the e.m.f. of the dry cell is 1.5 volts.

The concentration of reactants and products

The processes that occur at an electrode surface are reversible and are governed by the equilibrium law. If the concentration of a reactant or product is altered, the position of equilibrium also alters and therefore the potential changes because electrons are part of the equilibrium:

$$\text{oxidized form} \quad + \quad \text{electrons} \quad \rightleftharpoons \quad \text{reduced form}$$

$$Fe^{3+}_{(aq)} \quad + \quad e^- \quad \rightleftharpoons \quad Fe^{2+}_{(aq)}$$

The exact relationship between the electrode potential of a half-cell and the concentrations of reactants and products is given by the *Nernst equation* (derived from the second law of thermodynamics).

$$E = E^{\ominus} + \frac{RT}{nF} \log_e \left\{ \frac{[\text{oxidized form}]}{[\text{reduced form}]} \right\}$$

where E = the measured electrode potential relative to the standard hydrogen electrode
$E^{\ominus}$ = the standard electrode potential
R = the gas constant
T = the absolute temperature
n = the number of electrons transferred from the 'reduced form' to the 'oxidized form'.
F = the charge of a mole of electrons = 1 faraday (96 500 coulombs)

When the electrode in question is a metal or hydrogen gas, the 'reduced form' is a pure element.

$$Zn^{2+}_{(aq)} + 2e^- \rightleftharpoons Zn_{(s)}$$
$$2H_3O^+_{(aq)} + 2e^- \rightleftharpoons H_{2(g)} + 2H_2O_{(l)}$$

The activity of a pure element can be taken as unity and the Nernst equation for the system ions + electrons $\rightleftharpoons$ atoms then becomes

$$E = E^{\ominus} + \frac{RT}{nF} \log_e [\text{ions}]$$

But at $T = 298$ K, $\frac{RT}{nF} \log_e [\text{ions}] = \frac{0.0592}{n} \log_{10} [\text{ions}]$

$$\Rightarrow \quad \boxed{E = E^{\ominus} + \frac{0.0592}{n} \log_{10} [\text{ions}]}$$

For example, a standard silver electrode has a potential E given by

$$E = E^{\ominus} + \frac{0.0592}{1} \log_{10} (1.00); \qquad \text{but } \log_{10} 1 = 0$$

$$\Rightarrow \quad E = E^{\ominus} = 0.80 \text{ volts}$$

but if the solution contains only 0.001 mol dm^{-3} of Ag^+

$$E = E^{\ominus} + 0.0592 \log_{10} (0.001)$$
$$= 0.80 + [0.0592 \times (-3)] = 0.62 \text{ volts}$$

A cell can be set up by using the two half-cells described above. It is called a *concentration cell* because the e.m.f. of the cell depends on the difference in concentration in the two half-cells:

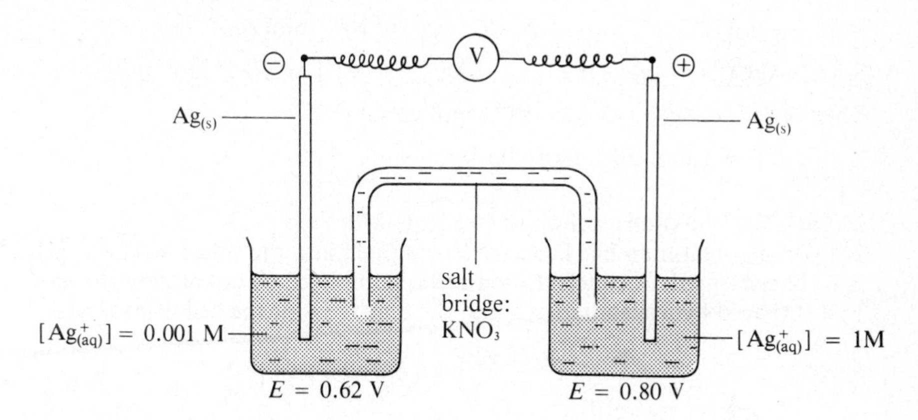

In the above cell, the electrode dipping into the more dilute solution is negative compared with the standard half-cell. Electrons therefore flow from left to right through the wire. Silver dissolves in the left-hand beaker and is deposited on the electrode of the right-hand beaker. The process continues until the concentration of silver ions in the two beakers becomes equal.

In a concentration cell of this sort, in which one half-cell is standard, the measured e.m.f. is given by the expression:

$$\text{e.m.f.} = E_R - E_L$$

where E_L = electrode potential of the left-hand half-cell
E_R = electrode potential of the right-hand half-cell

but $E_R = E^{\ominus}$ (standard)

and $E_L = E^{\ominus} + \frac{RT}{nF} \log_e [\text{ions}]$

$$\Rightarrow \quad \text{e.m.f.} = E^{\ominus} - (E^{\ominus} + \frac{RT}{nF} \log_e [\text{ions}])$$

$$= -\frac{0.0592}{n} \log_{10} [\text{ions}]$$

Concentration cells can be used to measure the concentration of ions in an unknown solution. This has two applications:

1. to calculate solubility products (where ionic concentrations are very low),
2. to measure hydrogen ion concentrations and therefore the pH of solutions.

1. Solubility products (see page 295)
The solubility product of silver chloride can be measured by making up a saturated solution of silver chloride at 298 K. By dipping a silver electrode into this solution, a concentration cell is made up with a standard silver electrode as the other half-cell. The measured e.m.f. is 0.29 volts.

$$\Rightarrow \frac{0.059}{1} \log_{10} [Ag^+] = -0.29$$

$$\Rightarrow \log_{10} [Ag^+] = -4.9 \quad \text{or} \quad Ag^+ = 10^{-4.9} \text{ mol dm}^{-3}$$

Since $AgCl_{(s)} \rightleftharpoons Ag^+_{(aq)} + Cl^-_{(aq)}$, $[Ag^+] = [Cl^-] = 10^{-4.9}$ mol dm^{-3}

Thus $K_{sp} = [Ag^+]\,[Cl^-] = 10^{-9.8}$ mol^2 dm^{-6}

$K_{sp} = 1.6 \times 10^{-10}$ mol^2 dm^{-6}

2. The pH of unknown solutions (see page 278)
A hydrogen-platinum black electrode is dipped into the solution whose $[H_3O^+]$ is to be measured. A concentration cell is completed by connecting the half-cell to a standard hydrogen half-cell and the e.m.f. E_{ob} of the cell is measured.

Since
$$E_{ob} = \frac{-0.0592}{1} \log_{10} [H_3O^+]$$

$$-\log_{10} [H_3O^+] = \frac{E_{ob}}{0.0592}$$

Thus
$$pH = 16.9\, E_{ob}$$

Temperature

The effect of temperature change on the electrode potential of a half-cell is also given by the Nernst equation.

$$E = E^\ominus + \frac{RT}{nF} \log_e [\text{ions}]$$

An increase in temperature results in the measured electrode potential becoming more positive compared with the standard hydrogen electrode. A more detailed treatment shows that the temperature coefficient of the e.m.f. of a cell is directly proportional to the entropy change involved in the reaction (see page 221). Important thermodynamic information is therefore available from the measurement of cell e.m.f.'s at different temperatures.

Questions

1. Explain the meaning of the following terms:
   a) oxidation b) oxidant c) standard redox potential d) oxidation number
   e) half-cell.

2. Describe how the standard electrode potential of an element is measured.

3. a) Discuss the factors that affect the magnitude of the electrode potential of an element.
   b) How can the direction of a redox reaction be predicted from a knowledge of standard electrode potentials? Explain any limitations that must be kept in mind.

4. Write down the oxidation numbers of the underlined atoms:
   a) $Na\underline{Cl}O_4$ b) $Na\underline{Cl}$ c) $Na\underline{Cl}O$ d) $K_2\underline{Cr}O_4$ e) $H_2\underline{C}O_3$ f) $\underline{Pb}O_2$ g) $\underline{N}_2O_4$
   h) $H_2\underline{P}O_4^-$ i) $\underline{P}Cl_6^+$ j) $\underline{Ti}Cl_4$ k) $\underline{Mn}O_4^{2-}$ l) $\underline{Cu}I$ m) $\underline{N}H_4^+$ n) $\underline{Hg}_2Cl_2$ o) $H_2\underline{P}O_2^-$

5. Write balanced half-equations for the following redox couples in aqueous acid solution.
   a) $MnO^-_{4(aq)}/MnO_{2(s)}$ b) $S_2O^{2-}_{3(aq)}/S_4O^{2-}_{6(aq)}$
   c) $HSO^-_{4(aq)}/H_2S_{(g)}$ d) $NO^-_{3(aq)}/N_2O_{(g)}$
   e) $ClO^-_{3(aq)}/Cl^-_{(aq)}$

6. Write balanced half-equations for the following redox couples in aqueous alkaline solution.
   a) $PbO_{2(s)}/Pb(OH)^-_{3(aq)}$ b) $CrO^{2-}_{4(aq)}/Cr(OH)_{3(s)}$
   c) $Al_{(s)}/Al(OH)^-_{4(aq)}$ d) $Cl^-_{(aq)}/ClO^-_{(aq)}$
   e) $FeO^{2-}_{4(aq)}/Fe(OH)_{3(s)}$

7. Write balanced equations for the redox reactions which occur between:
   a) I^- and H_2SO_4 to yield I_2 and SO_2
   b) Zn and HNO_3 to yield Zn^{2+} and NH_4^+
   c) MnO_4^- and Fe^{2+} in acid solution to yield Mn^{2+} and Fe^{3+}
   d) Cl_2 and OH^- to yield ClO_3^- and Cl^-

8. a) Given the following standard electrode potentials:
   Fe^{3+}/Fe^{2+} $E^{\ominus} = +0.77$ volts
   Fe^{2+}/Fe $E^{\ominus} = -0.44$ volts
   Zn^{2+}/Zn $E^{\ominus} = -0.76$ volts
   Predict any reactions that might occur between any of the oxidants and reductants above, and then write balanced equations for the reactions.

   b) Given the following standard electrode potentials
   MnO_4^-/Mn^{2+} $E^{\ominus} = +1.52$ volts
   Fe^{3+}/Fe^{2+} $E^{\ominus} = +0.77$ volts
   O_2/H_2O_2 $E^{\ominus} = +0.68$ volts
   H_2O_2/H_2O $E^{\ominus} = +1.77$ volts
   Predict any reactions that might occur between any of the oxidants and reductants above, and then write balanced equations for the reactions.
   Comment on the behaviour of H_2O_2.

9. Given the following standard electrode potentials
   Mg^{2+}/Mg $E^{\ominus} = -2.38$ volts
   Zn^{2+}/Zn $E^{\ominus} = -0.76$ volts
   Fe^{2+}/Fe $E^{\ominus} = -0.44$ volts
   Cu^{2+}/Cu $E^{\ominus} = +0.34$ volts
   a) Describe how you would set up cells with the following e.m.f.'s: i) 1.1 volts; ii) 1.62 volts; iii) 1.94 volts; iv) 0.78 volts and in each case mark in the polarity of the electrodes in each cell.

b) What is the largest e.m.f. obtainable from a combination of two of these half-cells under standard conditions? Which pair of half-cells would be used?

c) Using the same pair of half-cells as in b) above, describe ways of changing the conditions in the half-cells to increase the e.m.f. of the cell.

d) If only the components of one type of half-cell were available, would it be possible to set up a complete cell that would produce a current. Describe the cell and mark in the polarity of its electrodes.

10. a) What is the origin of the potential difference between a metal and a solution of its ions?

b) How is this potential difference related to the ionization energies of the metal?

Chapter 20

Electrochemistry

Electrochemistry is the study of the processes taking place during the passage of electricity through an *electrolyte*.

> An electrolyte is a conductor in which the flow of charge is carried by ions and not by electrons

The study is divided into two major areas: conduction and electrolysis. *Conduction* focuses attention on the processes occuring in the bulk of the electrolyte; *electrolysis* concerns those that take place at the surface of the electrodes. The differences are illustrated in the diagrams below.

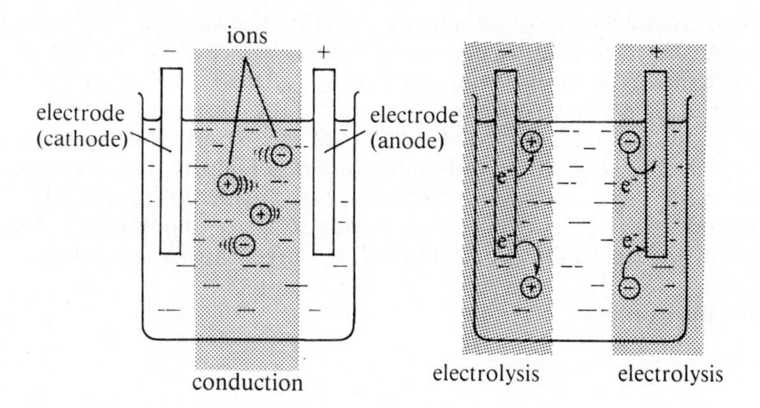

20.1 Conduction

Definitions

A metal normally conducts electricity in such a way that Ohm's Law is obeyed.

$$R = \frac{V}{i}$$

where V is the applied p.d. in volts (V)
i is the resulting current in amperes (A)
R is the resistance in ohms (Ω) which remains constant.

However, it is found that the resistance depends on

1. the nature of the metal
2. the length and cross-sectional area of the conductor
3. the temperature of the metal.

The behaviour of two different metals can be compared by means of their resistivity ρ (ohm metres) defined by

$$\rho = \frac{RA}{l}$$

where A is the cross-sectional area of the metal (m^2)
l is the length of the specimen (m)

This may also be put in terms of the conductivity κ of the metal, measured in $\Omega^{-1}\ m^{-1}$.

$$\kappa = \frac{1}{\rho} = \frac{l}{RA}$$

The conductivity is a measure of how easily the substance conducts electricity, irrespective of the geometry of the specimen.

Note that some authorities quote values of conductivity in siemens per metre ($S\ m^{-1}$), where $1\ S = 1\ \Omega^{-1}$

Conductivity of electrolyte solutions

The electrolytic conductivity of an aqueous solution is measured by placing the solution at constant temperature in a special silica glass cell that is made to form one arm of a Wheatstone bridge circuit. The plates of the cell are of known area and rigidly fixed a set distance apart. They are made of thick platinum black to minimize electrolytic effects at the surface. For the same reason, the Wheatstone bridge circuit makes use of a high frequency source of alternating current. The electrolyte solution must be made up from highly purified distilled water as its conductivity is very sensitive to impurities. Two typical electrolytic cells and a Wheatstone bridge circuit are shown below.

A fixed container cell

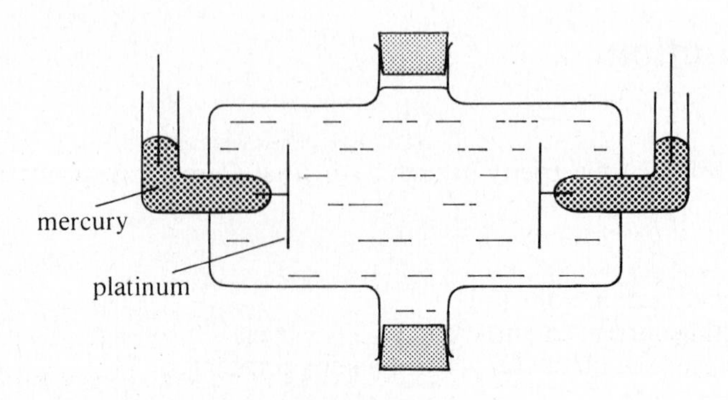

A dipping cell

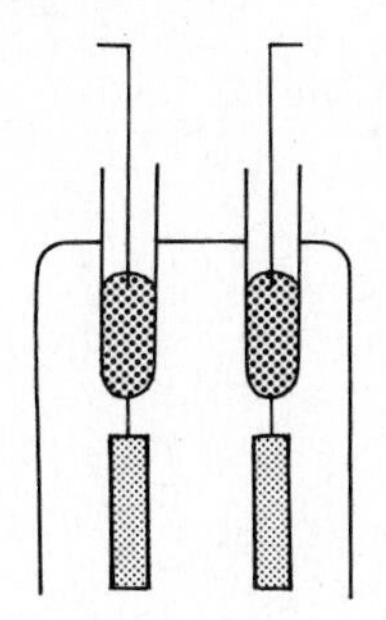

Wheatstone bridge circuit

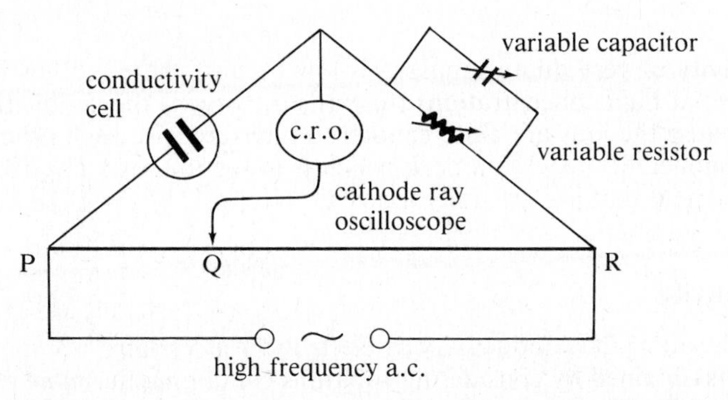

At the point of balance of the circuit, as shown by the oscilloscope,

$$\frac{\text{resistance of conductivity cell}}{\text{resistance of variable resistor}} = \frac{\text{PQ}}{\text{QR}}$$

The conductivity of the electrolyte in the cell is the reciprocal of the cell resistance multiplied by a cell constant (l/A) to allow for the geometry of the cell.

$$\kappa = \frac{1}{\text{measured resistance}} \times \text{cell constant}$$

Alternatively, the conductivity may be found by comparison with the resistance $R^{\ominus}$ of the same cell containing a standard 0.1 M solution of KCl whose electrolytic conductivity $\kappa^{\ominus}$ is known.

$$\kappa = \frac{R^{\ominus}}{R} \times \kappa^{\ominus}$$

The graph below shows the variation of conductivity of an electrolyte with dilution (1/concentration) for a strong electrolyte and for a weak electrolyte.

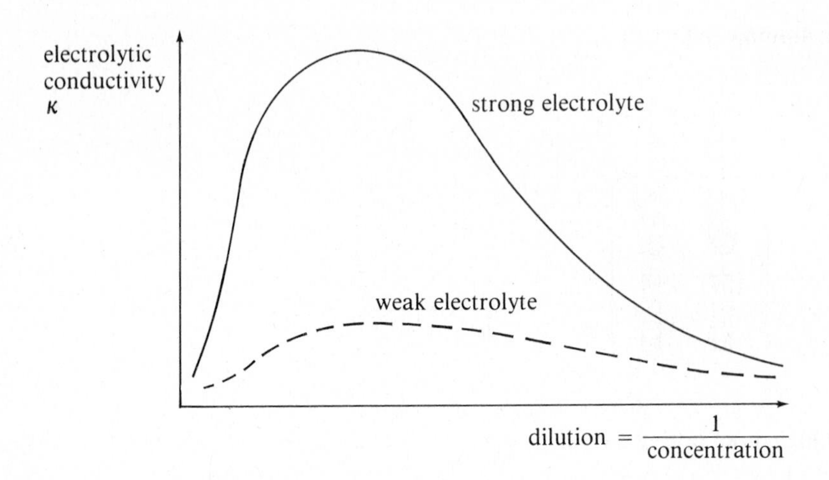

The conductivity of very dilute solutions is low because there are few ions to carry the charge; at high concentrations (towards the origin) the conductivity is equally low because the ions are close enough to interfere with each other. In between, the conductivity reaches a peak which is lower for weak electrolytes such as ethanoic acid that are not fully ionized.

Molar conductivity

A fairer comparison of the conductivity of electrolytes at various concentrations is obtained by considering solutions containing the *same* number of particles. For this we use the term molar conductivity, Λ.

> The molar conductivity Λ of a solution is the conductivity of that volume of the solution which contains 1 mole of electrolyte between plates 1 metre apart.

The plates of the cell are kept 1 metre apart, but it can be imagined that their area is varied so that the number of particles between the plates is the same for different concentrations of electrolyte. Thus for very dilute solutions, a large plate area is needed to contain the larger volume for the same number of particles.

$$\Lambda = \kappa V$$

where V is the volume in m^3 that contains one mole of the solution
Λ is measured in $\Omega^{-1}\ m^2\ mol^{-1}$
κ is measured in $\Omega^{-1}\ m^{-1}$

The relationship between κ and Λ can be brought out in a more visual way by considering an aqueous solution being poured into a cubic container of volume 1 m^3. If two of the sides of the container can act as electrodes, the conductivity of the system reaches the value of κ for the solution when the container is completely full.

0.01 mol dm^{-3} KCl

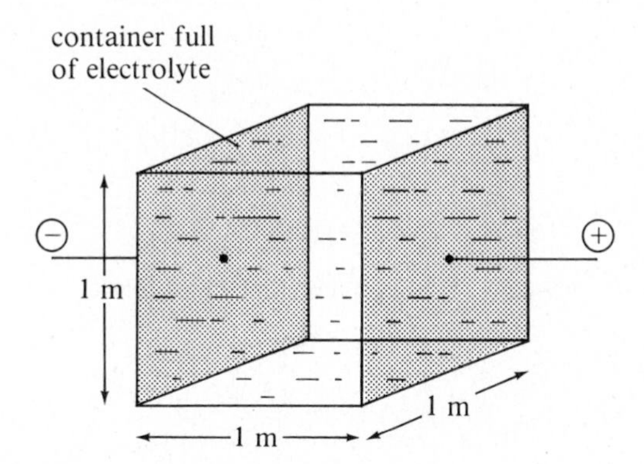

Conductivity = κ : the conductivity of a solution between plates 1 m^2 in area and 1 m apart

In the above case, there is more than one mole of electrolyte between the plates. In 1 m^3, there are 1000 dm^3, and each dm^3 contains 0.01 mole of electrolyte; in other words, there are 10 moles of electrolyte. When the container is 1/10 full, there is exactly one mole of electrolyte between the plates. The conductivity of the system is then equal to the value of Λ.

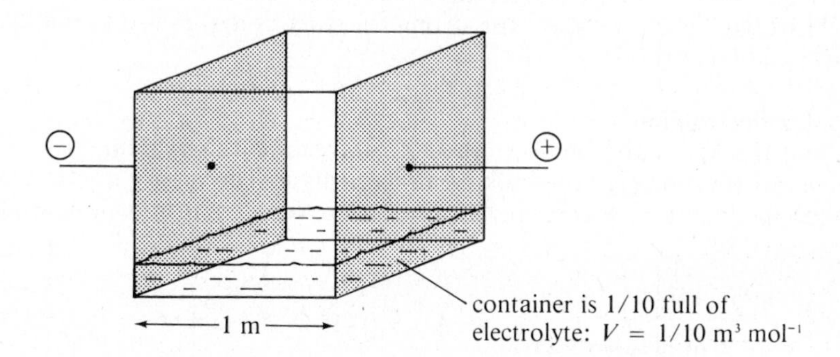

Conductivity = Λ : the conductivity of a solution containing 1 mole of electrolyte between plates 1 metre apart

$$\Lambda = \kappa \times 1/10$$

Using the apparatus described on page 333, a series of readings of κ for varying concentrations of electrolyte can be obtained. In each case, Λ can be calculated and the dependence of Λ on concentration compared. The graph overleaf illustrates this dependence for a strong electrolyte such as potassium chloride and a weak electrolyte such as ethanoic acid. A weak electrolyte is one in which the position of the dissociation equilibrium lies well to the left and so the molar conductivity falls sharply as concentration increases because most of the electrolyte is present as molecules.

$$XY_{(aq)} \rightleftharpoons X^+_{(aq)} + Y^-_{(aq)}$$

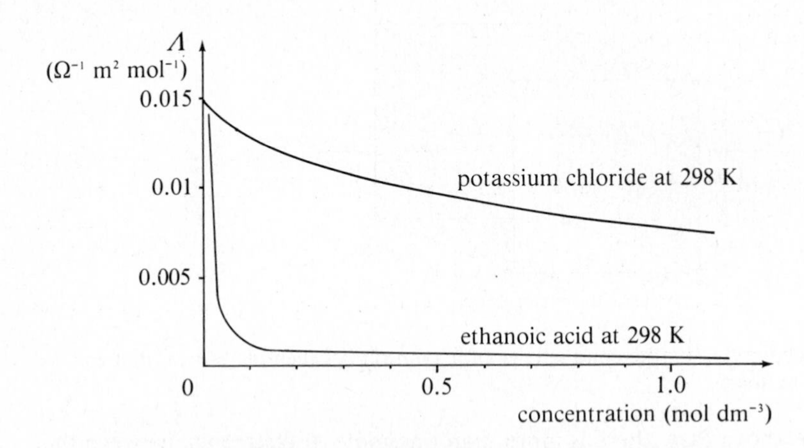

Molar conductivity at infinite dilution

The molar conductivity of a solution reaches a limiting value, Λ_∞, as the solution is progressively diluted. This value is of significance because it gives a measure of the ionic mobility when the ions are (in principle) completely free from the influence of one another.

A different practical procedure is necessary to obtain a value of Λ_∞ for a weak electrolyte. The procedure for strong electrolytes makes use of the theory of Debye and Hückel.

1. Strong electrolytes

Debye and Hückel established a mathematical treatment to explain the movement of ions through the bulk of an electrolyte. The theory predicts that molar conductivity at low concentrations is approximately proportional to the

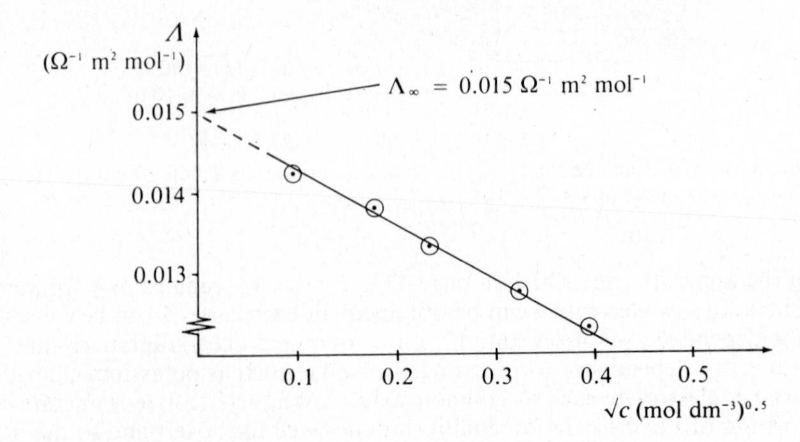

square-root of the concentration:

$$\Lambda \propto \sqrt{c}$$

When the measured values of Λ at varying concentrations of potassium chloride are plotted against $\sqrt{c}$, the graph shown at the bottom of page 336 is obtained. By extrapolating the 'best' straight line back to the vertical axis where $c = 0$, a value for Λ_∞ is obtained.

2. 'Weak' electrolytes

No meaningful result can be obtained from a similar plot for a weak electrolyte because the curve approaches the vertical axis asymptotically. However Kohlrausch in 1876 noticed that values of Λ_∞ for pairs of salts having the same anion showed a constant difference as in the tables below (the values are quoted in units of Ω^{-1} m^2 mol^{-1}).

	Λ_∞
KCl	0.0150
NaCl	0.0127
difference	0.0023

	Λ_∞
KBr	0.0152
NaBr	0.0129
difference	0.0023

	Λ_∞
KNO_3	0.0145
$NaNO_3$	0.0121
difference	0.0024

These results led him to put forward the law of independent ionic mobilities.

The conductivity of an electrolyte at infinite dilution is the sum of the ionic mobilities of the ions forming the electrolyte, and these mobilities are independent of the presence of other ions.

$$\Lambda_\infty = \lambda^+_\infty + \lambda^-_\infty$$

where λ^+_∞ is the ionic mobility of the cation
λ^-_∞ is the ionic mobility of the anion.

(an ionic mobility is the same as the molar conductivity of the particular ion under conditions of infinite dilution).

The constant differences observed by Kohlrausch reflect the difference between the mobility of potassium and sodium ions: $\lambda_\infty(K^+) - \lambda_\infty(Na^+) = 0.0023\ \Omega^{-1}\ m^2\ mol^{-1}$. The law can be used to determine Λ_∞ for a weak electrolyte by combining the separate values of the ionic mobilities concerned. At infinite dilution even a weak electrolyte is fully dissociated.

Example Λ_∞ for ethanoic acid is found from the measured values of Λ_∞ for hydrochloric acid, sodium chloride and sodium ethanoate.

$$\Lambda_\infty(HCl) = \lambda_\infty(H^+) + \lambda_\infty(Cl^-) = 0.0426\ \Omega^{-1}\ m^2\ mol^{-1}$$

$$\Lambda_\infty(NaCl) = \lambda_\infty(Na^+) + \lambda_\infty(Cl^-) = 0.0126\ \Omega^{-1}\ m^2\ mol^{-1}$$

$$\Lambda_\infty(CH_3COONa) = \lambda_\infty(Na^+) + \lambda_\infty(CH_3COO^-) = 0.0091\ \Omega^{-1}\ m^2\ mol^{-1}$$

HCl, NaCl and CH_3COONa are all strong electrolytes and the above values are all obtained by using the method described opposite.

Now, by Kohlrausch,

$$\Lambda_\infty(CH_3COOH) = \lambda_\infty(H^+) + \lambda_\infty(CH_3COO^-)$$

$$\begin{aligned}\Lambda_\infty(CH_3COOH) &= \Lambda_\infty(HCl) + \Lambda_\infty(CH_3COONa) - \Lambda_\infty(NaCl)\\ &= (0.0517 - 0.0126)\ \Omega^{-1}\ m^2\ mol^{-1}\\ &= 0.0391\ \Omega^{-1}\ m^2\ mol^{-1}.\end{aligned}$$

Ionic mobility and transport number

We have seen above that the mobility of an ion is the molar conductivity of the ion at infinite dilution. The value of an ionic mobility is determined from the fraction of the total charge transferred by the particular ion; this fraction is called the transport number t.

$$t^+ = \frac{\lambda^+_\infty}{\Lambda_\infty}\ ;\ t^- = \frac{\lambda^-_\infty}{\Lambda_\infty}\ ; \qquad \text{where } t^+ + t^- = 1$$

Hittorf developed a method of measuring transport numbers using the cell shown below. This is divided into three compartments for sampling the concentration of ions before and after a measured charge has been passed.

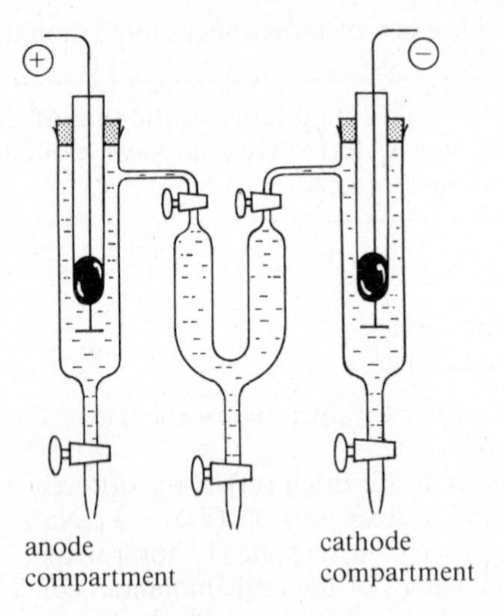

The net loss of electrolyte from the anode compartment can be shown to be proportional to the fraction t^+ of the total charge Q passed, i.e. the transport number of the cation.

In terms of moles:

$$\text{molar loss of electrolyte from the anode compartment} = \frac{Qt^+}{F} \text{ moles}$$

where F is the Faraday constant (= 96 500 C mol^{-1}).

Each transport number can be converted to ionic mobility using the equations above and the value of Λ_∞ for the electrolyte. The table on page 339 gives some accepted values for ionic mobility.

cation	*ionic radius* (m × 10^{-10})	λ^+_∞ (Ω^{-1} m^2 mol^{-1})	*anion*	λ^-_∞ (Ω^{-1} m^2 mol^{-1})
H_3O^+		0.0350	OH^-	0.0199
Li^+	0.60	0.0039	F^-	0.0055
Na^+	0.95	0.0050	Cl^-	0.0076
K^+	1.33	0.0074	Br^-	0.0078
			I^-	0.0077
Mg^{2+}	0.65	0.0106		
Ca^{2+}	0.99	0.0119	CO_3^{2-}	0.0119
			SO_4^{2-}	0.0160
Al^{3+}	0.50	0.0189		

The mobility of an ion is a measure of its speed of migration through the solution. In general, the smaller the ion the slower it travels because the greater charge density of the ion results in the attachment of more hydrating water molecules which act as a drag on the progress of the ion. Also, multiple-charged ions transport charge faster than single-charged ones, with the two exceptions of the hydroxonium ion H_3O^+ and hydroxide ion OH^-. These two ions owe their high mobilities to the proton transfer that takes place in their aqueous solutions:

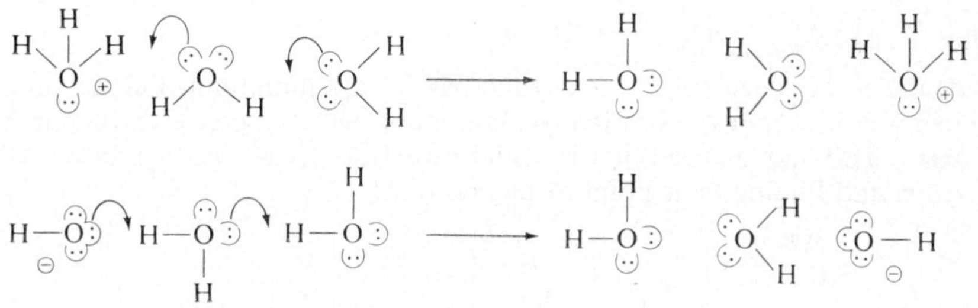

Uses of conductivity measurements

Conductivity values are useful in two major ways:

1. To calculate degrees of dissociation, α.
2. To follow the course of certain titrations, 'conductimetric' titrations.

1. Degree of dissociation

It is assumed that an electrolyte is fully dissociated at infinite dilution and so Λ_∞ is a measure of the conductivity of *all* its ions. At some other dilution, the molar conductivity is lower because a proportion of the molecules have not dissociated. The proportion that have dissociated is called the 'degree of dissociation', α, and is discussed for weak acids on page 279. It is related to molar conductivity by the expression below.

$$\alpha = \frac{\Lambda_c}{\Lambda_\infty}$$

where Λ_c = the molar conductivity at the concentration c.

2. Conductimetric titrations

During the titration of a strong acid by a strong base, the numbers of hydroxonium ions and hydroxide ions change. This causes the conductivity of the solution to change, as is shown on the following graph.

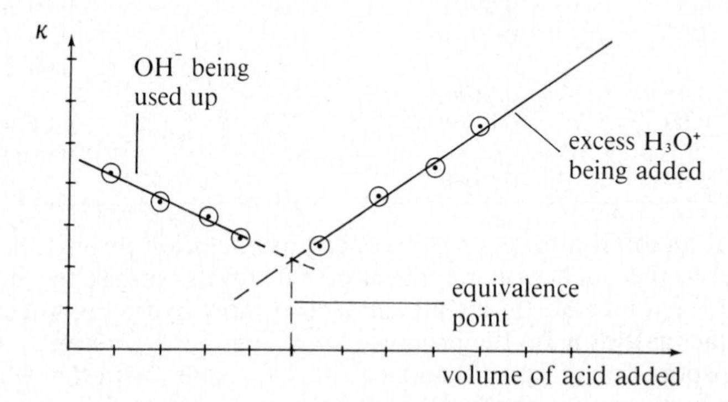

When hydroxonium ions are first added to the alkali, the reaction results in a decrease of hydroxide ion concentration.

$$H_3O^+_{(aq)} + OH^-_{(aq)} \longrightarrow 2H_2O_{(l)}$$

At the equivalence point, the conductivity is at a minimum. Further addition of acid produces an excess of hydroxonium ions which increase the conductivity again. The equivalence point is found by extrapolating the two halves of the graph and finding their point of intersection.

20.2 Electrolysis

Electrolysis is the process in which ions undergo electron transfer at the electrode surfaces. The products are often gases or metals deposited at the cathode. In many cases the electrolysis of an aqueous solution leads to the evolution of hydrogen at the cathode and oxygen at the anode because water contains a low concentration of hydroxonium and hydroxide ions.

$$2H_3O^+_{(aq)} + 2e^- \longrightarrow 2H_2O_{(l)} + H_{2(g)}$$

$$4OH^-_{(aq)} \longrightarrow 2H_2O_{(l)} + O_{2(g)} + 4e^-$$

There are two principal characteristics of electrolysis that require explanation:

1. the variation of the applied potential difference (p.d.) with the current flowing through an electrolysis cell,
2. the selective discharge of one ion in preference to another at the electrode surface.

Electrolysis and Ohm's law

Unlike metallic conductors, electrolytic conductors only obey Ohm's Law (page 331) beyond a particular value of applied potential difference. A typical set of results is illustrated below for the electrolysis of sulphuric acid between platinum electrodes.

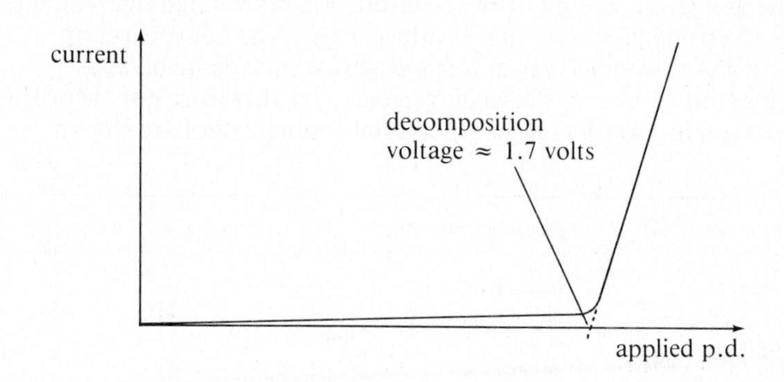

As the applied voltage is gradually increased, the current passing rises only slightly, and no electrolysis appears to take place at the electrodes. At the 'decomposition voltage', a sudden increase in current occurs and Ohm's Law is then obeyed for further increases in applied potential difference. Electrolysis now takes place at the electrodes and hydrogen and oxygen are produced according to the reactions shown below.

Anode: $6H_2O_{(l)} \longrightarrow 4H_3O^+_{(aq)} + O_{2(g)} + 4e^-$
(These are water ligands around the sulphate anions; they come into contact with the anode first)

Cathode: $2H_3O^+_{(aq)} + 2e^- \longrightarrow 2H_2O_{(l)} + H_{2(g)}$

The minimum voltage is necessary for two reasons:

1. The potential difference between the electrodes causes the ions in the system to move to their respective electrodes. The hydrogen and oxygen first formed coat the surfaces of the electrodes and the back reaction of the electrode processes can then take place:

$$6H_2O_{(l)} \rightleftharpoons 4H_3O^+_{(aq)} + O_{2(g)} + 4e^- \qquad E^\ominus = +1.23 \text{ volts}$$

$$2e^- + 2H_3O^+_{(aq)} \rightleftharpoons 2H_2O_{(l)} + H_{2(g)} \qquad E^\ominus = 0.00 \text{ volts}$$

The potential difference must therefore reach a value at least equal to the sum of the two electrode potentials to overcome the 'back e.m.f.' caused by the adsorption of the electrolysis products on the electrode surfaces. A *theoretical decomposition voltage* is therefore the sum of the two electrode potentials involved. This is 1.23 volts in the case of sulphuric acid using platinum electrodes. The observed decomposition voltage is over 1.7 volts, so there must be another factor as well.

2. The extra voltage required to discharge ions in an electrolysis cell is called the *overpotential* or *overvoltage*. Whereas the theoretical decomposition voltage concerns the thermodynamic feasibility of carrying out the electrode reaction, the overpotential is concerned with the kinetic requirements of the process. Even though a reaction may be energetically favourable, the activation energy may be so high that almost no reaction takes place. The overvoltage is a measure of the activation energy of the electrode reaction. High overvoltages occur in most electrode processes that produce a gas. The conversion of aqueous ions to gaseous molecules at a metal lattice surface involves a considerable amount of electron re-arrangement. It is therefore not surprising that the process might have a high overpotential. Some values are shown below:

gas being evolved	*electrode surface*	*overpotential (volts)*
hydrogen	clean platinum	0.03
oxygen	clean platinum	0.44
hydrogen	silver	0.15
oxygen	silver	0.45
hydrogen	mercury	0.78
oxygen	graphite	0.37
chlorine	clean platinum	0.70

To account for the total observed decomposition voltage (E_{ob}) of 1.70 volts, we now have the following expression:

$$E_{ob} = (E^{\ominus}_{anode} - E^{\ominus}_{cathode}) + \text{overpotentials at the two electrodes}$$

$$= (1.23 - 0.00) + 0.03 + 0.44 = 1.70 \text{ volts}$$

Selective discharge

In an aqueous solution of an electrolyte there is more than one type of cation and anion present. Water itself is weakly ionized:

$$2H_2O_{(l)} \rightleftharpoons H_3O^+ + OH^-$$

and the cations or anions of the electrolyte may be hydrolysed as well:

carbonate anions $\quad CO^{2-}_{3(aq)} + H_2O_{(l)} \rightleftharpoons HCO^-_{3(aq)} + OH^-_{(aq)}$

ammonium cations $\quad NH^+_{4(aq)} + H_2O_{(l)} \rightleftharpoons NH_{3(aq)} + H_3O^+_{(aq)}$

A competition for reaction at the electrode surface takes place between the ions of the electrolyte and the ions from water. The winner of the competition can be predicted from the relevant values of electrode potentials and overpotentials.

Example 1 Copper sulphate with graphite electrodes.
The relevant $E^{\ominus}$ values are shown on the scale below:

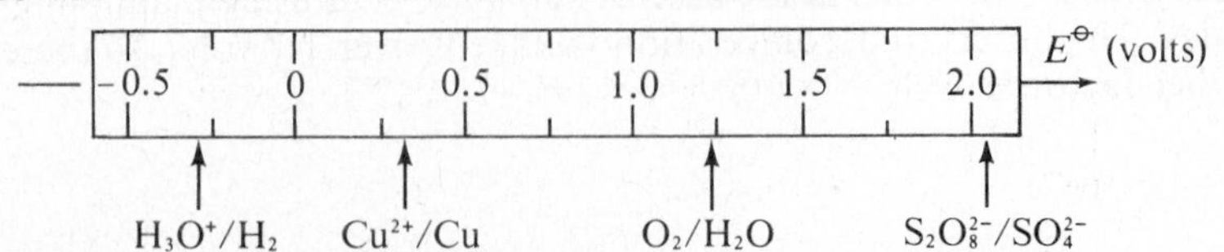

Notice that the hydrogen couple is below 0.00 volts because of the non-standard conditions. In the above solution $[H_3O^+] \approx 10^{-4}$ mol dm^{-3}.

At the anode the discharge of oxygen is more favourable than that of sulphate ions. Even after the addition of the overpotential for oxygen on graphite, the discharge potential for oxygen is still less than that for sulphate ions: (1.23 + 0.44) volt compared with 2.01 volt relative to the standard hydrogen couple:

$$6H_2O_{(l)} \longrightarrow 4H_3O^+_{(aq)} + O_{2(g)} + 4e^-: \qquad (E^\ominus + \text{overpotential}) = +1.67 \text{ volt}$$

At the cathode the discharge potential for copper is reached well before that required to discharge hydrogen and so copper is likely to be deposited on the cathode.

$$Cu^{2+}_{(aq)} + 2e^- \longrightarrow Cu_{(s)} \qquad E^\ominus = +0.34 \text{ volt}$$

The observed decomposition voltage is therefore approximately 1.33 volt.

Example 2 Copper sulphate with copper electrodes
When the electrodes are made of copper instead of graphite (as in example 1), the cathode reaction is unaltered but an alternative anode reaction competes:

$$Cu_{(s)} \longrightarrow Cu^{2+}_{(aq)} + 2e^-$$

The value of $E^\ominus$ for this process is the same as that for the cathode reaction because it is the reverse reaction. The decomposition voltage therefore drops to zero; copper dissolves from the anode and is deposited at the cathode. So an electrolysis cell containing copper sulphate between copper electrodes obeys Ohm's Law for any value of applied potential difference. Compare the variation of current flowing with applied voltage for the two examples discussed so far.

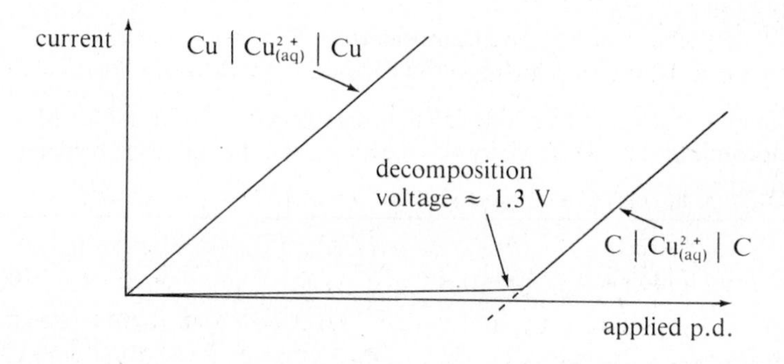

Example 3 Sodium chloride with graphite electrodes
The relevant values of $E^\ominus$ are shown on the scale below:

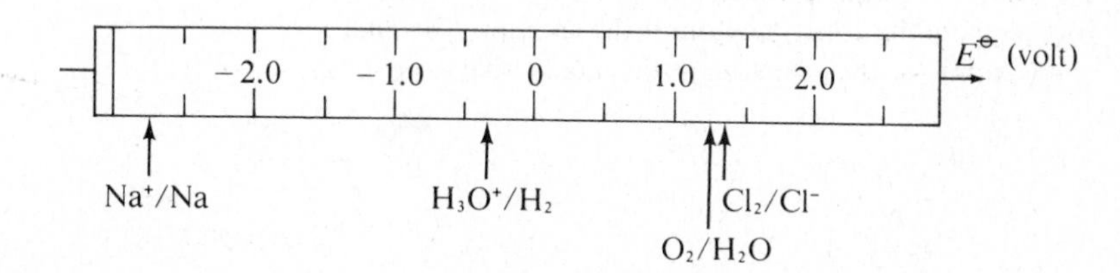

At the anode the overpotential of oxygen on graphite is about 0.4 volt whereas that for chlorine is only about 0.1 volt. This leads to discharge potentials of about 1.7 and 1.5 volt respectively compared with the standard hydrogen half-cell. In other words, chlorine is discharged in preference to oxygen despite the higher value of $E^\ominus$:

$$2Cl^-_{(aq)} \longrightarrow Cl_{2(g)} + 2e^-: \quad (E^\ominus + \text{overpotential}) = +1.48 \text{ volt}$$

At the cathode the discharge potential of hydrogen is reached well before that of sodium despite the low hydroxonium ion concentrations. Hydrogen is therefore discharged at the cathode. However, just as the oxygen and chlorine order is reversed by the effect of a high overpotential, it is possible to do the same with hydrogen and sodium. If a flowing mercury cathode is used (see page 628), the overpotential of hydrogen is as high as 0.8—0.9 volt; also, sodium forms an amalgam with mercury so that its discharge potential drops to about −1.0 volt. Under these conditions, sodium ions can be selectively discharged in preference to hydroxonium ions at the cathode.

Questions

1. Distinguish between the following:
   a) conduction and electrolysis
   b) resistance and conductivity
   c) conductivity and molar conductivity
   d) decomposition voltage and overvoltage.

2. Describe how the molar conductivity at infinite dilution is determined in the laboratory for:
   a) a strong electrolyte such as potassium chloride
   b) a weak electrolyte such as hydrogen fluoride.

3. Small volumes of sodium hydroxide solution were added to 50 cm^3 of 0.1 M hydrogen chloride. The conductivity was measured and the following readings obtained:

volume of NaOH (cm^3)	0	1	2	3	4	5	6	7	8	9	10
conductivity (Ω^{-1} m^{-1})	3.90	3.35	2.80	2.27	1.72	1.17	0.62	0.76	1.19	1.62	2.05

   a) What is the equivalence point of the above reaction?
   b) Calculate the concentration of the sodium hydroxide.
   c) Account for the variation in the conductivity.

4. The molar conductivity of ethanoic acid at two different temperatures and dilutions is shown below.

   At 18°C; $\Lambda_c = 8.5 \times 10^{-4}$, $\Lambda_\infty = 347 \times 10^{-4}\ \Omega^{-1}\ m^2\ mol^{-1}$

   At 100°C; $\Lambda_c = 14.7 \times 10^{-4}$, $\Lambda_\infty = 773 \times 10^{-4}\ \Omega^{-1}\ m^2\ mol^{-1}$

   a) Given that $c = 0.03\ mol\ dm^{-3}$, calculate the degree of dissociation of ethanoic acid and K_a at each of the two temperatures.
   b) Using this information, predict whether the dissociation of ethanoic acid is endothermic or exothermic.

5. The resistance of a conductivity cell containing 0.1 M lithium nitrate solution is 192 Ω. The resistance of the same cell containing 0.02 M potassium chloride solution is 620 Ω. The standard conductivity of 0.02 M potassium chloride is 0.242 $\Omega^{-1}\ m^{-1}$.
   Calculate a) the cell constant, b) the conductivity of the lithium nitrate solution, c) the molar conductivity of the solution.

6. a) Calculate the value of the molar conductivity at infinite dilution of hydrogen fluoride, given the following data:

strong electrolyte	Λ_∞ ($\Omega^{-1}\ m^2\ mol^{-2}$)
nitric acid	0.0421
potassium nitrate	0.0145
potassium fluoride	0.0129

   b) Calculate K_a for hydrogen fluoride given that a 0.1 M solution has a conductivity of $3.15 \times 10^{-3}\ \Omega^{-1}\ m^{-1}$.

7. Describe what takes place when a gradually increasing potential difference is applied across the following electrodes:
   a) copper electrodes in copper chloride solution
   b) carbon electrodes in copper chloride solution
   c) carbon electrodes in sodium chloride solution.

8. a) Why do most electrolytic conductors *not* obey Ohm's Law when the applied potential difference is small?
   b) Given an example of an electrolytic conductor that proves an exception to this generalization. How does it work?

9. Discuss how the process of 'selective discharge' might be used to design a method of separating a mixture of metallic compounds.

Part 3

Organic chemistry

Chapter 21
The chemistry of carbon

21.1 The molecular nature of carbon compounds

The chemistry of carbon is called 'organic chemistry' because so many of the first known carbon compounds were isolated from living systems. Today over four million different carbon compounds have been isolated. This number takes on particular significance when contrasted with the total number of all compounds found on Earth. About five million in all are known, which means that the proportion of the total that are carbon compounds is as high as 80 per cent or more. And yet the Earth contains less than 0.01 per cent by mass of carbon.

This huge discrepancy indicates that carbon atoms have some very unusual bonding properties. Before we can usefully tackle organic compounds, some preliminary discussion of these bonding properties is necessary.

The electronic configuration of carbon

Carbon atoms have the configuration $1s^2\ 2s^2\ 2p^2$. The outer shell has four electrons and so carbon atoms tend to form four covalent bonds. This bonding picture suggests that carbon compounds are likely to be molecular. The physical properties of carbon compounds (e.g. low melting point, low conductivity) support this suggestion.

One of the simplest carbon compounds is methane, CH_4. The valence-bond (V.B.) treatment for methane indicates a tetrahedral molecule:

H
|
C
H H
H

When we consider the atomic orbitals in which the outer electrons are found in an isolated carbon atom, it is not at first sight easy to see why carbon forms four bonds.

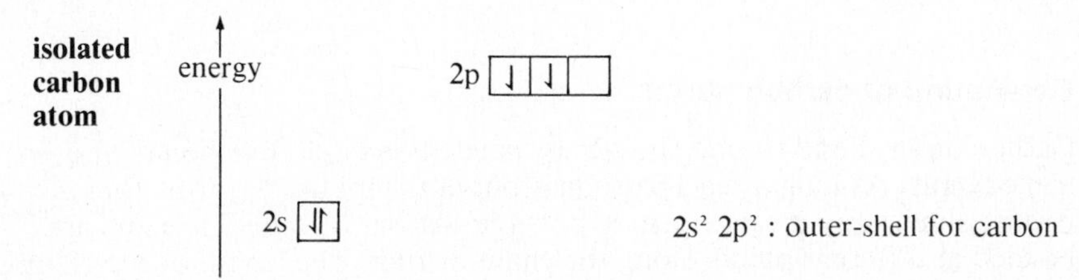

This might suggest that only two bonds can form as a result of pairing the single electrons in $2p_x$ and $2p_y$.

However, all the evidence shows that four bonds are formed and that they are identical. It is possible to account for this by considering the promotion of an electron from the 2s- into a 2p-orbital. The energy required for the promotion is more than offset by the energy released in forming two more bonds. The fact that the four bonds are the same is accounted for by the electrons occupying a set of hybrid orbitals deriving from an s-orbital (2s) and three p-orbitals ($2p_x$, $2p_y$ and $2p_z$) which lead to four goods σ-overlaps. The hybrid orbitals are called sp^3-hybrids (see page 58).

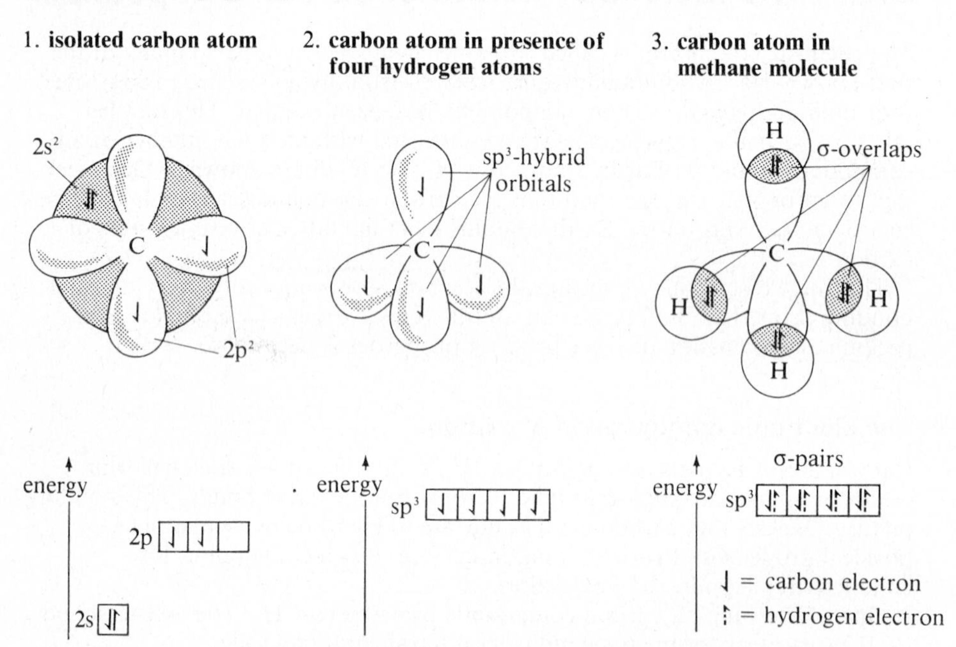

The four covalent bonds made to the hydrogen atoms are strong as a result of the extensive overlap between C-sp^3 and H-1s. Carbon atoms also form strong bonds to one another for the same reason. The facility with which carbon atoms can bond in this way explains why so many different carbon compounds exist and why one reaction so often leads to the formation of more than one set of products. All sizes of chains and interlinking rings of bonded carbon atoms may be produced onto which other atoms may be attached. The property of an atom to bond well to other atoms of the same sort is called *catenation*.

Catenation of carbon atoms

Carbon atoms bond to one another by single bonds, double bonds or even triple bonds. As long as each carbon atom in the molecule forms four covalent bonds, almost any arrangement is possible. Atoms of other elements are bonded at different points along the chain or ring. The length of the chains or construction of the rings, combined with the variation in different atoms bonded to the skeleton provided, determines the identity of the molecule.

The reactivity (or, conversely, inertness) of an organic molecule depends on its composition and structure. In particular, the type of electron pairs that it contains is of prime importance. In the following section, we shall look more closely at the organization of outer-shell electrons in some simple organic systems. You should find that this introduction will make it easier to understand the nature of organic reactions.

21.2 The electronic pairs of organic molecules

σ-pairs

A σ-pair of electrons is a pair of electrons trapped between two covalently-bonded nuclei. The electrons occupy a σ molecular orbital (M.O.), see page 54, as a result of a single overlap between the atomic orbitals of two atoms. For example, the pairs of electrons in the methane molecule are all σ-pairs. The shared pairs represent regions of electron density along the line of centres between the bonded nuclei. Any single bond between two atoms is in fact a σ-pair of electrons:

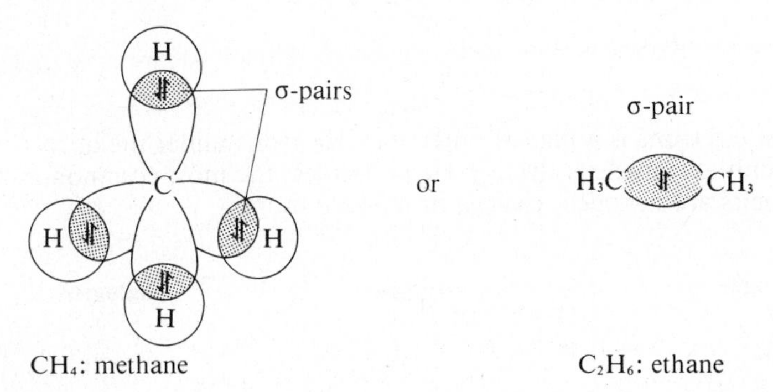

CH_4: methane

C_2H_6: ethane

π-pairs

A π-pair of electrons can only be shared between nuclei that are already bonded together by a σ-bond. π-electrons are found in a split region above and below the line of centres of the bonded nuclei. Once there is electron density between the nuclei (a σ-pair), then the other pair must split itself round the σ-pair. Electrons occupy a π-M.O. (see page 55) as a result of a double overlap between the atomic orbitals of two atoms. For example, in ethene, C_2H_4, the valence-bond model is

```
H       H
 \     /
  C = C
 /     \
H       H
```

Here the two carbon atoms share two pairs of electrons: one pair is a σ-pair, the other must be a π-pair.

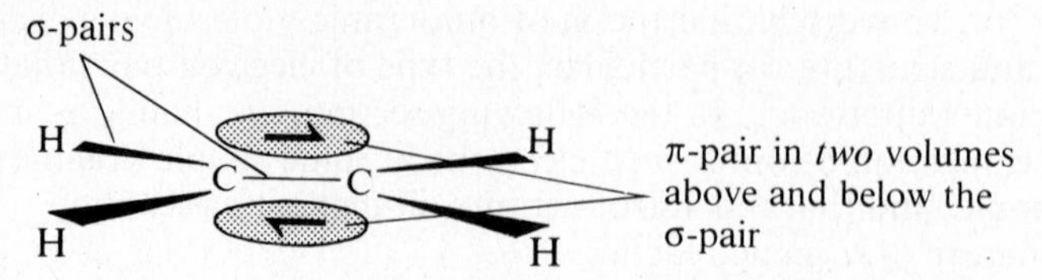

Another example of a π-pair occurs in any molecule containing the group shown below:

>C=O (the 'carbonyl' group), e.g. methanal: H_2CO

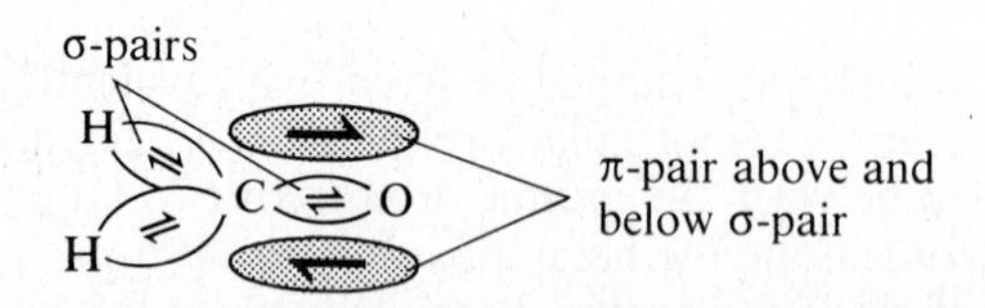

Although there are two distinct regions occupied by a π-pair, there is only *one* pair of electrons distributed through the two volumes.

Lone pairs

A lone pair of electrons is a pair of outer-shell electrons under the attractive influence of only one nucleus. In organic molecules, the most common atoms having lone pairs are nitrogen, oxygen or halogen atoms.

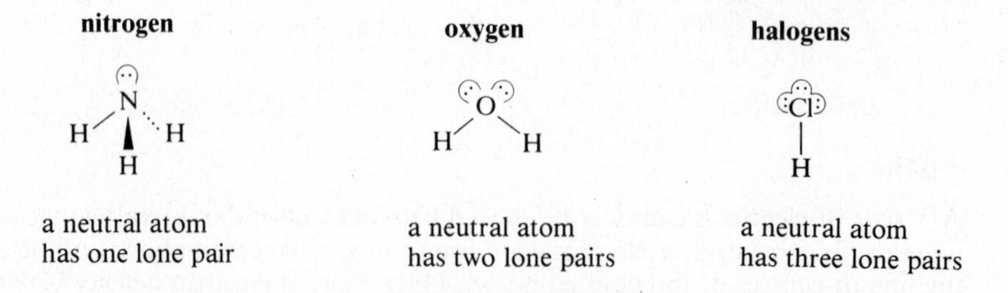

Notice that the neutrality of the different atoms bonded in the molecules has been emphasized. The number of lone pairs changes if proton transfer occurs: you should be familiar with this idea from the theory of acids and bases. Some typical examples are shown at the top of page 353.

+ charged		− charged	
NH_4^+	H_3O^+	OH^-	Cl^-
N with no lone pairs	O with one lone pair	O with three lone pairs	Cl with four lone pairs

In a reaction between two molecules, bonds are broken and made. This means that the types of electron pairs in the colliding molecules must change and, in particular, lone pairs often become σ-pairs in the process. As will be seen in 21.3, it is very important to know the relationship between the number of lone pairs on an atom and the charge of the whole particle.

The further a lone pair protrudes from its atomic core, the more easily it can be attracted away to form a σ-pair. The electronegativity of an atom is a measure of the attracting power that that atom has for its outer-shell electrons: the lower the electronegativity of an atom, the more prominent are its lone pairs. For the three cases considered earlier in this section, the nitrogen lone pair is likely to protrude the most because nitrogen has the lowest electronegativity.

In summary, therefore, there are three types of electron pairs found in organic molecules:

σ-pairs
π-pairs
lone pairs

In any reaction, there are bound to be changes in electron-pair type as bonds are broken and made.

In the next section, the general methods by which bonds make and break are discussed. These methods are usually represented by a series of particle drawings known as a *reaction mechanism*. An awareness of electron-pair type is essential to the understanding of mechanism.

21.3 The mechanism of organic reactions

Collision theory

The collision theory of reactions is described on page 231. Here is a reminder of the major points:

1. Two particles must collide in order to react.
2. Not every collision results in a reaction.
3. The particles must collide with a certain minimum energy to cause the reorganization of electron pairs.
4. An 'activated complex' is produced for a very short instant as a result of the merger of one particle with another during a collision. The 'activated

complex' breaks up to give products if there is sufficient energy in the collision. If not, the original particles reform after the collision.

5. Even a collision with sufficient energy sometimes does not lead to a reaction. The particles must collide correctly orientated towards each other.

It is the last of these factors that provides the key to an understanding of 'mechanism'. To make a new bond, a pair of electrons is needed. In most cases a pair of electrons from one particle comes into contact with a favourable site on the other particle during a successful collision. If the pair of electrons is not correctly lined up in a particular collision, the new bond is unlikely to form. The formation of a new bond usually leads to the breaking of an already existing one.

Electron availability and deficiency

In most successful collisions, we can usually distinguish:

1. a particle that has a pair of electrons able to start to form a new bond,
2. another particle that has an electron-deficient site which attracts the first particle's electrons.

$$A: \curvearrowright \quad \overset{\delta+}{B} - \overset{\delta-}{C}$$

Particle A is acting as an electron-pair donor or *Lewis Base*
Particle BC is acting as an electron-pair acceptor or *Lewis Acid*

Brønsted-Lowry acid-base theory (Chapter 17) focuses attention on proton transfer and this limits application of the theory. A broader acid-base theory called the *Lewis acid-base theory* concentrates on the movement of electron pairs.

> A Lewis base is an electron-pair donor; a Lewis acid is an electron-pair acceptor.

The Lewis basicity of a particle is governed by the type of electron pairs that it possesses:

σ-pairs — low basicity: they are held rigidly between two nuclei.
π-pairs — more basic than σ-pairs: they can be attracted away from their bonding position.
lone pairs — the most basic of all: the further the lone pair protrudes from the atomic core, the more basic the particle becomes.

The Lewis acidity of a particle usually depends on the electronegativity differences between its bonded atoms. The permanent polarization of a covalent bond as a result of an electronegativity difference is defined as an *inductive effect* (I). An atom that tends to attract electrons towards itself exerts a $-$I effect. An atom that tends to lose control of electrons exerts a $+$I effect.

$-$I effects are typically associated with atoms of high electronegativity such as oxygen or chlorine. For example, the chlorine atom in chloromethane exerts a $-$I effect on the carbon atom.

$$H_3C^{\delta+}-Cl^{\delta-}$$

carbon electronegativity = 2.5
chlorine electronegativity = 3.0

The electron-deficient site produced on the carbon atom leads the chloromethane molecule to act as a Lewis Acid.

On the other hand an alkyl group, like the methyl group, exerts a +I effect. A methyl group is less electronegative than a single hydrogen atom on account of the secondary inductive effect of the three hydrogen atoms bonded in the methyl group. These are all less electronegative than the carbon atom to which they are bonded and so they release electron density to this atom. This electron density is in turn transmitted to the atom bonded to the methyl group and is greater than the electron density withdrawn directly from a single hydrogen atom. The extra electron density can enhance the Lewis basicity of a neighbouring group. For example, methylamine is more basic than ammonia:

NH_3: $K_b = 1.7 \times 10^{-5}$ mol dm^{-3}

CH_3NH_2: $K_b = 2.29 \times 10^{-4}$ mol dm^{-3}

The methyl group releases a greater electron density to the nitrogen atom than do the hydrogen atoms. This increases the prominence of the lone pair of the methylamine molecule compared with that of the ammonia molecule.

Reactive sites

A reactive particle can now be recognized to have either

1. Lewis basicity: a pair of electrons able to be donated, or
2. Lewis acidity: an electron-deficient site able to accept electrons.

The centres of Lewis basicity or acidity are the *reactive sites* of a particle. A diagram of two particles that are about to react with each other is drawn so that the reactive sites are emphasized. The best way to do this is to use a system of shorthand for all unreactive parts of the particle. In this way attention is drawn to the reactive sites, and the unreactive parts do not clutter up the diagram. Some typical reactive and unreactive sites are shown in the table at the top of page 356. The reactive sites are divided into those that depend on Lewis acidity and those that depend on Lewis basicity.

unreactive sites	*reactive sites*	
	Lewis acidity	*Lewis basicity*
unpolarized σ-pairs C—H C—C σ-pairs have low basicity, and there is little charge separation in C—C or C—H	polarized σ-pairs $\overset{\delta+}{C} - \overset{\delta-}{O}$ $\overset{\delta+}{C} - \overset{\delta-}{N}$ $\overset{\delta+}{C} - \overset{\delta-}{Hal}$	prominent lone pairs  O N π-pairs C—O, C—C 

Some examples of the shorthand used for unreactive sites are shown below:

unreactive site	*name*	*shorthand*
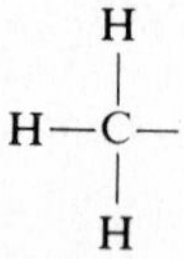 $H-CH_2-$ (H—C— with H above and below)	methyl group	Me —
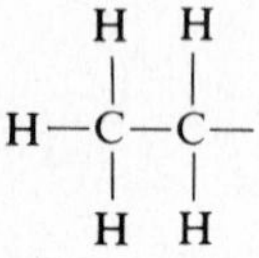 H—C—C— (each C bearing H above and below)	ethyl group	Et —
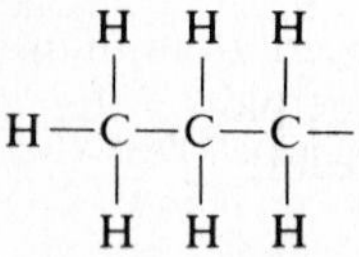 H—C—C—C— (each C bearing H above and below)	propyl group	Pr —

It might be helpful to work through a few examples of drawings that emphasize reactive sites. Try a few of your own afterwards to accustom yourself to this method of showing molecular structure.

name and structure	*reactive sites*	*particle drawing*
bromoethane 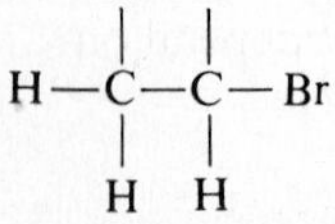 H—C—C—Br (each C bearing H above and below)	Lewis acidity $\overset{\delta+}{C} - \overset{\delta-}{Br}$	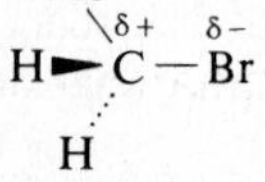 Me, H, H on $\overset{\delta+}{C} - \overset{\delta-}{Br}$

name and structure	*reactive sites*	*particle drawing*
ethanol 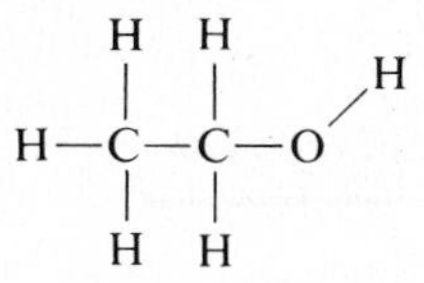 $H-C(H)(H)-C(H)(H)-O-H$	Lewis acidity $\overset{\delta+}{C}-\overset{\delta-}{O}$ Lewis basicity  O (lone pairs)	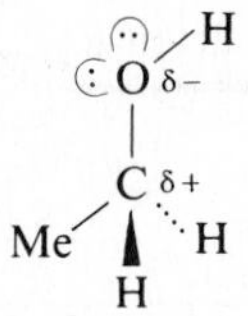 H, O δ−, C δ+, Me, H, H
propene $H-C(H)(H)-C(H)=C(H)(H)$ 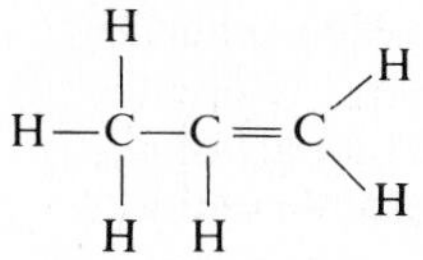	Lewis basicity C—C 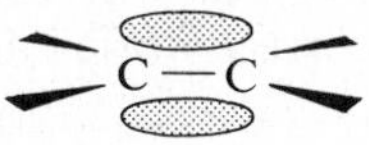	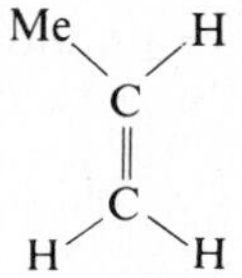 Me, H, C, C, H, H
ethanal $H-C(H)(H)-C(=O)-H$ 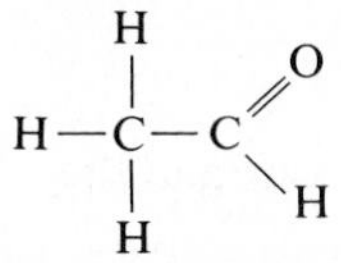	Lewis acidity $\overset{\delta+}{C}-\overset{\delta-}{O}$ Lewis basicity 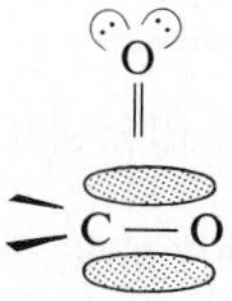 O (lone pairs) C—O	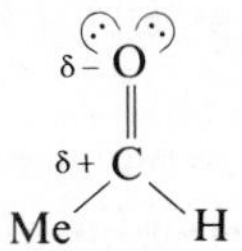 δ− O, δ+ C, Me, H

There are two other important terms that are used to describe the reactivity of particles. These are

1. *nucleophile* (symbol Nu:)
2. *electrophile* (symbol E)

A particle behaves as a nucleophile (most are negatively charged) when it has a lone pair of electrons and acts as a Lewis base. A mechanism that results from the action of a nucleophile is called a *nucleophilic reaction.*

A particle behaves as an electrophile (most are positively charged) when it has electron-deficient site and accepts a π-pair of electrons. Electrophiles act as Lewis acids, but note that the term electrophile can *only* be used for particles that attract π-pairs — *not* lone pairs. A mechanism that results from the action of an electrophile is called an *electrophilic reaction*. Some simple examples of nucleophilic and electrophilic reactions are discussed on pages 359 and 360.

Conventions of mechanism

1. The mechanism of a single-step reaction consists of three separate drawings: (i) the reactant particles, (ii) the activated complex formed, (iii) the product particles.

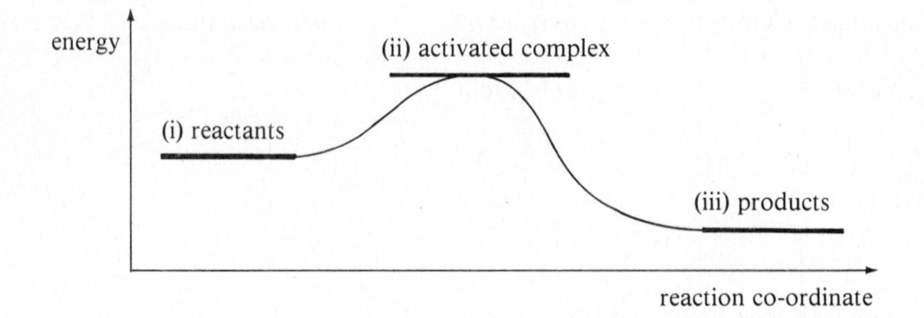

2. The reactant particles are drawn first with emphasis on their reactive sites. The reactive sites are shown correctly orientated towards each other.

3. A curly arrow, ↷ ,is used to represent the movement of a pair of electrons. For example, in a collision between a water molecule and a hydrogen molecule:

H_2O : H—Cl

The tail of the arrow marks the starting position of the electrons and the head shows where they finish.

4. The activated complex is drawn next and is shown in square brackets.

5. A localized charge sign is shown on any atom in the activated complex that has lost or gained control of an electron pair, for example:

⊕		⊖	
H_3O^+	NH_4^+	HO^-	[C]⊖ bond forming, bond breaking
an oxygen atom with three bonds and one lone pair	a nitrogen atom with four bonds and no lone pairs	an oxygen atom with one bond and three lone pairs	a carbon atom where one bond is forming and another is breaking

6. When a π-pair is attracted to an electrophile, the activated complex that forms contains the π-electrons shared between three atoms. This is shown

by a dotted triangle:

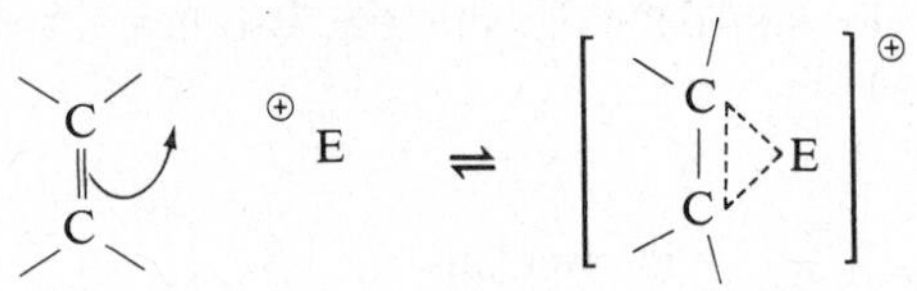

π-electrons shared between three atoms

7. The product particles are drawn finally.

Some simple examples of mechanisms

1. Nucleophilic substitution

The collision of a nucleophile with an electron-deficient, saturated carbon atom can lead to a substitution reaction. For example, bromoethane CH_3CH_2Br reacts when refluxed with aqueous alkali OH^-.

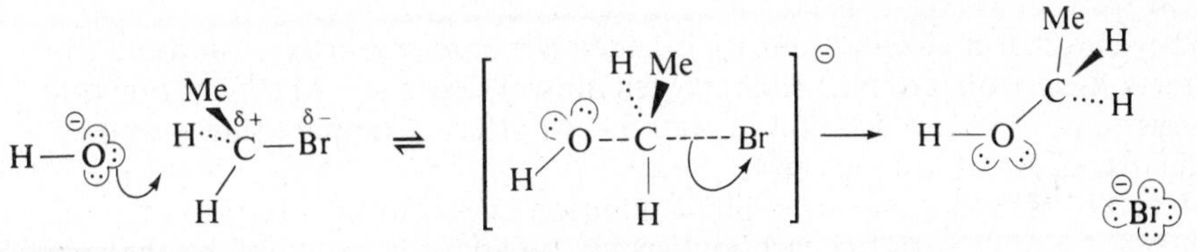

correct orientation needed to give collision of the two reactive sites

the three other pairs of electrons on the carbon atom have been repelled by the nucleophile's lone pair

a bromide ion has split out of the activated complex

The collision brings about the *substitution* of a bromine atom by a hydroxyl group. The bromide ion is called a 'leaving-group' because it leaves the activated complex when the products are formed.

A more general way of showing this type of nucleophilic substitution is to use:

$\overset{\ominus}{\text{Nu}}$: for any nucleophile
Le — for any leaving-group

Then

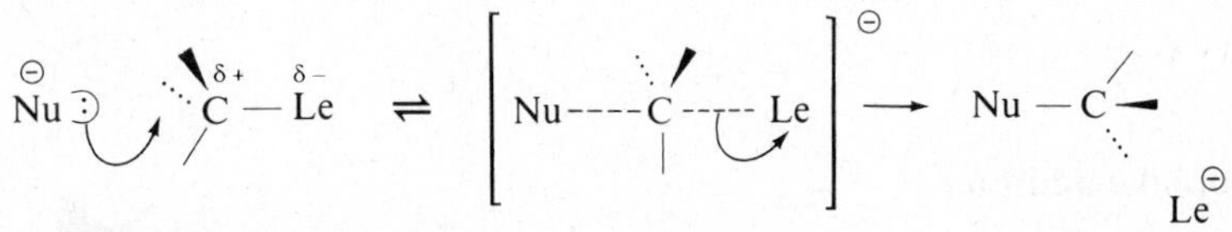

2. Nucleophilic addition

If a nucleophile collides with an electron-deficient site that also has a π-pair of electrons, 'addition' becomes possible.

For example,

a nucleophile: $Nu^{\ominus}$ an electron-deficient site with a π-pair $R_1R_2C^{\delta+}{=}O^{\delta-}$

$$Nu^{\ominus} + R_1R_2C^{\delta+}{=}O^{\delta-} \rightleftharpoons Nu{-}CR_1R_2{-}O^{\ominus}$$

the π-pair is repelled onto the oxygen atom by the nucleophile's pair of electrons

the intermediate is likely to be a strong base, like a hydroxide ion

A movement of π-electron density is called a *mesomeric effect*. An atom that attracts π-electrons to itself (like oxygen above) exerts a −M effect: one that releases electrons into a π-bond exerts a +M effect. Compare this with the inductive effect (I) on page 354.

A specific example of nucleophilic addition occurs in the reaction of hydrogen cyanide with the carbonyl group, an addition catalysed by the cyanide ion.

$$^{\ominus}C{\equiv}N + Me_2C^{\delta+}{=}O^{\delta-} \rightleftharpoons N{\equiv}C{-}CMe_2{-}O^{\ominus} \xrightarrow{H{-}C{\equiv}N} NC{-}CMe_2{-}O{-}H + {}^{\ominus}C{\equiv}N$$

cyanide ion regenerated i.e. a catalyst

Effectively H—CN is added across the double bond, $>C{=}O$. The result is

$$NC{-}C{-}OH$$

3. Electrophilic addition

The collision of an electrophile with a molecule possessing a π-pair and appreciable Lewis basicity can also lead to addition. On page 361, the reaction of electrophilic hydrogen bromide with ethene is illustrated.

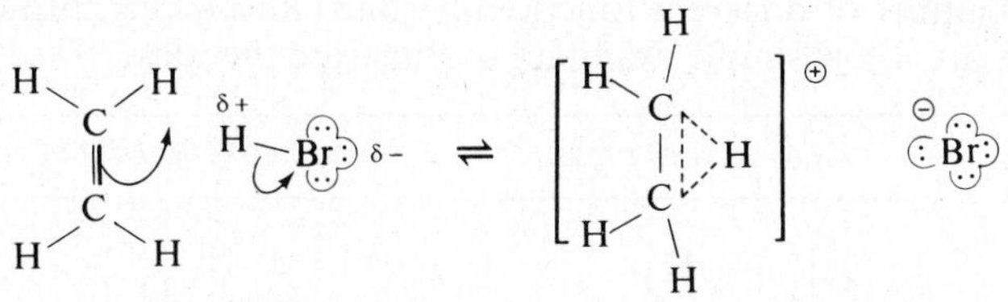

activated complex

If the activated complex collides with a bromide ion, two possible things could happen:

i) The trapped hydrogen could be removed again and the π-electrons returned to the carbon atoms.
ii) The bromide ion could collide with the other side of the activated complex. It would then act as a nucleophile and a new σ-bond would form. The π-pair in the activated complex would then become a σ-pair between one of the carbon atoms and the trapped hydrogen.

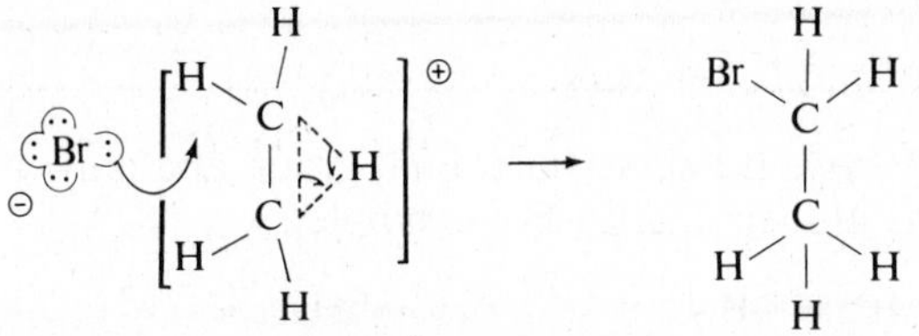

(⌒ indicates a shift of one electron)

Effectively, H—Br is added across the double bond, >C=C<. The result is

The use of reaction mechanisms to order and explain organic reactions has now become widespread. This is because mechanism is a powerful unifying technique which can save enormous amounts of repetition.

21.4 Functional groups

Dominant reactive sites

The chemical properties of an organic molecule are governed by its reactive sites. If two molecules have the same reactive sites, they are likely to have similar chemical properties. An atom, or group of atoms that constitutes a dominant reactive site in a molecule is called a *functional group*.

We shall meet a number of different functional groups in working through later chapters. Here are a few simple examples to illustrate the idea.

structure	*reactive sites present*	*name of functional group*
$-C-O-H$	$\overset{\delta+}{C}-\overset{\delta-}{O}-H$	hydroxyl group
$-C-NH_2$	$\overset{\delta+}{C}-\overset{\delta-}{N}H_2$	amino group
$>C=O$	$\underset{\delta+}{C}-\underset{\delta-}{O}$	carbonyl group
$-C(=O)-O-H$	combination of hydroxyl and carbonyl sites	carboxylic acid group

Two different molecules can have the same functional group and only differ in the number of carbon atoms in their chains. For example,

hydroxyl group: $H-CH_2-CH_2-CH_2-OH$ propan-1-ol; $H-CH_2-OH$ methanol

carboxylic acid group: $H-CH_2-C(=O)OH$ ethanoic acid; $H-CH_2-CH_2-C(=O)OH$ propanoic acid

For any functional group, there exists a whole family of organic compounds, each of which contains the particular group attached to a chain of different length. We can use the term *homologous series* to describe a family such as this, as long as its members differ only in the number of linking $—CH_2—$ units in their chains.

Names of organic compounds

The modern name of an organic compound depends on

1. the length (or ring structure) of its carbon chain,

2. the name of its functional group and substituents,
3. the position of the substituents on the chain.

A manual of naming practice has been brought out by IUPAC; for the present, we shall consider only the major principles outlined in it. Some homologous series have special names but discussion of these will be kept to later chapters. Sadly, the naming of organic compounds is complicated by the occasional use of old-fashioned 'trivial' names. There are no rules that govern the make-up of such names as acetic acid, formaldehyde, or lactic acid: either you know the structure or you don't. The present system of naming follows the pattern outlined below.

1. Count the number of carbon atoms in the longest chain present in the structure. If it is joined up as a ring, start with the prefix *cyclo*.

number of atoms	*use the first syllable*
1	meth
2	eth
3	prop
4	but
5	pent
6	hex

2. Check to see whether the chain is saturated or unsaturated.

saturation of chain	*use the second syllable*
fully saturated (all single bonds)	an
unsaturated (double bond)	en
unsaturated (triple bond)	yn

3. Identify the functional groups or substituents present in the structure. Substituents are indicated as prefixes before the start of the name. Functional groups are indicated as syllables added at the end of the name.

substituent	*name as prefix*
$-Br$	bromo
$-Cl$	chloro
$-I$	iodo
$-OH$	hydroxy
$-NH_2$	amino
$-CH_3$	methyl
$-C_2H_5$	ethyl

functional group	*name as added end syllable*
no functional group	e
$-OH$	ol
$-C(=O)H$	al
$-C(=O)-O-H$	oic acid
$-C(=O)NH_2$	amide

4. Sometimes a substituent occurs in the middle of a chain rather than at its end. To show this, a number is fitted into the name so that we can tell to which part of the chain the group is bonded. Count from the end of the chain nearest to the carbon atom that has the substituent: this number is included just before the name of the substituent itself.

The examples below illustrate these points and introduce a few others.

Example 1

```
   H  H  H
   |  |  |
H—C—C—C—H
   |  |  |
   H  H  H
```

propane

1. Three carbon atoms:	prop
2. Fully saturated chain:	an
3. No functional groups:	e

Example 2

a)

```
   H  H  H  H
   |  |  |  |
H—C—C—C—C—H
   |  |  |  |
   H  H  H  H
```

butane

1. Four carbons in the longest chain:	but
2. Fully saturated chain:	an
3. No functional groups:	e

b)

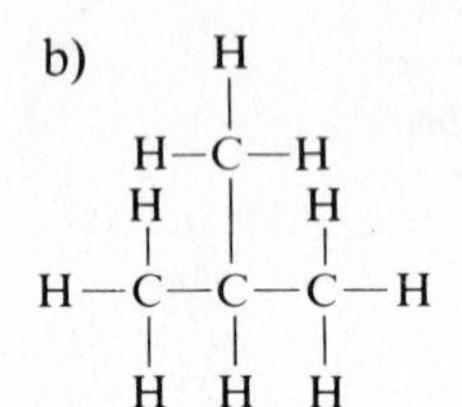

2-methylpropane

1. Three carbons in the longest chain:	prop
2. Fully saturated chain:	an
3. No functional groups:	e
4. A methyl group is present on carbon atom number 2:	2-methyl

Note that both these compounds have the molecular formula C_4H_{10}.

Example 3

```
H        Br
 \      /
  C=C
 /      \
H        C—H
        / \
       H   H
```

2-bromopropene

1. Three carbons in the longest chain:	prop
2. Unsaturated double bond:	en
3. No functional groups at the end:	e
4. Bromine atom on carbon no. 2:	2-bromo

Example 4

```
  Br     H     H    Br
    \    |     |   /
 H—C—C=C—C=C
   /       |       \
  H        H        H
```

1,5-dibromopent-1,3-diene

1. Five carbons in the longest chain:	pent
2. Unsaturated twice with 2 double-bonds on carbon atoms no. 1 and no. 3:	-1, 3-dien
3. No functional groups	e
4. Two bromine atoms on carbon atoms no.1 and no. 5:	1, 5-dibromo

Example 5

```
H     O
 \   //
H—C—C
 /    \
H      H
```

ethanal

1. Two carbons in the longest chain: eth
2. Fully saturated chain: an
3. Functional group: al

Example 6

```
  H  OH  H
  |  |   |
H—C—C—C—H
  |  |   |
  H  H   H
```

propan-2-ol or
2-hydroxypropane

1. Three carbons in the longest chain: prop
2. Fully saturated chain: an
3. -OH group on carbon atom no. 2: -2-ol or 2-hydroxy

Example 7

```
     O
    //
H—C
    \
     OH
```

methanoic acid

Example 8

```
      H
H     |    O
 \    |   //
H—C—C—C
 /    |   \
NH2   H    OH
```

3-aminopropanoic acid

Example 9

```
         Cl
         |
         C
        / \
     H2C   H
      |   CH
     H2C  ||
      \   CH
       H2C
```

3-chlorocyclohexene

Questions

1. Classify all the types of electron pairs found in the molecules of the following compounds: a) propane, b) propan-1-ol, c) water, d) carbon dioxide, e) propanonitrile, f) methyl ethanoate.
2. What evidence can you find to support the following suggestions?
   a) Carbon does *not* have the configuration $1s^2\ 2s^2\ 2p^2$ in organic molecules.
   b) Carbon is sp^3-hybridized in methane
3. Distinguish clearly the two acid-base theories of Brønsted-Lowry and Lewis. Show how a Brønsted-Lowry base can also be classified as a Lewis base.
4. Explain the meaning of the following terms: a) reactive site, b) nucleophile, c) electrophile, d) leaving group, e) reaction mechanism

5. Name the following compounds:

a) 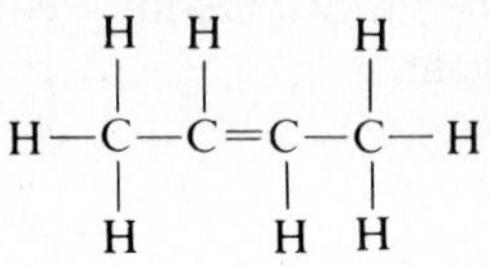
```
H   H       H
|   |       |
H—C—C=C—C—H
|     |   |
H     H   H
```

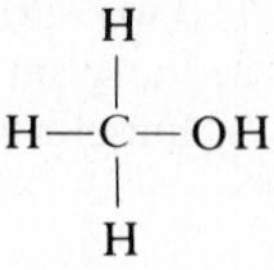

b)
```
   H
   |
H—C—OH
   |
   H
```

c)
```
  H NH₂ H
  |  |  |
H—C—C—C—H
  |  |  |
  H  H  H
```

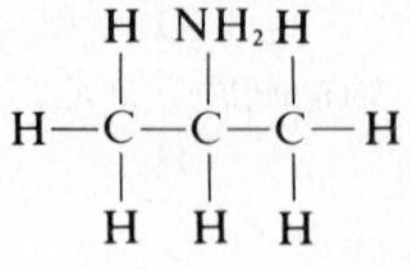

d) 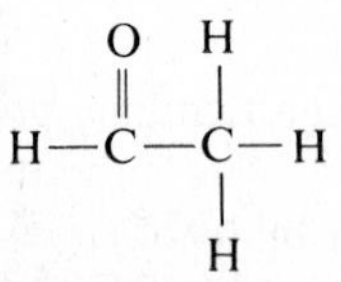
```
   O  H
   ‖  |
H—C—C—H
      |
      H
```

e) 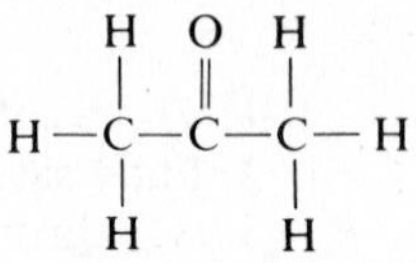
```
   H  O  H
   |  ‖  |
H—C—C—C—H
   |     |
   H     H
```

f) 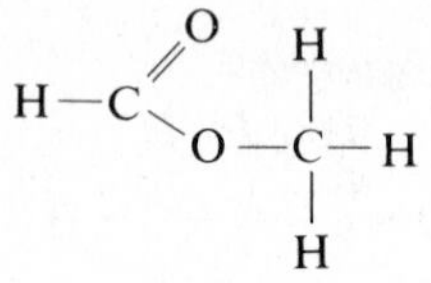
```
     O      H
    //      |
H—C        |
    \      |
     O —C—H
            |
            H
```

6. Redraw the structure of each of the above molecules so that their reactive sites are clearly emphasized.

7. Pick from the following list a) a nucleophile, b) an electrophile, c) a Lewis base, d) a Brønsted-Lowry acid, e) a Lewis acid.
i) H_2O ii) H^+ iii) HI iv) CH_4 v) NH_3 vi) NH_4^+
Explain your choice in each case.

8. Take the following reactions as examples to illustrate the distinction between nucleophilic and electrophilic addition. Give their mechanisms in full.
a) Hydrogen chloride and propene
b) Sodium hydroxide solution (very dilute) and methanal.

9. Copy out the reaction scheme below and fill in all the missing symbols:

```
          R₁                 R₁
           \           O    .
H—Ö:        C=O  ⇌  H/  \  C   —→  :Ö—C/R₁      H/O\R₂
          /                 :Ö/        \\O
         O                    \R₂
          \R₂
```

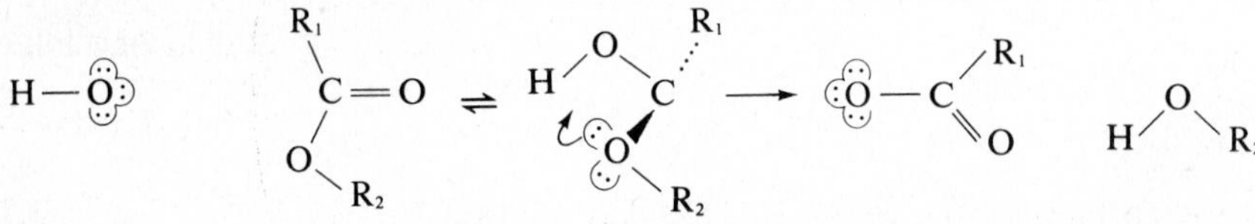

10. Copy out the electrophilic mechanism below and fill in all the missing symbols:

```
      O—H              O—H          :O—H
      |                |              \H
H  C=C  H   ⇌   [ H  C—C  H ]    —→   H   C—C=O  H
   |                |                  H   Br
   H   Br—Br        H         :Br:⊖             :Br:⊖
```

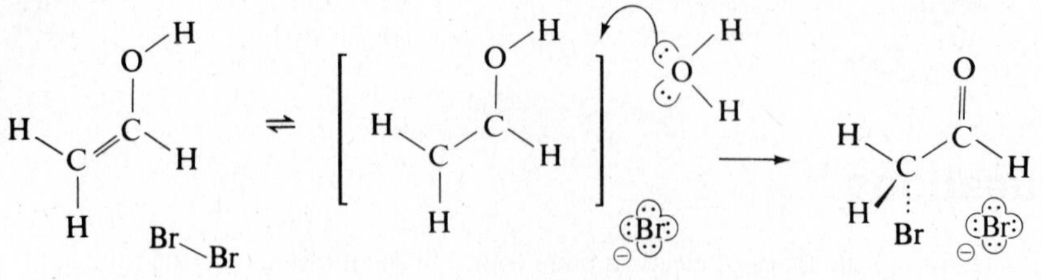

Chapter 22
Alkanes

22.1 Carbon-carbon single bonds

Functional group

The alkanes are a class of hydrocarbons: compounds made of only hydrogen and carbon. They are fully *saturated*, which means that each carbon atom bonds to the next by single bonds only. The first few members of the alkanes are:

methane 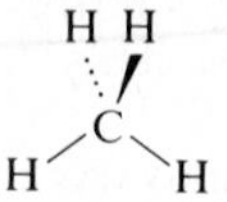	*ethane* 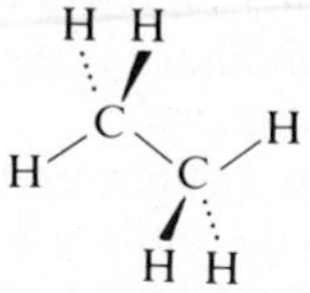	*propane* 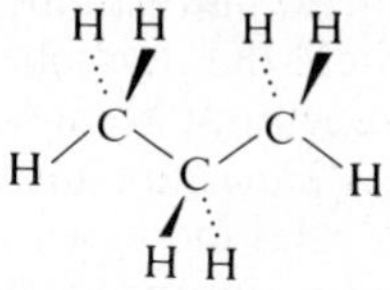	*butane* 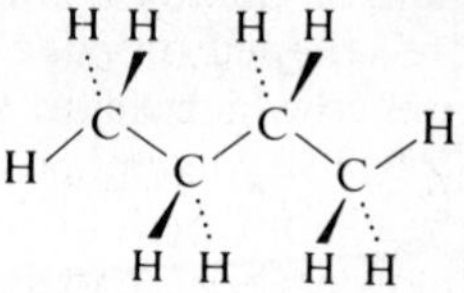

There are two patterns that can be noticed about the molecular structure of the alkanes

1. The molecules are made of zig-zag chains of carbon atoms. At right angles to the chain there are pairs of hydrogen atoms bonded to each carbon atom. The bond angles are all close to 109° 28′.
2. The general formula of any alkane is C_nH_{2n+2} where n can be any whole number.

Strictly speaking, there does not appear to be any functional group associated with the alkanes. We can write the general formula $CH_3\,(CH_2)_n\,CH_3$, where $n = 0, 1, 2 \ldots$, for any alkane except methane. From this formula the functional group of the alkanes can be thought of as $—CH_2—CH_2—$.

The concept of a functional group is not easily applicable to the alkanes because as a class of organic compounds, they are rather unreactive. The unreactive nature of saturated hydrocarbon chains has led to the use of a schematic way of drawing their structures:

propane butane
('straight-chain' hydrocarbons)

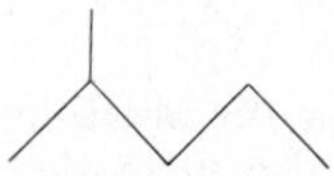

2–methylpentane
(a 'branched-chain' hydrocarbon)

Only the single bonds between the carbon atoms are shown. The rest of the molecule is not drawn in because it can be worked out: no hydrogen is shown at all. Where each bond 'links' to the next, there are two hydrogen atoms: where the bond finishes 'unattached', there are three hydrogen atoms. This ensures that each carbon atom has four bonds.

For naming practice, try to identify the structures below. The solutions are given in the answers section of this book.

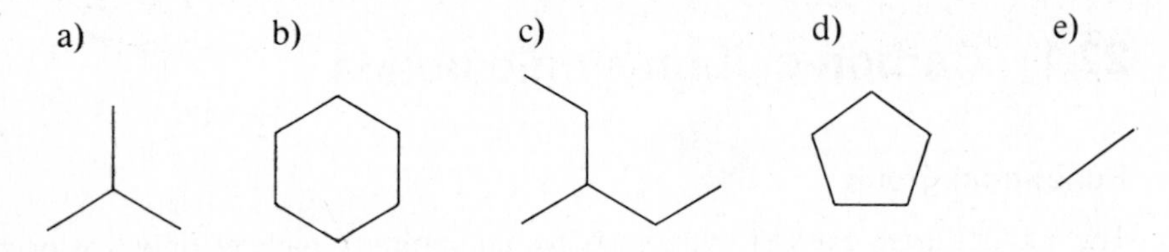

Source of alkanes

Large natural deposits of alkanes are quite widespread as fossil fuels. Coal, oil and natural gas all contain high proportions of different alkanes. Natural gas found in the North Sea is almost pure methane. The longer chain alkanes are isolated mainly from the fractional distillation of oil. The details of this process are outlined on page 584 where the petrochemical industry is discussed. The uses to which alkanes have been put are also described.

The first members of the alkanes are gases or low boiling-point liquids. As the chain length of the molecules increases so does the boiling point of the hydrocarbon. This is a direct consequence of the gradually increasing van der Waals' forces between the molecules: van der Waals' forces depend on the number of electrons in a molecule. The longer chain alkanes are therefore waxy solids.

There is little need to prepare samples of alkanes in industry or in the laboratory, because of the readily available natural sources. Where chemical methods are necessary to make a particular alkane, the most common strategies are as follows.

1. Saturate a suitable alkene

Ethane can be produced by passing ethene and hydrogen at $T \approx$ 370 K and $P \approx$ atmospheric pressure over a finely divided palladium catalyst

$$H_2C{=}CH_2 \xrightarrow[\text{finely divided Pd}]{H_2} H_3C{-}CH_3$$

2. Electrolyse the alkali metal salt of a suitable carboxylic acid

Ethane can also be made by electrolysing sodium ethanoate in methanol solution between platinium-foil electrodes:

$$2CH_3COO^- \xrightarrow[\text{at anode}]{} C_2H_6 + 2CO_2 + 2e^-$$

22.2 Reactive sites

Alkanes have no Lewis acidic character: there are no electron-deficient sites because carbon and hydrogen have about the same electronegativity.

Alkanes also have no Lewis basic character: the electron pairs are all tightly controlled σ-pairs.

This conspicuous lack of reactive sites is responsible for the comparative inertness of the alkanes. The only reactions that they undergo occur by a 'free radical' mechanism. The theory of free radical reactions (or homolytic processes) is introduced on page 248.

22.3 Reactions

With chlorine

An alkane reacts vigorously with chlorine when it is brought into contact with the gas in the presence of sunlight. The products are a mixture of substituted chloro-alkanes. Methane and chlorine, for example, combine explosively to give chloromethane as well as dichloro-, trichloro- and tetrachloromethane. The mechanism is similar to that of the hydrogen-chlorine reaction discussed in 15.5:

initiation $Cl{-}Cl + h\nu \longrightarrow Cl\cdot \quad \cdot Cl$

The breaking of a bond in this way, so that each fragment has one of the electrons from the pair that originally formed the bond, is called *homolysis* or *homolytic fission* of the bond.

propagation

1. $H_3C{-}H \;\; \cdot Cl \longrightarrow H_3C\cdot \quad H{-}Cl$

methyl radical

2. $Cl{-}Cl \;\; \cdot CH_3 \longrightarrow Cl{-}CH_3 \quad \cdot Cl$

3. $ClH_2C{-}H \;\; \cdot Cl \longrightarrow ClH_2C\cdot \quad H{-}Cl$

chloromethyl radical

termination

$R\cdot \;\; \cdot R \xrightarrow{\text{at the reaction vessel wall}} R{-}R$

(where R· is any free radical)

An initiation produces two chlorine atoms which are very reactive free radicals. When a chlorine atom collides with a methane molecule, electron reorganization occurs; a molecule of hydrogen chloride and a methyl radical are formed. The methyl radical propagates the chain reaction by colliding with a chlorine molecule. Once again a further radical is produced. The chain continues until two radicals meet and bond together at the vessel wall. The wall absorbs the excess energy from the bonding radicals which might otherwise split up again.

With excess chlorine, the reaction should produce tetrachloromethane

$$CH_4 + 4Cl_2 \xrightarrow{h\nu} CCl_4 + 4HCl$$

With oxygen: eudiometry

A similar reaction takes place between an alkane and oxygen. Alkanes burn to produce carbon dioxide and water. All hydrocarbons undergo this unspecific form of oxidation and the products are always carbon dioxide and water providing there is excess oxygen present; otherwise carbon monoxide or even carbon can be produced (as with a Bunsen with the air hole closed):

$$C_3H_8 + \underset{\text{excess}}{5O_2} \rightarrow 3CO_2 + 4H_2O$$

The actual pathway of these free-radical reactions is rather more complicated than that of the chlorine reaction. The initiation is probably the homolytic fission of the oxygen π-bond. A 'branched chain' reaction then occurs because more than one radical is produced in the major propagation step.

Apart from the use of the hydrocarbon-oxygen reaction in fuel technology, the technique of eudiometry is a well-known application. The equation for the reaction of *any* hydrocarbon with oxygen can be written in general form:

$$C_xH_y + (x + \frac{y}{4})O_2 \rightarrow xCO_2 + \frac{y}{2}H_2O$$

where x and y can be any whole number. It doesn't matter whether the hydrocarbon is an alkane or not. Eudiometry is a method which uses the above reaction to identify an unknown hydrocarbon.

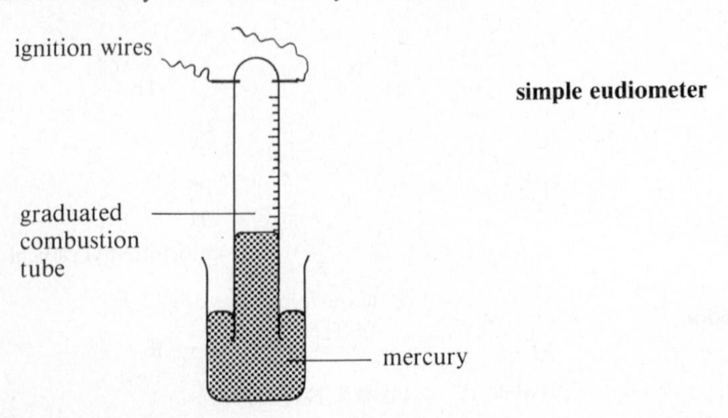

simple eudiometer

1. A volume (V cm^3) of the unknown hydrocarbon is bubbled into the tube.
2. A large volume of oxygen is bubbled in to make up a total volume (T cm^3).
3. The mixture is ignited and the hydrocarbon reacts quickly with the oxygen. Excess oxygen is left in the mixture.
4. When the temperature and pressure have returned to their initial values, the volume is measured (R cm^3) and then the residual gases are shaken with concentrated alkali which absorbs the carbon dioxide.
5. The final volume of gas remaining is measured (F cm^3).

By assuming that

i) all the hydrocarbon burns to carbon dioxide and water vapour,
ii) the water vapour condenses and therefore occupies negligible volume
iii) all the carbon dioxide is absorbed by the alkali:

$$CO_{2(g)} + HO^-_{(aq)} \longrightarrow HCO^-_{3(aq)}$$

then F cm^3 represents the excess oxygen used in the reaction. Oxygen should be the only gas that is left in the tube after the process. This means that the volume ratio of gases in the reaction can be calculated

$$[C_x\,H_y] : [O_2] : [CO_2] = V : (T-V-F) : (R-F)$$

Avogadro's law equates molar and volume ratios for gases under the same conditions of temperature and pressure. The calculated volume ratios can therefore be used to work out the stoichiometry of the unknown hydrocarbon.

Example 10 cm^3 of a hydrocarbon was made up to 100 cm^3 with oxygen. After ignition, the volume dropped to 70 cm^3 when measured under the same conditions. Shaking with alkali caused the volume to drop further to 30 cm^3.

Let the formula of the hydrocarbon be $C_x\,H_y$.
The volume of carbon dioxide absorbed in the alkali is
$70 - 30 = 40$ cm^3

Therefore, each volume of hydrocarbon produces 4 times the volume of carbon dioxide because 10 cm^3 produces 40 cm^3.

$$C_x\,H_y \rightarrow 4\,CO_2$$

Thus $x = 4$

Volume of oxygen used is $90 - 30 = 60$ cm^3. So the equation for combustion of the hydrocarbon is

$$C_4\,H_y + 6\,O_2 \longrightarrow 4\,CO_2 + 4\,H_2O$$

By balancing for oxygen,

$$y = 8$$

So the hydrocarbon is $C_4\,H_8$.

Questions

1. The reaction between methane and oxygen proceeds by a *homolytic mechanism*. Explain the meaning of the words in italic.
2. Give two examples of a branched-chain alkane and two examples of a straight-chain alkane. Name these compounds and explain the difference between them.
3. Why are alkanes so unreactive?
4. Comment on the following observations
   a) A mixture of methane and chlorine is stable in the absence of sunlight or thermal energy.
   b) Chloroethane reacts with sodium hydroxide solution but ethane does not.
   c) When octane passes over a very hot alumina catalyst, an appreciable quantity of butane is produced.
5. Give two methods by which a sample of propane could be prepared in the laboratory.
6. Calculate the molecular formulas of two unknown gaseous hydrocarbons from the following data:
   a) 20 cm^3 of X was made up to 300 cm^3 with pure oxygen. After ignition, 230 cm^3 of gas remained when measured under the same conditions of temperature and pressure. This volume was shaken with alkali and reduced to 110 cm^3.
   b) 5 cm^3 of Y was made up to 45 cm^3 with pure oxygen. After ignition (as before) the volume reduced to 35 cm^3. This was reduced further to 30 cm^3 on shaking with alkali.
7. What volume of butane burns in one kilogram of oxygen at s.t.p.? (Assume the products are carbon dioxide and water.)
8. Alkanes are often used as fuels. Compare the energy released when a kilogram of butane burns with that evolved when a kilogram of coke burns.

$\Delta H_f^\ominus$ kJ mol^{-1}	$CO_{2(g)}$	$H_2O_{(g)}$	$C_4H_{10(g)}$
	-394	-242	-125

9. When ethanal vapour $\left(H-\overset{H}{\underset{H}{C}}-C\overset{O}{\underset{H}{}} \right)$ is irradiated with ultra-violet light at 450°C, methane is produced. The equation for the reaction is

$$CH_3CHO_{(g)} \xrightarrow{hv} CH_{4(g)} + CO_{(g)}$$

   a) Compare this reaction with that of methane and chlorine in sunlight. What effect does sunlight have on the methane-chlorine mixture?
   b) What effect do you think that ultra-violet has on ethanal vapour?

bond energies (kJ mol^{-1})	C − H	C − C	C = O
	412	348	743

   c) Suggest a likely mechanism that accounts for the production of methane and carbon monoxide.
10. Cycloalkanes become more reactive as the number of carbon atoms in each molecule gets less. Comment on this observation.

Chapter 23

Alkenes

23.1 Carbon-carbon double bonds

Functional group

The alkenes are a class of hydrocarbons that differ from the alkanes. Instead of full saturation of the carbon chain, an alkene has a double bond in its chain and is described as *unsaturated*. Some typical alkenes are shown below.

mono-alkenes	*dienes or trienes*
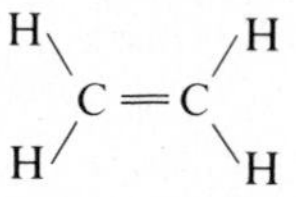 ethene	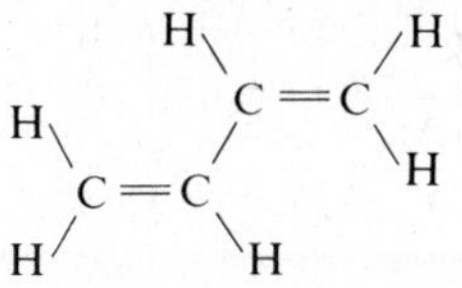 butadiene
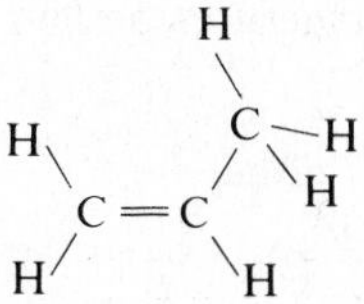 propene	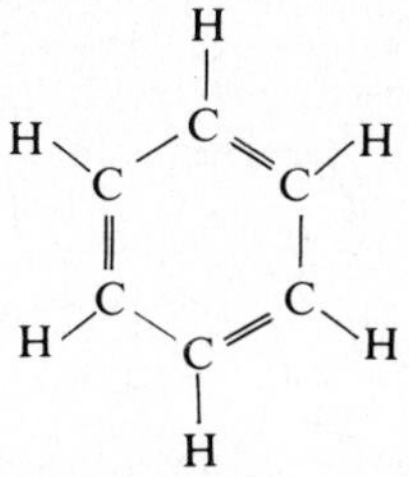 cyclohexatriene
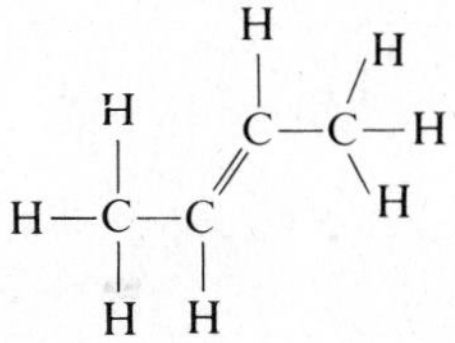 but-2-ene	

There are two patterns to notice about the molecular structure of the alkenes. Compare these with the alkane patterns described on page 367.

1. The carbon chain is planar about each double bond.
   sideways on:

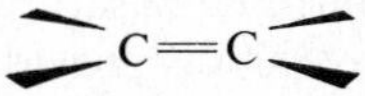

   C=C

   from above:

   C=C

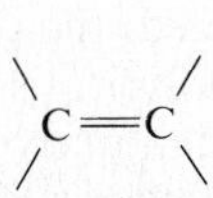

   The bond angles are all about 120°.

2. The general formula of a non-cyclic mono-alkene is $C_n H_{2n}$ where n = 2, 3, 4 . . . . There are two fewer hydrogen atoms in a monoalkene than in an alkane of the same chain length. This is a measure of the alkene's 'unsaturation': it has the ability to fit two more bonded atoms onto its carbon chain.

The functional group of an alkene is the carbon–carbon double bond $>C{=}C<$.

The shorthand method of writing alkane structures is also often used for writing alkene structures.

1. Leave out the letters for carbon and hydrogen atoms.
2. Assume that each carbon atom makes four bonds, so where there are bonds 'missing', hydrogen atoms are taken to be present.

For example

$H_2C{=}CH_2$ is ═══ ethene

$H_2C{=}CH{-}CH{=}CH_2$ is butadiene

It is useful to practise naming compounds. Try identifying the structures below. The solutions are given in the answers section of this book.

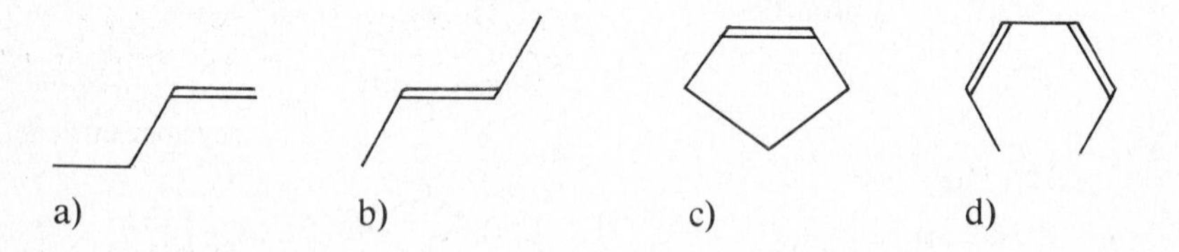

Source of alkenes: cracking alkanes

Ethene and propene are produced in huge amounts by a process called 'cracking'. The practical details of cracking are described in the discussion of the petrochemical industry (page 586) but some theoretical points can usefully be considered here.

The comparative inertness of an alkane has been explained so far in terms of the saturation of its carbon chain. An alkane's chemical properties depend only on free radical processes. However, if a source of unsaturation can be introduced into its chain, then a new reactive site is provided. *Cracking* is the process by which long-chain saturated hydrocarbons are turned into short-chain unsaturated hydrocarbons. As might be expected, the conditions for the reaction are severe and the mechanism is homolytic.

An alkane is passed over a catalyst of aluminium oxide and chromium (VI) oxide at a high temperature. Very many different reaction pathways are possible and a range of different alkenes, shorter-chain alkanes and some hydrogen is produced:

one possible cracking reaction

$$C_{16}H_{34} \xrightarrow[800\ K]{CrO_3,\ Al_2O_3} C_7H_{16} + C_3H_6 + 3C_2H_4$$

long-chain alkane — *shorter-chain alkane* — *mixture of short- chain alkenes*

The initiation occurs by the homolytic fission of one of the carbon-carbon single bonds in an alkane molecule. At the high temperatures over the catalyst, some molecules absorb enough energy for a bond to break.

initiation — absorbs enough thermal energy

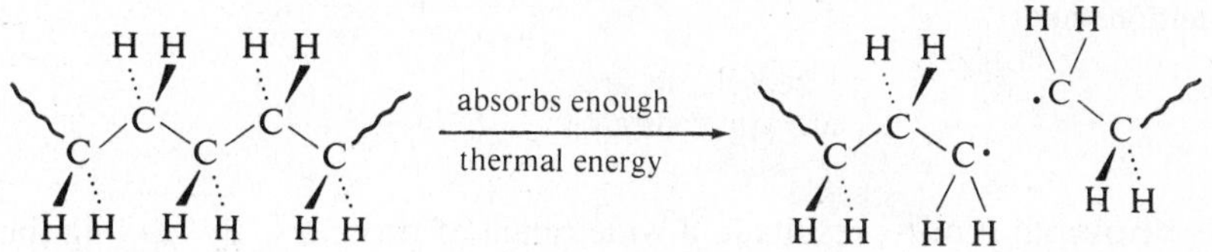

These unwieldy radicals undergo rapid disintegration until they collide with another particle. There are two main ways of disintegration.

i) $H\cdot$

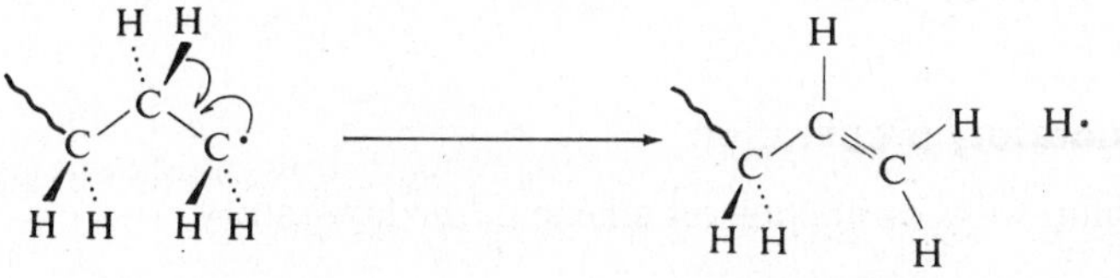

shorter-chain alkene product — hydrogen atom: a radical 'chain-carrier'

ii)

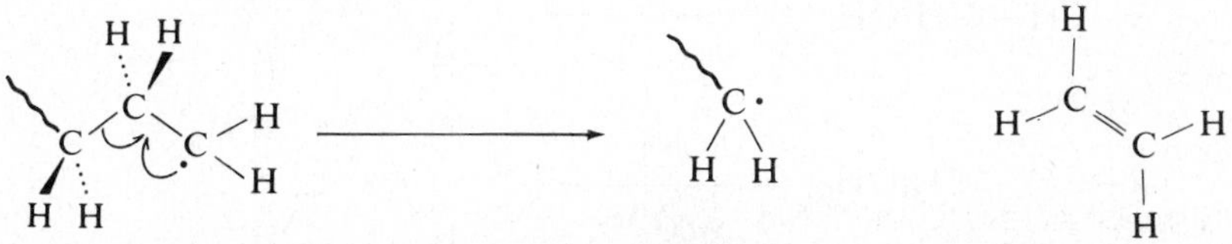

shorter-chain radical which disintegrates further — molecule of ethene

A collision with another particle can cause chain propagation via 'hydrogen abstraction':

propagation

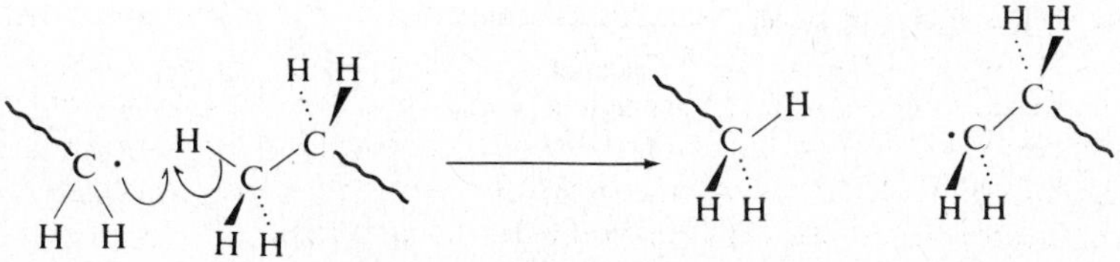

shorter-chain alkane product — new radical

or for disintegrations that have produced a hydrogen atom

$$H\cdot + H\text{–}CH_2\text{–}\sim \longrightarrow H\text{–}H + \cdot CH_2\text{–}\sim$$

hydrogen as product — new radical

The various possible chain reactions that are propagated by the presence of the radicals in the reaction-mixture are all terminated in the usual way:

termination $$R_1\cdot \;\; \cdot R_2 \xrightarrow[\text{vessel wall}]{\text{at the}} R_1\text{—}R_2$$

where R_1, R_2 are any two radicals

The overall process results in a wide range of products. By varying the conditions of the reaction and the time interval of the passage of vapour over the catalyst, some control is possible. The major aim of many cracking processes is the production from long-chain alkanes of octane-range alkanes (for use as petrol) and of ethene.

Methods of laboratory preparation

There are two main ways of making an alkene in the laboratory.

1. Elimination of water from an alcohol

(see page 412).

$$R\text{—}CH_2\text{—}CH_2OH \xrightarrow[(-H_2O)]{\text{conc } H_2SO_4} RHC{=}CH_2$$

or $$R\text{—}CH_2\text{—}CH_2OH_{(g)} \xrightarrow[\substack{Al_2O_3\\(-H_2O)}]{\text{heated}} RHC{=}CH_2$$

2. Elimination of HBr from a bromoalkane

(See page 404)

$$R\text{—}CH_2\text{—}CH_2Br \xrightarrow[\substack{\text{alcohol}\\\text{solution}\\(-HBr)}]{\text{KOH in}} RHC{=}CH_2$$

23.2 Reactive sites

Like the alkanes, the alkenes are formed from only carbon and hydrogen. This means that there is negligible charge separation in an alkene molecule and, therefore, no tendency for it to act as a Lewis acid.

Unlike the alkanes, the alkenes have π-electron density in their molecular chains. These π-electrons represent the reactive site of an alkene. They lead alkene molecules to be attracted to electrophiles.

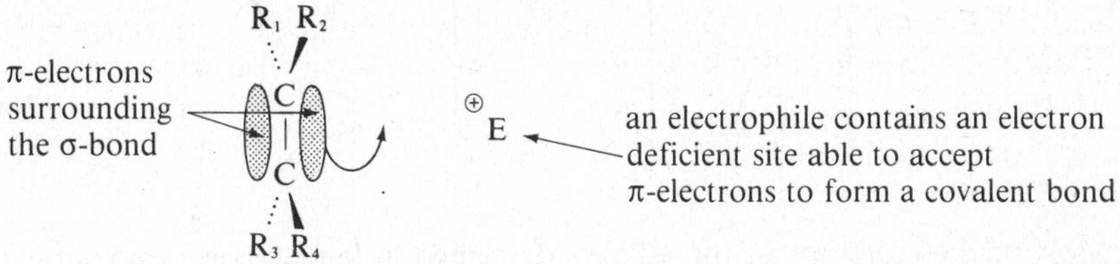

R_1 R_2 C C R_3 R_4

π-electrons surrounding the σ-bond

$E^{\oplus}$

an electrophile contains an electron deficient site able to accept π-electrons to form a covalent bond

The π-electrons become shared between three atoms: the two carbon atoms and the electron deficient site of the electrophile.

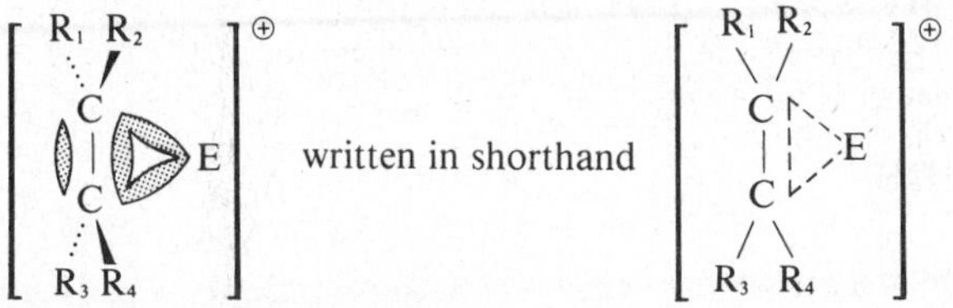

R_1 R_2 C C E R_3 R_4 ⊕

written in shorthand

R_1 R_2 C C E R_3 R_4 ⊕

This rather unstable activated complex either

1. quickly breaks up to reform the original particles, or
2. collides with a nucleophilic particle (Nu) in the reaction mixture. The carbon atoms have become electron-deficient because of the withdrawal of the π-electrons towards the electrophile.

Then:

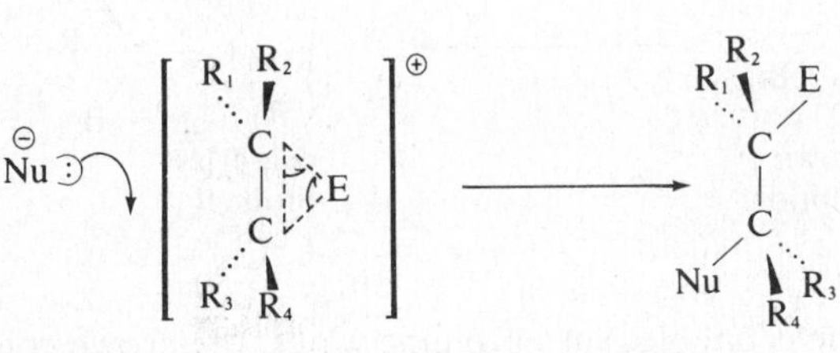

$Nu^{\ominus}$ R_1 R_2 C C E R_3 R_4 ⊕ ⟶ R_2 R_1 E C C Nu R_3 R_4

The incoming pair of electrons of the nucleophile repel the π-electrons away from one carbon atom. The nucleophile's lone pair becomes a σ-pair between Nu and one carbon atom; the π-pair becomes a σ-pair between E and the other carbon atom.

The whole mechanism is shown below. Note that the result of the process is addition. It is *electrophilic addition* because of the ability of the alkene π-electrons to interact with electrophiles.

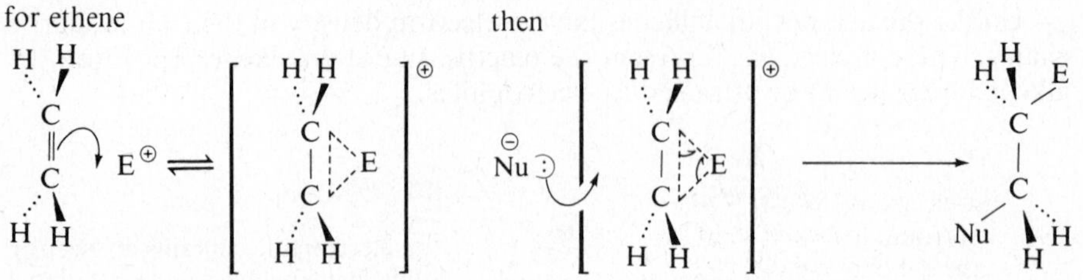

for ethene

then

Most of the reactions of the alkenes described in the next section follow the pattern outlined above. It is only the identity of the electrophile and the fate of the activated complex that change from one to the next.

23.3 Reactions

With halogens

The common laboratory test for an alkene depends on its reaction with bromine water. Rapid decolorization of the bromine occurs if a compound under test is an alkene. The reaction takes place at room temperature in aqueous conditions and in the absence of sunlight (compare alkanes' reaction). It is a very fast reaction.

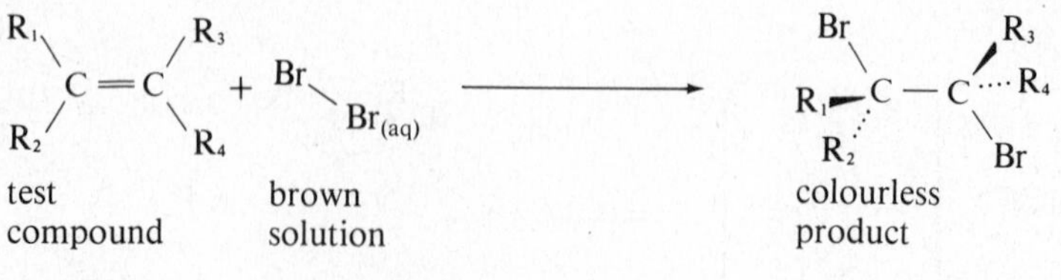

test compound

brown solution

colourless product

For example, when ethene is bubbled into bromine water, the brown colour disappears and 1,2-dibromoethane is produced. An interesting confirmation of the mechanism discussed is provided by the presence of 2-bromoethanol in the product. This seems to suggest the collision of a bromo-activated complex with the nucleophile, water:

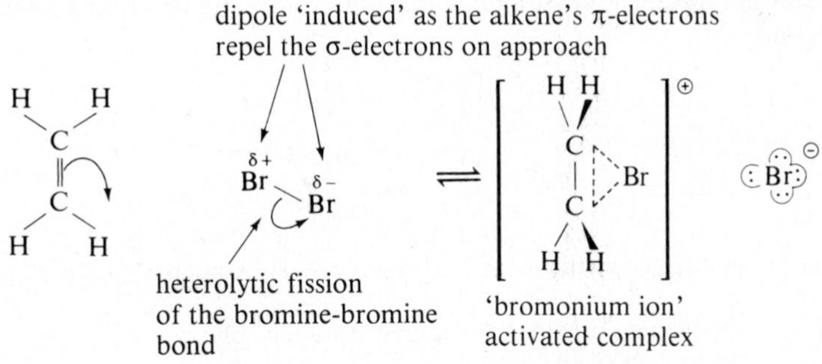

dipole 'induced' as the alkene's π-electrons repel the σ-electrons on approach

heterolytic fission of the bromine-bromine bond

'bromonium ion' activated complex

Notice that under the influence of the π-electrons from the alkene, the bond between the bromine atoms breaks unevenly. One bromine atom ends up with both electrons and hence a minus charge. The other, without either electron, is responsible for the positive charge on the bromonium ion.

This type of bond breaking is called *heterolysis* or *heterolytic fission* and is in contrast with the homolysis described on page 369.

i)

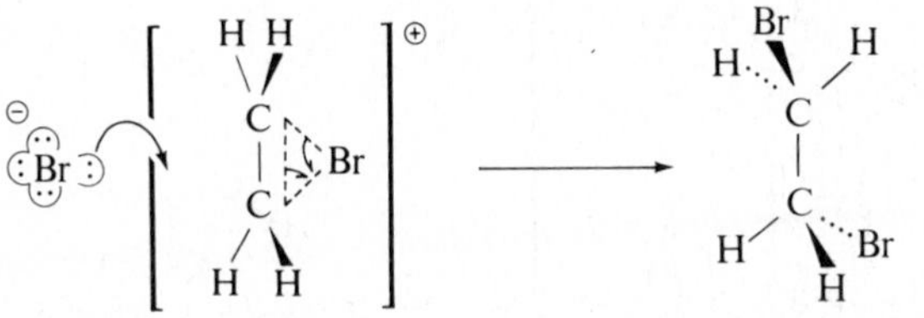

1,2-dibromoethane

or ii) if water is the nucleophile:

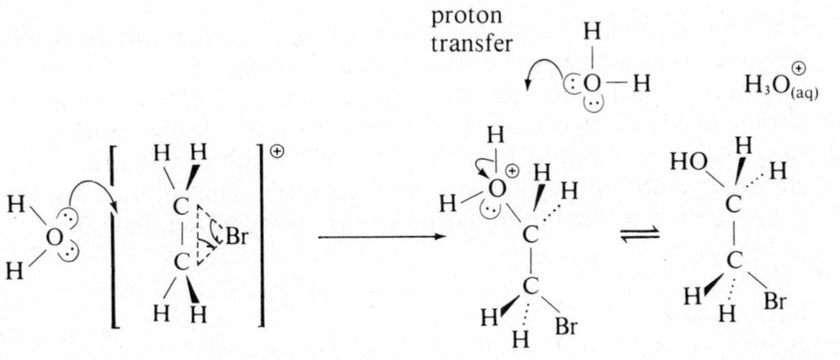

proton transfer

$H_3O^{\oplus}_{(aq)}$

2-bromoethanol

Alkenes react with chlorine and iodine in a similar way.

With hydrogen halides

Hydrogen halides add across an alkene double bond. The reaction pathway is again the same.

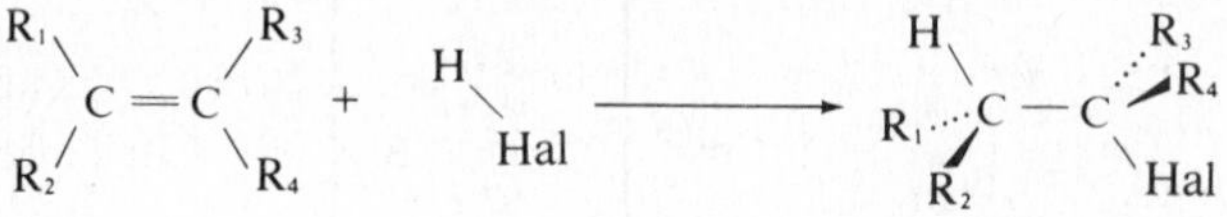

$$R_1R_2C{=}CR_3R_4 + H{-}Hal \longrightarrow H(R_1)(R_2)C{-}C(R_3)(R_4)Hal$$

For example, propene combines with hydrogen chloride to give 2-chloropropane

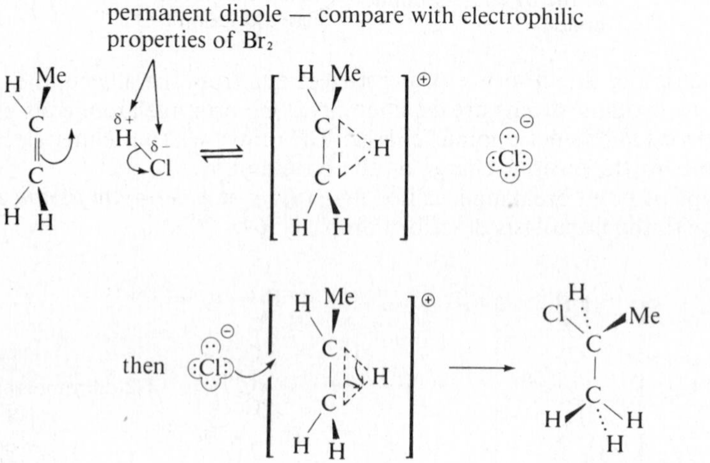

permanent dipole — compare with electrophilic properties of Br_2

$$H(Me)C{=}CH_2 + \overset{\delta+}{H}{-}\overset{\delta-}{Cl} \rightleftharpoons [\text{bridged H complex}]^{\oplus} + Cl^{\ominus}$$

then $Cl^{\ominus}$ attacks the complex $\longrightarrow$ 2-chloropropane

There are two important points to note about the above reaction.

1. The hydrogen halide acts as an electrophile because it has an electron-deficient hydrogen atom: the halogen atom exerts a $-I$ effect on hydrogen. This means that the hydrogen halide in effect is acting as a strong acid i.e. it is protonating the π-system of the alkene molecule.
2. In principle, two products are possible as a result of the attack of the chloride nucleophile on the activated complex. The chloride ion can attack either the top or the bottom atom shown in the above diagram.

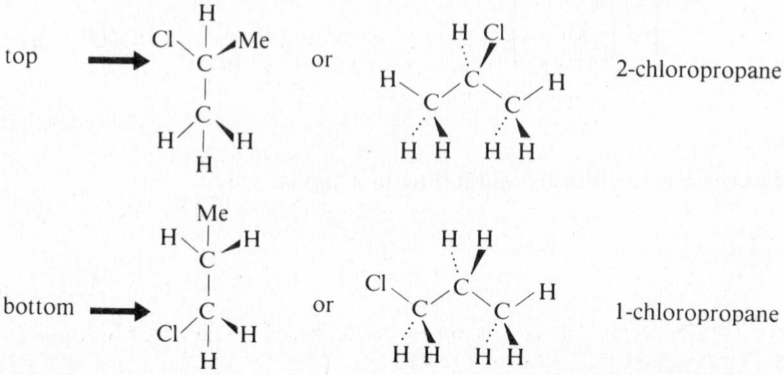

top $\longrightarrow$ $CH(Cl)(Me){-}CH_3$ or $CH_3{-}CHCl{-}CH_3$ 2-chloropropane

bottom $\longrightarrow$ $Me{-}CH_2{-}CH_2Cl$ or $ClCH_2{-}CH_2{-}CH_3$ 1-chloropropane

In practice, the actual product is almost entirely 2-chloropropane. The reasons for this are discussed on page 564, but a simple rule predicts the products of any alkene-hydrogen halide addition. It is named after Markovnikov, who first formulated it.

> Markovnikov's rule: When a molecule of H—X adds across a carbon-carbon double bond, the hydrogen atom adds onto the carbon atom which already has the most hydrogen atoms bonded to it.
>
> $$H_2C{=}CHR + H{-}X \longrightarrow H_3C{-}CHRX$$

Two further examples should help to make this clear:

$$\underset{\text{2-methylpropene}}{H_2\overset{1}{C}{=}\overset{2}{C}Me_2} + H{-}Br \longrightarrow \underset{\text{2-bromo-2-methylpropane}}{H_3C{-}CBrMe_2}$$

C_1 has more hydrogen atoms than C_2: 2 – 0

$$\text{1-ethylcyclohexene} + H{-}Cl \longrightarrow \text{1-chloro-1-ethylcyclohexane}$$

C_2 has more hydrogen atoms than C_1: 1 – 0

With sulphuric acid

Sulphuric acid is a strong acid like the hydrogen halides discussed above. When an alkene is passed into concentrated sulphuric acid, a similar reaction occurs. The addition obeys Markovnikov's rule:

$$H_2C{=}CHR + H{-}O{-}SO_2OH \longrightarrow H_3C{-}CHR{-}O{-}SO_2OH$$

The mechanism is the same as that for the hydrogen halide addition. So, for example, ethene gives ethyl hydrogen sulphate when bubbled through concentrated sulphuric acid. The reaction is reversible:

$$H_2C{=}CH_2 + H{-}O{-}SO_2{-}O{-}H \underset{\text{at } T>445\text{ K}}{\overset{\text{In cold}}{\rightleftharpoons}} H{-}CH_2{-}CH_2{-}O{-}SO_2{-}O{-}H \quad (\text{or } Et{-}O{-}SO_2{-}O{-}H)$$

ethyl hydrogen sulphate

Ethyl hydrogen sulphate decomposes at about 170°C. The reversibility of this reaction is discussed further on page 412.

With alkaline manganate (VII)

A solution of potassium manganate (VII) contains the ions $K^+_{(aq)}$ and $MnO^-_{4(aq)}$. Manganese is in a high oxidation state in $KMnO_4$ and this leads the compound to act as a strong oxidizing agent. The oxidizing strength of the solution actually depends on the pH: the lower the pH, the more powerful an oxidizing agent the solution is (see page 788).

An alkene is readily oxidized to a diol by alkaline manganate (VII). The reaction is another useful test for a carbon-carbon double bond: alkenes are among the few organic compounds that can be oxidized by manganate (VII) under the weaker alkaline conditions. The purple colour of the manganate solution rapidly turns to a brown suspension of manganese (IV) oxide:

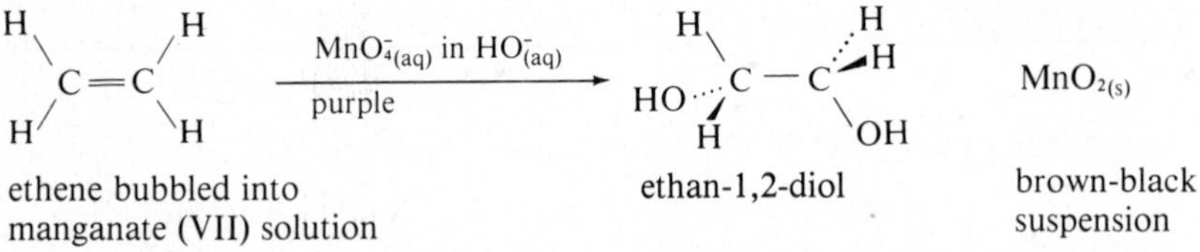

$$H_2C{=}CH_2 \xrightarrow[\text{purple}]{MnO^-_{4(aq)} \text{ in } HO^-_{(aq)}} HOCH_2{-}CH_2OH \quad MnO_{2(s)}$$

ethene bubbled into manganate (VII) solution → ethan-1,2-diol + brown-black suspension

The reaction pathway cannot be explained by the electrophilic mechanism that we have relied on so far. Like all redox reactions, the mechanism is likely to proceed by single electron transfer between the reacting particles. There is some evidence for a reaction intermediate of the 'bridge' type shown below:

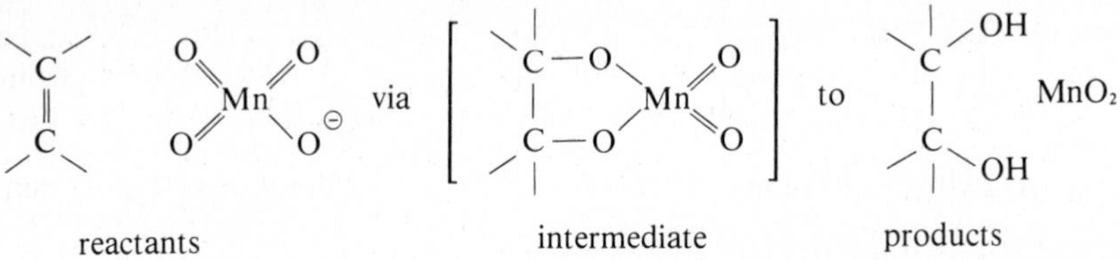

$$>C{=}C< \;+\; MnO_4^{\ominus} \quad \text{via} \quad \left[\begin{array}{c} -C-O \\ -C-O \end{array} > Mn(=O)_2 \right] \quad \text{to} \quad \begin{array}{c} >C-OH \\ >C-OH \end{array} \quad MnO_2$$

reactants — intermediate — products

With hydrogen

Alkenes are easily hydrogenated to alkanes. The reaction occurs rapidly when an alkene is passed over finely divided palladium with hydrogen at temperatures below 50°C and at atmospheric pressure:

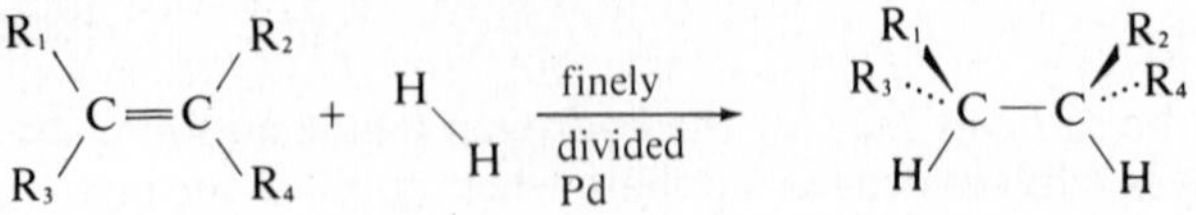

$$R_1R_3C{=}CR_2R_4 + H{-}H \xrightarrow{\text{finely divided Pd}} R_1R_3HC{-}CHR_2R_4$$

Palladium is a very expensive catalyst. A cheaper and more common catalyst is nickel. A form of nickel known as Raney nickel is usually used. It is prepared by treating a nickel-aluminium alloy with concentrated alkali:

$$2Ni-Al_{(s)} + 2OH^-_{(aq)} + 6H_2O_{(l)} \longrightarrow 2Ni_{(s)} + 3H_{2(g)} + 2\,[Al(OH)_4]^-_{(aq)}$$

Raney nickel

The nickel is deposited as a black suspension saturated with hydrogen. The sodium aluminate solution can be washed off the catalyst which is then dried and used to hydrogenate alkenes. Normally a temperature of about 100°C and pressures of about 3—5 atmospheres are necessary.

Ozonolysis

There is a third and more precise test that is used to help identify alkenes. Bromine water and alkaline manganate (VII) are able to show whether an unknown compound is or is not unsaturated; but ozonolysis provides a technique for pinpointing the exact location of any unsaturation in the carbon chain of the unknown compound.

Ozone (O_3) is an electrophilic reagent:

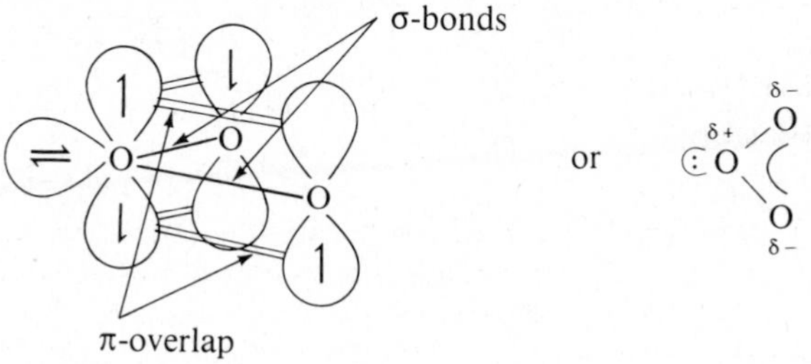

σ-bonds

or

π-overlap

The middle oxygen atom has one of its lone pairs in the $2p_y$-orbital. This overlaps with the $2p_y$-orbitals of the end two oxygen atoms. The effect is that electrons are donated by the middle atom to the end two: the middle atom is electron-deficient. The valence-bond model of the ozone molecule also emphasizes this:

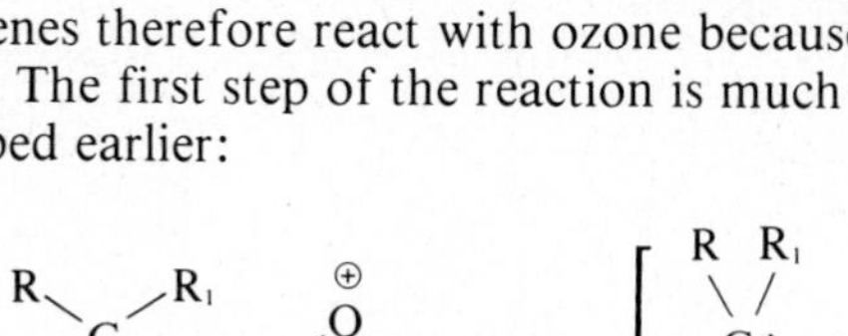

It shows the two extreme cases. The 'true' model lies between the two extremes.

Alkenes therefore react with ozone because of the electrophilic nature of ozone. The first step of the reaction is much the same as the general mechanism described earlier:

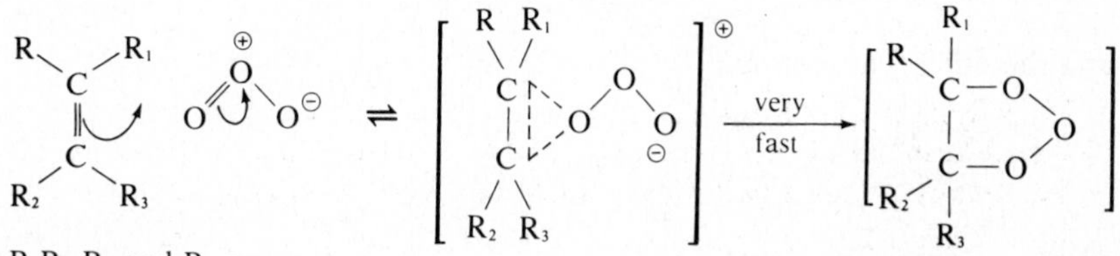

very fast

R R_1 R_2 and R_3 any carbon chains (or hydrogen)

activated complex

primary ozonide

This rather unstable 'primary ozonide' quickly undergoes reorganization to produce an ozonide which has the structure:

$$\begin{array}{c} R \quad R_1 \\ \diagdown\ \diagup \\ C{-}O \\ O \qquad | \\ C{-}O \\ \diagup\ \diagdown \\ R_2 \quad R_3 \end{array}$$

Notice that both the σ- and π-bonds between the carbon atoms of the alkene have been broken.

Ozonides are dangerously explosive and so are always hydrolysed at once in the presence of a reducing agent. Zinc in ethanoic acid is a common reagent which serves both purposes. A reducing agent is needed to decompose the hydrogen peroxide which is produced by the hydrolysis:

$$\text{ozonide} \xrightarrow[\text{CH}_3\text{COOH}]{\text{Zn in aqueous}} \underbrace{RR_1C{=}O \quad O{=}CR_2R_3}_{\text{hydrolysis products}} \quad H_2O$$

So, the overall reaction results in the breaking of the carbon–carbon double bond. Two fragments are produced, each containing a carbonyl group at the previous position of the double bond.

$$\text{overall}: \quad RR_1C{=}CR_2R_3 \xrightarrow[\text{2. Zn/CH}_3\text{COOH}]{\text{1. O}_3} RR_1C{=}O \quad O{=}CR_2R_3$$

We can separate the two carbonyl compounds from each other and then identify each by the standard method used for carbonyls (see page 438). Suppose one turned out to be *propanone*:

$$H_3C{-}C({=}O){-}CH_3$$

and the other was *ethanal*:

$$H_3C{-}C({=}O){-}H$$

then we would know that the original alkene was methylbut-2-ene

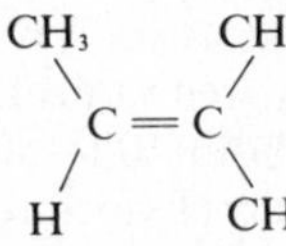

$(CH_3)_2C{=}CH(CH_3)$ methylbut-2-ene — deduced from ozonolysis products — CH_3CHO ethanal and $O{=}C(CH_3)_2$ propanone

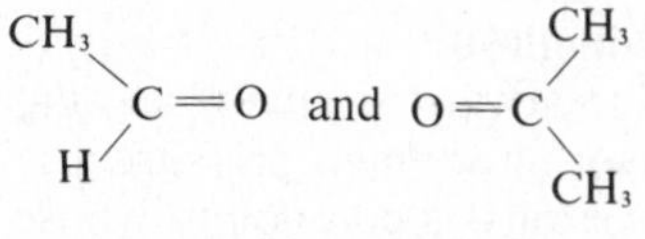

The reaction of an alkene with ozone is almost quantitative. By measuring the volume of ozone that a known mass of a hydrocarbon reacts with, we can obtain information about its structure. For example, an unknown hydrocarbon's formula is found by eudiometry to be C_5H_8; 0.3 g reacts with about 100 cm^3 of ozone at room temperature.

$$\Rightarrow \frac{0.3}{68} \text{ mole } C_5H_8 \text{ combines with about } \frac{100}{22\,400} \text{ mole ozone}$$

$$\Rightarrow 1 \text{ mole } C_5H_8 \text{ combines with } \left[\frac{68}{0.3} \times \frac{100}{22\,400}\right] \approx 1 \text{ mole of ozone}$$

⇒ there is only one double bond per molecule of C_5H_8 because one molecule of ozone is used for every double bond.

This means the hydrocarbon must be a cyclo-compound:

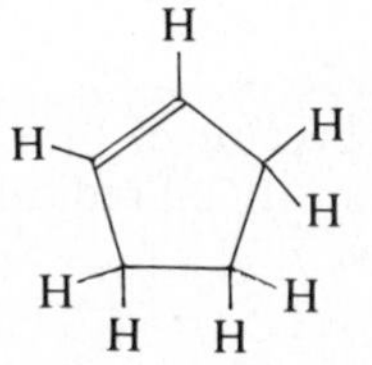

C_5H_8 : cyclopentene

One mole of a diene combines with two moles of ozone: straight-chain dienes give three ozonolysis products although some products turn out to be the same if the alkene is symmetrical, for example:

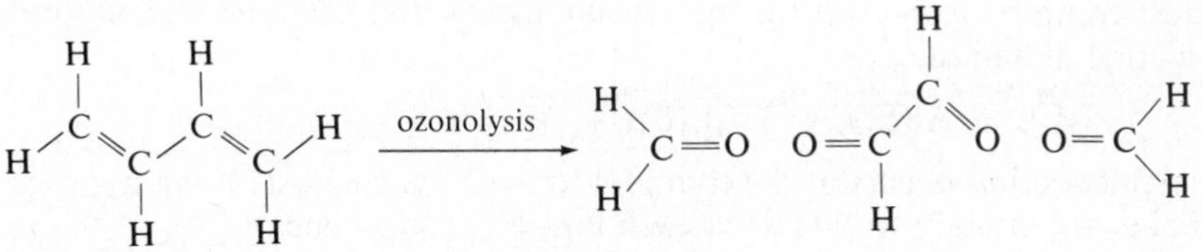

$CH_2{=}CH{-}CH{=}CH_2$ butadiene $\xrightarrow{\text{ozonolysis}}$ $H_2C{=}O$ methanal + $O{=}CH{-}CH{=}O$ ethandial + $O{=}CH_2$ methanal

Polymerization

Polymerization is a process in which the molecules of a compound add onto one another to produce a new macromolecular compound with the same empirical formula as the original compound. Alkenes can be polymerized in two different ways and the products are called 'plastics'. Some commercial plastics are discussed on page 587; most are now made by the second of the two methods below.

1. Free-radical method

The reverse of 'cracking' occurs if an alkene is mixed with a free radical initiator and kept under high pressure at a temperature of about 400—500 K. A free-radical initiator is a compound whose molecules are easily dissociated into radicals, e.g. a carboxylic acid peroxide

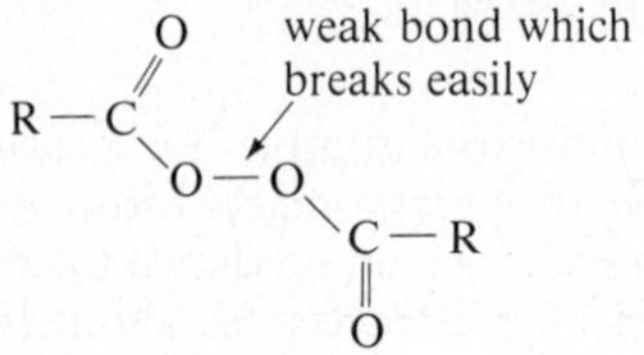

weak bond which breaks easily

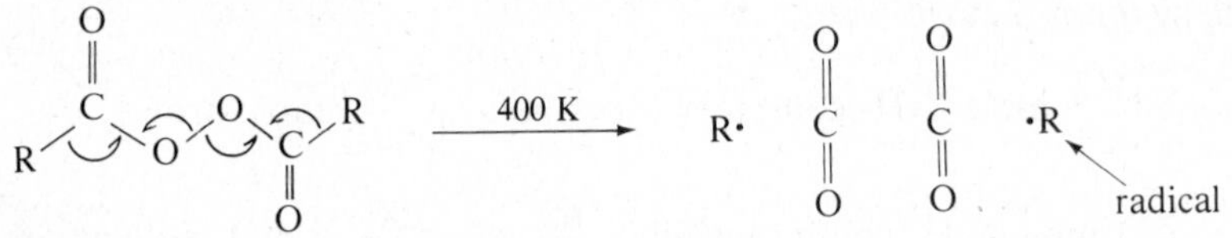

$$R{-}CO{-}O{-}O{-}CO{-}R \xrightarrow{400\ K} R\cdot \quad O{=}C{=}O \quad O{=}C{=}O \quad \cdot R$$

radical

A radical adds easily onto an alkene molecule and a chain reaction starts:

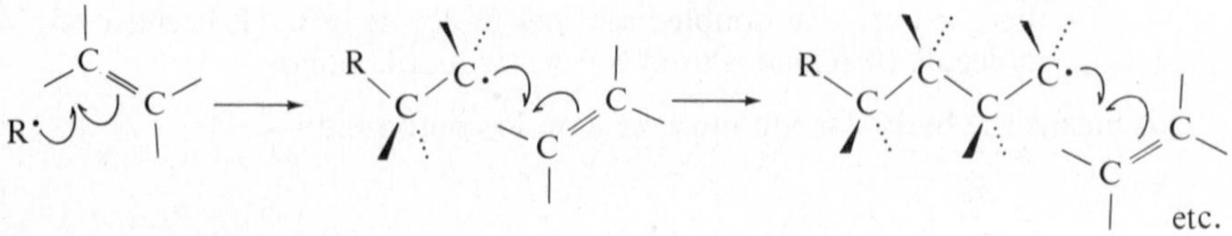

etc.

The pressure applied favours the polymerization process because of the reduction in volume that polymerization brings about. As with all radical processes, the reaction is hard to control and is unselective.

2. Electrophilic method

A far lower pressure is needed to polymerize an alkene if the reaction is carried out in the presence of a 'Ziegler' catalyst. A 'Ziegler' catalyst contains a source of electron-deficient particles able to interact with the alkene π-electrons. The most common catalyst is an equilibrium mixture of titanium (III) chloride and tri-ethyl aluminium.

$$TiCl_3 + Al(Et)_3 \rightleftharpoons TiCl_2(Et) + Al(Et)_2Cl$$

Polymerization occurs at the titanium ions of the catalyst. Long hydrocarbon chains are thought to 'grow' at each ion, e.g. for ethene

1.

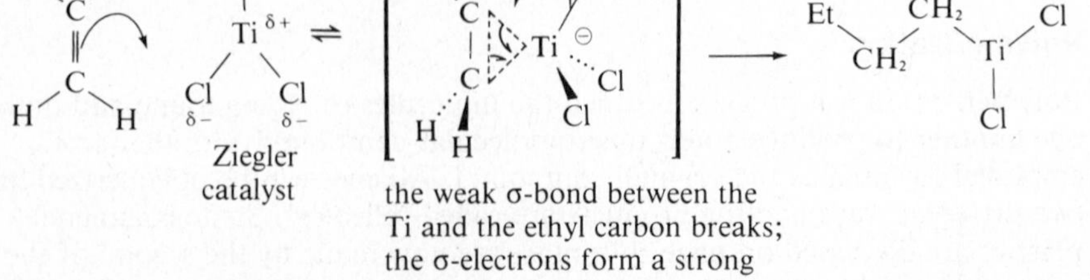

$Ti^{\delta+}$, $Cl^{\delta-}$

Ziegler catalyst

hydrocarbon chain 'growing'

Et–CH_2–CH_2–$TiCl_2$

the weak σ-bond between the Ti and the ethyl carbon breaks; the σ-electrons form a strong bond with a carbon atom from the trapped ethene molecule

2. Another ethene molecule is attached

Et–CH_2–CH_2–Ti(Cl_2) + $H_2C{=}CH_2$ → (after reorganization as before) Et–CH_2–CH_2–CH_2–CH_2–Ti(Cl_2)

longer chain

and so on

This heterolytic method of polymerization is much more specific. Unbranched chains are produced and the resulting plastic is stronger and more dense.

Substituted alkenes obviously give different polymerized products. Some common plastics are shown below.

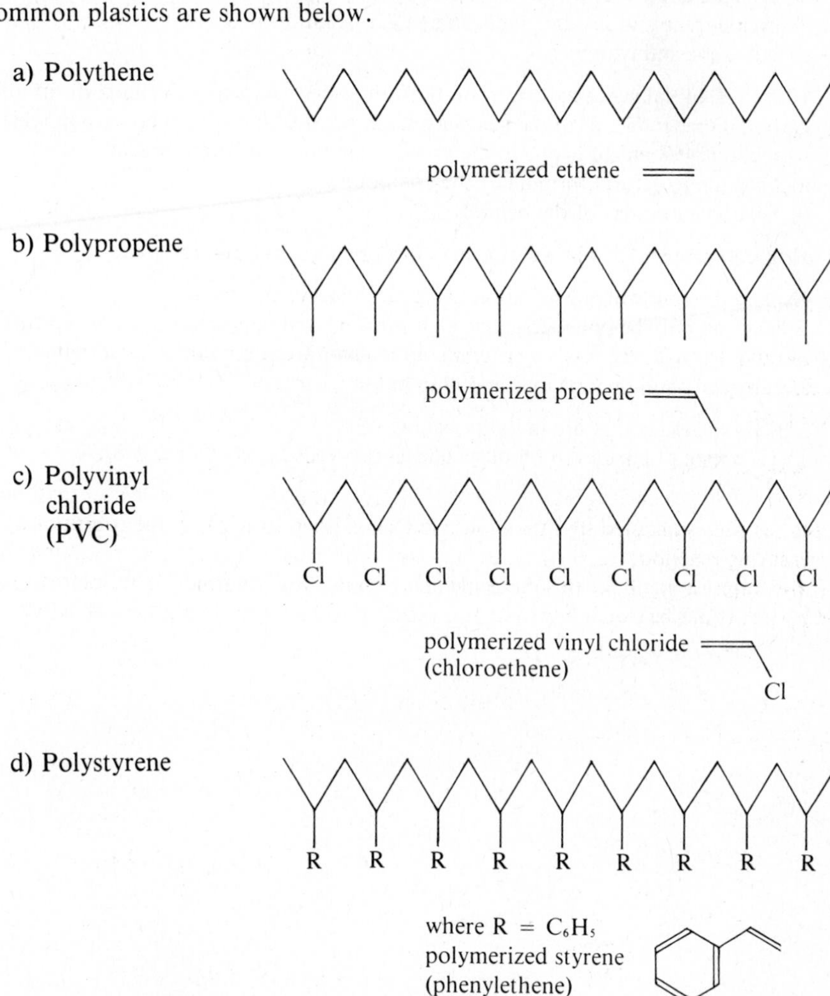

Questions

1. Why are alkenes more reactive than alkanes?
2. Compare and contrast the reaction of chlorine with ethane and ethene. Give the mechanisms of each.
3. Account for the presence of 2-bromoethanol among the products of the reaction of ethene with bromine water.
4. Describe two laboratory tests which could be used to show that a hydrocarbon is unsaturated. Give the equations for the two reactions.
5. Give the structural formula of the products of the following reactions
   a) but-1-ene and hydrogen chloride
   b) propene and concentrated sulphuric acid
   c) cyclohexene and alkaline manganate (VII) solution
   d) but-2-ene and ozone
6. 10 cm^3 of an unknown hydrocarbon burnt in excess oxygen to produce 40 cm^3 of carbon dioxide; 55 cm^3 of oxygen were required. 0.54 g of the hydrocarbon reacted with about 450 cm^3 of ozone in the presence of zinc and ethanoic acid.
   a) Give the structural formula of the compound.
   b) Give the products of the ozonolysis.
7. How can propene by obtained from a) propanol b) bromopropane?
8. Explain the following observation using mechanistic arguments:
   Ethene can only be polymerized at high pressure and temperature in the presence of ultra-violet light, but far lower temperatures and pressures can be used in the presence of titanium (III) ions and aluminium triethyl.
9. Explain the items that are in italics below:
   The *cracking* of alkanes to produce alkenes proceeds largely by a *homolytic* mechanism
10. It has been suggested that the molecule CH_2 exists transiently in the course of a cracking reaction.
    a) Suggest a name for this molecule and reasons why it has not been isolated.
    b) If it could be isolated, would you expect it to have typical alkene reactivity?

Chapter 24

Alkynes

24.1 Carbon-carbon triple bonds

Functional group

The alkynes are also hydrocarbons but differ yet again from the alkanes or the alkenes. While an alkane has a carbon chain that is fully saturated, and an alkene has a carbon chain that contains one or more carbon-carbon double bonds, an alkyne's carbon chain contains a triple bond. As with the alkenes, there may be more than one triple bond in the same chain. Some typical alkynes are shown below:

mono alkynes	*a diyne*
H—C≡C—H ethyne	
H—C≡C—C(H)(H)(H) propyne	H—C≡C—C≡C—H butadiyne

There are two patterns to notice about the molecular structure of the alkynes. Compare these with the alkane and alkene patterns described earlier.

1. The carbon chain is 'linear' at each triple bond. For example, in butadiyne, all six atoms form a straight line.
2. The general formula of any non-cyclic monoalkyne is $C_n H_{2n-2}$ where n = 2, 3, 4 . . . . There are four fewer hydrogen atoms than in the chain of an alkane of the same length. In other words, an alkyne has twice the unsaturation: it is able to fit four more bonded atoms onto its carbon chain.

   The functional group of an alkyne is the carbon-carbon triple bond —C≡C—. Again, shorthand methods are sometimes used to write alkyne structures:

   H—C≡C—H is ≡

   H—C≡C—C≡C—H is ≡—≡

Methods of preparation

Ethyne is very simply prepared. It is evolved during the hydrolysis of calcium dicarbide; this can be carried out by cold water.

$$CaC_{2(s)} + 2H_2O_{(l)} \longrightarrow Ca(OH)_{2(s)} + C_2H_{2(g)}$$

Calcium dicarbide is prepared by heating calcium oxide with carbon in an electric furnace at 2000°C. It is predominantly ionic, $Ca^{2+}\ [:\!C{\equiv}C\!:]^{2-}$, although the dicarbide ions C_2^{2-} are extensively polarized by the calcium ions.

The higher alkynes are generally prepared from the halogen derivatives of their corresponding alkenes. This is the same method as that used to make alkenes from halogenoalkanes (page 404). In the case of alkynes, a stronger base is usually needed: the most common one is sodium amide dissolved in liquid ammonia: $Na^+\ {}^-NH_2$ in $NH_{3(l)}$:

For example, propyne is prepared from 2-bromopropene using this reagent

$$H_2N\!:^{\ominus} + \begin{matrix} H & & Me \\ & C{=}C & \\ H & & Br \end{matrix} \xrightarrow[NH_{3(l)}]{in} \begin{matrix} NH_3 \\ H{-}C{\equiv}C{-}Me \\ Br^{\ominus} \end{matrix}$$

Higher alkynes can also be prepared from ethyne itself. Ethyne is a weak acid in the presence of sodamide and liquid ammonia (see page 394).

$$H{-}C{\equiv}C{-}H \underset{NH_{3(l)}}{\overset{NaNH_2}{\rightleftharpoons}} H{-}C{\equiv}C\!:^{\ominus} \xrightarrow{R-Br} H{-}C{\equiv}C{-}R \qquad Br^{\ominus}$$

If necessary the process can be repeated:

$$H{-}C{\equiv}C{-}R_1 \underset{NH_{3(l)}}{\overset{NaNH_2}{\rightleftharpoons}} {}^{\ominus}\!:C{\equiv}C{-}R_1 \xrightarrow{R_2-Br} R_1{-}C{\equiv}C{-}R_2 \qquad Br^{\ominus}$$

24.2 Reactive sites

The alkynes have π-electron density in two planes around the carbon chain.

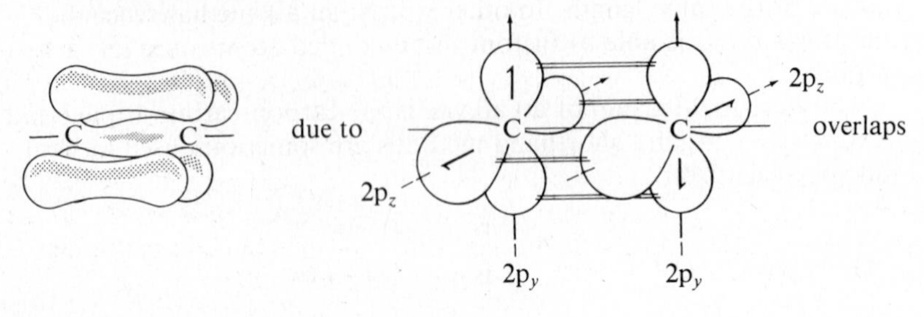

The double overlap has two important effects on the reactivity of the alkynes:

1. It pulls the nuclei closer together and makes it more difficult to draw the π-electrons away from the chain than from the corresponding alkene chain, i.e. alkynes have lower Lewis basicity than alkenes.
2. The acidity of any hydrogen atom bonded to the unsaturated group is significantly greater than that of a hydrogen bonded to an alkane or alkene chain. This is because the charge on the anion of the conjugate basic particle can be more easily delocalized:

$$R-C\equiv C-H + H_2O \rightleftharpoons R-C\equiv C^-_{(aq)} + H_3O^+_{(aq)} \quad pK_a = 26$$
$$Me-H + H_2O \rightleftharpoons Me^-_{(aq)} + H_3O^+_{(aq)} \quad pK_a = 54$$

The reactive sites of an alkyne are therefore not as immediately apparent as might be expected. Although alkynes show some tendency to undergo electrophilic addition like alkenes (page 378), more often than not a complex set of homolytic processes occurs between an alkyne and an electrophilic reagent. Carbon is a major product.

24.3 Reactions

With oxygen

All alkynes are endothermic compounds, and many are quite considerably unstable with respect to their elements, for example

ethyne: $\Delta H_f^\ominus = +227$ kJ mol^{-1}

Large amounts of energy are given out when an alkyne is burnt: oxy-acetylene welding equipment is a practical example utilizing the temperature of the acetylene flame (acetylene is the trivial name for ethyne):

$$2C_2H_2 + 5O_2 \longrightarrow 4CO_2 + 2H_2O \qquad \Delta H^\ominus_{c\,(ethyne)} = -1257 \text{ kJ mol}^{-1}$$

With electrophilic reagents

An alkyne tends to undergo a complex mixture of decomposition and substitution in the presence of electrophilic reagents like chlorine. Both the thermodynamic instability of these endothermic compounds and the comparatively low Lewis basicity of the π-electrons contribute to the effect. For example, ethyne and chlorine spontaneously ignite on mixing: flames and clouds of black soot are produced in a fumy atmosphere of hydrogen chloride. The reaction is quite spectacular and is best observed by generating both gases in the same conical flask: a mixture of calcium dicarbide and potassium manganate (VII) is added to 50% concentrated hydrochloric acid.

$$2MnO_{4(aq)}^{-} + 16H_3O_{(aq)}^{+} + 10Cl_{(aq)}^{-} \rightarrow 2Mn_{(aq)}^{2+} + 24H_2O_{(l)} + 5Cl_{2(g)}$$

$$CaC_{2(s)} + 2H_3O_{(aq)}^{+} \rightarrow Ca_{(aq)}^{2+} + 2H_2O_{(l)} + C_2H_{2(g)}$$

Then $$C_2H_{2(g)} + Cl_{2(g)} \xrightarrow{\text{mostly}} 2C_{(s)} + 2HCl_{(g)} \text{ (+ red flames)}$$

Alkynes *can* be made to undergo electrophilic addition under certain conditions. For example, they react with ozone by the same mechanism as alkenes (page 383) but more slowly: at about one-thousandth of the rate. The products are mostly carboxylic acids:

$$R_1-C\equiv C-R_2 \xrightarrow[\text{2. } H_2O]{\text{1. } O_3} R_1-C(=O)OH + HO(O=)C-R_2$$

One of the most important electrophilic reactions of an alkyne is its conversion to a carbonyl compound by the electrophilic addition of water. The reaction is carried out in an acidic solution of mercury (II) sulphate. The electrophile is likely to be a proton and the mercury (II) ions catalyse the reaction. For example, ethyne bubbled into dilute sulphuric acid, containing mercury (II) sulphate gives ethanal:

$$H_3O^+ + H-C\equiv C-H \rightleftharpoons [H-C=C-H \cdots H]^{\oplus} \rightarrow [H_2C=CH-\overset{\oplus}{O}H_2] \rightarrow H_3C-CH=\overset{\oplus}{O}H \rightarrow H_3C-CHO + H_3O^+$$

ethyne

protonated ethanal

ethanal

The mercury ions might also be acting as electrophiles via

$$H-C\equiv C-H + Hg^{2+} \rightleftharpoons [HC=CH \cdots Hg]^{2+}$$

or else they might simply be lowering the activation energy for the formation of the protonated alkyne activated complex shown in the first mechanism. This could come about through the interaction of the alkyne molecule with one of the solvating water molecules around the mercury ion:

$$H-C\equiv C-H \quad H_2\ddot{O}: \rightarrow Hg^{2+} \quad :\ddot{O}H_2$$

A good yield of ethanal can be obtained.

Ethyne undergoes addition with hydrogen bromide, but the reaction is much slower than the similar reaction of ethene. On the other hand, catalytic hydrogenation is more rapid for ethyne:

$$H-C\equiv C-H \xrightarrow[\text{slow}]{HBr} H_2C=CHBr \xrightarrow[\text{(Markovnikov)}]{HBr \text{ fast}} CH_3-CHBr_2$$

1,1 di-bromoethane

$$H-C\equiv C-H \xrightarrow[\text{Lindlar catalyst step 1}]{H_2} H_2C=CH_2 \xrightarrow[\text{Lindlar catalyst step 2}]{H_2} C_2H_6$$

A Lindlar catalyst employs palladium on a calcium carbonate support, partially poisoned by lead ethanoate. Step 1 proceeds much more rapidly than step 2 and often selective hydrogenation of an alkyne to an alkene can be achieved.

With ammoniacal silver (I) or copper (I)

If an alkyne that possesses the group $H-C\equiv C-$ is added to a solution of silver (I) ions (or copper (I) ions) complexed by ammonia, a precipitate is immediately produced. The solid is a salt resulting from the action of the alkyne as an acid. For example, the addition of ammonia solution to silver nitrate (V) produces a precipitate of silver oxide which redissolves on addition of excess ammonia:

$$NH_{3(aq)} + H_2O_{(l)} \rightleftharpoons NH^+_{4(aq)} + OH^-_{(aq)}$$
$$2Ag^+_{(aq)} + 2OH^-_{(aq)} \rightleftharpoons Ag_2O_{(s)} + H_2O_{(l)}$$
$$2Ag_2O_{(s)} + H_2O_{(l)} + 2NH_{3(aq)} \rightleftharpoons 2[Ag(NH_3)_2]^+ + 2OH^-_{(aq)}$$

When ethyne is bubbled into this mixture, silver dicarbide ('acetylide') precipitates

$$C_2H_{2(g)} + 2[Ag(NH_3)_2]^+ \longrightarrow Ag-C\equiv C-Ag_{(s)} + 2NH^+_{4(aq)} + 2NH_{3(aq)}$$

Copper (I) dicarbide can be produced in a similar way but both compounds are dangerously explosive if dried.

In the presence of liquid ammonia

Liquid ammonia is a useful solvent for performing a number of important reactions involving alkynes. Ammonia boils at 239.6 K and must be handled with great caution in vacuum-jacketted vessels. Proton transfer occurs in the liquid state and liquid ammonia resembles water in many of its solvent properties:

$$NH_{3(l)} + NH_{3(l)} \rightleftharpoons NH^+_{4(am)} + NH^-_{2(am)} \qquad K_{223\,K} = [NH_4^+][NH_2^-] = 10^{-30}\ mol^2\ dm^{-6}$$

$$H_2O_{(l)} + H_2O_{(l)} \rightleftharpoons H_3O^+_{(aq)} + OH^-_{(aq)} \qquad K_{298\,K} = [H_3O^+][OH^-] = 10^{-14}\ mol^2\ dm^{-6}$$

However, when sodium amide, $Na^+NH_2^-$, is dissolved in liquid ammonia, it is a far stronger base than its aqueous analogue sodium hydroxide in water, Na^+OH^-. It is this property which is used in connection with the alkynes. The weak acidity of the alkynes has already been described. In liquid ammonia containing sodium amide, the degree of dissociation of the alkynes is much more marked:

$$\underset{\text{acid}}{R{-}C{\equiv}C{-}H} + NH^-_{2(am)} \xrightleftharpoons{NH_{3(l)}} \underset{\text{conjugate base}}{R{-}C{\equiv}C\!:^{\ominus}} + NH_{3(l)}$$

The conjugate basic particle of an alkyne possesses a prominent lone pair of electrons. It can therefore act as a nucleophile if a suitably reactive compound is also dissolved in the liquid ammonia.

(i) With halogenoalkanes

Halogenoalkanes are susceptible to nucleophilic attacks. Propyne for example, reacts with bromoethane in the presence of sodium amide in liquid ammonia:

$$Me{-}C{\equiv}C{-}H + NH_2^{\ominus} \rightleftharpoons Me{-}C{\equiv}C\!:^{\ominus} + NH_3$$

$$Me{-}C{\equiv}C\!:^{\ominus} + H(Me)(H)C^{\delta+}{-}Br^{\delta-} \rightleftharpoons \left[Me{-}C{\equiv}C\text{----}C(H)(Me)(H)\text{----}Br\right]^{\ominus} \rightarrow Me{-}C{\equiv}C{-}Et \quad Br^{\ominus}$$

nucleophilic substitution

pent-2-yne

(ii) With carbonyl compounds
Similarly, aldehydes and ketones are susceptible to nucleophilic attack. Propyne reacts with methanal in the same reaction medium:

$$Me-C\equiv \overset{\ominus}{C}: \quad + \quad \underset{H}{\overset{H}{>}}\overset{\delta+}{C}=\overset{\delta-}{O} \quad \rightleftharpoons \quad \left[Me-C\equiv C-\underset{H}{\overset{H}{C}}-\overset{\ominus}{O} \right] \quad H-NH_2$$

$$\downarrow$$

$$Me-C\equiv C-CH_2OH$$

nucleophilic addition — but-2-yn-1-ol

Both these reactions result in the formation of new carbon-carbon bonds. This is an important process and we shall meet it in more detail in Chapter 34.

Questions

1. Why are the alkenes more reactive than either the alkanes or alkynes? The order of reactivity for aliphatic straight-chain hydrocarbons is: alkenes > alkynes > alkanes.
2. Compare and contrast the reactions of ethene and ethyne with a) chlorine and b) protonating reagents.
3. Give a reaction outline for each of the following conversions: a) ethene to ethyne, b) ethyne to propyne, c) ethyne to ethanol.
4. 20 cm^3 of an unknown hydrocarbon burnt in 90 cm^3 of oxygen to produce 80 cm^3 of carbon dioxide. Its rate of ozonolysis was about a thousand times slower than the rate of ozonolysis of ethene and the volume ratio appeared to be hydrocarbon: ozone = 1:4.
   Suggest a likely structural formula for the hydrocarbon and explain your reasons carefully.
5. When ethyne is passed into ammoniacal copper (I) solutions, a precipitate forms. No precipitates are given if the solution contains ammoniacal copper (II) or zinc (II). Comment.
6. a) Comment on the similarities and differences between the following particles: i) a dicarbide ion, ii) a cyanide ion, iii) a carbon monoxide molecule.
   b) How do these similarities and differences affect their reactivity?
7. When ethyne and carbon dioxide are passed into liquid ammonia in the presence of sodium amide, a small concentration of anions of empirical formula $(CO)_{2n}^{n-}$ can be detected.
   Suggest a likely value for n and account for the reaction.

8. Draw a bond energy diagram to illustrate why ethyne is classed as an endothermic compound. Comment on the relative magnitudes of the bond energies shown on your diagram.
9. Starting with the element as the only source of carbon, outline reaction schemes for the production of the following: a) ethene, b) bromoethane, c) butane, d) 1,2-dibromoethane, e) 1,1-dibromoethane.
10. In the synthesis of 6-6 nylon, the reactants are 'adipic acid' $HOOC — (CH_2)_4 — COOH$ and 'hexamethylenediamine' $H_2N — (CH_2)_6 — NH_2$.
    a) Name the reactants systematically.
    b) Both reactants are prepared from the diol, $HO — (CH_2)_4 — OH$. This is made from ethyne and methanal in the presence of an aqueous ammoniacal copper (I) solution. Give a possible reaction pathway for the production of the diol.

Chapter 25

Halogenoalkanes

25.1 Substituted alkanes

Functional group

The halogenoalkanes are a class of substituted alkanes. Each has a saturated hydrocarbon chain with a halogen atom (or atoms) attached at a particular position in place of a hydrogen atom. Some examples are shown below.

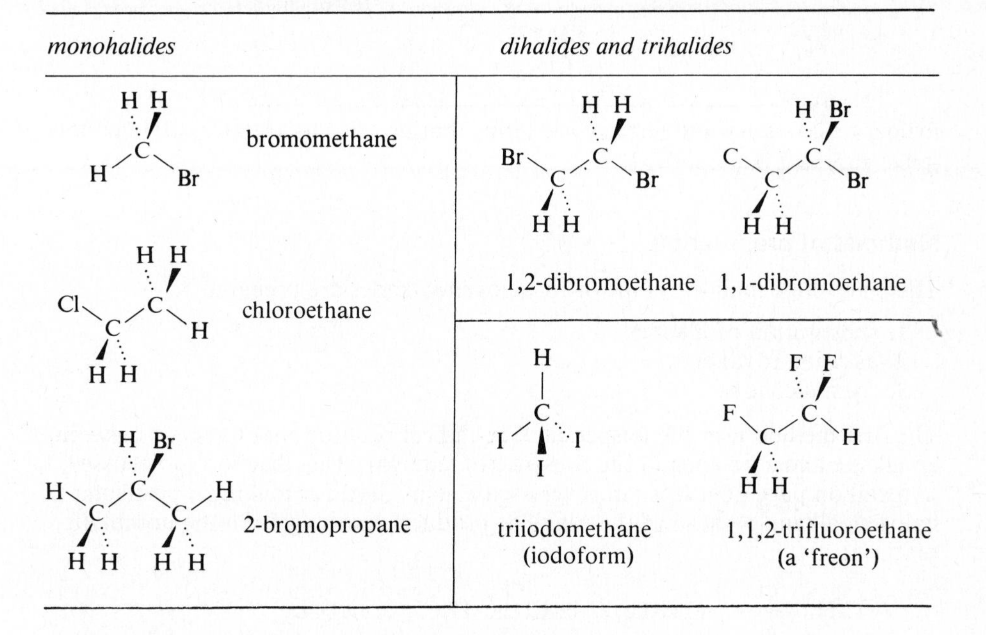

Like the alkanes, these molecules are zig-zag chains with pairs of atoms bonded to each carbon atom in the chain at 90° to the direction of the chain. All the bond angles are close to 109°. The functional group is the group of atoms containing the halogen atom and the saturated carbon atom to which it is bonded:

X = any halogen $\qquad R_1R_2R_3C{-}X$

and R_1, R_2, R_3 are alkyl groups or hydrogen atoms

The terms primary, secondary and tertiary are used to distinguish between the different structures resulting from differences in the identity of R_1, R_2 and R_3 above.

R_1, R_2 and R_3	*example*	*type*
one carbon chain two hydrogen atoms	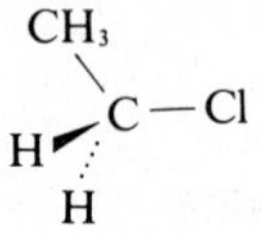 CH_3, H, H attached to C—Cl	*primary* (one chain)
two carbon chains one hydrogen atom	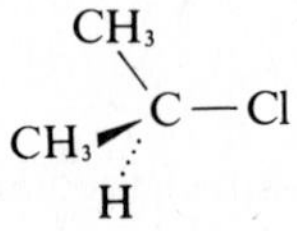 CH_3, CH_3, H attached to C—Cl	*secondary* (two chains)
three carbon chains	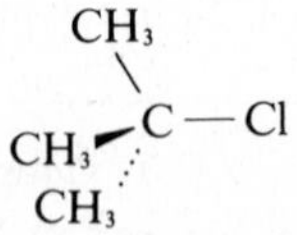 CH_3, CH_3, CH_3 attached to C—Cl	*tertiary* (three chains)

Primary, secondary and tertiary are terms that are also used to classify alcohols (page 407) and amines (419).

Methods of preparation

There are three main ways in which halogenoalkanes are prepared

1. substitution of alkanes
2. addition to alkenes
3. from alcohols.

The first method uses the unspecific free-radical reaction that happens between an alkane and a halogen in the presence of sunlight. This reaction is discussed in detail on page 369. It is not a very sensible method for making a particular halogenoalkane because of the range of products resulting from the homolytic reaction:

$$C_2H_6 \xrightarrow[h\nu]{Cl_2} C_2H_5Cl,\ C_2H_4Cl_2,\ C_2H_3Cl_3\ \ldots\ C_2Cl_6.$$

The second method is much more specific. Monohalogenoalkanes are produced by the addition of hydrogen halides; dihalogenoalkanes are produced by the addition of the halogen itself:

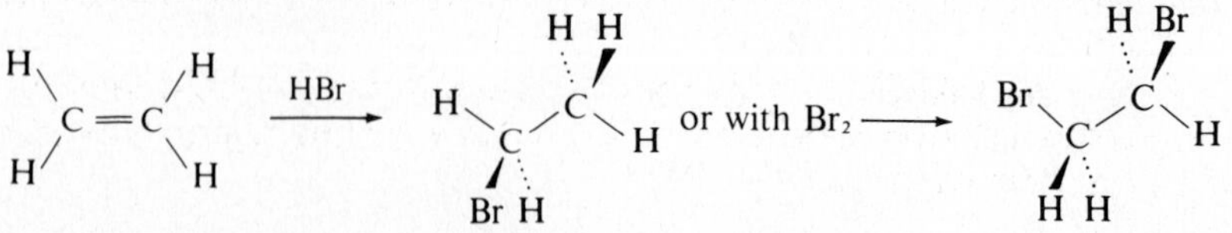

$$H_2C{=}CH_2 \xrightarrow{HBr} CH_3CH_2Br \quad \text{or with } Br_2 \longrightarrow CH_2BrCH_2Br$$

ethene — bromoethane — 1,2-dibromoethane

The third method uses the reaction between an alcohol and a strong hydrogen halide acid. The acid is usually prepared *in situ* from concentrated sulphuric acid and a halide salt.

$$R{-}OH \xrightarrow[\text{2. NaX, distil}]{\text{1. } H_2SO_4} R{-}X + H_2O$$

Chloroalkanes can be prepared from the specific chlorinating agents phosphorus (V) chloride or dichlorosulphur (IV) oxide (thionylchloride). These react with alcohols converting them to chloroalkanes directly:

$$C_3H_7OH \xrightarrow[\text{or } SOCl_2]{PCl_5} C_3H_7Cl$$

Of these two chlorinating agents, $SOCl_2$ is the more convenient to use because the by-products are all gaseous and can be driven off:

$$R{-}OH + SOCl_2 \rightarrow R{-}Cl + HCl_{(g)} + SO_{2(g)}$$

The schematic method of writing molecular structure is often used to draw the structures of the halogenoalkanes:

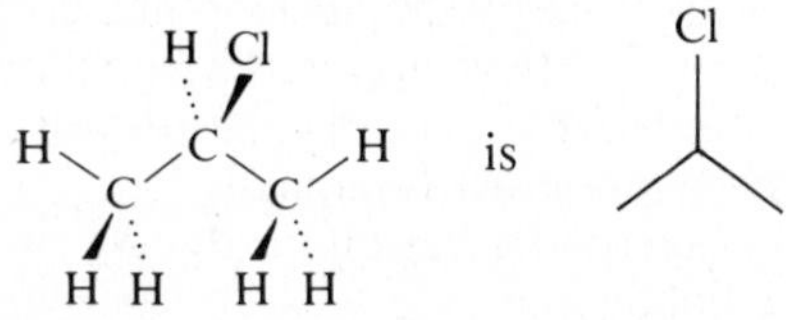

Now try naming the following; the solutions are given in the answers section of this book.

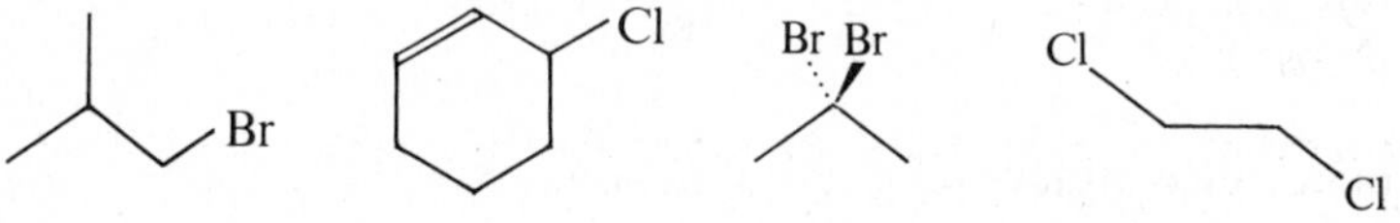

25.2 Reactive sites

The presence of a halogen atom in the saturated chain of a hydrocarbon causes Lewis acidity. This is because

1. the halogen atom is more electronegative than the carbon atom that it bonds to.
2. the halogen atom tends to act as a good leaving group owing to the comparative weakness of the C — Hal bond and the stability of the resulting halide ion formed.

These two effects counterbalance each other for the different halogens. The chlorine-carbon difference in electronegativity is large, but so is the bond strength when compared with the low carbon-iodine bond strength. There is little difference in electronegativity between carbon and iodine.

The two effects combine with the result that a halogen atom exerts a $-I$ effect on the carbon to which it is bonded:

$$\overset{\delta+}{C}-\overset{\delta-}{X}$$

This is the reactive site in any halogenoalkane molecule.

Although a halogen atom has three lone pairs of electrons, these show little Lewis basic character because they are largely withdrawn into the atomic core. Most of the reactivity of the halogenoalkanes can be explained in terms of the Lewis acidity of the halogen-bearing carbon atom.

Nucleophiles are able to attack the electron-deficient carbon atom.

$$Nu^{\ominus} + R_1R_2R_3\overset{\delta+}{C}-\overset{\delta-}{X} \rightleftharpoons \left[Nu\text{----}C(R_1R_2R_3)\text{----}X\right]^{\ominus} \longrightarrow Nu-C(R_1R_2R_3) + X^{\ominus}$$

The result is the substitution of the halogen atom by the nucleophilic group.

Under certain circumstances, the sequence of the processes shown above appears to alter. Instead of an attack on the carbon atom being followed by the leaving of the halide ion, sometimes the leaving process happens first.

The carbon-hydrogen bond undergoes heterolytic fission producing two ions, a positive carbonium ion and a negative halide ion.

$$R_1R_2R_3C-X \rightleftharpoons R_1R_2R_3C^{\oplus} + X^{\ominus}$$

The nucleophile then attacks the carbonium ion.

$$Nu^{\ominus} + R_1R_2R_3C^{\oplus} \longrightarrow Nu-C(R_1R_2R_3)$$

The distinction between these two mechanisms concerns the overall kinetics of the substitution reactions. This is described in detail on page 559 where the terms S_N2 and S_N1 are used as labels to distinguish between the two mechanisms.

25.3 Reactions

With aqueous alkali

Halogenoalkanes can be hydrolysed by the action of boiling aqueous alkali. An example is 1-chlorobutane.

$$\text{Bu}-\text{Cl} \xrightarrow[\text{under reflux}]{\text{OH}^-_{(aq)}\ \text{boil}} \text{Bu}-\text{OH} \quad \text{butan-1-ol}$$

The reaction is a slow one and the mixture must be boiled under reflux for half an hour to ensure complete hydrolysis.

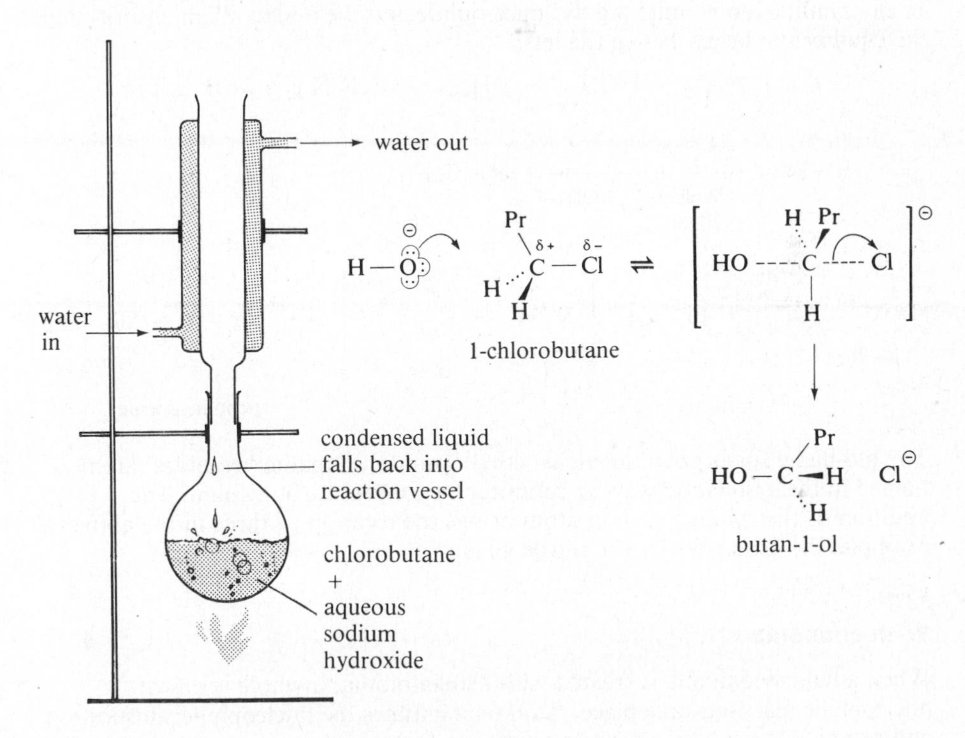

There are three general points that must be made about the above reaction for the different types of halogenoalkane.

1. The ease with which hydrolysis takes place is reduced when there is more than one halogen atom on the same carbon atom, e.g. tetrachloromethane, CCl_4, is inert to hydrolysis (see page 675). It becomes more difficult for the nucleophile to reach the electron-deficient carbon atom when it is surrounded by bulky halogen atoms: the nucleophile is said to be *sterically hindered*.
2. Iodoalkanes are more readily hydrolysed than bromoalkanes which are more readily hydrolysed than chloroalkanes. The sequence follows that of the quality of the leaving group: $I^- > Br^- > Cl^-$.

3. The rate of hydrolysis of a monohalogenoalkane depends on whether it is primary, secondary or tertiary (see page 398). This point concerns the distinction between the two possible mechanisms by which hydrolysis can take place (S_N1 and S_N2) and we shall return to it on page 559.

With alkaline cyanide

A similar reaction occurs when a halogenoalkane is refluxed with a solution of potassium cyanide containing a trace of alkali in aqueous alcohol. Halogeno-alkanes dissolve in alcohol but not in water, while the potassium cyanide is insoluble in alcohol but dissolves in water. To get them both to dissolve, the reaction is done in aqueous alcohol.

The cyanide ion is quite a good nucleophile and the added alkali ensures that the equilibrium below lies to the left:

$$CN^-_{(aq)} + H_2O_{(l)} \rightleftharpoons HCN_{(aq)} + OH^-_{(aq)} \qquad \text{(HCN is a weak acid)}$$

$$R-Br \xrightarrow[\text{dil alcoholic KOH) reflux}]{\text{KCN (in very}} R-CN$$

$$N{\equiv}C^{\ominus} + H(Me)(H)\overset{\delta+}{C}-\overset{\delta-}{Br} \rightleftharpoons \left[N{\equiv}C\text{----}C(H)(Me)(H)\text{----}Br\right]^{\ominus} \longrightarrow NC-C(H)(Me)(H) \quad Br^{\ominus}$$

bromoethane ... propanonitrile

The old name for propanonitrile is 'ethyl cyanide'. Organic 'cyanides' are now named in the systematic way by counting the longest carbon chain. The addition of the cyanide carbon atom brings the total up to three in the above example: the group N≡ is the nitrile group.

With ammonia

When a halogenoalkane is treated with 880-ammonia, a whole series of nucleophilic reactions take place. Ammonia initiates the nucleophilic substitution and produces a primary amine as a first product:

$$CH_3-I \xrightarrow{880\text{-}NH_3} CH_3-\ddot{N}H-H \qquad \text{methylamine}$$

But this primary amine is itself a strong nucleophile and may react with some unreacted iodomethane

$$CH_3-I \xrightarrow[\text{in mixture of } NH_3]{CH_3NH_2} CH_3-\ddot{N}(CH_3)-H \qquad \text{dimethylamine}$$

This is a secondary amine and it can also act as a nucleophile to produce a tertiary amine.

$$CH_3-I \xrightarrow[\text{in mixture of } NH_3]{(CH_3)_2NH} N(CH_3)_3 \quad \text{trimethylamine}$$

(The distinction between primary, secondary and tertiary amines is described on page 398).

The mechanism of the above reactions is shown below for the first two steps; the other step follows on in the same way.

nitrogen atom with four bonds: localized ⊕ charge

proton transfer

$$H_3N: + H_3C^{\delta+}-I^{\delta-} \rightleftharpoons [H_3N^{\oplus}\cdots CH_3\cdots I]^{\ominus} \rightarrow H_3N^{\oplus}-CH_3 \;\; I^{\ominus} + :NH_3 \rightleftharpoons H_2N-CH_3 + NH_4^{\oplus}$$

primary amine

Then a repeat of this process results in the following

$$CH_3NH_2: + H_3C^{\delta+}-I^{\delta-} \rightleftharpoons [CH_3NH_2^{\oplus}\cdots CH_3\cdots I]^{\ominus} \rightarrow CH_3NH_2^{\oplus}-CH_3 \;\; I^{\ominus} + :NH_3 \rightleftharpoons (CH_3)_2NH + NH_4^{\oplus}$$

secondary amine

If there is sufficient iodomethane present, a quaternary ammonium salt eventually forms after nucleophilic attack of the tertiary amine on an iodomethane molecule. There are now no protons left on the nitrogen to transfer: the ion produced resembles an ammonium ion

$$(CH_3)_3N: + H_3C^{\delta+}-I^{\delta-} \rightleftharpoons [(CH_3)_3N^{\oplus}\cdots CH_3\cdots I]^{\ominus} \rightarrow (CH_3)_4N^{\oplus} \; I^{\ominus}$$

tertiary amine → quaternary ammonium salt

In general, therefore, the reaction of the halogenoalkane R-X and ammonia gives a mixture of

$R\text{-}NH_2$	$(R)_2NH$	$(R)_3N$ and	$[(R)_4N^+]X$
primary amine	secondary amine	tertiary amine	quaternary ammonium halide salt

With alcoholic alkali

When potassium hydroxide is dissolved in ethanol instead of water, it has slightly different properties. The hydroxide ion has three lone pairs which give it the properties of either a) a nucleophile or b) a base. In aqueous conditions, the solvating particles around the ions are water molecules. In alcoholic conditions, however, the solvating particles are ethanol molecules. These larger molecules make the whole cluster much bulkier and thus restrict the ability of the hydroxide ion to act as a nucleophile.

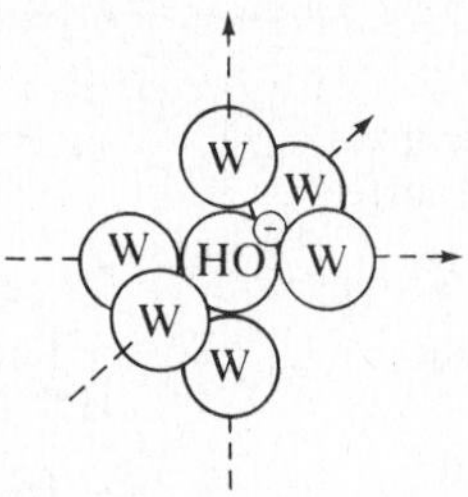

$HO^-_{(aq)}$ 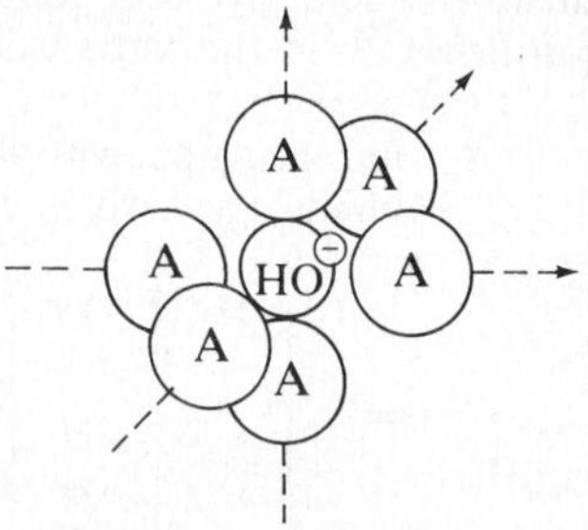 $\qquad\qquad$ $HO^-_{(eth)}$

A nucleophile must be able to reach the centre of electron deficiency in the molecule with which it reacts. Instead of acting as a nucleophile, $HO^-_{(eth)}$ acts predominantly as a base in its reaction with a halogenoalkane. It causes elimination of the elements of H-Hal from the compound and produces an alkene:

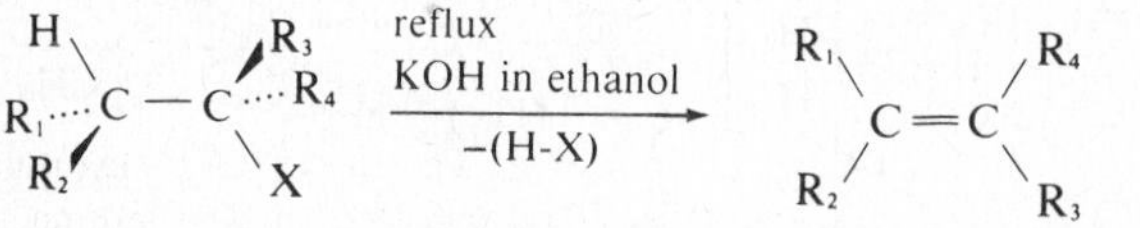

$$R_1R_2HC-CR_3R_4X \xrightarrow[-(H\text{-}X)]{\text{reflux, KOH in ethanol}} R_1R_2C=CR_3R_4$$

For example, ethene can be produced from bromoethane in low yield (~10%)

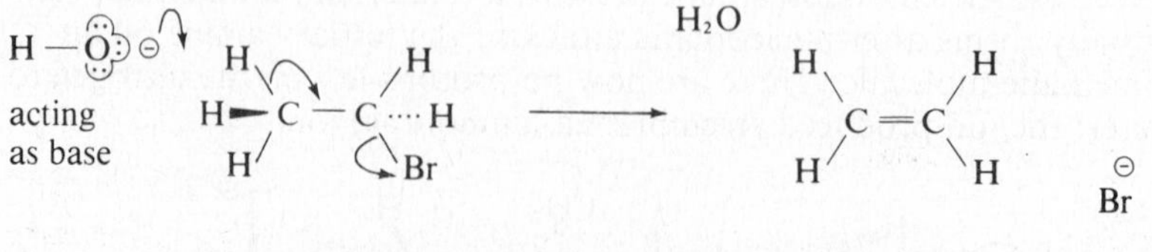

$$H-\ddot{O}:^{\ominus}\ (\text{acting as base}) + \underset{\text{bromoethane}}{H_3C-CH_2Br} \xrightarrow{} H_2O + \underset{\text{ethene}}{H_2C=CH_2} + Br^{\ominus}$$

but ethoxyethane is a major ‘substitution’ product. The mechanism is ‘concerted’: notice that there is a uniform shift of electron density through the structure of the bromoethane molecule.

> A ‘concerted’ mechanism is one in which all the changes in electron-pair type occur simultaneously.

In the elimination reaction:

1. a lone pair of electrons on the hydroxide ion becomes a σ-pair between the oxygen and a hydrogen atom at the same time as
2. the σ-pair of the hydrogen-carbon bond becomes a π-pair between the two carbon atoms, at the same time as
3. the σ-pair between the second carbon atom and the bromine atom become a lone pair on the bromine atom.

These shifts all happen as a result of a favourably oriented and energetic collision between the hydroxide ion and the bromoethane molecule.

The process of elimination competes with that of nucleophilic substitution. This is a theme we shall return to on page 570. In the case of the reaction described opposite, ethanol molecules act as nucleophiles when activated by the presence of hydroxide ions:

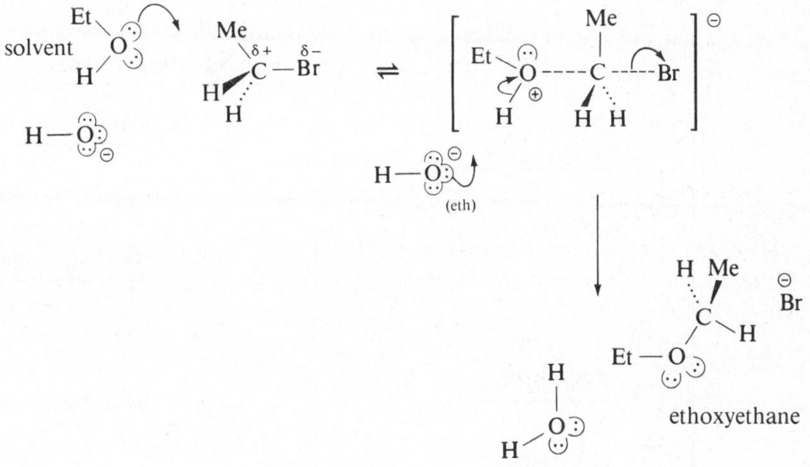

solvent Et–O–H + Me–CH₂–Br ($\delta+$ C, $\delta-$ Br) + H–O$^{\ominus}$ ⇌ [Et–O$^{\oplus}$(H)----C(Me)(H)(H)----Br]$^{\ominus}$ + H–O$^{\ominus}$ (eth)

→ Et–O–C(H)(Me)(H) ethoxyethane + $Br^{\ominus}$ + H–O–H

Questions

1. Account for the difference in reactivity between a halogenoalkane and its parent alkane.
2. Give the structural formulas of the products of bromoethane's reaction with a) aqueous alkali, b) alcoholic alkali, c) ammonia.
3. In the synthesis of chloropropane, why is propane not a good choice of reactant? What would be a more sensible starting-point?
4. Give an example of chloropropane reacting with a) a nucleophile, b) a base. Give the mechanisms in each case.
5. Give a reaction scheme to show how each of the following compounds can be prepared from unsaturated hydrocarbons: a) 2-chloropropane, b) 1,2-dichloropropane, c) 2, 2-dichloropropane.

6. Chloromethane is fairly readily hydrolysed by passing it into boiling alkali, dichloromethane is more readily hydrolysed still. However, trichloromethane ('chloroform') is much more resistant to hydrolysis and tetrachloromethane is inert to hydrolysis altogether. Explain these observations.
7. Give two possible reaction schemes by which ethoxyethane (ether) can be prepared from bromoethane.
8. Why does the reaction of ammonia and hydrogen chloride produce a single product while that of ammonia and 'methyl' chloride (chloromethane) produces a range of products?
   Explain your answer by outlining the likely mechanism of both reactions.
9. Discuss the factors responsible for making a bromide ion a better leaving group than a hydroxide ion.
10. Consider the energy diagram shown below; R is an alkyl group and X is a halogen atom.

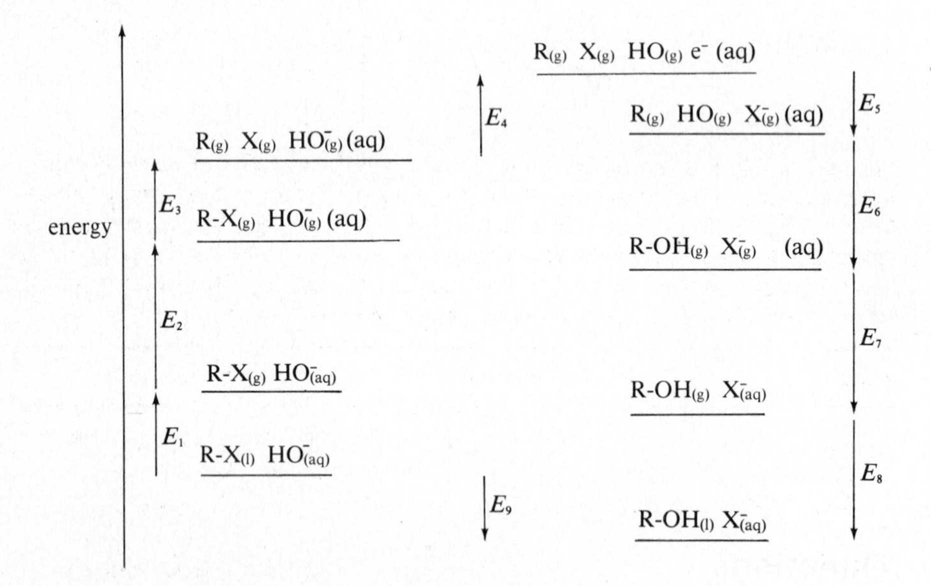

a) What process does the diagram outline in theory?
b) Name each energy term E_1 to E_9 from the diagram.
c) Which factors depend on the identity of X?
d) Discuss the change in magnitude of these factors when X is changed from fluorine to iodine.
e) How do you think that these changes might affect the reactivity of fluoroethane compared with that of iodoethane?

Chapter 26

Alcohols

26.1 Substituted water

Functional group

The alcohols have structures that show a marked similarity to that of water.

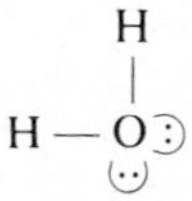

water	an alcohol	where R = a hydrocarbon chain
$H-\ddot{O}-H$	$R-\ddot{O}-H$	

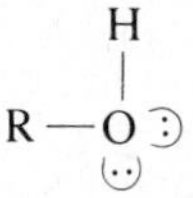

The hydrocarbon chain of an alcohol gives it many different properties from water, but the presence of the —OH group leads to some similar properties as well. Alcohols, like the halogenoalkanes described on page 397 can be primary, secondary or tertiary depending on how many carbon atoms are bonded to the functional group. The functional group is C_{sat}—OH where C_{sat} is a saturated carbon atom. Some simple examples are shown below:

primary	*secondary*	*tertiary*
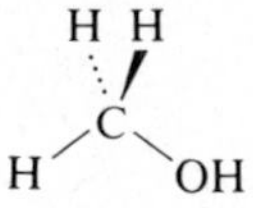 methanol	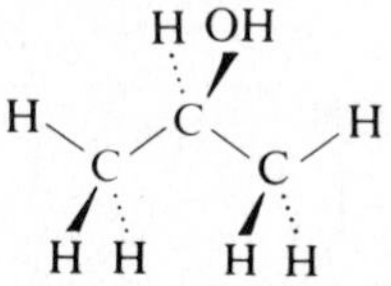 propan-2-ol	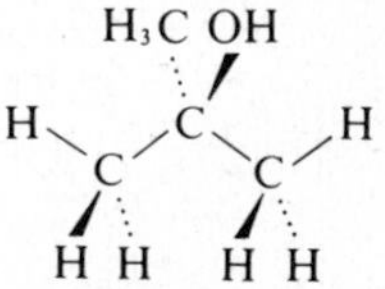 2-methyl-propan-2-ol
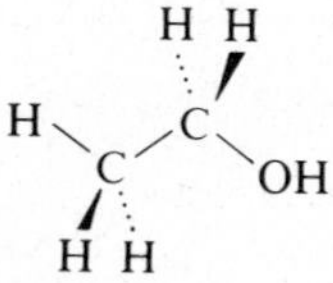 ethanol	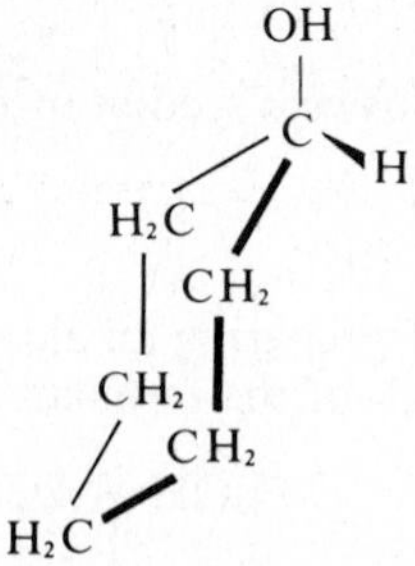 cyclohexanol	or
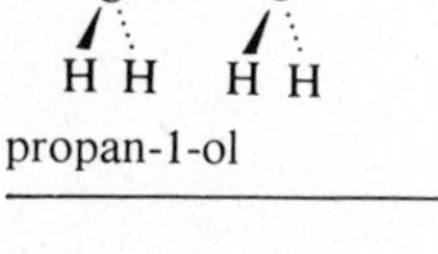 propan-1-ol		$C(Me)_3\,OH$

Diols and triols also exist. Be careful to distinguish these from secondary and tertiary alcohols: a common diol is 'glycol' or ethan-1,2-diol which is used as an anti-freeze; a common triol is 'glycerol' or propan-1,2,3-triol which is the alcohol on which many naturally occurring fats are based.

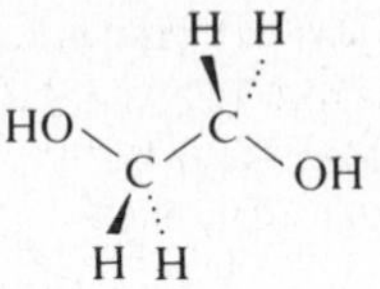

ethan-1,2-diol
(glycol)

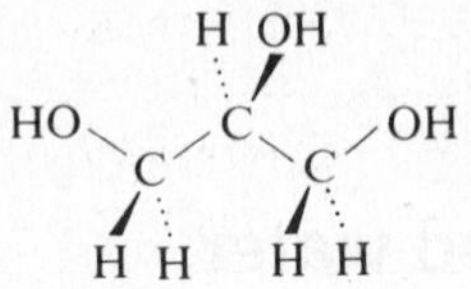

propan-1,2,3-triol
(glycerol)

The schematic method of drawing the structure of alcohols is often used. For example, the above two alcohols can be drawn:

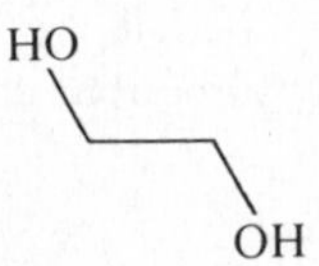

and

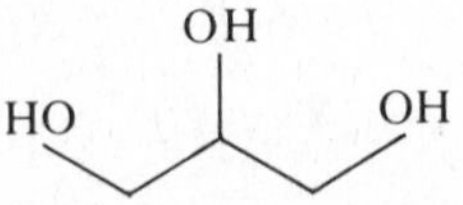

To practise naming compounds and to help recognition of alcohols, name the following alcohols. Decide whether they are primary, secondary, tertiary, mono, di or trialcohols:

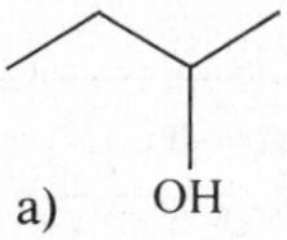

a) 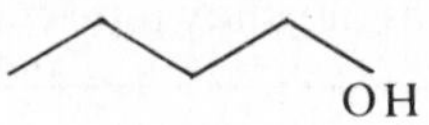 b) 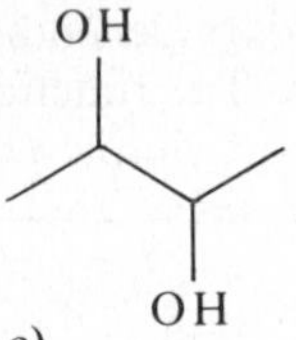 c)

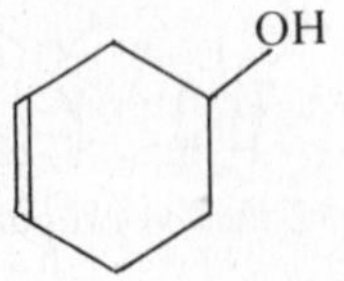

d)  e) 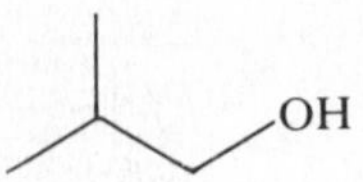 f)

The solutions are given in the answers section of this book.

Methods of preparation

There are three general ways of preparing an alcohol. Ethanol ('alcohol') can of course be produced as a result of the fermentation of sugars and starches:

$$C_6H_{12}O_6 \xrightarrow[\text{catalyst (an enzyme)}]{\text{zymase}} 2C_2H_5OH + 2CO_2$$

But the reaction is specific to ethanol.

1. Hydrolysis of an alkene

$$R_1R_2C{=}CR_3R_4 \xrightarrow{H_2O} R_1R_2C(H)-C(R_3)(R_4)OH$$

A wide variety of methods are used to bring about this reaction. Industrially, ethanol is produced from ethene by combining it directly with steam at high pressure over an acidic catalyst (see page 586). In the laboratory, the process is usually carried out in two stages. Either of the following two-stage reactions can be used:

1. Conversion to an alkyl hydrogen sulphate with concentrated H_2SO_4 followed by addition of water and distillation

$$R_1R_2C{=}CR_3R_4 \xrightarrow[H_2SO_4]{conc} R_1R_2C(H)-C(R_3)(R_4)OSO_2OH \xrightarrow{H_2O} R_1R_2C(H)-C(R_3)(R_4)OH$$

2. Conversion to an epoxide followed by reduction or hydrolysis

$$R_1R_2C{=}CR_3R_4 \xrightarrow[\text{or } O_2 \text{ and Ag catalyst 470 K: high pressure}]{\text{organic peracid}} \underset{\text{an epoxide}}{R_1R_2C\overset{O}{-}CR_3R_4}$$

$$\text{epoxide} \xrightarrow[H_2SO_4]{dilute} \underset{\text{a diol}}{HO(R_1)(R_2)C-C(R_3)(R_4)OH}$$

$$\text{epoxide} \xrightarrow[\text{with } LiAlH_4]{reduction} \underset{\text{a mono -ol}}{H(R_1)(R_2)C-C(R_3)(R_4)OH}$$

2. Hydrolysis of a halide

Refluxing a mixture of an organic halide with aqueous alkali gives a fair yield of alcohol. The reaction competes with an elimination process (page 404) which limits the yield:

$$R-X + OH^-_{(aq)} \xrightarrow{reflux} R-OH + X^-_{(aq)} \qquad (X = \text{any halogen})$$

3. Reduction of an aldehyde or ketone

Primary alcohols are prepared by reducing an aldehyde. Secondary alcohols are prepared by reducing a ketone. Lithium tetrahydrido-aluminate is the most common reducing agent

$$\underset{\text{aldehyde}}{RC(=O)H} \xrightarrow[2.\ H_2O]{1.\ LiAlH_4} \underset{\text{primary alcohol}}{R-C(OH)H_2}$$

$$\underset{\text{ketone}}{R_1C(=O)R_2} \xrightarrow[2.\ H_2O]{1.\ LiAlH_4} \underset{\text{secondary alcohol}}{R_1-C(OH)(H)R_2}$$

26.2 Reactive sites

Alcohols possess three different reactive sites.

1. They have *Lewis basicity* due to the prominence of the lone pairs on the oxygen atom:

R–Ö–H (lone pairs on O) — this can lead to action as a nucleophile

2. They have *Lewis acidity* due to the charge separation caused by the presence of the more electronegative atom, oxygen.

$\overset{\delta+}{\text{C}}-\overset{\delta-}{\text{OH}}$ — this can lead to attack by nucleophiles

However, the hydroxide group is a poor leaving group which reflects the strength of a hydroxide ion as a nucleophile: once it has become bonded to a carbon atom, it is less good at leaving. Therefore, in any reaction of an alcohol that involves a nucleophilic attack on its electron-deficient carbon atom, protonation of the hydroxide group is needed first. This is brought about by a strong acid, e.g. concentrated sulphuric acid.

$$R-\ddot{O}-H \quad H-O-SO_2OH \rightleftharpoons R-\overset{\oplus}{O}H_2 \quad HSO_4^{\ominus}$$

compare with H_3O^+

The protonation has two effects: a) it increases the electron-deficiency of the carbon atom and b) it turns the leaving-group into H_2O, a water molecule.

protonated alcohol

$$Nu^{\ominus} \quad C-\overset{\oplus}{O}H_2 \rightleftharpoons \left[Nu\text{----}C\text{----}\overset{\oplus}{O}H_2\right]^{\ominus} \longrightarrow Nu-C \quad H_2O$$

nucleophilic attack made easier due to increased electron-deficiency

water molecule as a leaving group: better than the hydroxide group

substitution product

3. An alcohol molecule has an *electron-deficient proton* in the same way that a water molecule has:

$$\overset{\delta-}{O}(H^{\delta+})(H^{\delta+}) \qquad \overset{\delta-}{O}(H^{\delta+})(R)$$

Alcohols are weakly acidic as a result of this. Compare

$$HO-H + H_2O \rightleftharpoons HO^- + H_3O^+$$

with

$$RO-H + H_2O \rightleftharpoons RO^- + H_3O^+$$

for example, the ethoxide ion is $Me(H)(H)C-O^{\ominus}$ and

$$K_a = \frac{[C_2H_5O^-][H_3O^+]}{[C_2H_5OH]} = 10^{-16}\ mol\ dm^{-3}$$

(cf. $K_w = 10^{-14}\ mol^2\ dm^{-6}$)

26.3 Reactions

With sodium

The weakly acidic nature of alcohols is demonstrated by their ability to react with sodium. The reaction is similar to the reaction of sodium with water but is slower in the case of an alcohol:

$$2R-OH_{(l)} + 2Na_{(s)} \longrightarrow 2R-O^-_{(alc)} + 2Na^+_{(alc)} + H_{2(g)}$$

For example, sodium in ethanol gives a solution of sodium ethoxide and hydrogen. Sodium ethoxide in ethanol is a more powerful base than sodium hydroxide in water and finds a number of uses in reactions called 'base-catalysed condensations'.

$$2\,CH_3-CH_2-\ddot{O}-H + 2Na \longrightarrow 2\,CH_3-CH_2-\ddot{O}^{\ominus}\ 2Na^{\oplus} + H_{2(g)}$$

sodium ethoxide

Although ethanol reacts with sodium, it does not react with calcium. This provides a good method for removing the last traces of water from the azeotropic mixture that results from the fractional distillation of ethanol-water solutions. A few small pieces of calcium metal are added to the flask. The calcium reacts with the water present.

With concentrated sulphuric acid

When sulphuric acid is gently and slowly added to some ethanol in a round-bottomed flask, the temperature rises. This effect is exactly the same as the effect of adding H_2SO_4 to water. After a while, however, the liquid starts to turn brown and remains brown even on cooling (unlike the water reaction). Clearly some new compound is formed. A cautious sniff at the reaction flask also indicates the presence of ether, which has a characteristic smell. Ether, however, is not brown, which suggest that at least two products have formed. If an attempt is made to extract the products from the solution, decomposition takes place (at ~ 445 K at atmospheric pressure) and ethene is produced in large quantities.

We can summarize these rather complex processes as below:

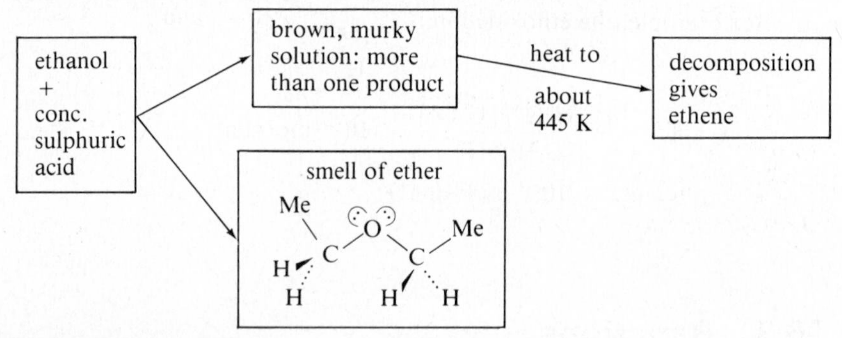

We have already met the reaction of ethene and concentrated sulphuric acid on page 381. Bubbling ethene into the acid produces a compound with the same characteristics as that made from ethanol and the acid. It has the structure:

ethyl hydrogen sulphate

This suggests that the initial reaction is

$$C_2H_5OH + H_2SO_4 \longrightarrow \underset{\text{ethyl hydrogen sulphate}}{C_2H_5OSO_2OH} + H_2O$$

and that ethyl hydrogen sulphate is unstable to heat.

$$CH_3CH_2-O-SO_2OH \xrightarrow{\text{heat}} \underset{\text{ethene}}{H_2C=CH_2} + HO-SO_2-OH$$

concerted shift of electron density in a circular movement

The presence of ether can be accounted for by considering the mechanism of the above reaction. The concentrated sulphuric acid initially protonates the ethanol as suggested on page 410. There are two possible nucleophiles that could attack the protonated molecules:

the hydrogen sulphate anion: $^{\ominus}\ddot{O}-SO_2OH$

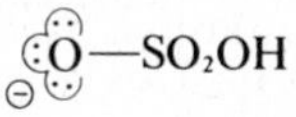

an unprotonated ethanol molecule: Et–$\ddot{O}$–H

Two possible pathways are then:

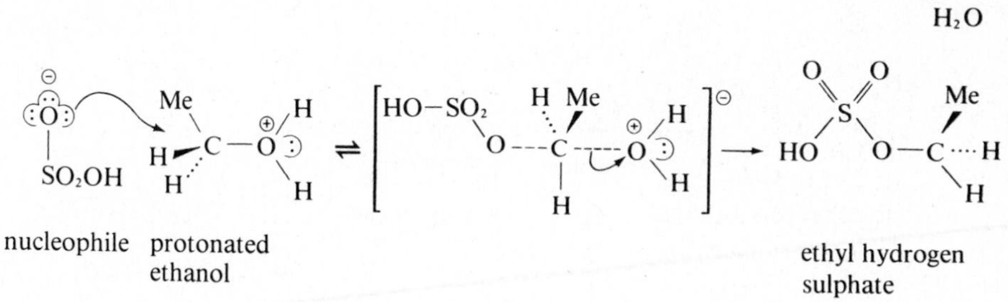

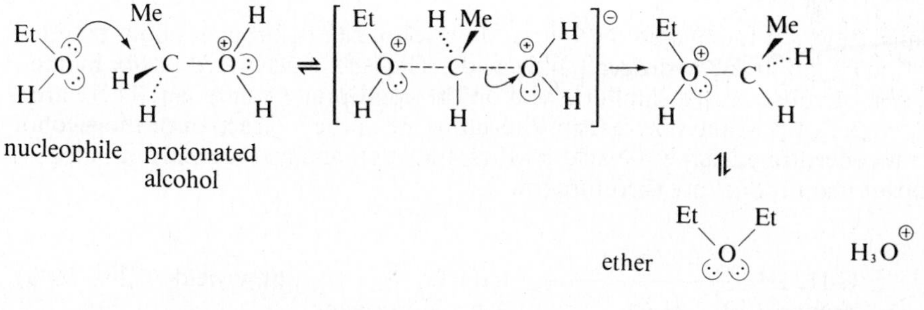

$$^{\ominus}\ddot{O}(SO_2OH) + H_3C{-}CH_2{-}\overset{\oplus}{O}H_2 \rightleftharpoons \left[HO{-}SO_2{-}O \cdots CH_2(Me) \cdots \overset{\oplus}{O}H_2\right]^{\ominus} \xrightarrow{} HO{-}SO_2{-}O{-}CH_2Me + H_2O$$

nucleophile protonated ethanol → ethyl hydrogen sulphate

$$Et{-}\ddot{O}{-}H + Me{-}CH_2{-}\overset{\oplus}{O}H_2 \rightleftharpoons \left[Et{-}\overset{\oplus}{O}(H) \cdots CH_2(Me) \cdots \overset{\oplus}{O}H_2\right]^{\ominus} \longrightarrow Et{-}\overset{\oplus}{O}(H){-}CH_2Me \rightleftharpoons Et{-}\ddot{O}{-}Et + H_3O^{\oplus}$$

nucleophile protonated alcohol → ether

If an excess of alcohol is used, the proportion of ether in the reaction products is greatly increased. This observation fits the mechanistic argument outlined above.

An overall equation summary of the effect of concentrated sulphuric acid on ethanol can now be given.

excess H_2SO_4 at $T < 440$ K $\quad C_2H_5OH \xrightarrow{H_2SO_4} C_2H_5\,OSO_2OH$ ethyl hydrogen sulphate

excess C_2H_5OH at $T < 440$ K $\quad 2C_2H_5OH \xrightarrow{H_2SO_4} H_5C_2-O-C_2H_5$ ethoxyethane (ether)

above ~ 445 K $\quad C_2H_5OH \xrightarrow[(-H_2O)\ \text{dehydration}]{H_2SO_4} C_2H_4$ ethene

Other primary alcohols behave in the same way when mixed with concentrated sulphuric acid, but secondary and tertiary alcohols differ slightly. Secondary alcohols are far more readily dehydrated than primary alcohols: only mild warming is necessary. Tertiary alcohols are even more easily dehydrated; no substitution products at all can be detected. Dehydration occurs at room temperature:

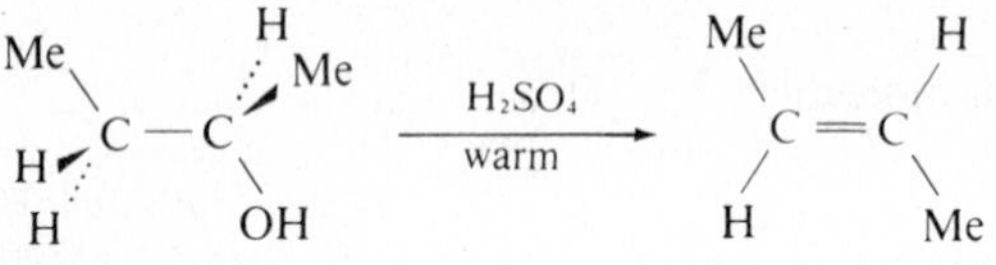

$$\text{butan-2-ol} \xrightarrow[\text{warm}]{H_2SO_4} \text{but-2-ene}$$

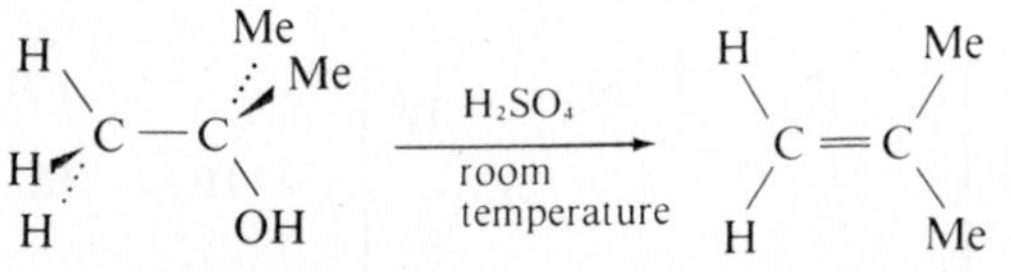

$$\text{methylpropan-2-ol} \xrightarrow[\text{room temperature}]{H_2SO_4} \text{methylpropene}$$

With hydrogen halides

Substitution of the hydroxyl group of an alcohol can be brought about by the action of the strong hydrogen halide acids. The acid is generated *in situ* by the action of concentrated sulphuric acid on the appropriate halide salt. There are a number of possible side reactions resulting from the interaction of the alcohol with concentrated sulphuric acid itself (see above), and so the yields of halogen-substituted product are therefore low:

$$\underset{\text{propan-1-ol}}{C_3H_7OH} \xrightarrow[\text{2. } H_2SO_4 \text{ added slowly, reflux}]{\text{1. KBr}} \underset{\text{1-bromopropane}}{C_3H_7Br} \quad (\text{low yield} \sim 20\text{—}30\%)$$

Again the mechanism is likely to be the same:

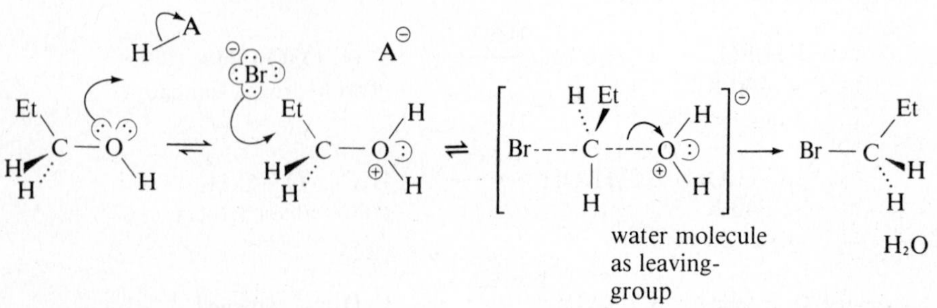

water molecule as leaving-group

H_2O

With chlorinating agents: PCl_5 and $SOCl_2$

Substitution of the hydroxyl group by a chlorine atom is readily brought about by the use of either phosphorus (V) chloride or dichlorosulphur (IV) oxide. The second of these two agents is preferred because the by-products are all gaseous and can be driven off the required product.

$$R-OH + PCl_5 \longrightarrow R-Cl + POCl_3 + HCl$$

$$R-OH + SOCl_2 \longrightarrow R-Cl + \underbrace{SO_2 + HCl}_{\text{gases}}$$

The mechanism probably involves the alcohol molecule acting as a nucleophile in an attack on the electron-deficient phosphorus or sulphur atom. Electron re-organization produces the halogenoalkane:

$$R-\ddot{O}-H + Cl_2\overset{\delta+}{S}=\overset{\delta-}{O} \rightleftharpoons \left[R-\overset{\oplus}{O}(H)-S(Cl)_2-O^{\ominus}\right] \longrightarrow R-Cl + O=S=O + H-Cl$$

With oxidizing agents

Common oxidizing agents used in organic chemistry include

potassium manganate (VII) in dilute acid: $KMnO_4/H_2SO_4$,
potassium dichromate (VI) in dilute acid: $K_2Cr_2O_7/H_2SO_4$.

Primary alcohols are oxidized first to aldehydes by these reagents and then to carboxylic acids. In practice, it is very difficult to stop the reaction at the half-way stage: the aldehyde.

$$\underset{\text{primary alcohol}}{R-CH_2OH} \xrightarrow{[O]} \underset{\text{aldehyde}}{R-C(=O)H} \xrightarrow{[O]} \underset{\text{carboxylic acid}}{R-C(=O)OH}$$

For example, ethanol is converted almost entirely to ethanoic acid when refluxed with acidic dichromate (VI) solution:

water out
ethanol
+
acidified
potassium
dichromate
water
in

$$\underset{\text{ethanol}}{CH_3CH_2OH} \xrightarrow[H_3O^+_{(aq)}]{Cr_2O_7^{2-}{}_{(aq)}} \underset{\text{ethanoic acid}}{CH_3C(=O)OH}$$

Secondary alcohols are oxidized to ketones by this process, but tertiary alcohols resist oxidation:

$$\underset{\text{propan-2-ol}}{(CH_3)_2CH(OH)} \xrightarrow[H_3O^+_{(aq)}]{Cr_2O_7^{2-}{}_{(aq)}} \underset{\text{propanone}}{(CH_3)_2C{=}O}$$

$$\text{but} \quad \underset{\text{methylpropan-2-ol}}{(CH_3)_3C(OH)} \longrightarrow \text{no reaction}$$

Sometimes it is necessary to prepare an aldehyde from a primary alcohol, i.e. to prevent the reaction from going through to the acid. Two common methods are usual:

1. Use the same oxidizing mixture, but take the following precautions
   a) Make up a mixture of alcohol and aqueous dichromate (just enough dichromate to oxidize the alcohol to the aldehyde but not to the acid)
   b) Using a tap funnel, drop the mixture into hot dilute acid which is just below the boiling point of the alcohol.

   The idea is that the aldehyde has a lower boiling point than the alcohol and vaporizes as soon as it is produced. The collecting vessel normally contains a mixture of alcohol and aldehyde with traces of acid.
2. A more satisfactory method is to use a different oxidizing agent. Chlorine has the property of being able to oxidize alcohols to aldehydes. The reaction also produces some chlorinated by-products, so again yields are often not high:

   $$RCH_2OH + Cl_2 \rightleftharpoons R{-}CHO + 2HCl$$

   For ethanal, further reactions result in the formation of chloroethanal and even di- and tri-chloroethanal. The mixture must be fractionally distilled to obtain ethanal.

   $$\underset{\text{chloroethanal}}{CH_2Cl{-}CHO} \qquad \underset{\text{dichloroethanal}}{CHCl_2{-}CHO} \qquad \underset{\text{trichloroethanal}}{CCl_3{-}CHO}$$

With acids and their derivatives

Alcohols are converted to esters by reaction with either a) a carboxylic acid or b) an acid anhydride or chloride. These reactions are discussed in detail in

Chapter 29 and so only the structural equations are given here:

$$\underset{\text{a carboxylic acid}}{R_1-C(=O)-O-H} + R_2-O-H \xrightleftharpoons[\text{reach equilibrium}]{\text{slow to}} \underset{\text{ester}}{R_1-C(=O)-O-R_2} + H_2O$$

$$\underset{\text{an acid chloride}}{R_1-C(=O)-Cl} + R_2-O-H \xrightarrow{\text{fast}} \underset{\text{ester}}{R_1-C(=O)-O-R_2} + HCl$$

In these reactions, an alcohol molecule acts as a nucleophile and attacks the electron-deficient carbon atom in an acid molecule.

Questions

1. Compare the reactivity of ethanol with the reactivity of a) water and b) chloroethane.

2. Give the structural formulas of the products of ethanol's reaction with a) phosphorus (V) chloride, b) ethanoic acid, c) concentrated sulphuric acid at temperatures greater than 170°C, d) sodium.

3. Bromoethane can be prepared by refluxing potassium bromide, concentrated sulphuric acid and ethanol. The corresponding cyanide compound, propanonitrile, cannot be prepared by refluxing potassium cyanide in the same reaction mixture.
   a) Explain this observation.
   b) Suggest a reaction scheme by which ethanol can be converted into propanonitrile. (pK_a of HBr < -7, pK_a of HCN > 3)

4. Give the equation for a reaction in which ethanol acts as a) an acid, b) a nucleophile, c) a substance undergoing nucleophilic attack.

5. a) Write down the structural formula of a primary alcohol, a secondary alcohol and a tertiary alcohol.
   b) Explain the difference in the structures.
   c) Suggest *two* methods by which they could be distinguished from one another in the laboratory. Give the equations for any reactions that you suggest.

6. Give *two* reaction schemes by which propan-2-ol can be obtained from propene.

7. How, and under what conditions, does ethanol react with concentrated sulphuric acid? Account for the detection of at least three organic substances among the products.

8. 1,2-dihydroxyethane (ethan-1,2-diol) is well known as an anti-freeze and is stable at quite high temperatures. 1,1-dihydroxyethane cannot be isolated and is rapidly dehydrated.
   a) Suggest the product of the dehydration.
   b) Account for the difference between the two stabilities by suggesting a likely mechanism for the dehydration.
   c) Under what conditions could 1,2-dihydroxyethane be dehydrated? What would the product be?

9. Explain why it is difficult to prepare a sample of an aldehyde from a primary alcohol.

10. How can the following compounds be prepared from propan-1-ol?
    a) propan-2-ol, b) methylpropanonitrile, c) 1,2-dichloropropane,
    d) propan-1,2-diol.

Chapter 27
Amines

27.1 Substituted ammonia

Functional group

An amine has a molecular structure that is similar to the structure of ammonia:

ammonia

primary amine

secondary amine

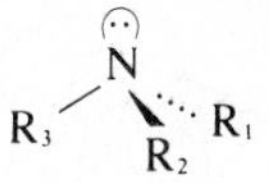
tertiary amine

The presence of the lone pair on the nitrogen atom of the amine group,  results in the amine having many properties in common with ammonia. The attached hydrocarbon chains R_1, R_2 and R_3 lead to some important differences.

In naming an amine, it is important to indicate how many carbon chains are attached to the nitrogen atom. Unlike an alcohol and a halogenoalkane, the names primary, secondary and tertiary are *not* used to describe 'branching' effects. For example, compare the structures below:

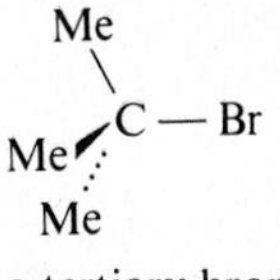
a tertiary bromoalkane

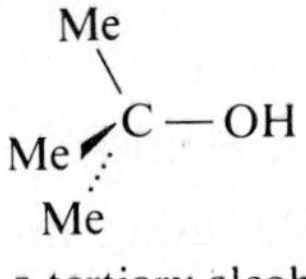
a tertiary alcohol

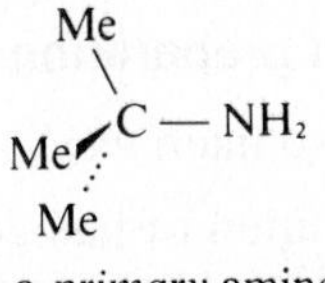
a *primary* amine

The amine is a primary one because there is only one carbon atom bonded to the nitrogen atom. If a second (or third) carbon chain is bonded to the nitrogen atom, its presence is indicated in the compound's name as an 'N' prefix: the 'N' in the name shows that the chain is bonded to the nitrogen atom.

primary	*secondary*	*tertiary*
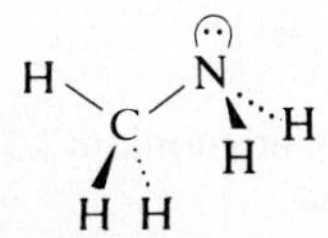	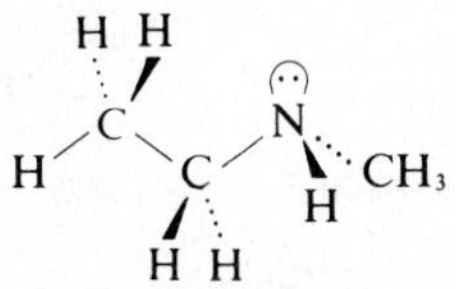	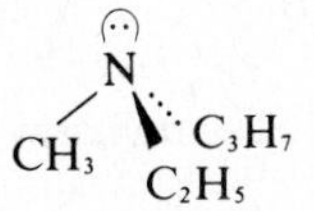
methylamine	N-methylethylamine	N-methyl-N-ethylpropylamine

Like the distinction made between a secondary alcohol and a diol (page 408), it is important to distinguish between a secondary amine and a diamine.

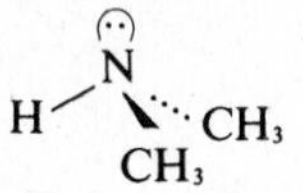

H N CH_3 CH_3

N-methyl methylamine
(or dimethylamine)

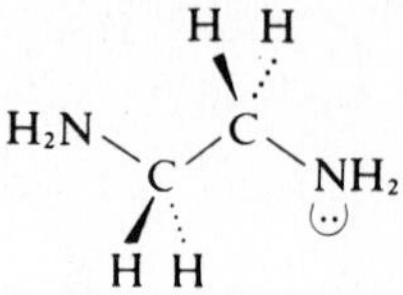

H H H_2N C C NH_2 H H

ethane-1-2-diamine

Here are some amines to practise naming (see answers section).

NH_2 NH H_2N NH_2

a)  b) 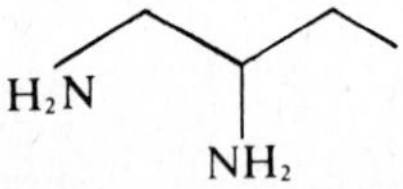 c)

N NH_2 N

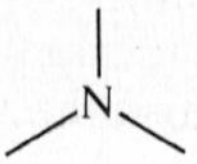

d) 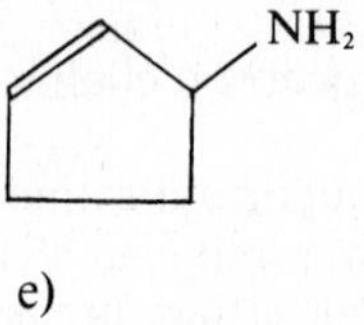 e) 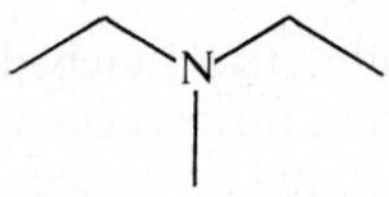 f)

Methods of preparation

There are two main ways of making amines:

1. substitution of halogenoalkanes using ammonia
2. reduction of unsaturated nitrogen compounds a) nitriles, b) amides, c) nitro-compounds.

The first method is described in detail on page 402. It is not a very satisfactory way of preparing a specific amine because a mixture of primary, secondary and tertiary amines is produced.

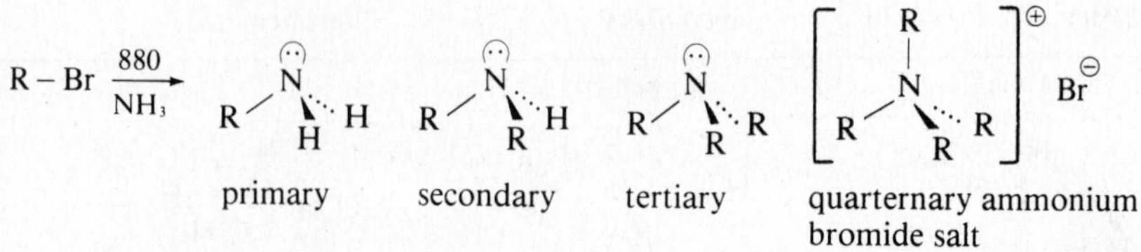

$$R-Br \xrightarrow[NH_3]{880}$$

R N H H — R N H R — R N R R — $[R_4N]^{\oplus} Br^{\ominus}$

primary secondary tertiary quarternary ammonium bromide salt

The second method is the most usual:

a) propylamine from propanonitrile:

$$H_3C{-}CH_2{-}C{\equiv}N \xrightarrow[\text{2. } H_2O]{\text{1. } LiAlH_4} H_3C{-}CH_2{-}CH_2{-}NH_2$$

b) ethylamine from ethanamide

$$H_3C{-}C({=}O){-}NH_2 \xrightarrow[\text{2. } H_2O]{\text{1. } LiAlH_4} H_3C{-}CH_2{-}NH_2$$

or, for aromatic amines, nitro-compounds are reduced

c) phenylamine from nitrobenzene

$$C_6H_5-NO_2 \xrightarrow[\text{2. NaOH}]{\text{1. Sn and conc. HCl}} C_6H_5-NH_2$$

27.2 Reactive sites

Compare the structures of chloromethane, hydroxymethane (methanol) and aminomethane (methylamine).

$$H_3\overset{\delta+}{C}-\overset{\delta-}{\ddot{Cl}}: \qquad H_3\overset{\delta+}{C}-\overset{\delta-}{\ddot{O}}-H \qquad H_3\overset{\delta+}{C}-\overset{\delta-}{\ddot{N}}H_2$$

There are two important points that can be identified by considering the similarities and differences between these structures.

similarities	*differences*
1. all three substituents exert a $-I$ effect on the carbon atom and make it open to attack by nucleophiles	1. the three substituents have markedly different leaving-group abilities $-Cl > -OH \gg -NH_2$
2. all three substituents have lone pairs of electrons which could lead to nucleophilic properties	2. the amount that these lone pairs protrude from the atomic core is markedly different $:N > \ddot{O} > :\ddot{Cl}$

The chemistry of the halogenoalkanes is dominated by their Lewis acidic character because the halogen group is a fairly good leaving group. Their tendency to show Lewis basic character via the halogen lone pairs is almost non-existent (see chapter 25).

The chemistry of the alcohols is more ambivalent. The hydroxide group causes both Lewis acid and Lewis base properties: nucleophilic attack is successful if the hydroxide group has first been protonated to improve its ability as a leaving group. Also alcohol molecules themselves act as nucleophiles as a result of the oxygen lone pairs (see chapter 26).

The amines provide the conclusion to this trend. The amino group is a very poor leaving group indeed. Even when protonated, it never acts as a leaving group in aqueous conditions because the protonated form (an ammonia molecule) is itself a more powerful nucleophile than any other likely in aqueous conditions. The chemistry of the amines is therefore dominated by the Lewis basic properties resulting from the presence of the nitrogen lone pair.

A particle with a prominent lone pair of electrons has three principal properties. It can act as:

1. *a base* — $B{:} \;\; H{-}A \rightleftharpoons B^{\oplus}{-}H \;\; A^{\ominus}$ (proton transfer)

2. *a ligand* — $L{:}\rightarrow M^{n+} \leftarrow{:}L$ (with two further L groups above and below): a complex ion. M^{n+} = a metal ion: particles with lone pairs are 'co-ordinated' around the ion

3. *a nucleophile* — $Nu{:} \rightarrow \overset{\delta+}{C}{-}\overset{\delta-}{Le}$ — the lone pair becomes a σ-pair bonded to an atom other than a hydrogen atom

It is important to be able to distinguish between these three different lone-pair properties. Amines (like ammonia) are capable of acting in all of these ways as we shall see in the next section.

27.3 Reactions

With water and mineral acids

Only the short-chain amines show appreciable solubility in water but all amines interact sufficiently with water to produce an alkaline solution.

$$\underset{\text{base}}{R\text{-}NH_2} + H_2O \rightleftharpoons R\text{-}\overset{\oplus}{N}H_{3(aq)} + OH^-_{(aq)} \qquad 7 < pH < 12$$

Their basic properties are better demonstrated in their ability to react with mineral acids. In many cases crystalline salts can be isolated as fairly stable compounds. Slow decomposition back into amine and acid is likely when these salts are in the solid form. For example, phenylamine (C_6H_5—NH_2) is very insoluble in water. Addition of some dilute hydrochloric acid brings about a large increase in solubility:

$$H_2N{-}C_6H_5 + H_3O^{\oplus} + Cl^{\ominus}_{(aq)} \rightleftharpoons H_3\overset{\oplus}{N}{-}C_6H_{5(aq)} + H_2O + Cl^{\ominus}_{(aq)}$$

solution of phenylammonium chloride

When a volatile amine (e.g. propylamine) is brought into contact with hydrogen chloride gas, a salt forms as 'smoke'. This is best demonstrated by putting the damp stoppers from a bottle of concentrated hydrochloric acid and a bottle of propylamine next to each other.

$$Pr\text{-}NH_{2(g)} + HCl_{(g)} \rightleftharpoons [Pr\text{-}\overset{+}{N}H_3][Cl^-]_{(s)}$$

With metal ion solutions

Amines are able to form a wide variety of complexes with transition-metal ions. The colours of these complexes are usually very similar to those that would be obtained if ammonia had been used. Significantly different colours, however, are produced for aromatic amines.

Addition of

a) ammonia
b) ethylamine
c) phenylamine (aromatic)

to copper (II) sulphate (VI) solution produces the following complex ions:

a) $[Cu(NH_3)_4]^{2+}$	b) $[Cu(EtNH_2)_4]^{2+}$	c) $[Cu(PhNH_2)_2]^{2+}$
intense deep blue	royal blue	lime green

For example:

$$[Cu^{2+}(:NH_3)_4] \quad (H_3N \text{ above, below, left and right of } Cu^{2+})$$

With halogenoalkanes

The reaction of an amine with a halogenoalkane is discussed in detail on page 402. The major points are summarized here.

1. Primary amines give a mixture of: secondary amines, tertiary amines and quaternary ammonium salts.

2. Secondary amines give tertiary amines and salts.
3. Tertiary amines give quaternary ammonium salts.

For example,

$$\underset{\text{primary}}{EtNH_2} \xrightarrow{EtBr} \underset{\text{secondary}}{(Et)_2NH} \xrightarrow{EtBr} \underset{\text{tertiary}}{(Et)_3N} \xrightarrow{EtBr} \underset{\text{quaternary ammonium salt}}{[(Et)_4\overset{\oplus}{N}][\overset{\ominus}{Br}]}$$

The amines are powerful nucleophiles and attack the electron-deficient carbon atom in the halogenoalkane.

$$R_1R_2R_3N\!: \curvearrowright \; Me(H)(H)\overset{\delta+}{C}-\overset{\delta-}{Br}$$

With nitrous (III) acid

Nitrous (III) acid has the formula HNO_2 and the structure:

$$H-O-\ddot{N}=O$$

It is unstable and decomposes to give a mixture of nitrogen oxides and water. Because of its instability, when required for a particular reaction, it is prepared *in situ*. The reactants are sodium nitrate (III) (sodium nitrite) and dilute acid:

$$NO_{2(aq)}^{-} + H_3O_{(aq)}^{+} \rightleftharpoons HO-NO_{(aq)} + H_2O_{(l)} \qquad \text{(it is a weak acid)}$$

Primary amines are converted to alcohols by the action of nitrous (III) acid. Nitrogen gas is evolved:

$$\underset{\text{primary amine}}{R-NH_2} + HONO \longrightarrow \underset{\text{alcohol}}{R-OH} + N_{2(g)} + H_2O$$

A secondary amine reacts slowly with nitrous (III) acid to produce a nitroso-compound. No nitrogen is evolved:

$$(R)_2NH + HONO \longrightarrow (R)_2N-\ddot{N}=O + H_2O$$

A tertiary amine reacts with nitrous (III) acid to form a salt (a nitrite).

$$(R)_3N + HONO \rightleftharpoons (R)_3\overset{\oplus}{N}H\ \overset{\ominus}{NO_2}$$

These salts are hydrolysed in water producing brown fumes of nitrogen (IV) oxide. The acid is therefore a useful test reagent to distinguish between primary, secondary and tertiary amines.

The mechanism involves the amine acting as a nucleophile in an attack on the electron-deficient nitrogen of the acid molecule

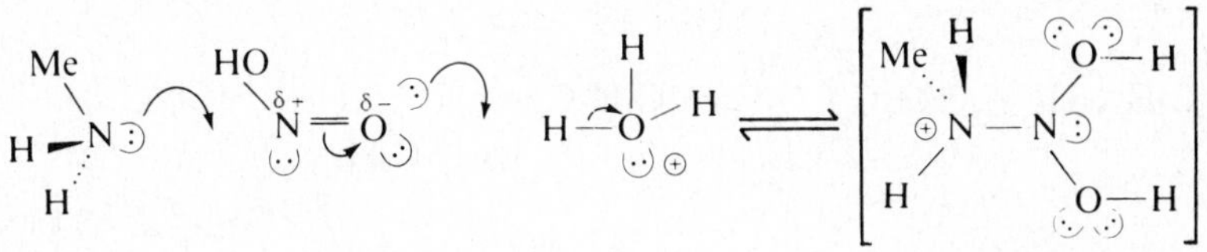

methylamine

two water molecules split out of this activated complex, as shown below

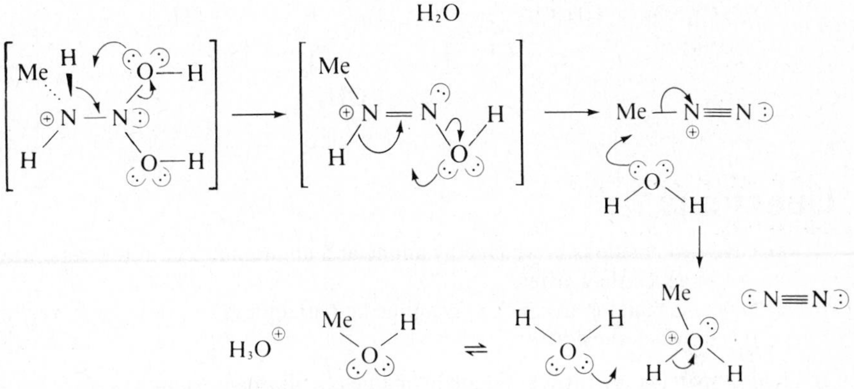

H_2O

methanol

With acids and their derivatives

Alcohols react with acid derivatives (the acid anhydrides and chlorides) to produce esters. Amines react in a similar way to produce amides. These reactions are considered in detail in Chapter 30 so only a summary is given below. Primary and secondary amines give substituted amides:

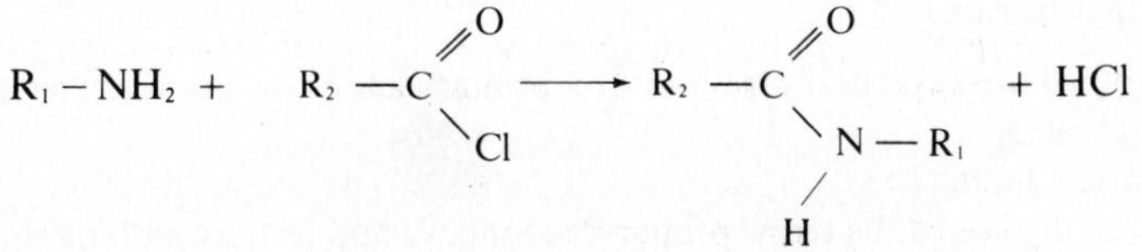

$$R_1-NH_2 + R_2-C(=O)Cl \longrightarrow R_2-C(=O)-N(H)-R_1 + HCl$$

primary amine　acid chloride　substituted amide

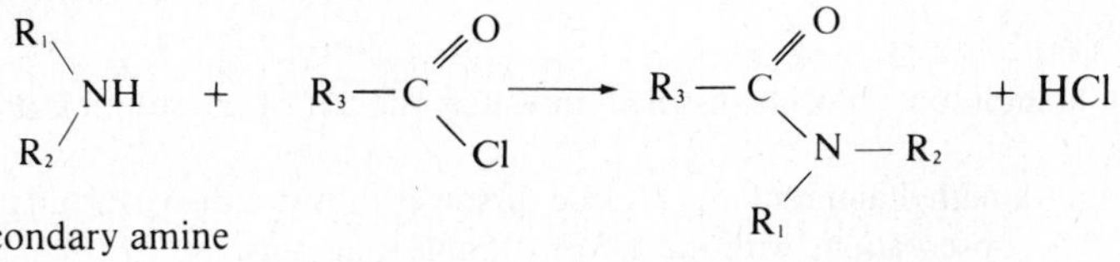

$$R_1R_2NH + R_3-C(=O)Cl \longrightarrow R_3-C(=O)-N(R_1)-R_2 + HCl$$

secondary amine

but tertiary amines do not react. For example, ethylamine and propanoyl chloride produce N-ethylpropanamide:

$$C_2H_5NH_2 + C_2H_5C\begin{matrix} \diagup O \\ \\ \diagdown Cl \end{matrix} \longrightarrow C_2H_5C\begin{matrix} \diagup O \\ \\ \diagdown N - C_2H_5 \\ \diagup \\ H \end{matrix} + HCl$$

dimethylamine and ethanoyl chloride produce N,N-dimethylethanamide

$$(CH_3)_2NH + CH_3C\begin{matrix} \diagup O \\ \\ \diagdown Cl \end{matrix} \longrightarrow CH_3C\begin{matrix} \diagup O \\ \\ \diagdown N - CH_3 \\ \diagup \\ CH_3 \end{matrix} + HCl$$

Questions

1. a) Give *two* reactions in which ethylamine and ammonia react in the same way, and *one* in which they differ.
   b) Give *one* reaction in which ethylamine and ethanol react in the same way, and *two* in which they differ.

2. Fluoromethane, methanol and methylamine are all *isoelectronic*
   a) Explain what this means.
   b) Are their proton structures the same?
   c) Compare their reactivity in view of your answers to a) and b).

3. Write equations, using structural formulas, for the reaction of propylamine with
   a) iodomethane, b) sulphuric (VI) acid, c) nitrous (III) acid, d) ethanoyl chloride.

4. Methylamine can act as a nucleophile, a base or a ligand.
   a) Explain the difference between the three properties.
   b) Give an example of each, with either an equation or a diagram to illustrate the important features.

5. a) Write down the structural formulas of a primary amine, a secondary amine and a tertiary amine.
   b) Name each amine.
   c) Compare the use of the terms primary, secondary and tertiary with their use to describe alcohols.

6. Give two different reaction schemes that outline the production of ethylamine from an alcohol.

7. a) When ammonium chloride dissolves in water, the pH of the solution drops below seven.
   b) When tetramethyl ammonium chloride dissolves in water the pH is unaffected.
   Explain these observations with the aid of suitable diagrams.

8. a) How does bromoethane react with aminoethane (ethylamine)?
   b) Why can more than one product be detected in the reaction mixture?

9. a) How can the following compounds be prepared from methylamine?
   i) Chloromethane, ii) ethylamine, iii) ethanoic acid, iv) N-ethyl ethanamide*.
   b) Outline means by which the following conversions could be carried out:
   i) ethene to butane-1,4-diamine, ii) ethylamine to ethyl ethanoate
   (*Hint: phosphorus (V) chloride reacts with an acid in a similar way to its reaction with an alcohol.)

10. a) For any base B in water, the equilibrium below is set up.

$$B_{(aq)} + H_2O_{(l)} \rightleftharpoons {}^{+}BH_{(aq)} + {}^{-}OH_{(aq)} \quad \text{and} \quad K_b = \frac{[BH^+][^-OH]}{[B]}$$

   Account for the different basic strengths of ammonia, methylamine and dimethylamine, given pK_b = 4.75, 3.36 and 3.26 respectively.
   b) Suggest an order of reactivity for the reactions between iodomethane and the three compounds mentioned above. Explain your suggestion.

Chapter 28

Aldehydes and ketones

28.1 The carbonyl group

Functional group

The structure of an aldehyde or ketone contains a hydrocarbon chain with a carbonyl group breaking the chain up at a particular point. If the carbonyl link occurs in the middle of the chain, the compound is a ketone; if the carbonyl group is at the end of the chain, the compound is an aldehyde. A carbonyl group has the structure shown below

carbonyl group

Compare the following molecules and the position of the carbonyl link in each.

aldehydes	*ketones*
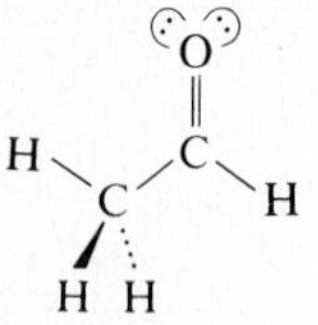 ethanal	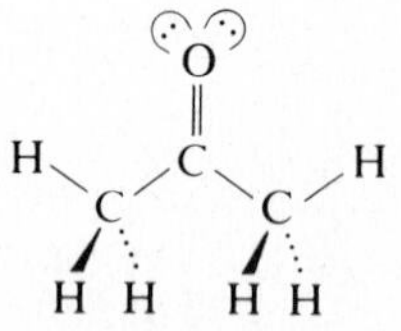 propanone
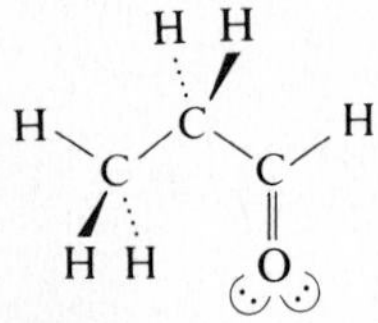 propanal	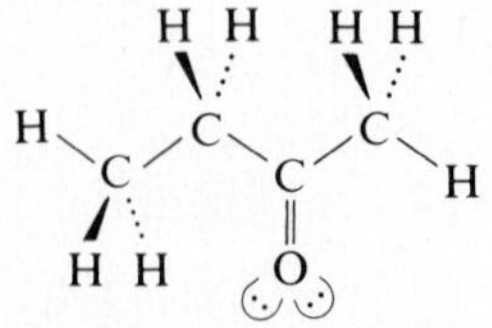 butanone

The names of aldehydes and ketones are based on the longest carbon chain present. The final syllables *-al* or *-one* are used for aldehydes or ketones respectively. For a ketone, it is sometimes necessary to specify the carbon atom that represents the link in the chain. Consider the examples below.

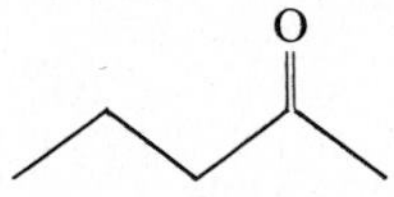

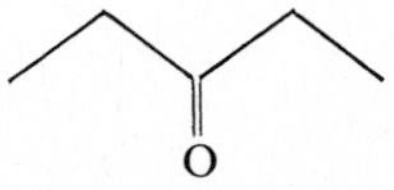

pentan-2-one pentan-3-one

The simplest aldehyde contains no carbon chain at all. Its structure is that of a hydrogen molecule broken up by a carbonyl link.

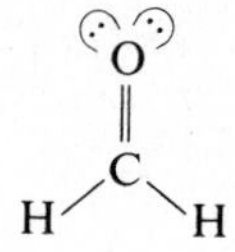

methanal

The functional groups of the two different types of compound have the following form (where R_1 and R_2 are carbon chains).

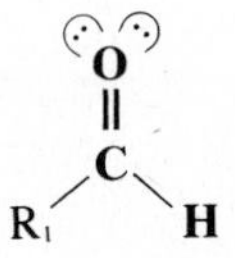

aldehyde 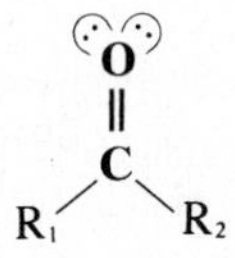 *ketone*

Many of the aldehydes and ketones that occur naturally are substituted compounds: there are other groups attached to the carbon chain as well as the carbonyl group. A most important class is that of the carbohydrates (see page 445). In naming a compound of this sort, it is sometimes best to show the carbonyl group as a prefix ('*oxy*'-) rather than as a final syllable ('-*al*' or -'*one*').

Example

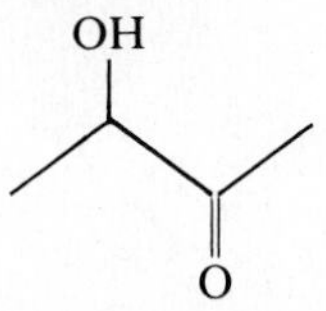

3-hydroxybutanone
or 3-oxybutane-2-ol

The choice of name depends on the requirements of emphasis: the hydroxyl group as a substituent or as a dominant functional group. For practice, name the schematic molecular drawings below. The solutions are given in the answers section of this book.

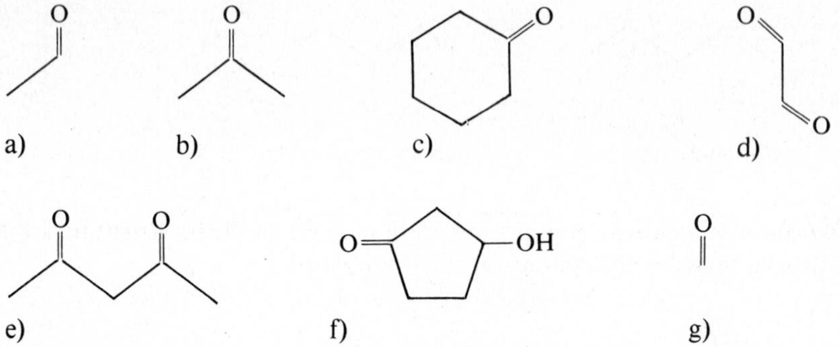

a) b) c) d)

e) f) g)

Methods of preparation

There are three main methods for preparing an aldehyde or a ketone.

1. Redox

An alcohol is oxidized by acidified dichromate (VI) to an aldehyde or ketone. Primary alcohols give aldehydes, secondary alcohols give ketones.

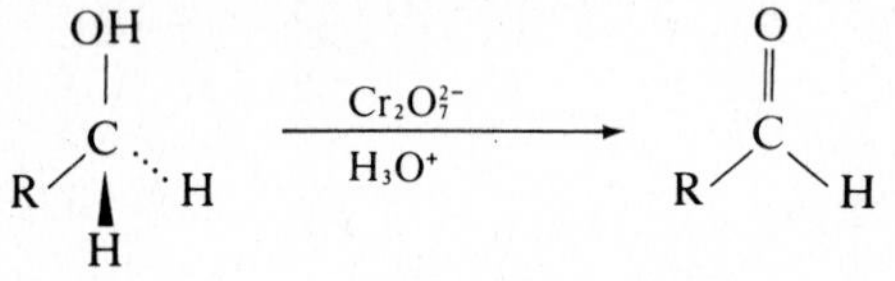

$$RCH_2OH \xrightarrow[H_3O^+]{Cr_2O_7^{2-}} RCHO \quad \text{(low yield)}$$

$$R_1CH(OH)R_2 \xrightarrow[H_3O^+]{Cr_2O_7^{2-}} R_1COR_2$$

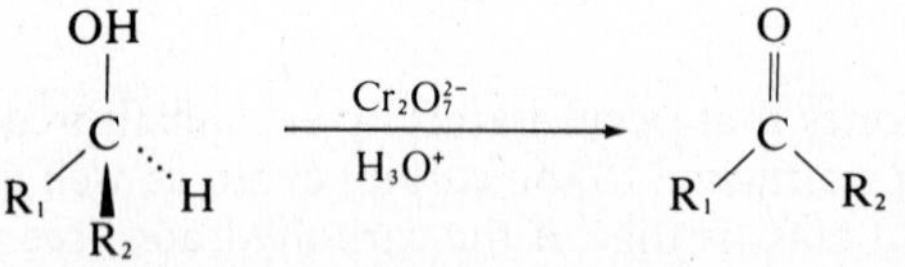

An aldehyde can also be prepared by reducing an acid or an acid derivative. There are no equivalent reactants that give a ketone as a product.

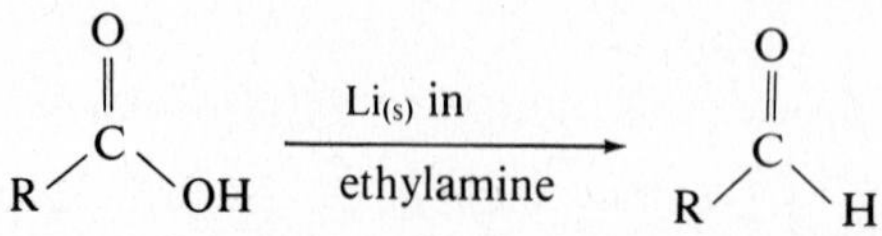

$$RCOOH \xrightarrow[\text{ethylamine}]{Li_{(s)}\text{ in}} RCHO \quad \text{(low yield)}$$

$$RCOCl \xrightarrow[\text{in xylene (reflux)}]{H_2-Pd/BaSO_4 \text{ catalyst}} RCHO \quad \text{(low yield)}$$

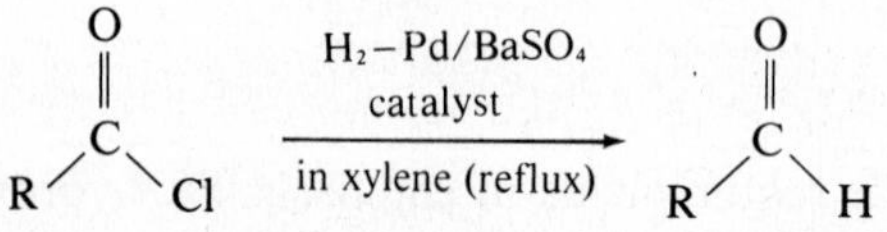

For aldehydes, the redox methods are not very satisfactory. The required aldehyde can itself be either oxidized or reduced: aldehydes are easily oxidized to acids or reduced to alcohols.

2. Hydrolysis
Alkynes are hydrolysed by dilute acid in the presence of mercury (II) ions. Alk-1-ynes give aldehydes, the others give ketones.

$$H-C{\equiv}C-R \xrightarrow{Hg^{2+}/H_3O^+} H-\overset{\overset{O}{\|}}{C}-CH_2-R$$

$$R_1-C{\equiv}C-R_2 \xrightarrow{Hg^{2+}/H_3O^+} R_1-\overset{\overset{O}{\|}}{C}-CH_2-R_2$$

Yields from this reaction are not high but ethanal can be produced satisfactorily from ethyne.

3. From a Grignard reagent
The treatment of a Grignard reagent (see page 528) with a nitrile gives a ketone after hydrolysis.

$$R_1-Br \underset{\text{dry ether}}{\overset{Mg_{(s)}}{\rightleftharpoons}} R_1-Mg-Br \xrightarrow[2.\ H_3O^+]{1.\ R_2-CN} R_1-\overset{\overset{O}{\|}}{C}-R_2$$

ketone

28.2 Reactive sites

Keto reactivity

The carbonyl group is often referred to as a keto group. It has three different reactive sites that are described overleaf.

C — O

π-pair drawn more onto the oxygen atom

sideways on

$O\ \delta-$

$C\ \delta+$

from above

1. The electron-deficient carbon atom

The oxygen atom is more electronegative than the carbon atom and therefore exerts a $-$I effect on it. This causes the carbon atom to become electron-deficient and susceptible to attack by nucleophiles. Nucleophilic attack is particularly favourable because the π-electrons are easily drawn away from the carbon atom to become a third lone pair on the oxygen atom. Compare the two mechanisms below:

nucleophilic attack on a carbonyl group		*nucleophilic attack on a saturated carbon atom*	
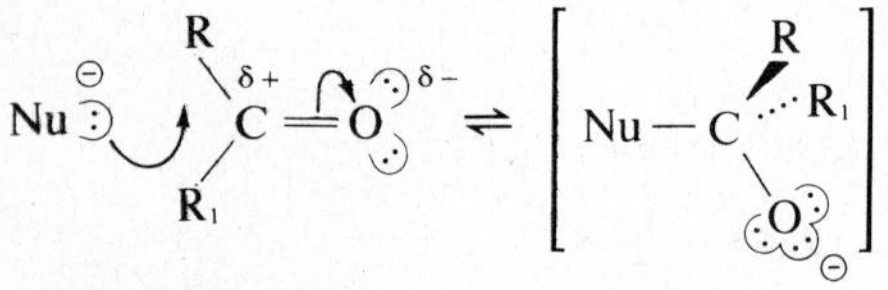 $Nu^{\ominus}$: + $R(R_1)\overset{\delta+}{C}{=}\overset{\delta-}{O}$ $\rightleftharpoons$	$[Nu{-}C(R)(R_1){-}O^{\ominus}]$	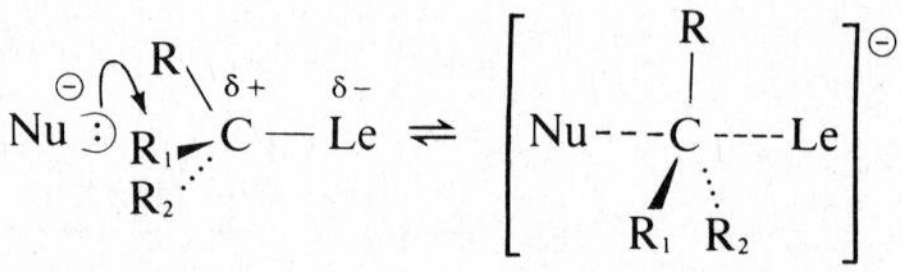 $Nu^{\ominus}$: + $R(R_1)(R_2)\overset{\delta+}{C}{-}\overset{\delta-}{Le}$ $\rightleftharpoons$	$[Nu{-}{-}{-}C(R)(R_1)(R_2){-}{-}{-}{-}Le]^{\ominus}$
the π-electrons are repelled by the approach of the nucleophile	4-bonded carbon atom	the three groups are pushed back by the approach of the nucleophile	carbon atom with 5 electron pairs around it

The activated complex deriving from the carbonyl is easier to form for two reasons:

1. Its formation depends only on the repulsion of π-electrons.
2. The complex has a structure based on a stable saturated carbon atom. It is not necessary for the central atom to be surrounded by five pairs of electrons.

Under most conditions, the complex acts as a base so that the overall reaction results in the addition of H — Nu across the carbonyl double bond.

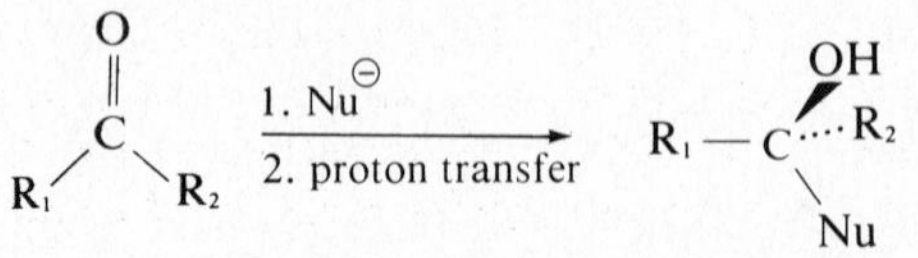

$$R_1{-}\underset{}{C}(=O){-}R_2 \xrightarrow[\text{2. proton transfer}]{\text{1. } Nu^{\ominus}} R_1{-}C(OH)(R_2){-}Nu$$

2. The carbonyl π-electrons

These are less easily drawn away than the π-electrons of an alkene because of the presence of the more electronegative oxygen atom in the carbonyl group. The oxygen atom exerts much more control over the π-electrons than do the atoms of an alkene molecule.

3. The oxygen lone pairs

These often interact with an electron-deficient site within the *same* molecule. Consider the geometry of a typical molecule having a carbonyl group: ethanal,

CH_3CHO. The π-overlap is responsible for holding four of its atoms and the two oxygen lone pairs in the same plane.

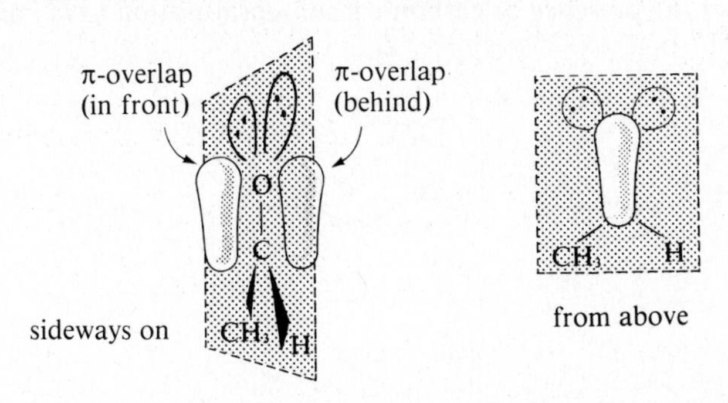

The methyl group, however, is free to rotate as shown below:

O
‖
H–C–C–H
H H

At a particular angle of rotation, one of the methyl hydrogen atoms lines up in the plane of the carbonyl group. In this position, the oxygen lone pair attracts the partially shielded proton of the hydrogen nucleus. A shift of electron density can then take place in the molecule with the following result.

O
‖
H–C–C–H
H H
⇌
H
O H
C=C
H H

The shifts in electron density are concerted and follow a uniform cyclic pattern: as the lone pair becomes a σ-pair, the C — H σ-pair becomes a π-pair and the C=O π-pair becomes a lone pair. The resulting molecule has a different structure from that of ethanal but can readily revert to ethanal.

H
O H
C=C
H H
⇌
O
‖
H–C–C–H
H H

The two different forms are called *tautomers* of ethanal and the effect itself is known as *tautomerism*.

> A compound exhibits tautomerism when it exists in two different structural isomeric forms that are in dynamic equilibrium with each other.

In the case of a carbonyl compound (such as ethanal above), the tautomers are known as *keto-* and *enol-* forms. *Keto-* signifies the carbonyl group typical of a ketone; *enol-* signifies the presence of carbon-carbon unsaturation (*-ene*) and an alcohol group, —OH (*-ol*)

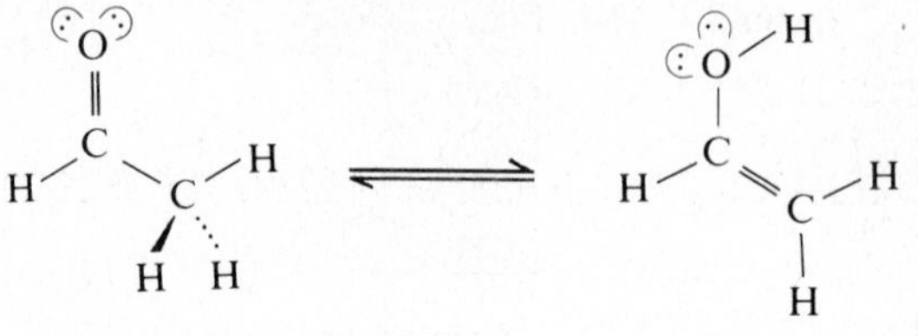

keto- ethanal ⇌ *enol-* ethanal

Enol reactivity

In most carbonyl compounds, the *keto-* tautomer dominates. However, although there may only be a trace of *enol-* form present, the reactivity of this tautomer is entirely different from that of the *keto-* tautomer. If the *enol-* form is used up in a reaction, the position of the *keto-enol* equilibrium is drawn over to replace the *enol-* form used up. The process can continue until all the carbonyl compound has reacted. In other words, the reactivity of the *enol-* tautomer provides an alternative to that of the *keto-* tautomer.

Only carbonyl compounds that have 'α-hydrogen atoms' are able to show *enol-* character. An 'α-hydrogen atom' is one that is bonded to a carbon atom next to the carbonyl group. Some examples are shown below.

compound	*molecular structure*	*number of α-hydrogen atoms*
propanone	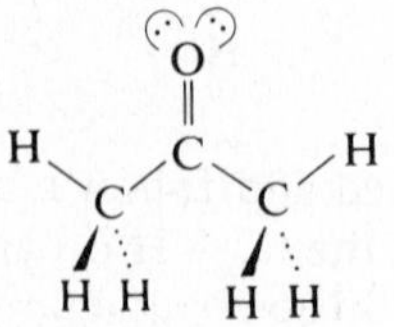	six
ethanal	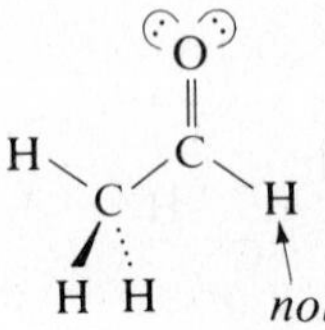 *not* α-hydrogen atoms	three
methanal	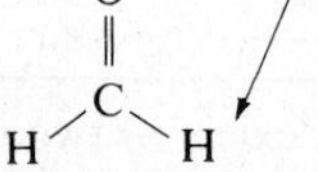	none

The reactive sites of an *enol-* structure are discussed in earlier chapters. They are those of an alkene (page 377) and of an alcohol (page 410). The dominant feature is the Lewis basicity of the carbon-carbon π-electrons. Carbonyl compounds interact with electrophiles through their *enol-* form.

28.3 Reactions common to aldehydes and ketones

With water and ammonia

Aldehydes and ketones undergo an addition reaction with water or ammonia. In the case of water, the addition is reversible and the addition compound is readily decomposed back to the original reactants.

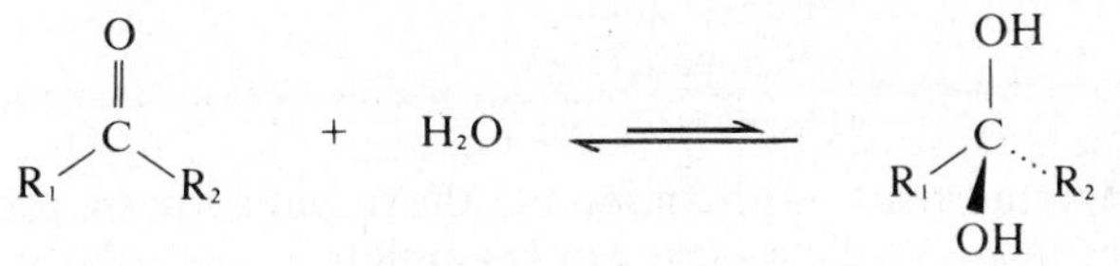

$$R_1C(=O)R_2 + H_2O \rightleftharpoons R_1C(OH)_2R_2$$

A similar addition product is formed when ammonia reacts with an aldehyde or ketone. Although this addition, in principle, is also reversible, in practice a complex set of polymerization reactions takes place so that the final product bears little resemblance to the initial addition product. It is only possible to isolate the simple addition product between an aldehyde and ammonia when the reaction is carried out in ether solution.

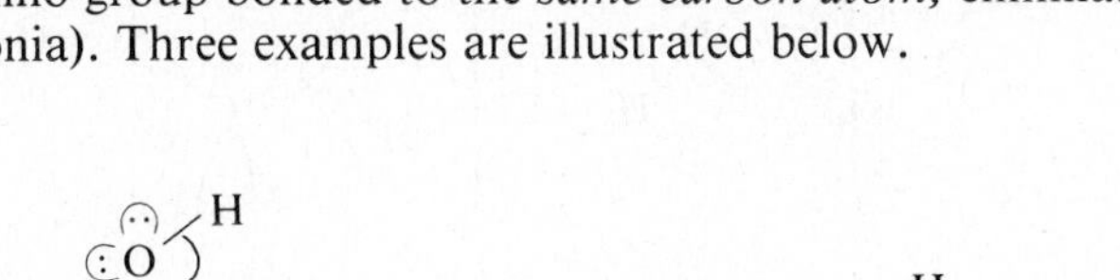

$$RC(=O)H + NH_3 \xrightleftharpoons{\text{ether}} RC(OH)(NH_2)H$$

As a general rule, a molecule that has two hydroxyl groups or a hydroxyl and an amino group bonded to the *same carbon atom,* eliminates water (or ammonia). Three examples are illustrated below.

a)

$$R_1C(OH)_2R_2 \longrightarrow R_1C(=O)R_2 + H_2O$$

two hydroxyl groups

loss of water

b)

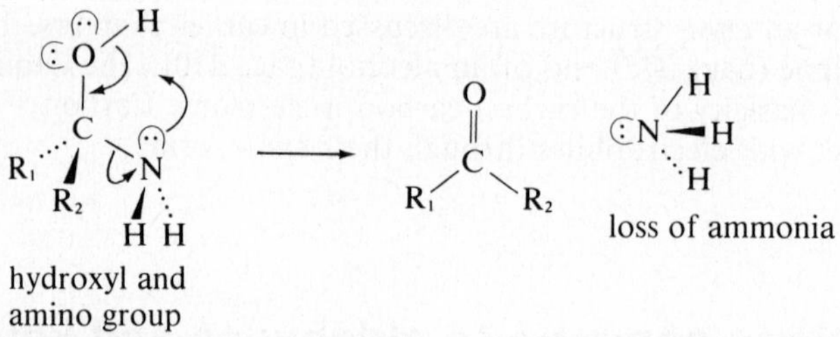

H
O
C
R_1 N
R_2
H H

hydroxyl and
amino group

$\longrightarrow$

O
C
R_1 R_2

H
N—H
H

loss of ammonia

c)

H
O
C
R_1 N H
R_2
H

hydroxyl and
amino group

$\longrightarrow$

R_1
C
R_2 N
H

unstable imine
which undergoes
polymerization

H
O—H

loss of water

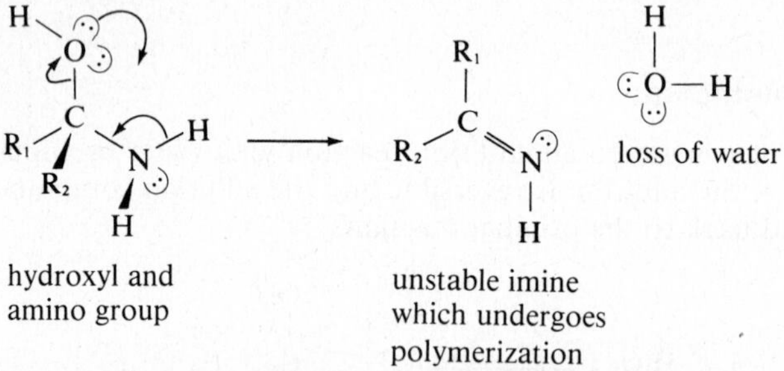

When a carbonyl compound reacts with ammonia, the initial addition product is formed as a result of nucleophilic attack. For example,

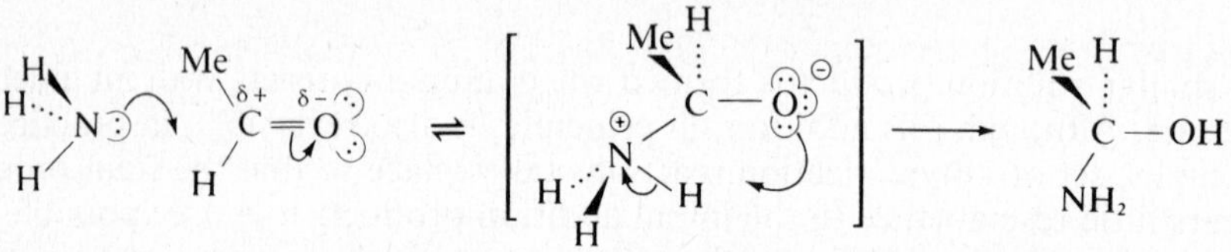

H
H N
H

Me
δ+ δ−
C=O
H

⇌

Me H
C—O ⊖
⊕
N
H H
H

$\longrightarrow$

Me H
C—OH
NH_2

Elimination of water from this addition product leads to the formation of an unstable imine by mechanism c) above. The imine polymerizes to give mostly a compound of formula $(CHMe.NH)_3$. The mechanism is likely to involve the nucleophilic attack of one imine molecule on another.

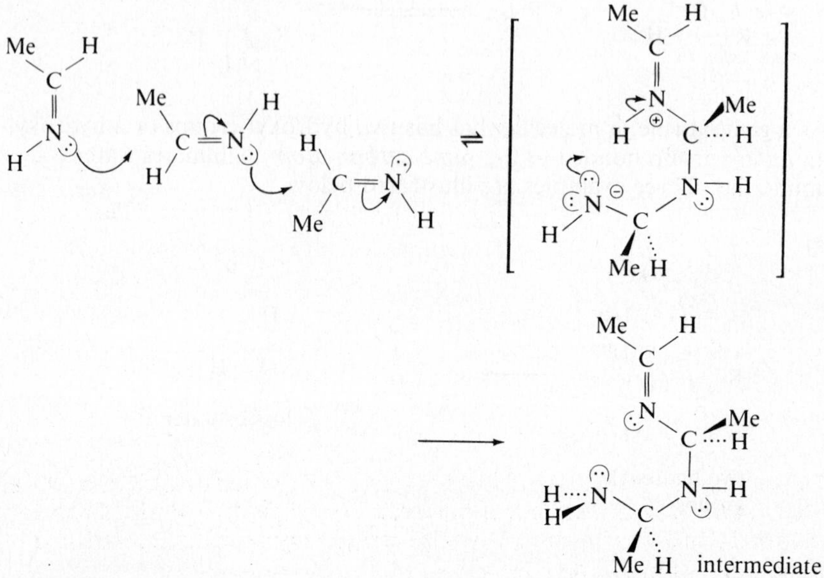

Me H
C
N
H

Me H
C=N
H

H
C=N
Me H

⇌

Me H
C
N ⊕ Me
H C H
⊖
N N—H
H C
Me H

$\longrightarrow$

Me H
C
N Me
C H
H N N—H
H C
Me H intermediate

This intermediate forms a cyclic product because one end of the chain has a nucleophilic nitrogen atom and the other has a carbon atom susceptible to nucleophilic attack.

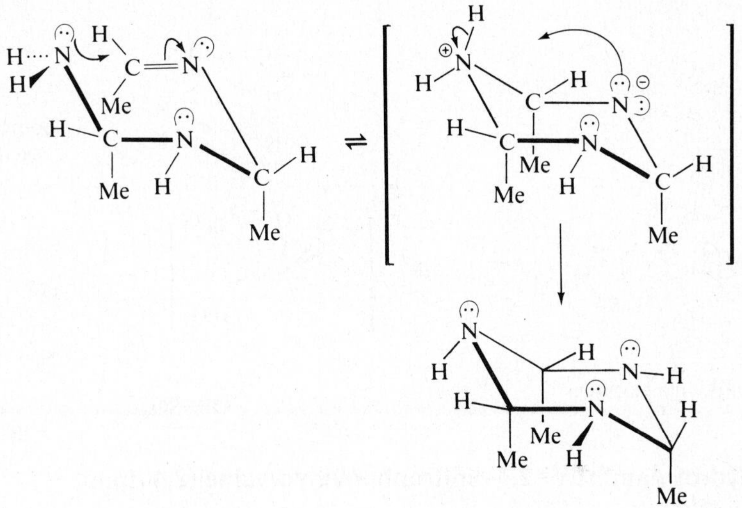

Ketones give polymeric products with ammonia in a similar manner. It is not possible to isolate a simple addition product even when the reaction is carried out in ethereal solution.

With hydrogen cyanide or sodium hydrogensulphate (IV)

A similar addition reaction occurs when either a ketone or an aldehyde is mixed with hydrogen cyanide or sodium hydrogensulphate (IV) in aqueous solution. The simple addition products can be isolated easily (unlike those of ammonia or water).

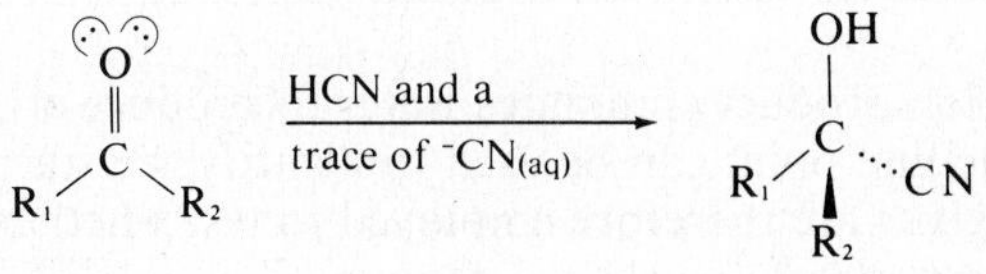

HCN and a trace of $^{-}CN_{(aq)}$

a hydroxynitrile (cyanohydrin)

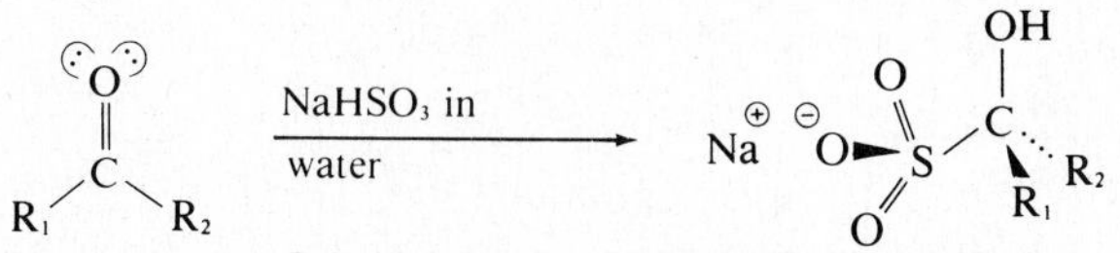

$NaHSO_3$ in water

crystalline salt (insoluble)

The reaction with sodium hydrogensulphate (IV) is slow if the groups R_1 and R_2 are bulky. The nucleophile is hindered ('steric hindrance') from reaching the electron-deficient carbon atom of the carbonyl group. Compare the two mechanisms with that on page 432.

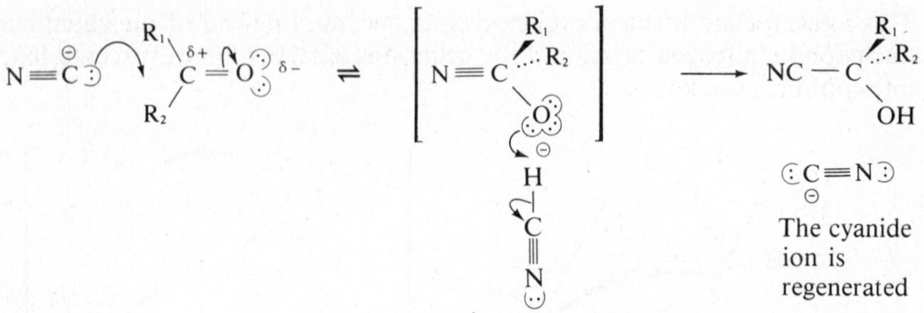

The cyanide ion is regenerated

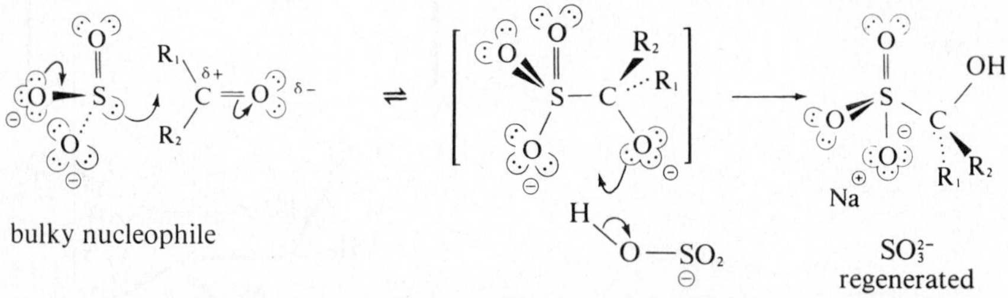

bulky nucleophile

SO_3^{2-} regenerated

With hydroxylamine or 2,4-dinitrophenylhydrazine (2,4 dnph)

The molecular structure of these two reagents is shown below. Both resemble ammonia in their nucleophilic properties and give addition products with carbonyl compounds.

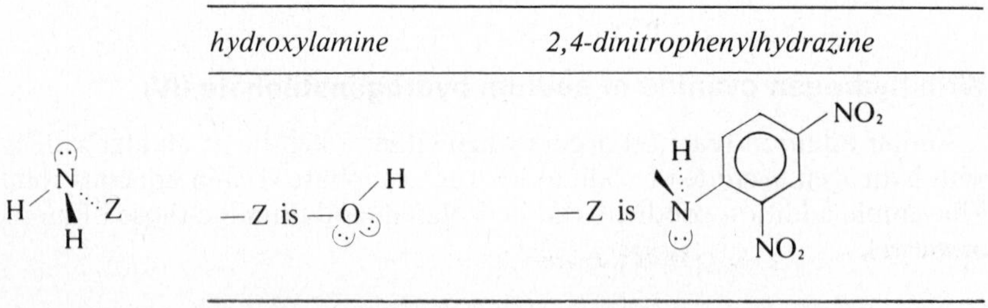

hydroxylamine	*2,4-dinitrophenylhydrazine*
—Z is —O—H	—Z is —N(H)—C$_6$H$_3$(NO$_2$)$_2$

In each case, the simple addition product eliminates water to produce an insoluble precipitate whose melting point can be used to identify the starting carbonyl compound. The reagents are therefore employed to test whether an unknown substance contains a particular carbonyl compound.

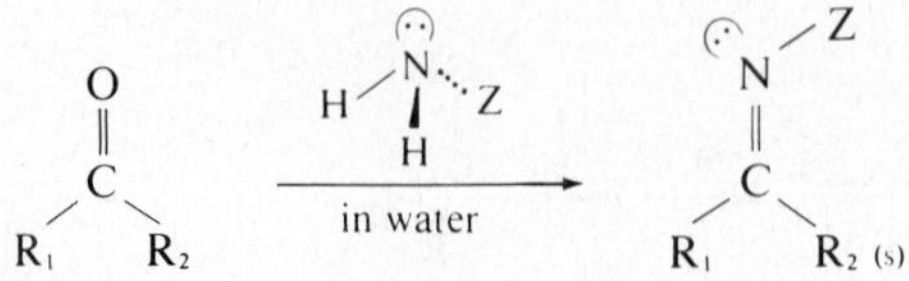

$$R_1R_2C{=}O + H_2N{-}Z \xrightarrow{\text{in water}} R_1R_2C{=}N{-}Z_{(s)}$$

The product is known as an oxime when hydroxylamine is used and a 2,4-dinitrophenylhydrazone when 2,4-dnph is used. Oximes are colourless but the 2,4-dinitrophenylhydrazones are orange.

The mechanism of the reaction is described as addition-elimination. Nucleophilic addition is the first step, but the addition product eliminates water before it can be isolated.

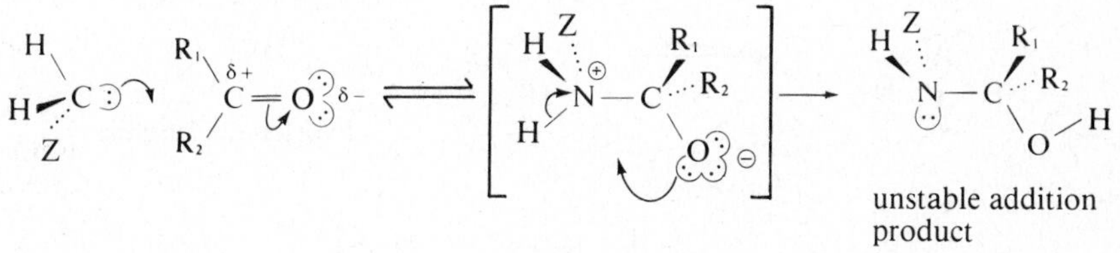

H H N Z R₁ R₂ C=O δ+ δ- ⇌ [H Z H N⊕—C R₁ R₂ O⊖] → H Z N—C R₁ R₂ O—H

unstable addition product

then

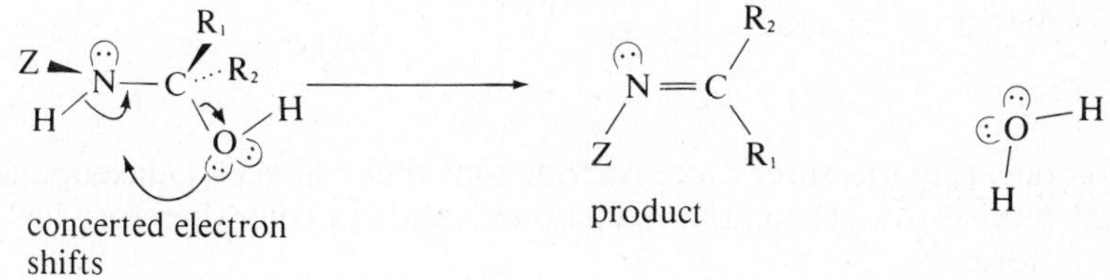

Z N H—C R₁ R₂ O H → N=C Z R₂ R₁ + H—O—H

concerted electron shifts

product

For example, propanone gives an orange precipitate of propanone-2,4-dinitrophenylhydrazone when two drops of the ketone are added to a freshly prepared solution of 2,4-dnph.

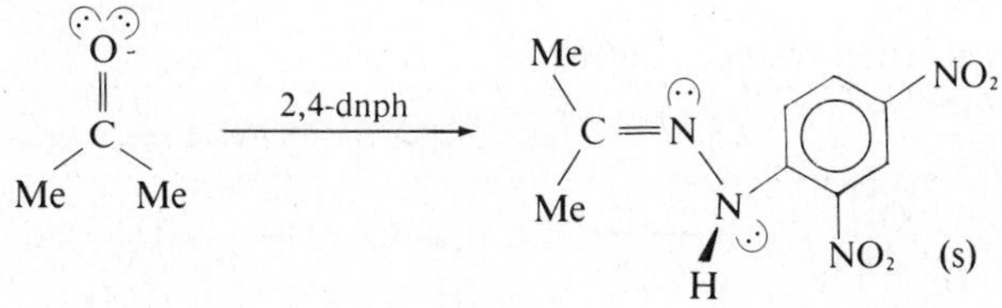

O=C Me Me —2,4-dnph→ Me Me C=N—N H (ring) NO_2 NO_2 (s)

With lithium tetrahydridoaluminate

Lithium tetrahydridoaluminate (lithium aluminium hydride, $LiAlH_4$) is a powerful reducing agent. It is violently decomposed by water and therefore is usually used in an ether solution. The reduction products are extracted from the ether solution by addition of water which hydrolyses the initial products. Aldehydes are reduced to primary alcohols and ketones to secondary alcohols.

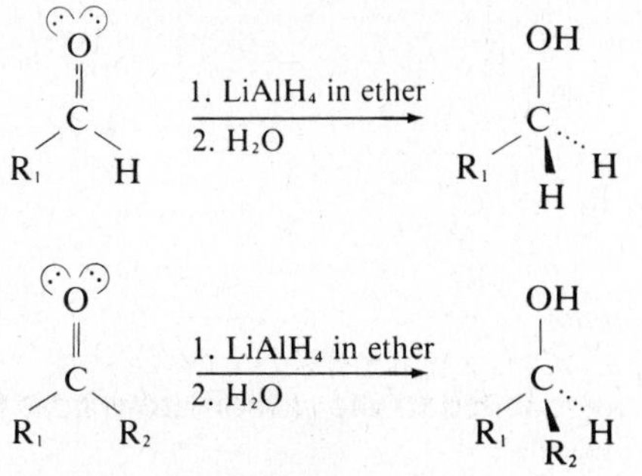

O=C R₁ H —1. $LiAlH_4$ in ether, 2. H_2O→ OH C R₁ H H

O=C R₁ R₂ —1. $LiAlH_4$ in ether, 2. H_2O→ OH C R₁ H R₂

Tetrahydridoaluminate ions are a source of hydride ions which in ethereal solution act as nucleophiles.

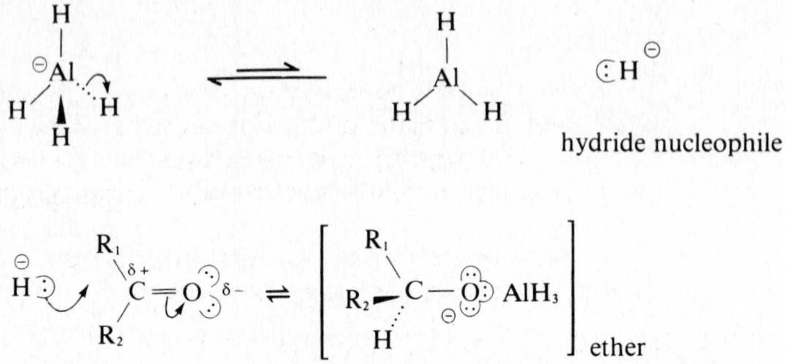

H H H H Al ⇌ H H H Al H

hydride nucleophile

H R1 C=O δ+ δ- R2 ⇌ [R1 R2 C—O AlH3 H] ether

This complex contains three more hydride ions that can act as nucleophiles by the same mechanism. The initial reduction product in ether therefore has the following formula.

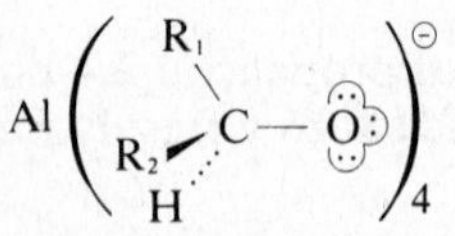

Al (R1 R2 C—O H)4

It is rapidly hydrolysed when water is added.

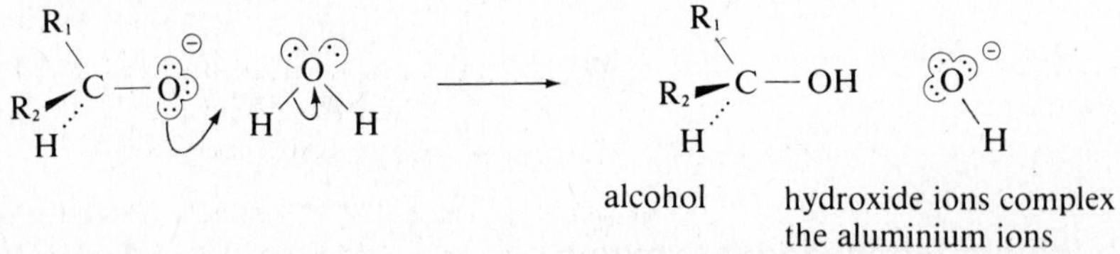

R1 R2 C—O H O H H ⟶ R1 R2 C—OH H O H

alcohol

hydroxide ions complex the aluminium ions

With aqueous halogen solutions

A carbonyl compound reacts with a halogen through its *enol-* form (see page 434). There is a strong tendency for a halogen molecule to add across a carbon-carbon double-bond, and any carbonyl that possesses α-hydrogen atoms can exhibit *enol-* character.

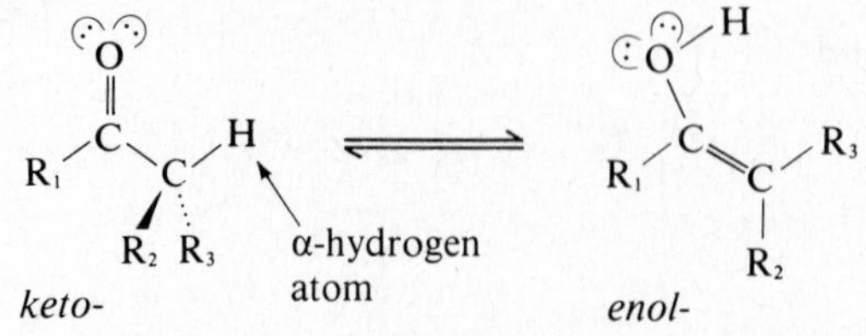

O R1 C C H R2 R3 ⇌ O H R1 C C R3 R2

α-hydrogen atom

keto-

enol-

(α-hydrogen atoms are hydrogen atoms bonded to the carbon atom next to the carbonyl group).

The reaction of a carbonyl compound and an aqueous halogen results in the replacement of the α-hydrogen atoms of the *keto*- form by halogen atoms. For example, propanal forms 2,2-dichloropropanal when shaken with an excess of chlorine water.

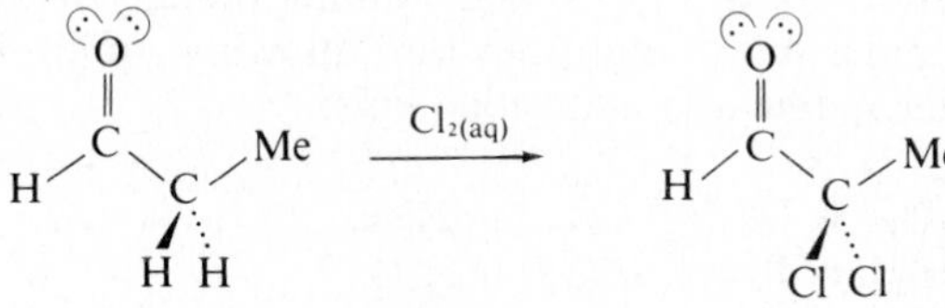

$$\mathrm{H{-}C({=}O){-}CH_2{-}Me} \xrightarrow{Cl_{2(aq)}} \mathrm{H{-}C({=}O){-}CCl_2{-}Me}$$

There is a particular laboratory test that makes use of the reactivity of iodine under alkaline conditions. It is known as the *iodoform test* and positive results are given for *any* carbonyl compound possessing three α-hydrogen atoms on the same carbon atom:

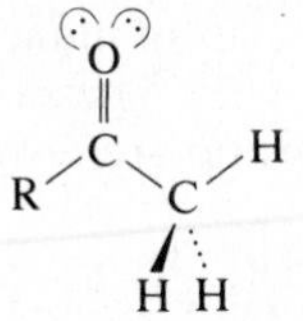

$$\mathrm{R{-}C({=}O){-}CH_3}$$

When a compound whose molecules contain the above group is treated with an alkaline solution of iodine, an insoluble precipitate of yellow iodoform (triiodomethane) is produced. The test is carried out as follows:

1. Take a few cm^3 of a solution of iodine dissolved in aqueous potassium iodide.
2. Add alkali drop by drop until the brown colour of iodine is just discharged.
3. Add two drops of the compound to be tested.
4. The appearance of a yellow precipitate indicates that the molecules of the compound contain the group $\mathrm{{-}C({=}O){-}CH_3}$

$$\mathrm{R{-}C({=}O){-}CH_3} \xrightarrow[\text{(or NaIO)}]{I_2 \text{ in } OH_{(aq)}^-} \mathrm{R{-}C({=}O){-}O^{\ominus}} \quad \mathrm{CHI_{3(s)}} \leftarrow \text{yellow iodoform}$$

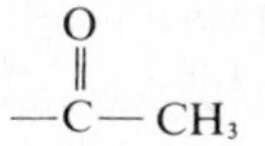

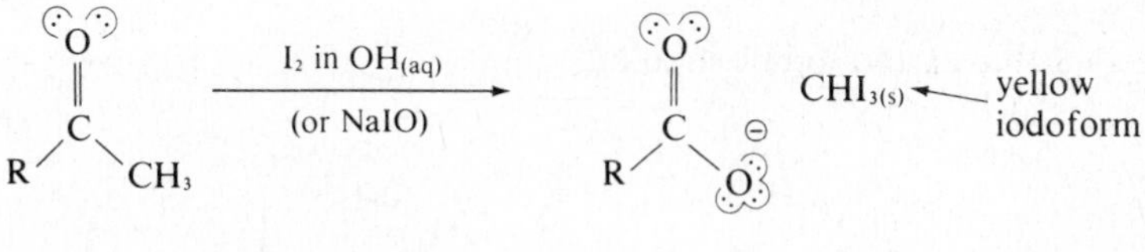

The test also works for alcohols of the following structure:

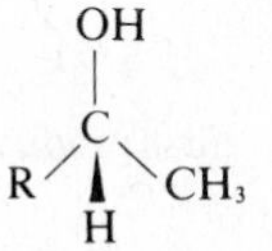

$$\mathrm{R{-}CH(OH){-}CH_3}$$

Under the oxidizing conditions of the reaction (iodine is a weak oxidant) the alcohol group is converted to the carbonyl group $R-\overset{\displaystyle O}{\overset{\|}{C}}-CH_3$

Sometimes a solution of sodium chlorate (I) and sodium iodide is used instead of iodine in alkali. Under these conditions the following equilibria provide a source of molecular iodine and hydroxide ions:

$$ClO^-_{(aq)} + 2I^-_{(aq)} + H_2O_{(l)} \rightleftharpoons I_{2(aq)} + Cl^-_{(aq)} + 2OH^-_{(aq)}$$
$$I_2 + 2OH^-_{(aq)} \rightleftharpoons I^-_{(aq)} + IO^-_{(aq)} + H_2O_{(l)}$$

The chlorate (I) is added for two main reasons:

1. to increase the concentration of molecular iodine in the testing reagent,
2. to oxidize alcohols more effectively and therefore improve the sensitivity of the test towards alcohols of the form described above.

The initial mechanism of the reaction of a carbonyl compound with an aqueous halogen is the same as that outlined on page 379. The *enol-* form of the compound reacts with a halogen molecule which behaves as an induced electrophile. This gives a π-complex that achieves stability by losing a proton from the hydroxyl group. The π-electrons of the carbonyl group are restored in the process.

Example 1 Propanal and chlorine.

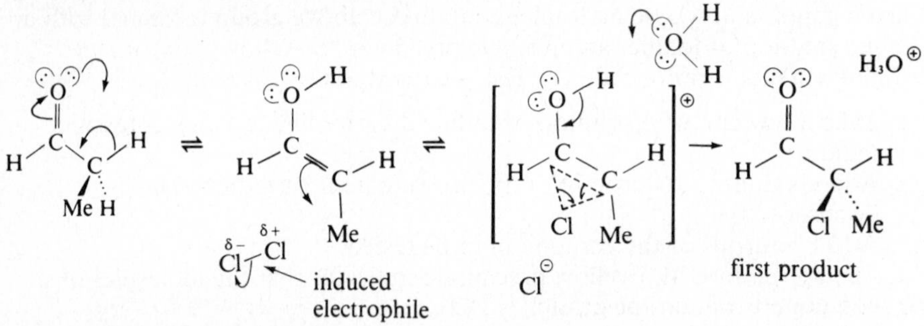

H
O
H
H
H₃O⊕
O
H
O
H
O
O
H
C
H
⇌
C
⇌
C
H
→
C
H
H
C
H
C
H
H
C
H
C
Me H
Me
Cl
Me
Cl Me
δ+
δ−
Cl
Cl
induced
electrophile
Cl⊖
first product

The first product also has an α-hydrogen atom and so the cycle repeats itself until the dichloro- derivative is produced.

Example 2 Ethanal and iodine in alkali

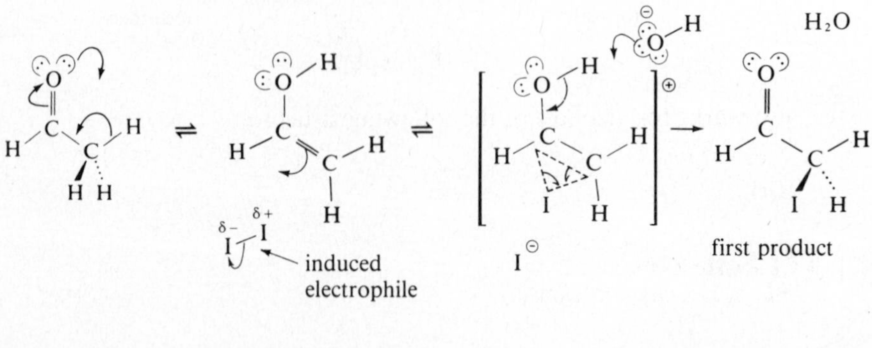

⊖
O
H
H₂O
O
O
H
O
H
O
C
H
⇌
C
H
⇌
C
H
→
C
H
H
C
H
C
H
C
H
C
H
H
H H
H
I
H
I H
δ+
δ−
I
I
induced
electrophile
I⊖
first product

The first product also has two α-hydrogen atoms and therefore a triiodo derivative is produced as an intermediate. In this intermediate, the σ-bond between the carbonyl group and the — CI_3 group is weak because of the three bulky iodine atoms. Also since — CI_3 can act as a leaving group (the negative charge can be delocalized towards the iodine atoms), the weakened C — C σ-bond is broken as a result of nucleophilic attack by the hydroxide present in the system.

CI_3 δ+ δ− H—O⁻ + C=O ⇌ [intermediate] → H—CI_3 + H—C(=O)O⁻

iodoform acid anion

28.4 Reactions specific to aldehydes

With alcohols

An aldehyde reacts with an alcohol in the presence of acid to form an acetal. The reaction is reversed by adding alkali.

$$R_1CHO + 2\,R_2OH \rightleftharpoons R_1CH(OR_2)_2 + H_2O$$

an acetal

The reaction is similar to the reaction of a carbonyl compound with water (see page 435). In the case of the formation of addition products with an alcohol, however, the product can be isolated. For example, ethanal and methanol give 1,1-dimethoxyethane in the presence of dilute sulphuric acid.

Me—OH + MeCHO (δ+ C=O δ−) + H_3O^+ ⇌ [intermediate] → MeCH(OH)OMe + H_3O^+

H_3O^+ ions assist the mesomeric shift of the carbonyl π-electrons

a hemi-acetal

The hemi-acetal is attacked by a similar mechanism.

1,1-dimethoxyethane

$H_3O^{\oplus}$

With oxidizing agents: the silver mirror test

Ketones are inert to oxidizing agents in solution but aldehydes can be oxidized to acids.

$$RCHO \xrightarrow{Cr_2O_7^{2-} \text{ in } H_3O^+} RCOOH$$

This difference in reactivity is used in the design of a laboratory test to distinguish between an aldehyde and a ketone. Silver (I) complexed by ammonia behaves as a weak oxidant: if an ammoniacal solution of silver nitrate is warmed with an aldehyde, a mirror of silver forms on the inside of the tube. No reaction occurs if a ketone is tested in this way.

$$RCHO \xrightarrow{Ag(NH_3)_2^+} RCOO^{\ominus} \quad Ag_{(s)}$$

Fehlings' solution is a similar reagent. It is a blue solution of complexed copper (II) that is reduced to a red precipitate of copper (I) oxide by an aldehyde, but not by a ketone.

With aqueous alkali

When an aldehyde is warmed with aqueous alkali, a set of polymerization and condensation reactions take place. These are known as base-catalysed condensations and are discussed in detail on pages 529 to 534. The distinction between polymerization and condensation is often disregarded in classifying these processes. *Polymerization* occurs when the molecules of a substance bond to one another to produce macromolecules; *condensation* is a polymerization process in which a small molecule is eliminated from the structure in each step (usually H_2O or NH_3). For example, ethanal readily undergoes base-catalysed condensation. In very dilute cold alkali a simple polymer is formed: *aldol*

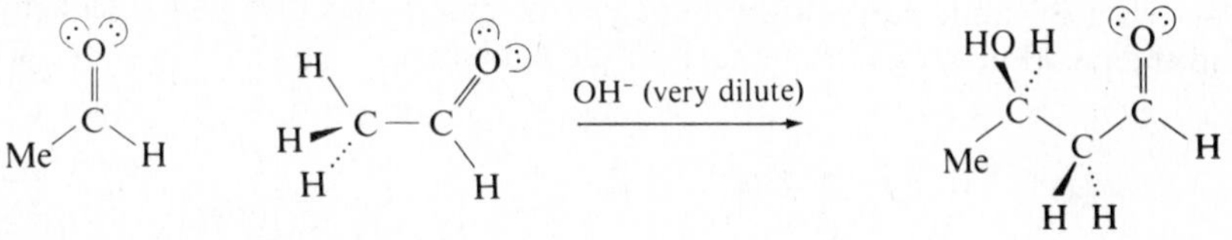

Me–C(=O)–H + H–CH₂–C(=O)–H —OH⁻ (very dilute)→ Me–CH(OH)–CH₂–C(=O)–H

aldol

It is called aldol because it contains an aldehyde group (ald-) and an alcohol group (-ol).

As the concentration of alkali is increased, further condensation occurs. Eventually a macromolecular yellow-brown resin that smells of rotten apples is deposited. The macromolecules have a structure based on the 'aldol' chain shown below.

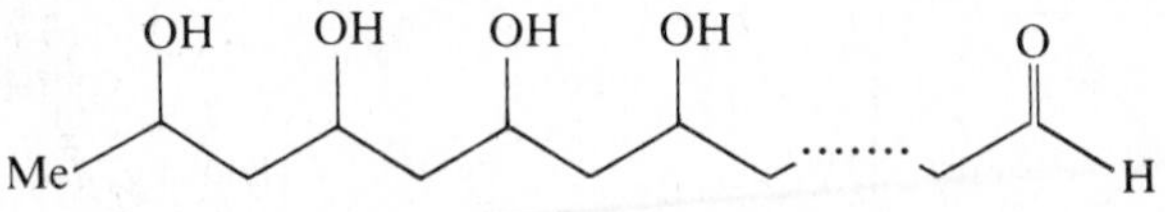

Me–CH(OH)–CH₂–CH(OH)–CH₂–CH(OH)–CH₂–CH(OH)–CH₂ .... CH₂–C(=O)–H

However there is considerable cross-linking between chains as well as the loss of water molecules from parts of the chain, for example:

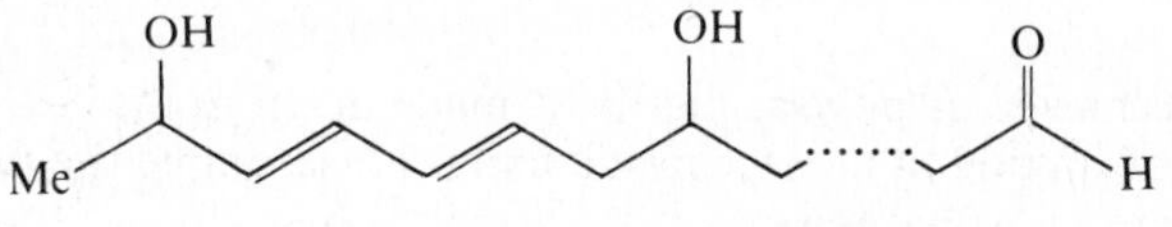

Me–CH(OH)–CH=CH–CH=CH–CH(OH)–CH₂ .... CH₂–C(=O)–H

28.5 Carbohydrates

A carbohydrate is a polyhydroxyaldehyde or polyhydroxyketone of the general formula $C_x(H_2O)_y$. Different carbohydrates are synthesized by a wide variety of organisms but the simplest all have a sweet taste and are known as sugars or saccharides. Sugar molecules show a marked tendency to condense with one another so that the most complex carbohydrates can be considered to have formed by the condensation of a number of smaller units. The smallest sugar molecules have $x = y = 5$ or 6 in the above general formula. These are called *monosaccharides* because they represent the 'monomers' from which the complex 'polymeric' carbohydrates such as starch can be considered to have formed:

$x = y = 5$ $C_5H_{10}O_5$ ribose
$x = y = 6$ $C_6H_{12}O_6$ glucose

Monosaccharides contain an aldehyde group (or sometimes a ketone group)

at one end of their chain and an alcohol group at the other. The chains therefore tend to form rings by an internal acetal-type reaction (see page 443). There is a dynamic equilibrium between the chain structure and ring structure, and the position of equilibrium is pH-dependent.

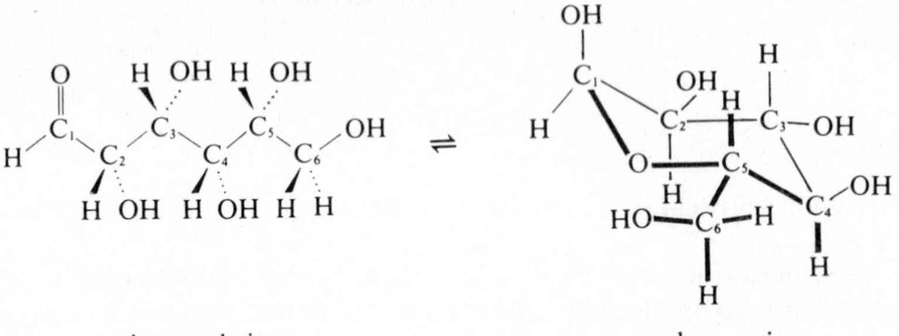

glucose chain ⇌ glucose ring

The ring forms as a result of the following interactions

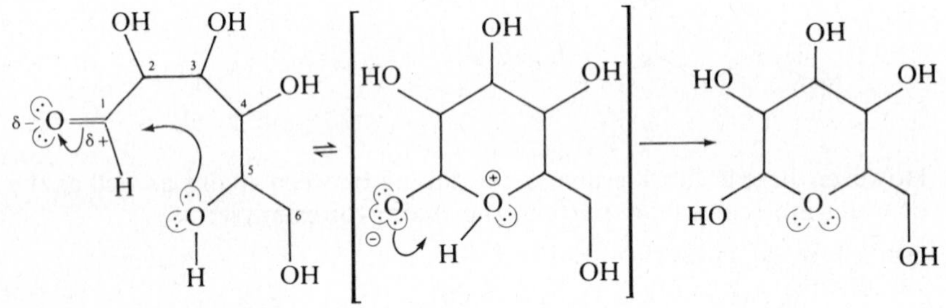

A disaccharide, trisaccharide or polysaccharide is made up from the appropriate number of linking monosaccharide units. For example the two sugars shown below are disaccharides.

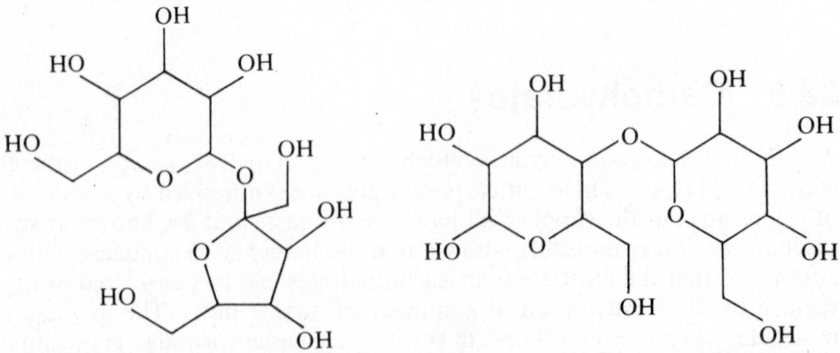

sucrose, $C_{12}H_{22}O_{11}$

maltose, $C_{12}H_{22}O_{11}$

If the chain form of a sugar molecule contains an aldehyde group, the sugar should be expected to give a positive result in the silver mirror test. As the chain form is used up in the reaction, the position of the chain-ring equilibrium is drawn over to replace the consumed chain form. Since certain sugar molecules do *not* have an aldehyde group, a distinction can be drawn between *reducing sugars* and *non-reducing sugars*. In order to possess an aldehyde group, the ring form must have the following link:

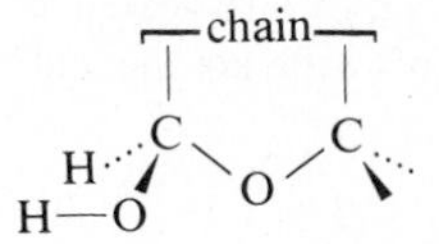

This link is the one that opens up to give a chain form by the mechanism below

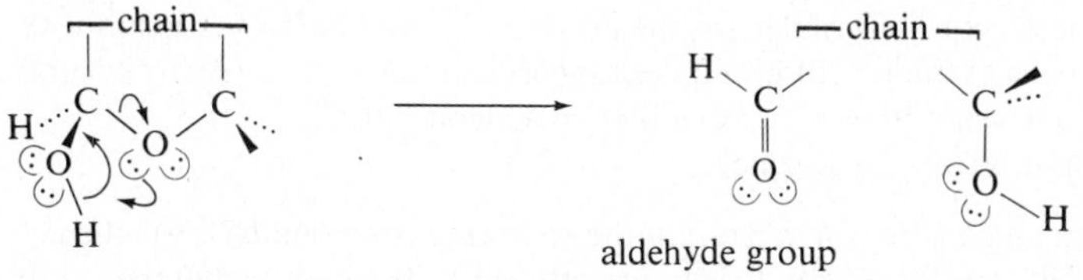

aldehyde group

By inspecting the structures of glucose, sucrose and maltose it can be seen that glucose and maltose have the link shown above but sucrose does not: glucose and maltose are reducing sugars and give a silver mirror slowly with ammoniacal silver nitrate; sucrose is a non-reducing sugar. All three give a precipitate with hydroxylamine because the chain forms of all three contain a carbonyl group.

Hydrolysis of polysaccharides.
All polysaccharides can be broken down into their linking monosaccharide units by hydrolysis under acidic conditions. For example, sucrose is hydrolysed to glucose and fructose when boiled in dilute acid.

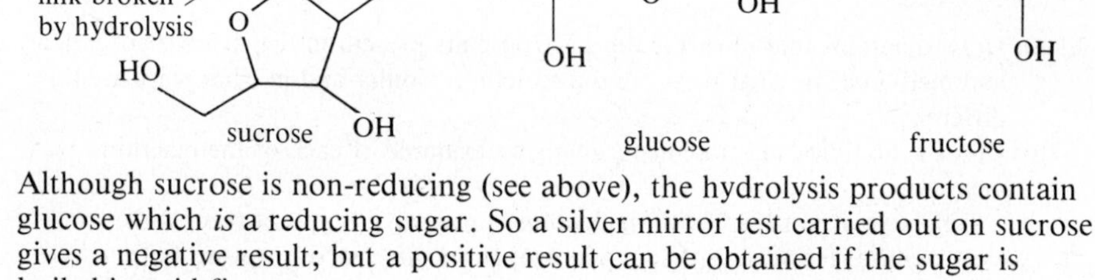

link broken by hydrolysis

sucrose $\xrightarrow[\text{boil}]{H_3O^+}$ glucose + fructose

Although sucrose is non-reducing (see above), the hydrolysis products contain glucose which *is* a reducing sugar. So a silver mirror test carried out on sucrose gives a negative result; but a positive result can be obtained if the sugar is boiled in acid first.

Glucose itself is manufactured by hydrolysing the polysaccharide starch.

$$C_x(H_2O)_y + (x-y)H_2O \xrightarrow[\text{heat under pressure}]{H_3O^+} \frac{x}{6}C_6H_{12}O_6$$

Questions

1. a) What is the characteristic functional group of i) a ketone, ii) an aldehyde?
   b) Give the structural formula of a typical ketone and that of a typical aldehyde. Give two reactions of the aldehyde that are also given by the ketone and two reactions *not* given by the ketone.
   c) Briefly account for the similarities and differences.

2. Discuss the difficulties that arise in most methods of preparing aldehydes. How are these usually overcome?
   Give *two* methods of preparing propanone from ethanol.

3. Give the structural formulas for the products of propanone's reaction with
   a) hydrogen cyanide, b) lithium tetrahydridoaluminate, c) hydroxylamine,
   d) bromine water in excess, e) iodine in aqueous alkali,
   f) 2,4 dinitrophenylhydrazine.

4. How can ethanal be converted into the following compounds? a) ethene, b)ethyl ethanoate, c) chloroform (trichloromethane), d) 3-hydroxybutanal, e) ethanoyl chloride.

5. How can propanone be converted into the following compounds? a) butan-2-ol, b) 1,2-dibromobutane, c) 2-hydroxymethylpropanonitrile.

6. a) Compare the reactivity of nucleophiles with i) propanone and ii) bromoethane.
   b) Compare the addition of H—X to a carbonyl group with the addition of H—X to an alkene group. Comment on the nature of X in each case.

7. a) The addition of hydrogen cyanide to a carbonyl compound is catalysed by cyanide ions. Explain this observation and account for the reversal of the reaction in the presence of hydroxide ions.
   b) Ethanol and ethanal both give positive results when treated with the 'iodoform test' reagents. Suggest a reason why ethanol reacts more slowly than ethanal. How could the rate of the reaction be increased without heating?
   c) A carbon-carbon bond is broken during the iodoform reaction. Comment on this and suggest a reason why the bond is so readily broken.

8. When dilute alkali is added drop by drop to a small sample of ethanal, the temperature rises. At first no other observation can be made, but gradually the solution darkens and a yellow-brown resin forms.
   Explain these observations by suggesting a likely mechanism for the reaction taking place.

9. a) Draw diagrams that illustrate the electron pairs present in the structure of ethene and methanal. In what ways are the structures similar and in what ways are they different?
   b) Discuss the following statement, giving an example of each of the reactions mentioned.
   Ethene does not react with nucleophiles; methanal does not react with electrophiles. However, ethanal and propanone are reactive toward both nucleophiles *and* electrophiles.

10. a) What is meant by the term 'reducing sugar'? Illustrate your answer with an example of a sugar that is 'reducing' and one that is not. Draw their structures.
    b) Give two tests that you would carry out to establish whether a particular sugar had 'reducing' properties.

Chapter 29

Acids

29.1 Substituted carbonic acid

Functional group

An organic acid has a molecular structure that is similar to the structure of carbonic acid, the weak acid resulting from carbon dioxide's reaction in water.

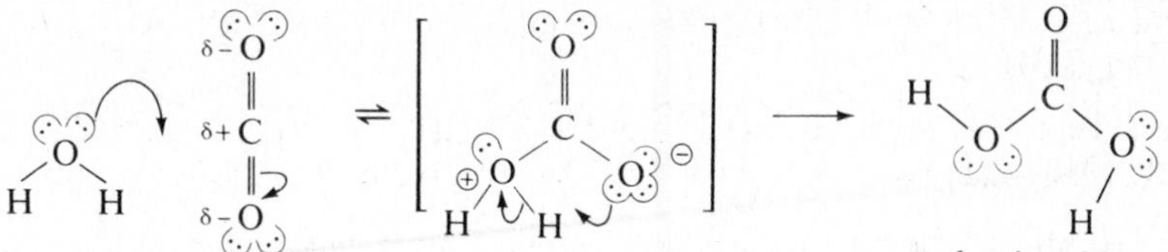

carbonic acid

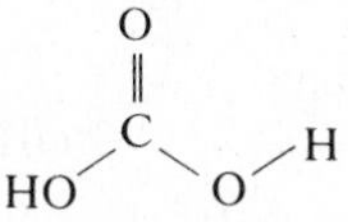

carbonic acid 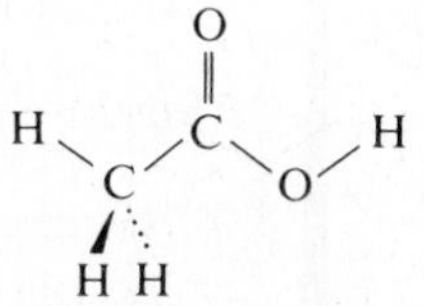   ethanoic acid 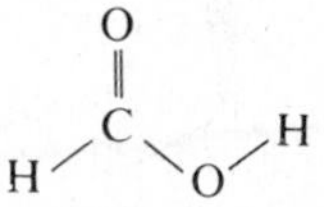   methanoic acid

The functional group of an acid is shown below:

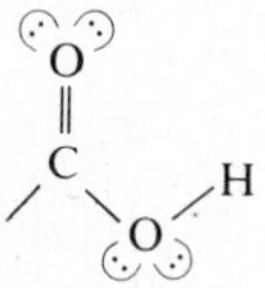

the carboxylic acid group

A carbon chain or ring is bonded to this group and the acid is named in the usual way by counting carbon atoms and adding the ending '-oic acid':

mono-acids		*di-acids*
H—COOH	methanoic acid	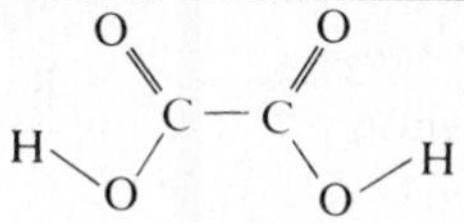 ethanedioic acid (oxalic acid)
Me—COOH	ethanoic acid	
Et—COOH	propanoic acid	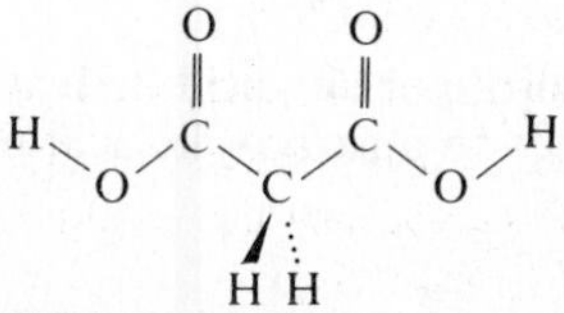 propanedioic acid (malonic acid)
Pr—COOH	butanoic acid	

The acid group is a *chain-ending* group: the chain is always numbered from the acid group backwards. The carbon atom of the functional group is taken as carbon atom number one:

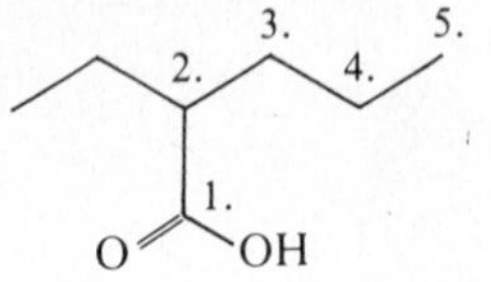

2-ethylpentanoic acid

Five is the longest carbon chain that also bears the acid group. A longer chain *does* exist, but it does not contain the acid group as a chain end.

Practise naming the following acids:

a) b) 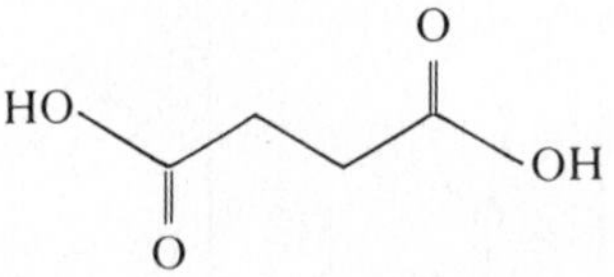 c)

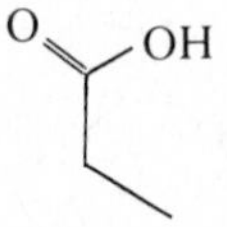

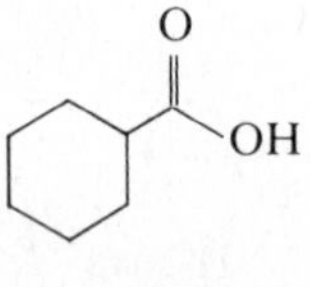

d) 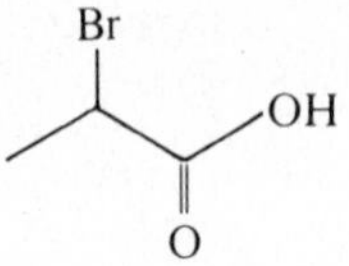 e) 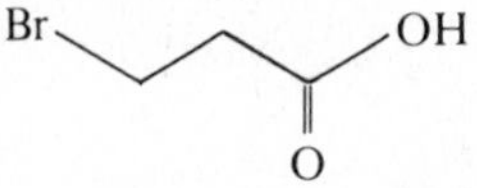 f)

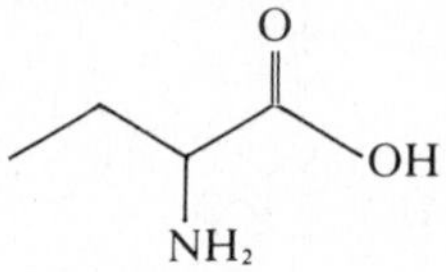

The solutions are given in the answers section of this book. Notice that d), e) and f) are substituted acids.

Methods of preparation

There are three main ways of preparing carboxylic acids:

1. Oxidation of an alcohol or an aldehyde

An alcohol or an aldehyde is readily oxidized to an acid by acidified potassium dichromate solution

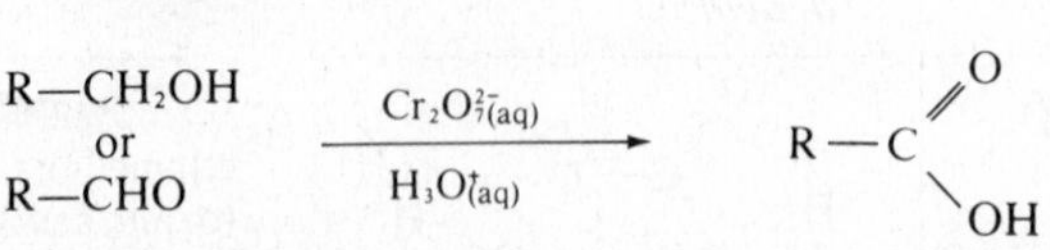

$$\begin{array}{c} R\text{—}CH_2OH \\ \text{or} \\ R\text{—}CHO \end{array} \xrightarrow[H_3O^+_{(aq)}]{Cr_2O_7^{2-}{}_{(aq)}} R\text{—}C \begin{array}{l} {=}O \\ \text{—}OH \end{array}$$

2. Hydrolysis of a nitrile or any acid derivative

A nitrile is hydrolysed to a carboxylic acid under either acidic or alkaline conditions

$$R{-}C{\equiv}N \xrightarrow[\text{reflux}]{OH^-_{(aq)}} R{-}C(=O){-}O^-_{(aq)} + NH_{3(g)}$$

$$R{-}C{\equiv}N \xrightarrow[\text{reflux}]{H_3O^+_{(aq)}} R{-}C(=O){-}O{-}H_{(aq)} + NH^+_{4(aq)}$$

Any acid derivative, e.g. ester, anhydride or acid chloride, can also be hydrolysed to give the parent acid.

3. From carbon dioxide with a Grignard reagent
A bromoalkane is converted to a carboxylic acid by reacting its Grignard derivative with carbon dioxide.

$$R{-}Br \underset{\text{dry ether}}{\overset{Mg_{(s)}}{\rightleftharpoons}} R{-}Mg{-}Br \xrightarrow[\text{2. } H_3O^+_{(aq)}]{\text{1. } CO_2} R{-}C(=O){-}OH$$

Grignard reagents are discussed on page 528.

29.2 Reactive sites

The functional group of an acid contains a carbonyl group and a hydroxide group. At first sight, the reactivity of an acid might be expected to combine the reactivity of a ketone or aldehyde with that of an alcohol.

carbonyl group → $\delta-$ O (=C) ; C $\delta+$; O $\delta-$; H $\delta+$ ← hydroxyl group

However, the two functional groups are so close together in the molecule that their reactivities are considerably modified. The electron deficiency of the 'carbonyl' carbon atom is reduced by the withdrawal of electron density from the 'hydroxyl' oxygen atom. This withdrawal has two effects on the hydroxyl group:

1. The lone pairs are less prominent and therefore the Lewis basicity of the group is reduced.
2. The bonding electrons between the oxygen and hydrogen atoms are attracted even more strongly towards the oxygen atom. The ability of the group to act as a proton donor is therefore more marked.

These effects can be illustrated using either valence-bond or orbital-overlap drawings. Both models suggest the likelihood of π-bonding between the carbon atom and *both* oxygen atoms.

Valence bond structures

The resonant hybrid of the above canonical forms can be shown as below

π-overlap drawing

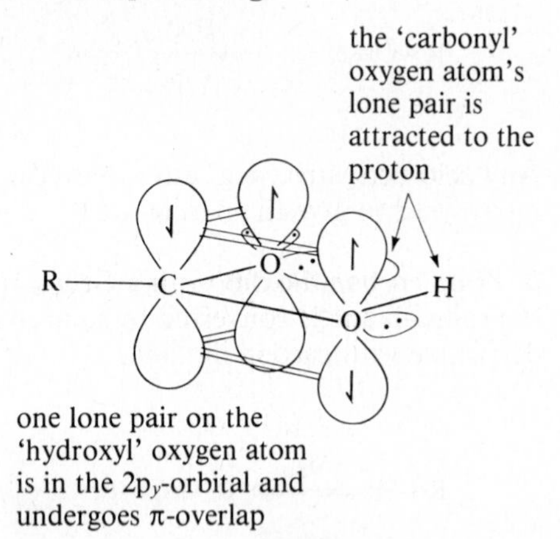
the 'carbonyl' oxygen atom's lone pair is attracted to the proton
R
C
O
H
one lone pair on the 'hydroxyl' oxygen atom is in the $2p_y$-orbital and undergoes π-overlap

The dominant reactive site of the acid group is the electron-deficient proton. Although the carbon atom is also electron deficient, its susceptibility to nucleophilic attack is very limited as a result of the acidic properties described. An attacking nucleophile is likely to be protonated before it can reach the carbon atom.

This deactivates the nucleophile and converts the acid molecule to an oxyanion that is stabilized by the effect of charge delocalization.

In summary, therefore, the reactivity of an acid is *not* simply the combined reactivity of a ketone and an alcohol, although there are *some* traces of the individual influence of the two groups which are described in the next section.

29.3 Reactions

With metals and inorganic bases

Most carboxylic acids are insoluble in water because of the hydrocarbon chains in their structures. However the short-chain acids are soluble enough to demonstrate the usual weak acid properties:

$$\underset{\text{acid}}{2CH_3COOH_{(aq)}} + \underset{\text{metal}}{Mg_{(s)}} \longrightarrow \underset{\text{salt solution}}{2\left[CH_3{-}C\begin{matrix}\nearrow O \\ \searrow O^{\ominus}\end{matrix}\right]_{(aq)}} + Mg^{2+}_{(aq)} + \underset{\text{hydrogen}}{H_{2(g)}}$$

$$\underset{\text{acid}}{2H\,COOH_{(aq)}} + \underset{\text{base}}{CaO_{(s)}} \longrightarrow \underset{\text{salt solution}}{2\left[H{-}C\begin{matrix}\nearrow O \\ \searrow O^{\ominus}\end{matrix}\right]_{(aq)}} + Ca^{2+}_{(aq)} + \underset{\text{water}}{H_2O_{(l)}}$$

The acidic nature of an insoluble carboxylic acid is best shown by using a soluble base:

Example 1 An insoluble carboxylic acid in sodium hydroxide solution:

1. suspension of acid in water: e.g. benzoic acid (C_6H_5COOH)
2. addition of sodium hydroxide dissolves the suspension
3. addition of sulphuric acid brings back the suspension

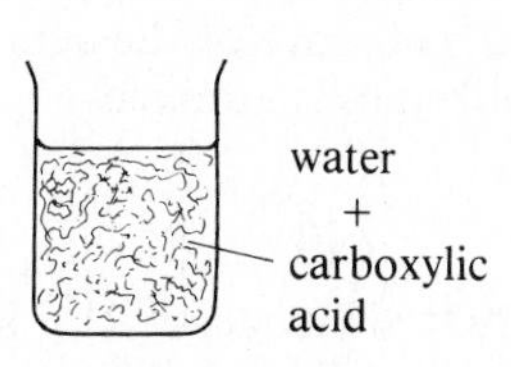

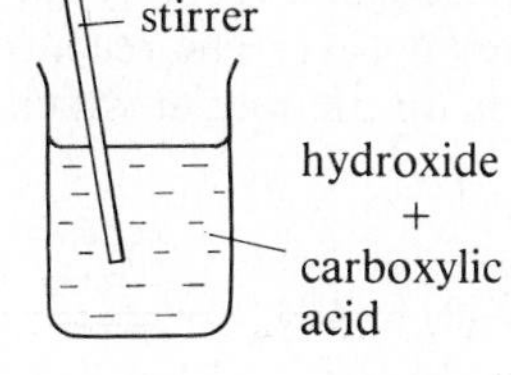

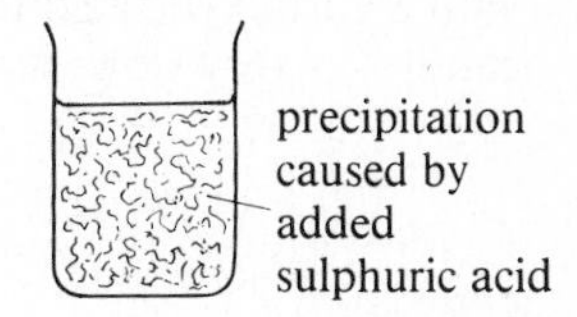

The acid dissolves in the hydroxide solution because the acid molecules protonate the hydroxide ions.

$$R{-}COOH_{(s)} + OH^-_{(aq)} \rightarrow R{-}COO^-_{(aq)} + H_2O$$

On adding sulphuric acid, the carboxylic acid anions are protonated by the hydroxonium ions and give back the insoluble acid.

$$R{-}COO^-_{(aq)} + H_3O^+_{(aq)} \rightarrow R{-}COOH_{(s)} + H_2O$$

Example 2 An insoluble carboxylic acid in sodium carbonate solution:
The acids are able to protonate carbonate and hydrogencarbonate ions, so producing evolution of carbon dioxide.

$$RCOOH_{(s)} + CO^{2-}_{3(aq)} \rightleftharpoons RCOO^-_{(aq)} + HCO^-_{3(aq)}$$

followed by

$$RCOOH_{(s)} + HCO^-_{3(aq)} \rightleftharpoons RCOO^-_{(aq)} + H_2CO_{3(aq)} \rightarrow H_2O_{(l)} + CO_{2(g)}$$

To test for the acidic character of a carboxylic acid, make up a suspension of the acid in water, add a little sodium carbonate solution, then look for the presence or absence of gas bubbles (CO_2).

With ammonia

As well as being a nucleophile, ammonia is also a base which reacts with a carboxylic acid to give an ammonium salt.

$$R{-}COOH + NH_3 \rightarrow R{-}COO^- \; {}^+NH_4$$

If an ammonium salt is heated strongly, it slowly decomposes. Water is driven off and an acid amide is produced:

$$Me{-}C(=O)O^{\ominus} \; {}^{\oplus}NH_4 \xrightarrow{\text{heat}} Me{-}C(=O)NH_{2(g)} + H_2O_{(g)}$$

ammonium ethanoate → ethanamide

With alcohols

In the presence of a concentrated acid (usually sulphuric), an alcohol reacts with a carboxylic acid to give an ester. The reaction in the absence of the acid catalyst is very slow: it speeds up the rate at which equilibrium is attained.

$$R_1{-}C(=O){-}O{-}H + R_2{-}OH \underset{}{\overset{\text{conc. } H_2SO_4}{\rightleftharpoons}} R_1{-}C(=O){-}O{-}R_2 + H{-}OH$$

This method is *not* usually used for preparing a sample of an ester in the laboratory. For the same reason, the preparation of an amide from ammonia and an acid is also not the best method. It is better to use an acid chloride instead of an acid in both cases (see page 452).

The mechanism of esterification is likely to proceed via the nucleophilic attack of an alcohol molecule on the carbonyl group of an acid molecule. In the presence of a concentrated acid catalyst, some of the alcohol molecules become protonated, as do some of the hydroxyl groups of carboxylic acid molecules. Esterification is favoured by the collision of an unprotonated alcohol molecule and a protonated carboxylic acid molecule.

Protonation

$$R_1-C(=O)-\ddot{O}-H \quad + \quad H-OSO_2OH \rightleftharpoons R_1-C(=O)-\overset{\oplus}{O}H_2 \quad {}^{\ominus}OSO_2OH$$

a protonated carboxylic acid molecule has a water molecule as a leaving group; compare page 410

Nucleophilic Attack

$$R_1-\overset{\delta+}{C}(=\overset{\delta-}{O})-\overset{\oplus}{O}H_2 + R_2-\ddot{O}-H \rightleftharpoons \left[R_1-C(O^{\ominus})(\overset{\oplus}{O}H_2)(\overset{\oplus}{O}(H)R_2) \right] \longrightarrow R_1-C(=O)-O-R_2 + H_3O^{\oplus}$$

the leaving water molecule is protonated as it leaves

Evidence that supports this mechanism has been provided by the use of a 'labelled' sample of alcohol. By synthesizing some alcohol that contains the oxygen isotope ^{18}O, the products can be separated by fractional distillation and then their R.M.M.'s found by mass spectrometry. When this is carried out in the ester-forming reaction, the ester product is found to be labelled, and not the water:

$$\underset{\text{ethanoic acid}}{Me-C(=O)-OH} + \underset{\text{labelled methanol}}{Me-\overset{*}{O}-H} \rightleftharpoons \underset{\text{labelled methyl ethanoate}}{Me-C(=O)-\overset{*}{O}-Me} + H_2O$$

This shows that the activated complex must contain the alcohol oxygen bonded to the carbonyl carbon of the acid. It was previously thought possible that the acid molecule might act as a nucleophile and attack the electron-deficient carbon atom of a protonated alcohol molecule:

$$\underset{\text{carboxylic acid molecule as nucleophile}}{Me-C(=O)-\ddot{O}-H} + \underset{\text{protonated alcohol molecule}}{H_3C-\overset{\oplus}{O}H_2} \rightleftharpoons \left[Me-C(=O)-\overset{\oplus}{O}(H)\text{----}CH_3\text{----}\overset{\oplus}{O}H_2 \right]^{\ominus}$$

labelled water leaving

As can be seen, this would result in the production of isotopically labelled water. There is very little labelled water in the products which suggests that the correct activated complex is the one shown below:

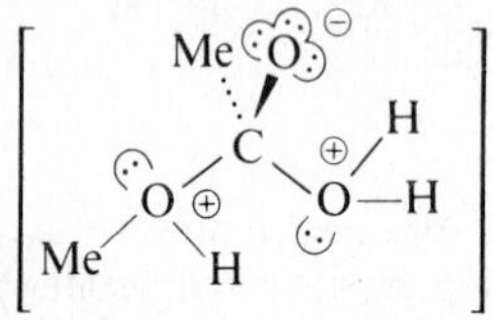

and *not* the 5-membered complex deriving from nucleophilic attack on the alcohol molecule.

With reducing agents

An acid is reduced to an alcohol by lithium tetrahydridoaluminate. The reaction is carried out in a solvent of dry ether (ethoxyethane) and the products in this solvent are hydrolysed to release the alcohol:

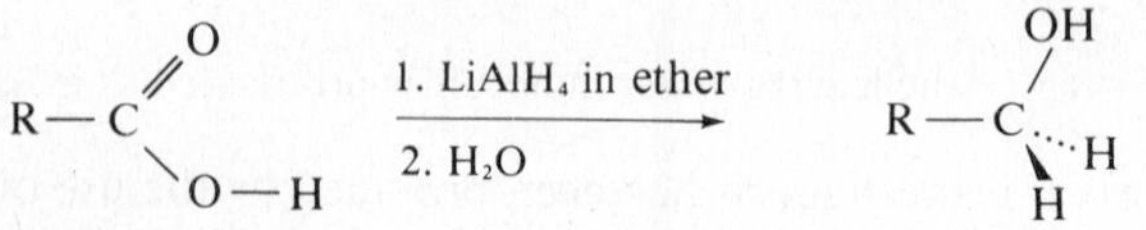

$$R-COOH \xrightarrow[2.\ H_2O]{1.\ LiAlH_4\ \text{in ether}} R-CH_2OH$$

The basis for the reactivity of lithium tetrahydridoaluminate (lithium aluminium hydride) is outlined on page 440. The reagent is a source of hydride ions which are powerful nucleophiles:

$$AlH_{4(ether)}^{\ominus} \rightleftharpoons AlH_{3(ether)} + H_{(ether)}^{\ominus}$$

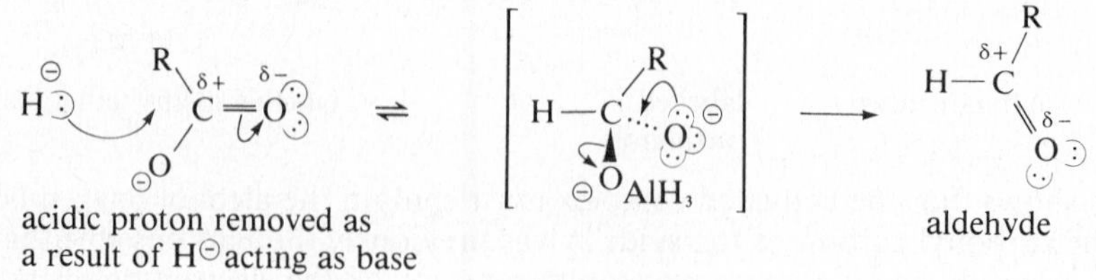

acidic proton removed as a result of $H^{\ominus}$ acting as base

aldehyde

The aldehyde is then reduced

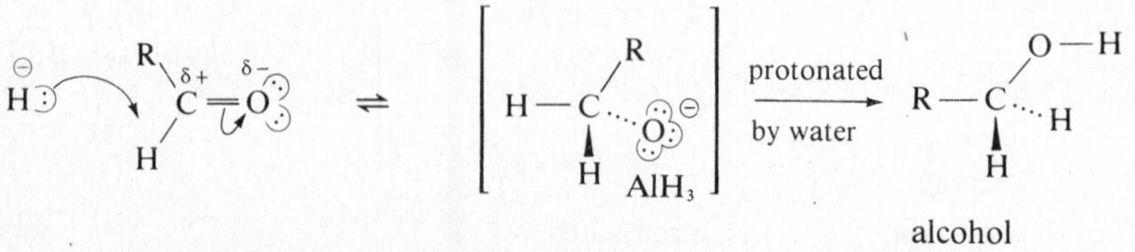

protonated by water

alcohol

It is not possible to isolate an aldehyde from this reaction mixture. Low yields of aldehyde can be produced from an acid by using the reducing agent lithium metal in a solvent of ethylamine. The mechanism is complex and involves single electron transfer

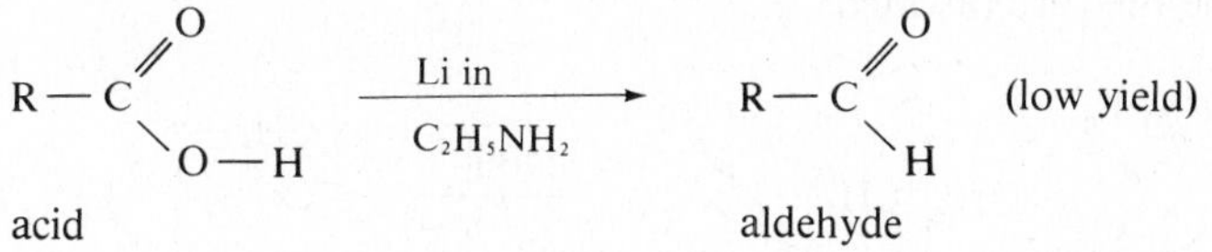

$$R-C(=O)-O-H \xrightarrow[C_2H_5NH_2]{\text{Li in}} R-C(=O)-H \quad \text{(low yield)}$$

acid → aldehyde

With chlorinating agents

Dichlorosulphur (IV) oxide and phosphorus (V) chloride both react with an acid to produce an acid chloride. The reaction is similar to the chlorination of an alcohol (page 415):

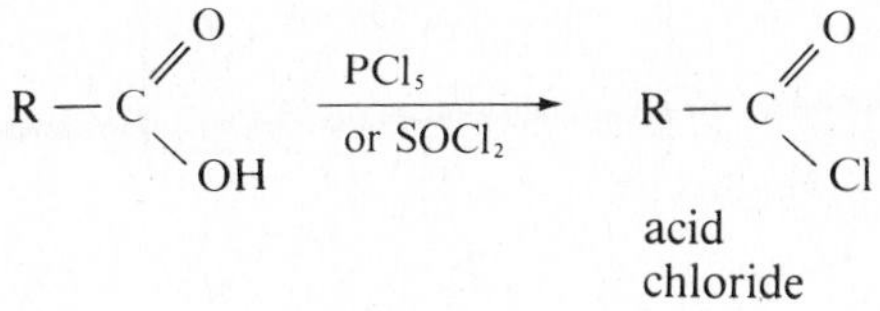

$$R-C(=O)-OH \xrightarrow[\text{or } SOCl_2]{PCl_5} R-C(=O)-Cl$$

acid chloride

Example

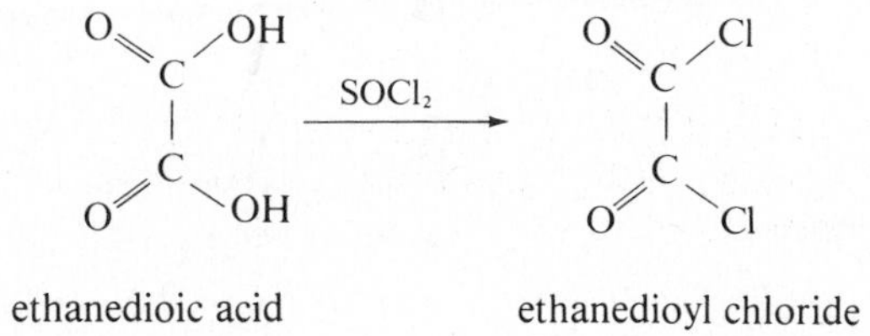

$$HOOC-COOH \xrightarrow{SOCl_2} ClOC-COCl$$

ethanedioic acid → ethanedioyl chloride

With dehydrating agents

A molecule of water can be eliminated between two molecules of a carboxylic acid under the action of a dehydrating agent. The best agent is phosphorus (V) oxide, P_4O_{10}. When an acid is distilled in the presence of this oxide, an acid anhydride is produced.

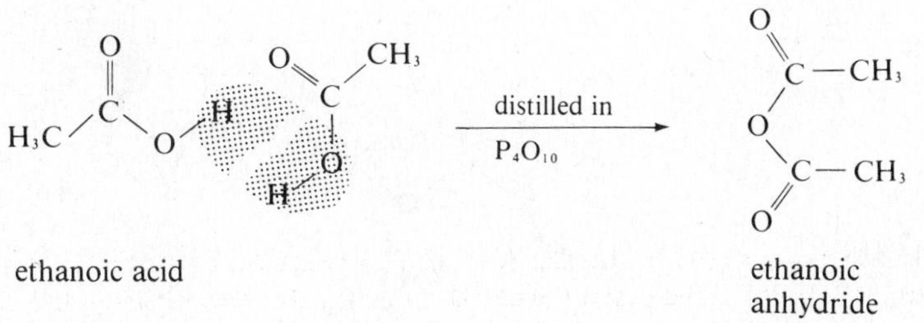

$$H_3C-C(=O)-O-H \;+\; H-O-C(=O)-CH_3 \xrightarrow[P_4O_{10}]{\text{distilled in}} CH_3-C(=O)-O-C(=O)-CH_3$$

ethanoic acid → ethanoic anhydride

A more satisfactory method of preparing an acid anhydride uses the sodium salt of the acid rather than the acid itself. The sodium salt is added to the acid chloride and the mixture distilled:

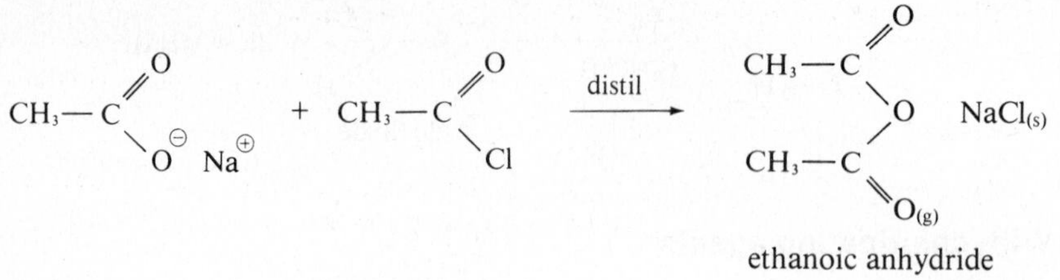

$$CH_3COO^{\ominus}Na^{\oplus} + CH_3COCl \xrightarrow{\text{distil}} (CH_3CO)_2O_{(g)} + NaCl_{(s)}$$

ethanoic anhydride

The ethanoate ions act as nucleophiles.

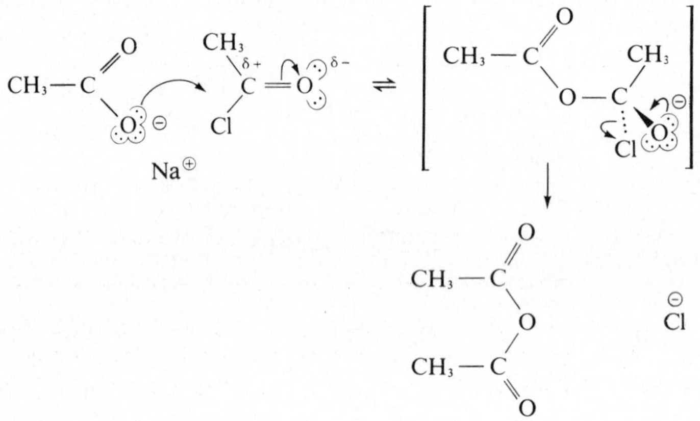

$$CH_3COO^{\ominus}\;Na^{\oplus} + CH_3C^{\delta+}(Cl){=}O^{\delta-} \rightleftharpoons \left[CH_3C(O)\text{–}O\text{–}C(CH_3)(Cl)O^{\ominus}\right] \longrightarrow CH_3C(O)\text{–}O\text{–}C(O)CH_3 + Cl^{\ominus}$$

29.4 Substituted acids

Structural characteristics

The substituted acids contain the usual acid functional group, but in the carbon chain bonded to the acid group, there are one or more groups substituted for hydrogen atoms. The most important substituents are:

1. halogen atoms
2. hydroxide groups
3. amine groups

for example

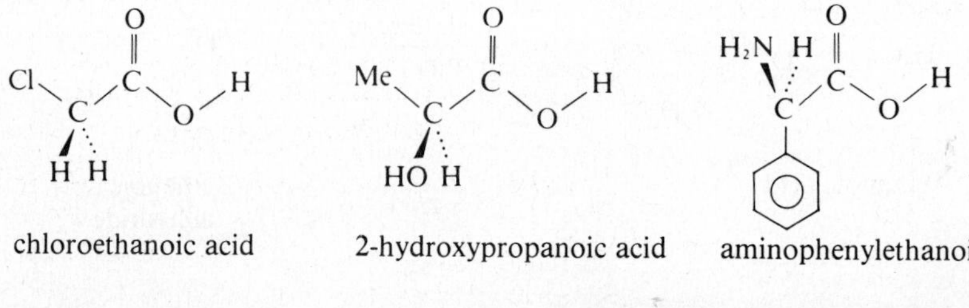

$ClCH_2COOH$ chloroethanoic acid

$CH_3CH(OH)COOH$ 2-hydroxypropanoic acid

$C_6H_5CH(NH_2)COOH$ aminophenylethanoic acid

All the above examples have substituents in the 'α-position'. In the discussion of keto-enol tautomerism on page 434, an α-carbon atom is defined as the atom next to a specified funtional group. In this case, it is the acid group. Substituents on this carbon atom are *α-substituents.*

α-amino acids	*β-amino acids*
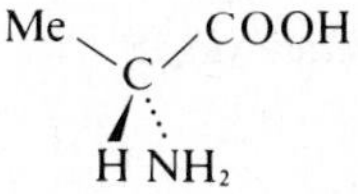 Me, COOH, H, NH_2 on C 2-aminopropanoic acid	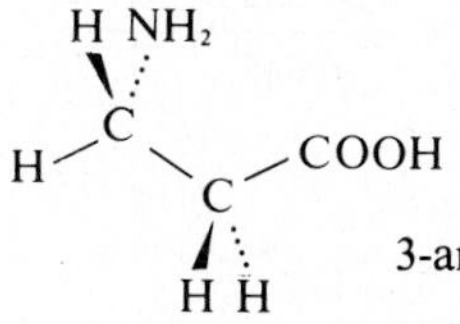 H, NH_2, H on C; C bearing COOH, H, H 3-aminopropanoic acid
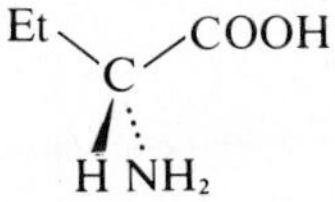 Et, COOH, H, NH_2 on C 2-aminobutanoic acid	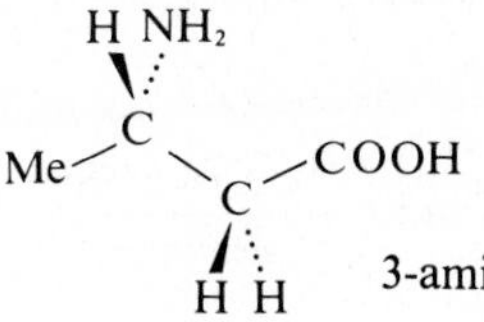 H, NH_2, Me on C; C bearing COOH, H, H 3-aminobutanoic acid

The α-substituted acids are the most common ones and we shall confine our attention to these. The reaction scheme below shows a possible route for synthesizing one each of the three types, starting from methanal and hydrogen cyanide.

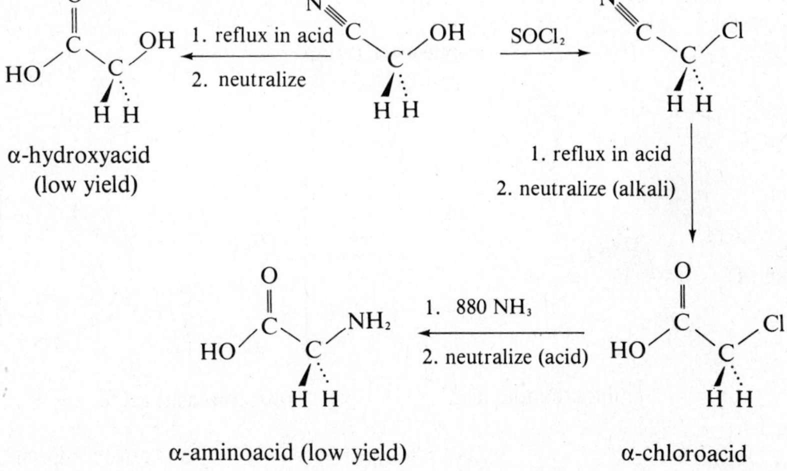

H, H, C=O

↓ HCN

N≡C, OH, H, H on C

← 1. reflux in acid 2. neutralize

HO, C=O, OH, H, H on C

α-hydroxyacid (low yield)

→ $SOCl_2$

N≡C, Cl, H, H on C

↓ 1. reflux in acid 2. neutralize (alkali)

HO, C=O, Cl, H, H on C

α-chloroacid

← 1. 880 NH_3 2. neutralize (acid)

HO, C=O, NH_2, H, H on C

α-aminoacid (low yield)

There are a number of better methods used for synthesizing these compounds. α-aminoacids, in particular, are important intermediates in the synthesis of proteins which contain α-aminoacids linked together.

α-bromoacids

Treat an acid with red phosphorus and bromine. The 'Hell-Volhard-Zelinsky' procedure:

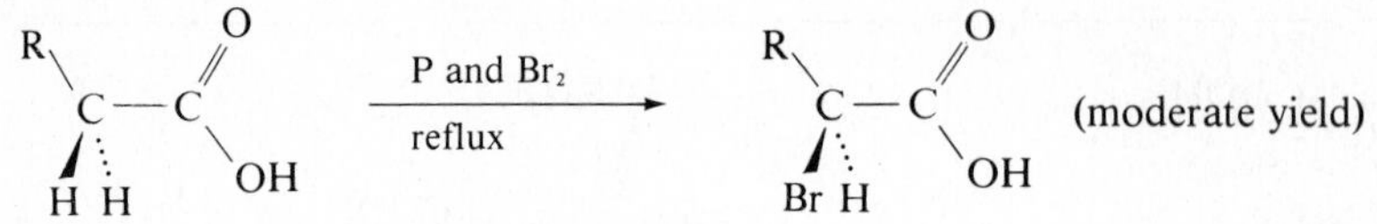

$$RCH_2COOH \xrightarrow[\text{reflux}]{\text{P and Br}_2} RCHBrCOOH$$

(moderate yield)

α-hydroxyacids

Hydrolyse the bromoacids (above) with alkali

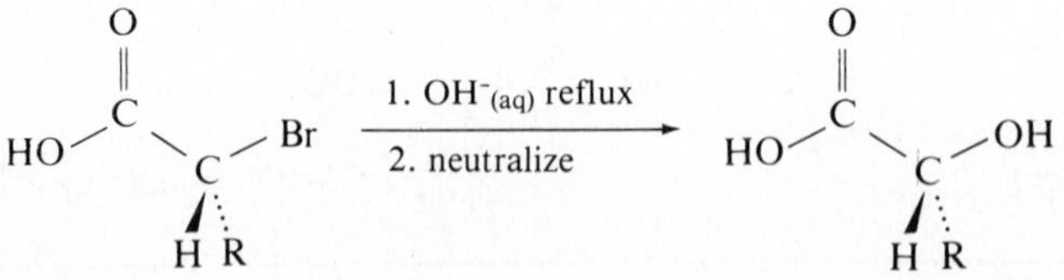

$$HOOC{-}CHRBr \xrightarrow[\text{2. neutralize}]{\text{1. OH}^-_{(aq)}\text{ reflux}} HOOC{-}CHR{-}OH$$

α-aminoacids

The mono-azide of a suitable di-acid is decomposed and hydrolysed to an α-aminoacid (Curtius method). For example, aminoethanoic acid can be prepared from diethylpropanedioate (a di-ester) as follows:

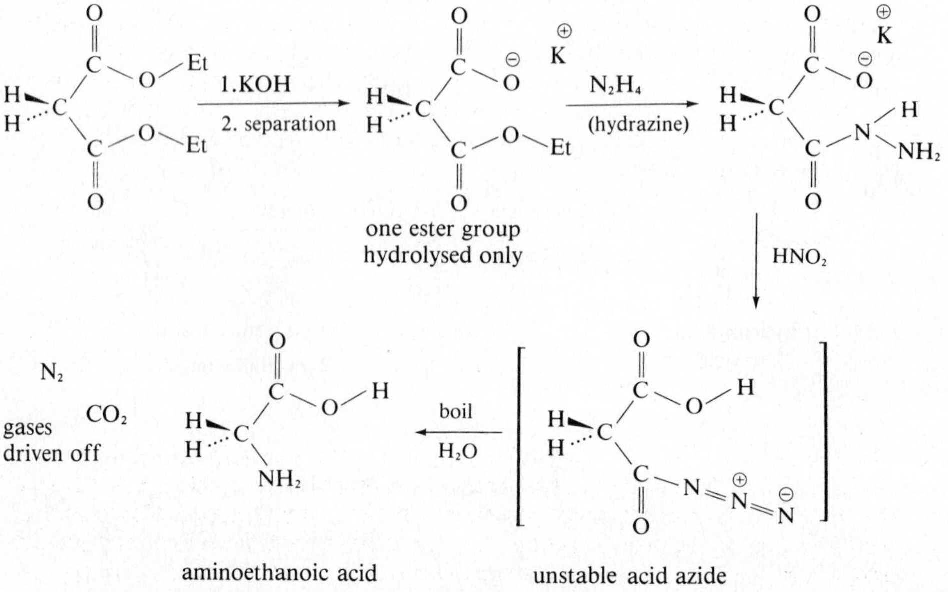

$$CH_2(COOEt)_2 \xrightarrow[\text{2. separation}]{\text{1.KOH}} CH_2(COO^-K^+)(COOEt) \xrightarrow[\text{(hydrazine)}]{N_2H_4} CH_2(COO^-K^+)(CONHNH_2)$$

one ester group hydrolysed only

$$\xrightarrow{HNO_2} [CH_2(COOH)(CO{-}N{=}\overset{\oplus}{N}{=}\overset{\ominus}{N})] \xrightarrow[H_2O]{\text{boil}} H_2C(NH_2)COOH$$

unstable acid azide

aminoethanoic acid

N_2 CO_2 gases driven off

The saturated carbon atom of the azide migrates to the electron-deficient nitrogen atom. The mechanism resembles those described on page 479.

Comparative properties

There are two major differences between a substituted acid and its parent acid: its acidity and its tendency to condense.

1. Acidity
The acid equilibrium is

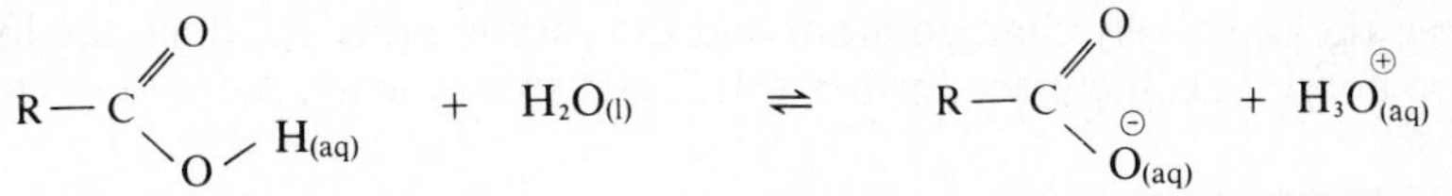

$$R{-}C(=O){-}O{-}H_{(aq)} + H_2O_{(l)} \rightleftharpoons R{-}C(=O){-}O^{\ominus}_{(aq)} + H_3O^{\oplus}_{(aq)}$$

$$K_a = \frac{[RCOO^-][H_3O^+]}{[RCOOH]}$$

An equilibrium constant depends on the free energy difference between the reactants and the products, for example:

$$\Delta G^{\ominus} = -RT\log_e K_p$$

As the thermodynamic stability of the products increases, so the free energy change for the reaction decreases (becomes a smaller positive value or a larger negative one). This leads to an increase in the value of the equilibrium constant. In comparing a set of different equilibrium constants, it is therefore fair to inspect the relative thermodynamic stabilities of their products. In the case of the acid reaction above, we shall therefore consider the relative stabilities of the anions.

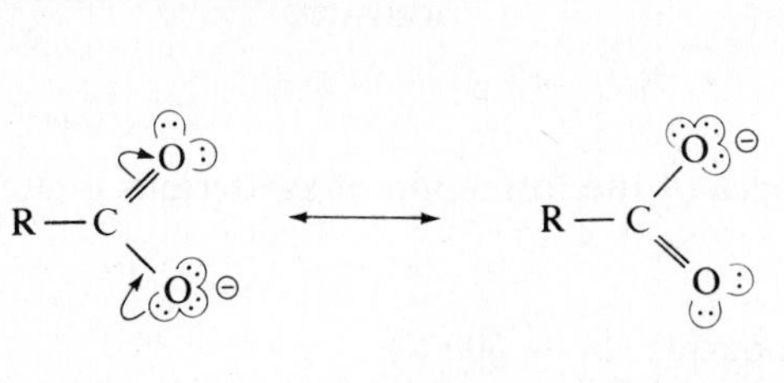

$$R{-}C(=O){-}O^{\ominus} \longleftrightarrow R{-}C(-O^{\ominus})=O$$

The identity of R— affects the ability of the structure to delocalize the anion charge further. The more the charge can be delocalized (or spread out) through the structure, the more thermodynamically stable the ion becomes. An empirical method of assessing delocalization uses the ideas of resonance and tautomerism and sometimes the relative electronegativities of the substituents:

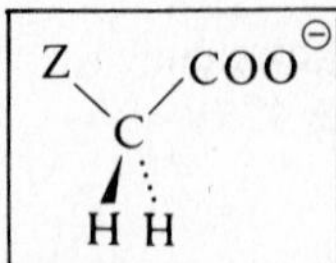 $ZCH_2COO^{\ominus}$	Z = H	Z = F	Z = Cl	Z = Br	Z = I	Z = OH
pK_a of the acid	4.75	2.66	2.86	2.86	3.12	3.83

The halogenoacids are stronger acids because the halogen atom is more electronegative than the hydrogen atom it has replaced. The negative charge of the anion is drawn towards the halogen atom through the σ-skeleton of the structure. This is a 'secondary' inductive effect because there is a σ-bond between the negatively charged atom and the carbon atom to which the halogen atom is bonded. If there are further α-halogen substituents, the acid strength increases more:

CCl_2COOH pK_a = 1.30
CCl_3COOH pK_a = 0.65

The acidity of the aminoacids is rather more of a complex issue. The amine group is capable of acting as a proton acceptor, and therefore the pure form of an aminoacid is likely to contain a proportion of *zwitterions* (molecular structures that possess oppositely charged ends):

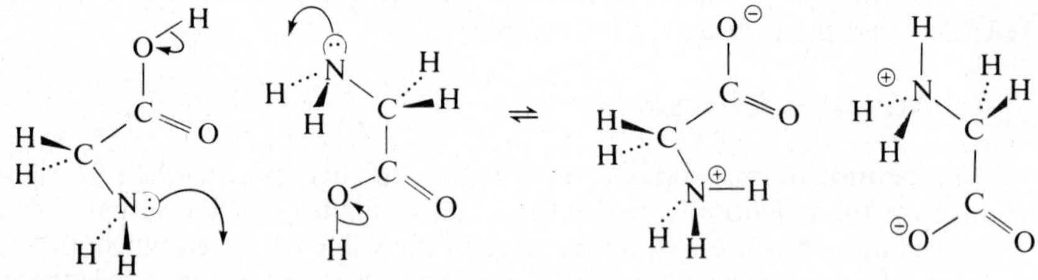

$$2NH_2.CH_2COOH \rightleftharpoons 2\overset{\oplus}{N}H_3.CH_2.CO\overset{\ominus}{O}$$

molecules ⇌ zwitterions

The evidence that supports the idea of the formation of zwitterions is supplied by the following data:

1. The high m.p.'s of α-aminoacids (> ~ 500 K)
2. The low solubility of α-aminoacids in non-polar solvents, but high solubility in polar solvents.
3. The very large dipole moments of α-aminoacids.

In aqueous solution, there is an equilibrium between the three forms shown below.

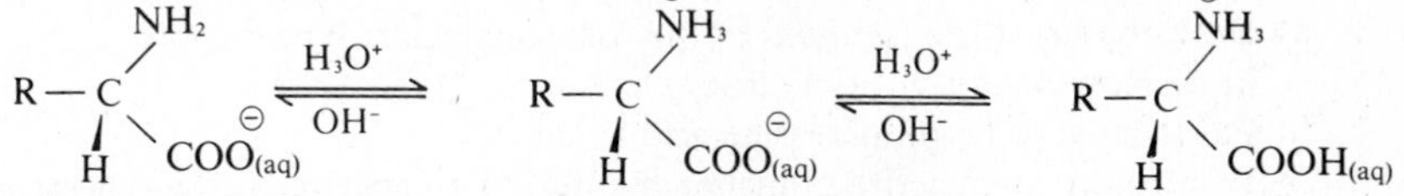

$$R-C(H)(NH_2)-COO^{\ominus}_{(aq)} \underset{OH^-}{\overset{H_3O^+}{\rightleftharpoons}} R-C(H)(\overset{\oplus}{N}H_3)-COO^{\ominus}_{(aq)} \underset{OH^-}{\overset{H_3O^+}{\rightleftharpoons}} R-C(H)(\overset{\oplus}{N}H_3)-COOH_{(aq)}$$

deprotonated form — 'neutral' form — protonated form

The position of the equilibrium is dependent on the pH of the solution. At low pH, an aminoacid could be collected as a cationic particle in an electrolysis cell. At high pH it could be collected as an anionic particle, and at a pH specific to the particular aminoacid, there would be no movement to either cathode or anode. Electrophoresis is a separational technique which is based on this principle.

2. Condensation

Aminoacids and hydroxyacids tend to condense. This is because they possess both a nucleophilic group and a group able to be attacked by a nucleophile in the same structure:

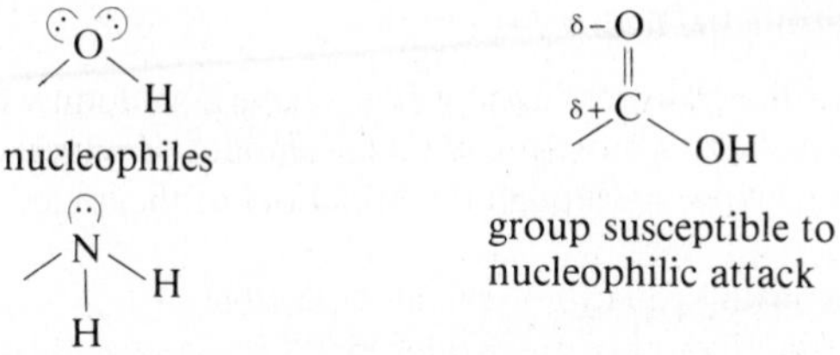

nucleophiles: $-\ddot{O}-H$, $-\ddot{N}H_2$

group susceptible to nucleophilic attack: $\overset{\delta+}{C}(=\overset{\delta-}{O})-OH$

The products of these condensations are giant molecular compounds whose chemistry is described in chapter 37. An outline of a typical condensation is shown below.

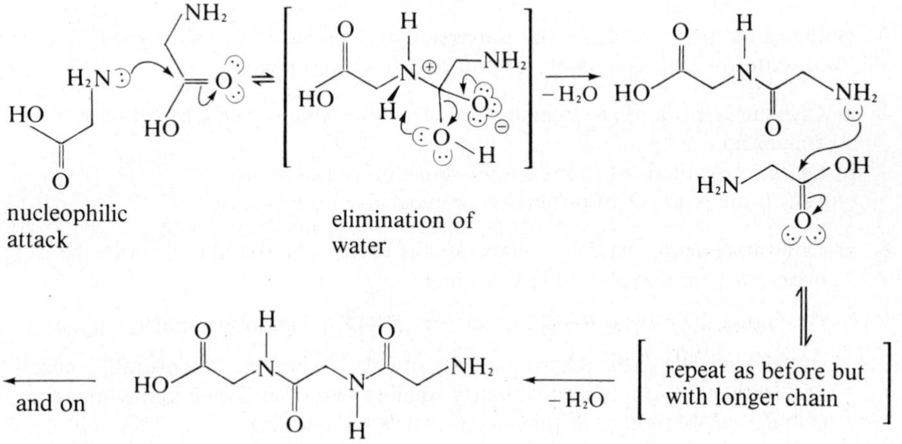

$$HOOC-CH_2-NH_2 + H_2N-CH_2-COOH \rightleftharpoons [HOOC-CH_2-\overset{\oplus}{N}H_2-C(O^{\ominus})(OH)-CH_2-NH_2] \xrightarrow{-H_2O} HOOC-CH_2-NH-CO-CH_2-NH_2$$

nucleophilic attack — elimination of water

$$HOOC-CH_2-NH-CO-CH_2-NH_2 + H_2N-CH_2-COOH \rightleftharpoons [\text{repeat as before but with longer chain}] \xrightarrow{-H_2O} HOOC-CH_2-NH-CO-CH_2-NH-CO-CH_2-NH_2 \longrightarrow \text{and on}$$

Questions

1. a) Draw the structural formulas of the following acids: i) propanoic ii) trichloroethanoic iii) hydroxyethanoic.
   b) Put them in order of increasing acid strength.
   c) What is the characteristic functional group of an acid? Give the general formula for an acid and an equation for the likely interaction between an acid molecule and a water molecule.

2. 'Organic acids contain a carbonyl and a hydroxyl group'. Criticize this statement by:
   a) giving *two* typical carbonyl reactions *not* given by acids
   b) giving *one* typical hydroxyl reaction *not* given by acids
   c) giving *any* reaction in which an acid does appear to act as either a typical carbonyl or a typical hydroxyl compound.

3. 'An acid can be prepared either as a result of oxidation or of hydrolysis'.
   a) Draw the structural formula of a compound that is readily oxidized to ethanoic acid.
   b) Draw the structural formula of a compound that is readily hydrolysed to ethanoic acid.
   c) Draw a mechanism to illustrate the hydrolysis reaction.

4. Z is a white crystalline solid. It dissolves readily in water to give a solution whose pH value lies between 9 and 10. Addition of an equimolar amount of mineral acid is unable to cause the pH to drop below 4 although the initial pH of the mineral acid is below 1.
   a) Explain these observations, giving equations wherever possible.
   b) Comment on the nature of the final mixture produced.
   c) Given that pure Z rapidly evolves carbon monoxide when added to concentrated sulphuric acid, suggest a possible structural formula for Z.

5. Draw the structural formula of the products of the reactions of propanoic acid with: a) phosphorus (V) oxide (distil) b) methanol c) aqueous ammonia d) dichlorosulphur (IV) oxide e) lithium tetrahydridoaluminate.

6. Outline a reaction scheme for the conversion of ethanoic acid to a) ethene b) chloroform c) propan-1-ol d) propan-2-ol e) propanone.

7. a) Give the formula of an 'α-aminoacid'. Describe what is meant by the term 'α-aminoacid'.
   b) Outline a synthesis of your example starting from ethanol.
   c) Why is the synthesis of different α-aminoacids of importance?

8. a) Draw an enthalpy diagram to illustrate the energy terms used to describe the dissociation of a weak acid HA in water.
   b) Given that the process $R{-}\ddot{\underset{\cdot\cdot}{O}}\cdot_{(g)} + e^- \rightarrow R{-}\ddot{\underset{\cdot\cdot}{O}}:^-_{(g)}$ has approximately the same enthalpy change as the electron affinity of chlorine and that the hydration energy of a methanoate ion is approximately equal to that of an iodide ion, estimate the enthalpy of solution of methanoic acid. (Use data-book.)

9. Discuss the factors that influence the strength of organic acids.

10. Some isotopically-labelled acid was prepared from deuterium (D = ^2_1H) with the intention of using it to provide evidence for the mechanism of esterification. Its structure is shown below:

$$CH_3-C(=O)-O-D$$

a) Give *two* reasons why an aqueous solution of this acid would be useless for the purpose required.
b) Suggest a better choice for a labelled acid giving the reasons for your choice.

Chapter 30
Acid derivatives

30.1 Acid chlorides and anhydrides

Derivatives

Functional group

All the acid derivatives have the structural form shown below.

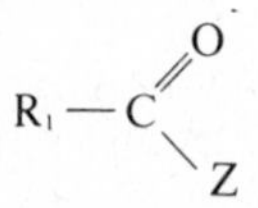

$$R_1-C(=O)-Z$$

$-Z$	$-Cl$	$-O-C(=O)-R_2$	$-O-R_2$	$-NH_2$	$-N(H)-R_2$
name of derivative	acid chloride	acid anhydride	ester	amide	N-substituted amide

Z takes the place of the hydroxyl group in the acid functional group. It is important to distinguish the type of substituted acid described on page 458 from an acid derivative. For example:

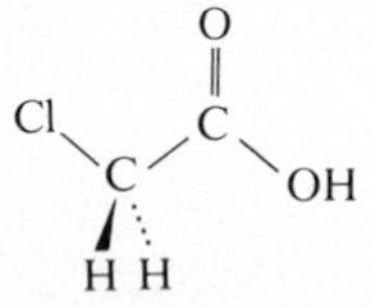

$Cl-CH_2-C(=O)-OH$ and $H-CH_2-C(=O)-Cl$

chloroethanoic acid   ethanoyl chloride

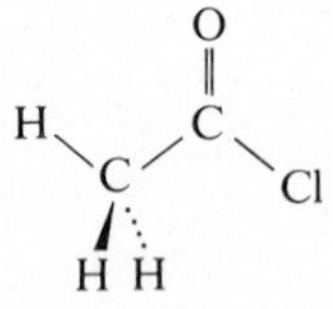

$Me-CH(NH_2)-C(=O)-OH$ and $Me-CH_2-C(=O)-NH_2$

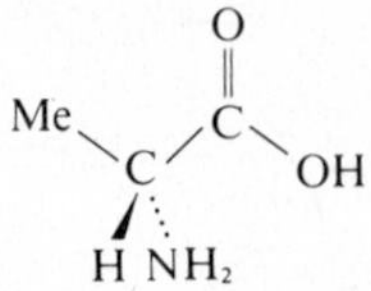

2-aminopropanoic acid   propanamide

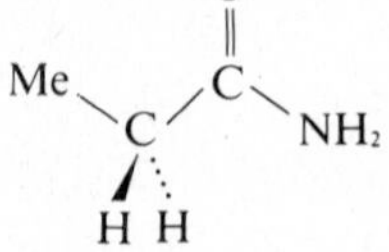

The first member of each pair is a substituted acid whereas the second member is an acid derivative in each case.

In the case of an acid chloride, the hydroxyl group is replaced by a chlorine atom.

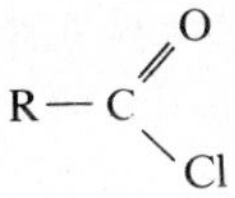

$$R-C\genfrac{}{}{0pt}{}{\nearrow O}{\searrow Cl} \quad \text{acid chloride}$$

It is named by using the ending -oyl chloride instead of -oic acid.

Acid anhydrides contain the carboxyl group in place of the hydroxyl group:

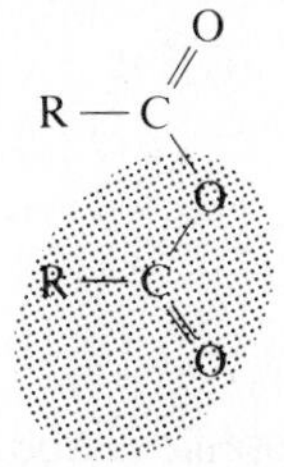

$$\begin{array}{l} R-C(=O) \\ \quad \backslash O \\ \quad / \\ R-C(=O) \end{array} \quad \text{acid anhydride}$$

The word anhydride is used instead of the word acid in the naming of these compounds. It is also possible to have a 'mixed' anhydride if the two carbon chains are not the same.

Methods of preparation

Acid chlorides are made by reacting an acid with a chlorinating agent

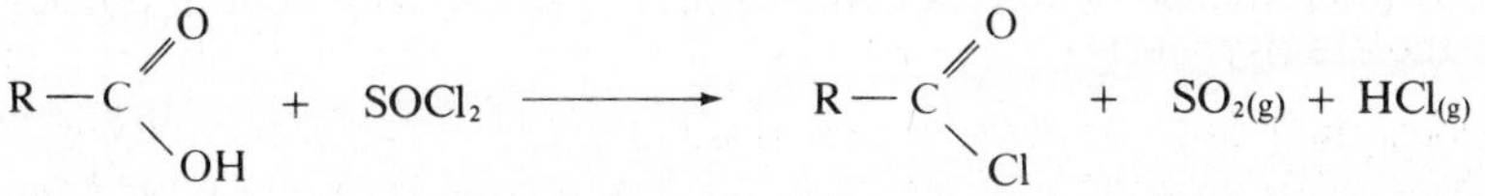

$$R-C(=O)OH + SOCl_2 \longrightarrow R-C(=O)Cl + SO_{2(g)} + HCl_{(g)}$$

An acid anhydride is prepared either by dehydrating the acid or by treating its sodium salt with an acid chloride:

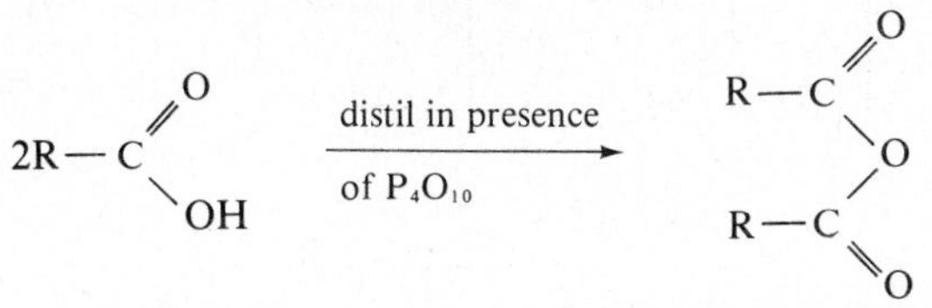

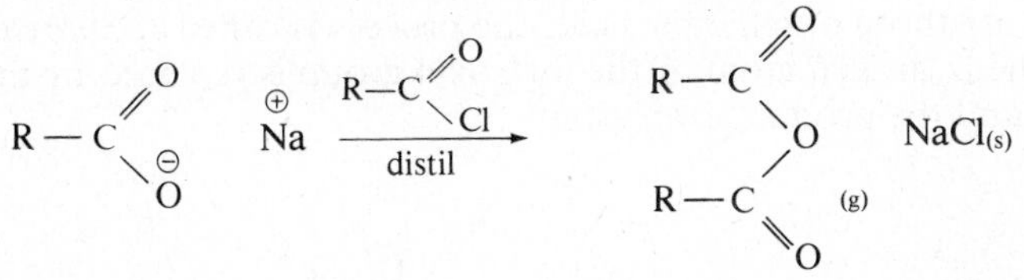

$$2R-C(=O)OH \xrightarrow[\text{of } P_4O_{10}]{\text{distil in presence}} (R-C(=O))_2O$$

$$R-C(=O)O^{\ominus} \; Na^{\oplus} \xrightarrow[\text{distil}]{R-C(=O)Cl} (R-C(=O))_2O_{(g)} \quad NaCl_{(s)}$$

Reactive sites

Acid chlorides and anhydrides are among the most reactive of organic compounds. Their functional group contains a carbonyl group bonded to a good leaving-group. This makes nucleophilic attack on the carbonyl successful and rapid:

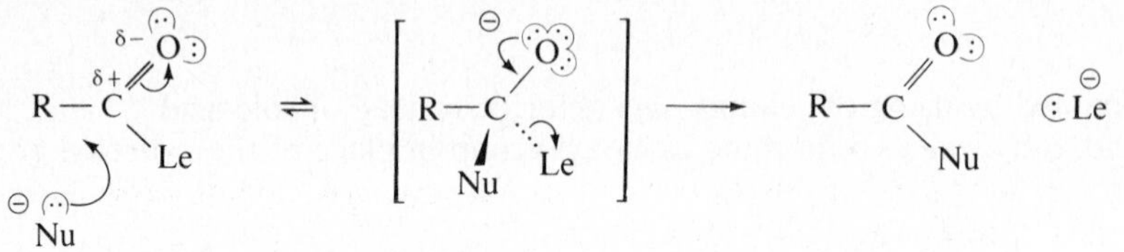

$$R{-}C(=O^{\delta-})^{\delta+}{-}Le + Nu^{-} \rightleftharpoons \left[R{-}C(O^{-})(Nu){-}Le \right] \longrightarrow R{-}C(=O){-}Nu + Le^{-}$$

where $-Le = -Cl$ in acid chlorides

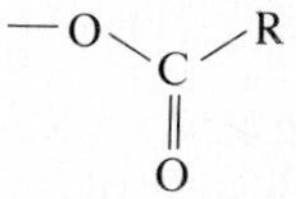

$= -O{-}C(=O){-}R$ in anhydrides

The mechanism above is called addition-elimination: a nucleophile is added as the leaving-group is eliminated.

Reactions

With water and alcohols

Acid chlorides and anhydrides are rapidly and violently hydrolysed by water to give the parent acids. These derivatives must therefore be kept in bottles sealed from the atmosphere:

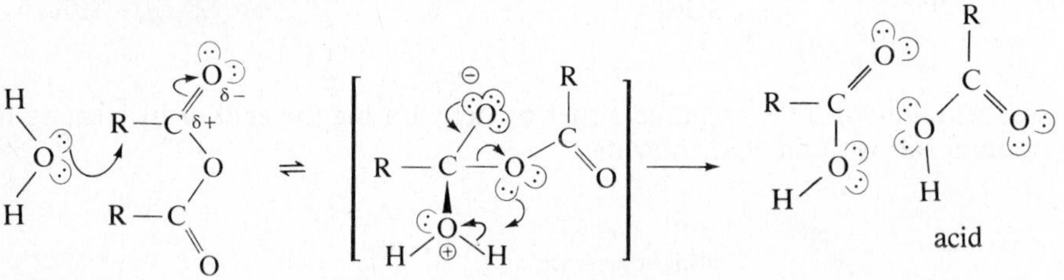

$$H_2O + R{-}C(=O^{\delta-})^{\delta+}{-}O{-}C(=O){-}R \rightleftharpoons \left[R{-}C(O^{-})(O^{+}H_2){-}O{-}C(=O){-}R \right] \longrightarrow R{-}C(=O){-}OH + R{-}C(=O){-}OH$$

acid

In a similar way, acid chlorides or anhydrides react with alcohols. When the acid derivatives are those of ethanoic acid, the process is called *ethanoylation* (acetylation). The hydrogen atom of the hydroxyl group is replaced by the ethanoyl group and the product is an ester:

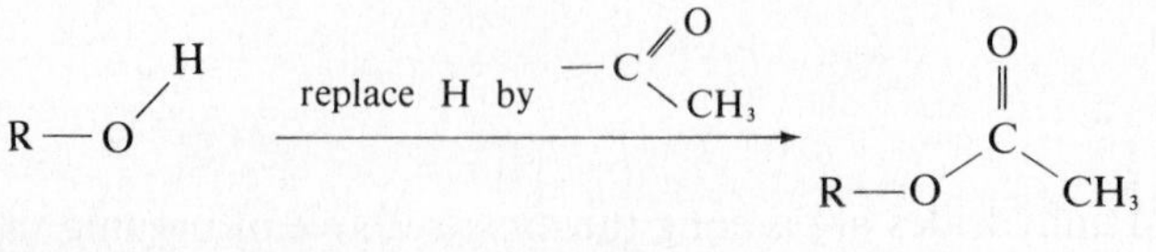

$$R{-}O{-}H \xrightarrow{\text{replace H by } -C(=O)CH_3} R{-}O{-}C(=O){-}CH_3$$

For example, propyl ethanoate is made from propanol and ethanoyl chloride:

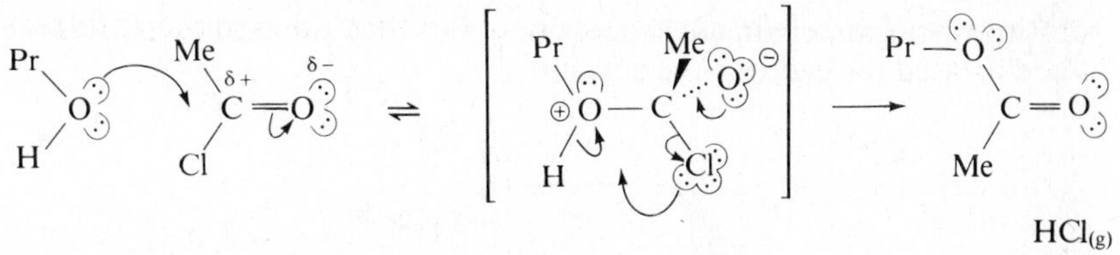

Pr, H, O, Me, δ+, C, O, δ−, Cl ⇌ [Pr, Me, ⊕O—C, O ⊖, H, Cl] ⟶ Pr—O, C=O, Me

$HCl_{(g)}$

Benzoylation is the term used to describe the same process when the acid derivatives are either of those shown below.

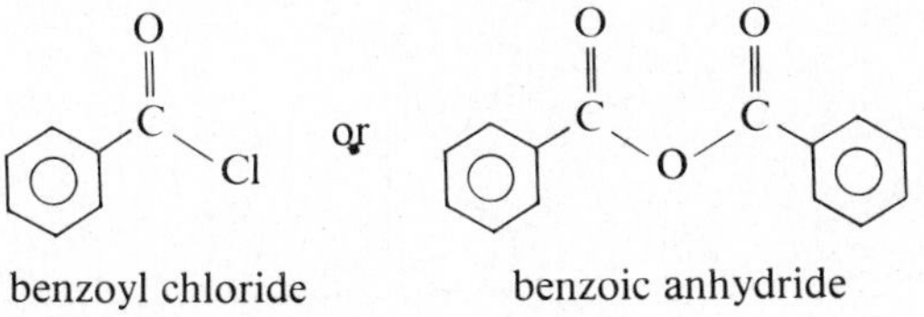

benzoyl chloride or benzoic anhydride

With ammonia and amines

Ammonia and amines react with the acid derivatives in a similar manner. The products are amides or N-substituted amides. For example, ammonia is ethanoylated to give ethanamide while methylamine is ethanoylated to give N-methylethanamide.

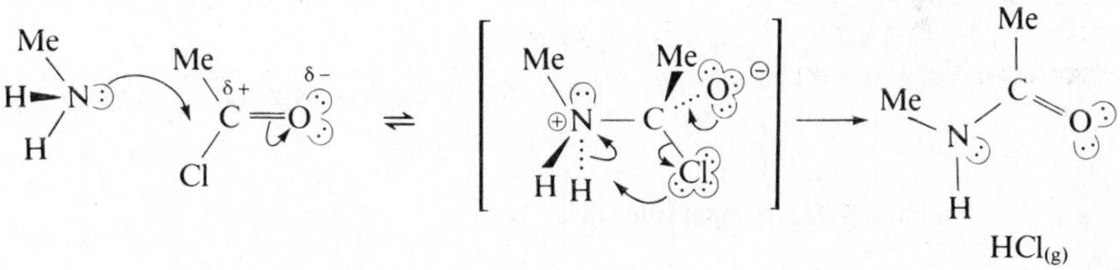

Me, H—N, H, Me, δ+, C, O, δ−, Cl ⇌ [Me, Me, ⊕N—C, O ⊖, H H, Cl] ⟶ Me, N, H, C, O, Me

$HCl_{(g)}$

30.2 Esters

The carboxyl link

Functional group

The molecular structure of an ester contains a hydrocarbon chain interrupted by a carboxyl link:

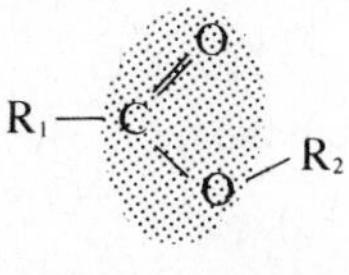

R_1—C(=O)—O—R_2

Esters are acid derivatives because they possess the characteristic acid structure introduced on page 466. An ester is named by reference to its parent acid. The substituent carbon chain (R_2) comes first and then the acid name follows: '-oate' is used instead of '-oic acid':

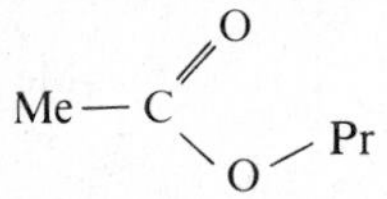

$$Me-C(=O)-O-Pr \qquad H-C(=O)-O-Et$$

propyl ethanoate 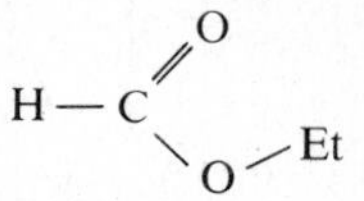 ethyl methanoate

Many esters contain more than one carboxyl link per molecule: these are known as polyesters.

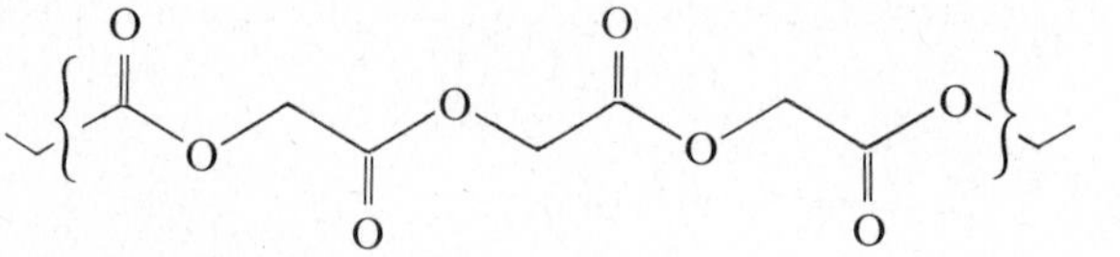

etc.

polyester polymer

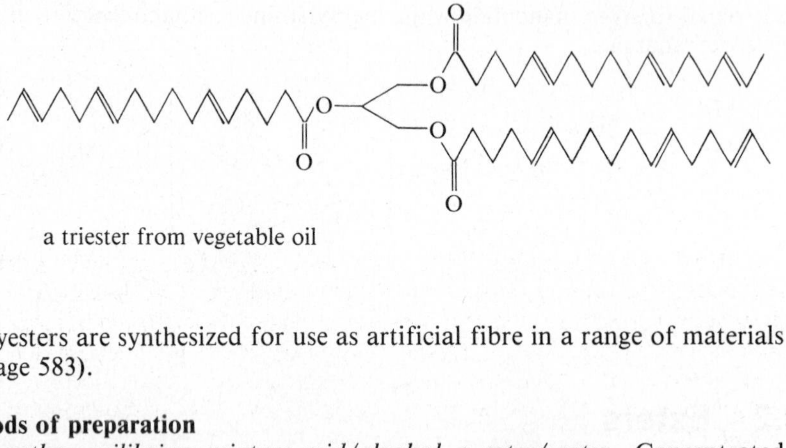

a triester from vegetable oil

Polyesters are synthesized for use as artificial fibre in a range of materials (see page 583).

Methods of preparation

From the equilibrium mixture acid/alcohol ⇌ ester/water Concentrated sulphuric acid is used to speed up the rate of attainment of equilibrium as well as reacting with one of the products (water), thus pulling the equilibrium position to the right. This is not the most satisfactory way of esterifying an alcohol:

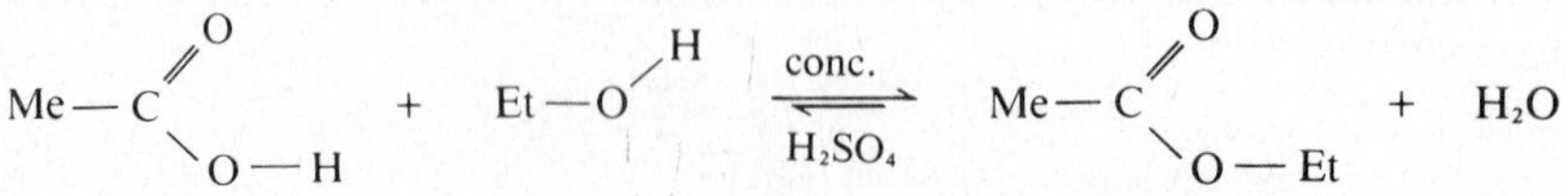

$$Me-C(=O)-O-H + Et-O-H \xrightleftharpoons[H_2SO_4]{conc.} Me-C(=O)-O-Et + H_2O$$

ethyl ethanoate

From acid chlorides or anhydrides The reaction between these acid derivatives and an alcohol is fast and goes to completion:

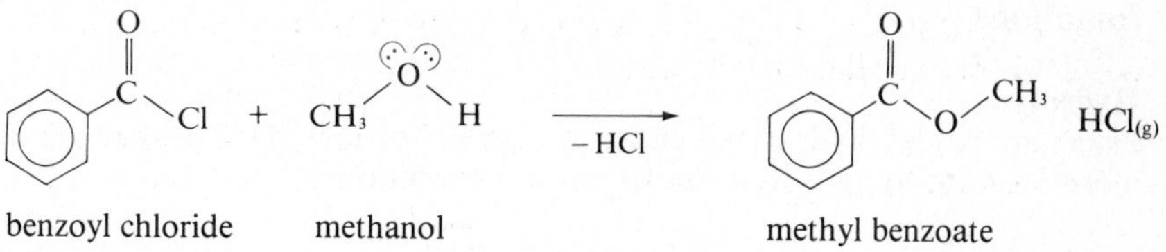

O || C Cl + CH₃ O H $\xrightarrow{-HCl}$ O || C O CH₃ $HCl_{(g)}$

benzoyl chloride methanol methyl benzoate

Reactive sites

When the reactive sites of an ester are compared with those of its parent acid, the significant difference is the presence of a carbon chain in place of the electron-deficient hydrogen atom. The carbonyl group, however, is modified in much the same way.

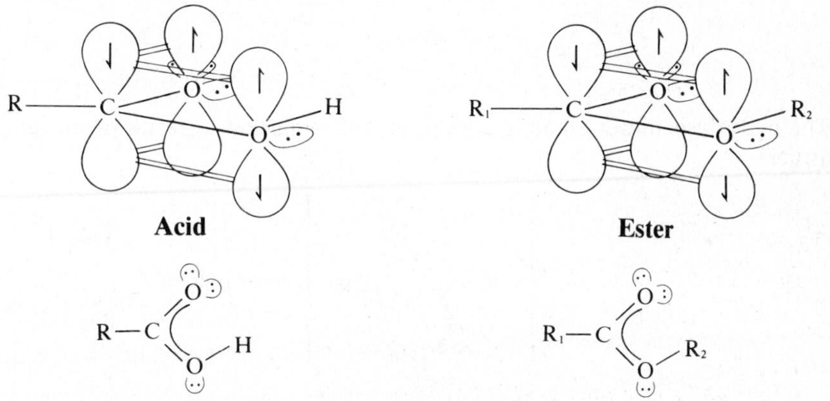

R — C O O H R₁ — C O O R₂

Acid **Ester**

R—C O O H R₁—C O O R₂

The modification of the carbonyl group makes esters (like acids) inert to attack by the nucleophilic testing reagents hydrogen cyanide, hydroxylamine and 2,4-dinitrophenylhydrazine (see page 438). Esters are, however, more readily attacked by nucleophiles than are acids. A nucleophile is likely to be protonated by an acid whereas this cannot occur when it attacks an ester. Compare the reaction of 880-ammonia and ethanoic acid with that of 880-ammonia and ethyl ethanoate:

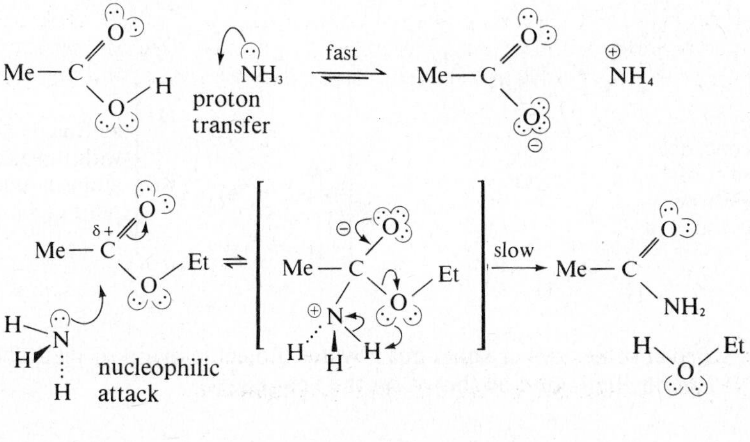

Me— C O O H + NH_3 (proton transfer) $\xrightarrow{fast}$ Me— C O $O^{\ominus}$ + $NH_4^{\oplus}$

Me— $C^{\delta+}$ O O Et + H—N(H)···H (nucleophilic attack) ⇌ [Me— C $O^{\ominus}$ O Et $^{\oplus}N$ H H H] $\xrightarrow{slow}$ Me— C O NH_2 + H O Et

Reactions

Hydrolysis

Esters are readily hydrolysed either at high pH or low pH; but they are much more resistant to hydrolysis under neutral conditions.

$$R_1-C(=O)-O-R_2 \xrightleftharpoons{OH^-_{(aq)}} R_1-C(=O)-O^\ominus + R_2-O-H$$

$$R_1-C(=O)-O-R_2 \xrightleftharpoons{H_3O^+_{(aq)}} R_1-C(=O)-O-H + R_2-O-H$$

The hydrolysis under alkaline conditions takes place as a result of nucleophilic attack:

$$R_1-\overset{\delta+}{C}(=\overset{\delta-}{O})-O-R_2 + H-O^\ominus \rightleftharpoons \left[R_1-C(O^\ominus)(OH)-O-R_2 \right] \longrightarrow R_1-C(=O)-O^\ominus + R_2-O-H$$

concerted shift of electron-density: an alcohol molecule leaves

Under acidic conditions, the situation is less obvious. Hydroxonium ions clearly speed up the rate of hydrolysis and therefore must assist the formation of an activated complex. They are most likely to promote the mesomeric shift of π-electrons onto the carbonyl oxygen; a water molecule acts as nucleophile.

$$R_1-\overset{\delta+}{C}(=O)-O-R_2 + H_3O^\oplus + H_2O \rightleftharpoons \left[R_1-C(OH)(\overset{\oplus}{O}H_2)-O-R_2 \right]$$

concerted shift of electron density

activated complex with three oxygen atoms bonded to the same carbon atom

The activated complex either splits out a water molecule again or an alcohol molecule can be eliminated as shown on the next page.

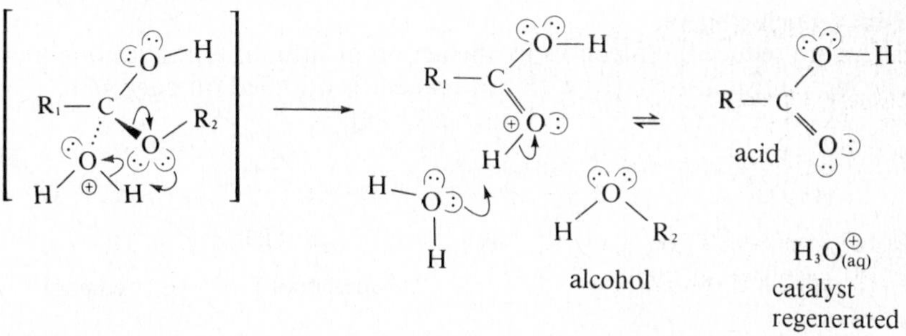

$$\left[R_1-C(-O-H)(\cdots O^{\oplus}H_2)(-O-R_2) \right] \longrightarrow R_1-C(-O-H)=O^{\oplus}-H \;+\; H-O-H \;\rightleftharpoons\; R-C(-O-H)=O \;+\; H-O-R_2 \;+\; H_3O^{\oplus}_{(aq)}$$

acid

alcohol

$H_3O^{\oplus}_{(aq)}$ catalyst regenerated

The reverse reaction of this equilibrium is discussed on page 455. The evidence obtained from 'labelling' one reactant with ^{18}O (prepared specially) is easily checked using the above reaction. In other words, by preparing some labelled water $H_2{}^{18}O$ for the hydrolysis, it is the acid that turns out to be labelled in the products:

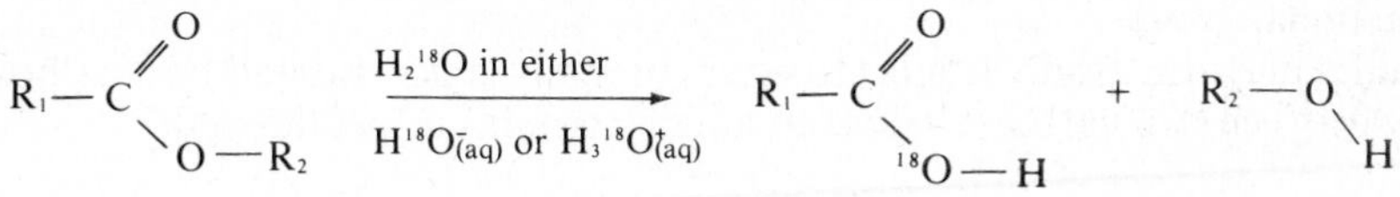

$$R_1-C(=O)-O-R_2 \xrightarrow[H^{18}O^-_{(aq)} \text{ or } H_3{}^{18}O^+_{(aq)}]{H_2{}^{18}O \text{ in either}} R_1-C(=O)-{}^{18}O-H + R_2-O-H$$

With ammonia

Esters react fairly slowly with 880 ammonia. Amides are produced.

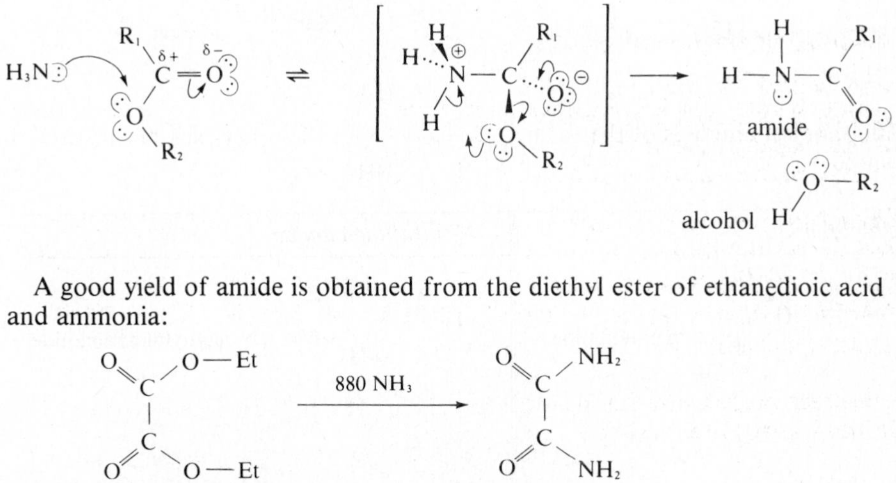

$$H_3N: + R_1-\overset{\delta+}{C}(-O-R_2)=\overset{\delta-}{O} \rightleftharpoons \left[H_3N^{\oplus}-C(R_1)(-O-R_2)-O^{\ominus} \right] \longrightarrow H-N(H)-C(R_1)=O + H-O-R_2$$

amide

alcohol

A good yield of amide is obtained from the diethyl ester of ethanedioic acid and ammonia:

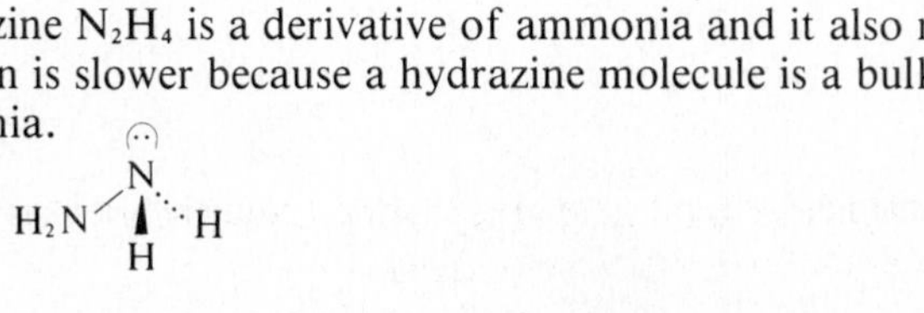

$$(COOEt)_2 \xrightarrow{880\ NH_3} (CONH_2)_2$$

diethylethanedioate ethanediamide

Hydrazine N_2H_4 is a derivative of ammonia and it also reacts with an ester. The reaction is slower because a hydrazine molecule is a bulkier nucleophile than ammonia.

$$H_2N-\ddot{N}H_2$$

With reducing agents
Esters are reduced to alcohols by the action of lithium tetrahydridoaluminate. The mechanism associated with this reagent is outlined on page 440.

$$\underset{\text{ethyl methanoate}}{H-C(=O)-O-Et} \xrightarrow[2.\ H_2O]{1.\ LiAlH_4\ \text{in ether}} \underset{\text{methanol}}{H-CH_2-OH} + \underset{\text{ethanol}}{Et-O-H}$$

30.3 Amides

The peptide link

Functional group
Amides are structurally similar to esters in many ways. Instead of a carboxylic link between two carbon chains, an amide contains a 'peptide link':

$R_1-C(=O)-N(H)-R_2$ (peptide link shaded) — an amide

$R_1-C(=O)-O-R_2$ (carboxyl link shaded) — an ester

The simplest amide is of the form: $R-C(=O)-NH_2$; the others are N-substituted amides.

simple amide		*N-substituted amide*	
$H-C(=O)-NH_2$	methanamide	$H-C(=O)-N(H)-Me$	N-methylmethanamide
$Me-C(=O)-NH_2$	ethanamide	$Me-C(=O)-N(H)-Et$	N-ethylethanamide
$Et-C(=O)-NH_2$	propanamide	$Et-C(=O)-N(H)-Me$	N-methylpropanamide

Polyamides, such as the protein structure shown below, contain many peptide links per molecule. They are often called polypeptides.

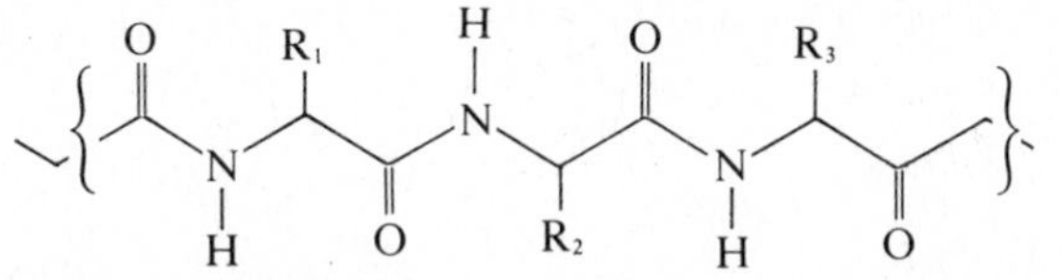

$$-[-CO-NH-CH(R_1)-CO-NH-CH(R_2)-CO-NH-CH(R_3)-CO-]- \text{ etc.}$$

R_1 R_2 and R_3 . . . R_n are all hydrocarbon rings or chains

Polypeptides of all sorts are described in more detail in Chapter 37.

Methods of preparation

1. Dehydration of the ammonium salt of an acid (low yield)

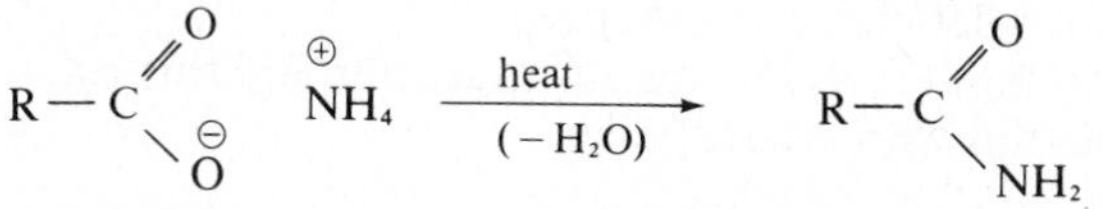

$$R-COO^{\ominus}\ \overset{\oplus}{N}H_4 \xrightarrow[(-H_2O)]{\text{heat}} R-CONH_2$$

2. From an acid chloride and ammonia (or an amine if a substituted amide is required)

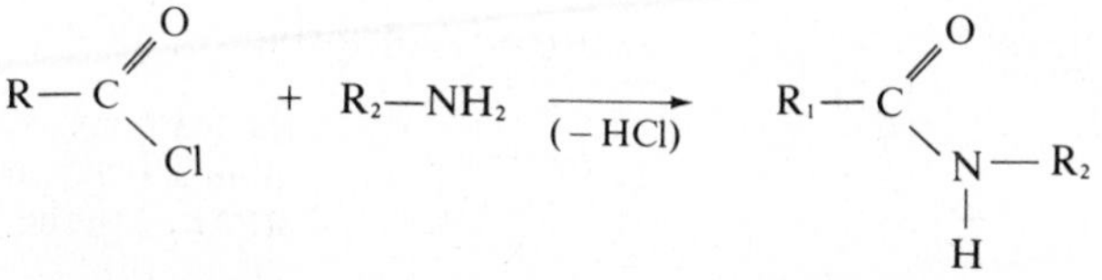

$$R-COCl + R_2-NH_2 \xrightarrow{(-HCl)} R_1-CO-NH-R_2$$

3. Oximes undergo rearrangement to give amides under acidic conditions (the Beckmann rearrangement, see page 480).

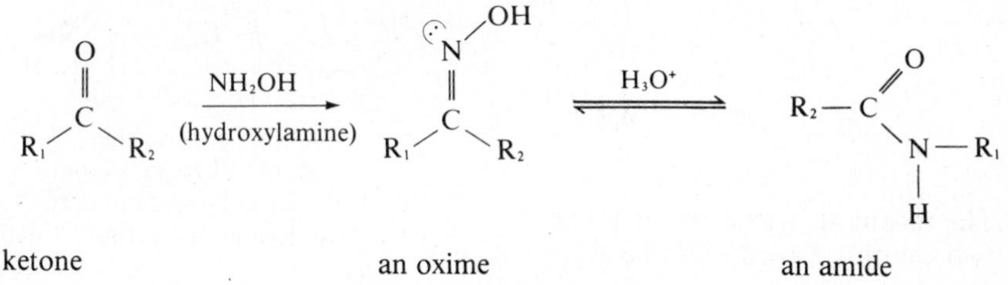

$$\underset{\text{ketone}}{R_1-CO-R_2} \xrightarrow[\text{(hydroxylamine)}]{NH_2OH} \underset{\text{an oxime}}{R_1-C(=N-OH)-R_2} \xrightleftharpoons{H_3O^+} \underset{\text{an amide}}{R_2-CO-NH-R_1}$$

The Beckmann rearrangement is a multi-step process. The mechanism involves the fission of a C—C bond.

Reactive sites

The functional group of an amide contains a carbonyl group adjacent to an amine group. As in a carboxylic acid molecule (page 452) the reactivities of the two groups become modified by the effect of the one on the other.

acid amide

In an amide, the lone pair on the nitrogen atom is drawn in towards the electron-deficient carbon atom. This has three major effects:

1. The reactivity associated with the lone pair of an amine is minimized: there are few basic or nucleophilic properties.
2. The susceptibility of the 'carbonyl' carbon to nucleophilic attack is reduced as its electron-deficiency is lessened.
3. The carbon-nitrogen bond receives some π-character and this leads to the possibility of interaction with electrophiles.

These effects can be illustrated using either valence-bond or orbital overlap drawings

Valence bond structures

π-overlap drawing

the 'carbonyl' oxygen atom's lone pair is attracted to the proton

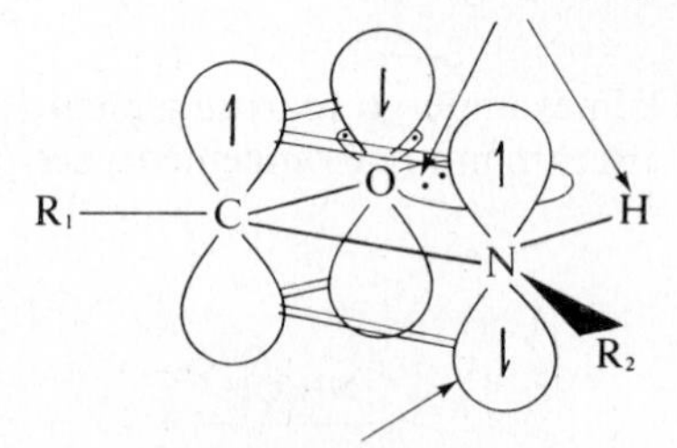

the lone pair on the 'amine' nitrogen atom is in the $2p_y$ orbital and undergoes π-overlap.

The resonant hybrid of the above two canonical forms can be shown as below.

Reactions

Hydrolysis and dehydration

An amide is related to a nitrile and to the ammonium salt of an acid by the hydrolysis pattern shown below:

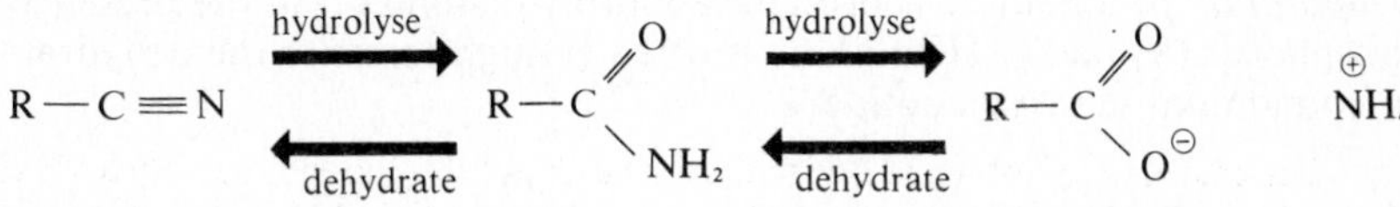

$$R-C{\equiv}N \underset{\text{dehydrate}}{\overset{\text{hydrolyse}}{\rightleftharpoons}} R-C(=O)NH_2 \underset{\text{dehydrate}}{\overset{\text{hydrolyse}}{\rightleftharpoons}} R-C(=O)O^{\ominus} \quad \overset{\oplus}{N}H_4$$

nitrile — amide — ammonium salt

Like esters, amides are hydrolysed to acid salts under either aqueous alkaline or acidic conditions. The mechanisms are similar, see page 472.

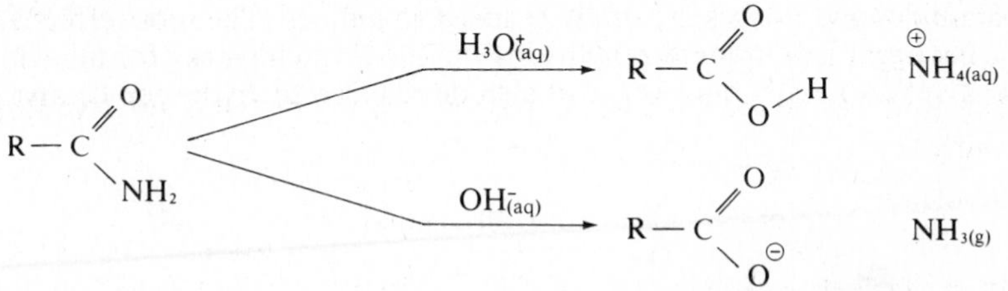

$$R-C(=O)NH_2 \xrightarrow{H_3O^+_{(aq)}} R-C(=O)O-H \quad \overset{\oplus}{N}H_{4(aq)}$$

$$R-C(=O)NH_2 \xrightarrow{OH^-_{(aq)}} R-C(=O)O^{\ominus} \quad NH_{3(g)}$$

alkaline hydrolysis

$$R-C^{\delta+}(=O^{\delta-})NH_2 + H-O^{\ominus} \rightleftharpoons \left[R-C(O^{\ominus})(OH)NH_2\right] \longrightarrow R-C(=O)O^{\ominus} \quad :NH_3$$

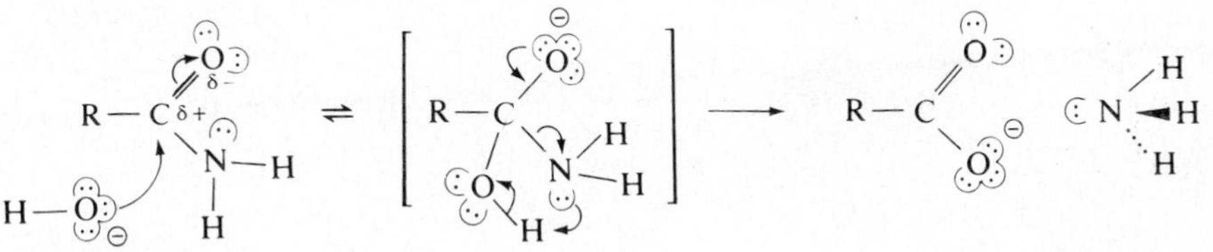

acid hydrolysis

An H_3O^+ ion assists the mesomeric shift of the π-electrons onto the oxygen atom after nucleophilic attack by a water molecule

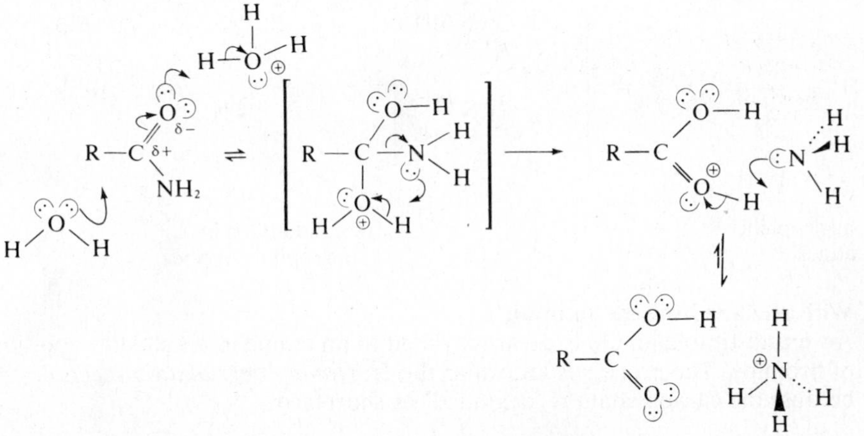

$$R-C^{\delta+}(=O^{\delta-})NH_2 + H_2O + H_3O^{\oplus} \rightleftharpoons \left[R-C(OH)(NH_2)(\overset{\oplus}{O}H_2)\right] \longrightarrow R-C(OH)=\overset{\oplus}{O}-H \quad :NH_3 \rightleftharpoons R-C(OH)=O \quad \overset{\oplus}{N}H_4$$

The common laboratory test for a simple amide makes use of its hydrolysis. If an unknown compound evolves ammonia on boiling in alkali, but not in cold alkali, it is likely to be a simple amide. The need for heat distinguishes an amide from an ammonium salt which liberates ammonia in the cold.

Dehydration of an amide is brought about by distilling it in the presence of phosphorus (V) oxide. Heat alone is often enough to cause the dehydration of an ammonium salt to an amide.

$$R-C(=O)O^{\ominus}\ \overset{\oplus}{N}H_4 \xrightarrow[-H_2O]{\text{heat}} R-C(=O)NH_2 \xrightarrow[-H_2O]{\text{distil} + P_4O_{10}} R-C\equiv N$$

Reduction
Both amides and nitriles are easily reduced to amines. The most effective reducing agent is lithium tetrahydridoaluminate which works for all acid derivatives. A nitrile, however, can also be reduced by hydrogen catalytically:

$$R-C\equiv N \text{ or } R-C(=O)NH_2 \xrightarrow[2.\ H_2O]{1.\ LiAlH_4 \text{ in ether}} R-CH_2-NH_2$$

$$R-C\equiv N \xrightarrow[P\approx 10\text{ atm},\ T\approx 400K]{H_2 \text{ over Raney Ni}} R-CH_2NH_2$$

Lithium tetrahydridoaluminate is a source of nucleophilic hydride ions:

$$\overset{\ominus}{Al}H_4 \rightleftharpoons AlH_3 + \overset{\ominus}{H}:$$

$$H:^{\ominus} + R_1-\overset{\delta+}{C}(=\overset{\delta-}{O})-NH_2 \rightleftharpoons [R_1-CH(-O^{\ominus}AlH_3)-NH_2] \longrightarrow H:^{\ominus} + R_1-CH=\overset{\oplus}{N}H_2 \ (+\ {}^{\ominus}O-AlH_3) \longrightarrow R_1-CH_2-NH_2$$

nucleophilic attack

second attack by nucleophilic hydride

With alkaline bromine solution
An unsubstituted amide is decarboxylated to an amine in an alkaline solution of bromine. The process is known as the *Hoffmann degradation reaction* because the carbon chain is 'degraded' or shortened.

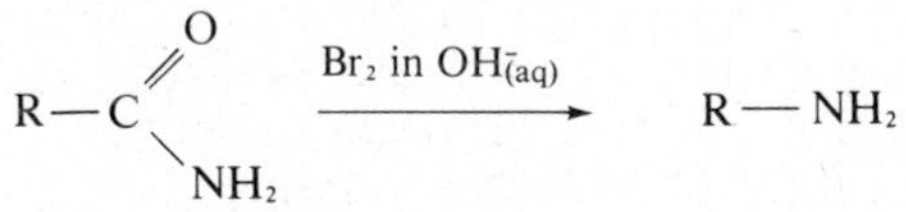

$$R{-}C(=O){-}NH_2 \xrightarrow{Br_2 \text{ in } OH^-_{(aq)}} R{-}NH_2$$

Bromine is an electrophilic reagent, and the mechanism is therefore likely to proceed in a similar manner to that of the reaction of ketones with alkaline iodine (the 'iodoform reaction'; see page 441). There is evidence of π-character between the carbon and nitrogen atoms of the amide's structure:

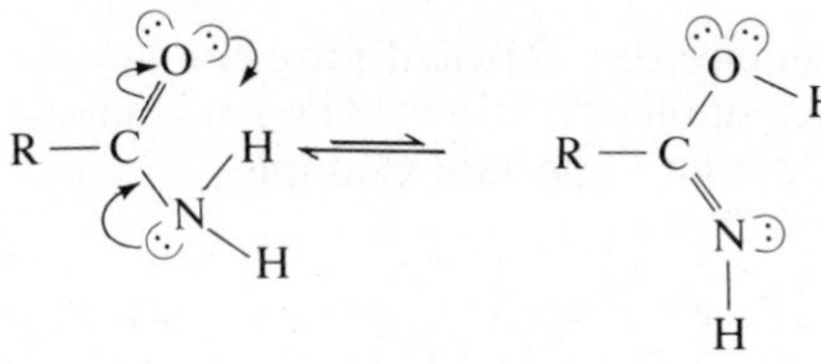

keto *enol*

The *enol* form of the amide reacts with bromine.

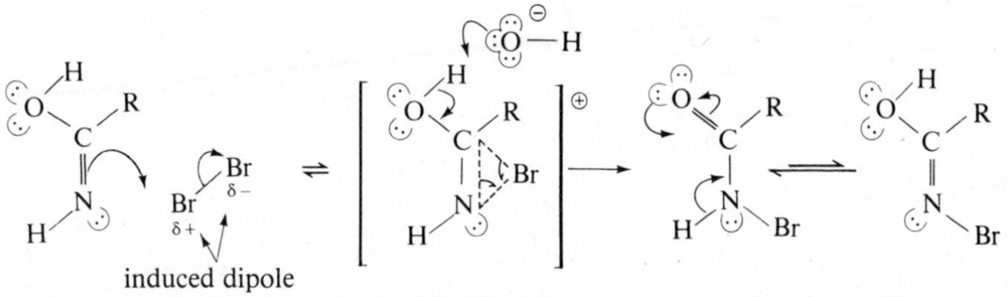

induced dipole

a hydroxide ion deprotonates the bromonium ion

N-bromoamide intermediate

Hydrogen bromide is eliminated from this intermediate via its *enol* form.

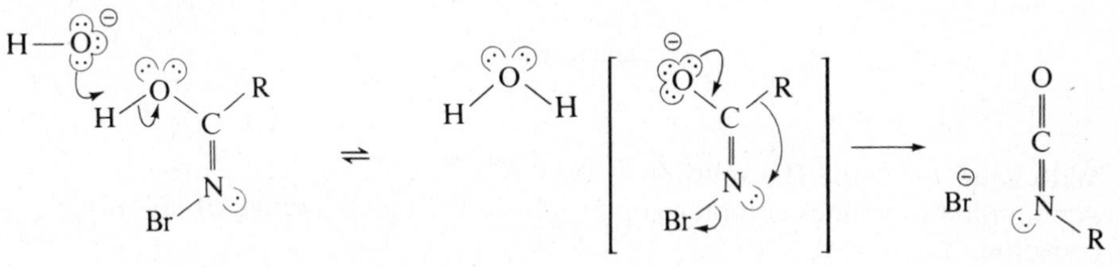

isocyanate intermediate

The elimination of the bromide ion and migration of the alkyl group are examples of a 'push-pull' mechanism. The combined effect of a bromide ion leaving and the lone pair of an oxygen atom becoming a π-pair provide the 'push' and 'pull' for the alkyl group to migrate. The resultant isocyanate is rapidly hydrolysed

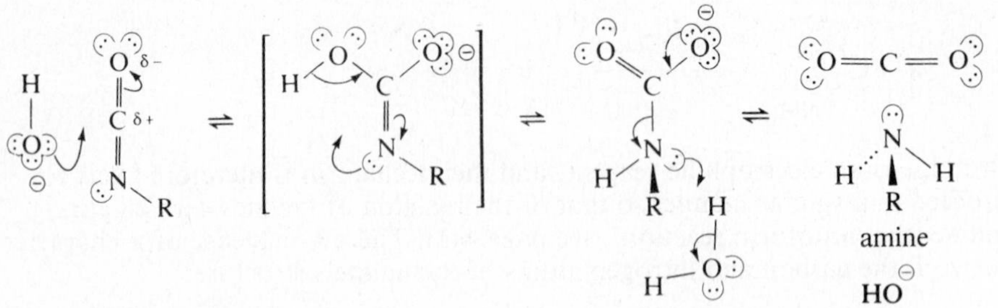

amine

HO

The migration of an alkyl group to an electron-deficient nitrogen atom is thought to occur in a number of other similar reactions. The Beckmann rearrangement of an oxime (page 475) is an important example.

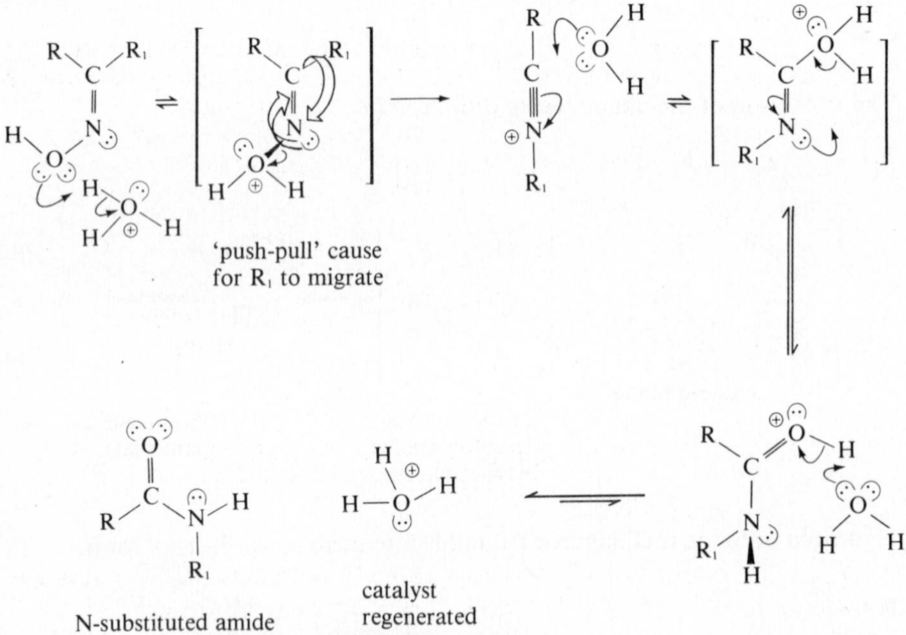

'push-pull' cause for R_1 to migrate

N-substituted amide

catalyst regenerated

With aqueous copper(II) solution

Ammonia and amines complex copper (II) ions to give a range of coloured coordination compounds

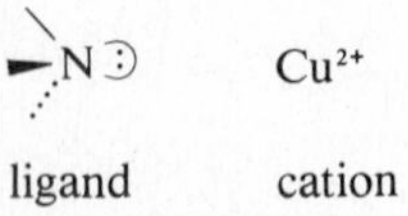

ligand $\quad$ Cu^{2+} cation

The delocalization of the lone pair of the nitrogen atom in an amide structure leads to the loss of most of its ability to complex. There is, however, one class of amides that gives a faint pink complex with a very dilute solution of copper

(II) in the presence of a trace of alkali. This class is that of the *conjugated* amides of the types shown below.

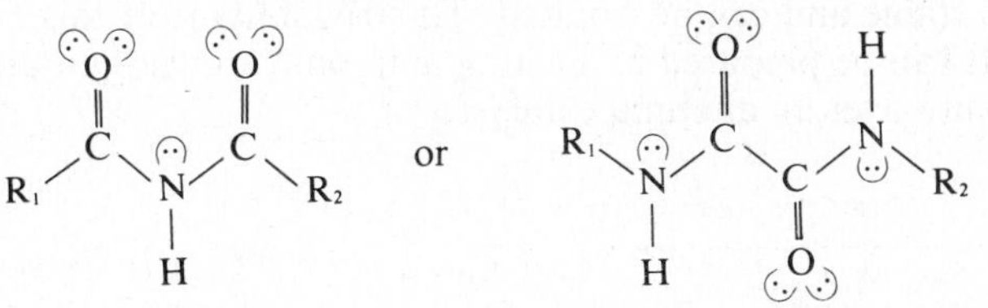

R_1–C(=O)–NH–C(=O)–R_2 or R_1–NH–C(=O)–C(=O)–NH–R_2

A conjugated structure contains one or more groups of π-bonded atoms that are themselves bonded together by a single σ-bond.

The term 'conjugation' is used to describe the merging of the separate π-overlaps that takes place because of their closeness. The increase in the extent of π-overlap leads to an increase in bond energy and therefore an increase in the thermodynamic stability of the structure. For example, the sideways view (below) of the first amide shown above illustrates the spread of the π-density.

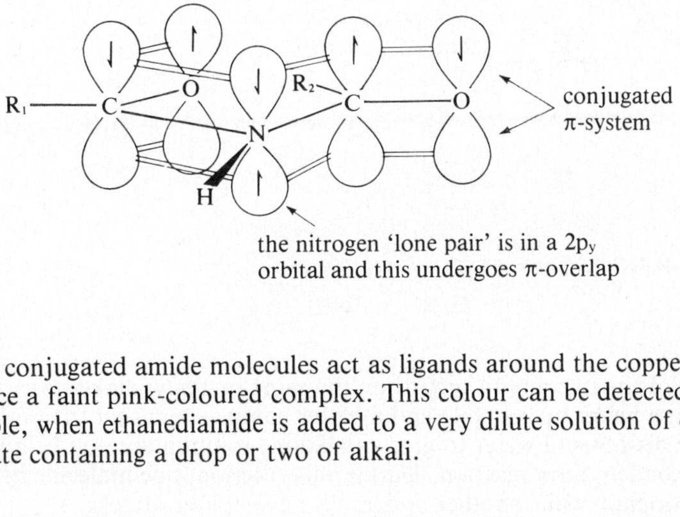

conjugated π-system

the nitrogen 'lone pair' is in a $2p_y$ orbital and this undergoes π-overlap

The conjugated amide molecules act as ligands around the copper (II) ions to produce a faint pink-coloured complex. This colour can be detected, for example, when ethanediamide is added to a very dilute solution of copper (II) sulphate containing a drop or two of alkali.

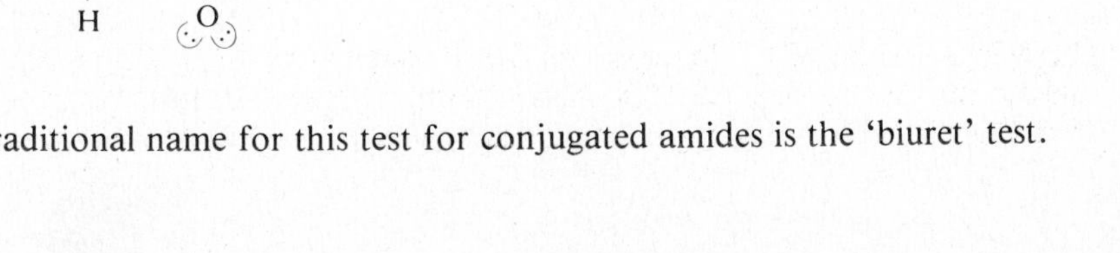

H_2N–C(=O)–C(=O)–NH_2 → (1. $Cu^{2+}_{(aq)}$ (very dilute); 2. 2 drops of $OH^-_{(aq)}$) → pink copper (II) complex

The traditional name for this test for conjugated amides is the 'biuret' test.

30.4 Urea: the diamide of carbonic acid

Although carbonic acid ($CO(OH)_2$) readily decomposes into carbon dioxide and water, its diamide is stable and can be isolated. The trivial name of this compound is urea. It can be prepared by heating ammonia and carbon dioxide together under pressure over an alumina catalyst.

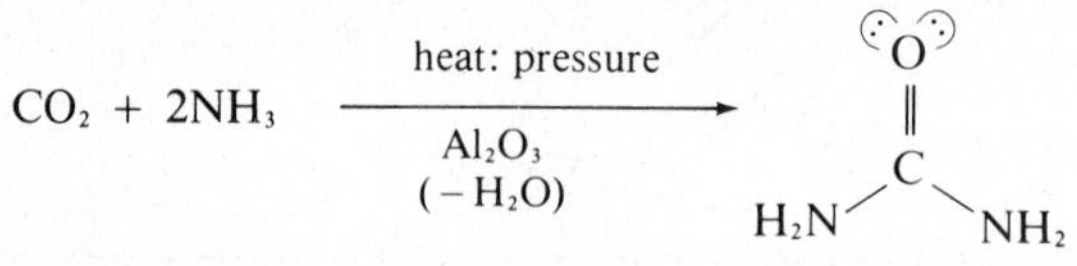

$$CO_2 + 2NH_3 \xrightarrow[Al_2O_3\ (-H_2O)]{\text{heat: pressure}} H_2N{-}C({=}O){-}NH_2 \quad \text{urea}$$

or by the decomposition of ammonium isocyanate

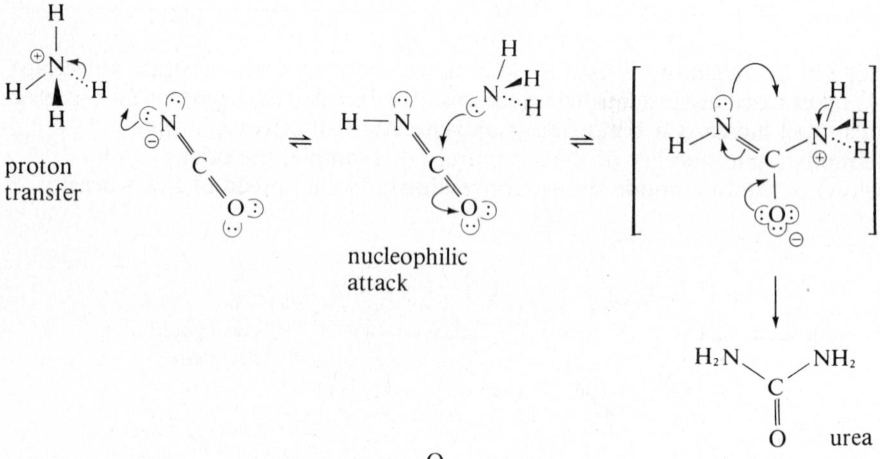

proton transfer

nucleophilic attack

urea

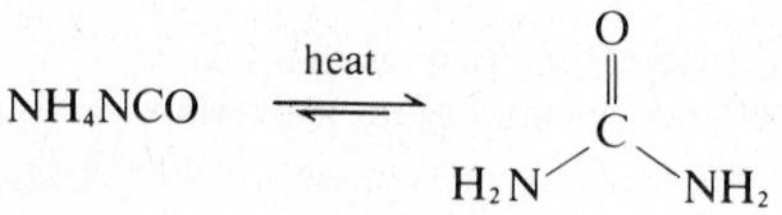

$$NH_4NCO \overset{\text{heat}}{\rightleftharpoons} H_2N{-}C({=}O){-}NH_2$$

The presence of the two amino groups on the same carbonyl makes urea a stronger base, a better nucleophile and a better complexing agent than other amides. Urea dissolves in water to give an alkaline solution and, on heating, undergoes a condensation reaction. During this reaction, one molecule of urea acts as a nucleophile while another undergoes nucleophilic attack.

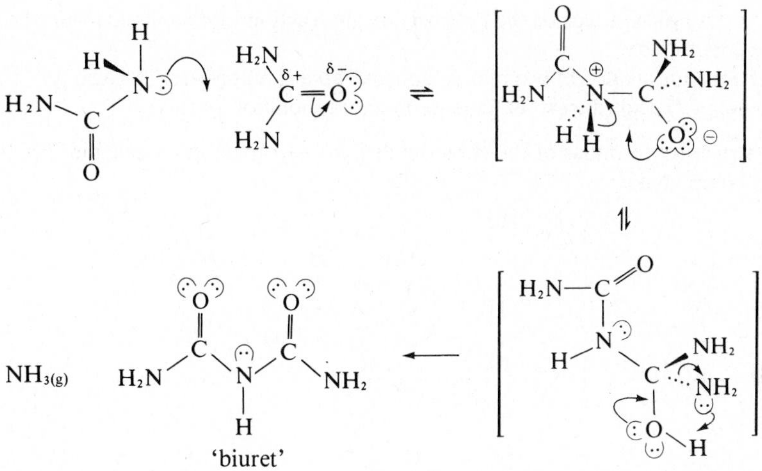

The condensation product ('biuret') is a conjugated amide which will complex with copper (II) as described on page 481.

Questions

1. Compare the reactivity of the following pairs of compounds.
   a) Ethanoic acid and ethanoic anhydride
   b) chloroethanoic acid and ethanoyl chloride
   c) aminoethanoic acid and ethanamide
2. a) How can methyl ethanoate be prepared in good yield from methanol?
   b) Why is ethanoyl chloride a more useful synthetic reagent than ethanoic acid for preparing esters?
3. The solubility of an ester in water is markedly dependent on pH. It is far more soluble in either boiling acid or alkali.
   a) Why does ethyl ethanoate tend to be rather immiscible with water? (Approximately 1 gram per dm^3 of water dissolves).
   b) Discuss the mechanism of the interaction between ethyl ethanoate and boiling alkali.
   c) Discuss the mechanism of the interaction between the ester and boiling acid.
4. a) What is meant by the term 'ethanoylation'?
   b) Give the structural formula of *two* powerful ethanoylating agents.
   c) Which of the following compounds can *not* be ethanoylated?
   i) methanol ii) ethanol iii) ethylamine iv) ammonia v) ethanamide
   d) Give the mechanism for a typical ethanoylation reaction.
5. 'Amides contain a carbonyl and an amino group'. Criticize this statement by
   a) giving *one* typical carbonyl reaction *not* given by amides
   b) giving *two* typical amine reactions *not* given by amides
   c) giving any reaction in which an amide does appear to act as either a typical carbonyl or amino compound.

6. Give the structural formulas of the products of the reaction of ethanamide with the following compounds:
   a) boiling aqueous alkali b) bromine in aqueous alkali c) nitrous (III) acid d) phosphoric (V) acid (distil) e) aqueous ammonia solution (reflux).

7. Give the structural formulas of the compounds that react together to give the following compounds:

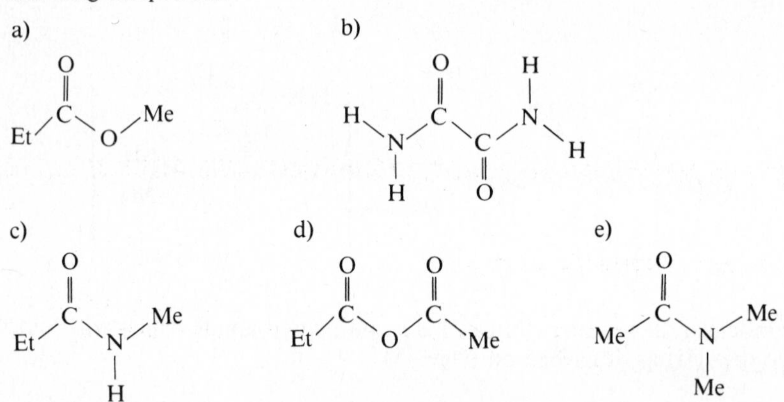

8. An unknown compound contains carbon, hydrogen, oxygen and chlorine. It reacts vigorously with water to give an acidic solution containing a white suspension that is soluble in alkali; 1.405 g of the compound requires 10 cm^3 of a molar sodium hydroxide solution to reach equivalence-point when methyl orange ($pK_a \approx 3.4$) is used as indicator. If phenolphthalein ($pK_a \approx 9.7$) is used, 20 cm^3 of the molar sodium hydroxide solution is needed.

   Suggest a likely formula for the compound and explain the reactions giving structural equations wherever possible.

9. Suggest a reason why the compound aminoethanoyl chloride has not been isolated although aminoethanoic acid is well known.

10. Give the structural formulas of the products of the reaction of methyl propanoate with the following reagents:
   a) dilute alkali (reflux) b) dilute acid (reflux) c) 880-ammonia d) lithium tetrahydridoaluminate.

Chapter 31

Aromaticity and aromatic substitution

31.1 Benzene: an unsaturated cyclohydrocarbon

Lack of typical alkene reactivity

When ethyne is passed backwards and forwards through a copper tube at temperatures above 650 K, a small amount of an unsaturated cyclic hydrocarbon is produced.

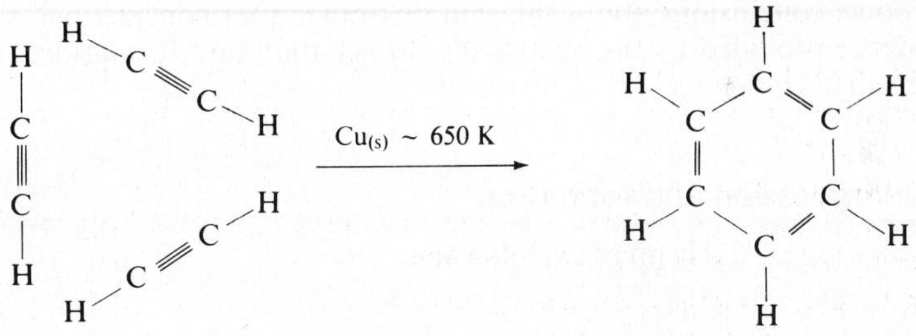

$$3C_2H_{2(g)} \xrightarrow{Cu_{(s)} \sim 650\text{ K}} C_6H_{6(g)} \quad \text{(very low yield)}$$

The formula of the hydrocarbon can be shown to be C_6H_6 by using the eudiometer described on page 370. By considering its molecular structure to result from the addition of the three ethyne molecules to one another, there are two canonical forms possible.

Based on this structure, its name is 'cyclohexatriene', and it should be expected to have the reactivity typical of an alkene. It should:

1. Decolorize bromine water

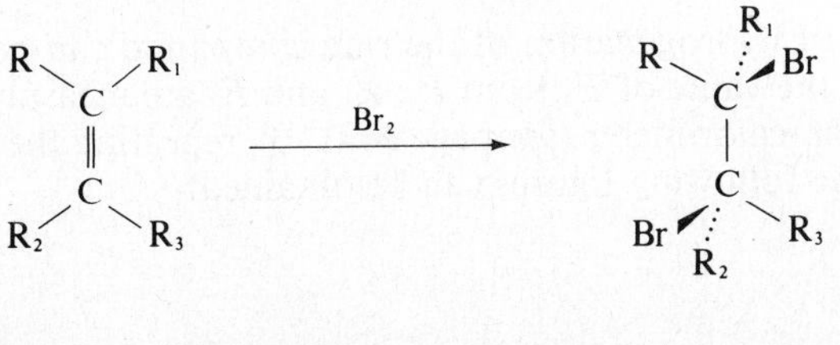

$$R(R_1)C{=}C(R_2)R_3 \xrightarrow{Br_2} R(R_1)(Br)C{-}C(Br)(R_2)R_3$$

2. Perform addition reactions with strong acids such as hydrogen bromide or sulphuric acid.

$$R(R_1)C{=}C(R_2)R_3 \xrightarrow{HBr} R(R_1)(Br)C{-}C(H)(R_2)R_3$$

3. Decolorize alkaline manganate (VII) solution

$$R(R_1)C{=}C(R_2)R_3 \xrightarrow[HO^-_{(aq)}]{MnO^-_{4(aq)}} R(R_1)(OH)C{-}C(OH)(R_2)R_3$$

In fact, it shows *none* of these properties. The 'triene' model of its molecular structure needs adapting in some way to account for the lack of alkene reactivity. Some conclusions about this will be drawn after considering some further evidence provided by the heats of hydrogenation and formation of the ring compound.

Heat of hydrogenation and formation

The hydrogenation of C_6H_6 gives cyclohexane:

$$C_6H_6 + 3H_2 \rightarrow C_6H_{12}$$

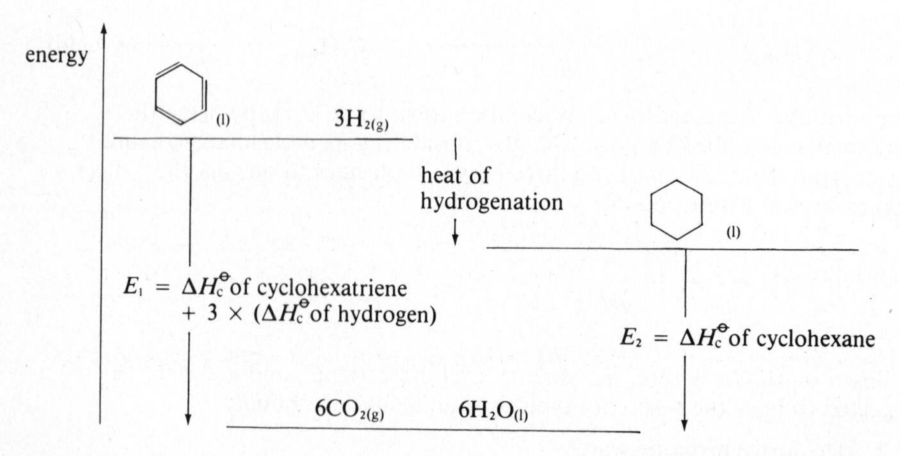

From Hess's law, the heat of hydrogenation of the ring compound can be worked out by subtracting the value of E_2 from E_1; E_1 and E_2 are measured experimentally using a bomb calorimeter (see page 212). By repeating the process for cyclohexene, the following figures can be obtained:

Cyclohexene

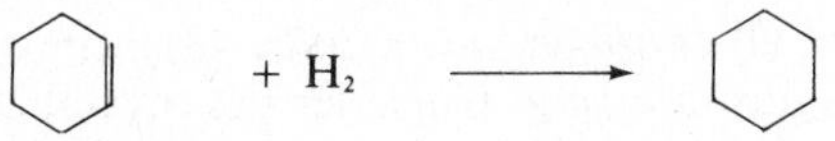

$$\text{C}_6\text{H}_{10} + H_2 \longrightarrow \text{C}_6\text{H}_{12} \quad : \quad \Delta H^{\ominus} = -122 \text{ kJ mol}^{-1}$$

hydrogenate
one double-bond

Cyclohexatriene

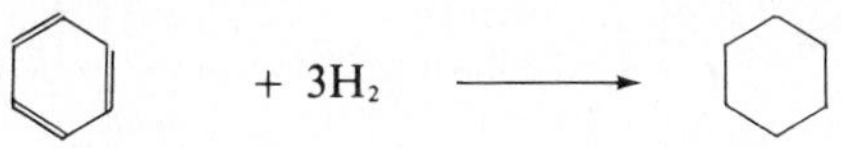

$$\text{C}_6\text{H}_6 + 3H_2 \longrightarrow \text{C}_6\text{H}_{12} \quad : \quad \Delta H^{\ominus} = -205 \text{ kJ mol}^{-1}$$

hydrogenate
three double-bonds

The heat of hydrogenation of cyclohexatriene ought to be three times that of cyclohexene if the three double-bonded model is accurate. In fact, 161 kJ mol^{-1} less energy is evolved than predicted. This means that the correct energy level for C_6H_6 is 161 kJ mol^{-1} lower than the one shown on the diagram. In other words, the actual molecular structure of the ring compound is 161 kJ mol^{-1} *more* thermodynamically stable than the structure shown below.

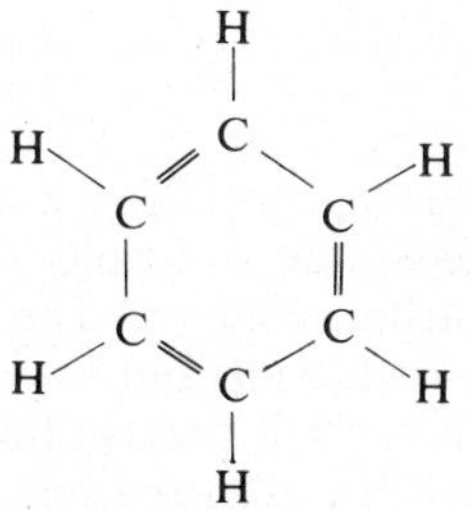

H
C
H C C H
C C
H C H
H

A similar conclusion is reached by examining the heat of formation of the compound. A '*theoretical*' value can be calculated by using bond energy data (see page 218). Assuming the three double-bonded structure, $\Delta H_f^{\ominus}$ depends on the formation of 3 [C—C], 3 [C=C] and 6 [C—H]. After subtracting the energy required to atomize 6 moles of carbon atoms and 6 moles of hydrogen atoms, the theoretical $\Delta H_f^{\ominus}$ = 212 kJ mol^{-1}. A '*measured*' value is readily calculated from combustion data, as in the case of the hydrogenation reaction. This measured value is 49 kJ mol^{-1} and is therefore 163 kJ mol^{-1} lower than the value predicted for the 'triene' structure.

All three pieces of evidence (chemical reactivity, heat of hydrogenation and heat of formation) suggest that the 'triene' structure is inadequate for the cyclohydrocarbon C_6H_6. It shows no alkene reactivity and appears to be thermodynamically much more stable than the trialkene structure.

The aromatic ring

The π-bonds of the ring structure are all *conjugated* (see page 481): the carbon atoms are connected by a set of σ-bonds in a ring. Consider the orbital overlap that is likely.

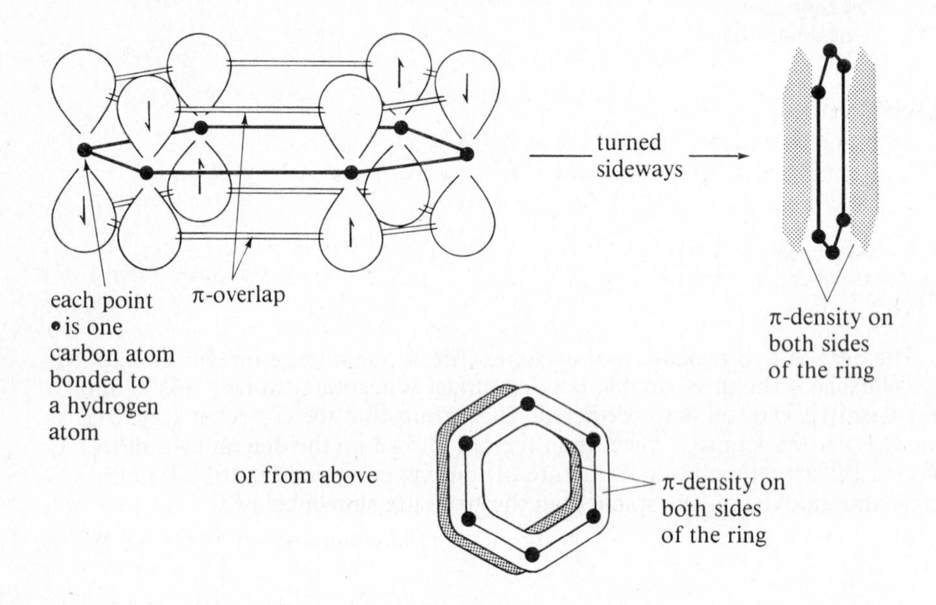

Instead of individual pairs of $2p_y$-orbitals overlapping to give discrete π-bonds, all the $2p_y$-orbitals overlap. The resulting π-system consists of two rings of negative charge, one above and the other below the carbon skeleton. The whole structure is called an *aromatic ring*. From the heat of formation and hydrogenation arguments, it can be concluded that the extra π-overlap lowers the energy level of the ring structure by ~ 160 kJ mol^{-1}. For this reason, the energy drop resulting from the conjugation is usually called the *stabilization* or *resonance energy*.

The conjugated hydrocarbon is called benzene and its structure is shown below (the circle denotes the conjugated π-electrons).

The absence of any discrete double bonds in the structure accounts for its inability to demonstrate alkene reactivity.

Other aromatic systems

The word 'aromatic' was first coined to describe the characteristic smell of the naturally occurring aromatic compounds. But any compound that contains a benzene ring is an 'aromatic' compound:

phenol phenylamine benzoic acid chlorobenzene

'Aromaticity' is associated with any unsaturated ring compound that has the following properties:

1. It burns with a smokey flame (high carbon content)
2. It does not undergo addition reactions.

An interesting discovery is that the hydrocarbon C_8H_8 is not aromatic! It quickly decolorizes bromine water. And yet, the ring structures shown below *are* aromatic.

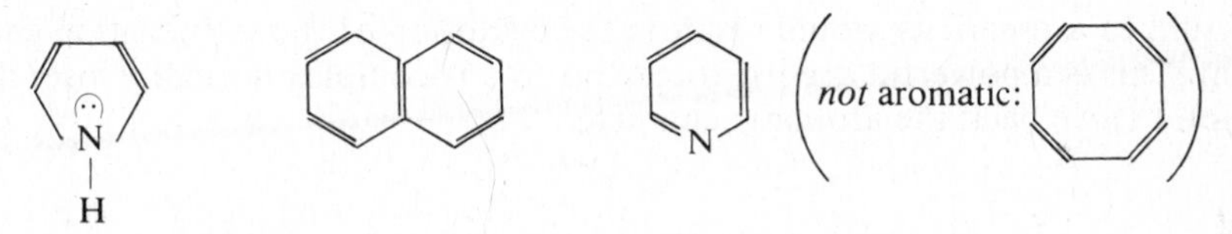

Non-aromatic compounds are called *aliphatic* compounds.

There is a simple rule that distinguishes an aliphatic ring compound from an aromatic compound:

> An aromatic ring contains a conjugated π-system in which there are a total of $(4n + 2)$ electrons; $n = 0, 1, 2, 3 \ldots$

31.2 Aromatic reactivity

Electrophilic substitution

The reactive site of an aromatic ring is the ring of π-electrons. The Lewis basicity of these electrons is quite high and they are readily withdrawn towards an electrophile.

π-clouds $\overset{\oplus}{E}$ ⇌ $[\ \overset{\oplus}{E}\ ]$

σ-skeleton π-complex

The unreactive nature of the aromatic ring can now be explained by considering the fate of the π-complex. If two electrons are to be removed from the aromatic ring to form a σ-bond with the electrophile, the aromatic ring must be broken up. To do this, the stabilization energy needs to be overcome. In other words, the activation energy for the process is high and therefore it proceeds very slowly. When a σ-complex does form, it has the structure shown below.

σ-bond: electrons taken from aromatic ring

$^{\oplus}$E ⇌ π-complex ⇌ [E, H]$^{\oplus}$

σ-complex

the broken ring signifies an aromatic ring that has lost two electrons

A return to aromaticity would result in the evolution of the stabilization energy again. This is a powerful driving force and so a σ-complex normally loses a proton to give back the aromatic character.

[E, H, $^{\ominus}$B]$^{\oplus}$ ⟶ E H—B

any basic particle in the reaction mixture

The overall reaction mechanism is usually given without reference to the initial π-complex.

$^{\oplus}$E ⇌ [E, H, $^{\ominus}$B]$^{\oplus}$ ⟶ E HB

The result is *substitution*. A hydrogen atom has been substituted by an electrophile.

The most readily reactive electrophiles are listed below:

1. $^{+}NO_2$: present in the equilibrium mixture of concentrated nitric and sulphuric acids (page 499).
2. $\overset{\delta+}{R}—\overset{\delta-}{X}$ in the presence of $AlCl_3$; X = halogen atom, R = carbon chain

The electrophilic reagents, grouped as type 2 above, are called *Friedel-Crafts* reagents. Aluminium chloride causes severe polarization of a molecule of R—X and this leads to the production of electrophiles as shown on the next page.

$CH_3Br + Al^{3+}(Cl^-)_3 \rightleftharpoons [AlCl_3Br]^{\ominus} \; {}^{\oplus}CH_3$

powerful electrophiles

or

$MeCOCl + Al^{3+}(Cl^-)_3 \rightleftharpoons Me{-}\overset{\oplus}{C}{=}O \; {}^{\ominus}[AlCl_4]$

So, for example, ethylbenzene is prepared by refluxing benzene and bromoethane over aluminium chloride.

$$C_6H_6 \xrightarrow[AlCl_3]{CH_3CH_2Br} C_6H_5CH_2CH_3$$

and phenylethanone is prepared by refluxing ethanoyl chloride and benzene

$$C_6H_6 \xrightarrow[AlCl_3]{MeCOCl} C_6H_5COMe$$

The mechanism is as before:

$$Me{-}Cl + AlCl_3 \rightleftharpoons \overset{\oplus}{Me} \; [\overset{\ominus}{AlCl_4}]$$

$$C_6H_6 \; {}^{\oplus}Me \rightleftharpoons \left[C_6H_6(Me) \; AlCl_4^{\ominus} \right]^{\oplus} \longrightarrow C_6H_5Me \quad AlCl_3 \quad HCl_{(g)}$$

A Friedel-Crafts catalyst is sometimes prepared *in situ*; iron is used instead of aluminium and the reactants are refluxed with chlorine over finely divided iron:

$$2Fe_{(s)} + 3Cl_{2(g)} \rightleftharpoons Fe_2Cl_{6(s)}$$

A strong acid can also interact with an aromatic ring but succeeds only in substituting a hydrogen atom for a hydrogen atom.

$$C_6H_6 \; H_1{-}A \rightleftharpoons \left[C_6H_6(H_1) \; A^{\ominus} \right]^{\oplus} \longrightarrow C_6H_5H_1 \quad H{-}A$$

The effect of substituents on the reactivity of the ring

A *substituent* is an atom or a group of bonded atoms that takes the place of a hydrogen atom on the ring. The presence of a substituent affects the reactivity of the ring because it either withdraws electron density from the ring or releases electron density into the ring. There are two ways by which this can occur (see pages 354 and 360).

1. as a result of an inductive effect (I)
2. as a result of a mesomeric effect (M).

Inductive effects describe the electron transfer that takes place when two atoms of different electronegativity are bonded. Mesomeric effects describe the electron transfer that can be transmitted through π-overlap in a structure. If electron density is withdrawn by a group, the group exerts a −I (or −M) effect because it increases in negative charge. Conversely, any group that releases electron density is exerting a +I (or +M) effect. Unfortunately, there are some groups that exert a −I but a +M effect: these groups are discussed later. At this stage it is helpful to classify substituents according to the dominant effect they have on the ring. Those that exert (+) effects *activate* the ring because their presence increases the electron density in the ring and makes electrophilic attack more likely. Conversely, substituents that exert (−) effects *deactivate* the ring by withdrawing electron density towards themselves. Some major examples are shown below.

activating groups		*deactivating groups*	
+I	+M	−I	−M
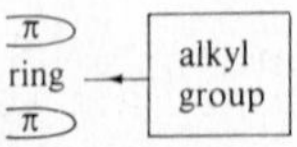 π ring → alkyl group π alkyl groups are *less* electronegative than the substituted hydrogen atom	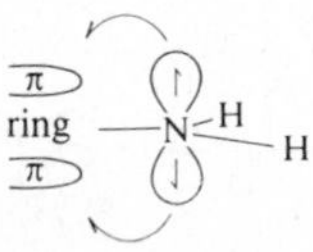 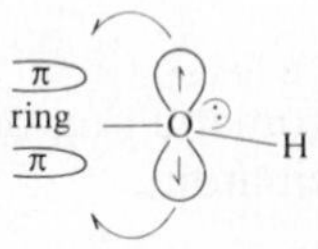 π ring —N—H H π π ring —O—H π	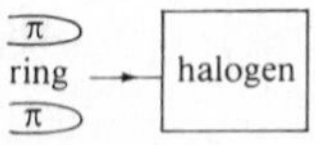 π ring → halogen π halogen atoms are *more* electronegative than the substituted hydrogen atom	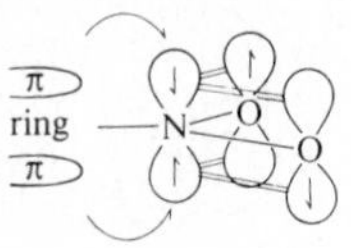 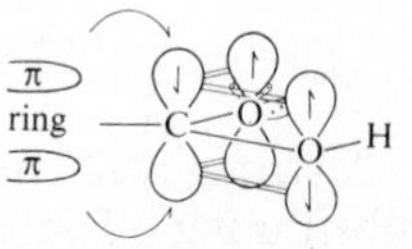 π ring —N O O π π ring —C O O—H π

Example 1 Chlorobenzene reacts more slowly with chloroethane and aluminium chloride than benzene. The chlorine atom exerts a −I effect on the rings and therefore deactivates it.

Cl $\xrightarrow[AlCl_3]{EtCl}$ Cl, Et much more slowly than $\xrightarrow[AlCl_3]{EtCl}$ Et

Although a chlorine atom has three lone pairs, these cannot be drawn into the ring because the outer chlorine shell is the third shell. The overlap of a chlorine $3p_y$-orbital and a carbon $2p_y$-orbital is not extensive because their sizes are not the same.

Example 2 Nitrobenzene reacts more slowly with a nitrating reagent than benzene. The nitro-group exerts a −M effect on the ring and therefore deactivates it (it also exerts a −I effect as well).

NO_2 $\xrightarrow[H_2SO_4]{conc.\ HNO_3}$ NO_2, NO_2 much more slowly than $\xrightarrow[H_2SO_4]{conc.\ HNO_3}$ NO_2

Example 3 Methylbenzene (toluene) reacts more quickly with a nitrating reagent than benzene. The methyl group exerts a +I effect compared with the substituted hydrogen atom.

Me $\xrightarrow[H_2SO_4]{conc.\ HNO_3}$ Me, NO_2 much more quickly than $\xrightarrow[H_2SO_4]{conc.\ HNO_3}$ NO_2

Example 4 Phenol and phenylamine contain powerfully activated aromatic rings: they react with cold bromine water (benzene only reacts with bromine in the presence of a Friedel-Crafts catalyst). Amino- and hydroxy- groups exert strong +M effects.

OH $\xrightarrow{Br_{2(aq)}}$ Br, OH, Br, Br and NH_2 $\xrightarrow{Br_{2(aq)}}$ Br, NH_2, Br, Br

Once a substituent is present on the ring, the remaining positions are no longer equivalent. The ring positions are labelled by counting round from the substituent which is taken to be on carbon atom number one. For example,

X is a 3-substituted phenol
Y is a 4-substituted phenol

Some examples of the naming of aromatic compounds are given below.

2-bromophenol

3-bromophenol

2,4,6-tribromophenylamine

4-hydroxyphenyl ethanoate

(Note that the group has the general name 'phenyl'.)

An older system of naming is still occasionally used. It gives special names to the positions of the ring.

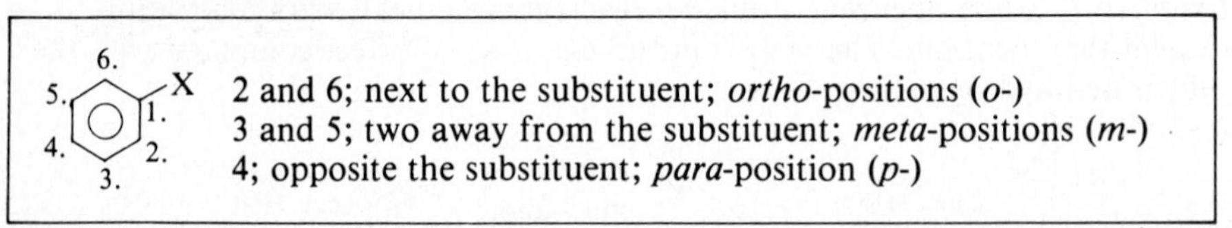

2 and 6; next to the substituent; *ortho*-positions (*o*-)
3 and 5; two away from the substituent; *meta*-positions (*m*-)
4; opposite the substituent; *para*-position (*p*-)

Examples

2-bromophenol is *o*-bromophenol.

4-bromophenol is *p*-bromophenol.

A substituted ring can therefore, in principle, give a number of different electrophilic susbtitution products depending on the position of the substitution and the number of hydrogen atoms substituted:

$$\xrightarrow[-H^{\oplus}]{E^{\oplus}}$$

2- or 3- or 4-

In practice, a simple rule is observed in most cases. A mechanistic explanation of the chemistry of this rule is described on page 561.

Saturated substituents direct electrophilic substitution to the 2, 4 and 6 positions (*o*- or *p*- directing).
Unsaturated substituents direct electrophilic substitution to the 3 and 5 positions (*m*- directing).

Examples

saturated substituents that direct 2-, 4- and 6-	*unsaturated substituents that direct 3- and 5-*
— alkyl — halogen — OH and — NH_2	— NO_2 — CN — COOH and — $C(=O)R$
a) Me → ($NO_2^{\oplus}$) Me, NO_2 , Me, NO_2 → ($NO_2^{\oplus}$) Me, NO_2, NO_2 → ($NO_2^{\oplus}$) Me, O_2N, NO_2, NO_2	b) NO_2 → ($NO_2^{\oplus}$) NO_2, NO_2 → ($NO_2^{\oplus}$) O_2N, NO_2, NO_2
c) Cl → (EtCl, $AlCl_3$) Cl, Et , Cl, Et	d) CN → (EtCl, $AlCl_3$) CN, Et
e) OH → ($Br_{2(aq)}$) OH, Br, Br, Br or NH_2 → ($Br_{2(aq)}$) NH_2, Br, Br, Br	f) $C(=O)Me$ → ($NO_2^{\oplus}$) $C(=O)Me$, NO_2 → ($NO_2^{\oplus}$) O_2N, $C(=O)Me$, NO_2

Questions

1. a) Draw the canonical forms of benzene and the resonant hybrid of these forms.
   b) What is meant by the term delocalization (or resonance) energy?
   c) How is this term used to explain the difference in the reactivity of bromine towards ethene and benzene?

2. Give the mechanism for the reaction of benzene with bromine in the presence of aluminium chloride.

3. Two drops of benzene are put into the bottom of a gas jar full of chlorine. The jar is sealed and left on a laboratory window sill for a few weeks. After this time, the inside of the jar contains a fine covering of colourless crystals.
   Explain this observation. Suggest a likely mechanism for the changes taking place.

4. Give an example of each of the following mechanisms. Draw each mechanism out in full.
   a) electrophilic substitution b) nucleophilic substitution c) electrophilic addition d) nucleophilic addition.

5. Although benzene does not react with bromine water, phenol and phenylamine do. Account for this observation.

6. a) What is meant by the expression 'ortho-para (or 2,4,6) directing'? Give an example of a substituent with these properties.
   b) What is meant by the expression 'meta (or 3,5) directing'? Give an example of a substituent with these properties.

7. a) Use the data book to find relevant bond energy data and thus calculate a theoretical value for the heat of combustion of phenol. State any assumptions that you make about the bonding in the phenol molecule.
   b) Would you expect your answer to agree with an experimentally determined value? Explain the reasons for your answer.

8. Butadiene has some features in common with benzene. It does not give a simple addition product with bromine, although an addition reaction *does* take place: the chief product is 1,4-dibromobutene.
   Discuss the similarities and differences apparent in the structures of butadiene and benzene as suggested by their reactivity towards bromine.

9. Write an essay on: The reactivity of unsaturated ring compounds.

Chapter 32

Phenylamine and diazonium salts

32.1 Preparation of phenylamine

Practical details

Phenylamine is prepared by nitrating benzene at a temperature below 40°C and then reducing the resulting nitrobenzene to phenylamine.

$$C_6H_6 \xrightarrow[\text{conc. } H_2SO_4]{\text{conc. } HNO_3} C_6H_5NO_2 \xrightarrow[\text{conc. HCl}]{\text{Sn and}} C_6H_5\overset{\oplus}{N}H_3\,\overset{\ominus}{Cl} \xrightarrow{\text{NaOH}} C_6H_5NH_2$$

benzene	nitrobenzene	phenylamine hydrogen chloride	phenylamine

stage 1   **stage 2**   **stage 3**

Stage 1
A 50/50 mixture of the two concentrated acids, nitric and sulphuric, is made up. This is added slowly to a calculated quantity of benzene in a round-bottomed flask. The reaction is performed in the fume cupboard and a cooling water bath is used to keep the temperature at about 40°C.

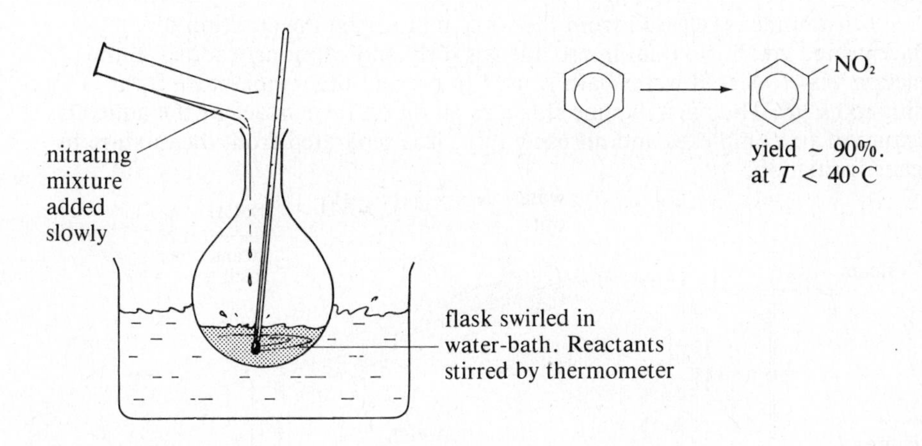

Stage 2
The nitrobenzene produced in stage 1 is extracted by fractional distillation. An excess of benzene in the reaction mixture ensures no further nitration during the

distillation. The nitrobenzene is then mixed with tin and concentrated hydrochloric acid and the reaction mixture is refluxed for a couple of hours to ensure complete reduction:

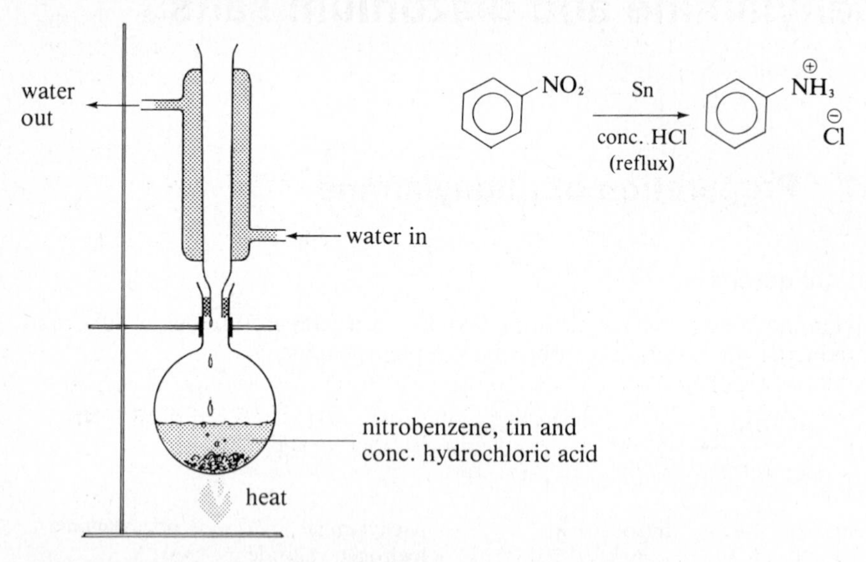

When the refluxing has been carried out, the reaction mixture is allowed to cool. It solidifies to a soft waxy mixture of complex salts of tin:

$$\left(C_6H_5-NH_3\right)_2 SnCl_6$$

phenyl ammonium hexachlorostannate (IV),

$$\left(C_6H_5-NH_3\right)_2 SnCl_4$$

phenyl ammonium tetrachlorostannate (II),

Stage 3

The phenylamine is released from the waxy mixture by the addition of concentrated alkali. Sodium hydroxide is slowly and cautiously added to the reaction vessel. A cold water bath is used to prevent the temperature from rising too high. The phenylamine floats as an oil on the surface of the aqueous mixture of tin complexes and unreacted tin. It is separated from the mixture by steam distillation:

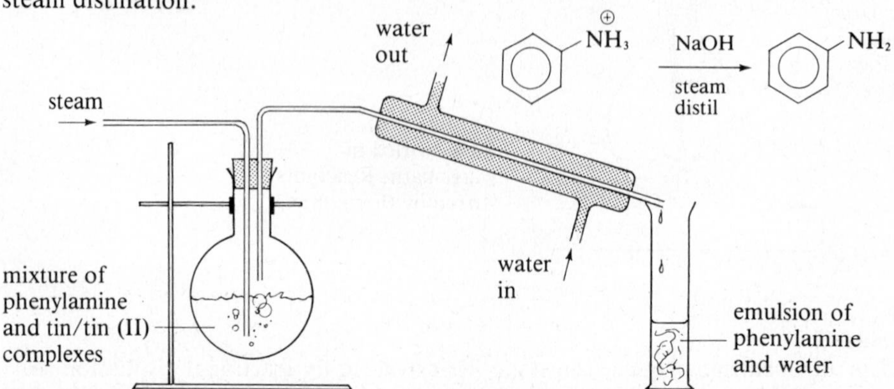

The mixture of phenylamine in water is shaken with solid sodium chloride which reduces the solubility of the phenylamine in the aqueous layer. Phenylamine is then extracted by shaking with several small portions of ether and dried using pellets of potassium hydroxide. The ether is distilled off using a beaker of hot water as a source of heat (no naked flames with ether). Finally the phenylamine is distilled using an air condenser and the fraction boiling between 181°C and 185°C is collected.

A detailed description of the conversion of benzene to phenylamine has been given because it illustrates a wide range of simple practical techniques used in synthesizing organic compounds and because nitration is an important process forming the first step towards producing the intermediate diazonium salts described overleaf.

Mechanism

Concentrated sulphuric acid and concentrated nitric acid exist together in a complex set of equilibria. Appreciable concentrations of the nitronium ion, $\overset{+}{N}O_2$ are present as a result of proton transfer and dehydration:

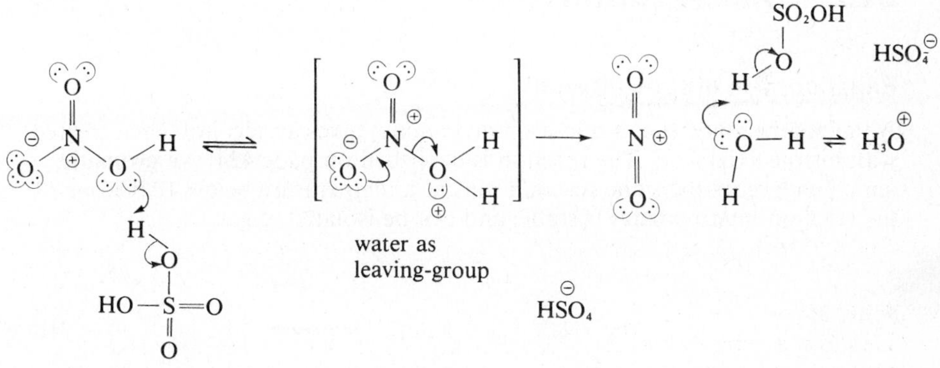

water as leaving-group

HSO_4^-

SO_2OH

HSO_4^-

H_3O^+

$$HNO_3 + 2H_2SO_4 \rightleftharpoons NO_2^+ + H_3O^+ + 2HSO_4^-$$

The nitronium ion is strongly electrophilic.

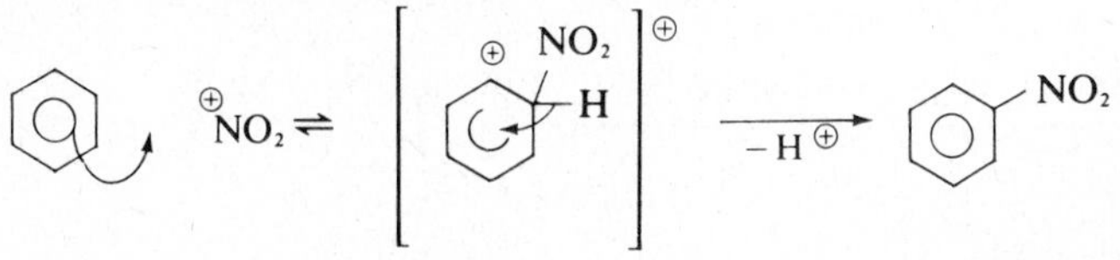

$-H^+$

The nitro-group is deactivating and so nitrobenzene is nitrated itself more slowly than benzene. At temperatures below 40°C, almost no disubstitution takes place. At 100°C, the product is almost entirely 1,3-dinitrobenzene.

Reduction in the presence of tin and concentrated hydrochloric acid occurs by a single electron transfer mechanism. Tin is converted mostly into complexed tin, e.g. $[SnCl_4]^{2-}$. The phenylamine is produced as its hydrochloride salt because it is a weak base and can be protonated by hydrochloric acid:

$$C_6H_5NH_2 + H_3O^+ \rightleftharpoons C_6H_5NH_3^+ + H_2O$$

Sodium hydroxide is added to deprotonate the cations of the salt.

$$C_6H_5NH_3^+ + OH^- \rightleftharpoons C_6H_5NH_2 + H_2O$$

32.2 Diazotization

Reaction with nitrous(III) acid

An aliphatic amine (i.e. a non-aromatic amine) gives an alcohol when treated with nitrous (III) acid. The reaction is described on page 424. An aromatic amine undergoes the same reaction but, at a temperature below 10°C, one of the reaction intermediates is stable and can be isolated, e.g.

$$C_6H_5NH_2 + HO-N{=}O_{(aq)} + H_3O^+_{(aq)} \rightleftharpoons C_6H_5N^+{\equiv}N_{(aq)} + 3H_2O_{(l)}$$

diazonium ion

Usually the reaction is carried out in dilute hydrochloric acid. For phenylamine, the product is benzenediazonium chloride.

$$C_6H_5NH_2 \xrightarrow[\text{temp} < 10^\circ C]{NaNO_{2(s)} + \text{hydrochloric acid}} C_6H_5N^+{\equiv}N_{(aq)} \quad Cl^-_{(aq)}$$

solution of benzenediazonium chloride

The practical details of this diazotization reaction are given in outline below:

1. Measure out a known volume of phenylamine into a boiling tube.
2. Add hydrochloric acid until the oil has dissolved; then add an excess.

3. Put the boiling tube into a freezing mixture and maintain a temperature between 0° and 8°C.
4. Add sodium nitrate (III) slowly, stirring the mixture all the while and keeping the temperature below 8°C.
5. Carry on adding sodium nitrate until no more reaction occurs.

If the temperature is allowed above 10°C, the diazonium salt decomposes and produces phenol. The phenol reacts with undecomposed diazonium salt to give a yellow dye. A good indication of the diazonium ion concentration can be gauged by the colour of the solution: benzenediazonium chloride is colourless.

elimination of two water molecules

The aromatic diazonium ion is more stable than its aliphatic analogue $Me-\overset{\oplus}{N}\equiv N$, because the positive charge is delocalized through the ring.

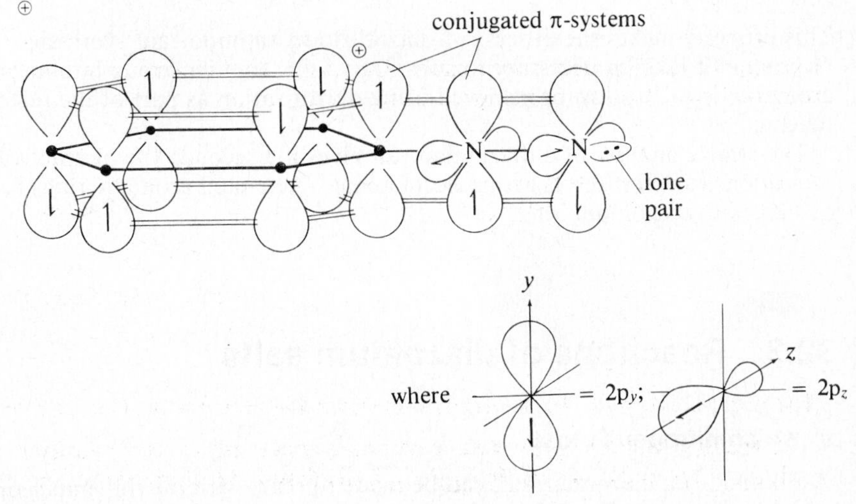

It is not possible to isolate $Me-\overset{\oplus}{N}\equiv N$ in aqueous solution.

The structure of the diazonium ion is that of a nitrogen molecule in which one of the nitrogen atoms has acted as a Lewis base to form a σ-bond to a carbon atom.

$$\overset{\oplus}{R}\ :N\equiv N: \rightleftharpoons R-\overset{\oplus}{N}\equiv N:$$

The reactive sites of a diazonium ion

Aromatic systems generally show very little reactivity toward nucleophiles. The π-electrons tend to deflect the approach of a nucleophile and also delocalize the effect of any charge separations. For example, bromobenzene is unaffected by alkali, but bromoethane is hydrolysed by alkali as shown below.

$$HO^{\ominus} \curvearrowright H_2(Me)\overset{\delta+}{C}-\overset{\delta-}{Br}$$

A diazonium ion, however, has an aromatic ring that is susceptible to nucleophilic attack. It is positively charged and tends to undergo dissociation in the presence of nucleophiles:

$$C_6H_5-\overset{\oplus}{N}\equiv N: \rightleftharpoons [C_6H_5^{\oplus}] + :N\equiv N: \xrightarrow{Nu^{\ominus}} C_6H_5-Nu$$

This property makes the process of diazotization an important synthetic technique: it is often necessary to substitute a nucleophilic group into an aromatic ring. This can be achieved using diazotization as part of the reaction scheme.

The ionic character of a diazonium salt gives it a second type of reactivity: the cations can themselves act as electrophiles. Activated aromatic rings react quickly with diazonium ions.

32.3 Reactions of diazonium salts

In which nitrogen is lost

A solution of a diazonium salt can be made to react with the three nucleophilic reagents shown opposite. In each case, nitrogen is evolved.

1. With water
The diazonium solution needs only warming above 10°C for the reaction with water to set in. The product is a phenol.

$$Ar-\overset{+}{N}\equiv N_{(aq)} + 2H_2O \rightarrow Ar-OH + N_{2(g)} + H_3O^+_{(aq)}$$

For example,

$$\text{2-methylbenzenediazonium} \xrightarrow[\text{above } 10°C]{H_2O} \text{2-methylphenol}$$

2-methylphenol

A synthetic route from methylbenzene to 2-methylphenol uses this reaction:

$$C_6H_5CH_3 \xrightarrow[HNO_3]{H_2SO_4} \text{2-nitromethylbenzene} \xrightarrow[\text{ConcHCl}]{Sn} \text{2-methylphenylamine}$$

nitration T < 30°C; reduction

$$\text{2-methylphenylamine} \xrightarrow[T < 10°C]{HONO} \text{2-methylbenzenediazonium ion}$$

diazotization

$$\text{2-methylbenzenediazonium ion} \xrightarrow[T > 10°C]{H_2O} \text{2-methylphenol}$$

hydrolysis

2. With chloride in the presence of copper (I) chloride
In the presence of copper (I) ions, a diazonium ion reacts with a chloride ion to give a chloro-substituted product. The mechanism for this reaction is unclear:

$$Ar-\overset{+}{N}\equiv N_{(aq)} + Cl^-_{(aq)} \xrightarrow[\text{catalyst}]{Cu^+_{(aq)}} Ar-Cl + N_{2(g)}$$

For example,

$$\text{4-chlorobenzenediazonium}_{(aq)} \xrightarrow[Cu^+_{(aq)}]{Cl^-_{(aq)}} \text{1,4-dichlorobenzene}$$

1, 4-dichlorobenzene

To convert chlorobenzene to 1,4-dichlorobenzene, the following route can be taken. The reagents needed are the same as before.

$$C_6H_5Cl \rightarrow O_2N\text{-}C_6H_4\text{-}Cl \rightarrow H_2N\text{-}C_6H_4\text{-}Cl \rightarrow N\equiv\overset{+}{N}\text{-}C_6H_4\text{-}Cl \rightarrow Cl\text{-}C_6H_4\text{-}Cl$$

Fractional distillation is necessary to isolate the 4-nitro derivative from the nitrating mixture: -Cl directs 2, 4, 6.

3. With cyanide in the presence of copper (I) cyanide

An exactly similar reaction takes place between diazonium ions and cyanide ions in the presence of copper (I). The product is an aromatic nitrile.

$$Ar-\overset{+}{N}\equiv N_{(aq)} + CN^-_{(aq)} \xrightarrow[\text{catalyst}]{Cu^+_{(aq)}} Ar-C\equiv N + N_{2(g)}$$

For example, in the synthesis of 2-methylbenzoic acid from methylbenzene, the following steps can be used.

$$C_6H_5CH_3 \xrightarrow[H_2SO_4-HNO_3,\ T<30°C]{conc} CH_3C_6H_4NO_2 \xrightarrow[\text{Conc HCl}]{Sn} CH_3C_6H_4NH_2$$

$$CH_3C_6H_4NH_2 \xrightarrow[T<10°C]{HONO} CH_3C_6H_4\overset{\oplus}{N}\equiv N_{(aq)}$$

$$CH_3C_6H_4\overset{\oplus}{N}\equiv N_{(aq)} \xrightarrow[Cu_{(aq)}]{CN^{\ominus}_{(aq)}} CH_3C_6H_4C\equiv N \xrightarrow[\text{reflux}]{H_3O^{\oplus}_{(aq)}} CH_3C_6H_4COOH$$

Nitriles are readily hydrolysed to carboxylic acids (see page 477).

In which nitrogen is retained

The reactions in which nitrogen is lost from a diazonium ion all involve interaction with nucleophiles. The diazonium ion itself, however, is positively charged and has electrophilic properties. It reacts with suitably activated aromatic rings in the presence of hydroxide ions. For example,

1. phenols: C_6H_5OH (and substituted derivatives)

2. phenylamines: $C_6H_5NH_2$ (and substituted derivatives)

Nitrogen is retained in the diazonium product and these products are highly coloured substances. They have found uses as dyestuffs.

$$HOC_6H_5 + C_6H_5\overset{\oplus}{N}\equiv N \rightleftharpoons \left[HOC_6H_5(H)N=NC_6H_5\right]^{\oplus} \xrightarrow{OH^{\ominus}} HOC_6H_4N=NC_6H_5 + H_2O$$

a 'diazo' dye

The reaction is often called a *'coupling'* reaction because the product contains one aromatic ring coupled to another by a double-bonded pair of nitrogen atoms. The substitution generally occurs at the 4- position of the phenolic or phenylamine ring. Hydroxyl and amino- groups both direct 2, 4, 6, but there is a considerable steric hindrance associated with substituting the bulky diazonium electrophile into the 2- or 6- position.

$$C_6H_5N^{\oplus}{\equiv}N_{(aq)} \xrightarrow[HO^{\ominus}_{(aq)}]{C_6H_5NMe_2} Me_2N{-}C_6H_4{-}N{=}N{-}C_6H_5$$

two aromatic rings coupled together

A diazonium salt can also be reduced in such a way that the nitrogen is retained. The most common reducing agent is either sodium sulphate (IV) or tin (II) chloride.

$$C_6H_5N^{\oplus}{\equiv}N_{(aq)} \xrightarrow[2.\ H_3O^{+}_{(aq)}]{1.\ SO_{3(aq)}^{2-}} C_6H_5{-}NH{-}NH_2$$

phenylhydrazine

It is also possible to reduce a diazonium salt so that the nitrogen is lost: a low yield of methylbenzene is produced when a methyl diazonium salt is added to phosphinic acid (hypophosphorous (I) acid).

$$2\text{-}CH_3C_6H_4N^{\oplus}{\equiv}N_{(aq)} \xrightarrow{H_3PO_2} C_6H_5CH_3 + N_{2(g)} \quad \text{(low yield)}$$

This reaction effectively brings about the substitution of a hydrogen atom for the $-N^{\oplus}{\equiv}N$ group, a useful process in some synthetic routes.

Example Synthesize 1,3,5-tribromobenzene from benzene

$$\text{benzene} \xrightarrow[H_2SO_4,\ T < 40°C]{HNO_3} C_6H_5NO_2 \xrightarrow[\text{conc HCl}]{Sn} C_6H_5NH_2 \xrightarrow{Br_{2(aq)}} 2,4,6\text{-}Br_3C_6H_2NH_2$$

$$2,4,6\text{-}Br_3C_6H_2NH_2 \xrightarrow[T < 10°C]{HONO} 2,4,6\text{-}Br_3C_6H_2N^{\oplus}{\equiv}N \xrightarrow{H_3PO_2} C_6H_3Br_3$$

1,3,5-tribromobenzene

Questions

1. Outline the practical methods used to carry out the conversion of benzene to benzene diazonium chloride. Give a series of diagrams to illustrate each stage of the process and mention any precautions necessary.

2. a) Compare the reactivity of phenylamine and benzene diazonium chloride towards i) electrophiles ii) nucleophiles.
   b) Why is benzene diazonium chloride such a useful synthetic reagent?

3. Explain the following observations.
   a) Phenylamine reacts with bromine water but benzene does not.
   b) Although benzene diazonium ions are stable in aqueous conditions below 8°C, ethyl diazonium ions are not.

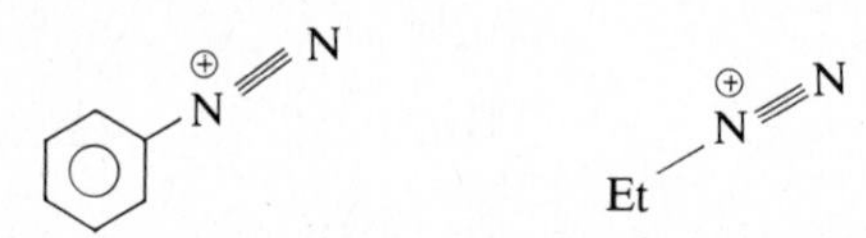

benzene diazonium ion ethyl diazonium ion

   c) A colourless solution of benzene diazonium chloride rapidly turns yellow when warmed above 10°C.

4. Give the reagents and reaction conditions that are needed to bring about the following conversions.

   a) benzene → $C_6H_5NH_2$ (NH₂)
   b) $C_6H_5NH_2$ → H_2N–C_6H_4–N=N–C_6H_5
   c) $C_6H_5NH_2$ → C_6H_5CN
   d) benzene → C_6H_5OH
   e) benzene → C_6H_5Cl
   f) $C_6H_5NH_2$ → 1,3,5-tribromobenzene (Br, Br, Br)

5. Give the equations for *three* reactions in which nitrogen is lost from a diazonium ion and *three* in which it is retained.

6. Draw mechanisms that illustrate the reaction of phenylamine with the following reagents:
   a) hydrogen chloride
   b) ethanoyl chloride
   c) copper (II) chloride solution.

7. Explain the two observations below
   a) Phenylamine is a weaker base than ethlyamine
   b) Ethylamine reacts with chloroethane but phenylamine does not.

8. a) Outline a synthesis of 2,4-dinitrophenylhydrazine starting from benzene.
   b) What use does this reagent have in the laboratory?
   c) Give an example of the use that you have mentioned in b) above.

9. Compare and contrast the reactions of nitrous (III) acid with the following nitrogen compounds.
   a) methylamine b) dimethylamine c) phenylamine.

10. Discuss the similarities and differences in the structure and chemical behaviour of phenylamine and ethanamide.

Chapter 33

Comparing aromatic and aliphatic compounds

33.1 Introduction

In this chapter we shall be comparing the properties of the two classes of compounds whose general structure is shown below.

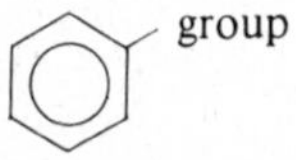

class (A): C_6H_5—group and class (B): >C—group

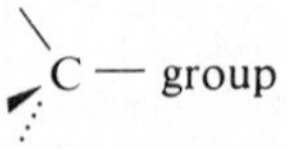

group is —Hal, $—NH_2$, —OH, —CHO, —COOH

Class (A) are obvious aromatic compounds. It is less obvious that class (B) can have aromatic character as well. Examine the structures shown below.

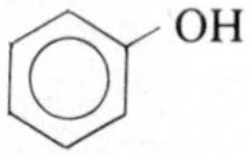

phenol (C_6H_5OH) — phenylmethanol ($C_6H_5CH_2OH$)

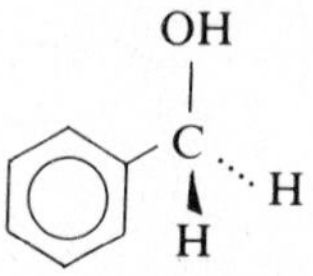

bromobenzene (C_6H_5Br) — 2-phenyl-1-bromoethane ($C_6H_5CH_2CH_2Br$)

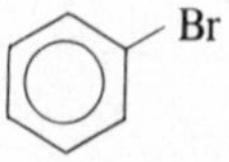

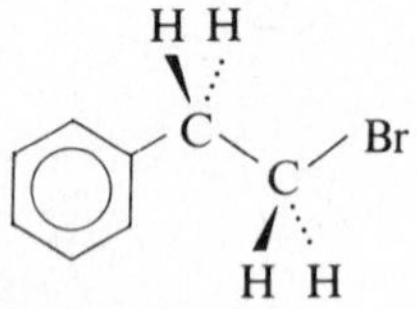

These are all aromatic. Those on the right, however, are class (B) compounds because they contain substituents bonded to a carbon chain and not directly to the aromatic ring. The aromatic rings themselves are substituents, as their names indicate.

In the sections that follow, simple examples are chosen to illustrate the different properties of the two classes of compound, e.g.

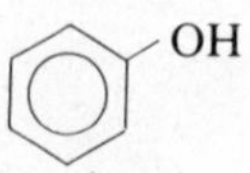

phenol (C_6H_5OH) and methanol (CH_3OH); phenylamine ($C_6H_5NH_2$) and ethylamine ($Et—NH_2$)

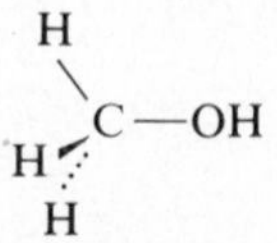

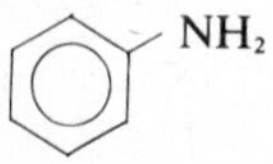

33.2 Halides

The halogen atom

A typical aromatic halide is chlorobenzene (PhCl); one of its aliphatic analogues is chloroethane (C_2H_5Cl). The methods for introducing a halogen atom onto a carbon chain are described on page 398, e.g.

$$R_3(R)C{=}C(R_2)R_1 \xrightarrow{HCl} R_3(R)(H)C{-}C(Cl)(R_2)R_1 \quad \text{or} \quad R{-}OH \xrightarrow{PCl_5} R{-}Cl$$

Neither of these methods can be used to introduce a chlorine atom onto an aromatic ring. There is no equivalent unsaturated hydrocarbon to undergo addition with hydrogen chloride, and phenol (PhOH) is inert to chlorinating agents.

A Friedel-Crafts catalyst of aluminium chloride activates benzene sufficiently to react with chlorine, but the yield of chlorobenzene is low and many side reactions occur.

$$C_6H_6 \xrightarrow[AlCl_3\text{: reflux}]{Cl{-}Cl \text{ with}} C_6H_5Cl \quad \text{(low yield)}$$

chlorobenzene

A more reliable method uses diazotization as an intermediate step.

$$C_6H_6 \xrightarrow[H_2SO_4]{HNO_3} C_6H_5NO_2 \xrightarrow[HCl]{Sn} C_6H_5NH_2 \xrightarrow[T<10°C]{HONO} C_6H_5N^{\oplus}{\equiv}N \xrightarrow[NaCl]{CuCl} C_6H_5Cl$$

The chemistry of the aliphatic halides is dominated by their ability to interact with nucleophiles and bases. Two typical examples are shown below.

1. Nucleophilic substitution

$$HO^{\ominus} + H_2(Me)C^{\delta+}{-}Cl^{\delta-} \rightleftharpoons [HO{\cdots}C(H)(H)(Me){\cdots}Cl]^{\ominus} \longrightarrow HO{-}C(H)(H)(Me) + Cl^{\ominus}$$

aqueous conditions

2. Base-induced elimination

$$HO^{\ominus} + H{-}CH_2{-}C^{\delta+}H_2{-}Cl^{\delta-} \longrightarrow H_2O + H_2C{=}CH_2 + Cl^{\ominus}$$

alcoholic conditions

Both the reactions shown on the previous page can be explained in terms of the electron-deficiency of the carbon atom bonded to the halogen atom. A halogen atom exerts a −I effect on a carbon atom.

An aromatic halide, on the other hand, does not react like this. The electron-deficiency caused by the presence of the halogen atom is delocalized around the aromatic ring. There is no suitable centre of Lewis acidity for a nucleophile to attack.

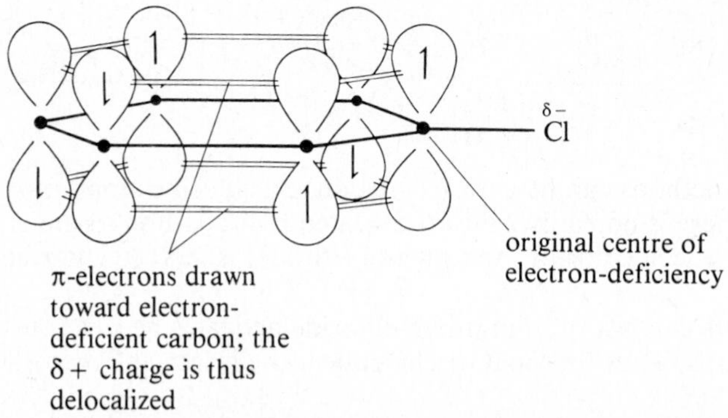

δ−
Cl

original centre of electron-deficiency

π-electrons drawn toward electron-deficient carbon; the δ+ charge is thus delocalized

Aromatic halides are rather unreactive compounds.

The aromatic ring

An aromatic halide not only possesses a halogen atom as a possible reactive site, it also has an aromatic ring. But just as the aromatic ring minimizes the reactivity of the C—halogen group, so the halogen atom's electron-withdrawing properties deactivate the ring. This idea is discussed on page 492. For example, chlorobenzene is nitrated by concentrated nitric and sulphuric acids four times more slowly than is benzene.

33.3 Phenols and alcohols

The hydroxyl group

An aromatic hydroxy-compound is called a 'phenol'. Phenol itself has the structure shown below: its name is derived from 'phenyl' and '-ol'.

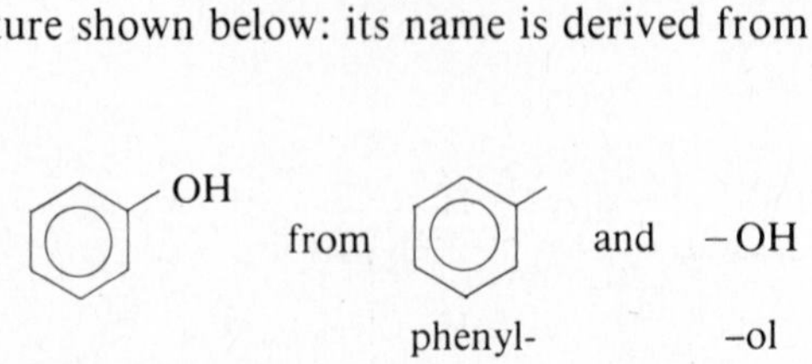

OH from and −OH

phenyl- −ol

Its aliphatic analogue is an alcohol such as ethanol, Et—OH. Alcohols are prepared by hydrating alkenes or hydrolysing halogenoalkanes.

$$\mathrm{RHC{=}CH_2} \xrightarrow[\text{2. } H_2O]{\text{1. } H_2SO_4} \mathrm{RCH_2{-}CH_2OH} \quad \text{or} \quad \mathrm{R{-}Br} \xrightarrow[\text{reflux}]{OH^-_{(aq)}} \mathrm{R{-}OH}$$

They can also be prepared by reducing acid derivatives with lithium tetrahydridoaluminate.

$$\mathrm{RCOOH};\ \mathrm{RCHO};\ \mathrm{RCOOR};\ \mathrm{RCOCl} \xrightarrow[\text{2. } H_2O]{\text{1. } LiAlH_4 \text{ in ether}} \mathrm{RCH_2OH}$$

None of these methods can be used to prepare a phenol. No equivalent alkene or acid exists from which a phenol could be produced, and aromatic halides are resistant to nucleophilic attack.

Phenol is prepared either by hydrolysing benzenesulphonic acid or via benzenediazonium chloride. (The *industrial* preparation is described on page 590.)

$$C_6H_6 \xrightarrow[T<40°C]{\text{fuming } H_2SO_4} \underset{\text{benzenesulphonic acid}}{C_6H_5SO_2{-}OH} \xrightarrow[\text{2. evaporate solvent}]{\text{1. NaOH}} C_6H_5SO_2O^{\ominus}\,Na^{\oplus}_{(s)} \xrightarrow[\text{2. neutralize } H_3O^+_{(aq)}]{\text{1. heat with NaOH}} C_6H_5OH$$

$$C_6H_6 \xrightarrow{\text{as on page 509}} C_6H_5N^{\oplus}{\equiv}N \xrightarrow[\text{above } 10°C]{\text{warm}} C_6H_5OH + N_{2(g)}$$

The mechanism of the sulphonation of benzene is shown below. Fuming sulphuric acid contains sulphur (VI) oxide dissolved in concentrated acid.

$$C_6H_6 + \overset{\delta+}{S}O_2{=}\overset{\delta-}{O} \rightleftharpoons \left[C_6H_5(H)SO_2O^{\ominus}\right]^{\oplus} \longrightarrow C_6H_5SO_2OH$$

When the sodium salt of this acid is heated with caustic soda, exchange of the sulphate (IV) ion occurs and the phenate ion is produced. This is protonated in acid solution to give phenol.

Phenol and ethanol both have a hydroxyl group. The reactive sites of this group and its associated chemical properties are discussed in detail on page 410. They are summarized below for ethanol:

1. The hydroxyl group exerts a $-$I effect on the carbon atom to which it is bonded. Nucleophilic attack can occur at this electron-deficient carbon atom. For example,

$$\text{Et—OH} \xrightarrow[\text{K}^+\ {}^-\text{Br}_{(s)}]{\text{conc H}_2\text{SO}_4} \text{Et—Br} \quad \text{bromoethane (low yield)}$$

2. The oxygen lone pairs of the hydroxyl group give it nucleophilic properties. For example,

$$\text{Et—OH} \xrightarrow[\text{Me—C(=O)—OH}]{\text{conc H}_2\text{SO}_4} \text{Me—C(=O)—O—Et} \quad \text{ethyl ethanoate}$$

3. The electron deficient proton of the hydroxyl group results in ethanol having weak acidic properties. For example,

$$\text{Et—OH} \xrightarrow{\text{Na}} \text{EtO}^{\ominus}\ \text{Na}^{\oplus} + \text{H}_{2(g)} \quad \text{sodium ethoxide}$$

The hydroxyl group in phenol has a rather different environment. For the same reason that bromobenzene is inert to attack by nucleophiles, phenol also does not interact with nucleophiles. The charge separation is delocalized by redistribution of π-electron density in the aromatic ring. The π-overlap extends to the oxygen atom which has one lone pair in the $2p_y$ orbital.

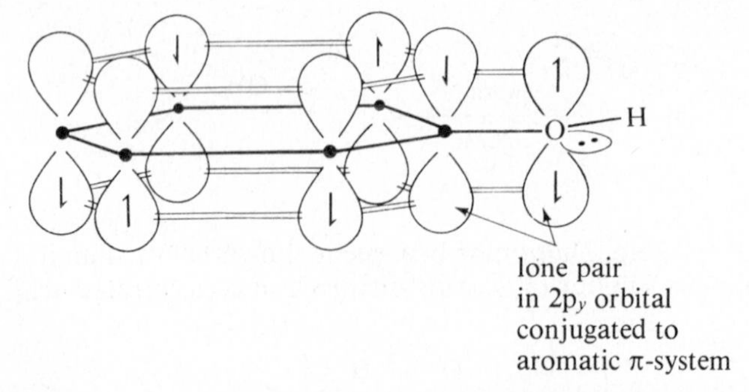

The conjugation of the oxygen lone pair in phenol has two major other effects:

1. There is a marked decrease in nucleophilic character: phenol does not react with an acid to give an ester. An acid chloride gives an ester with phenol, however, e.g.

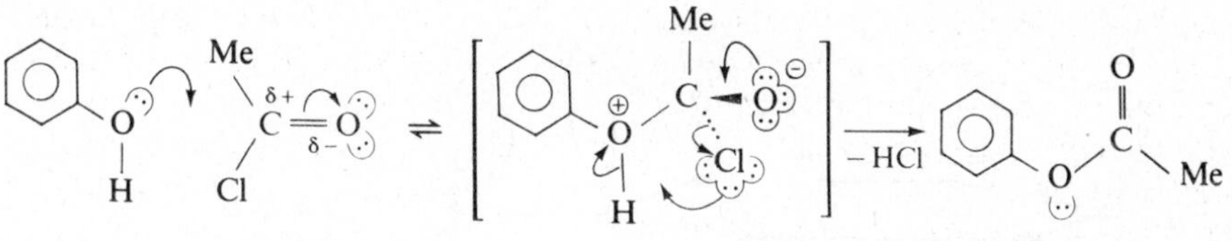

phenol + ethanoyl chloride ⇌ [intermediate] $\xrightarrow{-HCl}$ phenyl ethanoate

2. There is an increase in acid strength. Ethanol is a weaker acid than water but phenol is appreciably stronger than both.

$$H_2O_{(l)} + H_2O_{(l)} \rightleftharpoons HO^-_{(aq)} + H_3O^+_{(aq)} \qquad pK_a \approx 15.7$$

$$EtOH_{(aq)} + H_2O_{(l)} \rightleftharpoons EtO^-_{(aq)} + H_3O^+_{(aq)} \qquad pK_a \approx 16.0$$

$$PhOH_{(aq)} + H_2O_{(l)} \rightleftharpoons PhO^-_{(aq)} + H_3O^+_{(aq)} \qquad pK_a \approx 10.0$$

The more stable an acid anion, the stronger is the acid. The phenate ion is stabilized by delocalization of the negative charge through the aromatic ring. The canonical forms, using the valence bond model, show this effect best.

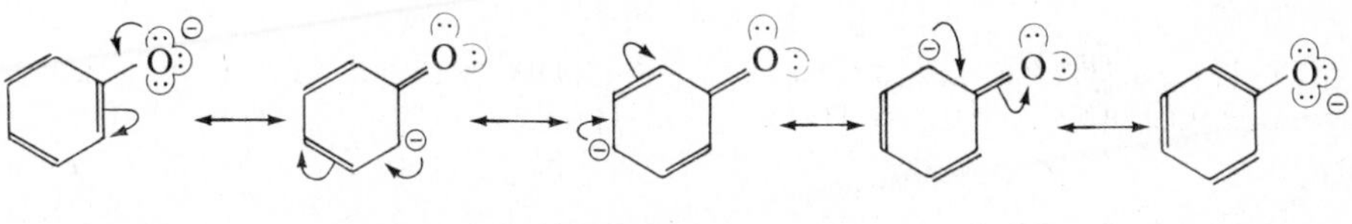

The stability of anions is discussed in more general terms for organic acids on page 461. Phenol is not as strong an acid as benzoic acid or ethanoic acid; it cannot liberate carbon dioxide from a solution of sodium carbonate whereas benzoic acid can:

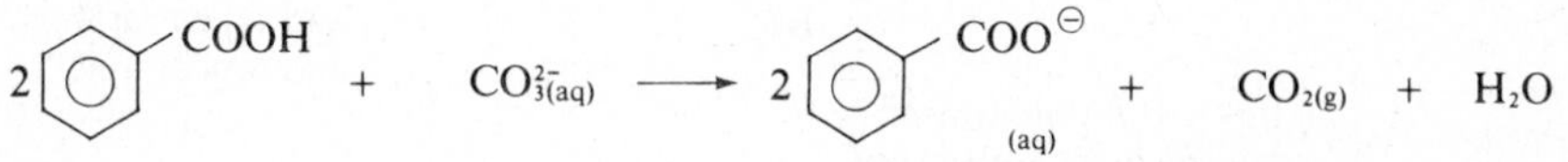

$$2\,C_6H_5COOH + CO_{3(aq)}^{2-} \longrightarrow 2\,C_6H_5COO^-_{(aq)} + CO_{2(g)} + H_2O$$

but

$$C_6H_5OH + CO_{3(aq)}^{2-} \rightleftharpoons C_6H_5O^-_{(aq)} + HCO^-_{3(aq)}$$

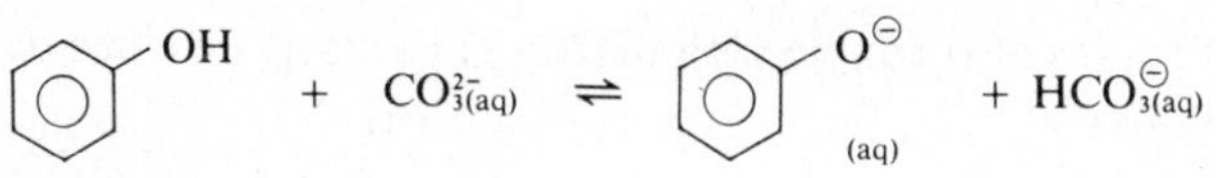

The aromatic ring

Phenols owe most of their reactivity to the presence of the aromatic ring in their molecular structure. The hydroxyl group activates the ring because of the extra electron density added by the conjugation of the oxygen lone-pair. This effect is illustrated in the diagram overleaf.

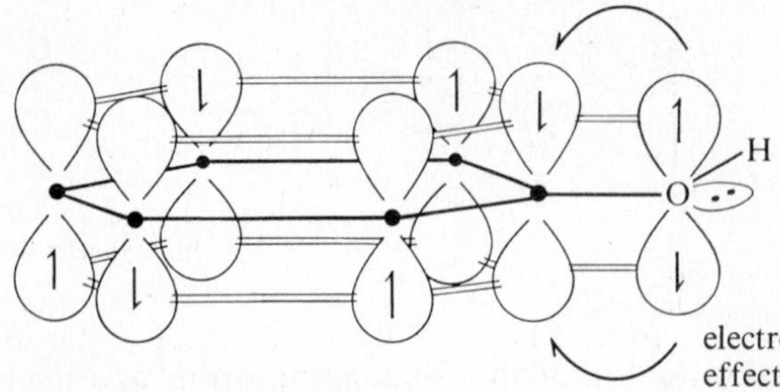

electron-releasing (+M) effect of conjugation of oxygen $2p_y$ with aromatic ring

It is interesting to note that, although the hydroxyl group is activating as a result of the +M effect illustrated above, the oxygen atom itself exerts a −I effect on the system. In other words, the electron density is likely to be at its highest in the region of the oxygen atom. Even so, the aromatic ring as a whole receives a sufficient increase in π-density for it to show activated properties.

Example 1 2,4,6-tribromophenol appears as a white suspension when sufficient bromine water is added to a phenol solution.

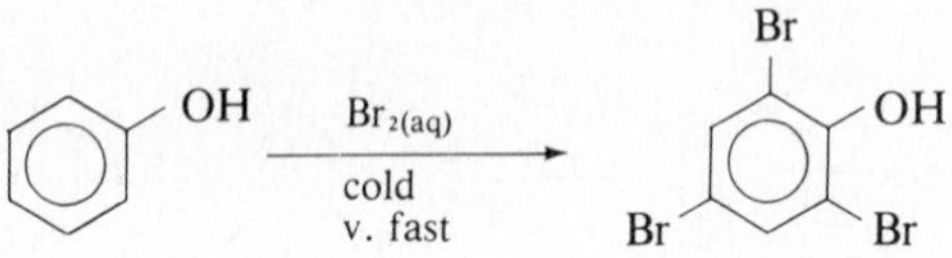

$Br_{2(aq)}$, cold, v. fast

via

H_2O

$H_3O^{\oplus}_{(aq)}$ $Br^{\ominus}_{(aq)}$

$Br^{\ominus}_{(aq)}$

and repeat at 4- and 6-positions

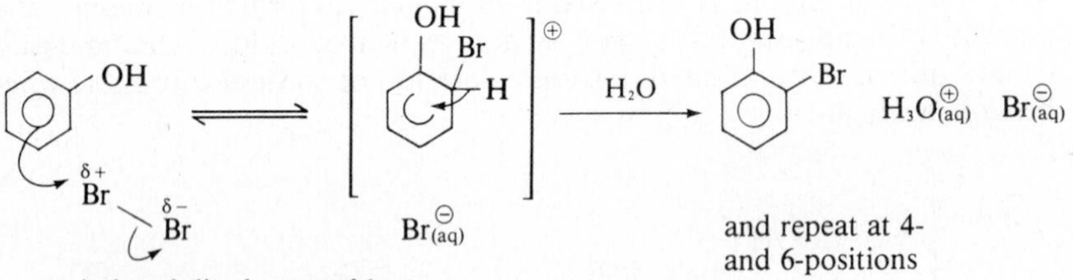

induced dipole caused by approach to π-electrons

Example 2 Phenols also couple with diazonium salts to produce dyes. Alkaline conditions are needed.

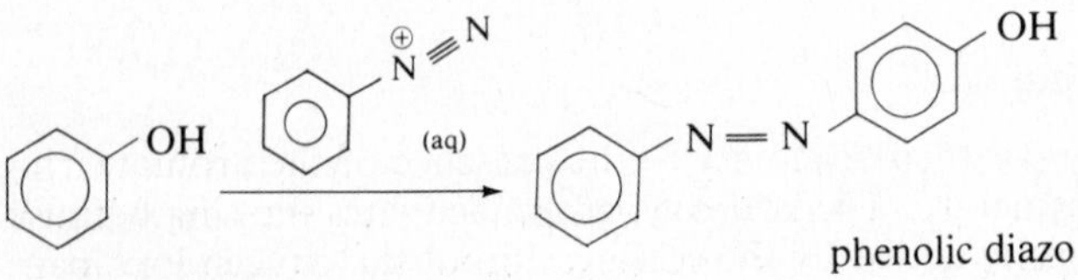

(aq)

phenolic diazo dye

Example 3 The hydroxyl group activates the ring sufficiently for nitration to be possible with dilute nitric (V) acid.

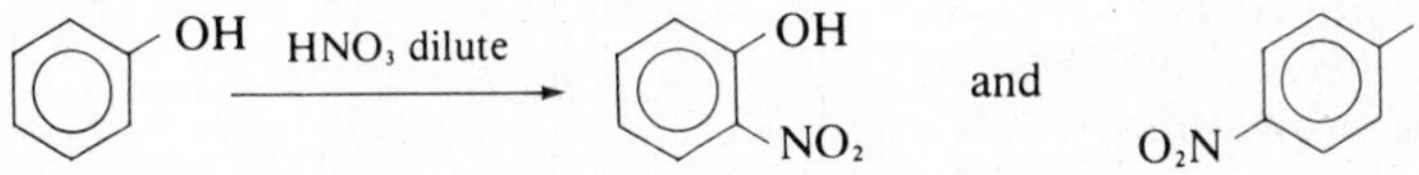

$$C_6H_5OH \xrightarrow{HNO_3\text{ dilute}} \text{2-}NO_2C_6H_4OH \text{ and } \text{4-}O_2NC_6H_4OH$$

It is thought that the electrophile responsible for the nitration is nitrous (III) acid which is present in small concentrations in dilute nitric (V) acid.

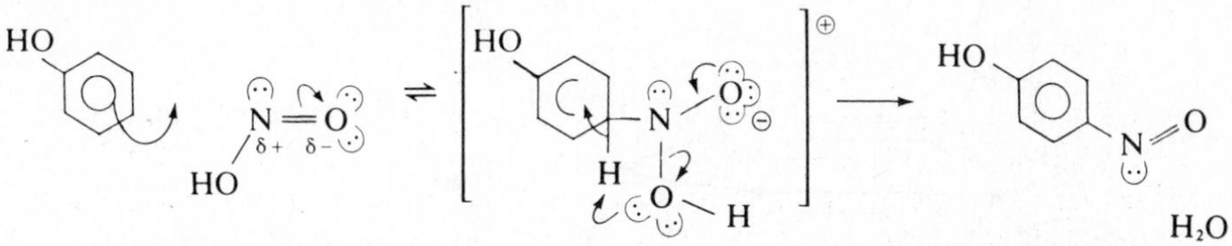

$$HOC_6H_5 + HO{-}N^{\delta+}{=}O^{\delta-} \rightleftharpoons [\,\text{intermediate}\,]^{\oplus} \longrightarrow HOC_6H_4N{=}O + H_2O$$

4-nitrosophenol

4-nitrosophenol is a reducing agent and is itself oxidized rapidly to nitrophenol by nitric (V) acid.

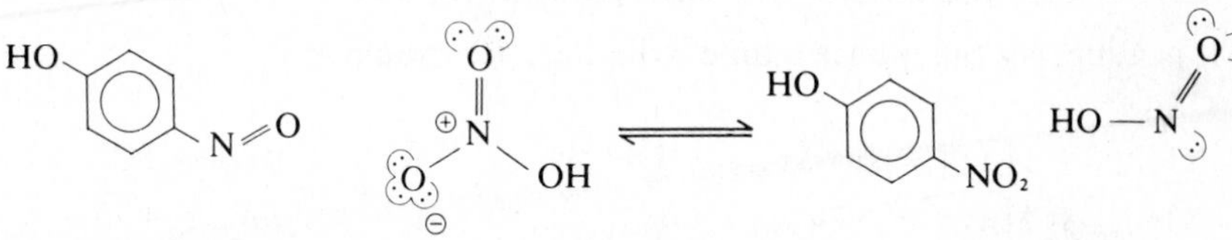

$$HOC_6H_4N{=}O + HNO_3 \rightleftharpoons HOC_6H_4NO_2 + HO{-}N{=}O$$

The regenerated nitrous (III) acid reacts with more phenol to continue the chain reaction.

33.4 Amines

The amino group

Phenylamine is an aromatic amine and one of its typical aliphatic analogues is ethylamine.

$C_6H_5NH_2$ — phenylamine

Me–CH_2–NH_2 — ethylamine

The amino groups of these two molecules differ in character in much the same way that the hydroxyl groups of a phenol and an alcohol do.

In aliphatic amines, the lone pair of electrons on the nitrogen atom of the amino group causes the group to possess:

1. basic properties
2. nucleophilic properties
3. ligand properties.

However, the nitrogen lone pair of a phenylamine molecule is extensively delocalized as shown below.

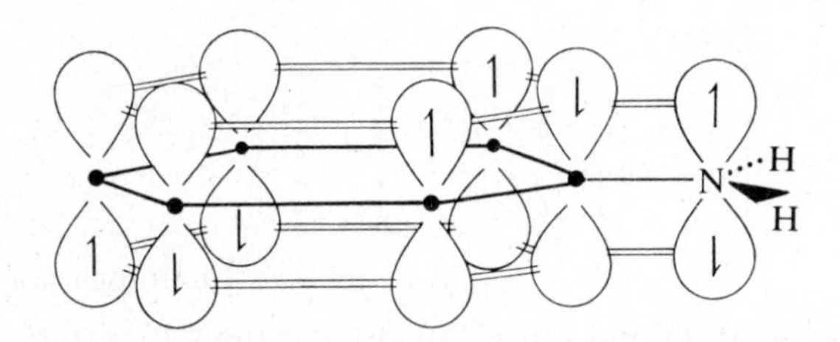

Phenylamine therefore differs from ethylamine as follows:

1. It is a much weaker base than ethylamine, for example:

$$Et{-}NH_{2(aq)} + H_2O_{(l)} \rightleftharpoons EtNH^+_{3(aq)} + OH^-_{(aq)} \quad : \quad pK_b = 3.27$$

$$NH_{3(aq)} + H_2O_{(l)} \rightleftharpoons NH^+_{4(aq)} + OH^-_{(aq)} \quad : \quad pK_b = 4.75$$

$$C_6H_5NH_{2(aq)} + H_2O_{(l)} \rightleftharpoons C_6H_5NH^+_{3(aq)} + OH^-_{(aq)} \quad : \quad pK_b = 9.38$$

where $B_{(aq)} + H_2O_{(l)} \rightleftharpoons BH^+_{(aq)} + OH^-_{(aq)}$ and $K_b = \dfrac{[BH^+]\,[OH^-]}{[B]}$

2. It is a poor nucleophile compared with ethylamine,
   e.g. phenylamine does not react with halogenoalkanes; it can, however, be ethanoylated.

$$C_6H_5NH_2 \xrightarrow{Me{-}COCl} C_6H_5{-}NH{-}C(=O){-}Me \qquad HCl_{(g)}$$

N-phenylethanamide

3. Ethylamine gives a blue complex with copper (II) like ammonia and other aliphatic primary amines. However, phenylamine gives a green complex.

$[Cu(PhNH_2)_2]^{2+}$ green $\qquad$ $[Cu(EtNH_2)_4]^{2+}$ blue

The delocalization of an aromatic amine's lone pair is rather similar to the delocalization of the nitrogen lone pair in an amide:

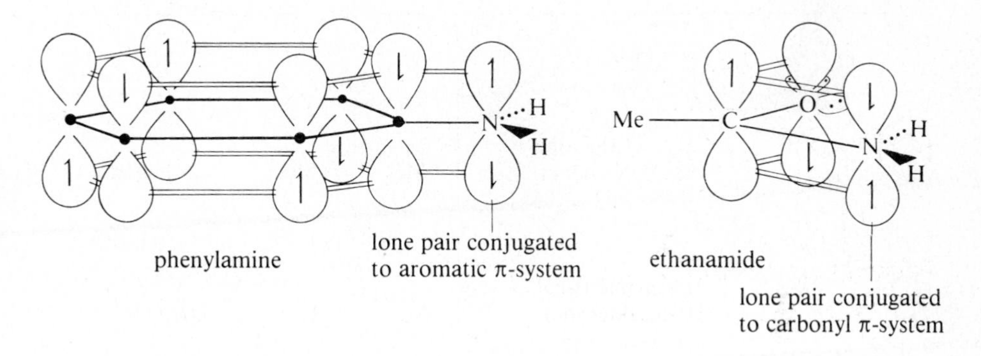

The aromatic ring

An amino group activates the aromatic ring in exactly the same way as a hydroxyl group. Both groups exert +M and −I effects on the aromatic system, but the overall result is activating.

Phenylamine reacts quickly with bromine water in the cold; it also couples with diazonium salts. It cannot, however, be nitrated by dilute nitric acid.

$$C_6H_5NH_2 + 3Br_{2(aq)} \xrightarrow{H_2O} C_6H_2Br_3NH_2 + 3H_3O^+_{(aq)} + 3Br^-_{(aq)}$$

2,4,6-tribromophenylamine

$$C_6H_5NH_2 \xrightarrow{C_6H_5N^{\oplus}\equiv N_{(aq)}} C_6H_5{-}N{=}N{-}C_6H_4{-}NH_2$$

a phenylamine diazo dye

33.5 Carbonyl compounds and acids

The carbonyl group

The carbonyl group is found in ketones, aldehydes, acids and acid derivatives. Some aromatic carbonyl compounds and their typical aliphatic analogues are shown in the table below.

aromatic		*aliphatic*	
C_6H_5–C(=O)–Me	phenylethanone (acetophenone)	Me–C(=O)–Me	propanone
C_6H_5–C(=O)–H	phenylmethanal (benzaldehyde)	Me–C(=O)–H	ethanal
C_6H_5–C(=O)–OH	phenlymethanoic acid (benzoic acid)	Me–C(=O)–OH	ethanoic acid
C_6H_5–C(=O)–O–Et	ethyl benzoate	Me–C(=O)–O–Et	ethyl ethanoate
C_6H_5–C(=O)–NH_2	benzamide	Me–C(=O)–NH_2	ethanamide

Benzoic acid and its derivatives are still most commonly given their 'trivial' names rather than those based on the name phenylmethanoic acid. Hence the amide is 'benzamide' instead of phenylmethanamide. The methods of preparing aromatic carbonyl compounds do not differ radically from those used to prepare their aliphatic analogues. Some examples are given opposite.

Ketones

1. Oxidize a secondary alcohol

   $C_6H_5CH(OH)Me \xrightarrow[H_3O^+]{MnO_4^-} C_6H_5C(=O)Me$

2. Use a Grignard reagent and a nitrile

   C_6H_5MgBr (ether) $\xrightarrow[2.\ H_3O^+]{1.\ MeCN} C_6H_5C(=O)Me$

Aldehydes

1. Oxidize a primary alcohol (low yield)

   $C_6H_5CH_2OH \xrightarrow[H_3O^+]{MnO_4^-} C_6H_5C(=O)H$

2. Hydrogenate an acid chloride over a palladium catalyst supported by barium sulphate.

   $C_6H_5C(=O)Cl \xrightarrow[\text{in xylene (reflux)}]{H_2 - Pd/BaSO_4 \text{ catalyst}} C_6H_5C(=O)H$

Acids

1. Oxidize a primary alcohol or aldehyde

   $C_6H_5CH_2OH \longrightarrow [C_6H_5C(=O)H] \rightarrow C_6H_5C(=O)OH$

2. Hydrolyse a nitrile under acidic or alkaline conditions

   $C_6H_5CN \xrightarrow[\text{or } HO^-]{\text{either } H_3O^+} C_6H_5C(=O)OH$

Acid derivatives

1. An acid chloride is first made from the corresponding acid

   $C_6H_5C(=O)OH \xrightarrow{PCl_5} C_6H_5C(=O)Cl$

2. Acid anhydrides are prepared from acid chlorides

   $C_6H_5C(=O)Cl \xrightarrow{C_6H_5COO^-Na^+} (C_6H_5C(=O))_2O$

3. Esters are prepared from alcohols

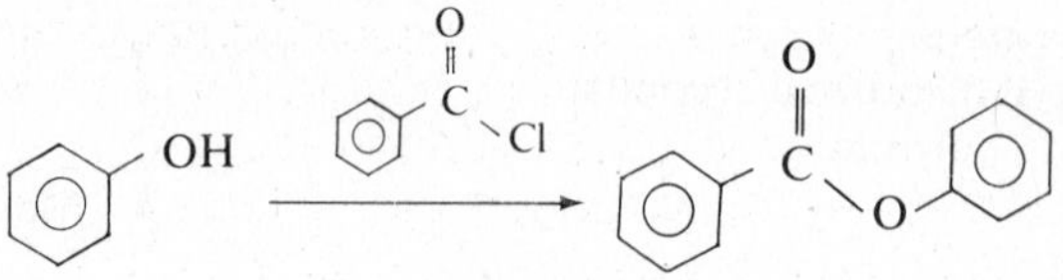

$$C_6H_5OH \xrightarrow{C_6H_5COCl} C_6H_5COOC_6H_5$$

4. Amides are prepared from amines

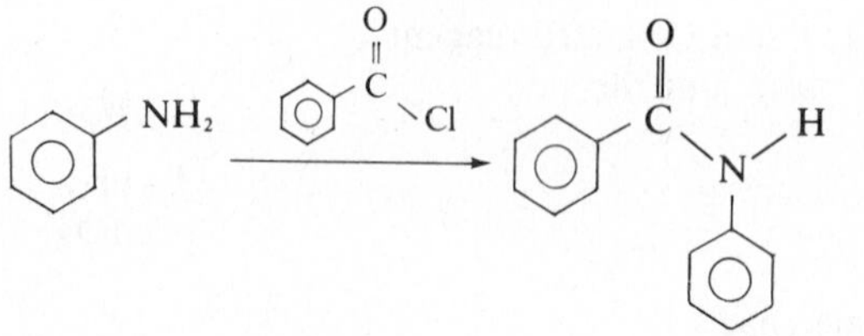

$$C_6H_5NH_2 \xrightarrow{C_6H_5COCl} C_6H_5CONHC_6H_5$$

There are two other useful methods by which an aromatic acid or aldehyde can be prepared. Both methods rely on the comparative inertness of the aromatic ring.

1. The oxidation of any carbon chain on an aromatic ring gives an acid if the oxidizing conditions are severe enough:

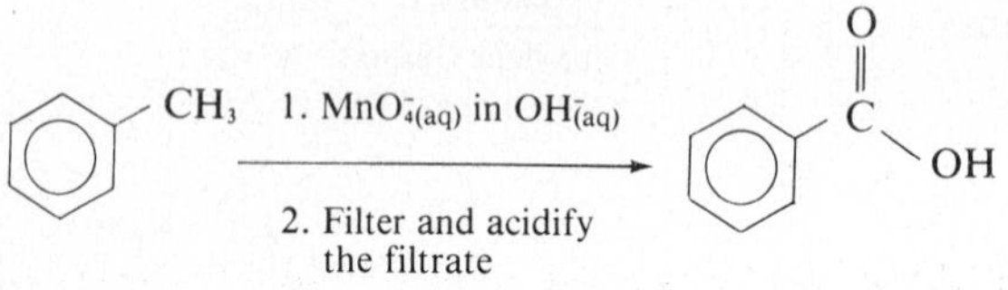

$$C_6H_5CH_3 \xrightarrow[\text{2. Filter and acidify the filtrate}]{\text{1. } MnO^-_{4(aq)} \text{ in } OH^-_{(aq)}} C_6H_5COOH$$

The aromatic ring is unaffected.

2. The selective free-radical chlorination of a carbon chain can be achieved by bubbling chlorine through an aromatic liquid in the presence of ultra-violet radiation. The aromatic ring resists chlorination unless the process is carried out in the gas phase:

gas phase

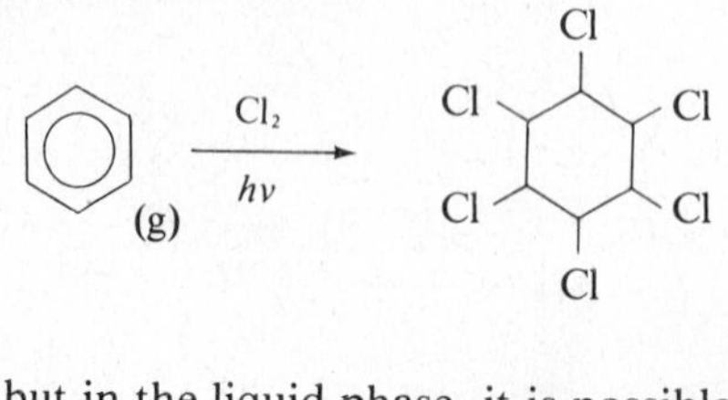

$$C_6H_{6(g)} \xrightarrow[h\nu]{Cl_2} C_6H_6Cl_6$$

1,2,3,4,5,6-hexachlorocyclohexane

but in the liquid phase, it is possible to obtain side-chain chlorinated compounds for methyl benzene in good yield:

liquid phase

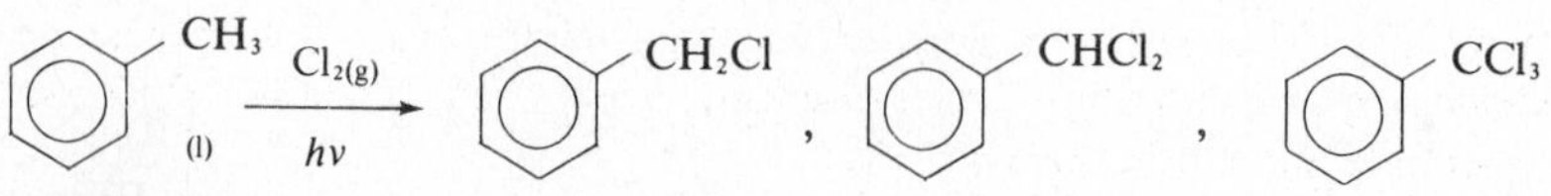

$$C_6H_5CH_{3(l)} \xrightarrow[h\nu]{Cl_{2(g)}} C_6H_5CH_2Cl, \quad C_6H_5CHCl_2, \quad C_6H_5CCl_3$$

At low temperatures, the monochloro derivative is the major product. At temperatures around 70°C, the trichloro derivative is produced almost quantitatively from methylbenzene.

By fractional distillation, each chloro-derivative can be isolated. The hydrolysis products of each are shown below.

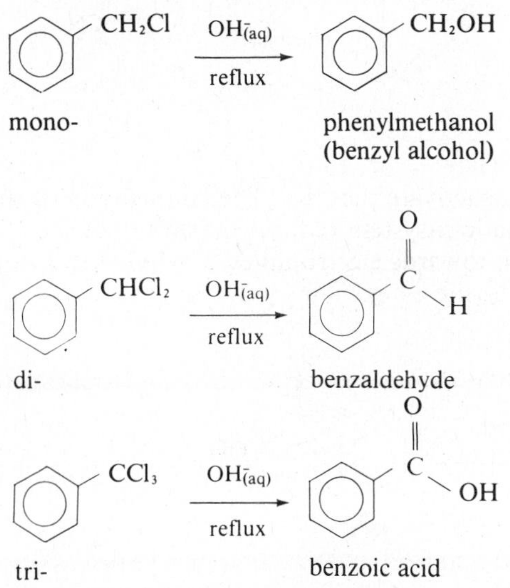

$C_6H_5CH_2Cl \xrightarrow[\text{reflux}]{OH^-_{(aq)}} C_6H_5CH_2OH$

mono- phenylmethanol (benzyl alcohol)

$C_6H_5CHCl_2 \xrightarrow[\text{reflux}]{OH^-_{(aq)}} C_6H_5CHO$

di- benzaldehyde

$C_6H_5CCl_3 \xrightarrow[\text{reflux}]{OH^-_{(aq)}} C_6H_5COOH$

tri- benzoic acid

This process is a good method for making benzaldehyde in particular. It is always difficult to use oxidation-reduction methods to give good yields of an aldehyde and so the above method is a useful alternative strategy.

The reactivity of aromatic carbonyl compounds is slightly less than that of their aliphatic analogues. They do however undergo the same types of reaction as the aliphatic carbonyl compounds. The reason for the reduced reactivity can again be attributed to conjugation effects. For example, benzaldehyde is illustrated below.

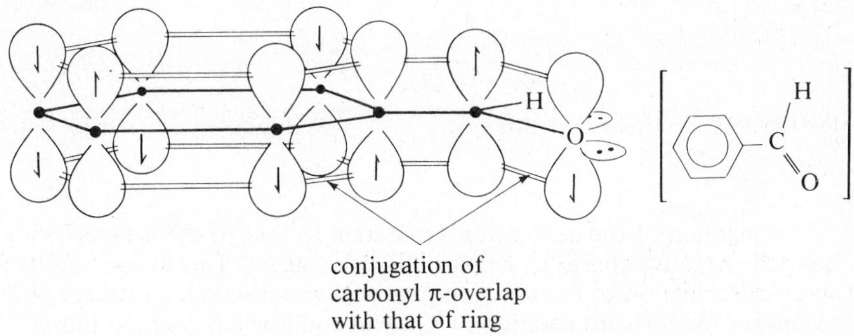

H
O
H
C
O

conjugation of carbonyl π-overlap with that of ring

The main effect of the conjugation is the delocalization of the electron-deficiency of the carbonyl carbon atom. This leads to reduced reactivity towards nucleophile reagents.

Example Benzaldehyde reacts very slowly with hydrogen cyanide

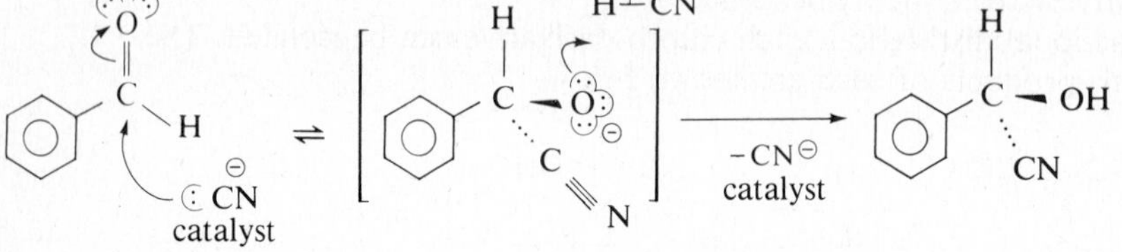

$$C_6H_5CHO \rightleftharpoons [C_6H_5CH(O^-)CN] \xrightarrow[\text{catalyst}]{-CN^-} C_6H_5CH(OH)CN$$

$\ominus$: CN catalyst; H—CN

The aromatic ring

A carbonyl substituent tends to deactivate the ring. The conjugation of the carbonyl π-system and the aromatic π-system reduces the ability of the aromatic π-electrons to be drawn towards electrophiles. Carbonyl substituents direct substitution to the 3- and 5- sites, e.g.

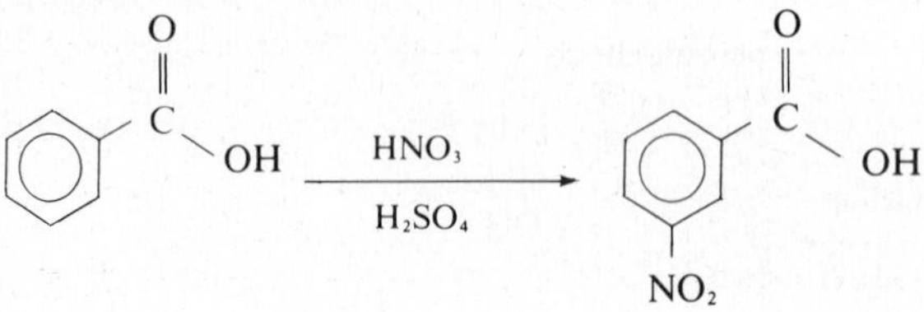

$$C_6H_5COOH \xrightarrow[H_2SO_4]{HNO_3} 3\text{-}NO_2C_6H_4COOH$$

Acidity

Aromatic acids vary in strength according to the nature of the aromatic ring to which the acid group is bonded. Benzoic acid itself is a stronger acid than its aliphatic analogue, ethanoic acid.

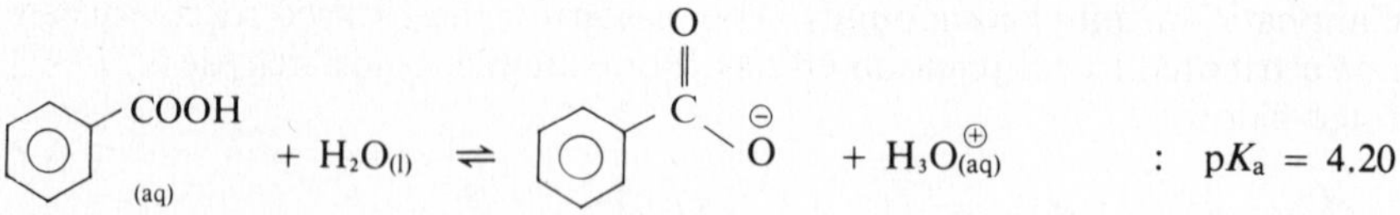

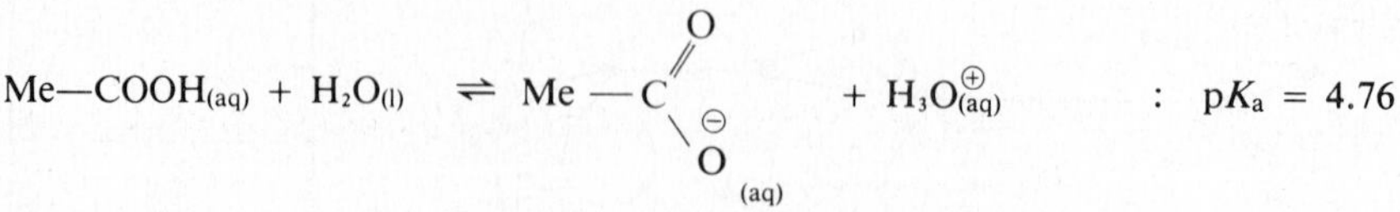

$$C_6H_5COOH_{(aq)} + H_2O_{(l)} \rightleftharpoons C_6H_5COO^- + H_3O^+_{(aq)} \quad : \ pK_a = 4.20$$

$$Me\text{—}COOH_{(aq)} + H_2O_{(l)} \rightleftharpoons Me\text{—}COO^-_{(aq)} + H_3O^+_{(aq)} \quad : \ pK_a = 4.76$$

The conjugation of the acid group's π-system to that of the aromatic ring allows the negative charge to be partially delocalized. This lowers the free energy of formation of the ion (compare with the discussion on page 461) and thus makes the forward reaction of the acid equilibrium reaction more favourable. The methyl group of the ethanoate ion is unable to behave in this way.

A substituent on the aromatic ring can further affect the acidity of either a phenolic or acidic group attached to the same ring. The figures for the pK_a values of some substituted phenols and aromatic acids are shown opposite.

aromatic acids		*phenols*	
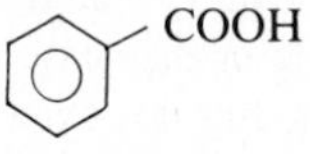 COOH	$pK_a = 4.20$	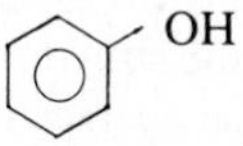 OH	$pK_a = 10.0$
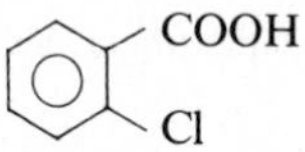 COOH, Cl	$pK_a = 2.94$	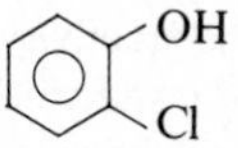 OH, Cl	$pK_a = 8.48$
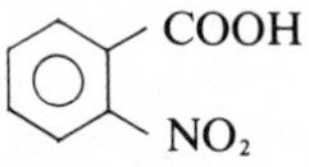 COOH, NO_2	$pK_a = 2.17$	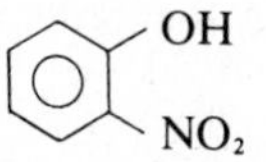 OH, NO_2	$pK_a = 7.21$

The chloro- and nitro- groups are electron-withdrawing in their properties. These groups therefore can stabilize the carboxylic acid anion group by their ability to draw the extra electron density towards themselves. The charge of the anion is more effectively delocalized and thus the anion more stable.

An interesting case is 2-hydroxybenzoic acid. It is able to act as a dibasic acid because it has both a phenolic and a carboxylic acid proton. The two pK_a figures are 2.97 and 13.40.

The first proton is very easily transferred but the second is even more difficult than would be expected for a phenolic proton. This is due to neighbouring group interaction:

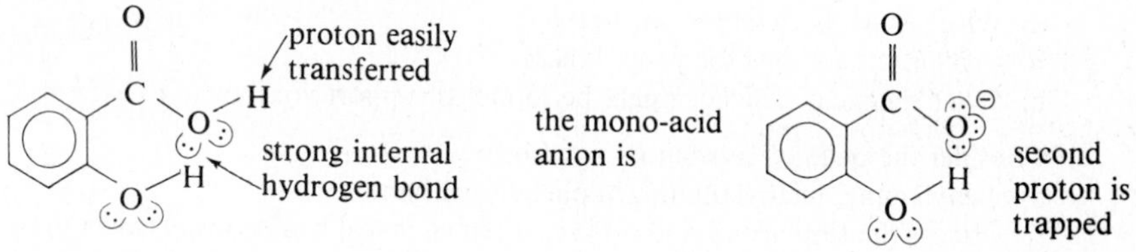

proton easily transferred

strong internal hydrogen bond

the mono-acid anion is

second proton is trapped

Questions

1. Illustrate the words 'aromatic' and 'aliphatic' by discussing the molecules whose structures are shown below.

a) 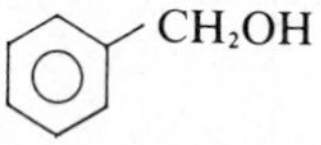 CH_2OH b) 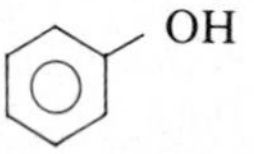 OH c) 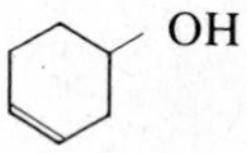 OH d) CH_3OH

2. i) Name the compounds whose structures are given in question 1.
   ii) Name a reagent that reacts with both b) and c). Give the structural formulas of the products in each case.
   iii) Name a reagent that reacts with both a) and d) but *not* with b). Explain why b) does not react.

3. Phenylamine is a weaker base than ethylamine but phenol is a stronger acid than ethanol. Explain these observations with the help of clear diagrams.

4. a) Give the equations for *two* reactions in which benzaldehyde and ethanal react in the same way. Draw the structural formulas of the products in each case.
   b) Give the equations for *two* reactions given by benzaldehyde but *not* by ethanal.
   c) Give the equations for *two* reactions given by ethanal but *not* by benzaldehyde.

5. Compare and contrast the reactivity of the following pairs of substances: benzene and chlorobenzene; ethane and chloroethane.
   Why does the substituent have such a different effect?

6. a) Put the following acids in order of acid strength.

   COOH (on benzene ring) $\quad$ CH_3—COOH $\quad$ H—COOH $\quad$ COOH (on benzene ring with two Cl)

   b) Give the reasons for the choice that you have made.

7. The strength of the carbon-carbon bond (marked *) in the following two molecules is *not* the same.

   (benzene ring)*—C(=O)—Me $\quad$ Me—*C(=O)—Me

   a) Which is likely to be the stronger?
   b) Explain the reasons for your choice.
   c) What chemical evidence might be found to support your answer?

8. Consider the pairs of substances listed below,
   a) phenylamine, methylamine b) phenol, methanol.
   i) Discuss the similarities and differences in chemical behaviour between the members of each pair.
   ii) In what ways are phenylamine and phenol similar?
   iii) In what ways are methylamine and methanol similar?

9. In Chapter 31, the following generalization was made: 'An aromatic molecule contains $(4n+2)$ delocalized π-electrons in a cyclic structure'. Discuss the aromaticity of the particles whose structures are shown below.

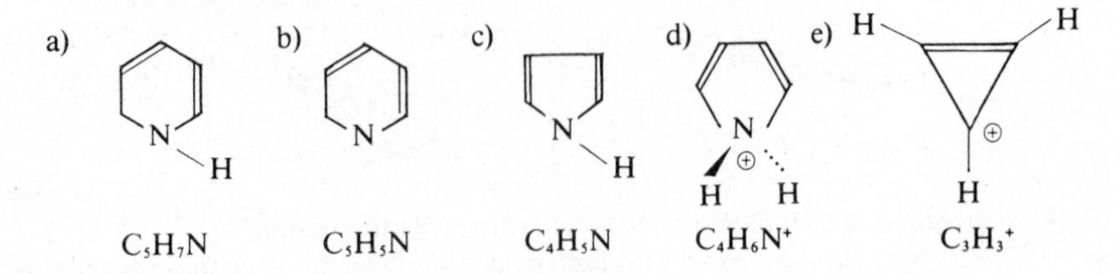

10. 'Aliphatic compounds are more reactive than aromatic compounds'. Discuss.

Chapter 34

Making carbon-carbon bonds

34.1 Introduction

The importance of carbon-carbon bonds

All living organisms synthesize a remarkable range of organic molecules often starting from the simplest of carbon chain sources. For plants the initial carbon source can be traced as far back as carbon dioxide, and for many animals to carbohydrates and fats. The molecular chains of these carbohydrates and fats are broken down and reformed in highly selective ways to generate the huge variety of organic molecules required to sustain the animal's life. In all these internal processes, carbon-carbon bonds are constantly both broken and made. An organism accomplishes this with the help of a number of extremely specific catalysts called enzymes. The mechanism of most enzyme-catalysed reactions is difficult to unravel and it is usually not possible to produce the equivalent conditions in a laboratory synthesis. In the drugs industry, where complex organic compounds need to be produced to high levels of purity, the chemical methods of making carbon-carbon bonds are particularly important. The majority of pharmaceutical processes are designed to synthesize a required drug from the cheapest and most readily available starting materials. Very rarely do these compounds possess the same molecular carbon framework as the final drug to be made. Ways must be found of bringing about the necessary alterations to the carbon framework and this involves both breaking and making carbon-carbon bonds. The laborious methods that often need to be used in the laboratory appear doubly frustrating to the research chemist when he knows that a particular organism may be able to employ a specific enzyme to achieve the conversion in one step: in the laboratory, as many as a dozen steps may be needed. And after each step, it may be necessary to separate a required intermediate from the reaction mixture.

Considerable research is being directed into the mechanisms of enzyme catalysis: a pharmaceutical company that finds the appropriate laboratory conditions to simulate the reaction conditions of a live organism, will undoubtedly have achieved a more efficient synthetic route.

To illustrate the significance of methods of making carbon-carbon bonds, a typical pharmaceutical process is shown in outline overleaf. Two carbon-carbon bonds are made per molecule during the synthesis of penicillin (V) from its major ingredients of valine, phenoxyethanoic acid and hydrogen sulphide.

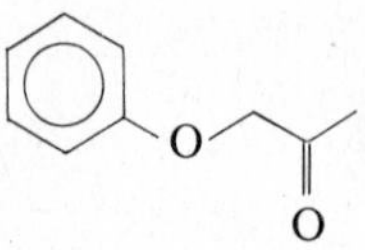

phenoxyethanoic acid 　　hydrogen sulphide 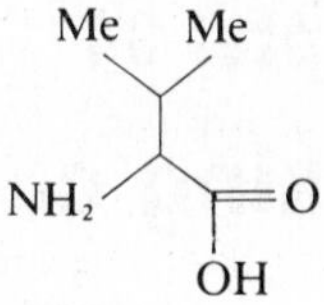　　valine

These ingredients are combined in a twelve-step process which employs a carbon-carbon bond formation strategy. The final product is pencillin (V).

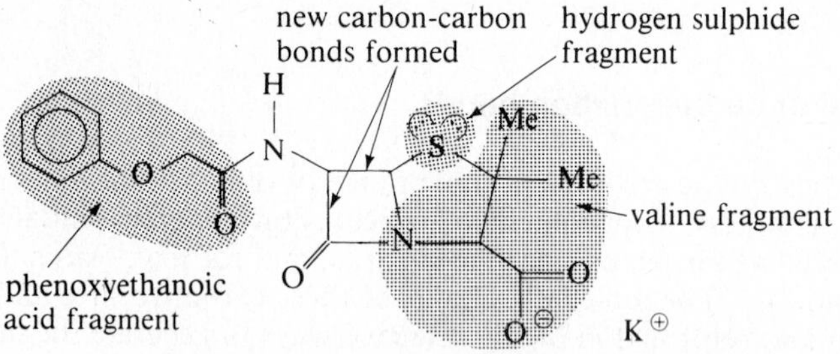

new carbon-carbon bonds formed　hydrogen sulphide fragment　valine fragment　phenoxyethanoic acid fragment

pencillin (V)

The difficulty in making carbon-carbon bonds

On page 354 the following characteristics were suggested as necessary for a collision between two particles to result in the redistribution of electron-pairs:

1. One particle should have an electron-deficient site (Lewis acidity).
2. The other particle should have a pair of electrons that can be easily transferred, e.g. a lone pair or a π-pair (Lewis basicity).

When these ideas are applied to the problem of forming carbon-carbon bonds, there is little difficulty in identifying molecules with electron-deficient carbon atoms, e.g. $\overset{\delta+}{\text{C}}-\text{Hal}$ or $\overset{\delta+}{\text{C}}=\overset{\delta-}{\text{O}}$.

The problem is to find a particle with sufficient Lewis basicity associated with a carbon atom. Carbon atoms tend not to have lone pairs: they are four-valent and form strong bonds, and although there are many molecules that possess carbon π-electrons, the Lewis basicity of these particles is often not sufficient to initiate successful attack on a source of electron-deficient carbon.

To overcome these difficulties, there are therefore two principal strategies.

1. Nucleophilic carbon is generated by producing particles that have a carbon atom with a certain amount of lone-pair character.
2. Unsaturated carbon is induced to attack a source of electron-deficient carbon i.e. electrophilic carbon.

34.2 Nucleophilic carbon

Cyanide ions

Carbonyl compounds react with hydrogen cyanide in the presence of a little potassium cyanide. The cyanide ions are a source of nucleophilic carbon.

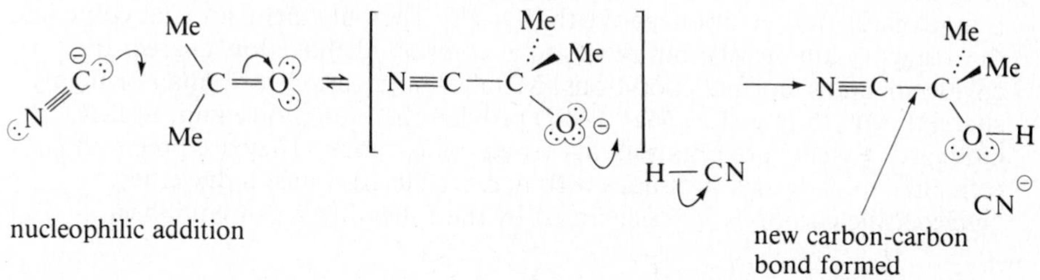

$$N{\equiv}C^{\ominus} + Me_2C{=}O \rightleftharpoons [N{\equiv}C{-}CMe_2{-}O^{\ominus}] \xrightarrow{H{-}CN} N{\equiv}C{-}CMe_2{-}O{-}H + CN^{\ominus}$$

nucleophilic addition

new carbon-carbon bond formed

Halogenoalkanes react when refluxed in aqueous potassium cyanide.

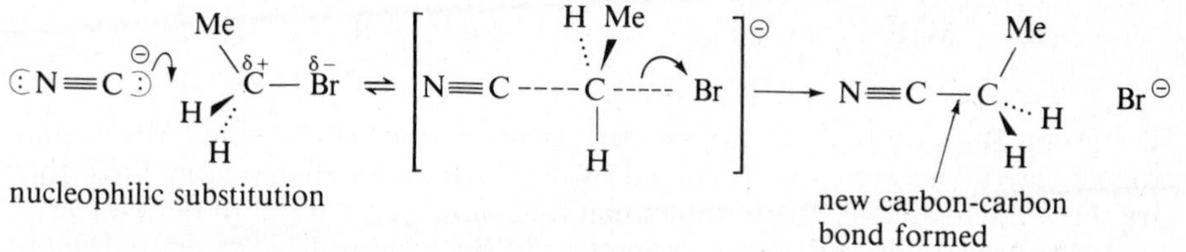

$$N{\equiv}C^{\ominus} + MeCH_2^{\delta+}{-}Br^{\delta-} \rightleftharpoons [N{\equiv}C\text{----}C(H)(Me)(H)\text{----}Br]^{\ominus} \longrightarrow N{\equiv}C{-}CH_2Me + Br^{\ominus}$$

nucleophilic substitution

new carbon-carbon bond formed

The nitriles produced are readily hydrolysed into acids and then the acids can be further converted to their derivatives if necessary.

Example Convert Et—Br to Et—C(=O)—N(Et)—Et

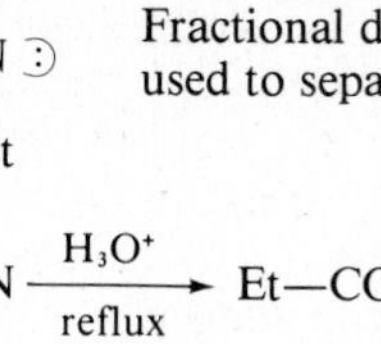

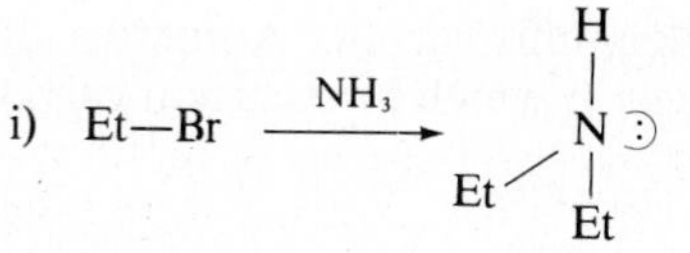

i) $Et{-}Br \xrightarrow{NH_3} Et_2NH$ and other amines. Fractional distillation is used to separate them.

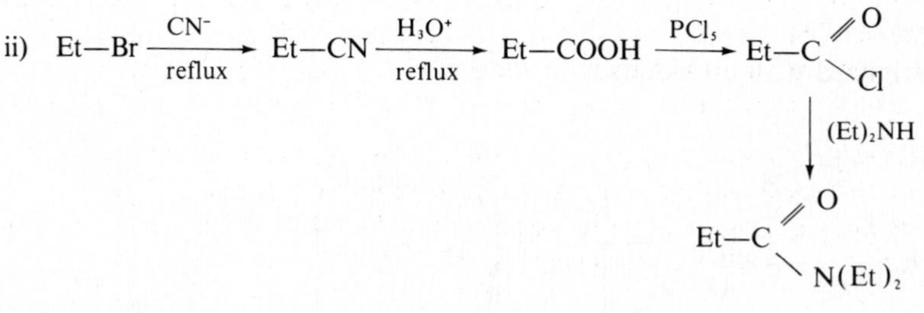

ii) $Et{-}Br \xrightarrow[\text{reflux}]{CN^-} Et{-}CN \xrightarrow[\text{reflux}]{H_3O^+} Et{-}COOH \xrightarrow{PCl_5} Et{-}C(=O){-}Cl \xrightarrow{(Et)_2NH} Et{-}C(=O){-}N(Et)_2$

Grignard reagents

In the majority of organic molecules, the carbon atoms are bonded either to more electronegative atoms such as oxygen, halogen or nitrogen atoms, or to other carbon or hydrogen atoms. In very few of these structures can carbon show any nucleophilic properties. The basic idea in forming a 'Grignard reagent' is to produce a structure in which carbon is bonded to an element that is appreciably *less* electronegative than itself. The only elements that come into this category are metals, but most metal compounds have low degrees of covalency under normal conditions. Metals whose cations are small or highly charged or both (e.g. Li^+, Mg^{2+}, Al^{3+}) provide the best choice and, in fact, Grignard reagents are organo-magnesium compounds. They are prepared by refluxing alkyl or aryl bromides with magnesium turnings in dry ether. Nucleophilic carbon $R\bar{:}$ is generated by the following set of equilibria.

$$R{-}Br + Mg \rightleftharpoons R{-}Mg{-}Br$$

$$R{-}Mg{-}Br + R{-}Mg{-}Br \rightleftharpoons R{-}Mg{-}R + MgBr_2$$

$$\updownarrow \qquad\qquad\qquad\qquad \updownarrow$$

$$R\bar{:} \quad {}^{+}MgBr \qquad\qquad R\bar{:} \quad Mg^{2+} \quad \bar{:}R$$

The group R— is usually an alkane derivative or an aromatic ring. Alkene and alkyne derivatives are easily prepared by exchange from alkane Grignards: they are thermodynamically more stable than the 'saturated' Grignard derivatives because of the ability of their π-systems to be delocalized toward the magnesium ions. For example, an ethyne Grignard is formed by bubbling ethyne into a refluxing mixture of bromoethane and magnesium in ether

$$Et{-}Br + Mg \xrightleftharpoons{\text{ether}} Et{-}Mg{-}Br$$

$$Et{-}Mg{-}Br + H{-}C{\equiv}C{-}H \underset{\text{exchange}}{\rightleftharpoons} Et{-}H + H{-}C{\equiv}C{-}Mg{-}Br$$

A Grignard reagent is a powerful source of nucleophilic carbon. A number of different classes of compounds can be synthesized in which new carbon-carbon bonds are made.

1. Alcohols

React a Grignard with an aldehyde or ketone

$$R_1{-}Mg{-}Br + R_2{-}\overset{H}{C}{=}\overset{\delta-}{O}\!: \;(C^{\delta+}) \rightleftharpoons \left[R_1{-}C(H)(R_2){-}O^{\ominus} \quad {}^{\oplus}MgBr \right] \xrightarrow{H_3O^{\oplus}(aq)} R_1{-}C(H)(R_2){-}O{-}H$$

2. Aldehydes and ketones
React a Grignard with a nitrile

$$R_1-MgBr + R_2-C\equiv N \rightleftharpoons \left[R_1-C(R_2)=N^{\ominus} \; {}^{\oplus}MgBr \right] \xrightarrow[(-NH_3)]{H_3O^{\oplus}_{(aq)}} R_1-C(=O)-R_2$$

3. Acids
React a Grignard with carbon dioxide

$$R-MgBr + O=C=O \rightleftharpoons \left[R-C(=O)-O^{\ominus} \; MgBr^{\oplus} \right] \xrightarrow{H_3O^{\oplus}_{(aq)}} R-C(=O)-OH$$

Base-catalysed condensation

Nucleophilic carbon can also be produced under conditions where the acidity of a carbon-hydrogen bond is of sufficient strength. Carbon-hydrogen bonds are not in general acidic. There are two structures, however, from which low concentrations of 'carbanions' are available (see pages 394 and 434).

1. Ethynes dissolved in liquid ammonia in the presence of sodium amide:

$$R-C\equiv C-H + {}^{\ominus}NH_2 \rightleftharpoons R-C\equiv C^{\ominus} + NH_3$$

carbanion

2. Keto-compounds that possess activated α-hydrogen atoms: an aqueous base or ethanolic base is used (i.e. NaOH in H_2O or NaOEt in EtOH).

$$R_1-C(=O)-CH_2-C(=O)-R_2 \rightleftharpoons R_1-C(=O)-CH=C(OH)-R_2 \rightleftharpoons \left[R_1-C(=O)-CH^{\ominus}-C(=O)-R_2 \right] + H-O-Et$$

keto-enol tautomerism; ${}^{\ominus}O-Et$; carbanion

This carbanion is stabilized by delocalization of the charge on to both oxygen atoms. The three canonical forms of the carbanion are:

$$R_1-C(=O)-\overset{\ominus}{C}H-C(=O)-R_2 \;\text{(drawn as } R_1-C(=O)-CH=C(-O^{\ominus})-R_2\text{)} \longleftrightarrow R_1-C(=O)-\overset{\ominus}{C}H-C(=O)-R_2 \longleftrightarrow R_1-C(-O^{\ominus})=CH-C(=O)-R_2$$

The two carbonyl groups adjacent to the α-carbon activate the α-hydrogen atoms because of the comparative stability of carbanion generated. Even so, the equilibrium lies to the left: low concentrations of carbanions are produced.

Either of these sources of carbanions can be used to form carbon-carbon bonds. They attack any molecules possessing electron-deficient sites.

Example 1 Add a bromoalkane to the alkyne-sodium amide mixture in ammonia.

$$R_1-C\equiv C:^{\ominus} + H_2\overset{\delta+}{C}(R)-\overset{\delta-}{Br} \rightleftharpoons \left[R_1-C\equiv C\text{----}CH_2(R)\text{-----}Br\right]^{\ominus} \rightarrow R_1-C\equiv C-CH_2R$$

longer-chain unsaturated hydrocarbon

Example 2 Add a carbonyl compound to the same reaction medium.

$$R_1-C\equiv C:^{\ominus} + R_2R_3\overset{\delta+}{C}=\overset{\delta-}{O} \rightleftharpoons \left[R_1-C\equiv C-C(R_2)(R_3)-O^{\ominus}\cdots H-NH_2\right] \longrightarrow R_1-C\equiv C-C(OH)(R_2)R_3$$

an unsaturated alcohol

The situation in the keto-enol mixture is rather more complex. There is already a source of electron-deficient carbon atoms present in the keto-compound itself. Condensation takes place which, in certain cases, prevents the possibility of any other reactions taking place. On page 444, the 'aldol' condensation of ethanal is described. This occurs very rapidly in aqueous alkali via the mechanism below.

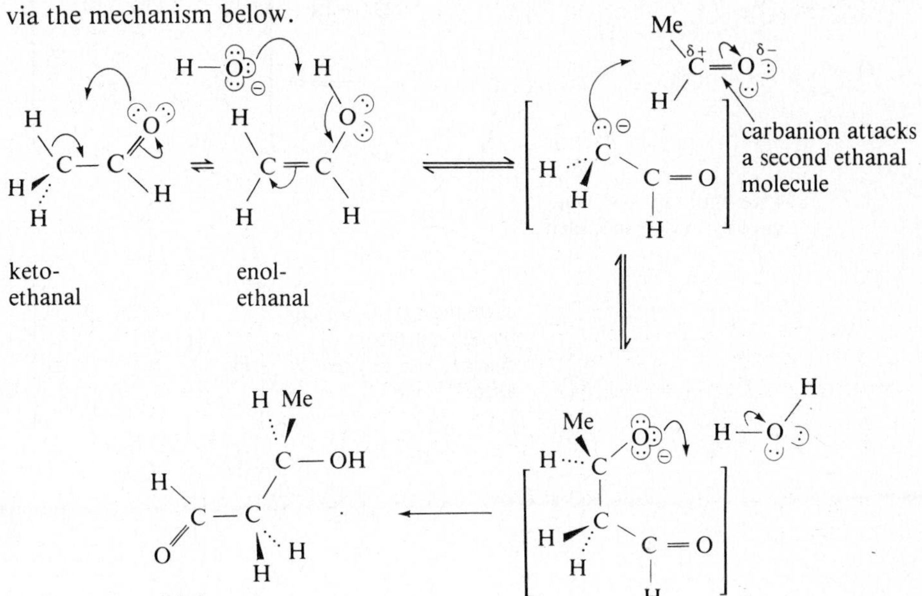

keto-ethanal ⇌ enol-ethanal ⟶ carbanion attacks a second ethanal molecule ⟶ aldol

Self-condensation therefore competes with any other reactions that are intended to occur by the addition of another reactant. As a synthetic reaction, a base-catalysed condensation is in this sense rather unselective. Yields of desired product are sometimes as low as 20%. Here are three main examples of the technique:

Example 1 The Claisen reaction This is the base-catalysed condensation of an ester (containing α-hydrogen) and an aldehyde that does not have α-hydrogen. It is usually carried out in ethanol and with sodium ethoxide as a base:

For example,

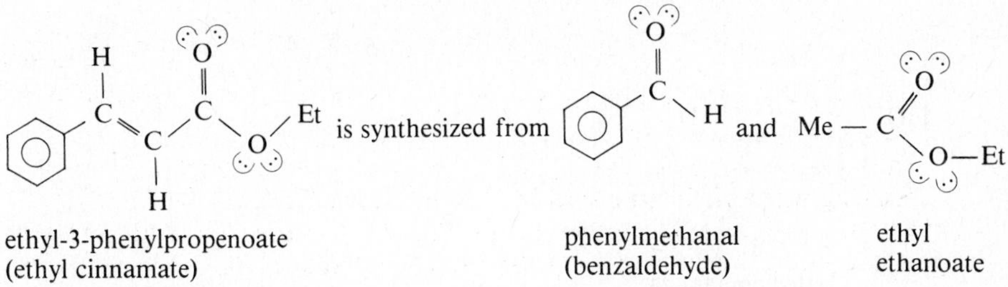

ethyl-3-phenylpropenoate (ethyl cinnamate) is synthesized from phenylmethanal (benzaldehyde) and ethyl ethanoate ($Me—COO—Et$)

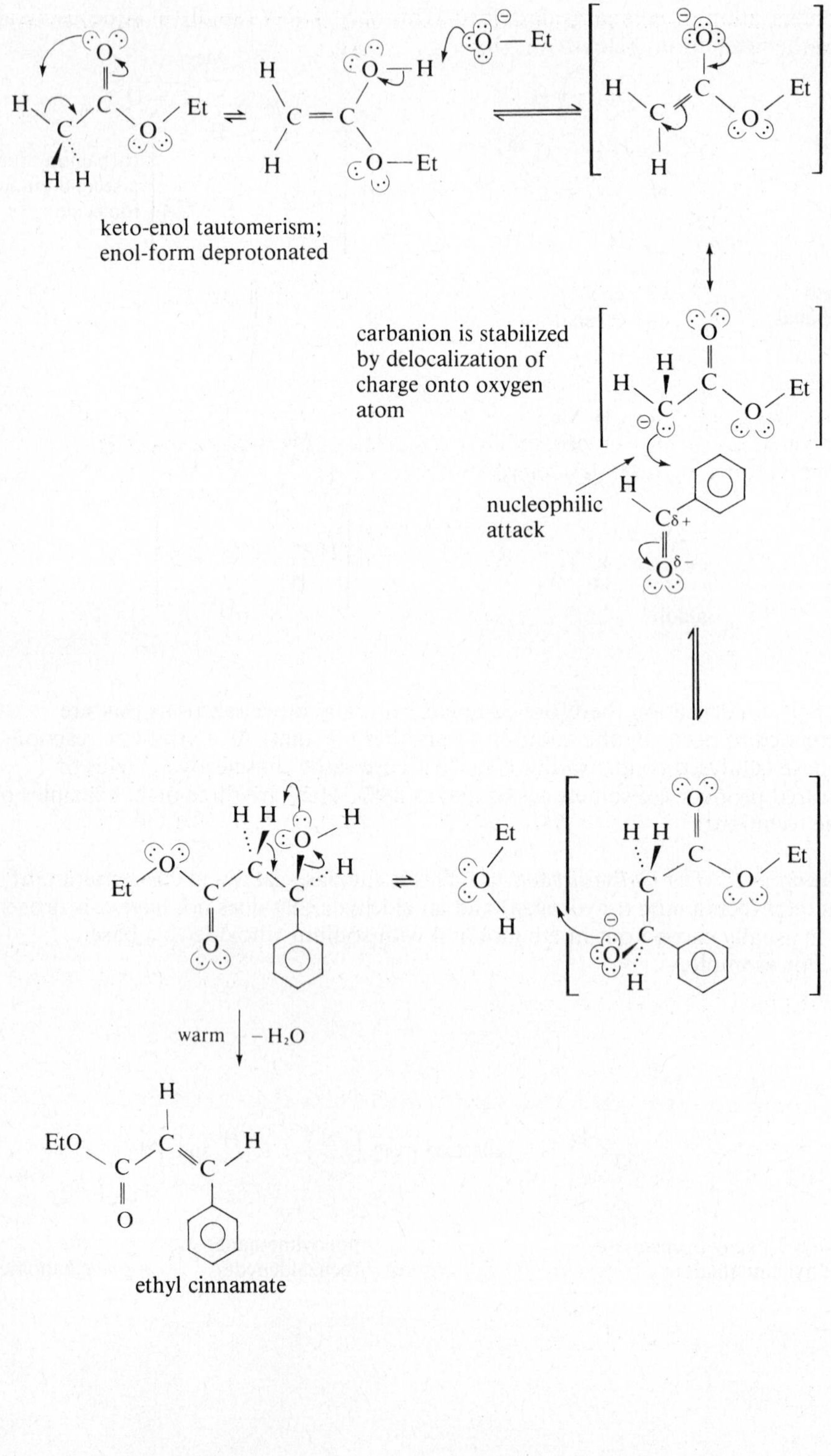

keto-enol tautomerism;
enol-form deprotonated

carbanion is stabilized
by delocalization of
charge onto oxygen
atom

nucleophilic
attack

warm $-H_2O$

ethyl cinnamate

The addition product is readily dehydrated to a more thermodynamically stable ester which is conjugated throughout its structure.

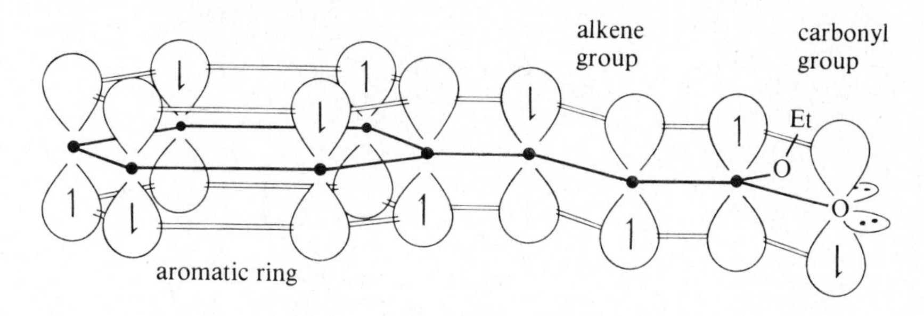

Example 2 The Knoevenagel reaction This reaction employs a doubly activated source of α-hydrogen atoms. The most common structure is diethylpropandioate (ethyl malonate).

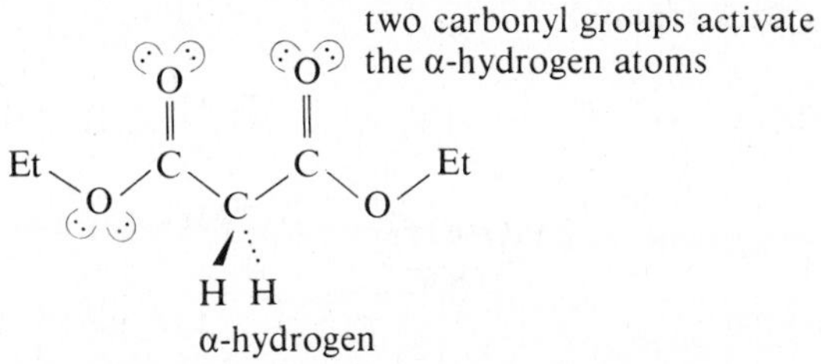

two carbonyl groups activate the α-hydrogen atoms

O O
C C
Et O C O Et
H H
α-hydrogen

The advantage of this structure is that a weaker base can be used which:

1. is still able to give a high enough concentration of carbanions for condensation to occur, but
2. minimizes the wastage due to self-condensation of the aldehyde or ketone reactant.

The resulting product is bifunctional. It is readily decarboxylated by hydrolysis and heat. For example, ethanal can be converted to 3-hydroxybutanoic acid, using piperidine as a weak base (B):

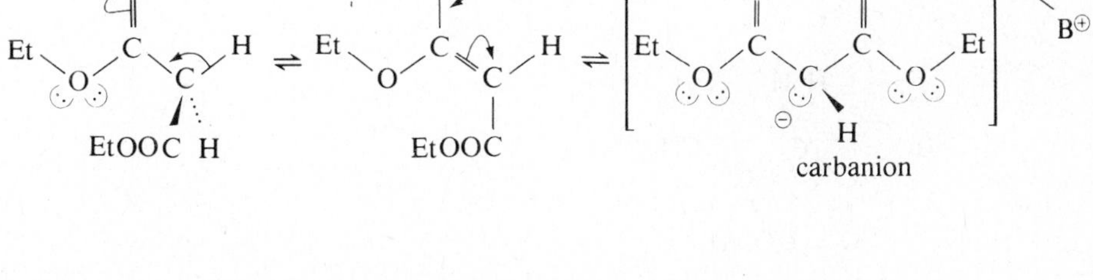

Et O C C H EtOOC H ⇌ Et O C C H EtOOC H B ⇌ [Et O C C C O Et H] H B⊕

carbanion

This carbanion is stabilized by the action of both carbonyls.

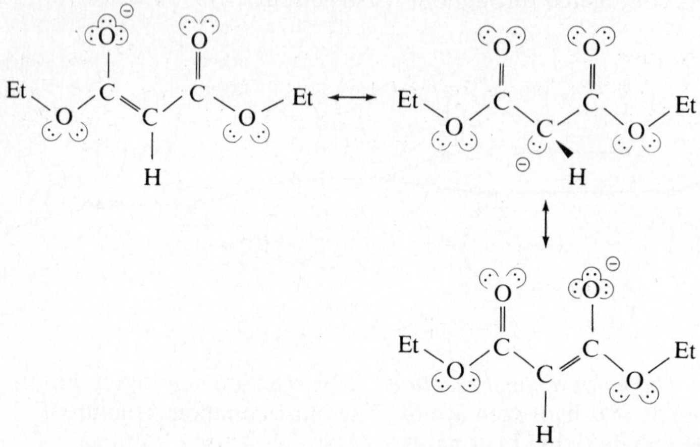

O Et C C O Et C O H ⟷ O O Et C C Et O C O H

↕

O O Et C C Et O C O H

It acts as a nucleophile to attack ethanal

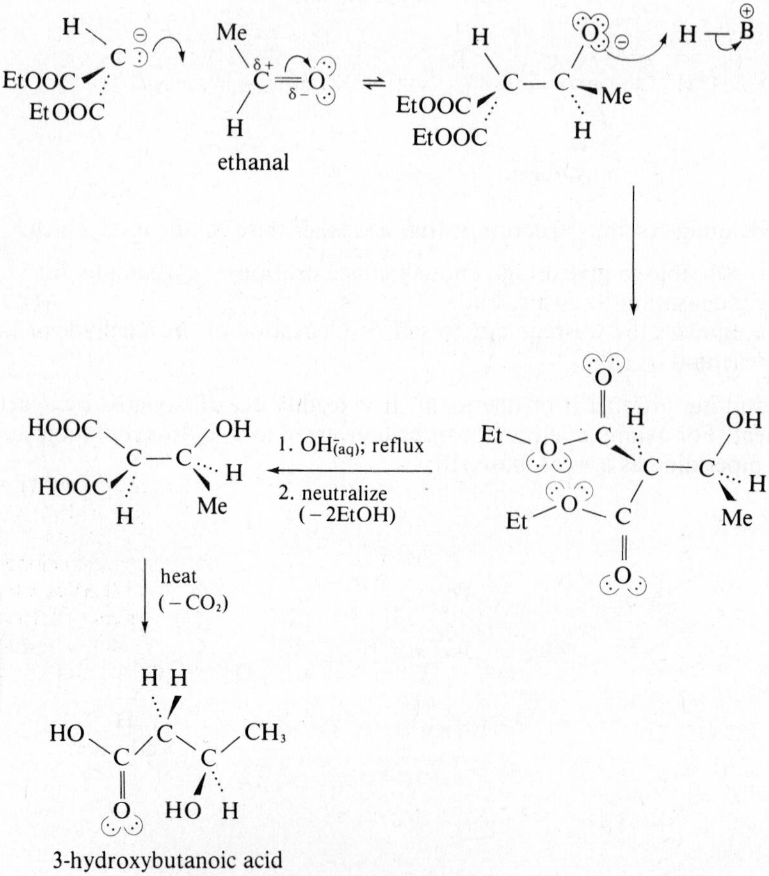

H C EtOOC EtOOC Me δ+ C O δ− H ⇌ H O H B EtOOC C C Me EtOOC H

ethanal

HOOC OH C C H HOOC H Me

1. $OH^-_{(aq)}$; reflux

2. neutralize

(−2EtOH)

O H OH Et O C C C H O Et C Me O

heat

$(-CO_2)$

H H HO C CH$_3$ C C O HO H

3-hydroxybutanoic acid

Example 3 Using halogen compounds In principle, it should be possible for a nucleophilic substitution to result in carbon-carbon bond formation, if, for example, a bromoalkane is added to a source of carbanions. In practice, the base present in the system often acts as a nucleophile itself and displaces the bromine from the bromoalkane. To overcome this, a basic particle that is as bulky as possible is used: its bulkiness hinders its nucleophilic properties. A typical example is the anion of the sodium salt of the tertiary alcohol, methylpropan-2-ol.

$(Me)_3C-O^{-}\ Na^{+}$ dissolved in $(Me)_3C-OH$

An interesting synthesis of cyclobutylmethanoic acid starts from diethyl malonate (the same bifunctional ester described on page 533); 1,3-dibromopropane is added to the reaction mixture.

$$CH_2(COOEt)_2 \xrightleftharpoons{(Me)_3C-O^{-}} {}^{-}CH(COOEt)_2 \quad \text{as before}$$

Br, δ−, Br, δ+, EtO, H, OEt, O, O ⇌ [Br, Br, H, H, EtO, H, OEt, O, O]⊖ → Br⊖, Br, EtO, H, OEt, O, O

$(Me)_3C-O^{-}$

2nd α-hydrogen that can be deprotonated by the base to give a new carbanion that undergoes internal nucleophilic attack. Cyclization results

[δ−, Br, δ+, OEt, EtO, O, O] → Br⊖, O, OEt, O, OEt

1. hydrolyse 2. heat (as before)

OH, O cyclobutylmethanoic acid

34.3 Electrophilic carbon

Friedel-Crafts catalysts

The π-electrons of an aromatic ring can be induced to attack the electron-deficient carbon atom of a halogen derivative by using a Friedel-Craft catalyst. These catalysts are described on page 490. They consist of either aluminium chloride, or iron filings in the presence of chlorine. The metal ions tend to polarize organic halogen derivatives such that the electron-deficiency of the halogen-bearing carbon atom is increased.

$$Al^{3+}(Cl^{\ominus})_3 \quad :Br{-}R \rightleftharpoons [AlCl_3Br]^{\ominus}\,[R^{\oplus}]$$

A carbon-carbon bond is made as a result of the attack of the aromatic π-electrons on the activated electrophile carbon source.

Example Benzene is converted to phenylethanone as shown below.

$$C_6H_6 + CH_3{-}C({=}O){-}Cl + AlCl_3 \rightleftharpoons \left[C_6H_5(H){-}C({=}O)CH_3 \quad :Cl{-}AlCl_3^{\ominus} \right]^{\oplus} \longrightarrow C_6H_5{-}C({=}O){-}CH_3 + HCl + AlCl_3$$

new carbon-carbon bond

phenylethanone AlCl$_3$

Friedel-Crafts catalysts can also be used to activate electrophiles for attack by alkene π-electrons. However, a number of side-reactions usually makes the process synthetically rather useless: rearrangement of both the alkene and the halogen derivative often occurs.

The Mannich reaction

The Mannich reaction is a more satisfactory method for inducing aliphatic carbon π-electrons to attack electrophilic carbon. The source of electrophilic carbon is produced by the action of a secondary amine (or sometimes a primary amine) on methanal. This reaction mixture is refluxed with an aldehyde or ketone that has α-hydrogen atoms. The whole process is catalysed by acid and the final product is liberated by treating the reaction mixture with alkali. It is called a Mannich base and contains a newly formed carbon-carbon bond.

A typical Mannich reaction is shown on the next page, first in outline and then with the mechanism:

1. In outline

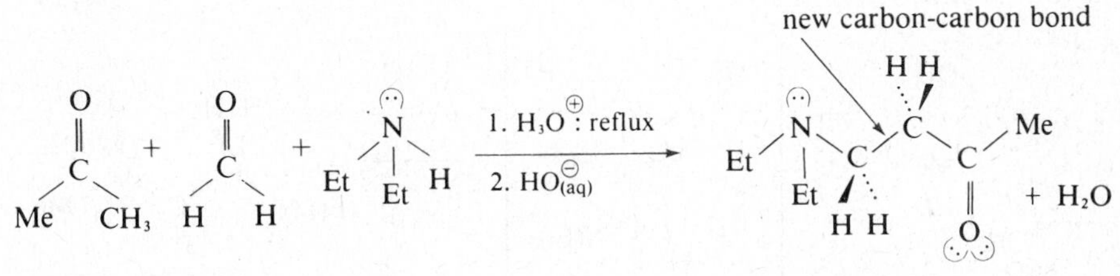

new carbon-carbon bond

1. $H_3O^{\oplus}$: reflux
2. $HO^{\ominus}_{(aq)}$

+ H_2O

propanone methanal diethylamine a Mannich base

2. Likely mechanism

Electrophilic carbon is generated as follows:

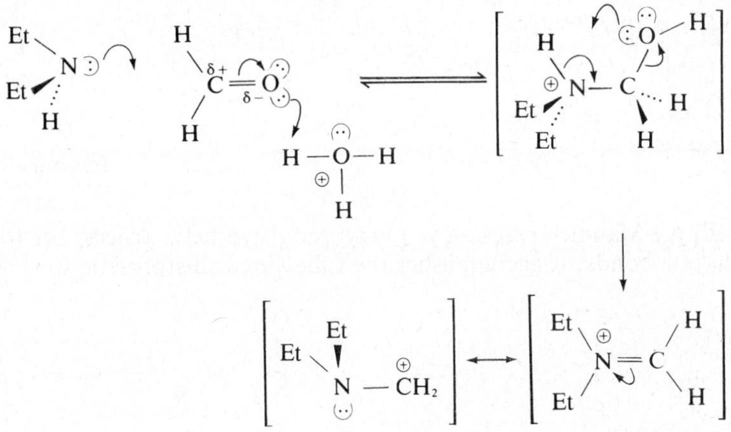

These electrophiles can interact with the π-electrons of the *enol*-form of propanone. The *keto*-form is converted more rapidly to *enol*-propanone in acid conditions.

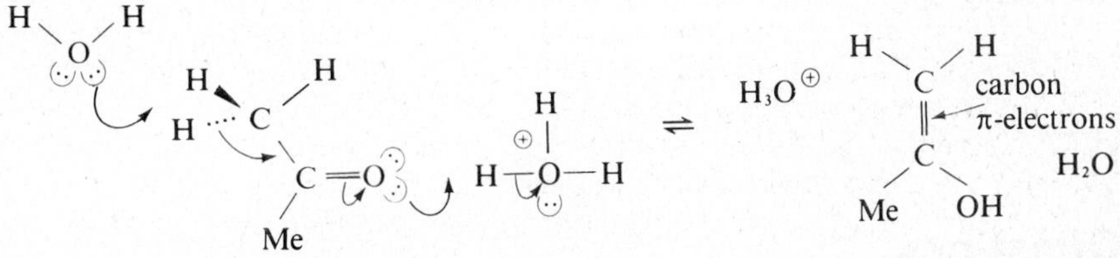

carbon π-electrons

keto *enol*

Electrophilic addition then takes place as follows.

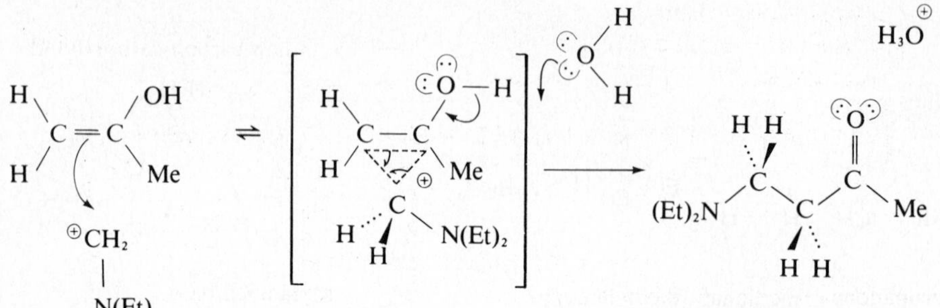

Mannich base

The hydrochloride salt of a Mannich base is readily decomposed on heating. The products are an amine hydrochloride and an α-unsaturated ketone which is stabilized by the conjugation of the two π-systems.

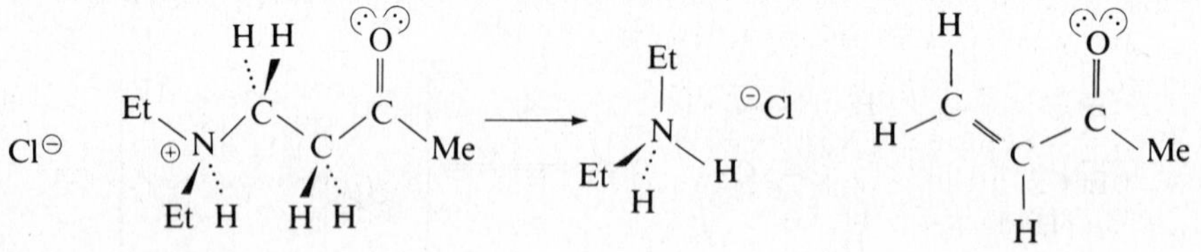

butenone

So, overall, the Mannich reaction is a very useful synthetic process for forming carbon-carbon bonds. It accomplishes the following transformation.

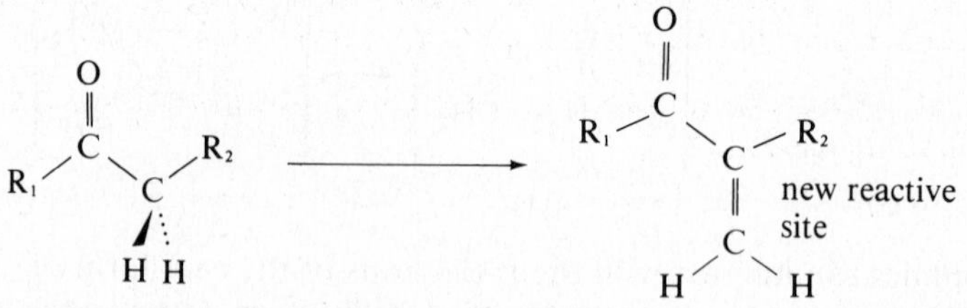

new reactive site

By introducing a new reactive site into the structure, further synthetic routes are opened up.

Questions

1. Compare and contrast the methods for making carbon-carbon bonds with those used for making carbon-nitrogen bonds.

2. a) How is an aqueous solution of potassium cyanide used in the formation of carbon-carbon bonds?
   b) Using this reagent, outline reaction schemes that bring about the following three conversions.

i) Me—OH → Et—OH ii) Et—OH → Me H COOH OH iii) Et—OH → N Et Pr H

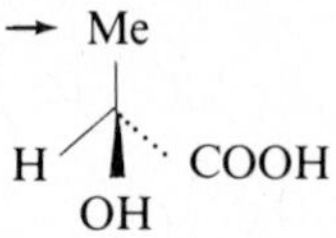

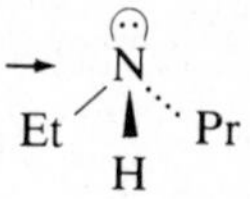

3. a) What is a Grignard reagent?
   b) Give an example of the formation of a Grignard reagent.
   c) Give *three* examples of the reactivity of such a reagent.

4. a) What is a base-catalysed condensation?
   b) Give an example of a reaction of this sort. Draw the structural formulas of the reactants and products.
   c) Give the mechanism of a typical base-catalysed condensation and emphasize newly-formed carbon-carbon bonds.

5. a) What is a Friedel-Crafts reaction?
   b) Give an example of a reaction of this sort.
   c) Why are Friedel-Crafts reactions more successful for aromatic compounds than for aliphatic compounds?

6. a) What is a Mannich base?
   b) Give an example of a base of this sort and give a reaction scheme by which it is formed.
   c) What points of mechanistic interest are there in the reaction described in your answer?
   d) Explain why Mannich bases readily undergo elimination. Give the elimination product derived from your example of a Mannich base.

7. Discuss the use of non-aqueous solvents in the practical methods available for forming carbon-carbon bonds.

8. Starting with propanone, outline syntheses of the following compounds:

a) O Me

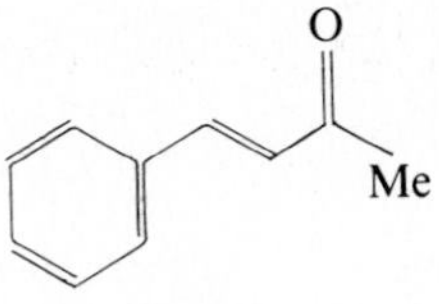

b) Me Me Me Me

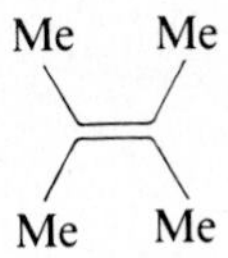

c) Cl H H Me Me Cl

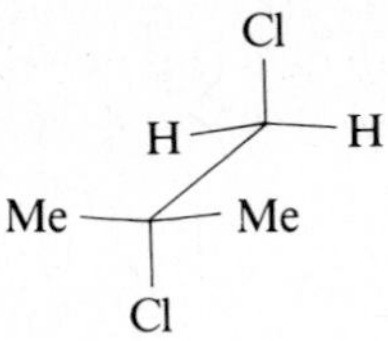

d) O Me Me Me

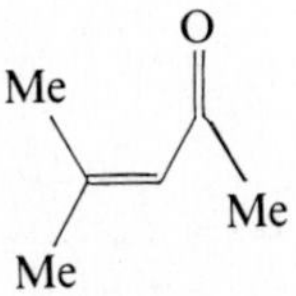

e) O Me Et

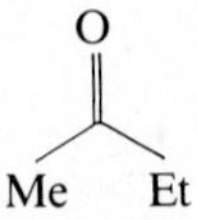

9. Starting with bromoethane, outline syntheses of the following compounds:

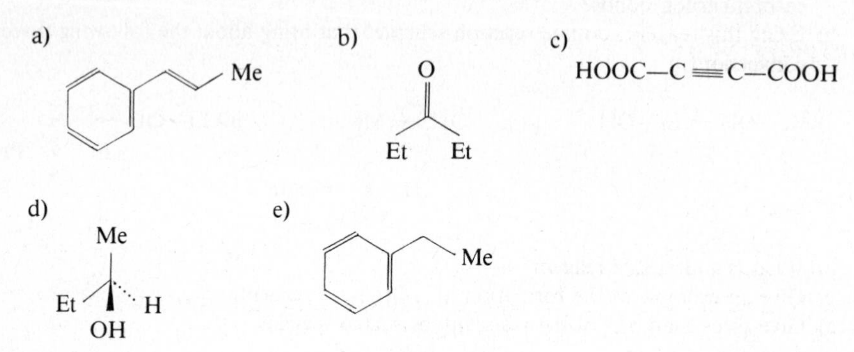

10. Outline syntheses of the following compounds:

a)

O

b)

O
Me

c)

Me
H COOH
H COOH
Me

d)

Me—C≡C—Et

Chapter 35

Isomerism

Molecular formula

The empirical formula of a compound shows the ratio of the number of atoms of each element in a compound. It can be found by analysing the relative amounts of each element in the compound. For example, determination of the amount of carbon dioxide and water obtained by burning a known amount of the compound gives the number of moles of carbon for every two moles of hydrogen in the molecule. Other techniques have been used for the determination of the ratio of nitrogen, sulphur, and the halogen atoms in the molecule, but all these methods have now been superseded by spectrographic analysis for finding the empirical formula (for example, see page 10).

The molecular formula of a compound gives the actual number of atoms of each element in the molecule; it is obtained from the empirical formula and from a determination of the R.M.M. of the compound.

name formula	*empirical formula*	*molecular*
ethane	CH_3	C_2H_6
ethene	CH_2	C_2H_4
ethyne	CH	C_2H_2
benzene	CH	C_6H_6
ethane-1,2-diol	CH_3O	$C_2H_6O_2$
ethanoic acid	CH_2O	$C_2H_4O_2$

However, even when the molecular formula is known, it gives little clue to the structure of an organic molecule because it is often possible for the same atoms to be arranged in more than one way in the molecule.

> Two molecules having the same molecular formula but different arrangements of their atoms are called isomers.

The difference in the arrangement of atoms in the isomers arises in two ways. Either the atoms are bonded together in a different order in each isomer in which case they are called *structural isomers*. Or, although the order of bonding in the isomers is the same, the arrangement of the atoms in space is different in each isomer; these are called *stereoisomers*.

The occurrence of isomers is commonest in carbon compounds because of the immense variety of ways in which carbon can form chains or rings, and because the four bonds around an unsaturated carbon atom are tetrahedrally arranged in space. However, isomerism is not confined to carbon compounds but is also found in the chemistry of other elements (see section 35.3 on page 552).

35.1 Structural isomerism

> Structural isomers have the same molecular formula but different structural formulas.

Structural isomerism can occur among compounds with the same functional group:

Example 1 Alcohols, e.g. C_3H_8O

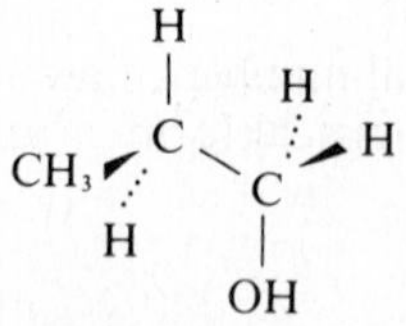

a primary alcohol

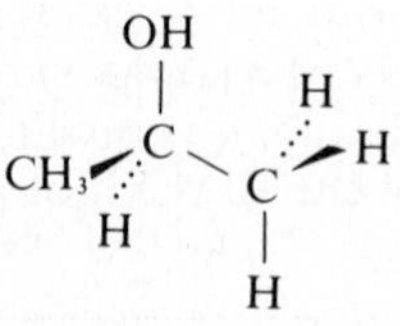

a secondary alcohol

Example 2 Aromatic compounds, e.g. $C_6H_4Cl_2$

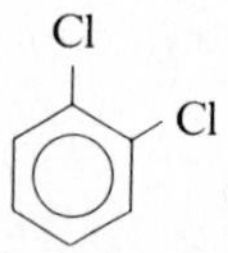

1,2 substituted (*ortho*)

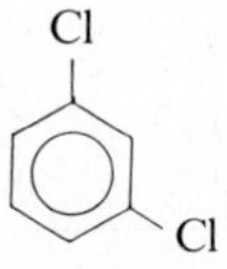

1,3 substituted (*meta*)

1,4 substituted (*para*)

Structural isomerism can also occur among compounds with different functional groups:

Example 1 C_3H_6O

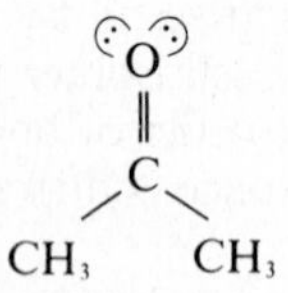

a ketone

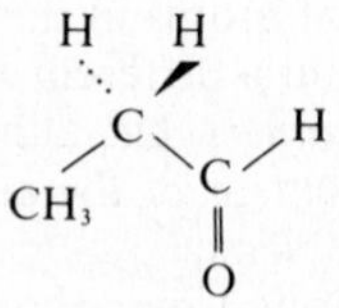

an aldehyde

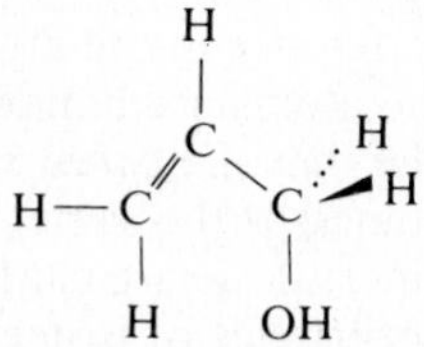

an unsaturated alcohol

Example 2 C_2H_6O

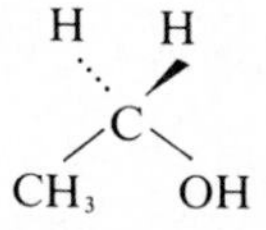

H H
C
CH_3 OH

a primary alcohol

O
CH_3 CH_3

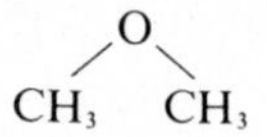

an ether

35.2 Stereoisomerism

Stereoisomers have the same molecular and structural formulas but the arrangement of their atoms in space is different.

There are two types of stereoisomerism; *geometrical* isomerism and *optical* isomerism.

Geometrical isomerism

Geometrical isomers occur in pairs. The members of the pair differ from each other in the positioning of two groups. One member has the two groups on the same side of the molecule and is called the *cis*-isomer from the latin meaning 'on the same side'. The other member of the pair has the two groups on opposite sides of the molecule and is called the *trans*-isomer from the latin meaning 'on the opposite side'.

Example 1 But-2-ene

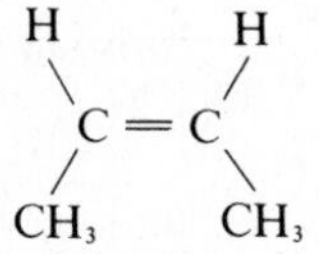

H H
C=C
CH_3 CH_3

cis isomer

CH_3 H
C=C
H CH_3

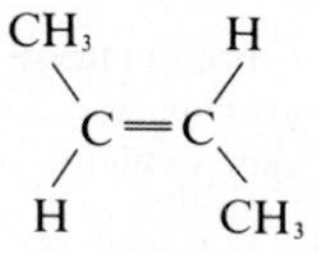

trans isomer

Example 2 Ethanal oxime

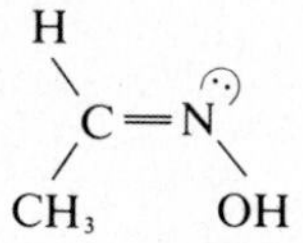

H
C=N
CH_3 OH

cis isomer

CH_3
C=N
H OH

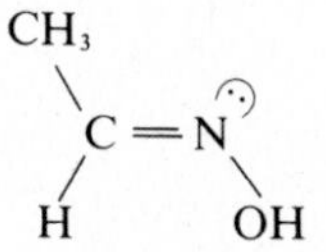

trans isomer

Example 3 1,2-dibromocyclopropane

H H
C—C
CH_2
Br Br

cis isomer

Br H
C—C
CH_2
H Br

trans isomer

Example 4 2-epoxybutane

H H
C—C
O
CH_3 CH_3

cis isomer

CH_3 H
C—C
O
H CH_3

trans isomer

Geometrical isomerism occurs when rotation about a carbon-carbon bond is prevented either by the presence of a π-bond or a ring structure including the carbon-carbon bond.

For example compare the structure of 1,2-dibromoethane with that of 1,2-dibromoethene.

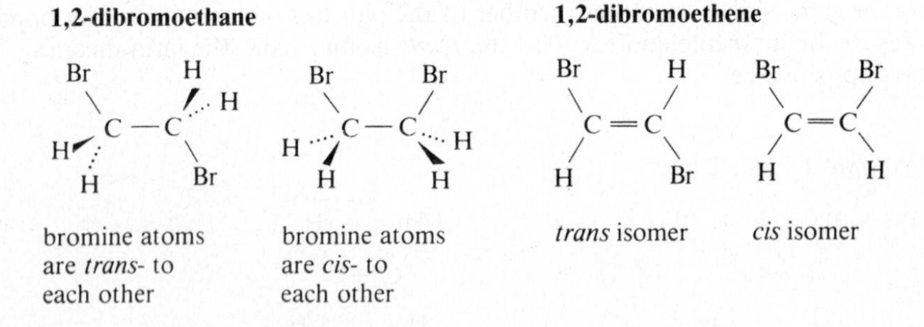

Only one form of 1,2-dibromoethane exists because the *trans*- and *cis*- forms are easily converted from one to the other by the rotation of the carbon-carbon single bond. Imagine looking from the left of the molecule along the line of its carbon-carbon bond.

A diagram showing this view is called a Newman projection.

bromine atoms are *trans-* here.

But rotation can readily occur about the single bond and the *cis-* position is immediately produced as shown below.

rotate front carbon atom through 180°

trans- position

cis- position

cis- and *trans-* positions are therefore temporary arrangements that occur during the rotational movements of a molecule of 1,2-dibromoethane. These are known as *conformations* of the structure of 1,2-dibromoethane.

> Two conformations of a given molecular structure are different spatial arrangements of the atoms that can be produced by rotation or twisting of the bonds in the structure.

Interconversion of *cis-* and *trans-* 1,2-dibromoethene does not happen in the same way: the presence of the π-overlap restricts rotation as shown in the following diagrams.

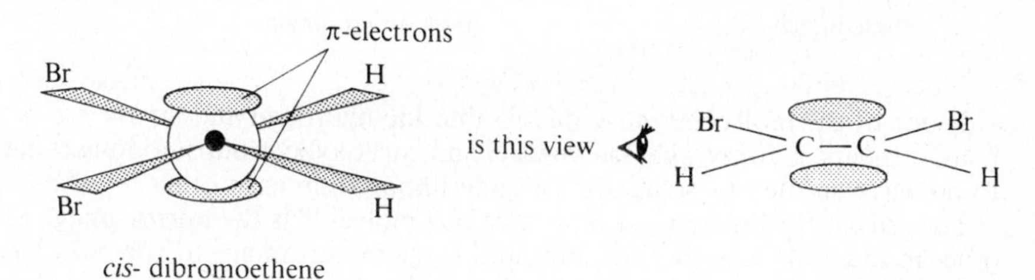

cis- dibromoethene

Rotation cannot take place unless the π-overlap is either broken or twisted. Neither is likely under normal conditions of temperature. Restriction on rotation also occurs in ring structures. There are two forms of 1,2-dibromocyclobutane and these can only be interconverted by breaking or twisting the ring.

Geometric isomers can therefore be defined as follows.

> Geometric isomers have the same structural formula but their molecules contain groups that are differently arranged in space as a result of restricted rotation within the molecular structure.

Optical isomerism

Symmetry

Optical isomers are stereoisomers which have the same structural formula but differ in their symmetry. Consider the two molecules of 2-aminopropanoic acid drawn below.

COOH | C — Me, H, NH_2 COOH | C — Me, NH_2, H

The same groups are bonded to the central carbon atom in each case, but the molecules are not the same!

Imagine putting one molecule on top of the other in order to try to match up the same groups. For example, the methyl and carboxylic groups can be made to match up; however, then the hydrogen and amino groups are found not to match as shown below.

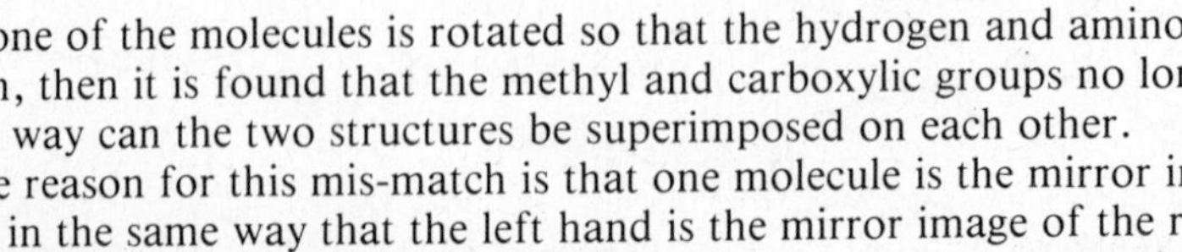

COOH | C — Me COOH | C — Me | C — H, NH_2 | C — NH_2, H

these match these do *not* match

If one of the molecules is rotated so that the hydrogen and amino groups do match, then it is found that the methyl and carboxylic groups no longer match. In no way can the two structures be superimposed on each other.

The reason for this mis-match is that one molecule is the mirror image of the other in the same way that the left hand is the mirror image of the right hand. This effect is called *chirality* (from the Greek word for 'hand') and the molecules are described as chiral molecules.

This can be seen in the diagram below where the right hand isomer must first be rotated through 180°.

COOH
C
Me H NH_2

COOH
C
Me H NH_2

rotate through 180°

COOH
C
Me NH_2 H

mirror

HOOC
C
H H_2N Me

mirror image

> Although optical isomers have the same structural formula, their molecules are mirror images which cannot be superimposed upon each other.

The reason for the non-superimposability is that the molecules possess no centre or plane of symmetry. The arrangement of four different groups tetrahedrally around a central carbon atom makes that carbon atom an *asymmetric centre*. The positions taken up by the four different groups can be represented using Newman Projections. Consider the two forms of 2-aminopropanoic acid being viewed along the carbon-carboxylic acid bond:

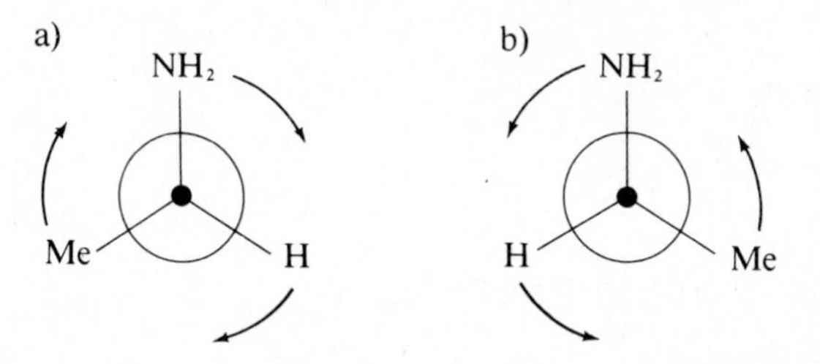

Follow the sequence of groups: amino → hydrogen → methyl
For a), rotation occurs in a right-handed circle
For b), rotation occurs in a left-handed circle.

These two chiral molecules are distinguished from each other by the prefixes D (from *dextro* meaning right) and L (from *laevo* meaning left). However these two labels only distinguish the two molecules from each other and have no absolute significance.

Notice that, if one of the groups in the molecules is changed so that it becomes the same as one of the others (e.g. change the methyl groups for hydrogen atoms), two things happen:

1. There is a plane of symmetry in the molecule that was not there before:

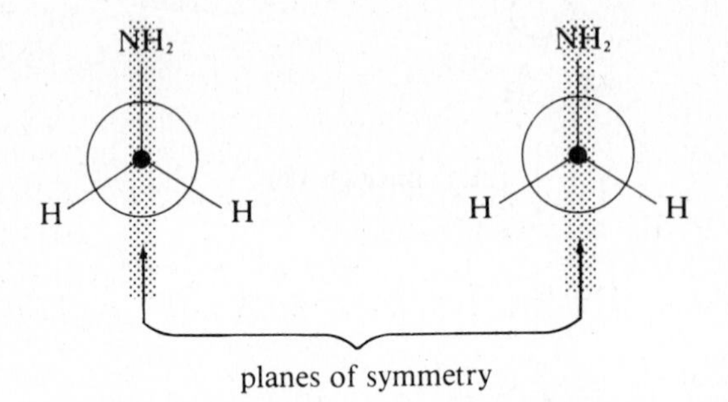

The left sides of these molecules are the mirror images of their right sides. They therefore possess symmetry.

2. Although they are mirror images of each other they are also superimposable, so they are no longer chiral. They are not optical isomers.

So, either a plane or centre of symmetry in a molecule is sufficient to indicate that the structure cannot give rise to optical isomerism.

To identify the likelihood of optical isomerism in a particular structure, first examine the structure for an asymmetric carbon atom. An asymmetric carbon atom is one that contains four different groups bonded to it. In the majority of cases, the presence of an asymmetric carbon atom gives the overall structure chirality and a lack of any planes or centres of symmetry.

Some examples are given below. By convention, they are drawn as mirror images of each other; the mirror drawing is a useful way of showing non-superimposability. In each case the asymmetric carbon atom is marked by an asterisk.

Example 1 A hydroxynitrile

CN, C*, Me, H, OH | NC, C*, HO, H, Me

mirror

Example 2 A secondary alcohol

OH, C*, H, Et, Me | HO, H, C*, Et, Me

mirror

Optical properties

There is a particular reason why this type of stereoisomerism became known as optical isomerism. If a pure sample is taken of one form of 2-aminopropanoic acid (described on page 546) and dissolved in water, the solution has an unusual property. It is able to interact with polarized light so that the plane of the polarization is rotated through an angle up to about 30°. This can be illustrated using two 'crossed' polaroids set up as shown below. The bottom polaroid allows the propagation of light in one plane only; the top polaroid is positioned in such a way that it only allows the propagation of light in a plane at 90° to the first plane. In this position the polaroids are 'crossed' and no light can shine through to the top.

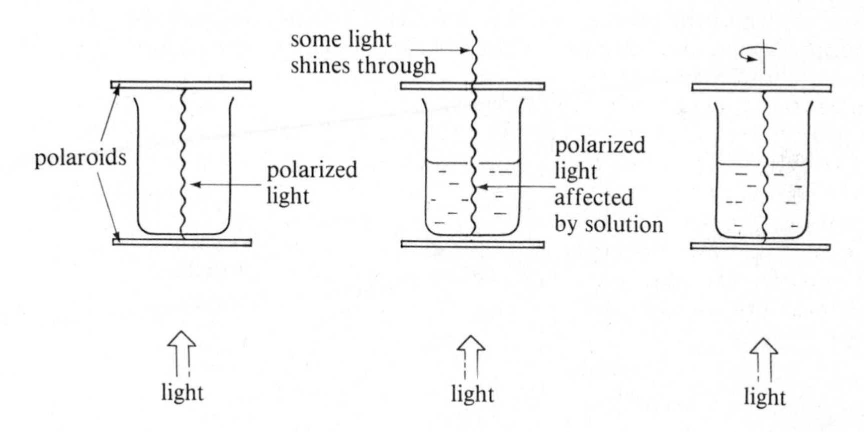
1. two 'crossed' polaroids prevent the passage of light through a beaker
2. addition of some of the solution causes light to shine through
3. the top polaroid needs to be rotated through a small angle to give back the darkness
some light shines through
polaroids
polarized light
polarized light affected by solution
light
light
light

The turning of the top polaroid represents the angle through which the plane of polarized light has been rotated by the solution. When the other form of the α-amino acid is used in this apparatus (sometimes called a polarimeter), it is found that the top polaroid needs to be turned in the opposite direction through an exactly equal angle. For this reason one optical isomer is labelled (+) and the other isomer is labelled (−): (+) stands for rotation to the right; (−) for rotation to the left.

Two optical isomers that rotate the plane of polarized light through equal angles, but in opposite directions, are called *enantiomers* (or enantiomorphs).

Two asymmetric centres

A molecule that has *two* asymmetric carbon atoms can exist in more than two stereoisomeric forms. A typical example of a molecule of this sort is shown below.

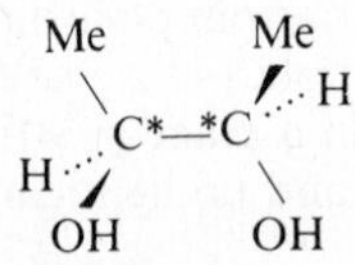

butane-2,3-diol

Although both the carbon atoms marked C* above are asymmetric, they have the *same* four different groups: —Me, —OH, —H and —C(OH).H.Me. A structure like this has three different stereoisomers, two of which, a) and b), are non-superimposable mirror-images (enantiomers) whilst the third c) is rather different. The first two are shown below using Newman projections and then using 'saw-horse' drawings.

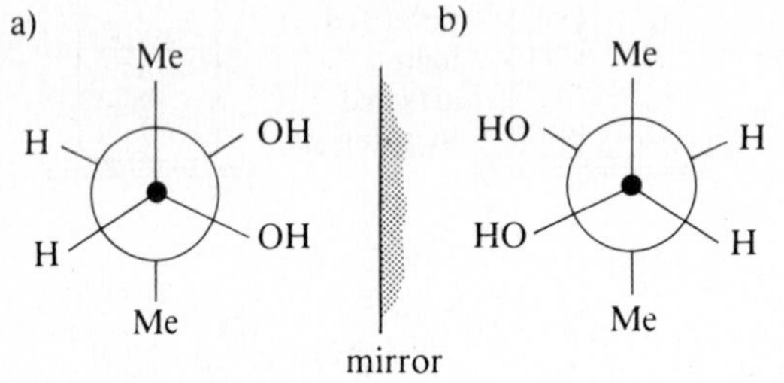

a) b)

mirror

Newman projections

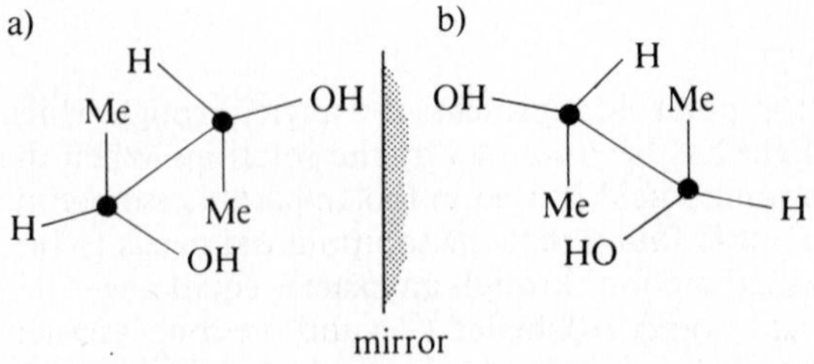

a) b)

mirror

'saw-horse' structures

The third stereoisomer c) is best shown in its 'eclipsed' conformation using a Newman projection. The saw-horse drawing beside the Newman projection on the next page is derived by rotating the rear carbon atom through 180°. It is a 'staggered' conformation because the bonded groups are in the gaps between those of the opposite carbon atom and are not eclipsed with one another as in the first conformation.

c) Me Me H H OH OH — or — c) HO Me H H Me OH

'eclipsed' conformation — 'staggered' conformation

These drawings bring out two major points:

1. a) and b) are non-superimposable mirror images of each other
2. c) is not the same as a) or b), but is also not a mirror image of either.

c) is therefore called a *diastereoisomer* of the enantiomeric pair a) and b).

When c) is dissolved in water, the resulting solution does not cause the rotation of the plane of polarized light passing through it: c) is *not* optically active. Its molecules possess a plane of symmetry and this prevents the possibility of optical activity. The front end of a molecule of c) is a mirror image of the rear end.

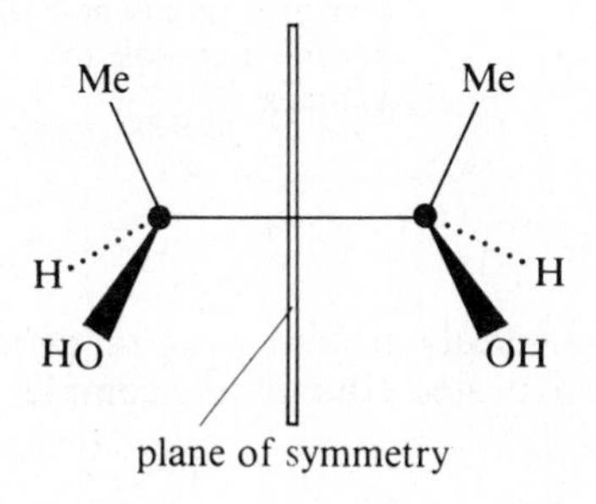

The asymmetry of each carbon atom in the above structure is said to be *internally compensated*. A stereoisomer of this type is called a *meso*-form.

Conversely, when either a) or b) are dissolved in water, the resulting solutions interact with polarized light. If a 50-50 mixture is made up, however, no effect is detected because one enantiomer rotates the plane in exactly the opposite direction to the other. The asymmetry of the carbon atoms in the structure is said to be *externally compensated* in a 50-50 mixture of this sort. The mixture is called a *racemic mixture* and is given the symbol ($\pm$) to illustrate the equal proportions of the (+) and (−) enantiomers. The term 'racemization' is applied to any process in which an optically active substance is converted into a racemic mixture.

35.3 Inorganic isomerism

Most inorganic isomers are complex salts of one form or another. Structural, geometric and optical isomerism are all in evidence in different types of complex.

Structural isomers

Three structural isomers of the compound $CrCl_3.6H_2O$ are known. They are structural isomers because the atoms are linked differently in the three forms. The first is violet in colour and, in solution, reacts with silver ions in the molar ratio of 1:3. The second is green and reacts with silver ions in the molar ratio 1:2. The third is dark green and gives a precipitate of silver chloride on a 1:1 molar ratio. The structures are therefore as below.

$[Cr^{III}(H_2O)_6]^{3+}\ 3Cl^-_{(s)}$

violet; three moles of chloride per mole of complex

$[Cr^{III}(H_2O)_5Cl]^{2+}\ 2Cl^-.H_2O_{(s)}$

green; two moles of chloride per mole of complex

$[Cr^{III}(H_2O)_4Cl_2]^{+}\ Cl^-.2H_2O_{(s)}$

dark green; one mole of chloride per mole of complex

Geometric isomers

The rigid arrangement of ligands placed octahedrally around a central cation can cause *cis-trans* effects. For example, the hydrated ethandioate complexes of chromium (III)

cis- and *trans-* $[Cr^{III}(H_2O)_2(\text{ethandioate})_2]^{\ominus}$

where the curved linkage is the ethandioate group ($^{\ominus}O-C(=O)-C(=O)-O^{\ominus}$)

Geometric isomerism is also common in square-planar arrangements. The nickel group tend to form square-planar complexes and so these compounds often exhibit geometric isomerism. For example, two forms of the di-(aminoethanoate) complex of Ni(II) are known and there are two compounds of formula $Pt(NH_3)_2\ Cl_2$:

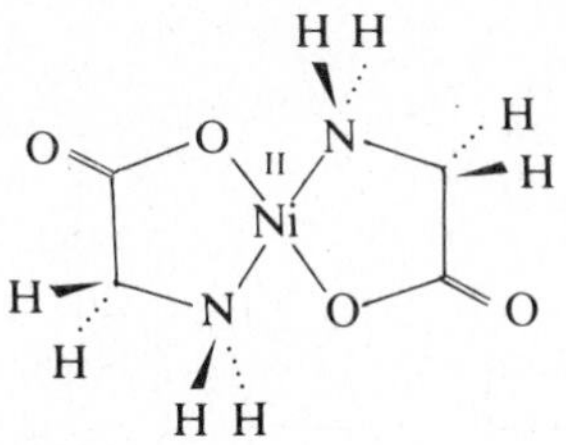

trans- 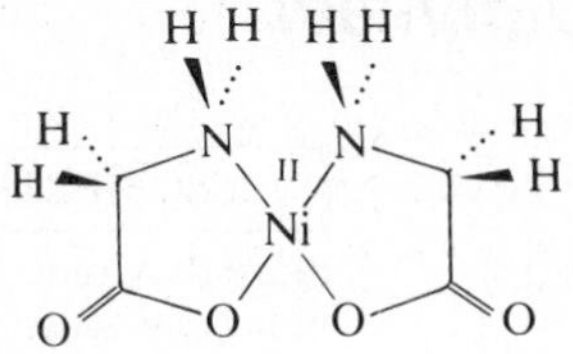 *cis-*

and

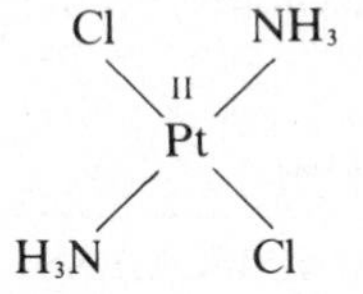

trans 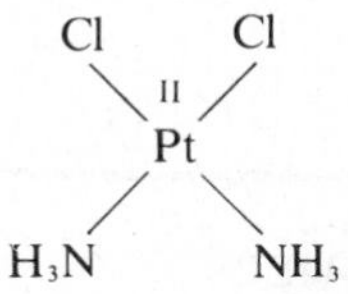 *cis*

Optical isomerism

Asymmetry can occur in certain octahedral and tetrahedral complexes. For example, the *cis-* ethandioate complex of chromium (III) shown above has a centre of asymmetry in the chromium ion. Two non-superimposable mirror images of this structure can be prepared and the two enantiomers do indeed rotate the plane of polarized light in opposite directions.

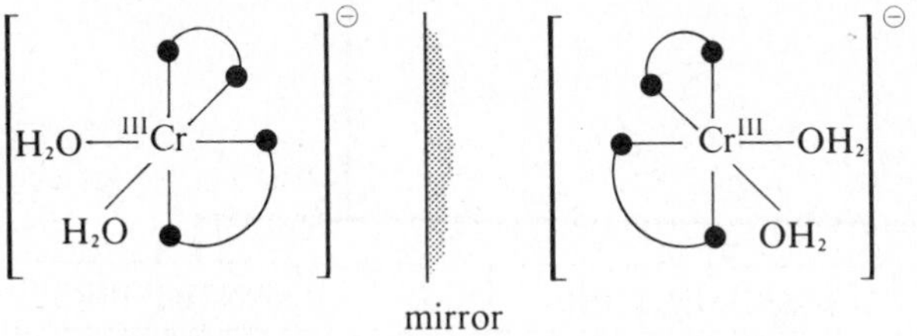

mirror

(+)form (−)form

where ⌒ is the ethandioate group shown on the previous page

The tri-ethandioate complexes of transition metal cations can also be prepared in different enantiomeric forms.

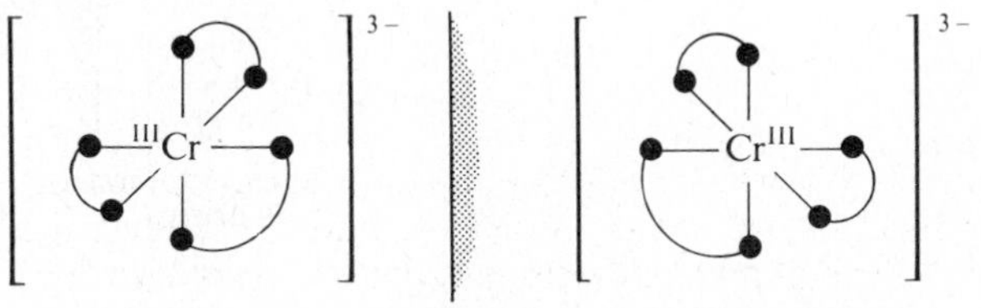

mirror

(+)form (−)form

35.4 Summary

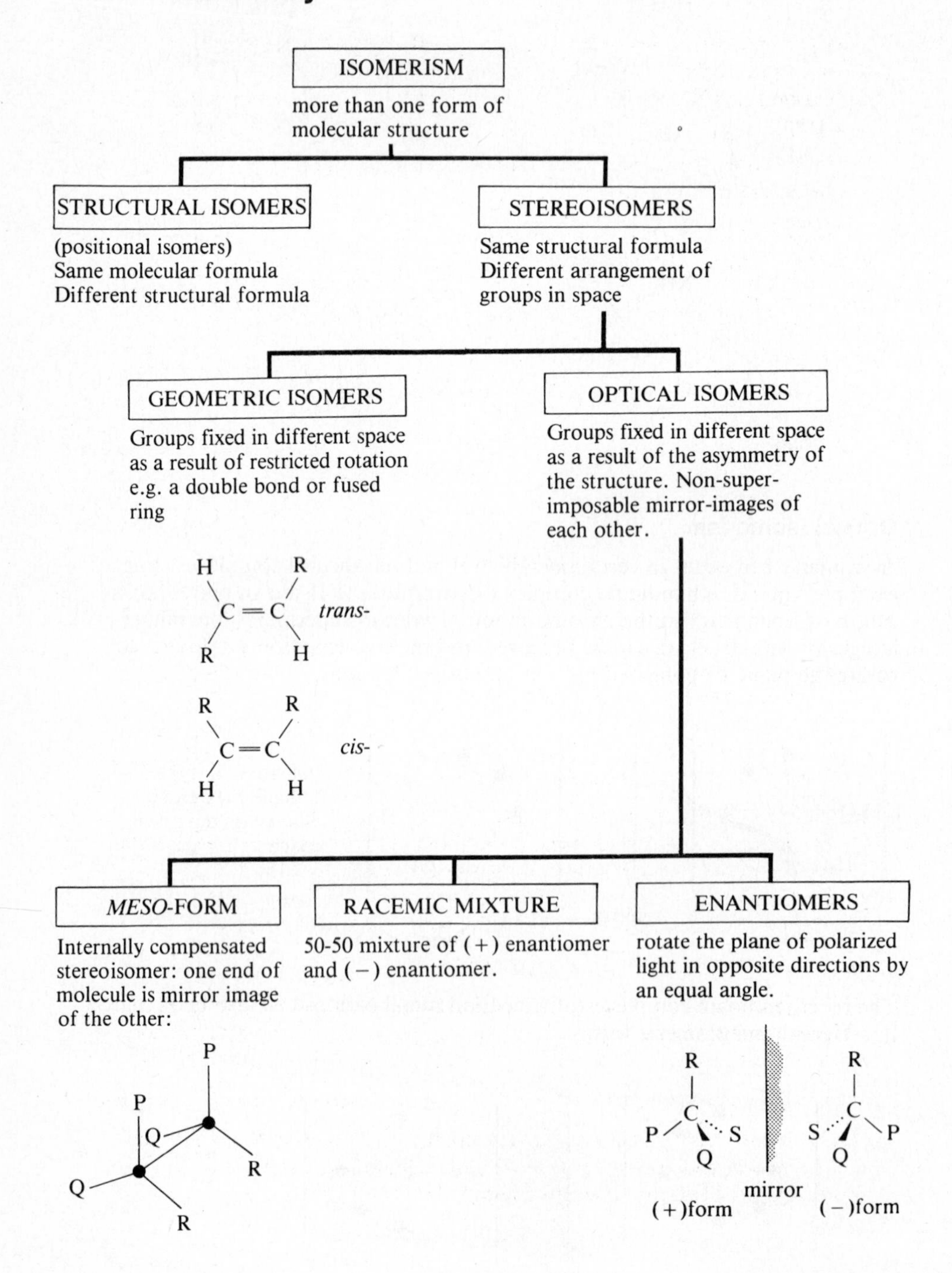

ISOMERISM
more than one form of molecular structure
STRUCTURAL ISOMERS
(positional isomers)
Same molecular formula
Different structural formula
STEREOISOMERS
Same structural formula
Different arrangement of groups in space
GEOMETRIC ISOMERS
Groups fixed in different space as a result of restricted rotation e.g. a double bond or fused ring
H R C=C R H trans-
R R C=C H H cis-
OPTICAL ISOMERS
Groups fixed in different space as a result of the asymmetry of the structure. Non-superimposable mirror-images of each other.
MESO-FORM
Internally compensated stereoisomer: one end of molecule is mirror image of the other:
P Q R P Q R
RACEMIC MIXTURE
50-50 mixture of (+) enantiomer and (−) enantiomer.
ENANTIOMERS
rotate the plane of polarized light in opposite directions by an equal angle.
R C P S Q
R C S P Q
mirror
(+)form
(−)form

Questions

1. a) Give all the structural isomers of formula C_3H_6O.
   b) Name each isomer.
   c) Explain the difference between structural isomerism and stereoisomerism.
2. a) Discuss the stereoisomers of 2-methylchlorocyclopropane.
   b) Using your answer to part a) as an example, explain the difference between optical and geometric stereoisomerism.
3. a) Why are optical isomers known as optical?
   b) Draw a diagram of a simple polarimeter and explain how it works.
4. In what stereoisomeric forms (if any) would you expect the following compounds to exist?

   a) MeCH = NOH

   b) H Br H H Br H

   c) chlorophenol

   d) 2-hydroxypropanonitrile

   e) but-2-ene

   f) MeCH — CHMe, with O bonded to both carbons

5. a) Define the following terms:
   i) enantiomer
   ii) diastereoisomer
   iii) meso-form.
   What is meant by racemization?
6. Account for the following observations.
   a) It is possible to isolate two forms of but-2-ene, but there are three forms of the product of the reaction between but-2-ene and bromine.
   b) It is possible to isolate two forms of 1,2-dichloroethene but only one form of the product of the reaction between 1,2-dichloroethene and hydrogen.
7. Describe the isomerism shown by $Co(NH_3)_4Cl_2$.
8. Two compounds of formula $Co(NH_3)_4Cl_2NO_2$ exist. A solution of one produces a white precipitate when added to silver nitrate solution, while the other does not. Account for these facts.
9. Discuss the isomerism of the compounds $Ni(en)_3SO_4$ where en = $NH_2CH_2CH_2NH_2$.
10. When the two ends of a bidentate chelate differ, e.g. $NH_2CH_2COO^-$, it is found that complexes of the form $Zn(chelate)_2$ exist in optical isomers while those of the form $Ni(chelate)_2$ exist in geometric isomers. Account for this.

Chapter 36

The stereochemistry of reactions

Mechanism

In Chapter 21, the ideas of 'reaction mechanism' are outlined. Two of the most important features of a mechanism are:

1. the orientation of the colliding particles in a collision
2. the energy level of the activated complex formed by the collision.

The outcome of a particular collision depends on these two factors. In many reactions, there is more than one way in which reacting particles collide, and more than one possible activated complex. But in spite of this, it is often found that only one product is formed and that somehow, none of the alternative reaction pathways are followed. Reactions of this sort are described by one of the following two terms.

Stereospecific — when one product is formed in complete preference to any other possible product

Stereoselective — when one product is formed in greater proportion than any other possible product.

An explanation for a stereospecific reaction is usually based on:

1. the ease with which a possible activated complex can form as a result of a correctly orientated collision,
2. the relative energy levels of the possible activated complexes.

These two factors are discussed in turn.

1. Steric hindrance

The first factor concerns the hindrance there might be to a correctly orientated collision. For example, it is easier for a nucleophile to collide with the electron-deficient carbon atom in chloromethane than with the electron-deficient carbon in 2-chloromethyl propane.

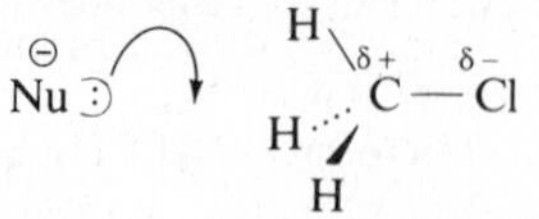

$\ominus$ Nu: → H, H, H–$C^{\delta+}$—$Cl^{\delta-}$ chloromethane

$\ominus$ Nu: → CH_3, CH_3, CH_3–$C^{\delta+}$—$Cl^{\delta-}$ 2-chloromethylpropane

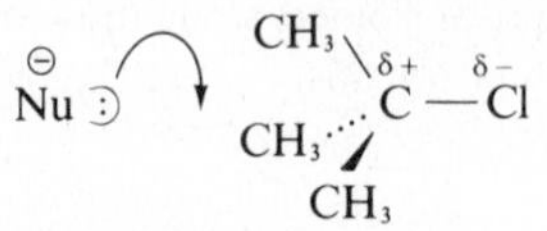

In the second case, the nucleophile's approach to the electron-deficient carbon is more hindered by the methyl groups than it was by the hydrogen atoms in the first case.

There are many occasions when a reaction produces a particular product because an alternative pathway involves a collision that is considerably more hindered.

2. The reaction pathway

The second factor is concerned with relative stabilities. Consider the energy diagram below for two hypothetical reactions that compete with each other: A→B or A→C. Two possible pathways exist.

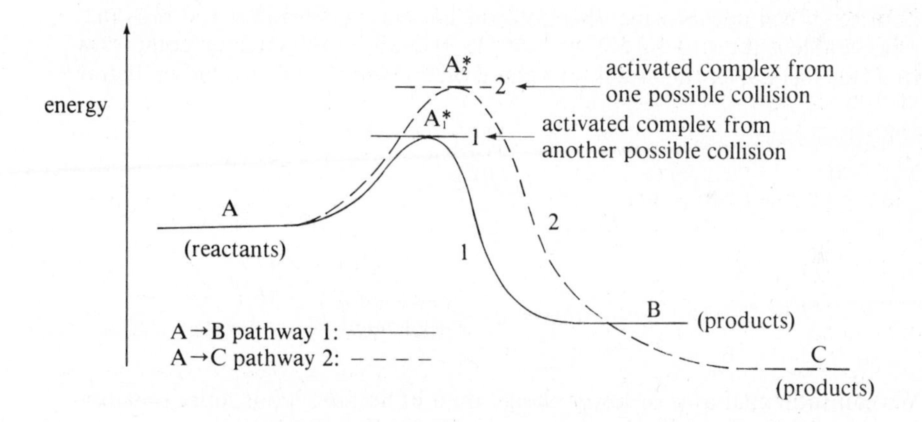

From this diagram the following points emerge.

i) The activation energy for pathway 1 is lower than that for pathway 2: $(A_1^* - A) < (A_2^* - A)$.
ii) The products of pathway 2 are thermodynamically more stable than those of pathway 1: (A—C) > (A—B).

Under nearly all conditions, it should be expected that the reaction would favour pathway 1 stereoselectively. The activation energy is lower and therefore this process occurs at a faster rate. For every one successful collision of the sort described by pathway 2, there might be a hundred successful collisions of type 1. On this reasoning, the products would be about 99% B and 1% C (despite C being more stable itself than B). A reaction controlled in this way is *kinetically controlled*: most organic reactions are kinetically controlled.

To increase the proportion of C in the products, it is necessary to find conditions under which the reverse reactions C → A and B → A are also favourable.

By refluxing the reaction mixture for a long period until it reaches equilibrium, a different sort of control operates. The position of equilibrium depends on the free energy difference between reactants and products and *not* on the activation energy. C would therefore tend to dominate any reaction mixture in equilibrium. Processes in which this strategy is used are called *thermodynamically controlled* reactions. They are not common, and we shall concentrate on the more typical of the two processes, kinetically controlled reactions.

The stability of an activated complex

Since many activated complexes contain localized charges when formally represented in bonding diagrams, a useful method of assessing their stability is to write down all possible canonical forms of the structure. The larger the number of canonical forms, the more the charge is delocalized and thus the more stable is the structure. For example, two different activated complexes that can form after the attack of phenol on an electrophile are shown below.

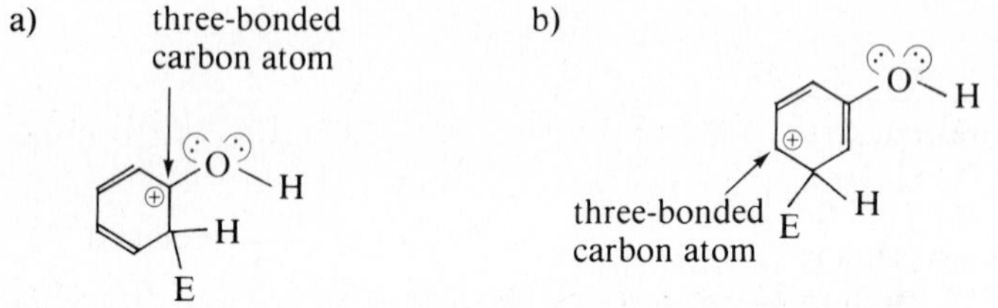

a) three-bonded carbon atom

b) three-bonded carbon atom

We can show that a) is of lower energy than b) because it has more possible canonical forms, i.e. the charge is more effectively delocalized.

a)

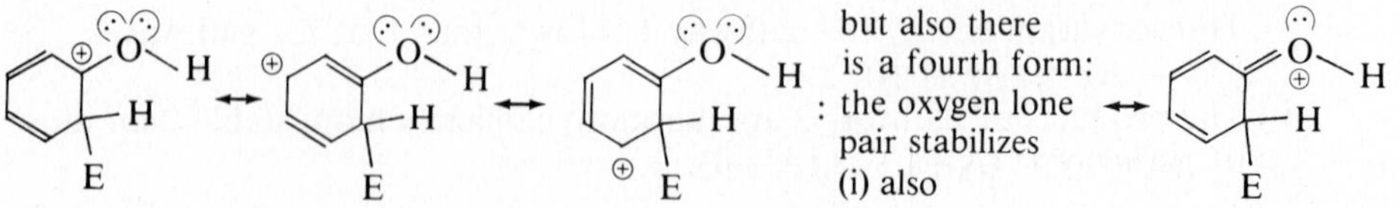

but also there is a fourth form: the oxygen lone pair stabilizes (i) also

b)

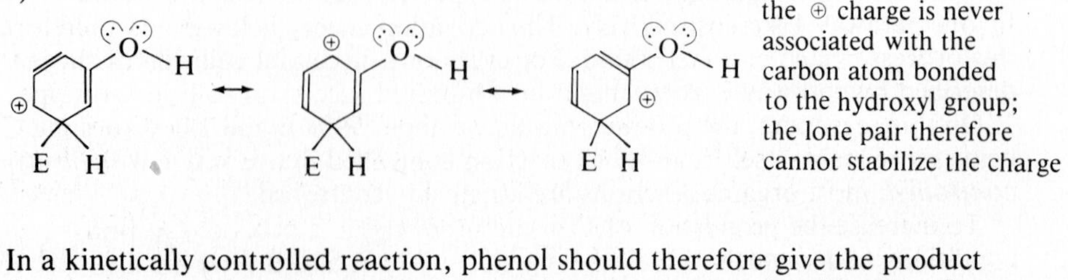

the ⊕ charge is never associated with the carbon atom bonded to the hydroxyl group; the lone pair therefore cannot stabilize the charge

In a kinetically controlled reaction, phenol should therefore give the product deriving from activated complex a) in much greater proportion than that from activated complex b). Pathway a) is of lower activation energy and therefore occurs faster.

Summary

In summary, two principal factors are usually responsible for stereospecificity:

1. the comparative steric hindrance of activated complex formation
2. the relative energy levels (stability) of the possible activated complexes.

> A reaction proceeds via the activated complex of lowest energy, providing that its formation is not severely sterically hindered.

36.1 Substitution

Nucleophilic substitution

There are two possible mechanisms of nucleophilic substitution:

S_N2

$$\overset{\ominus}{Nu}: + \overset{\delta+}{C}-\overset{\delta-}{Le} \rightleftharpoons \left[Nu\text{----}C\text{-----}Le \right]^{\ominus} \longrightarrow Nu-C \quad \overset{\ominus}{Le}$$

Here the formation of the activated complex depends on the collision of a nucleophile and a molecule of the carbon compound. The rate is therefore a function both of the nucleophile concentration and of the carbon compound concentration. It is therefore a second-order reaction and is labelled S_N2. This stands for **S**ubstitution, **N**ucleophilic, **2**nd order.

S_N1

$$\overset{\delta+}{C}-\overset{\delta-}{Le} \underset{}{\overset{\text{slow}}{\rightleftharpoons}} \left[C^{\oplus} \leftarrow \overset{\ominus}{Nu}: \right] \overset{\ominus}{Le} \xrightarrow{\text{very fast}} C-Nu \text{ and } Nu-C$$

In this case the activated complex is a *carbonium ion*, i.e. a positively charged ion derived from molecular carbon. The rate of formation of this ion depends only on the concentration of the carbon compound. The overall rate is *independent* of the concentration of nucleophiles because the interaction of nucleophile and carbonium ion is very fast indeed compared with the slow rate-determining dissociation of the carbon-compound. This reaction is therefore a first-order process and is labelled S_N1. This stands for **S**ubstitution, **N**ucleophilic, **1**st order.

In any nucleophilic substitution reaction, the two mechanisms compete. Since the order of the reaction can be found experimentally, its value gives a guide to the mechanism dominating. Some of the factors that favour each type of mechanism are listed in the table below.

favouring S_N2	*favouring S_N1*
1. the use of an unhindered nucleophile with a prominent lone pair of electrons	1. the presence of bulky groups on the carbon atom or a bulky nucleophile; this leads to considerable steric hindrance to S_N2.
2. the lack of any steric hindrance to the attack by the nucleophile: e.g. small groups bonded to the carbon atom.	2. the formation of a relatively stable carbonium ion.

Compare the hydrolyses of chloroethane and 1-phenyl-1-chloroethane.

$HO^{\ominus}$ — nucleophile

Me, H, H, $C^{\delta+}$—$Cl^{\delta-}$ — unhindered electron-deficient site

Me, C_6H_5, H, $C^{\delta+}$—$Cl^{\delta-}$ — hindered electron-deficient site

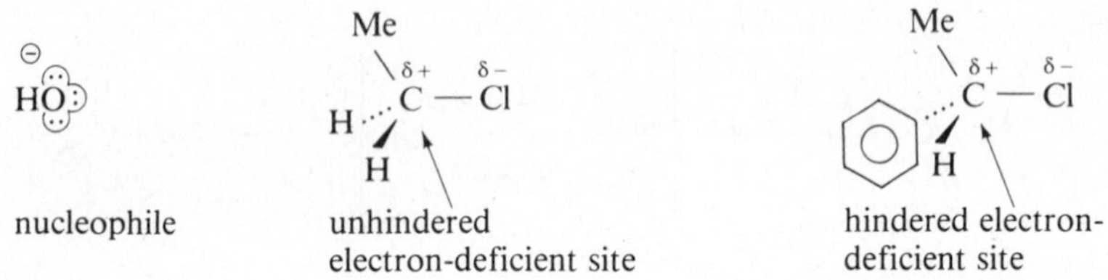

Applying the steric hindrance argument, it should be expected that the hydrolysis of chloroethane would be predominantly S_N2 whereas the hydrolysis of 1-phenyl-1-chloroethane seems likely to be S_N1.

The relative stability of the carbonium ions indicates the same conclusion.

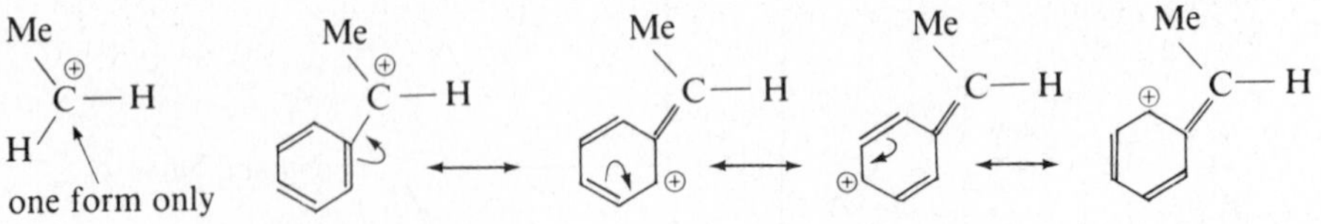

Me—$C^{\oplus}$—H, H — one form only

Me—$C^{\oplus}$—H (phenyl) ⟷ Me—C—H ⟷ Me—C—H ⟷ Me—C—H

There is a further stereochemical point that becomes clear if an optically active form of 1-phenyl-1-chloroethane is taken. The structure contains an asymmetric carbon atom.

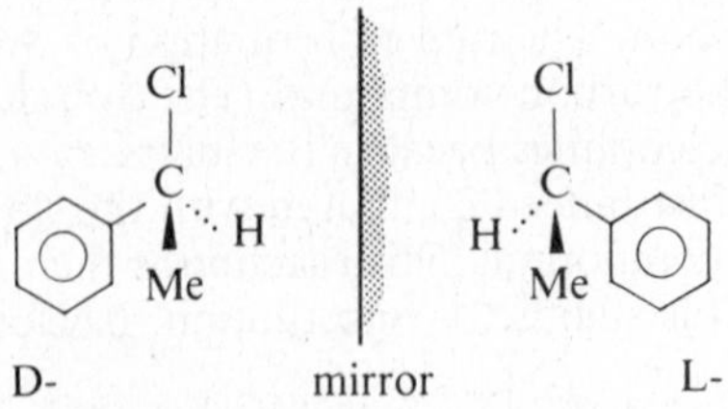

Cl, C, H, Me (D-) | mirror | Cl, C, H, Me (L-)

D- mirror L-

An S_N2 mechanism would result in an 'inversion' of the structure. The effect is the same as that of an umbrella being blown inside out by the wind:

D-form — the H, Ph and Me groups are pushed back until they become planar

the three groups are pushed over to the other side of the carbon atom

an L-form — still optically active, but with opposite symmetry

An S_N1 mechanism, however, would result in racemization because the carbonium ion has an equal chance of being attacked from either side.

either HO⁻ or HO⁻

D-form and L-form

(±) mixture or racemic mixture

The hydrolysis of optically active 1-phenyl-1-chloroethane does, in fact, result in a racemic mixture of products. This suggests that the reaction is dominated by an S_N1 mechanism.

Electrophilic substitution

On page 495, the directing properties of substituents on an aromatic ring are outlined. A saturated substituent (such as a hydroxyl group or an amine group) directs 2,4,6; an unsaturated substituent (such as a nitro group or a nitrile group) directs 3,5. These properties can be explained in terms of the relative stabilities of the possible activated complexes. For example, consider electrophilic substitution in the 3- and 4- positions for phenol and nitrobenzene. The canonical forms drawn for the 2- and 6- position activated complexes have the same charge distribution as that of the 4- position complex. Similarly the 5-position resembles the 3-position. Using the table on the next page, it can be concluded that the relative stabilities of the possible activated complexes *do* account for directing properties. The method can be applied to any substituent.

	electrophile position	*activated complexes involved*		*relative stability*
phenol	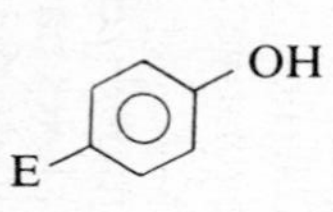 (4 −)		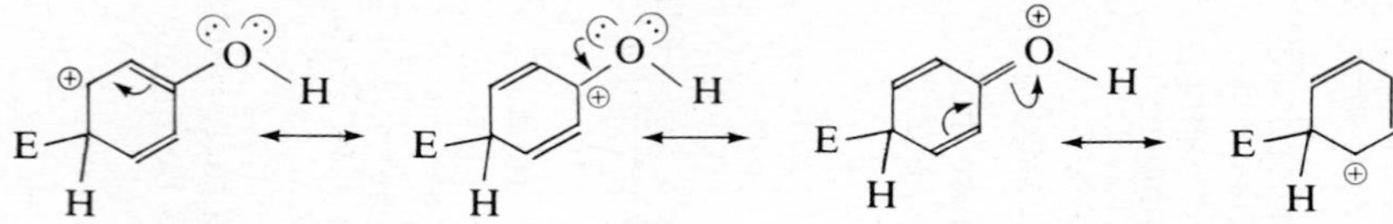 total = 4	more stable
	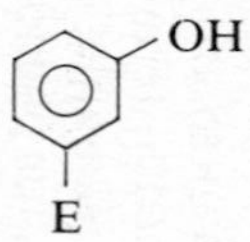 (3 −)		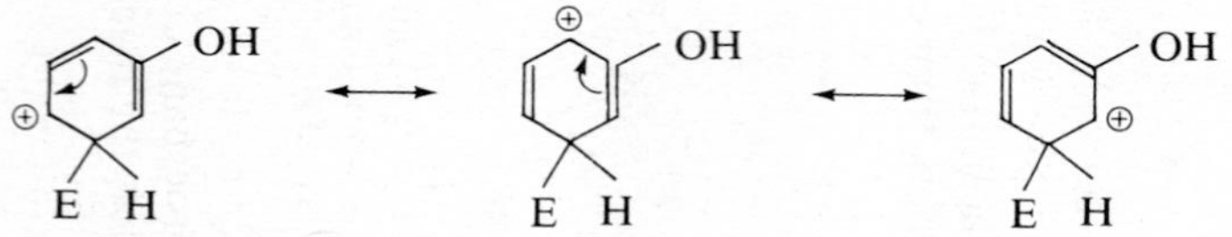 total = 3	less stable
nitro benzene	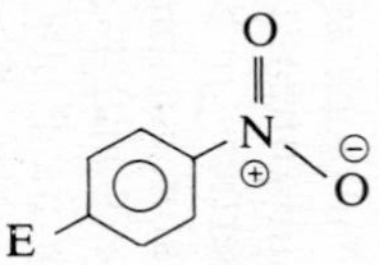 (4 −)	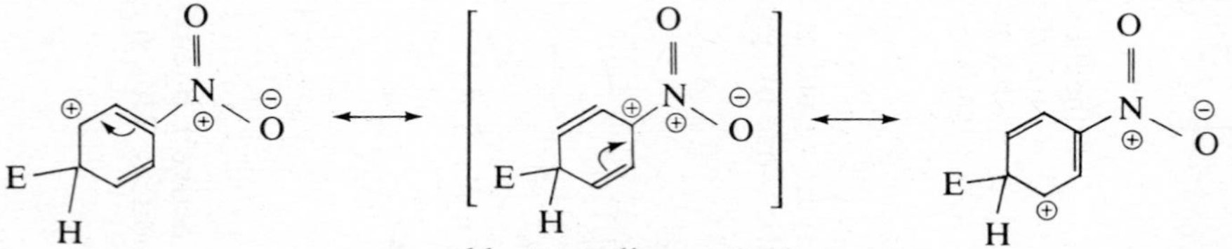 unstable: two adjacent ⊕ charges	total = 2	less stable
	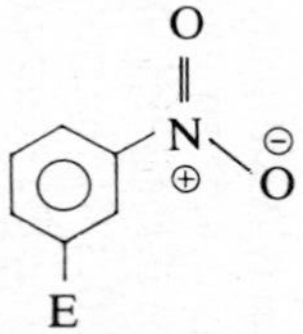 (3 −)		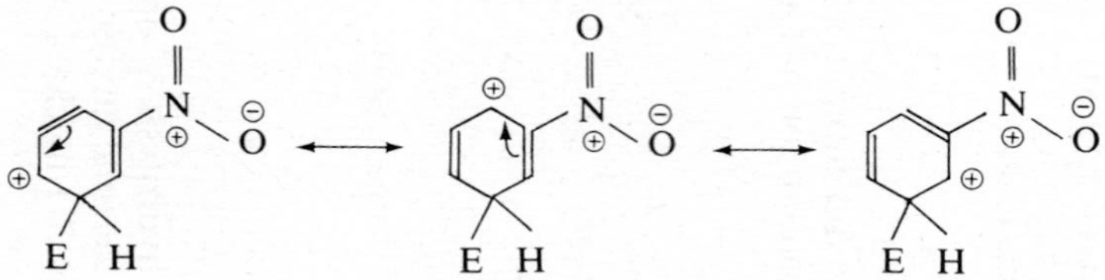 total = 3	more stable

36.2 Addition

Nucleophilic addition

The attack of a nucleophile on a carbonyl group can lead to an addition product with an asymmetric carbon atom. In principle, the product could therefore be optically active.

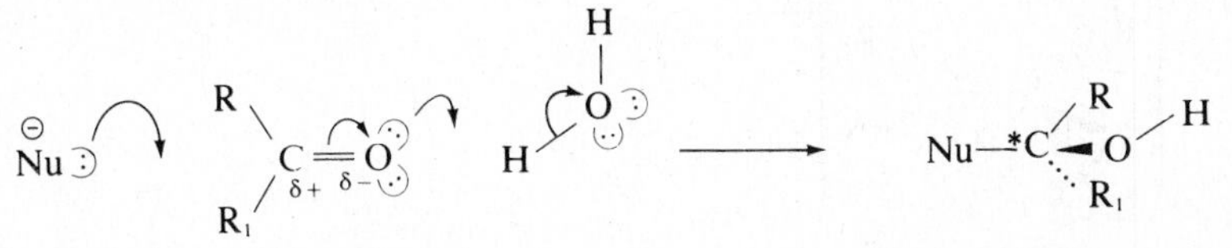

However, nucleophilic addition always results in the production of a racemic mixture. There is as much likelihood of the nucleophile attacking from beneath the carbonyl as there is of it attacking from above:

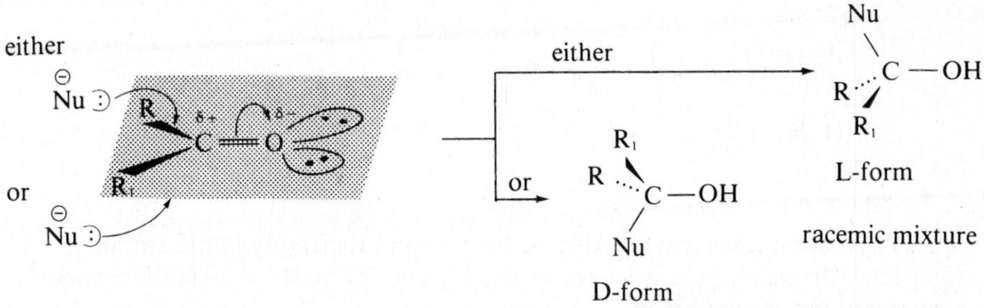

either

or

D-form

L-form

racemic mixture

An example of this is the formation of (±) 2-hydroxypropanonitrile from hydrogen cyanide and ethanal.

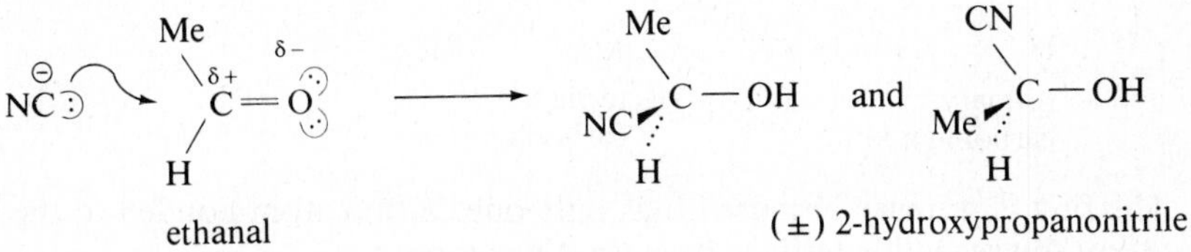

ethanal

(±) 2-hydroxypropanonitrile

Electrophilic addition

There are two principal types of stereospecificity associated with electrophilic addition:

1. Specific structural isomers are favoured on addition of H—X to an asymmetric alkene (Markovnikov's Rule, page 381).
2. The geometric isomers of an alkene give different optically active products after electrophilic addition.

1. Markovnikov's Rule

This example of stereospecificity can also be explained in terms of the relative stabilities of the possible activated complexes: the reaction pathway deriving from the most stable complex is the fastest, and therefore the products are controlled by this complex.

In the addition of H—X to an alkene, a π-complex is first formed:

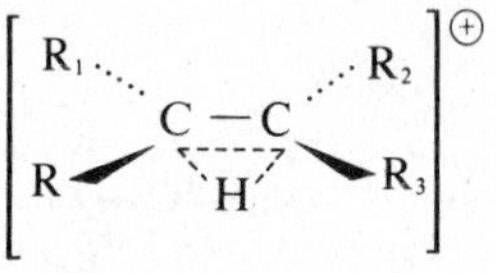

$$\left[\begin{array}{c} R_1 \cdots \quad \cdots R_2 \\ C-C \\ R \quad H \quad R_3 \end{array} \right]^{\oplus}$$

The π-complex is likely to break down into one of *two* possible σ-complexes under the action of the attacking nucleophile X^- :

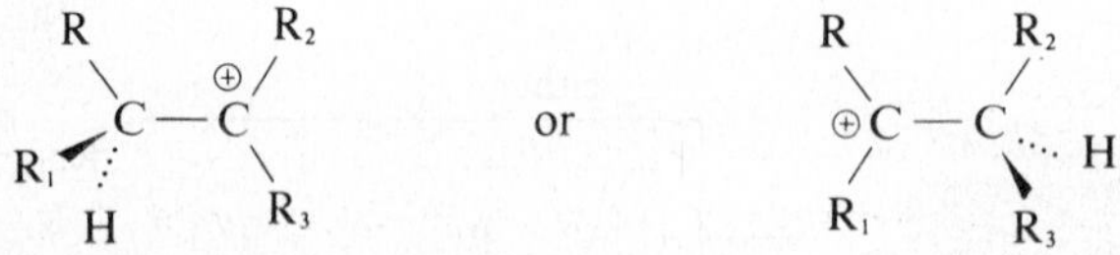

$$R(R_1)(H)C-\overset{\oplus}{C}(R_2)(R_3) \quad \text{or} \quad R(R_1)\overset{\oplus}{C}-C(R_2)(R_3)H$$

These two complexes are unlikely to be of equal thermodynamic stability. Consider the case where $R = R_1 = —CH_3$ and $R_2 = R_3 = —H$. The two complexes are shown below.

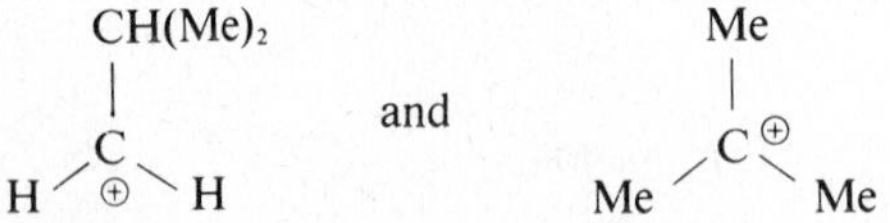

$$\underset{\text{primary carbonium ion}}{CH(Me)_2-\overset{\oplus}{C}H_2} \quad \text{and} \quad \underset{\text{tertiary carbonium ion}}{Me_3C^{\oplus}}$$

The first is primary because it has only one carbon atom bonded to the charged atom; the second is tertiary because it has three.

If we compare the inductive effect of a hydrogen atom with the overall inductive effect of an alkyl group, it can be shown that the alkyl group is electron-releasing with respect to a single hydrogen atom. This is because the alkyl group transmits the combined inductive effects of the hydrogen atoms bonded in the group, e.g.

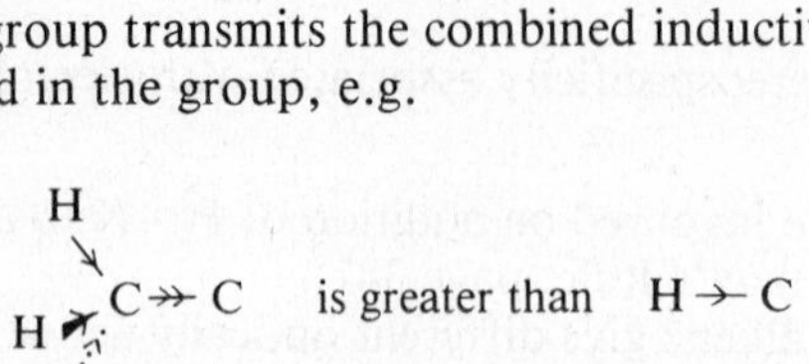

$$H_3C \twoheadrightarrow C \quad \text{is greater than} \quad H \rightarrow C$$

This suggests that tertiary carbonium ions (with three electron-releasing alkyl groups) are more stable than secondary ions which are in turn more stable than primary ones. The alkyl groups 'stabilize' the positive charge better than do the hydrogen atoms.

In applying these ideas to the addition reaction, therefore, we should expect the product to be largely the one derived from the tertiary complex.

overall (H)(H)C=C(Me)(Me) $\xrightarrow{\text{H—X}}$ (Me)(Me)(Me)C—X

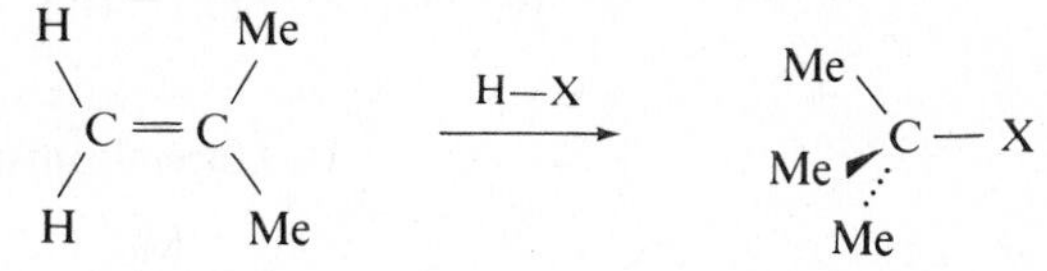

In other words, the hydrogen atom bonds to the carbon atom with the most hydrogen atoms there already. This happens because the reaction proceeds via the most substituted of carbonium ions which is the one of lowest energy.

2. Optical activity of addition products

The reactions of but-2-ene with potassium manganate (VII) and with benzoic peracid are stereospecific. Both reagents can oxidize but-2-ene to butane-2,3-diol but the stereochemistry of the products depends on the stereochemistry of the original alkene. The opposite effects of the two oxidizing reagents are illustrated in the table below and continued overleaf.

but-2-ene	*oxidized by*	*gives butane-2-3-diol*
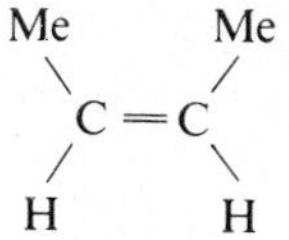 (Me)(H)C=C(Me)(H) *cis-*	alkaline $KMnO_4$	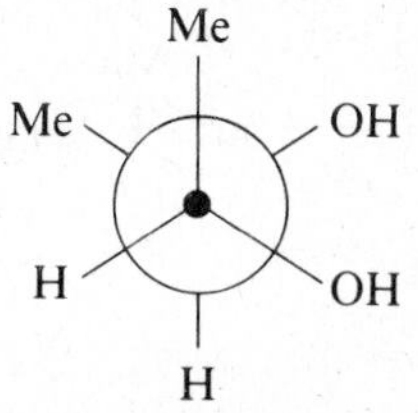 Newman projection (front: Me, OH, OH; back: Me, H, H) *meso*-form
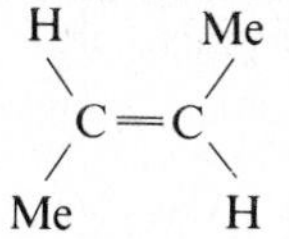 (H)(Me)C=C(Me)(H) *trans-*	alkaline $KMnO_4$	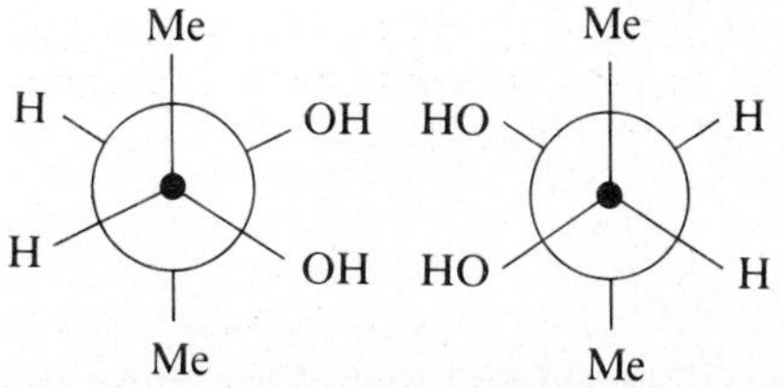 Newman projections: (Me, H, OH / H, OH, Me) and (Me, HO, H / HO, H, Me) (±) racemic mixture

but-2-ene	*oxidized by*	*gives butane-2-3-diol*
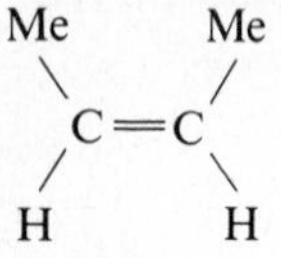 Me, Me / C=C / H, H *cis-*	1. peracid 2. hydrolyse the epoxide produced	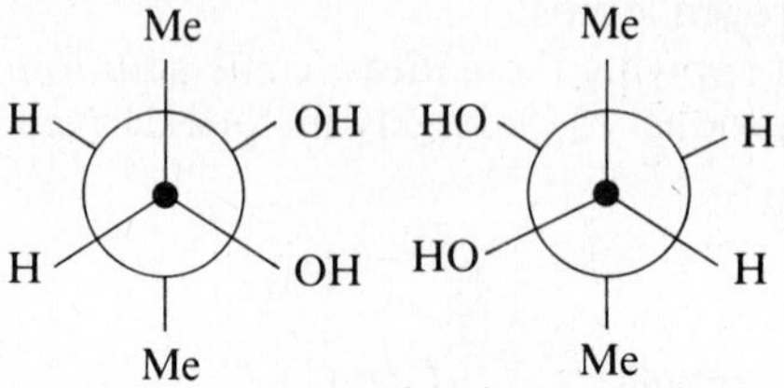 Me; H, OH; H, OH; Me and Me; HO, H; HO, H; Me ($\pm$) racemic mixture
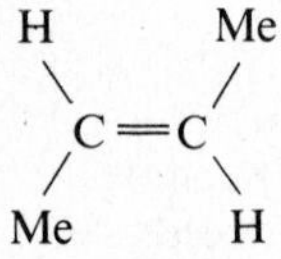 H, Me / C=C / Me, H *trans-*	1. peracid 2. hydrolyse the epoxide produced	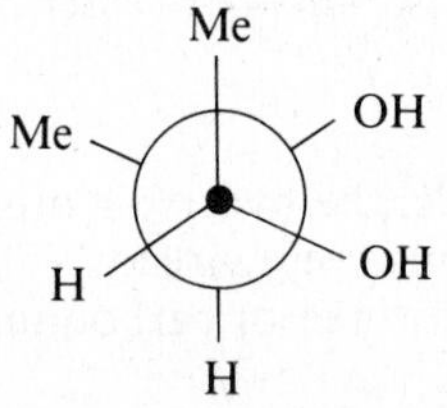 Me; Me, OH; H, OH; H *meso*-form

The stereospecificity of these reactions can be explained by considering the steric conditions of the collisions that form the products. Manganate (VII) causes *cis-* addition by the following mechanism.

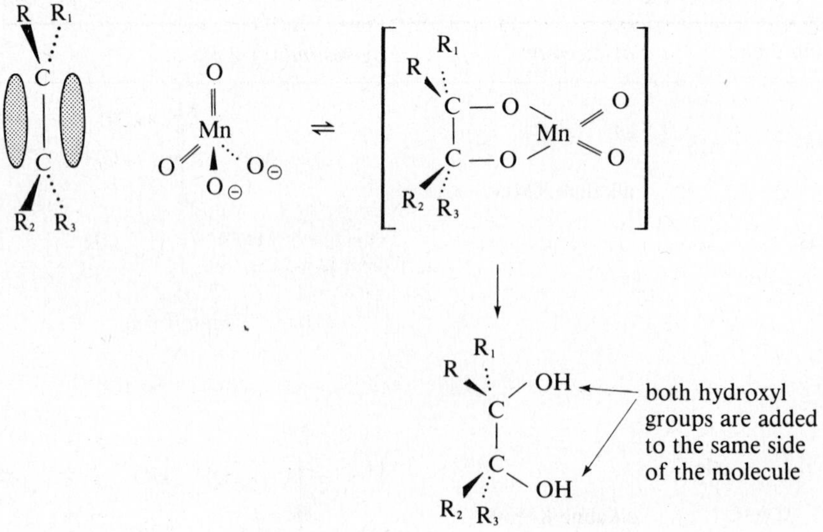

R R_1 C C R_2 R_3 O Mn O $O^{\ominus}$ $O^{\ominus}$ ⇌ [R_1 R C—O Mn O C—O O R_2 R_3]

↓

R_1 R OH C C OH R_2 R_3 ← both hydroxyl groups are added to the same side of the molecule

whereas peracid and hydrolysis cause *trans-* addition by the mechanism given at the top of the next page.

36.3 Elimination

Base-catalysed reactions

The most common type of elimination is catalysed by base. For example an alkene can be produced by the reaction of a halogenoalkane and alcoholic potassium hydroxide.

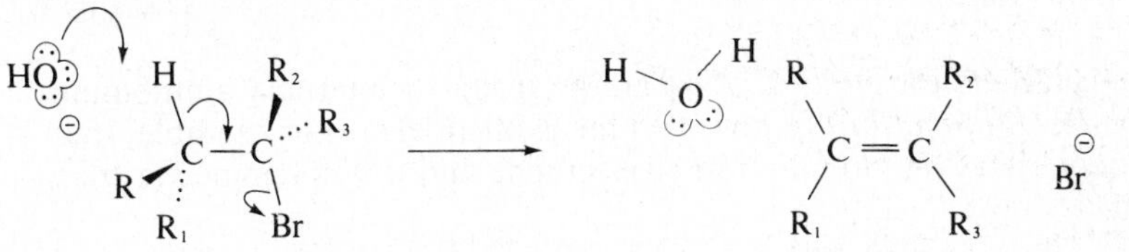

The reorganization of electron density is concerted and is likely to be most favourable when it occurs in a uniform direction. This is best achieved when the hydrogen atom and halogen atom are *trans*- to each other.

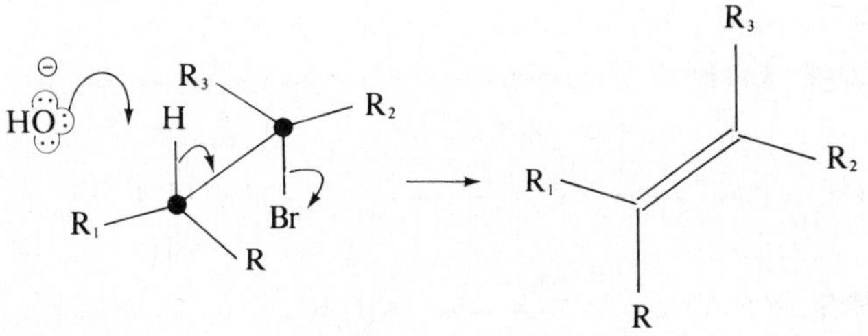

The *trans*- nature of this mechanism gives rise to the formation of stereospecific products when the groups R, R_1, R_2, and R_3 are different. The process is the exact reverse of the *trans*- addition of H—X discussed in the previous paragraphs. So, the first elimination products formed from 3,4-dibromobutane are different depending on whether the *meso*-form or an optically active form of the compound is used.

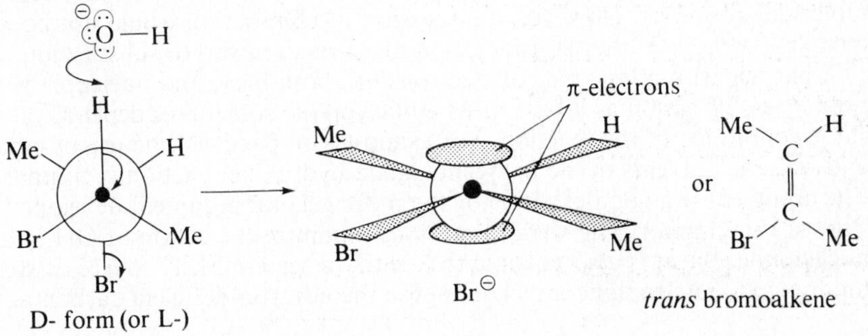

D- form (or L-)

trans bromoalkene

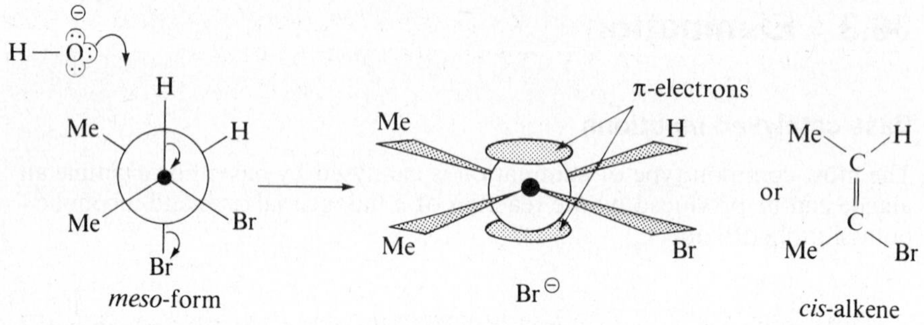

$$H-\ddot{O}:^{\ominus}$$

meso-form → π-electrons; or *cis*-alkene; $Br^{\ominus}$

The base-catalysed elimination of an amine from a quaternary ammonium salt (*Hoffmann Elimination*) follows this mechanism also. For example, the tetraethyl quaternary ammonium ion gives ethene and triethylamine via *trans*-elimination.

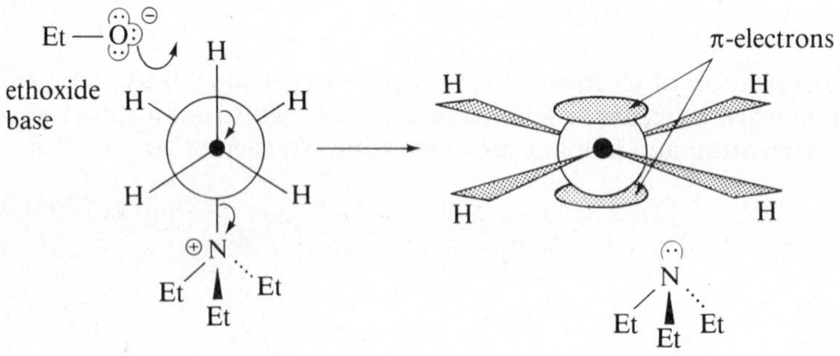

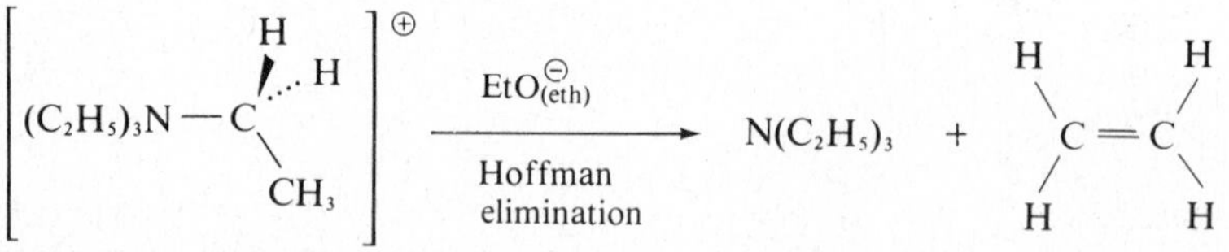

ethoxide base; π-electrons

$$\left[(C_2H_5)_3N-CH_2CH_3\right]^{\oplus} \xrightarrow[\text{Hoffman elimination}]{EtO^{\ominus}_{(eth)}} N(C_2H_5)_3 + H_2C{=}CH_2$$

Competition with substitution

Strictly speaking, the competition between substitution and elimination is not a stereospecificity conflict. The products are not isomers but have different molecular formulas. However, a discussion of elimination would not be complete without a brief consideration of its relationship to substitution.

A particle with a lone pair of electrons has both basic and nucleophilic properties. Its action as a base or as a nucleophile sometimes depends on the steric conditions of the reaction. For example, on page 404 the use of ethanol and water as solvents in the halogenoalkane-hydroxide reaction is compared. The argument is a good example of the more general argument developed at the start of this chapter. The steric conditions minimize the action of $OH^{-}_{(eth)}$ as a nucleophile. But it is also possible to identify a second likely source of steric hindrance to nucleophilic attack: suppose the electron-deficient carbon atom has bulky groups bonded to it. For example 2-methyl-2-bromobutane has an ethyl and two methyl groups bonded to the electron-deficient carbon atom.

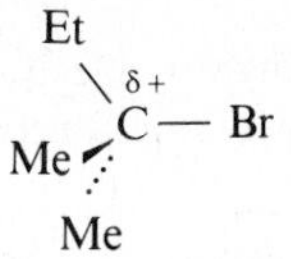

Et
δ+
C— Br
Me
Me

Nucleophilic attack on this atom is hindered, no matter what the nucleophile. Elimination is thus favoured over any S_N2 process. However, tertiary compounds like the one above are more likely to undergo substitution by an S_N1 mechanism (see page 559) because the carbonium ion is stabilized by the electron-releasing alkyl groups.

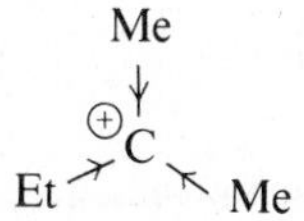

Me
⊕C
Et Me

Even so, the carbonium ion can readily lose a proton to an incoming particle with a lone pair and this effectively leads to elimination.

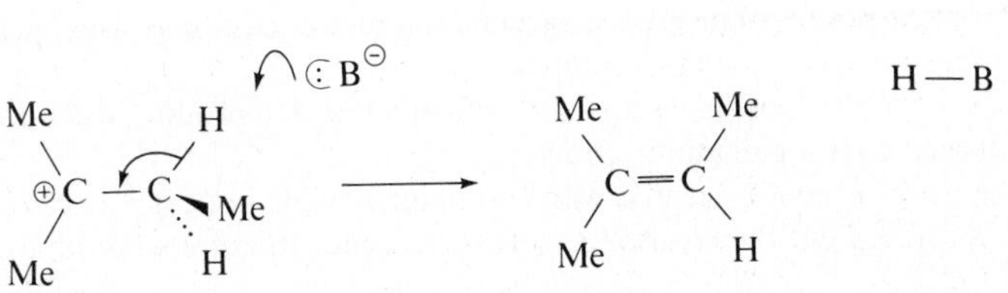

$:B^{\ominus}$ H—B

Me H Me Me
⊕C — C Me ⟶ C=C
Me H Me H

Elimination therefore competes also with S_N1. The whole mechanism of the above process is called *E*1 because it is first-order in the same way that S_N1 is first-order.

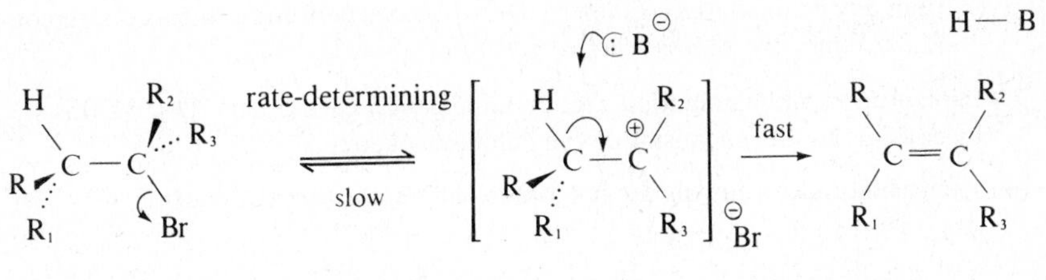

$:B^{\ominus}$ H—B

H R_2 rate-determining H R_2 R R_2
R_3 ⊕ fast
R C—C ⇌ R C—C ⟶ C=C
slow
R_1 Br R_1 R_3 $Br^{\ominus}$ R_1 R_3

In contrast, the concerted elimination mechanism discussed on page 569 is known as *E*2 because the rate depends on the concentration of both the base and the halide. Therefore there are four processes that can compete with one another in this reaction medium: S_N1, S_N2, *E*1 and *E*2.

Questions

1. a) Explain the difference between a reaction that is stereospecific and one that is stereoselective.
   b) Give an example of a stereospecific reaction and explain why it is specific.
   c) Give an example of a stereoselective reaction and explain why it is selective rather than specific.

2. a) What is meant by the terms S_N1 and S_N2 as applied to a reaction?
   b) Give an example of a reaction that proceeds largely by an S_N2 mechanism. Draw the mechanism and explain why it is favoured.
   c) Give an example of a reaction that proceeds largely by an S_N1 mechanism. Draw the mechanism and explain why it is favoured.

3. The rate at which the following bromoalkanes are hydrolysed by dilute alkali increases in the order shown below.

   $$MeCH_2Br < Me_2CHBr \approx CH_3Br \ll Me_3CBr$$

   Explain the observed order.

4. Account for the following observations
   a) Propene forms 2-bromopropane when treated with hydrogen bromide.
   b) Ethanal forms (±) butan-2-ol (a racemic mixture) when treated with the Grignard reagent derived from bromoethane.
   c) (+) 2-bromobutane racemizes in the presence of bromide ions.

5. Account for the following observations:
   a) *trans*- 2,3-dichlorobut-2-ene gives a racemic mixture of products when treated with hydrogen over a palladium catalyst.
   b) *cis*- 2,3-dichlorobut-2-ene gives a single optically inactive product when treated with hydrogen over a palladium catalyst.
   c) When the *trans*- isomer is treated with bromine, a single optically inactive product is formed but the *cis*- isomer yields a racemic mixture of products.

6. a) Nitrobenzene forms 1,3-dinitrobenzene when refluxed with concentrated nitric and sulphuric acids. Methylbenzene forms 2,4,6-trinitromethylbenzene under similar reaction conditions. Account for the differences.
   b) One diastereoisomer of benzene hexachloride (1,2,3,4,5,6-hexachlorocyclohexane) undergoes elimination of hydrogen chloride about a hundred times more slowly than any of its diastereoisomers. Draw the configuration of this diastereoisomer and account for its less reactive nature.

7. 'Substitution and elimination often compete with each other'. Discuss this statement by considering the chemistry of the halogenoalkanes.

8. 'Steric hindrance controls the outcome of most reactions'. Discuss.

Chapter 37

Applied organic chemistry

37.1 Natural polymers

Proteins and polypeptides

Proteins and polypeptides are long-chain polymers which occur in the cells of all living organisms. They constitute the connective tissue of the organism, including skin, muscle, tendons and ligaments. These polymers also take a vital part in the processes occurring within the cells of the organism, for they form enzymes and a whole range of metabolic intermediates.

The molecular mass of a protein or a polypeptide can be found from osmotic pressure measurements (see page 157). A very wide range of molecular masses is found with values ranging from below 100 to around 100 000. This range of molecular mass has been arbitrarily divided in two:

'polypeptides'	R.M.M. $< 10\,000$
'proteins'	R.M.M. $> 10\,000$

Hydrolysis of the naturally occurring proteins and polypeptides reveals the following:

1. The hydrolysis products are nearly all α-amino acids, i.e. the amino group is attached to the carbon atom next to the carboxylic acid group.

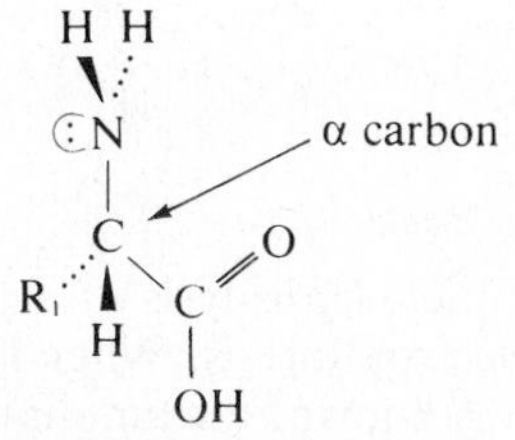

H H
N
α carbon
C
O
R
C
H
OH

2. Only about 20 different α-amino acids are produced as hydrolysis products despite the great range and variety of biological material that contains proteins.
3. All the α-amino acids contain an asymmetric carbon atom and they all have the same configuration — the one exception is aminoethanoic acid (glycine) which is the simplest of the α-amino acids.

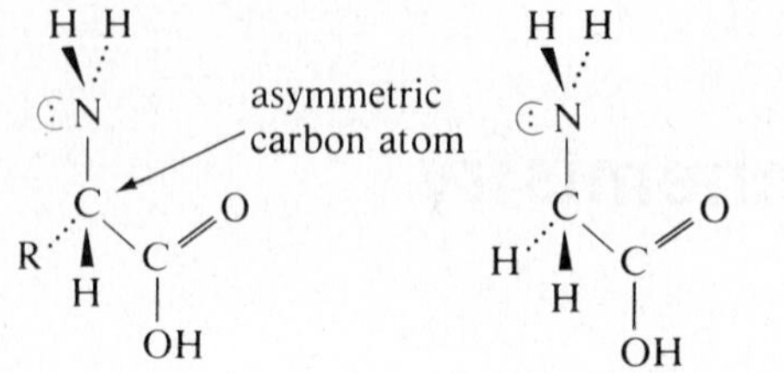

asymmetric carbon atom

aminoethanoic acid

Because the hydrolysis of a protein yields a series of α-amino acids, the protein itself must consist of these amino acids linked together in a long chain.

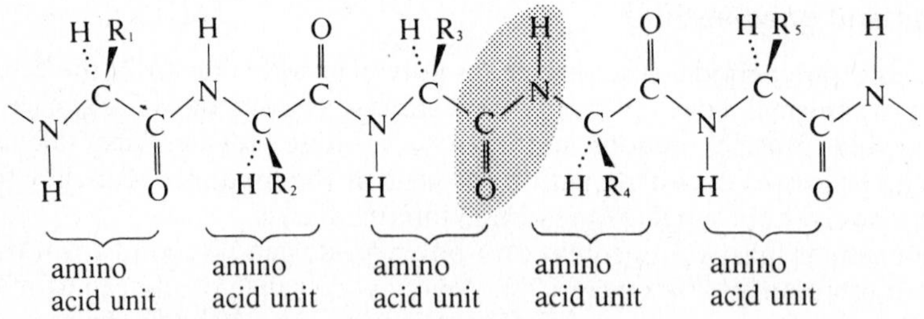

amino acid unit amino acid unit amino acid unit amino acid unit amino acid unit

Each pair of units is joined by the same peptide link. The peptide link is broken during hydrolysis just as it is with a substituted amide.

the peptide link (see page 474)

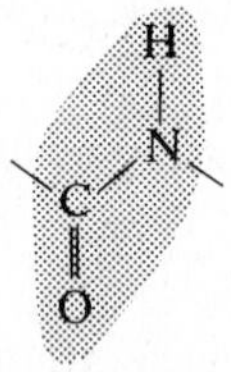

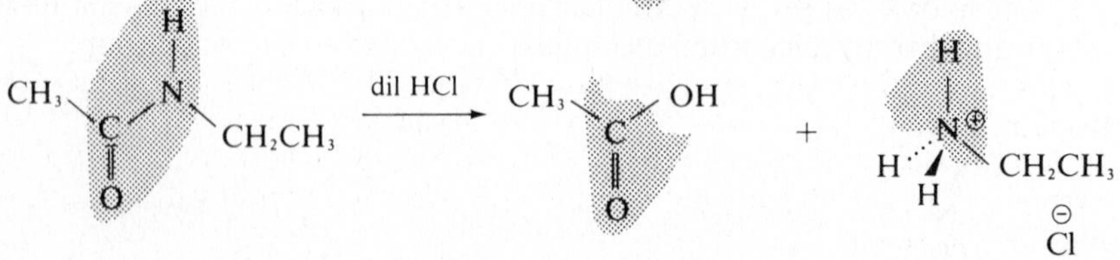

$$CH_3CONHCH_2CH_3 \xrightarrow{\text{dil HCl}} CH_3COOH + CH_3CH_2NH_3^+\,Cl^-$$

The collection of amino acids produced by the complete hydrolysis of a protein can be analysed by means of chromatography or electrophoresis. When the identity of the constituent amino acids in a protein is known, the amount of each amino acid can be determined using colorimetry. Each amino acid is reacted with a reagent to produce a coloured compound; the intensity of the colour is proportional to the amount of the amino acid as measured by means of a spectrophotometer.

Protein structure

Having determined the number and variety of amino acids in a protein, the next step is to determine the sequence of the amino acids in the molecule (known as its *primary structure*). This was first achieved for the protein insulin by Sanger in 1955.

The general procedure for working out this sequence is as follows.

1. Hydrolyse the protein selectively into some peptide fragments. This is done using enzymes that split particular peptide links in the chain.
2. Identify these peptide fragments using chromatography.
3. Identify the amino acid at one end of each of these fragments. The acid with the free amino group is called the *N-terminal acid*; it can be identified because it will react with 2,4-dinitrofluorobenzene.

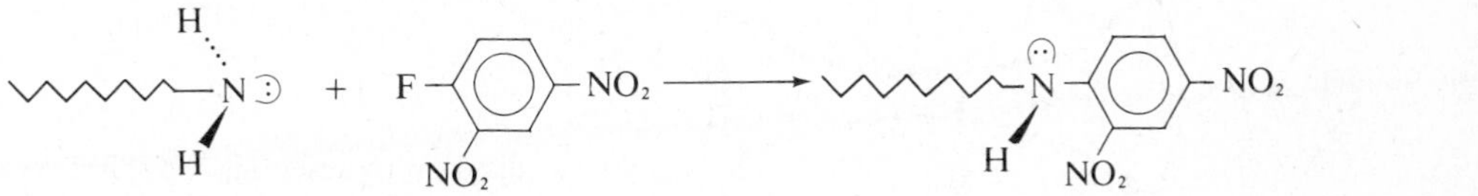

   After complete hydrolysis, the amino acid attached to the dinitrobenzene is identified using chromatography.
4. Identify the amino acid at the other end of each peptide fragment. This acid has a free carboxylic acid group and is called the *C-terminal acid*; it is identified by means of the enzyme carboxypeptidase. The enzyme breaks the peptide link only of the amino acid that has the free carboxylic acid group, so freeing it. In fact successive use of this enzyme allows the C-terminal acids to be picked off one at a time; they are then recognized using chromatography.

The above steps allow much of the amino acid sequence of the protein to be pieced together.

However, the sequence of α-amino acids in a protein is only part of the picture. The presence of electron-deficient hydrogen atoms in the amino groups and of lone pairs on the oxygen atoms of the carbonyl groups means that hydrogen bonds can form either between different portions of the chain or even between adjacent chains.

As a consequence, the chain can become folded into a spiral shape as shown on the next page. The actual conformation of the chain is known as the protein's *secondary structure*. The secondary structure is destroyed by heating above 70°C or by the addition of an acid or a base. When this occurs, the protein is said to be denatured.

The structure of the protein may be further complicated by the fact that the spiral may be folded back on itself in a loop. The loop is held in place by bonds between sulphur atoms that occur in some of the amino acids. This gives the protein a *tertiary structure*.

While the primary structure is determined largely by means of selective hydrolysis, the secondary and tertiary structure can only be deduced from X-ray analysis.

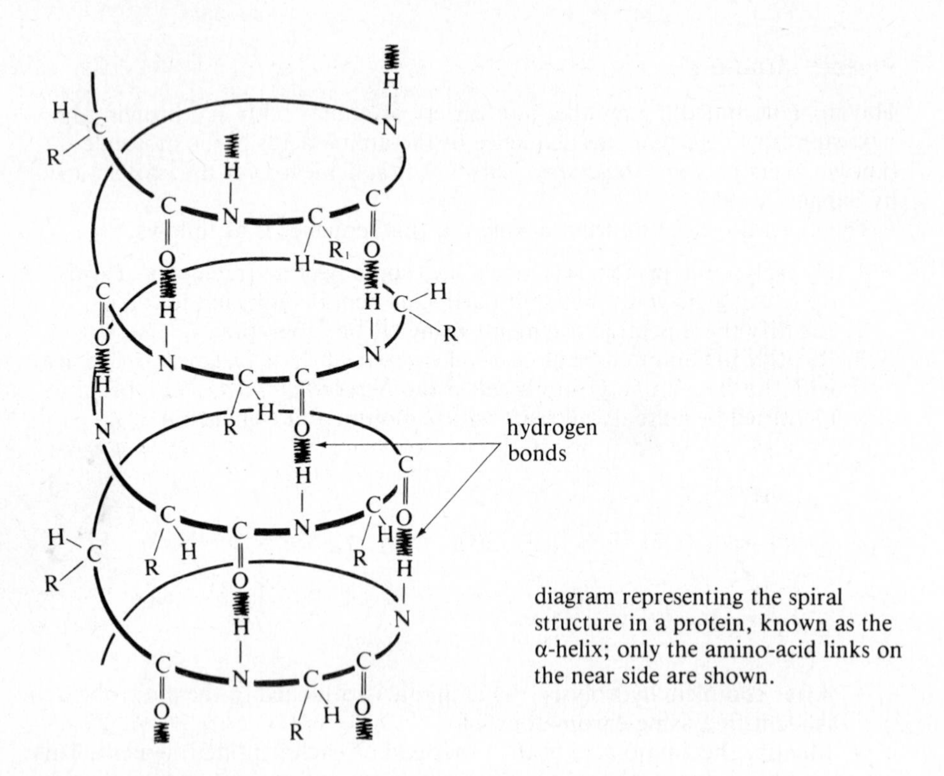

diagram representing the spiral structure in a protein, known as the α-helix; only the amino-acid links on the near side are shown.

Protein synthesis

In living cells proteins are synthesized by the stepwise addition of single amino acids to the carboxyl end of a growing chain. The sequence of amino acids added to the chain is controlled by a substance called messenger ribonucleic acid which contains genetic information of the organism. The formation of the new peptide links is catalysed by another protein, an enzyme called aminoacyl-t RNA synthetase.

In the laboratory it is possible, in principle, to synthesize proteins by successive addition-elimination reactions between the nucleophilic amine group of one amino-acid and the electron-deficient carbon of another:

$$\mathrm{H_2\ddot{N}}- \;+\; \overset{\delta+}{\mathrm{C}}(\mathrm{X})=\overset{\delta-}{\mathrm{O}} \longrightarrow \left[\, -\overset{\oplus}{\mathrm{N}}\mathrm{H_2}\cdots\mathrm{C}(\mathrm{X})-\mathrm{O}^{\ominus} \,\right] \xrightarrow{-\mathrm{HX}} \mathrm{H}-\mathrm{N}-\mathrm{C}=\mathrm{O}$$

new peptide link

In practice there are three problems that must be overcome before successful synthesis can be achieved.

a) There is a competing acid-base reaction between the carboxylic acid groups and the amino groups. This proton transfer reaction is much more rapid than the addition-elimination reaction. It removes the nucleophilic properties of the amino-group and produces the less reactive carboxylate ion.

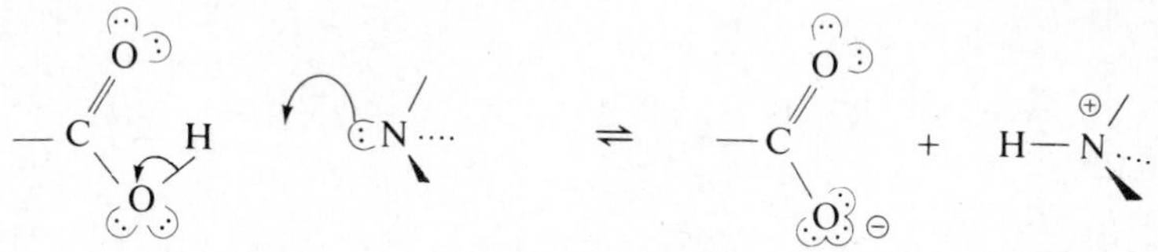

b) Amino acids are bifunctional. Each amino acid molecule contains both a nucleophilic nitrogen atom and an electron-deficient carbon atom. A reaction mixture of two amino acids produces a joint product, but also two other products as the result of reaction between two molecules of the same amino-acid, e.g.

mixture A + B $\longrightarrow$ AB but also AA and BB.

c) All the asymmetric carbon atoms in a natural polypeptide have the same configuration. Laboratory syntheses tend to produce racemic mixtures as products.

To overcome these problems, the following general procedure is adopted.

1. Two amino acids are selected
2. The carboxylic acid group of one amino acid is protected by reacting it with methyl propene.

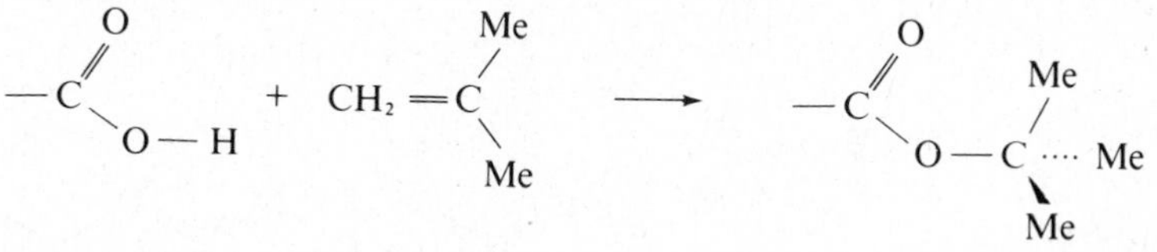

   Later this protecting group can be removed under mild hydrolysis which will not affect the newly formed peptide link.
3. The amino group of the second amino acid is protected by converting it into an acylamino group.

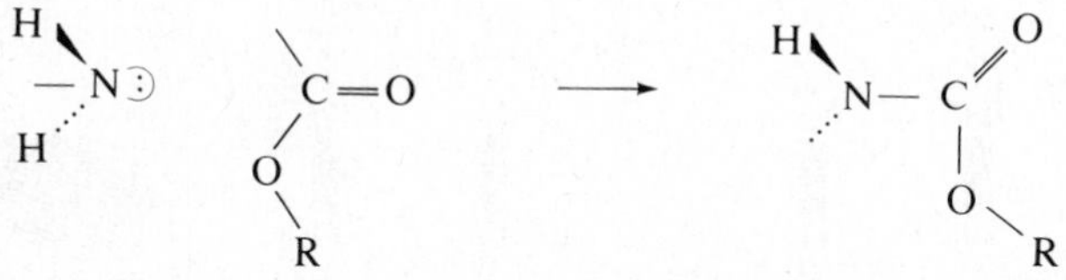

   This can also be removed later by mild hydrolysis.
4. The carboxylic group of the second amino acid is activated by converting it to an acid chloride.

5. The two modified amino acids are now reacted and the new peptide link is formed, e.g.

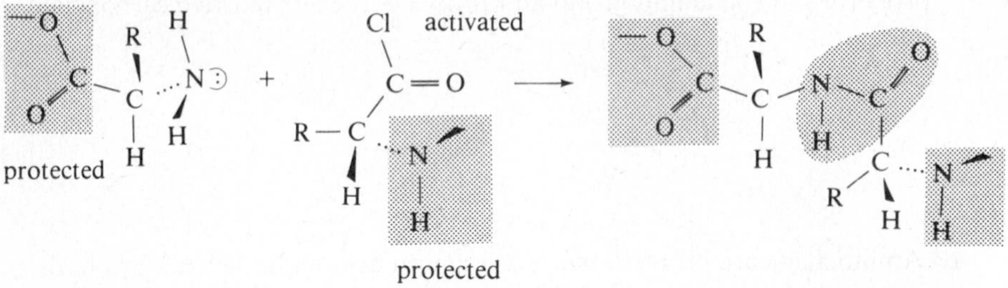

protected

activated

protected

6. The protecting group at one end of the embryo chain is removed and the whole sequence repeated successively until synthesis is complete.

Proteins and their uses

The properties and uses of proteins and polypeptides are closely related to their structure. Proteins can be divided into two broad groups.

1. Fibrous proteins

These make up much of the structure and connective tissue in an animal and fall into one of three different categories.

a) Proteins consisting of single chains lying side by side. Because these chains are parallel and run in opposite directions, there is the possibility of extensive hydrogen bonding between the chains.

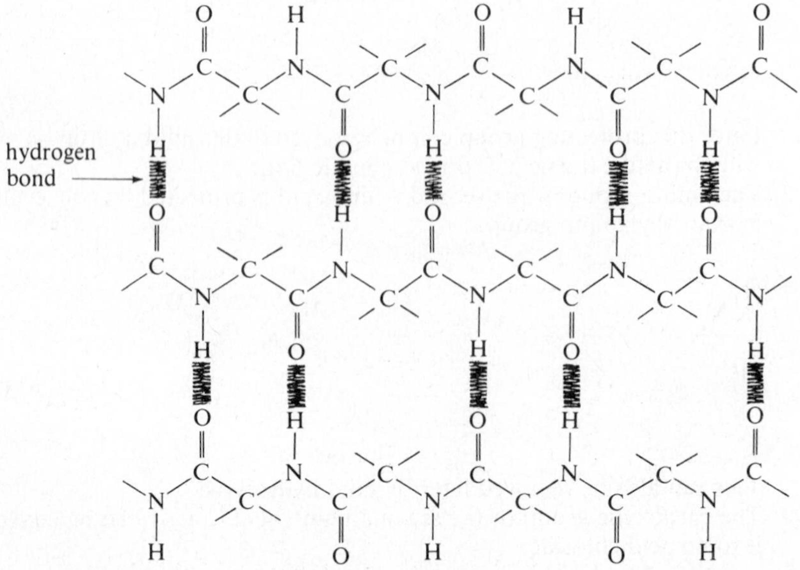

hydrogen bond

Rows of these chains produce a pleated sheet-like structure and form the supple and flexible fibres such as silk.

b) Proteins in which two or more chains are twisted around one another, producing a rope-like structure. Again there are hydrogen bonds between adjacent spiralling chains that hold them together. The flexibility and strength of the protein is dependent on the number of chains that are twisted together.

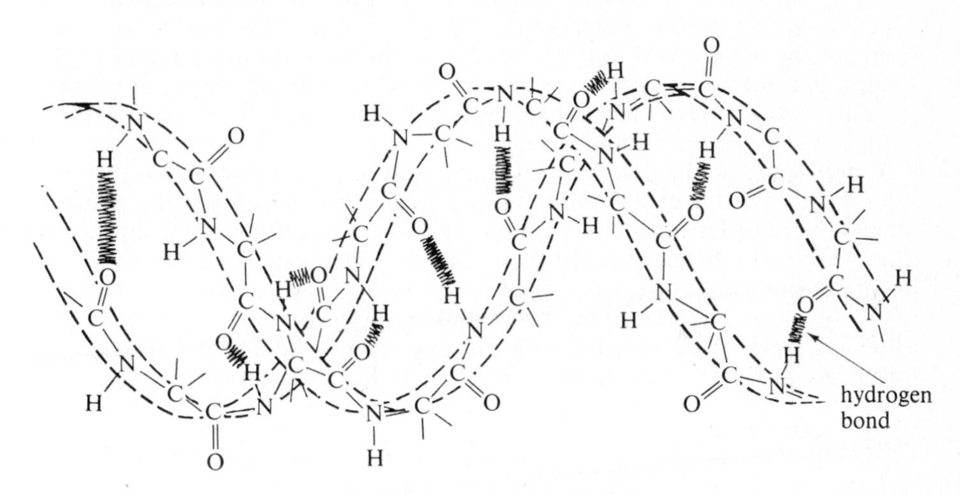

two adjacent spirals linked by hydrogen bonds.
(side groups are left off the asymmetric carbons for simplicity).

Proteins such as collagen are of this structure and form the tissue of muscles, tendons and ligaments.

c) Proteins consisting of a single chain that is wound in a tight spiral known as the α-helix. Successive coils of the spiral are hydrogen-bonded to each other, so producing a spring-like structure.

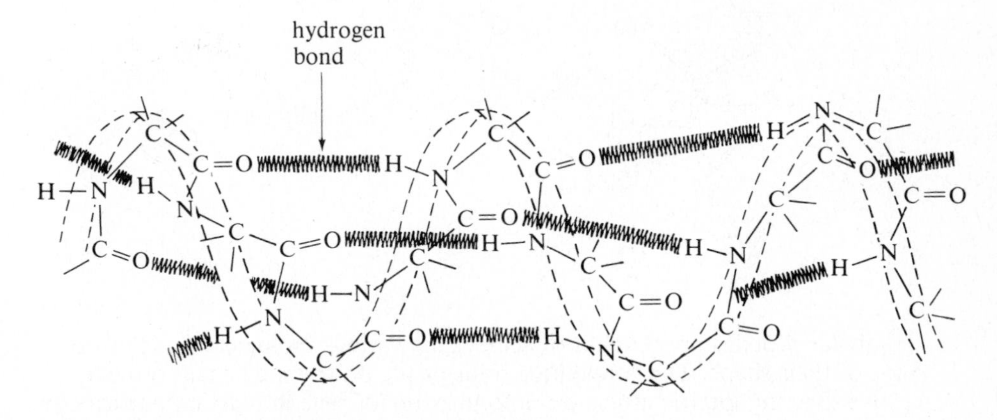

a portion of the α-helix (side chains omitted)

Proteins such as keratin and epidermin in hair and fingernails are of this type; they are flexible and elastic. Under stress the hydrogen bonds are broken and the spiral uncoils, but when the stress is removed the spiral reforms.

2. Globular proteins

The globular proteins constitute a group of proteins that serve a series of vital functions within the cell. For example, the group of proteins known as haemoglobin are proteins found in red blood cells that transport oxygen to the tissues. Enzymes also are proteins that act as catalysts to the processes within the cells. Likewise many hormones, such as insulin, are globular proteins that regulate the processes within the cells.

The structure of globular proteins is made up of polypeptide chains that are looped and folded back on themselves in complicated and apparently random tangles. Nevertheless, it is clear from recent studies of certain of these proteins that the complex arrangement of these chains is precise and specific. The diagram below shows the structure of the enzyme lysozyme. The chain consists of 129 α-amino acids, each placed at a kink in the chain. Notice the four sulphur bridges which, together with hydrogen bonds and Van der Waals' forces, are thought to hold this type of chain to its particular shape.

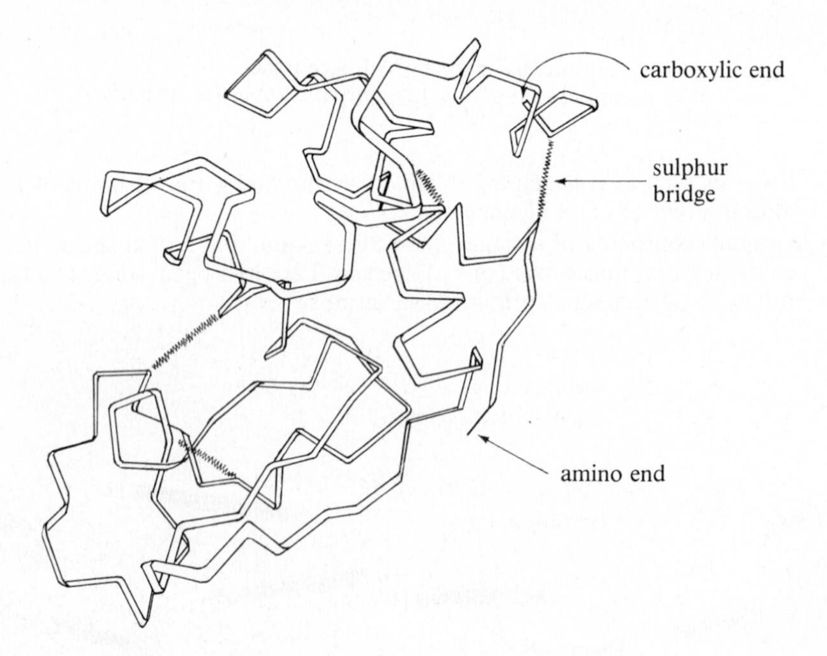

Globular proteins are thought to achieve the various biological functions by virtue of their shape. The convoluted folds of the polypeptide chain provide reactive sites for specific atoms or molecules, or for several reactant particles at the same time, as shown in the two schematic diagrams below.

oxygen transport in a globin

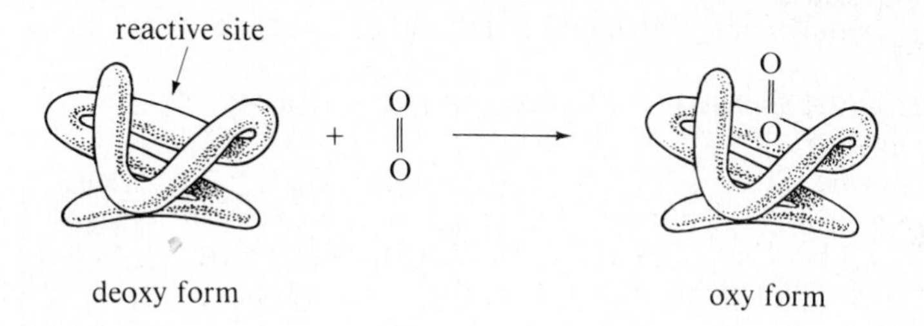

protein synthesis

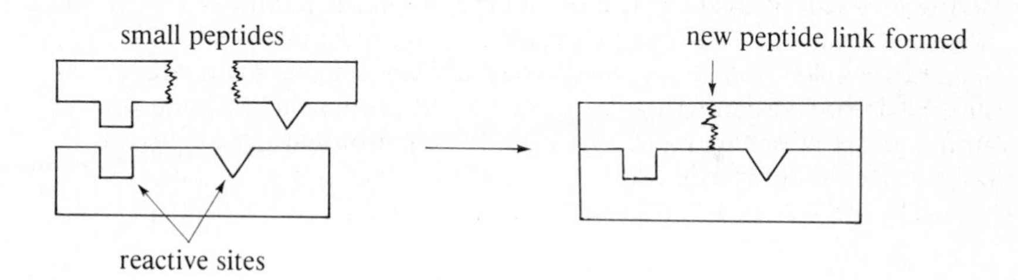

37.2 Synthetic polymers

Polyamides

Amino acids form the long chains that are proteins because the two ends of the molecule are different functional groups that will react together forming the peptide link. Synthetic polymers can be built using the same peptide link to join molecules, but in this case each of the two molecules has identical but complementary ends. The polymers formed are called nylon.

$$n\,[NH_2-(CH_2)_6-NH_2] + n\,[HOOC-(CH_2)_4-COOH]$$

$$\xrightarrow{\text{heat}} \sim\!\!\sim\left[(CH_2)_6-\underset{|}{\overset{H}{N}}-\underset{\|}{\overset{}{C}}(=O)-(CH_2)_4-\overset{O}{\overset{\|}{C}}-\underset{H}{\underset{|}{N}}\right]\sim\!\!\sim$$

Represented above is the synthesis of nylon 6.6 in which 1,6-diaminohexane having 6 carbon atoms is joined by a peptide link to hexanedioic acid, also having 6 carbon atoms. This was the first of the nylons to be widely used and was synthesized by Carothers in 1935.

Other nylons can be made from different starting materials. For example, nylon 6.10 is made from 1,6-diaminohexane and decanedioic acid.

$$n\,[NH_2-(CH_2)_6-NH_2] + n\,[HOOC-(CH_2)_8-COOH]$$

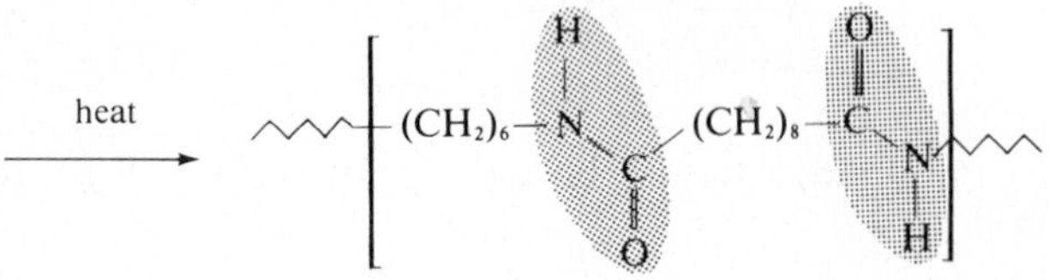

$$\xrightarrow{\text{heat}} \sim\!\!\left[(CH_2)_6-\underset{H}{N}-\overset{O}{\overset{\|}{C}}-(CH_2)_8-\overset{O}{\overset{\|}{C}}-\underset{H}{N}\right]\!\!\sim$$

Nylon 6.10 can be made at room temperature in the laboratory by substituting the more reactive decanedioyl chloride for the acid.

Another nylon, nylon 6, is made from a single starting material, caprolactam. This has a ring structure that is opened up to produce a molecule with an amino group at one end and a carboxylic acid group at the other end. It polymerizes to form nylon 6.

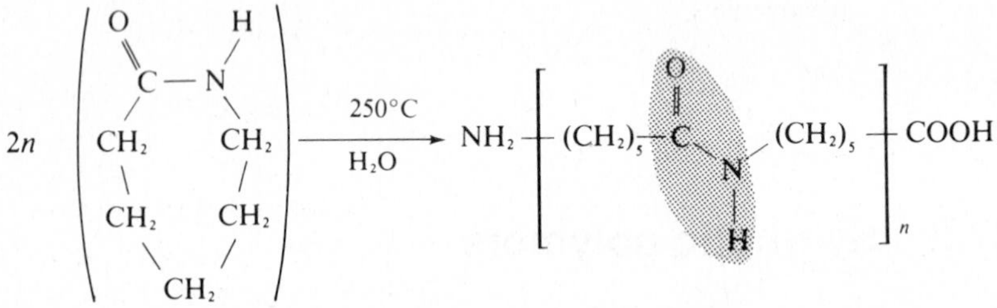

$$2n\,\text{(caprolactam)} \xrightarrow[H_2O]{250°C} NH_2-\left[(CH_2)_5-\overset{O}{\overset{\|}{C}}-\underset{H}{N}-(CH_2)_5\right]_n-COOH$$

caprolactam

Polyesters

Like polyamides, polyesters can be made from molecules having a functional group at each end. The polyester 'terylene' is made by the reaction between ethane-1,2-diol and an ester of benzene 1,4-dicarboxylic acid. An ester is used because the acid itself is insoluble in the diol.

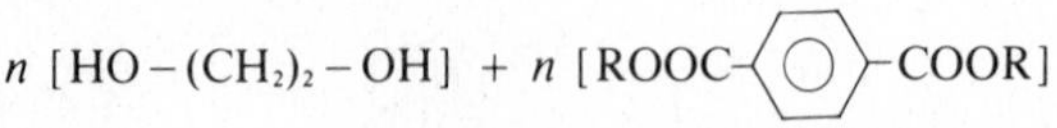

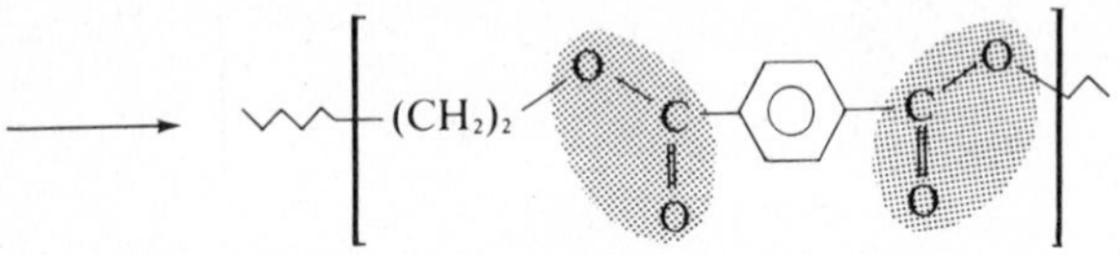

$$n\,[HO-(CH_2)_2-OH] + n\,[ROOC-C_6H_4-COOR]$$

$$\longrightarrow \sim\!\!\left[(CH_2)_2-O-\overset{O}{\overset{\|}{C}}-C_6H_4-\overset{O}{\overset{\|}{C}}-O\right]\!\!\sim$$

ester link

If a molecule with three functional groups is used in place of one with two for one of the reactants, the product is a cross-linked polymer and not a single long chain polymer.

$$n\,[HO-CH_2-CH(OH)-CH_2-OH] + 2n\,[ROOC-C_6H_4-COOR]$$

three functional groups

$$\longrightarrow HO-[CH_2-CH(O-CO-C_6H_4-COOR)-CH_2-O-CO-C_6H_4]-COOR$$

The product is called an alkyd resin and is used in paints.

Properties of the polymers

The physical properties of a synthetic polymer are influenced by:

1. Polymer chain length — this is controlled by the proportion of molecules with only one functional group added to the reacting mixture. When one of these molecules adds onto a chain it stops further growth.
2. Cross linking — this is controlled by the proportion of molecules with three functional groups added to the reacting mixture.

The appearance of the finished material depends on the treatment given to the crude polymer. After polymerization, it is melted and then may be cast or forced through fine holes producing fibres. These are spun or woven or used in thicker form as bristles in brushes. Objects made from polyamides and polyesters are strong and hard wearing.

Because both the peptide and ester links are susceptible to hydrolysis under acid or alkaline conditions, nylon and polyester objects should not be exposed to acid or alkaline conditions for long periods of time. Under these conditions, polymers based on polythene and its analogues should be used because they are not hydrolysed.

37.3 Petroleum

Natural sources

Petroleum is the hydrocarbon residue produced from the gradual decay of very small marine animals. It consists of a mixture of aliphatic and aromatic compounds whose proportions depend on the origin and history of the petroleum deposit. The physical properties of crude petroleum vary from those of natural gas to those of the thick waxy or tarry liquids which solidify as they are pumped from the ground.

Crude oil from the well is usually stripped of its gaseous components in a simple distillation tower and then fractionally distilled. Because the liquid portion of crude oil forms a nearly ideal mixture according to Raoult's Law, it can be effectively separated in a fractional distillation tower. Six main fractions are produced.

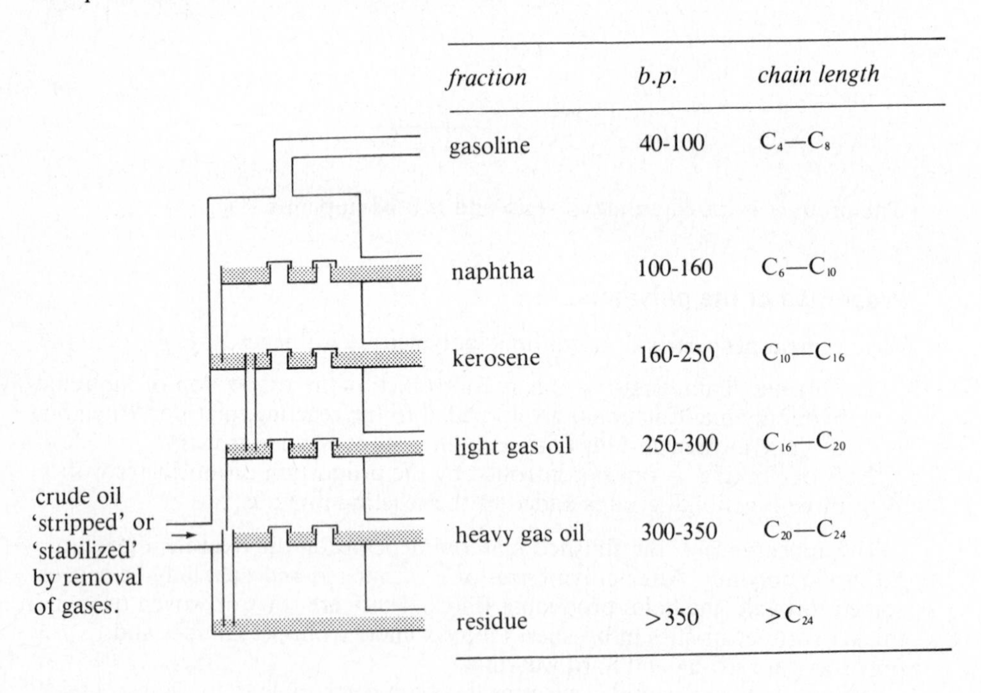

fraction	*b.p.*	*chain length*
gasoline	40-100	C_4—C_8
naphtha	100-160	C_6—C_{10}
kerosene	160-250	C_{10}—C_{16}
light gas oil	250-300	C_{16}—C_{20}
heavy gas oil	300-350	C_{20}—C_{24}
residue	>350	>C_{24}

Synthetic petroleum

As the two pie charts below show, there is an imbalance between the reserves and the consumption of the different fossil fuels.

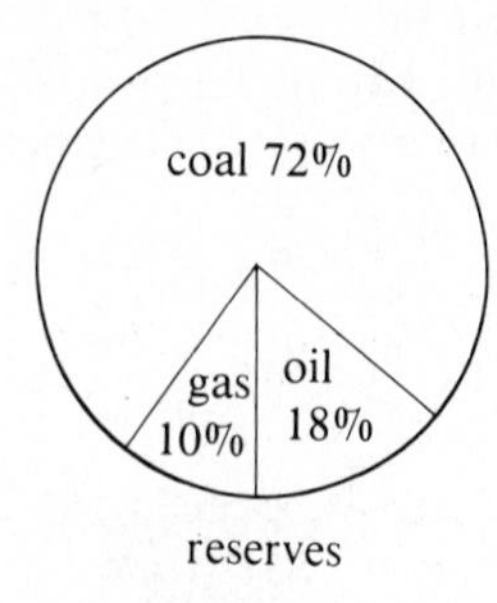

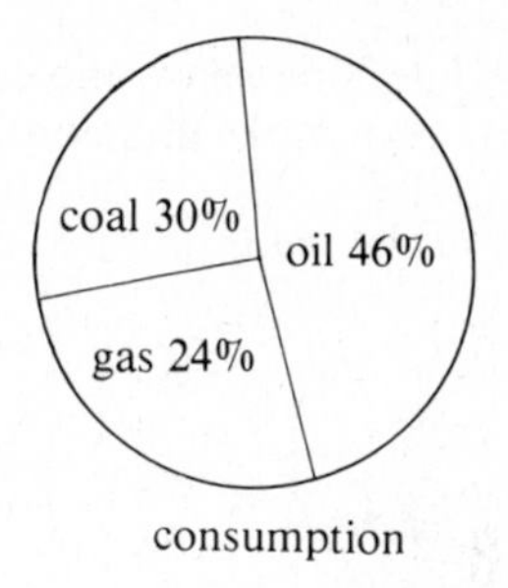

The need to match consumption of fuels more closely to available reserves has stimulated much research into methods of converting coal into liquid fuels. This process was pioneered in Germany in the 1930s, and South Africa has a large scale plant in operation producing 10 000 tons of liquid products a day.

There are four main approaches to the problem of converting coal to liquid fuel:

1. destructive distillation: coal is heated in the absence of air.
2. hydrogenation: coal is heated in a stream of hydrogen.
3. solvation: coal is dissolved in a range of organic solvents.
4. Fischer-Tropsch process: coal is heated in a stream of steam and oxygen.

All the methods have the aim of increasing the hydrogen to carbon ratio. Typical values of this ratio for coal and petroleum are:

	hydrogen : carbon
coal	0.8 : 1
petroleum	1.75 : 1

The molecules found in coal are very complex as the structure below indicates, and they must be broken down into smaller structures.

a typical 'coal' molecule

The flow diagrams below indicate in outline two of the processes for decomposing coal molecules and how they are integrated together.

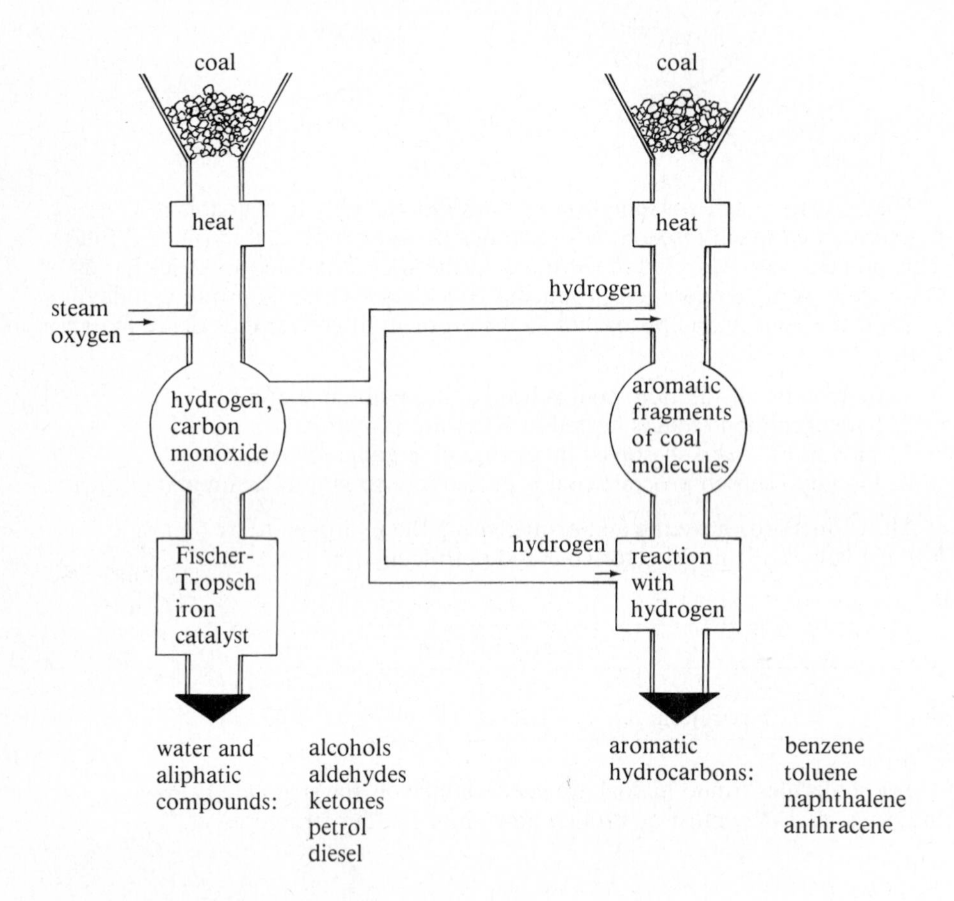

Cracking, reforming and isomerization

The fractions from the separation of crude oil or the products of the decomposition of coal are used as fuels. They are also subjected to a series of processes which modify them to meet the demands of the petrochemical industry. These processes include the following.

1. Cracking

This is the thermal decomposition of large saturated, hydrocarbon molecules in the presence of Lewis acidic catalysts. The more important products of cracking are unsaturated hydrocarbon molecules which are the starting points of several petrochemical processes. Cracking is a free-radical process and, like many free-radical reactions, produces a range of products. For example, the cracking of a C_{17} hydrocarbon might produce a significant proportion of saturated C_8 hydrocarbons for use as motor fuel, as well as large amounts of ethene and propene.

$$\text{e.g. } C_{17}H_{36} \xrightarrow[\sim 400°C]{Al_2O_3} C_8H_{18} + CH_2 = CHCH_3 + CH_2 = CH_2$$

among others

2. Reforming

This involves converting straight-chain hydrocarbons into cyclic ones, which are often dehydrogenated to give aromatic products, e.g.

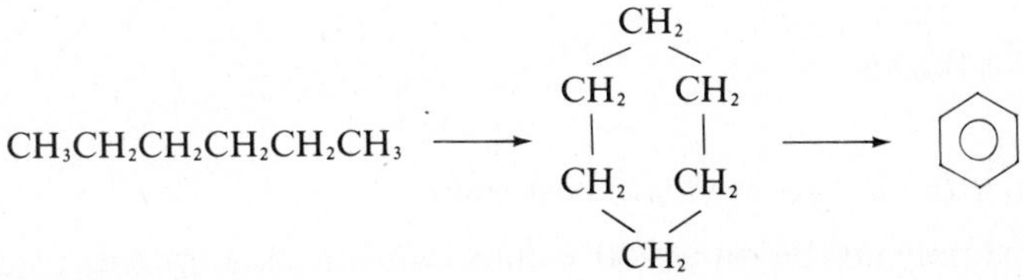

$$CH_3CH_2CH_2CH_2CH_2CH_3 \longrightarrow \text{cyclo-}(CH_2)_6 \longrightarrow C_6H_6$$

hexane — cyclohexane — benzene

3. Isomerization

This is the conversion of straight-chain hydrocarbons into branched-chain isomers. One reason for doing this is to improve the octane number of fuels for use in high compression engines. An example of an isomerization process is shown below. Notice the similarities it has with both simple Friedel-Crafts reactions and chain reactions discussed in earlier chapters.

initiation: $R{-}Cl + \overset{\delta+}{Al}\overset{\delta-}{Cl_3} \longrightarrow R^{\oplus} + AlCl_4^{\ominus}$

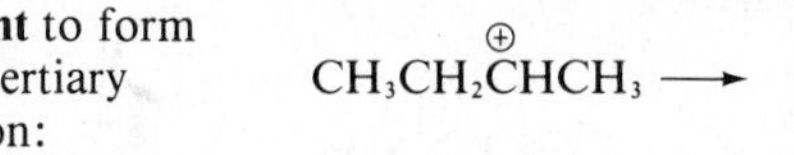

heterolytic cleavage

propagation: $R^{\oplus} + CH_3CH_2CH_2CH_3 \longrightarrow CH_3CH_2\overset{\oplus}{C}HCH_3 + RH$

rearrangement to form more stable tertiary carbonium ion: $CH_3CH_2\overset{\oplus}{C}HCH_3 \longrightarrow (CH_3)_2\overset{\oplus}{C}{-}CH_3$

$$(CH_3)_2\overset{\oplus}{C}{-}CH_3 + CH_3CH_2CH_2CH_3 \longrightarrow (CH_3)_2CH{-}CH_3 + CH_3CH_2\overset{\oplus}{C}HCH_3$$

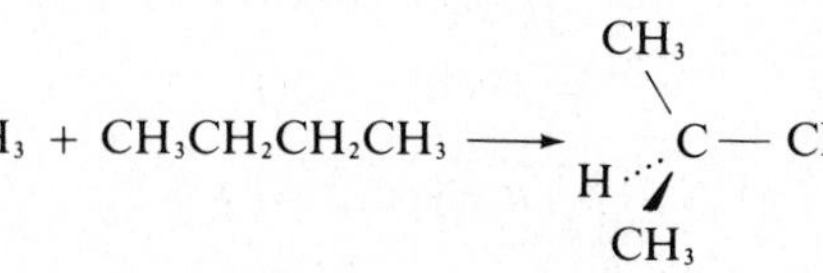

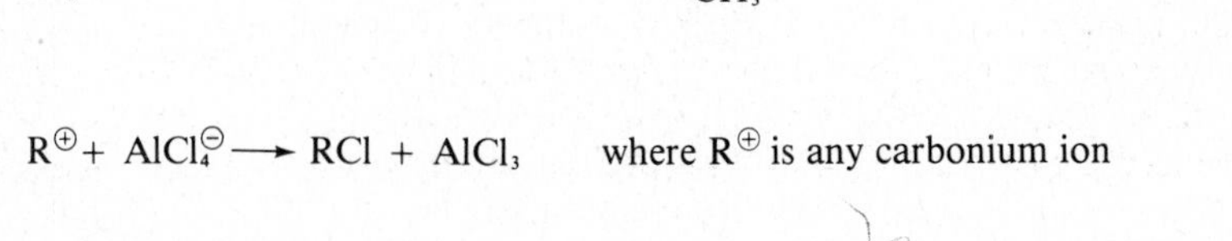

termination: $R^{\oplus} + AlCl_4^{\ominus} \longrightarrow RCl + AlCl_3$ where $R^{\oplus}$ is any carbonium ion

37.4 Petrochemical industry

Raw materials

Hydrocarbons are vital as the only major source of raw materials for the petrochemical industry. However they are also being consumed as fuels to satisfy the demand for energy. Unless a greater proportion of energy production is transferred to other fuels in the next two decades, the petrochemical industry will become critically short of crude oil, its basic raw material.

Below are listed the more important products made from two of the key compounds in the petrochemical industry.

Products from ethene

Polythene

Ethene is polymerized in two ways (see page 386).

1. Free radical polymerization at 200°C and 1500 atm in a trace of oxygen. This produces a low density polythene that softens considerably at 100°C.
2. Stereospecific heterolytic polymerization at 50°C and 3 atm in the presence of a Ziegler catalyst of $TiCl_3$ and $Al(Et)_3$. This produces a high density polythene that does not soften until about 125°C:

$$n\,[CH_2{=}CH_2] \longrightarrow \left[CH_2{-}CH_2\right]_n$$

polythene

Methylated spirits

Industrial ethanol is made by hydrating ethene over a catalyst of phosphoric acid on silica:

$$CH_2{=}CH_2 + H_2O \xrightarrow[\text{300°C and 70 atm}]{H_3PO_4} CH_3CH_2OH$$

ethanol

Terylene and antifreeze

Ethane-1,2-diol is the main constituent of antifreeze ('glycol'). It is also one of the reactants needed to make terylene (see page 582). It is made from epoxythene which is made from ethene:

$$2CH_2{=}CH_2 + O_2 \xrightarrow[\text{250°C}]{Ag} 2CH_2{-}CH_2 \text{ (with O bridging the two carbons)}$$

epoxyethene

$$CH_2{-}CH_2 \text{ (O-bridged)} + H_2O \xrightarrow[\text{200°C}]{} CH_2(OH){-}CH_2(OH)$$

ethane-1,2-diol

Polystyrene
Styrene, the monomer from which polystyrene is made, is produced from ethene and benzene. A Friedel-Crafts catalyst of $AlCl_3$ is used:

$$C_6H_6 + CH_2{=}CH_2 \xrightarrow[90°C]{AlCl_3} C_6H_5{-}CH_2{-}CH_3$$

ethyl benzene

$$C_6H_5{-}CH_2{-}CH_3 \xrightarrow[650°C \text{ and } 0.2 \text{ atm}]{ZnO} C_6H_5{-}CH{=}CH_2 + H_2$$

styrene

P.V.C. (polyvinylchloride)
Vinyl chloride, the monomer from which P.V.C. is made, is produced by chlorinating ethene and then decomposing the product catalytically. The catalyst is a silicate mineral:

$$CH_2{=}CH_2 + Cl_2 \xrightarrow[50°C]{ZnCl_2} CH_2Cl\,CH_2Cl$$

1,2-dichloroethane

$$CH_2ClCH_2Cl \xrightarrow[500°C \text{ and } 3 \text{ atm}]{\text{pumice}} CH_2{=}CHCl + HCl$$

vinylchloride (chloroethene)

Products from propene

Polypropylene
Propene is polymerized in the presence of a Ziegler catalyst. This catalyst promotes selective heterolytic polymerization and so the product has a regular structure. Polymers with a regular structure are described as *isotactic* while those with a random structure are said to be *atactic*:

$$n\,[CH_3CH{=}CH_2] \xrightarrow[60°C \text{ and } 5 \text{ atm}]{TiCl_3,\ Al(Et)_3} \left[\,\overset{H\ \ CH_3}{C}{-}CH_2\,\right]_n$$

isotactic polymer

Glycerol (propane-1,2,3-triol), propanone and hydrogen peroxide
Glycerol is used in many industries. It is also the starting material for making cellophane film. Propanone has wide use as a solvent. These two compounds and hydrogen peroxide are all made from propene in the same plant. The flow diagram on the next page shows in outline how the different reactions are related to each other in an integrated plant.

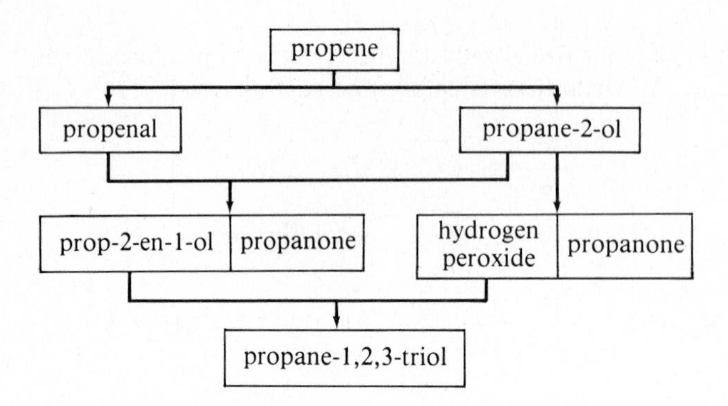

Propene is oxidized to propenal over a copper oxide catalyst:

$$CH_3CH{=}CH_2 + O_2 \xrightarrow[350^\circ C]{CuO} \underset{\text{propenal}}{CH_2{=}CHCHO}$$

Propene is hydrated directly to propane-2-ol over a catalyst of tungsten (VI) oxide and zinc oxide:

$$CH_3CH{=}CH_2 + H_2O \xrightarrow[200^\circ C \text{ and } 200 \text{ atm}]{WO_3,\ ZnO} \underset{\text{propane-2-ol}}{CH_3CHOHCH_3}$$

Propane-2-ol can be oxidized to produce propanone and hydrogen peroxide:

$$CH_3CHOHCH_3 + O_2 \xrightarrow[100^\circ C \text{ and } 4 \text{ atm}]{} \underset{\text{propanone}}{CH_3COCH_3} + \underset{\text{hydrogen peroxide}}{H_2O_2}$$

Propane-2-ol can also be reacted with propenal over a catalyst of magnesium oxide and zinc oxide:

$$\underset{\text{propane-2-ol}}{CH_3CHOHCH_3} + \underset{\text{propenal}}{CH_2{=}CHCHO} \xrightarrow[400^\circ C]{MgO,\ ZnO} \underset{\text{prop-2-en-1-ol}}{CH_2{=}CHCH_2OH} + \underset{\text{propanone}}{CH_3COCH_3}$$

The prop-2-en-1-ol is reacted with the hydrogen peroxide produced in the reaction above:

$$\underset{\text{prop-2-en-1-ol}}{CH_2{=}CHCH_2OH} + H_2O_2 \xrightarrow[60^\circ C]{WO_3} \underset{\text{propane-1,2,3-triol}}{CH_2OHCHOHCH_2OH}$$

Cumene and phenol

Cumene (2-phenylpropane) is an intermediate in the manufacture of phenol. Phenol is the starting material from which many resins and plastics are made. Both are derived from propene as follows:

$$C_6H_6 + CH_2{=}CHCH_3 \xrightarrow[250°C\ \text{and}\ 25\ \text{atm}]{AlCl_3} C_6H_5{-}CH(CH_3)_2$$

2-phenylpropane

$$C_6H_5{-}CH(CH_3)_2 + O_2 \xrightarrow[130°C]{Na_2CO_{3(aq)}} C_6H_5{-}C(CH_3)_2{-}O{-}O{-}H$$

$$C_6H_5{-}C(CH_3)_2{-}O{-}O{-}H \xrightarrow[100°C]{H_2SO_{4(aq)}} C_6H_5{-}OH + CH_3COCH_3$$

phenol propanone

Questions

1. Explain the meaning of: a) protein b) peptide link c) polyamide d) cracking e) reforming f) α-amino acid.
2. Distinguish between the primary, secondary and tertiary structure of a protein.
3. Outline, in your own words, the problems involved in attempting to synthesize proteins in the laboratory.
4. 2.5 g of a protein dissolved in 100 cm^3 of a suitable solvent at 20°C produced an osmotic pressure of 3.15 mm of mercury. Calculate its R.M.M.
5. Explain why R.M.M. determinations of proteins and polypeptides are carried out by measuring osmotic pressure and not any of the other colligative properties.
6. Draw a mechanism for the reaction between benzene and ethene in the presence of the Friedel-Crafts catalyst $AlCl_3$.
7. Compare the reactions of the two allotropes O_2 and O_3 with ethene.
8. Why does a mixture of alkanes obey Raoult's Law so closely?

Part 4

Inorganic chemistry

Chapter 38

Periodicity and chemical behaviour

38.1 Evidence of periodicity

Magazines and journals are often called 'periodicals' because they appear at regular intervals of time. Chemical periodicity refers to the regular occurrence of a set of properties of an element or its compounds in the Periodic Table. Because the Periodic Table lists the elements in order of atomic number, these periodic properties are related to the atomic number of the element.

1. Lithium, sodium and potassium
a) All burn vigorously in chlorine producing white salts

$$2Na_{(s)} + Cl_{2(g)} \rightarrow 2NaCl_{(s)}$$

b) All react exothermically and violently with water to produce alkaline solutions and hydrogen gas

$$2Li_{(s)} + 2H_2O_{(l)} \rightarrow 2Li^+_{(aq)} + 2OH^-_{(aq)} + H_{2(g)}$$

c) All tarnish very rapidly in the air. They are shiny and metallic in appearance when freshly cut.

2. Chlorine, bromine and iodine
a) All react exothermically with aluminium to give white solids that sublime easily

$$2Al_{(s)} + 3Cl_{2(g)} \rightarrow Al_2Cl_{6(g)}$$

b) The product of each element's reaction with aluminium dissolves in water and gives precipitates with silver nitrate solution

$Ag^+_{(aq)} + Cl^-_{(aq)} \rightarrow AgCl_{(s)}$ white

$Ag^+_{(aq)} + Br^-_{(aq)} \rightarrow AgBr_{(s)}$ cream

$Ag^+_{(aq)} + I^-_{(aq)} \rightarrow AgI_{(s)}$ yellow

c) Each element has a vapour that is pungent and irritating to smell. Damp blue litmus paper is turned red and then white by each element.

$$Cl_{2(g)} + 2H_2O_{(l)} \rightarrow H_3O^+_{(aq)} + Cl^-_{(aq)} + HOCl_{(aq)}$$

acidic properties ($H_3O^+_{(aq)}$); bleaching properties ($HOCl_{(aq)}$)

3. Helium, neon and argon
These are all inert gases. No stable compounds have been prepared from these elements.

The discovery of elements with similar properties led to the idea that there were families or *groups* of elements. In the last century, the Russian Mendeléef related the chemical properties of each group to those of the other groups. He drew up a table of elements based on his survey of chemical behaviour. Elements in the same group were placed in the same vertical column and the columns were arranged in the order that gave the most gradual and steady change in properties on moving across the table.

Mendeléef and his contemporaries, however, carried out their research at a time when little progress had been made on the theory of atomic structure. Their work took on new significance in the light of the idea that the atoms of each element contain the same sub-atomic particles. The structure of an atom can be imagined to result from the steady build up of protons and electrons. For example, starting with a lithium atom, a sodium atom is produced as a result of the addition of a fixed number of protons and electrons. This sodium atom has very similar properties to those of the lithium atom. Further addition to the sodium atom eventually results in the production of a potassium atom and, once again, the occurrence of a particular set of chemical properties. So the idea of periodicity became established and Mendeléef's table is now best known as the Periodic Table of the elements.

Ionization energies

A fuller understanding of chemical periodicity is reached by looking at the first ionization energies of the elements. The ease with which electrons can be removed from an atom is closely linked to its bonding capabilities and therefore also with the chemical properties of the element. An interesting pattern of periodicity emerges when the first ionization energies of the elements are plotted against atomic number:

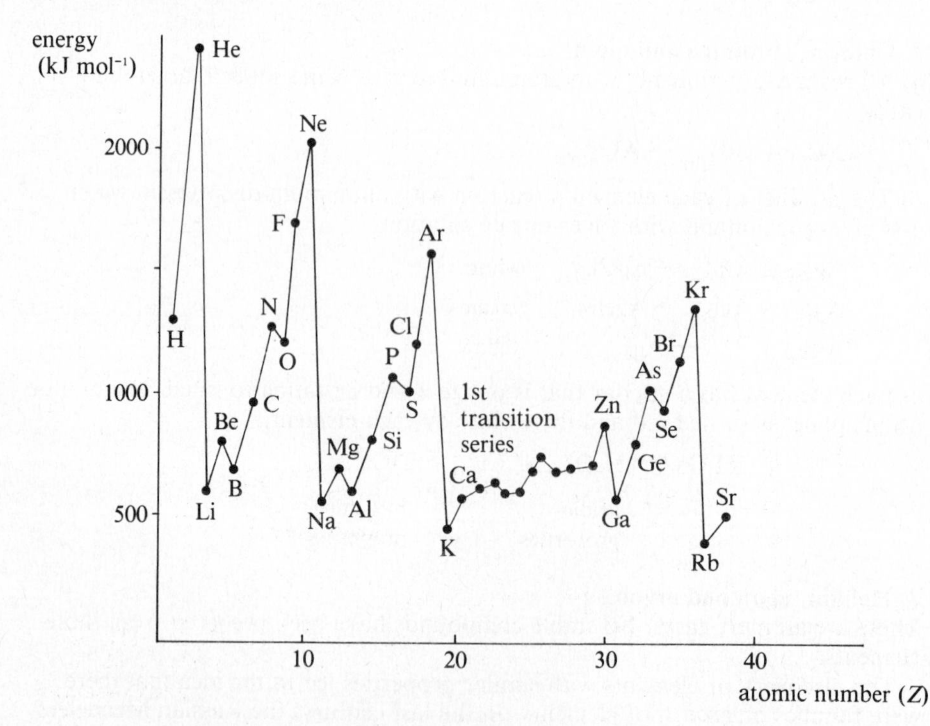

There are a number of distinctive features:

1. A set of peaks is reached at the elements He, Ne, Ar, Kr.
2. A set of troughs is reached at the elements Li, Na, K, Rb.
3. A set of interrupted increases occurs from trough to peak; the interruptions themselves form a regular pattern of their own.

The periodicity can now be seen to depend on the atomic number *Z*. The duration of a period, however, is not a constant. From the graph, the appearance of the peaks gives a measure of the duration of each period.

1st period: peak at $Z = 2$: duration = 2
2nd period: peak at $Z = 10$: duration = 8
3rd period: peak at $Z = 18$: duration = 8
4th period: peak at $Z = 36$: duration = 18

Certain physical properties also show periodic trends when plotted against atomic number.

Atomic volumes

In 1870 Lothar Meyer calculated the atomic volumes of the elements whose relative atomic masses were known from the measurements of their chemical combining weights. He used the formula:

$$\text{atomic volume} = \frac{\text{relative molar mass}}{\text{density}} \left(\frac{\text{g mol}^{-1}}{\text{g cm}^{-3}} \right)$$

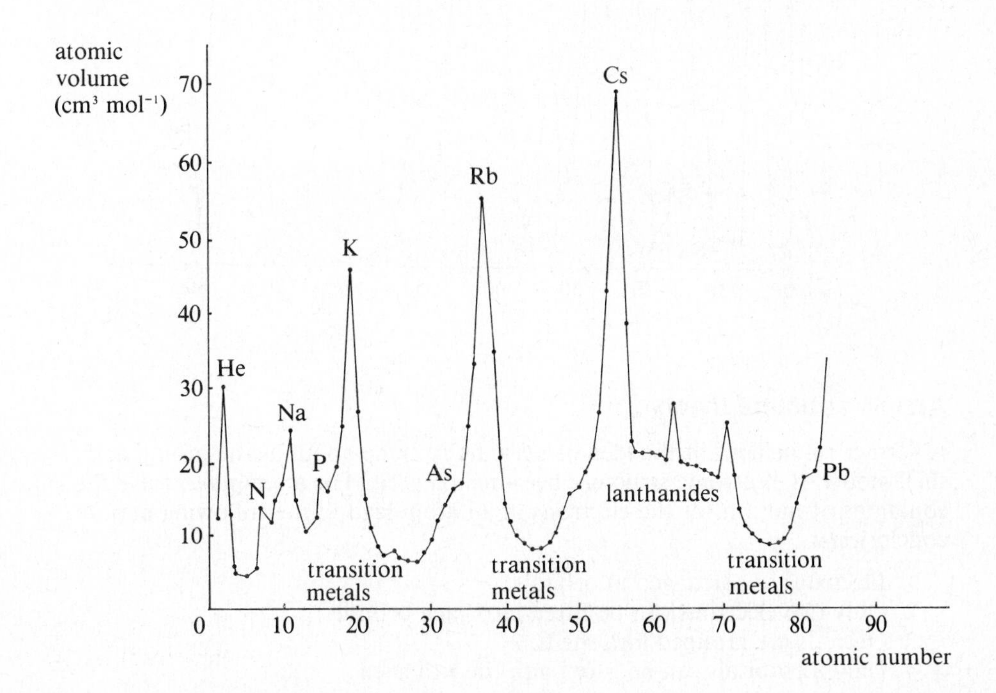

The result clearly indicates periodic behaviour. An atom's volume depends on three factors:

a) the identity of the outer shell: the more distant the outer shell the larger is its volume
b) the attraction of the shell to the nucleus: the higher the core charge the greater is the attraction
c) repulsion between electrons in the atom: the more electrons there are in the outer shell, the greater is the repulsive force on any particular electron.

Melting points

A plot of the melting points of the elements against atomic number shows clear indications of periodicity.

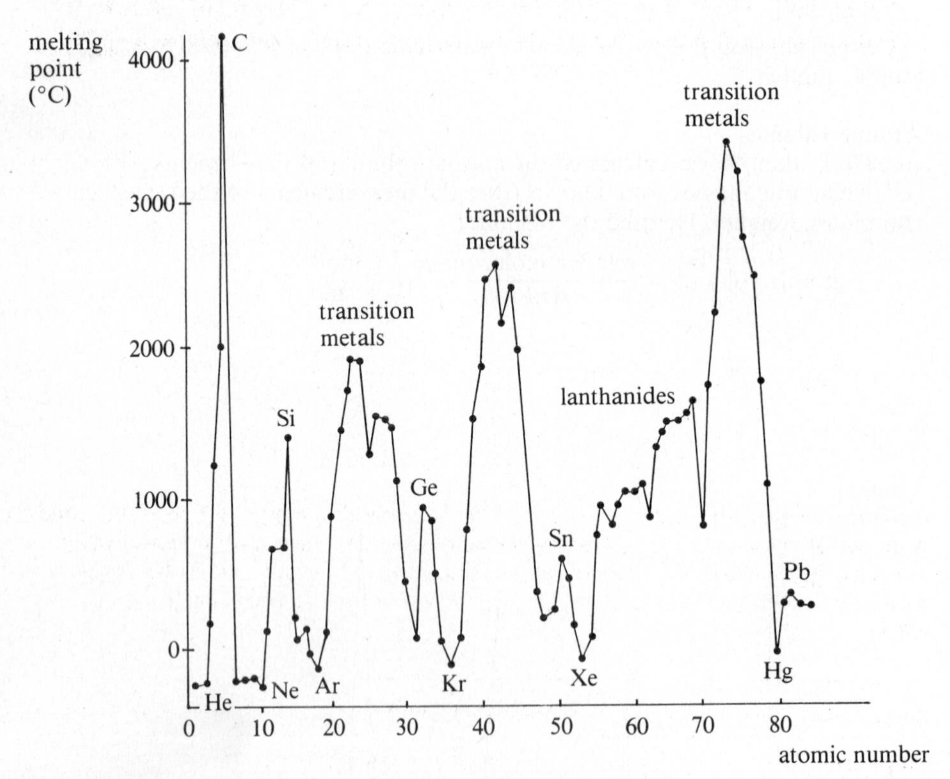

Atomic structure theory

A further refinement of the idea of periodicity came about as the orbital and shell model of electronic structure became accepted. The attempts to solve the equations of motion for the electrons in an atom lead to the following major conclusions:

1. Electrons are arranged in orbitals
2. Only two electrons can be fitted into each orbital
3. Orbitals are grouped into shells
4. Only n^2 orbitals can be fitted into the nth shell

5. There are different types of orbital, and each type has a different shape:
   a) s-orbital: spherical; one s-orbital per shell
   b) p-orbital: dumb-bell shaped; three p-orbitals per shell
   c) d-orbital: complex 4-leaf clover shaped; five d-orbitals per shell
6. A set of each type of orbital is often called a 'subshell' because the energies of the orbitals making it up are equal, e.g. the *third shell* ($n = 3$) contains 9 orbitals (n^2) in the three different subshells:

	an s-subshell	= 1 orbital
	a p-subshell	= 3 orbitals
	a d-subshell	= 5 orbitals
total:	3s, 3p, 3d	= 9 orbitals

7. The electrons in an atom are distributed into orbitals by filling successively the orbital of lowest energy (*the aufbau principle*). The order of orbital energies follows the diagram below:

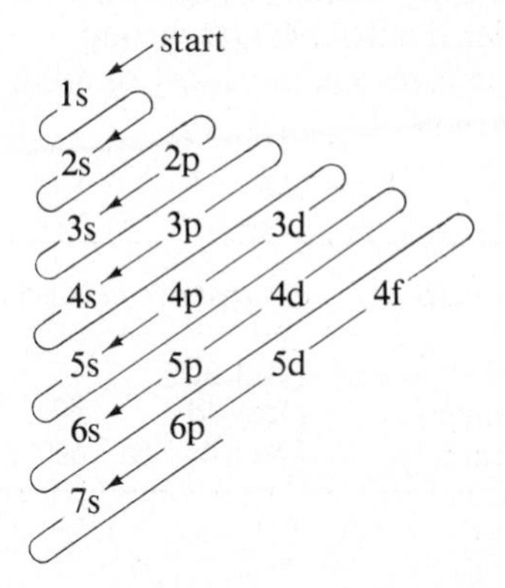

The recurrence of a set of events in chemical behaviour now can be understood in terms of the number of outer-shell electrons in each atom. Two elements in the same group have the same number of outer-shell electrons distributed in similar orbitals, as shown for group II and group VII elements in the tables below.

group II	electronic configuration		outer-shell electrons
beryllium	$1s^2$ $2s^2$	[He] $2s^2$	2
magnesium	$1s^2$ $2s^2$ $2p^6$ $3s^2$	[Ne] $3s^2$	2
calcium	$1s^2$ $2s^2$ $2p^6$ $3s^2$ $3p^6$ $4s^2$	[Ar] $4s^2$	2

group VII	electronic configuration		outer-shell electrons
fluorine	$1s^2$ $2s^2$ $2p^5$	[He] $2s^2$ $2p^5$	7
chlorine	$1s^2$ $2s^2$ $2p^6$ $3s^2$ $3p^5$	[Ne] $3s^2$ $3p^5$	7
bromine	$1s^2$ $2s^2$ $2p^6$ $3s^2$ $3p^6$ $3d^{10}$ $4s^2$ $4p^5$	[Ar] $4s^2$ $4p^5$	7

The number of outer-shell electrons and the number of vacancies left in an atom's outer shell determine many of the chemical properties of an element.

38.2 The Periodic Table

Main features

The Periodic Table is given at the back of the book. The elements on the left of the heavy black, stepped dividing line are metals; those on the right are non-metallic. Metallic character is associated with atoms whose nuclei have a low degree of electronic control, whereas non-metallic character is associated with atoms whose nuclei have a high degree of control of outer-shell electrons. The reasons for this are discussed on page 43.

The Periodic Table is classified by *periods* and *groups*. A period is a horizontal row of elements and a group is a vertical column of elements. Period numbers are written in Arabic numerals (usual numbers) but group numbers are always given as Roman numerals.

A letter, 'A' or 'B', is sometimes added to the group number to give information about the inner subshell structure of the atoms. 'A' group elements have atoms with empty, or partially empty, inner d-subshells; 'B' group elements have atoms whose inner d-subshell structure is complete. Compare groups IIA and IIB for the 4th, 5th and 6th periods.

group IIA	*electronic configuration*	*group IIB*	*electronic configuration*
calcium	[Ar] $4s^2$	zinc	[Ar] $3d^{10}$ $4s^2$
strontium	[Kr] $5s^2$	cadmium	[Kr] $4d^{10}$ $5s^2$
barium	[Xe] $6s^2$	mercury	[Xe] $4f^{14}$ $5d^{10}$ $6s^2$

The differences in the properties of the elements in the 'A' and 'B' groups are discussed in more detail in Chapter 50.

The terms s-block, p-block or d-block are used as an alternative method of describing the position of elements in the Periodic Table. An element is classified by the subshell that is being filled at that point in the table. Compare some elements from the 4th period.

s-block		*d-block*		*p-block*	
potassium	[Ar] $4s^1$	manganese	[Ar] $3d^5$ $4s^2$	germanium	[Ar] $3d^{10}$ $4s^2$ $4p^2$
calcium	[Ar] $4s^2$	iron	[Ar] $3d^6$ $4s^2$	bromine	[Ar] $3d^{10}$ $4s^2$ $4p^5$

Overleaf is an outline of the Periodic Table with the positions of the three blocks marked in.

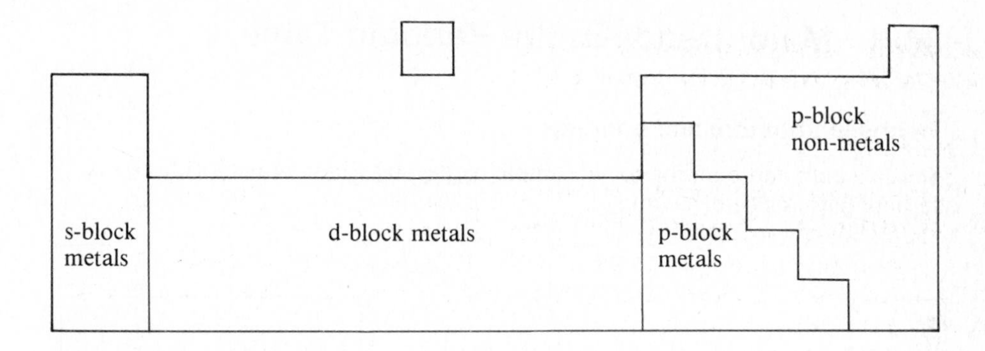

38.3 Using the Periodic Table

The inorganic chemistry of the elements may be studied by surveying the properties of different groups and periods in the Periodic Table.

Groups contain elements whose chemical properties are similar. By studying the group as a whole, repetition of information for each element in the group can be avoided. The common properties of a group of elements can be related to the electronic configuration of the outer shell, while trends within the group are related to the actual outer-shell number and hence the size and electronic control of the atom.

Periods contain elements whose character changes from metallic to non-metallic. The changing properties across a period are related to the changing size of the atoms and the different number of outer-shell electrons.

The major concepts of physical chemistry provide the context that gives order to the great number of inorganic facts, and for this reason the remainder of this section of the book is organized under key physical chemistry concepts. When each element and its compounds are discussed the important ideas which are emphasized are:

1. Structure and bonding
The way in which the particles are arranged and bonded in a given system accounts for many of the chemical properties of the system.

2. Acid-base properties
A particular set of properties is associated with the ability of the particles in a system to act as proton donors or acceptors.

3. Redox properties
Another set of properties is associated with the ability of the particles in a system to act as electron donors or acceptors.

4. Solubility and complexing properties
Many reactions take place in solution. The ability of the particles in a system to interact with solvent particles affects solubility and other solution properties.

The summaries on the next few pages show that there are distinct trends in these four classes of chemical behaviour within the Periodic Table.

38.4 Major trends in the Periodic Table

Trends in structure and bonding

The structure and bonding of the simple oxides, hydrides and chlorides show definite patterns of behaviour.

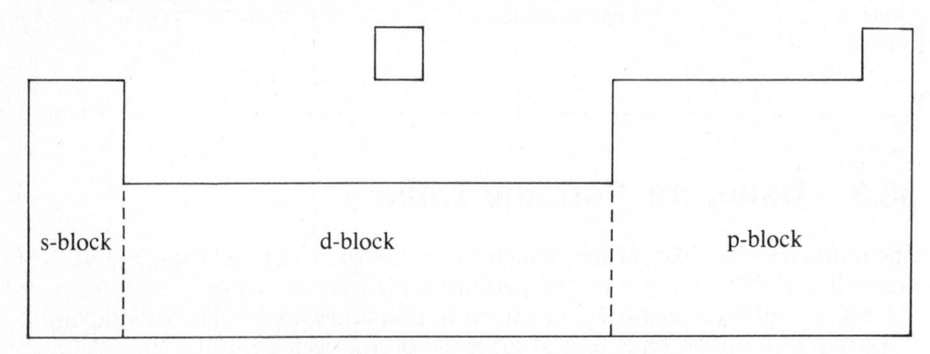

s-block
These elements form simple ionic compounds whose lattice types are indicated by radius ratio calculations, e.g.

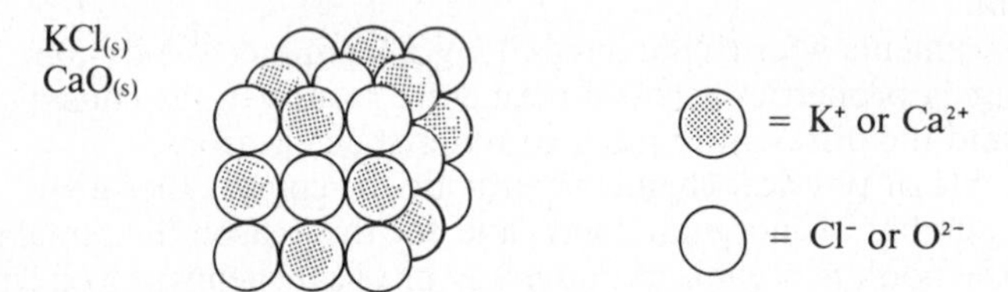

d-block
The oxides and chlorides of these elements are either ionic (with a high degree of covalent character) or covalent. The structures are either layer lattices, or chain lattices, or macromolecular.

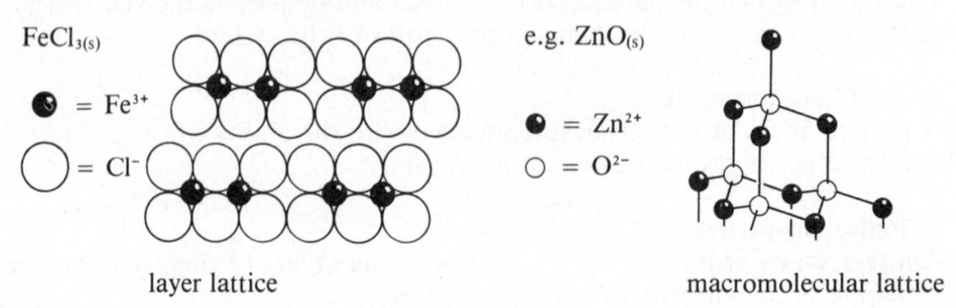

p-block
The oxides, hydrides and chlorides of these elements are covalently bonded molecular compounds, e.g.

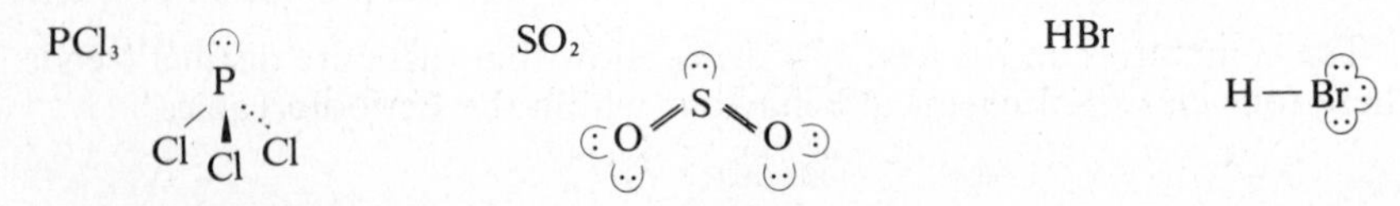

Trends in acid-base properties

The acid-base properties of the simple oxides, chlorides and hydrides also show definite patterns of behaviour.

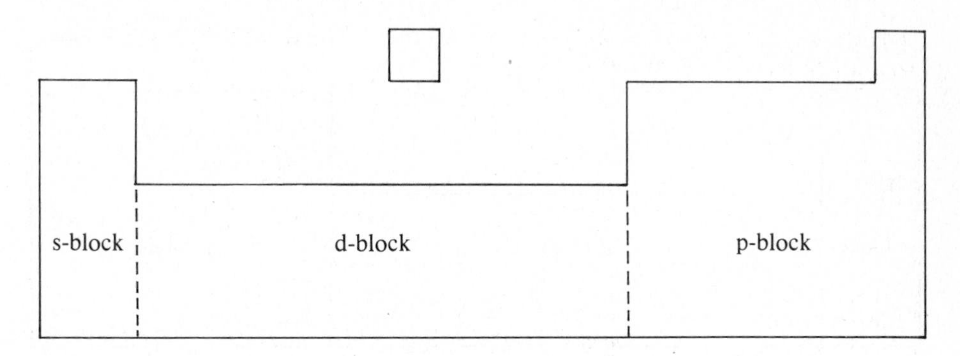

s-block

Oxides and hydrides are alkaline in solutions; the chlorides are neutral, e.g.

$K_2O_{(s)} + H_2O_{(l)} \rightarrow 2K^+_{(aq)} + 2OH^-_{(aq)}$ pH ≈ 14

$NaH_{(s)} + H_2O_{(l)} \rightarrow Na^+_{(aq)} + OH^-_{(aq)} + H_{2(g)}$ pH ≈ 14

$CaCl_{2(s)} + H_2O_{(l)} \rightarrow Ca^{2+}_{(aq)} + 2Cl^-_{(aq)}$ pH ≈ 7

d-block

Oxides and hydroxides are basic and insoluble but there is a tendency towards amphoteric behaviour. The chlorides hydrolyse to give acidic solutions, e.g.

$ZnO_{(s)} + 2H_3O^+_{(aq)} \rightarrow Zn^{2+}_{(aq)} + 3H_2O_{(l)}$ basic oxide

$Zn^{2+}_{(aq)} + 2OH^-_{(aq)} \rightarrow Zn(OH)_{2(s)}$ insoluble hydroxide

$Zn(OH)_{2(s)} + 2H_3O^+_{(aq)} \rightarrow Zn^{2+}_{(aq)} + 4H_2O_{(l)}$
$Zn(OH)_{2(s)} + 2OH^-_{(aq)} \rightarrow [Zn(OH)_4]^{2-}_{(aq)}$ } amphoteric hydroxide

$FeCl_{3(s)} + H_2O_{(l)} \rightarrow Fe(H_2O)^{3+}_{6(aq)} + 3Cl^-_{(aq)}$ hydrolysis of chloride

then $[Fe(H_2O)_6]^{3+}_{(aq)} + H_2O_{(l)} \rightarrow [Fe(H_2O)_5OH]^{2+}_{(aq)} + H_3O^+_{(aq)}$ pH ≈ 3

p-block

Oxides, chlorides and most hydrides are acidic, but there are some notable exceptions, e.g. NH_3 and CH_4.

$SO_{3(s)} + 2H_2O_{(l)} \rightarrow HSO^-_{4(aq)} + H_3O^+_{(aq)}$ pH ≈ 1

$SiCl_{4(l)} + 8H_2O_{(l)} \rightarrow Si(OH)_{4(s)} + 4Cl^-_{(aq)} + 4H_3O^+_{(aq)}$ pH ≈ 1

$HCl_{(g)} + H_2O_{(l)} \rightarrow H_3O^+_{(aq)} + Cl^-_{(aq)}$ pH ≈ 1

Trends in redox properties

The redox properties of the pure elements show definite patterns of behaviour.

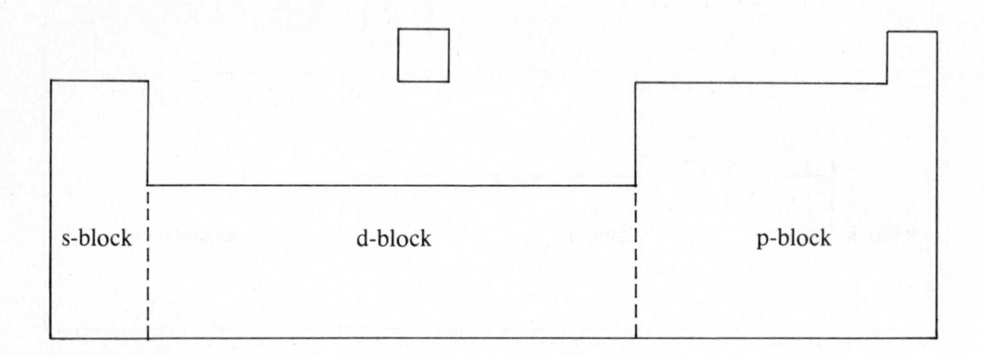

s-block
The elements are strong reductants. For example, sodium and calcium both reduce cold water to hydrogen.

$$2Na_{(s)} + 2\overset{I}{H}_2O_{(l)} \rightarrow 2Na^+_{(aq)} + 2OH^-_{(aq)} + \overset{0}{H}_{2(g)}$$

d-block
The elements are weak reductants. For example, iron is a weaker reductant than sodium or calcium; it reduces steam to hydrogen at red heat.

$$Fe_{(s)} + \overset{I}{H}_2O_{(g)} \rightarrow FeO_{(s)} + \overset{0}{H}_{2(g)}$$

p-block
Within this block there is a trend from weak reductant through weak oxidant to very strong oxidant. For example, nitrogen can act either as a weak oxidant in its reaction with hydrogen, or as a weak reductant in its reaction with oxygen.

$$N_{2(g)} + 3\overset{0}{H}_{2(g)} \rightleftharpoons 2N\overset{I}{H}_{3(g)}$$

$$N_{2(g)} + \overset{0}{O}_{2(g)} \rightleftharpoons 2N\overset{-II}{O}_{(g)}$$

Chlorine is a powerful oxidant. It will oxidize almost all metals.

$$2\overset{0}{Cr}_{(s)} + 3Cl_{2(g)} \rightarrow 2\overset{III}{Cr}Cl_{3(s)}$$

Trends in solubility and complexing properties

The precipitation and complexing properties of the elements do vary within the Periodic Table, but the patterns of behaviour are less clearly defined than with the other key types of behaviour.

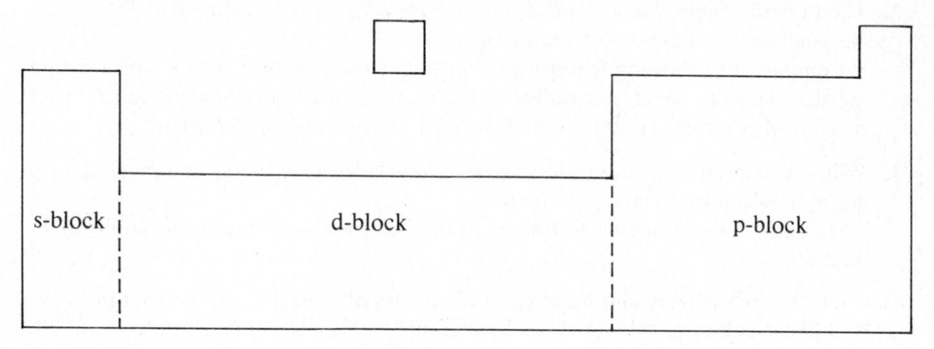

s-block
The cations of these elements have poor complexing properties. All group I compounds are soluble. The oxides and hydroxides of group II vary in their solubility.

$$KOH_{(s)} + H_2O_{(l)} \rightarrow K^+_{(aq)} + OH^-_{(aq)}$$

$$Ca(OH)_{2(s)} + H_2O_{(l)} \rightleftharpoons Ca^{2+}_{(aq)} + 2OH^-_{(aq)}$$

d-block
The cations of these elements form many complexes with a wide range of ligands. The oxides, hydroxides and even some of the chlorides of these elements are insoluble, e.g.

$[Cu(NH_3)_4]^{2+}$; $[Fe(CN)_6]^{3-}$ complexes

$$Mn^{2+}_{(aq)} + 2OH^-_{(aq)} \rightarrow Mn(OH)_{2(s)}$$

$$Ag^+_{(aq)} + Cl^-_{(aq)} \rightarrow AgCl_{(s)}$$

} precipitation reactions

p-block
The anions of these elements act as ligands, e.g.

$:\ddot{\underset{..}{F}}:^{\ominus}$ $\quad H—\ddot{\underset{..}{O}}:^{\ominus}$ $\quad H_3N:$ $\quad :C\equiv N:^{\ominus}$

These ligands form complexes such as $[CrF_6]^{3-}$; $Zn(OH)_4^{2-}$.

Questions

1. Explain the meaning of: a) periodicity b) period c) group d) ionization energy e) shell f) orbital g) p-block h) layer lattice.

2. The Periodic Table can be divided into two regions; one containing metallic elements and the other non-metallic elements.
   Compare the following features of a typical metallic element with a typical non-metallic element: a) atomic radius b) electronegativity c) degree of control over outer-shell electrons d) types of ion formed e) types of bond formed.

3. With some specific reactions as examples, show how the elements in a particular group have similar chemical properties.
   How can these properties be related to the electronic structure of the atoms of these elements?

4. Compare the structure and bonding of the oxides of typical elements from the s-, p- and d-blocks.

5. Compare the acid-base properties of the hydrides and chlorides of typical elements from the s-, p- and d-blocks.

6. Comment on the standard electrode potentials of the following three elements:

element	$E^\ominus$ volts
Na^+/Na	−2.71
Fe^{2+}/Fe	−0.44
Cl_2/Cl^-	+1.36

   Are they typical of elements from their respective blocks in the Periodic Table?

7. Relate the physical properties of the s-, p- and d-block elements to their structure and bonding.

8. Explain why the atomic volumes of the elements show periodic behaviour.

9. What information about the forces between the particles found in the lattices of elements can be acquired from a knowledge of the melting points of the elements?

10 Discuss the statement: 'Periodicity is the consequence of electronic structure'.

Chapter 39
Hydrogen

39.1 Structure of the element

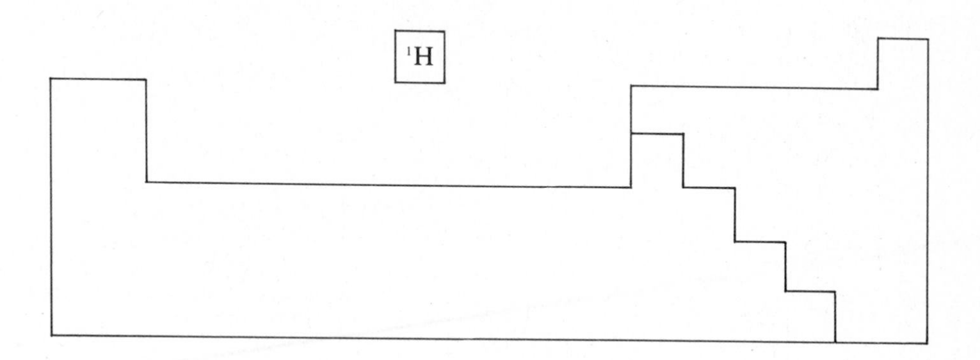

electronic configuration	*isotopes*	*radioactivity*	*percentage abundance*
	protium ^1H	stable	99.985
$1s^1$	deuterium ^2H	stable	0.015
	tritium ^3H	β^-emitter	$t_{1/2}$ = 12.3 years

Atomic structure

A hydrogen atom has the simplest of all atomic structures. Its nucleus contains only one proton and there is a single electron in the 1s-orbital. This gives a hydrogen atom two distinctive features:

1. There are no inner-shell electrons to shield the outer shell from the nuclear charge.
2. The outer shell is both half full and half empty at the same time.

The second of these two features explains the position assigned to hydrogen on the Periodic Table. With one electron in its outer shell, hydrogen resembles a group I element. At the same time, with one vacancy in its outer shell, hydrogen resembles a group VII element. For this reason, hydrogen is usually shown on its own half way between groups I and VII.

Physical structure

Hydrogen occurs as the molecule, H_2, and its physical properties are typical of a molecular substance. Its fixed points and bond energy are listed below.

Hydrogen gas has no colour or smell.

molecular structure	*R.M.M.*	*bond energy*	*m.p.*	*b.p.*
H—H	2	435 kJ mol^{-1}	−259°C	−253°C

The bond energy of the hydrogen molecule corresponds to the drop in the system's potential energy which results from the overlap of two atomic orbitals producing molecular orbitals.

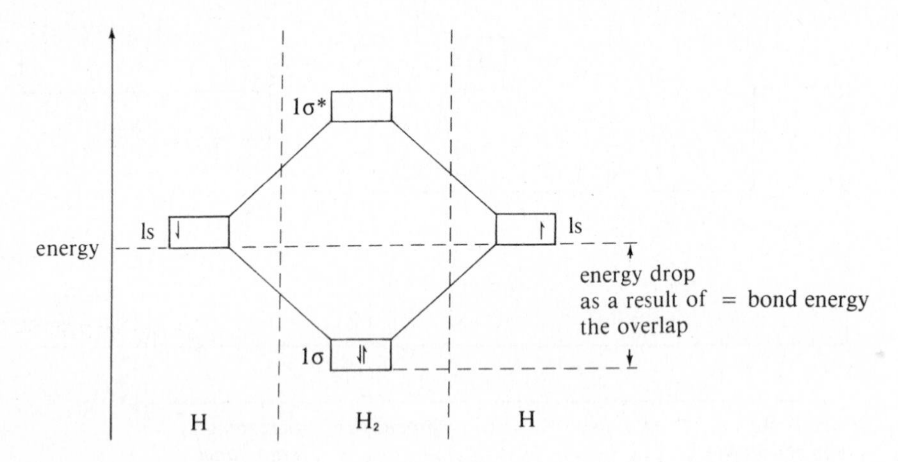

39.2 Compounds formed by the element

Hydrogen forms a wide range of compounds containing the element in oxidation state +I, 0 or −I. The electronegativity of hydrogen is 2.1. This value is between that for a reactive metal such as calcium (1.0) and a reactive non-metal such as chlorine (3.0). A hydrogen atom therefore either loses or gains control of an electron depending on the nature of the atom to which it is bonded.

The high ionization energy might suggest that positive oxidation state chemistry is unlikely. The value for hydrogen is the same as the first ionization energy of oxygen:

$$H_{(g)} \rightarrow H^{+}_{(g)} + e^{-} \quad \Delta H^{\ominus} = 1310 \text{ kJ mol}^{-1}$$

$$O_{(g)} \rightarrow O^{+}_{(g)} + e^{-} \quad \Delta H^{\ominus} = 1310 \text{ kJ mol}^{-1}$$

However, a hydrogen cation is simply a proton and is minute compared with the size of any other ion. The bonds formed to the tiny point charge are very strong and the energy released on the formation of these bonds more than compensates for the high ionization energy. Thus, the use of the ionic model to describe hydrogen (I) compounds is not very sensible: the polarizing power of a proton is enormous and high degrees of covalency result (see Fajans' rules, page 65).

Hydrogen has an exothermic, but fairly low, electron affinity. The energy released on the formation of a hydride ion H^- is often not enough to compensate for the bonds broken in the process. Hydrides are formed only with the more reactive metals.

$$H_{(g)} + e^- \rightarrow H^-_{(g)} \qquad \Delta H^\ominus = -72 \text{ kJ mol}^{-1}$$

The application of Fajans' rules suggests that these −I state hydrides should have high degrees of covalency. A hydride ion is large and has a high polarizability: its size is between that of a bromide and iodide ion.

ion	Br^-	H^-	I^-
ionic radius ($m \times 10^{-10}$)	1.95	2.08	2.16

The large size of the ion is the result of repulsion between the two 1s electrons.

39.3 Occurrence, extraction and uses

Hydrogen is not found in large quantities as the uncombined element. At sea level, the atmosphere contains about 0.0009% of free hydrogen. The molecules are so light and move so quickly that they can easily escape from the earth's gravitational field. However, the element occurs frequently in a combined state, e.g. as water, and in organic compounds such as hydrocarbons (fossil fuels) and carbohydrates, fats and proteins (living systems).

Hydrogen from steam

a) Steam is passed over white hot coke.

$$C_{(s)} + H_2O_{(g)} \rightarrow CO_{(g)} + H_{2(g)}$$

This reaction is called the *water gas process*; it is endothermic and causes the coke temperature to drop. A blast of air alternating with the steam restores the temperature because the oxidation of coke is exothermic.

b) Alternatively, an alkane such as methane is mixed with steam and passed over a nickel catalyst. The products are a mixture of carbon monoxide and hydrogen, known as *synthesis gas*.

$$CH_{4(g)} + H_2O_{(g)} \xrightarrow[1000°C\ 25\ atm]{Ni} CO_{(g)} + 3H_{2(g)}$$

The product gases from either of these reactions are mixed with more steam and passed over an iron oxide catalyst.

$$CO_{(g)} + H_2O_{(g)} \xrightarrow[450°C]{\text{iron oxide}} CO_{2(g)} + H_{2(g)}$$

The carbon dioxide is removed from the gas mixture by scrubbing with water under pressure.

$$CO_{2(g)} + H_2O_{(l)} \rightarrow H_2CO_{3(aq)}$$

Any residual carbon monoxide is reacted with hydrogen producing methane which is recycled.

$$CO_{(g)} + 3H_{2(g)} \rightarrow CH_{4(g)} + H_2O_{(g)}$$

Hydrogen is also produced
a) as a by-product from cracking hydrocarbons
b) as a by-product from the electrolysis of brine

Laboratory preparation
Hydrogen can be produced by the action of dilute acid on most metals, the action of concentrated alkali on zinc or aluminium, the electrolysis of dilute acid or the hydrolysis of metal hydrides. The equations for these reactions are shown below.

$$Zn_{(s)} + 2H_3O^+_{(aq)} \longrightarrow Zn^{2+}_{(aq)} + 2H_2O_{(l)} + H_{2(g)}$$

$$2Al_{(s)} + 2OH^-_{(aq)} + 6H_2O_{(l)} \longrightarrow 2Al(OH)^-_{4(aq)} + 3H_{2(g)}$$

$$2H_3O^+_{(aq)} + 2e^- \xrightarrow[\text{cathode}]{\text{at}} 2H_2O_{(l)} + H_{2(g)}$$

$$LiH_{(s)} + H_2O_{(l)} \longrightarrow Li^+_{(aq)} + OH^-_{(aq)} + H_{2(g)}$$

Uses of hydrogen
1. Manufacture of ammonia and nitric acid.
2. Manufacture of hydrogen chloride by burning a hydrogen jet in chlorine.
3. Manufacture of methanol from carbon monoxide and hydrogen. Methanol is a useful solvent as well as being a raw material for the manufacture of other organic compounds:

$$2H_{2(g)} + CO_{(g)} \xrightarrow{ZnO/Cr_2O_3} CH_3OH_{(g)} \qquad P = 300 \text{ Atm}, T = 400°C$$

4. Production of margarine by reacting an unsaturated vegetable oil with hydrogen over a nickel catalyst.

39.4 Compound structure

Hydrogen (I)

Hydrogen combines directly with many non-metals to give molecular hydrogen (I) compounds. With the more reactive non-metals, these reactions are explosively fast but, with the less reactive non-metals the position of equilibrium often favours the uncombined elements and the reactions are slow, e.g.

$$2H_{2(g)} + O_{2(g)} \xrightarrow{\text{spark}} 2H_2O_{(g)} \qquad \text{explosive}$$

$$H_{2(g)} + Cl_{2(g)} \xrightarrow{\text{sunlight}} 2HCl_{(g)} \qquad \text{explosive}$$

$$H_{2(g)} + I_{2(g)} \rightleftharpoons 2HI_{(g)} \qquad K_p = 45.6 \text{ at } 490°C$$

Hydrogen (I) compounds are usually prepared by protonating the appropriate anions. For example, hydrogen chloride gas is produced by the action of concentrated sulphuric acid on sodium chloride:

$$NaCl_{(s)} + H_2SO_{4(l)} \rightarrow NaHSO_4 + HCl_{(g)}$$

$$:\ddot{Cl}:^{\ominus} \quad H{-}\ddot{O}{-}SO_2OH \longrightarrow :\ddot{Cl}{-}H \quad {}^{\ominus}:\ddot{O}{-}SO_2OH$$

Hydrogen sulphide can be prepared using the weaker proton donors, hydroxonium ions, present in dilute acid. Iron (II) sulphide is a source of sulphide ions:

$$FeS_{(s)} + 2H_3O^+_{(aq)} \rightarrow Fe^{2+}_{(aq)} + 2H_2O_{(l)} + H_2S_{(g)}$$

There is a major group of hydrogen (I) compounds that contain hydrogen covalently bonded to oxygen. These are the *hydroxyl compounds* and are discussed on page 619. They are closely related to water, HOH, which can be considered as the hydroxyl compound of hydrogen itself.

The bonding in these molecular compounds is summarized below:

1. Strong covalent bonds hold the molecules together.

bond	*bond energy*		*bond*	*bond energy*
O—H	463 kJ mol^{-1}		O—O	146 kJ mol^{-1}
Cl—H	431 kJ mol^{-1}	Compare	Cl—Cl	242 kJ mol^{-1}
N—H	388 kJ mol^{-1}	with	N—N	163 kJ mol^{-1}
C—H	412 kJ mol^{-1}		C—C	348 kJ mol^{-1}

2. A charge separation exists along the bonds due to the electronegativity difference. This often results in polar molecules.

$\overset{\delta+}{H}{-}\overset{\delta-}{O}{-}\overset{\delta+}{H}$ $\quad$ $\overset{\delta-}{N}(\overset{\delta+}{H})_3$ $\quad$ $\overset{\delta-}{Cl}{-}\overset{\delta+}{H}$

3. The molecules attract one another by a combination of van der Waals' forces, dipole-dipole forces and hydrogen bonding where applicable.

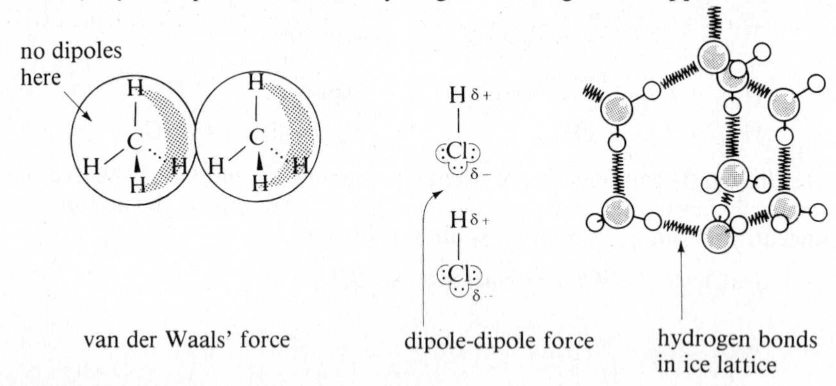

Hydrogen bonds are strongest between the hydrogen (I) compounds of the elements fluorine, oxygen and nitrogen.

Hydrogen (0)

In the pure element, the oxidation state of hydrogen is zero. Hydrogen also forms a number of *interstitial hydrides* with different transition metals. There is little evidence of electron transfer between the atoms in these compounds and so they can be viewed as hydrogen (0) compounds. Nearly all of them are non-stoichiometric and hydrogen can be extracted from them by reducing the pressure. It is sometimes better to think of these compounds as alloy phases rather than as chemical combinations. The hydrogen atoms fit into vacancies in the close-packed metal atom lattices. The uptake of hydrogen can vary from molar compositions as low as 10% to those as high as 67% in the well characterized hydride of zirconium, ZrH_2.

The presence of hydrogen atoms in a metal atom lattice leads to changes in conductivity, malleability and magnetic properties. The chemical activity of hydrogen is also very different in this environment.

The percentage of vacant sites occupied by hydrogen atoms in different metal lattices varies according to the metal and the temperature, for example:

hydride	*% of vacant sites occupied*
uranium	3.1 at 650°C
hafnium	15 at 250°C
palladium	30 at 160°C

Hydrogen (– I)

When hydrogen is passed over a molten lump of a reactive metal, a salt-like white solid forms. This saline hydride usually has the same structure as the corresponding metal halide. The size of a hydride ion is between that of a bromide ion and an iodide ion. With cations of low polarizing power, e.g. Na^+, K^+, simple ionic lattices are given, but with cations of higher polarizing power, layer lattice structures with higher degrees of covalency result. For example, NaH has the same structure as NaCl: each sodium ion is surrounded by six hydride ions and the co-ordination number of the hydride ion is also six. BeH_2 has a similar structure to that of $BeCl_2$. There are chains of atoms in which the hydrogen atoms act as bridging units:

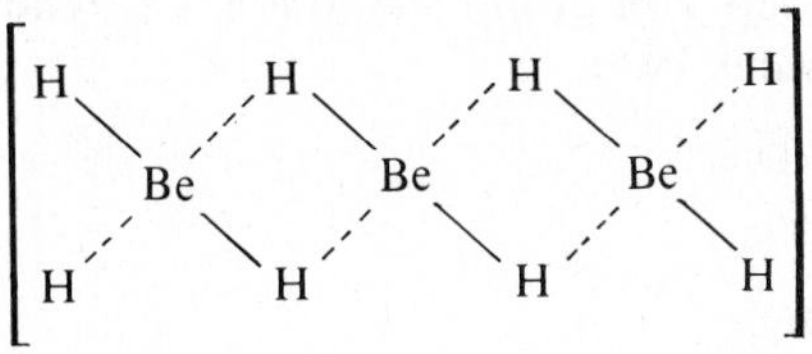

39.5 Acid-base properties

Hydrogen (I)

With the exception of the group IV hydrides and the hydroxyl compounds (see 39.8), hydrogen (I) compounds have molecules with

1. poorly shielded protons
2. lone pairs of electrons.

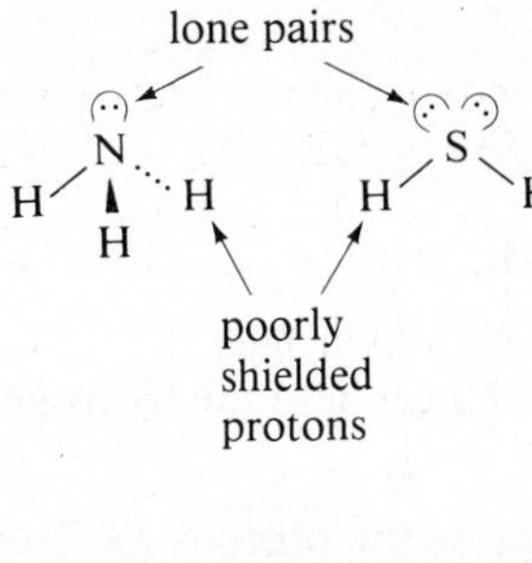

lone pairs

poorly shielded protons

These molecules therefore act as either proton donors (acids) or proton acceptors (bases), or as both (ampholytes).

There are two trends in the acid-base properties of the p-block hydrides.

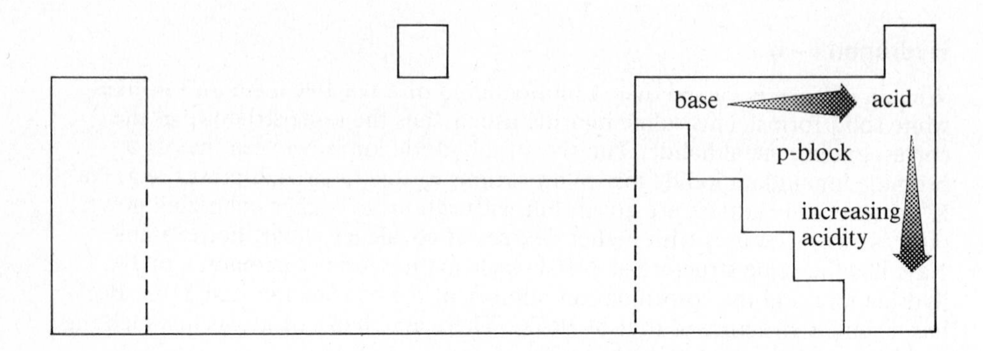

1. Going across the period from group V to group VII, there is a trend from base to acid, e.g. ammonia is a weak base.

   $NH_{3(g)} + H_2O_{(l)} \rightleftharpoons NH^+_{4(aq)} + OH^-_{(aq)}$

   water is an ampholyte

   $H_2O_{(l)} + H_2O_{(l)} \rightleftharpoons H_3O^+_{(aq)} + OH^-_{(aq)}$

   hydrogen fluoride is an acid

   $HF_{(g)} + H_2O_{(l)} \rightleftharpoons H_3O^+_{(aq)} + F^-_{(aq)}$

2. Going down a group, the acidity increases, e.g. water is an ampholyte but hydrogen sulphide is a weak acid.

   $H_2S_{(g)} + H_2O_{(l)} \rightleftharpoons H_3O^+_{(aq)} + HS^-_{(aq)}$

   the hydrogen halides are increasingly strong acids:

   $HX_{(g)} + H_2O_{(l)} \rightleftharpoons H_3O^+_{(aq)} + X^-_{(aq)}$

acid	K_a mol dm^{-3}
HF	5.6×10^{-4}
HCl	1×10^{7}
HBr	1×10^{9}
HI	1×10^{11}

There are two factors that account for these trends:

1. the electronegativity of the element combined with hydrogen
2. the strength of the bond between the non-metal atom and the hydrogen atom.

A high electronegativity favours acid properties because the protons receive less shielding, and also the non-metal atom's lone pairs become more withdrawn into the atomic core. Compare an ammonia molecule with a hydrogen fluoride molecule:

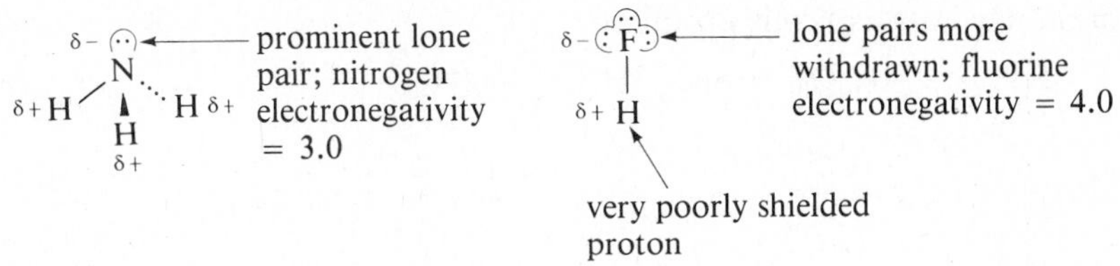

δ– prominent lone pair; nitrogen electronegativity = 3.0

δ– lone pairs more withdrawn; fluorine electronegativity = 4.0

δ+ H — very poorly shielded proton

hence

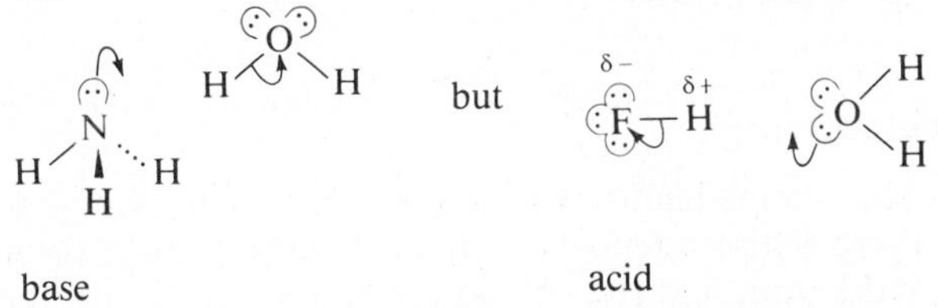

base — but — acid

A weak bond between the hydrogen atom and the non-metal atom favours acid properties as well. The proton is more readily lost. H_2S is a stronger acid than H_2O in spite of the decrease in electronegativity on moving from oxygen to sulphur. Similarly HI is stronger than HCl. Compare the bond strengths:

bond	*bond strength (kJ mol^{-1})*	*acid strength K_a (mol dm^{-3})*
H—O	463	$10^{-15.5}$
H—S	338	$10^{-7.1}$
H—Cl	431	10^{7}
H—I	299	10^{11}

Hydrogen (– I)

The ionic hydrides all act as strong bases. A hydride ion shows a marked tendency to act as a proton acceptor.

$$H^- + H_2O \longrightarrow H{-}H + OH^-$$

For example, sodium hydride is violently hydrolysed to sodium hydroxide solution and hydrogen gas.

$$NaH_{(s)} + H_2O_{(l)} \rightarrow Na^+_{(aq)} + OH^-_{(aq)} + H_{2(g)}$$

The group IV hydrides are an interesting class of compounds because the polarity of the hydrogen atoms in the molecules changes from positive to negative going down the group. Methane, CH_4, contains hydrogen nominally in oxidation state +I and shows no acid-base properties at all. Silane, SiH_4, contains hydrogen in oxidation state –I and is slowly hydrolysed like sodium hydride. Hydrated silica is produced:

$$SiH_{4(g)} + 4H_2O_{(l)} \rightarrow SiO_2.2H_2O_{(s)} + 4H_{2(g)}$$

39.6 Redox properties

Standard electrode potential

Hydrogen molecules show some tendency to lose electrons and become hydrogen ions and, similarly, there is also a tendency for the reverse process to occur. The first reaction is oxidation, and the reverse reaction is reduction; the two tendencies are compared in aqueous solution using a hydrogen electrode (see page 308). This electrode is taken as the standard for all other electrodes and a standard electrode potential $E^{\ominus}$ is the potential of an electrode compared with that of a hydrogen electrode. For example, $E^{\ominus}$ for $Zn^{2+}_{(aq)} + 2e^- \rightleftharpoons Zn_{(s)}$ is -0.76 volts.

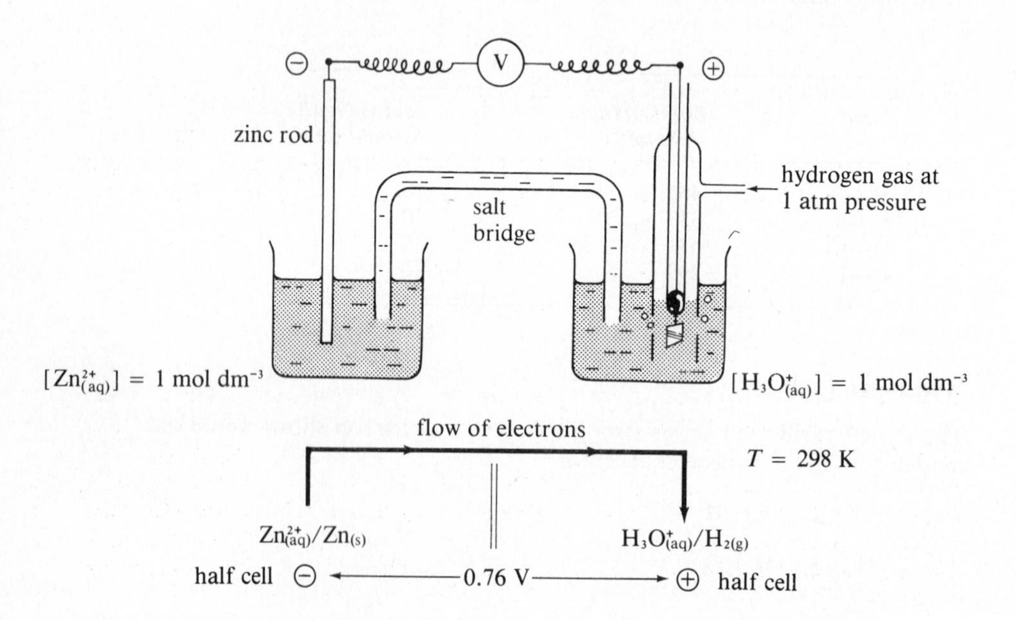

This suggests that zinc atoms lose electrons more readily than hydrogen molecules, and therefore that zinc should react with dilute acid to produce hydrogen. In other words, zinc reduces dilute acid to hydrogen while the dilute acid acts as an oxidant.

Although $E^{\ominus}$ values are useful guides for describing redox chemistry, many reactions do not take place under standard conditions, and so the values must be used with caution. In particular $E^{\ominus}$ values give no information about the rate of a reaction.

Hydrogen (I) and hydrogen (0) as oxidants

In the above paragraph, the oxidizing action of dilute acids on zinc is described. Dilute acids contain hydrogen (I) and oxidize any metals above hydrogen in the electrochemical series:

$$\overset{0}{Mg}_{(s)} + 2\overset{I}{H}_3O^+_{(aq)} \rightarrow \overset{II}{Mg}^{2+}_{(aq)} + 2H_2O_{(l)} + \overset{0}{H}_{2(g)}$$

In the gas phase, water and hydrogen halides can also act as oxidants. The industrial preparation of hydrogen relies on water oxidizing white hot coke:

$$\overset{0}{C}_{(s)} + \overset{I}{H}_2O_{(g)} \rightarrow \overset{II}{C}O_{(g)} + \overset{0}{H}_{2(g)} \qquad (1200°C)$$

Hydrogen halides oxidize most metals to the corresponding halides:

$$\overset{0}{Fe}_{(s)} + 2\overset{I}{H}Cl_{(g)} \xrightarrow{heat} \overset{II}{Fe}Cl_{2(s)} + \overset{0}{H}_{2(g)}$$

Hydrogen (0) itself acts as an oxidant in its action on reactive metals. The metal atoms lose electrons to hydrogen molecules as the ionic hydride lattice forms:

$$2\overset{0}{Na}_{(s)} + \overset{0}{H}_{2(g)} \xrightarrow{heat} 2\overset{I}{Na}\overset{-I}{H}_{(s)}$$

Hydrogen (0) and hydrogen (– I) as reductants

In most of the reactions of hydrogen, the products contain the element in oxidation state +I. In these reactions, hydrogen acts as a reductant: hydrogen molecules lose electrons either to non-metal atoms or to the cations of unreactive metals such as copper or silver:

$$\overset{0}{N}_{2(g)} + 3\overset{0}{H}_{2(g)} \rightleftharpoons 2\overset{-III}{N}\overset{I}{H}_{3(g)}$$

$$\overset{II}{Cu}O_{(s)} + \overset{0}{H}_{2(g)} \xrightarrow{heat} \overset{0}{Cu}_{(s)} + \overset{I}{H}_2O_{(g)}$$

Molecular hydrogen is not as strong a reductant as the element in the presence of a suitable transition metal such as nickel or palladium. The formation of a non-stoichiometric, interstitial hydride increases the reducing strength of hydrogen. In this environment, it is likely to be atomic and so less energy is required for a reduction process because the strong bonds holding the molecules together are already overcome. Hydrogen in the presence of palladium or nickel ('Raney nickel' see page 382) is used to reduce a number of organic compounds:

$$H_2C{=}CH_2 \xrightarrow{Ni/H_{2(g)}} H_3C{-}CH_3$$

alkene → alkane

$$R{-}C{\equiv}N \xrightarrow{Pd/H_{2(g)}} R{-}CH_2NH_2$$

nitrile → amine

Hydrogen (– I) compounds are amongst the most powerful reductants known. The hydride ions readily lose control of electrons and thus cause the reduction of many molecular substances. The hydride complexes discussed in the next section are the most useful source of hydride ions. They reduce a large number of organic compounds that cannot be reduced by hydrogen (0).

$$\underset{\text{acid}}{R(HO)C{=}O} \xrightarrow[\text{2. water}]{\text{1. LiAlH}_4} \underset{\text{alcohol}}{R(HO)CH_2}$$

$$\underset{\text{amide}}{R(H_2N)C{=}O} \xrightarrow[\text{2. water}]{\text{1. LiAlH}_4} \underset{\text{amine}}{R(H_2N)CH_2}$$

39.7 Solubility and complexing properties

A cationic complex of hydrogen is any particle that contains a proton acting as an electron-pair acceptor, e.g.

$H^+_{(aq)}$ (H_3O^+) $H^+_{(NH_3)}$ (NH_4^+)

Since the polarizing power of a hydrogen cation (one proton) is so enormous, it is not very useful to consider these particles in any detail as complexes. They are best viewed as molecular cations with no distinction drawn between any of the bonds within the particle.

Anionic complexes of hydrogen contain the hydride ion acting as a ligand. These complexes are stable only in anhydrous conditions and are formed predominantly by the group IIIB cations, B^{3+} and Al^{3+}. The degree of covalency of the hydridoborates and hydridoaluminates is high because the cations are both small and highly charged. Typical complexes are lithium tetrahydridoaluminate and sodium tetrahydridoborate: $Li^+ [AlH_4]^-$, $Na^+ [BH_4]^-$. In both these complexes, the anion structure can be interpreted in terms of a hypothetical equilibrium between the electron-deficient group III hydride and a hydride ion, e.g.

$$H^{\ominus} + AlH_3 \rightleftharpoons [H{-}AlH_3]^{\ominus}$$

Hydridocomplexes are sparingly soluble in non-polar solvents such as ether and are usually prepared in solution. For example, lithium tetrahydridoaluminate is prepared by refluxing aluminium chloride and lithium hydride in ether. An excess of aluminium chloride must be avoided otherwise aluminium hydride precipitates out of the ether solution in its polymeric form $(AlH_3)_n$:

$$4LiH + AlCl_3 \xrightleftharpoons[\text{ether}]{} LiAlH_4 + 3LiCl$$

The molecular hydrides of aluminium and boron are themselves unusual in structure and properties. Their chemistry is in some ways similar to that of the hydride complexes, and there is therefore some justification in treating them also as ionic complexes with high degrees of covalency. These hydrides are discussed in more detail in Chapter 42. The most common one is diborane, B_2H_6, whose structure is interpreted in terms of the ionic model as two tetrahedral arrangements of ions with one edge of each tetrahedron shared between the two:

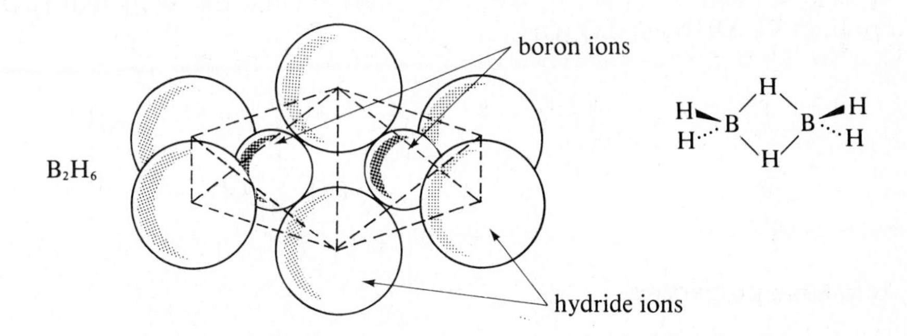

39.8 The hydroxyl group

Structure

The hydroxyl group contains one oxygen atom covalently bonded to a hydrogen atom:

$$\cdot\ddot{\underset{\cdot\cdot}{O}} \times H$$

It is therefore isoelectronic with a fluorine atom and is often referred to as a 'pseudohalide' for this reason (other pseudohalides include —CN, —SCN). A pseudohalide is any group with the three properties outlined below.

1. It forms a stable X_2 molecule: HO—OH like F—F
2. It forms ionic X^- compounds: $HO^-\ {}^+Na$ like $F^-\ {}^+Na$
3. It forms covalent —X compounds: HO—H like F—H.

The periodic trends in the bonding of the halides are described on page 602. The bonding in the hydroxyl compounds follows a similar trend.

Trends in bonding of hydroxyl compounds

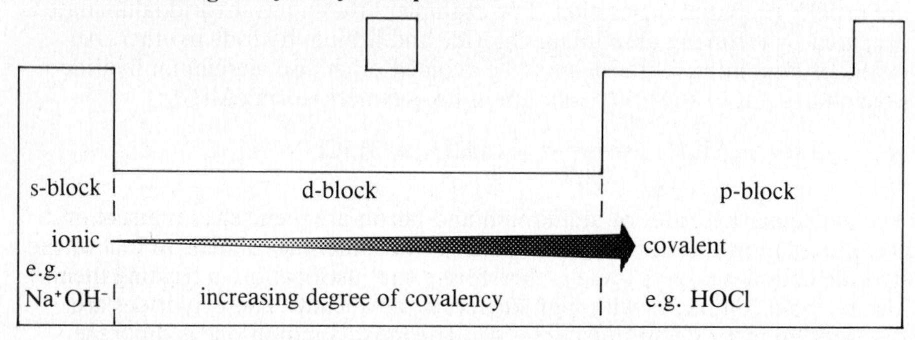

Unlike the halides, however, the hydroxyl group contains a poorly shielded proton. This leads to the elimination of water molecules from the molecules of hydroxy compounds containing more than three hydroxyl groups. For example, PF_5 and SF_6 both exist, but $P(OH)_5$ and $S(OH)_6$ do not. Elimination of H_2O produces $PO(OH)_3$ and $SO_2(OH)_2$.

PF_5 but $P(OH)_5 \longrightarrow PO(OH)_3 + H_2O$

Acid-base properties

The hydroxyl group has lone pairs of electrons as well as an electron-deficient proton.

$O^{\delta-}$ ← lone pairs

$H^{\delta+}$ ← electron-deficient proton

It is therefore capable of acting as a proton donor, a proton acceptor, or both. There is a marked trend in acid-base properties on going across the Periodic Table.

Trends in acid-base properties of hydroxyl compounds

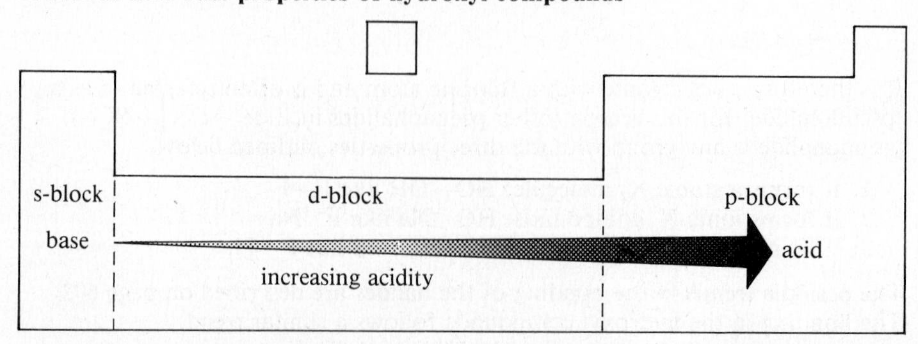

a) NaOH is alkaline; pH = 14.
b) $Fe(OH)_2$ is basic; it is insoluble but neutralizes an acid solution.
c) $Al(OH)_3$, $Zn(OH)_2$, $Pb(OH)_2$, $Sn(OH)_2$ are amphoteric; they dissolve in both acids and bases.
d) H_2SO_4, HNO_3 are acidic; pH = 1.

This trend can be interpreted by using the ionic model to discuss the structure of the hydroxyl compounds. The polarizing power of a cation increases steadily on moving across the Periodic Table: this is due to the increase in charge and decrease in size of the ions. When a hydroxide ion is in the presence of a cation with a low polarizing power, its three prominent lone pairs readily lead it to act as a proton acceptor:

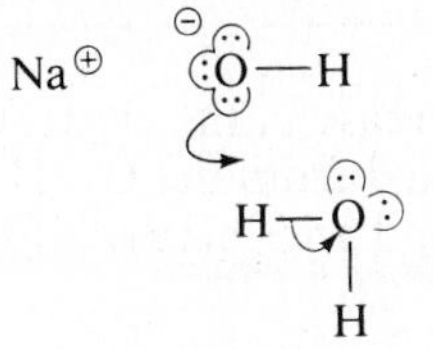

$Na^{\oplus}$ $^{\ominus}$:O—H → H—O(H)

NaOH dissolves well in water and is therefore an alkali

As the polarizing power of the cations increases, the lone pairs become more tightly held by the cations, and also the electron deficiency of the hydroxyl proton begins to increase. Aluminium hydroxide is therefore amphoteric: both proton loss and gain are able to take place:

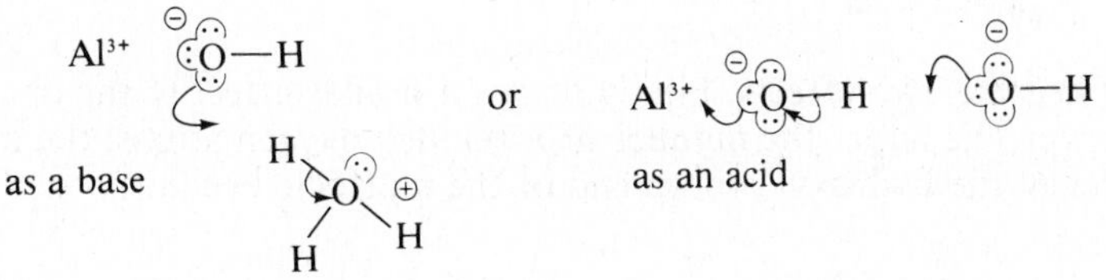

Al^{3+} $^{\ominus}$:O—H → $H_3O^{\oplus}$ as a base

or

Al^{3+} $^{\ominus}$:O—H → $^{\ominus}$:O—H as an acid

The ionic model is not the usual choice to discuss the hydroxyl compounds of sulphur but, as an extension of the above argument, the choice becomes worth-while. An S^{6+} ion would have an enormous polarizing power and would attract the hydroxide lone pairs very strongly. A high degree of covalency would result and the electron-deficient proton therefore dominates any acid-base interaction.

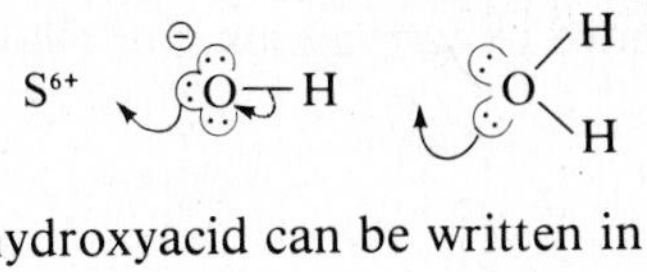

S^{6+} $^{\ominus}$:O—H :O(H)H strong acid

Any hydroxyacid can be written in the general form, $XO_p\,(OH)_q$.

element	*acid formulas*		*p*	*q*
phosphorus	H_3PO_4	$PO(OH)_3$	1	3
nitrogen	HNO_3	$NO_2(OH)$	2	1
sulphur	H_2SO_4	$SO_2(OH)_2$	2	2
chlorine	$HClO_4$	$ClO_3(OH)$	3	1

The strength of these acids increases with two main factors:

1. An increase in the value of p
2. An increase in the electronegativity of X.

Compare the acid strengths below.

acid	p	K_a (mol dm^{-3})	*acid*	*electronegativity*	K_a (mol dm^{-3})
$HClO$	0	$10^{-7.2}$	HIO	I = 2.5	$10^{-10.5}$
$HClO_2$	1	$10^{-2.0}$	$HBrO$	Br = 2.8	$10^{-8.7}$
$HClO_3$	2	$10^{1.0}$	$HClO$	Cl = 3.0	$10^{-7.2}$
$HClO_4$	3	$> 10^{10}$			

The increase in electronegativity of X leads to an increase in the electron deficiency of the hydroxyl proton. Electrons are attracted from the O—H bond towards the electronegative atom. This weakens the bond and increases the tendency for the molecule to act as a proton donor.

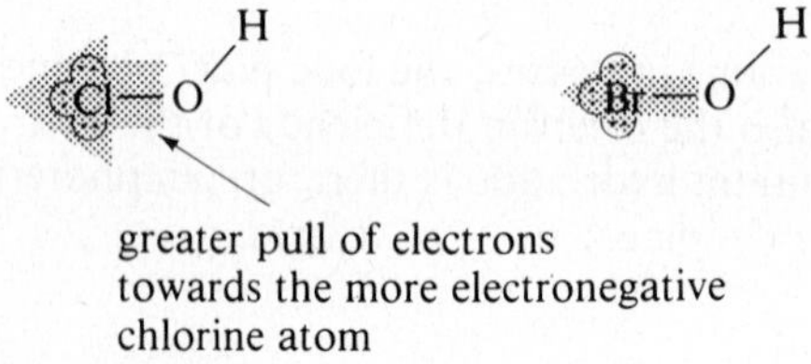

greater pull of electrons towards the more electronegative chlorine atom

The increase in p has two effects. Firstly, it has a similar effect to the one described above. The larger the number of π-bonded oxygen atoms, the more electron-deficient the hydroxyl proton end of the molecule becomes:

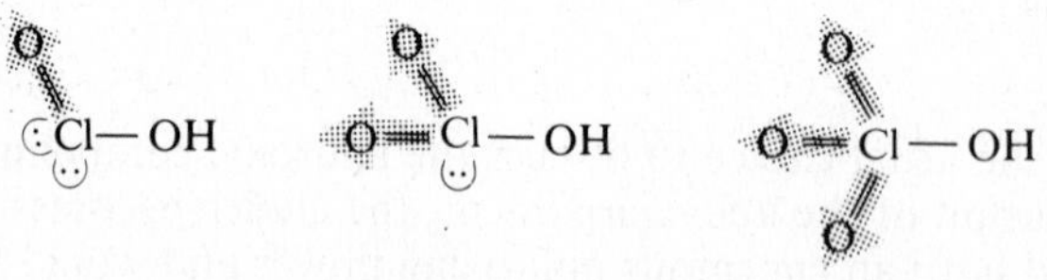

Secondly, it leads to an increase in the thermodynamic stability of the corresponding anions. There are more ways of delocalizing the charge on the anion when there is a larger number of π-bonded oxygen atoms. For chlorate (VII) ions, there are few canonical forms:

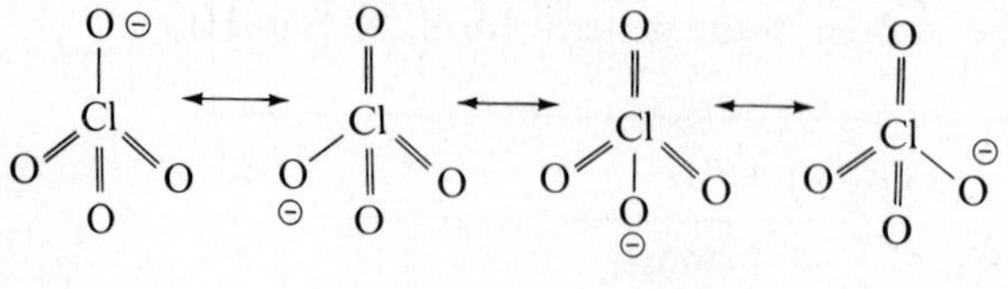

The hydroxide ion

The lone pairs of electrons on the hydroxide ion are responsible for its three major properties

1. as a base
2. as a ligand
3. as a nucleophile.

Its action as a base sometimes leads to precipitation taking place; for example, when sodium hydroxide solution is added to zinc (II) chloride in solution. The hydroxide ions deprotonate the zinc aquo-ions to produce an insoluble hydrated form of zinc hydroxide:

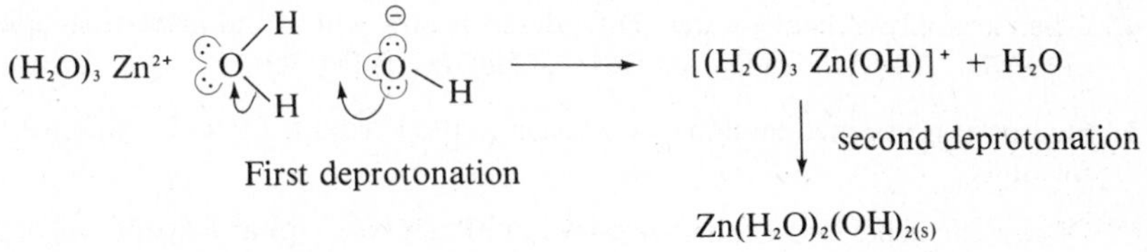

$$(H_2O)_3\,Zn^{2+}\,O(H)H + {}^{\ominus}OH \longrightarrow [(H_2O)_3\,Zn(OH)]^+ + H_2O$$

First deprotonation

second deprotonation

$$Zn(H_2O)_2(OH)_{2(s)}$$

In this system, an excess of hydroxide ions causes the zinc hydroxide to redissolve. Hydroxide ions act as ligands to complex the zinc cation.

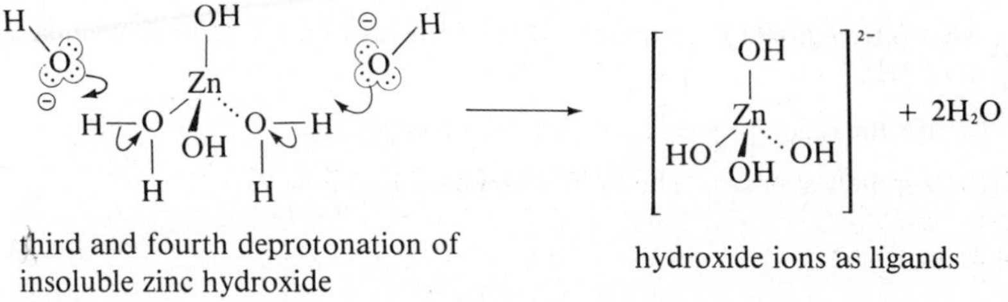

$$Zn(OH)_2(H_2O)_2 + 2\,{}^{\ominus}OH \longrightarrow [Zn(OH)_4]^{2-} + 2H_2O$$

third and fourth deprotonation of insoluble zinc hydroxide

hydroxide ions as ligands

The hydroxide ion's action as a nucleophile is outlined throughout Part 3 of this book. A typical example is shown below.

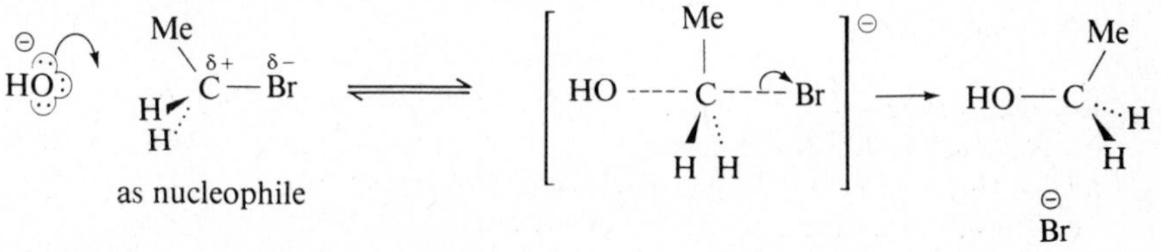

$$HO^{\ominus} + MeH_2C^{\delta+}\!-\!Br^{\delta-} \rightleftharpoons [HO\cdots C(Me)(H)(H)\cdots Br]^{\ominus} \longrightarrow HO\!-\!C(Me)(H)(H) + Br^{\ominus}$$

as nucleophile

In all of these reactions, the lone pairs of the hydroxide ions interact with electron-deficient parts of other particles:

1. Hydrogen atoms — as a base
2. Metal cations — as a ligand
3. Non-metal atoms — as a nucleophile.

Questions

1. Explain the meaning of: a) hydrogen bond b) interstitial hydride c) acid strength d) bond strength e) standard electrode potential f) reductant g) thermodynamic stability h) ligand.

2. Compare the reactions of the following hydrides with water: a) NaH b) CH_4 c) NH_3 d) HCl.

3. a) Describe one large scale method for extracting hydrogen from its compounds. Give three large scale uses of hydrogen.
   b) Give four different methods of producing hydrogen in the laboratory. In each case, classify the type of reaction taking place.

4. Given a supply of heavy water, D_2O, describe how you would prepare samples of: a) $LiAlD_4$ b) ND_3 c) D_2 d) DCl e) NaOD f) $Fe(OD)_3$.

5. Hydrogen is assigned an unusual position in the Periodic Table. Explain the reasons for this.

6. Discuss the behaviour of the hydroxyl ion a) as a base b) as a ligand c) as a nucleophile.

7. Describe a hydrogen electrode and explain its uses. Why is the standard electrode potential for hydrogen zero?

8. What factors affect the strengths of the oxyacids with the general formula $XO_p(OH)_q$?

9. Discuss the redox properties of hydrogen compounds.

10. Discuss the statement: 'The hydrogen proton is unique'.

Chapter 40

The alkali metals: group IA

40.1 Structure of the elements

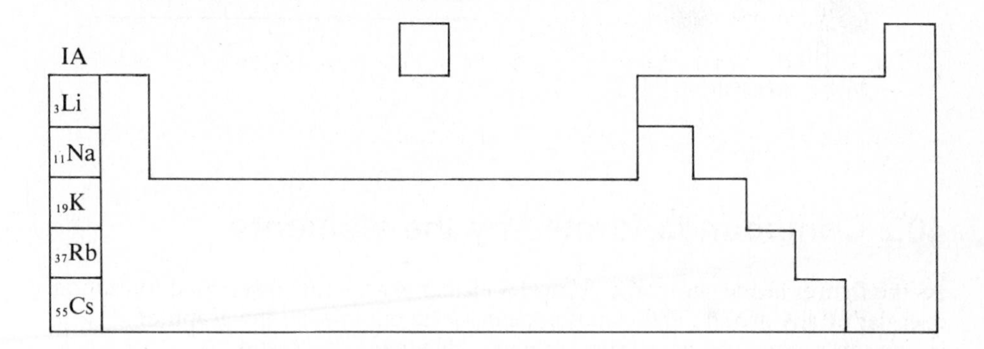

element	*electronic configuration*	*electro-negativity*	*atomic radius* (m × 10^{-10})	*ionic (M^+) radius* (m × 10^{-10})
lithium	[He] $2s^1$	1.0	1.23	0.60
sodium	[Ne] $3s^1$	0.9	1.57	0.95
potassium	[Ar] $4s^1$	0.8	2.03	1.33
rubidium	[Kr] $5s^1$	0.8	2.16	1.48
caesium	[Xe] $6s^1$	0.7	2.35	1.69

Atomic structure

A group IA atom has the following characteristic properties:

1. One outer-shell electron in an s-orbital.
2. A core charge of 1.
3. An empty inner d sub-shell, where applicable, e.g. lithium and sodium have no inner d sub-shells, but potassium has an empty 3d sub-shell and a single electron in the 4th shell.

Physical structure

All the group IA elements are metals with low melting points and boiling points. The low core charge of the atoms leads to very poor control of the outer shell electrons which are therefore easily lost from the atoms. The pure elements exist as lattices of single-charged cations in a cloud of delocalized outer-shell electrons. The ions adopt a body-centred cubic arrangement in these lattices. The low core charge and large size of the ions result in weak bonding within the lattice and therefore the elements have low fixed points and are fairly soft.

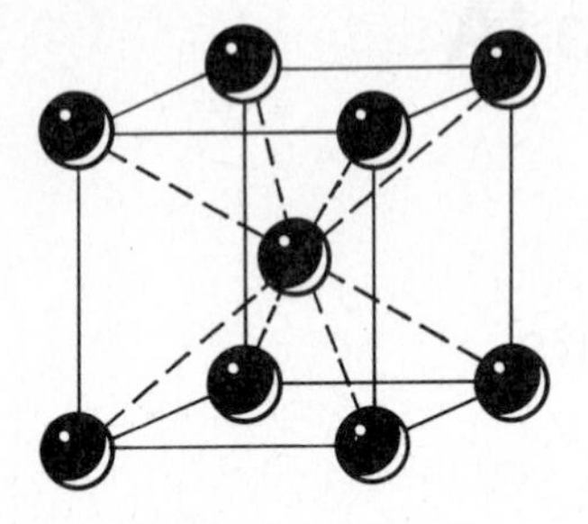

b.c.c. arrangement

element	*m.p. (°C)*	*b.p. (°C)*
lithium	180	1330
sodium	98	890
potassium	64	890
rubidium	39	688
caesium	29	690

40.2 Compounds formed by the elements

As the figures below show, the group IA elements have the lowest first ionization energies of any group. The values appear as the minima on the graph of ionization energies on page 596. Similarly, they have the lowest electronegativities of any group.

element	*first ionization energy* (kJ mol^{-1})
lithium	520
sodium	494
potassium	418
rubidium	403
caesium	374

Ionic compounds containing the elements in an oxidation state of +I are therefore readily formed. With the exception of those of lithium, these compounds have a high percentage of ionic character. The group IA cations have low polarizing powers as a result of their large sizes (except Li^+) and low charge. Lithium compounds show more covalent character than those of the other group IA elements because the small lithium ion causes a greater degree of polarization.

LiI m.p. 469°C NaI m.p. 660°C

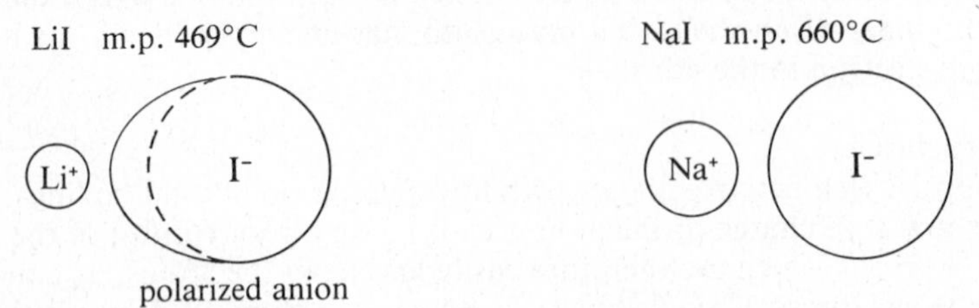

polarized anion

This is an example of the first member of a periodic group having atypical properties. This is discussed in detail in Chapter 50.

40.3 Occurrence, extraction and uses

Sodium and potassium are the sixth and seventh most abundant elements in the earth's crust (2.83% and 2.59%). The alkali metals are far too reactive to occur naturally, but are found in a number of mineral deposits combined as the chloride, nitrate or silicate, e.g.

lithium as spodumene $LiAl(SiO_3)_2$
sodium as rocksalt NaCl and Chile saltpetre $NaNO_3$
potassium as carnallite $KCl.MgCl_2.6H_2O$

The pure metals can be extracted from their compounds only by electrolysis. There are no chemical reductants strong enough to reduce a group IA metal salt to the metal. The most common industrial process for extracting sodium is outlined below. Sodium chloride is either mined as rock salt or separated from sea water which contains approximately 3% by mass of dissolved minerals, many of which are group IA compounds.

Extraction of sodium

The Downs cell

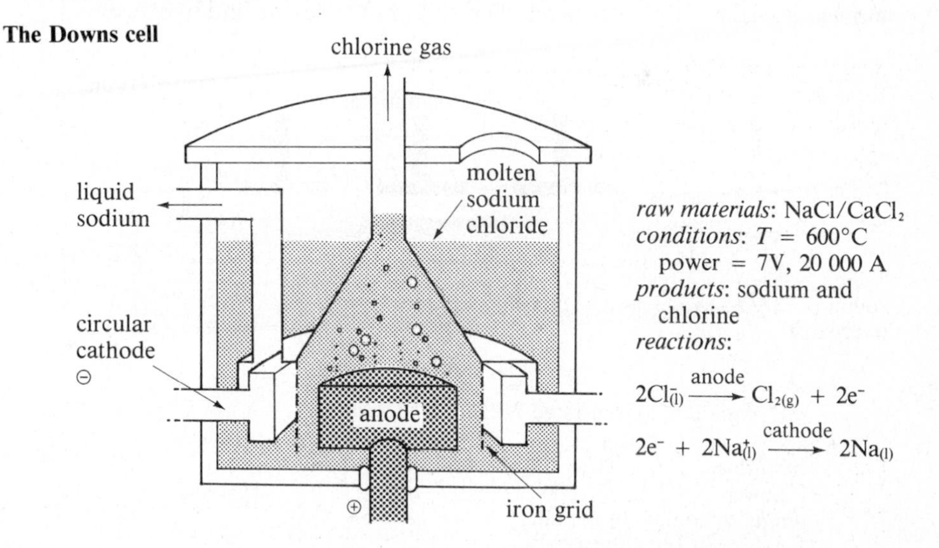

Molten sodium chloride is electrolysed in a circular cell between a central graphitic anode and a circular iron cathode. The heating effect of the very large current keeps the electrolyte molten.

The two important features about this process are:

1. The two products are extremely reactive and must be kept apart. The shape of the cell and an iron grid between the two electrodes are designed to achieve this.
2. The boiling point of sodium, 890°C, is close to the melting point of sodium chloride, 801°C. At temperatures above 800°C the high vapour pressure of liquid sodium causes problems in containing the sodium. These are avoided by adding enough calcium chloride (40% by mass) to the sodium chloride to produce a eutectic mixture whose melting point is only 600°C.

Sodium hydroxide and chlorine from brine

The electrolysis of aqueous sodium chloride involves the following electrode reactions:

cathode: $2HOH_{(l)} + 2e^- \rightarrow 2OH^-_{(aq)} + H_{2(g)}$
anode: $2Cl^- \rightarrow Cl_{2(g)} + 2e^-$

The two gaseous products, hydrogen and chlorine, must be prevented from mixing with each other, and the chlorine must be prevented from reacting with the hydroxide ions in solution

$$Cl_{2(g)} + 2OH^-_{(aq)} \rightarrow ClO^-_{(aq)} + Cl^-_{(aq)}$$

a) **The Castner-Kellner cell**
A long thin tank contains brine into which titanium anodes dip from above. The cathode is a thin layer of mercury which flows along the bottom of the tank. Chlorine gas is produced at the anodes and is piped off. Due to the high over-potential of hydrogen on a mercury electrode (0.8 volts), sodium is preferentially discharged at the cathode. The sodium forms an amalgam with the mercury and flows out of the cell into a second tank. Here the sodium amalgam reacts with water producing sodium hydroxide and hydrogen.

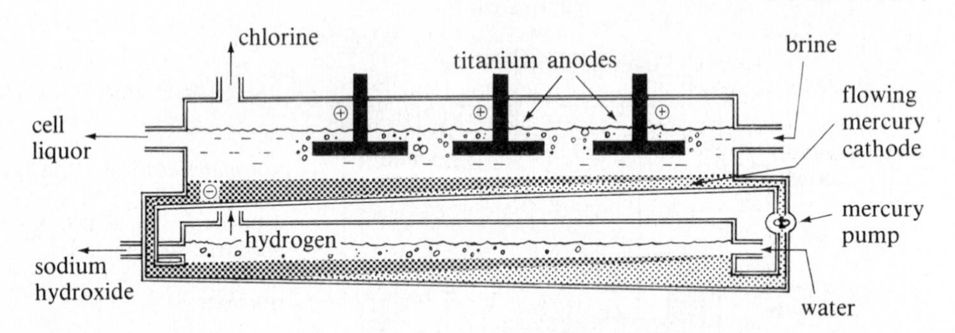

raw materials: concentrated brine
$Na^+_{(aq)}$ $Cl^-_{(aq)}$
conditions: $T = 20°C$
power = 4.3 V, 300 000 A
products: sodium hydroxide, hydrogen and chlorine

reactions:

$$2Cl^-_{(aq)} \xrightarrow{\text{anode}} Cl_{2(g)} + 2e^-$$

$$2Na^+_{(aq)} + 2e^- \xrightarrow{\text{Hg cathode}} 2Na/Hg$$

$$2Na/Hg + 2H_2O \longrightarrow 2NaOH + H_2 + Hg$$

b) **The diaphragm cell**
The brine flows into the top of the cell. About half the chloride ions are discharged at the anodes producing chlorine gas which is piped off. The remaining solution seeps through the asbestos diaphragm into the region of the cathodes. Here water molecules surrounding the sodium ions accept electrons at the cathodes yielding hydrogen molecules and hydroxyl ions. The resulting solution is piped from the cell and consists of 10% by mass of sodium hydroxide and 15% of sodium chloride. The solution is evaporated to 20% of its original volume causing the relatively insoluble sodium chloride to crystallize out. The resulting solution contains 50% by mass of sodium hydroxide and only 1% of sodium chloride.

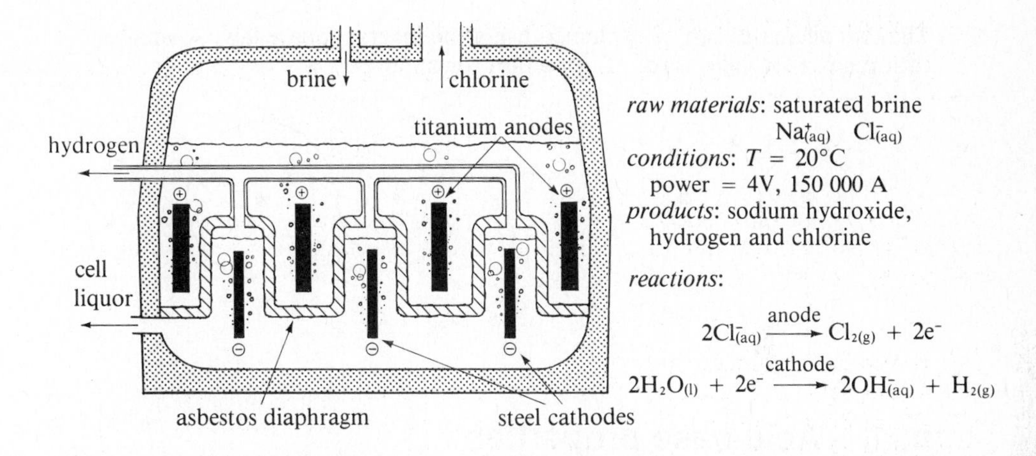

Uses of the alkali metals

Sodium is used in larger quantities than any of the other alkali metals and so only the uses of sodium are given below:

1. In sodium vapour lights, e.g. for road lighting.
2. In the manufacture of 'lead tetra-ethyl', an additive to petrol that acts as a chain inhibitor and controls the combustion of petrol in the engine cylinder.

$$\text{Pb—Na (alloy)} \xrightarrow{4C_2H_5Cl} Pb(C_2H_5)_4 + 4NaCl$$

3. As a coolant for nuclear reactor piles. Liquid sodium conducts heat extremely well.
4. In the manufacture of titanium: sodium displaces the metal from its chloride because sodium is a powerful reductant. $TiCl_4$ vapour is passed into molten sodium under pressure.

$$TiCl_4 + 4Na \rightarrow Ti + 4NaCl$$

5. In the manufacture of sodium cyanide: this is a useful synthetic reagent in a number of organic processes.

40.4 Compound structure

The ionic nature of the group IA compounds has already been described. The elements are always combined in the +I oxidation state and ionic lattices are formed with almost all anions, e.g. oxy-anions such as CO_3^{2-} and MnO_4^-, and simple atomic anions such as Cl^- and O^{2-}. The increase in ionic radius on going down the group from Li^+ to Cs^+ has the following effects:

1. an increase in cation co-ordination number, e.g. 6 in NaCl but 8 in CsCl.
2. an increase in percentage ionic character: a large cation has a smaller polarizing power.
3. a decrease in lattice energy, e.g. NaCl : 771 kJ mol^{-1}; CsCl : 645 kJ mol^{-1}.
4. a decrease in melting point, e.g. NaCl : 801°C; CsCl : 645°C.

The two decreases can be explained using the inverse square law. A smaller cation has a stronger force of attraction for an anion.

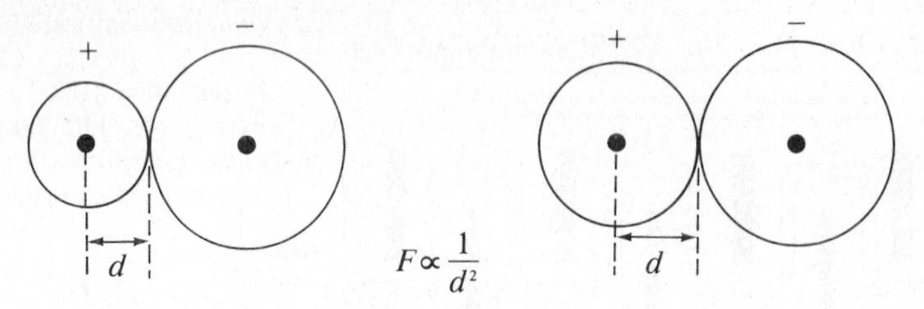

40.5 Acid-base properties

The metals, their oxides, hydrides and carbonates are all soluble bases and therefore alkalis. The metals and hydrides react vigorously with water to produce hydroxides and hydrogen, while the oxides and carbonates dissolve in water and the anions act as bases. For example, for sodium:

$$2Na_{(s)} + 2H_2O_{(l)} \rightarrow 2Na^+_{(aq)} + OH^-_{(aq)} + H_{2(g)}$$

$$NaH_{(s)} + H_2O_{(l)} \rightarrow Na^+_{(aq)} + OH^-_{(aq)} + H_{2(g)}$$

$$Na_2O_{(s)} + H_2O_{(l)} \rightarrow 2Na^+_{(aq)} + 2OH^-_{(aq)}$$

$$Na_2CO_{3(s)} \xrightarrow{(aq)} 2Na^+_{(aq)} + CO^{2-}_{3(aq)}$$

and $$CO^{2-}_{3(aq)} + H_2O_{(l)} \rightleftharpoons HCO^-_{3(aq)} + OH^-_{(aq)}$$

A wide range of neutral and acidic salts can be prepared from an alkaline solution. Group IA cations are sufficiently large and of a low enough polarizing power that they can form stable lattices with a number of acidic anions, e.g. $NaHCO_3$, $NaHSO_4$ and NaH_2PO_4.

For example, sodium dihydrogenphosphate (V) is prepared by titrating orthophosphoric (V) acid against sodium hydroxide using a methyl orange indicator. Methyl orange indicates the first equivalence point and shows the volume of sodium hydroxide that is required to carry out the following reaction:

$$NaOH + H_3PO_4 \rightarrow NaH_2PO_4 + H_2O$$

The salt is removed from the solution by evaporating off the solvent.

The cations themselves have little effect on the pH of a solution. The small lithium aquo-ion has a $K_a = 3 \times 10^{-14}$ mol dm^{-3}, and so a molar solution of lithium chloride has a pH of 6.98.

$$(H_2O)_3Li^+ \cdot OH_2 + H_2O \rightleftharpoons (H_2O)_3Li^+ \,{}^{\ominus}\!:\!O{-}H + H_3O^{\oplus}$$

Lithium aquo-ions are stronger proton donors than any of the other group IA aquo-ions, but the effect is still very limited, even for lithium.

40.6 Redox compounds

The pure elements are among the strongest reductants known. The atoms react readily with water yielding the corresponding aquo-ions:

$$M^{+}_{(aq)} + e^{-} \rightleftharpoons M_{(s)}$$

element	$E^{\ominus}$ *(volts)*
sodium	−2.71
potassium	−2.92
rubidium	−2.99

The elements cannot be used as reductants in aqueous conditions because they reduce water violently to hydrogen. Under non-aqueous conditions, they reduce most non-metals and also the compounds of less reactive metals. When dissolved in liquid ammonia, the elements provide a valuable and important means of reducing certain organic compounds.

Reducing a non-metal
When burnt in air, lithium reduces oxygen to the oxide ion, while sodium produces a mixture of sodium oxide and sodium peroxide.

When burnt in pure oxygen the members of group IA show the following trend in the type of oxide formed.

reaction	*product*	*anion*
$4\,Li_{(s)} + O_{2(g)} \longrightarrow 2\,Li_2O_{(s)}$	lithium oxide	$[:\ddot{\underset{..}{O}}:]^{2-}$
$2\,Na_{(s)} + O_{2(g)} \longrightarrow Na_2O_{2(s)}$	sodium peroxide	$^{\ominus}:\ddot{\underset{..}{O}}-\ddot{\underset{..}{O}}:^{\ominus}$
$K_{(s)} + O_{2(g)} \longrightarrow KO_{2(g)}$	the superoxide	$\cdot\ddot{\underset{..}{O}}-\ddot{\underset{..}{O}}:^{\ominus}$
$Rb_{(s)} + O_{2(g)} \longrightarrow RbO_{2(s)}$		unpaired electron

This trend is the result of the increasing size of the metal ions as the group is descended. The lithium ion is too small to accommodate enough of the larger peroxide ions around it to form a stable lattice.

Reducing a compound of a less reactive metal
The industrial preparation of titanium relies on sodium's reducing properties.

$$\overset{IV}{Ti}Cl_4 + 4\overset{0}{Na} \rightarrow \overset{0}{Ti} + 4\overset{I}{Na}Cl$$

Reducing an amide to an aldehyde
In liquid ammonia, sodium is partially dissociated into solvated ions and electrons

$$Na_{(am)} \rightleftharpoons Na^{+}_{(am)} + e^{-}_{(am)}$$

In the presence of a weak proton donor such as the ammonium salt of a weak acid, an amide is reduced to an aldehyde by this reagent:

$$R{-}C(=O)NH_2 \xrightarrow{+\,2e^{-}_{(am)}} \left[R{-}C(O^{\ominus})^{\ominus}NH_2 \right] \longrightarrow \left[R{-}C(O^{\ominus})(H)NH_2 \right] \xrightarrow{-NH_3} R{-}C(=O)H$$

amide — $H{-}NH_3^{\oplus}$ — $:NH_3$ — aldehyde

40.7 Solubility and complexing properties

Nearly all the compounds of the alkali metals are soluble in water, and some are extremely soluble indeed. The aquo-ions are easily formed and are very hard to precipitate from solution. The large size and low charge of the cations leads to low lattice energies, but also to low hydration energies as well. Compare the energy terms for sodium chloride dissolving in water with those for calcium chloride.

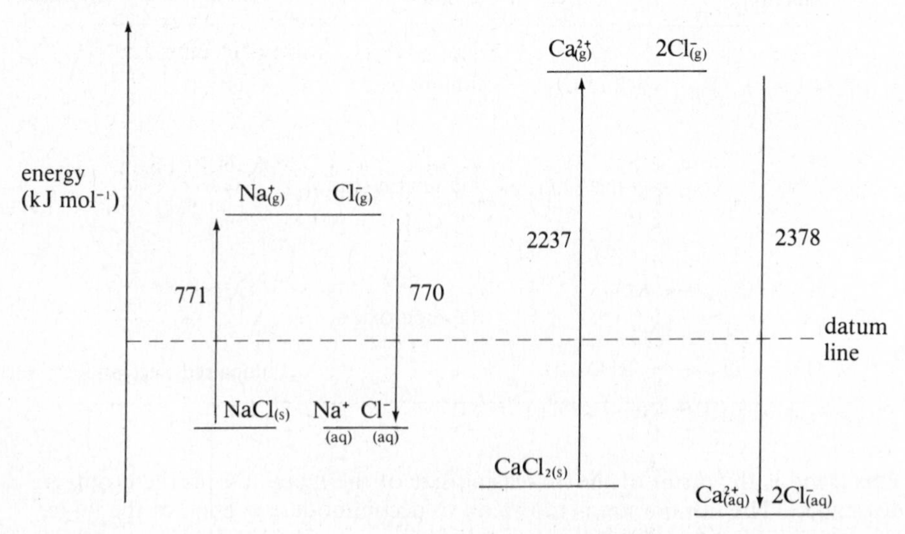

Both are soluble, but the lattice energy and hydration energy of calcium chloride are almost three times greater than those of sodium chloride.

Apart from forming loosely bound aquo-ions, the group IA metal cations have almost no other co-ordination chemistry. The energy given out when ligands co-ordinate to the rather large and low-charged ions is too small to favour complex formation. Lithium, once again, provides some exceptions to this pattern and these are described in more detail in Chapter 50.

Because alkali-metal compounds are all so soluble, it is difficult to confirm the presence of the group IA cations during the analysis of a substance. It is not possible to precipitate all the ions from solution, nor is it possible to find a characteristic complex that each ion forms. An alkali metal compound is therefore usually characterized by a flame test.

Under certain circumstances it is possible to precipitate group IA metal ions from solution. These rely on the use of the 'common ion effect' (see page 297). An important industrial application of the use of this effect can be seen in the Solvay Process for the manufacture of sodium carbonate.

The Solvay process

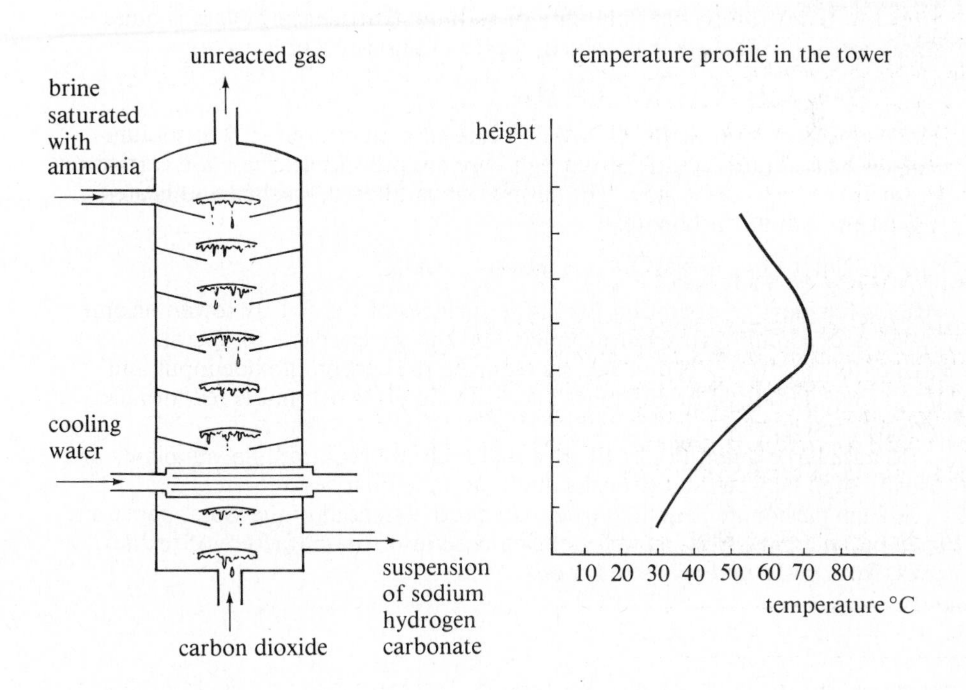

raw materials: calcium carbonate; sodium chloride; ammonia

products: sodium carbonate; ammonium chloride; sodium hydrogen carbonate; calcium chloride

Carbon dioxide is produced by heating calcium carbonate.

$$CaCO_{3(s)} \rightarrow CaO_{(s)} + CO_{2(g)}$$

The gas is pumped into the bottom of the Solvay tower and meets a descending stream of concentrated brine saturated with ammonia. As it ascends the tower, the carbon dioxide dissolves in and reacts with the descending water:

$$CO_{2(g)} + H_2O_{(l)} \rightleftharpoons H_2CO_{3(aq)}$$

$$H_2CO_{3(aq)} + H_2O_{(l)} \rightleftharpoons HCO^-_{3(aq)} + H_3O^+_{(aq)}$$

The presence of ammonia shifts the position of the second equilibrium to the right by removing $H_3O^+_{(aq)}$ ions:

$$H_3O^+_{(aq)} + NH_{3(aq)} \rightleftharpoons NH^+_{4(aq)} + H_2O_{(l)}$$

$$NH_{3(aq)} + H_2O_{(l)} \rightleftharpoons NH^+_{4(aq)} + OH^-_{(aq)}$$

$$OH^-_{(aq)} + H_3O^+_{(aq)} \rightleftharpoons 2H_2O_{(l)}$$

This series of reactions evolves heat and results in a high concentration of hydrogen carbonate ions, $HCO^-_{3(aq)}$. There is also present a high concentration of sodium ions from the brine.

At low temperatures the solubility of sodium hydrogencarbonate is quite small, and so, if the tower is cooled, a precipitation reaction occurs:

$$Na^+_{(aq)} + HCO^-_{3(aq)} \rightarrow NaHCO_{3(s)}$$

The temperature profile of the tower is carefully monitored so that sodium hydrogencarbonate crystals of the right size are produced. These are carried out of the tower as a suspension. The suspension is filtered, washed and heated producing sodium carbonate:

$$2NaHCO_{3(s)} \rightarrow Na_2CO_{3(s)} + H_2O_{(l)} + CO_{2(g)}$$

After a few days of operation the inside surfaces of the Solvay tower become scaled with sodium hydrogencarbonate and the tower ceases to operate efficiently. The tower is descaled by reducing the carbon dioxide input and increasing the rate of brine input. The scale dissolves off and is fed into an adjacent tower.

Because large quantities of the raw materials are required, Solvay plants are usually sited between salt deposits and limestone quarries.

Sodium carbonate is an important chemical, essential in the glass, paper and soap making industries, as well as being used in the manufacture of textiles, dyes, soft drinks and processed foods.

40.8 The thermal stability of the oxy-compounds

Although copper (II) carbonate decomposes below 300°C and calcium carbonate below 900°C, sodium carbonate does not decompose at all but melts at 851°C. The oxy-compounds of the group IA metals generally are very stable to the effect of heat. In cases where decomposition does take place, the products rarely include the metal oxide, unlike most other metal oxy-compounds. For example, lead (II) nitrate (V) decomposes easily to give the metal oxide, nitrogen (IV) oxide and oxygen. But sodium nitrate (V) decomposes with difficulty to give the nitrate (III):

$$2PO(NO_3)_{2(s)} \xrightarrow{\text{heat}} 2PbO_{(s)} + 4NO_{2(g)} + O_{2(g)}$$

$$2NaNO_{3(s)} \xrightarrow{\text{heat}} 2NaNO_{2(s)} + O_{2(g)}$$

The reason for the stability can once again be put down to the low polarizing power of the alkali metal cations resulting from their low charge and relatively large size. A cation of high polarizing power distorts the lone pairs of a neighbouring oxy-anion strongly and this leads to considerable covalent character:

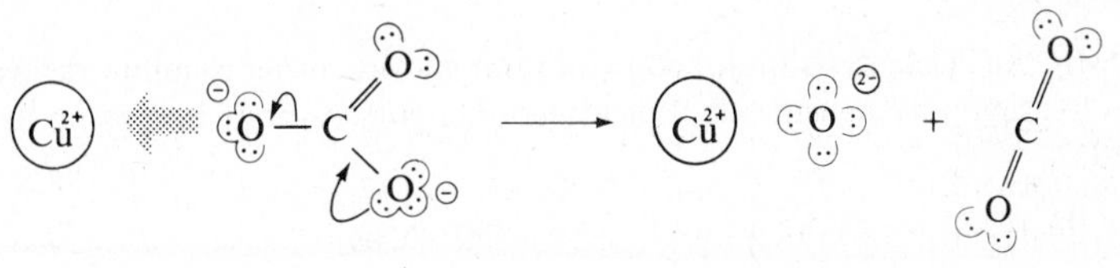

When heat energy is being supplied to the lattice, it becomes as likely for the covalent bonds in the anion to break as for the ions to separate.

Questions

1. Explain the meaning of: a) core charge b) subshell c) delocalized electrons d) body-centred cubic e) polarizing power f) lattice energy g) hydration energy h) electron affinity.

2. The atomic numbers of the group I elements are given below:

element	*atomic number*
Li	3
Na	11
K	19
Rb	37
Cs	55

   a) Write down their electronic configurations.
   b) Use these electronic configurations to account for similarities in chemical behaviour.
   c) Use these electronic configurations to account for differences in chemical behaviour as the group is descended.

3. Compare the lattice structures of sodium chloride and caesium chloride, and account for the differences.

4.

element	*standard electrode potential $E^\ominus$ volts*	*ionization energy*
Li	−3.04	+ 519 kJ mol^{-1}
Cs	−2.92	+ 376 kJ mol^{-1}

Account for the similarity in $E^\ominus$ values for these two elements when their ionization energies are so different.

5. On heating, lithium carbonate decomposes while sodium carbonate melts. Lithium hydroxide and lithium carbonate are only sparingly soluble in water, while sodium hydroxide and sodium carbonate are both very soluble. Account for these facts.
6. When sodium carbonate solution is added to a solution of copper sulphate the precipitate produced is a mixture known as a basic carbonate, $Cu(OH)_2.CuCO_3.xH_2O$. Account for this.
7. Explain why the electrolysis of aqueous sodium chloride using graphite electrodes produces hydrogen at the cathode, but when a mercury cathode is used no hydrogen is evolved.
8. Compare the reactions of the group I metals with oxygen.
9. Discuss the redox chemistry of group I metals and their compounds.

Chapter 41

The alkaline earth metals: group IIA

41.1 Structure of the elements

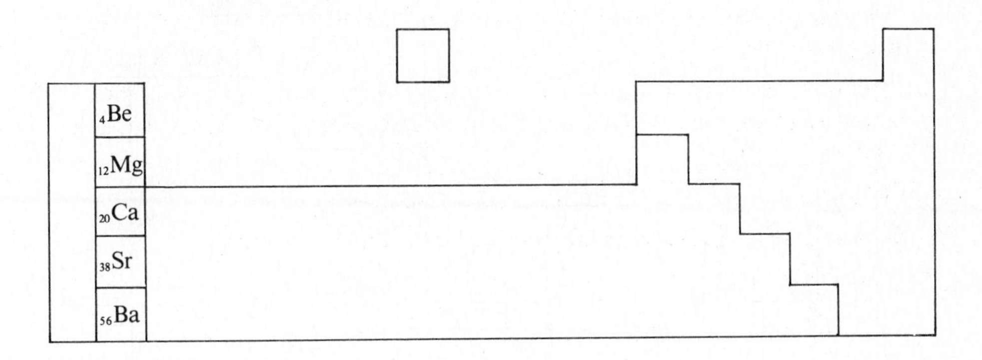

element	*electronic configuration*	*electro-negativity*	*atomic radius* (m × 10^{-10})	*ionic* (M^{2+}) *radius* (m × 10^{-10})
beryllium	[He] $2s^2$	1.5	0.89	0.31
magnesium	[Ne] $3s^2$	1.2	1.36	0.65
calcium	[Ar] $4s^2$	1.0	1.74	0.99
strontium	[Kr] $5s^2$	1.0	1.91	1.13
barium	[Xe] $6s^2$	0.9	1.98	1.35

The atomic and physical structures of the group IIA elements resemble those of the group IA elements. The major differences are summarized below.

1. There are two outer-shell electrons in each atom instead of one.
2. The atoms are smaller because they have a core charge of two.
3. The ions are also smaller and are double charged. Therefore they have higher polarizing powers.
4. The melting points and boiling points of the elements are greater as a result of the effects of increased core charge and decreased atomic radius.

	Be	Mg	Ca	Sr	Ba
melting point (°C)	1280	650	850	768	714
boiling point (°C)	2470	1100	1490	1380	1640

41.2 Compounds formed by the elements

The first ionization energies of the group IIA elements are low, but not as low as those of the group IA elements. Their second ionization energies are about twice the value of the first ionization energies. It might therefore be expected that the elements should form compounds in oxidation state +I rather than +II. In fact, only +II state compounds are formed and this is born out by consideration of the energy terms involved.

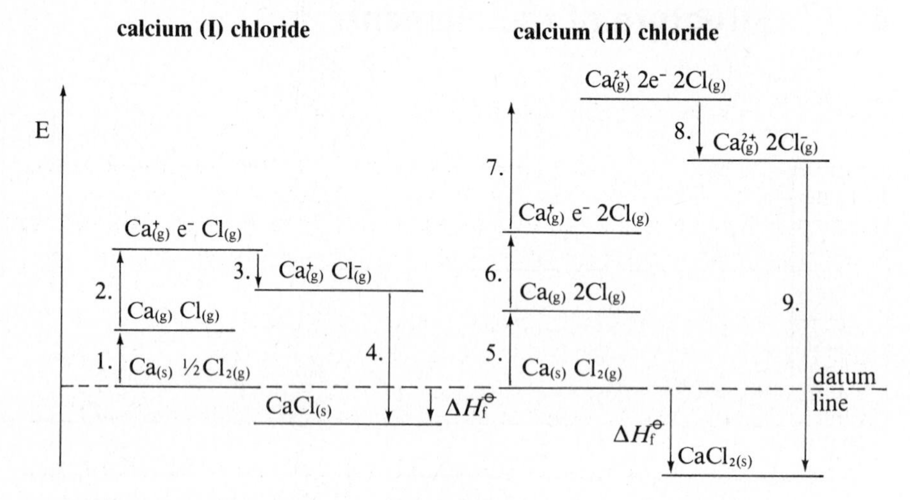

1. $\Delta H^{\ominus}_{at}$ of Ca + ½(bond dissociation energy of Cl_2) = 314 kJ mol⁻¹
2. first ionization energy of Ca = 590 kJ mol^{-1}
3. electron affinity of chlorine = −364 kJ mol^{-1}
4. lattice energy of Ca^+Cl^- = −720 kJ mol^{-1}

$\Delta H^{\ominus}_{f}$ of Ca^+Cl^- = (1. + 2.) + (3. + 4.) = 314 + 590 − 364 − 720
= −180 kJ mol^{-1}

5. $\Delta H^{\ominus}_{at}$ of Ca + (bond dissociation energy of Cl_2) = 435 kJ mol^{-1}
6. first ionization energy of Ca = 590 kJ mol^{-1}
7. second ionization energy of Ca = 1150 kJ mol^{-1}
8. 2 × (electron affinity of chlorine) = −728 kJ mol^{-1}
9. lattice energy of $Ca^{2+}(Cl^-)_2$ = −2237 kJ mol^{-1}

$\Delta H^{\ominus}_{f}$ of $CaCl_2$ = (5. + 6. + 7.) + (8. + 9.)
= 435 + 590 + 1150 − 728 − 2237 = −790 kJ mol^{-1}

More energy is required to form $Ca^{2+}_{(g)}$ and $Cl^{-}_{(g)}$ (1447 kJ mol^{-1}) than to form $Ca^{+}_{(g)}$ and $Cl^{-}_{(g)}$ (540 kJ mol^{-1}), but the large lattice energy of $CaCl_2$ makes the overall process of forming the +II state more favourable. A Ca^{2+} ion is smaller than a Ca^+ ion and has twice the charge; thus by the inverse square law, the attractive force between a Ca^{2+} ion and an anion is correspondingly greater than that of a Ca^+ ion. Hence the lattice energy of $CaCl_2$ is larger than the lattice energy of CaCl.

The energy terms involved in forming a Ca^{3+} salt can be considered in the same way, but in this case the third ionization energy of calcium proves prohibitively high. Even though the lattice energy of $CaCl_3$ is about three times greater than that of $CaCl_2$, it cannot offset the huge amount of energy needed to produce $Ca^{3+}_{(g)}$. The removal of the third electron from the inner shell of the calcium atom requires 4940 kJ mol^{-1}.

The group IIA elements therefore form compounds in which they are present as M^{2+} ions. There is more covalent character in group IIA compounds than their corresponding group IA compounds and as before, the degree of covalency is greatest for the first member of the group.

41.3 Occurrence, extraction and uses

Magnesium is the eighth most abundant element in the earth's crust (2.09%), and calcium the fifth most abundant (3.63%). Like the alkali metals, the alkaline earth metals are far too reactive to occur in the free state but they are widespread in the rocks and minerals that form the earth's surface. For example,

beryllium occurs in small quantities as beryl, $3BeO.Al_2O_3.6SiO_2$
magnesium is in dolomite $CaCO_3.MgCO_3$, in carnallite $KCl.MgCl_2.6H_2O$, and in magnesite $MgCO_3$
calcium is widespread as chalk or limestone $CaCO_3$, gypsum $CaSO_4.2H_2O$, and anhydrite $CaSO_4$
strontium and barium occur in small quantities as their carbonates

Of the group IIA elements, only magnesium is extracted in any quantity. Limestone deposits are used to produce lime (calcium oxide) in quantity, but the element itself is rarely needed.

Extraction of magnesium

1. By reduction using electrolysis
Either sea water or an aqueous solution of a magnesium mineral is the source. Magnesium hydroxide is precipitated using slaked lime.

$$Mg^{2+}_{(aq)} + 2OH^-_{(aq)} \rightarrow Mg(OH)_{2(s)}$$

The hydroxide is collected and converted to the chloride using hydrogen chloride gas

$$Mg(OH)_{2(s)} + 2HCl_{(g)} \rightarrow MgCl_{2(s)} + 2H_2O_{(g)}$$

The chloride is melted and electrolysed in a similar cell to the one used for the Downs process (see page 627).

$$Mg^{2+}_{(l)} + 2e^- \rightarrow Mg_{(l)}$$

2. By reduction using silicon
Magnesium ore is roasted in a furnace to make magnesium oxide.

$$MgCO_{3(s)} \rightarrow MgO_{(s)} + CO_{2(g)}$$

The oxide is then heated with a ferroalloy of silicon at very low pressures. Magnesium vapour is distilled out of the mixture because magnesium itself melts at 650°C.

$$2MgO_{(s)} + Si_{(s)} \xrightarrow[\text{low pressure}]{700°C} SiO_{2(s)} + 2Mg_{(g)}$$

Uses of the elements and their compounds

Magnesium
The metal is useful because of its lightness; it is often alloyed with aluminium to make aircraft and motor components. The reactivity of the metal is used to provide sacrificial protection to steel structures such as ships' hulls and oil drilling rigs at sea. When bolted on to the steel, it is preferentially corroded thus protecting the steel structure. The metal is also sometimes used in place of sodium as a reducing agent in the manufacture of titanium from its chloride.

Calcium compounds
Limestone $CaCO_3$ is the source of calcium oxide ('quicklime') which is formed by thermal decomposition at 1000°C, and of the hydroxide $Ca(OH)_2$ ('slaked lime') which is formed by the addition of water to CaO.

$$CaCO_{3(s)} \rightarrow CaO_{(s)} + CO_{2(g)}$$

$$CaO_{(s)} + H_2O_{(l)} \rightarrow Ca(OH)_{2(l)}$$

Calcium hydroxide or lime is used in agriculture to neutralize acid soils. When mixed with water and sand, it is also used to make mortar which hardens by the loss of water and the slow reaction with atmospheric carbon dioxide to form calcium carbonate.

$$Ca(OH)_{2(s)} + CO_{2(g)} \rightarrow CaCO_{3(s)} + H_2O_{(g)}$$

Limestone is also used in the extraction of metals from their ores as in the manufacture of iron in the blast furnace (page 776). The limestone decomposes to the oxide which converts the acidic earthy impurities in the ore to a slag that floats on the surface of the metal. Carbon dioxide produced from the limestone reacts with the coke in the charge to produce carbon monoxide which is the most important of the reductants in the process.

Limestone and the various natural forms of calcium sulphate are used in the manufacture of cement. Either limestone and clay are heated together in a furnace:

$$4CaCO_{3(s)} + \underset{\text{'clay'}}{Al_2Si_2O_{7(s)}} \rightarrow \underbrace{2CaSiO_{3(s)} + Ca_2Al_2O_{5(s)}}_{\text{cement}} + 4CO_{2(g)}$$

or calcium sulphate is heated with coke:

$$2CaSO_{4(s)} + C_{(s)} \rightarrow 2CaO_{(s)} + CO_{2(g)} + SO_{2(g)}$$

and the calcium oxide heated with clay:

$$4CaO_{(s)} + \underset{\text{'clay'}}{Al_2Si_2O_{7(s)}} \rightarrow \underbrace{2CaSiO_{3(s)} + Ca_2Al_2O_{5(s)}}_{\text{cement}}$$

Calcium hydroxide is used to form calcium hydrogensulphate (IV) which is important to the paper-making industry because it converts wood extract to cellulose. The hydroxide is reacted with sulphur dioxide.

$$Ca(OH)_{2(s)} + 2SO_{2(g)} \rightarrow Ca(HSO_3)_{2(s)}$$

Other uses of slaked lime include the manufacture of bleaching powder from chlorine.

Calcium sulphate in various forms is used to make plaster for the building trade. The hydrated form ('gypsum') is partially dehydrated at 125°C.

$$2CaSO_4.2H_2O_{(s)} \rightarrow \underset{\text{plaster}}{(CaSO_4)_2.H_2O_{(s)}} + 3H_2O_{(g)}$$

In a purer form, the product is known as Plaster of Paris which is used for plaster casts.

41.4 Similarities with group IA

Compound structure

The compounds of group IIA, like those of group IA, have considerable ionic character. The lattice arrangements of the compounds can be predicted by calculating the radius ratio of the ions involved. In general the lattice energies of group IIA compounds are much larger than the corresponding group IA compounds as the figures below show. This is a consequence of the larger charge on the cation.

formula	*cation radius* m × 10^{-10}	*cation charge*	*anion radius* m × 10^{-10}	*anion charge*	*lattice energy* kJ mol^{-1}
NaF	0.95	+1	1.36	−1	902
CaF_2	0.99	+2	1.36	−1	2609
Na_2O	0.95	+1	1.40	−2	2724
CaO	0.99	+2	1.40	−2	3536

Beryllium, like lithium, is the first member of its group and has unusual properties. Beryllium compounds have considerable covalent character because of the greater polarizing power of the small beryllium ion. They are discussed in more detail in Chapter 50.

Acid-base properties

The group IIA metals, oxides and hydrides, like their group IA counterparts, act as alkalis:

$$Ca_{(s)} + 2H_2O_{(l)} \rightarrow Ca^{2+}_{(aq)} + 2OH^-_{(aq)} + H_{2(g)}$$

$$CaH_{2(s)} + 2H_2O_{(l)} \rightarrow Ca^{2+}_{(aq)} + 2OH^-_{(aq)} + 2H_{2(g)}$$

$$CaO_{(s)} + H_2O_{(l)} \rightarrow Ca^{2+}_{(aq)} + 2OH^-_{(aq)}$$

However there are the following differences between the two groups.

The group IIA metals are less reactive than the corresponding group IA metals in the same period. Magnesium reacts rapidly with water only when it is heated in steam. Calcium reacts steadily with cold water.

The group IIA hydroxides are much less soluble than those of group IA. This results in the formation of precipitates in all the above reactions, and a lower pH than would be produced with a group IA metal hydroxide because the precipitation reaction removes hydroxyl ions from solution:

$$Ca^{2+}_{(aq)} + OH^-_{(aq)} \rightleftharpoons Ca(OH)_{2(s)} \qquad K_{sp} = 4 \times 10^{-6}\ mol^3\ dm^{-9}$$

The carbonates of group IIA are very insoluble in water. Consequently (unlike those of group IA) they do not yield the carbonate ion in solution and so they do not affect the pH of water. Like all other metal carbonates, they are basic and neutralize acids with the evolution of carbon dioxide:

$$MgCO_{3(s)} + 2H_3O^+_{(aq)} \rightarrow Mg^{2+}_{(aq)} + CO_{2(g)} + 3H_2O_{(l)}$$

The higher polarizing power of the group IIA cations leads to an increased degree of hydrolysis of the salts in solution. Beryllium salts, in particular, are appreciably acidic.

$$(H_2O)_3Be^{2+}\text{—}OH_2 + H_2O \rightleftharpoons [Be(H_2O)_3OH]^+_{(aq)} + H_3O^+_{(aq)}$$

Redox properties

The pure metals are powerful reductants like the group IA metals. Their reactions are very similar. Compare the standard electrode potentials of the group members.

group I $M^+_{(aq)} + e \rightleftharpoons M_{(s)}$	$E^\ominus$ volts	*group II* $M^{2+}_{(aq)} + 2e \rightleftharpoons M_{(s)}$	$E^\ominus$ volts
lithium	−3.04	beryllium	−1.85
sodium	−2.71	magnesium	−2.38
potassium	−2.92	calcium	−2.87
rubidium	−2.99	strontium	−2.89

41.5 Differences from group IA

There are three major areas in which the chemistries of groups IA and IIA differ:

1. The solubility of their compounds in water.
2. The thermal stability of their compounds.
3. Their tendency to form complexes.

Solubility

The increased charge of the cations leads to higher lattice energies, hydration energies and degrees of covalency in the group IIA compounds. These factors combine to lower the tendency of the ionic lattices to dissociate in the presence of water molecules. The oxides, hydroxides, carbonates, fluorides and sulphates of the alkaline earth metals are all insoluble or sparingly soluble (except those of beryllium and $MgSO_4$).

Thermal stability

Group IIA compounds containing oxy-anions are less stable than the corresponding group IA compounds. For example, magnesium hydroxide and carbonate both decompose on heating:

$$Mg(OH)_{2(s)} \xrightarrow{\text{heat}} MgO_{(s)} + H_2O_{(l)}$$

$$MgCO_{3(s)} \xrightarrow{\text{heat}} MgO_{(s)} + CO_{2(g)}$$

while sodium hydroxide and carbonate do not decompose but melt. As the figures below indicate, the thermal stability of the compounds increases down the group:

compound	$BeCO_3$	$MgCO_3$	$CaCO_3$	$SrCO_3$	$BaCO_3$
size of cation m × 10^{-10}	0.31	0.65	0.99	1.13	1.35
decomposition temp. °C.	~ 100	540	900	1290	1360

In both cases, the reason for the change in stability is the different polarizing powers of the cations. Smaller and highly charged cations distort the charge clouds of an oxy-anion considerably causing the anion to break up:

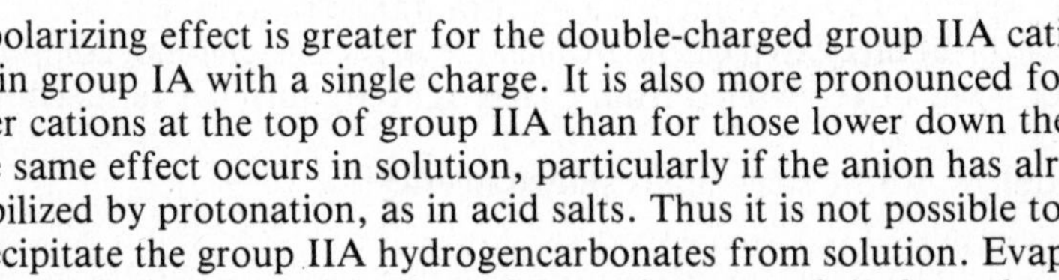

This polarizing effect is greater for the double-charged group IIA cations than those in group IA with a single charge. It is also more pronounced for the smaller cations at the top of group IIA than for those lower down the group.

The same effect occurs in solution, particularly if the anion has already been destabilized by protonation, as in acid salts. Thus it is not possible to crystallize or precipitate the group IIA hydrogencarbonates from solution. Evaporation or boiling results in the precipitation of the carbonate and the loss of carbon dioxide, as shown overleaf.

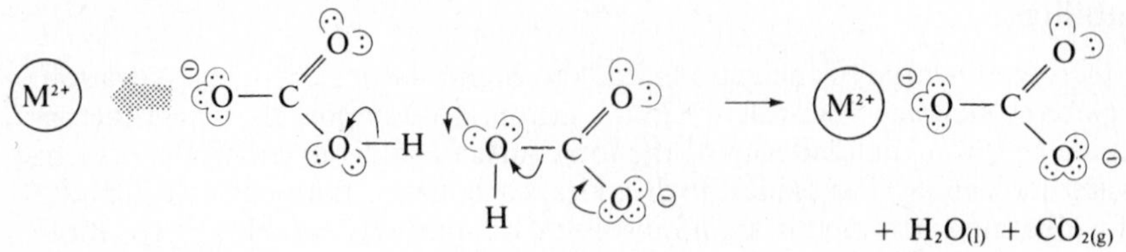

$$M^{2+} \Leftarrow {}^{\ominus}O{-}C({=}O){-}O{-}H \;\; H{-}O{-}C({=}O){-}O^{\ominus} \longrightarrow M^{2+} \; {}^{\ominus}O{-}C({=}O){-}O^{\ominus} + H_2O_{(l)} + CO_{2(g)}$$

Solutions of the group IIA hydrogencarbonates are responsible for temporary hardness in water. Weakly acidic rainwater (dilute carbonic acid) dissolves small amounts of the limestone rock that it trickles over. When the water is boiled, the reverse process takes place.

$$\underset{\text{limestone}}{CaCO_{3(s)}} + \underbrace{HCO^-_{3(aq)} + H_3O^+_{(aq)}}_{\text{rain}} \rightarrow \underbrace{Ca^{2+}_{(aq)} + 2HCO^-_{3(aq)}}_{\text{hard water}} + H_2O_{(l)}$$

$$\downarrow \text{boil}$$

$$CaCO_{3(s)} + CO_{2(g)} + H_2O_{(l)}$$

Complexing properties

The alkaline earth metal cations have a greater tendency to form complexes than alkali metal cations. They do not have an extensive co-ordination chemistry but there are one or two well-characterized complexes. The increased tendency toward complex formation can once again be explained in terms of the higher charge and smaller size of the group IIA cations. More energy is given out when a ligand bonds to the smaller, more highly-charged cation. Beryllium forms a particularly wide range of complexes for this reason. Some common examples of magnesium and calcium complexes are listed below.

1. $[Ca(NH_3)_8]^{2+}$ present in ammoniacal solutions of $CaCl_2$.
2. edta complexes of Mg^{2+} or Ca^{2+}; edta is ethylene diamine tetra acetic acid and its anion is a hexadentate ligand (see p. 303). With a standard solution of this acid, it is possible to analyse the concentration of a group IIA salt solution by using an indicator (called eriochrome black-T) which forms a weak complex with magnesium or calcium ions. The colour of this complex is red, but the uncomplexed indicator is blue. By adding the indicator to a volume of the solution to be analysed, a red colour is given. Standard edta is now run into the solution until it changes to blue, when all the Mg^{2+} or Ca^{2+} is complexed by edta. (mole ratio 1:1).
3. Grignard reagents (see page 528) which are organomagnesium compounds, e.g. Br—Mg—CH_3 formed from CH_3Br and Mg turnings refluxed in dry ether.
4. chlorophyll, whose structure is shown overleaf:

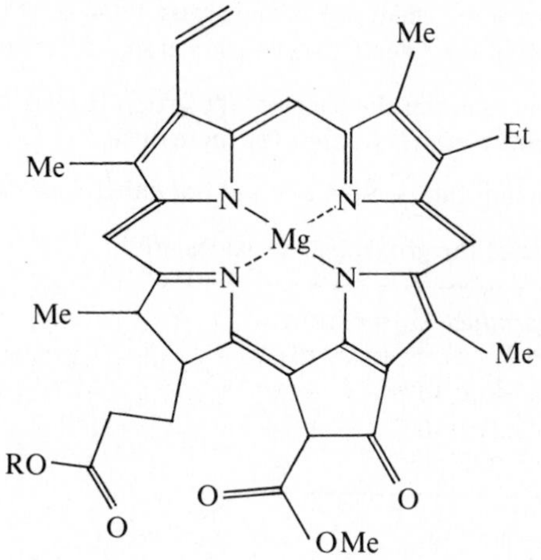

chlorophyll-a

Questions

1. Explain the meaning of: a) atomic radius b) oxidation state c) inverse square law d) ionic character e) co-ordination number f) precipitate g) solubility h) complex.

2.

element	*first ionization energy*	*second ionization energy*
sodium	$+444$ kJ mol^{-1}	$+4560$ kJ mol^{-1}
magnesium	$+736$ kJ mol^{-1}	$+1450$ kJ mol^{-1}

Explain why the first ionization energy of sodium is less than that of magnesium, but the second ionization energy of sodium is greater than that of magnesium.

3. The solubilities of some of the group (II) metal sulphates are given below:

sulphate	*solubility at 10°C in g/100 g*
$MgSO_4$	30.9
$CaSO_4$	0.192
$SrSO_4$	0.104
$BaSO_4$	0.00265

Account for the trend these figures show.

4. Explain why a solution of magnesium sulphate has a pH slightly less than 7 while a suspension of magnesium carbonate has a pH well above 7.

5. What similarities are there between the chemistry of the group I and of the group II elements? Why do these similarities exist when the elements are in different groups?

6. Point out the important differences in the chemistry of the group II metals compared with that of the group I metals. Account for these differences.

7. Explain why the formula of calcium fluoride is CaF_2 and not either CaF or CaF_3.

8. The solubility products for three of the group II hydroxides are:

hydroxide	*solubility product* K_{sp} $mol^3\ dm^{-9}$
$Ca(OH)_2$	5.5×10^{-6}
$Sr(OH)_2$	3.2×10^{-4}
$Ba(OH)_2$	5.0×10^{-3}

a) Calculate their solubilities in mol dm^{-3}.
b) How does the solubility change going down the group?
c) Account for this trend.

9. Compare the reactions of the following with water: a) $Ca_{(s)}$ b) $CaCl_{2(s)}$ c) $CaO_{(s)}$ d) $Ca(OH)_{2(s)}$ e) $CaH_{2(s)}$ f) $Ca(CH_3COO)_{2(s)}$.

10. Write an essay explaining the importance of limestone as a raw material.

Chapter 42

Boron and aluminium: group IIIB

42.1 Structure of the elements

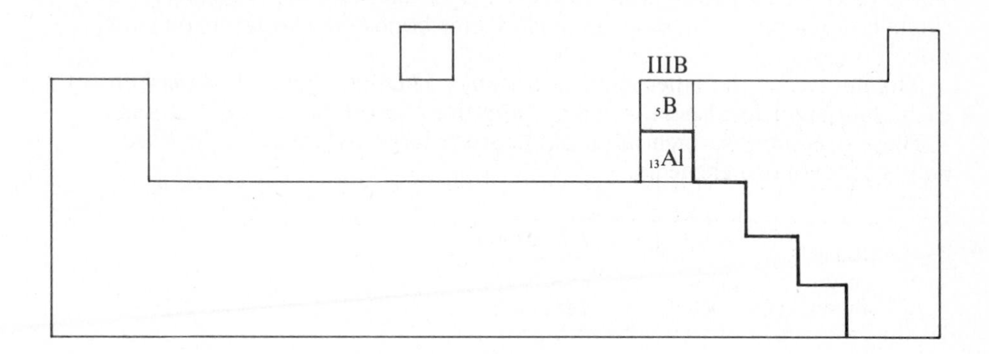

element	*electronic configuration*	*electro-negativity*	*atomic radius* (m × 10^{-10})	*ionic* (M^{3+}) *radius* (m × 10^{-10})
boron	[He] $2s^2$ $2p^1$	2.0	0.80	0.20
aluminium	[Ne] $3s^2$ $3p^1$	1.5	1.25	0.50

Atomic structure

Boron and aluminium atoms have three outer shell electrons each and are comparatively small in size. Their core charge is three and the outer-shell electrons are quite tightly held to the nucleus. The sum of the first three ionization energies is 6879 kJ mol^{-1} for boron and 5137 kJ mol^{-1} for aluminium.

Physical structure

The energy required for the delocalization of boron's outer-shell electrons is greater than that given out on the formation of a metallic lattice of close-packed B^{3+} ions. This gives boron the physical properties of a non-metal. Even though each boron atom can only reach six electrons in its valence shell by the formation of covalent bonds, crystalline boron is a macromolecular solid of high melting point and great hardness. It is extremely inert chemically and is difficult to prepare in a pure state. The structure of boron is complicated. It is thought to be made up from B_{12} molecular units each having the shape of an icosahedron.

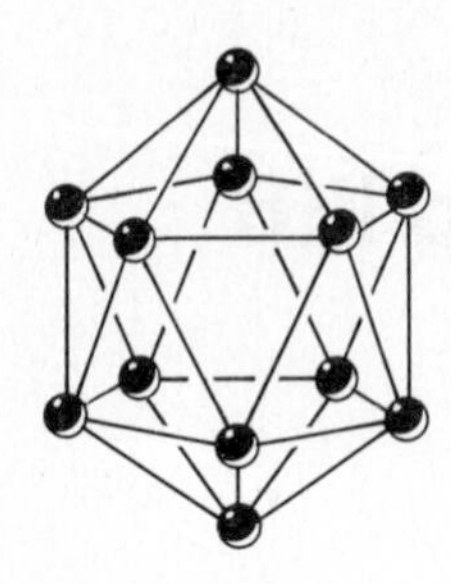

B_{12} icosahedron

These units are packed in layers and each layer alternates between a layer of single boron atoms. The layer of single atoms bonds the two layers on either side of it together.

Aluminium, on the other hand, is metallic. The three outer-shell electrons of each atom are delocalized throughout a lattice of close-packed Al^{3+} cations.

These structural differences account for the large difference in the fixed points of these two elements.

	m.p. (°C)	*b.p.* (°C)
boron	2030	3930
aluminium	660	2470

42.2 Compounds formed by the elements

The elements form compounds mostly with non-metals, and +III is the only stable oxidation state. Although it is sometimes useful to describe these compounds using the ionic model, the tiny size and high charge of the cations produce high degrees of covalency. In particular, boron compounds show very little ionic character at all. If these compounds are treated as molecular, the molecules have valence shells that are not full. This leads to a number of characteristic properties which are described later.

42.3 Occurrence, extraction and uses

Aluminium is the third most abundant element on earth and is the most abundant metal. It makes up 8.13% of the earth's crust.

Boron is found only in trace quantities and is not extracted by a large scale process.

Extraction of aluminium

Aluminium, like other reactive metals is extracted by electrolysis but first the ore must be purified. The most important ore of aluminium is bauxite,

$Al_2O_3.2H_2O$. This is purified by dissolving it in concentrated alkali:

$$Al_2O_3.2H_2O_{(s)} + 2OH^-_{(aq)} + H_2O \rightarrow 2Al(OH)^-_{4(aq)}$$

Impurities are then filtered off and the resulting filtrate is made to precipitate out by seeding it with aluminium hydroxide.

$$Al(OH)^-_{4(aq)} \rightarrow Al(OH)_{3(s)} + OH^-_{(aq)}$$

The solid aluminium hydroxide is now heated to yield pure aluminium oxide:

$$2Al(OH)_{3(s)} \rightarrow Al_2O_{3(s)} + 3H_2O_{(l)}$$

Pure aluminium oxide has a very high melting point, 2045°C, and to enable electrolysis to take place at a more manageable temperature, cryolite, Na_3AlF_6, is added to the aluminium oxide. This forms a eutectic mixture with a melting point of 950°C that can be electrolysed in the Hall cell.

The Hall cell

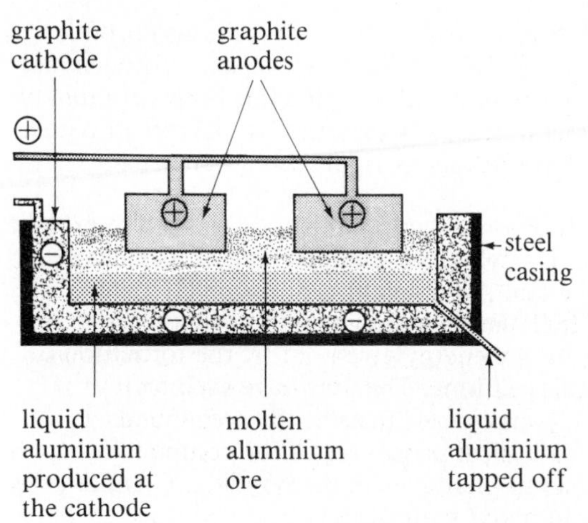

raw materials: Al_2O_3/Na_3AlF_6
conditions: T = 950°C
power = 4.5 V, 100 000 amps
products: Aluminium and oxygen (which combines with the carbon anode)
reactions: $Al^{3+}_{(l)} + 3e^- \rightarrow Al_{(l)}$
$2O^{2-}_{(l)} \rightarrow O_{2(g)} + 4e^-$
$C_{(s)} + O_{2(g)} \rightarrow CO_{2(g)}$

At the cathode, liquid aluminium is produced which is 99.5 – 99.8% pure. This collects at the bottom of the cell and is tapped off. Oxygen is produced at the graphite anodes and, at the cell temperature, reacts with them producing carbon dioxide. As the anodes are burnt away, more graphite in the form of a paste is added to the top of them and bakes hard. The cell is maintained at its operating temperature by the heating effect of the enormous current.

Uses of boron and aluminium

Boron

Boron itself has few uses although a form of boron is used as a moderator in some nuclear reactors. The oxy-compounds of boron (borates) are used in making glass and glass fibre. Boron nitride, which is isoelectronic with carbon, exists in two forms like carbon. The graphite form is used as a lubricant, and the very hard diamond form is used for tipping machine tools.

Aluminium
Aluminium is widely used as a structural metal. It is light, corrosion-resistant and strong when suitably alloyed and heat treated. It is used in the manufacture of cars, aircraft, ships and many other articles.

Aluminium also conducts heat and electricity well. Saucepans and high voltage transmission lines are manufactured from the metal.

Aluminium oxide (alumina) has a number of uses as well:

1. as 'corundum', a grinding material, made by heating amorphous alumina to about 2000°C
2. when dissolved in alkali, as a fixing agent in the dyeing industry
3. when finely powdered, as the stationary phase in chromatographic columns.

42.4 Compound structure

Boron and aluminium combine with most non-metals. Amorphous boron is very reactive and combines exothermically with chlorine, oxygen, nitrogen and sulphur. It also forms an unusual compound with magnesium, MgB_2. Aluminium appears less reactive because the metal is usually protected by a layer of oxide. When the layer is penetrated, the metal reacts as vigorously as amorphous boron.

The structure of the aluminium and boron compounds is often hard to explain in terms of only one bonding model. Each atom uses all three of its valence electrons for bonding and this results in the formation of boron (III) and aluminium (III) compounds. The high energy required to remove the three electrons from the atoms is offset by the energy given out by the formation of strong bonds to the small highly charged ions. The ions have such high polarizing powers, however, that it is often best to treat the compounds as molecular or macromolecular. The physical properties of the compounds certainly suggest that there is little ionic character in the bonding. Compare the melting points of the oxides, chlorides and hydrides:

compound	*boron*	*aluminium*	*sodium*	*silicon*	*sulphur*
oxide	460°C	2045°C	1275°C	1610°C	−73°C
chloride	−107°C	183°C	801°C	−70°C	−78°C
hydride	−166°C	decomposes (~ 150°C)	decomposes (~ 800°C)	−185°C	−86°C

The structure of aluminium chloride is discussed in some detail on page 91. Both oxides are macromolecular in character but yet molten alumina conducts electricity. It is unlikely that the melt contains simple Al^{3+} and O^{2-} ions; there is some evidence that suggests the following dissociation takes place:

$$Al_2O_{3(s)} \rightleftharpoons Al^{3+}_{(l)} + AlO^{3-}_{3(l)}$$

Boron forms a wide range of compounds with hydrogen e.g. B_9H_{15}, $B_{10}H_{14}$, $B_{20}H_{16}$. These compounds, known as boranes, have complex molecular structures.

Boranes

The simplest borane is diborane (B_2H_6), and was first prepared by the action of dilute acid on the compound formed when amorphous boron is fused with magnesium. This reaction produces a mixture of air-sensitive and inflammable compounds. Diborane is now usually prepared from lithium tetrahydrido-aluminate and boron chloride:

$$4BCl_3 + 3LiAlH_4 \xrightarrow[\text{ether}]{} 3LiCl + 3AlCl_3 + 2B_2H_{6(g)}$$

The molecular geometry of diborane is shown below:

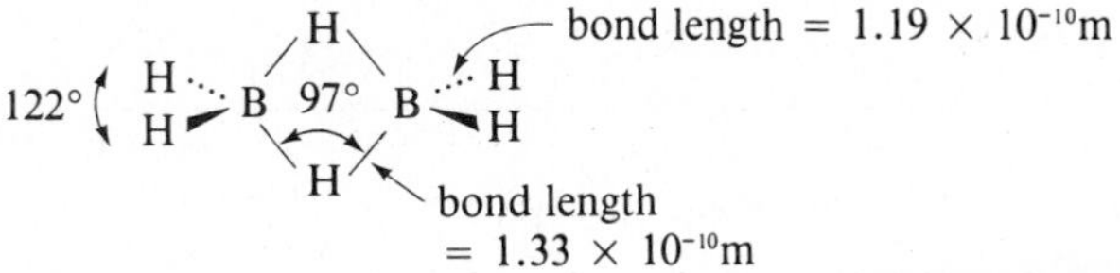

Its structure is described in terms of the ionic model ($2B^{3+}$, $6H^-$) on page 619. A more rewarding approach is to consider the unit to be molecular and discuss the formation of the molecular orbitals. The distorted tetrahedral arrangement of hydrogen atoms around each boron atom suggests that the boron atoms are sp^3-hybridized. The longer length of the bridging hydrogen-boron bonds suggests that these bonds are also electron-deficient. These observations can be explained by assuming that they form as a result of the overlap of the hydrogen 1s-orbital with sp^3-orbitals from each of two boron atoms. Each boron atom has only one electron available to be put into one of the two bridging units. This means that the bridging bonds are three-centred but have only two electrons each.

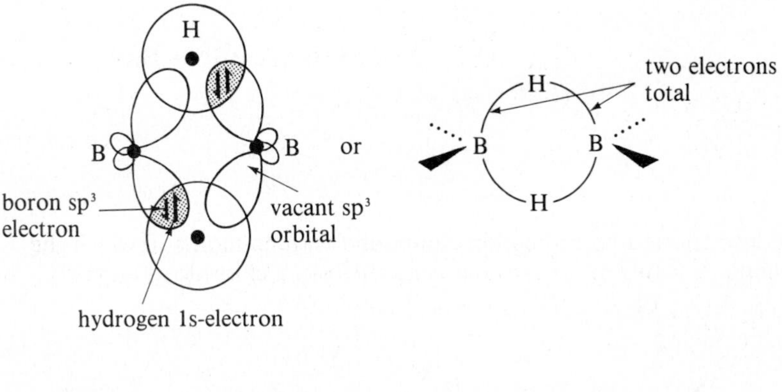

The effect can be described as an electron-deficient, multi-centred bond (sometimes known as a 'banana-bond'!)

Boron atoms themselves seem capable of forming multi-centred bonds without the presence of hydrogen atoms. The B_{12} icosahedral units making up the lattice of crystalline boron undoubtedly form a number of multi-centred bonds. A number of other boranes have been prepared showing these bonding characteristics and some electron-deficient chlorides have also been isolated. Two examples are shown below.

tetraborane B_4H_{10} 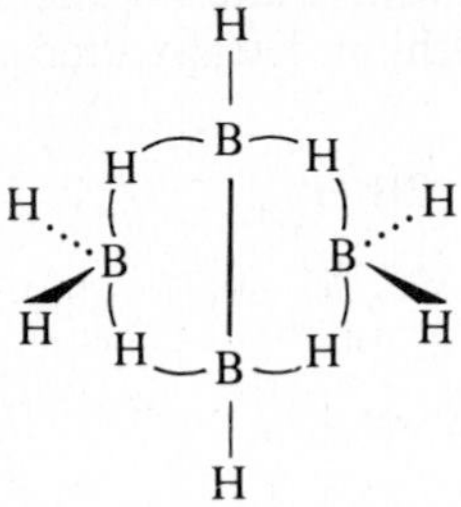	*tetraboron tetrachloride* B_4Cl_4 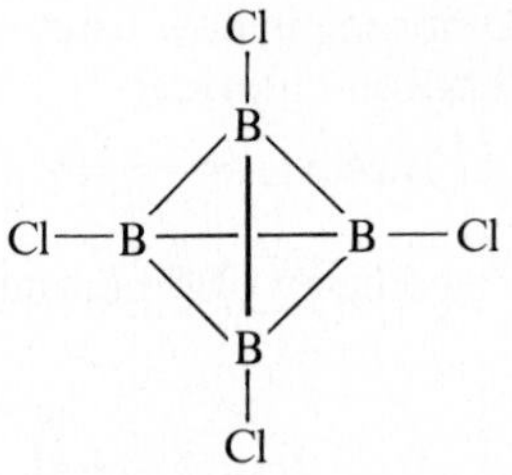

42.5 Acid-base properties

Acidity

The tiny highly charged cations of the IIIB elements act as powerful electron-pair acceptors. They are therefore acids in the Lewis sense (see page 354) and are often acidic in water as a result of accepting electron pairs from water molecules:

$$BCl_{3(l)} + 6H_2O_{(l)} \rightarrow B(OH)_{3(aq)} + 3H_3O^+_{(aq)} + 3Cl^-_{(aq)} \qquad pH \approx 1$$

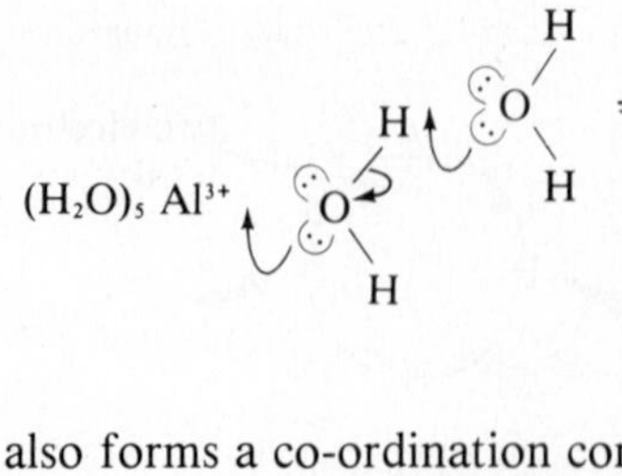

$$(H_2O)_5\,Al^{3+} + H_2O + H_2O \rightleftharpoons [(H_2O)_5Al(OH)]^{2+} + H_3O^+_{(aq)} \qquad pH \approx 3$$

BCl_3 also forms a co-ordination compound with ammonia in which the ammonia is acting as a Lewis base and the boron chloride as an acid:

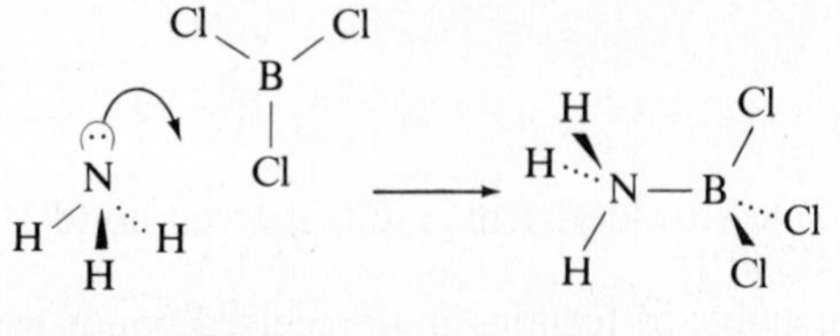

$$NH_3 + BCl_3 \longrightarrow H_3N{-}BCl_3$$

The oxides and hydroxides of aluminium and boron both have acidic properties and dissolve in alkali to give borate and aluminate salts.

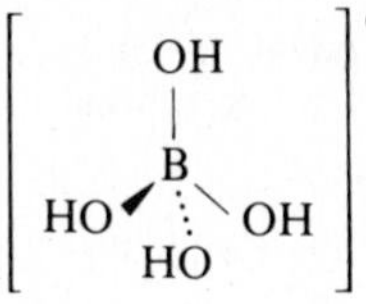

borate ion

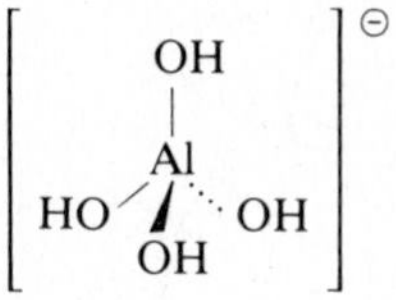

aluminate ion

In particular, boron hydroxide acts as a weak monobasic acid in solution and is usually called *orthoboric acid*. Unlike aluminium hydroxide, it is soluble in water and is acidic as a result of interaction with the lone pairs of the water molecules:

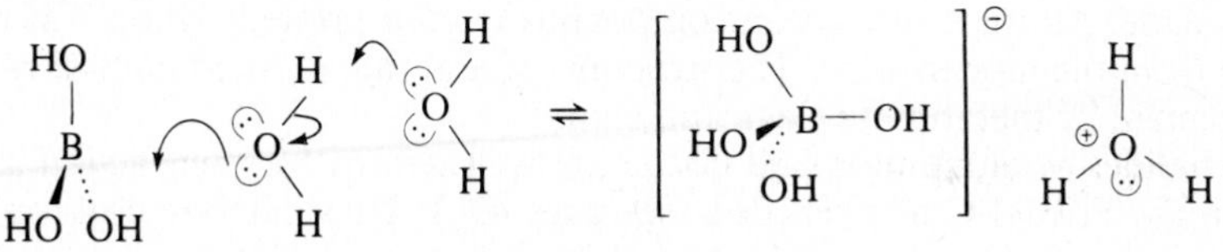

The hydroxy-borate anion, $[B(OH)_4]^-$, is rarely found in the crystal lattice of solid structures. It is so readily dehydrated that water molecules are eliminated between adjacent anions producing linked borate ions called 'metaborates':

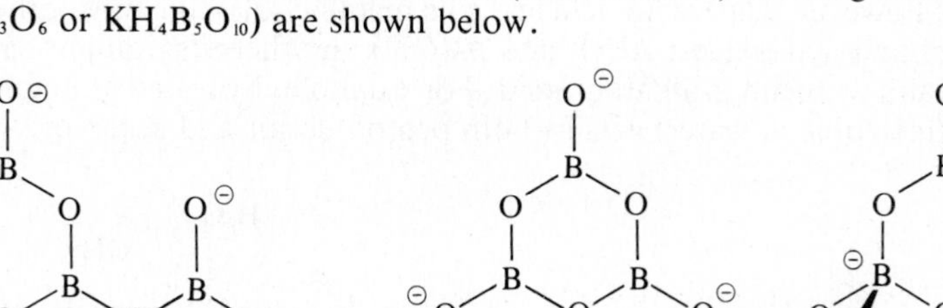

$+ H_2O$ etc.

Some examples of chain metaborates (as in CaB_2O_4) and ring metaborates (as in $K_3B_3O_6$ or $KH_4B_5O_{10}$) are shown below.

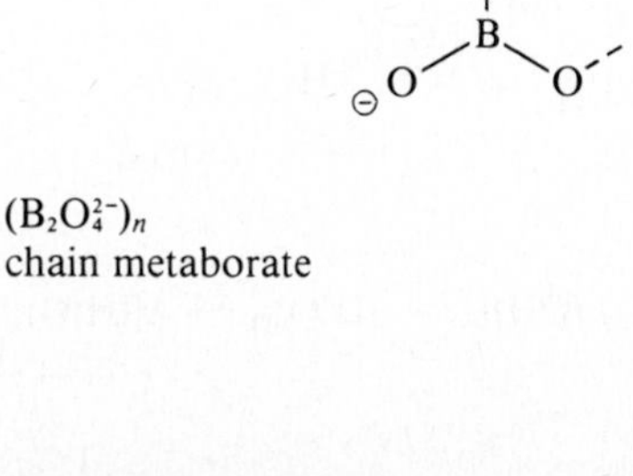

$(B_2O_4^{2-})_n$
chain metaborate

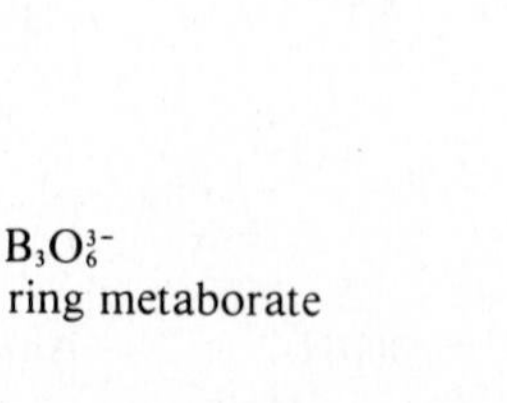

$B_3O_6^{3-}$
ring metaborate

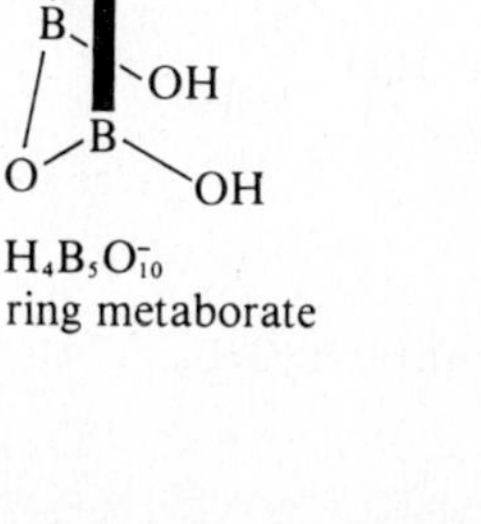

$H_4B_5O_{10}^-$
ring metaborate

The crystalline structure of orthoboric acid itself is rather interesting. There is extensive hydrogen bonding linking the molecular units together.

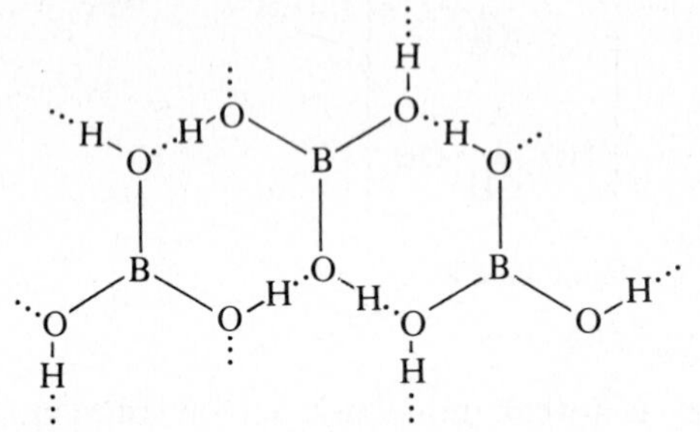

— = covalent bond
··· = hydrogen bond

The structure is easily dehydrated at about 100°C to a form similar to the chain metaborate above. Its empirical formula is HBO_2 and it is called *metaboric acid*. Heating to red heat brings about complete dehydration and the production of boron oxide. All these oxy-boron compounds tend to produce 'glasses' if rapidly cooled from the molten state. The structure of a borate glass contains a random arrangement of metaborate rings and chains.

The halides of aluminium and boron are well known for their action as Lewis acids in the Friedel-Crafts reaction (see page 490). They catalyse electrophilic substitution on an aromatic ring; e.g. chloromethane reacts with benzene in the presence of aluminium chloride.

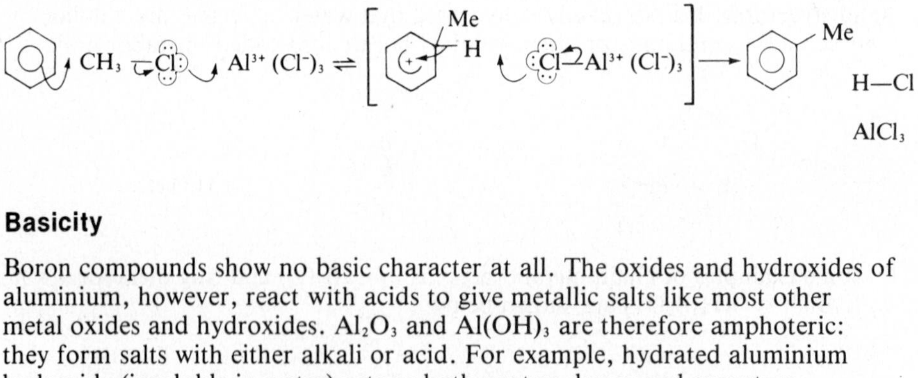

CH_3 Cl $Al^{3+}(Cl^-)_3$ ⇌ [Me H Cl $Al^{3+}(Cl^-)_3$] → Me H—Cl $AlCl_3$

Basicity

Boron compounds show no basic character at all. The oxides and hydroxides of aluminium, however, react with acids to give metallic salts like most other metal oxides and hydroxides. Al_2O_3 and $Al(OH)_3$ are therefore amphoteric: they form salts with either alkali or acid. For example, hydrated aluminium hydroxide (insoluble in water) acts as both proton donor and acceptor:

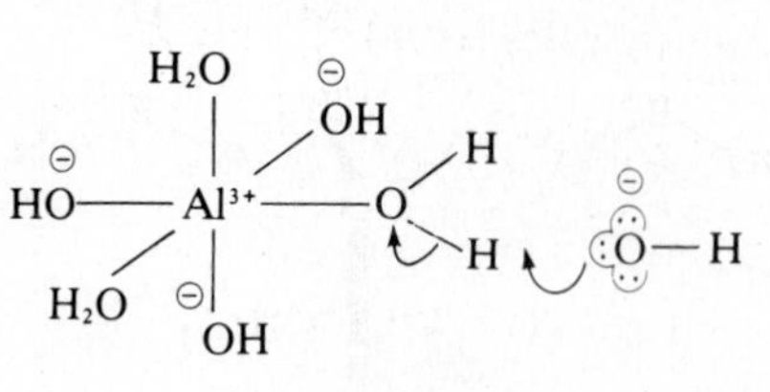

as acid

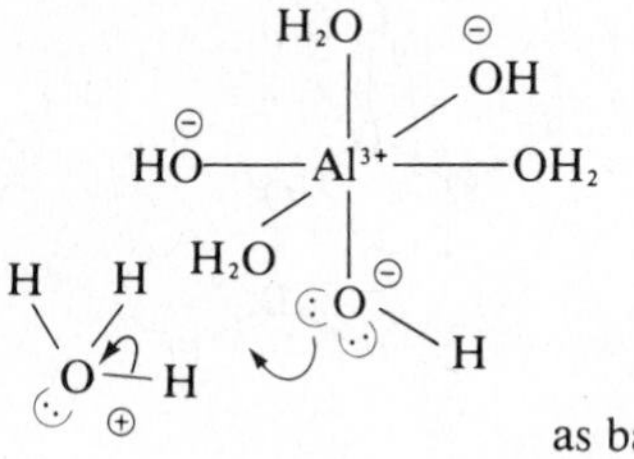

as base

$$Al(H_2O)_3(OH)_{3(s)} + HO^-_{(aq)} \rightarrow Al(OH)_4^- + 3H_2O_{(l)}$$

$$Al(H_2O)_3(OH)_{3(s)} + 3H_3O^+_{(aq)} \rightarrow Al(H_2O)_6^{3+} + 3H_2O_{(l)}$$

Aluminium has a tendency to form double sulphate salts with univalent metals or ammonium compounds. These are given the general name '*alums*'. Other metals that form trivalent cations share this tendency, e.g. chromium and iron. The structures of these alums are similar and are only stable when the univalent cations in the alum lattice are large, e.g. K^+, NH_4^+. For example, 'potash' alum crystallizes as large octahedral crystals when a solution of potassium sulphate and aluminium sulphate is evaporated: $KAl(SO_4)_2.12H_2O$, potash alum.

The general formula for an alum is $M_a^+ M_b^{3+} (SO_4^{2-})_2.12H_2O$, e.g. ammonium iron (III) alum, $NH_4Fe(SO_4)_2.12H_2O$.

42.6 Redox properties

The elements as reductants

Amorphous boron and metallic aluminium are both quite strong reductants. The oxide layer on aluminium gives a misleading impression of its strength as a reductant. Once this has been penetrated, the metal reacts vigorously with most non-metals and with the compounds of less reactive metals. It is less reactive than the groups IA and IIA metals but its standard electrode potential is still quite large and negative. Compare the values below:

$$Al^{3+}_{(aq)} + 3e^- \rightleftharpoons Al_{(s)} \qquad E^\ominus = -1.66 \text{ volts}$$

$$Mg^{2+}_{(aq)} + 2e^- \rightleftharpoons Mg_{(s)} \qquad E^\ominus = -2.37 \text{ volts}$$

Amorphous boron or aluminium reduce oxygen, sulphur or nitrogen.

$$4Al_{(s)} + 3O_{2(g)} \xrightarrow{\text{heat}} 2Al_2O_{3(s)}$$

$$16Al_{(s)} + 3S_{8(s)} \xrightarrow{\text{heat}} 8Al_2S_{3(s)}$$

$$2Al_{(s)} + N_{2(g)} \xrightarrow{\text{heat}} 2AlN_{(s)}$$

The presence of an oxide layer inhibits reactions between aluminium and solutions containing the ions of less reactive metals. However, if the oxide layer is scraped off in the solution, displacement reactions occur:

$$2Al_{(s)} + 3Cu^{2+}_{(aq)} \rightarrow 2Al^{3+}_{(aq)} + 3Cu_{(s)}$$

In the solid state, the oxide layer is sometimes removed under violent conditions, and then the true reactivity of aluminium is revealed. This is achieved in the 'thermite' reaction where a mixture of aluminium powder and iron oxide is activated using powdered magnesium:

$$2Al_{(s)} + Fe_2O_{3(s)} \rightarrow Al_2O_{3(s)} + 2Fe_{(l)}$$

This process yields molten iron and has been used to join railway lines.

The hydrides and hydrido-complexes as reductants

These compounds of boron and aluminium are powerful and useful reductants. They appear to act as a source of hydride ions and therefore provide a selective means of reducing polar groups such as carbonyls, nitriles and halides. They do not reduce non-polar groups such as alkenes and alkynes. Their action in organic chemistry is discussed on page 478.

They also reduce any inorganic polar group, e.g. NH_3. An interesting set of compounds is produced as a result of the reduction of ammonia by diborane B_2H_6. When the temperature is kept low and an excess of ammonia is used, a complex hydrido-borate forms with the following structure.

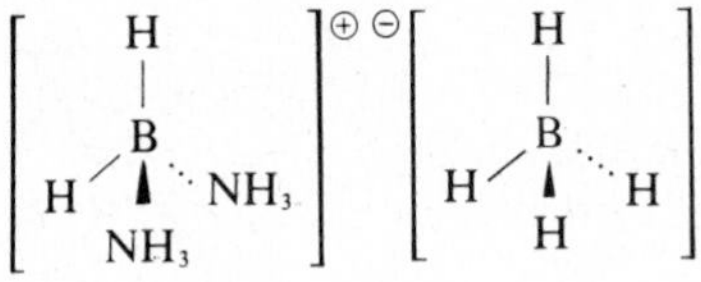

$$[BH_2(NH_3)_2]^{\oplus}[BH_4]^{\ominus}$$

When this complex is raised to a high temperature, it decomposes to give boron nitride $(BN)_n$ and hydrogen.

If the reaction between ammonia and diborane is carried out at a high temperature in the mole ratio $NH_3:B_2H_6 = 2:1$, a compound forms called borazole $B_3N_3H_6$. Both borazole and boron nitride have exact analogues in the chemistry of carbon. Boron nitride is isoelectronic with graphite and borazole is isoelectronic with benzene.

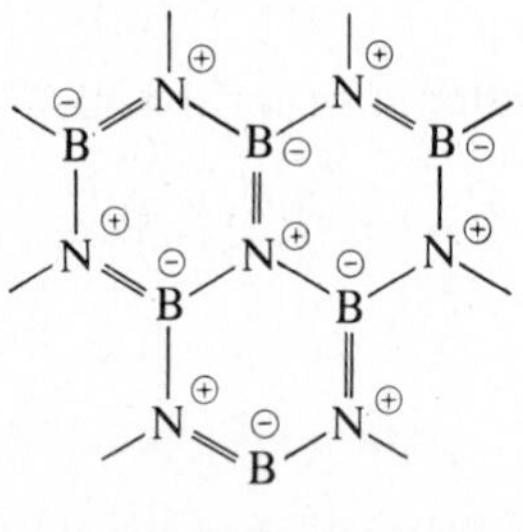

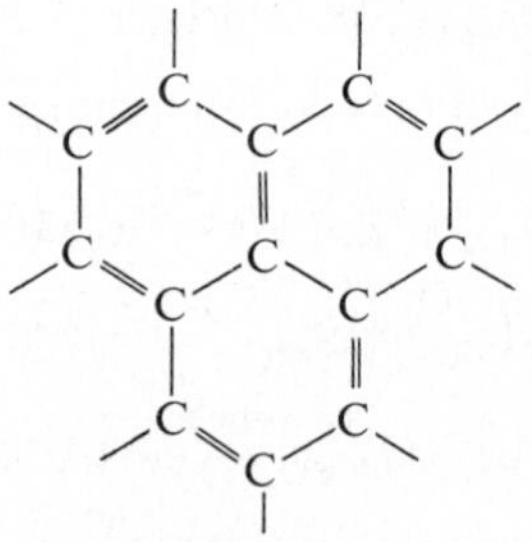

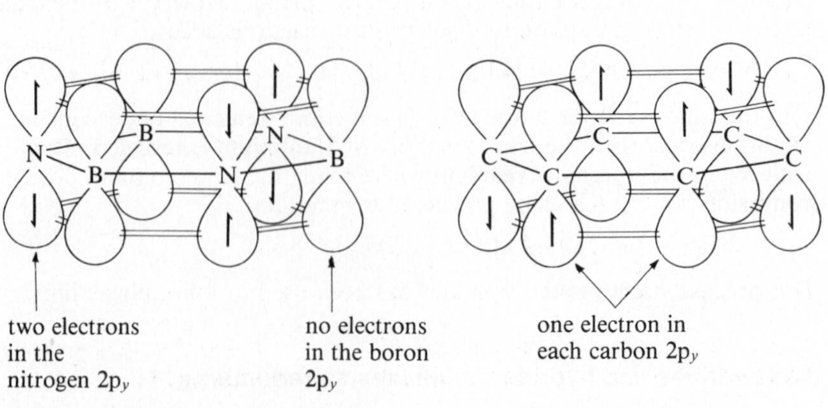

two electrons in the nitrogen $2p_y$

no electrons in the boron $2p_y$

one electron in each carbon $2p_y$

The charge separation in a borazole molecule gives it different properties from those of benzene. Borazole readily undergoes electrophilic addition.

42.7 Complexing properties

Both boron and aluminium have extensive co-ordination chemistries. The small, highly charged ions form strong bonds with a number of different ligands and there is a high degree of covalency in these bonds. The hydroxy-complexes, $Al(OH)_4^-$ and $B(OH)_4^-$, have already been described. Other important complexes are formed with:

1. halide ligands
2. hydride ligands
3. organic ligands.

Halide complexes

BF_4^- and BCl_4^- are both well characterized. Tetrafluoro-borate ions are readily prepared from orthoboric acid and hydrofluoric acid. The resulting strong acid is often called fluoroboric acid.

$$B(OH)_{3(s)} + 4HF_{(aq)} \rightarrow BF^-_{4(aq)} + H_3O^+_{(aq)} + 3H_2O_{(l)}$$

AlF_6^- and $AlCl_4^-$ both exist but the tetrachloro complex is very readily hydrolysed and so is rather unstable. Hexafluoro ions are present in cryolite, Na_3AlF_6, which is made by fusing aluminium fluoride with sodium hydroxide

$$6NaOH + 4AlF_3 \rightarrow 2Na_3AlF_6 + Al_2O_3 + 3H_2O$$

Hydride complexes

Tetrahydrido complex ions can be prepared by the action of group I hydrides on the halides of boron or aluminium. Hydrido-borate complexes are often made from the methyl ester of orthoboric acid instead. In both cases, hydride ligands replace those around the central boron or aluminium ions under anhydrous conditions

$$B(OMe)_3 + 4NaH \xrightarrow[\text{ether}]{} NaBH_4 + 3NaOMe$$

$$AlCl_3 + 4LiH \xrightarrow[\text{ether}]{} LiAlH_4 + 3LiCl$$

Organic complexes

The degree of covalency in these compounds is so high that they are best considered as covalent. Apart from the complexes formed by bidentate ligands such as ethanedioate ions, the most important class of complexes is the aluminium alkyls. Triethyl aluminium resembles aluminium chloride in its dimeric structure.

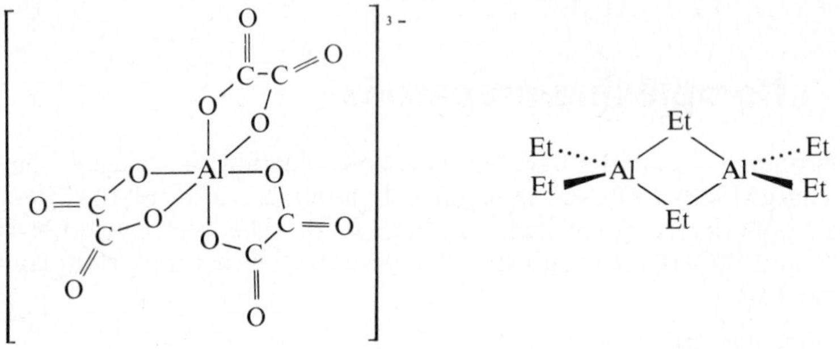

$Al^{3+} (C_2O_4^{2-})_3$ $Al_2(C_2H_5)_6$

Triethyl aluminium is used in the Ziegler-Natta process for polymerizing ethene (see page 386).

Questions

1. Explain the meaning of: a) covalent bond b) macromolecular c) isoelectronic d) ionic character e) Lewis acid f) co-ordination compound g) monobasic acid h) amphoteric.

2. Discuss the structure and bonding of: a) B_2H_6 b) BF_3 c) AlF_6^{3-} d) AlH_4^- e) $(BN)_x$ f) $B_3N_3H_6$.

3. Describe the extraction of aluminium from its ore, emphasizing the important chemical principles involved in the process.

4. Although they are in the same group, boron is a non-metal while aluminium is a metal. Taking their oxides and hydrides as examples, describe how their chemistries differ and reflect their non-metallic or metallic character.

5. Account for the following observations:
   a) Aluminium fluoride boils at 1270°C but aluminium chloride sublimes at 180°C.
   b) Molten aluminium chloride is a poor conductor but an aqueous solution of aluminium chloride conducts well.
   c) Boron melts at 2030°C but aluminium melts at 660°C.

6. Calculate the pH of a 0.1 mol dm^{-3} solution of aluminium nitrate if K_a for $Al(H_2O)_6^{3+}$ is 1.4×10^{-5} mol dm^{-3}.

7. If the standard heat of formation in kJ mol^{-1} for Al_2O_3 is -1669 and for Fe_2O_3 is -822, calculate the heat of reaction for:

$$2Al_{(s)} + Fe_2O_{3(s)} \rightarrow Al_2O_{3(s)} + 2Fe_{(s)}$$

8. The standard electrode potential for $Al^{3+}_{(aq)}/Al_{(s)}$ is -1.66 volts. From this information, comment on the use of aluminium a) as a reductant b) as a structural material.

9. Under what conditions does aluminium react with the following?
   a) dilute sulphuric acid
   b) dilute sodium hydroxide
   c) chlorine gas.

10. Compare the chemistry of beryllium with that of aluminium.

Chapter 43

Carbon to lead: group IVB

43.1 Structure of the elements

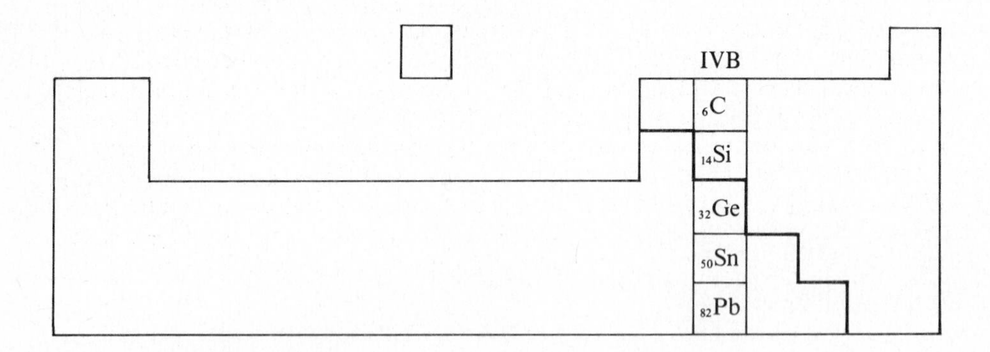

element	*electronic configuration*	*electro-negativity*	*atomic radius* ($m \times 10^{-10}$)	*stable oxidation states*
carbon	[He] $2s^2\ 2p^2$	2.5	0.77	II, **IV**
silicon	[Ne] $3s^2\ 3p^2$	1.8	1.17	**IV**
germanium	[Ar] $3d^{10}\ 4s^2\ 4p^2$	1.8	1.22	II, **IV**
tin	[Kr] $4d^{10}\ 5s^2\ 5p^2$	1.8	1.40	II, **IV**
lead	[Xe] $4f^{14}\ 5d^{10}\ 6s^2\ 6p^2$	1.8	1.54	**II**, IV

Atomic structure

A group IVB atom has four electrons in its outer shell and a full inner d-subshell (where applicable). The shielding of the nuclear charge by the inner-shell electrons is quite efficient for the first two members of the group, but it becomes increasingly less efficient for the lower members because of the increasing number of d- (and f-) electrons in their structure: d-electrons and f-electrons have poorer screening abilities than s- or p-electrons. The ineffectiveness of the shielding of the outer shell is felt more by the outer s-electrons (in their spherically symmetric orbitals) than by the angular p-electrons. This has the effect of withdrawing the outer s-electrons into the atomic core so that they begin to have the character of inner-shell electrons. The effect is often called the *inert-pair effect* and is much in evidence for a lead atom which, in many ways, behaves as if it had only two outer-shell electrons (see page 71).

Physical structure

The structure of the group IVB elements ranges from macromolecular non-metals through metalloids to metallic lattices of close-packed cations. The reasons for this can be traced to the inert-pair effect (described above) which becomes increasingly dominant going down the group, but is absent altogether in the structure of a carbon atom. Carbon atoms form four strong covalent bonds and, in doing so, achieve full valence shells. Carbon is known in two allotropic forms, diamond and graphite. Silicon atoms exist in a rather distorted diamond-type lattice which is also macromolecular and non-metallic in character. Germanium has unusual properties that suggest some metallic, and some non-metallic character. Its lattice structure is an even more distorted form of the diamond arrangement. but it conducts electricity under certain conditions. It is used (as is silicon) as a semiconductor in the construction of transistors. Its chemical properties, which are described in the next section, also show that it has some metallic character. Tin exhibits allotropy but, unlike carbon whose allotropes are monotropic, tin has enantiotropic forms. 'Grey' tin is stable below 13.2°C and 'white' tin above this temperature. 'White' tin is metallic and contains Sn^{2+} ions in a lattice of delocalized electrons; 'grey' tin has a similar structure to that of germanium. Lead exists only as a face-centred cubic lattice of Pb^{2+} ions in a delocalized electron charge cloud.

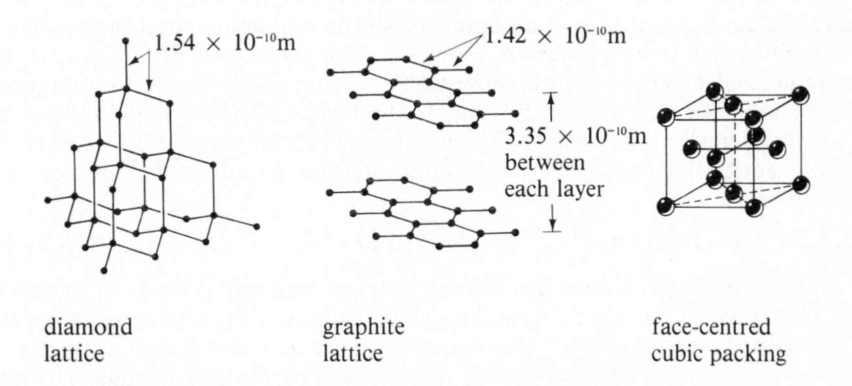

diamond lattice | graphite lattice | face-centred cubic packing

The fixed points and hardness of the elements reflect the gradual change in structure.

property	*carbon (diamond)*	*silicon*	*germanium*	*tin*	*lead*
m.p. (°C)	3730	1410	937	232	327
b.p. (°C)	4830	2360	2830	2270	1744
hardness (moh)	10.0	7.0	5.0	1.5 – 1.8	1.5

43.2 Compounds formed by the elements

The compounds of the group IVB elements further reflect the change from non-metallic to metallic character going down the group. Carbon forms molecular compounds with other non-metals, and anionic salts when heated with reactive metals. It has a remarkable tendency to form catenated compounds (catenation is the property of an element's atoms to bond together forming chains of atoms), and the whole study of organic chemistry is concerned with these catenated compounds. The electronic configuration of carbon suggests that the element is likely to be divalent:

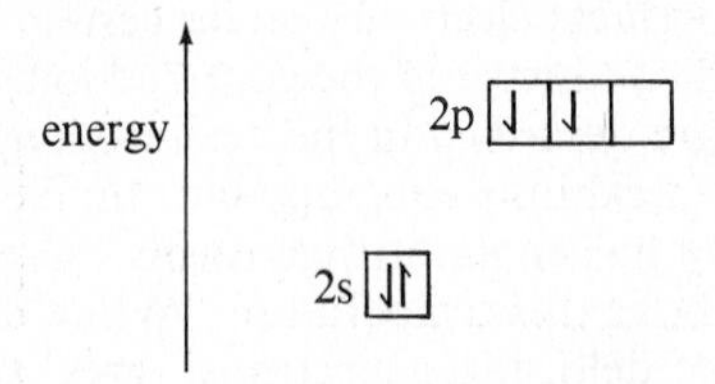

But, by promoting one of the 2s-electrons to the $2p_z$-orbital, four orbitals are half-filled and four covalent bonds can form (see page 350). The energy required for the promotion is offset by the large amount of energy released when the covalent bonds form. The compounds of silicon and germanium share some similarities with those of carbon. However there is less evidence of catenation and the divalent state becomes more stable. This is partly due to the emergence of the inert pair effect and partly due to the weaker bond strengths formed by silicon and germanium atoms. The energy required for promotion is not so readily got back by the evolution of bond energies. Compare the bond energies shown below.

C—C: 348 kJ mol^{-1} Si—Si: 176 kJ mol^{-1} Ge—Ge: 156 kJ mol^{-1}

Metallic character dominates the chemistry of lead and is much in evidence in the chemistry of tin. Lead (II) and tin (II) compounds are common; so are the IV state complexes $[MF_6]^{2-}$ whose properties are best interpreted using the ionic model: an M^{4+} ion surrounded octahedrally by fluoride ligands. The first two ionization energies of tin and lead compare favourably with those of magnesium which is a reactive metal:

element	*first ionization energy* (kJ mol^{-1})	*second ionization energy* (kJ mol^{-1})	*ionic (M^{2+}) radius* (m × 10^{-10})
magnesium	736	1450	0.65
tin	707	1410	1.12
lead	716	1450	1.20

43.3 Occurrence, extraction and uses

Occurrence and extraction of carbon

Carbon compounds make up the structure of all living matter and the element is also present as atmospheric carbon dioxide, carbonate rocks (limestone, chalk, dolomite) and in hydrocarbon fossil fuels. The amount of carbon dioxide in the atmosphere (0.03%) is kept constant by the effect of a natural cycle, called the carbon cycle.

The carbon cycle

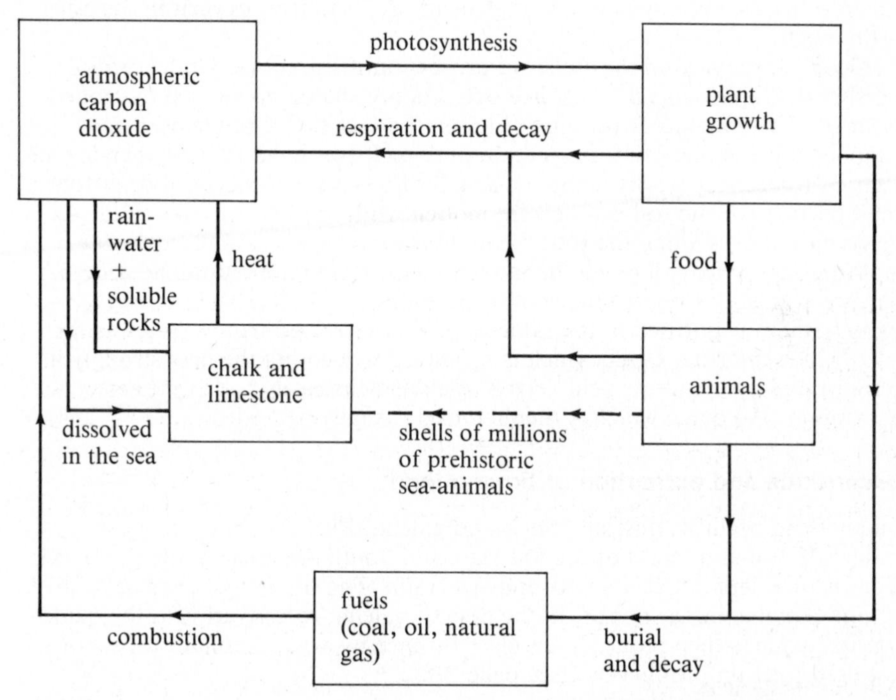

Carbon is found naturally as both diamond and graphite. Diamond is mined from igneous rocks that occur most widely in South Africa. Graphite is mined in parts of South America, the Alps and Siberia. The demand for these allotropes of carbon now cannot be met by mining alone.

The Acheson process

70% of the graphite required is produced by reacting coke (made from the destructive distillation of fossil fuels) with sand at very high temperatures in an electric arc furnace. This is known as the Acheson process:

$$C_nH_m \xrightarrow{1000°C} \underset{\text{coke}}{C} + \text{lower hydrocarbons} + \text{hydrogen}$$

Acheson process:

$$\underset{\text{coke}}{3C_{(s)}} + \underset{\text{sand}}{SiO_{2(s)}} \xrightarrow{2000°C} \underset{\text{graphite}}{C_{(s)}} + Si_{(s)} + 2CO_{(g)}$$

30% of the diamonds used in industry are made by heating graphite to about 3000°C under pressures of about 120 000 atmospheres in the presence of a rhodium catalyst.

Occurrence and extraction of silicon and germanium

Silicon occurs as silica, $(SiO_2)_n$, in sandstone and is present in many silicate minerals. These compounds are the most abundant in the earth's crust, and silicon is the second most abundant element, 27.7%, after oxygen in the crust of the earth.

Silicon is produced in the Acheson process outlined above. Under reduced pressure the liquid silicon is distilled off. Impure silicon is purified by 'zone refining'. This method of refining relies on the fact that a single pure component crystallizes first from an impure melt (see page 187). By passing an electric furnace very slowly along the length of a supported silicon rod, different small portions of the rod can be kept molten while the rest remains solid. As the furnace moves along the rod, the molten zone moves as well. Since the impurities are more soluble in the molten zone than in the crystalline zones, they are gradually swept to the end of the rod.

Germanium is purified in the same way. It is extracted from its ores as the volatile tetrachloride, $GeCl_4$ which is produced by heating the ores strongly in concentrated hydrochloric acid. The tetrachloride is easily hydrolysed to germanium (IV) oxide which is then reduced, usually by hydrogen.

Occurrence and extraction of tin and lead

Tin and lead occur as tinstone, SnO_2 and galena, PbS. Tin ores are most commonly found in parts of the Far East and South America, while the largest deposits of galena are in Canada and Australia. The ore is first roasted to drive off non-metallic impurities or, in the case of galena, to convert it to the oxide. Roasted oxide is then reduced in a blast furnace by a very similar process to that used for iron production, (see page 776).

Uses of elements and their compounds

element	*form*	*uses*
carbon	diamonds	tips for drills, saws and cutters; gemstones; styli for record-players
	graphite	furnace-lining; electrodes; lubricants; pencils; 'carbon black' for ink
	coke	as a reductant for metal extraction; fuel
	hydrocarbons	fuels; lubricants; the petrochemical industry; plastics; pharmaceutical manufacture.

element	*form*	*uses*
silicon	silicon	transistors; miniature circuit-boards for computers.
	silica	glass-making; cement manufacture (p. 641); phosphorus extraction (p. 682).
	silicones	protective materials; polishes; synthetic rubber.
germanium	germanium	transistors (like silicon)
tin	tin	protective coatings ('tin-plate')
	tin alloys	bronzes (Sn-Cu); solders (Sn-Pb)
lead	lead	roofing material; plumbing; cable-casing; accumulators; radioactivity shielding
	lead alloys	solders (Pb-Sn)
	lead salts	paint manufacture
	lead tetraethyl $Pb(C_2H_5)_4$	'anti-knock' for combustion engine fuel

43.4 Compound structure

The +II state

The table below lists examples of the +II oxidation state compounds for the group.

carbon	*silicon*	*germanium*	*tin*	*lead*
CO and carbenes e.g. $:CCl_2$ (not isolatable)	none	GeF_2 GeI_2	SnO $Sn(Hal)_2$ $SnCl_3^-$	PbO $PbO.PbO_2$ (Pb_3O_4) PbS $Pb(Hal)_2$ $Pb(C_2H_5)_2$

Carbon (II) compounds are entirely different in structure and character from those of lead (II). Carbon forms a +II oxide because of the ability of the small carbon atom to form pπ-bonds with the similarly sized oxygen atoms. Although carbon is nominally in an oxidation state of +II, it is likely that more than two electrons take part in the bonding:

V.B. model of CO	*M.O. model of CO*
 :C=O: ⟷ $\ominus$:C≡O:$\oplus$	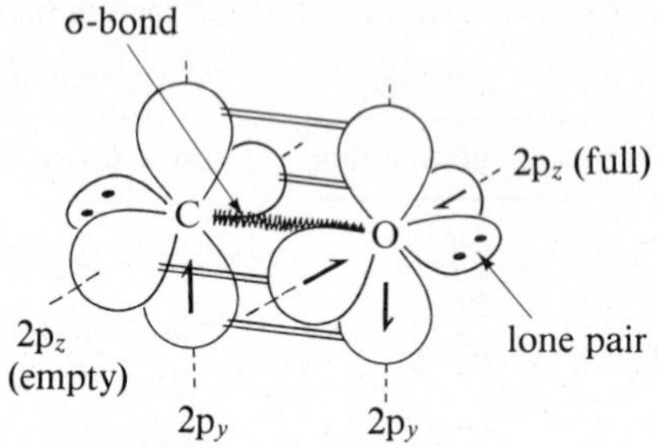 σ-bond; 2p$_z$ (full); lone pair; 2p$_z$ (empty); 2p$_y$; 2p$_y$

The molecule is stabilized by the overlap of the carbon atom's 2p$_y$-electrons with the oxygen atom's 2p$_y$-electrons, as well as by an overlap between the two atoms' 2p$_z$-orbitals. The tendency of carbon to form pπ-stabilized oxy-compounds is a feature particular to it in group IVB. Nitrogen is the only other element to form similar compounds (see Chapter 44).

Carbenes are molecules containing a simple σ-bonded structure and a lone pair on a carbon atom, e.g.

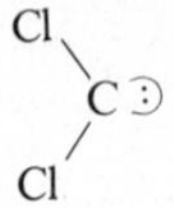

Cl_2C:

Although it has not proved possible to isolate a carbene in a pure state, there is spectroscopic evidence to suggest their existence as reaction intermediates in certain organic reactions, e.g. the Reimer-Tiemann reaction.

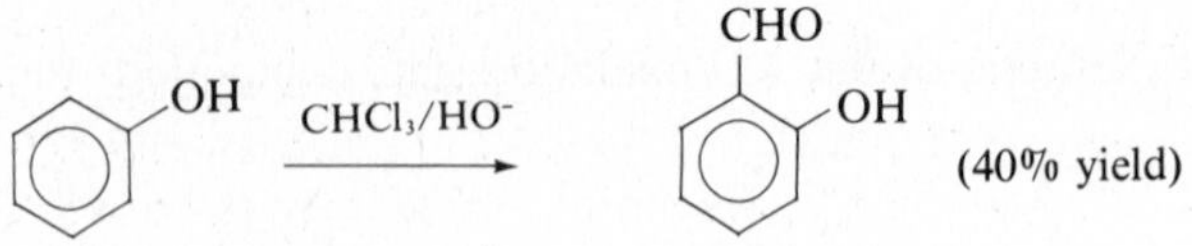

$$C_6H_5OH \xrightarrow{CHCl_3/HO^-} o\text{-}HOC_6H_4CHO \quad \text{(40\% yield)}$$

via

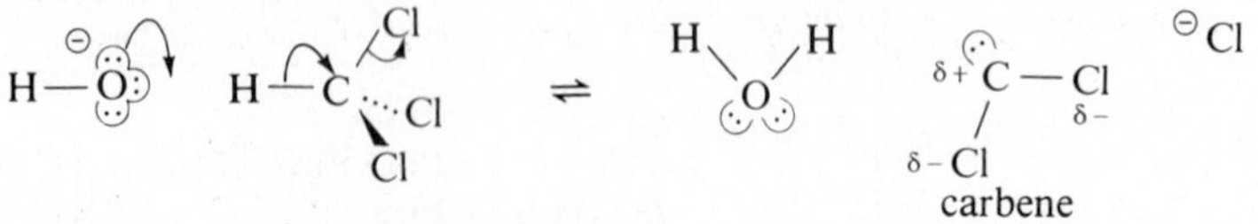

$$H{-}O^{\ominus} + H{-}CCl_3 \rightleftharpoons H_2O + {}^{\delta+}CCl_2^{\delta-} \text{ (carbene)} + {}^{\ominus}Cl$$

and

$$\text{phenoxide ion} + {}^{\delta+}C(Cl^{\delta-})_2 \rightleftharpoons [\text{intermediate}] \xrightarrow{H{-}OH} o\text{-}HOC_6H_4CHCl_2 + {}^{\ominus}OH$$

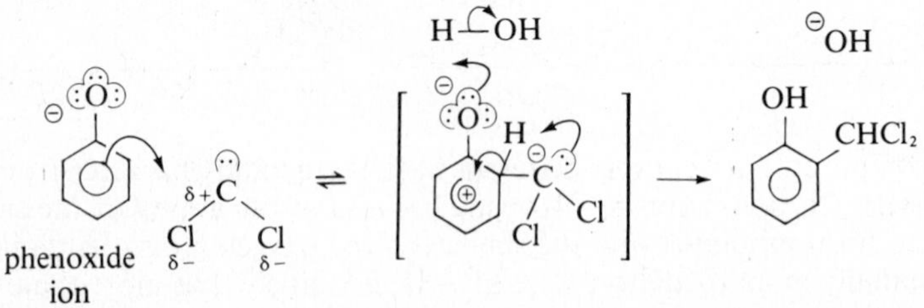

The dichloro side chain is hydrolysed readily to an aldehyde group (see page 521).

A carbene analogue has not been found for silicon, but the dihalides of germanium and tin resemble the carbene structure shown above. Tin also forms an ammonia-shaped trihalogenocomplex $[SnX_3]^-$ with chlorine and fluorine.

Sn Cl Cl 95°   Sn F F F

In contrast, there are no directional properties apparent for the lone pair of the lead (II) atom.

There is little evidence for the hydrated Sn^{2+} ion; in aqueous solution tin (II) is present largely as molecular SnX_2. Although most lead (II) compounds are insoluble in water, $Pb(NO_3)_2$ and $Pb(CH_3COO)_2$ are exceptions whose salts do contain Pb^{2+} ions.

These facts are examples of two important trends in group (IV) chemistry which are shown below.

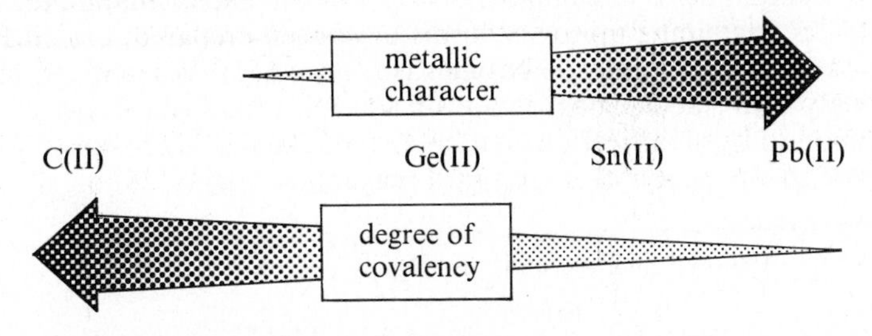

The +IV state

Going down the group the increase in ionic character of the elements' compounds is also reflected by the crystal structure of the M(IV) oxides, and by the stability of the hexahalide complexes $[MX_6]^{2-}$. For lead, these complexes correspond well to the model of a 4+ cation co-ordinated by six halide ligands.

Compare the structures of CO_2, SiO_2 and PbO_2. CO_2 (like CO) is molecular and pπ-stabilized; SiO_2 is macromolecular with little ionic character; PbO_2 is macromolecular with rather more ionic character. The lattice of PbO_2 has a co-ordination of 6:3, which is best accounted for in terms of a body-centred lattice of Pb^{4+} ions with O^{2-} ions occupying the trigonal holes (Rutile lattice).

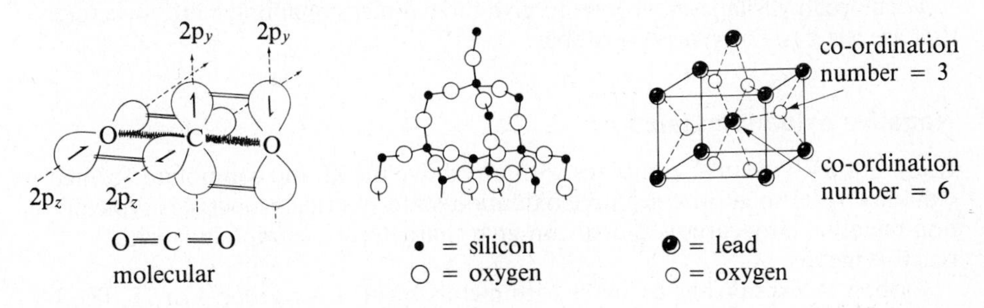

All the simple M(IV) halides and hydrides show typical molecular properties. Their fixed points are given below.

fixed point	CH_4	CCl_4	SiH_4	$SiCl_4$	GeH_4	$GeCl_4$
m.p. (°C)	−182	−23	−185	−70	−165	−50
b.p. (°C)	−160	77	−112	58	−89	84

fixed point	SnH_4	$SnCl_4$	PbH_4	$PbCl_4$
m.p. (°C)	−150	−33	unstable	−15
b.p. (°C)	−52	114		explodes at ~ 105

Silicon and germanium have some tendency to form catenated compounds like carbon. Molecules containing up to six atoms have been prepared, e.g. Si_6H_{14}. Silanes are made in the same way as boranes (see page 651). When magnesium is heated strongly with silicon, magnesium silicide forms and silanes are evolved from this compound by acidification. An important class of alkyl substituted silanes are made by the action of a Grignard reagent (see page 528) on silicon (IV) chloride:

$$SiCl_4 + R{-}Mg{-}Cl \rightleftharpoons R{-}SiCl_3 + MgCl_2$$

$$R{-}SiCl_3 + R{-}Mg{-}Cl \overset{\text{(ether)}}{\rightleftharpoons} R_2SiCl_2 + MgCl_2 \quad \text{etc.}$$

When a dichloroalkyl silane is hydrolysed, it might be expected that the silicon analogue of a ketone would be produced, i.e. $R_2Si{=}O$. But, instead, a polymer with silicon-oxygen single bonds forms. It is called a silicone.

$$nR_2SiCl_2 \xrightarrow{OH^-} {-}O{-}SiR_2{-}O{-}SiR_2{-}O{-}SiR_2{-}O{-}SiR_2{-}O{-}$$

silicone chain

This is analogous to the formation of a macromolecule by silica instead of existing in the molecular form O=Si=O.

Trichloroalkyl silanes hydrolyse to give three-dimensional polymeric structures that are the basis of synthetic rubber.

Negative oxidation states

If a group IVB element combines with a reactive metal, the compound formed contains the element in a negative oxidation state. As this property is typically non-metallic, only carbon and silicon give characteristic compounds with reactive metals.

Carbon forms salt-like carbides with metals from either group I or II. The IA and IIA carbides are ionic in character and contain the dicarbide ion $^-C{\equiv}C^-$;

the IB and IIB carbides (e.g. Ag_2C_2) have high degrees of covalency and are thermally unstable. Carbon forms a monocarbide with aluminium and beryllium. The Al^{3+} or Be^{2+} ions are too strongly polarizing for the dicarbide ion to be formed, e.g.

$$CaO_{(s)} + 3C_{(s)} \xrightarrow{2000°C} CaC_{2(s)} + CO_{(g)}$$

$$2Al_2O_{3(s)} + 6C_{(s)} \xrightarrow{2000°C} Al_4C_{3(s)} + 3CO_{(g)}$$

Silicon combines with metals from groups IA and IIA to give weakly ionic monosilicides, e.g.

$2Mg + Si \rightarrow Mg_2Si$ at $\sim 500°C$ in the absence of air.

43.5 Acid-base properties

The +II state

The oxides, hydroxides and aqueous halides of M(II) tend to be amphoteric in character. Carbon monoxide is an exception, even though it is the anhydride of methanoic acid:

$$Na^+ \; {}^{\ominus}O{-}C({=}O){-}H_{(s)} \xrightarrow[-H_2O]{conc.\ H_2SO_4} NaHSO_{4(s)} + \underset{\text{neutral oxide}}{CO_{(g)}}$$

There is a trend of increasing basicity going down the group which further reflects the increasing metallic character.

The dihalides are acidic by partial hydrolysis. For all of them (except lead (II) halides which are insoluble) the mechanism involves electron-pair acceptance rather than proton transfer from an aquo-ion.

$$H{-}\ddot{O}{:}^{\ominus} + SnCl_2 \rightleftharpoons [Sn(OH)Cl_2]^{\ominus}$$

and $2H_2O_{(l)} \rightleftharpoons H_3O^+_{(aq)} + HO^-_{(aq)}$
position of equilibrium shifted to the right

but for soluble lead (II) salts such as $Pb(NO_3)_2$ proton transfer does occur.

$$Pb(H_2O)_4^{2+} + H_2O \rightleftharpoons [Pb(H_2O)_3(OH)]^+ + H_3O^+_{(aq)}$$

Addition of alkali to a solution of M(II) leads to the precipitation of an insoluble hydroxide. This dissolves either in acid or base and hence shows amphoteric character. In excess alkali, the probable complex formed is $[M(OH)_6]^{4-}$. The oxides are less well characterized: SiO is not known and only a metastable form of GeO exists. SnO has a distorted octahedral lattice in which the co-ordination number appears to be five. PbO, on the other hand, is the most stable oxide of lead and behaves as a typical metal oxide. It dissolves in nitric acid to give a solution of $Pb(NO_3)_2$.

$$PbO_{(s)} + 2H_3O^+_{(aq)} \rightarrow Pb^{2+}_{(aq)} + 3H_2O_{(l)}$$

The +IV state

The acid-base properties of the C(IV) compounds are not typical of those of the rest of the group (see page 675). The oxides, halides and hydrides of M(IV) show a trend of decreasing acidity going down the group in much the same way as those of M(II). $SiCl_4$ and SiH_4 are violently hydrolysed in water, while $GeCl_4$ and GeH_4 are more resistant to hydrolysis, and the chlorides and hydrides of tin and lead are only partially hydrolysed in alkaline conditions, e.g.

$$SiCl_{4(l)} + 8H_2O_{(l)} \rightarrow SiO_2.2H_2O_{(s)} + 4H_3O^+_{(aq)} + 4Cl^-_{(aq)}$$

$$SiH_{4(g)} + 4H_2O_{(l)} \rightarrow SiO_2.2H_2O_{(s)} + 4H_{2(g)}$$

SiO_2 is an acidic oxide that forms a bewildering complexity of salts (the silicates, see below) while SnO_2 and PbO_2 are only weakly acidic and have a marked amphoteric character, e.g.

$$SnO_{2(s)} + 2H_2O_{(l)} + 2OH^-_{(aq)} \rightarrow [Sn(OH)_6]^{2-}$$

$$SnO_{2(s)} + 4H_3O^+_{(aq)} + 6Cl^-_{(aq)} \rightarrow [SnCl_6]^{2-} + 6H_2O_{(l)}$$

(~10M hydrochloric acid)

Silicates

In the lattice of silica, there are four oxygen atoms tetrahedrally arranged around each silicon. The same tetrahedral unit is present in all silicates:

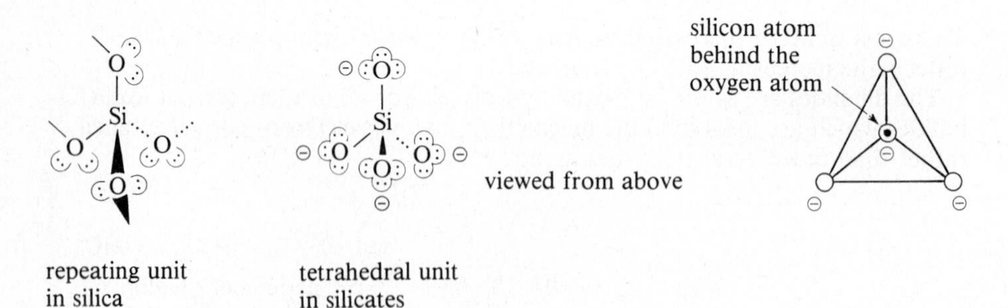

repeating unit in silica

tetrahedral unit in silicates

These units are rarely found as simple molecular ions. There is usually considerable cross-linking and this can take place in one direction to produce chain silicates or in two directions to produce cyclic or layer silicates. If all four oxygens of the SiO_4 tetrahedron cross-link with others, the silica lattice results. Sometimes there are metal atoms replacing the silicon atoms in this arrangement and these structures are three-dimensional silicates (e.g. zeolites).

Some examples of cyclic, chain and sheet silicates are shown below. Their formation can be interpreted in terms of the reaction between acidic silica and a basic metal oxide or carbonate, e.g.

$$SiO_{2(s)} + 2CO^{2-}_{3(s)} \xrightarrow{\text{heat}} [SiO_4]^{4-} + 2CO_{2(g)}$$

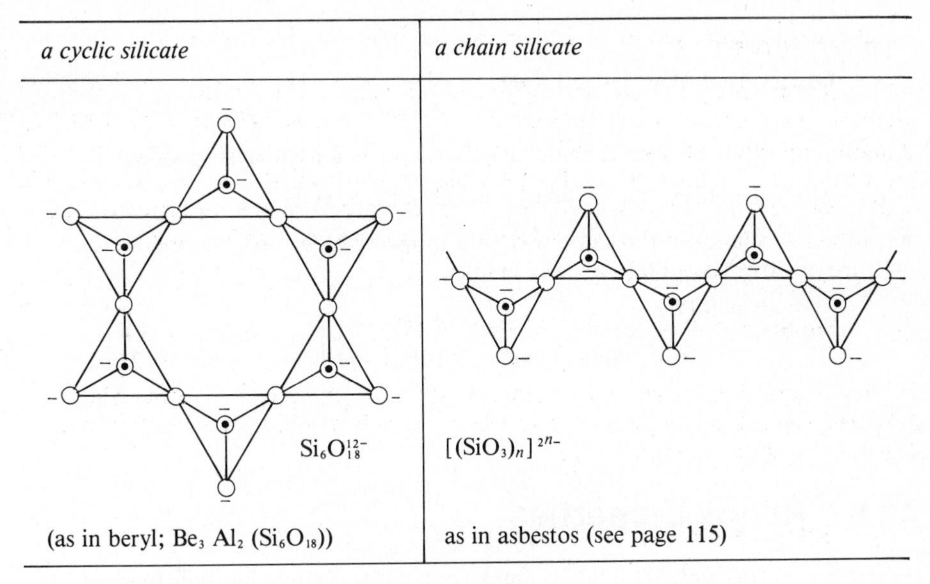

(as in beryl; $Be_3Al_2(Si_6O_{18})$)

as in asbestos (see page 115)

a sheet silicate; two-dimensional layer

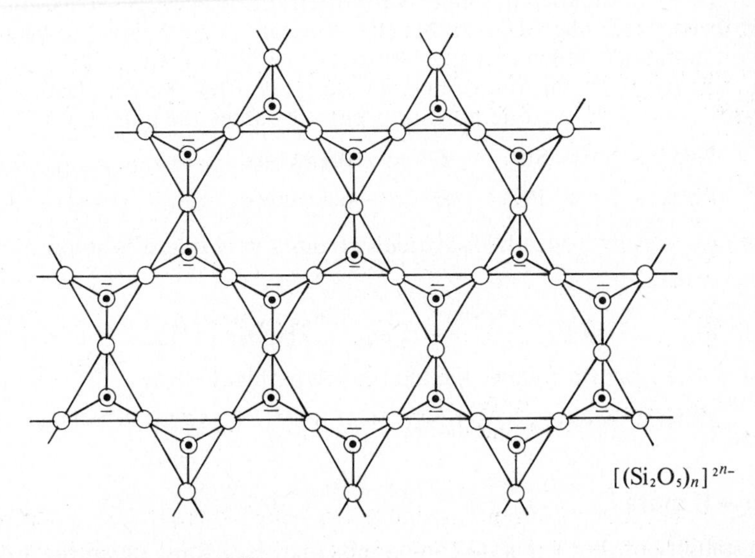

as in micas (see p. 114)

Negative oxidation states

The carbides and silicides are all strongly basic and are easily hydrolysed in neutral or acidic conditions. Dicarbides yield the hydrocarbon ethyne (acetylene) on hydrolysis, and used to be known as acetylides for this reason, e.g.

$$CaC_{2(s)} + 2H_2O_{(l)} \rightarrow \underset{pH \approx 12}{Ca(OH)_{2(s)}} + \underset{\text{ethyne}}{H{-}C{\equiv}C{-}H_{(g)}}$$

Aluminium and beryllium carbide give methane as a hydrolysis product

$$Al_4C_{3(s)} + 12H_3O^+_{(aq)} \rightarrow 4Al^{3+}_{(aq)} + 12H_2O_{(l)} + 3CH_{4(g)}$$

Silicides give silanes in the same way that monocarbides produce methane. A mixture of different silanes usually results:

$$Mg_2Si \xrightarrow{\text{dil } H_2SO_4} \underset{40\%}{SiH_4}, \quad \underset{30\%}{Si_2H_6}, \quad \underset{15\%}{Si_3H_8}, \quad \underset{10\%}{Si_4H_{10}}, \quad \underset{3\%}{Si_5H_{12}}, \quad \underset{2\%}{Si_6H_{14}}$$

43.6 Redox properties

The elements themselves all act as weak reductants, reflecting their tendency to form compounds in a positive oxidation state. Carbon is used to reduce many metal oxides to the metallic state. Very often this process is carried out in a blast furnace (see page 776) so that the reductant is quite likely to be gaseous carbon monoxide which is easily oxidized to carbon dioxide. Silicon is used as a reductant in the production of magnesium (page 639). Tin and lead both have negative standard electrode potentials but the values are quite low.

$$Sn^{2+}_{(aq)} + 2e^- \rightleftharpoons Sn_{(s)} \qquad E^\ominus = -0.14 \text{ volts}$$

$$Pb^{2+}_{(aq)} + 2e^- \rightleftharpoons Pb_{(s)} \qquad E^\ominus = -0.13 \text{ volts}$$

Lead displaces hydrogen with difficulty from a very dilute solution of nitric acid.

$$\overset{0}{Pb}_{(s)} + 2\overset{I}{H}_3O^+_{(aq)} \xrightarrow{\text{v. slow}} \overset{II}{Pb}^{2+}_{(aq)} + 2H_2O_{(l)} + \overset{0}{H}_{2(g)}$$

Most of the reaction follows the alternative pathway below

$$3Pb_{(s)} + 2NO^-_{3(aq)} + 8H_3O^+_{(aq)} \rightarrow 3Pb^{2+}_{(aq)} + 12H_2O_{(l)} + 2NO_{(g)}$$

The +II state

The stability of the +II state compounds increases going down the group as a result of the inert-pair effect. For the first few members of the group, M(II) compounds act as reductants and are readily air-oxidized or disproportionate in the absence of suitable compounds to reduce:

$$2\overset{II}{Ge}Cl_2 \rightleftharpoons \overset{0}{Ge} + \overset{IV}{Ge}Cl_4$$

An aqueous solution of tin (II) chloride is a useful reductant in alkaline conditions. The redox equilibrium involves the complexed tin (II) and tin (IV) hydroxides (stannate (II) and stannate (IV)).

$$Sn(OH)_6^{2-} + 2e^- \rightleftharpoons Sn(OH)_6^{4-} \qquad E^\ominus = -0.96 \text{ volts}$$

For example, iron (III) is reduced to iron (II) under these conditions

$$Fe(OH)_{3(s)} + e^- \rightleftharpoons Fe(OH)_{2(s)} + OH^-_{(aq)} \qquad E^\ominus = -0.56 \text{ volts}$$

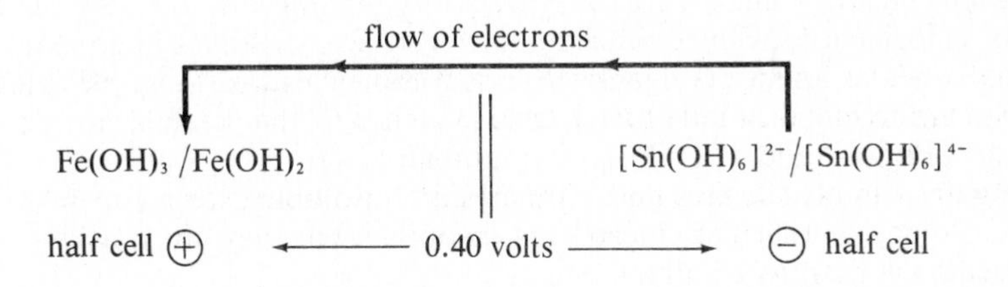

$$2\overset{III}{Fe}(OH)_{3(s)} + \overset{II}{Sn}(OH)^{4-}_{6(aq)} \rightarrow 2\overset{II}{Fe}(OH)_{2(s)} + \overset{IV}{Sn}(OH)^{2-}_{6(aq)} + 2OH^-_{(aq)}$$

The +IV state

The reverse of the above trend is in evidence for the M(IV) compounds. The +IV state compounds of silicon, germanium and tin have little tendency to show redox properties, but lead (IV) compounds are oxidants. The action of heat on lead (IV) oxide causes decomposition to lead (II) oxide:

$$2PbO_{2(s)} \xrightarrow{\text{heat}} 2PbO + O_{2(g)}$$

Under acidic conditions, the redox potential for Pb(IV) to Pb(II) is 1.46 volts.

$$PbO_{2(s)} + 4H_3O^+_{(aq)} + 2e^- \rightleftharpoons Pb^{2+}_{(aq)} + 6H_2O_{(l)} \qquad E^\ominus = 1.46 \text{ V}$$

e.g. lead (IV) oxide oxidizes hydrochloric acid to chlorine

$$PbO_2 + 4H_3O^+_{(aq)} + 6Cl^-_{(aq)} \rightarrow [PbCl_4]^{2-}_{(aq)} + 6H_2O_{(l)} + Cl_{2(g)}$$

43.7 Solubility and complexing properties

With the exception of carbon, the group IVB elements form halogeno-complexes and hydroxy-complexes in either the +II or +IV state. Silicon only forms one complex: the hexafluorosilicate ion $[SiF_6]^{2-}$. This is prepared by dissolving silica in hydrofluoric acid or by the partial hydrolysis of silicon (IV) fluoride.

$$SiO_{2(s)} + 6HF_{(aq)} \rightarrow [SiF_6]^{2-}_{(aq)} + 2H_3O^+_{(aq)}$$

Germanium and tin, however, are known in a number of complexes of the general formula $[M(X)_n(OH)_m]$ where $(n+m) = 6$. Many of these have already been described.

Lead (II) compounds are not very soluble in water and the easiest way to obtain a lead (II) solution is to prepare the tetrahalogeno complex. The insolubility of lead (II) chloride in low concentrations of chloride ions is a useful analytical test for the presence of lead ions in solution.

$$Pb^{2+}_{(aq)} + 2Cl^{-}_{(aq)} \rightarrow PbCl_{2(s)}; \qquad PbCl_{2(s)} + 2Cl^{-}_{(aq)} \rightleftharpoons [PbCl_4]^{2-}_{(aq)}$$

The precipitate redissolves in excess chloride solution.

Silver ions behave similarly if concentrated hydrochloric acid is used as a source of chloride ions. Silver chloride, however, also dissolves in aqueous ammonia, whereas ammonia ligands do not complex lead cations. Most lead complexes are tetrahedral but there is also evidence of the octahedral lead (IV) complexes $[PbF_6]^{2-}$ and $[Pb(OH)_6]^{2-}$ (plumbate (IV)).

Carbonates, simple silicates and stannates are insoluble except for those of group I. All macromolecular silicates are insoluble, but they are slowly hydrolysed by concentrated alkali.

43.8 The unusual properties of carbon

There are two main factors that account for the atypical properties of carbon as a group IVB element.

1. The small size of the atom.

atom	C	Si	Ge	Sn	Pb
atomic radius ($m \times 10^{-10}$)	0.77	1.17	1.22	1.40	1.54

2. The ability of a carbon atom to form a strong pπ-bond with an oxygen or a nitrogen atom.

Structure and bonding

The enormous range of organic structures can be explained by a combination of the two factors above. The small size of the carbon atom leads to the formation of strong covalent bonds, and the presence of delocalized electrons in the π-orbitals of many of these molecules produces added stability and combining power. The unsaturation present in a graphite lattice is responsible for a class of rather unusual compounds called 'graphitic compounds'. Different atoms can be fitted in between the layers of the carbon atoms to produce compounds of non-stoichiometric composition. For example, when the lattice contains atoms of a group IA metal such as potassium, a bronze-coloured solid forms whose electrical conductivity is much greater than that of pure graphite. Graphite oxide is prepared by the action of concentrated nitric acid on graphite. Its stoichiometric ratio C:O approaches 2:1 and the inter-layer spacing of the carbon atoms increases from 3.35×10^{-10}m to about twice this distance. The oxide does not conduct electricity and appears to have C—O—C links between the layers.

Acid-base properties

The molecular nature of carbon (IV) oxide is responsible for its much higher solubility in water than any of the other group IVB oxides. Aqueous solutions of carbon dioxide contain $CO_{2(aq)}$ as well as $H_3O^+_{(aq)}$ and $HCO^-_{3(aq)}$ by hydrolysis.

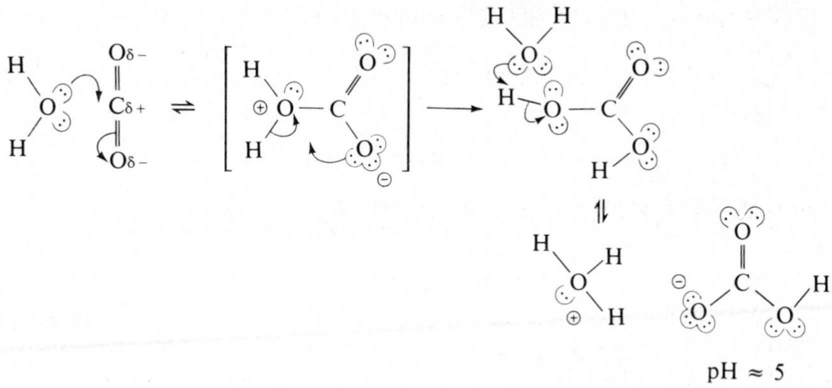

pH ≈ 5

The group IA hydrogen carbonates can be crystallized and are stable below about 300°C. They decompose to carbonates at this temperature

$$2NaHCO_{3(s)} \rightarrow Na_2CO_{3(s)} + H_2O_{(g)} + CO_{2(g)} \qquad T = 270°C$$

Although carbon (IV) oxide is hydrolysed, the chloride and hydride are not hydrolysed; again, unlike the chlorides and hydrides of the lower members. For example, silicon (IV) chloride reacts violently with water, but carbon (IV) chloride is immiscible and inert in water. The mechanism of hydrolysis is likely to proceed as follows

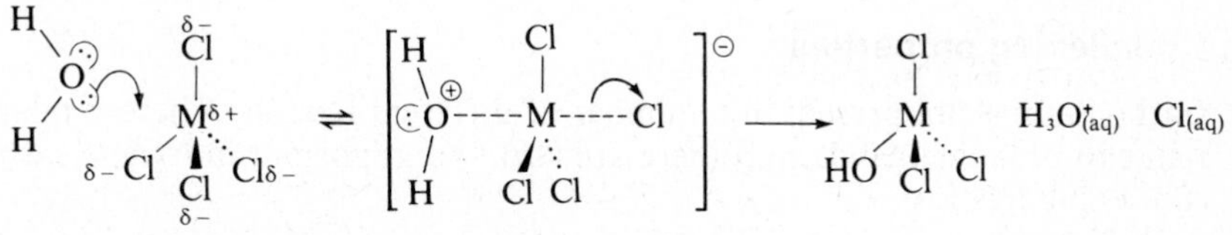

then

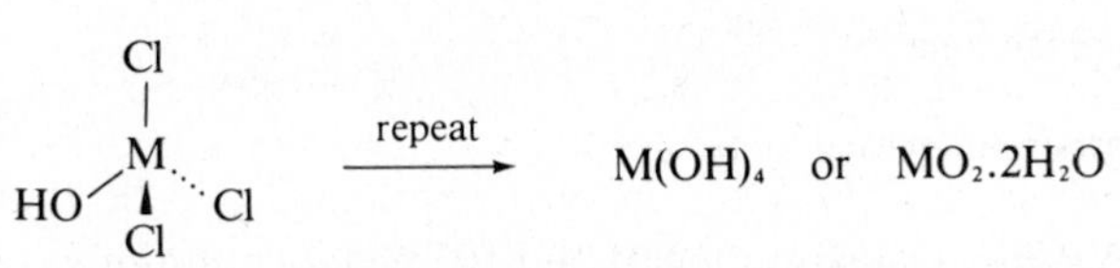

$$M(OH)Cl_3 \xrightarrow{\text{repeat}} M(OH)_4 \text{ or } MO_2.2H_2O$$

There are three reasons why the reaction proceeds so rapidly for $SiCl_4$ but not at all for CCl_4.

1. The small size of the carbon atom and the bulky chlorine atoms around it make it difficult for the attacking water nucleophile to reach the electron-deficient site.

2. The activated complex is a group IVB atom surrounded by five pairs of electrons. Silicon has vacant 3d-orbitals of low energy in which it can accommodate the extra pair of electrons. The outer shell of carbon is the second shell which has no d-orbitals.
3. The hydrolysis product for silicon is hydrated silica which is insoluble. A large amount of energy is released on the formation of this product which provides a driving force for the reaction.

The involvement of silicon 3d-orbitals is not confined only to the stabilizing of reaction intermediates. Compare the geometry of the tertiary amine, trimethylamine, with its silicon analogue trisilylamine.

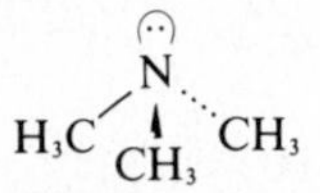

trigonal pyramidal
(like ammonia)

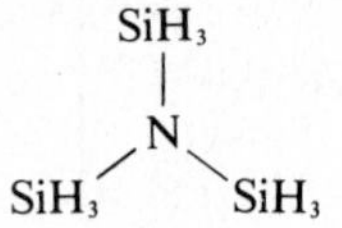

planar
(like boronfluoride)

The nitrogen lone pair appears to be in an unhybridized $2p_y$-orbital which overlaps with the vacant 3d-orbitals of the silicon atoms. This gives trisilylamine dπ-pπ-bonding character.

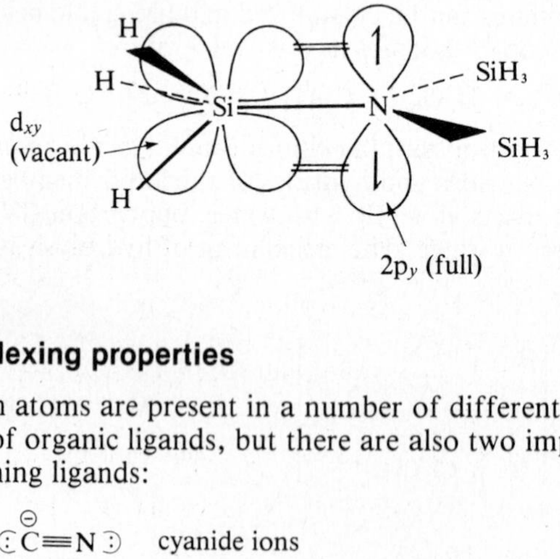

Complexing properties

Carbon atoms are present in a number of different ligands. There is a large range of organic ligands, but there are also two important inorganic carbon-containing ligands:

1. $\overset{\ominus}{:C}\equiv N:$ cyanide ions
2. $\overset{\ominus}{:C}\equiv \overset{\oplus}{O}:$ carbon monoxide molecules

Cyanide ions complex many transition metal and 'B' metal cations e.g. $[Fe(CN)_6]^{3-}$ and $[Zn(CN)_4]^{2-}$. They usually co-ordinate from the carbon atom end, but isocyanides are also known (NC).

Carbon monoxide acts as a ligand which is able to co-ordinate to transition metal atoms (zero oxidation state). The products are called carbonyls and are described on page 779.

Questions

1. Explain the meaning of: a) inert pair effect b) metalloid c) semiconductor d) catenation e) fossil fuel f) metallic character g) activated complex h) carbene.

2.

compound	*structure*	*m.p.* °C
carbon dioxide	molecular	−78.5 (sublimes)
silica	macromolecular	1883

Account for the different properties of these two oxides.

3. a) Contrast the reactions of the group IV halides with water
   b) Account for the anomalous behaviour of CCl_4.

4.

redox system	$E^\ominus$ volts
Ge (IV)/Ge^{2+}	−1.6
Sn (IV)/Sn^{2+}	+0.15
Pb (IV)/Pb^{2+}	+1.80

   a) What trend do these $E^\ominus$ values illustrate going down the group?
   b) Account for the change in $E^\ominus$ values.

5. a) How do the acid-base properties of the group IV oxides change going down the group?
   b) Account for the changes.

6. a) Describe with diagrams the structure of the two allotropic forms of carbon.
   b) Compare the physical properties of the two allotropes.
   c) Describe some of the uses of the two allotropes.

7. Account for the following observations:
   a) Addition of dilute hydrochloric acid to a solution of lead nitrate produces a white precipitate, but when concentrated hydrochloric acid is added the precipitate is only observed transiently, and then a clear solution results.
   b) Addition of sodium hydroxide solution to a solution of lead nitrate produces a white precipitate that dissolves in excess sodium hydroxide but not in excess aqueous ammonia.

8. How does the chemistry of carbon differ from that of the other group IV elements?

9. Why does the amount of carbon dioxide in the atmosphere remain constant?

10. Why are the differences in the properties of the group IV elements more pronounced than their similarities?

Chapter 44

Nitrogen and phosphorus: group VB

44.1 Structure of the elements

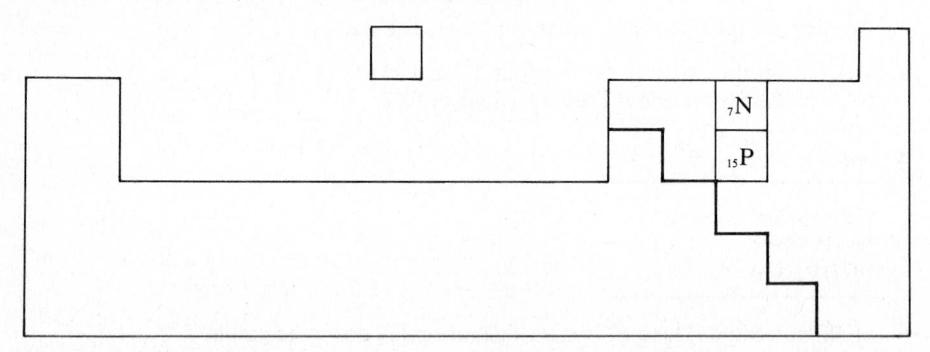

element	electronic configuration	electro-negativity	atomic radius (m × 10^{-10})	ionic (X^{3-}) radius (m × 10^{-10})	stable oxidation states
nitrogen	[He] $2s^2\ 2p^3$	3.0	0.74	1.71	– III, I, II, III, IV, V
phosphorus	[Ne] $3s^2\ 3p^3$	2.1	1.10	2.12	– III, I, III, V

Atomic structure

Nitrogen and phosphorus atoms each have five outer-shell electrons. They are distributed as shown below:

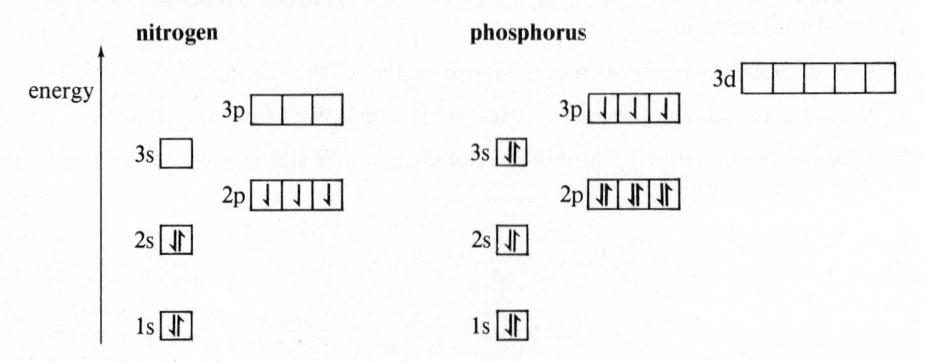

The atoms are quite small and have a core charge of five. Their degree of electronic control is therefore fairly high and they tend to form covalent bonds achieving fully paired p-orbitals as a consequence. Although the atoms are often trivalent as a result of this, each element has a different characteristic way of varying its valency.

1. Nitrogen
The atoms tend to donate their non-bonding pair of electrons and become tetravalent, for example:

The non-bonding pair can also be used to form pπ-bonds to atoms of oxygen, carbon or other nitrogen atoms. The small size of these three atoms and the ease with which their 2p-orbitals overlap favour the formation of these structures.

2. Phosphorus
The outer shell of a phosphorus atom can hold a maximum of eighteen electrons. The energy required to promote an electron into the vacant, energetically low-lying 3d-orbitals is often much less than the energy given out by the formation of two extra bonds.

phosphorus atom + promotion energy

In this configuration, a phosphorus atom can have a valency of five.

Physical structure

Both nitrogen and phosphorus are non-metallic in character. Nitrogen has typically molecular properties and phosphorus has three allotropic forms that show molecular or macromolecular properties. The different forms are monotropic and are discussed on page 125.

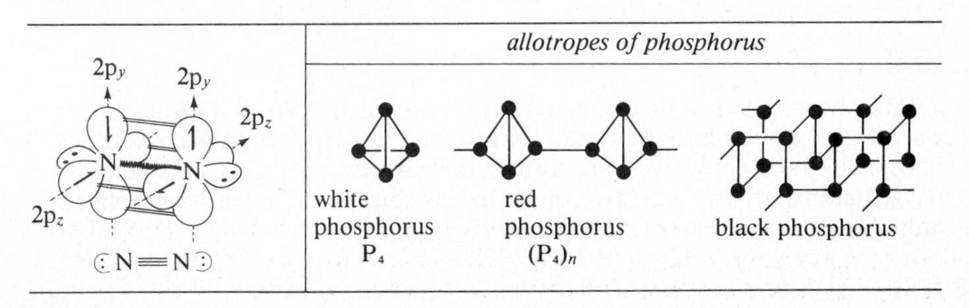

Notice that phosphorus atoms exist in σ-bonded chains, tetrahedra or lattices rather than as pπ-stabilized P_2 molecules. The π-bond energy resulting from the overlap of the phosphorus 3p-orbitals is not very high because the amount of overlap is small. The atomic radius of phosphorus is greater than that of nitrogen and it prevents an extensive π-overlap. Compare the bond energies and the bond lengths below.

bond	P—P	P≡P	N—N	N≡N
bond energy ($kJ\ mol^{-1}$)	172	492	163	944
bond length ($m \times 10^{-10}$)	2.21	2.03	1.46	1.10

44.2 Compounds formed by the elements

The compounds of nitrogen and phosphorus demonstrate the non-metallic nature of the two elements. They combine with the more reactive metals to give strongly basic nitrides and phosphides of some ionic character. The elements form molecular oxides and halides that are acidic, and molecular hydrides that have Lewis basic properties. They are combined in a wide range of oxy-compounds in which they are present as oxy-anions.

Nitrogen shows a marked tendency to form pπ-stabilized molecules and molecular ions, particularly with oxygen as a result of hydrolysis reactions. Nearly all inorganic nitrogen compounds are soluble in water and many are readily hydrolysed. Living systems contain a wide variety of different organic nitrogen compounds, and many of these contain pπ-bonded nitrogen. For example, in the two aromatic rings below (and their fused derivatives).

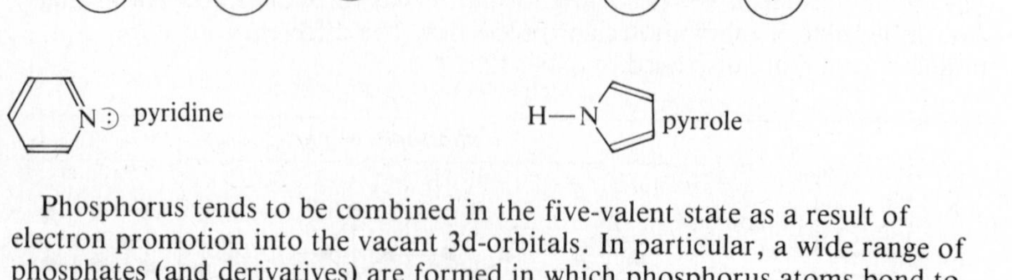

'lone-pair' in π-overlap

(4n + 2) delocalizable electrons see page 489

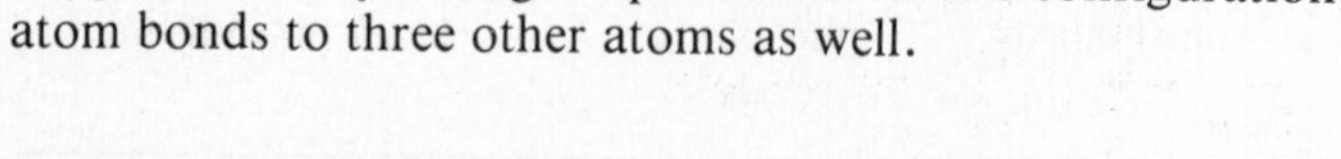

pyridine

H—N pyrrole

Phosphorus tends to be combined in the five-valent state as a result of electron promotion into the vacant 3d-orbitals. In particular, a wide range of phosphates (and derivatives) are formed in which phosphorus atoms bond to oxygen atoms by strong dπ-pπ-bonds. In this configuration, each phosphorus atom bonds to three other atoms as well.

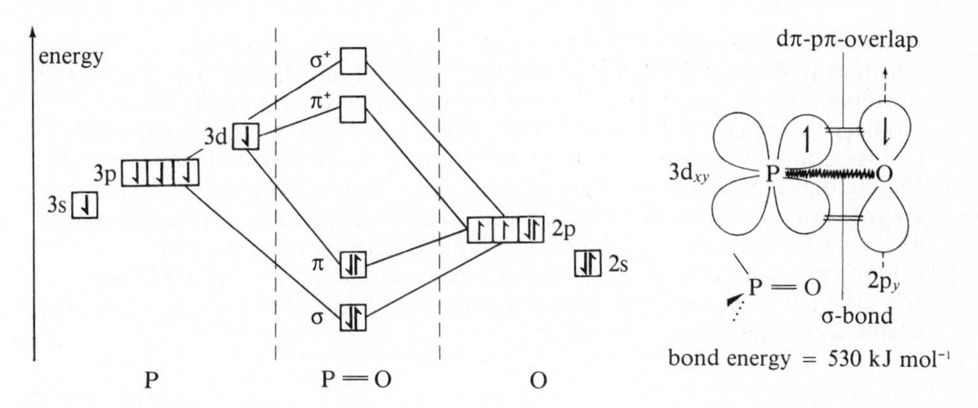

The energy given out on the formation of a P=O double bond is the driving force for many of the hydrolysis reactions of the phosphorus compounds.

44.3 Occurrence, extraction and uses

Nitrogen

Nitrogen occurs as the free element (~ 79% of the atmosphere), and is also found combined in proteins (and other naturally occurring organic compounds) and as nitrate deposits, e.g. Chile saltpetre $NaNO_3$. Ammonium salts are also present in the soil as a result of the decay of living systems and the action of bacteria. The proportion of nitrogen in the atmosphere remains approximately constant as a result of a natural cycle that is somewhat similar to the carbon cycle (see page 663). Man takes a more planned and active part in ensuring the nitrogen cycle operates efficiently by manufacturing 'fertilizers' and adding these to the soil. Nitrogenous fertilizers are ammonium salts or nitrate salts.

The nitrogen cycle

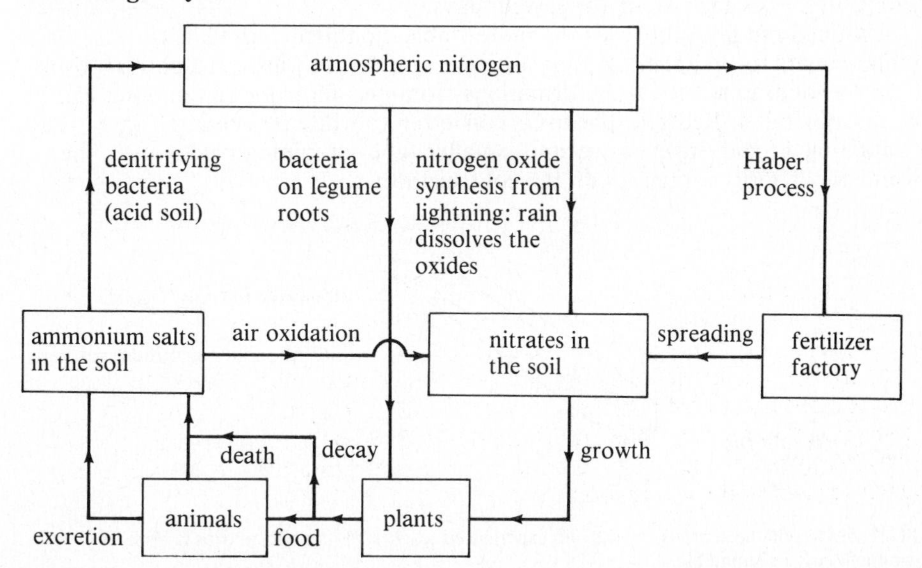

Extraction of nitrogen

Nitrogen is extracted from the atmosphere by liquefying air and then fractionally distilling the solution produced. Dust is removed by filtering and carbon dioxide and water are removed as they freeze out. When a gas is compressed, its temperature increases. If, after allowing it to cool to the temperature of its surroundings, it is suddenly allowed to expand, its temperature drops sharply. By repeating this cycle of compression – cooling – expansion, the temperature of the air is reduced until it liquefies, forming a nearly ideal solution. This solution is fractionally distilled and nitrogen vapour (b.p. $-196°C$) comes from the top of the column while liquid oxygen (b.p. $-183°C$) is run from the bottom (see page 166).

Phosphorus

Phosphorus is found in a number of phosphate minerals, e.g. apatite $CaF_2.3Ca_3(PO_4)_2$, and is present in the structure of all living matter. Bones contain about 50% calcium phosphate and phosphate ester links are present in nucleic acids and in carbohydrates that occur naturally. Phosphorus is the tenth most abundant element in the earth's crust (0.12%).

Extraction of phosphorus

Phosphorus is extracted from a phosphate ore by the high temperature reaction between the ore and silica in the presence of coke. Under the reaction conditions (about 1500°C in an electric furnace), a number of equilibria are set up and phosphorus vapour is removed in a stream of carbon monoxide from the top of the furnace. A silicate slag forms and is tapped off from time to time.

$$4PO^{3-}_{4(s)} + 6SiO_{2(s)} \rightleftharpoons 6SiO^{2-}_{3(l)} + P_4O_{10(g)}$$

$$P_4O_{10(g)} + 10C_{(s)} \rightleftharpoons 10CO_{(g)} + P_{4(g)} \rightleftharpoons 2P_{2(g)}\ (T > 800°C)$$

The vapour is condensed under water and white phosphorus forms. Red phosphorus is produced by keeping white phosphorus at a temperature of 270°C in the absence of air for several days.

Although red phosphorus is the more stable monotrope, the kinetic requirements for its formation must be met, otherwise white phosphorus forms.

In the laboratory† the transformations from one allotrope to the other can be demonstrated. Red phosphorus is converted to white by heating it in a tube containing a 'cold finger' or second, smaller tube containing cold water. The white phosphorus condenses on the cold surface:

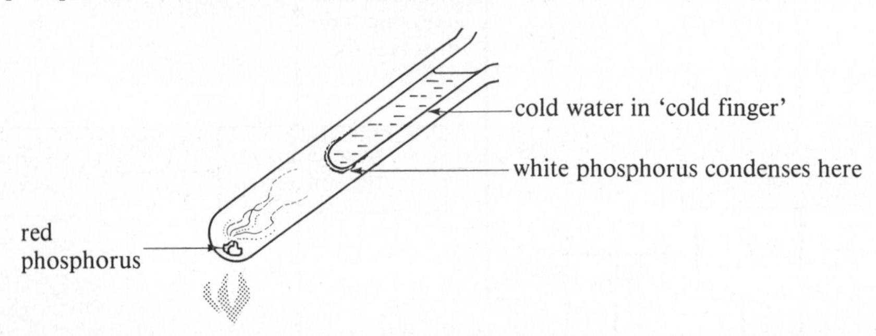

†This should only be demonstrated by an experienced teacher. White phosphorus is both toxic and spontaneously inflammable.

White phosphorus is converted to red by heating it in a tube with a crystal of iodine. To prevent the white phosphorus igniting in the air the tube must be flushed with carbon dioxide.

Black phosphorus can be produced by heating white phosphorus to about 200°C at very high pressure.

Uses of the elements and their compounds

nitrogen	in the Haber process to make ammonia (see p. 685). as an inert carrier gas.
ammonia	as an aqueous alkali. as a refrigerant and solvent (in the pure liquid phase). in the manufacture of nitric acid (see p. 694).
nitric acid	for making fertilizers and explosives (e.g. T.N.T.).
phosphorus	in the manufacture of orthophosphoric (V) acid and to make insecticides (halide and sulphide derivatives). phosphor bronzes.
orthophosphoric (V) acid and ortho-phosphates (V) e.g. $Ca_3(PO_4)_2$ from rocks	steel pickling and rust protection. as a dehydrating agent. in the manufacture of fertilizers e.g. 'Nitrophos': $Ca_3(PO_4)_2 + 4HNO_3 \rightarrow Ca(H_2PO_4)_2.2Ca(NO_3)_2$ 'triplesuperphosphate': $Ca_3(PO_4)_2 + 4H_3PO_4 \rightarrow 3Ca(H_2PO_4)_2$

44.4 Compound structure

The reactivity of the two elements is not quite what might be expected from their relative positions in the Periodic Table. For non-metallic elements, reactivity tends to decrease going down a group.

The comparative inertness of molecular nitrogen results from the stability of the triple-bonded molecules; 944 kJ mol^{-1} are required to cause disruption. The unexpectedly reactive nature of white phosphorus is due to the bond strain present in the tetrahedral molecules. There are three bonds at 60° to one another and the molecule has a strong tendency to open out.

The – III state

Both nitrogen and phosphorus combine with the more reactive metals to produce nitrides and phosphides. The degree of covalency in these compounds is high because the anions are large, highly charged and easily polarized:

$$3Mg_{(s)} + N_{2(g)} \xrightarrow[\text{strongly}]{\text{heat}} Mg_3N_{2(s)}$$

$$6Ca_{(s)} + P_{4(s)} \xrightarrow[\text{heat}]{} 2Ca_3P_{2(s)}$$

ionic radii

N^{3-} 1.71 P^{3-} 2.12 (m × 10^{-10})

They are hydrolysed rapidly to the hydride compounds, ammonia and phosphine. These are both gases at room temperature.

$$X^{3-}_{(s)} + 3H_2O_{(l)} \rightarrow XH_{3(g)} + 3OH^-_{(aq)}$$

The reaction is violent and not the usual method of preparing ammonia or phosphine. Ammonia is a base (see page 690) and forms crystalline ammonium salts from which the gas is easily obtained by using a stronger base (e.g. aqueous sodium hydroxide)

$$NH^+_{4(s)} + OH^-_{(aq)} \xrightarrow{\text{warm}} NH_{3(g)} + H_2O_{(l)}$$

Phosphine shows almost no basic properties and therefore cannot be prepared from 'phosphonium' salts. Instead, the gas is produced by the disproportionation of white phosphorus in concentrated, boiling alkali.

$$\overset{0}{P}_{4(s)} + 3H_2O_{(l)} + 3OH^-_{(aq)} \rightarrow 3H_2\overset{\text{I}}{P}O^-_{2(aq)} + \overset{-\text{III}}{P}H_{3(g)}$$

The hydroxide ions cause an induced dipole to be set up along the phosphorus-phosphorus σ-bonds.

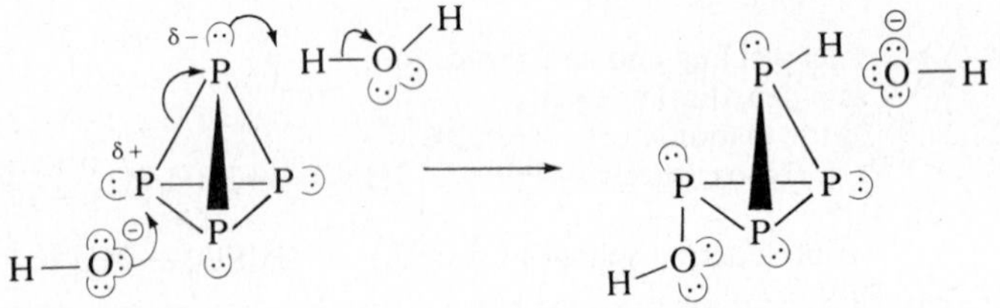

When this step is repeated five times, the products are as follows.

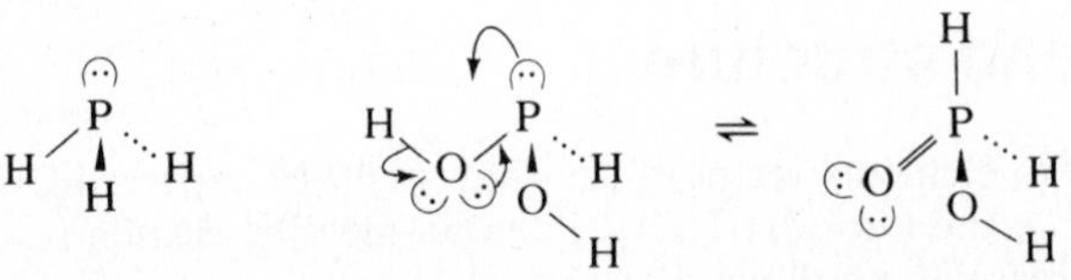

Under the alkaline conditions, the oxy-compound acts as an acid and is present as phosphinate (I) (or hypophosphite) anions. Diphosphene $H_2P—PH_2$ is produced as well in small quantities. This is a spontaneously inflammable gas which oxidizes instantly in air to phosphorus (V) oxide. The analogous nitrogen compound, hydrazine N_2H_4, is much more stable but still burns vigorously in air when ignited; it has been used as a rocket fuel for this reason. Compare the bond angles in the ammonia molecule with those in the phosphine molecule.

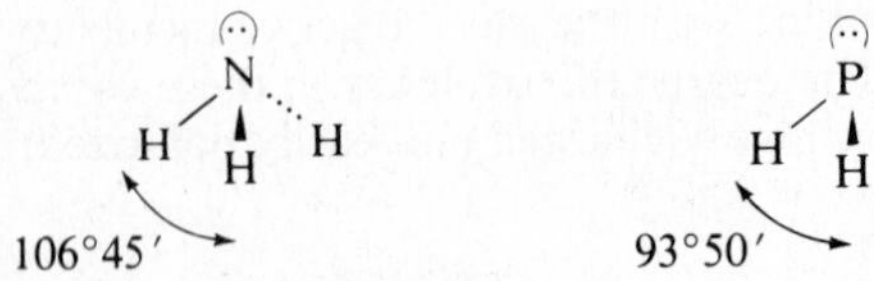

106°45′ 93°50′

The lone pair causes distortion of the tetrahedral arrangement of electron pairs (see page 98) and, for the less electronegative phosphorus atom, this distortion is considerable. The volume taken up by a P—H σ-pair is likely to be rather less than that taken up by an N—H σ-pair and this also leads to a reduction in bond angle.

The Haber process

Ammonia is produced on a very large scale industrially. Nitrogen is abundantly available from the air and hydrogen is obtained by the reduction of water using coke or hydrocarbons (see page 609). Nitrogen and hydrogen in the ratio of 1:3 are passed over a finely divided iron catalyst at a temperature of about 500°C and under very high pressure (from 200 – 1000 atm). The resulting gas mixture is rapidly cooled and the ammonia liquefies. The unreacted nitrogen and hydrogen are recycled.

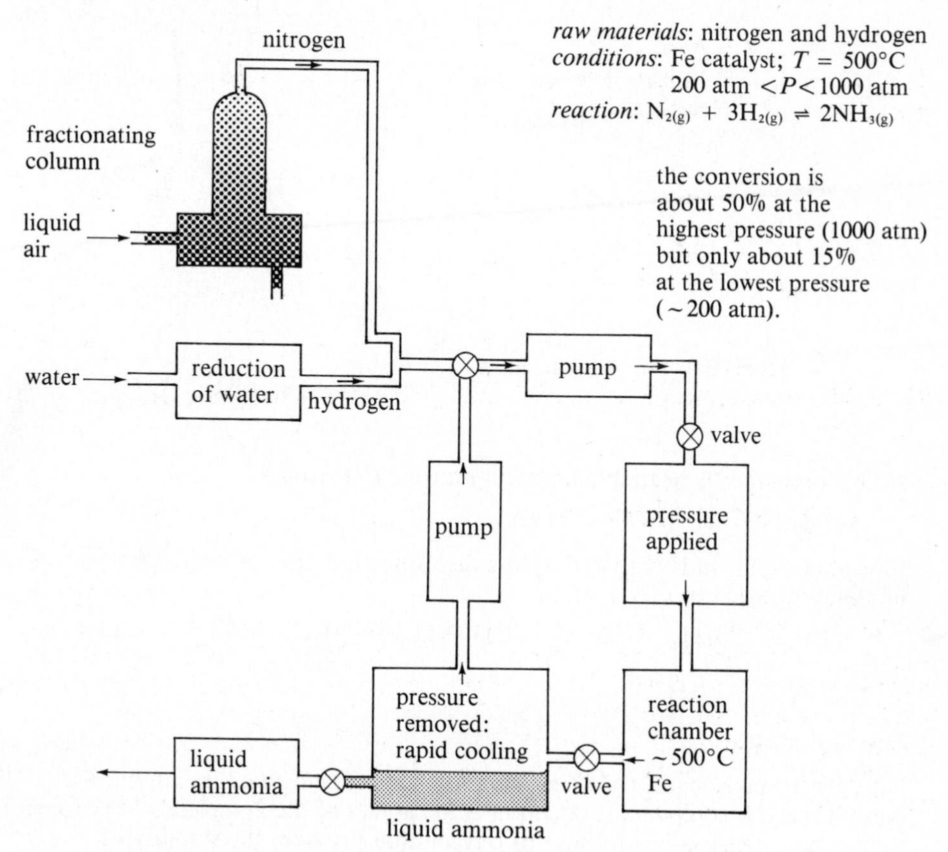

Maximum yield at equilibrium is produced by:

a) low temperature because the forward reaction is exothermic ($\Delta H_r^\ominus = -92$ kJ mol^{-1})

b) high pressures because the products occupy less volume (2 moles) than the reactants (4 moles).

In practice the reacting gases are not given time to reach equilibrium and are recycled continuously over the catalyst at 500°C which produces a faster rate of reaction. Removal of ammonia by sudden cooling prevents too much back reaction.

The +I state

The +I state compounds of nitrogen and phosphorus provide good examples of the different types of π-bond mentioned in section 44.2. Nitrogen (I) oxide contains pπ-stabilized molecules and is the only stable compound of nitrogen (I); phosphoric (I) acid (phosphinic) has dπ-pπ-stabilized molecules containing five-valent phosphorus.

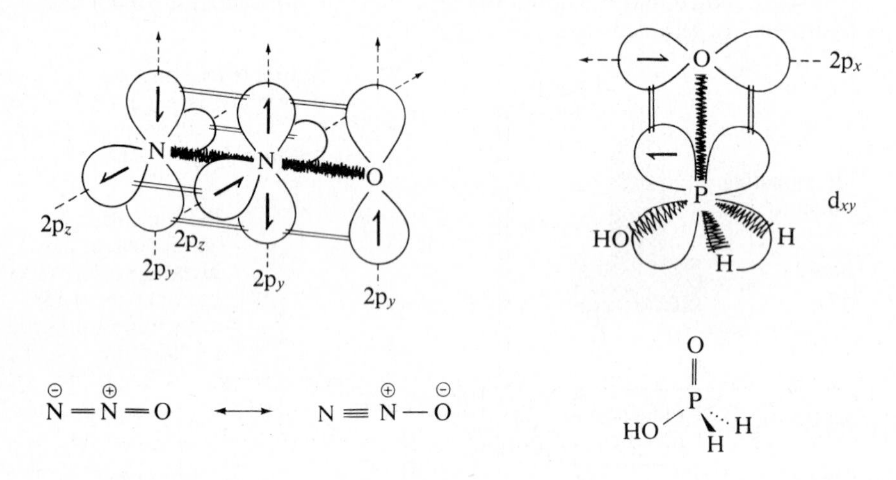

N_2O is prepared by heating ammonium nitrate (V) strongly

$$NH_4NO_{3(s)} \rightarrow N_2O_{(g)} + H_2O_{(g)}$$

Phosphinic (I) acid is prepared by the disproportionation of white phosphorus in alkali followed by protonation

$$P_{4(s)} + 2OH^-_{(aq)} + 3H_2O_{(l)} \rightarrow 3H_2PO^-_{2(aq)} + PH_{3(g)}$$

The +II state

Nitrogen forms a stable oxide in oxidation state II, but there are no known phosphorus (II) compounds. Again, it is the ability of the 2p-orbitals of nitrogen to overlap extensively with those of oxygen that provides the stability of nitrogen (II) oxide. It is prepared by passing an electric discharge through gaseous nitrogen and oxygen, or by the action of nitric acid (50%) on copper.

$$N_{2(g)} + O_{2(g)} \xrightarrow{\text{discharge}} 2NO_{(g)}$$

$$3Cu_{(s)} + 2NO^-_{3(aq)} + 8H_3O^+_{(aq)} \rightarrow 3Cu^{2+}_{(aq)} + 12H_2O_{(l)} + 2NO_{(g)}$$

Nitrogen (II) oxide is paramagnetic, which indicates that there are unpaired electrons present in the molecules. An NO molecule has the same orbital configuration as an N_2 molecule except that there is an extra electron to accommodate. This goes in an anti-bonding π-orbital, so that NO^+ is in fact more strongly bonded than NO. A molecule of NO is known as an 'odd electron' molecule because of this unpaired electron in its outer shell.

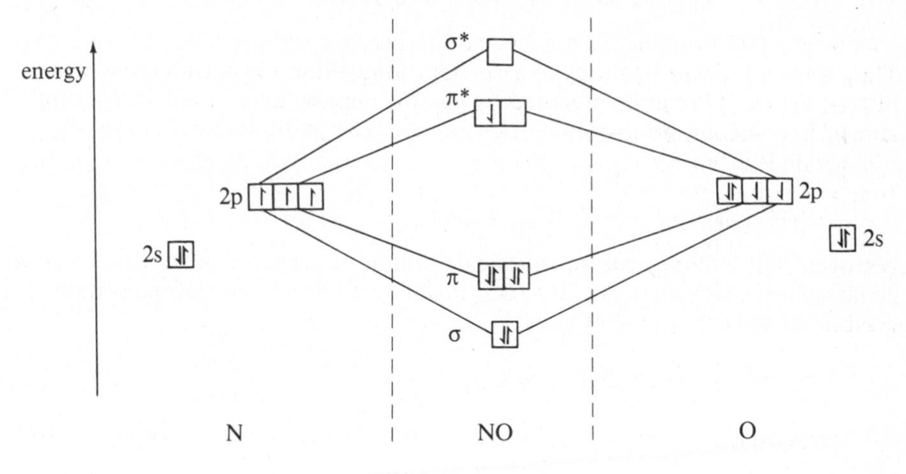

The +III state

Halides, oxides and oxyacids are known for both nitrogen (III) and phosphorus (III). Their structures again illustrate both the tendency of nitrogen to form $p\pi$-stabilized units and the reluctance of phosphorus to bond in this way.

element	*halide*	*oxide*	*hydroxyacid*
nitrogen	N, F, F, F	N_2O_3 is unstable and readily decomposes to NO and NO_2, but NOCl exists O=N–Cl	O=N–O–H pπ-pπ
phosphorus	P, F, F, F	P, O, O, O, P, O, O, P, O, O, P (cage)	O, dπ-pπ, P, HO, OH, H

Phosphorus (III) halides and phosphorus (III) oxide are easily prepared by reacting white phosphorus in a limited supply of a halogen or of oxygen. These compounds are rapidly hydrolysed to give the phosphorus (III) hydroxyacid, phosphonic acid.

$$P_{4(s)} + 3O_{2(g)} \rightarrow P_4O_{6(s)} \qquad P_4O_{6(s)} + 6H_2O_{(l)} \rightarrow 4H_3PO_{3(l)}$$

Nitrogen (III) compounds are less easy to prepare and are far less stable. They show a marked tendency to disproportionate into nitrogen (II) and nitrogen (IV). NF_3 can be prepared by passing ammonia and fluorine over hot copper in a stream of nitrogen carrier gas. It is thermally stable, but the trichloride is explosive.

$$2NH_{3(g)} + 3F_{2(g)} \xrightarrow{Cu_{(s)}} 2NF_{3(g)} + 3HF_{(g)}$$

Nitrogen (III) hydroxyacid (nitrous acid) cannot be prepared in the same way as its phosphorus (III) analogue. It is best made by acidifying the decomposition product of sodium nitrate (V):

$$2NaNO_{3(s)} \xrightarrow{\text{heat}} 2NaNO_{2(s)} + O_{2(g)}$$

$$2NaNO_{2(s)} + 2H_3O^+_{(aq)} \rightarrow 2Na^+_{(aq)} + 2HO{-}\overset{III}{N}O_{(aq)} \rightleftharpoons \overset{II}{N}O_{(g)} + \overset{IV}{N}O_{2(g)} + H_2O$$

The +IV state

NO_2 is the only common compound containing a group VB element in an oxidation state of IV. Like NO, it is paramagnetic and it contains 'odd electron' molecules. At low temperatures (or high pressures) these molecules tend to dimerize, which suggests that the single electron is not in an anti-bonding orbital.

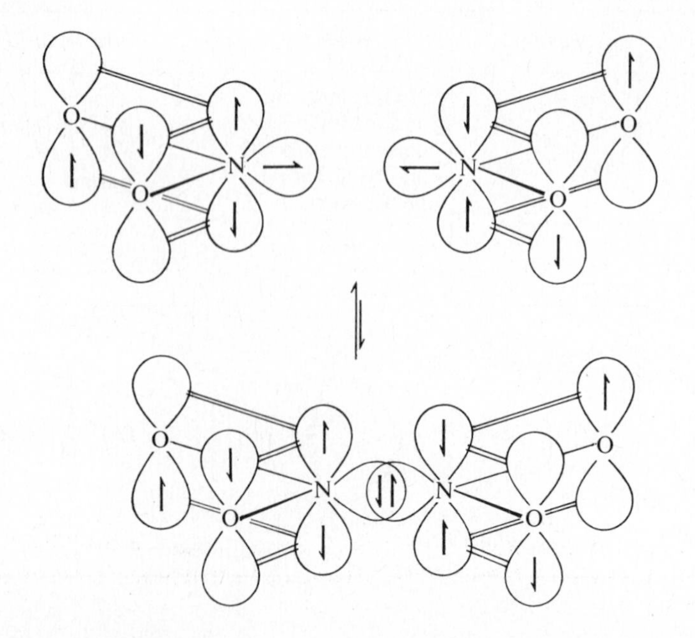

The N—N bond length in N_2O_4 is 1.75×10^{-10}m compared with an N—O bond length of 1.18×10^{-10}m. This suggests that the extent of π-overlap between the two dimerized molecules is small. The bond is certainly weak and N_2O_4 dissociates readily into $2NO_2$.

NO_2 is prepared by the action of concentrated nitric acid on copper or by the thermal decomposition of the nitrate (V) of any metal not in group I.

$$Cu_{(s)} + 4HNO_{3(l)} \rightarrow Cu(NO_3)_{2(s)} + 2NO_{2(g)} + 2H_2O_{(l)}$$

or

$$2Pb(NO_3)_{2(s)} \rightarrow 2PbO_{(s)} + 4NO_{2(g)} + O_{2(g)}$$

The +V state

Although nitrogen forms nitrogen (V) oxide and the nitrate (V) ion, it cannot combine with a halogen to give a V state halide. Five σ-bonds would be required, and this would mean ten electrons around the nitrogen atom. Nitrogen's outer shell is limited energetically to an octet. Phosphorus readily forms phosphorus (V) oxide, halides and hydroxy-derivatives by promoting an electron into the vacant 3d-orbitals.

element	*halide*	*oxide*	*hydroxyacid*
nitrogen	none formed	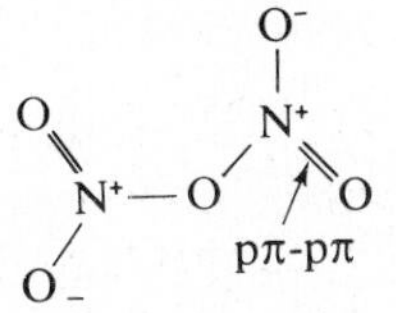 O^- , N^+, O; N^+—O; O, O_-; pπ-pπ	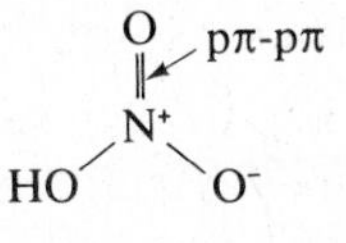 O, N^+, HO, O^-; pπ-pπ
phosphorus	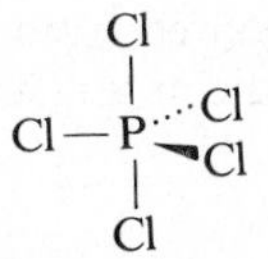 Cl, Cl—P, Cl, Cl, Cl	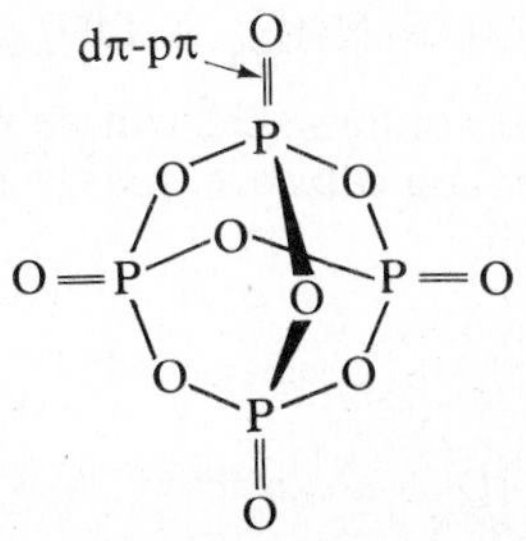 dπ-pπ; O, P, O, O, O=P, O, P=O, O, O, O, P, O	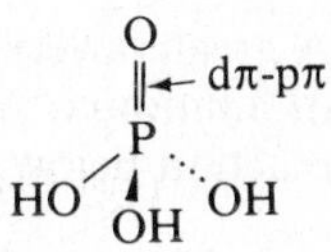 O, P, HO, OH, OH; dπ-pπ

The phosphorus (V) compounds are produced by the reaction of white phosphorus in excess oxygen or halogen:

$$P_{4(s)} + 5O_{2(g)} \rightarrow P_4O_{10(s)}$$

$$(\text{and } P_4O_{10(s)} + 6H_2O_{(l)} \rightarrow 4H_3PO_{4(l)})$$

$$P_{4(s)} + 10Cl_{2(g)} \rightarrow 4PCl_{5(s)}$$

The structure of P_4O_{10} is similar to the structure of P_4O_6. It is based on a tetrahedral arrangement of phosphorus atoms with oxygen atoms bridging the phosphorus atoms. In P_4O_{10}, there is also a terminal oxygen atom attached to each corner of the tetrahedron.

Although PCl_5 is molecular in the liquid state, in the solid state it exists as an ionic solid consisting of PCl_4^+ cations and PCl_6^- anions. It is not possible to fit five larger bromine atoms closely around a single phosphorus atom, and so the structure of phosphorus (V) bromide consists of the ions PBr_4^+ and Br^-. Iodine atoms are even larger and the substance PI_5 has not been isolated.

The nitrogen (V) oxy-compounds all contain pπ-stabilized molecules or molecular ions. Nitric acid is made by oxidizing ammonia (see page 694) and nitrogen (V) oxide is obtained as the acid anhydride by distilling with phosphorus (V) oxide.

44.5 Acid-base properties

Negative oxidation state

Metal nitrides and phosphides are strong bases and react with water (a weak acid) to give hydroxides and ammonia or phosphine:

$$X^{3-} + H\text{—}O\text{—}H \xrightarrow[\text{twice}]{\text{repeat}} XH_3 + 3\ {}^{\ominus}O\text{—}H$$

Ammonia is a weak base in water as well.

$$NH_{3(aq)} + H_2O_{(l)} \rightleftharpoons NH^+_{4(aq)} + OH^-_{(aq)} \qquad pK_b = 4.8$$

A great variety of ammonium salts can be prepared from an aqueous solution of ammonia. Ammonium salts are weakly acidic as a result of the hydrolysis reaction below.

$$H_3N^{\oplus}\text{—}H + H_2O \rightleftharpoons H_3N + H\text{—}\overset{\oplus}{O}\text{—}H \ (\text{with } H)$$

Phosphine, however, shows almost no basic character at all. Despite the more prominent lone pair of a phosphine molecule, phosphine does not behave like ammonia in water: it is almost insoluble. The reason for this is probably due to the comparatively weak P—H bond strength and the lower hydration energy of a bulky phosphonium ion. These two energy terms must counterbalance the breaking of a strong O—H bond in a water molecule during proton transfer. Phosphonium halides are thought to exist but $PH_4^+I^-$ is the only one that can be isolated at room temperature.

Positive oxidation states

The halides, oxides and hydroxyl compounds of nitrogen and phosphorus in a positive oxidation state all tend to be acidic. The +I and +II states for nitrogen are neutral, however. Neither N_2O nor NO shows any acidic properties at all and they are only sparingly soluble in water. The +III, +IV and +V states are all acidic, and compounds of nitrogen in these three states become increasingly acidic going from III to V.

Nitrogen (IV) oxide hydrolyses to give a mixture of nitrous (III) acid and nitric (V) acid:

$$2NO_{2(g)} + 3H_2O_{(l)} \rightleftharpoons 2H_3O^+_{(aq)} + NO^-_{2(aq)} + NO^-_{3(aq)}$$

Compare the K_a values of these two acids which illustrate the relationship between oxidation number and acid strength.

$$HO—\overset{III}{N}O_{(aq)} + H_2O_{(l)} \rightleftharpoons H_3O^+_{(aq)} + {}^-O—NO_{(aq)} \qquad K_a = 5 \times 10^{-4}\ mol\ dm^{-3}$$

$$HO—\overset{V}{N}O_{2(aq)} + H_2O_{(l)} \rightleftharpoons H_3O^+_{(aq)} + {}^-O—NO_{2(aq)} \qquad K_a = 1 \times 10^{2}\ mol\ dm^{-3}$$

Nitrogen (III) chloride is hydrolysed in a different way from phosphorus (III) chloride. The incoming nucleophilic water molecule induces a dipole along one of the chlorine-nitrogen bonds which are non-polar because nitrogen and chlorine have equal electronegativity. The nitrogen atom already has a full shell and so one of the chlorine atoms is attacked.

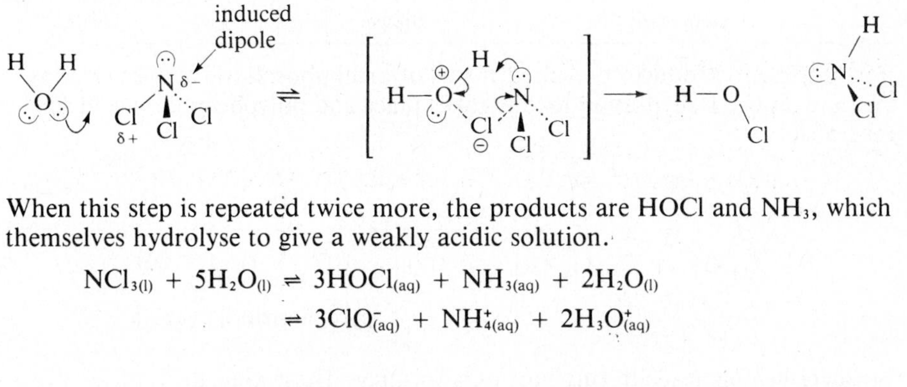

induced dipole

When this step is repeated twice more, the products are HOCl and NH_3, which themselves hydrolyse to give a weakly acidic solution.

$$NCl_{3(l)} + 5H_2O_{(l)} \rightleftharpoons 3HOCl_{(aq)} + NH_{3(aq)} + 2H_2O_{(l)}$$

$$\rightleftharpoons 3ClO^-_{(aq)} + NH^+_{4(aq)} + 2H_3O^+_{(aq)}$$

Phosphorus (III) chloride, on the other hand, is hydrolysed as a result of nucleophilic attack at the electron-deficient phosphorus atoms:

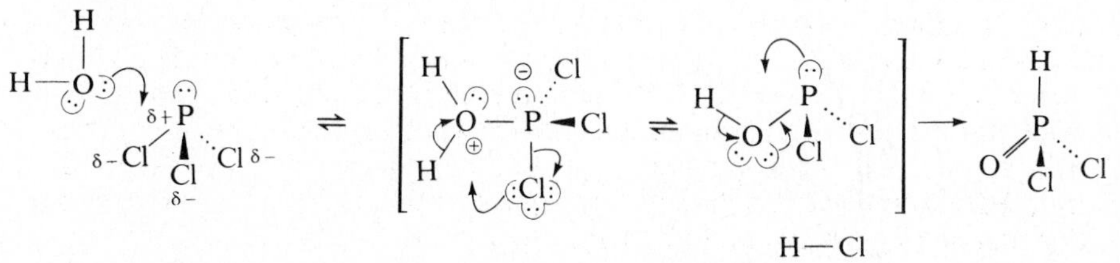

H—Cl

When this step is repeated twice more, the products are HCl and H_3PO_3, which again are themselves extensively hydrolysed:

$$H_3PO_{3(l)} + H_2O_{(l)} \rightleftharpoons H_3O^+_{(aq)} + HP(O)(OH)O^-$$

Phosphorus (III) oxide is similarly hydrolysed to give an acidic solution of H_3PO_3.

The halides and oxides of phosphorus (V) are even more violently hydrolysed than those of phosphorus (III). By an exactly analogous mechanism to that suggested above, the products are solutions of the phosphorus (V) acid, orthosphosphoric acid H_3PO_4. Phosphinic (I) acid is prepared by protonating the phosphate (I) anions produced when white phosphorus disproportionates in alkali (see page 684). The three monomeric hydroxyacids of phosphorus are shown below. Phosphinic (I) acid is monobasic, phosphonic (III) acid is dibasic and orthophosphoric (V) acid is tribasic; only the hydroxyl protons are acidic.

phosphinic (I) acid	*phosphonic (III) acid*	*orthophosphoric (V) acid*
$H_2P(O)(OH)$ monobasic	$HP(O)(OH)_2$ dibasic	$P(O)(OH)_3$ tribasic

It is possible to produce condensed forms of orthophosphoric acid by heating the pure liquid. Dehydration readily takes place and polyphosphoric acids are generated:

$$(HO)_2P(O)OH + HOP(O)(OH)_2 \xrightarrow[-H_2O]{\text{gentle heat}} (HO)_2P(O)\text{–O–}P(O)(OH)_2$$

pyrophosphoric (V) acid

Stronger heating leads to ring polymers forming. These are called metaphosphoric acids (cf metaborates page 653), e.g. trimetaphosphoric acid.

$$(HOPO_2)_3$$

Salts of the nitrogen and phosphorus acids

The only stable crystalline nitrate (III) salts are those of the group IA metals, but a wide range of nitrate (V) salts can be prepared by neutralizing nitric acid. Many of these are hydrated:

$$ZnO_{(s)} + 2H_3O^+_{(aq)} + 2NO^-_{3(aq)} + 3H_2O_{(l)} \xrightarrow{\text{evaporate}} Zn(NO_3)_2.6H_2O_{(s)}$$

An enormous variety of phosphate salts exists. These include the salts of the monomeric acids as well as those of the polyphosphoric and metaphosphoric acids. Some sodium salts are shown below.

Monomers

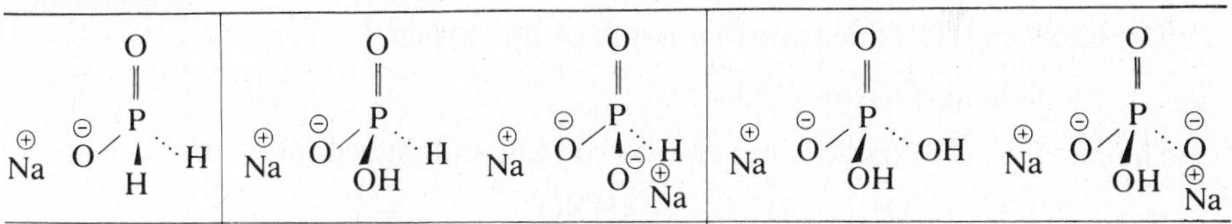

I	III	III	V	V
$Na[H_2PO_2]$	$Na[H_2PO_3]$	$Na_2[HPO_3]$	$Na[H_2PO_4]$	$Na_2[HPO_4]$

Polymers

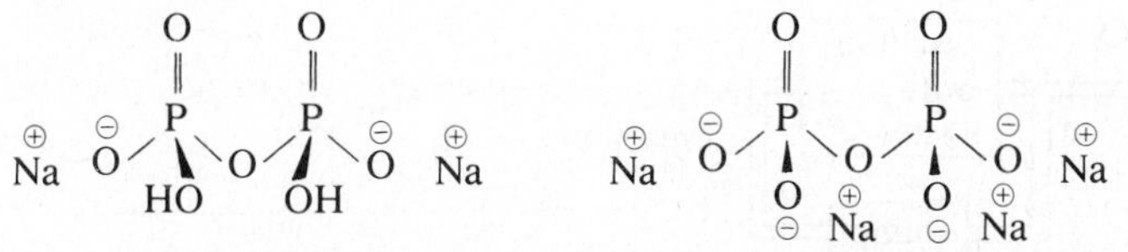

$Na_2[H_2P_2O_7]$	$Na_4[P_2O_7]$

$Na[H_2PO_3]$, $Na[H_2PO_4]$ and $Na_2[H_2P_2O_7]$ are acidic salts: there are still acidic protons present in the anions. Phosphate salts are used as fertilizers, e.g. 'Triple superphosphate' $Ca(H_2PO_4)_2$ and in ion exchange resins for water softeners (metaphosphates).

44.6 Redox properties

The – III state

Ammonia and phosphine both act as reductants. Phosphine is the stronger of the two and can reduce aqueous copper (II) ions or silver (I) ions to the metallic state.

$$8\overset{\text{I}}{Ag}{}^+_{(aq)} + \overset{-\text{III}}{P}H_{3(aq)} + 13H_2O_{(l)} \rightarrow 8\overset{0}{Ag}_{(s)} + H_2\overset{\text{V}}{P}O^-_{4(aq)} + 9H_3O^+_{(aq)}$$

Ammonia reduces copper (II) oxide to copper when passed over a heated sample of the solid:

$$3\overset{\text{II}}{\text{Cu}}\text{O}_{(s)} + 2\overset{-\text{III}}{\text{N}}\text{H}_{3(g)} \rightarrow 3\overset{0}{\text{Cu}}_{(s)} + \overset{0}{\text{N}}_{2(g)} + 3\text{H}_2\text{O}_{(g)}$$

The ammonia is oxidized to nitrogen during this reaction.

Manufacture of nitric acid

Ammonia can also be oxidized to nitrogen (II) oxide by oxygen over a platinum-rhodium catalyst at about 850°C. This reaction is the main step in the production of nitric acid from ammonia.

$$4\overset{-\text{III}}{\text{N}}\text{H}_{3(g)} + 5\overset{0}{\text{O}}_{2(g)} \xrightarrow[850^\circ\text{C}]{\text{Pt/Rh}} 4\overset{\text{II}}{\text{N}}\text{O}_{(g)} + 6\text{H}_2\overset{-\text{II}}{\text{O}}_{(g)}$$

The nitrogen (II) oxide is further oxidized by oxygen.

$$2\text{NO}_{(g)} + \text{O}_{2(g)} \rightarrow 2\text{NO}_{2(g)}$$

The nitrogen (IV) oxide is mixed with oxygen and passed into water

$$4\text{NO}_{2(g)} + \text{O}_{2(g)} + 2\text{H}_2\text{O}_{(l)} \rightleftharpoons 4\text{HNO}_{3(l)}$$

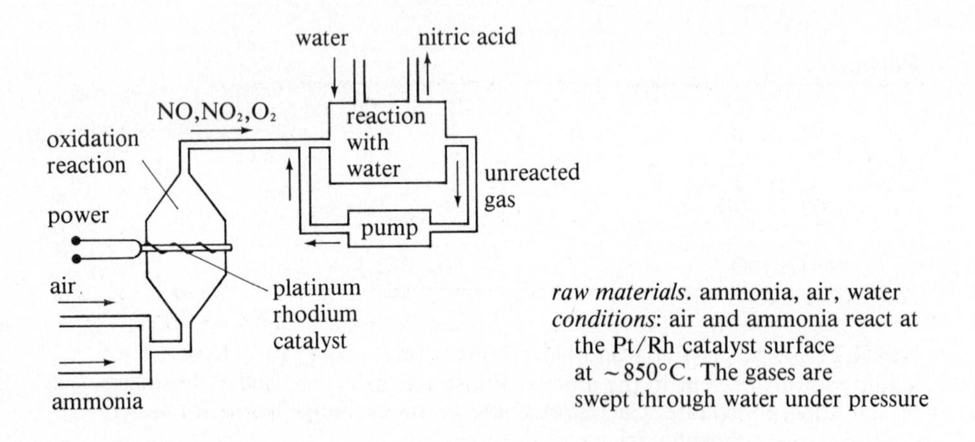

raw materials. ammonia, air, water
conditions: air and ammonia react at the Pt/Rh catalyst surface at ~850°C. The gases are swept through water under pressure

Once the reaction has started, no external heating is required because the first step is exothermic.

The nitric acid that is produced at first has a high concentration of dissolved nitrogen (IV) oxide. Excess NO_2 is removed by fractional distillation and reacted with alkali to give the useful salts sodium nitrate (III) and sodium nitrate (V). The nitric acid forms a 68% azeotrope with water and can only be obtained in a pure state by heating with phosphorus (V) oxide.

The nitrogen oxides

The redox relationship of the different nitrogen oxides shows a change from reducing character to oxidizing character as oxidation number increases.

Nitrogen (II) and nitrogen (III) show both oxidizing and reducing properties as the examples below show.

1. Nitrogen (I) oxide reduces sodium amide, $NaNH_2$, to sodium azide NaN_3. The lattice contains the pπ-stabilized N_3^- ion.

$$N_2O_{(g)} + \overset{\oplus\ \ominus}{NaNH_{2(s)}} \rightarrow \overset{\oplus\ \ominus}{NaN_{3(s)}} + H_2O_{(g)}$$

$$\overset{\ominus}{N}=\overset{\oplus}{N}=\overset{\ominus}{N}$$

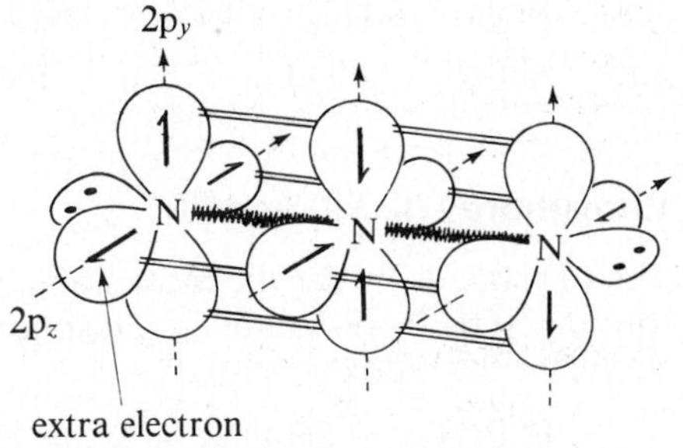

2p_y

2p_z

extra electron

2. Nitrogen (II) oxide reduces oxygen or chlorine in the presence of a catalyst of activated charcoal. It also reduces an aqueous solution of manganate (VII) ions.

$$2\overset{II}{N}O_{(g)} + \overset{0}{Cl}_{2(g)} \rightarrow 2\,\overset{III}{N}O\overset{-I}{Cl}$$

$$5\overset{II}{N}O_{(g)} + 3\overset{VII}{Mn}O^-_{4(aq)} + 4H_3O^+_{(aq)} \rightarrow 5\overset{V}{N}O^-_{3(aq)} + 3\overset{II}{Mn}{}^{2+}_{(aq)} + 6H_2O_{(l)}$$

but it oxidizes sulphur (IV) oxide in the presence of water to sulphuric (VI) acid:

$$\overset{IV}{S}O_{2(g)} + 2\overset{II}{N}O_{(g)} + H_2O_{(l)} \rightarrow H_2\overset{VI}{S}O_{4(l)} + \overset{I}{N}_2O_{(g)}$$

3. Nitrogen (III) hydroxyacid (nitrous acid) similarly, has both oxidizing and reducing properties. It is a stronger oxidant than a reductant and readily oxidizes iodide ions to iodine:

$$2\overset{-I}{I}{}^-_{(aq)} + 2\overset{III}{N}O^-_{2(aq)} + 4H_3O^+_{(aq)} \rightarrow \overset{0}{I}_{2(aq)} + 2\overset{II}{N}O_{(g)} + 2H_2O_{(l)}$$

But the acid also reduces manganate (VII).

4. Nitrogen (IV) oxide has its redox properties largely confined to its tendency to act as an oxidant. It can itself be oxidized to nitric acid by oxygen and water, but the mechanism is likely to proceed via the disproportionation of the oxide in water followed by the subsequent oxidation of the nitrogen (II) oxide produced:

disproportionation of nitrogen (IV)

a) $$2\overset{IV}{N}O_{2(g)} + 3H_2O_{(l)} \rightleftharpoons \overset{V}{N}O^-_{3(aq)} + \overset{III}{N}O^-_{2(aq)} + 2H_3O^+_{(aq)}$$

then disproportionation of nitrogen (III)

b) $$3\overset{III}{N}O^-_{2(aq)} + 2H_3O^+_{(aq)} \rightleftharpoons \overset{V}{N}O^-_{3(aq)} + 3H_2O_{(l)} + 2\overset{II}{N}O_{(aq)}$$

and oxidation of nitrogen (II) → (IV) and back to a)

c) $\overset{II}{2NO}_{(aq)} + O_{2(aq)} \rightarrow \overset{IV}{2NO}_{2(aq)}$

NO_2 oxidizes iodide ions or iron (II) ions in solution, and hydrogen sulphide or sulphur (IV) oxide in the gas phase:

$$Fe^{2+}_{(aq)} + NO_{2(g)} \rightarrow Fe^{3+}_{(aq)} + NO^-_{2(aq)}$$

and $8H_2S_{(g)} + 8NO_{2(g)} \rightarrow S_{8(s)} + 8NO_{(g)} + 8H_2O_{(l)}$

Phosphorus (I), (III) and (V)

The (V) state is by far the most stable oxidation state for phosphorus in solution. The P—H bonds in phosphorus (I) and (III) are strongly reducing. Compare the redox potentials.

$$H_3PO_{3(aq)} + 2H_3O^+_{(aq)} + 2e^- \rightleftharpoons H_3PO_{2(aq)} + 3H_2O_{(l)} \qquad E^\ominus = -0.50 \text{ volts}$$

$$H_3PO_{4(aq)} + 2H_3O^+_{(aq)} + 2e^- \rightleftharpoons H_3PO_{3(aq)} + 3H_2O_{(l)} \qquad E^\ominus = -0.49 \text{ volts}$$

Phosphates (I) and (III) reduce a wide variety of inorganic or organic compounds under aqueous conditions. A useful reduction is effected by phosphinic acid on diazonium ions in the presence of copper (I) ions (see page 505).

$$H_3O^+_{(aq)} + 2e^- + Ar{-}\overset{+}{N}{\equiv}N_{(aq)} \rightarrow Ar{-}H + N_{2(g)} + H_2O_{(l)}$$

Orthophosphoric (V) acid is unlike nitric (V) acid in that it shows almost no oxidizing character. The syrupy liquid is extensively hydrogen bonded as a result of the three hydroxyl groups on each molecule. One of its uses is as a drying agent.

Nitrogen (V)

Nitric (V) acid and nitrate (V) compounds are oxidants. The acid in particular is a powerful oxidant and reacts with both metals and non-metals. Dilute solutions of nitric acid do not evolve hydrogen when added to most metals; nitrogen oxides are evolved instead of hydrogen:

$$3\overset{0}{Mg}_{(s)} + 2\overset{V}{NO}{}^-_{3(aq)} + 8H_3O^+_{(aq)} \rightarrow 3\overset{II}{Mg}{}^{2+}_{(aq)} + 12H_2O_{(l)} + 2\overset{II}{NO}_{(g)}$$

$$\overset{0}{C}_{(s)} + 2H\overset{V}{N}O_{3(l)} \rightarrow \overset{IV}{C}O_{2(g)} + 2\overset{IV}{NO}_{2(g)} + H_2O_{(l)}$$

Nitrogen (V) is less stable than phosphorus (V) because of its pπ-character and the ability of nitrogen atoms to bond to oxygen atoms in a number of different arrangements. Nitric (V) acid is slowly decomposed in the presence of sunlight and the nitrogen (IV) oxide produced is responsible for the yellow colour of old samples of the concentrated acid.

$$4HNO_{3(l)} \xrightarrow{hv} 4NO_{2(g)} + 2H_2O_{(l)} + O_{2(g)}$$

In the pure liquid state, the acid is weakly conducting as a result of the ionization shown below:

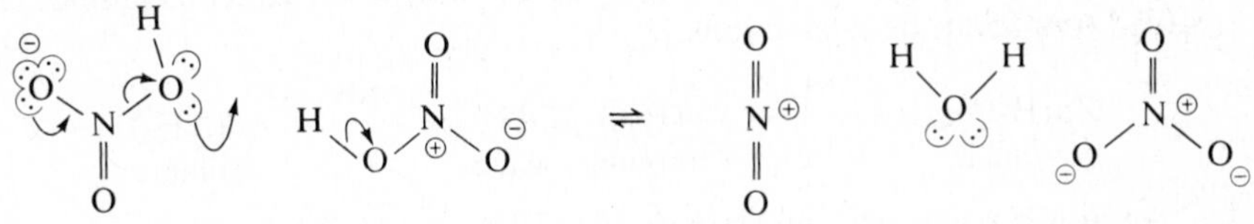

nitronium ion

This equilibrium is powerfully affected by the presence of concentrated sulphuric acid which reacts with the water produced and pulls the position of equilibrium over to the right.

$$2HNO_{3(l)} + H_2SO_{4(l)} \rightleftharpoons NO_2^+ + H_3O^+ + NO_3^- + HSO_4^-$$

A competing acid-base reaction, which also generates nitronium ions, is set up in the mixture as well, see page 499.

The mixture of the two concentrated acids is used as a nitrating agent for aromatic organic compounds because the π-electrons of the ring are attacked by the nitronium ions.

44.7 Solubility and complexing properties

Ammonia and phosphine

The molecules of these two compounds have lone pairs of electrons which can become co-ordinated to metal cations. Ammonia is a stronger complexing agent than phosphine whose lone pair character is limited.

Aqueous ammonia dissolves many insoluble metal salts and hydroxides as a result of complex formation. The number of ammonia molecules that are involved in the formation of the complex varies from two to six:

$$Ag(NH_3)_2^+ \qquad Zn(NH_3)_4^{2+} \qquad Cr(NH_3)_6^{2+}$$

The derivatives of ammonia and phosphine are also often able to act as ligands. Amines, amides and other amino-compounds all form metal complexes. The presence of electron-releasing groups attached to the nitrogen or phosphorus atoms can enhance their complexing properties by making the lone pairs on the atoms more prominent. Thus alkylated and arylated phosphines are quite good ligands, e.g. triphenylphosphine $(C_6H_5)_3P$ stabilizes transition metal cations in low oxidation states.

Ligand, base, or nucleophile

Sometimes a competition takes place between the tendency of ammonia to act as a ligand, as a base or as a nucleophile. Compare this with the discussion on page 623 about hydroxide ions acting in these three roles.

Ammonia acts first as a base, then as a ligand when aqueous ammonia is added to a solution of zinc ions.

$$\underset{\text{soluble}}{Zn(H_2O)^{2+}_{4(aq)}} \xrightarrow{2NH_3} \underset{\text{insoluble, white}}{Zn(H_2O)_2(OH)_{2(s)}} \xrightarrow{4NH_3} \underset{\text{soluble}}{Zn(NH_3)^{2+}_4} + 2H_2O$$

Ammonia reacts with an ester as a nucleophile. The lone pair on the ammonia molecule attacks the electron-deficient carbon atom of the carbonyl group.

$$H_3N\colon + \overset{\delta+}{C}(R)(OR')=\overset{\delta-}{O} \rightleftharpoons \left[H_3\overset{\oplus}{N}-C(R)(OR')-O^{\ominus}\right] \longrightarrow H-O-R' + H_2N-C(R)=O$$

alcohol

amide

However, ammonia reacts with an organic acid as a base, deprotonating the acid.

$$NH_{3(aq)} + RCOOH_{(aq)} \rightarrow RCOO^-_{(aq)} + NH^+_{4(aq)}$$

Nitrosyl complexes (NO^+)

The nitrosyl ion, NO^+, is isoelectronic with a carbon monoxide molecule and it acts as a ligand in a similar way.

The brown ring test

An important nitrosyl complex that is formed in solution and that is used analytically to test for nitrates is the brown-coloured iron complex $[Fe(H_2O)_5NO]^{2+}$. The 'brown-ring' test depends on the formation of this ion. If a solution contains nitrate (V) ions, positive results are obtained as follows:

1. Iron (II) sulphate is added in excess.
2. Concentrated sulphuric acid is poured slowly down the inside of the tube so that two separate layers form.
3. At the acid boundary, the nitrate (V) ions are reduced to nitrogen (II) oxide:

$$3Fe^{2+}_{(aq)} + NO^-_{3(aq)} + 4H_3O^+_{(aq)} \rightarrow NO_{(aq)} + 3Fe^{3+}_{(aq)} + 6H_2O_{(l)}$$

4. The nitrogen (II) oxide immediately combines with unreacted iron (II) ions to give the characteristic brown colour which is observed at the liquid phase boundary.

It is likely that electron transfer takes place from the neutral 'odd electron' molecule, NO, to the iron (II) ion. This means that the brown complex is strictly an iron (I) complex containing an NO^+ ligand.

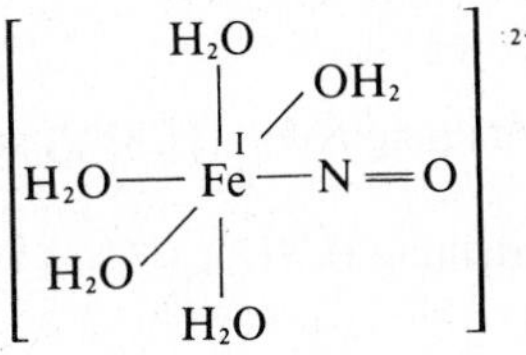

```
                                  2+
 [        H2O                    ]
 [         |   OH2               ]
 [         |I /                  ]
 [ H2O — Fe — N=O                ]
 [        /  |                   ]
 [   H2O     |                   ]
 [          H2O                  ]
```

NO^+ has a greater thermodynamic stability than NO (see page 687).

Questions

1. Explain the meaning of: a) monotropy b) π-overlap c) Lewis base d) oxy-anion e) hydrolysis f) induced dipole g) bond strain h) disproportionation.
2. a) Describe how you would prepare samples of the two group (V) hydrides, NH_3 and PH_3 in the laboratory.
   b) Explain why different methods are required for each.
3. a) Describe the industrial preparation of ammonia emphasizing the important chemical principles involved.
   b) Describe the industrial conversion of ammonia to nitric acid.
   c) What are the important uses of ammonia and nitric acid?
4.

element	*m.p. °C*	*b.p. °C*
nitrogen	−210	−196
phosphorus — red:	590	—
— white:	44.2	280

Account for these very different fixed points.

5. Comment on the following figures:

compound	pK_a
NH_3	9.25
CH_3NH_2	10.64
$(CH_3)_2NH$	10.72
$(CH_3)_3N$	9.80
$C_6H_5NH_2$	4.62
$C_6H_5NHCH_3$	4.40
$(C_6H_5)_2NH$	0.8

6. Compare and contrast the behaviour of NH_3 and OH^- as: a) a base b) a ligand c) a nucleophile.

7. Account for the following facts:
   a) NF_3 does not react with water
   b) hydrolysis of NCl_3 yields ultimately a solution containing $NH^+_{4(aq)}$, $ClO^-_{(aq)}$ and $H_3O^+_{(aq)}$
   c) hydrolysis of PCl_3 yields ultimately a solution containing $H_2PO^-_{2(aq)}$, $Cl^-_{(aq)}$ and $H_3O^+_{(aq)}$.

8. a) Explain why the pK_a figures for the following acids are similar.

acid	pK_a
H_3PO_2	2.0
H_3PO_3	1.8
H_3PO_4	2.1

   b) Why are there no acid salts of H_3PO_2, one for H_3PO_3 and two for H_3PO_4?

9. a) Explain why the ionic solid, PCl_5, does not conduct in the liquid phase.
   b) Explain why PCl_5 melts at 148°C while PBr_5 sublimes at 106°C.
   c) Explain why PI_5 has never been isolated.

10. Compare the acid-base chemistries of nitrogen and phosphorus.

Chapter 45

Oxygen and sulphur: group VIB

45.1 Structure of the elements

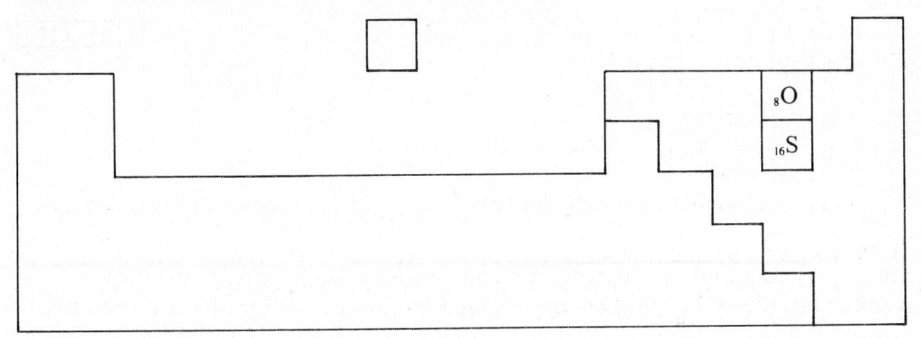

element	electronic configuration	electro-negativity	atomic radius (m × 10^{-10})	anionic (X^{2-}) radius (m × 10^{-10})	stable oxidation states
oxygen	[He] $2s^2$ $2p^4$	3.5	0.74	1.40	–II, –I
sulphur	[Ne] $3s^2$ $3p^4$	2.5	1.04	1.84	–II, II, IV, VI

Atomic structure

Oxygen and sulphur atoms each have six outer-shell electrons and a core charge of six. Their degree of electronic control is therefore high and only exceeded by atoms of elements in the halogen group. The electrons are distributed as shown below.

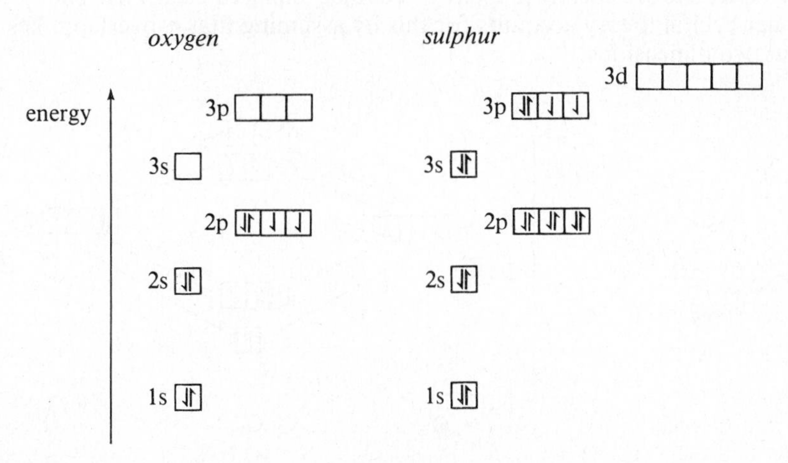

The atoms tend to gain control of two more electrons to fill the vacancies in their p-orbitals and therefore are often found in the −II state. However, oxygen and sulphur also share the bonding characteristics evident in the chemistries of nitrogen and phosphorus (see page 680). Oxygen atoms tend to form pπ-bonds whereas sulphur exhibits variable valency by use of the vacant, energetically low-lying 3d-orbitals.

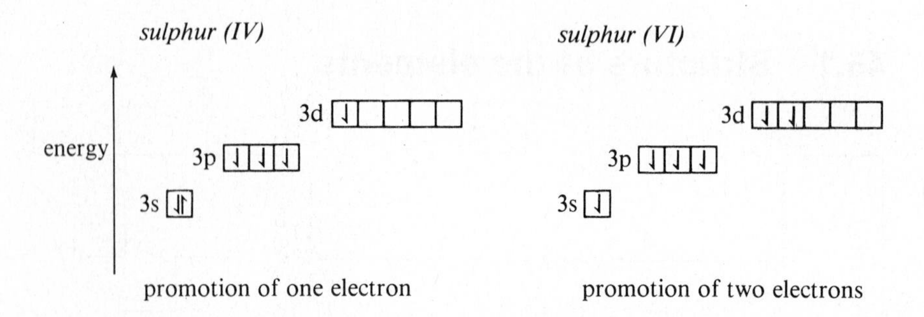

The energy required to promote electrons into the higher energy orbitals is recouped in the extra bond energy evolved by the formation of further bonds.

Physical structure

Both oxygen and sulphur are non-metallic in character and both exist in more than one allotropic form. Oxygen exhibits dynamic allotropy but sulphur is enantiotropic. The enantiotropes of sulphur and their interconversion are described on page 126. Oxygen's allotropes are diatomic oxygen, O_2, and triatomic oxygen, ozone, O_3. The diatomic form is the most common allotrope but ozone is formed in the upper atmosphere as a result of the effect of ultra-violet radiation on diatomic oxygen.

$$O_{2(g)} \overset{h\nu}{\rightleftharpoons} 2O_{(g)} \qquad O_{(g)} + O_{2(g)} \rightleftharpoons O_{3(g)}$$

Both oxygen and ozone contain pπ-bonded molecules. Oxygen is paramagnetic and its molecules are therefore likely to contain unpaired electrons. The molecular orbital theory accounts for this by assuming that π-overlap takes place in two dimensions.

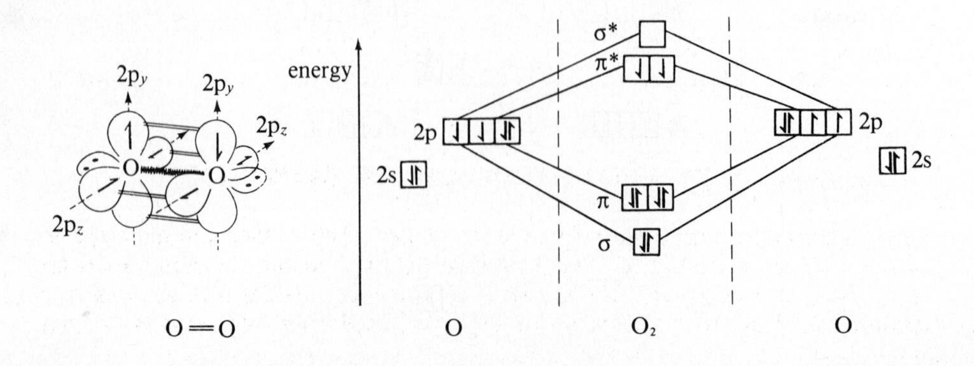

In O_2 there are three pairs of electrons in bonding molecular orbitals and two electrons in the anti-bonding π-molecular orbitals. The net number of bonds is therefore two but, by the Hund principle, the two electrons in the anti-bonding orbitals go separately into the two available π^* orbitals. These single electrons have unpaired spins and so oxygen is paramagnetic.

Ozone is diamagnetic and has the bent structure shown below.

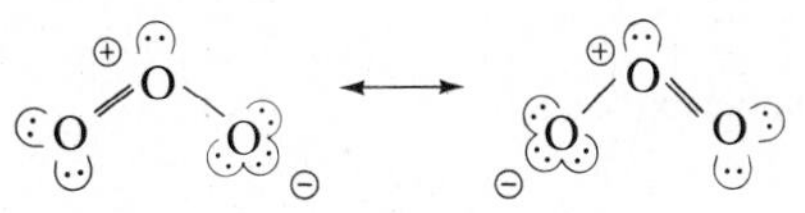

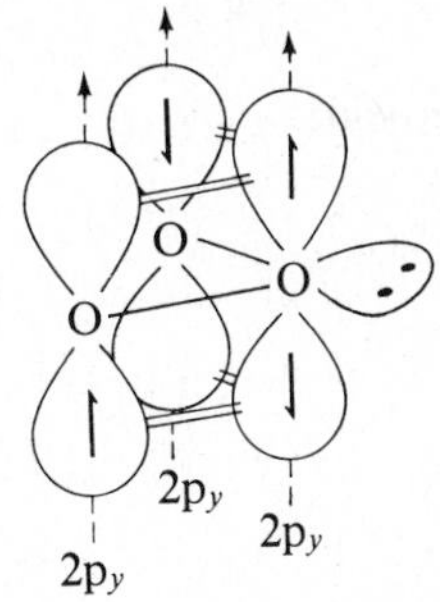

$2p_y$ $2p_y$ $2p_y$

Sulphur does not exist as $p\pi$-stabilized S_2 molecules at room temperature (in much the same way that phosphorus does not exist as P_2). Crown-shaped S_8 molecules are formed in which each sulphur atom is σ-bonded to two others.

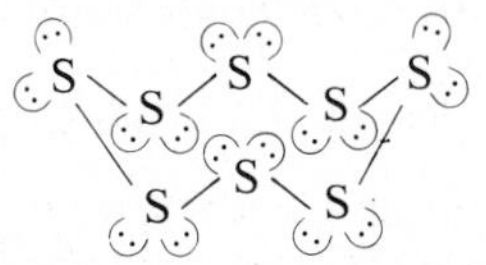

S_8

Some data concerning the elemental forms of oxygen and sulphur are shown in the table below.

form	*oxygen*	*ozone*	*rhombic sulphur*
molecule	O=O		
bond angle	180°	116°50′	107°50′
bond strength (kJ mol^{-1})	496	301	264
m.p. (°C)	−218	−193	113
b.p. (°C)	−183	−112	445

45.2 Compounds formed by the elements

Oxygen combines with almost every other element. With other non-metals, the compounds are molecular and tend to be acidic; with metals the compounds are ionic and tend to be basic. The reaction of basic metal oxides with acidic non-metal oxides leads to a wide range of oxy-salts. In all these compounds, oxygen is present in the −II oxidation state. A few catenated compounds exist, but these are not typical. For example,

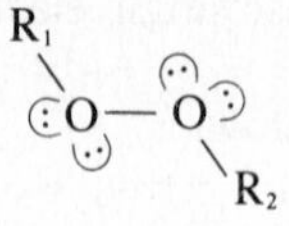

peroxides

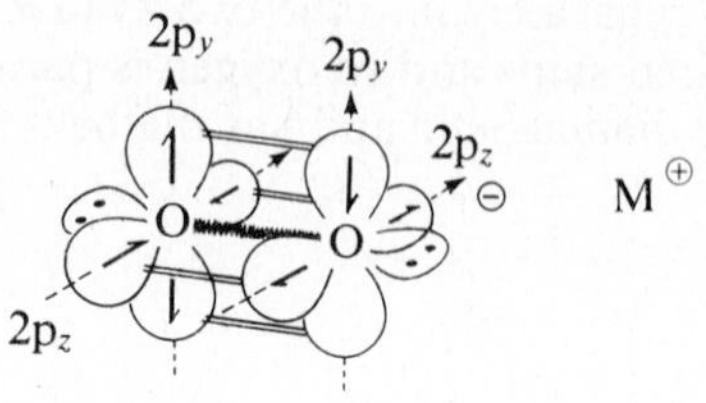

superoxides

Sulphur also forms compounds with nearly all the elements. In the −II state, these are less stable than their oxygen analogues because of sulphur's tendency to expand its octet and achieve higher valencies. Like phosphorus, sulphur's chemistry is dominated by the formation of dπ-pπ-bonded oxy-compounds. These are often produced as a result of air oxidation or hydrolysis, both processes involving oxygen.

For example, the structure of sulphur (IV) oxide is dπ-pπ-stabilized whereas the oxygen analogue, ozone, is pπ-pπ-stabilized.

45.3 Occurrence, extraction and uses

Oxygen

Oxygen is the most abundant element on earth. It makes up 46.6% of the crust and is present as its hydride, water, in the seas, lakes and rivers of the world. The earth's crust is largely made of silicate, carbonate and phosphate rocks, all of which contain combined oxygen. Oxygen also occurs as the free element in the atmosphere (≃21% by volume). The proportion of oxygen in the atmosphere is kept at this level as a result of plant photosynthesis: oxygen used for respiration and combustion releases carbon dioxide into the atmosphere; plants combine this carbon dioxide with water in the presence of sunlight to produce carbohydrates and to release oxygen again which is returned to the atmosphere. It is extracted from the atmosphere by the fractional distillation of liquid air (see page 165).

Laboratory preparation of oxygen

In the laboratory, oxygen can be produced by one of the following methods

1. Catalytic decomposition of hydrogen peroxide

$$2H_2O_{2(aq)} \xrightarrow{MnO_2} 2H_2O_{(l)} + O_{2(g)}$$

2. Thermal decomposition of oxides of unreactive metals or metals in high oxidation states.

$$2Ag_2O_{(s)} \rightarrow 4Ag_{(s)} + O_{2(g)}$$

or $$2PbO_{2(s)} \rightarrow 2PbO_{(s)} + O_{2(g)}$$

3. Thermal decomposition of oxy-salts of elements in high oxidation states.

$$2KMnO_{4(s)} \rightarrow K_2MnO_{4(s)} + MnO_{2(s)} + O_{2(g)}$$

$$KClO_{4(s)} \rightarrow KCl_{(s)} + 2O_{2(g)}$$

Oxy-salts should only be heated in very small quantities. They may explode.

4. Electrolysis of dilute acids and alkalis: at the anode, the water ligands around the anions give up electrons and oxygen forms (see page 341).

$$6H_2O_{(l)} \xrightarrow[\text{anode}]{\text{at}} 4H_3O^+_{(aq)} + O_{2(g)} + 4e^-$$

Ozone is prepared in the laboratory by passing an electric discharge through oxygen. The inner and outer surface of a water condenser are coated with aluminium foil and connected to an induction coil.

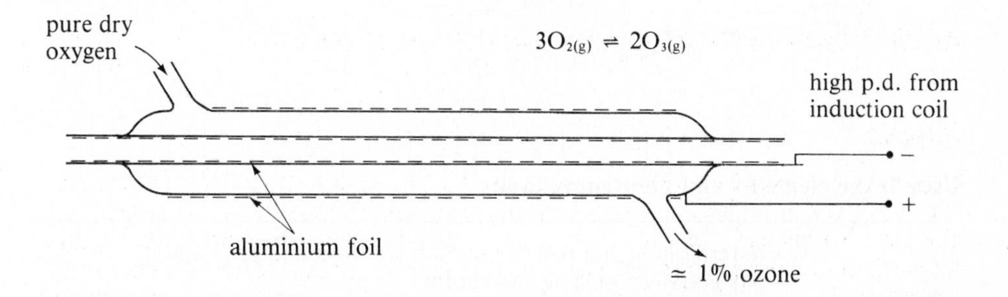

Sulphur

Sulphur occurs as the free element trapped between layers of volcanic rock underground, mainly in Japan and in the U.S.A. Sulphur also occurs in the combined state as sulphates such as gypsum, $CaSO_4.2H_2O$, and anhydrite, $CaSO_4$, and many sulphide ores such as iron pyrites, FeS_2. It is also found as hydrogen sulphide mixed with crude oil and natural gas.

Extraction of sulphur: the Frasch process

Sulphur is extracted from volcanic sulphur beds using a Frasch pump. This consists of a series of concentric pipes which penetrate down into the sulphur deposit. The sulphur is melted using steam and pumped to the surface. The details are shown below:

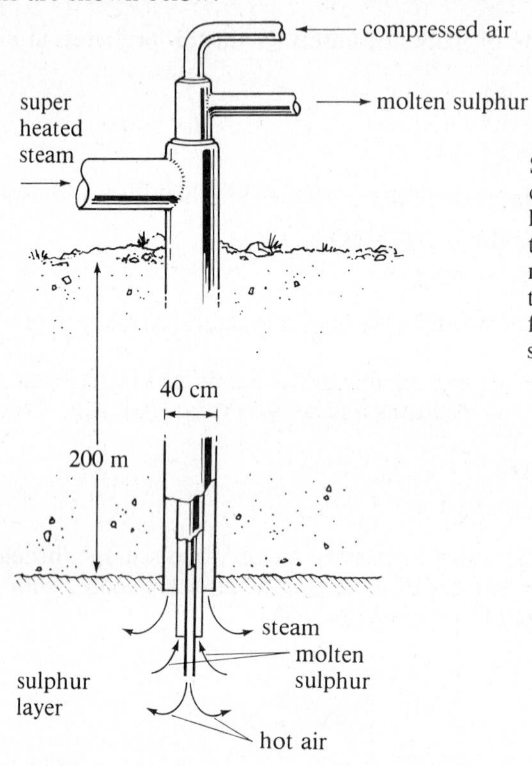

Super-heated steam and compressed air, at 170°C and high pressure, are blown into the sulphur layer. The sulphur melts and forms a froth with the steam and air. The froth is forced up the inner pipe to the surface.

The froth is passed into huge containers. The sulphur solidifies out of the air and water mixture. 99% pure sulphur is produced.

Uses of the elements and their compounds

oxygen	in respiration and fuel combustion (for heat and propulsion). oxy-acetylene welding and cutting steel manufacture hospitals, high altitude flying, diving
sulphur	most of the sulphur extracted is used to manufacture sulphuric acid. Some is used in the manufacture of gunpowder, for vulcanizing rubber and as a pesticide. Some sulphur dioxide is also used in the paper industry
sulphuric acid	in the manufacture of fertilizers ('superphosphate' and ammonium sulphate), paints, synthetic fibres, hydrofluoric acid, detergents, dyes, explosives and drugs.

45.4 Compound structure

The – II state

This oxidation state dominates the chemistry of oxygen compounds. Oxygen is the second most electronegative element known and can only exist in a positive oxidation state under unusual circumstances (see for example page 711) or when combined with fluorine. F_2O_2 and F_2O have both been prepared. The classification of the oxygen (– II) compounds, the oxides, in terms of their structures is shown below.

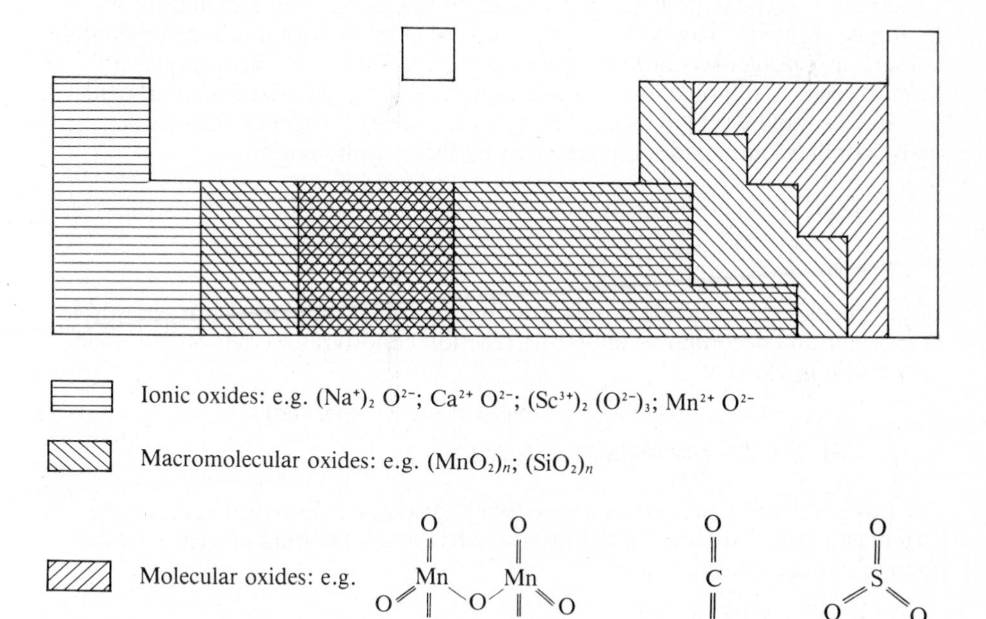

These structures illustrate the ability of an oxygen atom to bond in three different ways

1. As the oxide ion O^{2-}
2. As σ-bonded divalent oxygen: O
3. As σ/π bonded divalent oxygen: =O

Either pπ-pπ or dπ-pπ bonds form depending on the nature of the bonded atom. For example, carbon forms pπ bonds, but phosphorus forms dπ bonds as shown overleaf.

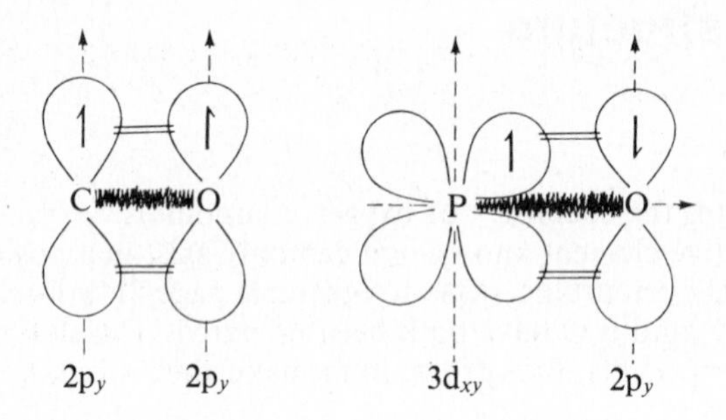

Sulphur (−II) compounds are confined to the metal sulphides and to hydrogen sulphide. Most other compounds of sulphur with a non-metal contain sulphur in a positive oxidation state because it has a lower electronegativity than oxygen. A sulphide ion is larger and more easily polarized than an oxide ion and so sulphides tend to have a higher degree of covalency than oxides. Most sulphides and oxides are prepared by direct combination:

$$8Zn_{(s)} + S_{8(s)} \xrightarrow{\text{heat}} 8ZnS_{(s)}$$

$$4Fe_{(s)} + 3O_{2(g)} \xrightarrow{\text{heat}} 2Fe_2O_{3(s)}$$

Both hydrides can be made in this way, but hydrogen sulphide is far less stable than water and decomposes under the reaction conditions so that only a small conversion is achieved.

$$2H_{2(g)} + O_{2(g)} \rightarrow 2H_2O_{(g)} \quad \text{(explosive when sparked)}$$

$$16H_{2(g)} + S_{8(l)} \underset{\text{heat}}{\rightleftharpoons} 8H_2S_{(g)}$$

The rather different properties of the two hydrides are described later. These differences are mostly due to the extensive hydrogen bonding present in water and the stronger O—H bonds:

O—H = 463 kJ mol^{-1}
S—H = 338 kJ mol^{-1}

compound	*m.p.* (°C)	*b.p.* (°C)	$\Delta H_f^{\ominus}$(kJ mol^{-1})	K_a (mol dm^{-3})
water	0	100	−286	2×10^{-16}
hydrogen sulphide	−85	−61	−20	1×10^{-7}

The −I state

Peroxide bonds
Oxygen combines with reactive metals to produce a mixture of simple oxides (O^{2-}), peroxides ($^-O—O^-$) and sometimes even superoxides (O_2^-). These reactions are described on page 631. The peroxide link occurs not only in ionic peroxides but also in molecular compounds such as hydrogen peroxide and the peroxyacids.

Hydrogen peroxide
Hydrogen peroxide is made in the laboratory by hydrolysing barium peroxide with ice-cold sulphuric acid. Barium sulphate is precipitated and filtered from the solution.

$$Ba^{2+}\ [O{-}O]^{2-}_{(s)} + 2H_3O^{+}_{(aq)} + SO^{2-}_{4(aq)} \rightarrow BaSO_{4(s)} + 2H_2O_{(l)} + H_2O_{2(aq)}$$

Industrially, it is prepared by the reaction of a quinol with oxygen. The quinol is produced *in situ* by the catalytic reduction of a quinone. The reaction of quinol with oxygen regenerates the quinone, which is therefore almost acting as a catalyst itself:

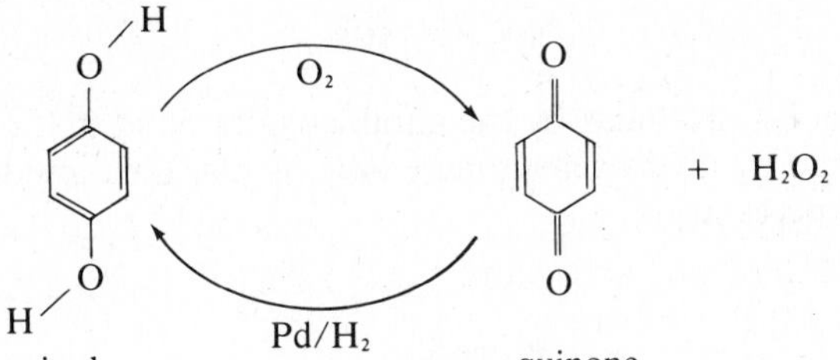

quinol $\xrightarrow{O_2}$ quinone + H_2O_2 ; quinone $\xrightarrow{Pd/H_2}$ quinol

The mechanism is homolytic and usually an anthraquinone derivative is used as catalyst:

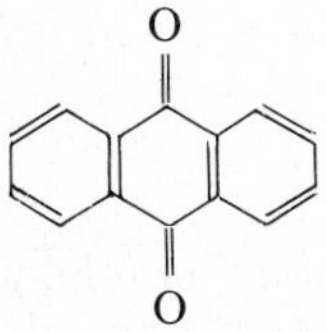

(anthraquinone)

The structure of hydrogen peroxide is shown below.

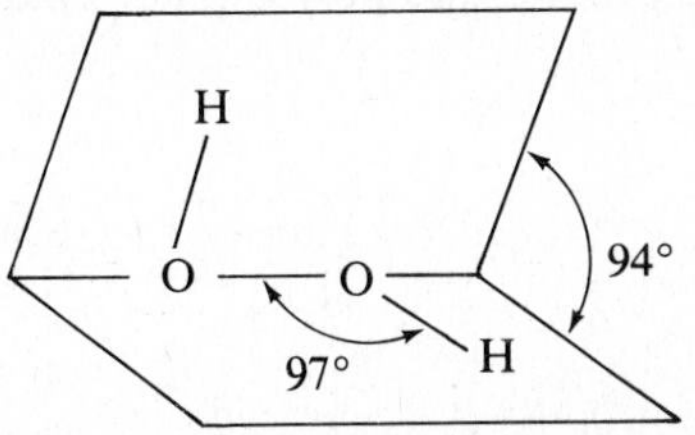

H, O—O, H; 97°; 94°

Peroxyacids
Two typical 'peracids' are formed by sulphur. Peroxymonosulphuric (VI) acid (Caro's acid), H_2SO_5, and peroxydisulphuric (VI) acid, $H_2S_2O_8$, both have peroxide links in their molecular structure.

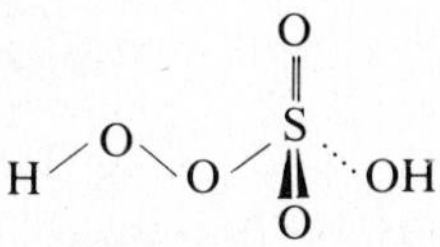

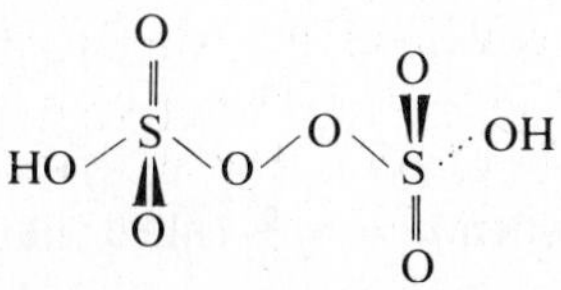

A number of organic peracids can also be prepared. The general method used for making these compounds is to combine an acid anhydride or chloride with hydrogen peroxide. For example, Caro's acid is made by reacting the acid monochloride of sulphuric acid with hydrogen peroxide:

$$HO{-}SO_2{-}Cl + H_2O_2 \longrightarrow HO{-}SO_2{-}O{-}O{-}H$$

Caro's acid

Peroxydisulphuric (VI) acid is produced by the anodic oxidation at 0°C of a saturated solution of potassium hydrogensulphate (VI). A platinum anode and a high charge density are necessary.

$$HO{-}SO_2{-}O^{\ominus} \quad {}^{\ominus}O{-}SO_2{-}OH \xrightarrow[-2e^-]{\text{Pt anode}} HO{-}SO_2{-}O{-}O{-}SO_2{-}OH$$

It is readily hydrolysed to Caro's acid.

Sulphur-sulphur bonds

Sulphur itself also tends to form catenated compounds containing sulphur-sulphur σ-bonds analogous to the peroxide link. When sulphur is boiled in aqueous alkali, the element disproportionates. The reaction mechanism is very similar to that described for phosphorus (page 684). However, not all the sulphur-sulphur σ-bonds are broken in the process and polysulphide chains are produced in the solution:

$$nS_8 \xrightarrow[\text{concentrated}]{OH^-_{(aq)}} \{S{-}S{-}S{-}S{-}S{-}S{-}S\}^{2-} + SO^{2-}_{3(aq)}$$

Insoluble hydrogen polysulphides are precipitated when the solution is neutralized. They smell foul. Acid derivatives of these polysulphides are also produced during the disproportionation. They have the general formula

$$HO{-}SO_2{-}[S{-}S]_n{-}SO_2{-}OH$$

The simplest, when $n = 1$, is called the tetrathionate and is the product of the oxidation of thiosulphate ions:

$$I_{2(aq)} + 2S_2O^{2-}_{3(aq)} \rightarrow 2I^-_{(aq)} + S_4O^{2-}_{6(aq)}$$

Two halides that also contain sulphur-sulphur bonds are described in the next section. They are:

hal—S—S—hal and $F_5S—SF_5$

(S_2hal_2) (S_2F_{10})

Positive oxidation states

Apart from the fluorides and some unusual complexes (e.g. $O_2^+[PtF_6]^-$) oxygen has no positive oxidation state chemistry. Sulphur, however, forms compounds over a range of positive oxidation states.

I and II	*IV*		*V and VI*	
Hal—S—S—Hal	SF_4	SCl_4	SF_6	$F_5S—SF_5$
$SHal_2$	SO_2	$SOCl_2$	SO_3	SO_2Cl_2
	H_2SO_3 and its salts		H_2SO_4 and its salts	

Sulphuric (IV) acid, H_2SO_3, is unstable and decomposes to SO_2 and water.

Sulphur reacts with fluorine to produce a complex mixture of the I, II, IV, V and VI state fluorides. In excess fluorine, SF_6 is the major product. This is a remarkably inert and stable compound and is used as an insulator for high voltage transformers. With chlorine, the I state dominates because both sulphur (II) and (IV) chloride readily decompose. Disulphur (I) dichloride is a foul-smelling liquid that is used as a vulcanizing agent in the rubber industry. Vulcanizing is the process of cross-linking hydrocarbon chains in rubber with —S—S— links. It hardens the rubber and gives it a wider range of uses.

Sulphur burns in oxygen to produce the dioxide; the trioxide is manufactured on a massive scale by catalytic oxidation of sulphur (IV) oxide (see page 720). Both oxides are extensively hydrolysed by water to give sulphates (IV) and (VI).

The oxy-chloro compounds can be prepared from the unhydrolysed oxides or from anhydrous sulphate salts. For example, dichlorosulphur (VI) oxide is obtained by reacting sulphur (IV) oxide and chlorine in the presence of a hot, activated charcoal catalyst.

$$SO_2 + Cl_2 \xrightarrow[C_{(s)}]{heat} SO_2Cl_2$$

Dichlorosulphur (IV) oxide is prepared by chlorinating anhydrous sodium sulphate (IV) with a non-oxidizing chlorinating agent e.g. trichlorophosphorus (V) oxide.

$$3Na_2SO_{3(s)} + 2\,POCl_{3(l)} \xrightarrow{distil} 2Na_3PO_{4(s)} + 3SOCl_{2(g)}$$

$SOCl_2$ and SO_2Cl_2 are chlorinating agents like PCl_5 (see page 415).

Thio compounds

Although pπ-stabilized S_2 is not known at room temperature, sulphur atoms do show some tendency to undergo pπ-overlap with one another and with other atoms. Sulphur atoms that bond in this way are acting more like oxygen atoms than as typical sulphur atoms. Any compound that contains sulphur atoms where there are usually oxygen atoms is called a 'thio' compound. Two important examples are shown below:

$(Na^+)_2$ $S_2O_3^{2-}$ (S=S with three O, two bearing ⊖)

sodium thiosulphate (VI)

$K^{\oplus}$ $^{\ominus}S{-}C{\equiv}N$

potassium thiocyanate

Sodium thiosulphate is obtained by boiling a solution of sodium sulphate (IV) in the presence of powdered sulphur:

$$8SO^{2-}_{3(aq)} + S_{8(s)} \rightarrow 8S_2O^{2-}_{3(aq)}$$

Potassium thiocyanate is produced by fusing potassium cyanide with sulphur.

$$8KCN_{(s)} + S_{8(s)} \xrightarrow{heat} 8K(SCN)_{(s)}$$

The oxygen analogues of the two thiocompounds are sodium sulphate (VI) (Na_2SO_4) and potassium cyanate (KCNO).

45.5 Acid-base properties

Oxides and sulphides (– II)

Metal oxides and sulphides
These are all insoluble bases (except for those of the group IA metals, which are soluble and alkaline).

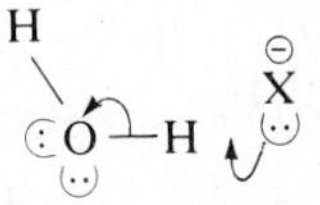

$$Na_2O_{(s)} + H_2O_{(l)} \rightarrow 2Na^+_{(aq)} + 2HO^-_{(aq)}$$
$$Na_2O_{2(s)} + 2H_2O_{(l)} \rightarrow 2Na^+_{(aq)} + 2HO^-_{(aq)} + H_2O_{2(aq)}$$
$$Na_2S_{(s)} + H_2O_{(l)} \rightarrow 2Na^+_{(aq)} + HS^-_{(aq)} + HO^-_{(aq)}$$

As the polarizing power of the metal cation increases in an oxide lattice, so its basic character decreases. The lone pairs of the oxide ion are held more closely to the cation and are thus less able to be donated to protons.

M^{n+} O^{2-}

strong polarizing power

The oxides of aluminium and beryllium are therefore amphoteric because the cations act as electron-pair acceptors for attacking hydroxide ions:

$$\underset{\text{acid}}{Al_2O_{3(s)}} + \underset{\text{base}}{2OH^-_{(aq)}} + 3H_2O_{(l)} \rightleftharpoons \underset{\text{salt}}{2[Al(OH)_4]^-}$$

$$\underset{\text{base}}{Al_2O_{3(s)}} + \underset{\text{acid}}{6H_3O^+_{(aq)}} \rightarrow \underset{\text{salt}}{2Al^{3+}_{(aq)}} + 9H_2O_{(l)}$$

The basic character of a metal sulphide is used in the preparation of hydrogen sulphide gas. Dilute hydrochloric acid is usually added to iron (II) sulphide:

$$FeS_{(s)} + 2H_3O^+_{(aq)} \rightarrow Fe^{2+}_{(aq)} + 2H_2O_{(l)} + H_2S_{(g)}$$

Hydrogen sulphide itself is a weak acid as a result of the large discrepancy between the bond energies of S—H and O—H.

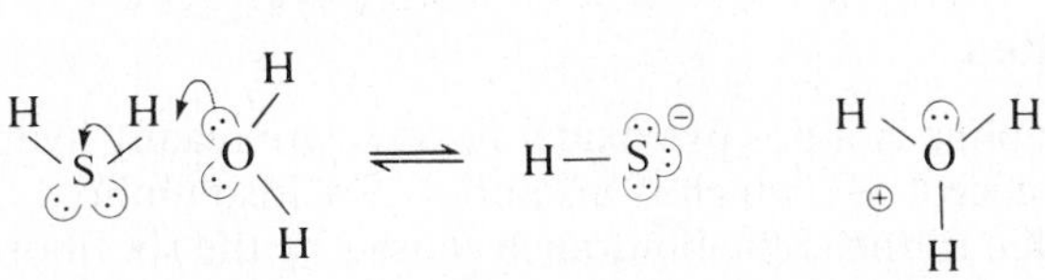

$$H_2S + H_2O \rightleftharpoons HS^- + H_3O^+ \qquad pK_a = 7$$

It is a stronger acid than water for this reason.

Non-metal oxides

In general these tend to be acidic. There are a few that are neutral (e.g. N_2O, NO and CO) and these contain a non-metal in a low oxidation state. This leads to less charge separation in the pπ-stabilized molecules and therefore less tendency for the molecules to accept electron pairs from water molecules. It is the electron deficiency of the non-metal atoms in the molecules that is responsible for the acidic nature of the oxides:

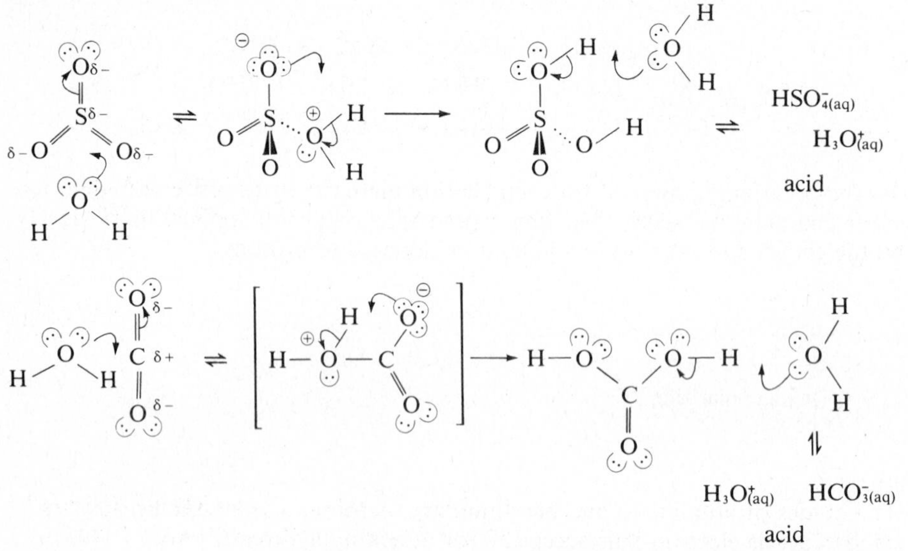

$HSO^-_{4(aq)}$ $H_3O^+_{(aq)}$ acid

$H_3O^+_{(aq)}$ $HCO^-_{3(aq)}$ acid

The non-metal anions are also stabilized by the delocalization of the charge through the π-system of the anion.

Hydrogen oxide is one of the most interesting of all the non-metal oxides. It is one of the very few that is not pπ-stabilized and it acts as both acid and base. The acidic action of one water molecule on another acting as a base leads to partial ionization in the liquid phase:

$$H_2O_{(l)} + H_2O_{(l)} \rightleftharpoons H_3O^+_{(aq)} + HO^-_{(aq)} \qquad pK_c = 17.4$$

All oxides are closely related to their hydroxides. This is inevitable because a hydroxide is simply the protonated form of an oxide. Compare the acid-base properties of the hydroxides discussed on page 620.

Positive oxidation states

All the compounds of sulphur in a positive oxidation state are readily hydrolysed to give acidic solutions, except SF_6 which is insoluble. The insolubility of this compound is probably due to the steric hindrance caused by the six fluoride groups around the S(VI) atom. The lower state sulphur compounds disproportionate in water while sulphur (IV) and sulphur (VI) generate the respective (IV) and (VI) acids. Hydrogen sulphate (IV) is not stable under acidic conditions and decomposes to give sulphur (IV) oxide and water. Compare the hydrolysis reactions below:

sulphur's oxidation state	*equation*
I	$2\overset{I}{S}_2Cl_{2(l)} + 6H_2O_{(l)} \rightarrow 3\overset{0}{S}_{(s)} + \overset{IV}{S}O_{2(g)} + 4Cl^-_{(aq)} + 4H_3O^+_{(aq)}$
IV	$SO_{2(g)} + 3H_2O_{(l)} \rightleftharpoons HSO^-_{3(aq)} + H_3O^+_{(aq)}$
IV	$SOCl_{2(l)} + 3H_2O_{(l)} \rightarrow SO_{2(g)} + 2Cl^-_{(aq)} + 2H_3O^+_{(aq)}$
VI	$SO_2Cl_{2(l)} + 5H_2O_{(l)} \rightarrow HSO^-_{4(aq)} + 2Cl^-_{(aq)} + 3H_3O^+_{(aq)}$

The lack of stability of H_2SO_3 is used as a laboratory method of preparing samples of sulphur (IV) oxide. Dilute hydrochloric acid is added to a metal sulphate (IV):

$$CaSO_{3(s)} + 2H_3O^+_{(aq)} \rightarrow Ca^{2+}_{(aq)} + 3H_2O_{(l)} + SO_{2(g)}$$

Thiosulphuric (VI) acid is also unstable and decomposes to sulphur and sulphur (IV) oxide. The sulphur produced is colloidal.

$$Na_2S_2O_{3(s)} + 2H_3O^+_{(aq)} \rightarrow 2Na^+_{(aq)} + 3H_2O_{(l)} + S_{(s)} + SO_{2(g)}$$

Sulphate salts

Although sulphuric (IV) acid is not stable, its salts are. Both sulphate (IV) and hydrogensulphate (IV) salts exist in solution, and sulphate (IV) salts of group (I) metals can be crystallized. Hydrogensulphate (IV) salts cannot be crystallized from solution because the neighbouring anions react, and water condenses out leaving a pyrosulphate (IV):

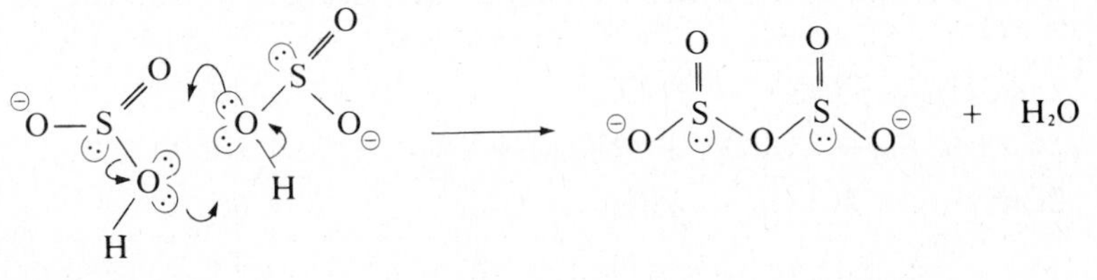

$$\text{2 HSO}_3^- \longrightarrow {}^-O{-}S(=O){-}O{-}S(=O){-}O^- + H_2O$$

Similarly, sulphate (VI) and hydrogensulphate (VI) salts are obtained by neutralizing sulphuric acid. Like many non-metal hydroxy-acids, sulphuric acid can be dehydrated to give pyrosulphuric acid:

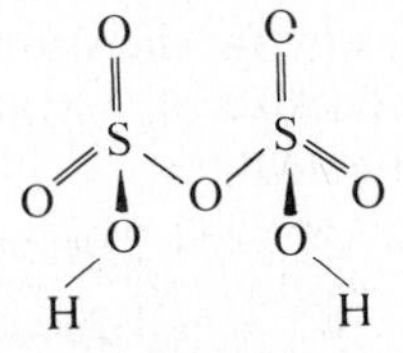

$$H_2S_2O_7$$

This acid may also be obtained by dissolving SO_3 in H_2SO_4.

Salts of these pyro-acids, as well as the salts of a wide range of catenated and thio-acids are also known. Some examples are shown below.

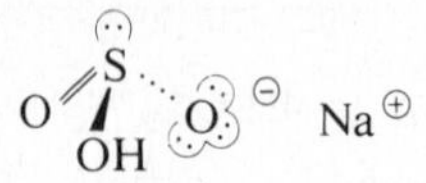

$NaHSO_3$ sodium hydrogen sulphate (IV) (an acid salt)

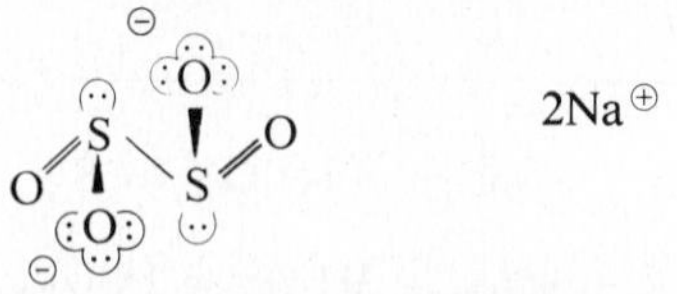

$Na_2S_2O_4$ sodium dithionate (IV) (made by reducing Na_2SO_3 with SO_2 and finely divided zinc)

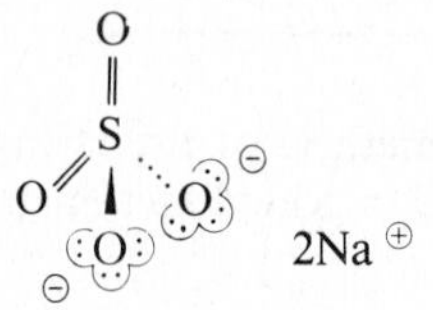

Na_2SO_4 sodium sulphate (VI)

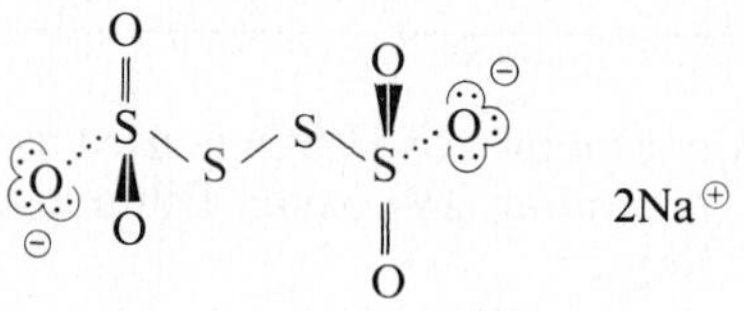

$Na_2S_4O_6$ sodium tetrathionate (VI)

The most common sulphates are sulphates (VI), and these salts occur naturally in certain minerals and rocks, e.g. gypsum $CaSO_4.2H_2O$ and Epsom Salts $MgSO_4.7H_2O$. They are mostly soluble in water with the important exceptions of lead (II) sulphate and barium (II) sulphate. A test for the presence of sulphate (VI) ions in solution is the addition of an acidified barium chloride solution. A precipitate of barium sulphate confirms their presence:

$$Ba^{2+}_{(aq)} + SO^{2-}_{4(aq)} \rightarrow BaSO_{4(s)}$$

They are more thermally stable than most oxy-salts. Compare the decomposition temperatures of copper (II) hydroxide, copper (II) nitrate (V), and copper (II) sulphate (VI).

$$Cu(OH)_{2(s)} \rightarrow CuO_{(s)} + H_2O_{(g)} \qquad T > 70°C$$

$$2Cu(NO_3)_{2(s)} \rightarrow 2CuO_{(s)} + 4NO_{2(g)} + O_{2(g)} \qquad T > 200°C$$

$$2CuSO_{4(s)} \rightarrow 2CuO_{(s)} + 2SO_{2(g)} + O_{2(g)} \qquad T > 650°C$$

45.6 Redox properties

The elements themselves tend to act as oxidants. Ozone is by far the strongest oxidant, followed by pure oxygen. Sulphur can itself be oxidized by oxygen and the halogens, but in reactions with metals, it acts as an oxidant.

$$O_{3(g)} + 6H_3O^+_{(aq)} + 6e^- \rightleftharpoons 9H_2O_{(l)} \qquad E^\ominus = +2.07 \text{ volts}$$

$$O_{2(g)} + 4H_3O^+_{(aq)} + 4e^- \rightleftharpoons 6H_2O_{(l)} \qquad E^\ominus = +1.23 \text{ volts}$$

$$S_{8(s)} + 16H_3O^+_{(aq)} + 16e^- \rightleftharpoons 8H_2S_{(g)} + 16H_2O_{(l)} \qquad E^\ominus = +0.14 \text{ volts}$$

Ozone is used for ozonolysis (see page 383) and is strong enough to oxidize solid lead (II) sulphide to the sulphate and to oxidize chloride ions in solution.

$$Pb\overset{-II}{S}_{(s)} + 4\overset{0}{O}_{3(g)} \rightarrow Pb\overset{IV}{S}\overset{-II}{O}_{4(s)} + 4O_{2(g)}$$

$$2\overset{-I}{Cl}^{-}_{(aq)} + \overset{0}{O}_{3(g)} + 2H_3O^{+}_{(aq)} \rightarrow \overset{0}{Cl}_{2(g)} + 3H_2\overset{-II}{O}_{(l)} + O_{2(g)}$$

The word 'oxidation' was first used for the reactions of oxygen with other elements and with organic compounds. Oxygen's oxidation state goes down from 0 to −II in these processes while the other element's oxidation state increases as its atoms lose control of electrons to the oxygen atoms:

$$2\overset{0}{Mg}_{(s)} + \overset{0}{O}_{2(g)} \rightarrow 2\overset{II}{Mg}\overset{-II}{O}_{(s)}$$

$$\overset{0}{P}_{4(s)} + 5\overset{0}{O}_{2(g)} \rightarrow \overset{V}{P}_4\overset{-II}{O}_{10(s)}$$

Peroxides and peracids

Peracids are used as oxidants in both inorganic and organic systems. The peroxide link contains oxygen in the −I state and this tends to drop to the more stable −II state, bringing about oxidation. For example,

Persulphate ions readily oxidize iodide ions or iron (II) ions in solution. In the presence of silver ions (catalysts) they even oxidize manganate (II) ions to manganate (VII)

$$5S_2O^{2-}_{8(aq)} + 2Mn^{2+}_{(aq)} + 24H_2O_{(l)} \rightarrow 10SO^{2-}_{4(aq)} + 2MnO^{-}_{4(aq)} + 16H_3O^{+}_{(aq)}$$

Compare the standard redox potentials for the two couples involved.

$$S_2O^{2-}_{8(aq)} + 2e^- \rightleftharpoons 2SO^{2-}_{4(aq)} \qquad E^{\ominus} = +2.01 \text{ volts}$$

$$MnO^{-}_{4(aq)} + 8H_3O^{+}_{(aq)} + 5e^- \rightleftharpoons Mn^{2+}_{(aq)} + 12H_2O_{(l)} \qquad E^{\ominus} = +1.52 \text{ volts}$$

Hydrogen peroxide

This is able to act as either an oxidant or as a reductant, and its properties depend on the nature of the system in which it is mixed.

as oxidant: $$H_2O_{2(aq)} + 2H_3O^{+}_{(aq)} + 2e^- \rightleftharpoons 4H_2O_{(l)}: \qquad E^{\ominus} = +1.77 \text{ volts}$$

as reductant: $$O_{2(aq)} + 2H_3O^{+}_{(aq)} + 2e^- \rightleftharpoons H_2O_{2(aq)} + 2H_2O_{(l)}: \qquad E^{\ominus} = +0.68 \text{ volts}$$

a) H_2O_2 oxidizes lead (II) sulphide to lead (II) sulphate (VI) and iron (II) to iron (III) or iodide (−I) to iodine in solution.

$$Pb\overset{-II}{S}_{(s)} + 4H_2\overset{-I}{O}_{2(aq)} \rightarrow Pb\overset{VI}{S}O_{4(s)} + 4H_2\overset{-II}{O}_{(l)}$$

$$2\overset{II}{Fe}^{2+}_{(aq)} + H_2\overset{-I}{O}_{2(aq)} + 2H_3O^{+}_{(aq)} \rightarrow 2\overset{III}{Fe}^{3+}_{(aq)} + 4H_2\overset{-II}{O}_{(l)}$$

The first reaction is useful to picture restorers. Old oil paintings contain lead-based pigments that have darkened with age. This is a result of the lead ions reacting with sulphides released from burning coal. Lead sulphide is black while lead sulphate, its oxidation product, is white.

b) H_2O_2 reduces manganate (VII) to manganate (II) and chlorate (V) to chloride (– I):

$$MnO^-_{4(aq)} + 5H_2O_{2(aq)} + 6H_3O^+_{(aq)} \rightarrow Mn^{2+}_{(aq)} + 14H_2O_{(l)} + 3O_{2(g)}$$
$$ClO^-_{3(aq)} + 3H_2O_{2(aq)} \rightarrow Cl^-_{(aq)} + 3H_2O_{(l)} + 3O_{2(g)}$$

Thio-compounds

Thio-compounds based on sulphur (IV) are usually strong reductants while those based on sulphur (VI) often show weak reducing properties. The two examples given below illustrate these points.

Sodium dithionate (IV)
This is used as an efficient reactant to remove oxygen from a mixture of gases. It reacts rapidly with oxygen in the presence of the β-sulphonate of anthraquinone which catalyses the reaction:

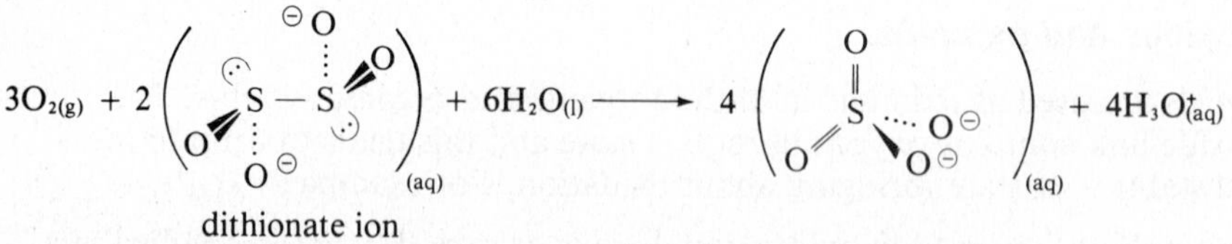

$3O_{2(g)}$ + 2 (dithionate ion)$_{(aq)}$ + $6H_2O_{(l)}$ ⟶ 4 (sulphate ion)$_{(aq)}$ + $4H_3O^+_{(aq)}$

dithionate ion

The catalyst has the structure

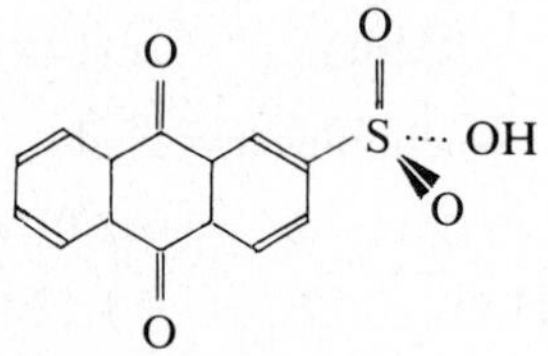

Sodium thiosulphate (VI)
This is used in the volumetric analysis of iodine. It reduces iodine to iodide ions while being oxidized to the tetrathionate salt which shows no appreciable redox properties:

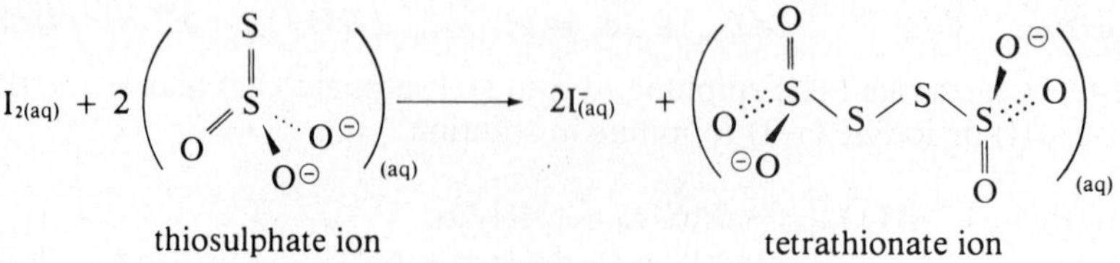

$I_{2(aq)}$ + 2 (thiosulphate ion)$_{(aq)}$ ⟶ $2I^-_{(aq)}$ + (tetrathionate ion)$_{(aq)}$

thiosulphate ion

tetrathionate ion

Hydride, halides and oxides of sulphur

Hydride
Any sulphur compound that contains the element in an oxidation state less than +IV is a reductant. So hydrogen sulphide acts as a reductant in solution or in the gas phase, e.g. with chlorine or moist sulphur dioxide.

$$8\overset{0}{Cl}_{2(g)} + 8H_2\overset{-II}{S}_{(g)} \rightarrow 16H\overset{-I}{Cl}_{(g)} + \overset{0}{S}_{8(s)}$$

$$8\,\overset{IV}{S}O_{2(g)} + 16H_2\overset{-II}{S}_{(g)} \rightarrow 3\overset{0}{S}_{8(s)} + 16H_2O_{(l)}$$

Hydrogen sulphide also reduces aqueous iron (III) solutions:

$$16\,\overset{III}{Fe}^{3+}_{(aq)} + 8H_2\overset{-II}{S}_{(g)} + 16H_2O_{(l)} \rightarrow 16\overset{II}{Fe}^{2+}_{(aq)} + 16H_3O^+_{(aq)} + \overset{0}{S}_{8(s)}$$

Halides of sulphur

The halides of sulphur in low oxidation states act as reductants, but their properties are largely due to the disproportionation products that form under aqueous reaction conditions.

Sulphur (IV) oxide

Sulphur (IV) compounds can act as either oxidants or reductants. They act as oxidants when they are reduced to sulphur themselves:

$$16\overset{0}{Mg}_{(s)} + 8\overset{IV}{S}O_{2(g)} \xrightarrow{\text{heat}} 16\overset{II}{Mg}O_{(s)} + \overset{0}{S}_{8(s)}$$

Also notice the reactions above with hydrogen sulphide.

However, more often sulphur (IV) compounds act as reductants. In these reactions they are converted to the more stable sulphur (VI) state. For example, sulphur (IV) oxide reduces chlorine over a hot charcoal catalyst.

$$\overset{IV}{S}O_{2(g)} + \overset{0}{Cl}_{2(g)} \rightarrow \overset{VI}{S}O_2\overset{-I}{Cl}_2$$

In solution sulphur (IV) oxide is a reductant either in its molecular form or as sulphate (IV) ions:

$$2\overset{VII}{Mn}O^-_{4(aq)} + 5\overset{IV}{S}O_{2(aq)} + 6H_2O_{(l)} \rightarrow 2\overset{II}{Mn}^{2+}_{(aq)} + 5\,\overset{VI}{S}O^{2-}_{4(aq)} + 4H_3O^+_{(aq)}$$

$$\overset{0}{I}_{2(aq)} + \overset{IV}{S}O^{2-}_{3(aq)} + 3H_2O_{(l)} \rightarrow 2\overset{-I}{I}^-_{(aq)} + \overset{VI}{S}O^{2-}_{4(aq)} + 2H_3O^+_{(aq)}$$

One of the most important reactions of sulphur (IV) oxide is its oxidation to sulphur (VI) oxide. This can be achieved in the laboratory as shown below:

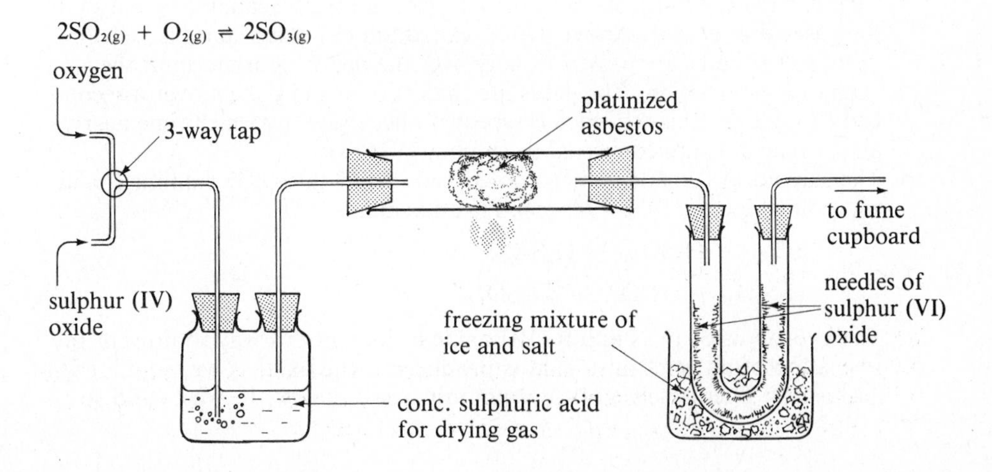

Sulphur (VI) oxide

The structure of sulphur (VI) oxide is complex, but there are at least two well-defined forms:

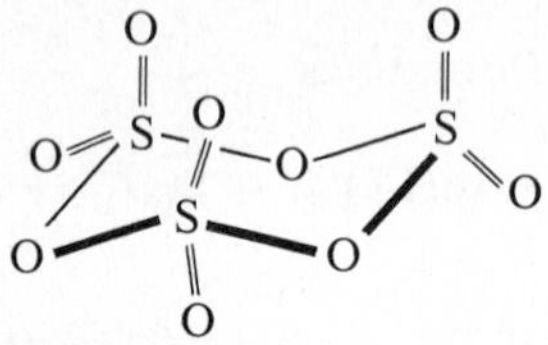

cyclic trimer 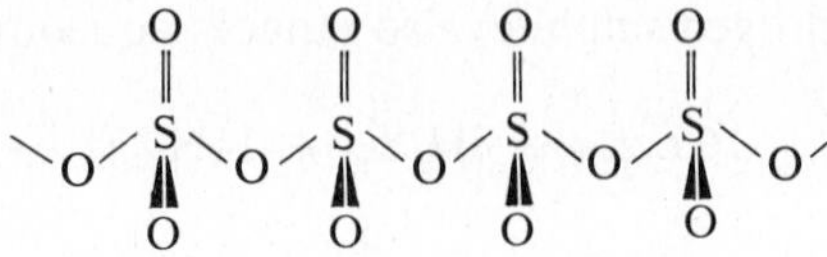   helical chains (asbestos-like)

The oxide is violently hydrolysed to sulphuric (VI) acid, H_2SO_4, a stable, oily liquid of b.p. 338°C.

Sulphuric acid

The contact process

Sulphuric acid is a vitally important industrial chemical and is manufactured in huge quantities by the contact process. The main steps in this process are:

1. Molten sulphur is burnt in air producing sulphur (IV) oxide.

$$S_{8(l)} + 8O_{2(g)} \rightarrow 8SO_{2(g)}$$

2. The sulphur (IV) oxide is purified. This consists of removing solid particle impurities electrostatically, washing out soluble impurities with hot water, and then drying the gas stream by passing it through a spray of concentrated sulphuric acid. This purification is necessary to prevent the catalyst being 'poisoned': i.e. becoming inactive because it has reacted with some impurity.
3. The sulphur (IV) oxide is then mixed with air and passed over a bed of vanadium pentoxide catalyst.

$$2SO_{2(g)} + O_{2(g)} \rightleftharpoons 2SO_{3(g)} \qquad \Delta H^{\ominus} = -98 \text{ kJ mol}^{-1}$$

The reaction is exothermic and the temperature of the catalyst increases as the gases pass over it. Above 450°C vanadium (V) oxide starts to decompose, so the gases are passed rapidly over the bed in order to limit the extent of the reaction. The gases are then cooled and passed over a second bed of catalyst. This sequence is repeated once more, by which time nearly all the reactants are converted to sulphur (VI) oxide.

4. The sulphur (VI) oxide is then cooled and passed into 98% sulphuric acid solution where the following reactions take place.

$$SO_{3(g)} + H_2SO_{4(l)} \rightarrow H_2S_2O_{7(l)}$$

$$H_2S_2O_{7(l)} + H_2O_{(l)} \rightarrow 2H_2SO_{4(l)}$$

The hydrolysis of the sulphur (VI) oxide is done in this way to prevent the formation of an acid 'mist' that is produced if the oxide is passed directly into water. The presence of the sulphuric acid 'dilutes' the water and so reduces the violence of the reaction. Also the boiling point of the 98% sulphuric acid mixture is much higher (~ 350°C) than that of water.

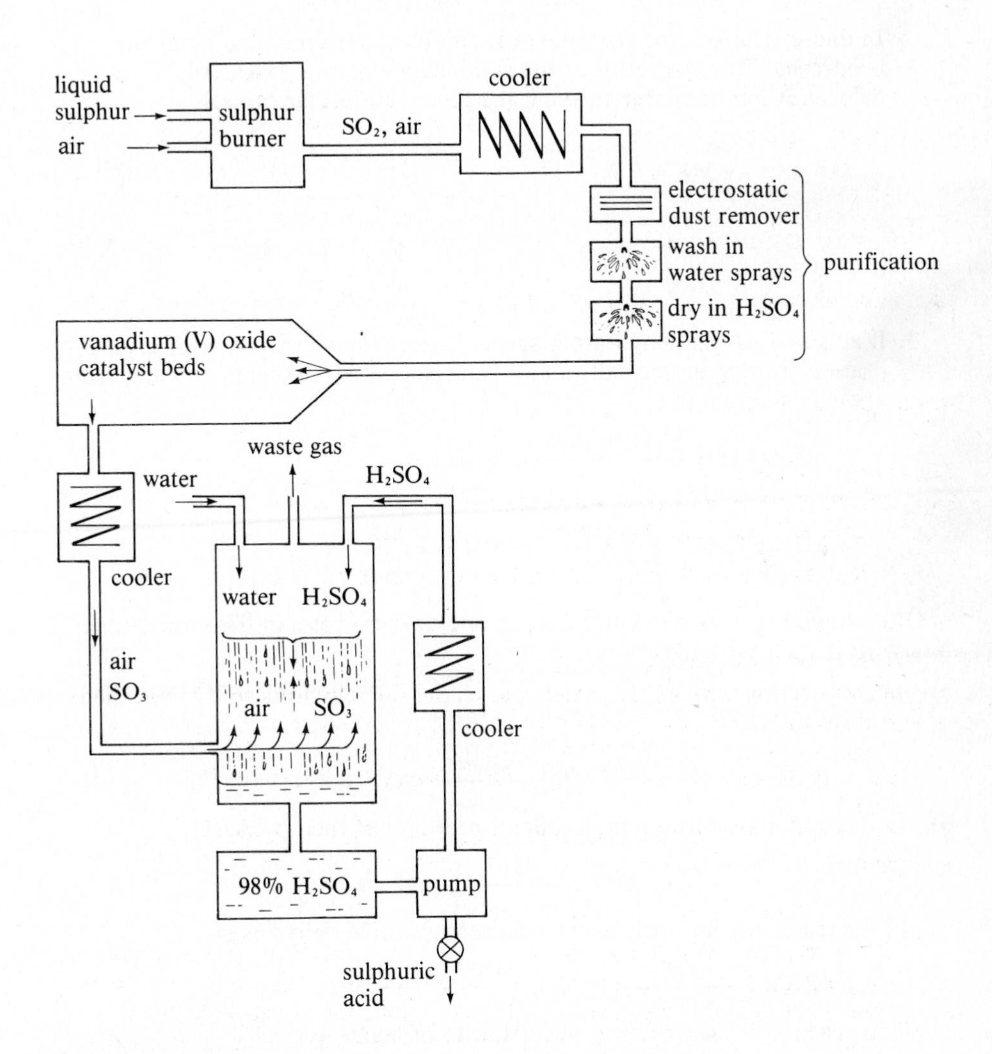

Properties of sulphuric acid
The many chemical properties of sulphuric acid can be categorized under three headings.

1. It is a strong acid:

$$H_2SO_{4(l)} + H_2O_{(l)} \rightarrow H_3O^+_{(aq)} + HSO^-_{4(aq)}$$

$$H_2SO_{4(l)} + NaCl_{(s)} \rightarrow HCl_{(g)} + NaHSO_{4(s)}$$

2. The concentrated acid is a strong oxidant. It oxidizes unreactive metals like copper and non-metals like carbon.

$$\overset{0}{Cu}_{(s)} + 2H_2\overset{VI}{S}O_{4(l)} \rightarrow \overset{II}{Cu}SO_4 + \overset{IV}{S}O_{2(g)} + 2H_2O_{(l)}$$

$$\overset{0}{C}_{(s)} + 2H_2\overset{IV}{S}O_{4(l)} \rightarrow \overset{IV}{C}O_{2(g)} + 2\overset{IV}{S}O_{2(g)} + 2H_2O_{(l)}$$

In dilute solution, the sulphate (VI) anions show almost no oxidizing properties. This is a result of the stabilizing effect of extensive delocalization of charge through the dπ-pπ molecular orbitals.

3. It acts as a strong dehydrating agent. It can remove the water of crystallization from hydrated salts and it can remove the elements of water from organic compounds:

$$\underset{\text{blue}}{CuSO_4.5H_2O} \xrightarrow{\text{conc. } H_2SO_4} \underset{\text{white}}{CuSO_4}$$

$$\underset{\text{ethanol}}{CH_3CH_2OH} \xrightarrow{\text{conc. } H_2SO_4} \underset{\text{ethene}}{CH_2{=}CH_2}$$

Often sulphuric acid acts in more than one of these roles in the same system, as shown in the examples below.

a) In the reaction with sugar, it dehydrates the carbohydrate to carbon, then oxidizes the carbon.

$$C_{12}H_{22}O_{11} \xrightarrow{\text{dehydrate}} [C] \xrightarrow{\text{oxidize}} CO_2,\ SO_2$$

b) In the reaction with sodium iodide, it protonates, then oxidizes.

$$NaI \xrightarrow{\text{protonate}} [HI] \xrightarrow{\text{oxidize}} I_2,\ SO_2,\ H_2S$$

c) In the reaction with nitric acid, it protonates, then dehydrates.

$$HNO_3 \xrightarrow{\text{protonate}} [H_2NO_3^+] \xrightarrow{\text{dehydrate}} NO_2^+$$

This reaction is important in the nitration of benzene.

45.7 Solvent and complexing properties

Solvents

Water is a polar solvent and dissolves many ionic compounds. The water molecules act as ligands to form complexes whose stability depends on the charge density of the ion to which the ligands are co-ordinated. The smaller and the more highly charged the ion, the more stable is the complex. A number of other ligands co-ordinate through a bonded oxygen atom. These include the important ligand OH^- and a range of organic ligands, three of which are shown below.

ligand type	*monodentate*	*bidentate*	
structure	Et(H)O:	$^-$O(=O)C—C(=O)O$^-$	Me–C(=O)–CH=C(–O$^-$)–Me 'enol' form
	ethanol	ethanedioate (oxalate)	pentane-2,4-dione (acetoacetone)
typical complex	$\overset{II}{Ca}(EtOH)_4^{2+}$	$\overset{III}{Cr}(C_2O_4)_3^{3-}$	$\overset{II}{Be}(C_5H_7O_2)_2$

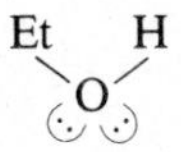

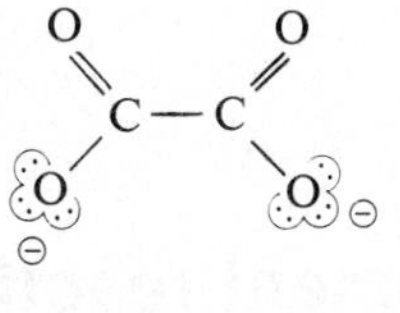

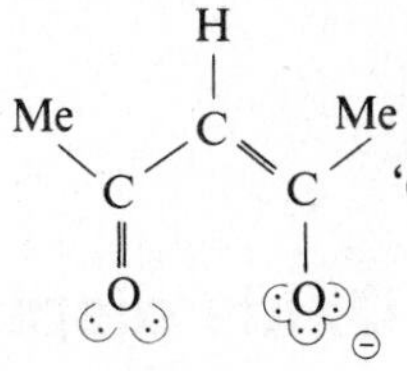

Many organo-sulphur compounds are useful solvents. The lower electronegativity of sulphur leads to less polar character in these solvents, and many are therefore good solvents for molecular compounds. Two examples of these are:

a) carbon disulphide (b.p. 46°C): S=C=S

b) dimethyl disulphide (b.p. 110°C): Me—S—S—Me

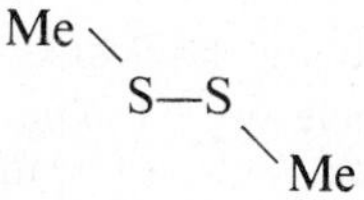

Carbon disulphide is an unusual compound because its molecules have no charge separation (the electronegativity of carbon is the same as the electronegativity of sulphur, 2.5) and also contain pπ-bonded sulphur atoms. Carbon disulphide can be prepared by strongly heating the powdered elements together.

Thio ligands

Thiosulphate ions and thiocyanate ions are able to form complexes with 'B' metal cations and transition metal cations. For example iron (III) forms a complex with thiocyanate ions. The blood-red colour given when a drop of potassium thiocyanate is added to an iron (III) salt solution is used as an analytical test for iron (III).

$$Fe^{3+}_{(aq)} + {}^-:\ddot{S}—C\equiv N \rightleftharpoons [Fe—S—C\equiv N]^{2+}_{(aq)}$$

blood red

When sodium thiosulphate is added to an iron (III) solution, a purple colour appears which is then rapidly discharged and replaced by a fine colloidal suspension of sulphur. The thiosulphate ions at first form a complex with the iron (III) ions but the complex breaks up as the iron (III) ions oxidize the thiosulphate ligands:

$$16Fe^{3+}_{(aq)} + 8S_2O^{2-}_{3(aq)} + 24H_2O_{(l)} \rightarrow 16Fe^{2+}_{(aq)} + 16H_3O^{+}_{(aq)} + 8SO^{2-}_{4(aq)} + S_{8(s)}$$

45.8 Oxide displacement reactions

The change in free energy involved in the formation of an oxide is temperature-dependent. The change is a function of the entropy change for the reaction:

$$\Delta G^{\ominus} = \Delta H^{\ominus} - T\Delta S^{\ominus}$$

Differentiate partially with respect to temperature at constant pressure.

$$\Rightarrow \left(\frac{\partial(\Delta G^{\ominus})}{\partial T}\right)_P = -\Delta S^{\ominus} \quad \text{since} \left(\frac{\partial(\Delta H^{\ominus})}{\partial T}\right)_P = 0$$

The formation of an oxide usually results in a decrease in the gaseous content of the system. Since the entropy of the gas phase far exceeds that of any other phase, this leads to a decrease in entropy i.e. $\Delta S^{\ominus}$ is negative for the reaction:

$$Pb_{(s)} + \tfrac{1}{2}O_{2(g)} \rightarrow PbO_{(s)} \qquad \Delta S^{\ominus} = -97 \text{ J K}^{-1} \text{ mol}^{-1}$$

1 mole of solid + ½ mole of gas → 1 mole of solid

When the free energy of formation is plotted as a function of temperature, the slope of the line is likely to be positive because, (from above) the gradient is $-\Delta S^{\ominus}$ and $\Delta S^{\ominus}$ itself is negative.

The formation of the gaseous carbon oxides, however, presents a different picture.

$$C_{(s)} + \tfrac{1}{2}O_{2(g)} \rightarrow CO_{(g)}$$

1 mole of solid + ½ mole of gas → 1 mole of gas

$\Delta S^{\ominus}$ is **positive**

$$C_{(s)} + O_{2(g)} \rightarrow CO_{2(g)}$$

1 mole of solid + 1 mole of gas → 1 mole of gas

$\Delta S^{\ominus}$ is $\simeq$ **zero**

In the first of these two reactions, the proportion of gas increases, so the free energy/temperature plot for the oxide has a negative slope. In the second reaction, the proportion of gas remains the same and the free energy/temperature plot is flat. When these free energy/temperature graphs are plotted with those of a number of other oxides on the same graph the result is called an *Ellingham diagram*.

Ellingham diagram for copper, iron, silicon and carbon

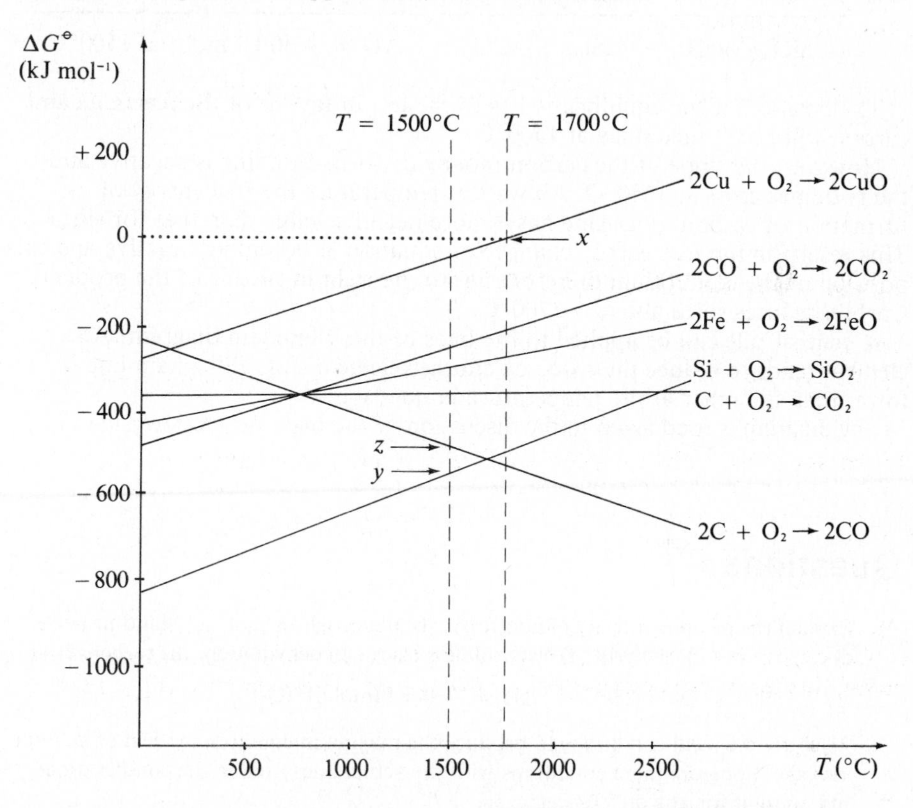

The diagram can be used to predict the position of equilibrium in an oxide displacement reaction. When $\Delta G^\ominus$ is positive for a process, it does not take place spontaneously (see page 223). For example, at 1700°C the free energy change x for the formulation of copper (II) oxide is zero. Above 1700°C, $\Delta G^\ominus$ becomes positive and this suggests that the position of equilibrium will move towards the reactants.

$$2Cu_{(s)} + O_{2(g)} \rightleftharpoons 2CuO_{(s)} \qquad \Delta G^\ominus \text{ is positive at } T > 1700°C$$

that is, by heating copper (II) oxide above 1700°C decomposition to the elements is predicted.

Consider the reaction below.

$$SiO_{2(s)} + 2C_{(s)} \rightleftharpoons 2Si_{(s)} + 2CO_{(g)}$$

From the diagram, it can be seen that the silica formation line y at 1500°C is below the carbon monoxide line z at 1500°C.

1. $Si_{(s)} + O_{2(g)} \rightarrow SiO_2 \qquad \Delta G^\ominus = -520 \text{ kJ mol}^{-1}$
2. $2C_{(s)} + O_{2(g)} \rightarrow 2CO_{(g)} \qquad \Delta G^\ominus = -490 \text{ kJ mol}^{-1}$

By subtracting equation 1. from equation 2. a new equation, 3. is produced.

2.	$2C_{(s)} + O_{2(g)} \rightarrow 2CO_{(g)}$	$\Delta G^{\circ} = -490\ kJ\ mol^{-1}$
1.	$Si_{(s)} + O_{2(g)} \rightarrow SiO_{2(s)}$	$\Delta G^{\circ} = -520\ kJ\ mol^{-1}$
3.	$SiO_{2(s)} + 2C_{(s)} \rightarrow 2Si_{(s)} + 2CO_{(g)}$	$\Delta G^{\circ} = +30\ kJ\ mol^{-1}$ at 1500°C

The position of the equilibrium lies to the left in favour of the reactants and carbon will not reduce silica at 1500°C.

However, the slope of the carbon monoxide formation line is negative and the two lines cross at 1650°C. Above this temperature the free energy of formation of carbon monoxide has a more negative value than that for silica. This results in the free energy change for equation 3. becoming negative and the position of the equilibrium therefore lies to the right in favour of the products. Carbon reduces silica above ~1700°C.

A general rule can be applied to the lines of the Ellingham diagram. An element tends to reduce the oxide of another element if its Ellingham line is lower than the other at the temperature in question.

The diagram is used again in the discussion of the blast furnace on page 777.

Questions

1. Explain the meaning of: a) enantiotropy b) molecular orbital c) Hund principle d) catenation e) catalysis f) electron deficiency g) dehydration h) carbohydrate.

2. $2SO_2 + O_2 \rightleftharpoons 2SO_3 \qquad \Delta H = -98\ kJ\ (mol\ of\ SO_3)^{-1}$

   What are the ideal conditions to produce the maximum equilibrium yield of sulphur trioxide? Compare these conditions with the actual ones used in the contact process and account for any differences.

3. Draw the structures of the following particles:
   a) O_3 b) S_8 c) H_2O_2 d) SO_3^{2-} e) $S_2O_3^{2-}$ f) SO_4^{2-} g) SO_2 h) SO_3 i) SF_6 j) SCl_4 k) $SOCl_2$ l) $S_4O_6^{2-}$

4.

oxidized form	*reduced form*	E° *volts*
H_2O_2	H_2O	+1.77
O_2	H_2O	+1.23
O_2	H_2O_2	+0.68
H_2SO_3	S	+0.45
SO_4^{2-}	H_2SO_3	+0.17
S	H_2S	+0.14

   a) Which oxygen species is the most powerful oxidant present in the list?
   b) Which oxygen species is the most powerful reductant present in the list?
   c) What information does this give you about the nature of the following reaction?

   $$2H_2O_2 \rightarrow 2H_2O + O_2$$

d) From the data above, is sulphur an oxidant or a reductant?
e) Predict whether reaction is likely to occur between the following species:
   i) H_2O_2 and S
   ii) H_2O_2 and SO_4^{2-}
   iii) H_2SO_3 and H_2S.

5. Classify the following equations as either acid-base or redox by completing and balancing them, and putting in oxidation numbers:
   a) $Na_2S_{(s)} + H_2O_{(l)} \rightarrow$
   b) $SO_{2(g)} + Cl_{2(g)} \rightarrow$
   c) $Ba^{2+}_{(aq)} + SO^{2-}_{4(aq)} \rightarrow$
   d) $PbS_{(s)} + 4O_{3(g)} \rightarrow$
   e) $2Cu(NO_3)_2 \xrightarrow{\text{heat}}$
   f) $I_{2(aq)} + S_2O^{2-}_{3(aq)} \rightarrow$

6. Account for the following:
   a) H_2SO_4 is a stronger acid than H_2SO_3
   b) Sulphate (VI) salts are more stable than sulphate (IV) salts
   c) SF_6 is a very unreactive compound.

7. Compare and contrast:
   a) H_2O and H_2S
   b) CO_2 and CS_2
   c) Na_2SO_4 and $Na_2S_2O_3$.

8. a) Give four methods of preparing oxygen in the laboratory. In each case give a balanced equation and mark in the oxidation states.
   b) Starting with a pure sample of ^{34}S, describe how you would prepare samples of the following: i) $H_2\,^{34}SO_4$

ii)

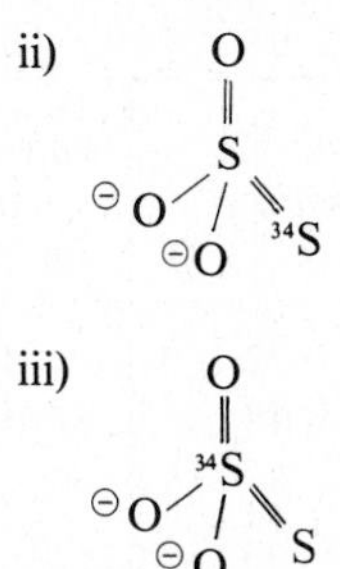

iii)

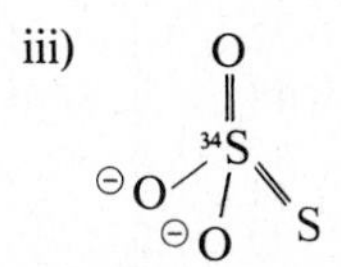

9. a) Outline the industrial extraction of oxygen.
   b) Outline the industrial extraction of sulphur.
   c) Outline the industrial manufacture of sulphuric acid.

10. Compare the bonding shown by oxygen and sulphur in their compounds.

Chapter 46

The halogens: group VIIB

46.1 Structure of the elements

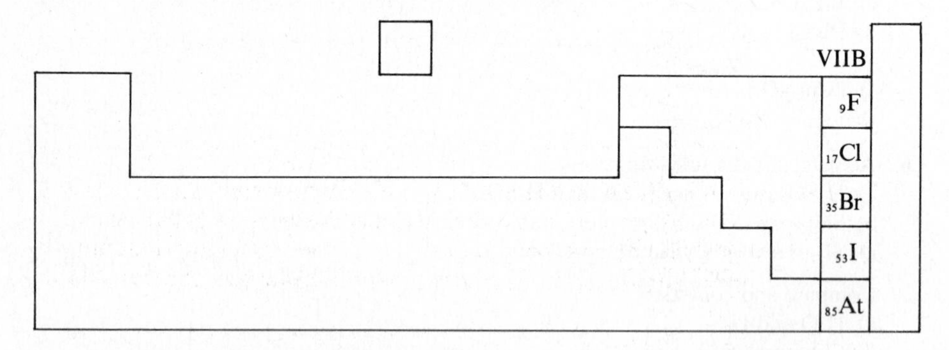

Astatine is radioactive and has been prepared only in trace quantities.

$$^{209}_{83}Bi + ^{4}_{2}He \xrightarrow[\text{(α-bombardment)}]{} ^{211}_{85}At + ^{1}_{0}n + ^{1}_{0}n \quad (t_{½} = 7.5 \text{ hours})$$

The limited evidence suggests that its properties are very similar to those of iodine.

element	*electronic configuration*	*electro-negativity*	*atomic radius* (m × 10^{-10})	*anionic* (X^-) *radius* (m × 10^{-10})	*stable oxidation states*
fluorine	[He] $2s^2$ $2p^5$	4.0	0.72	1.36	−I
chlorine	[Ne] $3s^2$ $3p^5$	3.0	0.99	1.81	−I, I, III, IV, V, VI, VII
bromine	[Ar] $3d^{10}$ $4s^2$ $4p^5$	2.8	1.14	1.95	−I, I, III, IV, V, VI, VII
iodine	[Kr] $4d^{10}$ $5s^2$ $5p^5$	2.5	1.33	2.16	−I, I, III, V, VII

Atomic structure

A halogen atom has seven outer-shell electrons and the largest possible core charge of seven. The halogens as a group are therefore the most electronegative of all the elements and their atoms tend to have a high degree of electronic control. With the exception of a fluorine atom, each halogen atom can also expand its octet of electrons by using the vacant, energetically low-lying d-orbitals present in its outer-shell. Compare the electronic structures of a fluorine atom and a chlorine atom.

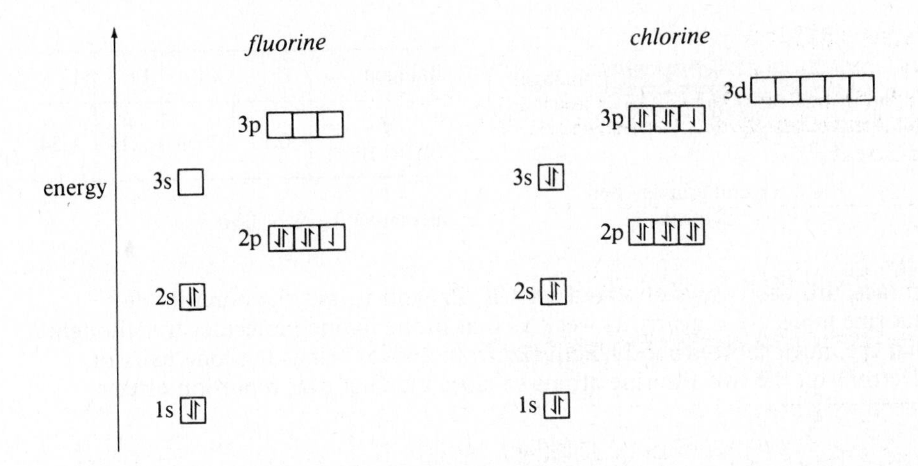

Fluorine cannot expand its octet because there are no 2d-orbitals and the energy required to promote electrons into the third shell is prohibitively high. Chlorine can expand its octet by making use of the vacant 3d-orbitals. By promoting electrons into these, a chlorine atom forms more bonds (energy is given out when bonds form).

The halogens are 'B' elements because their inner d-subshells (where they exist) are full.

Physical structure

All the halogens are non-metallic in character. They exist as diatomic molecules, formed by the overlap of the unpaired p-orbitals of two separate atoms.

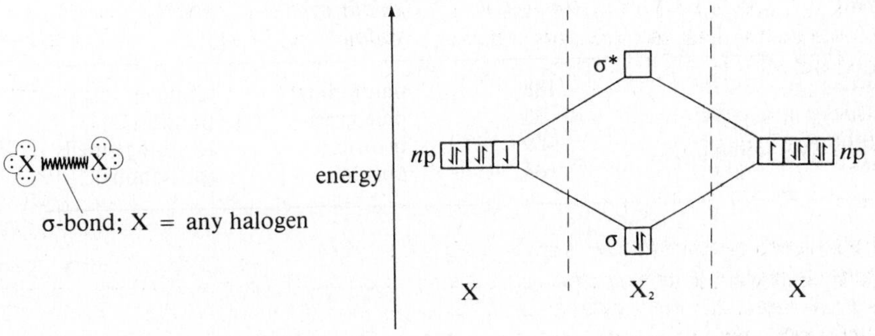

X X

σ-bond; X = any halogen

energy

σ*

np *np*

σ

X X_2 X

At room temperature, fluorine and chlorine are gaseous, bromine is a liquid and iodine is a solid. This reflects the increasing van der Waals' forces between the X_2 molecules as the number of electrons in each X_2 molecule increases. It might be expected that the bond dissociation energies of the X_2 molecules would decrease steadily going down the group because the distance of the nuclei from the bonding pair steadily increases and $F \propto 1/d^2$.

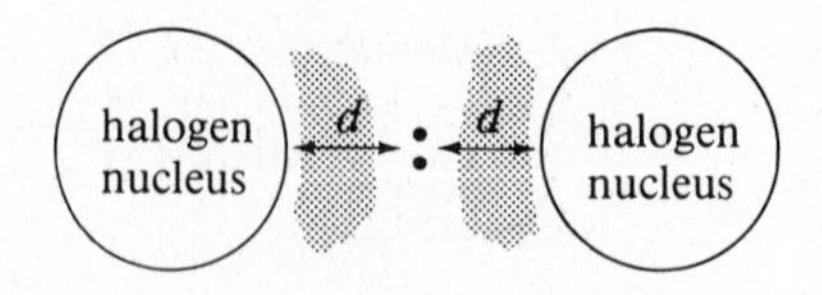

screening inner-shell electrons

halogen	F	Cl	Br	I
d (m × 10^{-10})	0.71	1.00	1.14	1.34

attractive force $\propto 1/d^2$

In fact, this decrease is observed for Cl_2, Br_2 and I_2, but the bond in the fluorine molecule is nearly as weak as that in the iodine molecule. It is thought that the much shorter bond length (2*d* from above) brings the lone pairs of electrons on the two fluorine atoms so close together that repulsion effects become significant.

F—F weakening of the fluorine-fluorine bond

molecule	F—F	Cl—Cl	Br—Br	I—I
bond dissociation energy (kJ mol^{-1})	158	242	193	151

A collection of some other physical properties of the halogens is given in the table below.

element	*m.p.* (°C)	*b.p.* (°C)	*colour of vapour*	*smell*
fluorine	−220	−188	pale yellow	all have
chlorine	−101	−35	pale green	pungent,
bromine	−7	59	brown	irritating smells
iodine	114	183	purple	(poisonous)

46.2 Compounds formed by the elements

The halogens combine with nearly every other element. Fluorine is the most reactive non-metal of all and forms univalent compounds in which the fluorine oxidation state is always −I. With non-metals, halogen compounds are molecular and tend to be acidic in water. With metals, they form either simple halide salts containing the X^- ion or oxy-salts containing the halogen in a positive oxidation state. The chemistry of the halogens is dominated by the stability of the halide ions. Only one electron is required to produce the electronic configuration of a full shell and the electron affinities of the halogens are the highest of any group.

$X_{(g)} + e^- \rightarrow X^-_{(g)}$: $\Delta H^{\ominus}$ = first electron affinity

element	*fluorine*	*chlorine*	*bromine*	*iodine*
first electron affinity (kJ mol^{-1})	−348	−364	−342	−314

The high hydration energy of these ions is also an important factor that favours their production in aqueous solution. Consider the Born-Haber cycle below for the enthalpy of formation of the aquo-ions from the molecular halogens. This energy change is a measure of the oxidizing strength of a halogen.

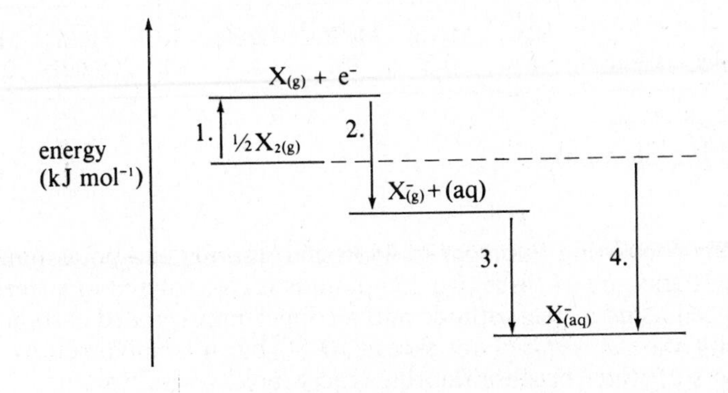

element	1. *half bond dissociation energy*	2. *first electron affinity*	3. $\Delta H^{\ominus}_{hyd}$	4. *total*
fluorine	+79	−348	−506	−775
chlorine	+121	−364	−364	−607
bromine	+97	−342	−335	−580
iodine	+76	−314	−293	−541

Notice the very large heat of hydration of the fluoride ion; a result of its small size. In many of the reactions of the halogens and their compounds in water, halide ions are produced as a result of their considerable stability.

46.3 Occurrence, extraction and uses

Occurrence of the elements

The halogens are found combined in a number of mineral deposits and in the salts dissolved in the sea. The most important minerals or salts are listed below. The typical composition of 100 grams of sea water is also given.

element	*sources*
fluorine	fluorspar CaF_2; cryolite Na_3AlF_6
chlorine	rock salt NaCl; carnallite $KCl.MgCl_2.6H_2O$ and in dissolved salts in the sea
bromine	only in dissolved salts in the sea
iodine	Chilean saltpetre (KNO_3) also contains small quantities of KIO_3; and in dissolved salts in the sea

Typical composition of sea water

	NaCl	$MgCl_2$	$MgSO_4$	$CaSO_4$	KCl	$MgBr_2$	MgI_2
mass (g per 100 g sea water)	2.6	0.3	0.2	0.1	0.1	0.001	0.0003

Extraction of the elements

Fluorine

It is extracted by converting fluorspar to hydrogen fluoride and potassium fluoride. A 50-50 mixture of these two compounds is electrolysed in a steel (or copper-nickel) cell using a steel cathode and a copper-impregnated carbon anode. The cell temperature need not exceed 100°C but it is important to exclude all traces of water because fluorine reacts rapidly with water.

$$CaF_{(s)} \xrightarrow{H_2SO_4} 2HF_{(g)} \xrightarrow[\text{electrolyse}]{KF} H_{2(g)} + F_{2(g)}$$

Chlorine

Its extraction is described on pages 627 to 629.

Bromine

The element is obtained from sea water using chlorine as an oxidant. Sea water is first acidified to repress the hydrolysis of molecular chlorine (see page 742) and then chlorine is bubbled through it.

$$1.\ Cl_{2(g)} + 2Br^-_{(aq)} \rightarrow 2Cl^-_{(aq)} + Br_{2(aq)}$$

The bromine produced is evaporated from the treated sea water by blowing a blast of air through the system. The air carrying trace quantities of bromine, is passed into sodium carbonate solution where the bromine is absorbed as bromate (V) and bromide. On acidification, pure bromine is obtained by distillation.

2. $3Br_{2(g)} + 6CO^{2-}_{3(aq)} + 3H_2O_{(l)} \rightarrow BrO^-_{3(aq)} + 5Br^-_{(aq)} + 6HCO^-_{3(aq)}$

3. $BrO^-_{3(aq)} + 5Br^-_{(aq)} + 6H_3O^+_{(aq)} \xrightarrow{\text{distil}} 3Br_{2(g)} + 9H_2O_{(l)}$

Iodine

The element is extracted from certain types of seaweed (e.g. *Laminaria*). The weeds concentrate the combined element in their cellular structure and are a useful source of the element. However, the majority of the world's iodine comes from the small concentrations of sodium iodate (V) present in Chile saltpetre. After fractional crystalization, the more soluble sodium iodate (V) is reduced with sodium hydrogen sulphate (IV). Iodine precipitates and is filtered. It is purified by sublimation in the presence of potassium iodide which reacts with any other halogen impurity.

$$2IO^-_{3(aq)} + 5HSO^-_{3(aq)} \rightarrow 2SO^{2-}_{4(aq)} + 3HSO^-_{4(aq)} + H_2O_{(l)} + I_{2(s)}$$

Laboratory preparation of the halogens

A laboratory sample of a halogen is best prepared by oxidizing a halide using manganese (IV) oxide. This method does not work for fluorine which is itself the most powerful oxidant known. For chlorine, the oxidation requires a high concentration of acid. Concentrated hydrochloric acid is therefore used:

$$MnO_{2(s)} + 4HCl_{(conc)} \xrightarrow{\text{heat}} MnCl_{2(s)} + 2H_2O_{(l)} + Cl_{2(g)}$$

For bromine and iodine, the hydrogen halides can be generated by mixing a potassium halide with concentrated sulphuric acid. The addition of manganese (IV) oxide and distillation produces a sample of the halogen.

Uses of the elements and their compounds

substance	*uses*
fluorine	manufacture of fluorocarbons (used as refrigerants, propellants for aerosols and insulators); for separating ^{235}U and ^{238}U by the different rates of the gaseous effusion of their volatile hexafluorides
hydrogen fluoride	fluorination; manufacture of cryolite; etching; solvent for silicates.
chlorine	manufacture of hydrogen chloride, organic solvents (e.g. trichloroethane), plastics (e.g. PVC), disinfectants (e.g. TCP), bleaching agents.
bromine	manufacture of petrol additives that volatilize the lead in the petrol as $PbBr_2$; silver bromide is used for photographic films.
iodine	manufacture of iodoform, alkyl halides and dyes; medicinal uses (e.g. tincture of iodine); silver iodide is also used in films.

46.4 Compound structure

The − I state: halides

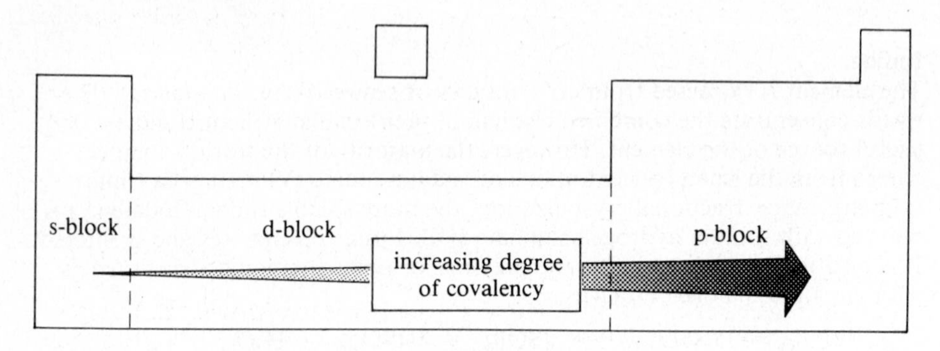

s-block
These halides are typical ionic compounds:

$K^+ X^-$; $Mg^{2+} (X^-)_2$

d-block
The low oxidation state halides are ionic:

$Mn^{2+} (X^-)_2$, $Zn^{2+} (X^-)_2$

The higher oxidation state halides are covalent molecular compounds:

F
F—V—F (F, F)
F

(m.p. 19°C)

p-block
The halides are covalent molecular compounds:

Cl — C — Cl, Cl, Cl P (F, F, F) Cl — Pb — Cl, Cl, Cl

Most halides can be prepared by the direct combination of the elements concerned. The reactivity of the halogens decreases going down the group and so it is more difficult to prepare iodides by this method:

$H_{2(g)} + F_{2(g)} \rightarrow 2HF_{(g)}$ explosively fast, even in the dark
but $H_{2(g)} + I_{2(g)} \rightleftharpoons 2HI_{(g)}$ incomplete conversion even at 450°C

Many metals and non-metals burn in a halogen atmosphere and others react when heated in a stream of halogen gas:

$C_{(s)} + 2Cl_{2(g)} \xrightarrow{\text{heat}} CCl_{4(g)}$ hot carbon glows brightly in chlorine

$P_{4(s)} + 6Br_{2(g)} \longrightarrow 4PBr_{3(s)}$ phosphorus ignites in bromine vapour

$2Na_{(s)} + I_{2(g)} \longrightarrow 2NaI_{(s)}$ sodium burns in iodine vapour

$2Al_{(s)} + 3Cl_{2(g)} \longrightarrow Al_2Cl_{6(g)}$ aluminium combines with chlorine when heated in a stream of the gas.

Fluorine often causes an element to combine in its highest possible oxidation state. For example, SF_6, PF_5 and VF_5 have no bromide analogues: S_2Br_2, PBr_3 and VBr_3 are formed when sulphur, phosphorus and vanadium react with bromine. There are a number of factors that are responsible for this:

1. Fluorine's oxidizing strength is greatest.
2. Fluorine atoms form the strongest bonds and more energy is given out when these bonds form.
3. The larger size of the lower halogen atoms makes it difficult to fit more than three or four around another atom.

Compare some typical halogen bond strengths given in kJ mol^{-1} in the table below. The trends follow the order suggested in the discussion on page 730.

atom (Z)	Z—F	Z—Cl	Z—Br	Z—I
hydrogen	562	431	366	299
carbon	484	338	276	238
silicon	542	361	290	215
phosphorus	462	332	273	215

The steady increase in covalent character observed in the halides going across a period is best interpreted in terms of the increasing polarizing power of the different cations as their size decreases and their charge increases. Similarly, fluorides show more ionic character than iodides because the iodide ion (being much larger) is more polarizable.

The +I state

With the exception of fluorine, all the halogens disproportionate in alkali. The disproportionation of an element in alkali is a characteristic of molecular non-metals that have σ-bonded molecules. Both white phosphorus and sulphur behave in a similar way (see pages 684 and 710). In the case of a halogen, hydroxide ions cause induced dipoles in the halogen molecules which are then attacked nucleophilically:

$$H{-}\ddot{O}:^{\ominus} + \overset{\delta+}{X}{-}\overset{\delta-}{X} \rightleftharpoons H{-}\overset{\ominus}{O}\cdots X{-}O{-}H \;\; :X:^{\ominus} \rightleftharpoons H{-}O{-}H \;\; {}^{\ominus}:O{-}X \;\; :X:^{\ominus}$$

$$\overset{0}{Cl_2} + 2OH^-_{(aq)} \rightleftharpoons \underset{\text{halide}}{\overset{-I}{Cl^-_{(aq)}}} + \underset{\text{halate (I)}}{\overset{I}{ClO^-_{(aq)}}} + H_2O_{(l)}$$

Disproportionation is a redox reaction in which a substance containing an element in one oxidation state reacts to form products containing the same element in two other oxidation states.

The halate (I) itself tends to disproportionate leading to the further production of halide ions:

$$3\overset{\mathrm{I}}{X}O^-_{(aq)} \rightleftharpoons \overset{\mathrm{V}}{X}O^-_{3(aq)} + 2\overset{-\mathrm{I}}{X}^-_{(aq)}$$

The halogen (I) oxides of both chlorine and bromine have been prepared by passing the halogens over mercury (II) oxide. They are dangerously explosive compounds:

$$2Br_{2(g)} + 2HgO_{(s)} \longrightarrow HgO.HgBr_{2(s)} + Br_2O_{(g)}$$

It is also possible to prepare interhalogen compounds containing one of the halogens in a +I oxidation state. ClF, BrF, BrCl, ICl and IBr have all been isolated from equimolar mixtures of two halogens:

$$Cl_{2(g)} + F_{2(g)} \rightleftharpoons 2ClF_{(g)} \qquad T \approx 200°C$$
$$I_{2(s)} + Cl_{2(l)} \rightleftharpoons 2ICl_{(l)} \qquad T \approx -50°C$$

The +III state

Chlorine, bromine and iodine can all increase their valency from one (as in halides (−I) and halates (+I)) to a maximum of seven. This is achieved by promotion of electrons into the vacant, energetically low-lying, outer-shell d-orbitals. The 'promotion energy' is recouped by the energy given out when the extra bonds form. Compare the electronic configuration of chlorine in the interhalogens $\overset{\mathrm{I}}{Cl}—F$ and $F—\overset{\mathrm{III}}{Cl}(—F)—F$ (T-shaped, see page 97)

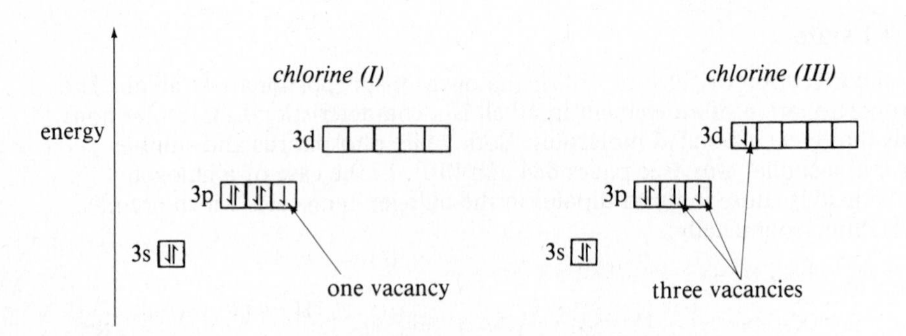

Interhalogens containing halogen (III) are prepared by adding an excess of the more reactive halogen to the less reactive one:

$$I_{2(s)} + 3Cl_{2(g)} \rightleftharpoons 2ICl_{3(s)}$$

The trifluorohalogens are useful fluorinating agents for either metals or their oxides:

$$6ZnO_{(s)} + 4BrF_{3(l)} \longrightarrow 6ZnF_{2(s)} + 2Br_{2(g)} + 3O_{2(g)}$$

There are no stable halogen (III) oxides, but it is possible to prepare a solution of a chlorate (III) by reducing chlorine (IV) oxide with an alkaline solution of hydrogen peroxide. The free acid, hydrogen chlorate (III), is not stable and, even under alkaline conditions, chlorate (III) disproportionates readily into chlorate (V) and chloride.

$$2ClO_{2(g)} + H_2O_{2(aq)} + 2OH^-_{(aq)} \longrightarrow 2ClO^-_{2(aq)} + 2H_2O_{(l)} + O_{2(g)}$$

The chlorate (III) ion is bent like a water molecule, and contains dπ-pπ-bonded chlorine and oxygen atoms:

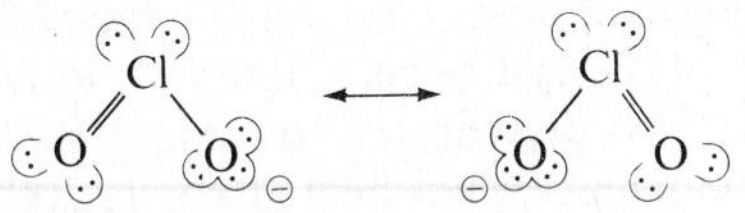

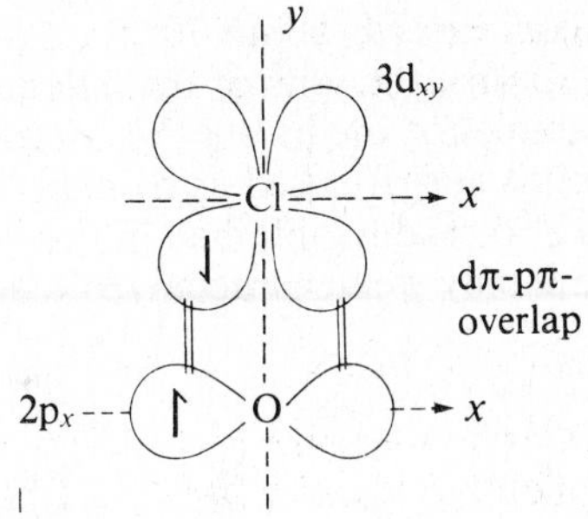

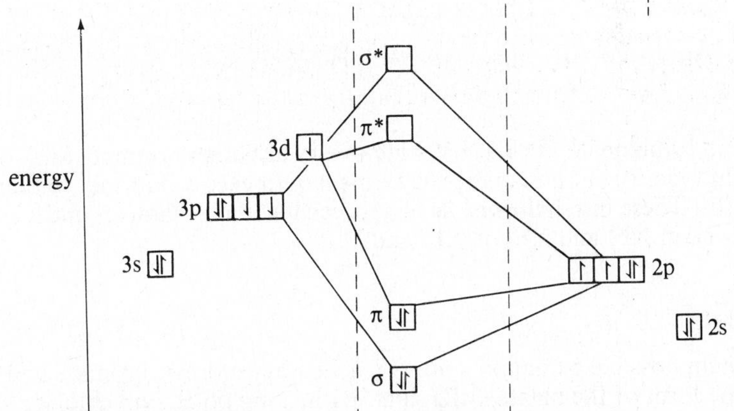

The +IV state and +VI state

Chlorine and bromine form pπ-dπ-stabilized oxides in the IV and VI states. These are dangerously explosive compounds, although chlorine (IV) oxide is produced commercially as a bleaching agent for flour. It is made by reducing sodium chlorate (V) with sulphur (IV) oxide in fairly concentrated sulphuric acid:

$$2NaClO_{3(s)} + H_2SO_{4(l)} + SO_{2(g)} \longrightarrow 2NaHSO_{4(s)} + 2ClO_{2(g)}$$

ClO_3 can be prepared by oxidizing ClO_2 with ozone. It is an oily red liquid at room temperature and its structure is not clearly defined. There is considerable evidence for association, and chlorine (VI) oxide is often known as dichlorine hexoxide Cl_2O_6. Both ClO_2 and ClO_3 exist as 'odd electron' molecules like NO_2.

The +V state

The halate (V) salts are produced by the disproportionation of the halate (I) salts. The tendency to disproportionate increases down the group. Thus iodate (I) is only transiently produced when iodine is treated with alkali; iodate (V) and iodide ions are produced instead.

$$3I_{2(s)} + 6OH^-_{(aq)} \rightarrow 5I^-_{(aq)} + IO^-_{3(aq)} + 3H_2O_{(l)}$$

Sodium chlorate (I) disproportionates like this when the temperature of the alkali exceeds about 70°C

Unlike the salts of the halogen (I) acids, it is possible to prepare crystalline samples of the halate (V) compounds. These are comparatively stable colourless solids and are used as oxidants (see page 743). Hydrogen iodate (V), iodic acid, is also readily obtained in the solid phase. It is dehydrated to iodine (V) oxide (the only well-characterized oxide of the element) when heated above 200°C.

$$2\ O{=}I({=}O)OH \xrightarrow{\text{heat}} (O{=})_2I{-}O{-}I({=}O)_{2\,(s)} + H_2O_{(g)}$$

Chlorine and bromine do not have V state oxides although bromine does form a V state fluoride, BrF_5 on reacting the element with excess fluorine. Iodine also gives IF_5. These interhalogens have appreciable ionic character and probably contain XF_4^+ and XF_6^- or XF_4^+ and F^-.

The VII state

The maximum possible oxidation state is reached by chlorine, bromine and iodine in the form of the halate (VII) salts. By heating potassium chlorate (V) gently, disproportionation occurs and potassium chlorate (VII) is produced.

$$4KClO_{3(s)} \longrightarrow 3KClO_{4(s)} + KCl_{(s)}$$

Stronger heating causes the chlorate (VII) to decompose with the evolution of oxygen. Chlorate (VII) salts are also prepared by electrolysing a chlorate (V) solution using bright platinum electrodes which have a high oxygen overpotential and thereby discourage the electrolysis of water (see page 342)

At anode: $ClO^-_{3(aq)} + 3H_2O_{(l)} \longrightarrow ClO^-_{4(aq)} + 2H_3O^+_{(aq)} + 2e^-$

At cathode: $2H_3O^+_{(aq)} + 2e^- \longrightarrow 2H_2O_{(l)} + H_{2(g)}$

Overall: $ClO^-_{3(aq)} + H_2O_{(l)} \xrightarrow[\text{Pt electrodes}]{\text{electrolysis}} ClO^-_{4(aq)} + H_{2(g)}$

Both the free acid and the salts of chlorine (VII) and iodine (VII) are well known. $HClO_4$ and HIO_4 can be prepared by protonating a halate (VII) salt in non-aqueous conditions:

$$NaClO_4 \xrightarrow{\text{conc. } H_2SO_4} HClO_4$$

The anhydride of hydrogen chlorate (VII), Cl_2O_7 is produced when the acid is dehydrated in the presence of phosphorus (V) oxide

$$2\ O_3Cl{-}OH \xrightarrow{P_4O_{10}} O_3Cl{-}O{-}ClO_3$$

I_2O_7 is not known, but a VII state fluoride results from the reaction of iodine with excess fluorine. IF_7 has a structure based on a pentagonal biprism.

IF_7

energy

iodine (VII)

5d ⇂ ⇂ ⇂ _ _

5p ⇂ ⇂ ⇂

5s ⇂

46.5 Acid-base properties

The halides

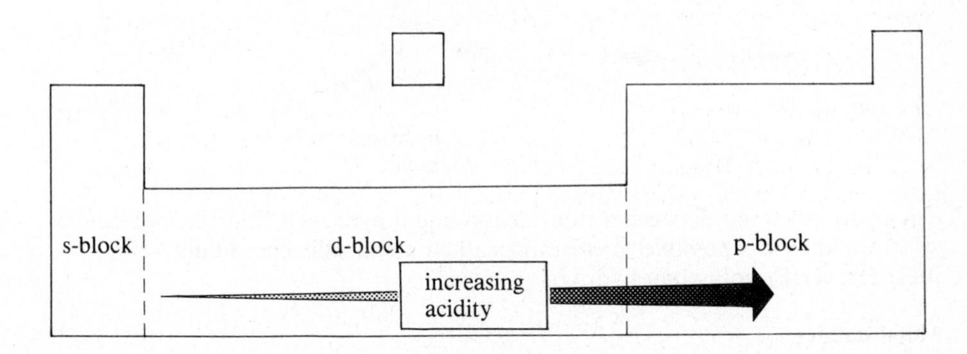

s-block halides

These dissolve in water producing neutral solutions:

$$NaX_{(s)} + H_2O_{(l)} \longrightarrow Na^+_{(aq)} + X^-_{(aq)} \qquad pH = 7$$

d-block halides

These dissolve in water producing acidic solutions as a result of the hydrolysis of the metal cation. The smaller and more highly charged the metal cation, the more acidic is the solution:

$$FeX_{3(s)} + H_2O_{(l)} \longrightarrow Fe^{3+}_{(aq)} + 3X^-_{(aq)}$$

$$[Fe(H_2O)_6]^{3+}_{(aq)} + H_2O_{(l)} \rightleftharpoons [Fe(H_2O)_5OH]^{2+} + H_3O^+_{(aq)} \qquad pH \simeq 3$$

p-block halides

These tend to react with water producing acidic solutions:

$$\left.\begin{array}{l} PCl_{3(l)} + 6H_2O_{(l)} \rightarrow 3H_3O^+_{(aq)} + 3Cl^-_{(aq)} + H_3PO_{3(aq)} \\ PCl_{5(s)} + 9H_2O_{(l)} \rightarrow 5H_3O^+_{(aq)} + 5Cl^-_{(aq)} + H_3PO_{4(aq)} \\ SiCl_{4(l)} + 6H_2O_{(l)} \rightarrow 4H_3O^+_{(aq)} + 4Cl^-_{(aq)} + SiO_{2(s)} \end{array}\right\} \quad pH \simeq 1$$

The acid strength of the hydrogen halides increases going down the group from HF to HI. This is in agreement with the decreasing bond strengths of the molecules (see the table on page 735).

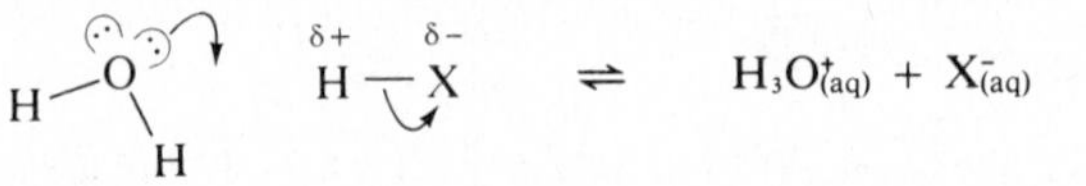

$$H_2O + \overset{\delta+}{H}-\overset{\delta-}{X} \rightleftharpoons H_3O^+_{(aq)} + X^-_{(aq)}$$

acid	K_a (mol dm^{-3})
HF	5.6×10^{-4}
HCl	1×10^{7}
HBr	1×10^{9}
HI	1×10^{11}

Hydrofluoric acid is much weaker than the others. The bond strength of H—F is in fact greater than the bond strength of H—O. Since the proton transfer involves breaking H—F and forming H—O, the process is unlikely to be very favourable. Molecular hydrogen fluoride is also stabilized by hydrogen bonding in aqueous solution and this further discourages ionization.

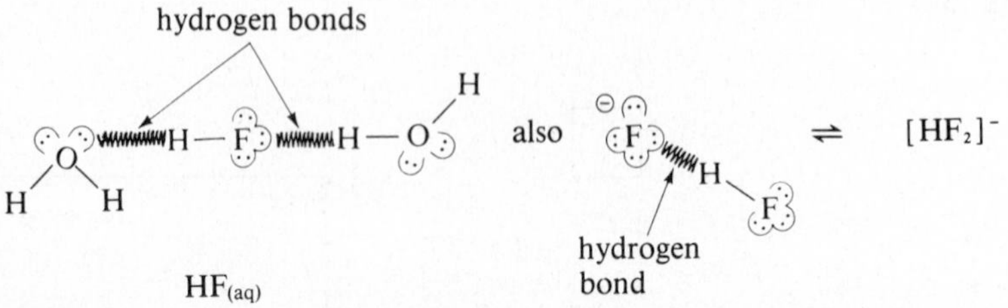

hydrogen bonds

$H_2O \cdots H—F \cdots H—O—H$ also $F^- \cdots H—F \rightleftharpoons [HF_2]^-$

$HF_{(aq)}$ hydrogen bond

The hydrogen bond between a fluoride ion and a hydrogen fluoride molecule is so strong that it is possible to crystallize alkali metal salts containing $[HF_2]^-$ ions: e.g. KHF_2 potassium hydrogen fluoride.

Metal halides

Most metal halides can be prepared from the aqueous hydrogen halide acids. Metal oxides or carbonates neutralize the acids and leave aqueous salt solutions.

The acids themselves are usually produced by the reverse reaction in the laboratory: an anhydrous metal halide is protonated under non-aqueous conditions:

$$NaCl_{(s)} + H_2SO_{4(l)} \rightarrow NaHSO_{4(s)} + HCl_{(g)}$$

Concentrated sulphuric acid cannot be used to give hydrogen bromide or iodide because of its oxidizing properties (see page 721). Phosphoric (V) acid is used instead. Industrially, hydrogen chloride is manufactured by burning a jet of hydrogen in chlorine.

Non-metal halides

Non-metal halides are nearly all strongly acidic in water. Hydrolysis occurs as a result of nucleophilic attack on the electron-deficient non-metal atoms:

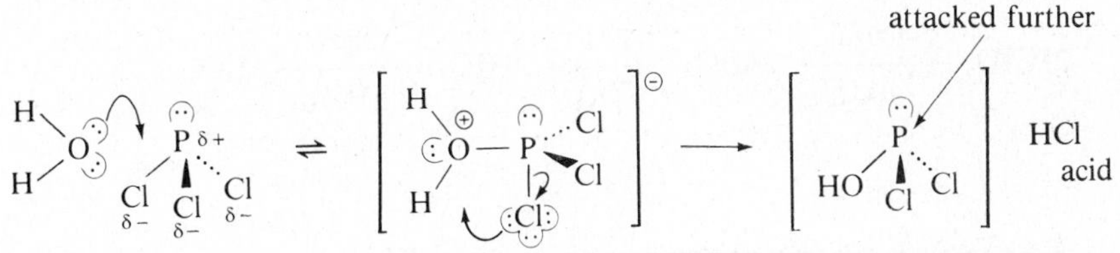

There are one or two exceptions (e.g. CCl_4 and SF_6) and their anomalous behaviour can usually be explained in terms of the steric hindrance of the attacking nucleophiles caused by the envelope of bonded halogen atoms.

Positive oxidation states

All the halogen compounds that contain the elements in a positive oxidation state hydrolyse to give acidic halate solutions. The three most common states in solution are I, V and VII, but chlorine shows some tendency to exist as chlorate (III) as well. Some typical hydrolysis reactions are those of the oxides and fluorides of chlorine.

oxidation state	*oxides*	*fluorides*
I	$Cl_2O_{(g)} + 3H_2O_{(l)} \rightarrow 2ClO^-_{(aq)} + 2H_3O^+_{(aq)}$	$ClF_{(g)} + 3H_2O_{(l)} \rightarrow ClO^-_{(aq)} + F^-_{(aq)} + 2H_3O^+_{(aq)}$
III		$ClF_{3(l)} + 6H_2O_{(l)} \rightarrow ClO^-_{2(aq)} + 3F^-_{(aq)} + 4H_3O^+_{(aq)}$
IV	$2ClO_{2(g)} + 3H_2O_{(l)} \rightarrow ClO^-_{2(aq)} + ClO^-_{3(aq)} + 2H_3O^+_{(aq)}$	
V		$(ClF_{5(s)} + 9H_2O_{(l)} \rightarrow ClO^-_{3(aq)} + 5F^-_{(aq)} + 6H_3O^+_{(aq)})$
VI	$Cl_2O_{6(l)} + 3H_2O_{(l)} \rightarrow ClO^-_{3(aq)} + ClO^-_{4(aq)} + 2H_3O^+_{(aq)}$	
VII	$Cl_2O_{7(l)} + 3H_2O_{(l)} \rightarrow 2ClO^-_{4(aq)} + 2H_3O^+_{(aq)}$	

The strength of the different acids increases as n increases in HXO_n (see page 622), and the only stable halate salts that can be crystallized are those containing halate (V) or (VII). Iodic (VII) acid can be prepared in a number of differently hydrated forms. The common form has the molecular formula H_5IO_6 but this can be dehydrated to HIO_4 by heating to about 100°C.

$$H_5IO_6 \ (\text{O=I(OH)}_5) \xrightarrow[(\sim 100°C)]{\text{heat}} HIO_{4(s)} \ (\text{IO}_3\text{OH}) + 2H_2O_{(g)}$$

Stronger heating leads to decomposition and the production of iodic (V) acid and oxygen, and eventually iodine (V) oxide

$$2H_5IO_6 \xrightarrow[-4H_2O_{(g)}]{100°C} 2HIO_4 \xrightarrow[-O_{2(g)}]{150°C} 2HIO_3 \xrightarrow[-H_2O_{(g)}]{200°C} I_2O_5$$

46.6 Redox properties

Disproportionation

The considerable thermodynamic stability of the −I state halides provides the driving force for a number of halogen disproportionation reactions.

The elements themselves disproportionate in alkali or water,

$$\overset{0}{X_2} + 2H_2O \rightleftharpoons \overset{-I}{X^-_{(aq)}} + H_3O^+_{(aq)} + HO\overset{I}{X}$$

and the weak halogen (I) acid tends to disproportionate further.

$$3HO\overset{I}{X}_{(aq)} + 3H_2O_{(l)} \rightleftharpoons \overset{V}{X}O^-_{3(aq)} + 2\overset{-I}{X^-_{(aq)}} + 3H_3O^+_{(aq)}$$

The action of heat on a solid halate (V) salt leads to one last disproportionation:

$$4K\overset{V}{Cl}O_3 \rightarrow 3K\overset{VII}{Cl}O_4 + K\overset{-I}{Cl}$$

Oxidants

The oxy-halogen compounds, the interhalogens and the halogens themselves all act as oxidants. Once again, it is the tendency of the elements to exist in the −I state that is responsible for the oxidizing strength of these compounds. This strength decreases going down the group as the electronegativity of the halogens gets less from fluorine to iodine.

Fluorine is the most powerful oxidant known; an important contributory factor in making this so is the large hydration energy of the fluoride ion.

The oxy-acids are useful laboratory oxidants. Compare the electrode potentials overleaf.

elements	$E^\ominus$ (volts)
$F_{2(g)} + 2e^- \rightleftharpoons 2F^-_{(aq)}$	2.87
$Cl_{2(g)} + 2e^- \rightleftharpoons 2Cl^-_{(aq)}$	1.36
$Br_{2(l)} + 2e^- \rightleftharpoons 2Br^-_{(aq)}$	1.06
$I_{2(s)} + 2e^- \rightleftharpoons 2I^-_{(aq)}$	0.54

halates	$E^\ominus$ (volts)
$ClO^-_{4(aq)} + 8H_3O^+_{(aq)} + 8e^- \rightleftharpoons Cl^-_{(aq)} + 12H_2O_{(l)}$	1.37
$ClO^-_{3(aq)} + 6H_3O^+_{(aq)} + 6e^- \rightleftharpoons Cl^-_{(aq)} + 9H_2O_{(l)}$	1.45
$ClO^-_{(aq)} + 2H_3O^+_{(aq)} + 2e^- \rightleftharpoons Cl^-_{(aq)} + 3H_2O_{(l)}$	1.49
$IO^-_{3(aq)} + 6H_3O^+_{(aq)} + 6e^- \rightleftharpoons I^-_{(aq)} + 9H_2O_{(l)}$	1.09
$IO^-_{(aq)} + 2H_3O^+_{(aq)} + 2e^- \rightleftharpoons I^-_{(aq)} + 3H_2O_{(l)}$	0.99

The elements oxidize metals, non-metals and aqueous solutions of compounds. For example, chlorine oxidizes aluminium, sulphur and iron (II) sulphate. In the gas phase, chlorine oxidizes hydrogen and hydrogen sulphide.

$$2\overset{0}{Al}_{(s)} + 3\overset{0}{Cl}_{2(g)} \rightarrow \overset{III}{Al}_2\overset{-I}{Cl}_{6(g)}$$

$$2\overset{0}{S}_{8(s)} + 8\overset{0}{Cl}_{2(g)} \rightarrow 8\overset{I}{S}_2\overset{-I}{Cl}_{2(l)}$$

$$2\overset{II}{Fe}{}^{2+}_{(aq)} + \overset{0}{Cl}_{2(g)} \rightarrow 2\overset{III}{Fe}{}^{3+}_{(aq)} + 2\overset{-I}{Cl}{}^-_{(aq)}$$

$$\overset{0}{H}_{2(g)} + \overset{0}{Cl}_{2(g)} \rightarrow 2\overset{I}{H}\overset{-I}{Cl}_{(g)}$$

$$8H_2\overset{-II}{S}_{(g)} + 8\overset{0}{Cl}_{2(g)} \rightarrow 16H\overset{-I}{Cl}_{(g)} + \overset{0}{S}_{8(s)}$$

The steady decrease in oxidizing strength of the aqueous chlorates as oxidation state increases suggests that the chlorate (VII) ion is the most stable. This agrees with the predictions of resonance theory. There are more ways in which the anionic charge can be delocalized by a chlorate (VII) ion than by any other chlorate ion:

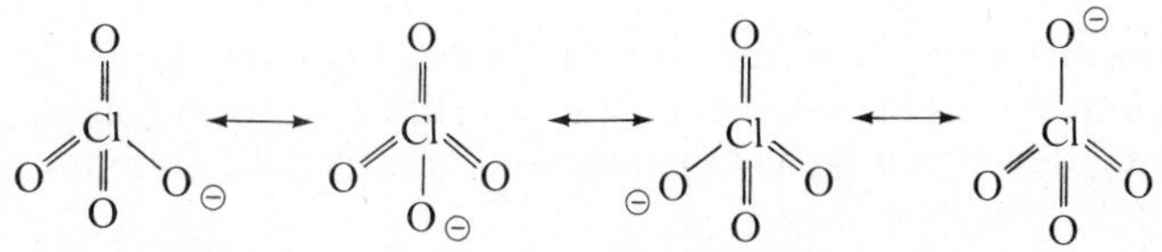

It is interesting to note that the acid strength of the hydrogen halates follows the opposite pattern: $HClO_4 > HClO_3 > HClO_2 > HClO$. Proton loss is increasingly favoured as the acid anion becomes more stable.

Iodates are useful redox reagents for quantitative work. They are reduced to iodine under acidic conditions and the iodine can be analysed by using sodium thiosulphate. For example, the purity of an impure sulphate (IV) salt can be estimated by treating it with excess acidified sodium iodate (V). The iodine liberated is then titrated against standard thiosulphate. Knowing the stoichiometries of the reactions, the concentration of sulphate (IV) in the sample is calculated.

1. $5SO_{3(aq)}^{2-} + 2IO_{3(aq)}^{-} + 2H_3O_{(aq)}^{+} \rightarrow 5SO_{4(aq)}^{2-} + I_{2(aq)} + 3H_2O_{(l)}$
2. $I_{2(aq)} + 2S_2O_{3(aq)}^{2-} \rightarrow 2I_{(aq)}^{-} + S_4O_{6(aq)}^{2-}$

In summary, $5[SO_3^{2-}] \equiv 2[S_2O_3^{2-}]$

A similar technique has been used to analyse gas mixtures for carbon monoxide. Iodine (V) oxide oxidizes carbon monoxide quantitatively to carbon dioxide.

$$I_2O_{5(s)} + 5CO_{(g)} \xrightarrow{\text{heat}} I_{2(g)} + 5CO_{2(g)}$$

Reductants

The decrease in oxidizing strength of the halogens going down the group is mirrored by an increase in the reducing strength of the corresponding halides. For example, although it is possible to prepare hydrogen chloride gas from sodium chloride and concentrated sulphuric acid, bromides and iodides reduce the sulphuric acid so that a mixture of free halogens, hydrogen halides and sulphur (IV) oxide results. Some hydrogen sulphide can also be detected. An aqueous solution of potassium iodide is a useful laboratory reagent because it is fairly readily oxidized to iodine. For example, it reduces a copper (II) salt solution to copper (I) iodide which precipitates as an insoluble solid:

$$2\overset{\text{II}}{Cu}{}_{(aq)}^{2+} + 4\overset{-\text{I}}{I}{}_{(aq)}^{-} \rightarrow 2\overset{\text{I}}{Cu}I_{(s)} + \overset{0}{I}{}_{2(aq)}$$

'Starch iodide' paper contains an iodide salt mixed with starch absorbed into the test paper. In the presence of oxidants, the paper turns blue because the iodide is oxidized to iodine which forms a blue complex with starch.

46.7 Solubility and complexing properties

Halide ions as ligands

The halide ions have a strong tendency to act as ligands and they form complexes with most 'B' metal and transition metal cations. They also form complexes with non-metal halides. Some typical metallic and non-metallic complexes are listed below.

metallic complexes	*non-metallic complexes*
$[AlF_6]^{3-}$	$[SiF_6]^{2-}$
$[PbCl_4]^{2-}$	$[PF_6]^{-}$
$[CoBr_4]^{2-}$	$[I_3]^{-}$, $[ICl_4]^{-}$, $[BrF_6]^{-}$

Metallic complex formation is often used in qualitative analysis to give information about the nature of the metal cations present. For example, when a few drops of hydrochloric acid are added to a solution of a lead (II) salt, a precipitate of insoluble lead (II) chloride forms.

$$Pb^{2+}_{(aq)} + 2Cl^{-}_{(aq)} \rightarrow PbCl_{2(s)}$$

The precipitate redissolves in concentrated hydrochloric acid because the soluble complex $[PbCl_4]^{2-}$ forms.

$$PbCl_{2(s)} + 2Cl^{-}_{(aq)} \rightarrow [PbCl_4]^{2-}_{(aq)}$$

The formation of non-metallic complexes usually takes place as a result of electron-pair donation by the halide ligands. For example, molecular iodine is not very soluble in water but dissolves quite readily in aqueous potassium iodide to give a brown-coloured solution. An iodide ion attacks an iodine molecule, and the valency of the attacked iodine atom is increased to two.

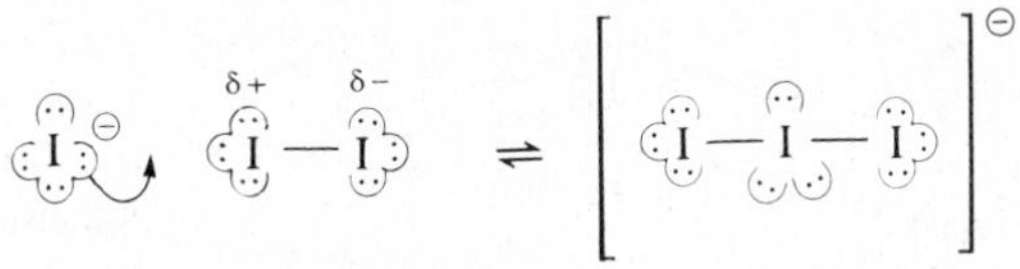

$$I^{\ominus} + \overset{\delta+}{I}-\overset{\delta-}{I} \rightleftharpoons [I-I-I]^{\ominus}$$

I^+ and I^{3+} complexes

When iodine and silver nitrate are treated with an excess of the complexing agent pyridine in a solvent of chloroform, the complex salt $[I(py)_2]^+ NO_3^-$ can be isolated. In all non-metallic groups, there is an increase in metallic character going down the group and the halogens are no exception. Iodine shows very few metallic properties, but the formation of a strongly co-ordinated +I state complex is more typical of a metal than a non-metal. The decrease in control of outer-shell electrons is responsible for the steady increase in metallic character down a group: the outer shell is getting progressively further away from the nucleus and therefore the force of attraction gets less and less.

There is also some evidence of an I^{3+} complex that is analogous to a complex of iron (III). When iodine is refluxed with ethanoic anhydride in fuming concentrated nitric acid, iodine (III) ethanoate is produced.

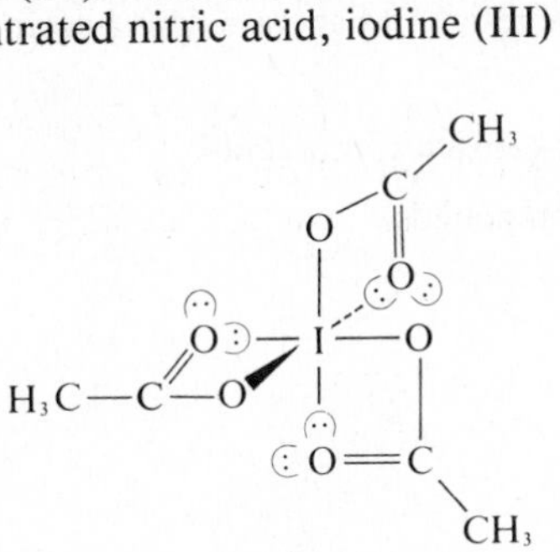

$$I(CH_3COO)_3$$

Questions

1. Explain the meaning of the following: a) 'odd electron' molecule b) stoichiometry c) oxidation state d) van der Waals' force e) Born-Haber cycle f) bond dissociation energy g) d-subshell h) nucleophilic attack.

2. Compare the reactions of fluorine and chlorine with: a) water b) sodium hydroxide c) chromium.

3. Compare and contrast the action of sodium hydroxide on: a) chlorine b) sulphur c) phosphorus.

4. Account for the trends in group VII properties shown in the table below:

hydride	*b.p.* °C	pK_a	*HX bond energy* (kJ mol^{-1})
HF	+19.5	3	562
HCl	−85	< -7	431
HBr	−67	< -9	366
HI	−35	< -10	299

5. a) Account for the difference in acid strength of the various chloric acids:
   HOCl, pK_a = 7; HOClO, pK_a = 2; $HOClO_2$, pK_a = −1; $HOClO_3$, pK_a = −10
   b) Account for the strengths of the various halic acids:
   HOCl, pK_a = 7; HOBr, pK_a = 9; HOI, pK_a = 10

6. Account for the difference in oxidizing strength of the halogens as indicated by the $E^{\ominus}$ values on page 743.

7. How does concentrated hydrochloric acid react with the following?
   a) aluminium
   b) manganese (IV) oxide
   c) copper (II) oxide
   d) lead (II) hydroxide.

8. Fluorine is the first member of the halogen group. It is often said that the first member of a periodic group differs from the rest. In what ways does fluorine fulfil this statement?

9. Why is disproportionation so common with halogen compounds?

10. Discuss the chemistry of the interhalogen compounds.

Chapter 47

The noble gases

47.1 Structure of the elements

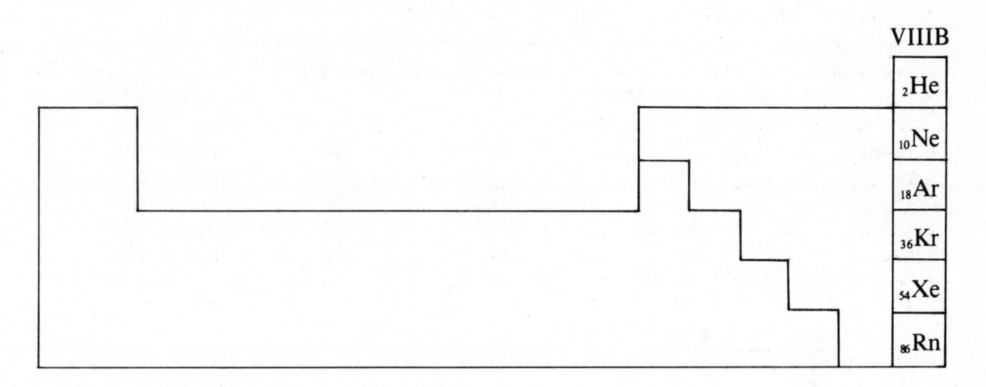

Note that radon is radioactive. $^{219}_{86}Rn \rightarrow ^{215}_{84}Po + ^{4}_{2}He$

element	*electronic configuration*	*first ionization energy* (kJ mol^{-1})	*promotion energy to next vacant orbital* (kJ mol^{-1})	*stable oxidation states*
helium	1s^2	2370	—	none
neon	[He] 2s^2 2p^6	2080	1600	none
argon	[Ne] 3s^2 3p^6	1520	1110	none
krypton	[Ar] 3d^{10} 4s^2 4p^6	1350	950	II, IV
xenon	[Kr] 4d^{10} 5s^2 5p^6	1170	800	II, IV, VI, VIII

Atomic structure

With the exception of helium, all the noble gas atoms have eight outer-shell electrons and hence the group number is usually VIIIB. Sometimes the group number is written as zero to indicate that there are in a sense no outer-shell electrons at all.

Compared with elements in the other periodic groups, the noble gases form very few compounds. In fact until 1962, no group VIIIB compounds had been isolated. This lack of reactivity is not surprising. Their very high ionization energies and low electron affinities (almost zero) indicate that electron loss or gain is unlikely. Helium and neon have the maximum number of electrons allowed in their outer shells and so cannot share any more. The other noble gas atoms can expand their outer shells and so covalent bonding with other atoms is feasible. The promotion energy in the table above is a measure of this. In fact, no compounds of helium, neon, or argon have ever been isolated, but a variety of compounds of krypton and xenon have been prepared.

Physical structure

The elements exist as monatomic gases at room temperature. The only forces between the atoms are van der Waals' forces and these increase with increasing number of electrons present.

element	*number of electrons per atom*	*m.p.* (°C)	*b.p.* (°C)	*% composition in the atmosphere*
helium	2	−270	−269	5.2×10^{-4}
neon	10	−249	−246	1.8×10^{-3}
argon	18	−189	−186	9.3×10^{-1}
krypton	36	−157	−152	1.1×10^{-3}
xenon	54	−112	−108	8.7×10^{-6}

47.2 Compound formation

The only compounds of the noble gases known before 1962 were 'clathrate compounds'. These substances can hardly be thought of as noble gas compounds because they do not contain chemically combined noble gas atoms. They consist of crystal lattices in which atoms of the group VIIIB elements have become trapped. The most common noble gas clathrates are those of β-quinol (1,4-dihydroxybenzene). When this substance is crystallized in the presence of argon, krypton or xenon under a pressure of about ten atmospheres, a clathrate structure forms. The lattice contains noble gas atoms, each trapped in a 'cage' made out of three quinol molecules in a roughly spherical arrangement. The quinol molecules are bonded to one another by hydrogen bonds and the volume that they enclose is about 4×10^{-10}m in diameter. This is approximately equal to the volume of a noble gas atom. The ratio (quinol : noble gas) = 3:1, is rarely achieved exactly because some of the cages are empty.

In 1962, Bartlett observed that the first ionization energy of xenon was almost identical to that of molecular oxygen. Since it was known that oxygen formed the complex $[O_2]^+[PtF_6]^-$ with the powerful oxidant, platinum (VI) fluoride, he reasoned that xenon might react similarly. By mixing xenon and platinum (VI) fluoride vapour under pressure, a yellow powder, xenon hexafluoro-platinate (V), was produced.

$$Xe_{(g)} + PtF_{6(g)} \rightarrow Xe^+[PtF_6]^-$$

Following this discovery, xenon (and krypton to a lesser extent) has been oxidized by other powerful oxidants, notably fluorine. A range of fluorides, oxyfluorides and aqueous oxyanions has been established.

47.3 Occurrence, extraction and uses

The percentage composition of the noble gases in the atmosphere is shown on page 748. Radon occurs as a result of the α-decay of radium-226 used in the treatment of cancerous cells.

$$^{226}_{88}Ra \rightarrow ^{222}_{86}Rn + ^{4}_{2}He$$

Radon-222 is the most stable isotope of the element but it still undergoes α-decay with a half life of 91.8 hours.

Helium occurs in fossil fuel deposits, probably as a result of natural α-decay processes. In some sources, up to seven per cent of natural gas is helium. It is extracted from these sources and from the atmosphere by fractional distillation of the liquefied gas mixtures. Argon, neon and krypton are similarly extracted from liquid air. The more important uses of the noble gases are listed below.

element	*uses*
helium	to provide an inert atmosphere for welding; to inflate balloons and sometimes even the tyres of large aircraft (the gas is very light and non-inflammable); mixed with oxygen in the breathing equipment of divers (replacing nitrogen which can cause the 'bends' in divers); as a liquid coolant for super-conductivity apparatus.
argon	welding operations (like helium): in light bulbs to prevent oxidation and evaporation of the filament.
neon	in discharge tubes for shop lights and street lamps (red).

47.4 Compound structure

The + II state

Both krypton and xenon form a II-state fluoride when the elements are irradiated by a mercury vapour lamp in the presence of fluorine. Xenon (II) tends to react further and produce xenon (IV) fluoride if the temperature is allowed to increase. XeF_2 can also be produced by treating solid xenon with difluorine oxide at $-120°C$.

$$Xe_{(s)} + F_2O_{2(g)} \rightleftharpoons XeF_{2(s)} + O_{2(g)}$$

XeF_2 and KrF_2 are linear molecules that are similar in electronic structure to the tri-iodide ion discussed on page 745.

F
|
(:)(:) Xe (:)
|
F

The +IV state

There is some evidence to suggest the existence of krypton (IV) fluoride, but this compound is far less stable than xenon (IV) fluoride which can be prepared by heating a 1:5 mixture of xenon and fluorine together at a pressure of about 6 atmospheres. Nickel catalyses this reaction.

$$Xe_{(g)} + 2F_{2(g)} \xrightarrow[6\ atm]{Ni_{(s)}} XeF_{4(g)} \text{ at } 400°C$$

XeF_4 has a square planar structure.

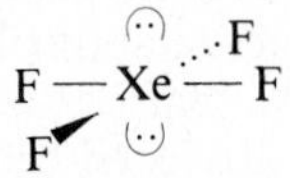

F — Xe — F
F F

Note that the m.p. of XeF_4 is only 114°C compared with that of XeF_2 (140°C). This suggests that there is some ionic character associated with these fluorides which decreases from Xe(II) to Xe(IV). It seems unlikely that $Xe^{2+}(F^-)_2$ exists, but the structure has sometimes been described as being a resonant hybrid of the two canonical forms below:

$$[F—X]^+ \, {}^-F \longleftrightarrow F^- \, [X—F]^+$$

The +VI state

Only xenon has been found to combine in this oxidation state. Xenon (VI) fluoride is produced by reacting the two elements together in a ratio of ($Xe:F_2$) = 1:20 at a pressure of about 50 atmospheres and a temperature of 300°C.

$$Xe_{(g)} + 3F_{2(g)} \xrightarrow{50\ atm} XeF_{6(g)} \text{ at } 300°C$$

Xenon (VI) fluoride has a distorted octahedral structure and exists at room temperature as colourless crystals of m.p. 48°C.

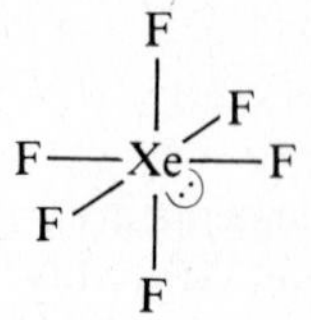

F
F — Xe — F
F F
F

It attacks silica to produce the most stable noble gas oxy-compound known, $XeOF_4$, tetrafluoroxenon (VI) oxide.

$$SiO_{2(s)} + 2XeF_{6(l)} \rightarrow SiF_{4(g)} + 2XeOF_{4(l)}$$

At room temperature this is a colourless liquid. The compound remains thermally stable in the vapour phase up to 500°C.

Xenon (VI) oxide has been prepared by the hydrolysis of xenon (VI) fluoride. It is a soluble molecular substance showing little tendency itself to hydrolyse like other VI-state oxides, e.g. CrO_3, SO_3. It is very explosive when extracted in the solid state as colourless crystals: its heat of formation shows it to be a very endothermic compound, $\Delta H_f^{\ominus} \approx +400\ \text{kJ mol}^{-1}$.

$$XeF_{6(s)} + 3H_2O_{(l)} \rightleftharpoons XeO_{3(aq)} + 6HF_{(aq)}$$

The structures of the two xenon oxy-compounds are shown below.

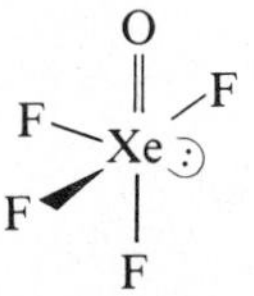

distorted square pyramidal

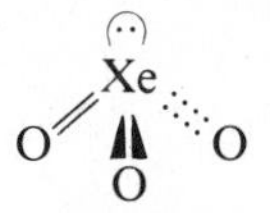

trigonal pyramidal

The +VIII state

Xe(VI) can be oxidized to Xe(VIII) by ozone in alkaline aqueous conditions. Xe(VIII) is only stable in solution; attempts to isolate XeO_4 from these solutions have shown that it is even more explosive than XeO_3. In solution, the majority of Xe(VIII) appears to exist as $H_3XeO_6^-$ although there is a strong tendency for Xe(VIII) to decompose into Xe(VI) and oxygen.

$$XeO_{3(aq)} + O_{3(g)} + 3H_2O_{(l)} \rightarrow [H_3XeO_6]^-_{(aq)} + H_3O^+_{(aq)} + O_{2(g)}$$

$$\rightleftharpoons$$

$$[HXeO_4]^-_{(aq)} + H_2O_{(l)} + \tfrac{1}{2}O_{2(g)}$$

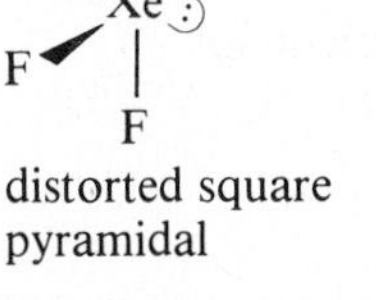

$H_3XeO_6^-$ [Xe(VIII)]

$HXeO_4^-$ [Xe(VI]

47.5 Reactivity

As might be expected, the reactivity of these group VIIIB compounds centre on their oxidizing and acid-base properties. Their tendency to decompose and occasionally disproportionate is also in evidence.

Oxidizing properties

Xe(II) and Xe(VIII) both oxidize water to oxygen. XeF_2 hydrolyses rapidly in alkaline conditions with the evolution of oxygen.

$$2\overset{\text{II}}{Xe}F_{2(s)} + 4\overset{-\text{II}}{O}H^-_{(aq)} \rightarrow 2\overset{0}{Xe}_{(g)} + 4F^-_{(aq)} + 2H_2O_{(l)} + \overset{0}{O}_{2(g)}$$

Xenon (IV) fluoride brings about the oxidation of water by disproportionating to xenon (0) and xenon (VI):

$$6\overset{IV}{Xe}F_{4(s)} + 12H_2\overset{-II}{O}_{(l)} \rightarrow 2\overset{VI}{Xe}O_{3(aq)} + 24HF_{(aq)} + 4\overset{0}{Xe}_{(g)} + 3\overset{0}{O}_{2(g)}$$

All the xenon compounds oxidize hydrogen quantitatively and are themselves reduced to xenon:

$$XeF_{2(s)} + H_{2(g)} \rightarrow Xe_{(g)} + 2HF_{(g)}$$

$$XeOF_{4(g)} + 3H_{2(g)} \rightarrow Xe_{(g)} + 4HF_{(g)} + H_2O_{(g)} \text{ (explosive)}$$

Acid-base properties

The presence of lone pairs as well as electron-deficiency in the molecules of the group VIIIB compounds gives them some unusual acid-base properties. Xenon (II) fluoride acts as a Lewis base:

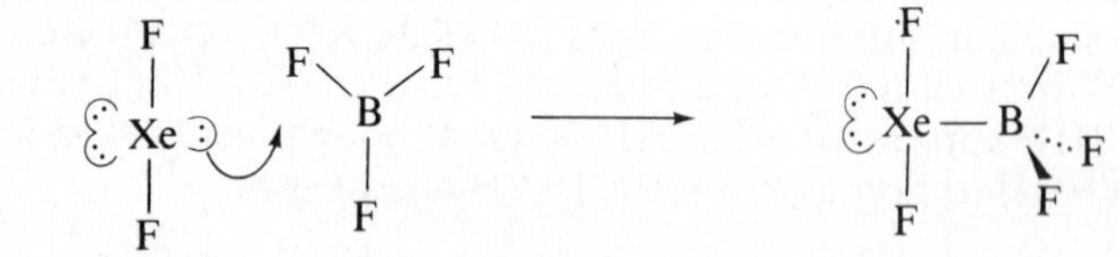

$$F-Xe-F + BF_3 \longrightarrow F_2Xe-BF_3$$

electron-pair donor (Lewis base) | electron-pair acceptor (Lewis acid) | salt

and yet XeF_2 is hydrolysed by water to give an acidic solution.

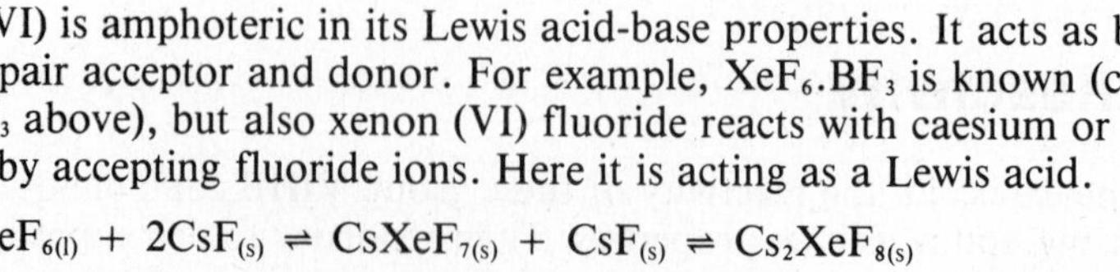

$$H_2O + F^{\delta-}-Xe^{\delta+}-F^{\delta-} \xrightarrow[\text{transfer}]{\text{electron}} Xe_{(g)} + \tfrac{1}{2}O_{2(g)} + 2HF_{(aq)}$$

weak acid

Xenon (VI) is amphoteric in its Lewis acid-base properties. It acts as both electron-pair acceptor and donor. For example, $XeF_6.BF_3$ is known (c.f. $XeF_2.BF_3$ above), but also xenon (VI) fluoride reacts with caesium or rubidium fluoride by accepting fluoride ions. Here it is acting as a Lewis acid.

$$XeF_{6(l)} + 2CsF_{(s)} \rightleftharpoons CsXeF_{7(s)} + CsF_{(s)} \rightleftharpoons Cs_2XeF_{8(s)}$$

The structure of the $[XeF_8]^{2-}$ ion is shown below.

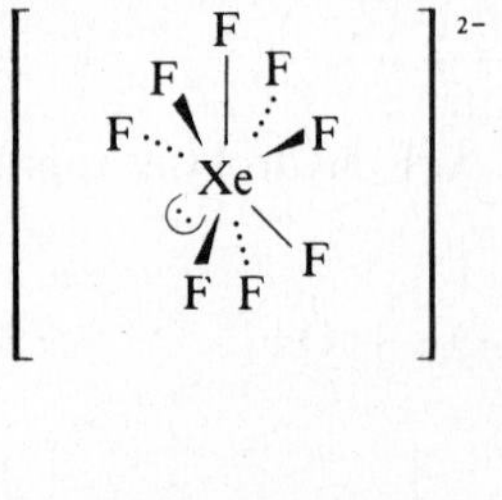

$$[XeF_8]^{2-}$$

Like XeF_2, XeF_4 and XeF_6 hydrolyse to give acidic solutions. Xenon (VI) oxide is weakly acidic in aqueous solution. It acts as a Lewis acid by accepting electron pairs from hydroxide ions. The resulting hydrogen xenate (VI) ions tend to disproportionate in the presence of group IA metal cations to give a precipitate of the metal xenate (VIII).

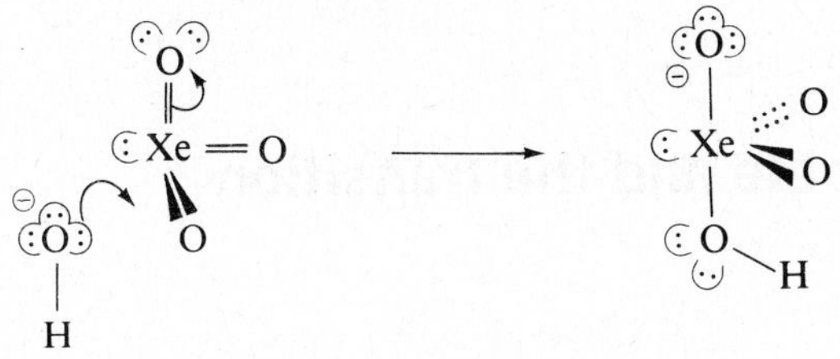

$[HXeO_4]^-$

and $2[H\overset{VI}{X}eO_4]^-_{(aq)} + 2OH^-_{(aq)} + 4Na^+_{(aq)} \rightarrow Na_4[\overset{VIII}{X}eO_6]_{(s)} + \overset{0}{X}e_{(g)} + O_{2(g)} + 2H_2O_{(l)}$

Sodium xenate (VIII) is one of the most insoluble sodium salts known. It is a powerful oxidant.

Questions

1. Explain the meaning of the following: a) weak acid b) oxidant c) thermal stability d) endothermic compound e) linear molecules f) melting point g) resonant hybrid h) canonical form.
2. Explain why the so called 'clathrate' compounds formed between the noble gases and β-quinol (1,4-dihydroxybenzene) are not classified as true compounds.
3. a) Why are the noble gases used in welding?
   b) Is the light from a discharge tube filled with neon always the same colour? Explain.
4. Discuss with examples the acid-base properties of: i) XeF_2 ii) XeF_6.
5. a) Write down the formula of five particles isoelectronic with the argon atom.
   b) Put the particles in decreasing order of size.
6. Calculate the empirical formulas of the following compounds:
   a) Kr 68.9%; F 31.1%.
   b) Kr 52.5%; F 47.5%.
   c) Xe 70.8%; O 8.7%; F 20.7%.
7. Explain why group VIII compounds are so reactive and susceptible to hydrolysis.

Chapter 48

The transition metals: a general survey

48.1 The Periodic Table and the transition elements

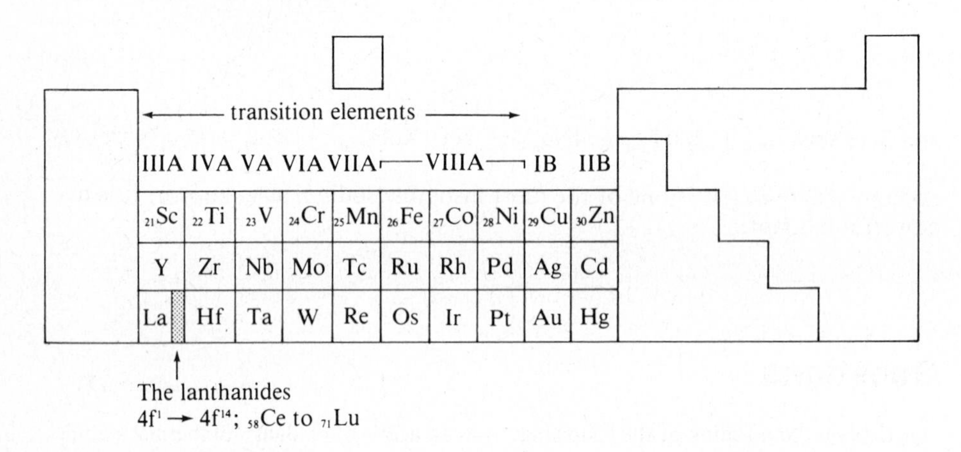

Orbital energies

The arrangement of the electrons in the structure of an atom is determined by the relative energies of the available vacant orbitals. Each electron occupies the orbital of lowest available energy according to the aufbau, Pauli and Hund principles (chapter 3). However, as the number of protons in the atom's nucleus increases, so the energy level of each orbital within the atom drops: more energy is required to remove an electron from the increased nuclear charge. In a hydrogen atom ($Z = 1$), the orbitals in a particular shell have the same energy, but as soon as Z begins to increase, each orbital's energy begins to change. The spherically symmetric s-orbitals are affected to a greater degree than are the more angular p- and d-orbitals, and so their energy drops by a larger amount. The energy drop of the orbitals as Z increases is shown opposite. At $Z = 16$, the 4s-orbital energy level has dropped below that of the 3d-orbitals and so, after the filling of the remaining 3p-orbitals, the fourth shell is started. However, as soon as an electron enters this outer orbital, the inner 3d-orbitals receive no shielding at all from the extra addition to the nuclear charge. This accounts for the sharp drop in their energy level from $Z = 18$ to $Z = 21$. At $Z = 21$, the energy levels of 4s and 3d are approximately equal and, with the filling of the inner d-subshell of electrons, the first transition series is produced. All the atoms have the electron-deficient fourth shell as their outer-shell and,

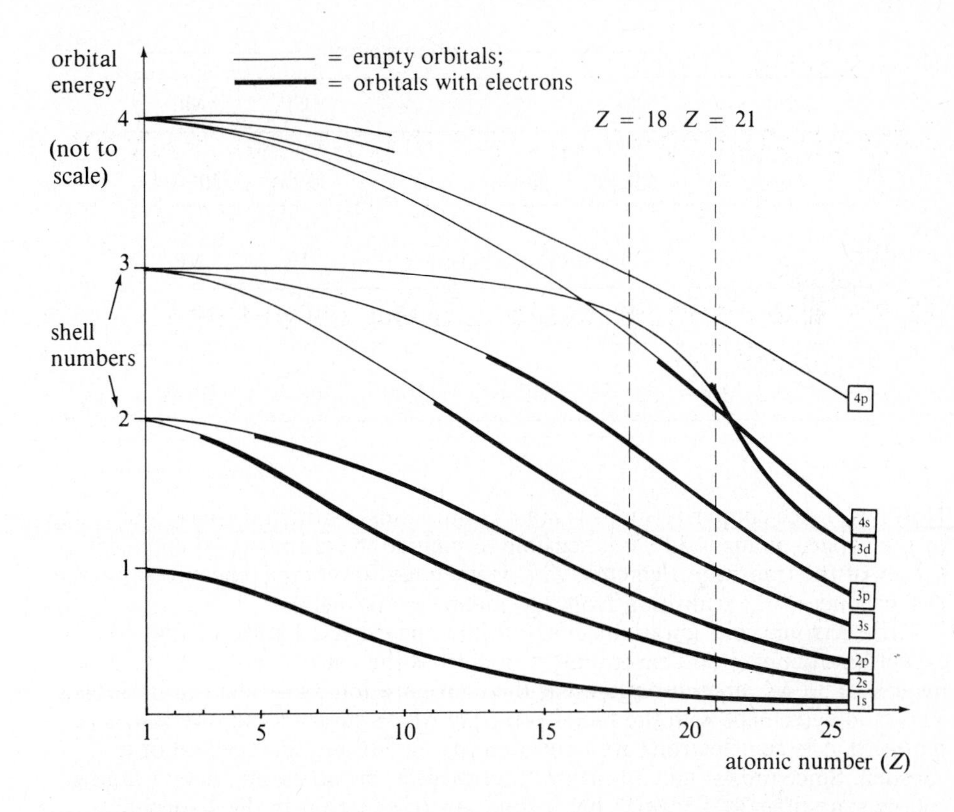

because the degree of nuclear control of these electrons is low, the elements show metallic properties.

A similar pattern repeats itself for the orbital energies of 5s and 4d. The filling of the inner 4d-subshell leads to the second transition series. At $Z = 57$ (lanthanum), the position becomes more complex still because the energy level of the 4f-orbitals drops below that of the 5d- and 6s-orbitals. The third transition series is therefore interrupted by the 'lanthanides' produced by the filling of an inner f-subshell. This pattern also repeats itself for the orbital energies of 5f, 6d and 7s. As the 5f-orbitals become occupied, the 'actinides' are produced. The heaviest named atom that has been synthesized in cyclotron bombardment experiments is an atom of lawrencium $_{103}Lw$. This is the last member of the actinide series and has the electronic configuration $[Rn]5f^{14}6d^{1}7s^{2}$. Its nuclear stability is extremely low and it decays with a half life of only a few seconds.

Electronic configuration of the first transition series

The characteristic property of a transition metal is that its atoms contain partially-filled inner d-subshells. In this sense, it is therefore incorrect to classify copper and zinc as transition metals. Their electronic configurations are shown overleaf.

	← transition elements				---
atom	Sc	Ti	V	Cr	Mn
inner shells $1s^2\ 2s^2\ 2p^6\ 3s^2\ 3p^6$	$3d^1\ 4s^2$	$3d^2\ 4s^2$	$3d^3\ 4s^2$	$3d^5\ 4s^1$	$3d^5\ 4s^2$

	--- transition elements →			IB	IIB
atom	Fe	Co	Ni	(Cu)	(Zn)
inner shells $1s^2\ 2s^2\ 2p^6\ 3s^2\ 3p^6$	$3d^6\ 4s^2$	$3d^7\ 4s^2$	$3d^8\ 4s^2$	$3d^{10}\ 4s^1$	$3d^{10}\ 4s^2$

However, since copper forms +II state compounds in which the electronic state of the copper atoms is $3d^9$, it is sensible to include the chemistry of copper in a survey of the transition elements. Zinc is included for completeness to show the end product of the transition from 'A' metals to 'B' metals.

The electronic configurations above follow the expected pattern with the exception of copper and chromium. Copper has the configuration $3d^{10}4s^1$ instead of $3d^9 4s^2$, and chromium has the configuration $3d^5 4s^1$ instead of $3d^4 4s^2$. This is in agreement with the Hund principle which suggests that the forces of repulsion between electrons are minimized by half-filling each orbital of a subshell. Since the 4s- and 3d-orbital energy levels are so close, $3d^5 4s^1$ is a lower energy state than $3d^4 4s^2$ which has a spin pair of electrons in the 4s-orbital. This is shown for chromium in the energy diagrams below.

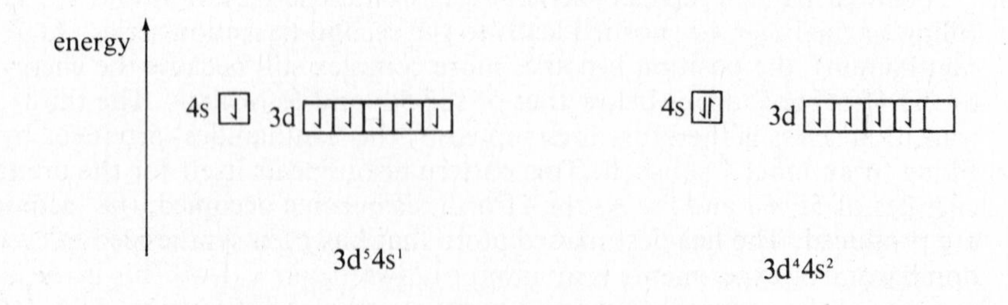

There are a number of similarities in the chemistries of the different transition elements. These are due principally to two characteristics that the atoms of the transition elements share.

1. They have the same outer-shell configurations ($4s^2$ or $4s^1$)
2. They have unpaired inner-shell d-electrons whose energy level is not much lower than that of the outer 4s-electrons.

The degree of shielding of the outer shell gradually decreases across the series because of the poor screening qualities of the d-electrons. This leads to a general increase in the ionization energies, but the increase is far less pronounced than the increase observed going across a p-block.

first ionization energies of the fourth period

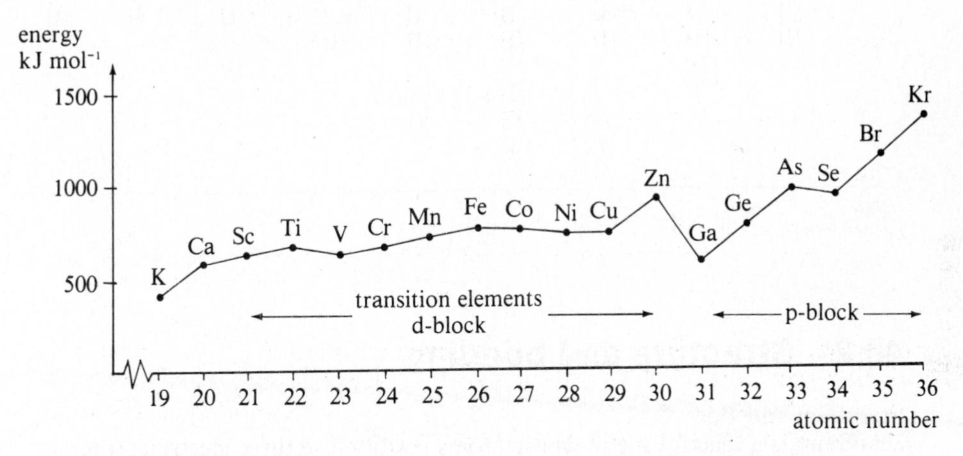

The graph of successive ionization energies for two transition metals is compared below with that for two s-block metals. In the former, there is no sudden increase indicating that these transition metals may exhibit a range of oxidation states.

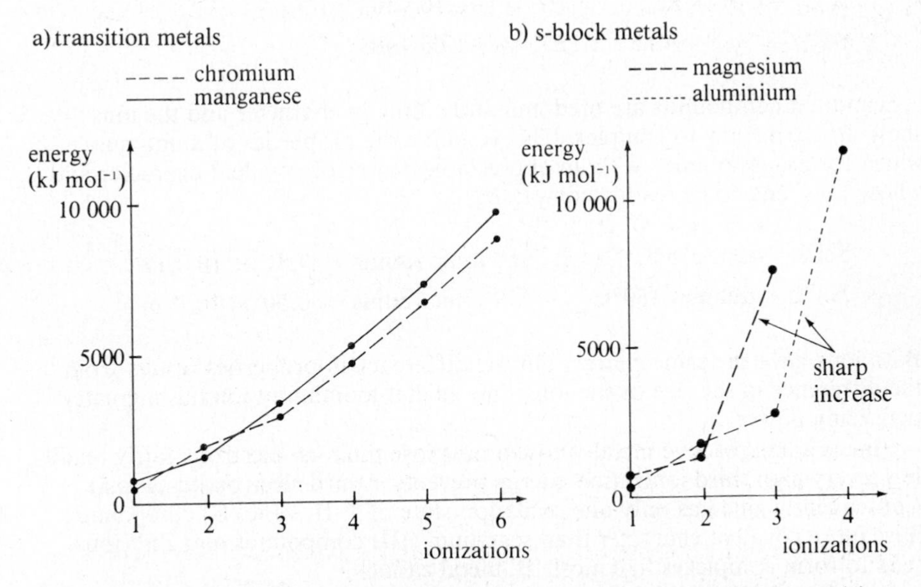

The table on the next page shows the stable oxidation states of the transition metals in the fourth period. The stability of a particular oxidation state is dependent on the nature of the bonded atoms or ligands as we shall see in the next section.

The stable oxidation states of transition metals

$_{21}Sc$	$_{22}Ti$	$_{23}V$	$_{24}Cr$	$_{25}Mn$	$_{26}Fe$	$_{27}Co$	$_{28}Ni$	$_{29}Cu$	$_{30}Zn$
								I	
	II	II	II	II	II	II	II	II	II
III	III	III	III	III	III	III			
	IV	IV		IV			IV		
		V							
			VI	VI	VI				
				VII					

48.2 Structure and bonding

Scandium and zinc

Scandium is a reactive metal whose atoms readily lose three electrons (the $3d^14s^2$-electrons) to form scandium (III) compounds. This is the only oxidation state for scandium. The redox potential for the element is not much lower than that for magnesium, but it is higher than that for aluminium.

$$Mg^{2+}_{(aq)} + 2e^- \rightleftharpoons Mg_{(s)} \qquad E^\ominus = -2.38 \text{ volts}$$
$$Sc^{3+}_{(aq)} + 3e^- \rightleftharpoons Sc_{(s)} \qquad E^\ominus = -2.10 \text{ volts}$$
$$Al^{3+}_{(aq)} + 3e^- \rightleftharpoons Al_{(s)} \qquad E^\ominus = -1.66 \text{ volts}$$

Scandium compounds are predominantly ionic in character and the ions show little tendency to complex. This is unlike the properties of aluminium which forms compounds with an appreciable degree of covalent character and whose ions tend to be readily complexed.

$ScCl_3$ m.p. 938°C Sc^{3+} ionic radius $= 0.81 \times 10^{-10}$ m

$AlCl_3$ sublimes 180°C Al^{3+} ionic radius $= 0.50 \times 10^{-10}$ m

Both ions have the same charge, but the difference in properties results from the difference in the size of the ions: the smaller aluminium ion has a greater polarizing power.

Zinc is a less reactive metal whose atoms lose their $4s^2$-electrons fairly readily, but a very high third ionization energy prevents a third electron being lost. Consequently zinc has only one oxidation state of +II. Zinc (II) compounds have more covalent character than scandium (III) compounds and Zn^{2+} ions tend to form complexes like most 'B' metal cations.

In their compounds, therefore, scandium and zinc are present with the electronic configurations d^0 and d^{10} respectively. Neither possesses typical transitional character but each represents the limits within which this character is defined.

Typical transition metals

The particular characteristic associated with the structure and bonding of transition metal compounds is the formation of complexes in a wide range of different oxidation states. The stability of a particular oxidation state of a transition metal is governed by the nature of the ligand present.

The formation of complexes

The tendency of a transition metal atom to form complex ions rather than simple ions is due to two main factors:

1. The atoms readily lose electrons to form small cations whose polarizing power increases with an increase in the number of electrons lost. Even the cations, M^{2+}, formed by losing only two electrons have appreciable polarizing powers, and therefore bonds between cation and ligand, which have quite a high degree of covalency, are favoured. Compare the radii of some M^{2+} ions from the d-block with the radius of a Ca^{2+} ion:

ion	Ca^{2+}	Mn^{2+}	Fe^{2+}	Co^{2+}	Ni^{2+}	Cu^{2+}	Zn^{2+}
ionic radius ($m \times 10^{-10}$)	1.00	0.80	0.76	0.78	0.76	0.69	0.74

2. The metal atoms have partly filled 3d-orbitals that can interact with the ligand orbitals. Electron density can either be transferred to the transition metal atom by orbital overlap or, occasionally, the d-electrons of the metal atom become delocalized onto the ligands as a result of the dπ-overlap.

Below is a table showing different oxidation state complexes stabilized by particular ligands.

	$:\bar{C}\equiv\overset{+}{O}:$	$:\bar{C}\equiv N:$	$:NH_3$	H_2O	$:F:^{-}$	$:O:^{2-}$
–II	$[Cr(CO)_5]^{2-}$					
–I	$[V(CO)_6]^{-}$					
0	$Ni(CO)_4$	$[Co(CN)_4]^{4-}$				
I	$[Mn(CO)_5]^{+}$	$[Cu(CN)_2]^{-}$	$[Cu(NH_3)_2]^{+}$			
II		$[Fe(CN)_6]^{4-}$	$[Co(NH_3)_6]^{2+}$	$[Mn(H_2O)_6]^{2+}$		
III		$[Cr(CN)_6]^{3-}$	$[V(NH_3)_6]^{3+}$	$[Fe(H_2O)_6]^{3+}$	$[CuF_6]^{3-}$	
IV					$[TiF_6]^{2-}$	$[Ti_2O_5]^{2-}$
V					$[VF_6]^{-}$	$[VO_3]^{-}$
VI						$[CrO_4]^{2-}$
VII						$[MnO_4]^{-}$

Type of ligand
The table above shows that certain ligands favour particular oxidation states although they may also form complexes with metal ions in oxidation states outside these ranges. Below is a diagram indicating the stabilizing effects of the different ligands.

ligand	*low oxidation state*	*stability*	*high oxidation state*
CO			
^{-}CN			
NH_3, H_2O			
F^-			
O^{2-}			

All ligands have a lone pair of electrons which can become co-ordinated to a metal atom by forming a σ-overlap with the vacant 3d-orbitals of the atom.

$M^n \xleftarrow{\sigma} :L$

L = ligand
M^n = a metal in oxidation state n

When the metal is in a low oxidation state, its atoms still have considerable electron density in the 3d-orbitals. Carbon monoxide molecules and cyanide ions are able to stabilize these low oxidation state atoms because they have vacant π-molecular orbitals that can accept electron density from the metal atoms.

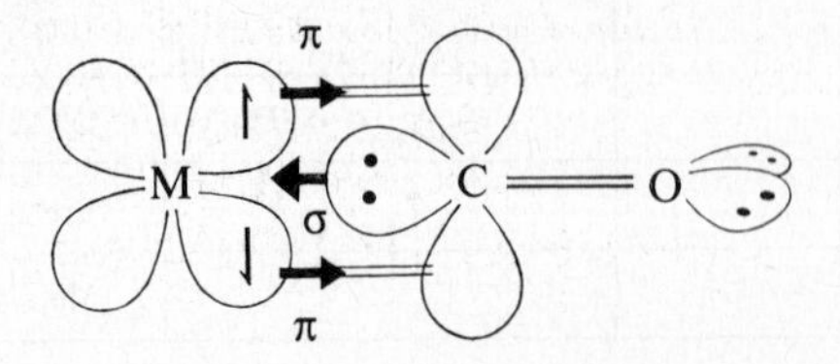

When the metal is in a high oxidation state, its atoms contain empty (or nearly empty) 3d-orbitals. Oxide ions stabilize these high oxidation state atoms because they have p-electrons that can be donated to the vacant 3d-orbitals by dπ-pπ-overlap.

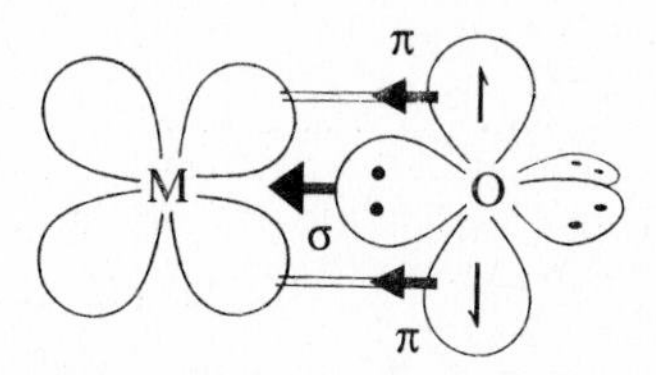

Some ligands are unable to interact with the metal atom by π-overlap. These tend to stabilize the middle value oxidation states, +II and +III. Fluoride ions form such strong bonds that their range extends to the +IV and +V states as well. Fluorine usually induces an element to combine in its highest possible oxidation state.

Summary of ligand type

All ligands have a lone pair of electrons that can be donated towards the metal atom, so they can be described as σ-electron donors. Some ligands are π-electron acceptors as well while others are π-electron donors. The table below, in which M represents the metal atom and L the ligand, summarizes these ideas.

low oxidation states	*medium oxidation states*	*high oxidation states*
$\mathrm{M} \underset{\sigma}{\overset{\pi}{\rightleftarrows}} \mathrm{L}$	$\mathrm{M} \underset{\sigma}{\longleftarrow} \mathrm{L}$	$\mathrm{M} \underset{\sigma}{\overset{\pi}{\leftleftarrows}} \mathrm{L}$
σ-electron donor π-electron acceptor e.g. CO; CN^-	σ-electron donor only e.g. NH_3; H_2O	σ-electron donor π-electron donor e.g. O^{2-}

48.3 Acid-base properties

Throughout the chemistry of the p-block elements, it can be seen that the acid-base properties of an element's oxide, hydroxide and halides depend to a great extent on the oxidation state of the element. The same pattern is true for the chemistry of the transition metals. The compounds of any particular element become less basic and more acidic as the oxidation state of the element increases. The examples below illustrate this trend.

Oxides

The black vanadium (III) oxide, V_2O_3, is basic. It is insoluble in water but dissolves in acids.

$$V_2O_{3(s)} + 6H_3O^+_{(aq)} \rightarrow 2V^{3+}_{(aq)} + 9H_2O_{(l)}$$
base acid

The orange vanadium (V) oxide, V_2O_5 is acidic. It is sparingly soluble in water giving a weakly acidic solution and it dissolves in alkalis giving vanadate (V) ions in solution.

$$V_2O_{5(s)} + 2OH^-_{(aq)} \rightarrow 2VO^-_{3(aq)} + H_2O_{(l)}$$
acid base

Hydroxides

Chromium (III) hydroxide is amphoteric. It dissolves in dilute acids and in excess alkali.

$$Cr(OH)_{3(s)} + 3H_3O^+_{(aq)} \rightarrow Cr^{3+}_{(aq)} + 6H_2O_{(l)}$$
base acid

$$Cr(OH)_{3(s)} + 3OH^-_{(aq)} \rightarrow Cr(OH)^{3-}_{6(aq)}$$
acid base

Aqueous solutions of chromium (VI) oxide contain the hydroxy-compound H_2CrO_4 which is quite a strong acid (c.f. H_2SO_4).

$$H_2CrO_{4(aq)} + H_2O_{(l)} \rightleftharpoons HCrO^-_{4(aq)} + H_3O^+_{(aq)} \qquad pK_a = -1$$

Chlorides

Both the chlorides of iron, $FeCl_2$ and $FeCl_3$ are soluble in water and hydrolysed by it. However the smaller size and greater charge of the Fe^{3+} ion results in a more acidic solution than that of the Fe^{2+} ion.

$$[Fe(H_2O)_6]^{2+} + H_2O \rightleftharpoons [Fe(H_2O)_5OH]^+ + H_3O^+ \qquad pK_a = 9.5$$

$$[Fe(H_2O)_6]^{3+} + H_2O \rightleftharpoons [Fe(H_2O)_5OH]^{2+} + H_3O^+ \qquad pK_a = 2.2$$

48.4 Redox properties

The redox trends evident in p-block chemistry are also found in the chemistry of the transition metals. Thus the stability of a wide range of different oxidation states leads to many redox reactions among transition metal compounds. These reactions show the following trends.

High oxidation states

Compounds containing transition elements in high oxidation states have oxidizing properties, e.g. M (V), (VI) and (VII) compounds tend to be oxidants as the standard redox potentials indicate.

V → IV	$VO^{+}_{2(aq)}/VO^{2+}_{2}$	$E^{\ominus}$ = +1.00 volts
VI → III	$Cr_2O^{2-}_{7(aq)}/Cr^{3+}_{(aq)}$	$E^{\ominus}$ = +1.33 volts
VII → II	$MnO^{-}_{4(aq)}/Mn^{2+}_{(aq)}$	$E^{\ominus}$ = +1.52 volts

Solutions of manganate (VII) or dichromate (VI) in acid are the most widely used laboratory oxidants. Either readily oxidizes iron (II) to iron (III):

$$5Fe^{2+}_{(aq)} + MnO^{-}_{4(aq)} + 8H_3O^{+}_{(aq)} \rightarrow 5Fe^{3+}_{(aq)} + Mn^{2+}_{(aq)} + 12H_2O_{(l)}$$

Low oxidation states

The metals themselves, that is oxidation state 0, are reductants, often quite strong ones. The strength of the metals as reductants decreases as the d-subshell fills across the series. The tendency for the metal atoms to lose their electrons decreases because the number of protons in the nucleus is getting bigger and the screening by the inner electrons is not very effective.

Compare the redox potentials for the equilibria

$$M^{2+}_{(aq)} + 2e^{-} \rightleftharpoons M_{(s)}$$

element	Ti	V	Cr	Mn	Fe	Co	Ni	Cu
$E^{\ominus}$ volts	−1.63	−1.20	−0.91	−1.03	−0.44	−0.28	−0.23	+0.34

Medium oxidation states

The medium oxidation states may lead to either oxidizing or reducing properties depending on the relative stabilities of the members of the conjugate redox pair.

Many of the middle oxidation states of the transition metals are unstable in aqueous solution. This is due to one of three reasons:

1. They disproportionate:

$$2Mn^{3+}_{(aq)} + 6H_2O_{(l)} \rightarrow Mn^{2+}_{(aq)} + MnO_{2(s)} + 4H_3O^{+}_{(aq)}$$

2. They are oxidized by water itself:

$$2Ti^{2+}_{(aq)} + 2H_2O_{(l)} \rightarrow 2Ti^{3+}_{(aq)} + H_{2(g)} + 2OH^{-}_{(aq)}$$

3. They are oxidized by dissolved oxygen from the air:

$$4Cr^{2+}_{(aq)} + O_{2(aq)} + 4H_3O^{+}_{(aq)} \rightarrow 4Cr^{3+}_{(aq)} + 6H_2O_{(l)}$$

However there are some that are fairly stable and can be used as laboratory reagents.

Iron (II) salts are weak reductants and can reduce strong oxidants:

$$2Fe^{2+}_{(aq)} + Cl_{2(aq)} \rightarrow 2Fe^{3+}_{(aq)} + 2Cl^{-}_{(aq)}$$

Iron (III) salts are oxidants and can oxidize quite weak reductants:

$$2Fe^{3+}_{(aq)} + Cu_{(s)} \rightarrow 2Fe^{2+}_{(aq)} + Cu^{2+}_{(aq)}$$

This reaction is used to etch copper in the manufacture of printed circuit boards in the electronics industry.

48.5 Typical transition element properties

Many of the chemical properties discussed so far in this chapter depend to some extent on the atoms of the transition elements having partially filled d-orbitals. The properties discussed already are:

1. variable oxidation state
2. complex formation.

Three further important properties typical of a transition metal compound are:

3. colour of the compounds
4. magnetic properties of the compounds
5. catalytic properties of the elements and their compounds.

Colour

The colour of each element in its common oxidation states in aqueous solution are shown below.

element	Ti	V	Cr	Mn	Fe	Co	Ni	Cu
oxidation state and colour	III violet	III,IV blue	III green	II pink	II green	II pink	II green	II blue
		V orange (yellow)	VI orange (yellow)	VI green	III yellow	III (ammonia complex) yellow		
				VII purple	VI violet			

The particles present in a coloured substance of any sort are able to interact with visible light. 'White' light is a mixture of electromagnetic rays of wavelength range 4 × 10^{-7} to 7 × 10^{-7} metres.

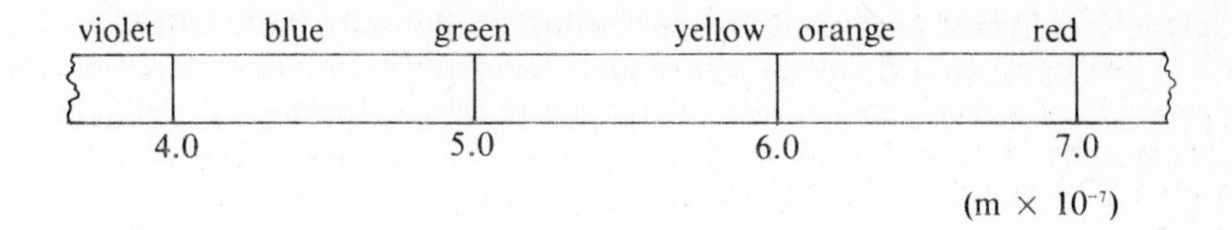

Certain atoms, particularly those of the transition elements, have two electronic orbitals whose energy levels differ by a value corresponding to the energy of part of the visible spectrum.

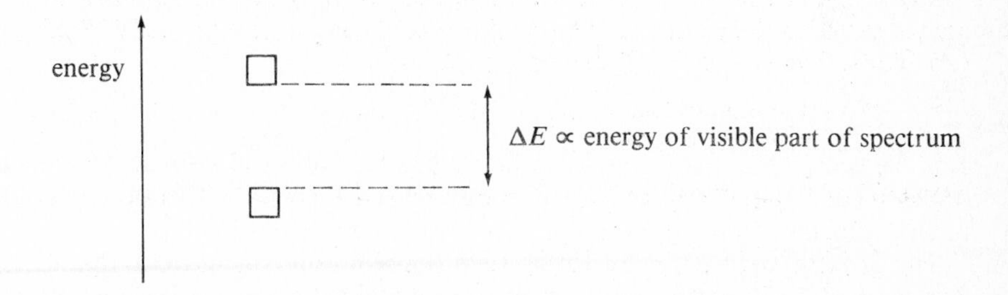

When white light falls on a substance containing these atoms, particular wavelengths are absorbed by the system as electrons are promoted from one orbital to another of higher energy. The observed colour of the substance is the visible spectrum minus these absorbed wavelengths.

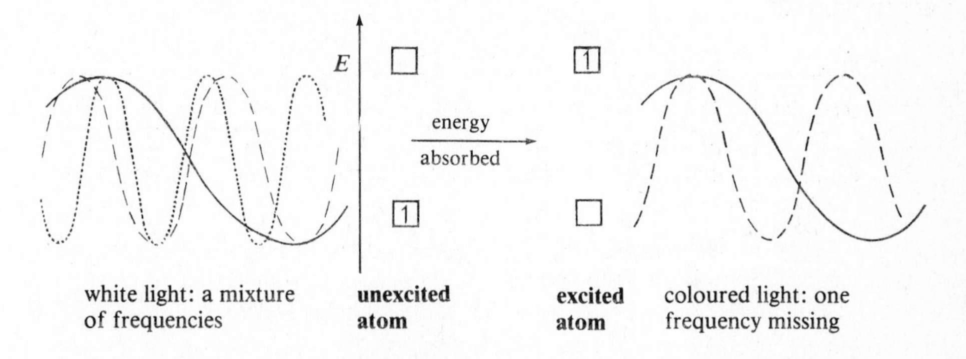

The excited atoms do not re-emit the absorbed frequencies, but lose their extra energy as kinetic energy in collisions.

To understand why different complexes of the transition metal atoms have different colours, it is necessary to consider the electronic structure of the atoms in the complexes. One explanation is provided by the crystal field theory.

Crystal field theory

This theory has its roots in the ionic model of bonding. It treats the ligands as point charges; the formation of the complex as a whole is justified in terms of the energy released as a result of the setting up of electrostatic forces of attraction between the central cation and the ligands; for example, in a complex consisting of a metal cation, M^{n+}, surrounded by six ligands, L^{m-}, arranged octahedrally:

$$M^{n+}_{(g)} + 6L^{m-}_{(g)} \longrightarrow [ML_6]^{(n-6m)}_{(g)}$$

energy

$M^{n+}_{(g)}$ $6L^{m-}_{(g)}$

E

$[ML_6]^{(n-6m)}_{(g)}$

E is the energy required to remove the six ligands from the attractive influence of the metal cation: this also is the energy released when the ligands co-ordinate to the central cation.

The ligands create an electrostatic field (known as a crystal field) that is symmetrically placed around the transition metal cation. In an isolated atom, the energy of the five 3d-orbitals is the same; but in the crystal field created by the ligands, the five 3d-orbitals do not all have the same energy. This is because the 3d-orbitals are arranged differently in the space around the metal atom and so interact in different ways with the crystal field.

Consider an octahedral complex whose ligand point charges are on the three Cartesian axes:

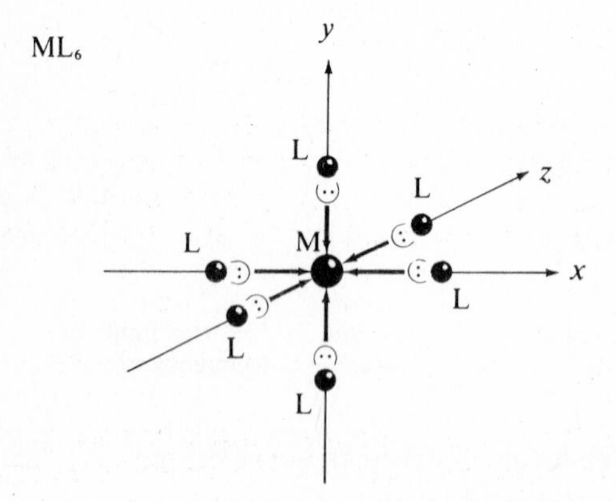

In this crystal field, the electrons in the $3d_{x^2-y^2}$ and $3d_{z^2}$ orbitals are repelled more than those in the $3d_{xy}$, $3d_{yz}$ and $3d_{xz}$ orbitals which lie between the axes. Thus the effect of the crystal field produced by the approach of the ligands is to raise the energy of all the d-orbitals but by different amounts so that two d-orbitals are raised more than the remaining three. This is shown in the diagram below and the effect is known as d-splitting.

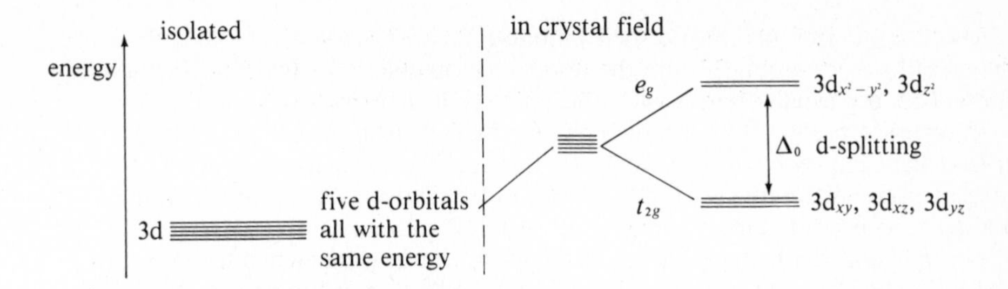

The size of the d-splitting is given the symbol Δ_0 and the two groups of orbitals produced as a result of the energy split are often named after their symmetric properties: t_{2g} and e_g are symbols derived from the symmetry of groups.

In a tetrahedral arrangement of ligands, the d-splitting is the reverse of that in an octahedral arrangement. The t_2 orbitals are all raised in energy by a greater amount than the e orbitals because the ligands are now directed towards the lobes of the $3d_{xy}$, $3d_{xz}$ and $3d_{yz}$ orbitals.

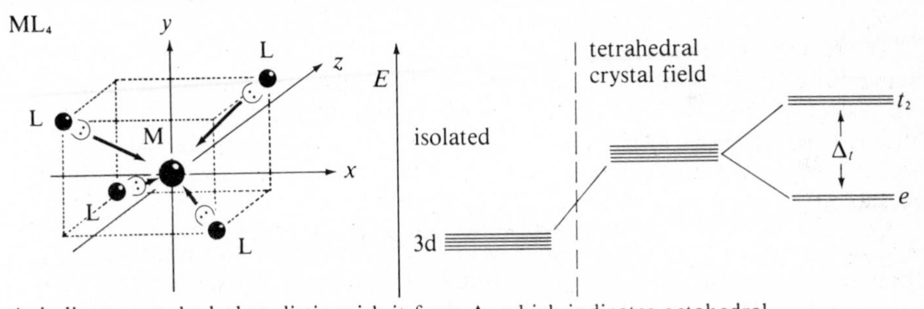

Δ_t indicates tetrahedral to distinguish it from Δ_0 which indicates octahedral.

The d-splitting caused by the crystal fields is often in the energy range of visible wavelengths of light. Consequently when light falls on a complex salt, a particular wavelength may be absorbed as electrons are promoted to the higher group of orbitals. Because this energy is not re-emitted as light, the complex will have a characteristic colour.

Magnetic properties

A spinning, charged object creates a magnetic moment. If a substance contains atoms with unpaired electrons in its structure, the spin effects of these unpaired electrons exert small magnetic moments that become aligned to one another. When this happens, the substance is able to interact with an external magnetic field.

> Paramagnetism is the property of a substance to line itself in the N—S direction of an external field.
>
> Diagmagnetic substances orientate themselves at 90° to the direction of the field.

Many transition metals and their compounds are paramagnetic and, in some, the effect is so pronounced that the term 'ferromagnetic' is used. Ferromagnetic substances are usually derivatives of iron, cobalt or nickel.

In order to predict the paramagnetism of a transition metal compound, crystal field theory is used to work out the number of electrons likely to be present in the different d-orbitals of the central cation. An example should make this procedure clear: consider an iron (II) salt containing d^6 ions, Fe^{2+}. In a strong ligand field, the electrons are paired, but in a weak ligand field, the unpaired state may be of lower energy because there is less electron repulsion when electrons are assigned to separate orbitals of a subshell.

iron (II) octahedral complexes

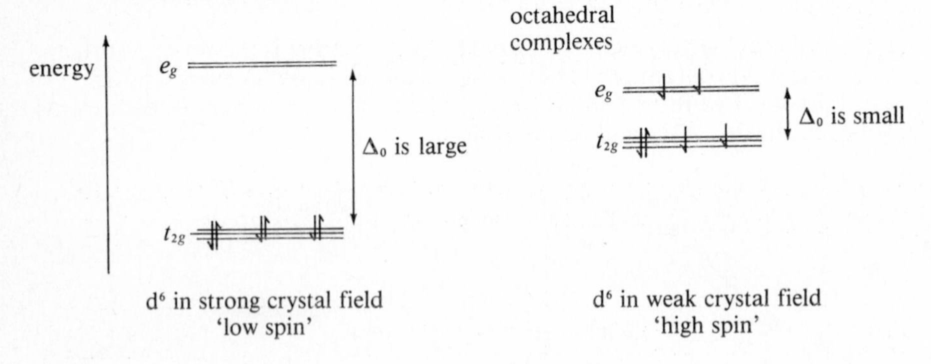

The 'low spin' d^6 complex is not paramagnetic because it has no unpaired spins; the 'high spin' d^6 complex is strongly paramagnetic. The strength of the crystal field is therefore all important in deciding the magnetic properties of a particular complex.

Spectrochemical series

The size of the d-splitting, Δ_o, has been measured for particular cation-ligand combinations and some values in kJ mol^{-1} are given below:

cation		Ti^{3+}	V^{3+}	Cr^{3+}	Mn^{2+}	Fe^{2+}	Fe^{3+}	Co^{2+}	Ni^{2+}	Cu^{2+}
ligand	H_2O	228	223	209	101	113	170	112	102	138
	NH_3			259				132	130	
	CN^-		281	319		406	420			
	F^-	210	193	181			168			
	Cl^-	156	155	166	86				86	

From these values, and others like them, it is evident that the strength of the crystal field, and hence the d-splitting, depends on the identity of the metal cation and the ligand.

For a given ligand, it is possible to list the cations in order of increasing d-splitting:

cation Δ_0	$Mn^{2+} < Ni^{2+} < Co^{2+} < Fe^{2+} < Fe^{3+} < Cr^{3+} < V^{3+} < Ti^{3+}$ →

Similarly, for a particular cation the ligands may be arranged in order of increasing d-splitting:

ligand Δ_0	$I^- < Br^- < Cl^- < F^- < OH^- < CH_3COO^- < H_2O < NH_3 < CN^-$ →

These lists are known as the spectrochemical series.

Catalytic activity

The transition elements and many of their compounds are useful heterogeneous catalysts for a variety of reactions. Some examples are given in the following table.

catalyst	*example*
iron	synthesis of ammonia in the Haber process $N_{2(g)} + 3H_{2(g)} \rightleftharpoons 2NH_{3(g)}$
vanadium V oxide, V_2O_5	oxidation of sulphur (IV) oxide in the contact process $2SO_{2(g)} + O_{2(g)} \rightleftharpoons 2SO_{3(g)}$
nickel, palladium or platinum	hydrogenation of an alkene — the manufacture of margarine $HRC{=}CHR^1 + H_2 \longrightarrow H_2RC{-}CR^1H_2$
titanium (III) chloride and aluminium triethyl	manufacture of polythene and polypropylene $nCH_2{=}CHR \longrightarrow [CH_2{-}CHR]_n$

The presence of partially filled d-orbitals has two possible effects that might influence the rate of a reaction:

1. Vacant orbitals can act as electron acceptors from reactant particles.
2. Full orbitals can act as electron donors to reactant particles.

In either case weak bonds are formed from reactant particles to the catalyst surface. This 'chemisorption' leads to an increase in reaction rate both because reactant particles are brought into close contact at the surface, and because the bonds in the reactant particles themselves are weakened as a result of the chemisorption.

If either or both of these assumptions is true, and they do seem to be, then it appears that the purely ionic model suggested by crystal field theory is an oversimplification. The evidence of catalytic activity indicates that some degree of electron transfer between ligands and cations does take place and it is possible to account for the bonding in complexes using molecular orbital theory. However both crystal field and molecular orbital theories represent extreme descriptions of the bonding in a complex; a more satisfactory approach is to apply a combination of the two approaches. This is called 'ligand field' theory and is beyond the scope of this chapter, but see for example *Physical Chemistry* by P.W. Atkins (O.U.P.).

Questions

1. Explain the meaning of the following: a) transition element b) aufbau principle c) dπ-overlap d) crystal field e) paramagnetism f) low spin complex g) d-splitting h) poor screening.

2. Explain why it is appropriate to consider the transition elements as a whole rather than by groups like the other elements in the s-block and the p-block.

3. a) Explain why the transition elements exhibit so many oxidation states.
   b) Choose the good oxidants and the good reductants from the list below.

system	$E^{\ominus}$ volts
MnO_2/Mn^{2+}	+1.23
Fe^{3+}/Fe	−0.05
Mn^{2+}/Mn	−1.18
MnO_4^-/MnO_2	+1.70
FeO_4^{2-}/Fe^{3+}	+2.20
$CrO_4^{2-}/Cr(OH)_3$	−0.13
Cr^{3+}/Cr^{2+}	−0.41

   c) Give the oxidation numbers of the underlined atoms in the following species: i) $\underline{Mn}O_4^{2-}$ ii) $\underline{Cr}(OH)_3$ iii) $\underline{V}O_4^{3-}$ iv) $\underline{V}O^{2+}$ v) $\underline{Ti}O_2$ vi) $Na_2\underline{Fe}O_4$.

4. a) Calculate the pH of a 0.01 mol dm^{-3} solution of iron (III) chloride given K_a for the hexaquo-iron (III) ion is 6.0×10^{-3} mol dm^{-3}.
   b) For a 0.05 mol dm^{-3} solution of chromium (III) nitrate, the pH is 2.5. Calculate K_a for the hexaquo-chromium (III) ion.
   c) The stability constant K_{stab} is 1×10^3 mol^{-1} dm^3 for the following complex equilibrium.

$$Fe^{3+}_{(aq)} + SCN^-_{(aq)} \rightleftharpoons FeSCN^{2+}_{(aq)}$$

   Find the concentration of the complex ion in a system in which $[Fe^{3+}] = 1.5 \times 10^{-4}$ mol dm^{-3} and $[SCN^-] = 2 \times 10^{-2}$ mol dm^{-3}.

5. a) Write out balanced equations for the following reactions that occur in acid conditions.

   i) $MnO^-_{4(aq)} + Cl^-_{(aq)} \rightarrow Mn^{2+}_{(aq)} + Cl_{2(g)}$

   ii) $Cr_2O^{2-}_{7(aq)} + Fe^{2+}_{(aq)} \rightarrow Cr^{3+}_{(aq)} + Fe^{3+}_{(aq)}$

   iii) $Fe^{3+}_{(aq)} + SO^{2-}_{3(aq)} \rightarrow Fe^{2+}_{(aq)} + SO^{2-}_{4(aq)}$

   b) Write out balanced equations for the following reactions that occur in alkaline conditions.

   i) $Cr(OH)^-_{4(aq)} + H_2O_{2(aq)} \rightarrow CrO^{2-}_{4(aq)} + H_2O_{(l)}$

   ii) $MnO^-_{4(aq)} + OH^-_{(aq)} \rightarrow MnO^{2-}_{4(aq)} + O_{2(g)}$

   iii) $Fe(OH)_{3(s)} + ClO^-_{(aq)} \rightarrow FeO^{2-}_{4(aq)} + Cl^-_{(aq)}$

6. Explain why so many transition element compounds are coloured.

7. Account for the magnetic properties of certain of the transition elements and their compounds.

8. The transition elements and their compounds are used as catalysts. Give specific examples and suggest reasons for their activity.

9. Discuss the possible relationship between acid-base properties and redox properties of the compounds of a transition metal.

10. Outline and account for the properties of a typical transition element.

Chapter 49

Chromium, manganese and iron

49.1 Structure of the elements

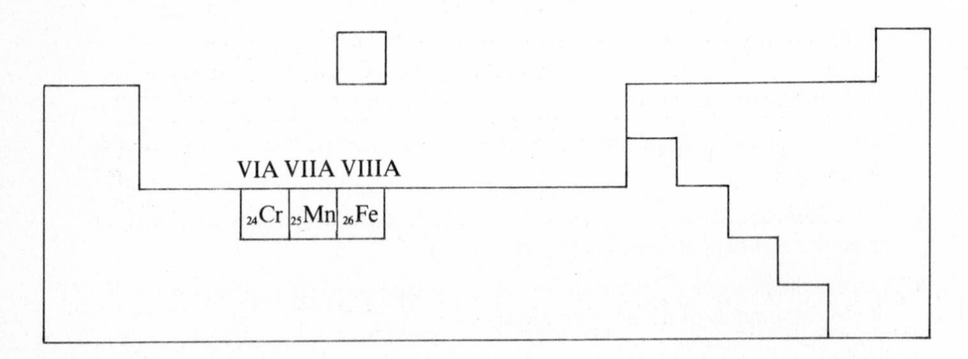

element	electronic configuration	electro-negativity	atomic radius ($m \times 10^{-10}$)	ionic (M^{2+}) radius ($m \times 10^{-10}$)	common oxidation states
Cr	[Ar] $3d^5$ $4s^1$	1.6	1.17	0.81	II, III, VI
Mn	[Ar] $3d^5$ $4s^2$	1.5	1.17	0.80	II, III, IV VI, VII
Fe	[Ar] $3d^6$ $4s^2$	1.8	1.16	0.76	II, III, VI

Atomic structure

Chromium, manganese and iron occupy the central positions in the first transition series. Their atoms have half-full inner 3d-subshells ($3d^5$ and $3d^6$) and an outer shell of one or two ($4s^1$ and $4s^2$). Like all transition metal atoms, their degree of electronic control is fairly low and their successive ionization energies increase more slowly than those of a non-transition element.

Physical structure

The low degree of electronic control leads the atoms to form lattices containing delocalized electrons. Both d- and s- bands are produced and their energy levels overlap. The lattice structure is body-centred cubic for all three metals at room temperature. However, a number of other arrangements become stable at higher temperatures (the allotropy of iron is briefly mentioned on page 78).

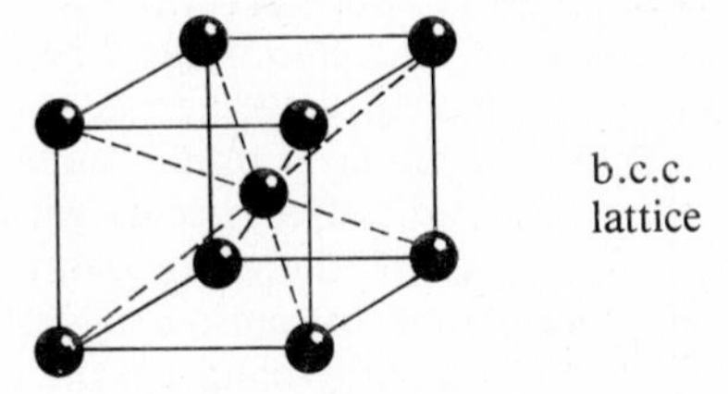

b.c.c. lattice

metal	*m.p.* (°C)	*b.p.* (°C)	*density* (kg m^{-3})	*conductivity* (Ω^{-1} m^{-1})
chromium	1890	2482	7190	7.87×10^6
manganese	1240	2100	7200	7.38×10^5
iron	1535	3000	7860	1.12×10^7

While iron is malleable and can be worked quite easily, both chromium and manganese are hard and brittle at room temperature.

49.2 Compounds formed by the elements

The three elements show the characteristic properties discussed in Chapter 48. They form compounds in a range of oxidation states and the stability of any particular state depends on the type of ligand bonded to the transition metal ion. Each complex has a characteristic colour as the table below shows:

metal	*oxidation state*	*ligand*	*complex*	*colour*
Cr	+VI	O^{2-}	CrO_4^{2-}	yellow
Cr	+III	H_2O	$Cr(H_2O)_6^{3+}$	violet
Cr	+II	CH_3COO^- and H_2O	$Cr(CH_3COO)_2H_2O$	red
Mn	+VII	O^{2-}	MnO_4^-	purple
Mn	+VI	O^{2-}	MnO_4^{2-}	green
Mn	+IV	O^{2-}	MnO_2	black
Mn	+III	F^-	MnF_3	magenta
Mn	+II	H_2O	$Mn(H_2O)_6^{2+}$	pink
Fe	+VI	O^{2-}	FeO_4^{2-}	red
Fe	+III	H_2O	$Fe(H_2O)_6^{3+}$	yellow
Fe	+II	H_2O	$Fe(H_2O)_6^{2+}$	green

The metals themselves are relatively reactive, although reactions with non-metals do not occur readily unless the temperature is raised. The activation energy for synthesizing binary compounds is quite high for most transition metals. Iron is the most reactive of the three in that it combines with non-metals at a lower temperature and with greater ease. Chromium metal has the unusual property of being extremely resistant to most corrosive agents at low temperatures. It appears that these form a thin oxidized layer on the surface of the metal which prevents any further reaction. The same effect occurs with iron but the oxidized layer tends to be hydrous, and flakes off therefore giving the metal no protection. The thermodynamics of compound formation suggest that the elements should be more reactive than they actually are (compare the redox potentials on page 763). A few non-aqueous reactions are given below to illustrate typical compound formation.

$$Cr_{(s)} + 2HCl_{(g)} \xrightarrow{700°C} CrCl_{2(s)} + H_{2(g)}$$

$$2Cr_{(s)} + 3Cl_{2(g)} \xrightarrow{500°C} 2CrCl_{3(s)} \text{ (sublimes above 600°C)}$$

$$3Mn_{(s)} + N_{2(g)} \xrightarrow{1200°C} Mn_3N_{2(s)}$$

$$8Mn_{(s)} + S_{8(s)} \xrightarrow{200°C} 8MnS_{(s)}$$

$$Fe_{(s)} + 2H_3O^+_{(aq)} \xrightarrow[\text{temperature}]{\text{room}} Fe^{2+}_{(aq)} + 2H_2O_{(l)} + H_{2(g)}$$

Rust

Rust is formed by the reaction between iron and moist air.

$$O_{2(g)} + 2H_2O_{(l)} + 4e^- \rightleftharpoons 4OH^-_{(aq)} \qquad E^\ominus = +0.40 \text{ volts}$$

$$Fe^{2+}_{(aq)} + 2e^- \rightleftharpoons Fe_{(s)} \qquad E^\ominus = -0.44 \text{ volts}$$

The electrode potentials above indicate that moist air forms a system that readily oxidizes iron according to the following equation:

$$2Fe_{(s)} + O_{2(aq)} + H_2O_{(l)} \rightarrow 2Fe^{2+}_{(aq)} + 4OH^-_{(aq)}$$

Iron (II) hydroxide is very insoluble and precipitates.

$$Fe^{2+}_{(aq)} + 2OH^-_{(aq)} \rightarrow Fe(OH)_{2(s)}$$

The solid is then rapidly oxidized by moist air to iron (III) hydroxide, some of which may lose water.

$$2Fe(OH)_{2(s)} + O_{2(aq)} + H_2O_{(l)} \rightarrow \underbrace{2Fe(OH)_3 \rightarrow Fe_2O_3 + 3H_2O}_{\text{rust}}$$

The hydrous iron (III) oxide is the basis of rust.

49.3 Occurrence, extraction and uses

Occurrence

The most important mineral ores of these metals and the abundance of the metals in the Earth's crust are shown in the following table.

element	*mineral sources*	*% abundance in Earth's crust*
chromium	chromite, $FeO.Cr_2O_3$	0.02
manganese	pyrolusite, MnO_2	0.10
iron	haematite; magnetite; siderite Fe_2O_3 Fe_3O_4 $FeCO_3$	5.00

Extraction of chromium

Chromium is required by industry in two forms, pure metallic chromium and the iron-chromium alloy called ferrochromium. Chromium metal is extracted from the ore by the following steps:

1. Convert chromium (III) in the ore to dichromate (VI). This removes the iron.

$$FeO.Cr_2O_{3(s)} + 4OH^-_{(aq)} + O_{2(g)} \rightarrow FeO_{(s)} + CrO^{2-}_{4(aq)} + 2H_2O_{(l)}$$

$$2CrO^{2-}_{4(aq)} \xrightarrow{\text{acidify}} Cr_2O^{2-}_{7(aq)}$$

2. Reduce the dichromate (VI) to chromium (III) oxide

$$Na_2Cr_2O_{7(s)} + 2C_{(s)} \xrightarrow{200°C} Cr_2O_{3(s)} + Na_2CO_{3(s)} + CO_{(g)}$$

3. Reduce the chromium (III) oxide using aluminium as a reductant in a thermite type reaction.

$$Cr_2O_{3(s)} + 2Al_{(s)} \rightarrow Al_2O_{3(s)} + 2Cr_{(s)}$$

This produces chromium metal which is 97 – 99% pure.

Ferrochromium is obtained from the ore by reducing it with coke or silicon in an electric arc furnace:

$$FeO.Cr_2O_{3(s)} + 4C_{(s)} \rightarrow \underset{\text{ferrochrome alloy}}{FeCr_{(s)}} + 4CO_{(g)}$$

Extraction of manganese

Nearly all the manganese extracted is in the form of the alloy ferro-manganese.

This is extracted like ferrochrome by reducing a suitable ore mixture in an electric arc furnace or a blast furnace with carbon.

Extraction of iron

The economic importance of iron is such that it is extracted on a vast scale. Very little iron is produced as the pure metal because it is almost always used in some alloyed form.

The blast furnace

Iron ores are reduced to iron by reacting them at high temperatures with coke and limestone (calcium carbonate) in a blast furnace. Iron ore, coke and limestone are crushed and mixed together to form what is known as 'the charge'. The coke is present as both a fuel for the furnace and a reductant, while the limestone is added to form a slag in which high melting point impurities can dissolve. The charge is introduced at the top of the furnace and falls down through a blast of preheated air that is injected near the base of the furnace.

the blast furnace

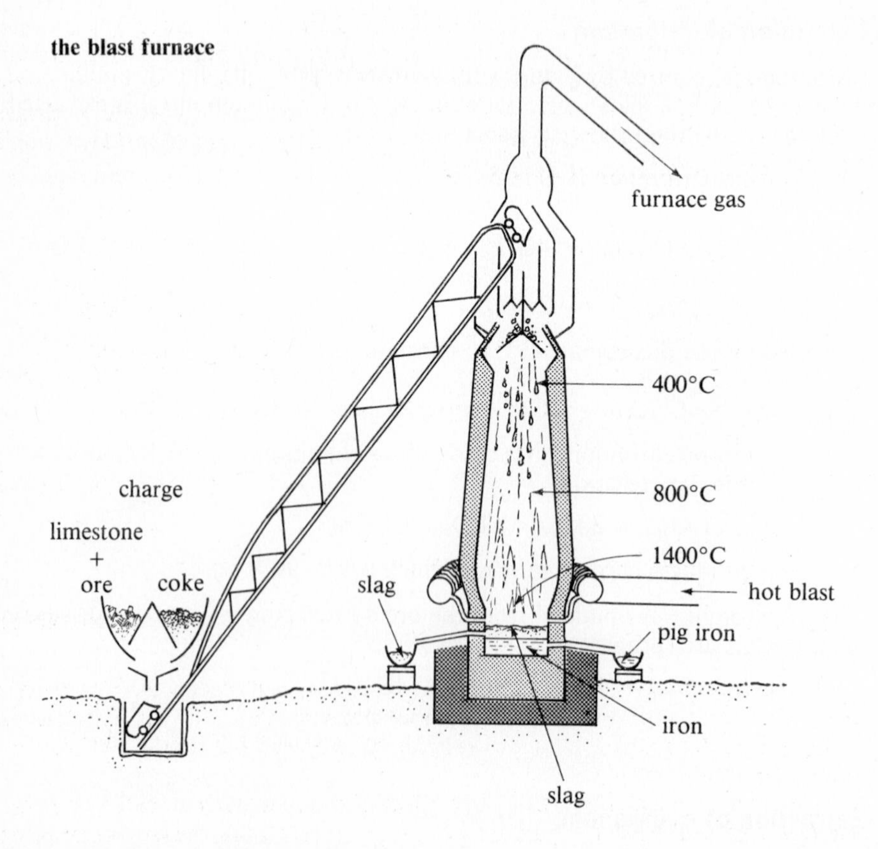

A large number of reactions take place in the blast furnace but the more important are as follows:

coke burns in the hot air blast

$$C_{(s)} + O_{2(g)} \rightarrow CO_{2(g)}$$

carbon dioxide is reduced by hot coke

$$C_{(s)} + CO_{2(g)} \rightleftharpoons 2CO_{(g)}$$

carbon monoxide reduces iron ore

$$Fe_2O_{3(s)} + 3CO_{(g)} \rightarrow 2Fe_{(l)} + 3CO_{2(g)}$$

more ore is reduced by carbon in the hotter, lower part of the furnace

$$Fe_2O_{3(s)} + 3C_{(s)} \rightarrow 2Fe_{(l)} + 3CO_{(g)}$$

limestone decomposes to produce calcium oxide

$$CaCO_{3(s)} \rightarrow CaO_{(s)} + CO_{2(g)}$$

calcium oxide reacts with non-metal oxide impurities to form a slag

$$CaO_{(s)} + SiO_{2(s)} \rightarrow CaSiO_{3(l)}$$

$$CaO_{(s)} + P_2O_{5(g)} \rightarrow Ca_3(PO_4)_{2(l)}$$

The chemical changes taking place in the blast furnace are extremely complex. The decomposition of limestone and the reaction of the basic calcium oxide with the acidic earthy impurities are fairly straightforward, but the reduction of iron oxide by the C/CO/CO_2 system occurs by a number of possible mechanisms. These are best interpreted using the Ellingham diagram on page 725, a small section of which is shown below.

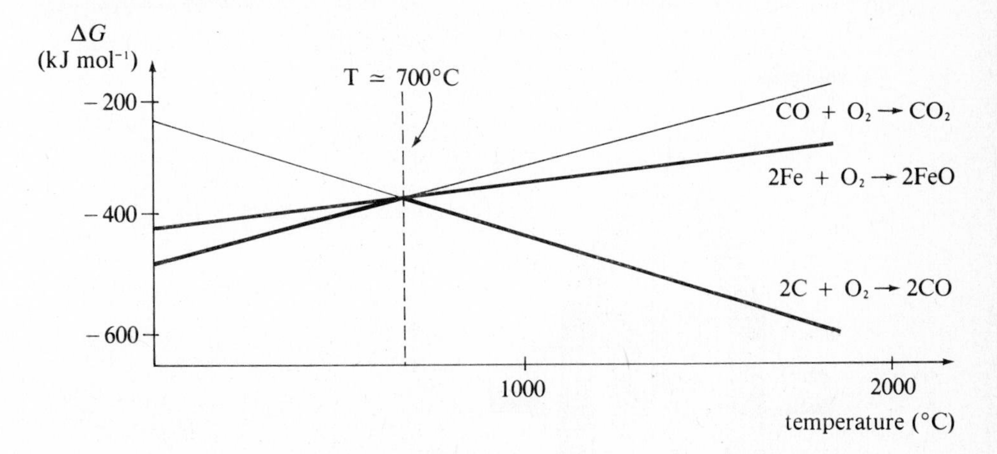

The diagram shows that below 700°C the oxidation of carbon monoxide has a greater negative free energy change than the oxidation of iron. This means that carbon monoxide reduces iron oxide at temperatures below 700°C, but not above that temperature.

The diagram also shows that above 700°C the oxidation of carbon to carbon monoxide has a greater negative free energy change than the oxidation of iron. This indicates that carbon reduces iron oxide at temperatures above 700°C, but not below that temperature.

Consequently, in the cooler top part of the furnace, the reductant is carbon monoxide and in the hotter, lower part of the furnace, carbon is the reductant.

Iron produced in the blast furnace has a number of impurities dissolved in it. The main ones are carbon, silicon, sulphur and phosphorus. These impurities give the iron a rather brittle nature and it is known as pig iron. The composition of pig iron varies depending on its source. Below are typical values for impurities in pig iron.

pig iron composition

element	%
C	2.5 – 4.0
Si	0.6 – 3.5
S	0.04 – 0.1
P	0.04 – 2.5

Steel

Steel is manufactured from pig iron by removing most of the impurities and then adding controlled amounts of carbon and other alloying elements to modify the properties of the iron.

The most common method of steel making consists of pouring molten pig iron into an enormous brick-lined bucket, known as a converter, and blowing oxygen through the molten metal.

converter

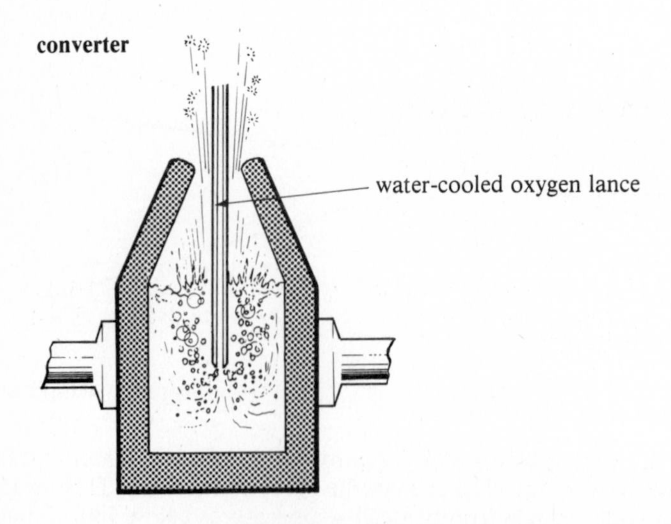

The carbon, sulphur and phosphorus are removed as their gaseous oxides, while the oxide of silicon forms a slag on top of the iron. This slag is scraped off and carbon, manganese and other alloying elements are then added to the molten iron. The amount of carbon in steel is much less than that in pig iron and ranges from 0.09 – 0.9%.

Uses of the metals

The composition of some alloy steels and their uses are given below.

type of steel	*composition*	*particular property and uses*	
mild steel	0.09 – 0.2%C 0.5 – 1.0%Mn 0.2 – 0.75%Si	easily shaped	— car bodies — nails — tubes
high carbon steel	0.4 – 0.9%C 0.5 – 1.0%Mn 0.2 – 0.75%Si	hard	— tools — masonry nails
stainless steel	0.2 – 0.4%C 18%Cr 8%Ni	resists corrosion	— cutlery — garden tools — boat fittings
manganese steel	0.4 – 0.9%C 13%Mn	strong and hard	— springs — drill bits
tungsten steel	0.4 – 0.9%C 5%W	very hard	— cutting tool edges

49.4 Compound structure

Low oxidation states

Carbonyl compounds
When finely divided iron is treated with carbon monoxide under pressure at about 200°C, iron pentacarbonyl is produced. At room conditions, this is a pale yellow liquid (b.p. 103°C).

$$Fe_{(s)} + 5CO_{(g)} \rightleftharpoons Fe(CO)_{5(l)}$$

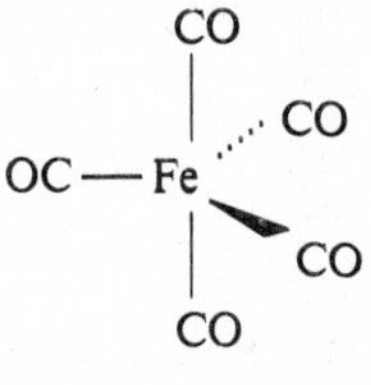

(Structure of $Fe(CO)_5$: Fe bonded to OC, CO, CO, CO, CO)

(The bonding in metal carbonyls is described on page 760.) Iron atoms in the pentacarbonyl have an oxidation state of zero. More violent conditions are necessary to produce the carbonyls of chromium and manganese:

$$CrCl_{3(s)} + \underset{\text{excess}}{6CO_{(g)}} + 3Na_{(s)} \xrightarrow[\substack{\text{tetrahydrofuran} \\ T\,=\,300°C \\ P\,=\,200\text{ atm.}}]{\text{solvent of}} Cr(CO)_{6(sol)} + 3NaCl_{(s)}$$

The simplest manganese carbonyl contains molecules with metal-metal bonds.

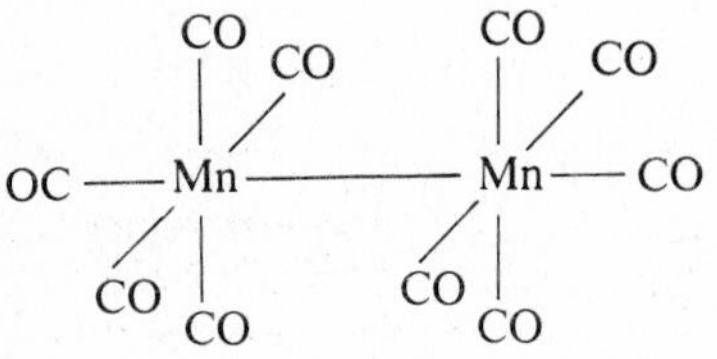

$Mn_2(CO)_{10}$ (compare S_2F_{10}, page 711)

The metal-metal bond is readily broken by treating the carbonyl with a reductant such as sodium in an organic solvent.

$$Mn_2(CO)_{10} + 2Na \xrightarrow[\text{tetrahydrofuran}]{\text{solvent of}} 2Mn(CO)_5^- + 2Na^+$$

The bond can also be broken by an oxidant such as a halogen.

$$Mn_2(CO)_{10} + Cl_2 \rightarrow 2Mn(CO)_5Cl$$

The carbonyl ligands can therefore stabilize Mn(−I), Mn(0) and Mn(+I). More complex carbonyls are produced as a result of thermal or photochemical reactions between molecules of simple carbonyls. In these ‘derived’ carbonyls, there are ketone-like bridging carbonyl units, as well as an increased tendency for metal-metal bonds to form. For example, iron enneacarbonyl ($Fe_2(CO)_9$) is produced when a solution of the pentacarbonyl is irradiated by ultra-violet radiation.

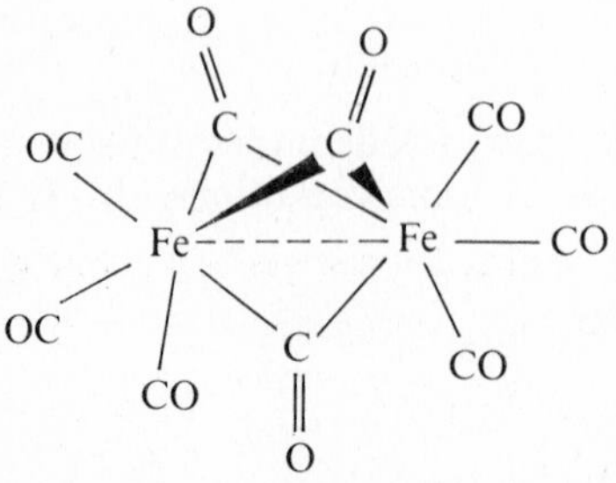

iron enneacarbonyl

$$2Fe(CO)_{5(sol)} \xrightarrow{h\nu} Fe_2(CO)_{9(sol)} + CO_{(g)}$$

Each iron atom has a distorted octahedral arrangement of CO molecules around it.

Sandwich compounds

Another class of compounds can be prepared from the carbonyls. They are known as 'sandwich compounds' and contain the atom of a transition metal in a low oxidation state complexed by aromatic particles. For example, when iron pentacarbonyl is refluxed with cyclopentadiene, C_5H_6, the carbonyl ligands are replaced by aromatic cyclopentadienyl anions π-$C_5H_5^-$ (all aromatic rings contain $(4n + 2)$ delocalizable electrons).

compare with benzene

π-$C_5H_5^-$ $(n = 1)$ $\qquad$ π-C_6H_6 $(n = 1)$

The π signifies that these ligands offer π-electrons towards the metal atom.

Two cyclopentadienyl anions sandwich each iron atom. The complex is neutral and molecular because iron is present as Fe(II).

$$\overset{0}{Fe}(CO)_5 + 2C_5H_6 \xrightarrow[T > 200°C]{\text{reflux}} \overset{II}{Fe}(\pi\text{-}C_5H_5)_2 + 4CO + HCHO$$

'ferrocene'

The 'sandwich' structure of ferrocene is responsible for the general name given to this class of compound.

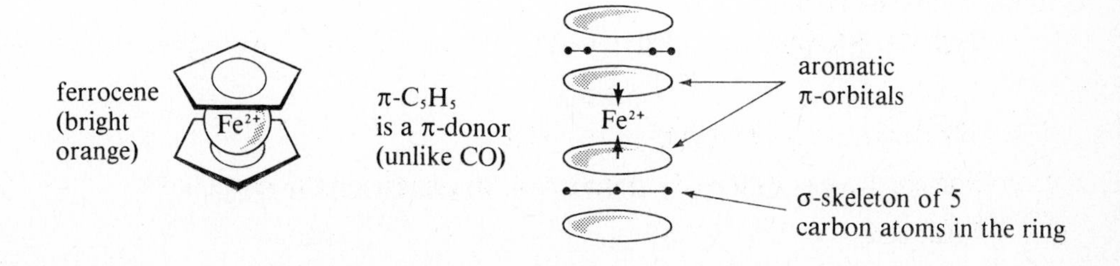

Chromium forms a sandwich compound with benzene:

$$\overset{III}{Cr}Cl_3 + Al + 2C_6H_6 \xrightarrow[150°C]{AlCl_3} \overset{0}{Cr}(\pi\text{-}C_6H_6)_2 + AlCl_3$$

dibenzenechromium

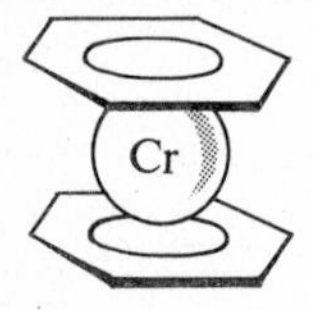

This crystalline compound (m.p. 284°C) is oxidized by the air to an oxide containing the cation $Cr(\pi\text{-}C_6H_6)_2^+$.

Many sandwich compounds have been prepared and studied. These compounds are often highly coloured and, with the exception of ferrocene, tend to be easily oxidized in air. Mixed carbonyl and sandwich complexes are also known:

$$C_6H_6 + Mn(CO)_5Cl + AlCl_3 \xrightleftharpoons{reflux} [(C_6H_6)Mn(CO)_3]^{\oplus} [AlCl_4]^{\ominus} + 2CO$$

The +II and +III oxidation states

The elements all react with dilute mineral acids to give M^{2+} ions and hydrogen. In aqueous solution, the stability of these ions is very different for the three metals. Mn^{2+} is by far the most stable state for manganese; Fe^{2+} is air-oxidized to Fe^{3+}, and Cr^{2+} is very readily converted to Cr^{3+}. In order to isolate a chromium (II) compound, it is necessary to precipitate it from solution. $Cr^{2+}_{(aq)}$ is a powerful reductant and is pale blue in colour.

The easiest method for obtaining stable chromium (II) is to prepare the insoluble, red ethanoate salt. When a freshly made solution of chromium (II) chloride is added to a concentrated sodium ethanoate solution, a hydrated precipitate forms. The structure of the complex contains metal-metal bonding. The bond between the two chromium atoms is a quadruple bond consisting of a σ-, two π- and a δ-bond. A δ-bond results from the overlap between the four lobes of two d-orbitals.

$$Cr_{(s)} + 2HCl_{(g)} \xrightarrow{700°C} CrCl_{2(s)} + H_{2(g)}$$

$$CrCl_{2(s)} \xrightarrow{(aq)\ in\ N_{2(g)}} Cr^{2+}_{(aq)}$$

$$2Cr^{2+}_{(aq)} + 4Me.COO^-_{(aq)} + 2H_2O_{(l)} \rightarrow [Cr_2(MeCOO)_4(H_2O)_2]_{(s)}$$

pale blue red

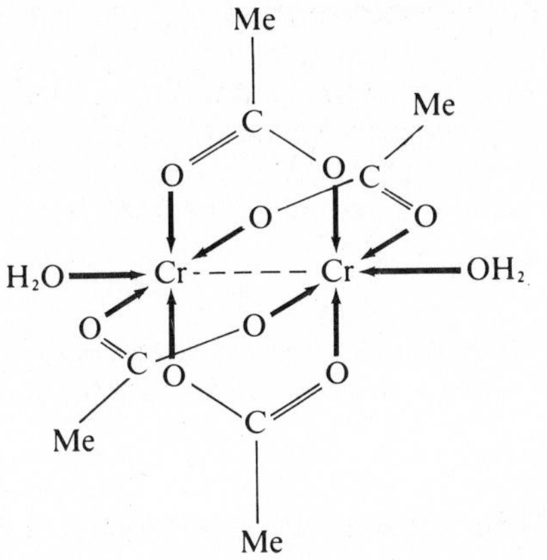

bond lengths (m × 10^{-10})	
Cr - - - - - Cr	2.46
Cr ⟵ ethanoate	1.97
Cr ⟵ water	2.20

The analogous Mn(III) complex exists; it is isoelectronic with the Cr^{2+} complex. Both are unstable under aqueous conditions. Whereas Cr^{2+} tends to form Cr^{3+}, Mn^{3+} either tends to act as an oxidant and form Mn^{2+}, or disproportionates to give $MnO_{2(s)}$ and $Mn^{2+}_{(aq)}$. Fe(III) and Cr(III) are stable and are easily prepared by the action of oxidants on either the metals or M(II) compounds:

$$2Fe^{2+}_{(aq)} + Cl_{2(g)} \longrightarrow 2Fe^{3+}_{(aq)} + 2Cl^{-}_{(aq)}$$

$$2Cr_{(s)} + 3O_{2(g)} \xrightarrow{600\,°C} Cr_2O_{3(s)}$$

The only simple Mn(III) compounds that have been prepared are MnF_3 and $MnPO_4$. Apart from the ethanoate complex, Mn(III) exists in fluoro-complexes such as MnF_4^- and MnF_5^{2-} but not in many other compounds unlike Cr(III) and Fe(III). These form a large range of complexes some of which are described in 49.7.

Higher oxidation states

Compounds of chromium, manganese and iron in oxidation states greater than III are either oxy-compounds or fluoro-compounds. Mn (IV) exists as $Mn(SO_4)_2$ but it is the only exception. The most common compounds are shown in the table below.

oxidation state	*chromium*	*manganese*	*iron*
IV	CrF_4	MnF_4 MnO_2	$SrFeO_3$, Ba_2FeO_4
VI	CrF_6 CrO_3 CrO_2Cl_2 K_2CrO_4 CrO_2F_2 $K_2Cr_2O_7$	K_2MnO_4	K_2FeO_4
VII	None	MnO_3F Mn_2O_7 $KMnO_4$	None

CrF_4 and MnF_4 are both obtained by direct combination of the elements above 300°C (chromium gives CrF_5 and CrF_6 at pressures greater than 400 atmospheres). Neither of the IV state fluorides is stable in moist air and both are violently hydrolysed in water. Manganese, however, does form a stable +IV state oxide. MnO_2 occurs as the mineral pyrolusite, and can be prepared in the laboratory by decomposing the nitrate salt of manganese (II).

$$\overset{II}{Mn}(\overset{V}{N}O_3)_2 \xrightarrow{T > 500°C} \overset{IV}{Mn}O_{2(s)} + 2\overset{IV}{N}O_{2(g)}$$

+VI state compounds are obtained by oxidation in concentrated alkaline conditions. Strong oxidants are often required in order to prepare them, as shown.

1. **Chromium VI:** (yellow)
$$2\overset{III}{Cr}(OH)^{3-}_{6(aq)} + 3H_2O_{2(aq)} \rightarrow 2\overset{VI}{Cr}O^{2-}_{4(aq)} + 8H_2O_{(l)} + 2HO^-_{(aq)}$$

2. **Manganese (VI):** (green)
$$3\overset{IV}{Mn}O_{2(s)} + ClO^-_{3(aq)} + 6HO^-_{(aq)} \rightarrow 3\overset{VI}{Mn}O^{2-}_{4(aq)} + Cl^-_{(aq)} + 3H_2O_{(l)}$$

3. **Iron (VI):** (purple)
Electrolyse concentrated alkali using a soft iron anode and a high current density. At the anode,
$$\overset{0}{Fe}_{(s)} + 8HO^-_{(aq)} \rightarrow \overset{VI}{Fe}O^{2-}_{4(aq)} + 4H_2O_{(l)} + 6e^-$$

The chemistry of chromium (VI) is more extensive than that of the other two. In basic solution, chromium (VI) is present as yellow chromate, but it is converted to the orange dichromate (VI) on acidification.

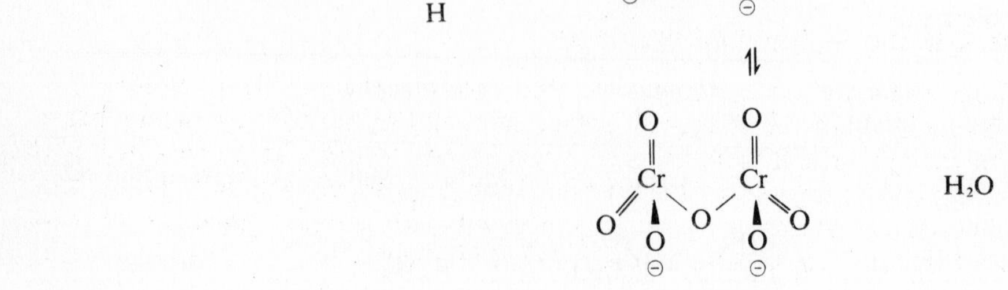

O O O O O O Cr Cr Cr Cr O O O O O O H H H H H H H
H_3O^+ $\rightleftharpoons$ H_2O H_2O H_2O

$$2CrO^{2-}_{4(aq)} + 2H_3O^+_{(aq)} \rightleftharpoons Cr_2O^{2-}_{7(aq)} + 3H_2O_{(l)}$$

yellow — orange

The potassium salts of both chromium (VI) compounds can be isolated, as can the potassium salts of manganese (VI) and of iron (VI). The only stable VI-state parent oxide that can be prepared is chromium (VI) oxide. (This preparation should not be attempted in a school laboratory: the reaction may be violent). Chromium (VI) oxide is precipitated as an orange-red solid when concentrated sulphuric acid is added to a concentrated solution of potassium dichromate. When this reaction is carried out in the presence of chloride ions, some dichlorochromium (VI) oxide is also formed. It is a deep-red liquid of boiling point 117°C.

$$Cr_2O_{7(aq)}^{2-} \xrightarrow{H_2SO_4} [H_2Cr_2O_7] \xrightarrow[(-2H_2O)]{H_2SO_4} 2CrO_{3(s)}$$

$$CrO_{3(s)} \xrightarrow{2HCl} CrO_2Cl_{2(l)} + H_2O$$

Manganese (VII) resembles chromium (VI) in some respects. In acid conditions, the manganate (VI) ion is not stable and disproportionates to give manganate (VII) and manganese (IV) oxide.

$$\underset{\text{green}}{3\overset{\text{VI}}{Mn}O_{4(aq)}^{2-}} + 4H_3O_{(aq)}^{+} \rightarrow \underset{\text{purple}}{2\overset{\text{VII}}{Mn}O_{4(aq)}^{-}} + \underset{\text{black}}{\overset{\text{IV}}{Mn}O_{2(s)}} + 6H_2O_{(l)}$$

An even more dangerously explosive oxide can be prepared by dehydrating potassium manganate (VII) with very cold, concentrated sulphuric acid

$$2KMnO_{4(s)} \xrightarrow[-H_2O]{2H_2SO_4} 2KHSO_{4(s)} + Mn_2O_{7(l)}$$

Manganese (VII) oxide readily decomposes to give manganese (IV) oxide and oxygen. Considerable energy is evolved.

$$2Mn_2O_{7(l)} \rightarrow 4MnO_{2(s)} + 3O_{2(g)} \qquad \Delta H^{\ominus} = -306 \text{ kJ (mol } Mn_2O_7)^{-1}$$

49.5 Acid-base properties

The +II and +III oxidation states

All the +II state oxides, hydroxides and carbonates are basic with little amphoteric character. The +III state oxides and hydroxides are mostly basic but are more strongly amphoteric. A wide range of salts can be obtained by the neutralization reactions of these bases. Iron (III) and chromium (III) also form double salts or alums (see page 655). Typical neutralization reactions are shown below: the acid anions could be Cl^-, SO_4^{2-}, NO_3^- etc.

$$Cr_2O_{3(s)} + 6H_3O_{(aq)}^{+} \rightarrow 2Cr_{(aq)}^{3+} + 9H_2O_{(l)}$$

$$MnCO_{3(s)} + 2H_3O_{(aq)}^{+} \rightarrow Mn_{(aq)}^{2+} + CO_{2(g)} + 3H_2O_{(l)}$$

$$Fe(OH)_{2(s)} + 2H_3O_{(aq)}^{+} \rightarrow Fe_{(aq)}^{2+} + 3H_2O_{(l)}$$

Solutions of these metal salts tend to be weakly acidic because the metal cations polarize the water ligands quite strongly.

$$(H_2O)_5M^{n+}\text{—}OH_2 + H_2O \rightleftharpoons (H_2O)_5M^{n+}\text{—}O^{-}\text{—}H + H_3O^+_{(aq)}$$

The addition of alkali gives a gelatinous precipitate that often redissolves in excess alkali because of the amphoteric nature of the hydroxides. Chromium (III) has the following solubility characteristics:

$$[Cr^{III}(H_2O)_6]^{3+} \underset{3H_3O^+}{\overset{3HO^-}{\rightleftharpoons}} [Cr^{III}(H_2O)_3(OH)_3] \underset{3H_3O^+}{\overset{3HO^-}{\rightleftharpoons}} [Cr^{III}(OH)_6]^{3-}$$

$$\underset{\text{soluble}}{Cr^{3+}_{(aq)}} \underset{\text{protonate}}{\overset{\text{deprotonate}}{\rightleftharpoons}} \underset{\text{insoluble}}{Cr(H_2O)_3(OH)_3} \underset{\text{protonate}}{\overset{\text{deprotonate}}{\rightleftharpoons}} \underset{\text{soluble}}{Cr(OH)_6^{3-}}$$

Higher oxidation states

With the exception of MnO_2, all the oxides and fluorides of the three metals in an oxidation state greater than three are acidic. The metals occur in some oxidation states only as the anions because the oxides and fluorides are too unstable. For example, FeO_4^{2-} and MnO_4^{2-} are both known but FeO_3 and MnO_3 do not exist, unlike CrO_3.

CrO_3 (O=Cr(=O)=O) resembles SO_3 (O=S(=O)=O)

$$CrO_{3(s)} + 2H_2O_{(l)} \rightarrow HCrO^-_{4(aq)} + H_3O^+_{(aq)}$$

The fluorides hydrolyse to give acidic solutions of fluoro-complexes or oxy-compounds:

$$MnF_{4(s)} + 2H_2O_{(l)} \rightarrow MnO_{2(s)} + 4HF_{(aq)}$$

49.6 Redox properties

The elements themselves act as reducing agents although the activation energy for these reactions is often high. They readily combine with non-metals at higher temperatures or when finely divided. Pure manganese in fact can displace hydrogen from hot water.

$$\overset{0}{Mn}_{(s)} + 2\overset{I}{H}_2O_{(l)} \rightarrow \overset{II}{Mn}(OH)_{2(s)} + \overset{0}{H}_{2(g)}$$

The compounds of the metals in lower oxidation state tend to be reductants while those in the higher states are often oxidants. The comparative strength of these reductants and oxidants is shown in the table below:

reductants	*oxidants*	*redox equilibrium*	$E^{\ominus}$ (volts)
Cr^{2+}	Cr^{3+}	$Cr^{3+}_{(aq)} + e^- \rightleftharpoons Cr^{2+}_{(aq)}$	−0.41
Fe^{2+}	Fe^{3+}	$Fe^{3+}_{(aq)} + e^- \rightleftharpoons Fe^{2+}_{(aq)}$	+0.76
Cr^{3+}	$H\overset{VI}{Cr}O_4^-$	$HCrO^-_{4(aq)} + 7H_3O^+_{(aq)} + 3e^- \rightleftharpoons Cr^{3+}_{(aq)} + 11H_2O_{(l)}$	+1.19
Mn^{2+}	$\overset{IV}{Mn}O_2$	$MnO_{2(s)} + 4H_3O^+_{(aq)} + 2e^- \rightleftharpoons Mn^{2+}_{(aq)} + 6H_2O_{(l)}$	+1.23
	$\overset{VII}{Mn}O_4^-$	$MnO^-_{4(aq)} + 8H_3O^+_{(aq)} + 5e^- \rightleftharpoons Mn^{2+}_{(aq)} + 12H_2O_{(l)}$	+1.52
Fe^{3+}	$H\overset{VI}{Fe}O_4^-$	$HFeO^-_{4(aq)} + 7H_3O^+_{(aq)} + 3e^- \rightleftharpoons Fe^{3+}_{(aq)} + 11H_2O_{(l)}$	+2.20

The nature of the ligands present in a redox system affects the value of the redox potential. There are two important examples of this.

1. Ligand replacement
For example in the Fe^{3+}/Fe^{2+} system:

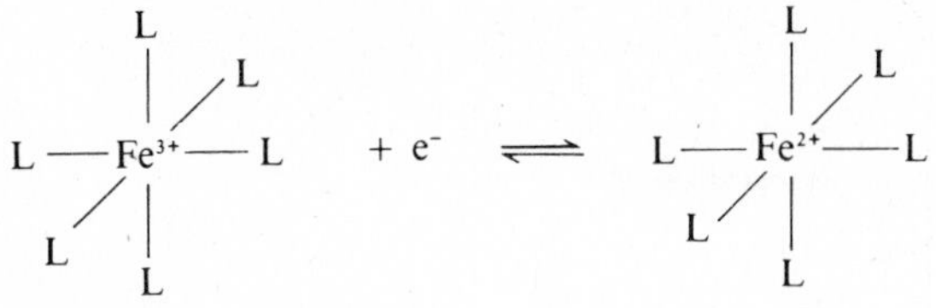

$$[Fe^{3+}L_6] + e^- \rightleftharpoons [Fe^{2+}L_6]$$

identity of L	$E^{\ominus}$ volts
dipyridyl (×3)	+0.96
water (×6)	+0.76
edta^{4-} (×1)	−0.10

Note that 2,2-dipyridyl ligands are bidentate, but that edta^{4-} is hexadentate (see page 303).

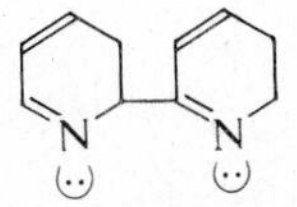

2,2-dipyridyl

The redox potentials fall with the increasing ability of the ligands to stabilize Fe^{3+} ions. The dipyridyl ligands are π-acceptors and therefore tend to stabilize the lower-charged Fe^{2+} in preference to the more electron-deficient Fe^{3+} ions; $edta^{4-}$ ions, on the other hand, are electron-excessive and are strong σ-donors thus stabilizing the Fe^{3+} ions with respect to the Fe^{2+} ions.

2. pH changes in aqueous redox equilibria

The table below shows the redox potentials of the manganese and chromium systems in acid and alkaline solution.

acidity	*redox equilibria*	$E^{\ominus}$ (volts)
alkaline	$\overset{VI}{Cr}O^{2-}_{4(aq)} + 4H_2O_{(l)} + 3e^- \rightleftharpoons [\overset{III}{Cr}(OH)_6]^{3-}_{(aq)} + 2OH^-_{(aq)}$	−0.12
acid	$H\overset{VI}{Cr}O^-_{4(aq)} + 7H_3O^+_{(aq)} + 3e^- \rightleftharpoons \overset{III}{Cr}^{3+}_{(aq)} + 11H_2O_{(l)}$	+1.19
alkaline	$\overset{VII}{Mn}O^-_{4(aq)} + 2H_2O_{(l)} + 3e^- \rightleftharpoons \overset{IV}{Mn}O_{2(s)} + 4OH^-_{(aq)}$	+0.58
acid	$\overset{VII}{Mn}O^-_{4(aq)} + 4H_3O^+_{(aq)} + 3e^- \rightleftharpoons \overset{IV}{Mn}O_{2(s)} + 6H_2O_{(l)}$	+1.68

The higher oxidation states are far more stable in alkaline conditions than in acid conditions. Chromate (VI) and manganate (VII) are both strong oxidants at pH $\approx$ 1. At pH $>$ 12, chromate (VI) ions are weaker oxidants than hydroxonium ions.

These observations can be explained by the different stabilizing effect of the ligands on the two possible oxidation states. High oxidation states are stabilized by strong π-donor ligands of which oxide ions are the prime example (see page 761). The medium oxidation states are stabilized by simple σ-coordinating ligands such as water molecules. In acidic conditions, there is an increased tendency for the O^{2-} ligands to be protonated and form water molecules. This favours the lower of the two oxidation states and therefore enhances the oxidizing strength of the higher oxidation state compounds. Similarly, in alkaline conditions the water ligands tend to be more readily deprotonated and this favours the formation of the π-donor O^{2-} ligands.

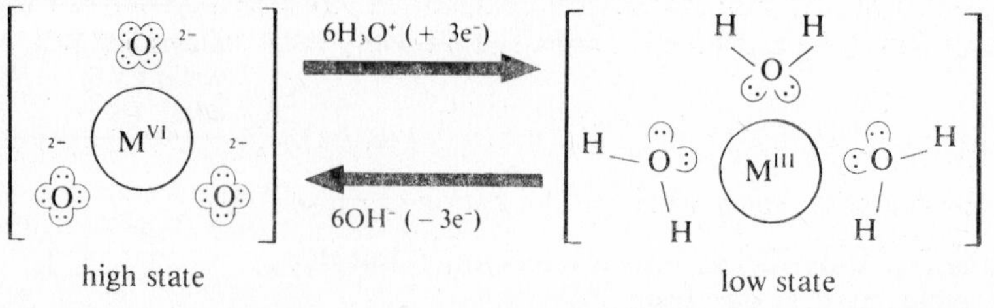

M^{VI} — $6H_3O^+$ (+ $3e^-$) → M^{III}

$6OH^-$ (− $3e^-$) ←

high state

low state

49.7 Complex formation

A large number of different ligands can stabilize the metals in an oxidation state of II or III. The blood-red iron (III) thiocyanate complex is used in the qualitative analysis of iron salts (page 723) as are the cyano-complexes. A very deep blue precipitate forms when an iron (III) solution is added to hexacyanoferrate (II) solution.

$$Fe^{3+}_{(aq)} + Fe(CN)_6^{4-} \rightarrow \overset{III}{Fe}.\overset{II}{Fe}(CN)_6^{-}$$

Some other typical II-state and III-state complexes are shown in the table below.

chromium	*manganese*	*iron*
$[\overset{III}{Cr}F_6]^{3-}$; $[\overset{III}{Cr}(NH_3)_6]^{3+}$	$[\overset{II}{Mn}Cl_4]^{2-}$	$[\overset{II}{Fe}Cl_4]^{2-}$; $[\overset{III}{Fe}Cl_4]^{-}$
$[\overset{III}{Cr}(C_2O_4)_2(H_2O)_2]^{-}$	$[\overset{II}{Mn}(SCN)_6]^{4-}$	$[\overset{III}{Fe}(C_2O_4)_3]^{3-}$

The diethanedioate complex of chromium (III) listed in the table has two isomeric forms.

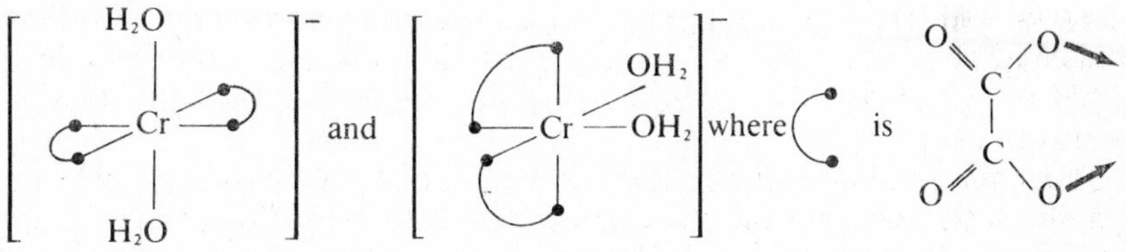

H_2O Cr H_2O — and — Cr OH_2 OH_2 — where (is $O=C(O^-)-C(O^-)=O$

trans- *cis-*

The isomerism of inorganic compounds is discussed in Chapter 35. The *trans*-isomer above is prepared by crushing crystals of potassium dichromate (VI) and ethanedioic acid together in an evaporating basin with a very little water. A molar ratio of about 1:10 should be used. The dichromate (VI) is reduced to chromium (III) when the reaction mixture is warmed and some of the ethanedioic acid is oxidized to water and carbon dioxide. The chromium (III) ions complex with the excess ethanedioate and the complex can be separated by adding ethanol. Ethanol reacts with or dissolves any excess reactants left but the chromium (III) complex is precipitated as very dark coloured crystals.

$$K_2Cr_2O_{7(s)} + 7H_2C_2O_{4(s)} \xrightarrow{\text{warm}} 2K[Cr(C_2O_4)_2(H_2O)_2]_{(s)} + 6CO_{2(g)} + 3H_2O_{(g)}$$

potassium dichromate (VI) — ethanedioic acid — *trans* ethanedioate complex

Peroxo-complexes of chromium

Chromium (VI) is unusual in that it forms a range of peroxo-complexes. The simplest has the structural formula CrO_5 and is shown below.

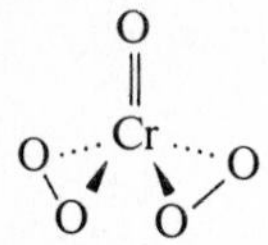

It can be prepared by the action of hydrogen peroxide on acidified dichromate solutions. When hydrogen peroxide is added to potassium dichromate solution in sulphuric acid, a deep blue colour rapidly appears, but as quickly fades with the evolution of oxygen from the solution. The blue colour is due to the formation of CrO_5 which then decomposes to give chromium (III) and oxygen.

$$4CrO_5 + 12H_3O^+_{(aq)} \rightarrow 4Cr^{3+}_{(aq)} + 18H_2O_{(l)} + 7O_{2(g)}$$

The peroxo-complex can be extracted if ether is also present in the reaction system. The complex is more stable in the organic solvent and a number of other derivatives of it have been prepared in ether solution, for example:

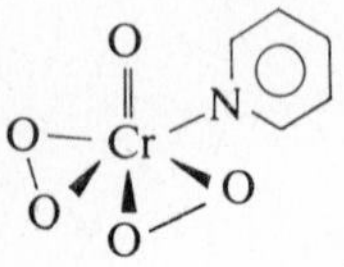

Questions

1. Explain the meaning of the following: a) binary compound b) isoelectronic c) alloy d) rusting e) slag f) sandwich compound.

2. a) Outline the extraction of iron from its ore.
   b) Describe how iron is converted into steel.

3.

Redox system	$E^{\ominus}$ volts
$Cr^{3+}_{(aq)}/Cr^{2+}_{(aq)}$	-0.41
$Cr_2O^{2-}_{7(aq)}/Cr^{3+}_{(aq)}$	$+1.33$
$Fe^{3+}_{(aq)}/Fe^{2+}_{(aq)}$	$+0.77$
$Mn^{2+}_{(aq)}/Mn_{(s)}$	-1.19
$MnO^{-}_{4(aq)}/Mn^{2+}_{(aq)}$	$+1.51$

   Predict the likelihood of a redox reaction when the following are mixed together, and write balanced redox equations for those cases where a reaction does take place.
   a) $Cr^{3+}_{(aq)} + Mn^{2+}_{(aq)}$
   b) $Cr^{3+}_{(aq)} + MnO^{-}_{4(aq)}$
   c) $Fe^{3+}_{(aq)} + Cr^{2+}_{(aq)}$
   d) $Cr^{2+}_{(aq)} + Mn^{2+}_{(aq)}$

4. a) Compare the bonding and oxidation numbers of the atoms in CrO_4^{2-} and MnO_4^-
   b) Compare the effect of dilute acid on i) CrO_4^{2-}; ii) MnO_4^-

5. Explain why oxy-anions of chromium and manganese in their highest oxidation states are prepared under alkaline conditions, but when these oxy-anions are used as oxidants, dilute acid is always added to them.

6. a) Draw the structure of the chromium (III) aquo-ion.
   b) Explain why a solution of chromium (III) ions in water has a pH lower than 7.
   c) Explain why addition of sodium hydroxide solution to a solution of chromium (III) ions first produces a greenish, gelatinous precipitate and then a clear green solution.

7. K_{sp} for $Fe(OH)_2 = 7.9 \times 10^{-16}$ mol^3 dm^{-9} at 25°C.
   K_{sp} for $Fe(OH)_3 = 2.0 \times 10^{-39}$ mol^4 dm^{-12} at 25°C.
   Calculate the solubilities of these two compounds at 25°C.

8. Discuss the formation of complex ions formed by the transition metals iron and chromium.

Chapter 50

The Periodic Table: further comparisons

The major trends in the properties of the elements and their compounds can be classified by period or by group.

As atomic number increases across a *period*, there is an increase in the number of electrons in the outer shell and a decrease in the size of the outer shell as the core charge increases. These factors lead to an increase in the electronic control of the atoms across the period which results in the trend from metallic to non-metallic elements.

The similar properties within a *group* of the Periodic Table are a consequence of the common outer-shell structure that each atom of the group has. Going down any group, the common group properties are modified by the increase in atomic size that accompanies the addition of each successive shell of electrons.

There are three other aspects that have not been considered yet. These are as follows.

1. The anomalous or unusual properties of the first member of a group in the Periodic Table.
2. The 'diagonal relationships' which refer to the similarity of certain pairs of elements which are diagonal neighbours in the Periodic Table.
3. The difference between the properties of metals in the 'A' and 'B' groups of the Periodic Table.

50.1 Anomalous properties of the first member of a group

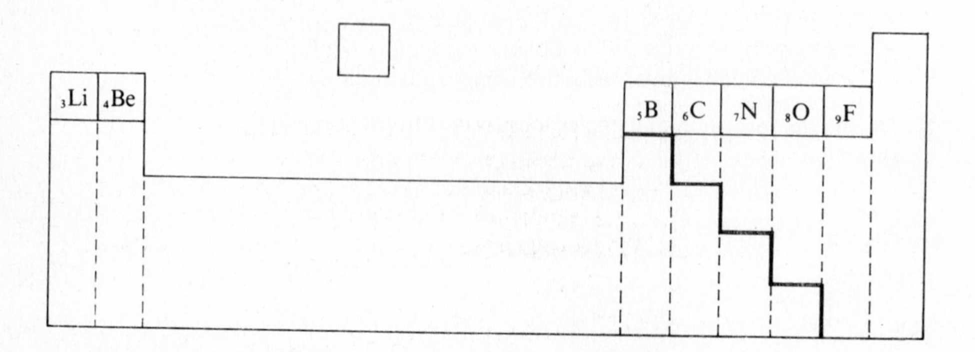

The atoms of the elements which are at the top of the groups in the Periodic Table differ from the atoms of the other elements within their groups in a number of ways:

Size
The atoms of these elements are the smallest in the group and consequently the ions they form are also smallest. This influences lattice energies, hydration energies, and (for non-metal atoms) the ability to undergo pπ-overlap.

Electronic shielding
The inner shell of the first member atoms consists of the helium structure $1s^2$ while other atoms have at least an octet of inner-shell electrons. This means that the nuclear charge is less well shielded and so attraction for electrons in the outer shell is greater. This is reflected in higher first ionization energies and higher electronegativities.

Octet limitation
The first members of groups I to VII are all in the second period of the table. Elements in the second period have a maximum number of electrons within the outer shell of $2n^2$ (where $n = 2$) $= 8$. This is often referred to by the statement that they cannot expand their outer shell.

These factors can be used to explain a number of anomalous properties of the first member elements. Some examples are given below.

Structure and bonding

Metallic groups
The properties of lithium and beryllium are atypical of their groups in three major respects:

1. the lattice arrangements of their compounds
2. their tendency to form organometallic compounds
3. the lower thermal stability of their oxy-salts.

1. Unlike the other members of group IA, lithium does not form 'alums': the lithium ions are too small to form a stable lattice with M^{3+} and sulphate ions. Also lithium salts are almost always hydrated (a few sodium and potassium salts are but no rubidium salts are).

   Beryllium salts are likewise heavily hydrated but, unlike the other group IIA elements, only four hydrating ligands fit round each cation. The small size of Be^{2+} is responsible for this and also for the formation of macromolecular layer-lattice structures instead of simple ionic lattices. For example, compare $BeCl_2$ and $CaCl_2$.

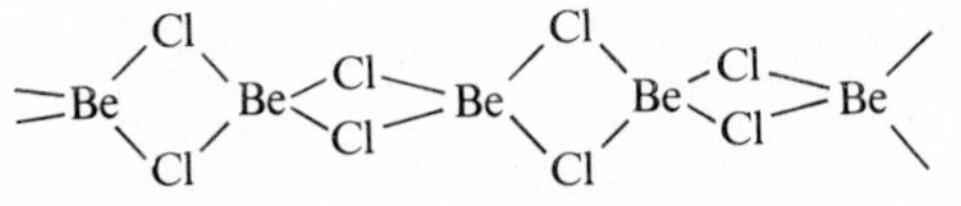

part of $BeCl_2$ macromolecular chain

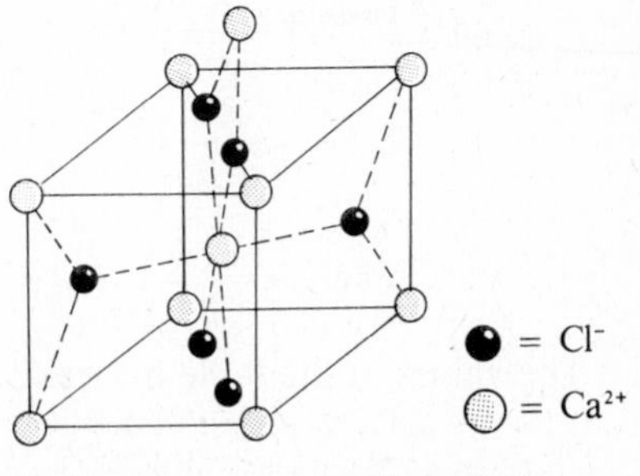

● = Cl^-
○ = Ca^{2+}

co-ordination = 6:3 (rutile lattice)

2. The variety of lithium and beryllium organic compounds results from the higher degrees of covalency of bonded lithium and beryllium atoms. These compounds resemble the Grignard reagents in their reactions.
3. Thermal stability of oxy-salts is discussed on page 643. The large polarizing power of small cations causes distortion in the oxy-anions and favours the formation of an oxide lattice.

Non-metallic groups

Among the first members of groups IV to VII there are four atypical properties.

1. The anomalous strength of the bonds formed.
2. The tendency of the atoms to undergo pπ-overlap.
3. Hydrogen bonding in hydrides whose molecules also have a lone pair.
4. The octet limitation in the outer shell.

1. Going down a group, the increase in atomic size leads to a decrease in bond strength between the atoms as the distance from the nucleus to the bonding pair grows. In group IV, carbon has a relatively strong bond energy which, with the high C—H bond energy, leads to the great number of catenated organic compounds. In the remaining groups, the bond energies are all weaker than would be expected. This is a result of repulsion between the closely adjacent lone pairs in such systems. The table below illustrates this.

bond dissociation energies (kJ mol^{-1})

group IV		*group V*		*group VI*		*group VII*	
C—C	+348	N—N	+163	O—O	+146	F—F	+158
Si—Si	+176	P—P	+172	S—S	+264	Cl—Cl	+242

2. The first members (except fluorine, see chapter 46) form a diversity of pπ-stabilized molecules and molecular ions. The π-overlap between $2p_y$-orbitals of σ-bonded atoms leads to strong bonds.

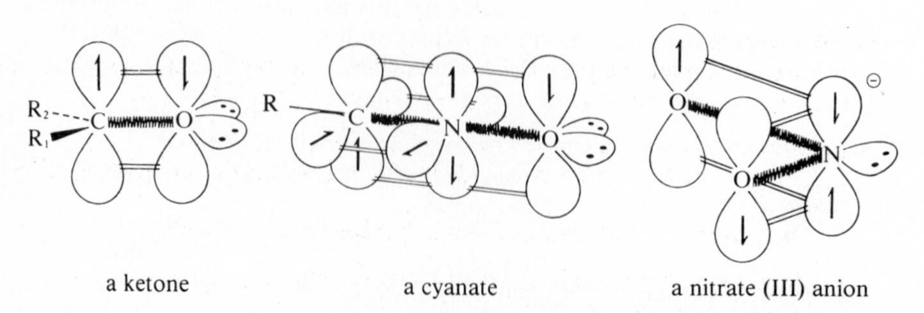

a ketone a cyanate a nitrate (III) anion

The values in the table overleaf illustrate the strength of the multiple bonds.

bond dissociation energies (kJ mol^{-1})

group IV		*group V*		*group VI*	
C—C	+348	N—N	+163	O—O	+146
C=C	+612	N=N	+409	O=O	+496
C≡C	+837	N≡N	+944		

3. Hydrogen bonding is described on page 108. It is responsible for the anomalously high fixed points of compounds containing molecules in which hydrogen is bonded to nitrogen, oxygen or fluorine.
4. The first members cannot expand their octet of bonding electrons because the atoms do not have d-orbitals in their outer shells, (there are no 2d-orbitals). The promotion energy required to fill the nearest vacant orbitals cannot be got back by forming strong enough bonds. The lower members tend to combine in higher valencies and to form dπ-pπ stabilized oxy-compounds.

   For example, there are no 'first-member' analogues of the following molecules.

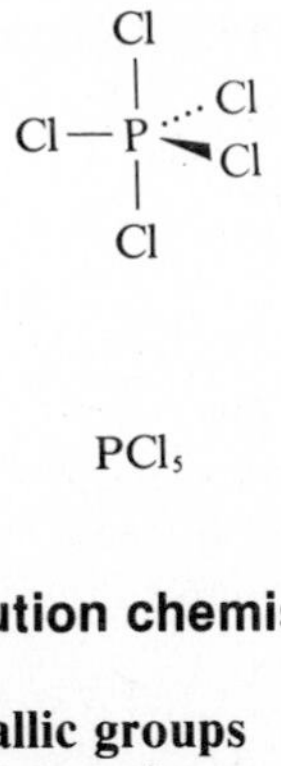

PCl_5 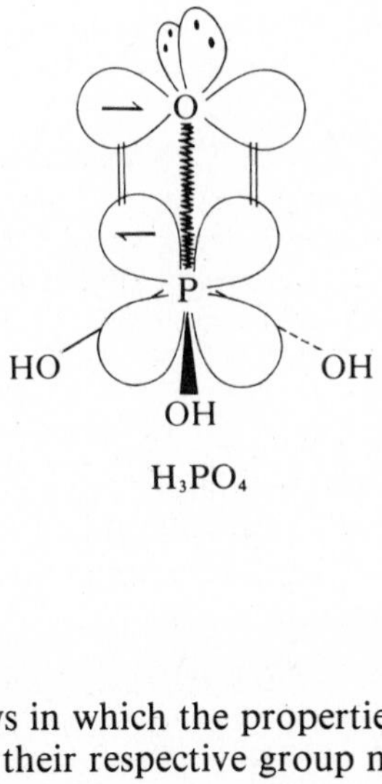 H_3PO_4 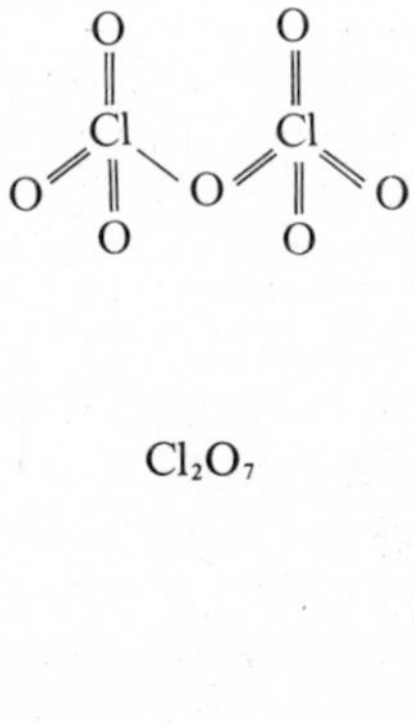 Cl_2O_7

Solution chemistry

Metallic groups

There are three main ways in which the properties of lithium and beryllium are at variance with those of their respective group members.

1. Lithium has an unusually low standard electrode potential
2. The salts of the metals either tend to be less soluble or are hydrolysed in water
3. The cations of both metals are much more readily complexed.

1. Redox potentials generally show an increase going down a group. For metals, the equilibrium concerned is:

$$M^{n+}_{(aq)} + ne^- \rightleftharpoons M_{(s)}$$

The energy terms involved include ionizing the element and then hydrating the ions produced. Both hydration energies and ionization energies decrease going down a group as the size of the atoms and the ions produced increases. However, the hydration energy of Li^+ is appreciably larger than that of the other group members and this makes up for the higher ionization energy:

	$E^\ominus$ (volts)		$E^\ominus$ (volts)
$Be^{2+}_{(aq)} + 2e^- \rightleftharpoons Be_{(s)}$	-1.85	$Li^+_{(aq)} + e^- \rightleftharpoons Li_{(s)}$	-3.04
$Mg^{2+}_{(aq)} + 2e^- \rightleftharpoons Mg_{(s)}$	-2.36	$Na^+_{(aq)} + e^- \rightleftharpoons Na_{(s)}$	-2.71
$Ca^{2+}_{(aq)} + 2e^- \rightleftharpoons Ca_{(s)}$	-2.87	$K^+_{(aq)} + e^- \rightleftharpoons K_{(s)}$	-2.92

2. The lower solubility or higher extent of hydrolysis of the salts are both functions of the smaller size of the ions. For a solid to dissolve, the lattice energy must be supplied while the hydration energy of the ions is given out. Even when it is thermodynamically favourable for this process to take place, there is often a high activation energy resulting from the increased degree of covalency in the lattice (e.g. the $BeCl_2$ lattice). On dissociation of the small cations into aqueous solution, hydrolysis takes place as a result of the large polarizing power of the cations.

$$(H_2O)_3\,Be^{2+}\;\;OH_2\;\;\;OH_2 \;\rightleftharpoons\; (H_2O)_3\,Be^{2+}\,\overset{\ominus}{O}{-}H \;\;\; H{-}\overset{\oplus}{O}H_2$$

3. The small size of the cations and their larger polarizing powers are also responsible for the ability of beryllium and lithium to form complexes. For example, beryllium and its oxide are amphoteric and beryllates are formed in strongly alkaline solutions:

$$Be_{(s)} + 2HO^-_{(aq)} + 2H_2O_{(l)} \rightarrow [Be(OH)_4]^{2-}_{(aq)} + H_{2(g)}$$

Non-metallic groups

The atypical properties in the solution chemistry of the non-metals are:

1. They have unusually high electrode potentials
2. The solubility of fluorides differs from that of the other halides.

1. The redox potentials for oxygen and fluorine are particularly high as the values below indicate. Even that for nitrogen is higher than other members of its group.

		$E^{\ominus}$ (volts)
Group V	$\frac{1}{2}N_{2(g)} + 4H_3O^+_{(aq)} + 3e^- \rightleftharpoons NH^+_{4(aq)}$ 2 $4H_2O_{(l)}$	+0.27
	$P_{(s)} + 3H_3O^+_{(aq)} + 3e^- \rightleftharpoons PH_{3(aq)} + 3H_2O_{(l)}$	−0.04
Group VI	$\frac{1}{2}O_{2(g)} + 2H_3O^+_{(aq)} + 2e^- \rightleftharpoons 3H_2O_{(l)}$	+1.23
	$S_{(s)} + 2H_3O^+_{(aq)} + 2e^- \rightleftharpoons H_2S_{(aq)} + 2H_2O_{(l)}$	+0.14
Group VII	$\frac{1}{2}F_{2(g)} + e^- \rightleftharpoons F^-_{(aq)}$	+2.87
	$\frac{1}{2}Cl_{2(g)} + e^- \rightleftharpoons Cl^-_{(aq)}$	+1.36

The main reason for these high values is the high hydration energy of the small ion produced in each case.

2. The small size of the fluoride ion (with its effect on lattice energies and hydration energies of fluorides) explains some anomalous solubilities. For example, all the sodium halides are very soluble except sodium fluoride which is only sparingly soluble. All the silver halides are insoluble except silver fluoride which is very soluble.

50.2 Diagonal relationships

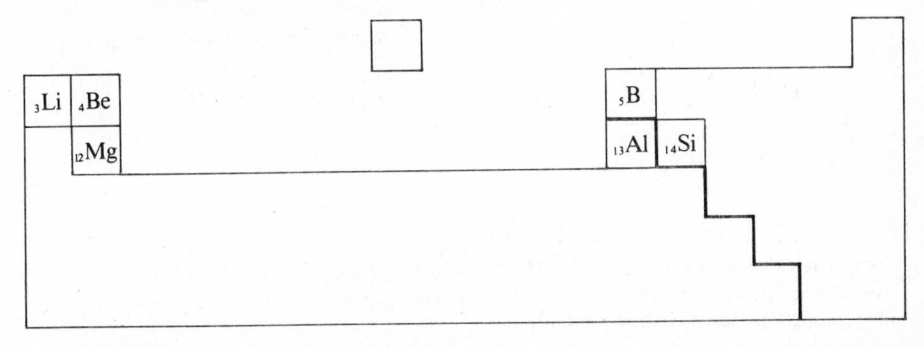

In the Periodic Table above, lithium's diagonal neighbour is magnesium, and boron has silicon diagonally next to it. Beryllium is the diagonal neighbour of aluminium, but this is less obvious because the transition elements have been fitted in between them in the period below.

As shown below, the elements in each of the three pairs have certain similarities which give rise to what are known as the diagonal relationships.

Lithium and magnesium

1. They both form organo-metallic compounds of similar reactivity.
2. Their carbonate, phosphate, and fluoride salts are insoluble.

3. They both form nitrides when heated in a stream of nitrogen (no other group IA metal does this).
4. Neither forms a peroxide when burnt in excess oxygen.

Beryllium and aluminium

1. They both dissolve in acid and in alkali with the evolution of hydrogen.
2. They both form a wide range of complexes.
3. In the metallic state, they are both protected by a persistent oxide layer. Strong oxidizing acids (e.g. concentrated nitric) render them passive.

Boron and silicon

1. Both have semiconducting properties and are macromolecular solids having high fixed points.
2. Both form a great variety of oxy-salts that contain chains and rings of bonded atoms.
3. Both tend to form 'glasses'.

These similarities can be accounted for by considering the change that occurs in atomic size and electronic control both going across a period and down a group.

Across a period, atomic size decreases, but electronic control increases. Going down a group, atomic size increases but electronic control decreases. Because these changes are in opposite directions, going diagonally they tend to cancel one another out as the diagram below shows.

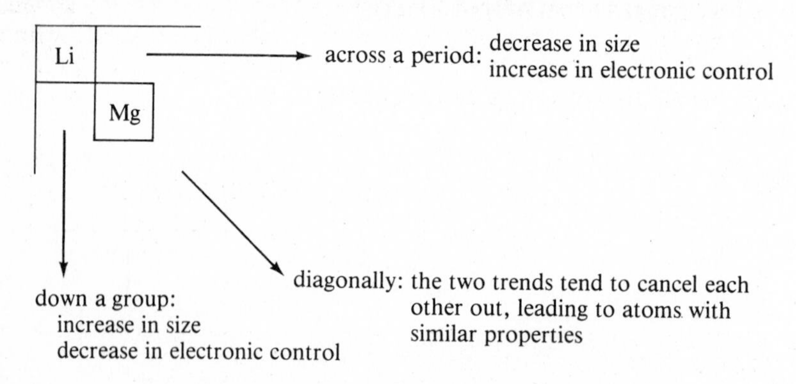

Another expression of the diagonal relationship is the stepped line in the Periodic Table that divides metallic from non-metallic elements.

50.3 'A' metals and 'B' metals

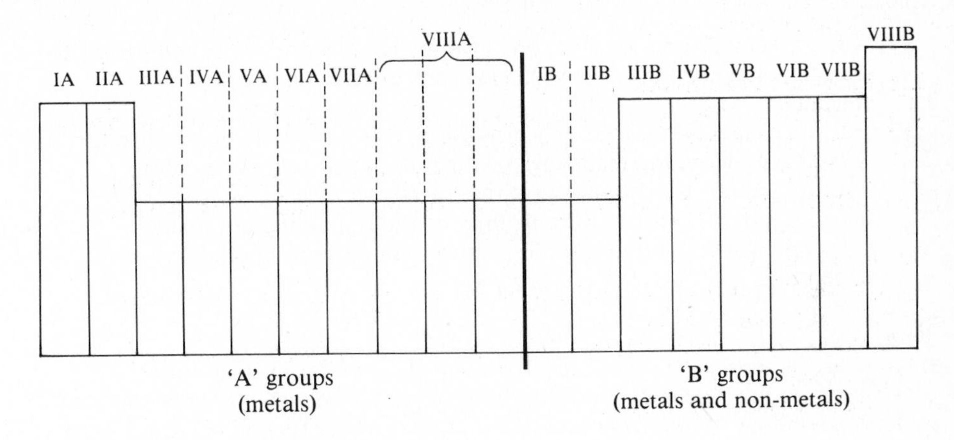

The Periodic Table above shows the division of the elements into eight 'A' groups and eight 'B' groups. This division represents an attempt to relate atomic structure to an octet of electrons, but it is only successful for the first five periods and by making group VIIIA consist of three columns of elements. The 'A' and 'B' labels are often useful to distinguish between different groups of metals.

All the 'A' groups consist of metals and these have already been described in previous chapters. Some of the 'B' groups contain metallic elements as shown below. In the following pages, the chemistry of the 'B' metals is compared with that of the metals already met in earlier chapters.

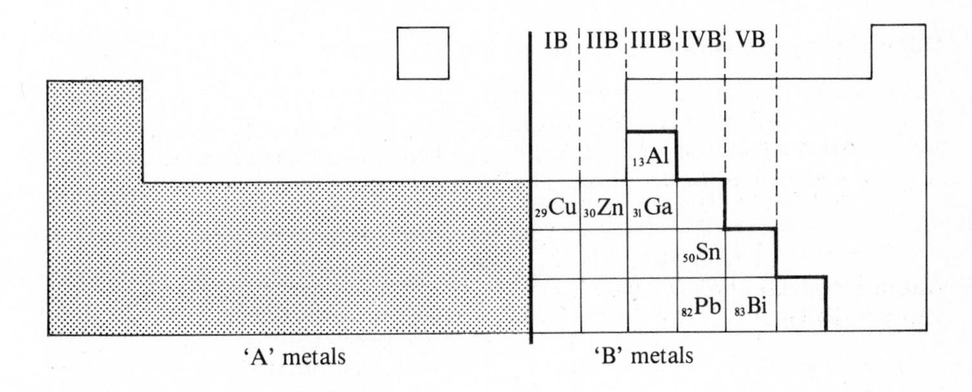

Character of the elements

Structure

'A' metals are found in the s-block and transition series of the elements. Their atoms have incomplete inner d-subshells and their nuclei have a low degree of control of the outer-shell electrons.

'B' metals have full inner d-subshells and their atoms have metallic characteristics for one of two reasons:

1. The outer shell has few electrons (groups IB, IIB and IIIB) and the core charge is low; so the outer-shell electrons are not tightly held to the nucleus.
2. For the elements tin, lead and bismuth at the bottom of the 'B' groups, there is an increasing tendency for the outer-shell s-electrons to be withdrawn into the atomic core: the 'inert pair effect' (see page 660). These electrons assist in the shielding of the remaining outer-shell electrons which are therefore less tightly bound than might have been expected.

Reactivity
In general, 'A' metals are more reactive than 'B' metals. The full inner d-subshells do not shield the nuclear charge as effectively as the inner shells of the corresponding 'A' metal atom. It is therefore easier for an 'A' metal atom to lose its outer-shell electrons than it is for a 'B' metal atom. Compare the first ionization energies, the electronegativities and the standard electrode potentials of the 'A' and 'B' metals of the third period in the table below.

element	*group IA* *potassium*	*group IB* *copper*	*group IIA* *calcium*	*group IIB* *zinc*
first ionization energy (kJ mol^{-1})	418	745	590	908
electronegativity	0.8	1.9	1.0	1.6
$E^{\ominus}$ (volts)	−2.92	+0.34	−2.87	−0.76

Occurrence, extraction and uses of the 'B' metals

The principal ores of the 'B' metals are mostly sulphides. The ores are usually found mixed with those of other metals. In their extraction, the different ores are concentrated together and separation of the different metals usually occurs late in the sequence of operations. The main ores are shown in the table below.

metal	*% by mass in Earth's crust*	*ore*
copper	8.5×10^{-3}	copper pyrites $CuFeS_2$
zinc	1.6×10^{-2}	sphalerite ZnS
gallium	1.8×10^{-3}	with sphalerite Ga_2S_3
tin	4.9×10^{-3}	cassiterite SnO_2
lead	1.9×10^{-3}	galena PbS
bismuth	2.4×10^{-5}	bismite Bi_2O_3
		bismuthite Bi_2S_3

It is interesting to note that the first metals to be discovered and exploited by man were the 'B' metals. In general these metals are easier to extract from their compounds, and some are found native as the metal. However, the metals are often difficult to obtain in a high degree of purity.

Extraction of copper

Copper pyrites is converted to copper sulphide by roasting it in a limited supply of air. The iron (II) oxide produced is removed as a slag by adding silica and reheating the mixture in the absence of air.

$$2\,CuFeS_{2(s)} + 4O_{2(g)} \longrightarrow Cu_2S_{(s)} + 2FeO_{(s)} + 3SO_{2(g)}$$

$$2FeO_{(s)} \xrightarrow{2SiO_2} 2FeSiO_{3(l)} \text{ slag}$$

Copper (I) sulphide is easily reduced to copper by heating in a limited supply of air

$$Cu_2S_{(s)} + O_{2(g)} \longrightarrow 2Cu_{(s)} + SO_{2(g)}$$

Impure copper from this process is purified by making it the anode of an electrolytic cell containing a pure copper cathode. The electrode processes are:

anode: $\underset{\text{impure}}{Cu_{(s)}} \longrightarrow Cu^{2+}_{(aq)} + 2e^-$

cathode: $Cu^{2+}_{(aq)} + 2e^- \longrightarrow \underset{\text{pure}}{Cu_{(s)}}$

Many precious and less reactive metals are found as impurities in the copper and fall to the bottom of the cell to be extracted from the 'anode sludge' that collects there.

Extraction of zinc

Zinc is extracted in a blast furnace very similar to the iron blast furnace. First the zinc ore is roasted in air, converting it to zinc oxide:

$$2ZnS_{(s)} + 3O_{2(g)} \longrightarrow 2ZnO_{(s)} + 2SO_{2(g)}$$

The roasted ore and coke are fed into the top of the furnace, and an air blast injected near the bottom. As with the iron furnace, a whole series of reactions takes place, the most important being:

$$C_{(s)} + O_{2(g)} \longrightarrow CO_{2(g)}$$
$$C_{(s)} + CO_{2(g)} \rightleftharpoons 2CO_{(g)}$$
$$ZnO_{(s)} + CO_{(g)} \rightleftharpoons Zn_{(g)} + CO_{2(g)}$$

However, unlike the iron furnace, here the zinc metal is produced in the vapour phase due to the much lower fixed points of the metal:

metal	*m.p. °C*	*b.p. °C*
iron	1540	3000
zinc	419	906

The zinc vapour is drawn off the top of the furnace with the other furnace gases and shock-cooled by a spray of molten lead. This prevents the zinc vapour re-oxidizing. Molten lead and zinc are only partially miscible in each other and so, by cooling the lead-zinc solution, a layer of zinc can be run off which is 99% pure.

Extraction of tin and lead

Both these metals are extracted in a blast furnace in processes very similar to the extraction of iron. The main reactions are as follows.

For tin:

$$SnO_{2(s)} + 2CO_{(g)} \rightarrow Sn_{(l)} + 2CO_{2(g)}$$
$$SnO_{2(s)} + 2C_{(s)} \rightarrow Sn_{(l)} + 2CO_{(g)}$$

For lead:

$$2PbS_{(s)} + 3O_{2(g)} \rightarrow 2PbO_{(s)} + 2SO_{2(g)}$$
$$PbO_{(s)} + CO_{(g)} \rightarrow Pb_{(l)} + CO_{2(g)}$$
$$2PbO_{(s)} + C_{(s)} \rightarrow 2Pb_{(l)} + CO_{2(g)}$$

Extraction of gallium and bismuth

These two metals are not extracted in any large scale process, and their specialized extraction from the ores of other metals is beyond the scope of this chapter.

Uses of the metals

Some of the important uses of the metals are listed below.

metal	*uses*
copper	electrical use as conductor pipes alloyed in brass and coinage metal
zinc	galvanizing castings alloyed in brass dry cell battery casings
gallium	semiconductors
tin	canning alloyed in solder
lead	batteries in cars sheet metal alloyed in solder lead tetraethyl in petrol
bismuth	bismuth compounds in pharmaceuticals and cosmetics

Chemical properties of the 'B' metals

Compound formation

With the exception of aluminium and zinc, the 'B' metals form compounds in two different oxidation states. The two valencies are usually N and $(N - 2)$, where N is the group number. The lower valency is usually the more stable as a result of the inert pair effect. Copper is not typical in that it shows transitional character by forming II state compounds containing $3d^9$ ions, and these compounds of copper have the properties associated with transition metals, i.e. colour, paramagnetism and tendency to complex (see chapter 48).

Pre-transitional 'A' metals have only one stable oxidation state, whereas the transition 'A' metals exhibit variable valency. However, unlike the 'B' metals, the transition 'A' metals have oxidation states that do not differ by 2.

The oxidation states of metals are compared in the table below.

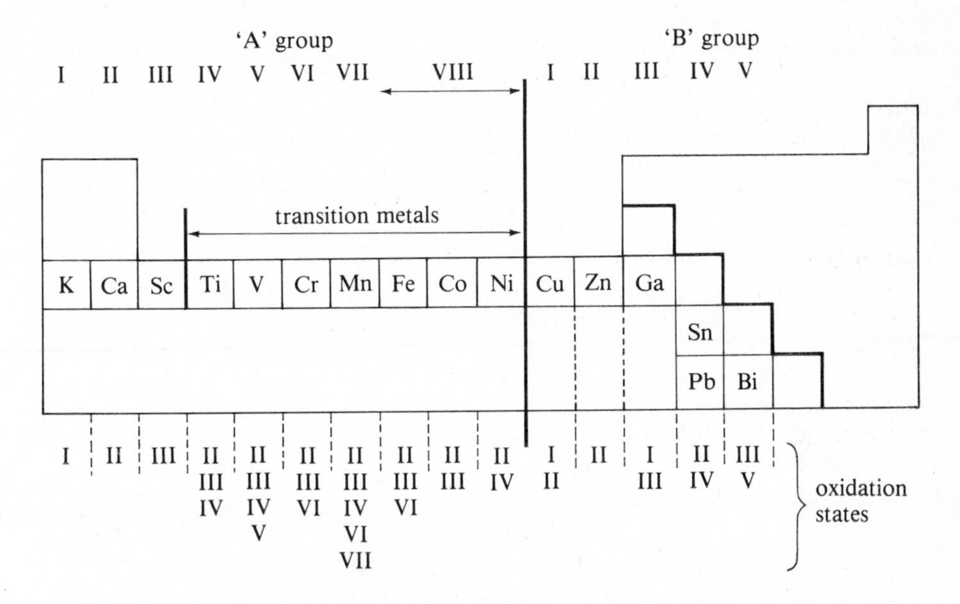

'B' metals combine when heated with oxygen, with sulphur, or with a halogen to give compounds that are essentially ionic in character. In the direct combination of the metal and the non-metal, usually the lower of the two 'B' metal oxidation states is produced unless the non-metal is fluorine.

$$2Pb_{(s)} + O_{2(g)} \rightarrow \overset{II}{2PbO_{(s)}}$$

$$8Zn_{(s)} + S_{8(s)} \rightarrow \overset{II}{8ZnS_{(s)}}$$

$$2Bi_{(s)} + 3Cl_{2(g)} \rightarrow \overset{III}{2BiCl_{3(s)}}$$

However, $2Ga_{(s)} + 3F_{2(g)} \rightarrow \overset{III}{2GaF_{3(s)}}$

Copper shows a greater tendency to form the higher state, however.

$$2Cu_{(s)} + O_{2(g)} \rightarrow \overset{II}{2CuO_{(s)}}$$

$$Cu_{(s)} + Cl_{2(g)} \rightarrow \overset{II}{CuCl_{2(s)}}$$

except $16Cu_{(s)} + S_{8(g)} \rightarrow \overset{I}{8Cu_2S_{(s)}}$

and when heated strongly, many of these copper (II) compounds decompose to give the thermodynamically more stable copper (I) compound.

$$4CuO_{(s)} \xrightarrow{1000°C} 2Cu_2O_{(s)} + O_{2(g)}$$

$$2CuCl_{2(s)} \longrightarrow 2CuCl_{(s)} + Cl_{2(g)}$$

In contrast, aqueous conditions favour the formation of copper (II) compounds because the copper (II) ion is smaller and more highly charged and so has a much higher solvation energy than the copper (I) compound.

The degree of covalency of these compounds is greater than in the corresponding 'A' metal compounds. The nuclear charge of a 'B' metal cation is more poorly screened than in an 'A' metal cation. For example, the structures of calcium oxide and zinc oxide illustrate this point. CaO has an ionic 6:6 lattice like that of NaCl; ZnO has a diamond-like lattice in which each zinc atom is surrounded tetrahedrally by four oxygen atoms and each oxygen atom has a similar arrangement of zinc atoms around it.

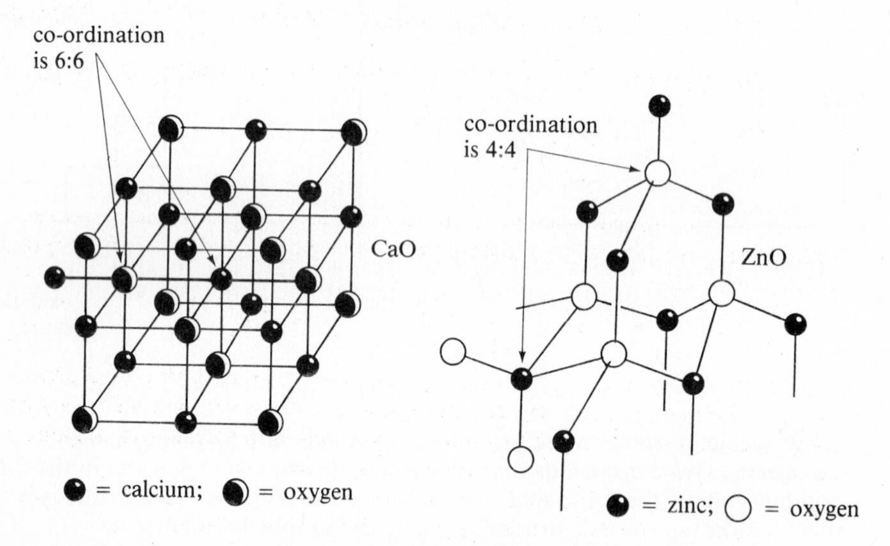

Acid-base properties

The lower state oxides and hydroxides of 'B' metals are basic with some amphoteric character. The higher state oxides and hydroxides (when formed) are all amphoteric, although copper again is atypical because the II-state oxide is basic with no amphoteric properties and the univalent ions disproportionate in aqueous solution as a result of the decrease in energy on formation of the more stable hydrated copper (II) ions.

Zinc, aluminium and lead (II) oxides and hydroxides all show typical 'B' metal characteristics.

$$ZnO_{(s)} + 2H_3O^+_{(aq)} \rightarrow Zn^{2+}_{(aq)} + 3H_2O_{(l)}$$

$$ZnO_{(s)} + 2HO^-_{(aq)} + H_2O_{(l)} \rightarrow [Zn(OH)_4]^{2-}_{(aq)}$$

The salts of the 'B' metals are acidic by hydrolysis. The effect of the polarizing power of the cations is not the only factor responsible for this. There is also a tendency for the simple cations to form dπ-pπ stabilized oxy-cations:

$$BiCl_{3(s)} + 3H_2O_{(l)} \rightarrow [BiO]^+_{(aq)} + 2H_3O^+_{(aq)} + 3Cl^-_{(aq)}$$

Redox properties

The lower electrode potentials of the 'B' metals are consistent with their weaker reducing strengths compared with 'A' metals. Aluminium is the most powerful reductant of all the 'B' metals and is used in the extraction of manganese and chromium from their ores. It is still weaker than the IIIA element, scandium.

In their higher oxidation states, the 'B' metal compounds tend to be oxidants; Bi(V), Pb(IV) and Tl(III) are all oxidizing, and even Cu(II) oxidizes iodide ions:

$$2\overset{\text{II}}{Cu}^{2+}_{(aq)} + 4\overset{-\text{I}}{I}^{-}_{(aq)} \rightarrow 2\overset{\text{I}}{Cu}I_{(s)} + \overset{0}{I}_{2(aq)}$$

The copper (I) produced is immediately precipitated and therefore does not have time to disproportionate. Sodium bismuthate (V) is a powerful oxidant in acid solution while lead (IV) oxide is a useful oxidant for preparing chlorine.

$$BiO^{-}_{3(aq)} + 4H_3O^{+}_{(aq)} + 2e^{-} \rightleftharpoons BiO^{+}_{(aq)} + 6H_2O_{(l)} \qquad E^{\ominus} = +2.10 \text{ volts}$$

$$2\overset{\text{II}}{Mn}^{2+}_{(aq)} + 5\overset{\text{V}}{Bi}O^{-}_{3(aq)} + 4H_3O^{+}_{(aq)} \rightarrow 2\overset{\text{VII}}{Mn}O^{-}_{4(aq)} + 5\overset{\text{III}}{Bi}O^{+}_{(aq)} + 6H_2O_{(l)}$$

$$\overset{\text{IV}}{Pb}O_{2(s)} + \underbrace{4H_3O^{+} + 6\overset{-\text{I}}{Cl}^{-}}_{\text{conc. HCl}} \rightarrow [\overset{\text{II}}{Pb}Cl_4]^{2-} + 6H_2O_{(l)} + \overset{0}{Cl}_{2(g)}$$

Sodium bismuthate is prepared in reasonable yield by heating together sodium peroxide and bismuth (III) oxide. The peroxide oxidizes Bi(III) to Bi(V) in the solid state.

$$2Na_2O_{2(s)} + Bi_2O_{3(s)} \xrightarrow[\text{heat}]{} 2NaBiO_3 + Na_2O_{(s)}$$

The stability of the lower oxidation state decreases on going up a group because the inert pair effect becomes less prominent. Tin and antimony are more stable in the (IV) and (V) state respectively and their lower oxidation state compounds have quite strong reducing properties:

$$[\overset{\text{IV}}{Sn}Cl_6]^{2-} + 2e^{-} \rightleftharpoons [\overset{\text{II}}{Sn}Cl_6]^{4-} \qquad E^{\ominus} = +0.15 \text{ volts}$$

Tin (II) chloride in hydrochloric acid readily reduces iodine to iodide or iron (III) to iron (II).

The tendency for the lower oxidation state compounds of copper to disproportionate in solution can be accounted for in terms of the two electrode processes shown below.

$$Cu^{+}_{(aq)} + e^{-} \rightleftharpoons Cu_{(s)}: \qquad E^{\ominus} = +0.52 \text{ volts}$$

$$Cu^{2+}_{(aq)} + e^{-} \rightleftharpoons Cu^{+}_{(aq)}: \qquad E^{\ominus} = +0.16 \text{ volts}$$

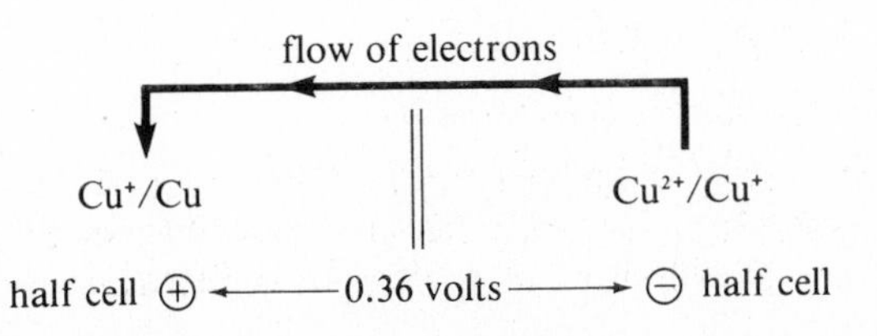

overall reaction is $2Cu^{+}_{(aq)} \rightarrow Cu^{2+}_{(aq)} + Cu_{(s)}$

Solubility and complexing properties

Like the transition metals, the 'B' metals form a wide range of complexes. Their ability to complex is due largely to the high polarizing power of their cations and the considerable covalent character associated with metal-ligand bonds. Both the upper and lower 'B' metal oxidation states are found in different complexes. Fluoride or oxide ligands stabilize the upper states whereas cyanide, ammonia or other halide ligands tend to stabilize the lower states (compare with the discussion on page 760).

lower state complexes		*upper state complexes*	
$[Ag(NH_3)_2]^+$	$[TlI_4]^{3-}$	$[CuF_4]^{2-}$	$[TlF_6]^{3-}$
$[PbCl_4]^{2-}$	$[BiCl_4]^-$	$[SbF_6]^-$	$[SnO_3.3H_2O]^{2-}$ $[BiO_3]^-$

The ability of copper (I) to form a dichloro-complex leads to the unusual property that the metal can displace hydrogen from concentrated hydrochloric acid. In dilute acid, copper is inert because the electrode potentials for copper (I) and for copper (II) are positive.

$$Cu^{2+}_{(aq)} + 2e^- \rightleftharpoons Cu_{(s)} \qquad E^\ominus = +0.34 \text{ volts}$$

$$Cu^{+}_{(aq)} + e^- \rightleftharpoons Cu_{(s)} \qquad E^\ominus = +0.52 \text{ volts}$$

whereas $2H_3O^+_{(aq)} + 2e^- \rightleftharpoons H_{2(g)} + 2H_2O_{(l)} \qquad E^\ominus = 0.00 \text{ volts}$

However, in concentrated acid, copper (I) is stabilized by the presence of the chloride ligands and the following reaction takes place.

$$2Cu + 2HCl + 2Cl^- \rightarrow 2[CuCl_2]^- + H_{2(g)}$$

The stereochemistry of the complexes of copper and zinc show a number of differences.

The zinc ion is surrounded by four ligands and these are arranged tetrahedrally around the central ion:

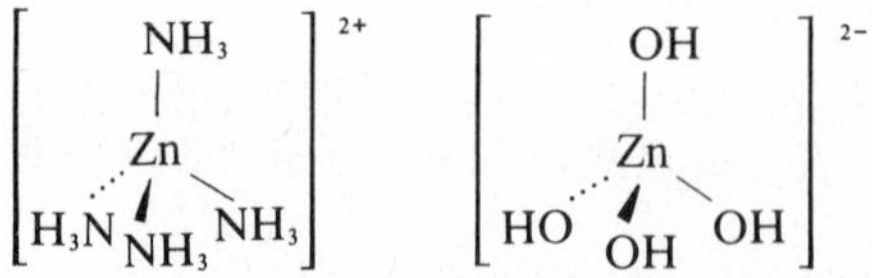

$$\left[\begin{array}{c} NH_3 \\ | \\ Zn \\ H_3N \quad NH_3 \quad NH_3 \end{array}\right]^{2+} \qquad \left[\begin{array}{c} OH \\ | \\ Zn \\ HO \quad OH \quad OH \end{array}\right]^{2-}$$

The copper ion is surrounded in water by six molecules, but these are placed around the copper ion in a distorted octahedral arrangement so that two are further away than the other four:

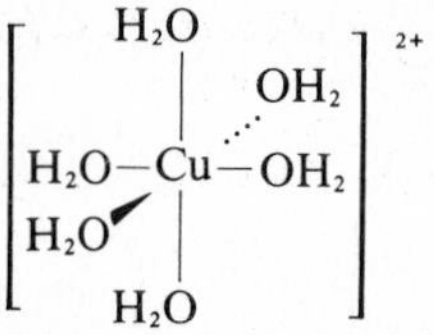

$$\left[\begin{array}{c} H_2O \quad\quad \\ | \quad OH_2 \\ H_2O-Cu-OH_2 \\ H_2O \quad | \quad\quad \\ H_2O \quad\quad \end{array}\right]^{2+}$$

This effect, known as the Jahn-Teller distortion, leads to the copper (II) ion having only four near-neighbours all in the same plane.

When other ligands are present in aqueous solution, they tend to occupy the close-in positions so producing planar complexes:

H_2O
NH_3
$H_3N-Cu-NH_3$
H_3N
H_2O
$2+$

the deep blue tetraammine copper (II) ion

H_2O
Cl
$Cl-Cu-Cl$
Cl
H_2O
$2-$

the yellow tetrachlorocuprate (II) ion

Questions

1. Explain the meaning of the following: a) cation radius b) cationic charge c) degree of covalency d) lattice energy e) organo-metallic compound f) oxy-salt g) thermal stability h) dissociation.
2. a) Outline the extraction of copper from its ore.
   b) Describe how copper is purified and explain why its purity is important.
   c) Explain the main uses of copper.
3. a) Contrast the extraction of zinc with that of iron.
   b) Explain why iron is frequently coated with zinc.
   c) Contrast the effect of corrosion on two pieces of iron, one coated with zinc and the other with chromium. What happens when the coating is scratched through?
4. Discuss diagonal relationships in the periodic table using as specific examples lithium and magnesium, and beryllium and aluminium.
5. Account for the following anomalies:
   a) Lithium carbonate is unstable to heat while other group I carbonates are stable.
   b) Lithium hydrogen carbonate cannot be crystallised out while other group I hydrogen carbonates can be.
   c) Beryllium oxide is amphoteric while the other group II oxides are basic.
   d) Beryllium halides fume in moist air while other group II halides do not.
6. Account for the following facts:
   a) Hydrogen fluoride is a weak acid while the other hydrogen halides are strong acids.
   b) Fluorine reacts with water yielding oxygen while the other halogens do not.
   c) Silver fluoride is soluble while the other silver halides are insoluble.
7. How do nitrogen and oxygen differ from the other members of their groups in the compounds they form?
8. Compare and contrast the chemistry of calcium and zinc.
9. In what situations are inter-group similarities more dominant than intra-group similarities?

Appendixes

Indexes

Periodic table

Appendix 1

Diagnostic tests

It is possible to obtain information about the nature of an unknown substance by carrying out a range of simple tests on it in the laboratory. In some tests a positive result is clearly visible (for example a colour change) while in others it may be necessary to identify gases evolved during the test.

> These diagnostic tests should be conducted *only* under the direct supervision of a qualified chemist who is aware of the necessary safety precautions. Always wear safety glasses.

For each test, use only a small quantity of the unknown substance: if inconclusive, the test can always be repeated. For a solid, 2–5 g is sufficient (about as much as will fit on the head of a drawing pin). For a liquid, use 10–20 drops (0.5–1 cm^3).

In general a test can be classified according to the action that it has on a substance. In this appendix, laboratory tests are given under the following headings.

A. **The action of heat**: does the substance change phase, decompose or burn?

B. **The action of solvents**: does the substance dissolve in or react with the solvent?

C. **The action of acids and bases**: does the substance have basic or acidic properties?

D. **The action of redox reagents**: does the substance have oxidizing or reducing properties?

E. **The action of complexing reagents**: does the substance give a complex of characteristic colour?

F. **The action of specific reagents**: for example, the use of bromine water to test for carbon-carbon unsaturation.

Organic compounds

A. Heat

1. Put a small portion of the compound on an inverted crucible lid on a pipe-clay triangle.
2. Start heating gently at one side and then gradually increase the heat.
3. Continue heating until no more change takes place.

observations	*conclusions*
a) the compound burns with an almost colourless flame	⇒ saturated, aliphatic compound e.g.: $CH_3CH_2OH + 3O_2 \rightarrow 2CO_2 + 3H_2O$
b) the compound burns with a yellow, luminous and smoky flame	⇒ high carbon content and therefore likely to be unsaturated or aromatic, e.g. $2C_6H_5COOH + 5O_2 \rightarrow 4CO_2 + 6H_2O + 10C$
c) the compound ignites with difficulty and a greyish residue is left	⇒ the residue is likely to be a metal oxide or carbonate and therefore the compound is probably a metal salt, e.g. $2CH_3COONa + 4O_2 \rightarrow Na_2CO_3 + 3CO_2 + 3H_2O$
d) the compound sublimes and produces a cloud of white smoke	⇒ ammonium salt, e.g. $NH_4X_{(s)} \rightleftharpoons NH_{3(g)} + HX_{(g)}$

B. Solvents

1. Pour a little of the solvent into a test tube.
2. Cautiously add a little of the compound and shake. Warm if necessary.
3. If the solvent is water, test the pH for evidence of hydrolysis.

observations	*conclusions*
a) the compound dissolves in water or is partially miscible with water	⇒ ionic or polar compound, e.g. $CH_3COO^-Na^+$, $CH_3\overset{\delta+}{C}(=O^{\delta-})CH_3$
b) the compound is insoluble in water (or is immiscible with water) but dissolves in organic solvents such as propanone, or 1,1,1-trichloroethane	⇒ molecular compound with minimal polar character, e.g. C_6H_5COOH, C_8H_{10}
c) the compound dissolves in water and changes the pH of the water	⇒ compound undergoes hydrolysis to produce H_3O^+ or OH^- in solution, e.g.
i) pH < 7	$C_6H_5NH_3^+Cl^-_{(s)} + H_2O_{(l)} \rightleftharpoons C_6H_5NH_{2(l)} + H_3O^+_{(aq)} + Cl^-_{(aq)}$
ii) pH > 7	$CH_3COO^-Na^+_{(s)} + H_2O_{(l)} \rightleftharpoons CH_3COOH_{(aq)} + OH^-_{(aq)} + Na^+_{(aq)}$

C. Acids and bases

Acidic and basic reagents are usually used in dilute aqueous solution. Under these conditions they provide a source of either hydroxonium or hydroxide ions to test the basic or acidic nature of a compound.

In an undiluted state, they are also useful testing reagents but their reactivity is more extensive. For example, concentrated sulphuric acid is a dehydrating agent as well as an acid while 880-ammonia is a powerful nucleophilic reagent as well as a base.

Dilute aqueous acids and bases: H_2SO_4, HCl; NaOH, Na_2CO_3, NH_3

1. Pour a little aqueous acid or base into a test-tube.
2. Cautiously add a small portion of the compound. Warm if necessary.
3. Add an excess amount of the compound if no change occurs.

observations	*conclusions*
a) the compound dissolves in aqueous acid but not in water	⇒ a basic compound, e.g. $C_6H_5NH_{2(l)} + H_3O^+_{(aq)} \rightleftharpoons C_6H_5NH^+_{3(aq)} + H_2O_{(l)}$
b) the addition of aqueous acid to a solution of the compound causes a crystalline precipitate to form	⇒ the compound is likely to be a salt of an insoluble, weak acid, e.g. benzoic acid $C_6H_5COO^-_{(aq)} + H_3O^+_{(aq)} \rightleftharpoons C_6H_5COOH_{(s)} + H_2O_{(l)}$
c) the compound dissolves in aqueous alkali	⇒ acidic compound, e.g. $C_6H_5COOH_{(s)} + OH^-_{(aq)} \rightleftharpoons C_6H_5COO^-_{(aq)} + H_2O_{(l)}$
d) the compound dissolves in cold alkali and liberates a gas turning red litmus blue	⇒ the gas is ammonia and therefore the compound is an ammonium salt, e.g. $NH_4X_{(s)} + OH^-_{(aq)} \rightarrow NH_{3(g)} + X^-_{(aq)} + H_2O_{(l)}$
e) the compound dissolves in aqueous alkali only after prolonged boiling; no gases can be detected	⇒ the compound is being hydrolysed; a hydroxide ion can act as a nucleophile to attack an ester or a halide, e.g. $CH_3COOR_{(s)} + OH^-_{(aq)} \xrightarrow{\text{boil}} CH_3COO^-_{(aq)} + ROH_{(aq)}$ $RBr_{(s)} + OH^-_{(aq)} \xrightarrow{\text{boil}} ROH_{(aq)} + Br^-_{(aq)}$
f) the compound dissolves as in e) above but a gas is liberated that turns red litmus blue	⇒ an amide is being hydrolysed, e.g. $C_6H_5CONH_{2(s)} + OH^-_{(aq)} \xrightarrow{\text{boil}} C_6H_5COO^-_{(aq)} + NH_{3(g)}$
g) the compound dissolves in aqueous sodium carbonate and a gas is evolved that turns lime-water cloudy	⇒ acidic compound strong enough to protonate hydrogen carbonate ions: i.e. a carboxylic acid but *not* a phenol, e.g. $2RCOOH_{(s)} + CO^{2-}_{3(aq)} \rightarrow 2RCOO^-_{(aq)} + CO_{2(g)} + H_2O_{(l)}$ but $C_6H_5OH_{(s)} + CO^{2-}_{3(aq)} \rightleftharpoons C_6H_5O^-_{(aq)} + HCO^-_{3(aq)}$

Concentrated acids and bases: H_2SO_4 and NH_3

1. Put a small portion of the organic compound into a test tube.
2. Cautiously and slowly add the concentrated liquid down the wall of the test-tube holding it at an angle of 45°; warm if necessary.

observations	*conclusions*
a) the compound effervesces on addition of concentrated sulphuric acid; a mixture of gases is evolved that will both burn at the mouth of the tube (blue flame) and turn lime-water cloudy	⇒ compound is likely to be an ethanedioate (oxalate): H_2SO_4 protonates the weakly basic ions and then dehydrates the acid produced. Carbon monoxide and carbon dioxide evolve $\begin{matrix} COO^- \\ \vert \\ COO^- \end{matrix} \xrightarrow{H_2SO_4} \begin{matrix} COOH \\ \vert \\ COOH \end{matrix} \xrightarrow[-H_2O]{} CO_2 + CO$
b) the compound behaves as in a) above, but the evolved gas does not turn lime-water cloudy	⇒ compound is likely to be a methanoate (formate). Only carbon monoxide is evolved $HCOO^- \xrightarrow{H_2SO_4} HCOOH \xrightarrow[-H_2O]{} CO$
c) a water-insoluble compound dissolves in 880-ammonia and, after warming, gives a product that dissolves in hot water	⇒ the compound is probably an ester undergoing nucleophilic attack to produce a more soluble amide, e.g. $C_6H_5COOR + NH_3 \rightarrow C_6H_5CONH_2 + ROH$

D. Redox reagents

Few organic compounds have oxidizing properties in aqueous solution but a number have reducing properties, for example aldehydes, alcohols and alkenes. Acidified dichromate (VI) or manganate (VII) can be used to show that a compound can act as a reductant but it is also possible to find specific testing reagents for an aldehyde because aldehydes are stronger reductants than alcohols or alkenes.

Acidified potassium dichromate (VI) or manganate (VII)

1. Add ten drops of test solution to about 2 cm^3 of dilute sulphuric acid.
2. Add a small sample of the compound to be tested.
3. Warm and boil gently if necessary.

observations	*conclusions*
a) the compound turns the orange dichromate to a green coloured solution, or the purple manganate to a colourless solution; no gases are evolved	⇒ compound is a reductant: $\overset{VI}{Cr} \rightarrow \overset{III}{Cr}$ or $\overset{VII}{Mn} \rightarrow \overset{II}{Mn}$ e.g. $\begin{matrix} RCH_2OH \\ RCHO \end{matrix} \xrightarrow[H_3O^+]{Cr_2O_7^{2-} \text{ or } MnO_4^-} RCOOH$
b) as in a) above but a gas is also evolved which turns lime-water cloudy	⇒ the gas is carbon dioxide and therefore the reductant is likely to be an ethanedioate (oxalate), e.g. $\begin{matrix} COO^- \\ \vert \\ COO^- \end{matrix} \rightarrow 2CO_2 + 2e^-$

Ammoniacal silver nitrate or Fehling's solution

Both these two test reagents are specific for aldehydes. They are prepared as follows

ammoniacal silver nitrate

1. Add three drops of alkali to about 1 cm^3 of silver nitrate solution
2. Add aqueous ammonia until the precipitate is dissolved.

Fehling's solution

1. Make up an alkaline solution of potassium 2,3-dihydroxybutanedioate
2. Add an equal volume of copper (II) sulphate solution.

observations	*conclusions*
a) on adding a small sample of the compound to ammoniacal silver nitrate, the solution darkens; after warming, a precipitate of silver is seen to coat the inside of the tube ('silver mirror')	⇒ only an aldehyde is a strong enough reductant to reduce complexed silver (I) to pure silver: $RCHO + 2Ag(NH_3)_2^+ + H_2O \longrightarrow RCOOH + Ag + 2NH_4^+$
b) on adding a small sample of the compound to Fehling's solution and warming, a red-brown precipitate forms	⇒ Fehling's solution contains complexed copper (II) and, as above, only an aldehyde is strong enough to reduce this to copper (I) oxide which is seen as an insoluble precipitate

E. Complexing reagents

If a compound forms a coloured complex when added to a solution of transition metal cations, it shows that the molecules of the compound can act as ligands. The transition metal salts most commonly used are iron (III) chloride and copper (II) sulphate.

Iron (III) chloride

1. Shake a small portion of the compound in 2 cm^3 of water.
2. Add iron (II) chloride solution drop by drop.

observations	*conclusion*
a) a brick-red colour is produced	⇒ aliphatic acid anions as ligands, e.g. $R{-}C(=O){-}\ddot{O}^{\ominus}$ → Fe^{III}
b) a buff colour is produced	⇒ aromatic acid anions as ligands, e.g. $C_6H_5C(=O){-}\ddot{O}^{\ominus}$ → Fe^{III}
c) a violet colour is produced	⇒ phenol molecules (or derivatives) as ligands, e.g. $C_6H_5\ddot{O}(H)$ → Fe^{III}

Copper (II) sulphate

This reagent is either used in the same way as iron (III) chloride described above or in very dilute alkaline conditions. In its second form, it is made up as follows:

1. Add five drops of dilute copper (II) sulphate solution to a few cm^3 of water.
2. Add two drops of sodium hydroxide solution.
3. Add a small sample of the compound to be tested.

observations	*conclusions*
a) a royal blue coloured complex is produced when a small portion of the compound in water is shaken with a few drops of copper (II) sulphate	⇒ aliphatic amine molecules acting as ligands, e.g. $R-\ddot{N}H_2 \rightarrow Cu^{II}$
b) a lime green complex (instead of a royal blue one) is produced as in a) above	⇒ phenylamine molecules (or derivatives) acting as ligands, e.g. $C_6H_5-\ddot{N}H_2 \rightarrow Cu^{II}$
c) a pale pink coloured complex is produced when a small sample of the compound is added to the very dilute alkaline solution of copper (II) sulphate (the 'biuret test')	⇒ conjugated amides of the type shown below acting as ligands: $>\ddot{N}- \rightarrow Cu^{II}$ R–CO–NH–CO–R , R–CO–NH–NH–CO–R e.g. $NH(CONH_2)_2$ biuret R–NH–CO–CO–NH–R e.g. $(CONH_2)_2$ ethanediamide

F. Specific reagents

Hydroxylamine or 2,4-dinitrophenylhydrazine

These reagents are used to confirm the presence of a carbonyl group in the structure of a compound. Precipitates of oximes or 2,4-dinitrophenylhydrazones are given by ketones and aldehydes, but *not* by acids or by their derivatives whose carbonyl groups are modified by the presence of other functional groups.

1. Add 2 cm^3 of the reagent to a test-tube.
2. Add a very small sample of the compound to be tested.

observations	*conclusions*
a) a bright orange precipitate is produced in the 2,4-dinitrophenylhydrazine solution	⇒ the compound is a ketone or aldehyde giving an insoluble 2,4-dinitrophenylhydrazone, e.g. $RCOR \rightarrow R_2C{=}N{-}NH{-}C_6H_3(NO_2)_{2(s)}$ (reagent: 2,4-dinitrophenylhydrazine, $H_2N{-}NH{-}C_6H_3(NO_2)_2$)
b) a white crystalline precipitate is produced in the solution of hydroxylamine	⇒ the compound is a ketone or aldehyde giving an insoluble oxime, e.g. $RCOR \xrightarrow{NH_2OH} R_2C{=}N{-}OH_{(s)}$

Bromine water

Bromine water is used to confirm the presence of carbon-carbon double bonds in the structure of a compound. It is also used to test for phenols and phenylamines, whose activated aromatic rings can attack the weakly electrophilic bromine molecules.

1. Shake a small sample of the compound in a few cm^3 of water.
2. Add bromine-water drop by drop until there is an excess present.

observations	*conclusions*
a) the brown colour of the bromine is rapidly discharged and a colourless suspension of water and immiscible droplets is produced	⇒ the compound contains carbon-carbon double bonds which become saturated to give a dibromo- addition product, e.g. $>C{=}C< \xrightarrow{Br_2} BrC{-}CBr$
b) the brown colour of the bromine is discharged and a white precipitate forms when excess bromine is added; the precipitate often smells of antiseptic	⇒ the compound is a phenol or phenylamine derivative. Electrophilic substitution occurs to produce a tribromo product, e.g. $C_6H_5OH \xrightarrow{3Br_2} C_6H_2Br_3OH_{(s)} + 3HBr$

Iodine in alkali: the iodoform test
This reagent can identify the presence of an ethanoyl group in ketones or aldehydes. It also gives positive results for an alcohol that can be oxidized to give an aldehyde or ketone containing the ethanoyl group. In other words, the reagent is a test for the following molecular groups:

$$R-\overset{\overset{O}{\|}}{C}-CH_3 \quad \text{or} \quad R-\overset{H\ \ OH}{C}-CH_3$$

1. Add 2 cm^3 of iodine in potassium iodide solution to a test-tube.
2. Add sodium hydroxide drop by drop until the brown colour is almost gone.
3. Add 2 drops of the compound (or solution) to be tested. Warm if necessary.

observations	*conclusions*
a) a fine yellow precipitate is produced when two drops of the compound are shaken with the test reagent	⇒ the compound's molecular structure contains one of the above two groups, electrophilic addition of iodine takes place and hydrolysis of the product gives insoluble iodoform, CHI_3. $R-\overset{\overset{O}{\|}}{C}-CH_3 \xrightarrow{I_2 \text{ in } OH^-} R-\overset{\overset{O}{\|}}{C}-O^- \quad CHI_{3(s)}$

Inorganic Compounds

A. Heat

Unlike organic compounds, very few inorganic compounds burn when heated in air. The purpose of heating them is to observe how they change phase or decompose. A wide range of different gases can be driven off as decomposition products so that, without some previous information about the substance under test, it is advisable to carry out a heating test twice.

On the first occasion, no chemical tests should be attempted but the heated substance should be carefully observed to obtain a general outline of the changes taking place. On the second occasion, it becomes possible to decide on some sensible tests to perform on any decomposition products thought likely as a result of the first heating. Most gases are recognised by one (or a combination) of the following properties: smell, colour, precipitating properties, acid-base properties or redox properties. For example, sulphur (IV) oxide is colourless but has an acrid, throat-catching smell; it turns damp blue litmus paper red and turns acidified dichromate paper from orange to green.

1. Put a small portion of the compound in a clean, dry hard-glass tube.
2. Cautiously heat gently at first and then more strongly.
3. Heat very strongly until no more changes are observed.

observations	*conclusions*
a) water vapour is seen at the top of the tube and steam is evolved	⇒ damp sample or hydrated salt, e.g. $CaSO_4.2H_2O_{(s)} \rightarrow CaSO_{4(s)} + 2H_2O_{(g)}$
b) the compound appears to melt, but steam pours off on stronger heating and eventually a solid is produced again; further decomposition of the solid may then occur	⇒ a hydrated salt dissolving in its own water of crystallization, e.g. $Na_2CO_3.10H_2O_{(s)} \rightarrow [2Na^+_{(aq)} + CO^{2-}_{3(aq)}] \downarrow Na_2CO_{3(s)} + 10H_2O_{(g)}$
c) no obvious changes appear to take place, but a glowing splint is relit at the mouth of the tube	⇒ a group IA metal nitrate (V) producing oxygen, e.g. $2NaNO_{3(s)} \rightarrow 2NaNO_{2(s)} + O_{2(g)}$ or an oxygen compound of an element in a high oxidation state $2PbO_{2(s)} \rightarrow 2PbO + O_{2(g)}$
d) no obvious changes appear to take place, but a gas is evolved which turns damp blue litmus paper red and lime-water cloudy; sometimes the solid changes colour during heating	⇒ a metal carbonate decomposing to give carbon dioxide, e.g. $CuCO_{3(s)} \rightarrow CuO_{(s)} + CO_{2(g)}$ $CO_{2(g)} + Ca^{2+}_{(aq)} + 2OH^-_{(aq)} \rightarrow CaCO_{3(s)} + H_2O_{(l)}$
e) as d) above but water vapour is driven off as well	⇒ a hydrogen carbonate salt decomposing, e.g. $2NaHCO_{3(s)} \rightarrow Na_2CO_{3(s)} + H_2O_{(g)} + CO_{2(g)}$
f) an acrid, throat-catching gas is evolved, damp blue litmus paper turns red and acidified dichromate paper goes from orange to green	⇒ a metal sulphate (IV) decomposing to give sulphur (IV) oxide, e.g. $ZnSO_{3(s)} \rightarrow ZnO_{(s)} + SO_{2(g)}$ $SO_{2(g)} + 2H_2O_{(l)} \rightarrow HSO^-_{3(aq)} + H_3O^+_{(aq)}$ S(IV) → S(VI) while Cr(VI) (orange) → Cr(III) (green)

observations	*conclusions*
g) as f) above but a yellow vapour is also driven off; this forms a layer of solid around the inside of the neck of the tube; the vapour may catch fire	⇒ a thiosulphate decomposing to give sulphur vapour and sulphur (VI) oxide, e.g. $8S_2O_3^{2-} \rightarrow S_8 + 8SO_2 + 8O^{2-}$
h) a brown, choking gas is driven off, damp copper turnings turn green in the gas and damp, blue litmus paper goes red, a glowing splint also glows more brightly (or is relit) at the mouth of the tube	⇒ a metal nitrate (V) decomposing to give nitrogen (IV) oxide (brown, acidic) and oxygen, e.g. $2Pb(NO_3)_{2(s)} \rightarrow 2PbO_{(s)} + 4NO_{2(g)} + O_{2(g)}$ $\underset{\text{brown}}{2NO_{2(g)}} + 3H_2O_{(l)} \rightarrow NO^-_{3(aq)} + NO^-_{2(aq)} + 2H_3O^+_{(aq)}$
i) the compound vapourizes and gives smoke which forms a layer of solid around the inside of the neck of the tube, no liquid can be detected	⇒ the compound sublimes, e.g. ammonium salts, iron (III) or aluminium halides, e.g. $NH_4Cl_{(s)} \rightleftharpoons NH_{3(g)} + HCl_{(g)}$ $2FeCl_{3(s)} \rightleftharpoons Fe_2Cl_{6(g)}$

B. Solvents

1. Pour a little of the solvent into a test-tube.
2. Add a small portion of the compound and shake. Warm if necessary.
3. If the solvent is water, test the pH for evidence of hydrolysis.

observations	*conclusions*
a) the compound dissolves in water to give a neutral solution; no hydrolysis (reaction with water) appears to take place	⇒ appreciable ionic character — probably the salt of a strong acid and strong base, e.g. $Ca(NO_3)_{2(s)} \rightarrow \underbrace{Ca^{2+}_{(aq)} + 2NO^-_{3(aq)}}_{\text{hydrated ions}}$
b) the compound dissolves in water to give a solution whose pH is less than seven	⇒ the salt of a strong acid and weak base; for example, an ammonium compound or an ionic compound whose cations have a high polarizing power, e.g. Al^{3+}, Fe^{3+} $NH^+_{4(aq)} + H_2O_{(l)} \rightleftharpoons NH_{3(aq)} + H_3O^+_{(aq)}$ $Al(H_2O)_6^{3+} + H_2O_{(l)} \rightleftharpoons Al(H_2O)_5OH^{2+} + H_3O^+_{(aq)}$

observations	*conclusions*
c) the compound dissolves in water to give a solution of pH greater than seven.	⇒ the salt of a weak acid and strong base, for example, a group I metal fluoride or sulphate (IV) $F^-_{(aq)} + H_2O_{(l)} \rightleftharpoons HF_{(aq)} + OH^-_{(aq)}$ $SO^{2-}_{3(aq)} + H_2O_{(l)} \rightleftharpoons HSO^-_{3(aq)} + OH^-_{(aq)}$
d) the compound is partially soluble in organic solvents of low (or zero) polarity, such as methyl benzene or 1,1,1-trichloroethane	⇒ appreciable covalent character: probably contains cations of high polarizing power, e.g. $2AlCl_{3(s)} \rightleftharpoons Al_2Cl_{6(sol)}$

C. Acids and bases

Dilute aqueous acids; HCl, H_2SO_4

A dilute aqueous acid is used as a source of hydroxonium ions to protonate the anions of a weak acid. If the compound to be tested is a metal salt of a weak acid, the weak acid itself is liberated into solution by the addition of dilute hydrochloric or sulphuric acid. Since many weak acids are unstable and decompose to give gaseous products, useful information can be obtained by identifying any gases evolved.

1. Put a small sample of the compound into a test-tube.
2. Add about 5 cm³ of dilute acid. Warm if necessary.
3. Test for evolved gas with litmus, dichromate paper or other specific tests.

observations	*conclusions*
a) the compound dissolves in acid but not in water; no gas can be detected	⇒ oxide, hydroxide or insoluble fluoride, e.g. $ZnO_{(s)} + 2H_3O^+_{(aq)} \rightarrow Zn^{2+}_{(aq)} + 3H_2O_{(l)}$ $CaF_{2(s)} + 2H_3O^+_{(aq)} \rightarrow Ca^{2+}_{(aq)} + 2HF_{(aq)} + 2H_2O_{(l)}$
b) the compound dissolves in acid and a gas is evolved which turns damp blue litmus paper pink and lime-water cloudy	⇒ metal carbonate or hydrogen carbonate, e.g. $FeCO_{3(s)} + 2H_3O^+_{(aq)} \rightarrow Fe^{2+}_{(aq)} + 3H_2O_{(l)} + CO_{2(g)}$ $CO_{2(g)} + 2OH^-_{(aq)} + Ca^{2+}_{(aq)} \rightarrow CaCO_{3(s)} + H_2O_{(l)}$
c) the compound dissolves in acid and a gas is evolved which has an acrid, throat-catching smell; damp blue litmus goes red and acidified dichromate paper turns from orange to green	⇒ metal sulphate (IV), e.g. $MgSO_{3(s)} + 2H_3O^+_{(aq)} \rightarrow Mg^{2+}_{(aq)} + 3H_2O_{(l)} + SO_{2(g)}$ $SO_{2(g)} + 2H_2O_{(l)} \rightleftharpoons HSO^-_{3(aq)} + H_3O^+_{(aq)}$ S(IV) → S(VI) while Cr(VI) → Cr(III) (orange → green)

observations	*conclusions*
d) as in c) above but a fine, yellow precipitate can also be seen in the mixture produced	⇒ metal thiosulphate, e.g. $Na_2S_2O_{3(s)} + 2H_3O^+_{(aq)} \rightarrow 2Na^+_{(aq)} + 3H_2O_{(l)} + SO_{2(g)} + S_{(s)}$
e) the compound dissolves in acid and a brown, choking gas is evolved; damp blue litmus is turned red and damp copper goes green in the gas	⇒ metal nitrate (III), e.g. $2KNO_{2(s)} + 2H_3O^+_{(s)} \rightarrow 2K^+_{(aq)} + 3H_2O_{(l)} + NO_{2(g)} + NO_{(g)}$ $\underset{\text{brown}}{2NO_{2(g)}} + 3H_2O_{(l)} \rightleftharpoons NO^-_{2(aq)} + NO^-_{3(aq)} + 2H_3O^+_{(aq)}$
f) the compound dissolves in acid and a gas smelling of bad eggs is evolved; damp blue litmus paper goes red and lead ethanoate paper is stained black	⇒ metal sulphide, e.g. $CuS_{(s)} + 2H_3O^+_{(aq)} \rightarrow Cu^{2+}_{(aq)} + 2H_2O_{(l)} + H_2S_{(g)}$ $H_2S_{(g)} + H_2O_{(l)} \rightleftharpoons HS^-_{(aq)} + H_3O^+_{(aq)}$ $Pb^{2+}_{(aq)} + H_2S_{(g)} + 2H_2O_{(l)} \rightarrow \underset{\text{black}}{PbS_{(s)}} + 2H_3O^+_{(aq)}$

Concentrated acids: HCl, H_2SO_4, HNO_3
Although these reagents are powerful proton donors, their reactivity is not confined solely to acid-base properties. Concentrated hydrochloric acid is the most convenient concentrated source of chloride ions when it is added to aqueous solution, and is therefore useful as a complexing reagent. Concentrated sulphuric and concentrated nitric acid are both oxidizing agents and most diagnostic tests make use of their redox properties, see pages 825 and 826. Concentrated sulphuric acid is also a dehydrating agent.

The page references given above contain all the diagnostic tests based on the concentrated acids, including the tests dependent on acid-base properties.

Dilute aqueous bases: NaOH, NH_3, Na_2CO_3
These reagents are used as a source of hydroxide ions to deprotonate the cations derived from weak bases. In the majority of cases, the cations are either hydrated metal ions or ammonium ions. An insoluble precipitate of a metal hydroxide is produced as a result of deprotonating hydrated metal ions unless the metal concerned comes from group IA, for example

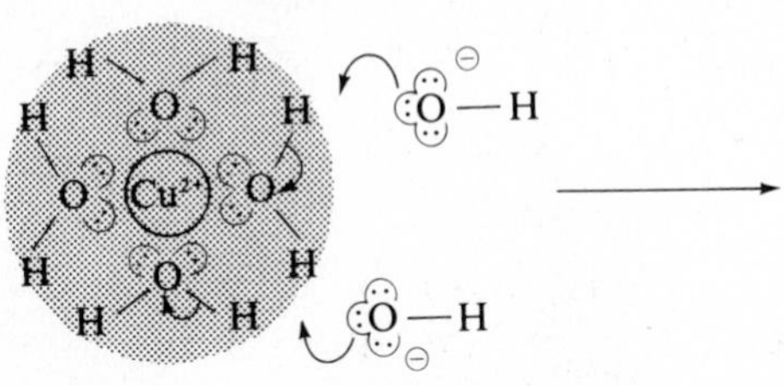

$$[Cu(H_2O)_6]^{2+} + 2\,{}^-OH \longrightarrow Cu(OH)_2(H_2O)_{2(s)} + 2H_2O_{(l)}$$

gelatinous,
pale blue precipitate

On addition of excess sodium hydroxide or ammonia, many insoluble precipitates redissolve. This occurs because of complex formation: a metal may form a soluble hydroxy- or amino-complex, e.g.

$Al(OH)_4^-$	$Cu(NH_3)_4^{2+}$
soluble, colourless	soluble, deep blue

1. Take a small portion of the compound and shake it in about 5 cm^3 of water.
2. Add soluble base dropwise, shaking the tube after each drop added.
3. Add an excess of soluble base.

observations	*conclusions*
a) when sodium hydroxide is added, no precipitate forms but, on warming, a pungent gas is evolved which turns damp red litmus blue	⇒ an ammonium salt — N.B. some ammonium salts are double salts, e.g. $Fe(NH_4)(SO_4)_2$ $NH_{4(aq)}^+ + OH_{(aq)}^- \underset{}{\overset{water}{\rightleftharpoons}} NH_{3(g)} + H_2O_{(l)}$
b) when sodium hydroxide or sodium carbonate is added to the solution of the compound, a gelatinous precipitate forms; the precipitate does *not* redissolve when excess soluble base is added	⇒ salt of a metal whose hydroxide is insoluble and not amphoteric, some of the cations that fit the above pattern are shown below, e.g. $M_{(aq)}^{n+} + nOH_{(aq)}^- \rightarrow M(OH)_{n(s)}$
c) on addition of sodium hydroxide, a gelatinous precipitate forms, the precipitate redissolves in excess alkali	⇒ salt of a metal whose hydroxide is amphoteric, e.g. $Zn_{(aq)}^{2+} + 2OH_{(aq)}^- \rightleftharpoons Zn(OH)_{2(s)}$ insoluble $Zn(OH)_{2(s)} + 2OH_{(aq)}^- \rightleftharpoons Zn(OH)_{4(aq)}^-$ soluble

Table for b):

colour of precipitate	*ion*
white	GpIIA
grey-green	Fe^{2+}
brown	Fe^{3+}
buff	Mn^{2+}
pale blue	Cu^{2+}
black	Ag^+
yellow	Hg^{2+}

Table for c):

colour of precipitate	*ion*
white	Al^{3+}, Zn^{2+}, Sn^{2+}, Pb^{2+}
green	Cr^{3+}

observations	*conclusions*
d) on addition of aqueous ammonia a gelatinous precipitate forms, the precipitate redissolves in excess aqueous ammonia	⇒ salt of a metal which forms an insoluble hydroxide but a soluble amino-complex, e.g. Cu^{2+}, Zn^{2+}, Ag^{+}, Ni^{2+} $Cu^{2+}_{(aq)} + 2NH_{3(aq)} + 2H_2O_{(l)} \rightleftharpoons \underset{\text{insoluble}}{Cu(OH)_{2(s)}} + 2NH^{+}_{4(aq)}$ $Cu(OH)_{2(s)} + 4NH_{3(aq)} \rightleftharpoons \underset{\text{soluble}}{Cu(NH_3)^{2+}_4} + 2OH^{-}_{(aq)}$

Concentrated base: 880-ammonia

880-ammonia has two main uses: either it acts as a concentrated source of ammonia ligands, for example in its effect on the different silver halides,

$$\underset{\text{white}}{AgCl_{(s)}} \xrightarrow{\text{aqueous ammonia}} \underset{\text{soluble}}{Ag(NH_3)^{+}_2} + Cl^{-}$$

$$\underset{\text{cream}}{AgBr_{(s)}} \xrightarrow{\text{aqueous ammonia}} \underset{\text{insoluble}}{AgBr_{(s)}} \xrightarrow{\text{880-ammonia}} \underset{\text{soluble}}{Ag(NH_3)^{+}_2} + Br^{-}$$

$$\underset{\text{yellow}}{AgI_{(s)}} \xrightarrow{\text{aqueous ammonia}} \underset{\text{insoluble}}{AgI_{(s)}} \xrightarrow{\text{880-ammonia}} \underset{\text{insoluble}}{AgI_{(s)}}$$

or the 880-ammonia on a damp bottle-stopper can be used to provide an ammonia-rich atmosphere in which acidic gases form smoke, for example:

$$HCl_{(g)} + NH_{3(g)} \rightarrow NH_4Cl_{(s)}$$

D. Redox reagents

Concentrated sulphuric acid

Concentrated sulphuric acid acts as acid, oxidant and dehydrating agent. Its action may depend on any of these properties.

1. Put a very small sample of the compound into a clean, dry test-tube.
2. Add concentrated sulphuric acid dropwise down the inside wall of the tube while holding it at an angle of about 45° [CARE].
3. Test for any gas being evolved. Warm very gently if necessary.

observations	*conclusions*
a) the compound effervesces and a colourless gas is evolved which fumes in the air; damp blue litmus is turned red and clouds of smoke are formed when the stopper of an 880-ammonia bottle is held at the mouth of the tube	⇒ metal chloride or complex containing chloride ligands; the chloride ions are protonated to give hydrogen chloride (but H_2SO_4 is not a strong enough oxidant to produce chlorine from a chloride) $Cl^{-}_{(s)} + H_2SO_{4(l)} \rightarrow HCl_{(g)} + HSO^{-}_{4(s)}$

observations	*conclusions*
b) as above, but the evolved gas is coloured yellow-brown; the damp blue litmus goes red but is then bleached; there is both an acrid and a pungent, irritating smell	⇒ metal bromide: a mixture of hydrogen bromide, bromine and sulphur dioxide is evolved $Br^-_{(s)} + H_2SO_{4(l)} \rightarrow HBr_{(g)}\ 2\ HSO^-_{4(s)}$ $HBr_{(g)} \rightarrow \underset{\text{brown}}{Br_{2(g)}}$ while $H_2SO_{4(l)} \rightarrow SO_{2(g)}$
c) also as above, but the evolved gas is purple and a black crystalline deposit forms on the inside of the tube; combined with the irritating smell is the smell of rotten eggs as well	⇒ metal iodide: a mixture of hydrogen iodide (very little), iodine, sulphur dioxide and hydrogen sulphide is evolved $I^-_{(s)} + H_2SO_{4(l)} \rightarrow HI_{(g)} + HSO^-_{4(s)}$ $H\overset{-I}{I}_{(g)} \rightarrow \underset{\text{purple}}{\overset{0}{I}_{2(g)}} \rightarrow \underset{\text{black}}{\overset{0}{I}_{2(s)}}$ while $H_2\overset{VI}{S}O_{4(l)} \rightarrow \overset{IV}{S}O_{2(g)}$ and $H_2\overset{-II}{S}_{(g)}$
d) a coloured compound loses its colour on addition of concentrated sulphuric acid	⇒ protonation of ligands, e.g. $\underset{\text{green}}{CuCl^{2-}_{4(s)}} + H_2SO_{4(l)} \rightarrow \underset{\text{white}}{CuSO_{4(s)}} + 2Cl^-_{(s)} + 2HCl_{(g)}$
i) no gas is evolved	⇒ probably dehydration of a hydrated salt: H_2O ligands → H_3O^+ ions $\underset{\text{blue}}{CuSO_4.5H_2O_{(s)}} \xrightarrow{H_2SO_4} \underset{\text{white}}{CuSO_{4(s)}} + 5H_3O^+$
ii) gas is evolved as described in a), b) or c) above	⇒ Cl^-, Br^- or I^- as ligands
iii) a mixture of gas is evolved: it turns lime-water cloudy and burns with a blue flame when ignited at the mouth of the tube	⇒ ethandioate ions as ligands $^{\ominus}O(O{=})C{-}C({=}O)O^{\ominus} \xrightarrow{H_2SO_4} HO(O{=})C{-}C({=}O)OH \xrightarrow[-H_2O]{H_2SO_4} CO_{2(g)} + CO_{(g)}$

Concentrated sulphuric acid is also used in 'ring' tests, see page 829.

Concentrated nitric acid

1. Put a small sample of the compound in a clean, dry test-tube.
2. Add concentrated nitric acid dropwise to the tube. Warm if necessary.

observations	*conclusions*
clouds of brown, choking gas are evolved. Damp blue litmus is turned red and a piece of damp copper goes green in the gas	⇒ compound can act as a reductant, e.g. $\overset{II}{Fe^{2+}} \rightarrow \overset{III}{Fe^{3+}}$ or $\overset{-I}{I^-} \rightarrow \overset{0}{I_2}$ while $H\overset{V}{N}O_3 \rightarrow \overset{IV}{N}O_2$ brown, acidic

Aqueous oxidizing agents: $K_2Cr_2O_7$ and $KMnO_4$

1. Shake a small sample of the compound in a few cm^3 of water.
2. Add 1 cm^3 of acidified dichromate (VI) or manganate (VII). Warm.

observations	*conclusions*
a) on warming the solution containing dichromate (VI), the original orange colour becomes green	⇒ compound can act as a reductant to reduce Cr(VI) → Cr(III) (orange → green)
b) on warming the solution containing manganate (VII), the original purple colour disappears; no gas is evolved	⇒ compound can act as a reductant to reduce Mn(VII) → Mn(II) (purple → colourless) e.g. Fe^{2+} salts, I^- salts, SO_3^{2-} salts
c) as b) above but bubbles of gas can be seen; if enough is collected, lime-water turns cloudy	⇒ compound contains ethandioate ions; in acid solution, these are oxidized to carbon dioxide $(\overset{III}{C}OO)_2^{2-} \rightarrow 2\overset{IV}{C}O_{2(g)}$ while $\overset{VII}{Mn}O_4^- \rightarrow \overset{II}{Mn^{2+}}$

Aqueous reducing agents: KI and $FeSO_4$

These two reagents are used to test for the ability of a compound to act as an oxidant. On addition of an oxidant to potassium iodide in acid solution, there is an immediate darkening of the colour of the solution as iodine is produced, e.g.

$$\underset{\text{an oxidant}}{Na\overset{V}{Cl}O_3} \rightarrow Na\overset{-I}{Cl} \quad \text{while} \quad \underset{\text{pale yellow}}{K\overset{-I}{I}} \rightarrow \underset{\text{dark brown}}{\overset{0}{I_2}}$$

On addition of an oxidant to iron (II) in acid solution, iron (III) is produced. Since iron (II) is green and iron (III) brown, the change can be shown by adding sodium hydroxide and precipitating the iron as its insoluble hydroxide.

E. Precipitating or complexing reagents

1. Shake a small sample of the compound in about 5 cm^3 of water.
2. Add a few drops of the complexing reagent.

observations	*conclusions*		
a) dilute nitric acid and silver nitrate			
i) an off-white or yellow precipitate forms which is soluble in:			
aqueous ammonia	⇒ a metal chloride	$AgCl_{(s)}$	white
880-ammonia	⇒ a metal bromide	$AgBr_{(s)}$	cream
neither	⇒ a metal iodide	$AgI_{(s)}$	yellow
ii) a red-brown precipitate forms which is soluble in excess nitric acid	⇒ a metal chromate	$Ag_2CrO_{4(s)}$	red-brown
b) dilute nitric acid and barium chloride			
i) a white precipitate forms	⇒ a metal sulphate	$BaSO_{4(s)}$	white
ii) when barium chloride is added *first* to the solution of the compound, a yellow precipitate forms. On addition of nitric acid an orange solution is produced as the yellow solid dissolves	⇒ a metal chromate in acid solution, the soluble barium dichromate (orange) is formed via $\underset{\text{yellow}}{2BaCrO_{4(s)}} + 2H_3O^+_{(aq)} \rightleftharpoons 2Ba^{2+}_{(aq)} + \underset{\text{orange}}{Cr_2O^{2-}_{7(aq)}} + 3H_2O_{(l)}$	$BaCrO_{4(s)}$	yellow
c) potassium hexacyanoferrate (III)			
i) a very dark blue precipitate forms	⇒ an iron (II) compound	$K\overset{II}{Fe}(\overset{III}{Fe}[CN]_6)_{(s)}$	blue
ii) a green precipitate forms	⇒ a copper (II) compound	$\overset{II}{Cu}_3(\overset{III}{Fe}[CN]_6)_{2(s)}$	green
d) potassium hexacyanoferrate (II)			
a very dark blue precipitate forms	⇒ an iron (III) compound	$K\overset{III}{Fe}(\overset{II}{Fe}[CN]_6)_{(s)}$	blue
e) potassium thiocyanate			
i) a blood-red colour is produced	⇒ an iron (III) compound	$\overset{III}{Fe}(CNS)^{2+}_{(aq)}$	red
ii) a green colour is visible at first and then a black precipitate begins to form	⇒ a copper (II) compound $\underset{\text{blue}}{Cu^{2+}_{(aq)}} \rightleftharpoons \underset{\text{green}}{Cu(CNS)^+_{(aq)}} \rightleftharpoons \underset{\text{black}}{Cu(CNS)_{2(s)}}$		

F. Specific tests

Flame test

1. Take a piece of nichrome wire and bend a small loop at the end of it.
2. Dip the looped end in concentrated hydrochloric acid and put it into the hottest part of a bunsen flame until it no longer colours the flame.
3. Remove from the flame and dip in concentrated hydrochloric acid again.
4. Now dip the hot wire into some of the solid to be tested.
5. Put the 'loaded' wire into the edge of the flame and observe the colour given.

observations		*conclusions*
colour:	yellow	⇒ sodium salt
	lilac	potassium salt
	brick red	calcium salt
	apple green	barium salt
	pale blue-green	copper salt

Brown ring test for nitrates (V)

1. Shake a small sample of the compound with about 5 cm^3 water. Add an equal amount of iron (II) sulphate.
2. Holding the test-tube at an angle of 45°, pour concentrated sulphuric acid cautiously down the inside wall of the tube.
3. Do not allow the liquids to mix but ensure that two layers are formed.

observations	*conclusions*	
a brown ring is seen to form at the aqueous solution-sulphuric acid junction	⇒ a nitrate (V) salt; e.g. $Na\overset{V}{N}O_3$	at the liquid junction $Fe(NO)^{2+}_{(aq)}$ brown

Blue ring test for thiosulphates

1. Add about 3 cm^3 of concentrated sulphuric acid to a test-tube.
2. In a second test-tube, shake a small sample of the compound in about 5 cm^3 of ammonium molybdate solution.
3. Hold the first tube at an angle of 45° and CAREFULLY pour in some liquid from the second tube so that two liquid layers form.

observations	*conclusions*	
a deep blue ring is seen to form at the sulphuric acid-aqueous solution junction	⇒ a thiosulphate salt: e.g. $K_2S_2O_3$	at the liquid junction Mo(V)/Mo(VI) oxide is blue

Appendix 2

Organic reaction routes

A. Aliphatic compounds

The map below outlines the major routes between different classes of aliphatic compounds. The three succeeding diagrams supply some of the details of this map; R is any alkyl group and X is any halide group.

map A

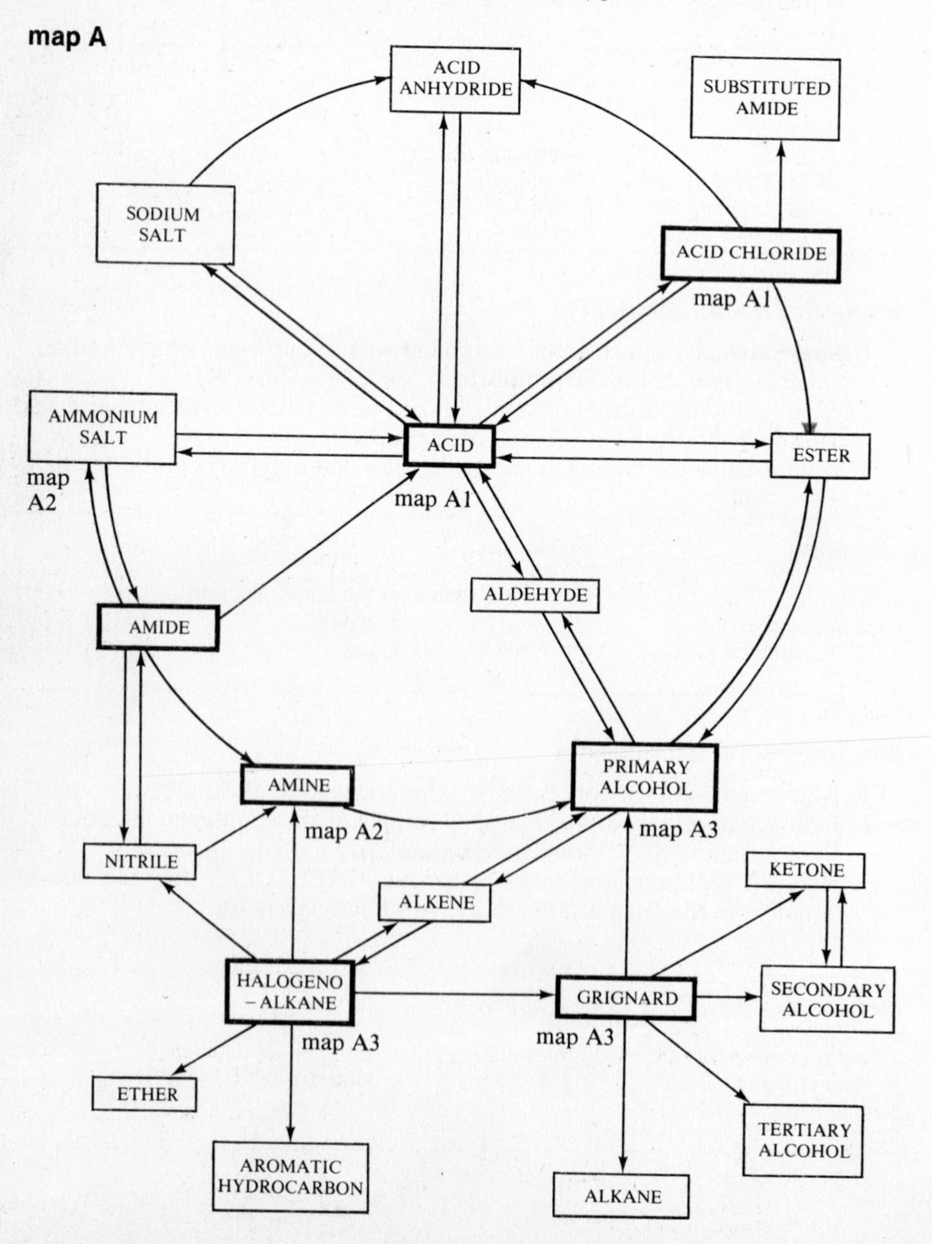

map A1

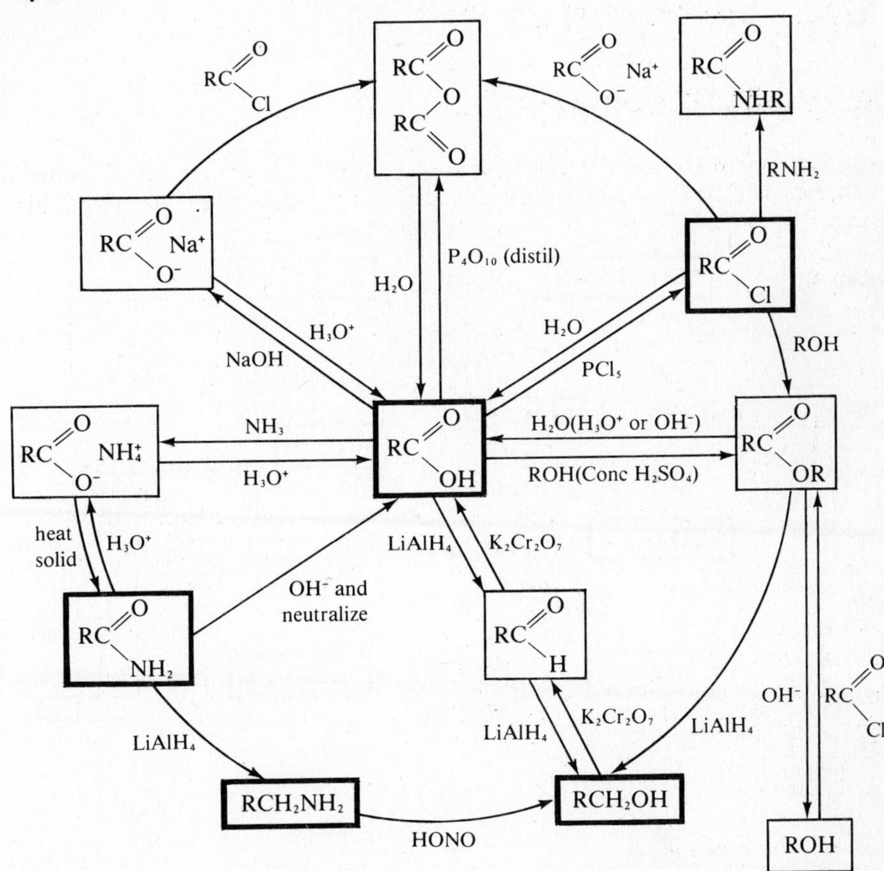
RCOCl
RCOOCOR
RCOO⁻ Na⁺
RCONHR
RNH₂
RCOO⁻ Na⁺
P₄O₁₀ (distil)
H₂O
RCOCl
H₃O⁺
NaOH
H₂O
PCl₅
ROH
RCOO⁻ NH₄⁺
NH₃
H₃O⁺
RCOOH
H₂O(H₃O⁺ or OH⁻)
ROH(Conc H₂SO₄)
RCOOR
heat
solid
H₃O⁺
LiAlH₄
K₂Cr₂O₇
OH⁻ and
neutralize
RCONH₂
RCHO
OH⁻
RCOCl
LiAlH₄
K₂Cr₂O₇
LiAlH₄
LiAlH₄
RCH₂NH₂
RCH₂OH
HONO
ROH

map A2

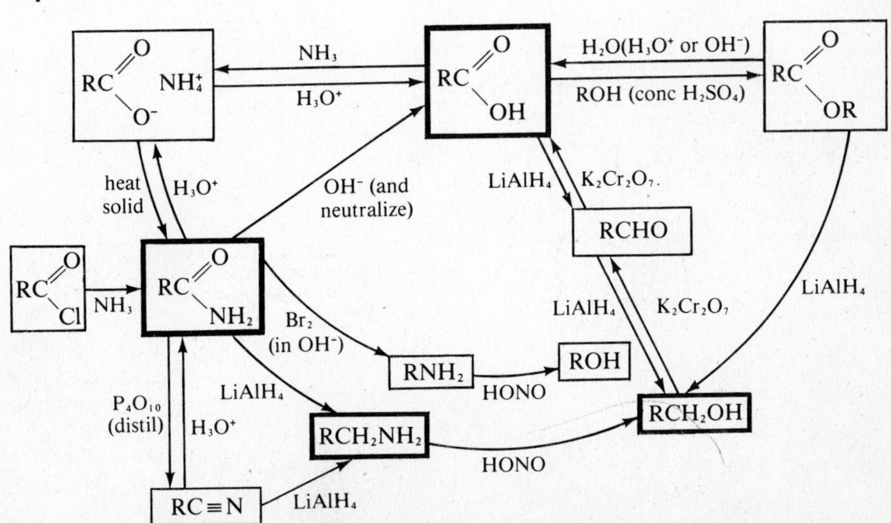
RCOO⁻ NH₄⁺
NH₃
H₃O⁺
RCOOH
H₂O(H₃O⁺ or OH⁻)
ROH (conc H₂SO₄)
RCOOR
heat
solid
H₃O⁺
OH⁻ (and
neutralize)
LiAlH₄
K₂Cr₂O₇.
RCHO
RCOCl
NH₃
RCONH₂
LiAlH₄
K₂Cr₂O₇
LiAlH₄
Br₂
(in OH⁻)
RNH₂
ROH
HONO
LiAlH₄
P₄O₁₀
(distil)
H₃O⁺
RCH₂NH₂
RCH₂OH
HONO
RC≡N
LiAlH₄

map A3

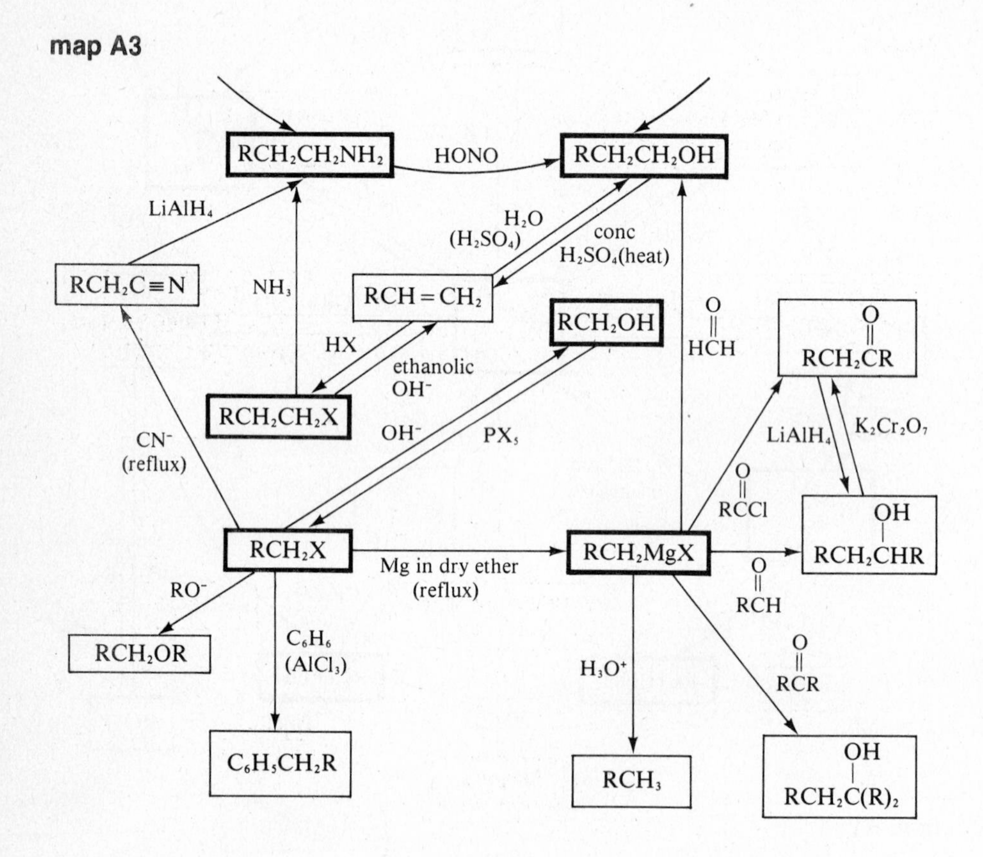
RCH₂CH₂NH₂
HONO
RCH₂CH₂OH
LiAlH₄
H₂O
(H₂SO₄)
conc
H₂SO₄(heat)
RCH₂C≡N
NH₃
RCH=CH₂
RCH₂OH
HCH
RCH₂CR
HX
ethanolic
OH⁻
RCH₂CH₂X
OH⁻
PX₅
LiAlH₄
K₂Cr₂O₇
CN⁻
(reflux)
RCCl
OH
RCH₂X
Mg in dry ether
(reflux)
RCH₂MgX
RCH₂CHR
RO⁻
RCH
RCH₂OR
C₆H₆
(AlCl₃)
H₃O⁺
RCR
C₆H₅CH₂R
RCH₃
RCH₂C(R)₂

B. Aromatic compounds

The map below outlines the major routes between different classes of aromatic compounds. The two succeeding diagrams supply some of the details of this map.

map B

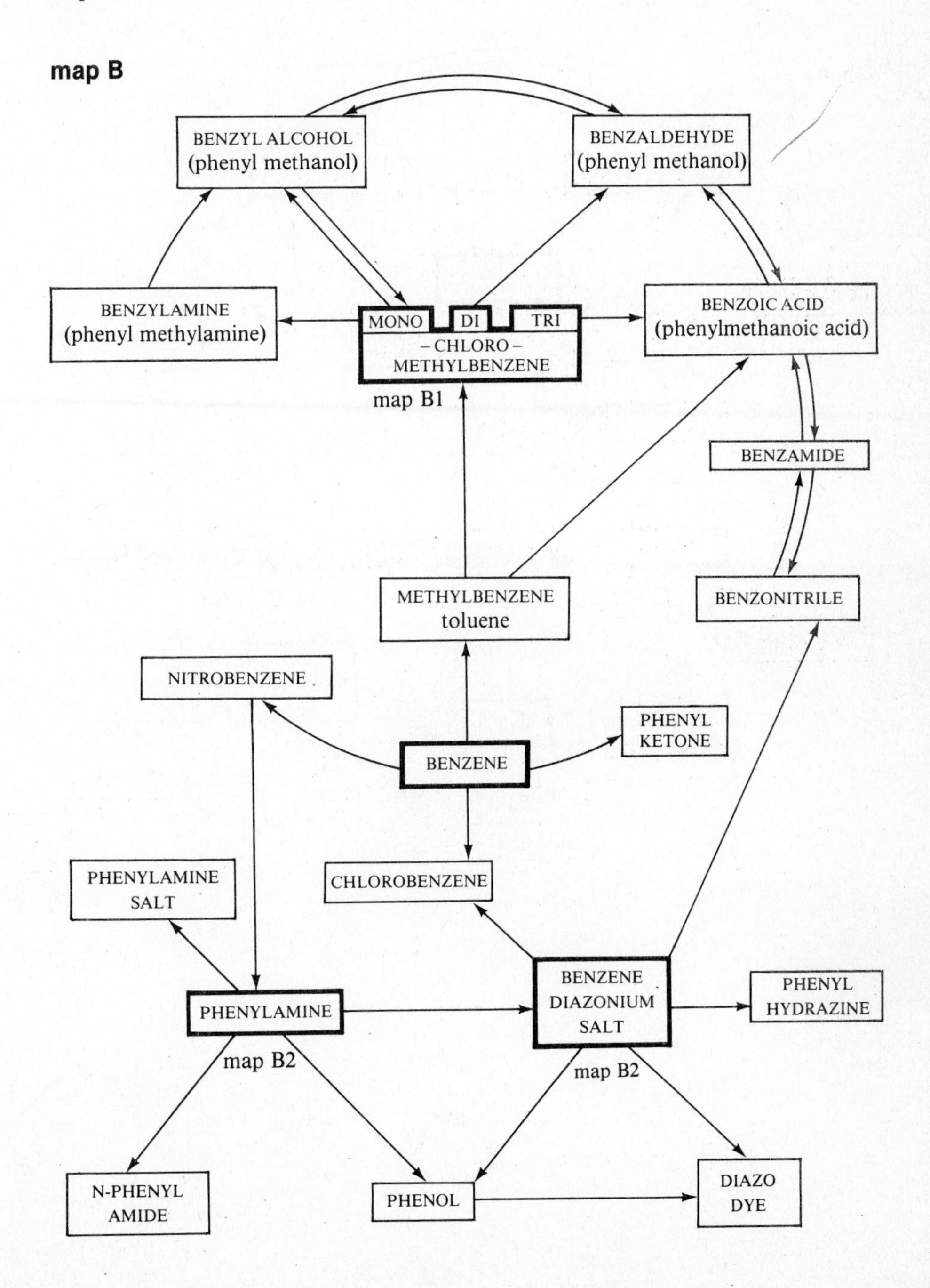

map B1

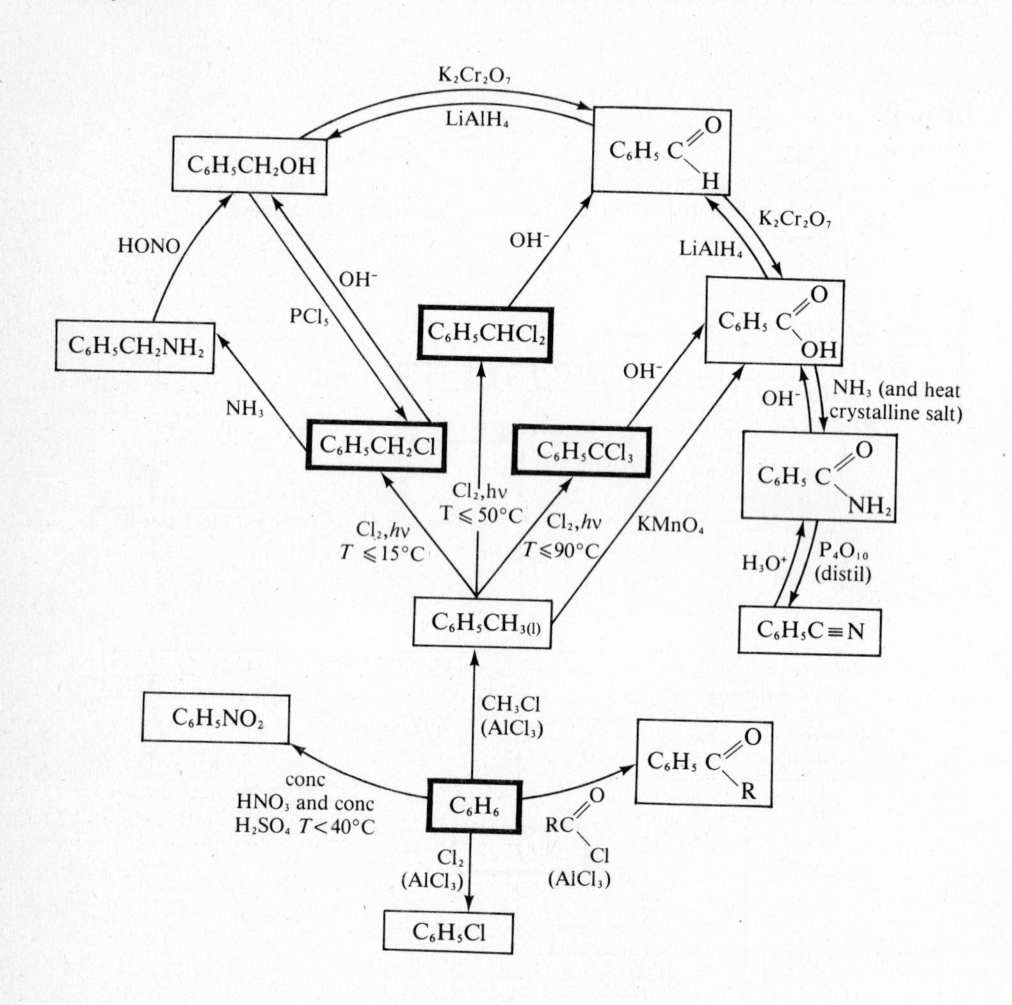

$K_2Cr_2O_7$
$LiAlH_4$
$C_6H_5CH_2OH$
$C_6H_5C(=O)H$
HONO
OH^-
PCl_5
OH^-
$K_2Cr_2O_7$
$LiAlH_4$
$C_6H_5CH_2NH_2$
$C_6H_5CHCl_2$
$C_6H_5C(=O)OH$
OH^-
NH_3
OH^-
NH_3 (and heat crystalline salt)
$C_6H_5CH_2Cl$
$C_6H_5CCl_3$
$C_6H_5C(=O)NH_2$
Cl_2,hν $T \leqslant 50°C$
Cl_2,*hν* $T \leqslant 15°C$
Cl_2,*hν* $T \leqslant 90°C$
$KMnO_4$
H_3O^+
P_4O_{10} (distil)
$C_6H_5CH_{3(l)}$
$C_6H_5C \equiv N$
$C_6H_5NO_2$
CH_3Cl ($AlCl_3$)
$C_6H_5C(=O)R$
conc HNO_3 and conc H_2SO_4 $T < 40°C$
C_6H_6
$RC(=O)Cl$ ($AlCl_3$)
Cl_2 ($AlCl_3$)
C_6H_5Cl

map B2

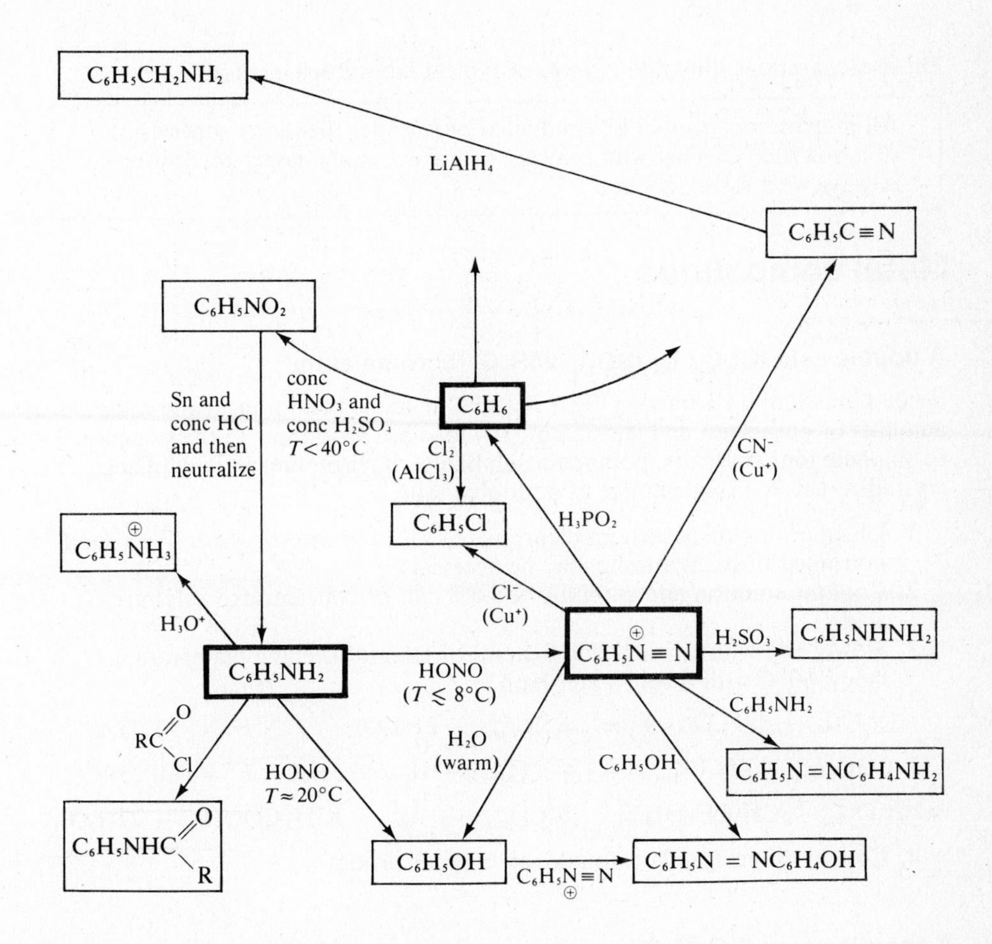

C6H5CH2NH2
LiAlH4
C6H5C≡N
C6H5NO2
conc
HNO3 and
conc H2SO4
T<40°C
Sn and
conc HCl
and then
neutralize
C6H6
Cl2
(AlCl3)
C6H5Cl
H3PO2
CN-
(Cu+)
C6H5NH3 ⊕
H3O+
Cl-
(Cu+)
C6H5N≡N ⊕
H2SO3
C6H5NHNH2
C6H5NH2
HONO
(T ≤ 8°C)
C6H5NH2
C6H5N=NC6H4NH2
RC(=O)Cl
HONO
T≈20°C
H2O
(warm)
C6H5OH
C6H5NHC(=O)R
C6H5OH
C6H5N≡N ⊕
C6H5N = NC6H4OH

Appendix 3

Inorganic preparations

These preparations illustrate a range of typical laboratory methods.

> All preparations should be conducted *only* under the direct supervision of a qualified chemist who is aware of the necessary safety precautions. Always wear safety glasses.

Metal compounds

A double salt: $K_2SO_4.\ Cr_2(SO_4)_3.\ 24H_2O$, 'chrome alum'

When potassium dichromate (VI) is reduced in acidic conditions, equimolar amounts of potassium and chromium (III) ions are produced. In the presence of sulphate ions two salts, potassium sulphate and chromium (III) sulphate, crystallize out in a single lattice as a 'double salt'.

1. Dissolve 10 g of potassium dichromate (VI) in 85 cm^3 of water in a round-bottomed flask. Warming may be necessary.
2. Cool the solution and carefully add 8.5 cm^3 of concentrated sulphuric acid.
3. Slowly and with stirring, add 8.5 cm^3 of ethanol keeping temperature below 60°C with a cold water bath.

$$Cr_2O^{2-}_{7(aq)} + 14H_3O^+_{(aq)} + 6e^- \rightleftharpoons 2Cr^{3+}_{(aq)} + 21H_2O \qquad E^\ominus = +1.33\ V$$

$$CH_3\ COOH_{(aq)} + 4H_3O^+_{(aq)} + 4e^- \rightleftharpoons CH_3CH_2OH_{(aq)} + 5H_2O_{(l)} \qquad E^\ominus = +0.25\ V$$

$$\Rightarrow 2Cr_2O^{2-}_{7(aq)} + 3CH_3CH_2OH_{(aq)} + 16H_3O^+_{(aq)} \rightarrow 4Cr^{3+}_{(aq)} + 3CH_3\ COOH_{(aq)} + 27H_2O_{(l)}$$

4. Cool in the fridge. The double salt crystallizes out

A complex salt: $K_3Cr(C_2O_4)_3$

The reduction of dichromate (VI) ions by an excess of ethanedioate ions produces complexed Cr^{3+} ions.

$Cr(C_2O_4)_3^{3-}$

Each chromium (III) ion is surrounded by three ethanedioate (oxalate) ligands.

1. Dissolve 6.5 g of potassium ethanedioate monohydrate and 15 g of ethanedioic acid dihydrate in 170 cm³ of water.
2. Add 5 g of potassium dichromate (VI) slowly, stirring after each small addition. Allow all the carbon dioxide to escape.

$$C_2O_{7(aq)}^{2-} + 14H_3O_{(aq)}^{+} + 6e \rightleftharpoons 2Cr_{(aq)}^{3+} + 21H_2O_{(l)} \qquad E^\ominus = +1.33 \text{ volts}$$

$$CO_{2(g)} + 2H_3O_{(aq)}^{+} + 2e \rightleftharpoons (COOH)_{2(aq)} + 2H_2O_{(l)} \quad E^\ominus = -0.49 \text{ volts}$$

$$\Rightarrow Cr_2O_{7(aq)}^{2-} + 3(COOH)_2 + 8H_3O_{(aq)}^{+} \rightarrow 2Cr_{(aq)}^{3+} + 6CO_{2(g)} + 15H_2O_{(l)}$$

3. Evaporate the solution to about half its volume and add 17 cm³ of ethanol because the complex salt is less soluble in ethanol.
4. Cool in the fridge. The complex salt crystallizes out.

A compound containing a transition metal in its highest oxidation state: $K_2Cr_2O_7$

Chromium (III) hydroxide is amphoteric and dissolves in alkali to give a tetrahydroxychromate (III) complex. Hydrogen peroxide is a strong enough oxidant under these conditions to convert chromium (III) to chromium (VI)

1. Dissolve 25 g of chromium (III) chloride hexahydrate in 70 cm³ of water.
2. Carefully dissolve 30 g of potassium hydroxide in 70 cm³ of water and add it to the first solution. Warm.

$$Cr_{(aq)}^{3+} + 4OH_{(aq)}^{-} \rightarrow Cr(OH)_{3(s)} + OH_{(aq)}^{-} \rightleftharpoons Cr(OH)_{4(aq)}^{-}$$

3. While stirring slowly, add 100 cm³ of 20-volume hydrogen peroxide. Boil for a few minutes to decompose excess hydrogen peroxide.

$$H_2O_{2(aq)} + 2e^- \rightleftharpoons 2OH_{(aq)}^{-} \qquad E^\ominus = +0.87 \text{ volts}$$

$$CrO_{4(aq)}^{2-} + 4H_2O_{(l)} + 3e^- \rightleftharpoons Cr(OH_4^-)_{(aq)} + 4OH_{(aq)}^{-} \quad E^\ominus = -0.13 \text{ volts}$$

$$\Rightarrow 2Cr(OH)_{4(aq)}^{-} + 2OH^-_{(aq)} + 3H_2O_{2(aq)} \rightarrow 2CrO_{4(aq)}^{2-} + 8H_2O_{(l)}$$

4. Filter while hot through a Buchner funnel into an evaporating basin and evaporate the solution to half its volume.
5. Carefully add 10 cm³ of pure ethanoic acid to the hot solution. Under acidic conditions, the chromate (VI) ions form dichromate (VI) ions and potassium dichromate (VI) is not very soluble in ethanoic acid,

$$2CrO_{4(aq)}^{2-} + 2H_3O_{(aq)}^{+} \rightleftharpoons Cr_2O_{7(aq)}^{2-} + 3H_2O_{(l)}$$

6. Cool in the fridge. The orange potassium dichromate crystallizes out.

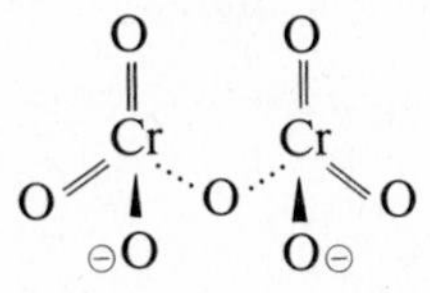

O O
Cr Cr
O O O
⊖O O⊖

A compound containing a transition metal in its lowest oxidation state: CuCl

Copper (II) chloride can be reduced to copper (I) chloride by sulphate (IV) ions. The copper (I) is rapidly oxidized by air and should be kept in an atmosphere of nitrogen.

1. Dissolve 12 g of copper (II) chloride dihydrate in 15 cm^3 of water. Stir and heat.
2. Make up a solution of 12 g of sodium sulphate (IV) hexahydrate in 60 cm^3 of water and add it to the first solution.
3. The solution darkens in colour and a white precipitate of copper (I) chloride forms. Gradually the solution colour lightens and the reaction is complete when no further change takes place.

$$Cu^{2+}_{(aq)} + Cl^-_{(aq)} + e^- \rightleftharpoons CuCl_{(s)} \qquad E^\ominus = +0.54 \text{ volts}$$

$$SO^{2-}_{4(aq)} + 2H_3O^+_{(aq)} + 2e^- \rightleftharpoons SO^{2-}_{3(aq)} + 3H_2O_{(l)} \qquad E^\ominus = +0.17 \text{ volts}$$

$$\Rightarrow 2Cu^{2+}_{(aq)} + 2Cl^-_{(aq)} + SO^{2-}_{3(aq)} + 3H_2O_{(l)} \rightarrow 2CuCl_{(s)} + SO^{2-}_{4(aq)} + 2H_3O^+_{(aq)}$$

4. Filter off the copper (I) chloride at a Buchner funnel and wash it with ethanol.

Non-metal compounds

A non-metal oxyacid: HIO_3

Iodine is oxidized under reflux by fuming nitric acid to produce iodic (V) acid. The relevant standard electrode potentials suggest that, under standard conditions, nitrogen (V) is not a strong enough oxidant to oxidize iodine to iodate (V). However, the reaction conditions are far removed from standard conditions and the reaction proceeds at about 80°C.

1. Put 0.5 g of iodine into a 50 cm^3 quickfit flask
2. Set up the flask for refluxing in a fume cupboard (as shown on the next page). Turn the cooling water on.
3. Slowly pour 15 cm^3 of fuming, concentrated nitric acid (CARE) down the reflux condenser.
4. Heat the flask to about 80°C using a water bath. Continue until no more brown fumes appear.

$$I_2 + 10HNO_3 \longrightarrow 2HIO_3 + 10NO_2 + 4H_2O$$

5. Pour the mixture into an evaporating basin and evaporate to dryness over a water bath.

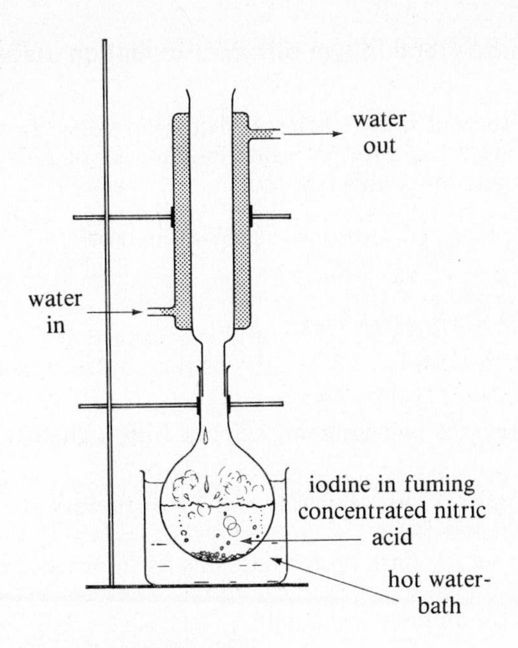

A non-metal halide: PCl_3

Red phosphorus combines directly with chlorine to produce phosphorus chloride.

1. Put 2 g of red phosphorus into a dry 150 cm³ quickfit flask. Ensure that the phosphorus is also dry.
2. Set up the apparatus in a fume cupboard as shown below. Flush it out with dry chlorine.
3. Pass chlorine over the phosphorus heated to 100°C by boiling water.
4. Collect the phosphorus (III) chloride in a small round-bottomed flask connected to the water condenser (b.p. PCl_3 = 75.5°C).

$$2P_{red} + 3Cl_{2(g)} \xrightarrow[100°C]{} 2PCl_{3(g)}$$

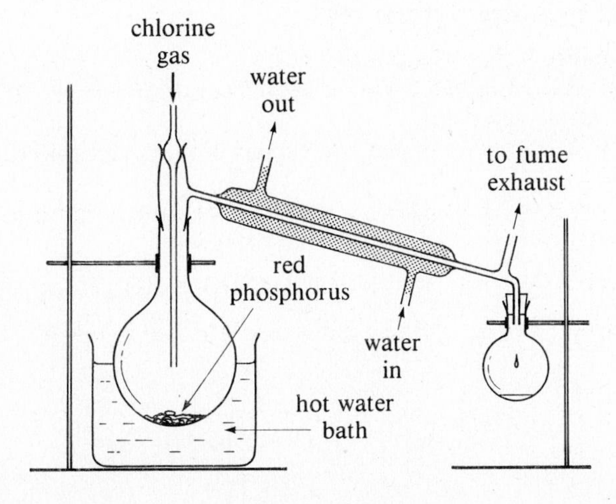

A compound containing a non-metal in two different oxidation states: $Na_2S_2O_3$

Sodium thiosulphate contains sulphur in two different oxidation states. In a thiosulphate ion, one sulphur atom has a formal oxidation number of zero while the other has a formal oxidation number of four.

S ← oxidation number = 0: no electrons transferred as a result of bonding

S ← oxidation number = 4: 4 electrons partially lost to the more electronegative oxygen atoms.

O ⊖ O ⊖ O

Sodium thiosulphate can be prepared by combining sulphur with sodium sulphate (IV).

1. Put 15 g of sodium sulphate (IV) heptahydrate and 2 g of flowers of sulphur into a 150 cm^3 quickfit flask.
2. Add 75 cm^3 of water and set the flask up for refluxing as shown at the top of the previous page.
3. Boil the mixture steadily for an hour and a half.

$$8SO^{2-}_{3(aq)} + S_{8(s)} \rightleftharpoons 8S_2O^{2-}_{3(aq)}$$

4. Filter off the excess sulphur and evaporate the remaining solution over a water-bath to about 15 cm^3.
5. Cool the solution and scratch the sides of the evaporating-basin vigorously under the liquid level. Sodium thiosulphate crystallizes out.

Answers

Chapter 1

6. b) ^{35}Cl:^{37}Cl = 3:1
8. a) 10.8; b) 5, 137.5
9. a) 35.5; b) 75% ^{63}Cu, 25% ^{65}Cu

Chapter 2

8. a) 15.3 h; b) 4.54×10^{-2} h^{-1}
9. a) 1980 years b) 15 600 years

Chapter 10

3. 400 cm^3; 0.2 atm; 48.1 cm^3; 10.4 *l*; 148 cm^3
4. 2.73 dm^3; 401 K; 322 K (49°C); 5.47 *l*; 1.3 dm^3
5. 1.5 atm; 2.5 dm^3; 140 kN m^{-2}; 273 K; 11.1 dm^3; 153 cm^3; 323 K; 700 K
6. 120; 60 7. 8.67 atm
8. c) 89.2 cm; d) 103 cm

Chapter 11

3. a) 79; b) 59; c) 65 g
4. a) 120; b) apparent R.M.M. = $\frac{58.5}{2}$; c) 1.56 K mol^{-1} 1000 g^{-1}
5. a) 14 100; b) 5980; c) 234 000
8. a) 0.0460 mol dm^{-3}; b) 736

Chapter 12

5. 2.22×10^{-10} m 6. 4°
7. a) 7.09 cm^3; b) 1.92 cm; c) 2.28×10^{-8} cm; d) 30°
8. a) 2.25×10^{-11} m radius; b) 4.14×10^{-11} m radius.

Chapter 13

1. a) 2.00 mol; b) 0.091 mol; c) 3.00 mol; d) 0.125 mol; e) 0.003 mol; f) 1.50 mol; g) 1.71 mol; h) 0.223 mol; i) 0.0112 mol; j) 0.005 70 mol
2. a) 22.0 g; b) 128 g; c) 3.65 g; d) 25.6 g; e) 24.5 g; f) 143 g; g) 2.09 g;
3. a) 49.0 g; b) 18.3 g; c) 20.0 g; d) 29.3 g; e) 53.0 g; f) 85.0 g; g) 26.8 g; h) 66.0 g; i) 171 g
4. a) 4.25 g; b) 0.913 g; c) 2.45 g; d) 1.00 g; e) 3.95 g; f) 6.20 g
5. a) 0.05 mol; b) 0.04 mol; c) 0.0025 mol.
6. a) 30.0 g; b) 1.38 g; c) 4.74 g
7. a) 6.0 g; b) 2.4 g
8. a) 4.48 dm^3; b) 5.56 g and 2.24 dm^3.
9. a) 0.24 g; b) 0.24 g; c) 2.54 g; d) 0.24 g
10. a) 0.5 mol; b) 53.0 g; c) 1 dm^3; d) 61.8 g; e) 1.25 dm^3; f) 0.8 mol dm^{-3}

Chapter 14

1 a) 24°C; b) 5040 J; c) 19°C; d) 3990 J; e) 1050 J; f) 19°C; g) 55.3 J $°C^{-1}$
3. b) 388.7 J c) 3.89 kJ mol^{-1}
4. 68.0 kJ mol^{-1}
5. − 196 kJ 6. − 541 kJ
7. − 103 kJ 8. − 10 kJ

Chapter 15

3. a) first order; b) k = 0.0460 h^{-1}
4. a) second order; b) k at 25°C = 4.86 mol^{-1} dm^3 min^{-1}, k at 60°C = 17.9 mol^{-1} dm^3 min^{-1}; c) 30 kJ mol^{-1}
5. a) first order; b) $k = 4.60 \times 10^{-3}$ min^{-1}
6. a) second order; b) $k = 5.00 \times 10^{-5}$ kPa^{-1} s^{-1}
7. a) first order; b) k = 0.043 min^{-1}
8. 70 kJ mol^{-1}

Chapter 16

4. a) 15.0 dm^3 mol^{-1}; b) 0.242 dm^3 mol^{-1}; c) 10.0 mol dm^3
5. a) p_Y = 2 atm; P_T = 10 atm; b) 7.5 mol dm^{-3}
6. a) (i) 0.100; (ii) 0.316; (iii) 3.16; b) (i) $[H_2] = [Br_2]$ = 2.5 mol dm^{-3}; $[HBr]$ = 0.791 mol dm^{-3}; (ii) 938 g
7. a) 4; b) 37.2 g
8. 176 g

Chapter 17

5. a) 0; b) 3; c) 5; d) 6.96; e) −0.0969; f) 3.89; g) 2.52; h) −1; i) −1.30; j) 14; k) 12; l) 11.2; m) 9.60; n) 8.84; o) 15
6. a) 0.01; b) 1×10^{-6}; c) 1×10^{-14}; d) 1; e) 10; f) 0.631; g) 0.235; h) 5.01×10^{-13}; i) 5.01×10^{-6} j) 2.09×10^{-13}; k) 1.56×10^{-10}; l) 1.29×10^{-13}; m) 3.98×10^{-8}; n) 2.89; o) 1.29
7. a) 3.0×10^{-3} mol dm^{-3}; b) 4.64; c) 4.11
8. a) 4.18×10^{-2} mol dm^{-3}; b) 6.31×10^{10}; c) 1.66×10^{-5}
9. a) (i) 2.37; 0.424%; (ii) 2.87; 1.34%; (iii) 3.87; 13.4%
10. a) 4.94; b) 10.4; c) 1.56:1

Chapter 18

2. 1.00×10^{-5} mol^2 dm^{-6}
3. 7.83×10^{-16} mol^3 dm^{-9}
4. 1.00×10^{-33} mol^4 dm^{-12}
5. 2.66×10^{-4} g 100 g^{-1}
6. Ag Br < $SrCO_3$ < $Co(OH)_2$ < Ag_2 CrO_4
9. 1.0×10^6 dm^6 mol^{-2}

Chapter 20

3. a) 6.35 cm^3; b) 0.787 mol dm^{-3}
4. a) $\alpha_{18} = 0.0245$; $\alpha_{100} = 0.0190$; K_a at 18°C = 1.80×10^{-5} mol dm^{-3}; K_a at 100°C = 1.08×10^{-5} mol dm^{-3}
5. a) 150 m^{-1}; b) 0.781 Ω^{-1} m^{-1}; c) 7.81 $\times 10^{-3}$ mol^{-1} Ω^{-1} m^2
6. a) 0.0405 Ω^{-1} m^2 mol^{-2}; b) 6.05×10^{-14} mol dm^{-3}

Chapter 21

5. a) but-2-ene b) methanol c) 2-aminopropane d) ethanal e) propanone f) methyl methanoate
7. a) NH_3 or H_2O b) H^+ or HI c) NH_3 (or H_2O) d) NH_4^+ or HI e) H^+, HI or NH_4^+

Chapter 22

p.368 a) methylpropane b) cyclohexane c) 3-methylpentane d) cyclopentane e) ethane

6. a) C_6H_{10} b) CH_4
7. 107.7 dm^3
8. $C_4H_{10(g)} \Rightarrow$ 45 880 kJ kg^{-1}; $C_{(s)} \Rightarrow$ 32 830 kJ kg^{-1}

Chapter 23

p.374 a) but-1-ene b) but-2-ene c) cyclopentene d) hexa-2,4-diene

6. a) $CH_2{=}C(H){-}C(C){=}CH_2$ (structure: CH_2=C with H on the first C; C=CH_2 with C on the second C)

   b) CH_2O and $(CHO)_2$

Chapter 24

4. H−C≡C−C≡C−H
7. $n = 2$

Chapter 25

p. 399 a) methyl-1-bromopropane b) 3-chlorocyclohexene c) 2, 2-dibromopropane d) 1,2-dichlorethane

Chapter 26

p. 408 a) butan-2-ol b) butan-1-ol c) butan-2,3-diol d) cyclohexen-4-ol e) cyclopropanol f) methylpropan-1-ol

Chapter 27

p. 420 a) propane-1-amine b) *N*-methylethylamine c) butane-1,2-diamine d) trimethylamine e) 3-aminocyclopentene f) *N*-methyldiethylamine

10. b) $(Me)_2NH$ > $MeNH_2$ > NH_3

Chapter 28

p.430 a) ethanal b) propanone c) cyclohexanone d) ethanedial e) pentan-2,4-dione f) 3-hydroxy-cyclopentanone g) methanal

Chapter 29

p. 450 a) propanoic acid b) butanedioic acid c) cyclohexylmethanoic acid d) 2-bromopropanoic acid e) 3-bromopropanoic acid

f) 2-aminobutanoic acid (an α-aminoacid)

4. $HCOO^- Na^+$ or $HCOO^- K^+$

Chapter 30

8. $C_6H_5C(=O)Cl$

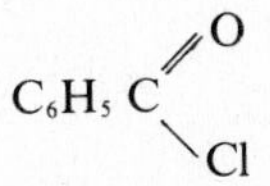

Chapter 33

2. i) a) phenylmethanol (benzyl alcohol)
 b) phenol c) 4-hydroxy-cyclohexene d) methanol

 ii) bromine water; 2,4,6-tribromophenol (Br, OH, Br, Br on ring)

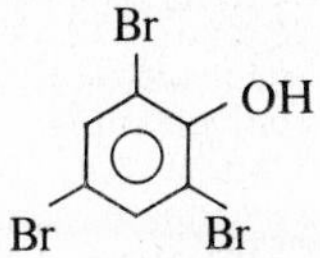

 and 3,4-dibromophenol (OH, Br, Br on ring)

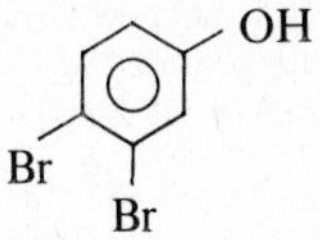

 iii) phosphorus (V) chloride

Chapter 37

4. 145 000

Chapter 41

8. a) 0.0111 mol dm^{-3};
 0.0431 mol dm^{-3}; 0.108 mol dm^{-3}

Chapter 42

6. 2.93 7. − 847 kJ

Chapter 47

6. a) KrF_2; b) KrF_4; c) $XeOF_2$

Chapter 48

4. a) 2.28; b) 2.14×10^{-4} mol dm^{-3};
 c) 3×10^{-3} mol dm^{-3}

Chapter 49

7. 5.82×10^{-6} mol dm^{-3};
 9.28×10^{-11} mol dm^{-3}

1 General index

2 Organic compound index

In this index, the chain length of the molecules is shown on the left and the abbreviations R and X are used to indicate a carbon chain and a halogen atom respectively.

3 Inorganic compound index

In this index each reference is followed by a letter to indicate the property under discussion: **s** = structure (and/or bonding) **p** = preparation **a** = acid-base properties **r** = redox properties **c** = complexing properties (or solubility)

THE PERIODIC TABLE

GORSEINON COLLEGE LIBRARY

RELATIVE ATOMIC MASS → 1
1 H hydrogen
ATOMIC NUMBER → 1

Transition Elements

	I	II											III	IV	V	VI	VII	Noble Gases
1																		4 He helium 2
2	7 Li lithium 3	9 Be beryllium 4											11 B boron 5	12 C carbon 6	14 N nitrogen 7	16 O oxygen 8	19 F fluorine 9	20 Ne neon 10
3	23 Na sodium 11	24 Mg magnesium 12											27 Al aluminium 13	28 Si silicon 14	31 P phosphorus 15	32 S sulphur 16	35·5 Cl chlorine 17	40 Ar argon 18
4	39 K potassium 19	40 Ca calcium 20	45 Sc scandium 21	48 Ti titanium 22	51 V vanadium 23	52 Cr chromium 24	55 Mn manganese 25	56 Fe iron 26	59 Co cobalt 27	59 Ni nickel 28	63·5 Cu copper 29	65 Zn zinc 30	70 Ga gallium 31	72·5 Ge germanium 32	75 As arsenic 33	79 Se selenium 34	80 Br bromine 35	84 Kr krypton 36
5	85·5 Rb rubidium 37	88 Sr strontium 38	89 Y yttrium 39	91 Zr zirconium 40	93 Nb niobium 41	96 Mo molybdenum 42	99 Tc technetium 43	101 Ru ruthenium 44	103 Rh rhodium 45	106 Pd palladium 46	108 Ag silver 47	112 Cd cadmium 48	115 In indium 49	119 Sn tin 50	122 Sb antimony 51	128 Te tellurium 52	127 I iodine 53	131 Xe xenon 54
6	133 Cs caesium 55	137 Ba barium 56	137 La lanthanum 57	* 178·5 Hf hafnium 72	181 Ta tantalum 73	184 W tungsten 74	186 Re rhenium 75	190 Os osmium 76	192 Ir iridium 77	195 Pt platinum 78	197 Au gold 79	201 Hg mercury 80	204 Tl thallium 81	207 Pb lead 82	209 Bi bismuth 83	209 Po polonium 84	210 At astatine 85	222 Rn radon 86
7	223 Fr francium 87	226 Ra radium 88	227 Ac actinium 89	*														

*														
140 Ce 58	141 Pr 59	144 Nd 60	147 Pm 61	150 Sm 62	152 Eu 63	157 Gd 64	159 Tb 65	162·5 Dy 66	165 Ho 67	167 Er 68	169 Tm 69	173 Yb 70	175 Lu 71	

*														
232 Th 90	231 Pa 91	238 U 92	237 Np 93	244 Pu 94	243 Am 95	247 Cm 96	247 Bk 97	251 Cf 98	254 Es 99	257 Fm 100	256 Md 101	256 No 102	257 Lw 103	